ADVANCED SOLUTIONS IN POWER SYSTEMS

IEEE Press
445 Hoes Lane
Piscataway, NJ 08854

ADVANCED SOLUTIONS IN POWER SYSTEMS

HVDC, FACTS, and Artificial Intelligence

Edited by

MIRCEA EREMIA
CHEN-CHING LIU
ABDEL-ATY EDRIS

IEEE PRESS SERIES ON POWER ENGINEERING

IEEE PRESS

WILEY

Published by John Wiley & Sons, Inc., Hoboken, New Jersey.
Published simultaneously in Canada

Library of Congress Cataloging-in-Publication Data is available.

ISBN: 978-1-119-03569-5

Printed in the United States of America

10 9 8 7 6 5 4 3 2 1

CONTENTS

CONTRIBUTORS xxi

FOREWORD xxiii

ACKNOWLEDGMENTS xxv

CHAPTER 1 *INTRODUCTION* 1
Mircea Eremia, Chen-Ching Liu, and Abdel-Aty Edris

PART I *HVDC TRANSMISSION*
Mircea Eremia

CHAPTER 2 *POWER SEMICONDUCTOR DEVICES FOR HVDC AND FACTS SYSTEMS* 11
Remus Teodorescu and Mircea Eremia

2.1 Power Semiconductor Overview **12**
- 2.1.1 Not-Controllable Power Semiconductor Devices **13**
- 2.1.2 Semicontrollable Power Semiconductor Devices **13**
- 2.1.3 Fully Controllable Power Semiconductor Devices **17**
  - 2.1.3.1 Gate Turn-Off Thyristor **18**
  - 2.1.3.2 Integrated Gate-Commutated Thyristor **18**
  - 2.1.3.3 Isolated Gate Bipolar Transistor **18**
- 2.1.4 Power Semiconductor Parameters **20**
  - 2.1.4.1 Steady-State Parameters **20**
  - 2.1.4.2 Switching Characteristics **20**
- 2.1.5 Future Power Semiconductor Devices **21**

2.2 Converter Types **21**

2.3 HVDC Evolution **23**
- 2.3.1 Line-Commutated HVDC Converters (LCC/CSC–HVDC) **24**
- 2.3.2 Capacitor-Commutated Converter (CCC–HVDC) **26**
- 2.3.3 Voltage Source Converter VSC–HVDC **28**
  - 2.3.3.1 VSC–HVDC Based on Two-Level Converters **29**
  - 2.3.3.2 VSC–HVDC Based on Multilevel Converters **29**
  - 2.3.3.3 Limitations of VSC Transmission **30**

2.4 FACTS Evolution **30**

References **33**

CHAPTER 3 *CSC–HVDC TRANSMISSION* **35**
Mircea Eremia and Constantin Bulac

3.1 Structure and Configurations **35**
3.1.1 Structure of HVDC Links **35**
3.1.2 HVDC Configurations **40**
3.2 Converter Bridge Modeling **47**
3.2.1 Rectifier Equations **47**
3.2.1.1 Ideal Converter Bridge Operation **47**
3.2.1.2 Commutation Process or Overlap **52**
3.2.1.3 Equivalent Circuit of the Rectifier **56**
3.2.2 Inverter Equations **57**
3.3 Control of CSC–HVDC Transmission **59**
3.3.1 Equivalent Circuit and Control Characteristics **59**
3.3.1.1 Equivalent Circuit of DC Transmission Link **59**
3.3.1.2 Voltage–Current Characteristics **62**
3.3.2 HVDC Control Principles **64**
3.3.2.1 State Variables of a HVDC Link **64**
3.3.2.2 Basic Control Principles of the DC Voltage and DC Current **65**
3.3.2.3 Control Modes **67**
3.3.3 HVDC Control Strategies **69**
3.3.3.1 Rectifier Control Strategy **69**
3.3.3.2 Inverter Control Strategy **71**
3.3.4 Hierarchical Control of a HVDC Link **72**
3.3.4.1 Master Control **72**
3.3.4.2 Pole Control **74**
3.3.4.3 Firing (Valve) Control **78**
3.3.4.4 Telecommunications **78**
3.3.4.5 Measurement Transducers **78**
3.4 Reactive Power and Harmonics **78**
3.4.1 Reactive Power Requirements and Sources **78**
3.4.2 Harmonics and Filters **83**
3.4.2.1 The Source of AC Harmonic Currents **83**
3.4.2.2 The Effect of Y/Δ Transformation on AC Harmonic Current **85**
3.4.2.3 Higher Pulse Operation Using Multiple Bridges and Transformers **86**
3.4.2.4 Elimination of Harmonics **86**
3.5 Load Flow in Mixed HVAC/HVDC-CSC Systems **91**
3.5.1 Steady-State Model **91**
3.5.1.1 The Extended Variables Method **93**
3.5.1.2 The Sequential Method **94**
3.5.1.3 The Eliminated Variables Method **94**
3.6 Interaction Between AC and DC Systems **96**
3.6.1 AC Systems Stabilization **96**
3.6.2 Influence of AC System Short-Circuit Ratio **96**
3.6.3 Effective Inertia Constant **99**
3.6.4 Reactive Power and the Strength of the AC System **100**
3.7 Comparison Between DC and AC Transmission **101**
3.8 Application on a CSC–HVDC Link **109**
3.8.1 Solution **111**

Appendix 3.1 CSC–HVDC Systems in the World 118
References 123

CHAPTER 4 *VSC–HVDC TRANSMISSION* 125
Mircea Eremia, José Antonio Jardini, Guangfu Tang, and Lucian Toma

4.1 VSC Converter Structures 126
4.1.1 Half-Bridge VSC or Two-Level Pole 126
4.1.2 Full-Bridge Single-Phase VSC 128
4.1.3 Three-Phase Two-Level VSC 128
4.1.4 Three-Level Pole VSC 129
4.1.5 Multimodule VSC Systems 131
4.1.6 Multilevel VSC Systems 132
4.1.7 Modular Multilevel Converter 138
4.1.7.1 Half-Bridge Modular Multilevel Converter 140
4.1.7.2 Full-Bridge Modular Multilevel Converter 143
4.1.7.3 The MMC–HVDC INELFE Project 144
4.1.8 Cascaded Two-Level Converters 147
4.2 Modulation Techniques 151
4.2.1 PWM Techniques 151
4.2.1.1 PWM Principle 151
4.2.1.2 PWM Strategy Control of a Half-Bridge Converter 155
4.2.1.3 Three-Phase Bridge Inverter with Sinusoidal PWM 159
4.2.2 Modulation Techniques for Multilevel Converters 163
4.2.2.1 PWM Algorithms for Multilevel Converters 163
4.2.2.2 Space Vector Modulation Algorithms 165
4.2.2.3 Other Modulation and Control Algorithms for Multilevel Converters 165
4.3 DC/AC Converter Analysis 166
4.3.1 Operation Modes of the Switched-Inductor Cell 166
4.3.2 Ideal DC/AC Half-Bridge Converter 168
4.3.3 Averaging Models 175
4.3.3.1 Circuit/Switch Averaging of DC–DC Converters 176
4.3.3.2 State-Space Averaging of DC–DC Converters 177
4.3.3.3 AVM of DC–AC Converters 178
4.3.4 Detailed and Averaged Models for MMC–HVDC Systems 180
4.3.4.1 Detailed Equivalent Models 181
4.3.4.2 AVM of MMC–HVDC Using Voltage- and Current-Controlled Sources 183
4.4 VSC Transmission Scheme and Operation 188
4.4.1 Power Equipment 188
4.4.2 Principles of Active and Reactive Power Control 192
4.4.3 VSC Transmission Control 196
4.4.3.1 VSC Converter Control Using the Vector Control Strategy 196
4.4.3.2 Levels of Control 199
4.4.3.3 Coordination of Controls 200
4.5 Multiterminal VSC–HVDC Systems and HVDC Grids 203
4.5.1 On the Conventional Multiterminal HVDC Configurations 203
4.5.2 Multiterminal HVDC Grid Configurations 204
4.5.3 Meshed HVDC Grid Configurations 209

4.5.4 Need for Fast and Low Loss HVDC Breakers 211
4.5.4.1 Preconditions 211
4.5.4.2 Schemes for the Current Zero Formation 212
4.5.4.3 Types of DC Circuit Breakers 214
4.5.5 HVDC Grid Protection 218
4.6 Load Flow and Stability Analysis 221
4.6.1 Load Flow in Meshed AC/DC Grids 221
4.6.1.1 Generalities 221
4.6.1.2 Load Flow Calculation in a DC Grid 223
4.6.1.3 Application 227
4.6.2 Dynamic Stability in Meshed AC/DC Grids 231
4.6.2.1 Generalities 231
4.6.2.2 Description of the VSC Model for Stability Analysis 233
4.6.2.3 Control Models 235
4.6.2.4 P–V Droop Control 237
4.6.2.5 Current and Voltage Limits 237
4.6.2.6 RMS Model Testing 238
4.6.2.7 Simulations on an AC/DC Meshed Grid 239
4.7 Comparison of CSC–HVDC Versus VSC–HVDC Transmission 246
4.7.1 Differences Resulting from the Commutation Principle 246
4.7.2 Differences Resulting from the Converter Type 248
4.8 Forward to Supergrid 249
4.8.1 Challenges and Solutions for Developing Supergrid 249
4.8.1.1 Connecting Renewable Energy Sources and Increased Transmission System Capacity 250
4.8.1.2 Compensating Reactive Power 250
4.8.1.3 Maintaining System Stability 252
4.8.2 Hybrid AC and DC Systems 252
4.8.3 Supernodes 254
4.8.4 Stepwise Development of the European Supergrid 255
4.8.5 Steps Toward a Planetary Supergrid 258
4.8.6 VSC Multiterminal in China 260
Appendix 4.1 VSC–HVDC Projects Around the World 261
Appendix 4.2 Examples of VSC–HVDC One-Line Diagrams 263
References 263

PART II *FACTS TECHNOLOGIES*
Abdel-Aty Edris and Mircea Eremia

CHAPTER 5 *STATIC VAr COMPENSATOR (SVC)* 271
Mircea Eremia, Aniruddha Gole, and Lucian Toma

5.1 Generalities 271
5.2 Thyristor-Controlled Reactor 273
5.3 Thyristor-Switched Capacitor 284
5.4 Configurations of SVC 287
5.4.1 Fixed Capacitor and Thyristor-Controlled Reactor 287
5.4.2 The SVC Device (TSC–TCR) 289
5.4.2.1 V–I Characteristics 289
5.4.2.2 Operating Domain 290

5.5 Control of SVC Operation 294
5.5.1 The Voltage Regulator 294
5.5.2 Gate Pulse Generator 296
5.6 SVC Modeling 296
5.6.1 Steady-State SVC Modeling 296
5.6.1.1 Modeling of an SVC That Operates Within or Outside the Linear Control Domain 297
5.6.1.2 Improved Models for SVC Representation 299
5.6.1.3 Newton–Raphson Modified Algorithm to Include the SVCs 305
5.6.2 SVC Dynamic Modeling 307
5.6.2.1 The Basic Dynamic Model 307
5.6.2.2 First-Order Dynamic Model 308
5.6.2.3 Complex SVC Dynamic Models 309
5.7 Placement of SVC 312
5.8 Applications of SVC 314
5.8.1 Maintaining the Voltage Level of a Bus or into an Area 315
5.8.2 Increasing the Transmission Capacity 315
5.8.3 Static and Transient Stability Reserve Improvement 317
5.8.4 Oscillations Damping 322
5.8.5 Reducing the Transient Overvoltages 323
5.9 SVC Installations Worldwide 324
5.9.1 SVC at Hagby, in Sweden 326
5.9.2 SVC at Forbes, in United States 327
5.9.3 SVC in Temascal, Mexico 328
5.9.4 Complex Compensation Scheme in Argentina 329
5.9.5 SVC in the 735 kV Transmission System in Canada 329
5.9.6 SVC at Auas, in Namibia 330
5.9.7 SVC at the Channel Tunnel Rail Link 333
5.9.8 SVC at Harker, in United Kingdom 334
5.9.9 Relocatable SVCs 336
References 337

CHAPTER 6 *SERIES CAPACITIVE COMPENSATION* 339
Mircea Eremia and Stig Nilsson

6.1 Generalities 339
6.2 Mechanical Commutation-Based Series Devices 339
6.3 Static-Controlled Series Capacitive Compensation 342
6.3.1 GTO-Controlled Series Capacitor 342
6.3.2 Thyristor-Switched Series Capacitor 345
6.3.3 Thyristor-Controlled Series Capacitor 348
6.3.3.1 Basic Structure 349
6.3.3.2 Operating Principles of TCSC. Steady-State Approach and Synchronous Voltage Reversal 351
6.3.3.3 Operation Modes and the Characteristics of the TCSC 357
6.3.3.4 Capability Characteristics of the TCSC 362
6.4 Control Schemes for the TCSC 365
6.4.1 Open Loop Impedance Control 365
6.4.2 Closed Loop Control 366
6.5 TCSC Modeling 370

6.5.1 Steady-State Modeling of TCSC 370
6.5.1.1 TCSC Modeling Through Series Variable Impedance 370
6.5.1.2 TCSC Impedance Modeling as a Function of the Firing Angle 374
6.5.2 TCSC Dynamic Models 376
6.5.2.1 Transient Stability Model 376
6.5.2.2 Long-Term Stability Model 379
6.6 Applications of TSSC/TCSC Installations 382
6.7 Series Capacitors Worldwide 387
6.7.1 Kanawha River Mechanically Switched Series Capacitor in United States 387
6.7.2 Kayenta TCSC in United States 389
6.7.3 Slatt TCSC in United States 392
6.7.4 Stöde TCSC in Sweden 396
6.7.5 Imperatriz-Serra da Mesa TCSC in Brazil 397
6.7.6 Purnea and Gorakhpur TCSC/FSC in India 400
6.7.7 Series-Compensated 500 kV Power Transmission Corridors in Argentina 402
Appendix 6.1 TCSC Systems Around the World 404
References 405

CHAPTER 7 *PHASE SHIFTING TRANSFORMER: MECHANICAL AND STATIC DEVICES* 409
Mylavarapu Ramamoorty and Lucian Toma

7.1 Introduction 409
7.2 Mechanical Phase Shifting Transformer 410
7.2.1 Principle of Operation of the PST 410
7.2.2 PST Topology 412
7.2.2.1 Direct-Type Asymmetrical PSTs 412
7.2.2.2 Direct-Type Symmetrical PSTs 414
7.2.2.3 Indirect-Type Asymmetrical and Symmetrical PSTs 416
7.2.2.4 Comparison of the Topologies 417
7.2.3 Steady-State Model of a Mechanical Phase Shifter 418
7.2.4 Equivalent Series Reactance as a Function of the Phase Shift Angle 420
7.2.4.1 Symmetrical Phase Shifter 420
7.2.4.2 Quadrature Booster 424
7.2.4.3 Asymmetrical Phase Shifter 425
7.2.4.4 In-Phase Transformer and Symmetrical/Asymmetrical Phase Shifter 426
7.3 Thyristor-Controlled Phase Shifting Transformer 428
7.3.1 Configurations of the Static Phase Shifter 428
7.3.1.1 Substitution of Mechanical Tap Changer by Electronic Switches 429
7.3.1.2 Thyristor-Controlled Quadrature Voltage Injection 429
7.3.1.3 Pulse-Width Modulation AC Controller 432
7.3.1.4 Delay-Angle Controlled AC-AC Bridge Converter 433
7.3.1.5 Discrete-Step Controlled AC-AC Bridge Converter 434
7.3.1.6 PWM Voltage Source Converter 434
7.3.2 Modeling of TCPST 436
7.3.2.1 Model of a Transmission System with a TCPST 436

7.3.2.2 Line Model with Thyristor-Controlled Phase Angle Regulator 437
7.3.2.3 The Dynamic Model of the Phase Shifter 439
7.4 Applications of the Phase Shifting Transformers 439
7.4.1 Power Flow Control by Phase Angle Regulators 440
7.4.2 Real and Reactive Loop Power Flow Control 442
7.4.3 Improvement of Transient Stability with PST 444
7.4.4 Power Oscillation Damping with PST 446
7.4.4.1 Application to Damp Power Oscillations 448
7.5 Phase Shifting Transformer Projects Around the World 450
References 456

CHAPTER 8 *STATIC SYNCHRONOUS COMPENSATOR – STATCOM* 459
Rafael Mihalic, Mircea Eremia, and Bostjan Blazic

8.1 Principles and Topologies of Voltage Source Converter 459
8.1.1 Basic Considerations 459
8.1.2 Converter Topologies 464
8.1.2.1 Two-Level Topologies 464
8.1.2.2 Multilevel Topologies 469
8.1.2.3 PWM Converter 471
8.1.3 Switching Function 472
8.2 STATCOM Operation 473
8.3 STATCOM Modeling 476
8.3.1 STATCOM Model for Steady-State Analysis 476
8.3.1.1 Basic Load Flow Equations 478
8.3.1.2 The Single-Phase Voltage-Based Model 480
8.3.1.3 The Single-Phase Current-Based Model 482
8.3.1.4 Three-Phase Voltage-Based Model 484
8.3.1.5 Three-Phase Current-Based Model 487
8.3.2 Dynamic Models of STATCOM 492
8.3.2.1 Simplified Dynamic Model 492
8.3.2.2 Detailed Dynamic Model 494
8.3.3 Control Algorithm 499
8.3.4 STATCOM Model for Unbalanced Operation 501
8.4 STATCOM Applications 506
8.4.1 Fast Voltage Control and Maintaining Voltage Levels of a Bus or an Area 506
8.4.2 Flicker Compensation 506
8.4.3 Improvement of the Network Transmission Capability 509
8.4.4 Improvement of Static and Transient Stability Reserve 512
8.4.5 Oscillations Damping 514
8.5 STATCOM Installations in Operation 515
8.5.1 ± 80 MVAr STATCOM in Japan 515
8.5.2 ± 100 MVAr STATCOM at Sullivan, in United States 516
8.5.3 +225/–52 MVAr TSC and STATCOM Mixed System at East Claydon, in Great Britain 520
8.5.4 +133/–41 MVAr STATCOM at Essex, in United States 520
8.5.5 STATCOM (+80/–110 MVAr) and Mechanic-Switched Capacitor (–93 MVAr) Mixed System, at Holly, in United States 521
8.5.6 ±100 MVAr STATCOM at Talega, in United States 522
References 524

CHAPTER 9 *STATIC SYNCHRONOUS SERIES COMPENSATOR (SSSC)* 527
Laszlo Gyugyi, Abded-Aty Edris, and Mircea Eremia

9.1 Introduction 527
9.2 Architecture and Operating Principles 528
9.2.1 The Basic Structure and Principles of Operation 528
9.2.2 Operating Modes of SSSC 530
9.2.3 The P_q-δ Characteristic of SSSC 532
9.3 Comparison of SSSC with Other Technologies 533
9.3.1 Comparison with Fixed Series Capacitor 533
9.3.2 Comparison with Fixed Series Reactor 534
9.3.3 Comparison with Phase Angle Regulator 534
9.3.4 Comparison with Thyristor-Controlled Series Capacitor 535
9.3.5 Comparison with Gate-Controlled Series Capacitor 538
9.3.6 Dynamic Flow Controller 540
9.4 Components of an SSSC 540
9.4.1 Overview of the Functional SSSC Components 540
9.4.2 Control 542
9.4.3 Protection 545
9.5 SSSC Modeling 546
9.5.1 Steady-State SSSC Model 546
9.5.1.1 VSC Controller Load Flow Models 546
9.5.1.2 Newton–Raphson Load Flow Solution 547
9.5.2 SSSC Dynamic Model 549
9.6 Applications 551
9.7 SSSC Installation 552
9.7.1 SSSC in Operation 552
9.7.2 SSSC for Power Flow Control: A Project in Spain 553
9.7.2.1 Project Overview 553
9.7.2.2 Components of the SSSC 554
9.7.2.3 Location Selection for Prototype Installation 555
References 556

CHAPTER 10 *UNIFIED POWER FLOW CONTROLLER (UPFC)* 559
Laszlo Gyugyi

10.1 Introduction 559
10.1.1 UPFC as the Functional Combination of Conventional Transmission Controllers 559
10.1.2 UPFC Directly Providing Line Current Forcing Function 566
10.2 Basic Characteristics of the UPFC 567
10.3 UPFC Versus Conventional Power Flow Controllers 571
10.3.1 UPFC versus Series Reactive Compensators 571
10.3.2 UPFC versus Phase Shifters 573
10.4 UPFC Control System 575
10.4.1 Functional Control of the Shunt Converter 578
10.4.2 Functional Control of the Series Converter 579
10.4.3 Stand-Alone Shunt and Series Compensation 580
10.4.4 Basic Control Structure for the Series and Shunt Converters 580
10.4.5 Practical Control Considerations 583

10.5 Equipment Structural and Rating Considerations 584
10.5.1 Circuit Structural Considerations 586
10.5.2 Rating Considerations for Series and Shunt Converters 588
10.5.2.1 Series Converter Rating to Meet Line Compensation Requirements 588
10.5.2.2 Shunt Converter Rating to Meet UPFC Operation Requirements 592
10.5.3 UPFC Rating Optimization by Combined Compensation 594
10.6 Protection Considerations 596
10.6.1 Protection of the Series Converter 596
10.6.2 Protection of the Shunt Converter 600
10.7 Application Example: UPFC at AEP's INEZ Station 600
10.7.1 Background and Planning Information at the Time of Installation 601
10.7.2 UPFC Operation Strategy 603
10.7.3 Description of the UPFC 604
10.7.4 Performance of the UPFC 607
10.7.5 Importance of Results and Possible Future Trends 613
10.8 Modeling of the UPFC Device 613
10.8.1 The Steady-State Model of UPFC 613
10.8.2 Power Flow and Active Power Balance Restrictions 616
10.8.3 Implementing the UPFC Model in the Newton–Raphson Method 618
10.8.4 The Dynamic Model of UPFC 623
References 627

CHAPTER 11 *INTERLINE POWER FLOW CONTROLLER (IPFC)* 629
Laszlo Gyugyi

11.1 Generalities 629
11.2 Basic Operating Principles and Characteristics of the IPFC 630
11.3 Generalized Interline Power Flow Controller for Multiline Systems 636
11.4 Basic Control System 638
11.5 Equipment Structural and Rating Considerations 640
11.6 Protection Considerations 642
11.7 Application Example: IPFC at NYPA's Marcy Substation 643
11.7.1 Background Information, System, and Equipment Requirements 643
11.7.2 Description of the CSC/IPFC 644
11.7.3 Importance of the NYPA Installation 645
References 649

CHAPTER 12 *SEN TRANSFORMER: A POWER REGULATING TRANSFORMER* 651
Kalyan K. Sen

12.1 Background 651
12.1.1 Traditional Power Flow Controllers 652
12.1.2 Essential Control Parameters and Their Implementations 655
12.2 The Sen Transformer Concept 656
12.2.1 Shunt-Series Configuration for ST 657
12.2.2 Principle of Operation of ST 658

12.2.3 Reduction in Number of Taps **660**
12.2.4 ST Circuit **663**
12.2.5 Operating Points of ST **665**
12.2.6 Control of ST **665**
12.2.6.1 Impedance Emulation **666**
12.2.6.2 Closed Loop Power Flow Control **667**
12.2.6.3 Open Loop Power Flow Control **668**
12.2.7 Simulation Results **669**
12.2.8 Limited Angle Operation of ST **671**
12.2.9 ST Configuration for Very High Voltage Applications **672**
12.2.10 ST Configuration for Voltage Matching and Power Flow Control **674**
12.2.11 Case-In Point for Using ST **675**
12.2.12 Optimization of Transformer and LTC Ratings **678**
12.2.13 Concluding Remarks **678**
References **679**

CHAPTER 13 *MEDIUM VOLTAGE POWER ELECTRONICS DEVICES FOR DISTRIBUTION GRIDS* **681**
Ion Etxeberria-Otadui, David Frey, Seddik Bacha, and Bertrand Raison

13.1 Introduction **681**
13.2 High Power Switching Valves: Association of Semiconductor Components **683**
13.2.1 Series Association **684**
13.2.1.1 Load Side Solutions **685**
13.2.1.2 Gate Side Solutions **687**
13.2.2 Parallel Association **689**
13.3 Topologies Used in High Power Converters **694**
13.3.1 Series H-Bridge Multilevel Inverter (Cascaded Topology) **694**
13.3.2 Structure NPC and MPC (Neutral Point Clamped—Multiple Point Clamped) **695**
13.3.3 Flying Capacitor Structure **696**
13.3.4 Some Industrial Examples **697**
13.4 Power Electronic Converter Control **697**
13.4.1 Modulation Techniques **698**
13.4.1.1 Full Wave Modulation **698**
13.4.1.2 Vertically Shifted Carrier Signals **699**
13.4.1.3 Horizontal Shifted Carrier Signals **700**
13.4.1.4 "Hybrid" Converters Control **701**
13.4.2 Inner Current Control Loop **703**
13.4.2.1 Modeling **703**
13.4.2.2 Current Control Techniques **705**
13.4.3 Inner Voltage Control Loop **709**
13.4.3.1 Modeling **710**
13.4.3.2 Voltage Control Techniques **711**
References **717**

PART III *ARTIFICIAL INTELLIGENCE TECHNIQUES*

Chen-Ching Liu and Mircea Eremia

CHAPTER 14 *ARTIFICIAL INTELLIGENCE AND COMPUTATIONAL INTELLIGENCE: A CHALLENGE FOR POWER SYSTEM ENGINEERS* 721
Chen-Ching Liu, Alexandru Stefanov, and Junho Hong

References 729

CHAPTER 15 *EXPERT SYSTEMS* 731
Mircea Eremia, Kevin Tomsovic, and Gheorghe Cârțină

15.1 Fundamental Concepts 731
15.1.1 Definitions 731
15.1.2 Expert System Characteristics 732
15.1.3 Artificial Expertise or Human Expertise? 734
15.2 Architecture of Expert Systems 735
15.2.1 Knowledge Base 735
15.2.1.1 Knowledge Representation 735
15.2.1.2 Facts Database 737
15.2.1.3 Quality of Knowledge Base 738
15.2.2 Inference Engine 738
15.2.2.1 Inference Mechanisms 739
15.2.2.2 Inference Engine Components 742
15.2.2.3 Other Modules 743
15.3 Expert Systems Application 745
15.3.1 Expert System for V–Q Control in Power Systems 745
15.3.1.1 The Structure of the Expert System 745
15.3.1.2 The Algorithmic Calculation Module 746
15.3.1.3 The Heuristic Module 748
15.3.1.4 Case Study 750
15.3.2 Other Applications of Expert Systems 751
15.3.2.1 Expert System for Distribution Networks Reconfiguration 751
15.3.2.2 Expert System Power System Restoration After Blackouts 753
References 753

CHAPTER 16 *NEURAL NETWORKS* 755
Dagmar Niebur, Ganesh Kumar Venayagamoorthy, and Ekrem Gursoy

16.1 Introduction 755
16.2 Neural Network Architectures 755
16.2.1 Feedforward Neural Networks 756
16.2.1.1 The Multi-Layer Perceptron 756
16.2.1.2 Radial Basis Function Networks 757
16.2.2 Feedback (Recurrent) Neural Networks 758
16.3 Adaptive Critic Designs 759
16.4 Independent Component Analysis 760
16.5 Learning Algorithms: The Determination of Weights 760
16.5.1 Supervised Learning Objectives 761

16.5.2 Unsupervised ICA Learning Objectives 761
16.5.2.1 Off-Line ICA Learning 762
16.5.2.2 On-Line (Adaptive) ICA Learning 762
16.6 Examples of Neural Network Applications for Power System Monitoring and Control 763
16.6.1 On-Line Estimation of Electric Power System Active Loads 763
16.6.2 Harmonic Source Identification Using Off-Line ICA 767
16.6.3 ICA-Based Harmonic Source Identification Case Study 769
16.6.4 Wind Speed Forecasting 772
16.6.5 Optimal Control of Grid Independent Photovoltaic System 773
16.6.6 Adaptive Neurocontrol of a FACTS Device: The Unified Power Flow Controller 776
16.6.7 Wide-Area Monitoring and Control 780
References 781

CHAPTER 17 *FUZZY SYSTEMS* 785
Germano Lambert-Torres, Luiz Eduardo Borges da Silva, Carlos Henrique Valerio de Moraes, and Yvo Marcelo Chiaradia Masselli

17.1 Introduction 785
17.2 Fundamental Notions 787
17.2.1 Classical Sets 787
17.2.2 Fuzzy Sets 788
17.2.2.1 Operations on Fuzzy Sets 788
17.2.2.2 Properties of Fuzzy Sets 789
17.2.3 Linguistic Values 790
17.2.4 Fuzzy Statements 793
17.2.5 Fuzzy Conditional Statements 793
17.2.6 Ordinary and Fuzzy Relations 796
17.3 Fuzzy Logic 797
17.3.1 Fuzzy Control 798
17.3.2 Fuzzy Controller 799
17.3.3 Fuzzy Inference Process 801
17.4 Fuzzy Model 801
17.4.1 Problem Formulation 802
17.4.2 The Algorithm to Solve for the Vector Θ 802
17.4.3 A Multiple Input/Output Decision System 804
17.4.4 Illustrative Example 807
17.5 An Application of Fuzzy Logic in Control System 811
17.5.1 Control Strategy 813
17.5.2 Production System 814
17.6 Final Remarks 816
Acknowledgments 817
References 817

CHAPTER 18 *DECISION TREES* 819
Constantin Bulac and Adrian Bulac

18.1 Introduction 819
18.2 Decision Trees 820

18.2.1 Decision Tree Construction 821
18.2.2 Decision Tree Pruning 824
18.2.2.1 Reduced Error Pruning 825
18.2.2.2 Pessimistic Error Pruning 826
18.2.2.3 Minimum Error Pruning 827
18.2.2.4 Critical Value Pruning 827
18.2.2.5 Cost-Complexity Pruning 828
18.2.2.6 Error-Based Pruning 829
18.3 Oblique Decision Trees 829
18.3.1 Recursive Least Squares Procedure 830
18.3.2 The Thermal Training Procedure 831
18.3.3 OC1 Algorithm 831
18.4 Applications of Decision Trees in Power Systems 833
18.5 Case Study 836
References 843

CHAPTER 19 *GENETIC ALGORITHMS* 845
Anastasios Bakirtzis and Spyros Kazarlis

19.1 Introduction to Evolutionary Computation 845
19.1.1 Taxonomy 846
19.1.2 Initial Inspiration and Basic Principles 846
19.1.3 On the Evolution Theory 848
19.1.4 DNA-Like Solution Encoding 849
19.1.5 Solution Evaluation 851
19.1.6 Genetic Information Recombination 852
19.1.7 The Circle of Evolution 853
19.1.8 Evolutionary Algorithms as Global Optimizers 853
19.1.9 Evolutionary Computation Paradigms 854
19.1.10 Application Areas 857
19.1.11 Advantages and Disadvantages 858
19.2 Genetic Algorithms 859
19.2.1 Basic GA Principles 860
19.2.2 GA Flow Diagram 862
19.2.3 Solution Encoding 863
19.2.4 Fitness Function 869
19.2.5 Parent Selection Methods 870
19.2.6 Basic Genetic Operators 873
19.2.6.1 The Crossover Operator 873
19.2.6.2 Mutation 877
19.2.7 Elitism 878
19.2.8 Other Genetic Operators 879
19.2.9 Hill-Climbing Operators 880
19.2.10 Parent Replacement Methods 883
19.2.11 Fitness Scaling 884
19.2.12 GA Control Parameters Determination 887
19.2.13 Niche and Species 888
19.2.14 Diversity Enhancement 893
19.2.15 Constrained Optimization with GAs 894

19.3 On The Optimal Location and Operation of FACTS Devices by Genetic Algorithms **897**
References **898**

CHAPTER 20 *MULTIAGENT SYSTEMS* **903**
Nan-Peng Yu and Chen-Ching Liu

20.1 Overview **903**
20.1.1 What is an Agent? What is a Multiagent System? **903**
20.1.2 Why Multiagent Systems? **904**
20.1.3 Applications of Multiagent Technology **904**
20.1.3.1 Industrial Applications **905**
20.1.3.2 Commercial Applications **906**
20.1.3.3 Medical Applications **907**
20.1.3.4 Entertainment Applications **907**
20.1.4 Challenges and Future of Multiagent Technology **908**
20.1.4.1 Design Methodologies for Software Development of Agent-Based Systems **908**
20.1.4.2 Ensure User Confidence and Trust in Agent-Based Systems **908**
20.1.4.3 Enable Agent Adaptation in Artificial System **908**
20.1.4.4 Promote Interoperability in an Open Environment **909**
20.1.4.5 Develop Semantic Infrastructure and Common Ontology for Agent Communication and Information Management **909**
20.1.4.6 Enhance Reasoning Capabilities for Agents in Open Environment **909**
20.2 Multiagent Technology Overview **909**
20.2.1 Architectures for Intelligent Agents **909**
20.2.1.1 Logic-Based Architectures **910**
20.2.1.2 Reactive Architectures **910**
20.2.1.3 Belief-Desire-Intention Architectures **911**
20.2.1.4 Layered (Hybrid) Architectures **911**
20.2.2 Multiagent Systems and Societies of Agents **912**
20.2.2.1 Communication **912**
20.2.2.2 Negotiation **913**
20.2.2.3 Coordination **913**
20.2.3 Programming Languages, Tools, and Frameworks for Multiagent Systems **914**
20.2.3.1 Programming Languages for Multiagent Systems **914**
20.2.3.2 Integrated Development Environment **915**
20.2.3.3 Frameworks for Multiagent Systems Development **915**
20.2.4 Multiagent System-Related Standards **915**
20.2.4.1 The Foundation for Intelligent Physical Agents **915**
20.2.4.2 The Object Management Group **917**
20.3 Applications of Multiagent Systems in Power Engineering **917**
20.3.1 Modeling and Simulation **917**
20.3.2 Monitoring and Diagnostics **918**
20.3.3 Restoration and Reconfiguration **919**
20.3.4 Distributed Control **919**
20.4 Electricity Markets Modeling and Simulation with Multiagent Systems **920**
20.4.1 Why Multiagent System? **921**

20.4.2 Literature on Multiagent-Based Modeling of Electricity Markets 921
20.4.3 Multiagent System Design for Electricity Market Modeling and Simulation 922
20.4.3.1 Purpose 922
20.4.3.2 MAS Structure 923
20.4.3.3 Agents 924
References 927

CHAPTER 21 *HEURISTIC OPTIMIZATION TECHNIQUES* 931
Kwang Y. Lee, Malihe M. Farsangi, Jong-Bae Park, and John G. Vlachogiannis

21.1 Introduction 931
21.2 Evolutionary Algorithms for Reactive Power Planning 932
21.2.1 Evolutionary Algorithms 932
21.2.1.1 Evolutionary Programming 932
21.2.1.2 Evolutionary Strategy 933
21.2.1.3 Genetic Algorithm 934
21.2.2 Optimal Reactive Power Planning Problem 935
21.2.2.1 Objective Functions 935
21.2.2.2 P–Q Decomposition 936
21.2.3 Case Studies 937
21.3 Genetic Algorithm for Generation Planning 943
21.3.1 Generation Expansion Planning Problem 943
21.3.2 Improved GA for the Least-Cost GEP 945
21.3.2.1 Overview of Genetic Algorithm 945
21.3.2.2 String Structure 945
21.3.2.3 Fitness Function 945
21.3.2.4 Creation of an Artificial Initial Population 946
21.3.2.5 Stochastic Crossover, Elitism, and Mutation 947
21.3.3 Case Studies 948
21.3.3.1 Test Systems Description 948
21.3.3.2 Parameters for GEP and IGA 949
21.3.3.3 Numerical Results 950
21.4 Particle Swarm Optimization for Economic Dispatch 951
21.4.1 Formulation of Economic Dispatch Problem 952
21.4.1.1 ED Problem with Smooth Cost Functions 952
21.4.2 Implementation of PSO for ED Problems 954
21.4.2.1 Overview of the PSO 954
21.4.2.2 Modified PSO for ED Problems 955
21.4.2.3 Case Studies 958
21.4.2.4 ED Problem with Smooth Cost Functions 959
21.4.2.5 ED Problem with Nonsmooth Cost Functions Considering Valve-Point Effects 959
21.5 Ant Colony System for Constrained Load Flow Problem 961
21.5.1 Formulation of Constrained Load Flow Problem 961
21.5.2 Development of Ant Colony System for the Constrained Load Flow Problem 962
21.5.3 Results 965

21.6 Immune Algorithm for Damping of Interarea Oscillation **968**
21.6.1 Study System and Problem Formulation **969**
21.6.2 Designing of Supplementary Controller **971**
21.7 Simulated Annealing and Tabu Search for Optimal Allocation of Static VAr Compensators **974**
21.7.1 Voltage Stability Analysis **974**
21.7.2 Simulated Annealing **975**
21.7.3 Tabu Search **975**
21.7.4 Study System and Optimal Allocation of SVCs **976**
21.7.4.1 A 5-Area-16-Machine System **976**
21.7.4.2 Optimal Allocation of SVCs **976**
21.8 Conclusions **980**
References **981**

CHAPTER 22 *UNSUPERVISED LEARNING AND HYBRID METHODS* **985**
Nikos Hatziargyriou and Manolis Voumvoulakis

22.1 Generalities **985**
22.2 Supervised Learning Methods **988**
22.2.1 Decision Trees **988**
22.2.2 Neuro-Fuzzy Decision Trees **990**
22.2.3 Radial Basis Function Neural Networks **992**
22.3 Unsupervised Learning Methods **996**
22.3.1 Self-Organized Maps **996**
22.4 Som Variants **1000**
22.4.1 Evolving SOM **1001**
22.4.2 Growing Hierarchical Self-Organized Map **1002**
22.4.3 Growing Neural Gas **1004**
22.4.4 Variable Local Topology-Self-Organized Map **1006**
22.5 Combined Use of Unsupervised with Supervised Learning Methods **1007**
22.6 Applications to Power Systems **1007**
22.6.1 Description of the Power System **1007**
22.6.1.1 RBFNN for DSA **1009**
22.6.1.2 Decision Trees for DSA **1010**
22.6.1.3 Decision Trees Application for Load Shedding **1011**
22.6.1.4 Genetic Algorithm Aided DTs for Load Shedding **1012**
22.6.1.5 Neuro-Fuzzy Decision Trees for DSA **1013**
22.6.1.6 SOM Application for Load Shedding **1014**
22.6.1.7 Decision Trees Aided SOM for Load Shedding **1016**
22.6.2 Preventive Security Control **1018**
22.6.2.1 Study Case System **1019**
22.6.2.2 Decision Trees for Security Constrained Economic Dispatch **1020**
22.6.3 Power System-Controlled Islanding **1022**
22.6.3.1 Application of the Method on the IEEE 30 Bus Test System **1027**
22.6.3.2 Application of the Method on the IEEE 118 Bus Test System **1028**
References **1030**

INDEX **1033**

CONTRIBUTORS

Seddik Bacha, G2Elab, Grenoble-Alpes University, Grenoble, France

Anastasios Bakirtzis, Power Systems Laboratory, School of Electrical and Computer Engineering, Aristotle University of Thessaloniki, Thessaloniki, Greece

Bostjan Blazic, Faculty of Electrical Engineering, University of Ljubljana, Ljubljana, Slovenia

Adrian Bulac, Department of Electrical Power Systems, University Politehnica of Bucharest, Bucharest, Romania

Constantin Bulac, Department of Electrical Power Systems, University Politehnica of Bucharest, Bucharest, Romania

Gheorghe Cârţină, Department of Power Systems, Gheorghe Asachi Technical University of Iaşi (retired), Iași, Romania

Abdel-Aty Edris, Santa Clara University, Santa Clara, CA, USA

Mircea Eremia, Department of Electrical Power Systems, University Politehnica of Bucharest, Bucharest, Romania

Ion Etxeberria-Otadui, IKERLAN, Arrasate-Mondragón, Spain

Malihe M. Farsangi, Department of Electrical Engineering, Shahid Bahonar University of Kerman, Kerman, Iran

David Frey, G2Elab, Grenoble-Alpes University, Grenoble, France

Aniruddha Gole, Department of Electrical and Computer Engineering, University of Manitoba, Manitoba, Canada

Ekrem Gursoy, Department of Electrical and Computer Engineering, Drexel University, Philadelphia, PA, USA

Laszlo Gyugyi, Westinghouse Electric Corporation (retired), Pittsburgh, PA, USA

Nikos D. Hatziargyriou, Division of Electric Power, School of Electrical and Computer Engineering, National Technical University of Athens, Athens, Greece

Junho Hong, Energy Systems Innovation Center, School of Electrical Engineering and Computer Science, Washington State University, Pullman, WA, USA

José Antonio Jardini, University of Sao Paulo, Sao Paulo, Brasil

Spyros Kazarlis, Department of Informatics and Communications, Technological Educational Institute of Serres, Serres, Greece

Germano Lambert-Torres, PS Solutions, Itajuba, Brazil

Kwang Y. Lee, Department of Electrical and Computer Engineering, Baylor University, Waco, TX, USA

Chen-Ching Liu, Energy Systems Innovation Center, School of Electrical Engineering and Computer Science, Washington State University, Pullman, WA, USA

Yvo Marcelo Chiaradia Masselli, INATEL – National Institute of Telecommunications, Santa Rita do Sapucai, Brazil

Rafael Mihalič, Faculty of Electrical Engineering, University of Ljubljana, Ljubljana, Slovenia

Carlos Henrique Valério de Moraes, Electrical Engineering Institute, Itajuba Federal University, Itajuba, Brazil

Dagmar Niebur, Department of Electrical and Computer Engineering, Drexel University, Philadelphia, PA, USA

Stig Nilsson, Exponent, Inc., Phoenix, AZ, USA

Jong-Bae Park, Department of Electrical Engineering, Konkuk University, Seoul, Republic of Korea

Bertrand Raison, G2Elab, Grenoble-Alpes University, Grenoble, France

Mylavarapu Ramamoorty, Koneru Lakshmaiah University, Guntur, Andhra Pradesh, India

Kalyan K. Sen, Sen Engineering Solutions, Inc., Monroeville, PA, USA

Luiz Eduardo Borges da Silva, Electrical Engineering Institute, Itajuba Federal University, Itajuba, Brazil

Alexandru Stefanov, ESB Networks, Dublin, Ireland

Guangfu Tang, C-EPRI Electrical Engineering Co. Ltd., Beijing, China

Remus Teodorescu, Department of Energy Technology, Aalborg University, Aalborg East, Denmark

Lucian Toma, Department of Electrical Power Systems, University Politehnica of Bucharest, Bucharest, Romania

Kevin Tomsovic, Department of Electrical Engineering and Computer Science, University of Tennessee, Knoxville, TN, USA

Ganesh Kumar Venayagamoorthy, Department of Electrical and Computer Engineering, Clemson University, Clemson, SC, USA

John G. Vlachogiannis, Industrial and Energy Informatics Lab, Lamia, Greece

Manolis Voumvoulakis, Division of Electric Power, School of Electrical and Computer Engineering, National Technical University of Athens, Athens, Greece

Nan-Peng Yu, Department of Electrical and Computer Engineering, Iowa State University, Ames, IA, USA

FOREWORD

THE ELECTRICITY has fascinated the humankind since ancient times, before the recorded history begun. While the first important theories about electric and magnetic phenomena were defined in the seventeenth century, the history of the modern electricity era begun at the end of the nineteenth century with the well-known "war of the currents." Engineers had to decide between using AC transmission or DC transmission. At that time, the first technology was the only solution for creating large-scale interconnected networks and allowed large amounts of power to be transferred from generators to loads.

Electrical energy has been the main ingredient for evolution of the modern civilization. Other technologies emerged as a result of the wide access to electrical energy, such as electronics, telecommunications, and computers. The technology has evolved exponentially in the last 50 years thanks also to the simulation capability. The more powerful the computers become and the faster the telecommunication infrastructure was achieved, the more flexible and reliable power grids have been developed. At the same time, the modern power electronics was developed and successfully employed in power system applications. The easy communication by Internet or other web resources has been essential in helping the engineers around the world to cooperate for the benefit of civilization.

The power electronics applications gradually broke the technological barriers faced by the AC technology. The current source converters (CSC) based high voltage direct current (HVDC) transmission systems are used for long-distance transmission, for sea crossing, or to asynchronously interconnect AC-operating power systems. The VSC–HVDC technology is about to break the technological barriers faced by the CSC–HVDC systems, that of creating interconnected HVDC networks. Developing the HVDC technology is essential for creating the already projected supergrids that will allow the human society to integrate at larger scale wind and solar power plants thereby to rely on clean energy to a greater extent.

FACTS devices are widely used as modern means of control of the AC power systems. Furthermore, under the fast-changing power flows caused by intermittent wind and solar power generation, the FACTS devices are mandatory for ensuring safe operation of the AC power systems.

The artificial intelligence techniques have been successfully employed in a large number of power systems applications, from the design stage to the real-time operation. Their application has been extended also due to the computers' capabilities, from both the hardware and software point of view.

International Council on Large Electric Systems (CIGRE), either through its working groups or together with IEEE, by contribution of great engineers, has supported the creation of reference technical guides related to both the power electronics and artificial intelligence techniques application in power systems.

All the three topics, well-denoted advanced solutions, are carefully covered in this handbook. Detailed theory of modeling the HVDC and FACTS systems, as well as implementation algorithms of the techniques, can be found in each chapter. Theoretical simulation examples and physical installations in the world are also presented, thus providing the reader a complete document.

I would like to congratulate the authors for their effort in preparing a comprehensive and thoughtful technical guide that will help the engineers, from the undergraduate level to the expert level, to understand and employ various theories of the power electronics and artificial intelligence based solutions in power systems design.

Philippe Adam
Secretary General of CIGRE

ACKNOWLEDGMENTS

FOR SOME CHAPTERS, the authors benefited from the kindness of some institutions or companies, which permitted reproducing or adapting figures, equations, or excerpts. Special thanks are thus addressed to Institute of Electrical and Electronics Engineers (IEEE), International Council on Large Electric Systems (CIGRE), as well as John Wiley & Sons, Inc. for their reproducing permission and support. The authors are also indebted to ABB, Alstom/General Electric, Siemens, and Infineon for granting the permission of reproducing pictures of equipment or excerpts describing various installations. The authors were also inspired from the publications of other companies like Mitsubishi Electric or Toshiba, to which they extend their gratitude.

Prof. Mircea Eremia is very grateful to several personalities for their support during his membership within the 14/B4 CIGRE Study Committee, among which Dr. Narain Hingorany and Dr. Mark Reynolds (Bonneville Power Administration), Prof. Williams (Bill) Long (Madison University), Prof. Dusan Povh (Siemens, Erlangen), Dr. Marcio Szechtmann (Brazil), Dr. Bjarne Andersen (Andersen Power Electronic Solutions), former chairman and/or secretary of the 14/B4 CIGRE Study Committee. Prof. Eremia extends his gratitude for the support or inspiration to Dr. Philippe Adam (Electricité de France), Dr. Ambra Sannino and Rolf Grünbaum (ABB Vasteras), Prof. Dieter Retzmann (Siemens, Erlangen), Denis Woodford (Electranix, Canada), Gilles Prud'Homme and Colin Davidson (General Electric), Dr. John Paserba (Mitsubishi Electric), and Dr. Rambabu Adapa (EPRI). Special acknowledgements are addressed to Acad. Erich Uhlmann (Sweden) and Dr. Alain Le Du (Electricite de France).

The authors of Chapter 3 would like to take this opportunity to thank Dr. Rashwan Mohamad (TransGrid Solutions, Canada) for reading and providing valuable inputs and comments.

The origins of this book reside also in the collaboration with Prof. Emeritus Jacques Trecat (Faculté Polytechnique de Mons) and the regretted Prof. Alain Germond (Ecole Polytechnique de Lausanne), which resulted in a book *Réseaux Électriques: Aspects Actuels* (2000) that contains chapters on FACTS devices and A.I. techniques. For this reason, the third part of this new book is dedicated to our beloved friend Prof. Alain Germond, who left too early the academic and scientific community.

Special thanks are addressed by the authors from University Politehnica of Bucharest to their colleagues Prof. Nouredine Hadjsaid and Prof. Yvon Besanger from INP Grenoble for the long collaboration within the Socrates/Erasmus Programme or by various PhD theses in the field of FACTS devices and A.I. techniques.

Writing a book is a complex work. The authors would like to extend their gratitude to Dr. Dragoş Petricică (GE, Romania), Dr. Daniel Radu (Schneider

Electric, Grenoble), Dr. Cristian Cristea (Next Energy, Romania), Dr. Petre Răzuși (Teletrans, Romania), Dr. Valeriu Presadă (GE, Romania), as well as Dr. Alexandru Mandiş, Andreea Neagoe-Ştefana, and Alisa Manoloiu from University Politehnica of Bucharest, for their contribution in drawing figures or editing text and equations.

The authors gratefully acknowledge the excellent collaboration with the IEEE Press and John Wiley & Sons, Inc. and address many thanks to Mary Hatcher, Brady Chin, Danielle LaCourciere and Shikha Sharma for their patience and professionalism in carrying out this printed book.

Mircea Eremia
Chen-Ching Liu
Abdel-Aty Edris

CHAPTER 1

INTRODUCTION

Mircea Eremia, Chen-Ching Liu, and Abdel-Aty Edris

POWER SYSTEM RELIABILITY is a primary concern for power system engineers in planning and operation of the power grids to ensure adequate and secure electricity service to consumers. As an electrical network, a power system should be operated in such a way that the electrical quantities, for example, bus voltages and line currents, will be maintained within an acceptable range in an operating condition. Power system security is a criterion for planning and operation of a power grid. To meet the system security standards, various control devices and tools are needed.

As policies and technologies evolve, power systems have become more complex and difficult to plan and operate. These major changes include the creation of electricity markets, large-scale integration of renewable energy sources, and increasing demand response programs on the customer side. Due to the intermittency of wind and solar generation resources, large and sudden changes in power flow may be experienced, causing the power system to be operated closer to its capability limits. Under these conditions, voltage control becomes a significant challenge for power system operators.

Major progress in technology for control, automation, protection, sensing, and communication has been achieved. New facilities are being added to replace the aging power infrastructures. Further investment in new transmission lines is important to upgrade transmission capacities to meet new requirements. Recent major events affected large parts of the interconnected power systems of Europe (the Italian blackout in September 2003, the UCTE (Union for the Co-ordination of Transmission of Electricity) event in November 2006), and the Northeast United States in August 2003. A root cause of these blackouts is the insufficient transmission capacity to serve the increasing load demand while meeting the N-1 security requirement.

Reliable and secure operation of power systems is fundamental to support the continuing development of civilization and provide the social and economic foundations. Power system engineers must be innovative in order to ensure highly reliable and cost-effective electric energy supply to the end users. A power system is expected to operate efficiently by supporting a well-designed market and achieve sustainable use of natural resources.

Power system operators need efficient solutions and tools to operate the power systems in order to meet the economic and regulatory requirements. Due to the difficulties regarding construction of new transmission lines and need for fast and

Advanced Solutions in Power Systems: HVDC, FACTS, and Artificial Intelligence, First Edition.
Edited by Mircea Eremia, Chen-Ching Liu, and Abdel-Aty Edris.

robust voltage and power flow control, the power electronic technology is a critical solution. Power electronics–based technology has shown excellent performance since its first use in direct current transmission in early 1960s and provide solutions for some limitations of the alternating current (AC) transmission systems. As technology advances, applications are also developed and deployed at the distribution system level.

Economic efficiency targets should be met from design to operation. Power system optimization is an important part of the literature. Optimal planning and operation as well as adaptation to constantly changing operating conditions can be achieved by well-designed tools for operation and decision support. Artificial intelligence (AI) techniques have been deployed in a range of applications due to the availability of powerful and versatile techniques. Application of power electronics and AI techniques help power systems to advance toward a "smart grid." Power electronic and AI techniques are among the critical tools available to modernize the power grids. As part of the vision for a smart grid, renewable energy sources and distributed generations have been integrated in large scale. Automation, protection, sensing, and other information and communication technologies have also advanced significantly.

Significant work has been done by authors to provide guidelines and techniques regarding the application of power electronics in power systems. We have benefited greatly from the prior work, including

- Adamson, C., and Hingorani, N. G., *High voltage Direct Current Power Transmission*, 1960
- Kimbark, E. W., *Direct Current Transmission*, 1971.
- Uhlmann, E., *Power Transmission by Direct Current*, 1975.
- Arrillaga, J., *High Voltage Direct Current Transmission*, 1983.
- Padiyar, K. R., *HVDC Power Transmission Systems. Technology and System Interactions*, 1990.
- Song, Y. H., and Johns, A. T. (Eds.), *Flexible AC Transmission Systems (FACTS)*, 1999.
- Hingorani, N. G., and Gyugyi, L., *Understanding FACTS. CONCEPTS and Technologies of Flexible AC Transmission Systems*, 2000.
- Mathur, R. M., and Varma, R. K., *Thyristor Based FACTS Controllers for Electrical Transmission Systems*, 2002.
- Sood, V. K., *HVDC and FACTS Controllers: Application of Static Converters in Power Systems*, 2004.
- Zhang, X. P., Rehtanz, C., and Pal, B., *Flexible AC Transmission Systems: Modelling and Control*, 2006.
- Sen, K. K., and Sen, M. L., *Introduction to FACTS Controllers. Theory, Modeling and Applications*, 2009.
- Yazdani, A., and Iravani, R., *Voltage Sourced Converters in Power Systems. Modeling, Control and Applications*, 2010.
- Jovcic, D., and Ahmed, K., *High-Voltage Direct-Current Transmission: Converters, Systems, and DC Grids*, 2015.

AI techniques were developed as complementary techniques to traditional methods that are based on rigorous mathematical foundations. AI techniques have been extensively applied to power system problems, such as genetic algorithms, artificial neural networks, expert systems, fuzzy logic, and decision trees. More recent applications are under development such as intelligent agents or particle swarm optimization. Genetic algorithms are good additions to the suite of tools including traditional optimization techniques. Expert systems can be used to support the power system operators in dispatching centers or substations in an online environment. Among the AI applications in power systems, rule- or logic-based technologies have been developed and deployed as decision support tools for distribution systems in an online environment. Artificial neural networks for load forecasting in power systems have been in practical use. Fuzzy logic is successfully applied in industrial controllers in power systems.

A significant amount of work has been done for development of AI applications in power systems, and further work is needed as the technology is continuously advancing. We acknowledge the following contributions:

- Nilsson, N.J., Learning machines, 1965.
- Zadeh, L. A., Fuzzy sets. Information and control, 1965.
- Kaufmann, A., Introduction to the theory of fuzzy sets, 1975.
- Quinlan, J. R., Introduction of decision trees, 1986.
- Barr, A., and Feigenbaum, A., Le manuel de l'intelligence artificielle, 1986
- Goldberg, D. E., Genetic algorithms in search, optimization and machine learning, 1989.
- Dillon, T. S., and Laughton, M.A., Expert systems applications in power systems, 1990.
- Zimmerman, H.J., Fuzzy set theory, 1990.
- El-Sharkawi, M., and Niebur, D. (Eds.), Artificial Neural Networks with applications to power systems, 1996.
- Tsoukalas, L. H., Uhrig, R. E., and Zadeh, L. A., Fuzzy and neural approaches in engineering, 1997.
- El-Hawary, M.E., Electric power applications of fuzzy systems, 1998.
- Jennings, N., and Wooldridge, M. (Eds.), Agent technology: Foundations, applications, and markets, 1998.
- Wehenkel, L., Automatic learning techniques in power systems, 1998.
- Lee, K. Y, and El-Sharkawi, M.A. (Eds.), Modern heuristic optimization techniques. Theory and applications to power systems, 2008.

The idea of this project was conceived as a comprehensive handbook on high voltage direct current (HVDC)/flexible alternating current transmission systems (FACTS) and AI applications for power engineering professionals and students. These subjects are already embedded in the academic curricula around the world. This is the case of the master program in electrical power systems at the University "Politehnica" of Bucharest (UPB), which includes courses on "high voltage direct current transmission" and "advanced technologies in power systems: FACTS and AI." Several international courses have been organized at UPB under the title "Advanced technologies in power systems: FACTS and AI," with participants from European countries. The support from various European programs (e.g., Erasmus, Tempus), the activities organized under Institute of Electrical and Electronics Engineers (IEEE) and International Council on Large Electric Systems / Conseil International des Grands Réseaux Électriques (CIGRE), and other opportunities have allowed the development of linkages among universities and industry from many countries, including Brazil, Canada, China, Denmark, France, Greece, Korea, India, Iran, Ireland, Romania, Slovenia, Spain and United States to carry out the project of this book. Topics related to the application of power electronics and AI techniques in power systems have been integrated in the academic curricula and extensively studied in PhD research in many universities around the world, for example, North America, South America, Europe, and Asia.

This book on "Advanced solutions in power systems: HVDC, FACTS and AI" is complementary to the book on "Electrical Power System Dynamics: Modeling, Stability and Control." The previous book was focused on providing dynamic models for the classical components of a power system, methods for stability assessment, strategies for voltage and frequency control, and analysis of power system blackouts. This book, on the other hand, presents advanced technologies and tools that are solutions to improve the performance of the power systems by enhancing the stability reserves and the transmission capacity, by improving the voltage control, by providing decision support tools for power system control, by improving the flexibility in operation and so on.

This book is organized into three parts, each dealing with one of the three main topics, that is, HVDC, FACTS, and AI. Each chapter is founded on the valuable knowledge and experience of its contributor(s).

The power electronic systems deployed in power systems include two types of installation: HVDC transmission links and FACTS devices. In both cases, two classes of converters exist, the *current source converters* (CSC) that are based on the conventional thyristors (with no intrinsic turn-off ability) and the *voltage source converters* (VSC) that are based on self-commutated devices. The advent of power electronic technology has removed several barriers in power transmission as regards the voltage level, power, and distance.

The first two parts are devoted to applications of power electronics. Each chapter is intended to guide the reader through the state-of-the art, principles of operation, modeling for steady-state and dynamic simulations, case studies, and installations in operation around the world.

The first part of the book is concerned with the theory of HVDC transmission. This part begins with a comprehensive description of the semiconductor devices and power electronic converters in Chapter 2 with a focus on their architectures and functionalities. The power electronic–based technologies are rapidly progressing as new power semiconductors are developed, reaching higher rated voltages, currents, or commutation frequencies. However, challenges remain in developing models and algorithms for both static and dynamic operation.

The next two chapters of the first part present the theory of CSC–HVDC and VSC–HVDC technologies. The CSC–HVDC technology employed for overhead transmission lines is mature today, reaching ultra high voltage levels and very high transmission capacities. However, in the absence of a commercial breaker, multi-terminal systems are limited in practice to three terminals only. CSC–HVDC links provide good power flow control under normal operating conditions, whereas severe events occurring in the AC system may affect it significantly. It is worth noticing that the CSC–HVDC links have reached the distance of 2400 km and transmission capacity of 8000 MW; these goals are difficult to achieve with AC lines. There are high expectations for the VSC–HVDC technology, which is fast developing due to the ability to eliminate problems associated with CSC–HVDC. Relative to the CSC–HVDC links, the VSC–HVDC technology is expensive as it is more suited for cable lines. In order to allow integration of renewable energy sources, mainly offshore, and face the unexpected critical events that may damage the power system, the vision for developing the transmission grids, *supergrid/highway*, is to adopt a hybrid AC–DC power system. The reader may find an extended presentation of both types of HVDC transmission links and be able to understand how power system performances can be improved.

Another class of power electronic applications, presented in the second part of the book, is FACTS devices. They can be series, shunt, or series–shunt connected and are designed to control various parameters of the AC power system in a wide range of operating conditions. For this reason, FACTS devices are also called controllers. These sophisticated controls are the modern version of the breaker-switched connected capacitors and reactors and conventional (mechanical) tap-changing transformers with a much faster response. Similar to the HVDC links, FACTS devices may be categorized into two classes depending on the type of converters.

The first class of controllers includes the static VAr controller (SVC), thyristor-controlled series capacitor (TCSC), and thyristor-controlled phase shifter, which employ conventional thyristors with no intrinsic turn-off capability. Depending on the connection type, these controllers may act on one of the three parameters that influence the power flow, that is, voltage (SVC), line impedance (TCSC), and phase angle (phase-shifter). While TCSC and the phase-shifter are designed to control the power flow on transmission lines, the main purpose of an SVC is to control the bus voltage.

The TCSC is inserted as variable capacitive impedance in series with the line inductive impedance at a distance calculated to achieve maximum efficiency. The device acts by developing a compensating voltage based on line voltage and thus affects the line current. The SVC is the most important FACTS device used in power systems, with at least one thousand installations in operation in the world in various configurations. An SVC is inserted into the electrical network as a variable shunt admittance and thus it acts by exchanging reactive power, which depends on the bus voltage. When operating in the normal domain, the thyristor-controlled devices demonstrate an outstanding performance. However, when large disturbances occur which significantly affect the line current or voltage, the SVC and TCSC are forced to operate outside their normal control range and are seen as fixed elements. Under these circumstances, the thyristor-controlled devices are no longer efficient and the power system cannot count on them. It is important for the reader to understand not only the operating principles of a specific device but also its importance and performance limits when the device is integrated in the power system as explained in Chapters from 5 to 7. This is important for engineers in order to select the best solutions to strengthen the power system.

The drawbacks of thyristor-controlled devices can be overcome by the second type of FACTS devices, based on self-commutated voltage-source switching converters, including the static synchronous compensator (STATCOM), static synchronous series compensator (SSSC), unified power flow controller (UPFC), interline power flow controller (ILPF), and the convertible static compensator (CSC). The VSC-based FACTS devices are similar to the synchronous machine as they are able to exchange active and reactive power while providing an almost instantaneous speed of response and control characteristics. A DC capacitor is used as the voltage source for the VSC, which may be able to generate or absorb reactive power with the AC system as the VSC voltage is greater or smaller than the voltage at the AC bus. The VSC technology has advanced considerably to reduce the active power losses in the converters and handle greater powers at higher voltages.

The VSC-based FACTS devices can perform significantly better than the thyristor-based FACTS devices, enhancing system stability, voltage control, and power flow control. However, due to the high cost of the VSC converters, a large number of STATCOM units have been implemented, whereas the UPFCs are primarily pilot projects. So far, the SSSC can be found only in the UPFC structure. A guide through the architecture, operation principles, modeling, and example installations is provided in Chapters 8–11.

A special class of flexible devices is the Sen transformer, presented in Chapter 12, that can be used to perform independent power flow control similar to FACTS devices.

Initially FACTS devices were intended to be installed in the transmission system to achieve various objectives. As the power electronics technology advances, FACTS applications have also been deployed in distribution networks, called D-FACTS, which is discussed in Chapter 13. They are designed for power quality improvement to mitigate voltage dips, flickers, and phase unbalance.

The third part of the book, consisting of Chapters 17–22, is devoted to applications of AI and computational intelligence (CI) techniques to power systems. The chapters provide a comprehensive overview of the AI and CI techniques that help realize the vision of a smart grid. These techniques include expert systems, artificial neural networks, fuzzy systems, decision trees, genetic algorithms, multiagent systems, heuristic optimization, and unsupervised learning.

Although AI and CI techniques emerged in mid-1950s as a computer science field, power engineers have been conducting research and development in practical applications to power systems since early 1980s. At that time, computers became more accessible for researchers around the world and the computing power has increased significantly, enabling conventional mathematical approaches to be utilized, such as linear programming, nonlinear programming, dynamic programming, and Pontryagin maximum principle, etc. However, these mathematical approaches have shortcomings in their applications as the power systems are becoming increasingly complex, large-scale, nonlinear, and stochastic. AI and CI techniques are complementary to the more rigorous mathematical techniques in many fields, such as operations research, control theory, and numerical analysis.

Electric power systems are constantly adapted to meet the technical and economic objectives. With the advent of computer and communication technologies, power systems are provided with more intelligence at all levels of operation, control, forecasting, and scheduling activities. The range of solutions to the increasingly complex problems in power system engineering is expanded to incorporate logic reasoning, heuristic search, perception, and the abilities to handle uncertainties. AI and CI open new opportunities for developing the intelligence of the future smart grid.

AI and CI techniques are applicable in a wide range of power system problems, including stability assessment/enhancement, power system control, security assessment, load forecasting, reactive power planning and control, state estimation, fault diagnosis, and behavior classification.

Successful applications of AI and CI techniques are also found in problems that involve HVDC and FACTS devices. Genetic algorithms are extensively used in optimization problems, such as placement of shunt or series FACTS devices, and voltage–VAr planning involving FACTS compensators. Decision threes and artificial neural networks are applied with good results for stability studies, in which increased attention is paid to power electronic applications. The number of AI and CI applications in power systems that include power electronic–based devices is increasing at the same time the computation power enables complex simulations.

The book is intended to provide insights into promising technologies and tools for application in power system operation and planning in such a way that the gap between theory and application can be bridged. This book is suitable for readers working in the fields of power systems, power electronics, computer applications, and industry applications. The content is addressed to students, faculty, researchers, engineers, consultants, utilities, and others.

PART I

HVDC TRANSMISSION

Advanced Solutions in Power Systems: HVDC, FACTS, and Artificial Intelligence, First Edition.
Edited by Mircea Eremia, Chen-Ching Liu, and Abdel-Aty Edris.

2. Power Semiconductor Devices for HVDC and FACTS Systems
Remus Teodorescu and Mircea Eremia

3. CSC–HVDC Transmission
Mircea Eremia and Constantin Bulac

4. VSC–HVDC Transmission
Mircea Eremia, José Antonio Jardini, Guangfu Tang, and Lucian Toma

CHAPTER 2

POWER SEMICONDUCTOR DEVICES FOR HVDC AND FACTS SYSTEMS

Remus Teodorescu and Mircea Eremia

THE FIRST STEPS TOWARD the modern power electronics as it is known today were made during 1940–1945, with the appearance of ignitron and excitron, along with silicium and germanium devices development [1]. The idea for a thyristor (THY) was initiated by W.B. Shockley in 1950, by his bipolar transistor with a *p-n* hook-collector. The operation mechanism of the THY was analyzed by Ebers in 1952, and then the switching mechanism was further investigated in 1956 by J.L. Moll. The first THY, known as silicon-controlled rectifier (SCR), was produced in 1957 by General Electric Company. The power semiconductor devices subsequently developed, that is, gate turn-off thyristor (GTO) in 1960 and the insulated gate controlled thyristor (IGCT) in 1968, were provided with turn-off capability, which is an essential feature for power control. The first transistor was developed at Bell Laboratories in 1948, and even nowadays it is a power device of great importance for very high power levels. The metal oxide semiconductor technology progress leaded to the appearance of metal oxide semiconductor field effect transistor (MOSFET) in 1975. In the early 1980s, a successful device with superior characteristic, which is very widely used in power electronics applications, was developed. This device, called isolated gate bipolar transistor (IGBT), cumulates the advantages of both bipolar transistor and MOSFET.

Power semiconductor devices (or switches) are the key components in power converters for High Voltage Direct Current Transmission (HVDC) and Flexible Alternating Current Transmission Systems (FACTS). They have quite complex semiconductor structure and can behave as unipolar and bipolar conducting or blocking devices in response to the gate signal and thus the flow of current in the circuit can be controlled. Thus, electrical energy flow through HVDC and FACTS systems can be accurately controlled. The diode is the only semiconductor device that does not have a gate signal and the conduction or blocking state is solely determined by the polarity of the voltage across.

Advanced Solutions in Power Systems: HVDC, FACTS, and Artificial Intelligence, First Edition.
Edited by Mircea Eremia, Chen-Ching Liu, and Abdel-Aty Edris.

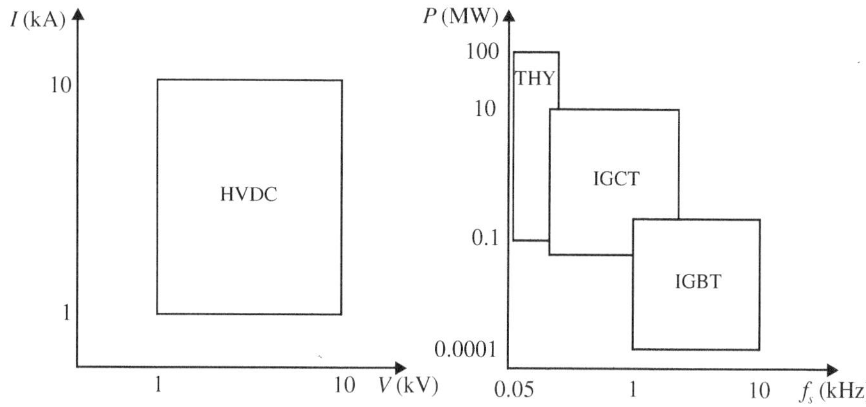

Figure 2.1 Si-based power devices evolution and ratings.

Since 1957, when Si-based semiconductors have been displaced by the Ge-based ones, two main device technologies have been developed [2]:

- bipolar—resulting in THY family;
- unipolar metal oxide semiconductor (MOS)—resulting in IGBT family.

Today, the most representative power devices used for HVDC are THY (also called phase-controlled thyristor [PCT] or silicon controlled rectifier [SCR]), IGCT, and IGBT. Figure 2.1 shows the current and power rating of these devices especially for HVDC applications. Due to the large rating of HVDC transmission applications, the MOS devices are not practical and therefore will not be covered further.

Since 1960, when the development of THY was started, silicon has been the material of choice for further development and ever since high efforts have been undertaken for reducing the conduction losses and increasing the voltage blocking capability, and switching speed.

Due to the high power handling capability required in HVDC, the power devices are typically packed in ceramic pressed-packs with double-sided cooling and power modules with plastic case and single-side cooling.

Currently, in excess of 8000 MW of HVDC transmission is installed worldwide with a very promising outlook for the future when intense urbanization and integrating of more renewable energy are calling for increased need of transmission. The highest installed voltage level today is ±800 kV (so-called Ultra High Voltage Direct Current [UHVDC]) with several installations recently commissioned in China. The modern THYs are rated 8.5 kV/4 kA and build in ceramic press-pack as shown in Figure 2.2. The main HVDC manufacturers are ABB, Siemens, and Alstom.

2.1 POWER SEMICONDUCTOR OVERVIEW

Power semiconductor devices can be classified depending on the number of degrees of freedom of controllability as follows.

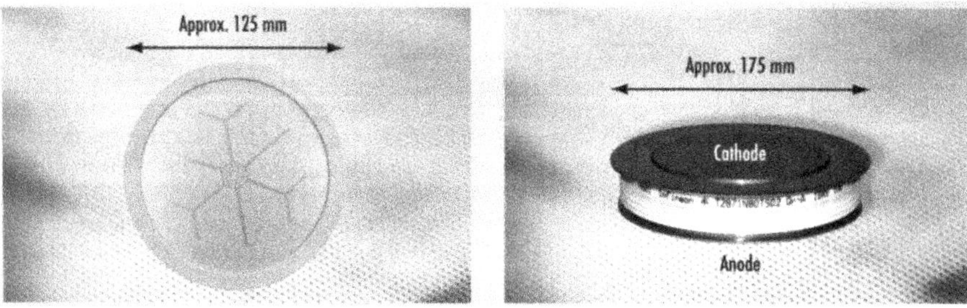

(a) (b)

Figure 2.2 Modern 8.5 kV 125 mm thyristor: (a) silicon wafer and (b) complete capsule. Reproduced with permission of Infineon Technologies AG.

2.1.1 Not-Controllable Power Semiconductor Devices

The only power semiconductor switch that does not need control is the power diode. Most of the power diodes consist of two silicon-based semiconductor layers forming a *p-n* junction. Depending on the polarity of the voltage across the terminals, the *p-n* junction can allow current conduction (forward operation for positive anode–cathode voltage) or current blocking. In forward operation, there is a voltage drop across the device in the range of few volts that adds to the conduction losses. The most common applications of power diodes are the rectification or the clamping in power conversion and overvoltage protection. In voltage-sourced converter (VSC)–HVDC applications, beside rectification, the power diodes are used as free-wheeling devices for the IGBTs, which are unidirectional power devices.

Snubber diodes are mainly used in applications with GTOs, or with IGBTs in series-connected stacks. A high demand of these diodes can be noticed because converters with GTOs are still built in high quantity.

2.1.2 Semicontrollable Power Semiconductor Devices

The most important switch in this category is the THY, also called SCR. Typically, THY structure is a four-layer of alternating *p*- and *n*-type material, respectively, three *p-n* junctions, as shown in Figure 2.3a. As a bipolar device, THY operates with charge carriers of both polarities (holes and electrons) and thus high current densities are possible (in the range of kA) The anode (A) and the cathode (K) terminals of the THY handle high potentials, while the gate (G) terminal is used for control.

THY can be turned on by applying a short (μs) positive current pulse (in the range of A) at the gate when the A–K voltage is positive and remains in the on-state (latched) until the next current zero crossing takes place when the THY resets and turns-off. Therefore, only one switching per half-cycle is possible, which limits its controllability.

The basic THY structure has been modified in order to deal with high ratings of HVDC. In order to reduce the power rating of gate driver, an amplifier has been realized using a low power "pilot-THY" as shown in Figure 2.3b. This allows the user to have to turn-on only the pilot THY, which in turn will turn-on the main THY. This

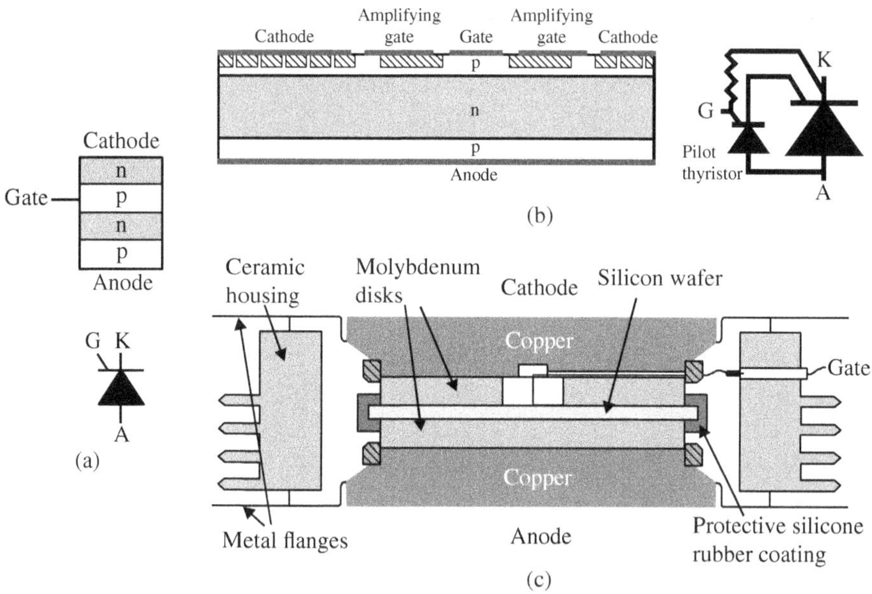

Figure 2.3 (a) Four-layer *p-n-p-n* structure; (b) thyristor with amplifying gate and cathode shorts; and (c) cross-section through an encapsulated thyristor. *Source*: ALSTOM Grid 2011 [3]. Adapted with permission of GE.

amplified gate driver allows large THYs to be turned on with small current pulses reducing the complexity of the gate drivers.

Another modification is related to the sensitivity to d*v*/d*t*. As voltage levels in the range of hundreds of kV are used in HVDC, great care has to be taken on the parasitic turn-on of THY due to induced currents in the gate circuit due to the high d*v*/d*t* rates. This problem can be alleviated by changing the structure of THY in the cathode area by creating internal "resistors" from gate to cathode as shown in Figure 2.3b. These resistors will quickly sink large currents induced by d*v*/d*t* and thus avoid this current to flow in the gate.

Finally, in order to increase the reliability, ceramic press-pack cases are used as shown in Figure 2.3c. The Si device is integrated in a "sandwich" structure where molybdenum (Mo) layers are included between the Si device and the copper electrodes. Normally, the press-pack THYs are stacked on top of each other in order to deal with high voltage (HV) levels. In order to reduce contact resistance, a controlled pressure is applied at the ends of the stacks using screws. Mo is highly conductive material, which from a mechanical point of view is very "soft," thus allowing a good pressure distribution at the surface of the Si device, which is critical for reliability.

This package is short-circuit proof, that is, during a short circuit, the device will safely fail into short-circuit condition, which is very important feature considering the fact that in HVDC many THYs are series connected and fault-redundant operation is required. Moreover, this package is explosion-proof, that is, in the case of short circuit, failure will not explode due to the high mechanical robustness of the package.

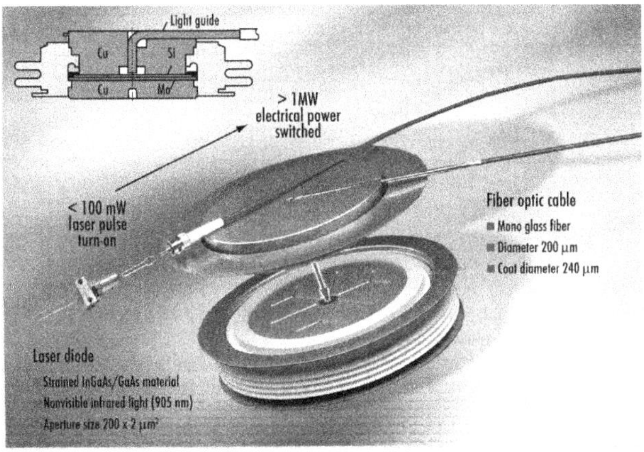

Figure 2.4 Infineon light-triggered thyristor. Reproduced with permission of Infineon Technologies AG.

The THY can be triggered by electrical gate signals (electrical-triggered thyristor [ETT]) or by light signals (light-triggered thyristor [LTT]).

In ETT, the gate current is provided by a current source grounded to the cathode of the THY, which is floating at high levels during operation. Therefore, supplying this current source is very challenging due to the high isolation demand. Additionally, the high count of components in the drive circuit can challenge the system reliability.

LTT is an improved solution using optical fibers to transmit the gate signal in the form of infrared light pulse directly to a light-sensitive region of the silicon around the gate (as shown in Figure 2.4).

The light-sensitive area has the ability to behave in the same way as the gate signal in the ETT, but more sensitive amplification is required. In LTT, the optical fiber, which easily can exhibit isolation levels of hundreds of kV, requires power supply only on the sender side usually grounded at the controller ground and thus simplifying the circuit. The LTT performance of noise rejection is superior to the ETT, but the lack of local protection functions at gate driver level, higher cost, and reduced number of manufacturers have prevented LTT from being largely used in HVDC. Recently, dv/dt protection has been demonstrated in LTT, but forward recovery protection is still a challenging issue.

With all these improvements, THYs have become the real "horse power" of line-commutated HVDC with excellent track in terms of reliability and robustness.

In HVDC, due to the HV level, typically hundreds of THYs are series connected in order to increase the blocking voltage in so-called THY valves. The term originates from the first mercury-arc vacuum tubes technology used in the 1950s for the first HVDC systems. The THY valves behave in essence in the same way as a very large THY but are complex circuits including cooling systems and ancillary circuits (gate drivers, di/dt snubbers, etc.). They are used practically like the building blocks of HVDC.

In Figure 2.5, the different stages of HVDC valve are shown.

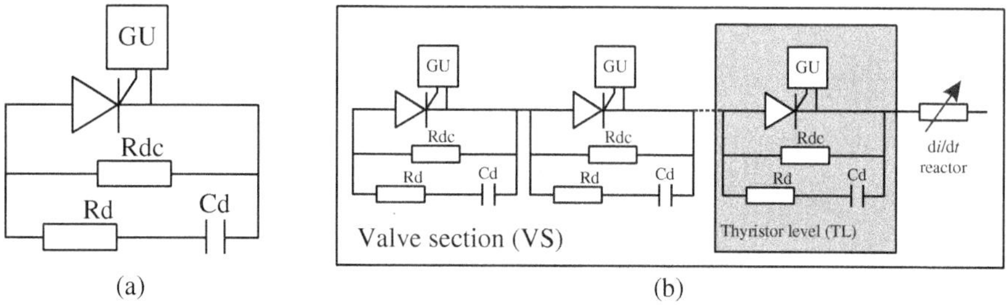

Figure 2.5 (a) Basic (minimal) circuit of one thyristor level, with electrically triggered thyristors and (b) thyristor valve section with lumped d*i*/d*t* reactors. *Source*: ALSTOM Grid 2011 [3]. Adapted with permission of GE.

First, the THY levels consist of THY with gate unit (GU), RC damping circuit for the limitation of the overvoltage during negative recovery (turn-off), and a voltage drop equalizer resistor (DC grading resistor). Second, lumped d*i*/d*t* reactors are added to the series-connected stack to form a valve section. Finally, the valve sections are used like building blocks to form the valve.

In order to ensure fault-tolerant operation, redundant THY levels are added to the valve section that can be activated when a THY is failing in a short.

An H400 valve module that contains 12 THY levels of 8.5 kV THYs is shown in Figure 2.6 [3]. Each valve module consists of two series-connected valve sections.

The THYs are still the devices for applications with the highest voltage and power levels. They are part of the mostly used FACTS devices up to the biggest HVDC transmissions with a voltage level above ±800 kV and power above 8000 MW. THYs are used as switches for capacities or inductances, in converters for reactive power compensators (static VAr compensator [SVC] and thyristor-controlled series-capacitor [TCSC]).

Bidirectional control thyristor (BCT) is built by the integration at wafer level of two antiparallel-connected THYs with two independent gates (Figure 2.7).

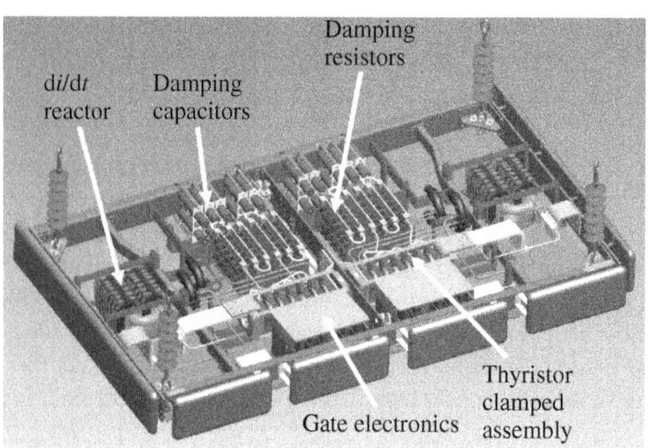

Figure 2.6 An H400 valve module. Reproduced with permission of GE from the book “HVDC – connecting to the future” [3].

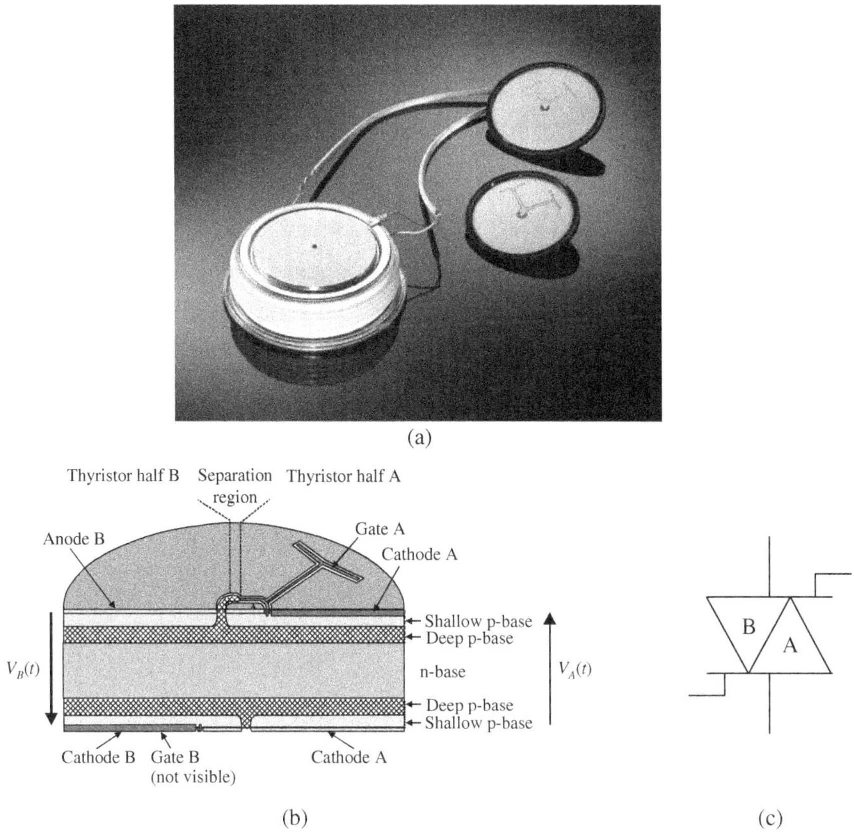

Figure 2.7 (a) Photograph of BCT element and wafer; (b) schematic cross-section of a BCT wafer showing A and B thyristor-halves and defining the two forward voltage directions $V_A(t)$ and $V_B(t)$; and (c) circuit diagram. *Source*: Thomas et al. 1999 [4]. Reproduced with permission of ABB Power Systems, Sweden.

The basic product philosophy is the same as for the PCTs. Each THY-half performs like the corresponding full-wafer THY in respect to its static and dynamic properties.

The BCT wafer has anode and cathode regions on each face (Figure 2.7b). The A and B THYs are identified on the wafer by letters A and B on the central gate metallization.

The advantage of such a device is lower footprint of the device resulting in more compact valve design for FACTS applications such as SVC and TCSC. BCTs are available on the market for blocking voltages up to 6.5 kV [4, 5].

2.1.3 Fully Controllable Power Semiconductor Devices

Fully controllable power devices allow both turn-on and turn-off process to be realized by gate signal command. A short review of the most widely used in HVDC fully controllable power devices is given in the following.

2.1.3.1 Gate Turn-Off Thyristor

The GTO has been developed in the late 1970s as an "improved" THY with turn-off controllability. The GTO has a similar structure as THY and it turns on with a short positive current pulse. For turn "off," it requires a large negative current pulse (in the range of hundreds of A). GTOs suffer from long turn-off times, which limit their applications to only 1 kHz switching frequency. Additionally, external snubber circuits are required to limit the d*i*/d*t* and commutation overvoltage (d*v*/d*t*). Nowadays, in more demanding applications, the GTOs are replaced by IGCTs that reduce the complexity of driving to a minimum.

- *Emitter turn-off (ETO) THY* is another new derivative of the GTO, focusing on drastically reduced gate drive requirements and, also, increased switching speed: optical control interface; turn-off time delay reduced from about 22 μs to about 1 μs; forward voltage drop (2 V); and increased switching speed and capable for > 1 kHz operation [5].
- *Super-GTO (S-GTO)* is also a GTO-type device with dramatically finer cell structure (160,000 vs. 50 cell/cm^2) than the conventional GTO. This is expected to result in considerable improvements over the conventional GTO in: turn-off time delay reduced from about 22 μs to about 1 μs; switching speed (3 times); current density and turn-off current density (2 and 10 times); and forward voltage drop (0.7 times) [5].

2.1.3.2 Integrated Gate-Commutated Thyristor

The IGCT is a press-pack device. It is pressed with a relatively high force onto heatsinks that also serve as electrical contacts to the power terminals. The IGCT is an improved version of GTO with reduced forward voltage and shorter commutation times. Additionally, as the name suggests, IGCTs have integrated gate units that ensure better control of the injection/extraction of the gate currents and more uniform current distribution inside the structure. They are capable of withstanding higher d*v*/d*t* with lower snubber requirements.

IGCT offers significant improvements over the GTO in the important areas of [4]: turn-off time delay reduced to <5 μs, reduced forward voltage drop, increased peak turn-off current, increased switching speed (~500 Hz), and lower conduction losses. The thermal losses are the only limitation regarding the upper switching frequency. Short on-off pulse bursts with switching frequencies of up to 40 kHz can be used. d*i*/d*t* protection is required during turn-on, which can be provided by an inductor, but the turn-off protection is not absolutely necessary, as in the GTO case.

Today, IGCTs are used in high power (MW) drive, FACTS, and traction applications but are not a preferred solution for VSC–HVDC, which is dominated by IGBTs.

2.1.3.3 Isolated Gate Bipolar Transistor

The IGBT has been developed in the early 1980s as an attempt to combine the benefits of the MOS high impedance low power gate input with the power handling capacity of normal bipolar transistors and THYs. The MOS transistors allow a high impedance control of the current flow through the device requiring extremely small amounts of power supplied to the control gate. The ability to withstand HVs and currents is

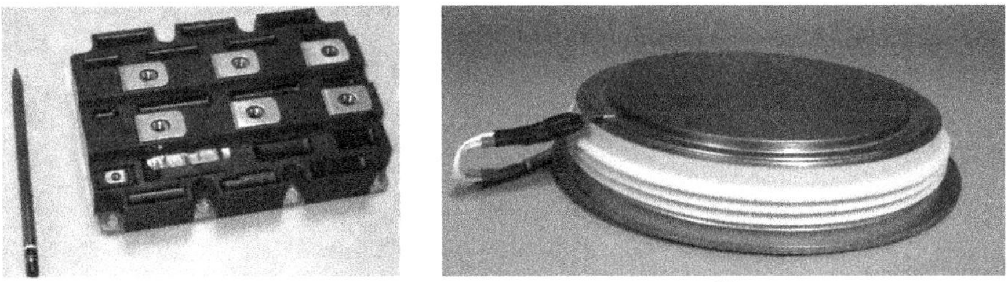

(a) (b)

Figure 2.8 High voltage IGBT: (a) power module and (b) press-pack.

provided by the vertical part of the device, comprising a bipolar transistor structure. In addition, the vertical transistor effects are crucial to enhancing the conductivity of the semiconductor material, and hence, to reducing excess voltage drop over the device in the conducting stage. The turn-on process requires a positive voltage on the gate, and turn-off can be achieved with either null or negative gate voltage. This can be achieved with a very simple drive circuit, typically in the form of input circuit (IC). IGBTs can be easily paralleled in high power low cost power modules and are optimized for either low forward voltage or high switching frequency, depending on the application. The conduction and switching losses have been reduced in the last decades through a number of technological generations (soft punch-through and field-stop).

Since the last 20 years, IGBTs have dominated the medium and high power applications such as drives, wind and high power PV, and traction. Due to the mass production and robotized manufacturing of power modules, IGBT is the most cost-effective device technology today. IGBTs can achieve 6.5 kV blocking voltage and in excess of 2 kA and can be found in both power module package and as a press-pack device, where IGBT chips are typically paralleled with free-wheeling diode (FWD) chips to achieve current bidirectional flow (Figure 2.8). The reversed voltage protection, in conjunction with a path provision for the current in opposite direction, is ensured by the antiparallel FWD integrated in the same package. Similar to IGBTs, the FWD normally consists of a number of chips in parallel.

Except for forward voltage drop and corresponding losses, the capabilities of present IGBTs are generally comparable to those of advanced GTO-type devices, and superior to most of them in the area of gate drive requirements. Moreover, IGBTs have a major, inherent advantage over the GTO-type devices in that they have a transistor-like characteristic of not latching into a conducting state and thus, their interval turn-off process is fully controllable from the gate.

The low voltage (LV) ranges of IGBT that can reach 1700 V/2 kA are packed in practical power modules and are used in the majority of grid connection applications (400 or 690 V for wind turbines). The HV ranges include 2.5, 3.3, 4.5, and 6.5 kV at currents in excess of 2 kA and are packed in both plastic power modules and ceramic press-packs for demanding applications. Today, VSC–HVDC sector is dominated by IGBTs, where both LV 1700 V IGBT power modules and HV 6.5 kV IGBTs are used.

Some manufacturers have made *injection-enhanced gate transistor* (*IEGT*) available. It represents an improved version of the IGBT for medium voltage applications. By optimizing the gate structure, it is claimed that the IEGT achieves a reduction in both on-state voltage and switching loss. A 42-chip 4.5 and 5.5 kV press-pack IEGT was reported. The IEGT is manufactured in both press-pack and single-sided modular version [6].

2.1.4 Power Semiconductor Parameters[1]

Power semiconductors are the most critical elements influencing the cost, reliability, and performance of converter-based transmission controllers. We consider the key parameters of power semiconductors and their main effects on converter design [7]:

2.1.4.1 Steady-State Parameters

- *Blocking voltage and R.M.S/average current* basically determine the number of devices the converter will need for a given MVA rating. Consequently, the number of devices largely determines the cost because the number of all converter auxiliary components (heat sink, gating circuit, snubber, communication fiber optics, and mechanical hardware) and assembly labor change in proportion with the number of power semiconductors.
- *Forward voltage drop* is an important parameter that determines the operating losses and cooling requirements, and influences the physical size of the device as well as that of the whole equipment.
- *Surge current* capability may be important to implement low cost protection schemes, particularly for converters operated in series with the transmission line.

2.1.4.2 Switching Characteristics

- *Peak current turn-off* is the maximum *instantaneous* current that the device can turn-off without failure. This parameter largely determines the derating required to accommodate harmonic ripple current and the safety margins needed for overcurrent protection; thereby, it is a critical factor for the high utilization of the converter MVA rating.
- *Turn-off time delay* (t_{off}) is the maximum delay time from the application of the turn-off signal to the completion of turn-off process. This parameter is crucial for reliable overcurrent protection without large safety margins and, thereby, another significant factor to achieve high converter utilization and low equipment cost.
- *Switching speed and losses* are the maximum time required for the device to transit from "on" state to "off" state, and vice versa, and the corresponding internal losses in the device. These determine the applicable range of switching frequency and the corresponding operating losses and cooling requirements. The capability to operate the devices at relatively high switching frequencies is important for the implementation of pulse-width-modulation techniques for output waveform synthesis that requires small output filters, and thereby allows

[1] Material reprinted with permission from CIGRE "*Static Synchronous Series Compensator - SSSC*, CIGRE Brochure No. 371, SC B4, WG B4-40, Edris, A.A. (convener) et al.", © Copyright 2009.

the high utilization of converter MVA. It is also important for providing relatively wide operating frequency band for the converter necessary to sustain *reliable* operation, with rigorous current limitation, under abnormal operating condition causing transformer saturation.

In Table 2.1, the main parameters of the power semiconductor devices used in HVDC and FACTS are shown.

2.1.5 Future Power Semiconductor Devices

Reverse-Conduction power devices. Reverse-conducting switch is realized with a unidirectional switch that can be either unipolar (asymmetrical) or bipolar (symmetrical), connected with an antiparallel diode. Thus, this switch can be regarded as a unipolar switch whose reverse breakdown voltage is approximately equal to the forward voltage drop of a diode, which typically is a few volts [8].

HVDC is a space demanding application, especially for offshore applications, and the footprint of the converters can be reduced by using the concept of hybrid power devices. With this concept, more devices can be hybrid-integrated on the same semiconductor structure. An example is the reverse conducting IGBT developed by ABB, where the FWD and the IGBT are integrated into a so-called BI-mode IGBT in order to obtain higher power capability for same footprint of power module.

The VSC requires reverse-conducting switches.

Silicon Carbide (SiC) power devices. SiC has been proved to be a superior wide-bandgap semiconductor material than Si as it can achieve higher (around 10 times) blocking voltage and significant switching losses. All types of devices, including THY, IGBT, GTO, and MOSFET, have been demonstrated by Cree Inc. [9]. The cost is still prohibitive, but it is expected to come down within next few years. Today, 1200 and 1700 V SiC MOSFET IGBT modules with current capabilities in excess of 300 A have started to displace conventional Si IGBT in applications like traction, solar inverters, and special drives. Ten kilovolts SiC MOSFET are also available for medium voltage applications. For HVDC applications, 20 kV SiC-GTOs have been demonstrated [9] showing the potential for significant footprint reduction.

2.2 CONVERTER TYPES

Depending on *the type of power device* used, the converters can be categorized as:

Line-commutated converter (LCC/CSC) (or naturally commutated). This type of converter is build using semicontrollable power devices like THY. In this case, the commutation is dictated by the polarity change of the grid voltage. It is typically used in the full-bridge topology for the LCC–HVDC applications.

Forced-commutated converter (or self-commutated). This type of converter uses fully controllable power devices such as IGCTs and IGBTs and can control both turn-on and turn-off commutation. It is largely used as a

TABLE 2.1 Characteristics of power semiconductor devices for HVDC and FACTS

Semiconductors type	Symbol	Characteristics: Max. blocking voltage (kV)	Max. direct current (kA)	Typical power rating	Typical switching frequency (Hz)	t_{off} (μs)	Main manufacturers	Comments
Diode	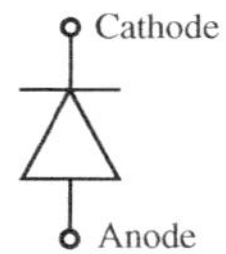	4.8 kV	2.4 kA	MW			Semikron, Infineon, IXYS Westcode, ABB, Toshiba, and Powerx	Packaged in either power module or press-pack
THY	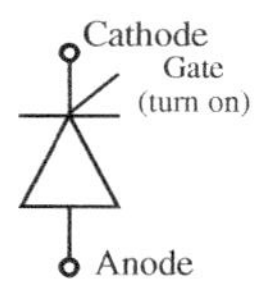	8.5 kV	4.5 kA	kW to GW	50/60 Hz	50	Semikron, Infineon, IXYS Westcode, ABB, and Powerx	Packaged in press-pack
GTO	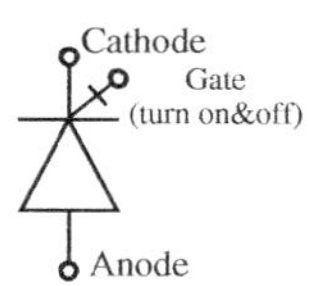	4.5 kV	4 kA	MW	< 500 Hz	10	IXYS Westcode and ABB	Packaged in press-pack
IGCT	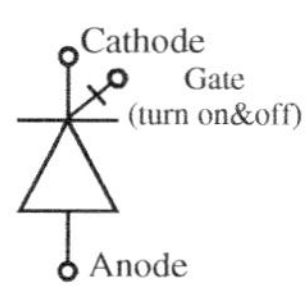	4.5 kV	1.5–4 kA	MW	< 1 kHz	5	ABB and Mitsubishi	Packaged in press-pack and integrated driver unit
IGBT	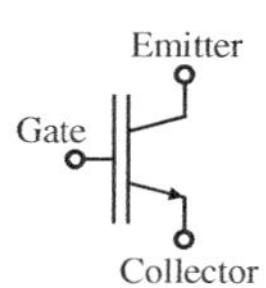	6.5 kV	2.4 kA	MW	< 1 kHz	2	Semikron, Mitsubishi, Infineon, ABB, Toshiba, Powerex, and Westcode	Packaged in either power module or press-pack

full-bridge topology in VSC–HVDC using pulse width modulation (PWM) technique.

Depending on the *type of the input source*, two types of converters exist:

Current-sourced converter (CSC/LCC) uses a constant DC current source implemented by a large inductance at the input. A typical example is the six-pulse, THY-based full bridge largely used in LCC–HVDC. In order to change the power flow direction, the converter is controlled in such a way that the polarity of the voltage at the output is reversed. HVDC systems based on CSC are typically called LCC–HVDC as they are commutated by the line.

VSC uses a constant DC voltage source implemented by a large capacitor at the input. A typical example is the PWM controlled full bridge topology largely used in VSC–HVDC and FACTS applications. This converter can reverse the voltage polarity by different switching schemes, and the flow of current is bidirectional.

2.3 HVDC EVOLUTION

HVDC power transmission technology evolution is directly dependent on the progress of semiconductor power devices. The technological advancement of THY and IGBT over the last 30 years has accelerated the growth of LCC/CSC–HVDC and VSC–HVDC systems, respectively.

HVDC transmission is advantageous compared with conventional AC transmission due to the following reasons:

- no synchronization requirements for the two AC areas;
- no need for reactive power (VAr) compensation along the line;
- lower cable cross-section requirement;
- lines can be easily buried underground saving the environment;
- cheaper for distances in excess of 500 km (OHL) or 40 km (underground cable);
- converter stations can provide ancillary services to the grid-like voltage regulation, damping oscillations, and synthetic inertia (only for VSC–HVDC).

Due to these advantages, HVDC is preferred in the following areas of application:

- long lines;
- cables crossing bodies of water wider than 30 km;
- interconnecting AC systems having different frequencies or where asynchronous operation is desired;
- congested urban areas or elsewhere where it is difficult to acquire right of way for overhead lines and where lengths involved make AC cables impracticable;
- DC grids (multiterminal DC).

HVDC systems are usually configured as monopolar or bipolar.

- In monopole configuration, one of the terminals of the rectifier station is grounded and the other terminal carrying either positive or negative potential is connected to the transmission line (overhead or cable).
- In bipolar configuration, two ground-isolated lines are used, leading to higher cost than monopole, but exhibiting advantages such as better fault management and redundancy due to double power injection.

2.3.1 Line-Commutated HVDC Converters (LCC/CSC–HVDC)

The first commercial HVDC link using conventional LCC/CSC technology was a 96 km 20 MW 100 kV submarine cable installed in 1954 between the island Gotland and the Swedish mainland. This system and all the other HVDCs commissioned until mid-1970s were based on mercury arc valves. A significant technical advancement came with the introduction of THYs (1972), although they only support the LCC concept. Until today, over 200 GW of LCC–HVDC has been installed worldwide. Power systems interconnection, along with bulk power transmission over long distance, was the purpose for installing the technology in conventional approach. Examples of long distance HVDC links are (see Appendix 3.1):

- Cross Channel DC Link (France–England) 71 km with rating of 2000 MW at a DC voltage $2 \times \pm 270$ kV.
- 1345 km Pacific Intertie DC link (United States) with rating of 3100 MW at a DC voltage of ±500 kV.
- The ±500 kV, 4000 MW double bipolar Nelson River HVDC system in Canada. It provides a bulk transmission link from remote generation to the load center.
- The 6300 MW Itaipu HVDC link—two bipoles each rated 3150 MW at a DC voltage ±600 kV—in Brazil (1984–1987). It plays a key role in the Brazilian power system.
- The ±450 kV, 2000 MW Quebec–New England link, which is a three-terminal HVDC system.
- The most powerful HVDC submarine cable to date is rated 600 MW at 450 kV. Typical examples are the 230 km Baltic Cable link between Sweden and Germany, and the 260 km cable for the SwePol link between Sweden and Poland.
- The 230 km Baltic cable link between Sweden and Germany, and the 260 km cable for the link between Sweden and Poland (SwePol) are currently embedding the most powerful HVDC submarine cable, which is rated for 600 MW and 450 kV.
- A major portion of the total generating power (18,200 MW), from Three Gorges hydroelectric power plant, was carried to China's industrialized coastal areas in Shanghai and Shenzheng by means of four HVDC links.
- The longest HVDC link is the Rio Madeira, Brazil, 2×3150 MW, ±600 kV, 2375 km installed in 2014.

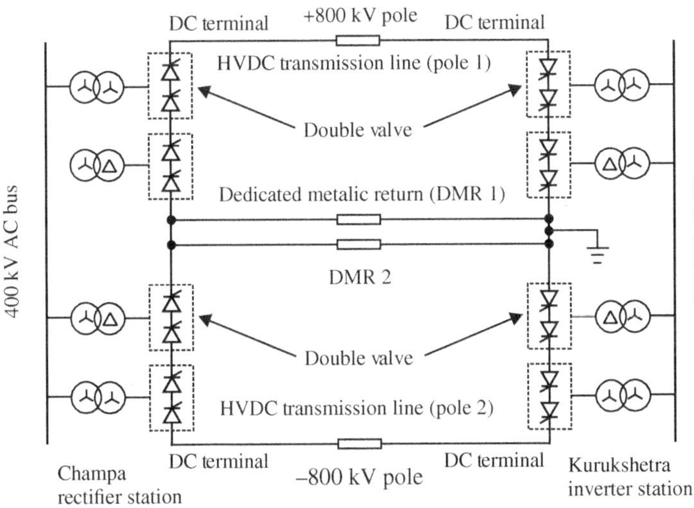

Figure 2.9 The single line diagram of the ±800 kV scheme with DME [10].

- The Champa-Kurukshetra scheme, 3000 MW, ±800 kV in India, 2015 (Figure 2.9) comprises four 12-pulse LCCs.

This scheme is the first application based on H400 THY valve technology (4 inch, 8.5 kV ETT) for such a voltage level. An unusual future for the ±800 kV transmission scheme is a dedicated metallic return, which is firmly earthed at Kurukshetra Station abandoning the requirement of grounding electrodes. This method has a main advantage of eliminating corrosion by buried pipelines [10].

Planned large LCC–HVDC projects:

- 2 × 4000 MW, ±800 kV Xingu–Minas–Rio in Brazil, 2018 (Figure 2.10). Here, a solution with two separated bipoles was adopted, as separated power injections in the southeast region network enhanced system performance, the short circuit ratio at the converter's point of connection, and reduced transmission costs [11].

A shortage of generation during the dry months and a significant and increasing power demand by the northern part of the grid imposed a power reversal transmission requirement for the new link. It will operate most of the time transmitting from north

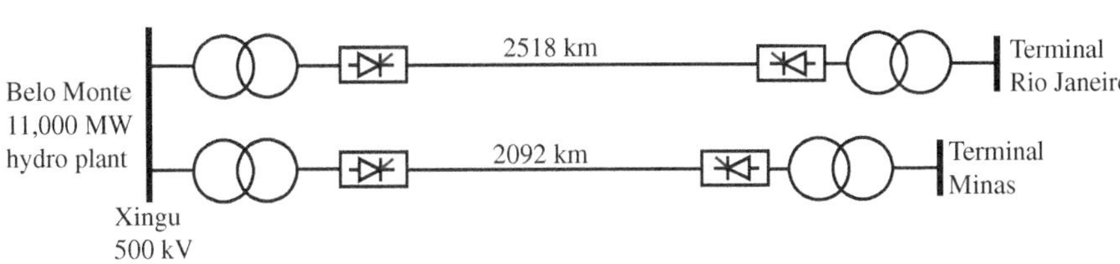

Figure 2.10 Two separated bipoles ± 800 kV.

to south, and during some months in the opposite direction. The ±800 kV HVDC link between Xingu and Minas is going to be the first embedded in a 500 kV AC network. "With such a connection, it can perform not only the basic function of bulk power transmission, but also, importantly, some additional control function within the AC network such as power flow control, voltage control, system stability improvement and the mitigation of system cascading failure" [12].

- An ±800 kV multiterminal HVDC scheme is planned for full commissioning in 2016, in India, and will bring 6000 MW from the North East to Agra (near Delhi).
- ±1100 kV HVDC schemes are being developed in China and are planned for commissioning in 2016/2017. The first ±1100 kV scheme, which will connect Zhundang and Chengdu, will have a power rating of 11,000 MW.

2.3.2 Capacitor-Commutated Converter (CCC–HVDC)

The CCC–HVDC scheme is characterized by the use of capacitors in series with the converter transformer. These capacitors are referred to as commutation capacitors and can either be located between the converter transformer and the THY valves or between the converter transformer and the AC filters, as shown in Figures 2.11a and 2.11b, respectively [3].

When the capacitors are connected between the transformer and AC filters, the capacitors may include parallel reactors and TCSC THY valves and are known as a controlled series capacitor converter.

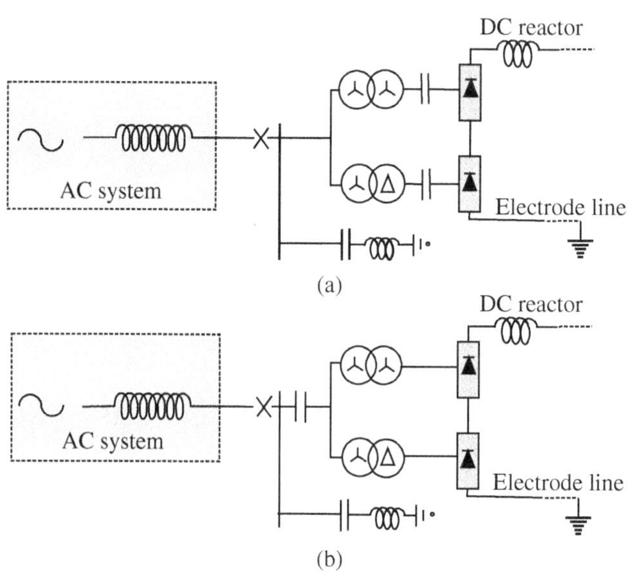

Figure 2.11 Configurations for applying series capacitors at HVDC substations: (a) with commutation capacitors on the valve side and (b) on the AC system side.

When using a CCC, the series capacitor generates reactive power proportional to the converter load current. The polarity of the voltage generated across the commutation capacitor is such that the voltage experienced by the THY valves is bigger than, and has a phase shift with regard to, the AC source voltage. The phase shift is such that the reactive power absorbed by a CCC is always lower compared to a conventional converter.

Once the DC conditions are fixed, the reactive power absorbed by the converter at any power level is determined by the capacitance of the commutation capacitor. Therefore, the size of the commutation capacitor can be chosen so that at full load, the reactive power consumption of the converter is small and can be compensated for by the reactive power generated by one small AC filter. Therefore, the filtering and reactive power supply functions are nearly fully separated and the need for switchable shunt capacitor banks for reactive power consumption is reduced. Another advantage of CCC is that temporary overvoltages (TOVs) are lower because the converter requires much less shunt reactive power compensation and the load rejection TOV is mainly determined by the amount of shunt reactive power connected.

With respect to LCC, CCC has also the following disadvantages:

- For conventional HVDC schemes, the only series component that can impact the overall availability and reliability of a back-to-back is the converter transformer, whereas in a CCC there is the additional capacitor bank that will directly impact the failure rate of the converter station.
- The series capacitor bank is under DC stress and therefore, if mounted externally, is subjected to additional pollution risk.
- The additional exposed conductors associated with the capacitor bank will increase the radio-frequency interference (RFI) radiated from the converter station.
- In order to avoid the risk of pollution build-up on the capacitor insulation and increased RFI generation, the capacitor can be located indoors. However, the additional civil works will have an associated additional cost and as the capacitor will be oil-filled, special protection methods must be employed.
- The series capacitor forces the converter to operate at higher firing angles, thereby increasing the harmonic current generated by the converter.
- The large amount of series capacitive energy can cause severe damage to the converter valve in the event of bushing flashover; hence, a large earthing resistor must be incorporated into the converter DC circuit.

The first commercial installation of the CCC type was at the 2200 MW, ±70 kV Garabi back-to-back (BtB) interconnection between the 500 kV systems of Argentina and Brazil in 2002 [13]. Since the Argentinean system is at 50 Hz, while the Brazilian system is at 60 Hz, a BtB frequency converter installation was necessary. Furthermore, since the short-circuit levels at the converters were low, a CCC option was selected to provide the enhanced stability for the AC system due to the forced-commutated converters. The second project was a 200 MW, ±13 kV back-to-back system installed in 2003 in the Rapid City, on the border between Nebraska and South Dakota.

The principal reasons for choosing CCC option were:

- Switchable shunt filter banks were not required and replaced by the series capacitor bank to compensate for the reactive power.
- The CCC alternative provided a dynamically more stable operation with the weak AC systems.
- Since the series capacitor's impedance is typically several times greater than the transformer's leakage impedance, it reduced the valve currents during DC side short circuits allowing the optimization of the transformer and valves.
- The power factor, seen from the AC bus, could be kept close to unity, or ever become positive during certain operating ranges.

Despite its advantages, CCC has remained only a niche application because of the advent of VSCs that completely eliminate the need for an extinction (turn-off) time.

2.3.3 Voltage Source Converter VSC–HVDC

VSC–HVDC technology is capturing more and more attention during the last decade. This new technology has only become possible due to important advances in HV IGBTs.

VSC–HVDC technology presents the following advantages compared with CSC–HVDC technology:

- Fully controlled power devices result in no need for an active commutation voltage like in the case of THY.
- Active and reactive power can be controlled independently.
- The direction of power flow can be easily changed by reversing direct current. No need of DC voltage reversal of polarity is enabling cheap crossed-linked polyethylene (XLPE) cable technology to be used.
- No minimum short circuit ratio requirement for the AC grids. Can interconnect weak AC grids or offshore wind power plants.
- Can provide black start of tripped grid (required DC source) as the DC-AC converter terminal can simulate a synchronous generator functionality by appropriate control.
- Faster dynamics and improved behavior during faults and disturbances. Better suited for meshed DC grids.Lower footprint due to significant reduction in harmonic filters making VSC–HVDC a viable solution for offshore deployment.
- No minimum level requirement for power transfer.

The first VSC–HVDC test system was installed by ABB in Sweden in 1997 (Helljön–Grangesberg ±10 kV, 3 MW). The first commercial link was 50 MW, ±80 kV linking the Swedish island of Gotland to the mainland in 1999. The main reason for choosing VSC–HVDC was to provide voltage support to the weak grid for the large amount of wind power installed in the South of Gotland.

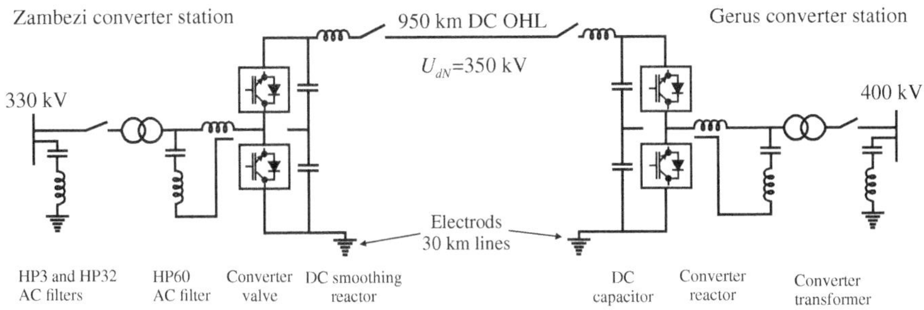

Figure 2.12 Caprivi Link [14].

2.3.3.1 VSC–HVDC Based on Two-Level Converters

Until few years ago, two-level converter technology was used and IGBT valves consisting of a large number of series-connected IGBTs in a similar mechanical stack arrangement to THY were used.

As IGBTs are fully controllable power devices, PWM technique has been used for control with switching frequencies in the range of 1–2 kHz. Due to the fast development of the offshore wind power, VSC–HVDC started to be used recently. The largest two-level VSC–HVDC link has been installed in 2012 in the North Sea (BorWin1 400 MW, ±150 kV, 75 km underground and 125 km submarine cable).

Another interesting example of using two-level VSC–HVDC for connecting two weak AC grid areas is Caprivi Link, monopole, 300 MVA, 350 kV in Africa using overhead lines with a distance of 952 km (Figure 2.12) [14].

2.3.3.2 VSC–HVDC Based on Multilevel Converters

The complexity problem of stacking a large number of IGBTs in two-level converters can be alleviated by using the concept of multilevel converter. The most acknowledged technology is the modular multilevel converter (MMC), where instead of series connection of IGBT, so-called submodules (SMs) typically in a half-bridge topology and a large storage capacitor are series connected. Each submodule can be inserted or bypassed, thus allowing the output voltage to be synthesized by a large number of levels with very low harmonic distortion. In comparison with two-level converters, MMC brings the following advantages:

- High level of modularity and upscaling possibilities. The voltage range is no longer limited by the blocking voltage of the devices, but by the number of SMs.
- Virtually, no grid connection filtering requirement due to the high number of voltage levels.
- Lower switching losses as the SMs are switched with very low frequency, typically 100–200 Hz.

The first MMC VSC–HVDC was installed in United States by Siemens in 2010 (Trans Bay Cable 400 MW, ±200 kV). It is important to mention that due to the advantages of MMC, LV–IGBT power modules with blocking voltage of 1700 V have been used.

ABB MMC approach for the SM is to use a combination of two series connections of eight IGBTs (4.5 kV press-pack) as an IGBT valve. The largest link with this technology is the offshore platform DolWin1, 800 MW, ±320 kV, 75 km underground and 90 km submarine cable commissioned in 2015 by ABB.

In recent years, significant progress has been made to reduce power losses in the converter stations. The new generation of VSC–HVDC based on MMC technology has reported losses per station down to 1.0%, comparable with the LCC–HVDC (0.8%).

Even though by June 2014, only 3370 MW of VSC–HVDC has been installed, new links for 11,770 MW are under actual construction (6430 MW for offshore wind power).

If in the United States and China there is a considerable interest for multiterminal VSC–HVDC systems, in Europe there is a high level of interest in building HVDC grids [15].

2.3.3.3 Limitations of VSC Transmission

- *Power rating.* Due to the reduced current rating of IGBT (about 2 kA) in comparison with THY (about 4 kA), the power rating of VSC–HVDC is limited to 1000 MW at the date of publication of this book. But, multipole solution can alleviate this shortcoming.
- *DC breaker.* The biggest limitation of VSC–HVDC is the need of HVDC breaker, which was recently overcome by ABB for a reduced voltage, but more time is required for maturing this technology. In point-to-point HVDC, DC breakers are not necessary as AC breakers but can be used in case of DC faults, while for multiterminal DC grids isolation of DC faults will require DC breakers.

2.4 FACTS EVOLUTION

The Institute of Electrical and Electronics Engineering (IEEE) and the International Council on Large Electric Systems (CIGRE) define FACTS devices as "Alternating Current Transmission Systems incorporating power electronic-based and other static controllers to enhance controllability and increase power transfer capability" [16].

It is the muscle by means of which an intelligent power system is able to act on at least one of the following system quantities: voltage magnitude, current magnitude, voltage phase angle, current phase angle, impedance, or admittance. For example, a transformer equipped with mechanical tap changers, that it is correctly controlled by the energy management system of a flexible transmission system, can be considered a FACTS controller because of its ability to act on voltage magnitude and avoid voltage collapse.

Various connections of mechanically switched capacitors and reactors in shunt and/or series with a transmission line allow the verification of the line voltage and, in the process, the power flow in the transmission line changes.

The mechanical switches of shunt-connected devices have been started to be replaced by static THY switches from the mid-1970s. This process delivered a new class of compensation devices called FACTS.

TABLE 2.2 Type of FACTS devices

	Controlled parameters		
Type of connection	P	Q	P and Q
Shunt connected		SVC and STATCOM	
Series connected	TSSC, TCSC, and SSSC		
Combined shunt and series connected	TCPST and TCPAR	TCVR	UPFC and IPFC

The types of FACTS controllers currently available can be categorized based on the groups of devices that control certain electrical parameters and have different types of connection to the controlled AC transmission system (Table 2.2).

The full names of FACTS devices mentioned in Table 2.2 and Figures 2.13 and 2.14 are listed below:

SVC : Static VAr Compensator
STATCOM : Static Synchronous Compensator
TSSC : Thyristor Switched Series Capacitor
TCSC : Thyristor Controlled Series Capacitor
SSSC : Static Synchronous Series Compensator
TCPST : Thyristor Controlled Phase Shifting Transformer
TCPAR : Thyristor Controlled Phase Angle Regulator
TCVR : Thyristor Controlled Voltage Regulator
UPFC : Unified Power Flow Controller
IPFC : Interline Power Flow Controller

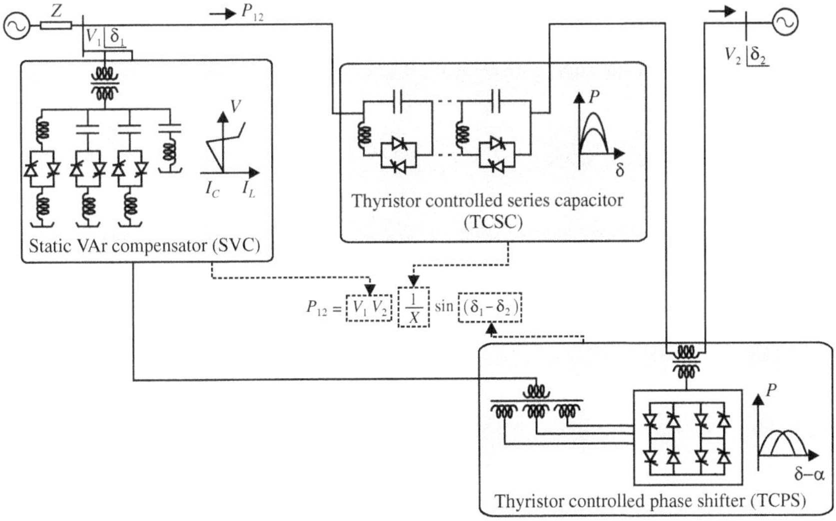

Figure 2.13 Conventional thyristor-based FACTS controllers [17].

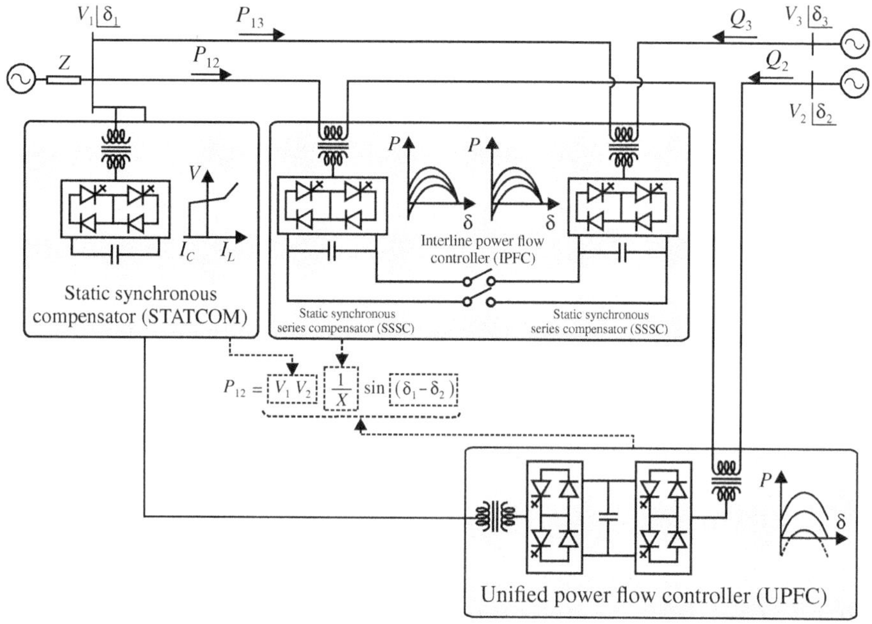

Figure 2.14 Family of VSC-based FACTS controllers [18].

At the present time, the shunt-connected FACTS devices and in particularly the SVC and static synchronous compensator (STATCOM) are widely used in power systems. The installation of the series compensation in uncompensated circuit was a viable solution of power delivery problems for more than eight decades. The modern series compensation devices employ many advanced technologies such as FACTS (thyristor switched series capacitor, TCSC, and static synchronous series compensator [SSSC]). The group of combined FACTS devices includes controllers that have both shunt and series connections to the controlled transmission system. This group is divided into two subgroups: the subgroup of thyristor-controlled phase shifting transformers or thyristor-controlled phase angle regulator and thyristor-controlled voltage regulators, and the subgroup of VSC-based devices (STATCOM, SSSC, unified power flow controller [UPFC], and interline power flow controller [IPFC]).

Special arrangement of two separated shunt- and series-connected VSCs, such as the STATCOM and the SSSC, controlled in a coordinated manner, results in the UPFC device [17], which represents the third generation of FACTS devices.

IPFC represents a last concept of the group, with the objective of providing a flexible power flow control scheme for a multiline power system, in which two or more lines employ an SSSC for series compensation.

The idea of FACTS for power flow control is explained in Figures 2.13 and 2.14, which illustrate how different types of FACTS devices can influence one or several electrical parameters from the equation $P_{12} = V_1 V_2 \sin(\delta_1 - \delta_2)/X_{12}$, and affect the active power flow in the transmission line.

The first group of controllers, the SVC, TCSC, and phase-shifter, employs conventional THYs (i.e., those having no intrinsic turn-off ability) in circuit arrangements that are similar to breaker-switched capacitors and reactors and conventional (mechanical) tap-changing transformers, but have much faster response and are operated by sophistical controls, each of these controllers can act on one of the three parameters determining power transmission, voltage (SVC), transmission impedance (TCSC), and transmission angle (phase-shifter), as illustrated in Figure 2.13 [17].

The second group of controllers employing VSCs, the STATCOM, the SSSC, the UPFC, and IPFC, is shown in Figure 2.14 together with the functional control objective [18].

The STATCOM controls transmission voltage by reactive shunt compensation. The SSSC provides series compensation by directly controlling the voltage across the series impedance, and thereby controlling the transmitted electric power. The UPFC can control individually or in combination, all three transmission parameters (voltage, impedance, and angle) or directly the real and reactive power flow in the line. The IPFC is able to transfer real power between lines, in addition to providing reactive series compensation, and thereby can facilitate a comprehensive overall real and reactive power management for a multiline transmission system.

These controllers are able to handle practically all voltage support, power flow, and stability problems, and, due to their unique capability of controlling both reactive and real power, can achieve the maximum utilization of power lines and other system assets. In addition, the technology can be used to tie asynchronous systems together and to connect remote-generating sources to the transmission network. The converter-based technology has also been instrumental to the practical solution of power quality problems in distribution systems. The capability of distribution STATCOM, and particularly the dynamic voltage restorer, together with active filters, for solving quality problems such as voltage sags, voltage flicker, and harmonics, has been successfully demonstrated by a growing number of installations.

REFERENCES

[1] Dijkhuizen, F. Multilevel converters: review, form, function and motivation. *Ecological Vehicles and Renewable Energies – EVER 2012*, Monaco, October 10–13, 2012.

[2] Rahimo, M. T. Ultra high voltage semiconductor power devices for grid applications. *2010 IEEE International Electron Devices Meeting (IEDM)*, San Francisco, December 6–8, 2010.

[3] ALSTOM Grid. *HVDC: Connecting to the Future*, 2nd edition. 2011.

[4] Thomas, K. M., Backlund, B., Toker, O., and Thorvaldsson, B. *The Bidirectional Control Thyristor (BCT)*. ABB Semiconductors AG Switzerland and ABB Power Systems AB, Sweden, 1999.

[5] Westermann, D. (convener), Davidson, C., Fu, Y., Muller, L., Nielson, S., Steimer, P., Takasaki, M., Weinhold, M. Increased system efficiency by use of new generation of power semiconductors. CIGRE Joint Working Group B4-43/A3/B3, Technical Brochure no. 337, December 2007.

[6] Jovcic, D., and Admed, K. *High Voltage Direct Current Transmission: Converters, Systems and DC Grids*. John Wiley & Sons, Chichester, United Kingdom, 2015.

[7] Edris, A. A. (Convener), Chow, J., Watanabe, E., Barbosa, P., Halvarsson, P., Angquist, L., Fardenesh, B., Uzunovic, E., Huang, A., Tyll, H., Batacharia, S., Enslin, J., Ivakin, V., and Wiik, J. Static synchronous series compensator - SSSC. CIGRE Brochure No. 371, SC B4, WG B4-40, February 2009.

[8] Yazdani, A., and Iravani, R. *Voltage-Sourced Converters in Power Systems. Modeling, Control, and Applications.* IEEE, John Wiley & Sons, Hoboken, New Jersey, 2010.

[9] Cheng, L., Agarwal, A. K., Capell, C., O'Loughlin, M., Lam, K., Richmond, J., Van Brunt, E., Burk, A., Palmour, J. W., O'Brien, H., Ogunniyi, A., and Scozzie, C. 20 kV, 2 cm^2, 4H-SiC gate turn-off thyristors for advanced pulsed power applications. *19th IEEE Pulsed Power Conference (PPC)*, June 2013.

[10] Jodeyri, M. H., and Dzus, A. Thyristor valve for the 12-pulse converter for the Champa-Kurukshetra HVDC transmission scheme. *IEEE ISGT Asia Conference*, 2013.

[11] Carvalho Jr., D. S., Souza, D. F., and Esmeraldo, P. C. V. Planning conceptions for a ± 800 kV HVDC transmission system in Brazil. *SC B4 Colloquium on HVDC and Power Electronics to Boost Network Performance*, Brasilia, 2013.

[12] Henry, S. (Convenor), Despouys, O., Adapa, R., Barthold, L., Bayfield, C., Bell, K., Binard, J. L., Edris, A., Egrot, P., Hung, W., Irwin, G., Karady, G., L'Abbate, A., Li, M., Manchen, M., Messner, J., Migliavacca, G., Muttik, P., Nourizian, M., Peard, A., Rauhala, T., Sakai, T., Sanz, S., Song, Z., Temtem, S., Van Hertem, D., Zhang, L. Influence of embedded HVDC transmission on system security and AC network performance. CIGRE Joint Working Group C4/B4/C1, Brochure No. 536, 2013.

[13] Graham, J., Jonsson, B., and Moni, R. S. *The Garabi 2000 MW Interconnection Back-to-Back HVDC to Connect Weak AC Systems.* ABB Utilities AB, Ludvika, Sweden.

[14] Magg, T., Machen, M., Krige, E., Wasborg, J., and Sundin, J. Connecting networks with VSC-HVDC in Africa: Caprivi Link interconnector. *IEEE PES Power Africa 2012 Conference,* Johannesburg, South Africa, July 9–13, 2012.

[15] Andersen, B. HVDC and power electronics. Annual Report 2014, CIGRE SC B4, ELECTRA, No. 274, June 2014.

[16] Edris, A. A., Adapa, R., Baker, M. H., Bohmann, L., Clark, K., Habashi, K., Gyugyi, L., Lemay, J., Mehraban, S., Meyers, A. K., Reeve, J., Sener, F., Torgerson, D. R., and Wood, R. R. Proposed terms and definitions for FACTS. *IEEE Transactions on Power Delivery*, Vol. 12, no. 4, pp. 1848–1853, October 1997.

[17] Hingorani, N. G., and Gyugyi, L. *Understanding FACTS. Concepts and Technology of Flexible AC Transmission Systems.* IEEE Press, New York, 2000.

[18] Song, Y. H., and Johns, A. T. (Eds.) *Flexible AC Transmission Systems (FACTS).* IEE Power and Energy Series, London, UK, 1999.

CHAPTER 3

CSC–HVDC TRANSMISSION

Mircea Eremia and Constantin Bulac

3.1 STRUCTURE AND CONFIGURATIONS

3.1.1 Structure of HVDC Links

An HVDC link can be either long distance or back-to-back (BTB). In a long-distance HVDC system, the rectifier and inverter stations are connected via an overhead line or cable or a combination of both. In a BTB system, there is no HVDC line or cable. The power flow is from a rectifier to the inverter. There is no reactive power exchange between the two stations. The components of an HVDC link are shown in Figure 3.1 [1].

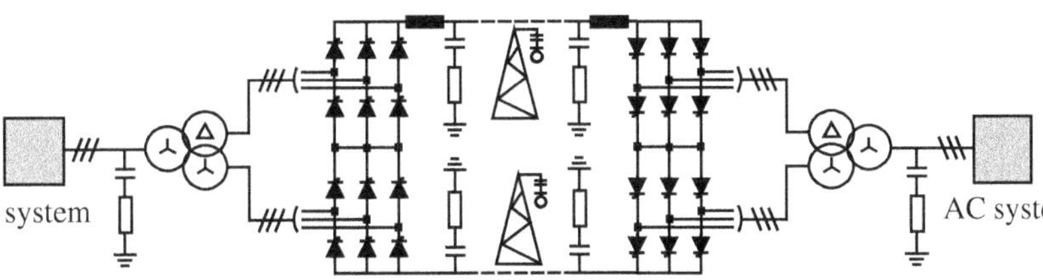

Figure 3.1 The main components of an HVDC link.

HVDC CONVERTER UNITS. The basic structure of a HVDC converter is the six-pulse bridge, consisting of three pairs of thyristor valves. The converter performs the conversion from AC to DC and DC to AC. A six-pulse bridge is connected to the AC system via a converter transformer (Figure 3.2a).

Two six-pulse bridges can be connected in series on the DC side and in parallel on the AC side through their respective converter transformers. If one bridge is connected to a Y/Y and the second bridge is connected to a Y/Δ transformer, and due to the phase shift of 30° between the Y and Δ, a 12-pulse operation is obtained. This configuration is referred to as 12-pulse converter and is employed in most of the HVDC systems (Figure 3.2b) [2, 3].

A 24-pulse converter unit consists of four six-pulse bridges connected in series on the DC side with their respective transformers, producing four three-phase systems with 15° phase shift between bridges (Figure 3.2c). Figure 3.2d shows the 12-pulse

Advanced Solutions in Power Systems: HVDC, FACTS, and Artificial Intelligence, First Edition.
Edited by Mircea Eremia, Chen-Ching Liu, and Abdel-Aty Edris.

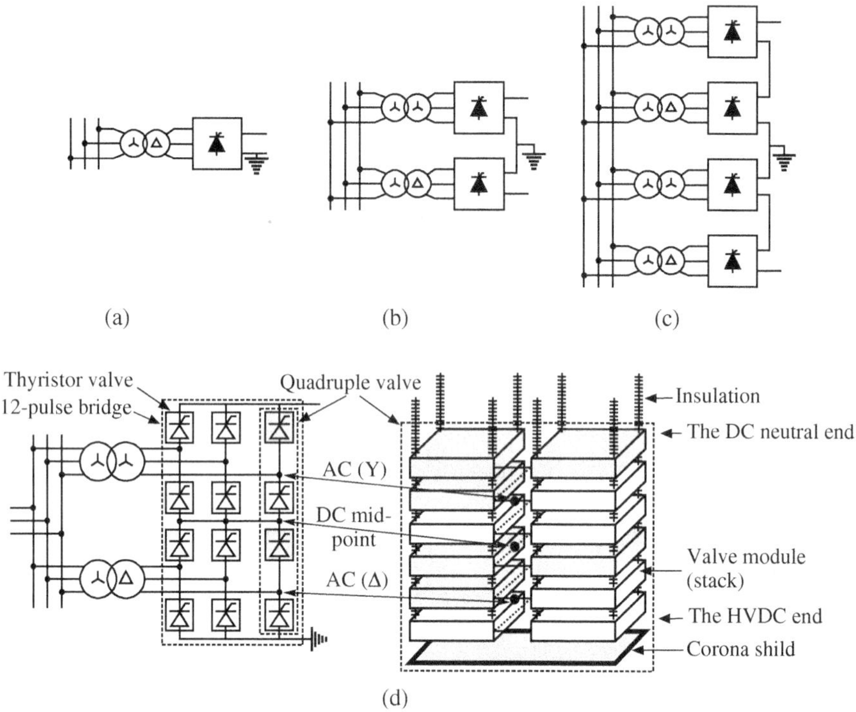

Figure 3.2 6, 12, and 24-pulse converter units (a), (b), (c), respectively, and the compositions of an HVDC valve tower with suspended 12-pulse series bridge converter (d).

converter and the arrangement of a HVDC quadruple valve tower suspended from the valve hall ceiling. One thyristor valve is made up of several valve modules (stack) [4,5]. Suspended designs are common today because of the lower cost of the structural components.

CONVERTER TRANSFORMER [6]. The converter transformer is designed to provide an ungrounded three-phase voltage source of the appropriate level to the bridge. It may have any winding connection, as long as the valve side winding is ungrounded. This allows the DC system to establish its own reference to ground, usually by grounding the positive or negative end of the converter.

Converter transformers are equipped with on-load tap changer (OLTC) to allow adequate control over DC voltage levels as AC voltages and DC currents vary. A typical range for the OLTC would be ±15% in 1–2% steps. Most will operate within 5 s of the occurrence of a change to restore desired conditions. The operation of the tap changer is closely coordinated with both valve-firing controls and switching of reactive power sources in the converter station.

Converter transformers are exposed to cyclic conduction of DC in their windings, which approximate rectangular current pulses of fundamental frequency. These current pulses may be thought of as being composed of the superposition of fundamental frequency and several harmonic frequency components. The harmonic frequency components increase losses. These effects combined with DC insulations requirements, large reactive power demands and OLTC mechanisms tend to make converter transformers much larger than conventional transformers of the same rating.

CONVERTER VALVES [6] The converter valves perform the switching to sequentially provide line-to-line voltages supplied by the converter transformer to the positive and negative terminals of the valve bridge. By controlling the timing of the switching operations, the level of DC voltage and its polarity can be varied from nearly the positive value of the peak line-to-line AC voltage to nearly the negative value of that voltage. The polarity and level of the resultant direct voltage relative to the direct current determines the direction and level of power transfer on the DC link. Operation of the converter will be discussed in Section 3.2.

Physically, a valve bridge consists of six high voltage valves. The six valves act as switches. It is usually convenient to mechanically stack two or four valves on top of one another with AC and DC connections at intermediate levels. These are called double valves or quadruple valves, respectively (Figure 3.2d) [5]. The valves are constructed using individual thyristor levels. Each valve consists of thyristor levels connected in series to attain the desired voltage level. Parallel connection of thyristor levels is not required. The thyristors in a valve are arranged into modules. Each module typically contains a gate unit for each thyristor. The gate unit controls the firing of each thyristor as well as monitors the thyristor-level components. In addition to the thyristors levels, the module contains series saturable reactors to limit current rate of change as well as grading capacitances to limit voltage rate of change.

Valves require a cooling system. Modern thyristor valves are water-cooled. Thermal time constants are on the order of seconds; therefore, any overload of the valves must be specified.

HARMONIC FILTERS. Converters generate voltage and current harmonic on both AC and DC sides. These harmonics can cause overheating of capacitors and nearby generators, and interferences with communication systems.

There are three types of filters used:

(i) *AC filters*. These are passive circuits used to provide low impedance, shunt paths for AC harmonic currents. Both tuned and damped filter arrangements are used.

(ii) *DC filters*. These are similar to the AC filters and are used for the filtering of harmonics present on the DC side.

(iii) *High frequency filters*. These are connected between the converter transformer and the station AC bus to suppress any high frequency currents that may interact with power line carrier communication used in the AC system. Sometimes such filters are provided on high-voltage DC bus converter between the DC filter and DC line and also on the neutral side.

REACTIVE POWER COMPENSATION. The converters require reactive power supply, which depends on the active power loading. Under steady-state conditions, the reactive power consumed is about 50–60% of active power transferred. Reactive power compensation is therefore provided near the converters, by shunt capacitors or Flexible Alternating Current Transmission System (FACTS) devices (static VAr compensator – SVC, static synchronous compensator – STATCOM). When sizing the compensation device, the reactive power produced by the AC filters must be taken into account.

Smoothing Reactor. A series reactor is used on the DC side to smooth direct current and also serves as a buffer between the converters and the DC line. The sizing of the reactor depends on various requirements: smooth the ripple in the direct current in order to prevent the current from becoming discontinuous at light loads, protect the converters against fast front transients originated in the lightning strikes occurring on the DC line, to reduce the incidence of commutation failure in inverters caused by dips in the AC voltage at the converter bus, and limit the crest current in the rectifier during short circuit on the DC line.

DC Lines/Cables. Selection of DC line/cable voltage is made only in conjunction with converter rating selection. Going to higher voltages will increase converter costs and decrease line losses for a given power level. Optimization of system voltage level is highly dependent upon both converter costs and line/cable costs. Thus DC lines, unlike AC lines, may not be designed at an optimum voltage level for a given power transfer. Converter cost as a function of voltage and current must be considered before DC voltage is selected. The evaluation should consider different voltage levels.

Electrode Systems. Bipolar DC projects are designed to use the earth as a neutral conductor for at least brief periods of time unless they are designed as two independent monopoles. The connection to the earth requires a large surface area conductor to minimize current densities and surface voltage gradients.

The conductor is termed an *electrode*. The electrode is usually located several kilometers from the converter station so that direct currents will not stray into the station's ground mat or into AC transmission system. The line connecting the converter station to the electrode is termed an *electrode line*. In some designs if ground currents are not allowed at all, a metallic return conductor is provided, referred to as dedicated metallic return (DMR). In bipolar systems, during monopolar operation, one of the pole conductors can be used for metallic return. This requires a metallic transfer breaker (MRTB) at the low voltage end of one station. This breaker has to commutate the DC current from the electrode line to the metallic return.

These three elements, electrode, electrode line, and metallic return transfer breaker, constitute the electrode system. Each has unique characteristics which are discussed in turn.

(i) *Electrode* [6]. Two very different types of electrodes need to be considered: anode and cathode. The anode is the point at which conventional current (not electron flow) enters the earth. The cathode is where it leaves the earth. At both types of electrodes, conduction is largely by ion flow in an electrolyte of water and salts. The major difference between the types of electrodes is that at the anode metal ions assume positive charges and are carried away from the electrode. At the cathode, metal ions are deposited. This causes severe corrosion of the anode and erosion of the metal. The anode is effectively "sacrificed" to allow current conduction. The cathode may increase in metallic mass, but that is not a problem. Anodes are more expensive to maintain because of the corrosion effect. To reduce costs, they are usually constructed with low cost conducting material surrounding the metallic conductors (coke and graphite are frequently used). Cathodes may consist of just a buried array of metallic conductors, since they are not affected by corrosion, if at least low currents flow most of time.

In practice, it is usually not possible to predict which electrode will be the anode and which will be the cathode, because the earth return usually carries the unbalanced current of bipolar system. Electrode construction varies with the resistivity and moisture content of the earth.

The most compact electrodes are constructed in ocean locations where saltwater provides a good conductor and unlimited cooling and moisture. Larger electrodes are required when the electrode is created on land since thousands of meters of buried conductors may be required. For instance, the electrode at the northern end of the Pacific Intertie is constructed as a large conducting circle with a diameter of 1036 m [6]. An alternative to the surface electrode is the deep electrode. If low resistivity rock or mineral formations are located below the surface, it is sometimes advantageous to construct several vertical shafts to connect with those formations.

A major problem with electrodes is that DC earth currents follow paths of least resistance. When located in vicinity of conductive man-made facilities, such as gas or water pipelines or even transmission lines, the leakage currents may produce corrosion where they leave metallic objects. Remote from the electrode, this will probably be deep within the conducting mantle of the earth. In the vicinity of the electrode, it may include following low resistance man-made facilities, such as gas or water pipelines or even transmission lines. Currents in these facilities may create problems, such as saturation of transformer cores.

(ii) *Electrode line* [6]. The electrode line requires insulation levels adequate to withstand the voltage drop along its length. That is usually only a small fraction of the line-to-ground voltage of the DC system. The electrode line may be carried as an insulated shield wire on one of the transmission lines leaving the converter station, as long as it has adequate capacity.

The electrode line has some unique protection problems. Faults are hard to detect since there will be no change in current, and voltages are very low all of the time. The effects of grounding the electrode line to a tower or the earth may be very adverse, however. In either case, sever corrosion can occur if high currents are present. Corrosion of the tower footing can destroy the tower; corrosion of the electrode line can cause loss of conducting capacity and failure of the electrode line.

Any break in the electrode line can cause DC arcing and overvoltages on neutral end of the pole. DC arcs can be much longer and stable than AC arcs. Any tendency to extinguish is countered by the inductances of the DC system, and recovery voltages which, on monopolar systems, can be even higher than the pole-to-ground voltage.

(iii) *Metallic return transfer breaker.* The MRTB [6] is designed to either force the current flowing in the earth to take a higher resistance metallic path to the other end of the link or to momentarily ground the metallic return conductor to help extinguish faults on that conductor. In either case, only one MRTB is required in a DC system, since one end of the link will always be grounded in both earth and metallic return modes.

In the transfer mode, the metallic return conductor is first connected between the neutrals of the converters at each end of the link. Only a small portion of the DC return current will flow through it in this connection. The MRTB is located in series with the much larger earth component of the return current. The MRTB then interrupts the earth component of the current (see Figure 3.5c).

The fault-clearing mode of operation of the MRTB could be used if a fault occurred on the metallic return, although such faults are very rare because the metallic return conductor is insulated for full pole voltage while being used for very low voltage during metallic return. Closing the MRTB would bring voltages and currents on the metallic return to nearly zero and may extinguish the fault. Opening of the MRTB after this represents another transfer to metallic return mode.

Insulation Coordination [6]. The insulation coordination of HVDC converter stations requires a specific knowledge of the behavior of the converter. The protection is afforded by ZnO arresters. The location, protective levels, and energy requirements are derived through insulation coordination studies. IEEE, IEC, and CIGRE have guidelines for the insulation coordination of HVDC converter stations.

3.1.2 HVDC Configurations

The main configurations of HVDC transmission systems are bipolar lines, monopolar lines, and BTB links (Figure 3.3).

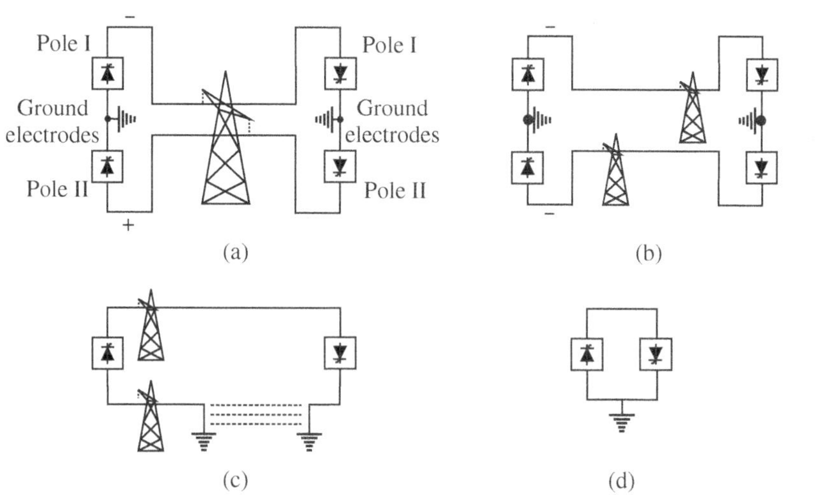

Figure 3.3 Types of HVDC transmission systems: (a) bipolar line; (b) two monopolar transmission lines forming a bipolar system; (c) monopolar transmission line with ground or sea return; (d) BTB [7].

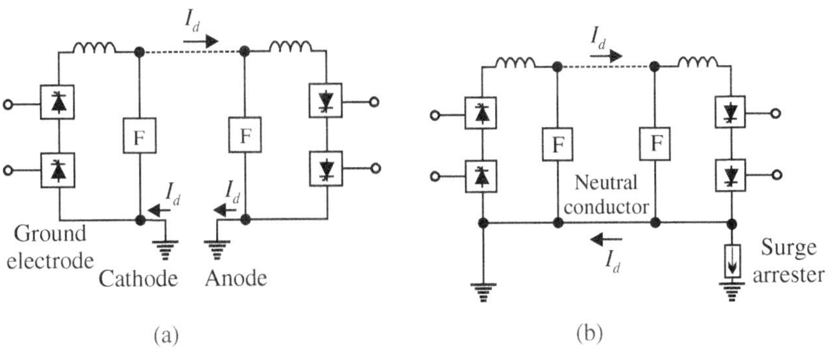

Figure 3.4 Monopolar transmission systems: (a) monopolar with ground return; (b) monopolar with metallic return. *Source:* IEEE Std. 2000 [8]. Reproduced with permission of IEEE.

MONOPOLAR HVDC SYSTEMS can be of two types: monopolar systems with ground return and monopolar systems with metallic return.

(a) *A monopolar HVDC system with ground return* consists of one or more 12-pulse converter units in series or parallel at each end, a single conductor and return through the ground (Figure 3.4a) [8] or seawater (Figure 3.3c).

It could be a cost-effective solution for an HVDC cable transmission and/or the first stage of a bipolar scheme. At each end of the line, it requires an electrode line and a ground or sea electrode built for continuous operation. Thus, possible interference with other structures, installations, or systems and magnetic field effects may be considered.

(b) *A monopolar HVDC system with metallic return* consists of one high voltage and one low voltage conductor, with the neutral tied at the one converter station to the station grounding grid or to the associated ground electrode. The other terminal is not grounded. It is generally used either as the first stage of a bipolar scheme, when ground currents are not allowed during the intermediate period, or when for environmental reasons ground current is not allowed, or there is no possibility of constructing an electrode due to location.

Both configurations require one or more DC smoothing reactors at each end of the HVDC line, usually located on the high voltage side or even on the low voltage side, if the resulting performance is acceptable, and DC filters (F), if the line is overhead [9].

BIPOLAR HVDC SYSTEMS. A bipolar HVDC system consists of two poles, each of one or more 12-pulse converter units, in series with electrode lines and ground electrodes or DMR. For power flow in the other direction, the two conductors reverse their polarities. It is a combination of two monopolar schemes with ground return, as Figure 3.5a presents. With both poles in operation, the imbalance current flow in the ground path can be held to a very low value.

This is a very common arrangement with the following operational capabilities [9]:

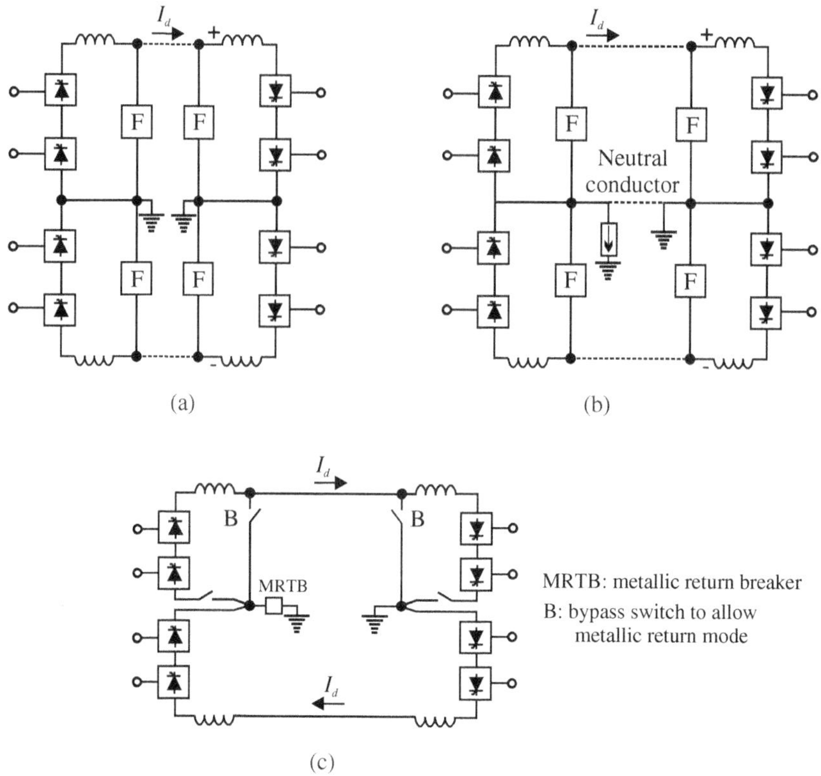

Figure 3.5 Bipolar HVDC systems: (a) bipolar system; (b) bipolar with metallic neutral; (c) bipolar, grounded midpoint configuration (common tower). *Source:* IEEE Std. 2000 [8]. Reproduced with permission of IEEE.

- during an outage of one pole, the other could be operated continuously with ground return;
- when one pole cannot be operated with full load current, the two poles of the bipolar scheme could be operated with different currents;
- when construction of electrode lines and ground electrodes constitutes an uneconomical solution due to shortness of the DC line or the high value of earth resistivity, a third conductor can be added to give a DMR, as Figure 3.5b presents. This conductor carries unbalanced current during bipolar operation and serves as return path, when a pole is out of service. The neutral of one terminal could be grounded, whereas the other would float or be tied to the grounding grid through an arrester or a capacitor.

The *bipolar grounded midpoint configuration* (Figure 3.5c) is one most widely used on existing HVDC systems [6, 8]. It is basically two monopolar ground return circuits of opposite polarity operating between the same terminals on the same tower. In this way, the ground return currents can be made to largely cancel each other under normal conditions, greatly decreasing electrode problems and communication problems that either one individually would have. The fact that two power circuits are present provides a substantial improvement in redundancy to cope with line or converter contingencies. If converter pole paralleling switches are provided along with bypass switching of subpole valve groups, substantially higher fractions of normal power capabilities can continue during contingencies.

POINT-TO-POINT HVDC SYSTEMS.[1] *Point-to-point HVDC overland transmission* is a very common application and a well-justified choice in case of power transmission over long distances or asynchronous interconnection of two power networks. For an evaluation process, the following parts, consisting of a "point-to-point overland system," may be considered [9]:

- converter stations at the two ends of the transmission;
- HVDC overhead line;
- electrode lines, which could be medium voltage insulated overhead lines, connecting the neutral point of the converter stations with the associated electrodes. It is noted that, if the route is the same for a certain length, HVDC line and electrode line could be on the same towers;
- electrode stations, which could be located in the ground or sea.

Point-to-point HVDC submarine systems. For sea crossings longer than certain distance, that is 50 km, the use of conventional AC cable or 100 km if cross-linked polyethylene (XLPE) or other new cable is use, the use of AC is virtually impossible, as the capacitive currents become too high. Thus, in these cases the most common application worldwide is the HVDC submarine cable transmission. Another similar application is the use of long HVDC underground cables in densely populated urban areas. In these areas, construction of overhead lines becomes very expensive and may face opposition due to environmental issues. The choice of HVDC underground cables versus the HVAC alternative has the advantage of low losses and the disadvantage of the space required for the converters stations though this may be reduced in case of voltage-sourced converters.

Main possible parts of a point-to-point HVDC submarine system are [9]

- converter stations at the two ends of the transmission;
- overhead line sections, linking converter stations with the overhead line/land cable junctions or/and land cable sections, linking the overhead line/land cable junctions with the land/sea cable joint;
- submarine cable section, linking the two land/sea cable joints;
- electrode lines;
- two marine electrodes, one anode and one cathode.

BTB HVDC LINKS are HVDC interconnections, where there is no DC transmission line, and both converters are located at the same site. For economic reasons, each converter is usually a 12-pulse converter unit. The valves for both converters

[1] Material reprinted with permission from CIGRE [9].

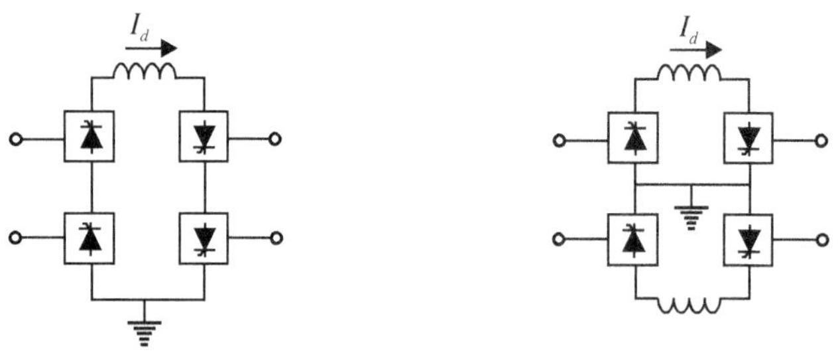

(a) (b)

Figure 3.6 BTB systems: (a) one DC smoothing reactor and one DC system ground; (b) two DC smoothing reactors and one DC system ground.

may be located in one valve hall, and similarly other items for the two converters, such as control system, cooling equipment, and auxiliary systems, may be integrated into configurations common to the two converters.

In Figure 3.6, two different circuit configurations, regarding locations of smoothing reactor and DC system ground are presented.

Generally, the DC voltage rating is typically low and thyristor valve current rating is high in comparison with HVDC interconnections with overhead lines and cables. The main reason is that, on the one hand, valve cost is much more voltage dependent, as the higher voltage increases the number of thyristors and thus the cost and, on the other hand, the highest possible current adds very little extra cost in the price of an individual thyristor. This also permits a low voltage tertiary winding of the converter transformer to which AC filters and compensation may be connected. Smaller reactive power switching steps can thus be achieved.

A large BTB HVDC system could be composed of two or more BTB links and thus loss of one converter unit will not cause loss of full power capability. Considering an HVDC transmission between two asynchronous power systems, the optimum location of the terminals should be considered. Either an HVDC intertie with overhead lines or a BTB link at the border of two existing grids should be used. BTB links certainly have a cost benefit regarding terminal costs only. Decision will depend upon the capacity of the existing grids to handle the new load. If reinforcements are necessary, the possibility of using a BTB link with a long AC feeder must be examined, too.

AC AND DC IN PARALLEL. Environmental restrictions are being imposed on the expansion of power systems and especially on the construction of new transmission lines. As an alternative to new lines, the enhancement of power transmission capability of the existing lines may be considered. One way of achieving this is by converting existing AC lines to DC, with possible alterations in insulators on the existing towers.

For double circuit AC lines, conversion of one AC circuit to a DC leads to a parallel operation of AC and DC lines on the same transmission towers. Studies performed for the close coupling of two parallel AC and DC lines running on conventional AC towers show no adverse effect on the operation of the DC link [9].

MULTITERMINAL HVDC SYSTEMS. Multiterminal HVDC (MTDC) refers to an HVDC system that consists of three or more transforming stations (Figure 3.7). Its architecture is more complex as compared to that of a two terminal point-to-point

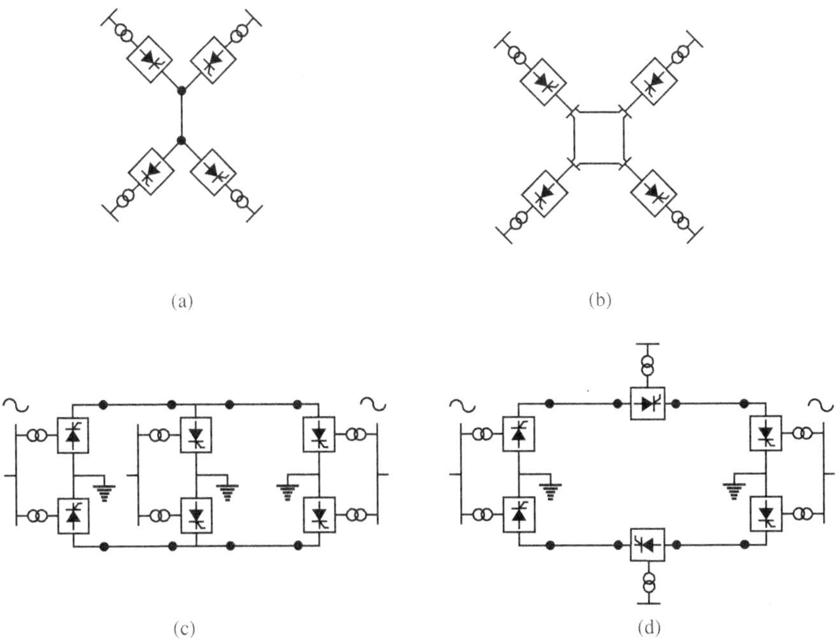

Figure 3.7 Multiterminal HVDC systems: (a) radial connections; (b) mesh or ring connection; (c) terminal in parallel; (d) terminal in series.

system. It requires a significant complexity to facilitate communication and control between each transforming station. The main possible configurations are mentioned, the first three referred to parallel connected 12-pulse converter units, rated for full direct voltage and the last one to series connected 12-pulse converter units, rated for full direct current.

The following classification is given according to Figure 3.7 [9]:

- *Radial MTDC system*, where each converter station is connected to a single DC line and for same of the distance the energy flows through a common DC line. No part of the DC system can be disconnected without significant energy flow change between the AC systems.
- *Meshed or ring MTDC system*, where each converter station is connected to more than one DC line. Any part of the DC system can be disconnected without energy flow change between the AC systems.
- *Series-connected MTDC system*, where all the converter stations are connected in series to a common ring shaped DC line. No part of the common DC line can be disconnected without interrupting energy exchange between the AC systems.

The first three configurations refer to parallel-connected 12-pulse converter units, rated for full direct voltage and that last one to series-connected 12-pulse converter units, rated for full current.

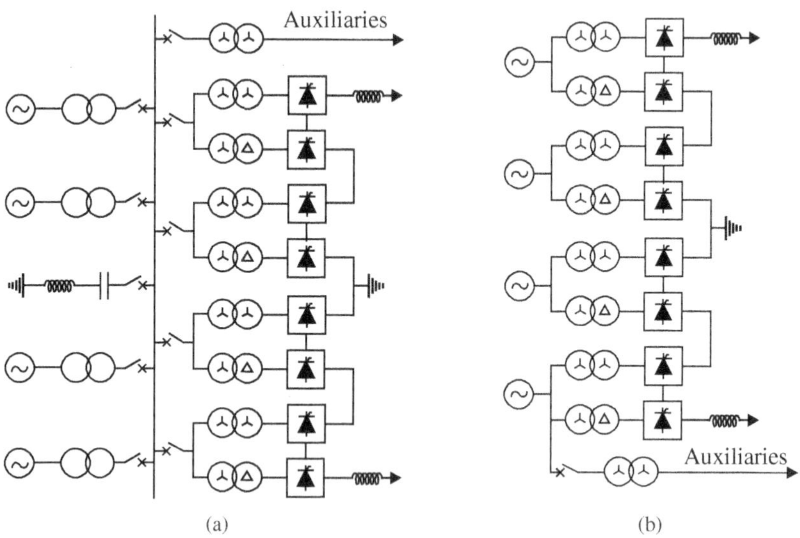

Figure 3.8 Schemes for generators connected to HVDC converters: (a) conventional scheme; (b) unit connection. *Source:* CIGRE Working Group 2001 [9]. Adapted with permission of CIGRE.

HVDC Unit Connected Generators. Several technical and economic reasons strongly suggest that in certain HVDC applications it may be advantageous to simplify the rectifier station, via a direct connection of each machine set to a separate converter group with series–parallel combinations made on the DC side. This arrangement is usually referred to in the HVDC literature as unit connection. Unit connection schemes could be very attractive solutions for applications as electrical generation from remote sources of power such as hydro and low grade coal fields, pump storage schemes, wind power station, new large generating stations near their load center, etc.

The arrangement traditionally used is shown in Figure 3.8a and termed a *conventional scheme* [9]. The basic feature is that the generating units feed a common AC busbar at the secondary side of the unit transformers. This busbar also takes the AC harmonic filters. The modified arrangement is indicated in Figure 3.8b and termed a *unit connection scheme.* Here the generator transformer and the AC busbar have been removed. The generators are directly connected to the converter transformers, and, if needed, series–parallel combinations of units are done at the DC side.

The small harmonic currents produced by the 12-pulse unit connected scheme are to be absorbed by the generator, so that the need for AC filters is eliminated. It is only expected that voltage control can be exercised entirely by the generator excitation. In this case, the transformer on-load top changers would no longer be needed.

Switching Possibilities on DC Side. For bipolar systems, several possible DC switching arrangements may be provided, which increase the availability of the overall HVDC interconnections [9]:

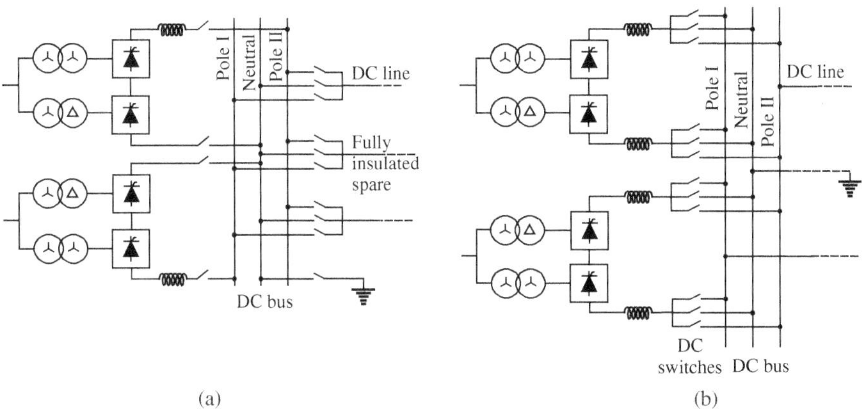

Figure 3.9 Switching possibilities on DC side: (a) switching of DC line conductors; (b) switching of converter poles. *Source:* IEEE Std. 2000 [8]. Reproduced with permission of IEEE.

- capability of connecting any line to any converter pole or to neutral through an arrangement with three DC buses—two pole buses and a neutral bus, as Figure 3.9a presents;
- capability of connecting the two valve groups in parallel (Figure 3.9b). It is noted that the 12-pulse converter unit of each group would need to be insulated for full line voltage;
- capability at the junction of overhead and cable sections of a bipolar DC line to connect any of the overhead line poles to any cable.

3.2 CONVERTER BRIDGE MODELING

3.2.1 Rectifier Equations

3.2.1.1 Ideal Converter Bridge Operation

The most used rectifier within an AC–DC conversion at high voltages, high powers, and industrial frequency is the Graetz bridge or the six-pulse converters (Figure 3.10) [2, 10, 11, 14].

This rectifier is considered as being composed by two three-phase connections, with median point, with outputs series connected. The cathodes of the upper group of valves T_1, T_3, T_5 are connected to the anodes of the lower group. The common potential of the cathodes of these valves is equal to the most positive anode voltage. The common potential of the anodes of valves T_2, T_4, T_6 is equal to the most negative cathode voltage.

The same transformer secondary windings feed two groups of three thyristor valves each. The thyristors are in conduction two by two, depending on the voltage variation in the secondary. With no grid control, conduction will take place between the cathode and the anode of highest potential. For example, during t_0–t_1

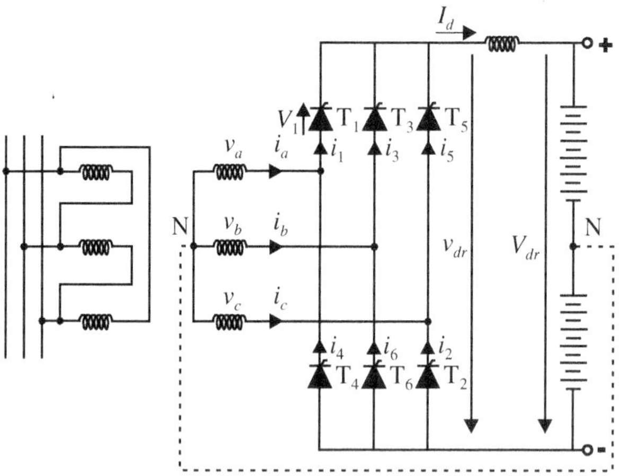

Figure 3.10 Three-phase bridge arrangement of valves, or Graetz rectifier circuit [2, 11].

(Figure 3.11a), among the thyristors belonging to the upper group (T_1, T_3, T_5), T_1 will be in conduction because it has the most positive anode; among the thyristors belonging to the lower group (T_2, T_4, T_6), T_6 will be in conduction because it has the most negative cathode. As a consequence, T_1 will also remain in conduction during t_1–t_2 interval (i.e., as long as $v_a > v_b$).

At the beginning of the t_2–t_3 period, T_3 will enter the conducting state. Similarly, at the beginning of t_1–t_2, the thyristor T_2, having the most negative cathode among the inferior group of thyristors, is conducting, while T_6 is blocked (T_2 remains in conduction during the entire interval t_1–t_3). The transformer secondary line-to-neutral voltages are shown in Figure 3.11a.

These are also the voltages of the anodes of the lower group of valves and the cathodes of the upper group, all with respect to neutral point *N*. The difference in ordinates between the upper and lower envelopes is the instantaneous direct voltage on the valve side of the smoothing reactor. This is shown in Figure 3.11b as the envelope of the line-to-line voltages. It is immediately to be noted that the ripple of the direct voltage is of frequency 6*f* and the magnitude of ripple is smaller. Each thyristor is in conduction a *T*/3 period, the shape of current wave being rectangular (Figures 3.11c, 3.11d and 3.11e).

The inverse voltage V_1 across valve 1, appearing at the ends of T_1 while T_3 is in conducting state, is u_{ab}; during the conduction period of T_5 the inverse voltage is u_{ac} (Figure 3.11f).

The maximal inverse voltage at thyristor ends is

$$V_{\max} = \sqrt{3}\hat{V}_s$$

where the line-to-neutral peak magnitude voltage is $\hat{V}_s = \sqrt{2}V_s$ (where V_s is the rms line-to-neutral alternating voltage) and the line-to-line rms voltage is $U_s = \sqrt{3}\,V_s$.

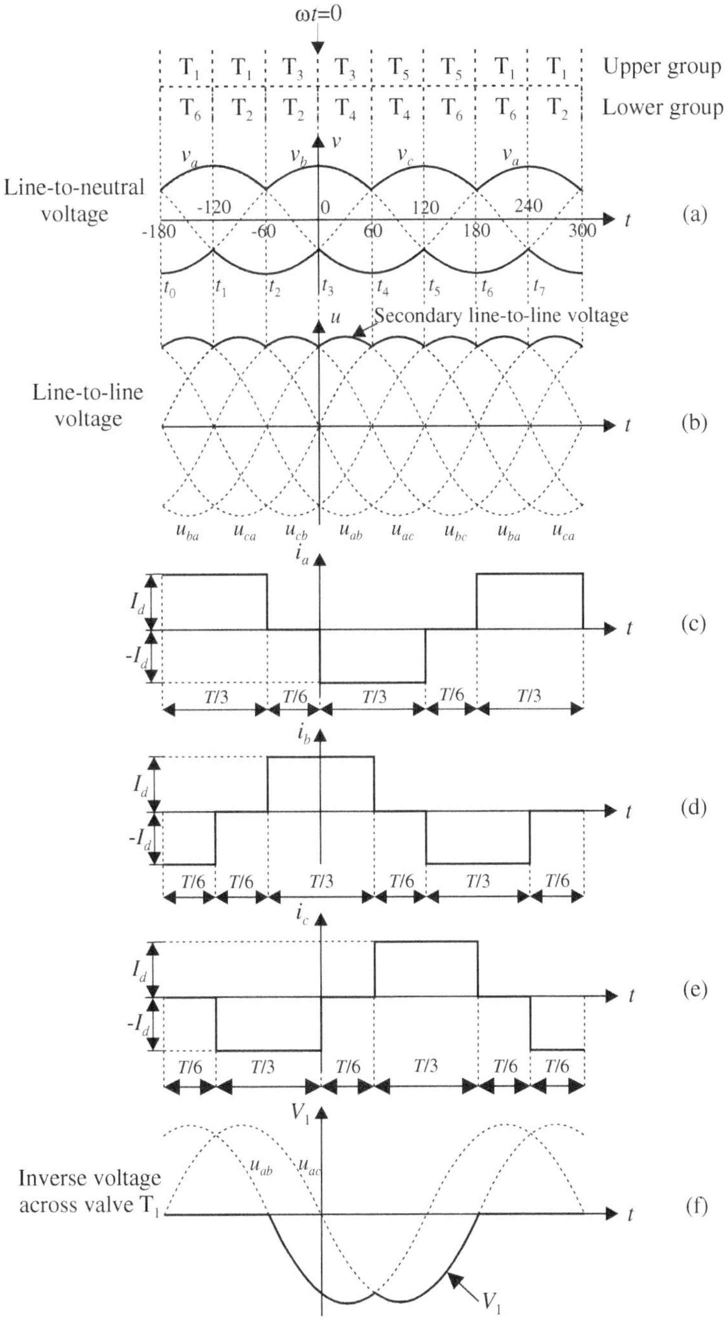

Figure 3.11 Voltage and current waveforms for uncontrolled rectifier. Adapted from Kimbark 1971[2] and Eremia et al. 2006 [14].

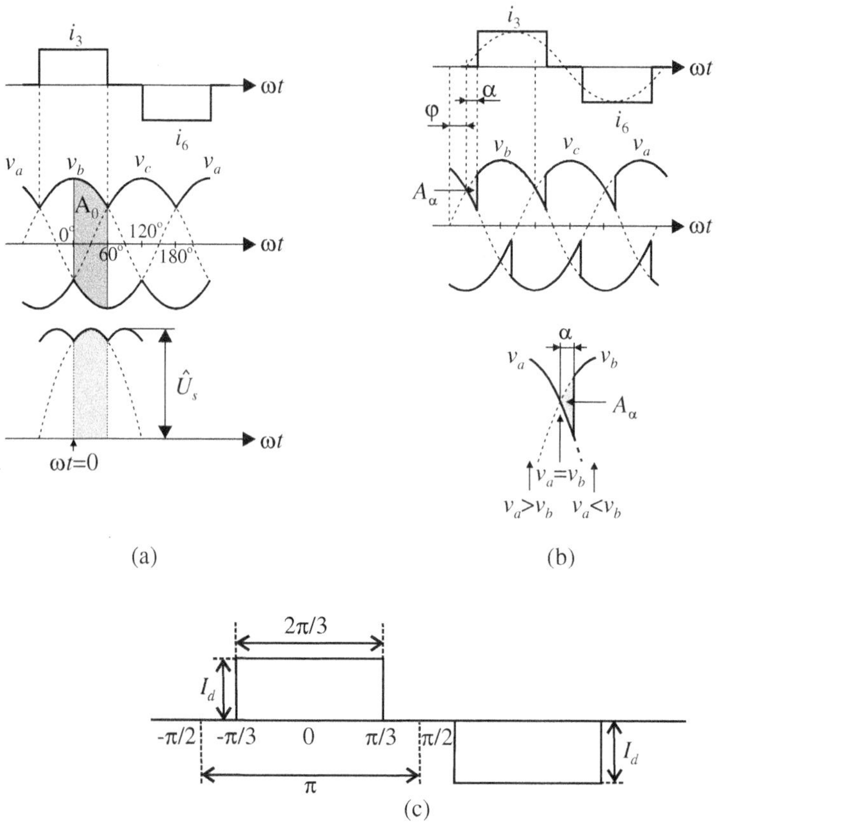

Figure 3.12 Voltage and current waveforms for the uncontrolled rectifier (a) and controlled rectifier (b) and (c) [12].

We consider the three-phase voltages system from the secondary windings of the transformer supplying the three-phase conversion bridge:

$$v_a = \hat{V}_s \cos(\omega t + 120°); \; v_b = \hat{V}_s \cos(\omega t); \; v_c = \hat{V}_s \cos(\omega t - 120°)$$

where the instant ($\omega t = 0$) when the b-phase voltage reaches its maximal value (Figure 3.11a) has been chosen as time origin. Two cases will be examined: the uncontrolled and the controlled rectifier.

THE CASE OF UNCONTROLLED CONVERTER. Let us consider the sequence with the thyristor T_3 in conduction (Figure 3.12a); the voltage from transformer's secondary windings is [2, 12]

$$v_b - v_a = \hat{V}_s[\cos(\omega t) - \cos(\omega t + 120°)] = 2\hat{V}_s \sin(\omega t + 60°) \sin 60°$$

respectively:

$$u_{ba} = v_b - v_a = \sqrt{3}\hat{V}_s \sin(\omega t + 60°) = \hat{U}_s \sin(\omega t + 60°) \tag{3.1}$$

By integrating u_{ba} voltage from equation (3.1) on a pulse width ($\pi/3$), we obtain the surface A_0 (Figure 3.12a):

$$A_0 = \int_0^{\pi/3} u_{ba} d(\omega t)$$

If one divides this value by the pulse width, one obtains a value that approximates good enough the value of the ideal no-load direct voltage:

$$V_{d0} = \frac{3}{\pi} A_0 = \frac{3}{\pi} \int_0^{60^\circ} \sqrt{3}\hat{V}_s \sin(\omega t + 60^\circ) d(\omega t) = \frac{3\sqrt{3}}{\pi} \hat{V}_s [-\cos(\omega t + 60^\circ)] \Big|_0^{60^\circ}$$

One obtains

$$V_{d0} = \frac{3\sqrt{3}}{\pi} \hat{V}_s = 1.653\, \hat{V}_s \tag{3.2}$$

where V_{d0} is the maximum DC voltage that a six-pulse thyristor converter can achieve.

THE CASE OF CONTROLLED RECTIFIER (with ignition delay). The gate control can be used to delay the ignition of the valves. As the six-pulse bridge operates cyclically, it is common to refer to this time delay in terms of electrical degrees with 0° being the point on wave where the ideal diode bridge (uncontrolled converter) would conduct. This delay angle is referred as the "firing" or "trigger" angle and is denoted by α (Figure 3.12b); it corresponds to time delay of α/ω seconds. The delay angle is limited to 180°. If α exceeds 180°, the valve fails to ignite. Figure 3.12b presents in detail the transfer of current from a phase to another.

The average direct voltage is calculated as for the uncontrolled rectifier:

$$V_d = \frac{3}{\pi} \int_{\alpha}^{60^\circ+\alpha} \sqrt{3}\hat{V}_s \sin(\omega t + 60^\circ) d(\omega t) = \frac{3\sqrt{3}}{\pi} \hat{V}_s [-\cos(\omega t + 60^\circ)] \Big|_{\alpha}^{60^\circ+\alpha}$$

Taking into account equation (3.2), results

$$V_d = V_{d0} \cos\alpha \tag{3.3}$$

The first effect of α angle is that the direct voltage is decreased by the factor $\cos\alpha$. Since α can range from 0° to 180°, $\cos\alpha$ can range from 1 to −1. As a consequence, the voltage V_d varies between $+V_{d0}$ and $-V_{d0}$. A firing with an angle α between 0 and $\pi/2$ determines the operation as the rectifier. For angles between $\pi/2$ and π, the polarity of V_d is negative, and because the current I_d does not change the direction, one obtains a power transfer in the opposite direction, from DC toward AC side; the converter acts like an inverter.

For $\alpha = 90^\circ$, the voltage waveform is symmetrical and consequently its value is zero. For $\alpha = 0^\circ$, the fundamental component of the alternating current from the secondary winding of the transformer is in phase with the voltage waveform on that phase. If the firing is delayed by α, the current pulses of fundamental component will be dephased with $\varphi = \alpha$ as regards the voltage waveform (they are not centered anymore on the voltage peak value).

The second effect is that the more α increases, the more increases φ angle between current and voltage waveforms. If the power losses inside the converter are neglected, then the power entering in the converter on the AC side should be equal to the power getting out on the DC side, meaning:

$$\sqrt{3}U_s I_{s1} \cos\varphi \cong V_d I_d = V_{d0} I_d \cos\alpha \tag{3.4}$$

where

I_{s1} is the effective value of the fundamental frequency component of alternating line current;

U_s – line-to-line voltage;

$\cos\varphi$ – the power factor,

all defined on the AC side of the converter.

The fundamental frequency component of the alternating line current can be determined by Fourier analysis of the current wave shape shown in Figure 3.12c. The peak value of the alternating line current is [2, 12]

$$\sqrt{2}I_{sM} = \frac{2}{\pi}\int_{-\frac{\pi}{3}}^{\frac{\pi}{3}} I_d \cos\theta \, d\theta = \frac{2}{\pi} I_d \sin\theta \Big|_{-60^\circ}^{60^\circ} = \frac{2}{\pi} I_d[\sin 60^\circ - \sin(-60^\circ)] = \frac{2}{\pi}\sqrt{3}I_d$$

The rms value is

$$I_{s1} = \frac{I_{sM}}{\sqrt{2}} = \frac{2\sqrt{3}}{\pi\sqrt{2}} I_d = \frac{\sqrt{6}}{\pi} I_d = 0.78 I_d \tag{3.5}$$

By substituting in (3.4), the values I_{s1} and V_{d0} can be obtained:

$$\cos\varphi \cong \cos\alpha$$

meaning that, the closer is α to $\pi/2$, the smaller is the power factor. As a consequence, the *rectifier will consume more reactive power from the AC system*.

3.2.1.2 Commutation Process or Overlap

Due to the leakage inductance of the converter transformers and the impedance in supply network, the current in a valve cannot change suddenly and thus commutation from one valve to the next in the same row cannot be instantaneous. As a consequence, due to the electromagnetic inertia, the I_d value at the beginning of commutation process, as well as the decrease to zero while blocking are not instantaneous, the time required is called *the overlap* or *commutation time*. The overlap angle is noted λ, and consequently the overlap time is λ/ω seconds. In normal operation $\lambda < 60^\circ$ [2, 13].

For example, when valve 3 is fired (Figure 3.13a), the current transfer from valve 1 to valve 3 takes a finite period λ during which both valves are conducting. Each interval of the period of supply can be divided into two subintervals. In first subinterval, during commutation, three valves conduct simultaneously (1,2,3), and in

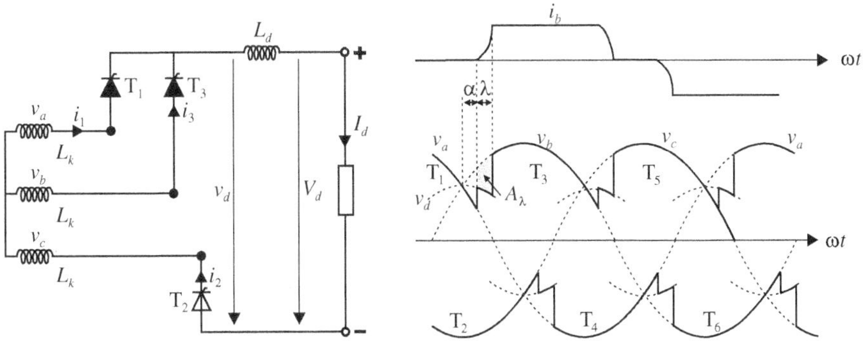

(a) (b)

Figure 3.13 Electric circuit for the commutation interval of T_1 with T_3 [2].

the second subinterval, between commutations, only two valves are conducting. A new commutation begins every 60° and lasts for an angle λ. Thus the angular interval when two valves conduct is (60° − λ). If λ > 60, unpleasant phenomenon may occur (i.e., the current increases to value which may destroy the junctions).

Summarizing, one can write the following conduction sequence of thyristors: 12, 123, 23, 234, 34, 345, 45, 456, 56, 561, 61, 612, etc.

The interval in which valves 1 and 2 conduct ends at $\omega t = \alpha$. Let us suppose that at the instant $\omega t = \alpha$, T_1 starts to block and T_3 enters in conduction. Within the lower group, T_2 is in conduction, saturated. So, at the instant $\omega t = \alpha$ commutation one has $i_1 = I_d$ and $i_3 = 0$, while at the instant $\omega t = \alpha + \lambda$ commutation results $i_1 = 0$ and $i_3 = I_d$. During this subinterval, the direct current is transferred from thyristor 1 to thyristor 3. The end of the interval is at $\omega t = \delta$ where δ is called the extinction angle and is given by $\delta = \alpha + \lambda$. For the considered case, the following relations could be written as $i_a = i_1$, $i_b = i_3$, $i_c = i_2$.

Because of the current variations during λ subinterval, at thyristor's ends appear voltage drops:

$$v_d - v_b = -L_k \frac{di_b}{dt} \tag{3.6a}$$

$$v_d - v_a = -L_k \frac{di_a}{dt} \tag{3.6b}$$

where L_k is the transformer secondary inductance.

Taking into account that during the interval λ exists the relation $i_1 + i_3 = I_d =$ ct. Differentiating both sides, we obtain:

$$\frac{di_a}{dt} + \frac{di_b}{dt} = \frac{dI_d}{dt} = 0$$

Using this result according to the relations (3.6a) and (3.6b), we obtain:

$$2v_d - (v_a + v_b) = -L_k \left(\frac{di_b}{dt} + \frac{di_a}{dt} \right) = 0 \Rightarrow v_d = \frac{v_a + v_b}{2} \tag{3.7}$$

Consequently, during λ interval, the unfiltered direct voltage v_d becomes equal to the half-sum of secondary voltages of the phases in commutation (the dotted line from Figure 3.13b). This situation stops once the overlap (anodic superposition) disappears. The corresponding diminishing of the direct voltage caused by the commutation is called *commutation voltage drop*. Within λ interval, the phases a and b of transformer's secondary windings are in short circuit. If in equation (3.6b) one replaces v_d from equation (3.7), it results:

$$v_b - v_a = -2L_k \frac{di_a}{dt} = 2L_k \frac{di_b}{dt} = 2X_k \frac{di_b}{d(\omega t)}$$

where $X_k = \omega L_k$ is the commutation reactance.

Therefore, the commutation voltage becomes

$$\sqrt{3}\hat{V}_s \sin \omega t = 2X_k \frac{di_b}{d(\omega t)}$$

respectively, if $I_{sc2} = \sqrt{3}\hat{V}_s / 2X_k$ is the current in a line-to-line short circuit on the AC source, then

$$di_b = I_{sc2} \sin \omega t \, d(\omega t)$$

By integrating the last equation, one obtains

$$\int di_b = \int I_{sc2} \sin \omega t \, d(\omega t)$$

thus

$$i_b(t) = I_{sc2}(-\cos \omega t) + C = i_3(t)$$

The integration constant, C, is obtained from the boundary condition at the beginning of commutation interval: $\omega t = \alpha$ and $i_3 = 0$. The result is

$$C = I_{sc2} \cos \alpha = \frac{\sqrt{3}}{2X_k} \hat{V}_s \cos \alpha$$

Substituting in the expression of $i_b = i_3(t)$, it results

$$i_3(t) = I_{sc2}(\cos \alpha - \cos \omega t) \tag{3.8}$$

The current i_3 of the incoming valve during commutation consists of a constant term ($I_{sc2} \cos \alpha$) and a sinusoidal term ($-I_{sc2} \cos \omega t$) lagging the commutating voltage by 90°, which is the characteristic of a purely inductive circuit and has a peak value I_{sc2}. The constant term of i_3 depends on α; it serves to make $i_3 = 0$ at the beginning of commutation. For $\alpha = 0°$, it shifts the sine wave upward by its peak value.

Imposing the conditions at the end of commutation interval: $\omega t = \alpha + \lambda$ and $i_b = i_3 = I_d$, we obtain from (3.8) the expression of *direct current* as function of angles α and λ:

$$I_d = I_{sc2}[\cos \alpha - \cos(\alpha + \lambda)] = I_{sc2}(\cos \alpha - \cos \delta) \tag{3.9}$$

During commutation, the shape of i_1 satisfies $i_1(t) = I_d - i_3(t)$, hence

$$i_1(t) = I_d - I_{sc2}(\cos \alpha - \cos \omega t) = I_{sc2}(\cos \omega t - \cos \delta) \tag{3.10}$$

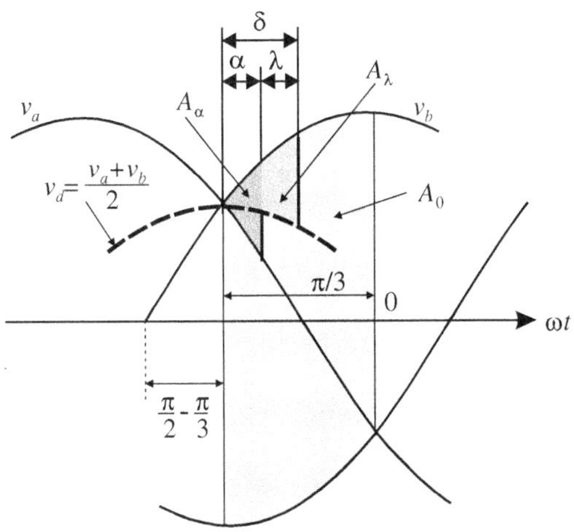

Figure 3.14 Figure explaining the voltage drop caused by overlap during commutation from valve 1 to valve 3. Adapted from Kimbark [2].

The current i_1 of the outgoing valve has a sine term of the same amplitude as that of i_3 but of opposite phase, and its constant term makes $i_1 = I_d$ at the beginning.

For α nearly equal to 0° (or 180°), the commutation period or the overlap is the greatest. The overlap is the shortest when $\alpha = 90°$, since i_3 is associated with the segment of the sine wave, which is nearly linear. Also, if the source voltage $\hat{V}_s$ is lowered or if I_d is increased, the overlap increases [12].

Commutation Process Effects. Voltage reduction is due to commutation overlap. The effect of the overlap on the voltage could be compared to the subtraction of an area A_λ from an area A_0 ($A_0 = V_{d0}\pi/3$) at every sixth of a cycle ($\pi/3$ rad), as shown in Figure 3.14.

$$A_\lambda = \int_{\alpha}^{\delta=\alpha+\lambda} \left(v_b - \frac{v_a + v_b}{2} \right) d(\omega t) = \frac{\sqrt{3}}{2} \int_{\alpha}^{\alpha+\lambda} \hat{V}_s \sin \omega t \; d(\omega t)$$

By integrating, we obtain

$$A_\lambda = \frac{\sqrt{3}}{2} \hat{V}_s (-\cos \omega t) \Big|_{\alpha}^{\alpha+\lambda} = \frac{\sqrt{3}}{2} \hat{V}_s [\cos \alpha - \cos(\alpha + \lambda)]$$

The average voltage drop caused by the commutation overlap is obtained from

$$\Delta V_d = \frac{3A_\lambda}{\pi} = \frac{V_{d0}}{2} [\cos \alpha - \cos(\alpha + \lambda)] \tag{3.11}$$

where V_{d0} is the ideal no-load voltage given by equation (3.2).

In the case when the commutation process and ignition delay is considered, the reduction in direct voltage is represented by areas A_α and A_λ; the direct voltage is given by [15]

$$V_d = V_{d0}\cos\alpha - \Delta V_d = \frac{V_{d0}}{2}[\cos\alpha + \cos(\alpha+\lambda)] \tag{3.12}$$

With no overlap ($\delta = \alpha$), it results $V_d = V_{d0}\cos\alpha$, which is the same as before.

(a) Taking into account that $\alpha + \lambda < 180°$ and the fact that λ could reach values up to 20°–30°, we obtain *a maximal value of* $\alpha < 150°$, fact which reduces α's range of values in inverter operation mode. As a consequence, the control characteristic for a self-commutated rectifier is operational only up to this value.

(b) The commutation process also modifies the shapes of current and voltage waveforms and consequently their *harmonics values*, as well as the *power factor* [2, 11, 13]:

$$\cos\varphi = \frac{1}{2k}[\cos\alpha + \cos(\alpha+\lambda)] \tag{3.13a}$$

where

$$k = \frac{\sqrt{[\cos 2\alpha - \cos 2(\alpha+\lambda)]^2 + [2\lambda + \sin 2\alpha - \sin 2(\alpha+\lambda)]^2}}{4[\cos\alpha - \cos(\alpha+\lambda)]} \tag{3.13b}$$

For most practical power flow cases, k is close to 1 and can be assumed to be constant. When α is varied between 5 and 20° and λ between 8 and 22°, k varies between 0.994 and 0.999.

(c) As the phase shift φ is positive (AC current is lagging), the reactive power Q_c absorbed by the converter is positive, that is Q_c is absorbed both on the rectifier side and on the inverter side. For converter bridges operating in normal mode (i.e., $\alpha = 15°$) and for a commutation reactance $X_k = 15\%$, *the consumed reactive power is about* 50% *of the transmitted active power.*

3.2.1.3 *Equivalent Circuit of the Rectifier*

If in equation (3.12), we replace the expressions of biphase short-circuit current I_{sc2} and we take into account (3.2), results a new form of the direct current:

$$I_d = \frac{\pi}{6X_k}V_{d0}[\cos\alpha - \cos(\alpha+\lambda)] \tag{3.14a}$$

If in expression (3.12) of the direct voltage, we replace $\cos(\alpha+\lambda)$ from (3.14a), we obtain

$$V_{dr} = V_{d0,r}\cos\alpha - \frac{3X_{kr}I_d}{\pi} = V_{d0,r}\cos\alpha - R_{cr}I_d \tag{3.15}$$

where $R_{cr} = 3X_{kr}/\pi = 6fL_k$ is the *equivalent commutation resistance on the rectifier side* (represents the equivalent in DC of transformer's leakage reactance). Note that the overlap angle has been eliminated and its place we have R_c. Based on equation (3.15), an equivalent circuit of the bridge rectifier may be conceived (Figure 3.15). The direct voltages and current in this circuit are the average without ripple.

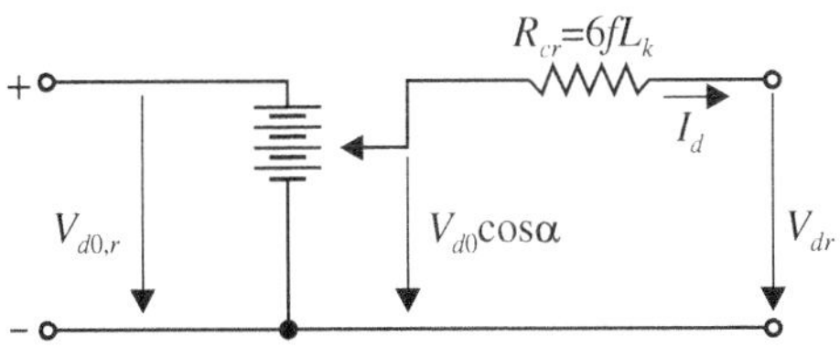

Figure 3.15 Equivalent circuit of the rectifier bridge [2].

3.2.2 Inverter Equations

With rectifier operation, the output current I_d and output voltage V_d are such that power is absorbed by a load. For inverter operation, it is required to transfer power from the direct current to the alternating current systems and as current can only flow from anode to cathode (i.e., in the same direction as with rectification) the direction of the associated voltage must be reversed. An alternating voltage system must exist on the primary side of the transformer, and grid control of the converters is essential.

If the bridge rectifier is given progressively great delay, the output voltage decreases becoming zero when α is 90°. With further delay, the average direct voltage becomes negative and the applied direct voltage, from the rectifier, forces current through the valves against this negative or back voltage. The converter thus receives power and inverts.

In the hypothesis that the commutation process is neglected, the polarity of direct voltage V_d could change when $\alpha = 90°$. The direct voltage becomes negative for $90° < \alpha < 180°$, and the converter operates in the inversion mode. The valve current cannot be reversed since conduction occurs in only one direction. Thus reversal of the direct voltage V_d implies a reversal of power.

In reality, due to the commutation process, the angle α_{tr}, for which the switch to inverter operation mode takes place, is smaller. In order to demonstrate this affirmation, in equation (3.12) the sum expression will be transformed:

$$V_d = \frac{V_{d0}}{2}[\cos\alpha + \cos(\alpha+\lambda)] = V_{d0}\cos\left(-\frac{\lambda}{2}\right)\cos\left(\alpha+\frac{\lambda}{2}\right) \tag{3.16}$$

It can be noticed that, for $\alpha + \lambda/2 < \pi/2$, the converter operates as the rectifier, and for $\alpha + \lambda/2 > \pi/2$ as the inverter. The limit case $\alpha_{tr} + \lambda/2 = \pi/2$ corresponds to the situation when the converter takes from the network only reactive power. It results that the delay angle corresponding to start of inversion is $\alpha_{tr} = 90° - \lambda/2$, always smaller than 90°.

In Figure 3.16b, the voltage waveforms for the inversion mode are presented.

In order to obtain the equations describing inverter's operation, we use (3.12) and (3.13a), where α is the delay angle of the thyristors while operating in the rectifier mode and $\delta = \alpha + \lambda$ is the extinction angle. Both angles are measured by the delay from the instant at which the commutating voltage is zero and increasing ($\omega t = 0°$).

Although angles α and δ could have been used in inverter's theory, in order to make a difference, other symbols will be employed. These angles are defined by their advance with respect to the instant (ωt = 180° for ignition of valve 3 and extinction of valve 1) when the commutating voltage is zero and decreasing, as

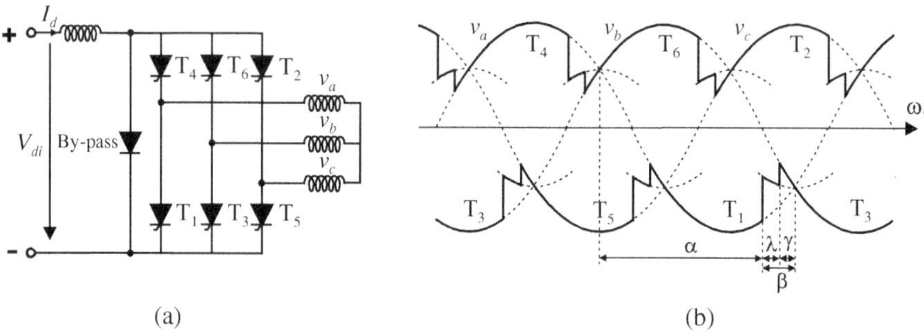

Figure 3.16 Bridge connection – inverter operation (a); voltage waveform (b) [2].

shown in Figure 3.17 [16]. From the figure, we see that $\beta = 180° - \alpha$ for the ignition advance angle, and $\gamma = \beta - \lambda = 180° - \delta$ for the extinction advance angle.

Back to rectifier's equation (3.12), we use opposite polarity for direct voltage and replaces $\alpha = 180° - \beta$ and $\lambda = \beta - \gamma$:

$$V_{di} = -\frac{V_{d0,i}}{2}[\cos\alpha + \cos(\alpha + \lambda)] = -\frac{V_{d0,i}}{2}[\cos(180° - \beta) + \cos(180° - \gamma)]$$

Taking into account that

$$\cos\alpha = \cos(180° - \beta) = -\cos\beta$$
$$\cos\delta = \cos(180° - \gamma) = -\cos\gamma$$

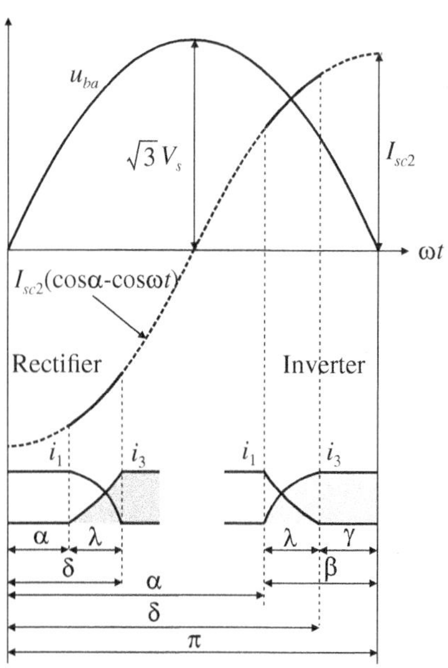

Figure 3.17 Relationships between the angles used in converter theory and why the curvature of the front of a current pulse of inverter differs from that of a rectifier. *Source:* IEEE Std. 1992 [16]. Reproduced with permission of IEEE.

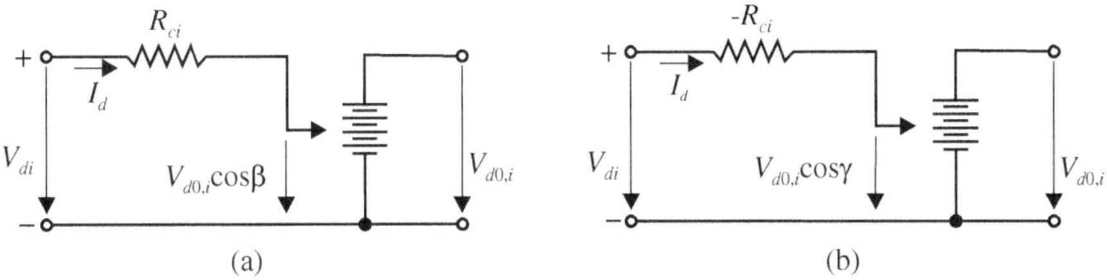

Figure 3.18 Equivalent circuits of the inverter bridge [2].

we obtain:

$$V_{di} = \frac{V_{d0,i}}{2}(\cos\beta + \cos\gamma) \tag{3.17}$$

Similarly, equation (3.14a) becomes

$$I_d = I_{sc2}(\cos\gamma - \cos\beta) \tag{3.18}$$

If one replaces $\cos\alpha = -\cos\beta$ in equation (3.17), we obtain a first equation for converter's operation in inversion mode:

$$V_{di} = V_{d0,i}\cos\beta + R_{ci}I_d \tag{3.19a}$$

Because inverters are commonly controlled so as to operate at constant extinction advanced angle γ, it is useful to have the relations between V_d and I_d for this condition. If we replaces $V_{d0,i}\cos\beta$ from equation (3.17) in equation (3.19a), results another form of the equation:

$$V_{di} = V_{d0,i}\cos\gamma - R_{ci}I_d \tag{3.19b}$$

However, it is to be noted that while α is directly controllable, γ is not.

Accordingly to the operation equations (3.19a) and (3.19b), two possible equivalent circuits could be built for the inverter operation (Figures 3.18a and 3.18b).

HVDC LINK EQUATIONS. Table 3.1 shows the equations necessary to calculate the parameters of an HVDC link for both rectifier and inverter.

Observation. The alternating voltages are phase-to-phase voltages, $N_{ik,r}$ and $N_{ik,i}$ are the transformation ratio of the transformers supplying rectifier, respectively, inverter bridges.

3.3 CONTROL OF CSC–HVDC TRANSMISSION

3.3.1 Equivalent Circuit and Control Characteristics

3.3.1.1 Equivalent Circuit of DC Transmission Link

A DC transmission link with two ends, composed by two transformers, rectifier, inverter and the electric line might be represented through an equivalent circuit, where the baseline for all elements is the DC part (Figure 3.19). Subscripts r and i refer to rectifier and inverter, respectively.

TABLE 3.1 HVDC link equations

	Rectifier	Inverter
Ideal no-load direct voltage	$V_{d0,r} = \frac{3\sqrt{2}}{\pi} N_{ik,r} V_{kr}$	$V_{d0,i} = \frac{3\sqrt{2}}{\pi} N_{ik,i} V_{ki}$
Direct voltage with commutation overlap and ignition delay	$V_{dr} = V_{d0,r} \cos\alpha - \frac{3}{\pi} X_{kr} I_d$	$V_{di} = V_{d0,i} \cos\gamma - \frac{3}{\pi} X_{ki} I_d$
Current in the transformer secondary	$I_{sr} \cong \frac{\sqrt{6}}{\pi} I_d$ $I_{pr} = N_{ik,r} I_{sr}$	$I_{si} \cong \frac{\sqrt{6}}{\pi} I_d$ $I_{pi} = N_{ik,i} I_{si}$
Active power on DC line	$P_{dr} = V_{dr} I_d$	$P_{di} = V_{di} I_d$
Apparent power at AC system terminal bus	$S_{kr} = \sqrt{3} V_{kr} I_{pr}$	$S_{ki} = \sqrt{3} V_{ki} I_{pi}$
Active power at AC system terminal bus	$P_{kr} \cong P_{dr}$	$P_{ki} \cong P_{di}$
Reactive power at AC system terminal bus	$Q_{kr} = \sqrt{S_{kr}^2 - P_{kr}^2}$	$Q_{ki} = \sqrt{S_{ki}^2 - P_{ki}^2}$
DC line equation	$V_{dr} = V_{di} + R_L I_d$	

When the scheme is build, several issues should be considered:

- the transformers have variable transforming ratios. The effect of leakage reactance on DC voltage was included through the commutation resistances R_{cr} on the rectifier side, respectively, R_{ci} on the inverter side;
- the DC overhead line is represented only through resistance R_L; its capacity and reactance have been neglected;
- harmonics filters and the elements for reactive power generation (capacitor banks, static or synchronous compensators) have not been included.

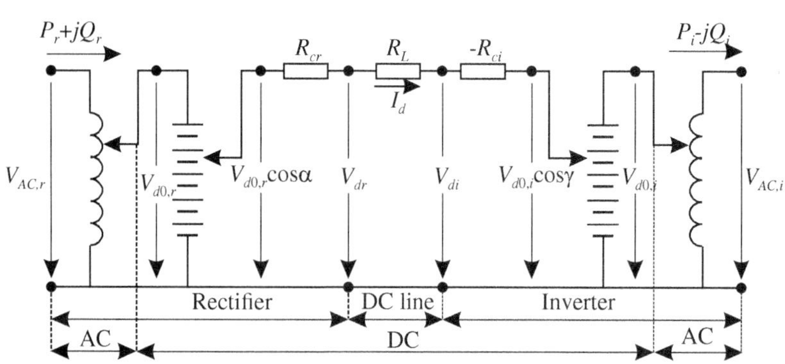

Figure 3.19 Equivalent circuit of the HVDC link valid for average currents and voltages in the steady state [2].

A converter can be used to either convert AC power to DC power or vice versa. Only the relative DC voltage and current polarities determine the direction of power flow. A terminal, which supplies power to DC link, is termed the rectifier terminal; the terminal, which takes power from the DC line, is termed the inverter terminal. The direction of power flow and therefore the terminology for the terminals can change in less than a second if the converter voltage levels are changed by firing angle control.

Power flow on a DC link is always from the terminal with the greater positive direct voltage to the lesser positive voltage or from the more negative terminal to the less negative terminal. However, a power direction reversal does not require a current direction reversal.

The direct current, I_d, flowing through the line from the rectifier to inverter is

$$I_d = \frac{V_{d0,r}\cos\alpha - V_{d0,i}\cos\gamma}{R_{cr} + R_L - R_{ci}} \tag{3.20a}$$

where:

$$R_{cr} = \frac{3X_{kr}}{\pi};\ R_{ci} = \frac{3X_{ki}}{\pi} \tag{3.21}$$

DC power is simply the product of the current and the voltage at the particular location. Power transfer on the DC link can be increased by either increasing the rectifier voltage or decreasing the inverter voltage. Either of these increases the current and power in the DC link.

The powers at the rectifier and inverter terminals are

$$P_{dr} = V_{dr} I_d \tag{3.22a}$$

$$P_{di} = V_{di} I_d = P_{dr} - (R_L + R_{cr} - R_{ci}) I_d^2 \tag{3.22b}$$

An essential characteristic of the transmission at the direct voltage is the possibility to rigorously control the transmitted active power, in terms of magnitude and direction.

The normal mode of power control on a DC link is to hold the inverter voltage constant and to control the current by changing the rectifier voltage level. The rectifier direct voltage will usually be between 0% (no load) and 10% higher than the inverter DC voltage, depending upon line losses and loading level.

The value of the direct current, I_d, could be controlled by the change of $V_{d0,r}$ and $V_{d0,i}$ values or by the change of α or γ angles:

- the values of $V_{d0,r}$ and $V_{d0,i}$ could be regulated by changing the converter transformer turn ratio N_{ik} with a slow acting control;
- the firing angle could be rapidly controlled by gate control. Usually, both the rectifier and the inverter are operating with $\alpha = 12° - 15°$, respectively, $\gamma_{\min} = 18°$.

As a consequence, through firing angles control, a converter could operate as the rectifier or as the inverter; the power flow direction could be changed by reversal of polarity of the direct voltages at the both ends.

The responsibilities for voltage regulation and current regulation are kept distinct and are assigned to separate terminals. Generally, in a DC transmission link, the inverter substations controls the direct voltage V_d, keeping it at a constant value and rigorously dependent on the voltage on the AC side. At the other terminal, the rectifier substation regulates the direct voltage so that I_d current corresponds to the necessary active power P_d.

3.3.1.2 Voltage–Current Characteristics

The *ideal voltage–current characteristics* are presented in Figure 3.20 [12]. The voltage V_d and the current I_d forming the coordinates may be measured at the same common point on the DC line. The rectifier and inverter characteristics are both measured at the rectifier. The inverter characteristic thus includes the voltage drop across the line. With the rectifier maintaining constant current (CC), its V_d–I_d characteristic is a vertical line.

Based on the inverter equation seen from the rectifier

$$V_d = V_{d0,i} \cos \gamma + (R_L - R_{ci})I_d \tag{3.23}$$

if the current is constant, also the voltage will be constant.

This gives the inverter characteristic, with γ maintained at a fixed value. If the commutating resistance R_{ci} is slightly larger than the line resistance R_L, the characteristic of the inverter has a small negative slope. Since an operating condition has to satisfy both rectifier and inverter characteristics, it is defined by the intersection of the two characteristics.

The rectifier characteristic can be shifted horizontally by adjusting the "current command" or "current order." If measured current is less than the command, the regulator advances the firing by decreasing α. The inverter characteristic can be raised or lowered by means of its transformer tap changer. When the tap changer is moved, the constant extinction angle (CEA) regulator quickly restores the desired γ. As a result, the direct current changes, which is then quickly restored by the current regulator of the rectifier. The rectifier tap changer acts to bring α into the desired range (10°–20°) to ensure a high power factor and adequate room for control [12].

The rectifier maintains CC by changing α. However, α cannot be less than its minimum value (α_{min}). Once α_{min} is reached, no further voltage increase is possible and the rectifier will operate at constant ignition angle (CIA).

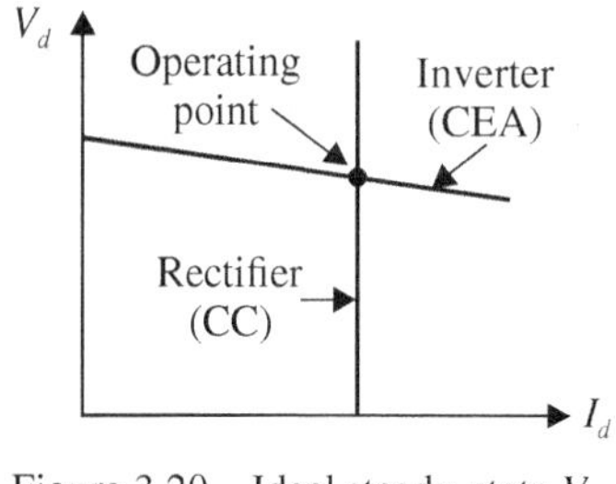

Figure 3.20 Ideal steady-state V_d–I_d characteristic seen from the rectifier terminal.

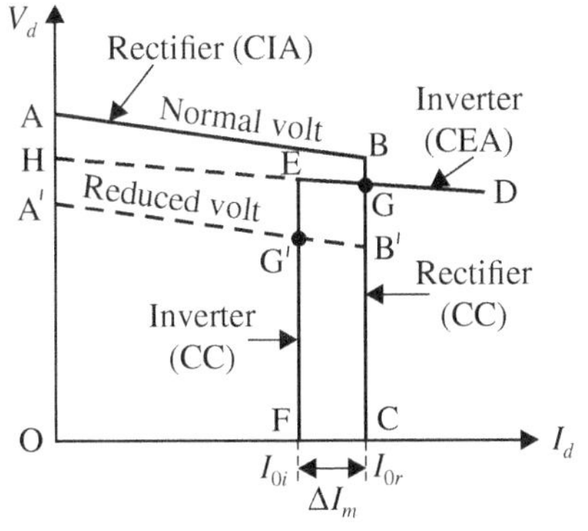

Figure 3.21 Actual converter control steady-state characteristics [6, 13].

The *actual steady-state characteristics*, based on the above description, are shown in Figure 3.21:

(i) The rectifier characteristic consists of the two segments (AB and BC). The segment AB corresponds to minimum ignition angle and represents the CIA control mode; the segment BC represents the normal CC control mode. The complete rectifier characteristic at the normal voltage is defined by ABC. At a reduced voltage it shifts, as indicated by A′B′C;

(ii) the inverter characteristic consists of two segments (DE and EF). The CEA characteristic of the inverter intersects the rectifier characteristic at G for the normal voltage. However, the inverter CEA characteristic (HD) does not intersect the rectifier characteristic at a reduced voltage represented by A′B′C. Therefore, a big reduction in a rectifier voltage would cause the current and power to be reduced to zero after a short time depending on the presence of the smoothing inductance. To avoid this situation, the inverter is equipped with a current regulator, whose reference I_{0i} is smaller than the reference I_{0r} of the rectifier.

The variables I_{0r} and I_{0i} are called rectifier current order and inverter current order, respectively, while their difference

$$\Delta I_m = I_{0r} - I_{0i}$$

is called current margin and its usual value is $\Delta I_m = 0.1 - 0.15$ p.u. from the rated current.

Under normal operating conditions, the operating point is G, the rectifier controls the direct current and the inverter the direct voltage; this is CCR (current control performed by the rectifier) operating mode. With a reduced rectifier voltage, the operating condition is represented by the intersection point G″. The inverter takes current control, and the rectifier established the voltage; this is CCI (current control performed by the inverter) operating mode. The change from a mode to another is referred to as a mode shift [12].

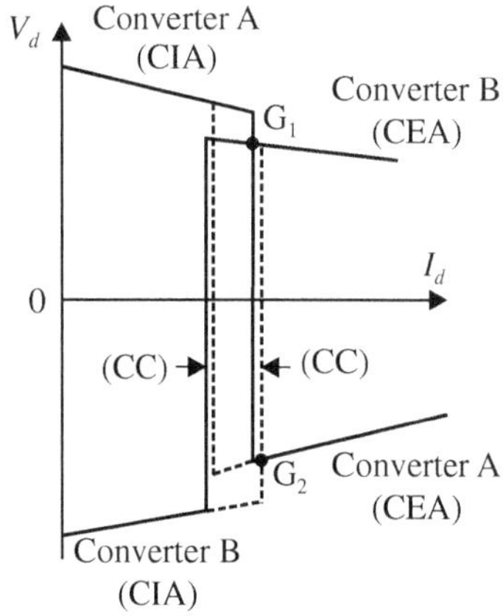

Figure 3.22 Combined characteristics.

In most HVDC systems, each converter is required to function as a rectifier as well as an inverter. Consequently, each converter is provided with a *combined characteristic* as shown in Figure 3.22.

The characteristic of each converter consists of three segments: CIA corresponding to α_{min}, CC, and CEA.

The power transfer is from converter A to converter B, when the characteristics are as shown in Figure 3.22 by solid lines. The operating condition in this mode of operation is represented by point G_1. The power flow is reversed when the characteristics are as shown by the dotted lines. This is achieved by reversing the "margin setting," that is by making the current order setting of converter B exceed that of converter A. The operating condition is now represented by G_2 in Figure 3.22; the current I_d is the same as before, but the voltage polarity has changed [12].

3.3.2 HVDC Control Principles

3.3.2.1 State Variables of a HVDC Link

The operating state of a DC link is determined by means of the electrical quantities associated with the converter station, called state variables, which are grouped into a vector [X]. In a simple approach, the one-line diagram from Figure 3.23 is used.

The following notations are used in Figure 3.23:

$V_k\angle\phi$ is the alternating voltage at the converter station bus: $V_{kr}\angle\phi_r$ (rectifier) and $V_{ki}\angle\phi_i$ (inverter);

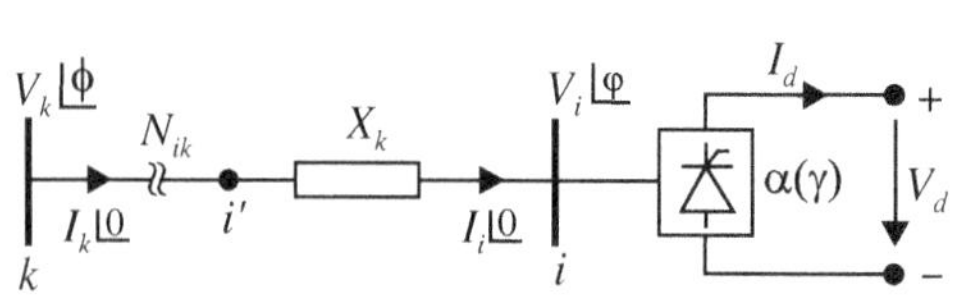

Figure 3.23 The one-line diagram of the converter station.

$V_i\angle\varphi$	–	the fundamental component of the alternating voltage on the secondary winding (valve winding) of the converter transformer: $V_{ir}\angle\varphi_r$ (rectifier) and $V_{ii}\angle\varphi_i$ (inverter);
I_p, I_s	–	the fundamental component of the alternating current on the primary and secondary windings of the transformer;
α, γ	–	the firing angle and extinction angle, respectively;
N_{ik}	–	the transformation ratio, with $N_{ik} = V_i/[V_k(1 + \text{Tap})]$;
Tap	–	the actual tap position;
V_d	–	the direct voltage at each converter;
I_d	–	the direct current.

In order to simplify the expressions of the mathematical model and to improve the performances of the computation algorithms, the alternating current at the secondary winding of the transformer is taken as reference.

The 10 variables defined above, of which nine are associated with the converter, and the voltage V_k form a possible choice of the vector $[X]$ of the quantities associated with the DC system.

In order to solve the power flow problem for a two-terminal HVDC system, two out of the 10 variables are sufficient to be chosen as independent variables (the others, determined in terms of these two independent variables, form the set of dependent or output variables). This choice leads to more complicated expressions of the other state variables, making difficult their implementation into traditional load flow programs. Thereby, we can use either a vector of nine independent variables

$$[X] = [V_{dr}, \phi_r, N_{ik,r}, \cos\alpha, V_{di}, \phi_i, N_{ik,i}, \cos\gamma, I_d]^t \tag{3.24a}$$

or a vector of seven independent variables

$$[X] = [V_{dr}, N_{ik,r}, \cos\alpha, V_{di}, N_{ik,i}, \cos\gamma, I_d]^t \tag{3.24b}$$

If the vector of seven components is chosen, besides the system of independent equations corresponding to the rectifier, the inverter and the DC line, another four equations are added, modeling the control strategy of the link by specifying the values of four independent variables and also the limit values (lower and upper) in terms of which commutation to a specific operation mode is performed.

3.3.2.2 Basic Control Principles of the DC Voltage and DC Current

Figure 3.19 illustrates the equivalent circuit of a two-terminal HVDC link for steady-state analysis, whereas Figure 3.23 emphasizes the state variables of a converter. In normal operation, the rectifier aims maintaining the DC current at a constant value, and the inverter aims maintaining constant the extinction angle, while controlling the DC voltage. The rectifier normally controls the DC current flowing in the circuit by adjusting its output DC voltage to give a current flow.

The rectifier DC voltage can be expressed as

$$V_{dr} = V_{d0,r}\cos\alpha - R_{cr}I_d \tag{3.15}$$

and inverter DC voltage as

$$V_{di} = V_{d0,i} \cos\gamma - R_{ci} I_d \tag{3.19b}$$

where $V_{d0,r} = (3\sqrt{2}/\pi)N_{ik,r}V_{kr}$ and $V_{d0,i} = (3\sqrt{2}/\pi)N_{ik,i}V_{ki}$ are the ideal no-load direct voltages. Depending on the AC voltages V_{kr} and V_{ki} and the desired current I_d, the DC voltages can, therefore, be controlled either by adjusting the converter angle (α or γ) or by adjusting the transformation ratio N_{ik} by means of the tap changer.

(i) *Firing angle control.* At no load, the DC voltage is proportional to the cosine of the firing angle. Hence, by adjusting the firing angle from 0° to 90°, the DC output voltage can be varied between 1.0 and 0.0 p.u., whereas for values of the firing angle between 90° and 180° the DC terminal voltage reverses polarity taking values between 0.0 and –1.0 p.u. In practice, however, a converter is never operated at an angle as low as 0° or as high as 180°, a much tighter domain being defined [4].

(ii) *Tap changer control* (TCC). The converter transformers, on both the rectifier and inverter sides, are typically equipped with an OLTC in order to provide voltage control on the DC side when the firing or extinction angle exceeds the normal operating range. The transformers provide discrete voltage control since a tap changer typically adds or takes off several turns when changing up or down one step, that account for a change in the valve side voltage of about 1–1.5% of the rated value. The maximum number of taps available on a transformer is usually around 32, which accounts for a maximum voltage change of around 40% [4]. The voltage control by a tap changer is much slower than the firing angle control as one tap changer step takes few seconds to execute. Thereby, there is a natural delay between the two control functions.

(iii) *Combining firing angle and tap changer control.* Normal control of a HVDC converter combines both firing angle and tap changer controls. The first control provides rapid responses to sudden system conditions changes or load demands, whereas the second control provides steady-state adjustments or replaces the firing angle control.

Current can flow through a thyristor from the anode terminal to the cathode terminal only. For this reason, in a two- or multiterminal HVDC system, the cathode terminal of the sending-end converter(s) must be connected to the anode terminal of the receiving-end converter(s), as shown in Figure 3.24.

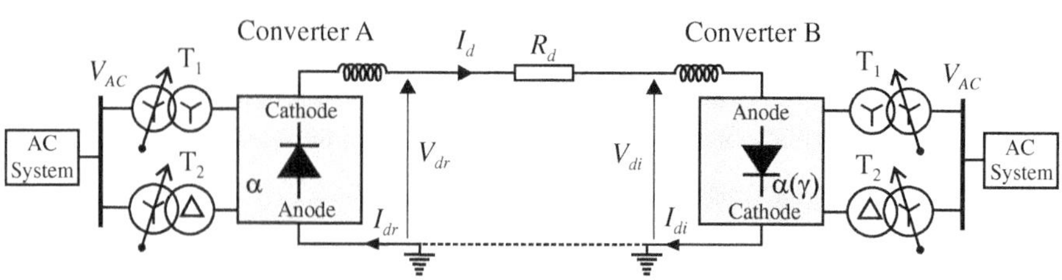

Figure 3.24 Simplified monopole HVDC circuit.

Considering the direction of current flow in Figure 3.24, the DC side resistance can be represented as a single lumped resistor:

$$R_d = R_{cr} + R_L - R_{ci} \tag{3.20b}$$

A return path has been represented in Figure 3.24 for both converter A and converter B, and an earth can be placed on either side. In this example, the converter B's cathode and the converter A's anode are connected to the return path. As the current flows from the cathode to anode, the current returns from converter B to converter A.

The power flow from converter A to converter B is associated with positive DC voltage, whereas the power flow from converter B to converter A is associated with negative DC voltage. The rectifier voltage and the inverter voltage are controlled independently, and hence there is always a voltage drop across the resistance R_d. As long as the rectifier voltage is larger than the inverter voltage, there will be a DC current flow given by

$$I_d = \frac{V_{dr} - V_{di}}{R_d} \tag{3.20c}$$

Because of the unidirectional nature of the converter valve, the current cannot become negative on the DC side. Thus, the direction of power flow is given by the polarity of the DC voltage. When polarity changes, the rectifier becomes inverter, and the inverter becomes rectifier.

In HVDC transmission lines, the converters can be hundreds or even thousands of miles apart from each other, and their measurement systems cannot be synchronized. As it results from equation (3.20c), the inverter voltage (V_{di}) is lower than the rectifier voltage (V_{dr}) by the amount of voltage drop across the DC circuit resistance R_d. Therefore, if the voltage measurement point—called a compounding point—is set at the rectifier, then the DC voltage measured at the inverter is compounded with the voltage drop calculated in terms of the measured current at the inverter. Hence, the control system on either a converter of a HVDC link can determine the voltage at the other converter or at a particular point in the circuit.

The compounding point is most commonly set at the rectifier, but in some cases it is located on the DC line. For example, the compounding point for the IFA 2000 Channel HVDC interconnection is located midway between the two converter stations that interconnect England and France [4].

3.3.2.3 Control Modes

A three position code is used in [17] to identify the control mode. The first position indicates whether constant power, denoted by *P*, or constant current, denoted by *I*, is used. The second position indicates if the current control is in the rectifier (*R*) or in the inverter (*I*). The index in the third position shows in which converter station the transformation ratio N_{ik} is fixed: *PR*1 (no station); *PR*2 (rectifier); *PR*3 (inverter); *PR*4 (both stations).

Taking into account the four main strategies *PR*, *PI*, *IR*, *II*, which are functions of the four variables chosen for the model, it results a total of 16 possible operation modes (Table 3.2). There are, therefore, four control strategies for each control mode.

TABLE 3.2 Possible control modes of a HVDC link

Control mode	Specified variables				Control mode	Specified variables			
*PR*1	α	γ	V_{di}	P_{di}	*IR*1	α	γ	V_{di}	I_d
*PR*2	$N_{ik,r}$	γ	V_{di}	P_{di}	*IR*2	$N_{ik,r}$	γ	V_{di}	I_d
*PR*3	α	γ	$N_{ik,i}$	P_{di}	*IR*3	α	γ	$N_{ik,i}$	I_d
*PR*4	$N_{ik,r}$	γ	$N_{ik,i}$	P_{di}	*IR*4	$N_{ik,r}$	γ	$N_{ik,i}$	I_d
*PI*1	α	γ	V_{dr}	P_{di}	*II*1	α	γ	V_{dr}	I_d
*PI*2	α	γ	$N_{ik,r}$	P_{di}	*II*2	α	γ	$N_{ik,r}$	I_d
*PI*3	α	$N_{ik,i}$	V_{dr}	P_{di}	*II*3	α	$N_{ik,i}$	V_{dr}	I_d
*PI*4	α	$N_{ik,i}$	$N_{ik,r}$	P_{di}	*II*4	α	$N_{ik,i}$	$N_{ik,r}$	I_d

The control mode with constant power at the rectifier, *PR*, is shown in Figure 3.25. The variables in boxes are those that are specified (to be maintained constant) for the respective control mode.

The *PR*1 control mode corresponds to normal control for V_{di} and α. The direct voltage V_{di} is controlled by the tap changer at the inverter; when the control range of the transformation ratio $N_{ik,i}$ is exceeded, commutation to a control mode with constant $N_{ik,i}$, set at the limit reached, is performed (from *PR*1 to *PR*3, or from *PR*2 to *PR*4). The firing angle α is controlled by the tap changer at the rectifier; when the control range of $N_{ik,r}$ is exceeded, commutation to a control mode with constant $N_{ik,r}$, set at the limit reached, is performed (from *PR*1 to *PR*2, or from *PR*3 to *PR*4). Thus, the control modes *PR*2, *PR*3, and *PR*4 correspond to the situation in which at least one of the variables $N_{ik,r}$ and $N_{ik,i}$ exceeds their control limits, a case in which commutation from one operation mode to another is performed (solid line in Figure 3.25).

One can see that modeling the control strategy of a HVDC link for steady-state calculations is similar with treating the generator buses.

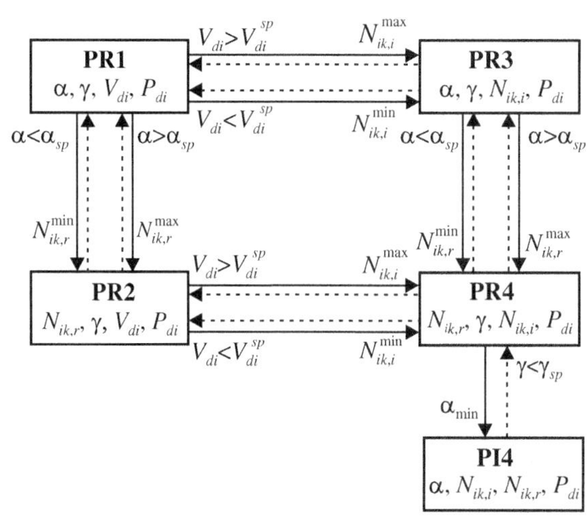

Figure 3.25 Transitions between control modes for a *PR* strategy [17].

3.3.3 HVDC Control Strategies

As shown in Table 3.3, there are two basic types of control associated with the rectifier and three with the inverter, for both normal and emergency situations. For the normal operation, various control strategies can be defined.

3.3.3.1 Rectifier Control Strategy

Under normal operating conditions, the rectifier is assigned to control the DC current, which can be done by varying the firing angle α or by controlling the AC commutation voltage. The control strategy chosen is the result of economical and technical optimization.

(i) *Constant valve winding voltage control.* This control strategy aims maintaining constant the ideal no-load direct voltage, $V_{dr,0}$, mainly by means of the transformer tap changer. As it results from the expression given in Table 3.1, depending on the AC system voltage, the voltage $V_{dr,0}$ can be controlled by changing the transformation ratio $N_{ik,r}$. By maintaining the valve winding voltage, V_{ir}, at almost constant value, and thus the ideal no-load direct voltage, since the DC voltage V_d is given by the inverter, the DC current I_d can be maintained constant at the reference value, provided by the master control, by a control of the firing angle α within a technically acceptable operation range.

As it is impractical to measure, the voltage on the valve winding side is determined in terms of the measured AC system side voltage and the turns ratio, that is $V_{kr}N_{ik,r}$. The turns ratio $N_{ik,r}$ can be calculated since the tap position can be identified by direct means.

The tap changer is typically located on the primary winding side of the converter transformer. If the valve voltage increases above the maximum limit $V_{ir,\max}$, the turns ratio must be decreased, thereby the tap changer must switch on an upper position. On the contrary, if the valve voltage decreases below the minimum limit $V_{ir,\min}$, the turns ratio must be increased, thereby the tap must be lowered. If the required tap position is outside the physical limits ($Tap_{\min}$ and $Tap_{\max}$), other voltage control means of additional changes in the HVDC system should considered. A simplified control block diagram is shown in Figure 3.26.

The tap changer is a mechanical device, and the number of operations it has to perform should be minimized in order to minimize the required maintenance. The voltage control performed by a tap-changing transformer is discontinuous, and thereby, in order to avoid unnecessary operations, a deadband of

TABLE 3.3 The basic control schemes of the converters

Rectifier control	Inverter control
• DC current control (main control mode)	• DC voltage control (main control mode)
• DC voltage limitation control (backup control mode)	• DC current control (backup control mode)
	• Extinction angle control (backup control mode)

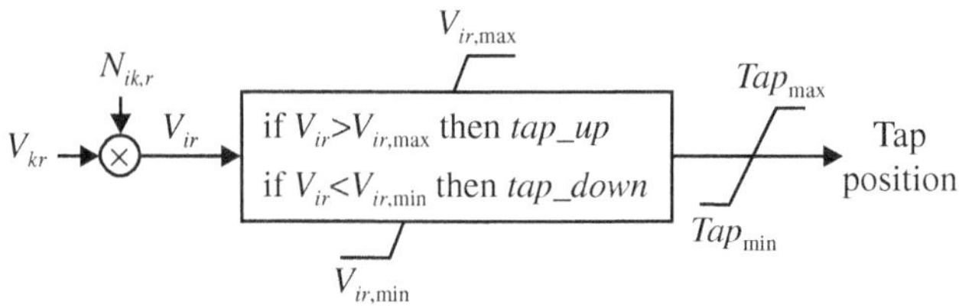

Figure 3.26 Tap changer controller with valve winding voltage limits.

the valve winding voltage between an upper and a lower limit has to be defined. This deadband should be larger than the voltage difference achieved by one tap step (typically $1^1/_2$) in order to ensure that when moving one tap up or down, return to the initial tap is not requested shortly afterward [4].

This control strategy may be used with good results at a power transfer around 1 p.u. If the power transfer order decreases below 1 p.u., the firing angle α must be increased, which in turn results in increased reactive power absorption, increased harmonic level, and higher converter valve losses. Thus, there are several side effects when protecting the tap changer from performing too many tap commutations.

(ii) *Firing angle limit control.* An optimal control range of the firing angle, typically in between 12° and 15°, is defined in this control strategy in order to keep the absorbed reactive power, the harmonic level, and converter losses al low values and avoid commutation failure. Figure 3.27 shows a simplified control block diagram, in which the output value is determined in terms of the firing angle limits.

Variations in the AC voltage will cause variations in the valve voltage. In order to keep the current constant at the reference value, changes in the firing angle are first required because it is faster than the tap changing control. If the resulted angle will exceed one of the predefined steady-state limits, the tap changing control is subsequently activated to bring the firing angle within the predefined steady-state operating range ($\alpha_{r,\min}$ and $\alpha_{r,\max}$). In this control strategy, the control range of the firing angle is tighter and defined around lower values, as compared to the constant valve winding voltage control. Under these conditions, in order to compensate for larger variations in the AC voltage, a larger number of tap steps is required than in the case of constant valve winding voltage control.

As it results from equation (3.15), the firing angle α is directly proportional to the no-load direct voltage $V_{dr,0}$ and thus to the valve winding voltage V_{ri}. When the

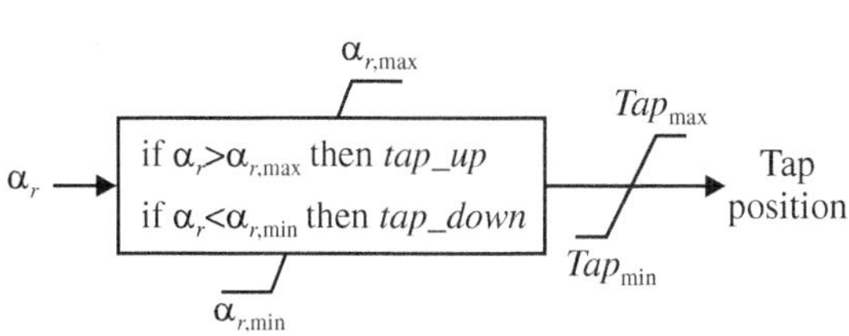

Figure 3.27 Tap changer controller with firing angle (α) limits.

required angle α_r exceeds the upper limit $\alpha_{r,\max}$, V_{ri} must be decreased thus the tap changer must switch on an upper tap position. On the other hand, when the required angle α_r exceeds the lower limit $\alpha_{r,\min}$, V_{ri} must be increased thus the tap changer must switch on a lower tap position.

In order to avoid excessive tap changes when the firing angle is around one of the limits, a deadband between the upper and the lower limits has to be defined. Given the minimum selected band limit, $\alpha_{r,\min}$, the maximum band limit can be defined by [4]

$$\alpha_{r,\max} = \cos^{-1}\left(\frac{\cos \alpha_{r,\min}}{1 + \mathrm{DB}}\right) \tag{3.25}$$

where DB is the deadband in p.u.

A special case is when the HVDC converter is connected to the end of a long AC line that produces too much reactive power. Thereby, either additional reactive power compensation devices (e.g., switched shunt reactors) should be installed in the converter station, or an increased operating angle that results in increased reactive power being absorbed by the converter should be accepted [4].

3.3.3.2 Inverter Control Strategy

Since the DC current is given by the rectifier, the inverter, the inverter is assigned to control the DC voltage. Depending on the internal and external characteristics of the HVDC link, three control strategies may be defined:

(i) *Constant valve winding voltage control.* Similar to the rectifier, in this strategy the ideal no-load direct voltage, $V_{di,0}$, is maintained around a reference value by tap changing. For a given value of the DC current, which is controlled by the rectifier, in order to keep the voltage at a reference value the extinction angle is adjusted, according to equation (3.23), but within an acceptable range. The inverter is not allowed to operate in the rectifier region in order to avoid power reversal. For this reason, the extinction angle γ_i is controlled such that the firing angle γ_i will not decrease under the a minimum value, around 100°–110° [18]. A block diagram for the tap position control similar to that illustrated in Figure 3.26 can be defined here.

(ii) *Extinction angle limit control.* This control strategy assumes that the extinction angle is maintained within an optimal range, narrower than for the constant valve winding voltage control, usually between 15° and 18°. Thereby, the no-load direct voltage is no longer maintained constant, and thus the tap changer will experience more tap movements. A controller similar to that illustrated in Figure 3.33 can be defined here for the tap position control.

(iii) *Constant extinction angle control (CEA control).* When maintaining the extinction angle at a constant value in steady state, the DC voltage is allowed to vary when the power transfer or the AC voltage changes. This strategy allows the inverter to operate at the minimum value of the extinction angle, $\gamma_{\min}$, so that to minimize the reactive power demand, power losses, and harmonic level; additionally, the valve winding voltage is lower and thus a lower number of thyristor levels in each valve are required.

Since the extinction angle γ is maintained constant, as it results from equation (3.19b), the DC voltage V_d is controlled by controlling the no-load direct voltage, which in turn is controlled by the inverter tap changer. However, since voltage control by tap changing is not continuous, a deadband for the DC voltage control (DC-VC) range equivalent to $1\frac{1}{2}$ tap steps must be defined.

When the DC voltage at the inverter increases, the rectifier must have enough firing angle to allow maintaining the DC current at a constant value. Also, when DC voltage decreases, the DC current increases, which leads to requirements for greater equipment rating.

Under CEA control, when operating within a weak power system, a disturbance in the AC system that leads to the reduction in the converter bus AC voltage will result in the DC voltage reduction. Under these circumstances, the power transmitted through the HVDC link can be maintained at the desired value by increasing the DC current. As the DC current is increased, the reactive power absorbed at the two ends will increase accordingly, which in turn may further reduce the converter AC voltage, that will require larger DC current to keep the DC power at the desired value and so on. In such cases, a stable operating point may not be found and the converter AC voltage may collapse. Therefore, the solution would be to optimize the cost of converter for normal operating conditions and accept a reduction in the transmitted power imposed by the short-circuit strength conditions [4,19]. Additional voltage control means/investments should be taken in the weak power system to cope with the temporary contingency conditions.

(iv) *Cascaded DC voltage/CEA control* [4]. CEA control is typically employed at the inverter. Any change in the AC voltage is reflected in the DC voltage, and thus in the DC current, which, on the other hand, on the reactive power absorbed by the converter. Changes in the reactive power absorbed by the converter leads to changes in the AC voltage. In weak power systems, with low short-circuit ratio (SCR), in order to deal with the large excursions in the AC voltage, a cascaded DC voltage/CEA control scheme can be used (see section 3.3.4.2).

3.3.4 Hierarchical Control of a HVDC Link

Figure 3.28 shows the general arrangement of control for a bipole HVDC link. In terms of the time constants, the DC side controls are hierarchized into three levels [7]:

(i) the master control, with time constants up to 10 s;

(ii) the pole control, in the range from 10 to 500 ms;

(iii) the valve electronics control, in the range up to 1 ms.

3.3.4.1 Master Control

This control level is interfacing the HVDC link control system with the AC national/regional dispatch center (NDC). The power-flow set-point value ($P_{d,0}$) is the result of the power-flow order ($P_{d,ord}$), provided manually by the dispatch center operator, and *power modulation* orders, such as from the system frequency control,

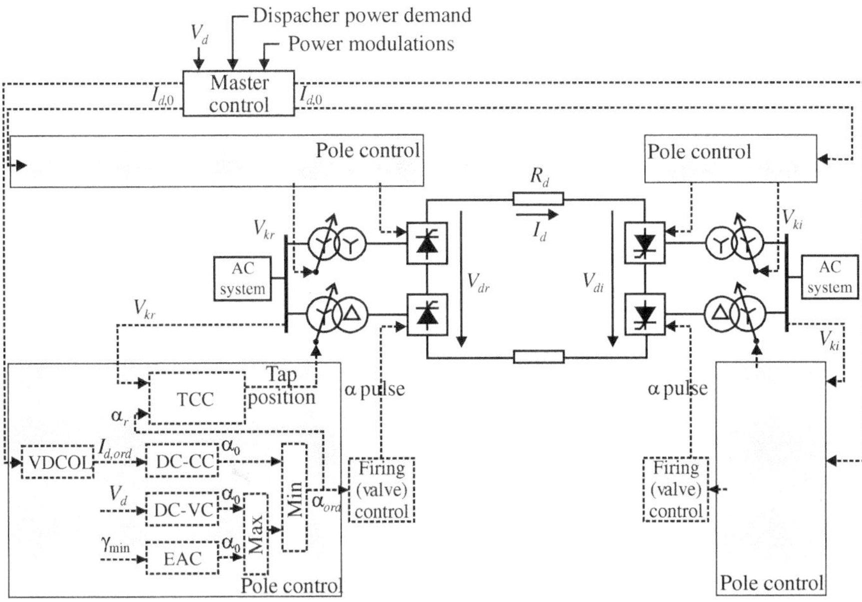

Figure 3.28 The hierarchical control scheme of a bipolar CSC–HVDC link.

frequency stabilization, power oscillation damping control (Figure 3.29). Besides the power demand, the dispatcher provides also the ramp speed in MW/min.

FREQUENCY CONTROL (FC). An HVDC link can be integrated into the frequency control of the AC system. When detecting frequency variations, the HVDC link frequency regulator ($G_{FC}(s)$) determines automatically the adjustment (ΔP_{FC}) of the power absorbed from or delivered into the AC system to ensure generation-load balance.

When the rectifier and inverter are in asynchronous power systems, the slow responding controller can transfer power from one system to the other to assist in

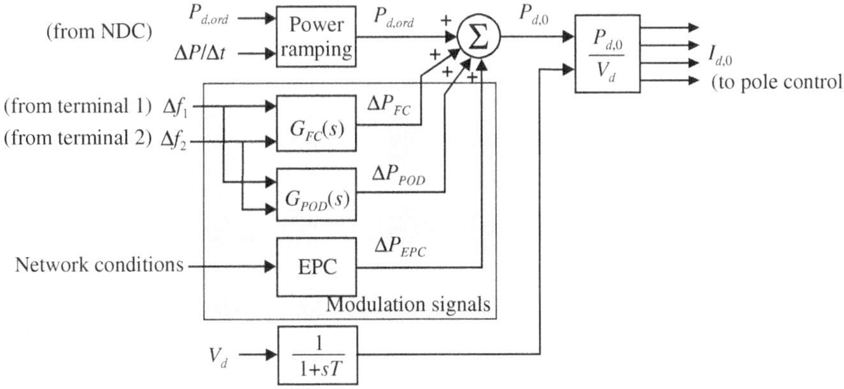

Figure 3.29 The master control scheme [20].

frequency stabilization of each [21]. Fast power control using an HVDC link is very important for a small power system to reduce the underfrequency or overfrequency.

Frequency thresholds may be applied to the power control function. For instance, if the frequency is to low in the sending-end AC system, the power absorbed by the HVDC link will be limited in order to avoid a severe frequency disturbance. The power controllability of an HVDC scheme is very important in case of an AC system consisting of islands.

POWER OSCILLATIONS DAMPING (POD). An AC power system is subject to power swings due to electromechanical oscillations. Based on frequency or voltage phase angle measurements at one or both ends of the HVDC link, the POD regulator ($G_{POD}(s)$) automatically modulates the DC power order by a value (ΔP_{POD}), typically in the range from 0.8 to 5 Hz, to provide damping to low frequency power oscillations in either of the AC power systems interconnected by the HVDC link [20].

EMERGENCY POWER CONTROL (EPC). The HVDC link can assist an AC power system to recover from a critical perturbation by rapidly increasing or decreasing the power transferred through it, or even to change the power flow direction, by a power signal (ΔP_{EPC}). In case of a generation unit or a line outage in the AC power system, an appropriate power adjustment signal is sent to the HVDC power control to compensate the loss. This function helps maintaining the AC system stability and limit the disturbance to spread over a wide area [21].

Based on the power order provided by the control center ($P_{d,ord}$) and other information relating to the DC link operation, such as the DC voltage (V_d) and the filter AC bus voltage, the master control unit, which is located at one station only, determines the desired current order ($I_{d,0}$), which is sent to the individual pole control.

The pole control level can include also the AC voltage control, reactive power control, and filter switching control, etc. For instance, when undervoltage voltage is detected into the sending-end AC system, reducing the active power transfer, and thus the reactive power demand, can remove the undervoltage stress on the AC system helping thereby the voltage to restore to normal values.

3.3.4.2 Pole Control

The operating values of the DC voltage and DC current in a HVDC link are determined by state quantities such as the scheduled DC power and the converter AC terminal voltages, and circuit parameters such as the DC reactance. The pole control is responsible for maintaining the transmitted DC power to the order value by appropriate combination of the DC voltage and DC current. The pole control coordinates the control of bridges, including the conversion of current order to a firing angle order and TCC. Certain protection sequences are also included in this control level, such as the coordination of starting up, deblocking, and balancing of bridge controls.

DC CURRENT CONTROLLER (DC–CC). A DC current order ($I_{d,ord}$) is sent to both the rectifier and inverter. The measured DC current (I_d) is compared with the DC current order, and the difference between them will cause an error, I_{err} (Figure 3.30). This current error is fed into a regulator that determines the firing angle order α_{ord} to reduce I_{err} to zero. According to equation (3.15), the DC voltage increases by decreasing α, or decreases by increasing α, in such a way to adjust the current I_d so that to reach the set-point $I_{d,ord}$.

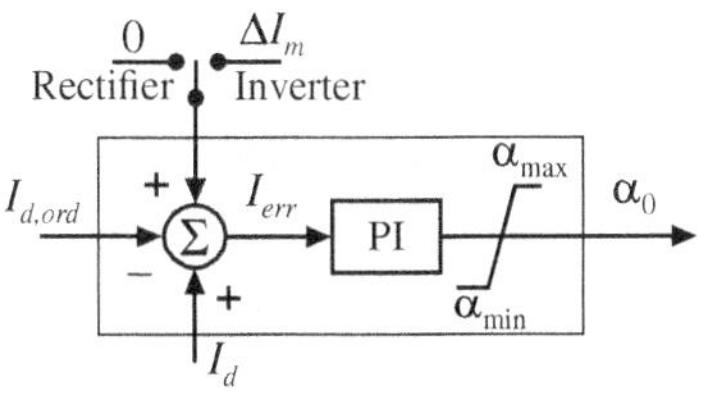

Figure 3.30 DC current controller.

This control loop is implemented in both the rectifier and inverter.

A current margin order ΔI_m is additionally introduced at the summation junction between the measured and reference currents. The current margin order function becomes active to the inverter only, whereas the current controller at the rectifier overrides it as the rectifier is responsible for controlling the DC current at the reference value $I_{d,ord}$. However, the current control at the inverter becomes active when the delay angle α is forced to drop below the minimum safety limit. The constant angle control is described by characteristic A-B of Figure 3.21.

For long transmission lines, the power losses become significant so that they must be considered when defining the reference current.

The direct current control loop is designed in terms of the circuit parameters, that is the DC line resistance, the DC reactor impedance, and the total impedance of the AC system and converter transformer.

Voltage-Dependent Current Order Limiter (VDCOL) is a function introduced in the pole control scheme at both terminals to provide an adjusted current order for more stability when disturbances in the AC system occurs, to define a fast and controlled restart after clearance of AC or DC faults, to avoid stress on the thyristors at continuous commutation failure, and to suppress the probability of consecutive failures at recovery [20, 22].

The steady-state V_d–I_d control characteristic of a HVDC link, modified to integrate the VDCOL function, is shown in Figure 3.31. The operating point A is given by the intersection between the DC current control characteristic at the rectifier and the

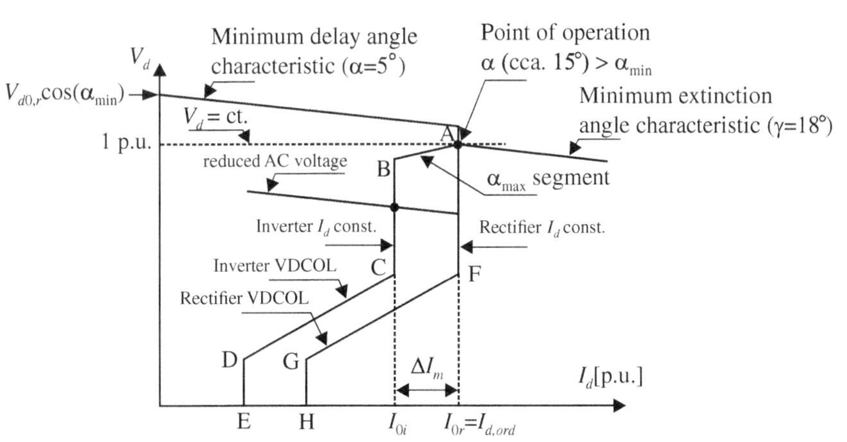

Figure 3.31 Modified V_d–I_d characteristic, with VDCOL.

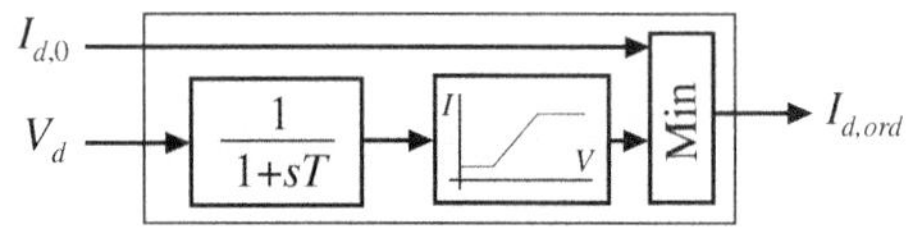

Figure 3.32 VDCOL controller.

DC voltage control characteristic at the inverter. At the inverter, the operating point is intentionally chosen as the crossover point between the DC voltage and extinction angle control (EAC) modes of the inverter [5]. To avoid coincidence between minimum firing angle and minimum extinction angle operation during unfavorable AC voltage conditions, for example, parallel characteristic, a positive slope (A-B) is added to the inverter characteristic, as α_{max} segment [20].

During AC system faults on the inverter side, there is a big risk for commutation failure. When the fault causes a too large DC voltage drop, the inverter may not be capable of recovering by itself. In order to reduce the stress on the inverter valves, a VDCOL function is activated at the rectifier which reduces the DC current order $I_{d,ord}$, according to the characteristic shown in Figure 3.31. The break points C and F are typically between 0.7 and 0.3 p.u. of the DC voltage.

Figure 3.32 shows the VDCOL controller. Its output is the minimum between the current order received from the master power control and the VDCOL limiter.

Extinction Angle Control. The firing angle at the inverter is controlled such that the extinction angle is maintained as small as possible. This will help minimizing the losses, reactive power demand, rating of converter transformers, and rating of converter valves, and thus reducing the capital costs. However, as the extinction angle becomes too small the probability for commutation failure will increase. Therefore, the angle γ_i is normally maintained in the range 15°–18°.

The simplified block diagram of the CEA controller is shown in Figure 3.33. An extinction angle order is introduced in the CEA controller to achieve a certain DC voltage. This order is compared with the operation value γ_i and determines the required firing angle. When changing to CC control is required, a current error control (CEC) signal is additionally introduced in the controller for a smoother transition from one control mode to the other. The extinction angle cannot be measured directly. It can be derived either from a measurement of the actual valve voltage or by a predictive method [18].

DC Voltage Control. When operating as the inverter, the converter is responsible for controlling the DC voltage. In some specific situations, there may be more effective to control the DC voltage by means of a DC voltage controller. At constant DC voltage control, the V_d–I_d characteristic is flat as compared to the CEA control that has a small negative slope characteristic. Furthermore, the DC voltage control

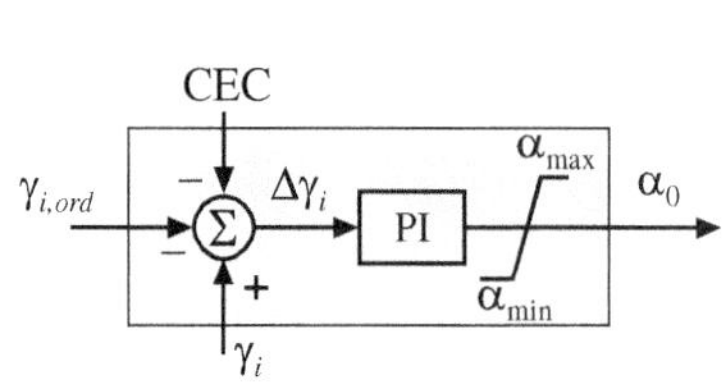

Figure 3.33 Inverter CEA controller.

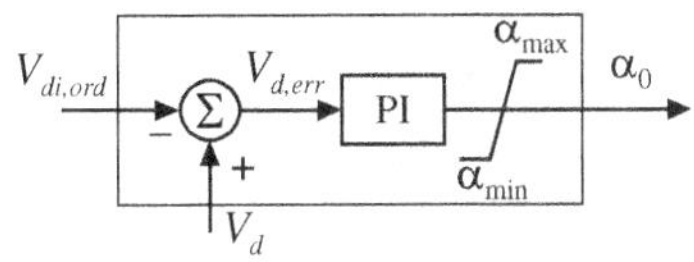

Figure 3.34 DC voltage controller.

has a slightly higher value of the extinction angle, which reduces the risk for commutation failure. A simplified block diagram of the DC voltage controller is shown in Figure 3.34.

With cascaded DC voltage/CEA control employed (Figure 3.35), in the event of a step reduction in the inverter busbar AC voltage, the extinction angle may temporarily be reduced toward the minimum steady-state value (γ_{min}) [4]. This scheme allows the inverter to maintain the DC voltage at a constant value for a given step reduction in inverter AC busbar voltage, thereby not changing the DC power. Furthermore, the demand for reactive power of the converter is limited, thus assisting the AC system in its recovery from the disturbance.

The DC voltage controller has the highest priority, and thus it is the fastest loop operating within a cycle. The extinction angle controller that aims keeping the angle γ at constant value is a much slower loop, operating with a time constant of around 1 s. By controlling the extinction angle at a desired value in steady state, savings in terms of station rating be achieved. However, by extending the control margin to deal with voltage disturbances in the AC system these savings are reduced.

Tap Changer Control. The HVDC link operation is sensitive to the AC system voltage. As the AC system voltage changes continuously on both sides of the HVDC link, the DC voltages V_{dr} and V_{di} may change accordingly. Under steady-state conditions, the AC voltage varies typically over the range ±5% to ±10%. In order to keep the transmitted power on the HVDC link at the desired value at the same time, the converter is maintained within its normal operation; the transformer OLTC is used.

At the rectifier station, the TCC aims maintaining the delay angle α within a certain range, usually between 12° and 15°. If α drops below or rises above the threshold limits, the tap changing control is activated and the AC voltage is increased or decreased accordingly.

At the inverter station, the TCC aims maintaining the DC voltage equal to the reference value. Since the EAC is typically maintained constant for safe commutation reasons, any change in the DC voltage is controlled by the tap changer.

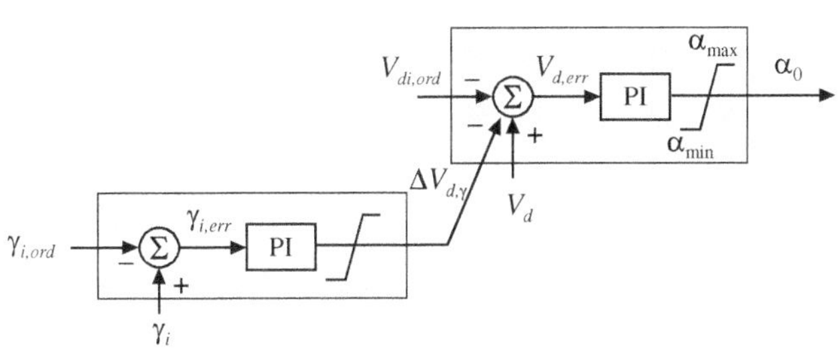

Figure 3.35 Cascaded DC voltage/CEA control.

The tap changer is slow as compared to the firing angle control and thus there is no risk for inadvertent interaction. Instead, they are complementary; when the first control takes places, the second is still in the previous steady-state conditions. Conversely, when the second control occurs, the first has already reached the new steady-state conditions [4].

3.3.4.3 Firing (Valve) Control

A phase-locked loop (PLL)-based firing system is employed at the valve control level to determine the time instants for thyristors gating according to the firing angle order received from the current controller. The firing pulses are synchronized with the AC commutation voltage. This has the fastest response within the control hierarchy.

3.3.4.4 Telecommunications

The overall control of an HVDC link requires adequate communication with the AC system national/regional control center. With the advent of telecommunication technology, the fiber optic became the best choice as transmission media, although other media could be used, such as direct wires via private lines or telephone networks, power-line carrier, or microwave systems. The master control of the HVDC link requires synchronization between its terminal stations, which can be performed via high-speed communication. In the case of rapid changes in the power transfer level ordered at the upper level, that is the overall system control, fast and reliable communication is required to maintain consistent current setting at the two terminals, whereas in case of change of power direction appropriate communication is requited to transfer the current margin setting from one terminal to the other [12]. Adequate communication is also required in case of transient or fault conditions.

3.3.4.5 Measurement Transducers

In a HVDC scheme, voltage and current transducers are installed on both the AC and the DC sides.

The line AC voltage is traditionally measured on the HV side of the converter transformer by a capacitive voltage divider (CVD), whereas the voltage used by the HVDC control system for determining the firing and extinction angles are obtained from the converter transformer line winding bushings, which are effectively capacitive dividers [4].

DC voltage measurement is made by either DC resistive voltage divider or an optical voltage divider. DC current measurement for both control and protection requires an electronic processing system. The DC current transducers can be zero flux current transducers or optical current measurements based on the Faraday effect. The DC power is then calculated as a product of the measured DC current and the DC voltage.

3.4 REACTIVE POWER AND HARMONICS

3.4.1 Reactive Power Requirements and Sources

Subsection 3.2.1 introduced the concept that the reactive power demands of a converter are directly related to the mode of operation of the converter, for example, an operation with a large extinction angle at the inverter means higher reactive power consumption. However, operation of the converter at the typical delay angle of 12°–15° at the rectifier and the extinction angle of 15°–20° at the inverter results in a reactive power demand at each terminal of typically 50%–60% of the active power transfers. Reactive power sources that are used vary from switched capacitors to static VAr systems. The requirements of voltage control and the costs dictate the choice of the speed of response of the reactive power control under dynamic conditions.

The active power supplied on the DC voltage side of the converter is $P_d = V_d I_d$, as given by expressions (3.22). Neglecting the power losses into the converter, it can be considered that the three-phase power on the AC side of the converter is equal to the active power, that is $P_{AC} = P_d$, where $P_{AC} = 3V_s I_1 \cos\varphi$.

The active and reactive fundamental components of the current are defined as

$$I_{1a} = I_{s1}\cos\varphi \tag{3.26a}$$

$$I_{1r} = I_{s1}\sin\varphi \tag{3.26b}$$

where I_{s1} is the rms value of the fundamental component of the AC line current.

Therefore,

$$3V_s I_{1a} = V_d I_d$$

$$I_{1a} = \frac{V_d I_d}{3V_s} \tag{3.27}$$

By substituting I_d from (3.5) and V_d from (3.12) into (3.27), and taking into account that $\hat{V}_s = \sqrt{2}V_s$, we obtain

$$I_{1a} = I_{s1}\frac{\cos\alpha + \cos(\alpha+\lambda)}{2} = \rho(\alpha,\lambda)I_{s1} \tag{3.28a}$$

where

$$\rho(\alpha,\lambda) = \frac{\cos\alpha + \cos(\alpha+\lambda)}{2} \tag{3.28b}$$

The power factor of the conversion stations can be calculated as follows:

$$\cos\varphi_r = \frac{\cos\alpha + \cos(\alpha+\lambda)}{2}$$
$$\cos\varphi_i = \frac{\cos\gamma + \cos(\gamma+\lambda)}{2} \tag{3.28c}$$

In order to obtain a similar expression as (3.28b), for the reactive fundamental component of the current, we express I_{s1} from (3.5), where the following expression of I_d is used:

$$I_d = \frac{\sqrt{6}V_s}{2X_k}(\cos\alpha - \cos\delta) = \frac{\sqrt{6}V_s}{X_k}\sin\frac{\alpha+\delta}{2}\sin\frac{\delta-\alpha}{2} \tag{3.14b}$$

where $\delta = \alpha + \lambda$ and X_k is the commutation reactance.

Then,

$$I_{s1} = \frac{\sqrt{6}}{\pi} I_d = \frac{6V_s}{\pi X_k}\left(\sin\frac{\lambda}{2}\sin\frac{2\alpha+\lambda}{2}\right) \tag{3.29a}$$

Using the expressions (3.5), (3.14b), and (3.28a), it results:

$$\begin{aligned} I_{1a} &= \frac{\sqrt{6}}{\pi}\frac{\sqrt{6}V_s}{2X_k}(\cos\alpha - \cos\delta)\frac{\cos\alpha + \cos\delta}{2} = \frac{3V_s}{2\pi X_k}(\cos^2\alpha - \cos^2\delta) \\ &= \frac{3V_s}{4\pi X_k}[\cos 2\alpha - \cos 2(\alpha+\lambda)] = \frac{3V_s}{2\pi X_k}\sin\lambda\sin(2\alpha+\lambda) \end{aligned} \tag{3.30a}$$

If in expressions (3.29a) and (3.30a), the following notations are used:

$$A' = \frac{3V_s}{2\pi X_k};\ \theta = \frac{2\alpha+\lambda}{2};\ \psi = \frac{\lambda}{2} \tag{3.31}$$

it results:

$$I_{s1} = 4A'\sin\theta\sin\psi \tag{3.29b}$$

$$I_{1a} = A'\sin 2\theta\sin 2\psi \tag{3.30b}$$

According to (3.26), (3.29b), and (3.30b), it can be written as

$$\cos\varphi = \cos\theta\cos\psi \tag{3.32a}$$

If the equality $\sin\varphi = \sqrt{1-\cos^2\varphi}$ is used:

$$\sin\varphi = \cos\psi\sqrt{\frac{1}{\cos^2\psi} - \cos^2\theta} \approx \cos\psi\sqrt{1-\cos^2\theta} \approx \sin\theta\cos\psi \tag{3.32b}$$

where it has been considered that $1/\cos^2\psi \approx 1$.

Taking into account (3.32b) and (3.29b), the expression (3.26b) becomes

$$\begin{aligned} I_{1r} &\cong I_{s1}\sin\theta\cos\psi = 4A'\sin^2\theta\sin\psi\cos\psi = 4A'(1-\cos^2\theta)\sin\psi\cos\psi \\ &= A'\sin 2\psi(1-\cos 2\theta) \end{aligned}$$

By using the notations from (3.31) and approximating $\sin\lambda \approx \lambda$, one obtains:

$$I_{1r} = \frac{3V_s}{2\pi X_k}[\lambda - \sin\lambda\cos(2\alpha+\lambda)] = \frac{3V_s}{2\pi X_k}\frac{2\lambda + \sin 2\alpha - \sin 2(\alpha+\lambda)}{2} \tag{3.33}$$

Expressing V_s from (3.14b) and taking into account (3.2) and (3.5), it results:

$$V_s = \frac{\pi X_k}{3[\cos\alpha - \cos(\alpha+\lambda)]} I_{s1}$$

By substituting in (3.33), one obtains:

$$I_{1r} = \frac{1}{4}\frac{2\lambda + \sin 2\alpha - \sin 2(\alpha+\lambda)}{\cos\alpha - \cos(\alpha+\lambda)} I_{s1} = \chi(\alpha,\lambda) I_{s1} \tag{3.34a}$$

where

$$\chi(\alpha, \lambda) = \frac{1}{4}\frac{2\lambda + \sin 2\alpha - \sin 2(\alpha + \lambda)}{\cos\alpha - \cos(\alpha + \lambda)} \tag{3.34b}$$

Using the expressions of $\rho(\alpha, \lambda)$ from (3.28b) and $\chi(\alpha, \lambda)$ from (3.34b), it results:

$$\tan\varphi = \chi(\alpha, \lambda)/\rho(\alpha, \lambda)$$

or taking into account the equality $\cos^2 x = \frac{1}{2} + \frac{1}{2}\cos 2x$, one obtains:

$$\tan\varphi = \frac{2\lambda + \sin 2\alpha - \sin 2(\alpha + \lambda)}{\cos 2\alpha - \cos 2(\alpha + \lambda)} \tag{3.35}$$

Therefore, using the expression (3.3), $\cos\varphi = V_d/V_{d0}$, the reactive power absorbed by the converter is

$$Q_{conv} = P_d \tan\varphi = P_d\sqrt{\frac{1}{\cos^2\varphi} - 1} \cong P_d\sqrt{\left(\frac{V_{d0}}{V_d}\right)^2 - 1} \tag{3.36}$$

For usual values of $\alpha = 15°$and $\lambda = 20°$ from (3.28c), it yields:

$$\cos\varphi = \frac{V_d}{V_{d0}} = \frac{\cos 15° + \cos(15° + 20°)}{2} = 0.8925$$

Substituting in (3.36), it results:

$$Q_{conv} \approx 0.505\, P_d$$

Figure 3.36 shows the reactive power absorbed by a HVDC converter as a function of the transmitted power and different delay angles α. It can be seen that any change in the power transmitted by the converter results in a corresponding change in the reactive power. This change in reactive power must be taken into consideration during the system design. The dashed line curve in Figure 3.36 shows the reactive power requirements for a constant power and different values of the delay angle α. One can use the delay angle α to control the reactive power especially at low DC power values.

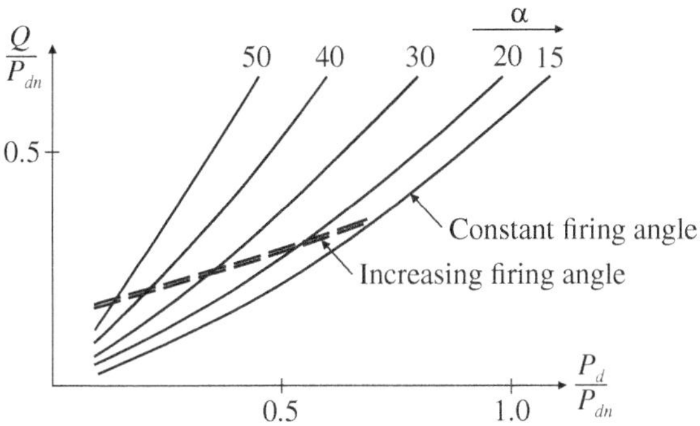

Figure 3.36 Reactive power demand of HVDC station.

The Uhlmann Approximation for Reactive Power. A commonly used method for the calculation of the reactive power absorbed by a HVDC converter is known as the Uhlmann approximation [4]. This approximation assumes that the converter line winding currents are trapezoidal, that is, that the DC side series reactance is infinite. This approximation provides a reasonable estimate of the reactive power absorbed at high DC current, close to 1.0 p.u., but can become significantly inaccurate if the AC voltage is low or if the overlap angle is large.

The Uhlmann approximation assumes that the real power (P) and the apparent power (S) can be defined, neglecting converter and transformer losses. This gives [4]

$$\cos\varphi = \frac{P_d}{\sqrt{3}U_s I_{s1}} = \frac{V_d I_d}{\sqrt{3}U_s \frac{\sqrt{6}}{\pi} I_d} = \frac{V_d}{V_{d0}}$$

and rearranging equation (3.12) the same ratio is

$$\frac{V_d}{V_{d0}} = \frac{1}{2}[\cos\alpha + \cos(\alpha+\lambda)] \rightarrow \cos\varphi = \frac{1}{2}[\cos\alpha + \cos(\alpha+\lambda)] \quad (3.28b)$$

The reactive power loading of the converter can then be found with the expression (3.36) above.

Sources of Reactive Power. The reactive power requirements of the converter are met by one or more of the following sources: AC system; AC filters (F); switched shunt capacitors (SSC); synchronous condensers (SC); static VAr compensator (SVC) or STATCOM (Figure 3.37). From voltage regulation, losses, and stability considerations, it is not desirable to draw reactive power from the system except at low loads or weak AC system [23].

The voltage regulation at the converter bus is desirable not only from the voltage control viewpoint but also from the minimization of loss and stability considerations. This requires adjustable reactive power source, which can provide variable reactive power.

The least expensive sources of reactive power for converters are *static capacitor banks* by themselves or as components of filters. Filters typically supply half or more of the reactive power requirements of a converter (see section 3.4.2). If the AC system is inherently strong enough to stand use of additional switched capacitors, they may

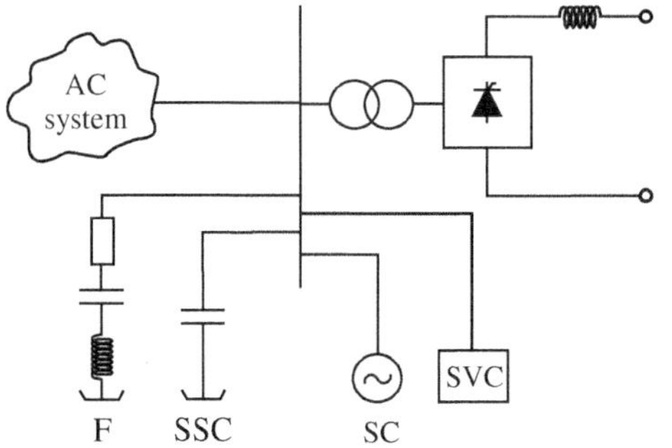

Figure 3.37 Reactive power sources.

be the preferred solution. The preferred location for a DC terminal is at a point in the system with a high short-circuit level. SC can be used to increase the short circuit levels and to provide dynamic voltage control. SC have been used on some DC links to obtain adequate short-circuit ratio. For very weak systems, such as islands, only rotating machines can supply an adequate short-circuit level. The drawback to use of SC is that they have high operating losses and maintenance requirements.

The static VAr systems—SVC and STATCOM—provide the fastest response following a disturbance.

3.4.2 Harmonics and Filters

Converters generate harmonic voltages and currents on both AC and DC sides. AC filters are invariably used to filter out AC current harmonics, which are critical. These filters are of band pass (tuned) or high-pass type and also supply reactive power. DC smoothing reactor along with DC filters perform the function of filtering DC harmonics.

In this respect, we consider two types of harmonics [23]:

- The characteristic harmonics, these are harmonics of those orders that are always present under "ideal" operation: balanced AC voltages, symmetric three-phase network, acceptable tolerances between components and equidistant firing pulses;
- the noncharacteristic harmonics—of the order other than the characteristic harmonics—are due to (i) unbalance between the angles of the two bridges forming a 12-phase converter, (ii) firing angle errors, (iii) unbalance and distortion in AC voltages, and (iv) unequal transformer impedances.

3.4.2.1 The Source of AC Harmonic Currents

Figure 3.38a shows the wave shape of the alternating current under the "ideal" condition with no commutation overlap, ripple-free direct current, balanced purely sinusoidal commutating voltages, and equally spaced converter-firing pulses. Each pulse would be considered as having 120° duration followed by 60° dead time. One positive pulse, one negative pulse, and two dead times constitute one full cycle.

A Fourier analysis of the AC current wave shape would reveal that it is composed of one fundamental component and higher order components, which are multiples of the fundamental frequency.

Each positive or negative series of rectangular current pulses can be shown to contain a DC component, a fundamental component, and components of all harmonic frequencies. Because there is a negative pulse for each positive pulse, the DC components cancel each other. Although there are two pulses per cycle, they are of opposite polarity and equally spaced. This forces the second all even harmonic components to cancel. Similarly, it can be shown that because the pulse width is one third of a cycle, third and all triple harmonics cancel as well [6].

Neglecting the effect of the commutating reactance, the current for a Y-Y connected converter transformer feeding a six-pulse group consists of rectangular pulses as shown in Figure 3.38a.

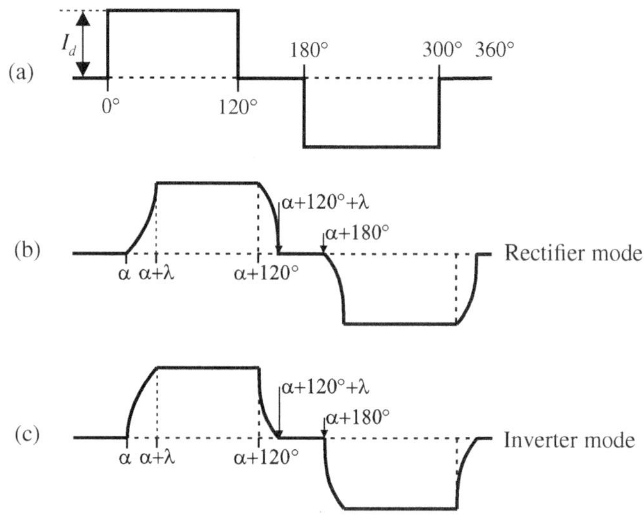

Figure 3.38 The curve of alternating current: (a) case of ideal converters; (b) and (c) cases with commutating reactance.

The Fourier series expansion for the alternating current is [15]:

$$i = \frac{2\sqrt{3}}{\pi} I_d \left(\cos \omega t - \frac{1}{5} \cos 5\omega t + \frac{1}{7} \cos 7\omega t - \frac{1}{11} \cos 11\omega t + \frac{1}{13} \cos 13\omega t - \ldots \right) \tag{3.37a}$$

The rms value of any term in (3.37a) is given by

$$I_n = \frac{\sqrt{6}}{n\pi} I_d$$

With $n = 1$, the rms value of the fundamental is given by

$$I_1 = \frac{\sqrt{6}}{\pi} I_d = 0.78\, I_d$$

For a Y–Δ transformer connection feeding a six-pulse group, from the Fourier analysis, the current is

$$i = \frac{2\sqrt{3}}{\pi} I_d \left(\cos \omega t + \frac{1}{5} \cos 5\omega t - \frac{1}{7} \cos 7\omega t - \frac{1}{11} \cos 11\omega t + \frac{1}{13} \cos 13\omega t - \ldots \right) \tag{3.37b}$$

The remaining harmonics are on the order of $n = 6q \pm 1$, where q is any positive integer, and $p = 6$ is the pulse number. The lowest of these "characteristic" harmonic are 5, 7, 11, 13, 17, 19, etc. times the fundamental. They are termed "characteristic" because all six-pulse converters produce them.

In a 12-pulse converter, there are two six-pulse bridges with two transformers, one with Y–Y connection and the other with Y–Δ connection. The transformers are

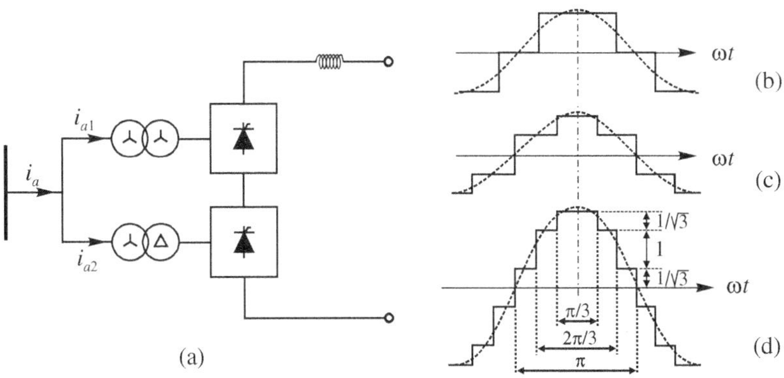

Figure 3.39 The scheme of a 12-pulse converter (a); AC current waveform for a why–why system (b); AC current waveform for a why–delta system (c); an idealized phase AC current waveform for a 12-pulse converter (d).

connected in parallel on the AC system and connected in series on the DC system (Figure 3.39). The harmonics of odd values of q cancel out. It results hence

$$i = \frac{2\sqrt{3}}{\pi} 2I_d \left(\cos \omega t - \frac{1}{11} \cos 11\omega t + \frac{1}{13} \cos 13\omega t - \frac{1}{23} \cos 23\omega t + \ldots \right) \tag{3.38}$$

The remaining harmonics that have the order $12q \pm 1$ (i.e., 11th, 13th, 23rd, 25th) flow into the AC system.

When the commutating reactance is considered, the overlap angle during commutation rounds off the square edges of the current waves, and this reduces the magnitude of harmonic components.

Figure 3.38b shows a typical current wave shape in the AC system before filtering. It consists essentially of series of alternating polarity rectangular current pulses, with significant transition times caused by commutation reactance.

The magnitude of the AC current harmonics produced can easily be predicted for the case of the ideal (instantaneous switching) converter. By Fourier analysis, it can be shown that the magnitudes of the fundamental and all characteristic harmonics are related by the following equation:

$$I_n = \frac{I_1}{n}$$

where I_n is the magnitude of the nth harmonic current and I_1 is the magnitude of the fundamental current, which is also proportional to the DC power. For example, the fifth harmonic is 20% of the value of the fundamental, but the 11th would be only 9% of the fundamental.

3.4.2.2 *The Effect of* Y/Δ *Transformation on AC Harmonic Current*

A Y/Δ transformer, when no voltage level change is involved, yields line currents on the Δ side, which are the difference of two line currents on the Y side, and vice versa. A series of ideal plus and minus current pulses in one phase of the value side of the converter transformer are subtracted from those of another phase by Y/Δ transformer to produce the current pattern shown in Figures 3.39b and 3.39c. Analysis of the harmonic content of this current wave shape reveals the same harmonic frequencies and magnitudes on both side of the Y–Δ transformer. However, the phase relationships of some of the harmonics have been shifted 180°. Specifically the harmonics of order 5, 7, 17, 19, 29, 31, etc. have been shifted totally out of phase with their relationship on the other side of the transformer [6, 15].

3.4.2.3 *Higher Pulse Operation Using Multiple Bridges and Transformers*

The fact that some harmonics are phase shifted differently in passing through a Y/Δ transformer than in a Y/Y transformer provides a means of canceling on those harmonics as far as the AC system is concerned. Figure 3.39d shows the effect of adding equal amount of two current wave shapes shown in Figures 3.39b and 3.39c. Positive and negative pulses generated on a phase of the transformer are deducted from the series generated on the adjacent phase, resulting a current that has the form as the one presented in Figure 3.39d.

The resulting wave shapes provide a better approximation to a sine wave, because the lowest harmonic frequencies cancel each other. The fifth and seventh harmonic components are not present if cancelation is complete. In fact, the only characteristic harmonics present are those which would result from a converter operating on a six-phase AC system, providing 12 pulse converter operation. These characteristic harmonics are determined by the $n = 12q \pm 1$, where n is the harmonic order and q is any positive integer.

This also introduces the concept of increasing the pulse number to reduce harmonics. The general equation to describe the characteristic harmonic orders produced by any converter is $n = pq \pm 1$, where p is the pulse number.

If the pulse number were increased to 24, one would expect the lowest harmonic to be the 23rd. Thus increasing pulse number is a very effective method of suppressing harmonics. The remaining harmonics have the same magnitude relative to the fundamental, which they have when present in a six-pulse operation [6].

3.4.2.4 *Elimination of Harmonics*

Unless measures are taken to limit the amplitude of harmonics entering the AC network and the DC line, some of the following undesirable effects may occur: overheating of capacitors and generators, instability of the converter control, and interference with telecommunication systems, especially noise on telephone line. These effects may not be confined to the vicinity of the converter station but may be propagated over large distances. The most difficult of these to eliminate is the telephone interference.

The principal method of eliminating the harmonic output of converters is the installation of filters. Filters are always used on the AC side of converters. AC filters serve the dual purpose of eliminating AC harmonics and supplying reactive power.

On the DC side, the reactor diminishes harmonics and, in many converters, especially those connected to DC cables, no additional filtering is required on the DC side. DC filters are required, however, on the same overhead DC lines [2].

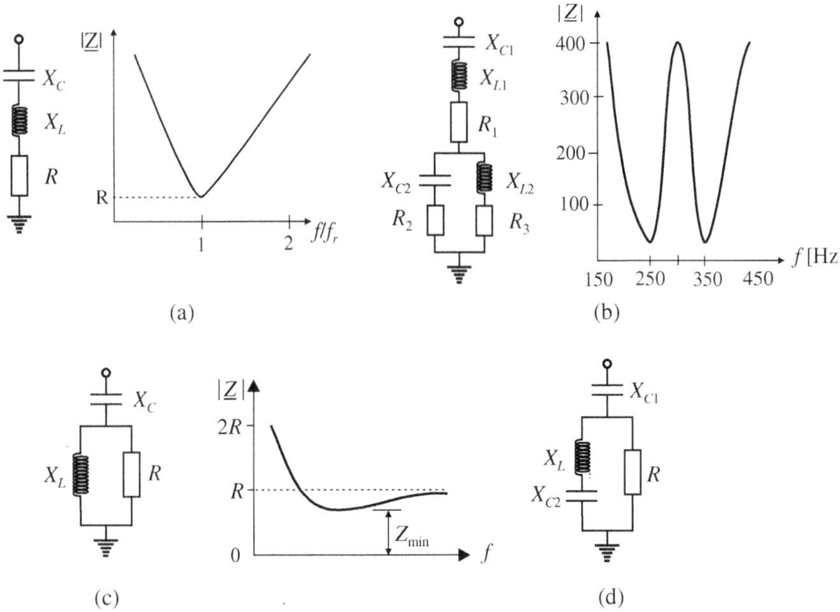

Figure 3.40 Configuration and impedance characteristics of filters: (a) series-resonant; (b) double tuned; (c) second-order high-pass; (d) high-pass "C" type.

AC HARMONIC FILTERS.[2] Harmonic filters are required at the AC terminals of the HVDC converter to limit harmonic distortion on the AC transmission system. Two main filter types are used today [4]:

- The *tuned filter* or band-pass filter, which is sharply tuned for one or several harmonic frequencies; they have low damping. Examples are single (e.g., 11th), double (e.g., 11/13th), and triple (e.g., 3/11/13th) tuned types.
  - The single-tuned band-pass filter consists of an RLC series circuit that is tuned at one harmonic frequency (in general, low order characteristic harmonics). Its impedance is $\underline{Z} = R + j(\omega L - 1/\omega C)$; the impedance at the resonance frequency consists of a low resistance. A tuned filter provides efficient suppression of individually selected harmonics but provides little damping at other harmonics (Figure 3.40a);
  - A double-tuned band-pass filter (Figure 3.40b) has the equivalent function of two single-tuned filters. The 11th and 13th filters are designed as a double-tuned filter to minimize cost. At high system voltage, it is easier to optimize one larger main capacitor than two small ones. Only one reactor is exposed to the full impulse voltage as the lower reactor is connected in parallel with the capacitor.
- The *damped filter* or high-pass filter offers a low impedance over a broad band of frequencies. The high-pass filters are designed to damp more than one

[2] Reproduced with permission of GE [4]

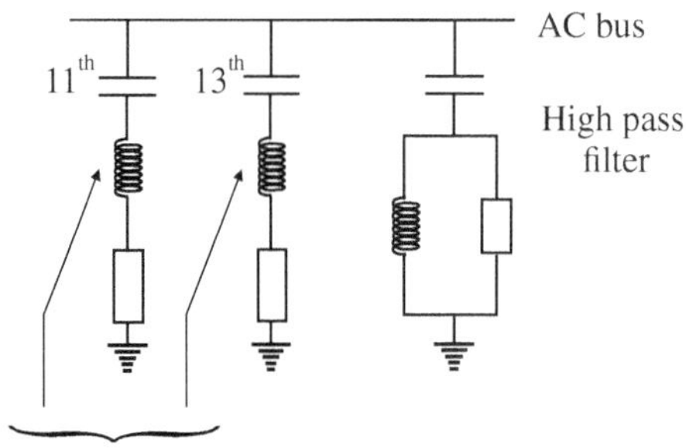

Figure 3.41 Typical filter system configuration.

harmonic; for example, a filter tuned at 24th harmonic will give low impedance for both 23rd and 25th harmonic and for most higher order harmonics. The losses are higher than tuned filters.

- Normally, a second-order high-pass filter is used (Figure 3.40c). The impedance of such a filter is given by

$$\underline{Z} = \frac{1}{j\omega C} + \left[1 \Big/ \left(\frac{1}{R} + \frac{1}{j\omega L}\right)\right]$$

- *C-type high-pass filter circuit.* An auxiliary capacitor C_2 is connected in series with the reactor and is tuned to form a fundamental frequency bypass of the resistor (then negligible fundamental frequency loss in a resistor). By creating a tuned filter C_2-L within a second-order filter, virtually all fundamental current is excluded from the resistor. At frequencies above the fundamental, a harmonic current flows through R, thus achieving the desired damping.

A typical filter system for a 12-pulse converter terminal is shown in Figure 3.41. The filter impedance is minimum at the 11th and 13th harmonics resulting from the two single-tuned branches. The high-pass filter maintains a low impedance for higher harmonic frequencies [12].

All the filter branches appear capacitive at fundamental frequency and supply reactive power. The fundamental frequency capacitive reactive power generated by the energized AC harmonic filters is defined by [4]:

$$Q_{filter} = Q_{filter0} \frac{V_{AC}^2}{V_{AC0}^2} \cdot \frac{f}{f_0}$$

where

$Q_{filter0}$ = $Q_{cap} \cdot n^2/(n^2 - 1)$ is the reactive power supplied by the filters at 1.0 p.u. voltage and frequency;

Q_{cap} – the reactive power supplied by the filter's main capacitor at 1.0 p.u. voltage and frequency;

n – a harmonic number to which the filter is tuned;

TABLE 3.4 Typical AC filters systems

HVDC systems	Filters rated power (MVAr)				
	5	7	11	13	High-pass
Skagerrak (500 MW):					
Denmark	—	—	2 × 20	2 × 20	1 × 80
Norway	—	—	2 × 20	2 × 20	2 × 45
Inga-Shaba (560 MW):					
Inga	—	—	2 × 10.1	2 × 7.2	2 × 19.9
Kolwezi	—	—	2 × 28.1	2 × 20.4	2 × 17
Pacific Intertie (1350 MW):					
Oregon	1 × 55.8	1 × 27.8	2 × 28.2	2 × 21	2 × 80
California	2 × 30	2 × 14.6	2 × 28.2	2 × 20.1	2 × 92
Konti–Skan (250 MW):					
Denmark	1 × 19	1 × 19	1 × 17	1 × 12	1 × 13
Sweden	1 × 13	1 × 7	1 × 12	1 × 9	1 × 40

V_{AC} – the actual HVDC station busbar voltage;
V_{AC0} – the 1.0 p.u. value of AC system voltage;
f – the actual HVDC station busbar frequency;
f_0 – the 1.0 p.u. value of AC system frequency.

In Table 3.4, the rated powers for harmonic filters installed on the AC side are given for some HVDC systems. As common features, the equipments operating with six pulses require filters for harmonics 5, 7, 11, 13 and a high-pass filter for the 17th harmonic and above. The HVDC equipments operating with 12 pulses require only filters for the 11th and 13th harmonics, respectively, a high-pass filter for the 23rd harmonic and above.

Some systems like Cross-Channel link are designed to operate only with high-pass filters. Although HVDC systems, given in Table 3.4, do not operate with six pulses, reduced size filters are yet used for the fifth and seventh harmonics during the unavailability of one conversion bridge of the two series connected.

At Chandrapur–Padghe HVDC Bipole transmission (India, 1500 MW; ± 500 kV; 752 km), there are four equally sized 200 MVAr *AC filter* banks. There are two banks consisting of one 120 MVAr HP12 filter and one 80 MVAr HP 24/36 filter, and two banks consisting of one 120 MVAr HP12 filter and one 80 MVAr HP3 filter [24].

The number and size of the reactive power compensation equipments in the HVDC systems are high rated in comparison with the transferred active power. Two examples are considered:

- For the *AC–DC Itaipu (Brazil) system*, 6300 MW rated, the following equipments were designed:

- harmonic filters provide 1451 MVAr at 500 kV on the rectifier's AC side and 2483 MVAr at 345 kV on the inverter's AC side;
- capacitor banks, with 588 MVAr installed capacity, on the 345 kV AC side;
- SC, with 1200 MVA installed capacity, in the inversion station.

- For the Cross-Channel link between France and England, having a maximum rated power of 2000 MW:
  - eight harmonic filters (11, 12, 13, 14, 21, 22, 23, 24) of 150 MVAr apiece in the 400 kV Mandarins (France) station;
  - eight harmonic filters of 120 MVAr apiece and two static compensators in the 400 kV Sellindge (England) station.

DC HARMONIC FILTERS. The harmonics in the DC voltage across the converter contain both characteristic and noncharacteristic orders. These harmonics result in current harmonics in the DC lines and cause noise in the telephone circuits. The harmonic current generated in the line can be computed if the harmonic voltage source at the converters, smoothing reactor, DC filter, and line parameters is known. The harmonic current varies with the distance, from the converter station along the line.

The DC filters are also of single- or double-tuned type to filter out 6th and 12th harmonics and a high-pass filter for higher order harmonics.

In [25], a modern system—with the active filter—for harmonics attenuation generated by the converter is presented (Figure 3.42). The basic idea of series active filters (Figure 3.42a) is to generate a harmonic voltage, equal to, but in phase opposition with the harmonic voltage generated by the converter. The selection of passive or active filters is performed by switching the circuit breakers 1 or 2.

A prototype active filter was installed, in 1991, on the Konti–Skan 2 link (Sweden–Denmark). Figure 3.42b illustrates the simplified scheme for hybrid

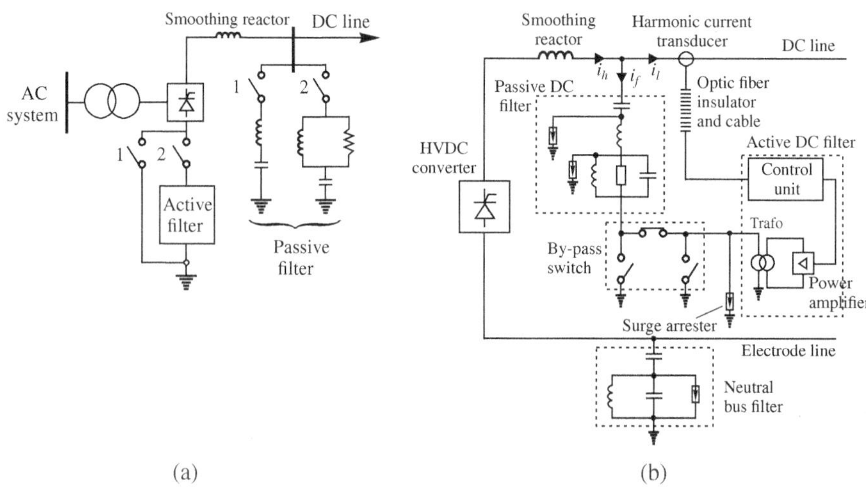

Figure 3.42 Active filters for harmonics attenuation. *Source:* Zhang and Asplund 1994 [25]. Reproduced with permission of IEEE.

solution—passive and active filters—on the DC side of the converter station from Lindome (Sweden).

A smoothing reactor is always installed on the DC side to reduce the ripple in the direct current i_h and a shunt-connected DC filter is used to reduce the level of harmonic currents i_l in the DC overhead line. The active filters will eliminate the reminder of harmonic currents. In this regard, by means of a transducer and an optic fiber system, a signal proportional to the harmonic current i_l is transmitted toward the control system (for signal processing). The active part of the active filter consists of a high-frequency transformer and a power amplifier. The power amplifier controlled by means of the control system acts as a harmonic voltage source. This generates harmonic currents in phase opposition with the harmonics generated by the conversion valve bridge, compensating on this way the harmonic currents in the AC line.

At the Chandrapur–Padghe converter stations, the *DC filters* are arranged as double-tuned filter branches and active DC filters. The double-tuned filters are provided as a single branch for each DC pole of the two converter stations to cancel the 2/6th and 12/24th harmonics. Due to the fact that the DC line length is close to quarter wavelength, second harmonic resonance was foreseen 2/6th harmonic filter is provided to reduce the effect of second harmonic resonance. Besides passive filters the HVDC bipole transmission is provided with an active filter to further minimize possible disturbances to nearly telecommunication lines.

A similar solution is applied at the TNB/EGAT HVDC transmission system, between Thailand and Malaysia, where the DC filters consist of a conventional passive filter, tuned for the 12th and 24th harmonics, connected in series with an active filter of 200 kVA rating [26].

3.5 LOAD FLOW IN MIXED HVAC/HVDC-CSC SYSTEMS

3.5.1 Steady-State Model

The steady-state mathematical model of an AC system consists of the system of equations expressing the balance of nodal powers [14,27,28]:

$$\begin{aligned} f_{P_i}([V],[\theta]) &= P_{gi} - P_{ci} - P_i = P_i^{sp} - P_i = 0, \qquad i \neq e \\ f_{Q_i}([V],[\theta]) &= Q_{gi} - Q_{ci} - Q_i = Q_i^{sp} - Q_i = 0, \qquad i \in n_c \end{aligned} \tag{3.39}$$

where P_i^{sp} and Q_i^{sp} are the specified active and reactive powers.

The load flow problem is solved by applying the iterative Newton–Raphson method, which consists of linearization of the system of equations (3.39) obtaining:

$$\begin{bmatrix} \Delta P \\ \Delta Q \end{bmatrix} = [J] \cdot \begin{bmatrix} \Delta\theta \\ \Delta V/V \end{bmatrix} \tag{3.40}$$

where

$\Delta P_i = P_i^{sp} - P_i$ is the active power mismatch;

$\Delta Q_i = Q_i^{sp} - Q_i$ – the reactive power mismatch;

$[J] = \begin{bmatrix} H & N \\ M & L \end{bmatrix}$ – the Jacobian matrix.

When the AC system includes a HVDC link, the equations of power balance at the terminal buses k of the AC system are modified by including the powers at the converter stations. Taking into account that P_{kr}^{dc}, Q_{kr}^{dc}, Q_{ki}^{dc} are consumed powers and P_{ki}^{dc} is injected power, the equations of the power mismatches at the terminal buses, given also in (3.39), get the following form:

$$\begin{aligned}
\Delta P_{kr} &= P_{kr}^{sp} - P_{kr}^{ac}([\theta],[V]) - P_{kr}^{dc} V_{kr}, V_{ki}, [X]) \\
\Delta P_{ki} &= P_{ki}^{sp} - P_{ki}^{ac}([\theta],[V]) + P_{ki}^{dc}(V_{kr}, V_{ki}, [X]) \\
\Delta Q_{kr} &= Q_{kr}^{sp} - Q_{kr}^{ac}([\theta],[V]) - Q_{kr}^{dc}(V_{kr}, V_{ki}, [X]) \\
\Delta Q_{ki} &= Q_{ki}^{sp} - Q_{ki}^{ac}([\theta],[V]) - Q_{ki}^{dc}(V_{kr}, V_{ki}, [X])
\end{aligned} \tag{3.41}$$

where:

P_{kr}^{sp}, Q_{kr}^{sp}, P_{ki}^{sp}, and Q_{ki}^{sp} are the specified active and reactive powers at the terminal buses corresponding to the rectifier and inverter, respectively;

P_{kr}^{ac}, Q_{kr}^{ac}, P_{ki}^{ac}, and Q_{ki}^{ac} are the powers transferred from/to the AC system through the terminal buses (rectifier or inverter) given by expressions:

$$P_i^{ac} = \sum_{k=1}^{n} V_i V_k [G_{ik} \cos(\theta_i - \theta_k) + B_{ik} \sin(\theta_i - \theta_k)]$$

$$Q_i^{ac} = \sum_{k=1}^{n} V_i V_k [G_{ik} \sin(\theta_i - \theta_k) - B_{ik} \cos(\theta_i - \theta_k)]$$

V_{kr} and V_{ki} are the voltage magnitudes at the rectifier and inverter buses;

$[X]$ is the vector of independent variables of the HVDC system.

Taking into account the introduction of the vector of unknowns $[X]$ into the system of equations (3.40), supplementary equations describing the DC link operation and control strategy are added:

$$R_i(V_{kr}, V_{ki}, [X]) = 0, \quad i = 1,\ 2,\ \ldots,\ n_x \tag{3.42}$$

where n_x denotes the number of independent variables chosen for the HVDC system, either nine or seven. The system of equations (3.42) consists of rectifier, inverter, and line equations as well as the equations corresponding to control strategy.

The methods for load flow calculation of a mixed AC–DC system can be grouped on the following categories [17]:

- the simultaneous solving method of the nonlinear system of equations that defines the operation of the AC system and the DC link, known as the extended variable method;
- the sequential solving method based on the diakoptics principle where the DC system is treated separately by the AC one;
- the eliminated variables method of the DC system.

3.5.1.1 The Extended Variables Method

In the frame of this method, the classical Newton–Raphson algorithm, for simultaneous solving of the two systems of equations (3.39) and (3.41) of the mathematical model, and the vector of independent variables $[X]$ having nine components are used. Vector $[X]$ is added to the vector of independent variables associated with the AC system (magnitudes and angles of nodal voltages) resulting in a set of extended variables [27].

In order to have a complete system of equations, to relations (3.43a) that describe the operation of rectifier, inverter, and DC line:

$$\begin{aligned} R_1 &= V_{dr} - \frac{3\sqrt{2}}{\pi} N_{ik,r} V_{kr} \cos\alpha + \frac{3}{\pi} X_{kr} I_d = 0 \\ R_2 &= V_{di} - \frac{3\sqrt{2}}{\pi} N_{ik,i} V_{ki} \cos\gamma + \frac{3}{\pi} X_{ki} I_d = 0 \\ R_3 &= V_{dr} - V_{di} - R_L I_d = 0 \end{aligned} \tag{3.43a}$$

two more equations are added

$$\begin{aligned} R_4 &= V_{dr} - k\frac{3\sqrt{2}}{\pi} N_{ik,r} V_{kr} \cos\phi_r = 0 \\ R_5 &= V_{di} - k\frac{3\sqrt{2}}{\pi} N_{ik,i} V_{ki} \cos\phi_i = 0 \end{aligned} \tag{3.43b}$$

corresponding to the independent variables ϕ_r and ϕ_i as well as a set of four equations describing the control strategy.

Relationships (3.43b) have been obtained by substituting the expression $\frac{1}{2}[\cos\alpha + \cos(\alpha + \lambda)]$ from relation (3.12) into relation (3.13a).

The reactive powers withdrawn at the converter stations are given by

$$Q_{kr} = k\frac{3\sqrt{2}}{\pi} N_{ik,r} I_d V_{kr} \cos\phi_r; \; Q_{ki} = k\frac{3\sqrt{2}}{\pi} N_{ik,i} I_d V_{ki} \cos\phi_i$$

Concluding, the nonlinear systems of equations use in the load flow calculation of mixed systems by using the extended variable method are

$$\begin{aligned} &P_i^{sp} - P_i([V],[\theta]) = 0 && i \in n\backslash t,\ i \neq e \\ &P_i^{sp} - P_i([V],[\theta],[X]) = 0 && i \in t \\ &Q_i^{sp} - Q_i([V],[\theta]) = 0 && i \in n_c\backslash t \\ &Q_i^{sp} - Q_i([V],[\theta],[X]) = 0 && i \in t \\ &R_i([V_t],[X]) = 0 && i = 1,\ 2,\ \ldots,9 \end{aligned} \tag{3.44}$$

where

$[V_t]$ = $[V_{kr}, V_{ki}]^T$ is the vector of nodal voltages at the rectifier and inverter, respectively, and T stands for transpose;

t is an assembly of pairs of two elements that contains the indices of the buses of the terminal stations;

e – the slack bus;

n_c – load buses.

Based on the system of equations (3.44), the classical Newton–Raphson algorithm involves the iterative solving of the matrix equations:

$$\begin{bmatrix} [\Delta P] \\ [\Delta P_t] \\ \hdashline [\Delta Q] \\ [\Delta Q_t] \\ \hdashline [\Delta R] \end{bmatrix} = \begin{bmatrix} H & N & \begin{matrix} 0 \\ A \end{matrix} \\ \hdashline M & L & \begin{matrix} 0 \\ C \end{matrix} \\ \hdashline 0\;0 & 0\;D & E \end{bmatrix} \cdot \begin{bmatrix} [\Delta\theta] \\ [\Delta\theta_t] \\ \hdashline [\Delta V/V] \\ [\Delta V_t/V_t] \\ \hdashline [\Delta X] \end{bmatrix} \tag{3.45}$$

The extended variable method represents a purely mathematical approach for solving a nonlinear system of equations corresponding to a steady state of a mixed AC–DC system. It presents the disadvantage that the expanding of the Jacobian matrix (which modifies its structure substantially) makes difficult its introduction into existing load flow programs for pure AC systems.

3.5.1.2 The Sequential Method

This constitutes a solving approach based on the diakoptics method, which allows the analysis of complex systems by the analysis of independent subsystems, corresponding to isolated operation by each other. In order to apply this method, the mixed system is divided into an AC subsystem and a DC subsystem. The separation of the DC link is modeled by introducing at the terminal buses of the AC system consumed or injected powers.

After each iteration, the operating state of the DC link is calculated to decide possible commutation from one operating mode to another, if the independent variables of the vector $[X]$ exceed their specified limits.

This method presents the advantage that does not necessitate supplementary steady-state calculation of the DC link after each AC iteration. The implementation into existent programs is easy, but its disadvantage results from the fact that, in certain situations, convergence problems can appear [27].

3.5.1.3 The Eliminated Variables Method

A way of keeping the modularity and the advantages of the sequential method without affecting the mathematical accuracy is to eliminate the vector of DC variables $[X]$, having seven independent variables (3.24b), from the system of equations (3.41) [17]. In this regard, from the system of equations (3.46a) that describes the DC link operation:

$$R_i([V_t],[X]) = 0 \quad i = 1,\ 2,\ \ldots,\ 7 \tag{3.46a}$$

the vector $[X]$ is determined in terms of the terminal voltage magnitude $[V_t]$, resulting

$$[X] = f([V_t]) = f(V_{kr}, V_{ki}) \tag{3.46b}$$

By physical considerations, one solution of (3.46a) type exists for each from the control strategies (Table 3.2). Taking into account this aspect and that the nodal voltage angles of the AC system do not appear in the expression of the active and reactive powers consumed, respectively, injected through the converter station at the terminal buses it results

$$\begin{aligned}
P_{kr} &= P_{kr}(V_{kr}, V_{ki}) \\
Q_{kr} &= Q_{kr}(V_{kr}, V_{ki}) \\
P_{ki} &= P_{ki}(V_{kr}, V_{ki}) \\
Q_{ki} &= Q_{ki}(V_{kr}, V_{ki})
\end{aligned}$$

and equations (3.41) expressing the nodal power mismatches at the terminal stations becomes

$$\begin{aligned}
\Delta P_{kr} &= P_{kr}^{sp} - P_{kr}^{ac}([\theta], [V]) - P_{kr}^{dc}(V_{kr}, V_{ki}) \\
\Delta P_{ki} &= P_{ki}^{sp} - P_{ki}^{ac}([\theta], [V]) + P_{ki}^{dc}(V_{kr}, V_{ki}) \\
\Delta Q_{kr} &= Q_{kr}^{sp} - Q_{kr}^{ac}([\theta], [V]) - Q_{kr}^{dc}(V_{kr}, V_{ki}) \\
\Delta Q_{ki} &= Q_{ki}^{sp} - Q_{ki}^{ac}([\theta], [V]) - Q_{ki}^{dc}(V_{kr}, V_{ki})
\end{aligned}$$

Thereby, under steady-state conditions, the conversion stations are seen as loads for the AC system, having the static characteristics dependent on the voltage of the terminal buses.

Then, the application of the classical Newton–Raphson algorithm for the steady-state calculation is simply reduced to solving, at each iteration, the system of linear equations:

$$\begin{bmatrix} \Delta P \\ \Delta Q \end{bmatrix} = \begin{bmatrix} H & N' \\ M & L' \end{bmatrix} \cdot \begin{bmatrix} \Delta\theta \\ \Delta V/V \end{bmatrix}$$

In this matrix equation, the Jacobian is the matrix from relationship (3.40) but with a modification of the terms that contains the partial derivatives of the powers with respect to voltage magnitudes corresponding to the terminal station (rectifier or inverter).

Concluding, by the analytical elimination of the DC variables from $[X]$, the implementation of a DC link into an existent power flow program based on the Newton–Raphson method requires a minimal computation effort, given by the modification of more eight elements of the Jacobian matrix. Moreover, if the modification is performed only on the submatrices $[N]$ and $[L]$ of the Jacobian matrix, the method can be easily adapted to fast decoupled–based power flow programs. In this case, the matrix $[B']$ remains unchanged, like in the case of pure AC system, only the matrix $[B'']$ is changed.

After each iteration, the values of the DC variables form $[X]$ are calculated to verify the appearance of possible commutation between the possible operating modes of the DC link.

3.6 INTERACTION BETWEEN AC AND DC SYSTEMS

3.6.1 AC Systems Stabilization

Significant improvements of the power systems operation are provided by DC links, especially in terms of controlling the active power flow to stabilize the interconnected AC system. A DC link behaves, mainly, as a load for the sending terminal and as a power source for the receiving terminal. The transferred power on the DC link can be modulated or changed based on the AC systems requirements. The DC link can be used, for example, to transfer active power toward any of the two AC systems with lack of generation, after a generator outage or changing of the transferred power between the AC systems so that to damp power oscillations into the AC systems.

The capability of the DC links to transfer power independent of the frequency of the interconnected AC systems has certain advantages. The DC link operating into AC meshed systems can influence the power flow on the AC lines operating in parallel with it. The propagation of the disturbances effects occurring in different parts of the AC interconnected systems can be limited by the DC link connecting them. By using a suitable controlling system, the power flow on the DC link can be controlled during the disturbances, so that to improve the stability of the AC system and to diminish the risk of overloading the AC interconnection lines.

Furthermore, the stability is improved by means of automatic control of the power flow on the DC link, counteracting and damping the low frequency electromechanic oscillations from the AC system.

The stability of the interconnected AC systems can also be improved by voltage control at the inverter station when the extinction angle γ regulation is set as a voltage control strategy. In this case, the reactive power exchange between the converter station and the AC system is performed continuously by maintaining an optimum voltage level. The reactive power changing by means of angle γ regulation can be performed until the limit of $\pm 5\%$ from the transferred active power through the converter station is reached.

DC links have, generally, the advantage of no increase the short-circuit currents in the AC system. Exceptions from this rule are the transformers from the converter station, having the windings delta connected, in case of short circuits into the AC systems.

3.6.2 Influence of AC System Short-Circuit Ratio

The interaction between the AC and DC systems is dependent on the strength of the AC system (SCR) relative to the power rating P_d of the DC link. The AC system can be considered as weak in two cases, such as when the short-circuit impedance is high and the mechanical inertia is low, respectively.

The short-circuit level, SCL (MVA) is calculated by using a Thevenin equivalent impedance of the AC system (Figure 3.43) [5].

$$\mathrm{SCL} = \frac{V_{AC}^2}{Z_{Th}}$$

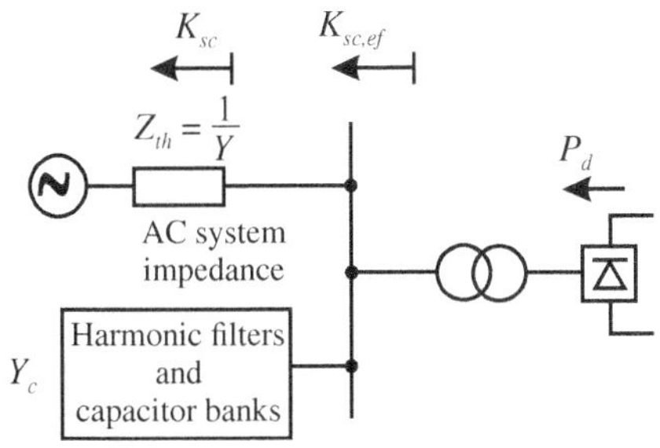

Figure 3.43 Definition the short-circuit ratios.

Then, the SCR is given by

$$\text{SCR} = \frac{\text{SCL(MVA)}}{P_d\text{(MW)}} = \frac{V_{AC}^2}{P_d Z_{Th}}$$

From the HVDC operating point of view is required that, in calculation of short-circuit ratio, the influence of equipments on the AC side associated with the DC link: harmonic filters, shunt capacitor banks, etc. to be considered. The harmonic filters, at fundamental frequency, behave practically as shunt capacitors (injecting reactive power). The capacitance of the harmonic filters and shunt capacitors will increase the equivalent impedance at fundamental frequency of the AC system. Consequently, the *effective short-circuit ratio* (ESCR) is calculated by [7]

$$\text{ESCR} = \frac{\text{SCL} - (Q_f + Q_{cb})}{P_d}$$

where

Q_f is the reactive power provided by the harmonic filters, at fundamental frequency, MVAr;

Q_{cb} – the capacitive reactive power of any additional shunt capacitors connected to the converter station terminals, MVAr.

In the initial stages of planning, the utility may know only the SCR of the system and the required HVDC capacity (MW).

The following values of the short-circuit ratio can be used to indicate approximately the strength of an AC system relative to the DC power infeed [7]:

- strong *system*, SCR > 3; no major problems are to be expected;
- *intermediate system*, 2 < SCR < 3; voltage support may have to be provided at the AC terminals of the converter station, by, for example, static VAr compensators;
- *weak system*, SCR < 2; SC or static VAr compensators may have to be added to strengthen of the AC system.

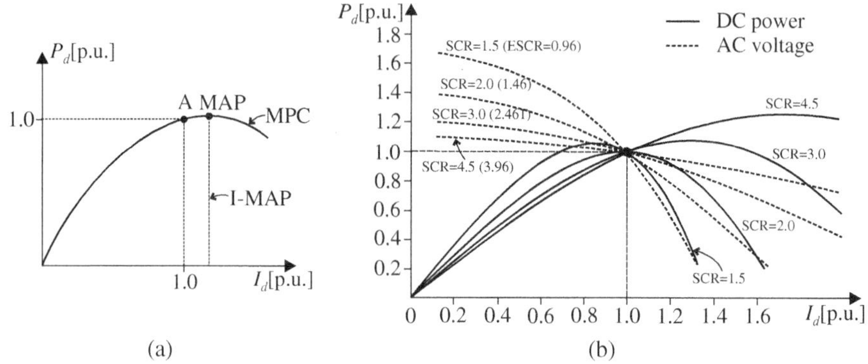

Figure 3.44 DC power-DC current characteristic (γ = constant) (a); variation of power and AC voltages with I_d for $X_C = 0.15$ p.u. and $\gamma = 18°$, $Q_c = Q_d = 0.54P_{dn}$, $V_{AC} = 1$ p.u. (b) [29].

- ***Maximum Available Power (MAP).*** Figure 3.43 represents the basic power circuit of an inverter feeding power into an AC system. The DC power-current P_d-I_d characteristic is shown in Figure 3.44a [29].

Most schemes in service are designed to operate normally at a point lower than the MAP, say at point A. These schemes operate in constant power control mode at a constant commutation margin angle γ having a minimum permissible value—normally between 15° and 18°.

The control strategy includes a DC current limit determined mainly by the convertor valve rating. Such arrangements have the lowest cost for a given power, as the ratings of inverter valves and transformers are the lowest possible and inverter VAr consumption is at a minimum.

Assume that in Figure 3.44a, the rated DC power and current at point A correspond to an AC busbar voltage $V_{AC} = 1.0$ p.u. The remainder of the P_d-I_d curve is obtained by varying I_d while keeping system emf constant. Each point of the curve is calculated by steady-state equations. As no action is taken to control AC voltage, the change of current will cause the corresponding changes in the power and the AC voltage. This quasi-steady-state curve is useful to understand the operation of HVDC.

For a given system impedance, system e.m.f. and other data of Figure 3.44a, there will be a unique P_d–I_d characteristic which will represent the maximum power curve (MPC). By increasing γ any power can be achieved below MPC but not above this curve, unless system conditions are changed.

- **Voltage/power instability**. The conditions required for voltage instability are met when the voltage sensitivity factors

$$V_{sf} = \partial V / \partial Q$$

or

$$\Psi = \partial Q / \partial V$$

are negative [29].

A negative V_{sf} or negative Ψ would indicate that an attempted operation in power control mode along the MPC curve at currents higher than the current corresponding to MAP, I-MAP, would be unstable; an increase in power demand would request a higher current, but this higher current would lead to an AC voltage reduction large enough to cause power to fall.

In CC control, $\delta P/\delta I_d$ can be negative without the operation being unstable.

It can be seen from Figure 3.44a that when operating at point A, power could be increased up to MAP by increasing current without changing AC system conditions. $\Delta P = P_{MAP} - P_{dA}$ represents the amount of power available without delay for overload or to be able to maintain P_d for some reduction of V_{AC}.

- ***AC variations and temporary overvoltages.*** An important factor to be taken into consideration is the AC voltage variation due to DC power changes. For Figure 3.44b, it is assumed that the same inverter is connected in turn to AC systems having different strengths [29]:
  - It can be seen that for a SCR = 2.5, the rated current is lower than I-MAP (i.e. 2.5) and voltage regulation is relatively modest;
  - for SCR = 2, the rated DC current corresponds to I-MAP (i.e. 2.0);
  - for SCR = 1.5, the voltage regulation is quite steep.

When the AC system has very high impedance, in addition to making sure to avoid voltage instability at normal operation steps must be taken to avoid large overvoltages.

Such classification of the AC system strength gives a preliminary evaluation of potential interactions between AC and DC systems. Also, the AC–DC interaction is influenced by the phase angle of the Thevenin equivalent (of the impedance $\underline{Z}_{Th}$); this is also called damping angle, and its value has significant influence on the DC system control stability.

3.6.3 Effective Inertia Constant

The capability of the AC system to maintain the required voltage and frequency depends on its rotational inertia. The mechanical inertia of an AC system is normally sufficiently large that a temporary reduction of power infeed would not appreciably reduce the system frequency.

However, if the local generation is small or nonexistent the interruption of the DC power infeed due to, for example, a temporary AC or DC system fault can cause unacceptable frequency reduction. Such systems need to be "strengthened" by synchronous compensators or other machines to provide additional inertia, and so avoid the need for load shedding to arrest the frequency decays.

The effective inertia constant, H_{dc}, gives a measure of the mechanical weakness of the AC systems [30]:

$$H_{dc} = H \cdot \frac{\text{MVA rating of the machine}}{\text{MW rating of the DC system}}$$

The relationship between change of machine frequency (df), mechanical power input (P_m), and electrical power output (P_e), both on machine ratings, for small changes of frequency can be represented by

$$df = \frac{(P_m - P_e)f_0 \cdot dt}{2H}$$

From the last two equations:

$$df = \frac{p \cdot dt \cdot f_0}{2H_{dc}}$$

where

H is the mechanical inertia constant, in MV-s/MVA;

f_0 – the nominal frequency;

p – per unit machine accelerating power to the base of DC power.

3.6.4 Reactive Power and the Strength of the AC System

It should be noted that because the AC systems are largely inductive, the reactive power exchange is mainly responsible for the effect of converter behavior on the AC network voltage side. Many schemes in the past were designed with transformer reactance of the order of 20% or even more to limit the thyristors fault currents. On the other hand, the reduction of the transformer reactance has some advantages, such as

- the reduction of the consumed reactive power at the converter Q_c;
- the AC system filters and any additional shunt capacitors are normally designed to supply at least all converter reactive power. By reducing the total reactive power $Q_f + Q_{cb}$ the cost of the equipments decreases while the ESCR increases;
- temporary overvoltages will be reduced, due to smaller shunt capacitors;

The bigger the reactive power consumed by any converter is, the transferred power increase. As it has been shown, for angles α and γ in the range of 15°–18° and a commutating reactance $X_k = 15\%$, a converter consumes 50%–60% reactive power. The reactive power necessary for converters operation is mainly provided by the capacitors from the filters constitution and by the capacitor banks. Since the consumed reactive power varies with the transferred DC power P_d, capacitors must be provided in appropriate sizes of switchable banks, so that the voltage is maintained in acceptable range at all load levels. The voltage level is also influenced, by the SCR of the AC system. Generators, if located near the DC terminals, can provide some of the required reactive power to maintain the voltage in an acceptable range.

For weak AC systems, it may be necessary to provide reactive power by means of static VAr compensators or SC.

Furthermore, at the operation of the HVDC system when connected to a weak AC system, other problems appear [12]:

- *Dynamic overvoltage*, when sudden interruption of the transferred power P_d through the DC link occurs. With a sudden decrease to zero of consumed reactive power of a converter station, the AC voltage increases suddenly due to shunt capacitors and harmonic filters.
- *Reduction of voltage stability reserve*. For a HVDC link connected to a weak AC system, the alternating as well as direct voltages, especially on the inverter side are very sensitive to changes in DC line loading. Therefore, an increase in the direct current is accompanied by a decrease in the alternating voltage. In such cases, the voltage control and recovery after faults presents problems; the DC system behavior can contribute to reduction in stability reserve or even to AC system collapse.
- *Voltage flickering* is due to the temporary switching off the capacitors and reactors. Also, the harmonic resonance at low frequency appearance is due to parallel resonance between capacitors and harmonic filters, on the one hand, and the inductive components from the AC system; consequently, *dangerous overvoltages* can occur.

3.7 COMPARISON BETWEEN DC AND AC TRANSMISSION

Normally, with bulk power transmission interconnections, there is a choice between AC and DC, and the determination in most cases be a matter of economics. In the case of DC, a large investment is required in terminal equipments for conversion, and this is mostly independent of the length of transmission. However, the DC overhead lines are cheaper than AC lines for the same power transfer and DC lines losses are less than those for AC lines.

In many countries, rising demand for power can go hand in hand with increasing environmental objections to the construction of new overhead lines. Converting one or more existing AC lines to HVDC lines can increase the transmission capacity within the same corridor [31].

By modifying the conductors, insulators and, for maximum benefit, the tower cross-arm structure, significantly more electrical power can be transmitted through the same corridor or right of way.

For a conventional three-phase AC transmission system (Figure 3.45a), each conductor must be insulated for the peak of the AC voltage to earth. Also, to avoid corona discharge, its equivalent outside diameter, that is the equivalent diameter created by a bundled conductor, must also be sized based on the peak AC voltage. For the DC transmission line, the voltage remains unchanged (Figure 3.45b) thus the insulation level and the equivalent outside conductor radius required for the conductor are proportional to the direct voltage applied to the conductor.

Generally, the construction cost of AC lines is higher than for the DC lines. For the same transmission capacity, the AC line requires either greater insulation level and/or greater conductor cross-sectional area. However, when terminal stations are taken into account, the line length is the main factor to determine the cost difference between the two technologies [9, 13, 31, 32].

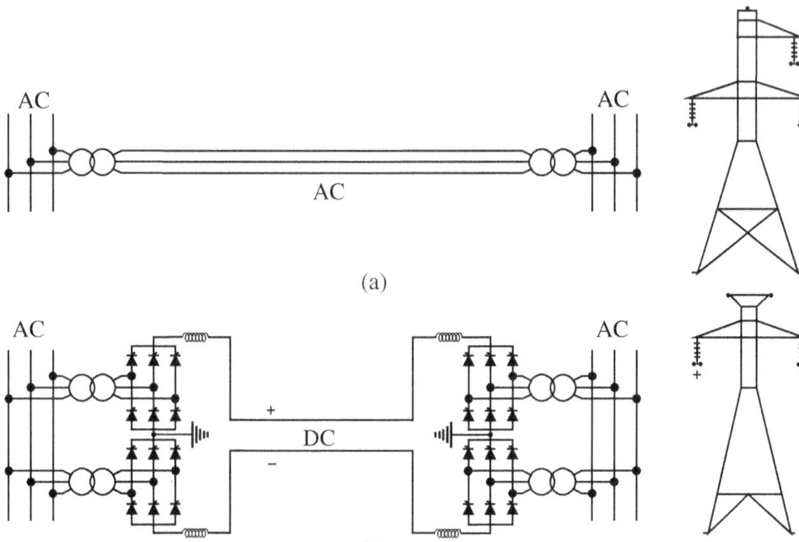

(b)

Figure 3.45 One line diagrams of a single-circuit three-phase line (a) and of a bipolar DC line (b).

Case I Single-circuit three-phase line versus bipolar DC line (Figure 3.45).

(a) For the same transmitted power and the same peak voltage to neutral, the ratio of power losses at AC to the ones at DC is 1.33.

The active power on the three-phase line is

$$P_{AC} = 3\,V_s I_{AC}$$

where $V_s = V_{\max,s}/\sqrt{2}$ ($V_{\max,s} = \hat{V}_s$ being the peak voltage to neutral of the AC line); the power factor $\cos\varphi = 1$ has been considered.

The active power on the bipolar DC line is

$$P_d = V_d I_d$$

where $V_d/2 = V_{\max,d}$ for the DC line.

Assuming that $P_{AC} = P_d$, we obtain

$$3\,V_s I_{AC} = V_d I_d \tag{3.47a}$$

or in other form

$$\left(3/\sqrt{2}\right)\ V_{\max,s} I_{AC} = 2V_{\max,d} I_d \tag{3.47b}$$

Assume also the insulators withstand the same peak voltage to neutral in both cases, $V_{\max,s} = V_{\max,d}$, it results

$$I_d = (3/2\sqrt{2})\,I_{AC} \tag{3.48}$$

Power losses for the two cases are given by

$$\Delta P_{AC} = 3I_{AC}^2 R; \quad \Delta P_d = 2I_d^2 R$$

where $R = R_d = R_{AC}$ is the ohm resistance of one phase.

Taking into account (3.48), it results the ratio of power losses as

$$\frac{\Delta P_{AC}}{\Delta P_d} = \frac{3}{2}\left(\frac{1}{1.06}\right)^2 = 1.33$$

(b) For *the same transmitted power and considering the same power losses and the same conductor cross-sectional area, the insulation level at the direct current is only 87% with respect to the one at the alternating current.*

The power losses for the two cases are

$$\Delta P_{AC} = 3I_{AC}^2 R; \quad \Delta P_d = 2I_d^2 R$$

Equating the expressions of power losses, we obtain

$$I_d = (\sqrt{3}/\sqrt{2})\, I_{AC} = 1.225\, I_{AC} \tag{3.49}$$

From (3.47a) and (3.49), it results

$$V_d = \sqrt{3} \cdot \sqrt{2}\, V_s \tag{3.50}$$

Assuming the discharge voltage of DC insulators is equal to the peak value of the alternating voltage that generates the discharging, it results in the following: the insulation level in the AC line is $k_1 \cdot \sqrt{2}\, V_s$ and $k_2 \cdot (V_d/2)$ in the DC line, respectively, where k_1 and k_2 are multiplication factors [32].

To simplify, assume $k_1 = k_2$ the ratio of insulation levels in DC and AC will be

$$\frac{\text{DC insulation level}}{\text{AC insulation level}} = \frac{V_d/2}{\sqrt{2}V_s} \tag{3.51}$$

For the chosen case, taking into account (3.50), it results

$$\frac{\text{DC insulation level}}{\text{AC insulation level}} = \frac{\sqrt{3}}{2} = 0.87$$

We can conclude that *the DC line* is more economical, besides having only two conductors as compared with the AC line that has three conductors and presents an insulation level of 87% from the AC one.

(c) If y stands for the ratio of DC and AC power losses:

$$y = \Delta P_d / \Delta P_{AC}$$

for $R_d = R_{AC}$, we obtain

$$\frac{I_d}{I_{AC}} = \sqrt{\frac{3y}{2}}$$

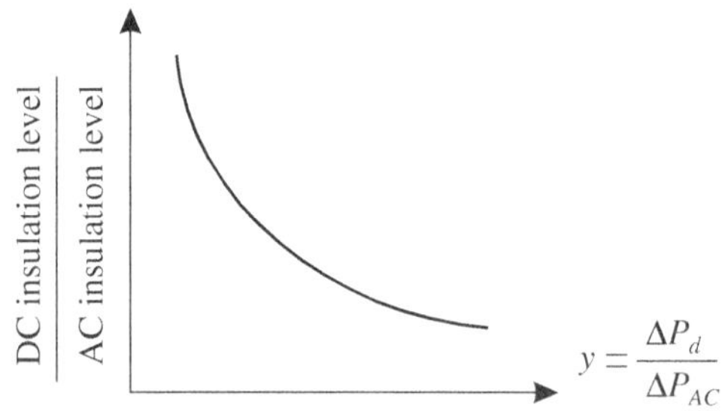

Figure 3.46 Dependency between the ratio of insulation levels and the ratio of power losses for the same transmitted power.

For $\cos\varphi = 1$ and assuming the transmitted power is the same for DC and AC as well, we can write

$$3\,V_s I_{AC} = V_d I_d$$

By compounding the last two equations, it results:

$$\frac{V_d}{V_s} = \frac{\sqrt{3}\cdot\sqrt{2}}{\sqrt{y}}$$

The ratio of the two voltages given by relationship (3.51) becomes

$$\frac{\text{DC insulation level}}{\text{AC insulation level}} = \frac{V_d/2}{\sqrt{2}V_s} = \frac{\sqrt{3}}{2\sqrt{y}} = \frac{0.87}{\sqrt{y}} \tag{3.52}$$

This latter dependency is provided graphically in Figure 3.46.

Case II Double-circuit three-phase line transformed into three DC circuits, having the same insulation level.

The transmitted power through the double-circuit AC line ($\cos\varphi = 1$) is

$$P_{AC} = 2\cdot 3V_s I_{AC}$$

The double-circuit AC line is transformed into three DC circuits, each one having two conductors on the polarities (+) and (–), respectively, and the potential $V_d/2$ referred to ground. The power transmitted through the three DC circuits is

$$P_d = 3\,V_d I_d$$

(a) For *the case* $I_{AC} = I_d$ *and considering the same voltage level* $V_d/2 = \sqrt{2}V_s$, the ratio of transmitted powers should be

$$\frac{P_d}{P_{AC}} = \frac{3\,V_d I_d}{2\cdot 3\,V_s I_{AC}} = \frac{2\sqrt{2}\,V_s}{2\,V_s} = \sqrt{2}$$

The ratio of percentage powers is

$$\frac{\text{DC power losses (\%)}}{\text{AC power losses (\%)}} = \frac{\Delta P_d}{P_d}\frac{P_{AC}}{\Delta P_{AC}}$$

or taking into account the above mentioned

$$\frac{\Delta P_d\,(\%)}{\Delta P_{AC}\,(\%)} = \frac{3 \cdot 2RI_d^2}{2 \cdot 3RI_{AC}^2}\frac{1}{\sqrt{2}} = 0.71 \rightarrow \Delta P_d\,(\%) = 0.71\ \Delta P_{AC}\,(\%)$$

(b) If one considers *the same percentage power losses and the same insulation level*

$$\frac{\Delta P_d}{P_d} = \frac{\Delta P_{AC}}{P_{AC}}; \quad \frac{V_d}{2} = \sqrt{2}\,V_s$$

it results

$$\frac{3 \cdot 2RI_d^2}{3U_dI_d} = \frac{2 \cdot 3RI_{AC}^2}{2 \cdot 3V_sI_{AC}} \text{ or } \frac{2I_d}{I_{AC}} = \frac{V_d}{V_s}$$

$$I_d = \sqrt{2}\,I_{AC}$$

In this case, the ratio of the transmitted powers will be

$$\frac{P_d}{P_{AC}} = \frac{3V_dI_d}{2 \cdot 3V_sI_{AC}} = 2 \rightarrow P_d = 2\,P_{AC}$$

From the above, it results that for the same rated voltage the DC lines require lower investments and softer constructions than the AC ones, thus with smaller conductors and insulators and softer electric towers (Figure 3.47).

For efficient utilization of the area occupied, a line operating at alternating current rated for 220 kV and transmission capacity of 480 MVA (Figure 3.48a) can be transformed into a line operating at DC transmission rated for ±380 kV, obtaining on this way a triple transmission capacity. The AC three-phase line with a double conductor per phase is converted into a DC bipolar line with a triple conductor per pole, with insulator rearrangement (Figure 3.48b).

A more interesting modification is shown in Figure 3.48c, where the number of conductors in the center bundle has been doubled; it acts as one pole, with the outer bundles serving in parallel as the other pole. As previously mentioned, insulator changes will be needed and structural upgrading may be required to support the 33% increase in weight. Thermal capacity is doubled [30].

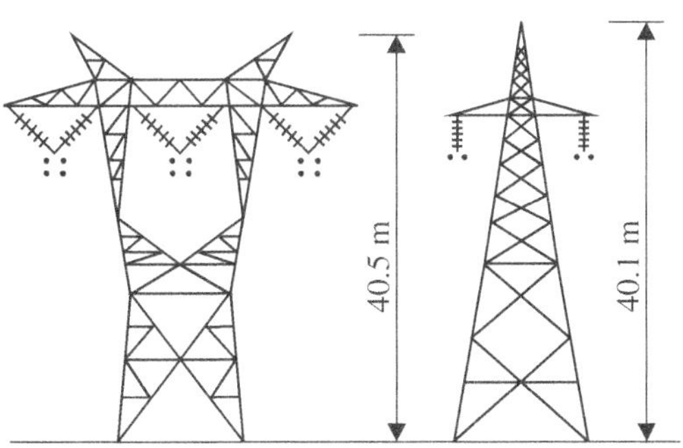

Figure 3.47 Comparison between towers sizes of 800 kV AC and ±500 kV DC, having the same transmission capacity (2000 MW).

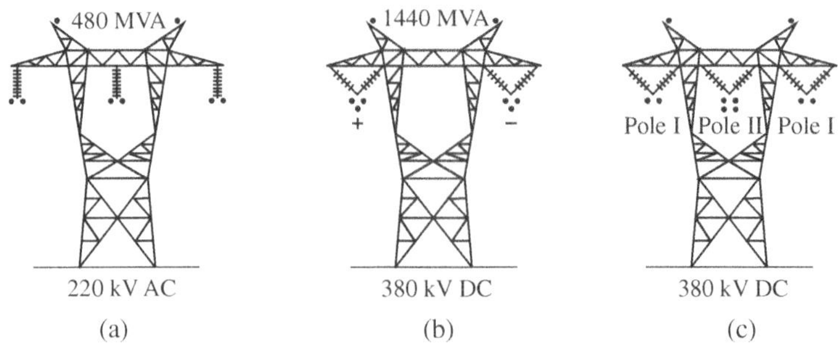

Figure 3.48 AC tower converted to bipole.

The ground area—also known as the right-of-way (RoW)—occupied by a DC power transmission is less than that required for the equivalent AC power transmission. For example, the RoW for 5000 MW, for different overhead lines (OHLs) or cable are shown in Figure 3.49.

The DC transmission lines have a bigger transmission capacity for the same RoW (Figure 3.50).

In Figure 3.51, (1) illustrates the initial cost for HVAC power transmission and (2) illustrates the initial cost of HVDC power transmission with a bigger initial cost due to a higher valve cost for HVDC transmission. In addition, (3) and (5) represent the cost for transmission line construction in HVAC and HVDC power transmission, respectively, and they demonstrate that HVDC power transmission has a lower cost for transmission line construction.

In the case of HVAC power transmission, a shunt reactor must be installed typically at every 100 or 200 km because of its electrostatic capacity. In other words, the increase in the total cost for power transmission lines is accompanied by additional costs due to the shunt capacitor [5].

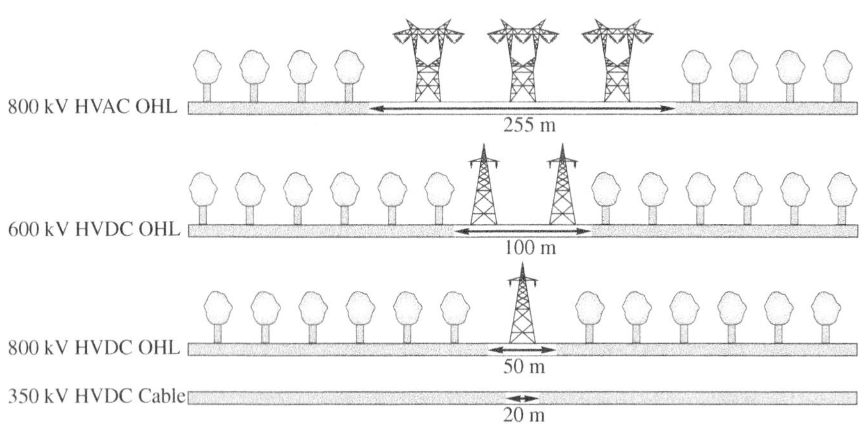

Figure 3.49 The need for RoW when comparing HVDC and HVAC for a power transmission of 5000 MW.

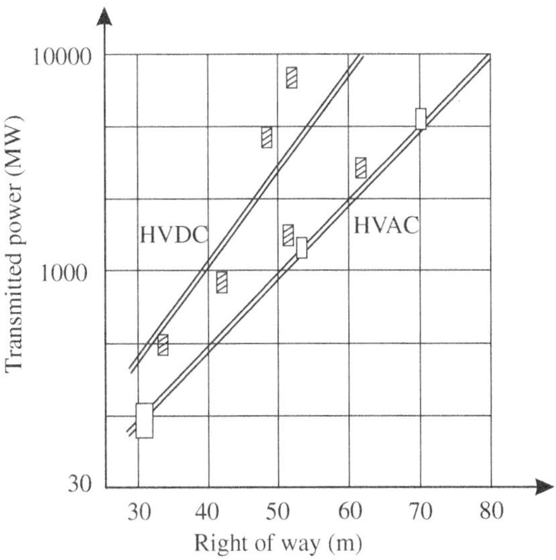

Figure 3.50 Transmission power versus RoW for HVDC and HVAC.

Also in Figure 3.51, (6) and (7) illustrate losses for HVAC and HVDC systems during power transmission. It is shown that the HVDC system has a smaller loss if the same amount of electric power is delivered. Therefore, even though the capital cost of the HVDC terminal station can be greater than that of an equivalent AC transmission substation, the per kilometer distance of the transmission circuit will be lower for DC

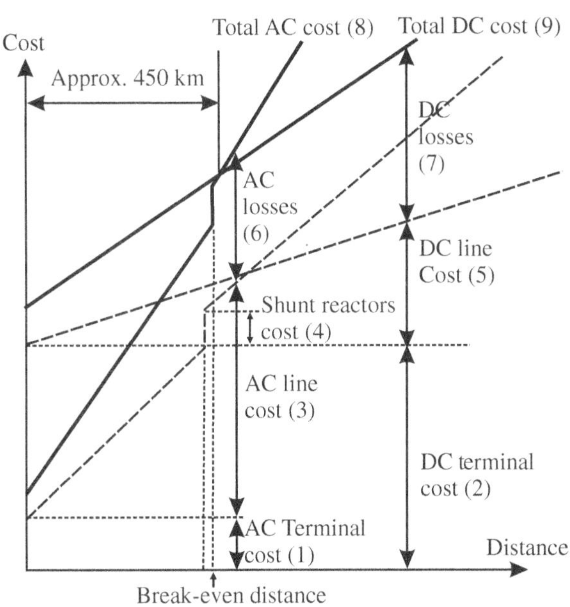

Figure 3.51 Transmission distance and investments costs AC and DC power transmission lines. *Source:* Kim et al. 2009 [5]. Reproduced with permission of IEEE.

TABLE 3.5 Transmission capacities of some AC and DC lines [33]

Economic loading of a AC line		Equivalent DC line			
		At the same insulation level		At the same right of way	
(kV)	(MW)	(kV)	(MW)	(kV)	(MW)
230	240	±200	400	±300	900
345	580	±300	900	±500	2500
500	1280	±400	1600	±700	4500
765	2700	±600	3600	±1000	8000

than for AC. Hence, a break-even distance can be defined beyond which it is more economical to transmit bulk real power as DC instead of AC. This break-even distance varies depending on the economic factors influencing each case. The HVAC transmission is favorable for distance less than about 450 km, and HVDC transmission is favorable for distance exceeding 450 km [5]. In case of undersea cables where the intersections of the bold lines are located at a relatively short distance (Figure 3.51), the DC system is much more economical.

A comparison between the transmission capacities of some AC and DC lines, based on technical-economic considerations, is given in Table 3.5 [33].

The 177 km long Murraylink underground high voltage interconnection, which uses HVDC light technology, is the world's longest [34]. The project of ABB connects the electricity grids in the states of Victoria and South Australia, allowing power to be traded directly between the two states. Underground cables were used because a large portion of terrain between the two states is made up of national parks with sensitive wildlife, as well as large privately owned agricultural areas. In addition to the visual and environmental impact, underground cables offer protection against Australia's traditional causes of power outages, such as lightning and damage caused by wildlife or bush fires.

In Table 3.6, the cost values, given in year 2000 in US$/kW/bipole (for both ends), for one valve group per pole, are presented.

These costs can be used to explore development options, but confirmatory figures obviously need to be obtained from manufacturers. Each power system is different with respect to voltage, system strength, and harmonic and reactive power limits. Each owner has different requirements concerning overloads, availability, reliability, etc. Each HVDC scheme is therefore unique, and caution is needed when utilizing the DC turnkey costs and additional facility cost variations discussed above for competing options. It is extremely important to consider all options on the same relative cost basis and also on approximately the same system scope basis (same capacity, dynamic performance, reliability, loss analysis, etc.).

It obviously must be treated with caution, as quoted costs of DC stations are subject, like anything else, to the vagaries of the marketplace. Though they have recently been dropping, it cannot be predicted whether such conditions will continue or reverse [6,9].

In Table 3.6, the "total" level gives typical turnkey costs of the vendor's HVDC supply and installations. These costs cover both terminals of a two-terminal scheme

TABLE 3.6 Historical HVDC turnkey cost division [9]

	Back-to-back		Monopole 500 kV	Bipole ±500 kV	Bipole ±500 kV	Bipole ±600 kV
	200 MW (%)	500 MW (%)	500 MW (%)	1000 MW (%)	2000 MW (%)	3000 MW (%)
Valve groups	19	19	21	21	22	22
Converter transformers	22.5	22.5	21	22	22	22
DC switchyard and filtering	3	3	6	6	6	6
AC switchyard & filtering	11	11	10	9.5	9	9
Control/Prot/Comm.	8.5	8.5	8	8	8	8
Civil/Mech. Works	13	13	14	14	13.5	13.5
Auxiliary power	2	2	2.5	2.5	2.5	2.5
Project eng. and admin.	21	21	17.5	17	17	17
Total, per kW	\$130	\$90	\$180	\$170	\$145	\$150

*Material reprinted with permission from CIGRE Working Group 14.20 [9].

and are based on some simplified assumptions: The DC bipole is made up of one valve group per pole; no special measures are required for reactive power and/or voltage control to incorporate a DC scheme into a weak AC system. These costs do not also include any costs by the purchaser entity itself, taxes, interest during construction, or other money borrowing costs. In certain applications, the purchaser's costs can be substantial.

3.8 APPLICATION ON A CSC–HVDC LINK

Let us consider the DC bipolar line from Figure 3.52, having the following characteristics [14,28]:

- The rated power and voltage values are 1200 MW and ±300 kV, respectively;
- the conversion stations consists of four bridges six-pulse configurations, each of them having the commutation reactance $X_{kB,r} = X_{kB,i} = 6\ \Omega$;
- the total resistance of the line is $R_L = 15\ \Omega$;
- the converter transformers are equipped with on-load tap change mechanisms to provide an appropriate level of the three-phase voltage to the valve bridge. This aims to restore, after disturbances, the values of the angles α and γ in intervals appropriate to the normal operating state ($\alpha \in [15°, 21°]$ and $\gamma \in [18°, 21°]$, respectively). The rated transformation ratio has the same value at both the rectifier and inverter stations ($N_{ik,r}^{rated} = N_{ik,i}^{rated} = 0.320$) and may vary in the interval from $N_{ik,r}^{\min} = N_{ik,i}^{\min} = 0.256$ to $N_{ik,r}^{\max} = N_{ik,i}^{\max} = 0.384$, corresponding to the

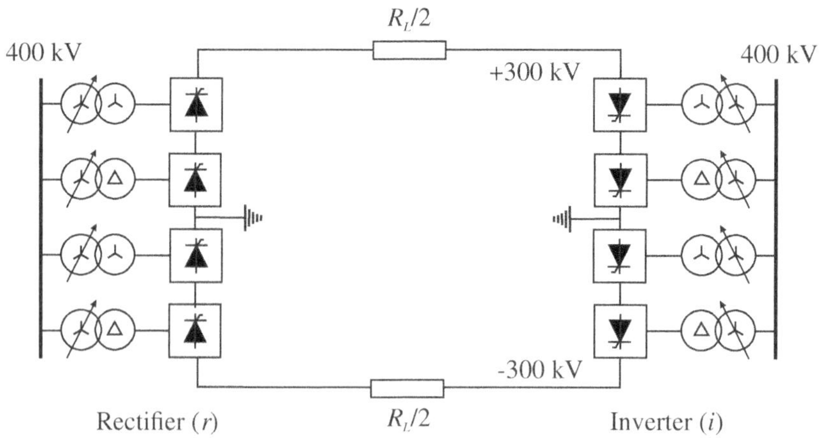

Figure 3.52 Bipolar line.

interval (0.8–1.2), in per unit, with increasing step of 0.01 p.u., in order to maintain the ignition angle α in appropriate range in a normal operating state and the inverter voltage in the range $\pm 2.5\%$ of the rated voltage ($U_{di} \in [585, 615]$ kV).

- the minimum value of the ignition delay angle is $\alpha_{\min} = 6°$;
- the current margin is set to $\Delta I_m = 15\%$;
- under normal operating conditions, the link operates in control mode 1, so that
  - the rectifier controls the current (mode CC), operates with $\alpha = 18°$, and provides the power at terminals $P_{dr} = 1260$ MW;
  - the inverter controls the voltage level, operates with CEA (mode CEA) $\gamma = 19°$ and has the terminal voltage $U_{di} = 600$ kV;
  - the transformation ratio values are $N_{ik,r} = N_{ik,i} = 0.32$.

I. *Determine for normal operation:*

(i) rectifier voltage V_{dr} and direct current I_d;

(ii) overlap angles λ_r and λ_i;

(iii) phase-to-phase voltages values V_{kr} and V_{ki}, and currents I_{kr} and I_{ki}, respectively, at high voltage terminals of the conversion stations;

(iv) active, reactive, and apparent powers as well as the power factor at high voltage terminals of the conversion stations (S_{kr}, P_{kr}, Q_{kr}, and $\cos \phi_r$, respectively, S_{ki}, P_{ki}, Q_{ki}, and $\cos \phi_i$).

II. *Considering that the AC voltage at rectifier decrease by $\Delta V = 10\%$ with respect to the voltage in the normal operating state and the AC voltage at the inverter remains unchanged, and the voltage control by means of tap changing is inactive, determine:*

(i) control strategy and the new value of the direct current I_d;

(ii) rectifier and inverter voltages V_{dr} and V_{di};

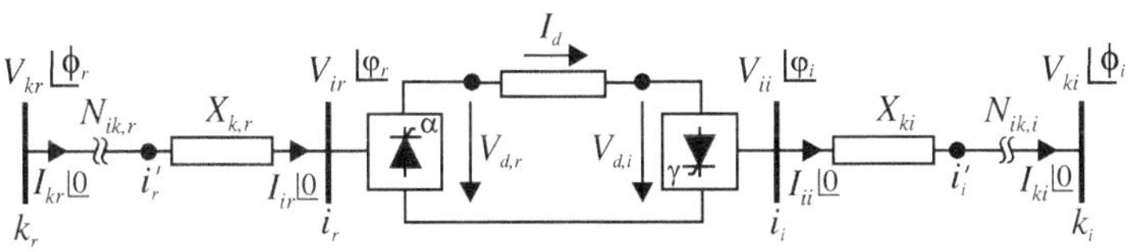

Figure 3.53 Equivalent circuit of the bipolar line.

(iii) extinction angle γ and the overlap angles λ_r and λ_i after the grid control system acts;

(iv) active, reactive, and apparent powers as well as the power factor at high voltage terminals of the conversion stations (S_{kr}, P_{kr}, Q_{kr}, and $\cos\phi_r$, respectively, S_{ki}, P_{ki}, Q_{ki}, and $\cos\phi_i$) before the control system of the tap-change position at transformer is activated;

(v) transformation ratio value $N_{ik,r}$ necessary to restore the normal operating conditions.

III. *Assuming that, with respect to the normal operating state, the AC voltage at inverter, increase by 2.5%, and the AC voltage at rectifier remain unchanged, under the hypothesis that the voltage control by means of tap changing is activated, determine:*

(i) rectifier and inverter voltages V_{dr} and V_{di} and the direct current I_d before the grid control system acts;

(ii) ignition delay α after the grid control system acts;

(iii) transformation ratio values $N_{ik,r}$ and $N_{ik,i}$ after the action of the control system of tap-changer position;

(iv) active, reactive, and apparent powers as well as the power factor at high voltage terminals of the conversion stations (S_{kr}, P_{kr}, Q_{kr}, and $\cos\phi_r$, respectively, S_{ki}, P_{ki}, Q_{ki}, and $\cos\phi_i$) under the new operating conditions.

Generally, for the operating state analysis of a bipolar DC link, the equivalent monopolar circuit is used (Figure 3.53). Let us denote by n_{Br} the number of rectifier bridges, and by n_{Bi} the number of inverter bridges and knowing that these are series connected on the DC side and parallel connected on the AC side, then the calculation relationships of the monopolar link (Table 3.1) get the form given in Table 3.7.

3.8.1 Solution

I. *Normal operating state*

(i) Calculation of rectifier voltage V_{dr} and direct current I_d.
From relation (A6), we express the direct current as $I_d = P_{dr}/V_{dr}$, which is substituted in relation (A9) resulting the second-order equation:

$$V_{dr}^2 - V_{di}V_{dr} - R_L P_{dr} = 0 \rightarrow V_{dr} = \frac{V_{di} \pm \sqrt{V_{di}^2 + 4R_L P_{dr}}}{2} \quad \text{(A10)}$$

TABLE 3.7 Operating equations of a bipolar line equivalated through a monopolar line

Rectifier	Inverter	
$V_{d0,r} = \frac{3\sqrt{2}}{\pi} n_{Br} N_{ik,r} V_{kr}$	$V_{d0,i} = \frac{3\sqrt{2}}{\pi} n_{Bi} N_{ik,i} V_{ki}$	(A1)
$V_{dr} = V_{d0,r} \cos\alpha - \frac{3}{\pi} n_{Br} X_{kB,r} I_d$	$V_{di} = V_{d0,i} \cos\gamma - \frac{3}{\pi} n_{Bi} X_{kB,i} I_d$	(A2)
$V_{dr} = \frac{V_{d0,r}}{2}[\cos\alpha + \cos(\alpha + \lambda_r)]$	$V_{di} = \frac{V_{d0,i}}{2}[\cos\gamma + \cos(\gamma + \lambda_i)]$	(A3)
$I_{kr} \cong \frac{\sqrt{6}}{\pi} n_{Br} N_{ik,r} I_d$	$I_{ki} \cong \frac{\sqrt{6}}{\pi} n_{Bi} N_{ik,i} I_d$	(A4)
$S_{kr} = \sqrt{3} V_{kr} I_{kr}$	$S_{ki} = \sqrt{3} V_{ki} I_{ki}$	(A5)
$P_{kr} \cong P_{dr} = V_{dr} I_d$	$P_{ki} \cong P_{di} = V_{di} I_d$	(A6)
$Q_{kr} = \sqrt{S_{kr}^2 - P_{kr}^2}$	$Q_{ki} = \sqrt{S_{ki}^2 - P_{ki}^2}$	(A7)
$\cos\phi_r = \frac{P_{kr}}{S_{kr}}$	$\cos\phi_i = \frac{P_{ki}}{S_{ki}}$	(A8)
$V_{dr} = V_{di} + R_L I_d$		(A9)

Of the two solutions of equation (A10), the one with positive "+" sign is kept because for the no-load conditions ($I_d = 0$ and $P_{dr} = 0$, respectively) this provides the value $V_{dr} = V_{di}$ which is physically correct.

Thus

$$V_{dr} = \frac{600 + \sqrt{600^2 + 4 \cdot 15 \cdot 1260}}{2} = 630 \text{ kV} \Rightarrow I_d = \frac{P_{dr}}{V_{dr}} = \frac{1260}{630} = 2 \text{ kA}$$

(ii) Calculation of the overlap angles.
From relation (A2), we determine

$$V_{d0,r} = \frac{V_{dr} + \frac{3}{\pi} n_{Br} X_{kB,r} I_d}{\cos\alpha} = \frac{630 + \frac{3}{\pi} \cdot 4 \cdot 6 \cdot 2}{\cos 18°} = 710.62 \text{ kV}$$

$$V_{d0,i} = \frac{V_{di} + \frac{3}{\pi} n_{Bi} X_{kB,i} I_d}{\cos\gamma} = \frac{600 + \frac{3}{\pi} \cdot 4 \cdot 6 \cdot 2}{\cos 19°} = 683.05 \text{ kV}$$

Taking into account relation (A3), it results

$$\cos(\alpha + \lambda_r) = \frac{2\,V_{dr}}{V_{d0,r}} - \cos\alpha = \frac{2 \cdot 630}{710.62} - \cos 18° = 0.822$$

$$\cos(\gamma + \lambda_i) = \frac{2\,V_{di}}{V_{d0,i}} - \cos\gamma = \frac{2 \cdot 600}{683.05} - \cos 19° = 0.811$$

and

$$\lambda_r = \text{acos}0.822 - 18° = 16.71°$$

$$\lambda_i = \text{acos}0.811 - 19° = 16.81°$$

(iii) The AC phase-to-phase voltages V_{kr} and V_{ki} at the terminal buses are determined using relation (A1), from which it results

$$V_{kr} = \frac{\pi V_{d0,r}}{3\sqrt{2}\, n_{Br} N_{ik,r}} = \frac{\pi \cdot 710.62}{3\sqrt{2} \cdot 4 \cdot 0.32} = 411.09 \text{ kV}$$

$$V_{ki} = \frac{\pi V_{d0,i}}{3\sqrt{2}\, n_{Bi} N_{ik,i}} = \frac{\pi \cdot 683.05}{3\sqrt{2} \cdot 4 \cdot 0.32} = 395.14 \text{ kV}$$

From relation (A4), it results

$$I_{kr} = \frac{\sqrt{6}}{\pi} n_{Br} N_{ik,r} I_d = \frac{\sqrt{6}}{\pi} \cdot 4 \cdot 0.32 \cdot 2 = 1.996 \text{ kA}$$

$$I_{ki} = \frac{\sqrt{6}}{\pi} n_{Bi} N_{ik,i} I_d = \frac{\sqrt{6}}{\pi} \cdot 4 \cdot 0.32 \cdot 2 = 1.996 \text{ kA}$$

(iv) The active, reactive, and apparent powers as well as the power factor on the AC sides are determined in the following:

Using relations (A6)–(A9), we proceed to the calculation of

- apparent powers:

$$S_{kr} = \sqrt{3}\, V_{kr} I_{kr} = \sqrt{3} \cdot 411.09 \cdot 1.996 = 1421.21 \text{ MVA}$$

$$S_{ki} = \sqrt{3}\, V_{ki} I_{ki} = \sqrt{3} \cdot 395.14 \cdot 1.996 = 1366.07 \text{ MVA}$$

- active powers:

$$P_{kr} \cong P_{dr} = V_{dr} I_d = 630 \cdot 2 = 1260 \text{ MW}$$

$$P_{ki} \cong P_{di} = V_{di} I_d = 600 \cdot 2 = 1200 \text{ MW}$$

- power losses on the DC line:

$$\Delta P_{DC} = P_{dr} - P_{di} = R_L I_d^2 = 15 \cdot 2^2 = 60 \text{ MW}$$

- reactive powers absorbed at the terminal buses:

$$Q_{kr} = \sqrt{S_{kr}^2 - P_{kr}^2} = 657.45 \text{ MVAr}$$

$$Q_{ki} = \sqrt{S_{ki}^2 - P_{ki}^2} = 652.80 \text{ MVAr}$$

- power factors:

$$\cos\phi_r = \frac{P_{kr}}{S_{kr}} = 0.887; \quad \cos\phi_i = \frac{P_{ki}}{S_{ki}} = 0.878$$

II. *Analysis of the disturbed state caused by the decrease in voltage by $\Delta V = 10\%$ on the AC side of the rectifier terminal.*

(i) The control strategy and direct current I_d.

Due to the decrease in voltage at the terminal bus, the rectifier voltage decreases. In order to maintain the current at the specified value, the control system of the link will trigger, in the first stage, the decrease of the ignition angle α in order to increase the voltage V_{dr}. Subsequently, the angle α will be brought in the normal operating range by changing the transformation ratio (shifting the tap-changer position).

In this section, the operation of the HVDC link will be analyzed for the time interval before to activation of the control system of the tap-changer position.

Assume that the HVDC link will keep on operation in normal conditions with: the rectifier on CC control with $I_d = 2$ kA and the inverter on CEA control with $V_{di} = 600$ kV.

Therefore, the ignition angle α should be changed so that:

$$V_{dr} = V_{di} + R_L I_d = 600 + 15 \cdot 2 = 630 \text{ kV}$$

From relation (A2) written under the form

$$V_{dr} = \frac{3\sqrt{2}}{\pi} n_{Br} N_{ik,r} \left(1 + \frac{\Delta V}{100}\right) V_{kr} \cos\alpha - \frac{3}{\pi} n_{Br} X_{kB,r} I_d$$

where ΔV stands for the percent voltage variation at the terminal bus with respect to the value corresponding to the normal operating state, we determine

$$\cos\alpha = \frac{V_{dr} + \frac{3}{\pi} n_{Br} X_{kB,r} I_d}{\frac{3\sqrt{2}}{\pi} n_{Br} N_{ik,r} \left(1 + \frac{\Delta V}{100}\right) V_{kr}}$$

For $\Delta V = -10\%$, it results

$$\cos\alpha = \frac{630 + \frac{3}{\pi} \cdot 4 \cdot 6 \cdot 2}{\frac{3\sqrt{2}}{\pi} \cdot 4 \cdot 0.32 \cdot (1 - 0.1) \cdot 411.09} = 1.057 > 1$$

Therefore, the normal operating conditions cannot be fulfilled, and the shift in the control mode is chosen: the rectifier on CIA control mode with $\alpha = \alpha_{min} = 6°$ and the inverter on CC control mode with

$$I_d = I_d^{sp} - \Delta I_m = (1 - 0.15) I_d^{sp} = 0.85 \cdot 2 = 1.7 \text{ kA}$$

where $I_d^{sp} = 2$ kA is the direct current value under normal operating conditions.

(ii) The rectifier and inverter direct voltages now are

$$\begin{aligned} V_{dr} &= \frac{3\sqrt{2}}{\pi} n_{Br} N_{ik,r} \cdot \left(1 + \frac{\Delta V}{100}\right) V_{kr} \cos\alpha_{min} - \frac{3}{\pi} n_{Br} X_{kB,r} I_d \\ &= \frac{3\sqrt{2}}{\pi} \cdot 4 \cdot 0.32 \cdot 0.9 \cdot 411.09 \cdot \cos 6° - \frac{3}{\pi} \cdot 4 \cdot 6 \cdot 1.7 = 597.09 \text{ kV} \end{aligned}$$

$$V_{di} = V_{dr} - R_L I_d = 597.09 - 15 \cdot 1.7 = 571.59 \text{ kV}$$

(iii) The angles γ, λ_r, and λ_i

Taking into account that the new AC rectifier voltage value is $V_{kr} = 0.9 \cdot 411.09$ kV (reduced by 10%) and that the AC inverter voltage remained unchanged ($V_{ki} = 395.14$ kV), from relation (A1) it results

$$V_{d0,r} = \frac{3\sqrt{2}}{\pi} n_{Br} N_{ik,r} V_{kr} = \frac{3\sqrt{2}}{\pi} \cdot 4 \cdot 0.32 \cdot 0.9 \cdot 411.09 = 639.55 \text{ kV}$$

which is 0.9 of the normal operating state value, whereas the inverter voltage remained unchanged, $V_{d0,i} = 683.05$ kV.

Next, from relations (A2) and (A3) we determine

$$\cos\gamma = \frac{V_{di} + \frac{3}{\pi} n_{Bi} X_{kB,i} I_d}{V_{d0,i}} = \frac{571.59 + \frac{3}{\pi} \cdot 4 \cdot 6 \cdot 1.7}{683.05} = 0.894 \Rightarrow \gamma = 26.64°$$

$$\cos(\alpha_{\min} + \lambda_r) = \frac{2\,V_{dr}}{V_{d0,r}} - \cos\alpha_{\min} = \frac{2 \cdot 597.09}{639.55} - \cos 6° = 0.873$$

$$\cos(\gamma + \lambda_i) = \frac{2\,V_{di}}{V_{d0,i}} - \cos\gamma = \frac{2 \cdot 571.59}{683.05} - \cos 26.64° = 0.780$$

and

$$\lambda_r = \text{acos}\, 0.873 - 6° = 23.191°$$
$$\lambda_i = \text{acos}\, 0.780 - 26.64° = 12.10°$$

(iv) The active, reactive, and apparent powers as well as the power factors at the terminal buses are determined in a similar manner like at the point (iv) from the case I. It results:

- currents at the terminal buses:

$$I_{kr} = \frac{\sqrt{6}}{\pi} n_{Br} N_{ik,r} I_d = \frac{\sqrt{6}}{\pi} \cdot 4 \cdot 0.32 \cdot 1.7 = 1.697 \text{ kA}$$
$$I_{ki} = \frac{\sqrt{6}}{\pi} n_{Bi} N_{ik,i} I_d = 1.697 \text{ kA}$$

- apparent powers:

$$S_{kr} = \sqrt{3}\, V_{kr} I_{kr} = \sqrt{3} \cdot 0.9 \cdot 411.09 \cdot 1.697 = 1087.48 \text{ MVA}$$
$$S_{ki} = \sqrt{3}\, V_{ki} I_{ki} = \sqrt{3} \cdot 395.14 \cdot 1.697 = 1161.43 \text{ MVA}$$

- active powers:

$$P_{kr} \cong P_{dr} = V_{dr} I_d = 597.09 \cdot 1.7 = 1015.05 \text{ MW}$$
$$P_{ki} \cong P_{di} = V_{di} I_d = 571.59 \cdot 1.7 = 971.70 \text{ MW}$$

- reactive powers:

$$Q_{kr} = \sqrt{S_{kr}^2 - P_{kr}^2} = 390.24 \text{ MVAr}$$
$$Q_{ki} = \sqrt{S_{ki}^2 - P_{ki}^2} = 636.18 \text{ MVAr}$$

- power factors:

$$\cos\phi_r = \frac{P_{kr}}{S_{kr}} = 0.933; \quad \cos\phi_i = \frac{P_{ki}}{S_{ki}} = 0.837$$

Observations:

(a) The transmitted power (the power at the inverter terminal) decreases by

$$\Delta P = \frac{1200 - 971.70}{1200} \cdot 100 \cong 19\%$$

with respect to the power transmitted in the normal operating state.

(b) The decrease in absorbed reactive power and the increase in power factor at rectifier terminal are due to the decrease of the ignition delay angle up to the value $\alpha_{\min} = 6°$.

(c) The reduction of the power factor at the inverter terminal bus is due to the increase of the extinction angle γ.

(d) In order to restore the normal operating conditions, the change of the transformation ratio $N_{ik,r}$ is required, so that for $\alpha = 18°$ the voltage V_{dr} to restore to the value $V_{dr}^{(n)} = 630$ kV.

From relations (A1) and (A2), it results that the normal operating state is restored if

$$V_{d0,r}^{(n)} = \frac{3\sqrt{2}}{\pi} n_{Br} N_{ik,r} \cdot 0.9 \cdot V_{kr}^{(n)}$$

where the superscript (n) designate the values corresponding to the normal operating state.

The new value of the transformation ratios is

$$N_{ik,r} = \frac{\pi V_{d0,r}^{(n)}}{3\sqrt{2}\, n_{Br} \cdot 0.9 \cdot V_{kr}^{(n)}} = \frac{\pi \cdot 710.62}{3 \cdot \sqrt{2} \cdot 4 \cdot 0.9 \cdot 411.09} = 0.356$$

or in per unit,

$$N_{ik,r} = \frac{0.356}{0.320} = 1.1125 \text{ p.u.}$$

III. *Analysis of the disturbed state caused by the increase in voltage by $\Delta V = 2.5\%$ on the AC side of the inverter terminal.*

(i) Because the inverter operates in mode CEA ($\gamma =$ ct.) and the control system of the tap-changer is not activated (this will act after a certain period of time which is of the second order), the increase in AC voltage at the inverter terminal bus will determine an increase in the inverter voltage V_{di}.

Therefore,

$$\begin{aligned} V_{d0,i} &= \frac{3\sqrt{2}}{\pi} n_{Bi} N_{ik,i} \left(1 + \frac{\Delta V}{100}\right) V_{ki}^{(n)} \\ &= \frac{3\sqrt{2}}{\pi} 4 \cdot 0.32 \cdot 1.025 \cdot 395.14 \text{ kV} = 700.12 \text{ kV} \end{aligned}$$

and thus

$$V_{di} = V_{d0,i}\cos\gamma - \frac{3}{\pi} n_{Bi} X_{kB,i} I_d$$
$$= 700.12 \cdot \cos 19^\circ - \frac{3}{\pi} \cdot 4 \cdot 6 \cdot 2 = 616.14\ \text{kV}$$

This change in the voltage V_{di} will result in change of the direct current I_d because the operating conditions at the rectifier remained unchanged ($V_{dr} = V_{dr}^{(n)} = 630$ kV).

The new value of the direct current is

$$I_d = \frac{V_{dr}^{(n)} - V_{di}}{R_L} = \frac{630 - 616.14}{15} = 0.924\ \text{kA}$$

(ii) Detecting this decrease in current, the grid control system will trigger the decrease of the ignition angle α so that the current will restore to the normal operating state value $I_d^{(n)} = 2$ kA.

The new value of the angle α is determined as it follows:

- The voltage necessary at rectifier is calculated as

$$V_{dr} = V_{di} + R_L I_d^{(n)} = 616.14 + 15 \cdot 2 = 646.14\ \text{kV}$$

- From relation (A2), it results

$$\cos\alpha = \frac{V_{dr} + \frac{3}{\pi} n_{Br} \cdot X_{kB,r} I_d}{V_{d0,r}^{(n)}} = \frac{646.14 + \frac{3}{\pi} \cdot 4 \cdot 6 \cdot 2}{710.62} = 0.974 \Rightarrow \alpha = 13.09^\circ$$

(iii) After a period of time (seconds) from the disturbance occurrence, the control system of the transformer tap-changer will be activated. In this way, the rectifier transformer tap-changer acts to hold α between 15° and 21° whereas the inverter transformer tap-changer acts to hold V_{di} between 585 and 615 kV.

Maintaining the current on the DC line at the value $I_d = I_d^{(n)} = 2$ kA, according to relations (A1), (A2), and (A9), the voltages V_{dr} and V_{di} meet the following relations:

$$V_{dr} = \frac{3\sqrt{2}}{\pi} n_{Br} N_{ik,r} V_{kr}^{(n)} \cos\alpha - \frac{3}{\pi} n_{Br} X_{kB,r} I_d = \tag{A11}$$
$$= \frac{3\sqrt{2}}{\pi} \cdot 4 \cdot N_{ik,r} \cdot 411.09 \cdot \cos\alpha - \frac{3}{\pi} \cdot 4 \cdot 6 \cdot 2 = 2220.67\, N_{ik,r} \cos\alpha - 45.84$$

$$V_{di} = \frac{3\sqrt{2}}{\pi} n_{Bi} N_{ik,i} \cdot 1.025 \cdot V_{ki}^{(n)} \cos\gamma - \frac{3}{\pi} n_{Bi} X_{kB,i} I_d = \tag{A12}$$
$$= \frac{3\sqrt{2}}{\pi} \cdot 4 \cdot 1.025 \cdot 395.14 \cdot N_{ik,i} \cos 19^\circ - \frac{3}{\pi} \cdot 4 \cdot 6 \cdot 2 = 2068.67\, N_{ik,i} - 45.84$$

$$V_{dr} - V_{di} = R_L I_d = 30\ \text{kV} \tag{A13}$$

TABLE 3.8 Voltages V_{dr} and V_{di}, angle α, variations in terms of the transformation ratio

$N_{ik,r}$	$N_{ik,i}$	V_{dr} (kV)	V_{di} (kV)	α (°)
0.320	0.320	646.13	616.13	13.15
0.323	0.317	639.51	609.51	17.27
0.326	0.314	632.89	602.89	20.54
0.330	0.310	626.28	596.28	23.32
…	…	…	…	…
0.374	0.266	533.60	503.60	45.82
0.378	0.262	526.98	496.98	46.91
0.381	0.259	520.36	490.36	47.97

Analyzing the expressions (A11)–(A13), we see that in order to restore the normal operating conditions, the increasing of the ratio $N_{ik,r}$ and angle α, and the decrease of ratio $N_{ik,i}$, respectively, are necessary.

In Table 3.8, the ignition angle α values and the voltages V_{dr} and V_{di} for the analyzed case are presented, where the transformation ratio $N_{ik,r}$ increases from $N_{ik,r}^{rated}$ to $N_{ik,r}^{\max}$ and the transformation ratio $N_{ik,i}$ decreases from $N_{ik,i}^{\text{rated}}$ to $N_{ik,i}^{\min}$.

We see that the shift in tap-changers with one position in the increasing direction of $N_{ik,r}$ and decreasing direction of $N_{ik,i}$, respectively, will restore the operation of the link with values of α and V_{di} in an admissible range, that is, $\alpha = 17.17° \in [15°, 21°]$ and $U_{di} = 609,51$ kV $\in [585, 615]$ kV.

APPENDIX 3.1 CSC–HVDC SYSTEMS IN THE WORLD

Line	Year	Power (MW)	Voltage (kV)]	Length (km) Over head	Under sea	Main reasons for choosing the HVDC system
a. Mercury-arc valve systems						
Kashira–Moscow	1950	30	±200	115	—	—
Gotland, Sweden I	1954–1970	20–30	100–150	—	96	Long sea crossing; frequency control
France–Great Britain I	1961	160	±100	—	7 + 50 + 8	Sea crossing; asynchronous link 50/60 Hz, out of service since 1984
Volgograd–Donbass	1964	750	±400	475	—	Long distance
Sakuma (Japan)	1965	300	2 × 125	—	—	Asynchronous link, rapid control, low losses (ASEA)
Konti–Skan (Denmark–Sweden)	1965/2006	250/380	±250/±285	25	87	Sea crossing; asynchronous link. Replacement of mercury-arc valves of Pole I with H400 thyristor valves (2006) (ASEA/Alstom)

Line	Year	Power (MW)	Voltage (kV)]	Length (km) Over head	Under sea	Main reasons for choosing the HVDC system
New Zeeland	1965	600	±250	575	42	Long line including sea crossing
Sardinia–Corsica–Italy (SACOI-1)	1967	200	200	290	116	The first multiterminal link (English Electric, ASEA)
Vancouver, Pole I (Canada)	1968/ 1969	312	+260	41	28	The first line operating in parallel with an AC circuit (ASEA)
Pacific - Intertie I, (Columbia River, Los Angeles)	1970 1985 1989	1600+ 400+ 1100	±500	1360	—	Long line in parallel with two AC lines; fast control (General Electric, ASEA)
Nelson River Bipole I (Canada)	1973– 1977	1620	±450	890	—	Long distance; stability (GEC Alstom)
b. Thyristor valve systems						
Eel River (Canada)	1972	320	2×80	—	—	BTB link
Skagerrak I–III (Norway–Denmark)	1974/ 1977/ 1993	500– 1500	±250	85 + 28	127	Sea crossing; DC active filter (ASEA, ABB)
Square–Butte (USA)	1977	500	±250	749	—	Long distance; stability
Cabora Bassa–Apollo (Mozambique–South Africa)	1977– 1979	1920	±533	1420	—	World's first long distance (oil-cooled thyristors) (AEG, BBC, Siemens)
Skin–Shinano (Japan)	1977	300	2 × 125	—	—	Frequency Converter 50 Hz/60 Hz
Vancouver Pole II (Canada)	1978	792	±260	41	32	—
Nelson River Bipole II and Bipole III (Canada)	1978– 1992/ 1997	900– 2000 1500	+250 ±500 ±500	940 930 930	—	Long distance (Water-cooled thyristors) (ABB/Siemens/AEG)
Hokkaido–Honshu (Japan)	1979 1993	150 600	125±250	27+97	44	Sea crossing (Hitachi)
Acaray (Paraguay–Brazil)	1981	50	26	—	—	Asynchronous link 50 Hz/60 Hz
Vyborg (Russia–Finland)	1982	1070	3 × ±85	—	—	BTB link
Inga Shaba (Republic of Congo)	1982	560	±500	1700	—	Long distance, finally, 1120 MW
Dürnrohr (Austria–Czech Republic)	1983	550	±145	—	—	BTB link, 50/50 Hz, out of service (1990)
Gotland - Sweden II/III	1983– 1987	130	150	7	96	Undersea cable, asynchronous link
Itaipu, Bipole I and II (Brazil)	1985– 1987	3150+ 3150	2× ±600	785+805	—	The highest transmitted DC power at long distance (18,432 thyristors), (ASEA)
Chateauguay (Canada)	1984	2 × 500	2 × 140.6	—	—	BTB link (ABB /Siemens)
Blackwater (USA)	1985	200	56.8	—	—	BTB link (ABB)

(continued)

Line	Year	Power (MW)	Voltage (kV)]	Length (km) Over head	Under sea	Main reasons for choosing the HVDC system
Highgate (USA)	1985	200	57	–	–	BTB link (ABB)
IFA 2000 Cross Channel (Great Britain–France II)	1986/ 2011	2 × 1000	2 × ±270	26	46	Undersea cable; asynchronous link 60 /50Hz; peak load (air-cooled thyristors/ H400 thyristor valves) (CGEE Alstom / GEC/Areva)
Intermountain (USA)	1986	1920	±500	785	—	Long distance (ASEA)
Ekibastuz-Centre (Russia)	1987	6000	±750	2414	—	The biggest overhead DC line (abandoned in 1990)
Fenno–Skan (Finland–Sweden)	1989/ 2010	500/ 800	±400 500	33	200	Monopolar line, sea crossing (ASEA)
Vindhyachal (India)	1989/ 1990	2 × 250	2 × 70	–	–	BTB link between North and West subsystems (ASEA)
Sileru–Barsoor (India)	1989	400	±200	196	—	HVDC transmission
Gezhouba–Shanghai (China)	1991	1200	±500	1046	—	Long distance; stability benefits (ABB/Siemens)
New Zeeland DC Hybrid link	1991/ 1992	560	−350	575	42	Long distance, sea crossing (ABB)
Quèbec–New England (Canada)	1990– 1992	2×2000 + 2×690	±450	1480	–	Long distance, asynchronous, with 3 terminals (ABB)
Rihand–Delhi (India)	1992	1500	±500	814		Long distance; stability (ABB)
South Vienna (Austria)	1992	550	145	—	—	BTB link, out of service (1996).
Etzenricht (Germany – Tchéquie)	1993	600	160	—	—	BTB link, out of service (1996). (Siemens)
Sakuma FC (Japan)	1965/ 1993	300	125	—	—	World's first frequency converter (FC); LTT valves (1993) (TOSHIBA)
Baltic Cable (Sweden–Germany)	1994	600	450	12	250	The second longest high voltage undersea link of high capacity for a single cable (ABB)
Kontek (Denmark–Germany)	1995	600	400	—	170	Sea crossing; asynchronous system (ABB)
Chandrapur–Padge (India)	1998	1500	±500	753	—	DC link in parallel with an AC line; stability (ABB)
Haenam - Cheju Island (South Korea)	1998	300	±180	—	101	Undersea cable; asynchronous link (GEC Alstom)
Chandrapur (India)	1998	2×500	2×205	—	–	BTB link between Western and Southern systems (Alstom)

Line	Year	Power (MW)	Voltage (kV)]	Length (km) Over head	Under sea	Main reasons for choosing the HVDC system
Visakhapatnan (India)	1999	500	205	—	—	BTB link between Eastern and Northern systems. (single monopole). (Alstom)
Leyte–Luzon (Philippine)	1999	880	±350	22	440	Long distance, undersea cable (ABB)
Shin–Shinano (Japan)	1999	600	125	—	—	Three terminal BTB 60/50/50 Hz frequency converter
Minami–Fukumitsu (Japan)	1999	300	±125	—	—	Asynchronous link (BTB)
Malaysia–Thailand	2000	300/ 600	±300	110	—	Thyristor 8 kV, 1 kA (Siemens)
Kii Channell (Japan)	2000	2800	2×±500	51	51	Thyristor 8 kV, 3500 A
Tian–Guang (China)	2000	1800	±500	986	—	DC link in parallel with an AC line (Siemens)
Higashi–Shimizu (Japan)	2001	300/ 600	±125	—	—	Asynchronous link 50/60 Hz
Greece–Italy	2001	500	400	105	163	Undersea cable; asynchronous line (ABB)
Moyle Interconnector Ecosse-North Ireland	2001	2×250	250	—	63.5	Undersea cable
Sasaram (India)	2002	500	205	—	—	BTB link between Eastern and Northern systems
Eurokabel (Norway–Germany)	2002	2×600	600	—	540	Undersea cable; asynchronous line
Vikingcable (Norway-Germany)	2003	2×800	500	—		Undersea cable; asynchronous line
Argentina–Brazil	2003	1000	±70	—	—	BTB link (ABB)
SWEPOL (Sweden–Poland)	2003	600	±450	–	230	Long distance and sea crossing (ABB)
Three Gorges–Changzhou (China)	2003	3000	±500	860	—	Long distance (ABB and Siemens)
Sarawak - Malaysia	2003	1500	2× ±500 4× +400	660	670	Undersea DC link (ABB) (Canceled)
Bangalore–Talcher (India)	2003	2000	±500	1450	—	East–South Interconnector (Siemens)
Three Gorges–Guangdong (China)	2004	3000	±500	940	—	Long distance (ABB)
Basslink (Australia–Tasmania)	2006	500	400	72	295	Monopolar with metallic return (Siemens)
Gulf Interconnector Al Fadhili (Saudit Arabia)	2006/ 2008	3×600	222	—	—	Three BTB converters 60/ 50 Hz (Alstom)
Neptun Link (USA)	2007	660	500	–	105	World's first HVDC 500 kV cable (Siemens)

(continued)

Line	Year	Power (MW)	Voltage (kV)]	Length (km) Over head	Under sea	Main reasons for choosing the HVDC system
NorNedkabel Link (Norway–Netherland)	2008	600/ 800	±500	—	580	Undersea cable; asynchronous line (ABB)
Yunnan-Guangdong (China)	2009	5000	±800	1418	—	World's first 800 kV DC with 5″ LTT valves (Siemens)
Ballia–Bhiwadi (India)	2010	2500	±500	800	—	(Siemens)
Xiangjiaba–Shanghai (China)	2010	6400	±800	2071	—	World's first 800 kV DC with 6″ ETT valves; two converters per pole (Siemens)
Ningdong–Shandong (China)	2010	2×2000	±660	1335	—	The world's first HVDC link at ±660 kV (Areva T&D)
Melo (Uruguay–Brazil)	2011	500	±79	—	—	BTB 50/60 Hz (Areva T&D)
BritNed (British–Netherland)	2011	1000	±450	—	260	Sea crossing (Siemens and ABB)
JinPing–SuNan (China)	2012	7200	±800	2090	—	With 6″ ETT valves (Siemens)
COMETA (Spain-Mallorca)	2012	400	±250	—	250	DC interconnector (Siemens)
Hudson Transmission Ridgefield (New Jersey, USA)	2013	660	170	—	-	BTB close to New York (Siemens)
Hami–Zhengzhou (China)	2013	8000	800	2400	—	Pole 2 of world's biggest and longest UHV project with 6″ ETT valves (Siemens)
Jindo–Cheju 2 (South Korea) Obs. The second HVDC link using a third cable as metallic return conductor.	2013	400	±250	—	120	A new control system to allow the power on the both HVDC links (1+2) to be automatically coordinated. (Areva T&D)
Rio Madeira (Brazil)	2014	3150	±600	2375	—	Longest HVDC link (ABB, Alstom, Toshiba)
Inter-Island Connector Pole 3 (New Zeeland)	2014	700	±350	649	40	HVDC link and STATCOM in parallel operation. (Siemens)
Western HVDC link (UK)	2015	2200	±600	—	420	Power exchange; world's first HVDC with 600 kV cable; HVDC and STATCOM in parallel operation.
NEA 800 Biswanath–Alipurduar–Agra (India)	2016	6000	±800	1728	—	First UHVDC multiterminal largest ever (ABB)

Line	Year	Power (MW)	Voltage (kV)]	Length (km) Over head	Under sea	Main reasons for choosing the HVDC system
c. Systems HVDC in construction or in designing						
Champa–Kurukshetra (India)	2015	3000	±800	1365	—	The first application of the H400 thyristor valve (4″, 8.5 kV) for such a voltage level (Alstom)
Belo Monte–Minas (Brazil) Bipole I	2018	4000	±800	2092		Very long distance.UHVDC system.
Belo Monte - Rio (Brazil) Bipole II		4000	±800	2518		
IbValley–Jaipur (India)		3000	±600	1500	—	Long distance
Karamsad–Korba (India)		3000	±600	1450	—	Long distance
Dehang-Bareilly (India)		5000	±600	1500	—	Long distance
Balipara–Ballabhgarh (India)		5000	±600	1500	—	Long distance
Pancheswar (Nepal)–Vadodra (India)		5000	±600	1100	—	Multiterminal system
Karnali–Vadodra (India)		5000	±600	1000		
Vadodra– Pune (India)		3000	±600	900		
Pune–Madras (India)		3000	±600	1050		
East-West High Power Link Smolensk–Berlin (Russia–Germany)		4000	±500	1800	—	Studies for long distance, asynchronous, with five terminals (Siemens)

REFERENCES

[1] Eremia, M., Trecat, J., and Germond, A. *Réseaux électriques. Aspects Actuels*. Editura Tehnică, Bucureşti, Romania, 2000 (in French).

[2] Kimbark, E. W. *Direct Current Transmission*, vol. I. John Wiley & Sons, Inc., New York, 1971.

[3] Sood, V. K., HVDC and FACTS Controllers: Applications of Static Converters in Power Systems, Springer-Verlag, 2004.

[4] Alstom, Grid. *HVDC: Connecting to the Future*. Alstom Grid, 2011.

[5] Kim, C. K., Sood, V. K., Jang, G. S., Lim, S. J., and Lee, S. J. *HVDC Transmission Power Conversion Applications in Power Systems*. IEEE Press and John Wiley & Sons (Asia) Pte Ltd, New Yor, 2009.

[6] EPRI Report El-3004. Methodology for integration of HVDC links in large AC systems, Phase 1: reference manual. Prepared by Ebasco Services Incorporated, New York, March 1983.

[7] Laughton, M. A., and Say, M. G. HVDC transmission. In: *Electrical Engineer's Reference Book*, A. Gavrilovic, Ed., 14th edition. Butterworths Heinemann, 1985.

[8] IEEE Std. P1030.1-2000, IEEE Guide for Specification of HVDC Systems, 2000.

[9] CIGRE Working Group 14.20. Economic assessment of HVDC links. Technical Brochure No. 186, June 2001.

[10] Weedy, B. M., and Cory, B. J. *Electric Power Systems*, 4th edition. John Wiley & Sons, Ltd., Chichester, UK, 1999.

[11] Uhlmann, E. *Power Transmission by Direct Current*. Springer-Verlag, Berlin, 1975.

[12] Kundur, P. *Power Systems Stability and Control*. McGraw-Hill, New York, 1994.

[13] Arrillaga, J. *High Voltage Direct Current Transmission*. Peter Peregrinus, London, 1983.

[14] Eremia, M., Ed. *Electric Power Systems. Electric Networks*. Publishing House of the Romanian Academy, Bucharest, Romania, 2006.

[15] Hingorani, N. G., and Gyugyi, L. *Understanding FACTS: Concepts and Technology of Flexible AC Transmission Systems*. IEEE Press, New York, 2000.

[16] IEEE Std. 519-1992, IEEE Guide for Harmonic Control and Reactive Compensation of Static Power Converters, 1992.

[17] Smed, T. Interaction between high voltage AC and DC systems. Ph.D. Thesis, Royal Institute of Technology, Stockholm, Sweden, 1991.

[18] Rashid, M. HVDC transmission. In: *Power Electronics Handbook: Devices, Circuits and Applications*, 3rd edition, V. K. Sood, Ed. Elsevier, Amsterdam, 2011.

[19] Andersen, B. R., Monkhouse, D. R., Whitehouse, R. S., Williams, J. D. G., Prasher, V. K, and Kumar, D. Commissioning the 1000 MW Back to Back HVDC Link at Chandrapur, India, Paper 14-114, CIGRÉ, Paris, 1998.

[20] Lindén, K. Overview of the HVDC control system. Technical Report, ABB, 2001.

[21] Woodford, D. *HVDC Transmission*, White paper written for Manitoba Research Center, Winnipeg, Canada, 1998.

[22] Nordström, B. Converter firing control. Technical Report, ABB, 2001.

[23] Padiyar, K. R. *HVDC Power Transmission Systems. Technology and Systems Interactions*, John Wiley & Sons, Inc., New York, 1991.

[24] Adhikari, T., Isacsson, G., and Ambekar, V. D. The Chandrapur–Padghe HVDC bipole transmission. In: *CIGRE Symposium of Power System Issues in Rapidly Industrializing Countries*, Kuala Lumpur, Malaysia, September 1999, pp. 20–23.

[25] Zhang, W., and Asplund, G. Active DC filter for HVDC systems. *IEEE Computer Applications in Power*, vol. 7, no. 1, pp. 40–44, January 1994.

[26] Sakdhnagool, S., Ab'Llah, Saleh., Wong, C. H., Bartzsch, C., Landgraf, E., Sadek, K., Schwegmann, M., and Zink, W. 300 MW TNB/EGAT HVDC interconnection project. In: *CIGRE Symposium of Power System Issues in Rapidly Industrializing Countries*, Kuala Lumpur, Malaysia, September 1999, pp. 20-23.

[27] Arrillaga, J., Arnold, C. P., and Haker, B. J. *Computer Modeling of Electrical Power Systems*, John Wiley and Sons, Inc., New York, 1983.

[28] Jardini, J. A., and Gole, A. (conveners), Rauhala, T. Woodford, D., Wachal, R., Dennetiere, S., Nguefeu, S., Barker, C., Eremia, M., Tsfasman, G., Manchen, M., Kell, D., and Bartzsch, C. Modelling and simulation studies to be performed during the lifecycle of HVDC systems, CIGRÉ WG 34-38, Brochure No. 563, December 2013.

[29] Gavrilovic, A. Krishnayya, P. C. S., Ainsworth, J. D., Bowles, J. P., Breuer, G. D., Hammad, A., Liss, G., Peixoto, C. O. A., Povh, D., and Thio, C. D. Interaction between DC and AC systems. In: *CIGRÉ Symposium on AC/DC Transmission. Interactions and Comparisons*, Boston, MA, September 28–30, 1987.

[30] IEEE Std. 1204-1997, IEEE Guide for planning DC links terminating at AC locations having low short-circuit capacities.

[31] Long, W. F., and Stovall J. P. Comparison of costs and benefits for DC and AC transmission. In: *CIGRE Symposium on AC/DC Transmission. Interactions and Comparison*, Boston, MA, September 28–30, 1987.

[32] Gonen, T., *Electric Power Transmission System Engineering. Analysis and Design*. A Wiley-Interscience Publication, John Wiley & Sons, Inc., New York, 1988.

[33] Comparison between AC and DC transmission systems. Report CIGRE no. 37-94, WG 12-06, June 1994.

[34] HVDC technology helps make the Murraylink project an environmental and engineering success. ABB Review, No. 52, April 2003.

CHAPTER 4

VSC–HVDC TRANSMISSION

Mircea Eremia, José Antonio Jardini, Guangfu Tang, and Lucian Toma

THE FORCED SWITCHING converters are a new version of the high voltage direct current (HVDC) technology. They are using voltage source converters (VSC), which are based mainly on insulated gate bipolar transistors (IGBT). VSCs utilize a power electronic valve with both turn-on and turn-off capability. The converters work by using pulse width modulation (PWM), making it possible to alter both the phase and amplitude, which allows the independent regulation of active and reactive power, as well as the voltage and frequency control. From the AC transmission grid point of view, the VSC–HVDC systems act similar to a motor or generator without inertia, which is able to control the active and/or reactive power. The VSC–HVDC technology is known as HVDC Light (ABB), HVDC Plus (Siemens), or MonSin (GE/Alstom).

Although relatively new to the power transmission industry, VSC technology was used in industrial application for several years prior to 1991. Since its introduction for HVDC transmission in 1997, VSC ratings for both voltage and power have increased steadily; ratings of up to ±320 kV (500 kV in monopolar applications) and 1000 MW are currently available.

The first generations of VSC developed were based on two- or three-level VSCs using the PWM technique with high switching frequencies (1–2 kHz). This resulted in large converter losses that could reach as high as 2–3%. Existing two-level or three-level converters using PWM have to switch the full DC voltage in large steps. This causes high harmonic distortion and high transient stresses, resulting in high frequency noise.

Recently, the industry moved to a new generation of modular multilevel VSCs, with the first commercial project Trans Bay Cable in the United States, at 400 MW and ±200 kV—going into commercial operation in 2010. The modular multilevel converter (MMC) provided certain technical advantages, mainly lower losses (between 0.9% and 1%) and lower or no filtering requirements [1].

Like line commutated converter (LCC) HVDC or current source converter (CSC) HVDC, a VSC transmission scheme enables reliable and controllable power transfer between networks. Additionally, VSC transmission converters (rectifier and inverter) provide independent control of the reactive power at the two ends and

Advanced Solutions in Power Systems: HVDC, FACTS, and Artificial Intelligence, First Edition.
Edited by Mircea Eremia, Chen-Ching Liu, and Abdel-Aty Edris.

independently of the active power transfer over the DC transmission. This enables a VSC transmission scheme to be connected to a very weak or even passive AC network, since it is self-commutating and can provide AC voltage support and/or voltage reference.

4.1 VSC CONVERTER STRUCTURES

The output voltage waveform of the converter must closely approximate a sine wave in order to eliminate or minimize the need for harmonic filtering. Since the converter is just an array, or matrix, of solid-state switches, the AC output voltage can only be produced by connecting the DC input voltage periodically, for predetermined intervals, to the output.

The switch-matrix of a VSC may be of different complexity depending on the number of DC voltage levels that are to be produced at the AC output of each constituent converter pole. The pole, composed of two or more solid-state switches or valves (usually comprising a number of switching elements directly connected in series), is the basic building block of the converter. The number of levels of a converter can be defined as the number of steps or constant voltage values that can be generated by the converter between the output terminal and any arbitrary internal reference node within the converter. Typically, it is a DC-link node, and it is usually called neutral. With an increasing number of voltage levels, the approximation to the desired sinusoidal output becomes closer, at the expense of increasing circuit complexity.

The multilevel power converters are considered state-of-the-art power conversion systems for high-power and power quality demanding applications. The term multilevel starts with three-level inverter introduced by Nabae et al. [2]. By increasing the number of levels in the converters, the output voltages have more steps generating staircase waveform, which has a reduced harmonic distortion. However, a high number of levels increases the control complexity and introduces voltages imbalance problems.

Four main topologies are known for multilevel converters: diode-clamped type (neutral point clamped (NPC) and up with DC-link capacitors), capacitor-clamped type (with flying capacitors (FCs) and DC-link capacitors), cascade H-bridge type (with separate DC sources, without common DC link), and modular converter type (without separate DC sources, with common DC link, but without DC-link capacitors) [3,4].

4.1.1 Half-Bridge VSC or Two-Level Pole

The half-bridge VSC or two-level pole is the simplest switching arrangement capable of producing AC output from a DC source in the form of a simple square wave, as illustrated in Figure 4.1.

The half-bridge VSC consists of an upper switch cell (S_1) and a lower switch cell (S_4). Each cell is composed of a fully controllable, unidirectional switch in

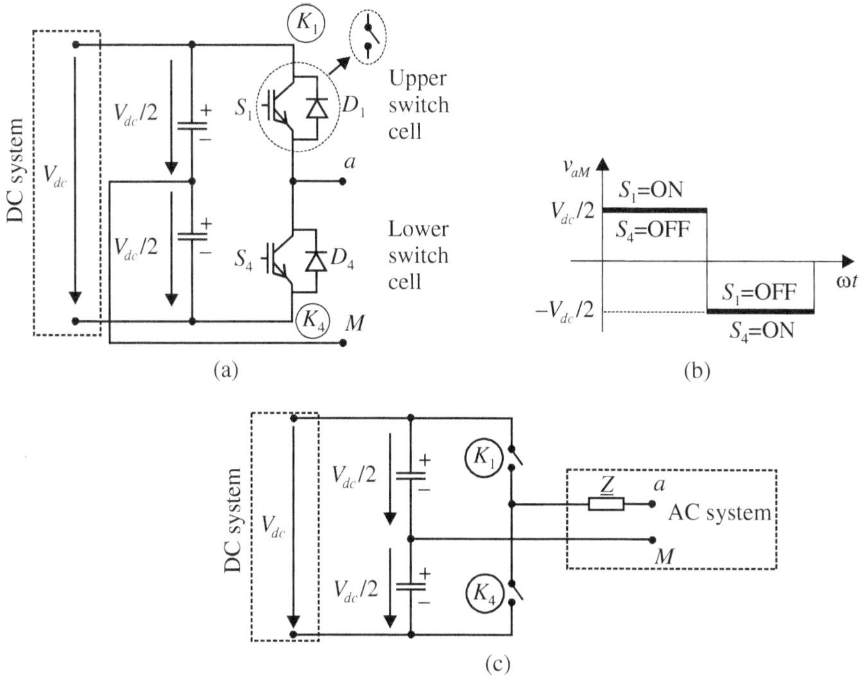

Figure 4.1 The configuration of a VSC with one pole and two levels, (a); AC voltage waveform (b); simplified scheme, (c) *Source:* Andersen et al. 2005 [6]. Adapted with permission of CIGRE.

antiparallel connection with a diode. This switch configuration constitutes a reverse conducting switch that is readily available in the form of IGBT and integrated gate commutated thyristor (IGCT) [5].

Such a configuration can be considered as a VSC with "two levels" and "one pole." The notion of "pole" must be perceived as a single phase of the converter.

When the S_1 is switched on and the S_4 valve is blocked, between the neutral point (M) of the converter and its output terminal (a), there will be a voltage of $V_{dc}/2$. When the two valves change their state, the potential of the output terminal with regard to the M point changes its polarity, reaching a value of $-V_{dc}/2$ (Figure 4.1b).

The switching state of a converter is a set of signal used to control each switching device of the power circuit. They can modify its conduction state and the way the load is connected to the different nodes of the DC side circuit. Hence, a particular switching state generates a corresponding output voltage level. The gate signal is of binary nature, representing by 0 the OFF state of switch and by 1 the ON state.

We note that $sw_1(t) + sw_4(t) \equiv 1$. The half-bridge converter based on the alternate switching of S_1 and S_4. The turn on/off command of S_1 and S_4 is issued through a PWM strategy.

The half-bridge or two-level pole has been the traditional building block of all types of power converters, usually in the so-called Graetz-bridge configuration. Apart from the poor (square wave) output voltage waveform, the main disadvantage of the

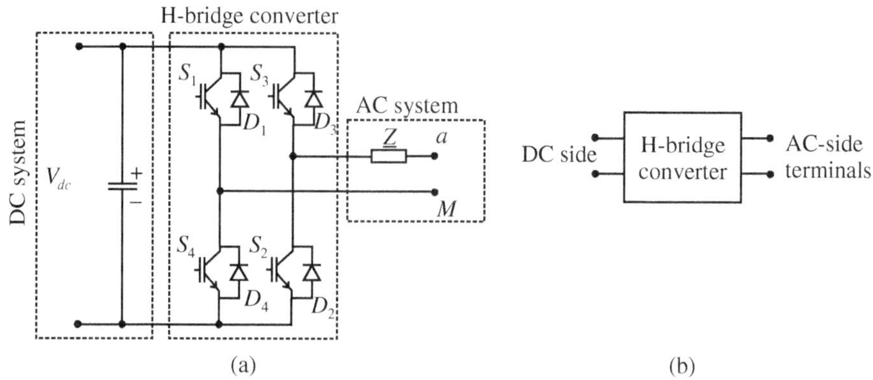

Figure 4.2 (a) Diagram of the full-bridge, single-phase, two-level VSC; (b) symbolic representation of the H-bridge converter.

two-level pole is its inability to provide zero output to facilitate direct control of the amplitude of its fundamental output voltage. This limitation can be overcome by the application of PWM or some indirect techniques (e.g., control of the DC voltage).

The advantages of the two-level topology are a simple circuit, small DC capacitors, a small footprint, and all semiconductor switches have the same duty. On the other hand, the disadvantages of the two-level topology are large blocking voltage of semiconductor switches, crude basic AC waveforms, and high values of converter switching losses due to the high switching frequency used.

4.1.2 Full-Bridge Single-Phase VSC

The full-bridge single-phase VSC (H-bridge converter) of Figure 4.2 consists of two half-bridge VSCs that are connected in parallel through their DC sides; the AC system can be interfaced with the AC-side terminals of the two half-bridge converters.

One advantage is that, for a given DC voltage, the synthesized AC voltage of the full-bridge VSC is twice larger in comparison with the half-bridge VSC, which corresponds to a more efficient utilization of the DC voltage and switch cells [5].

4.1.3 Three-Phase Two-Level VSC

The three-phase two-level, VSC (Figure 4.3a) is also an extension of the half-bridge VSC of Figure 4.1.

In power system applications, the three-phase VSC is interfaced with the AC system, typically through a three-phase transformer, based on a three-wire connection. This arrangement is capable of producing AC output from a DC source in the form of a simple square wave.

Figure 4.3c shows the three pole voltages (v_{aM}, v_{bM}, v_{cM}) that are 120° apart from each other and their fundamental components (v_{aM1}, v_{bM1}, v_{cM1}). The neutral voltage is the voltage at the midpoint of the DC capacitor. Due to the square-wave output voltage waveform, the two-level phase unit is unable to facilitate direct control

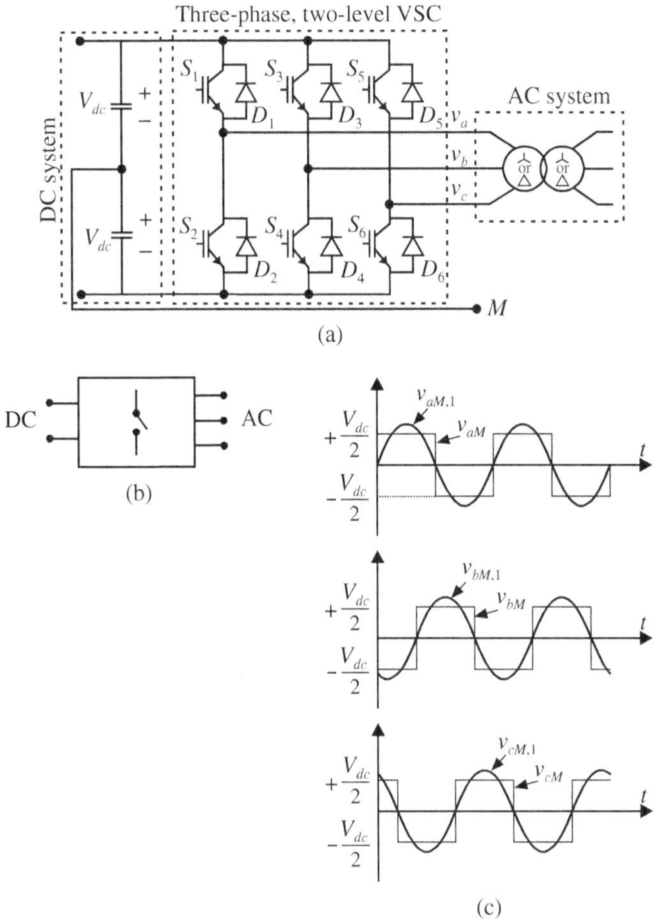

Figure 4.3 (a) Diagram of the three-wire, three-phase, two-level VSC; (b) symbolic representation of the three-phase VSC; (c) pole output to neutral square wave voltages with their fundamental components.

of the amplitude of its fundamental output voltage without the application of PWM (see section 4.2).

4.1.4 Three-Level Pole VSC

A three-level neutral point clamped pole, illustrated with its output voltage waveform in Figure 4.4, has the capability of controlling directly the output voltage of the converter.

As can be seen in Figure 4.4a, the negative bar of the upper converter and the positive bar of the lower one are joined together to form the new phase output *a*, while the original phase outputs are connected via two clamped diodes (D_1' and D_2') to form the neutral point *M*, dividing the DC-link voltage in two. The first half-bridge

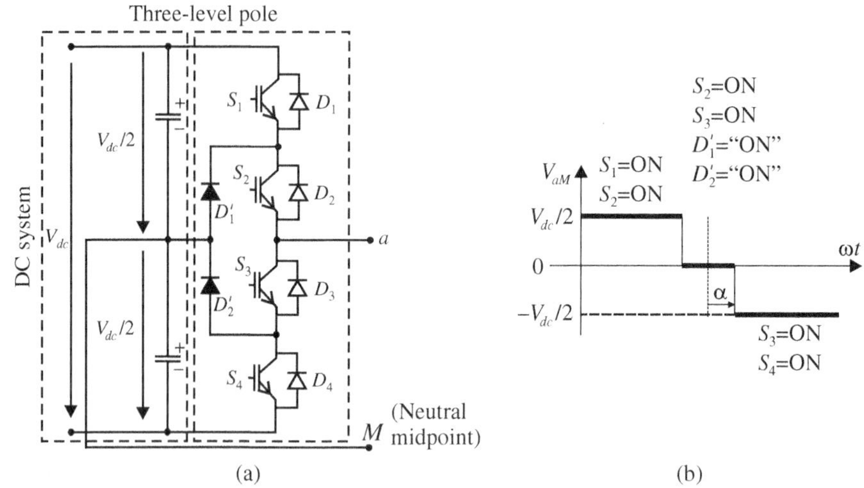

Figure 4.4 The configuration of the three-level pole neutral point clamped (a) and associated AC voltage waveform (b).

converter is composed of the switch cells S_1/D_1 and S_3/D_3, and the second one consists of the switch cells S_2/D_2 and S_4/D_4. All voltages in Figure 4.4a are expressed with the reference to the DC-side midpoint. Now, each power device has to block only half of the total converter voltage; hence, with the same semiconductor technology, the power rating of the converter can be doubled. In addition, the neutral point enables the generation of a zero voltage level, obtaining a total of the three different voltage levels [7]. As for the two-level half-bridge converter, the switching function for a switch cell is 0 for the OFF state and 1 for the ON state. The gating commands of S_1 and S_3 must be complementary, that is, $sw_1 + sw_3 \equiv 1$. Similarly, the gating commands of S_4 and S_2 are complementary, that is, $sw_4 + sw_2 \equiv 1$.

As seen, there are twice as many valves used as in the two-level phase unit (pole), and additional diodes (D'_1, D'_2) are also required to connect to the DC supply center-tap (which is reference zero potential). However, with identical valve terminal-to-terminal voltage rating, the total DC supply voltage can be doubled so that the output voltage per valve remains the same [6].

The AC waveform shown in Figure 4.4b is the phase-to-neutral voltage (V_{aM}), assuming fundamental frequency switching of the valves. The neutral voltage is the voltage at the midpoint of the DC capacitor. As illustrated in Figure 4.4b, the output voltage of the three-level pole can be positive, negative, or zero. Positive output ($+V_{dc}/2$) is produced by gating on both upper valves in the pole, while negative output ($-V_{dc}/2$) is produced by gating on both lower valves. Zero output is produced when the upper and lower middle valves, connecting the center tap of the DC supply via the two diodes (D'_1, D'_2) to the output, are gated on. At zero output, positive current is conducted by the upper-middle controllable device and the upper center-tap diode, and the negative current is conducted by the lower-middle controllable device and the lower center-tap diode.

As indicated in Figure 4.4b, the relative duration of the positive and negative output voltage with respect to the duration of the zero output is a function of control parameter "α," which defines the conduction interval of the top-upper, and the bottom-lower valves. Evidently, the magnitude of the fundamental frequency component of the output voltage produced by the pole is a function of parameter α: at $\alpha = 0°$ it is maximum, while at $\alpha = 90°$ it is zero. Thus, one advantage of the three-level pole is that it has an internal capability to control the magnitude of the output voltage without changing the number of valve switching per cycle. The other advantage is that with judicious choice of α, selected harmonic components of the output waveform can be eliminated.

In order to further reduce the harmonic content of the AC output voltage, the basic three-level pole can be conceptually extended to a multilevel, $2n + 1$ pole ($n = 1, 2, 3, \ldots$) configuration.

4.1.5 Multimodule VSC Systems

In high voltage, high power VSCs, the switch cell of Figure 4.1, which is composed of a fully controllable, unidirectional switch and a diode, may not be able to handle the voltage/current requirements. To overcome this limitation, the switch cells are connected in series and/or in parallel and form a composite switch structure, which is called a valve.

In most applications, the existing power semiconductor switches meet current handling requirements. However, in some applications, series-connected switch cells are inevitably required to satisfy the voltage requirements.

Due to various practical limitations including the unacceptable form factor, unequal off-state voltage distribution, and simultaneous-gating requirements, the number of series connected switch cells within a valve is limited. Thus, a two-level VSC unit cannot be constructed for any voltage level, and an upper voltage limit applies [5].

The maximum permissible voltage limit of a VSC system can be increased by a series connection of identical, three-phase, two-level VSC modules, to form a multimodule VSC. Figure 4.5a illustrates a scheme of an n-module VSC in which n identical two-level VSC modules are connected in series and parallel, respectively, at their AC and DC ports. Thus, the VSC modules share the same DC bus capacitor. Figure 4.5b illustrates an alternative configuration of an n-module VSC in which the two-level VSC modules are connected in series at both the AC and DC sides. In both cases, the VSC modules share the same DC-bus capacitor. In both configurations of Figures 4.5a and 4.5b, the AC-side voltages of the VSC modules are added up by the corresponding open-winding transformers, to achieve the desired voltage level (and waveform) for a connection to the AC system. One of the salient features of the n-module VSC configurations of Figures 4.5a and 4.5b is their modularity, as all the VSC modules and transformers are identical. Modularity is a desired feature that reduces manufacturing costs, facilitates maintenance, and permits provisions for spare parts [5].

The multimodule converter of Figure 4.5b can be further enhanced to acquire the AC-side voltage harmonic reduction capability of a multipulse configuration

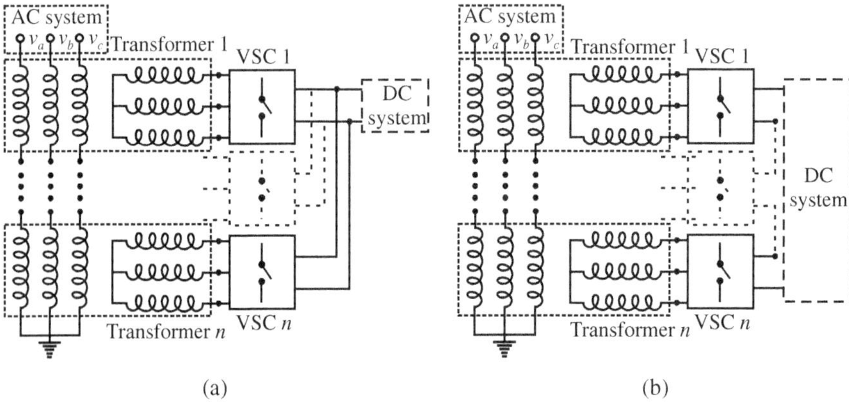

Figure 4.5 Schemes of a multimodule VSC composed of n two-level VSC modules connected in (a) paralleled from their DC sides; (b) in series with respect to their DC sides. *Source:* Yazdani and Iravani 2010 [5]. Reproduced with permission of IEEE.

while its modularity is preserved. This is achieved through appropriate phase shift in switching patterns of the constituent VSC modules, such that a prespecified set of voltage harmonics is canceled/minimized when added up by the open-winding transformers. The harmonic minimization enables operation of the multimodule converter at low switching frequencies, which, in turn, results in lower switching losses; it also mitigates the need for low-frequency harmonic filters at the converter AC side.

The *concept of multipulse conversion* is another technique employed to minimize low-frequency harmonic components of the synthesized AC voltage of a VSC, and thus minimize the associated filtering requirements.

4.1.6 Multilevel VSC Systems

A multilevel converter is a type of converter that can synthesize from DC voltage an approach sinusoidal waveform (Figure 4.6) [7, 8].

In general, the multilevel converters can be obtained by connecting several simple converters in different configurations. If two simple switching cells (Figure 4.7a) are connected, then a double cell is obtained (Figure 4.7b), which has the capability to generate three voltage levels, including zero. By connecting in series double switching cells, a chain of cells resembling with a dynamically controllable AC voltage source is obtained (Figure 4.7c).

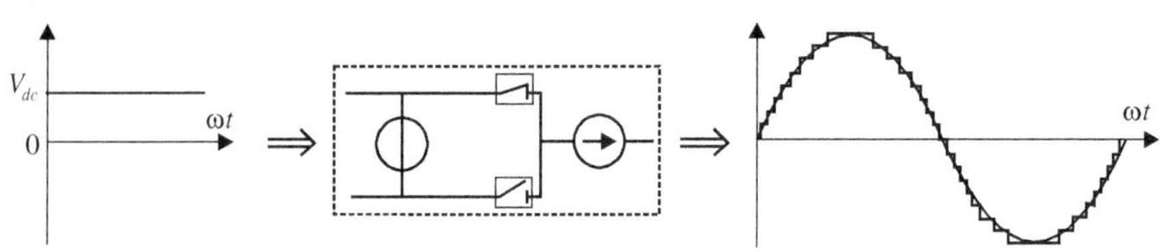

Figure 4.6 Static DC to AC conversion (voltage synthesis) using a multilevel converter.

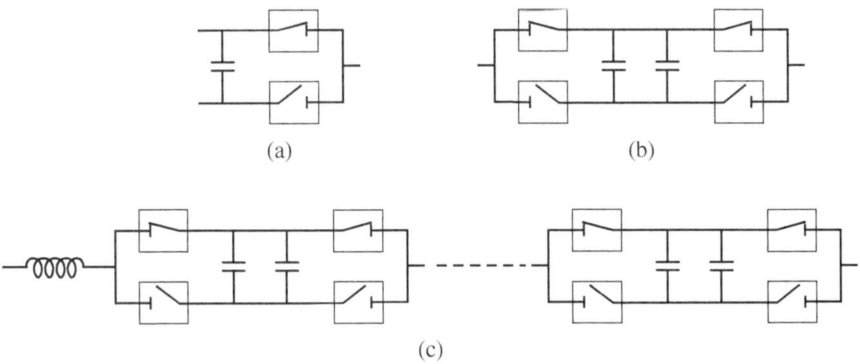

Figure 4.7 Simple switching cell(a); double switching cell (b); series connection of double switching cells (c). Adapted from Dijkhuizen 2012 [8].

Another possibility of arrangement of double cell is by connecting them into a stack (Figure 4.8). Compared to the double cell in Figure 4.7, that is able to generate three voltage levels including zero, the stack switching cell in Figure 4.8 is able to generate multiple voltage levels.

Because of the high number of series-connected IGBTs, the converter valve must be divided into several stacks that each other contains a manageable number of series-connected devices. Furthermore, the stacks must be very compact to ensure low stray inductance of the commutation loops.

Figure 4.9 shows a possible arrangement of a VSC valve. Shields to equalize the electric field around the stack may surround each other semiconductor component, which is clamped between heat sinks, and the entire IGBT level, including the gate control unit. This very compact design requires IGBTs of limited height that are well adapted for stack mounting. Their mechanical robustness must be sufficient to withstand the high mounting force needed to ensure a good thermal contact and the mechanical stability of the whole assembly. Furthermore, the IGBT housing should be designed to prevent fire or any other severe damage in the converter if the valve, by accident, should be subjected to a heavy short circuit.

Considering the switching cells from Figure 4.7 to emphasize that they can be used as building blocks to develop further circuit topologies, two cascade

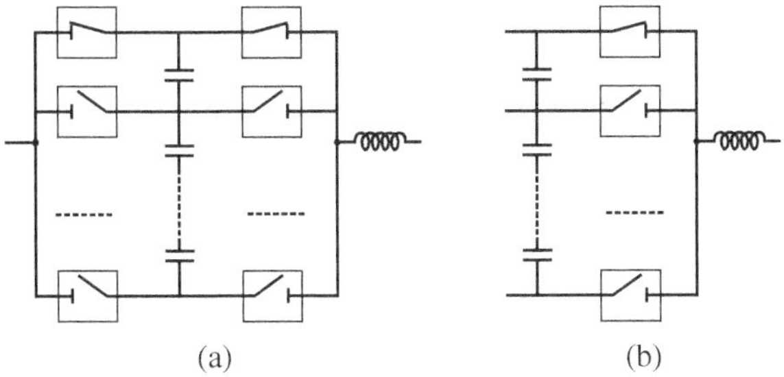

Figure 4.8 Switching cells stack. Adapted from Dijkhuizen 2012 [8].

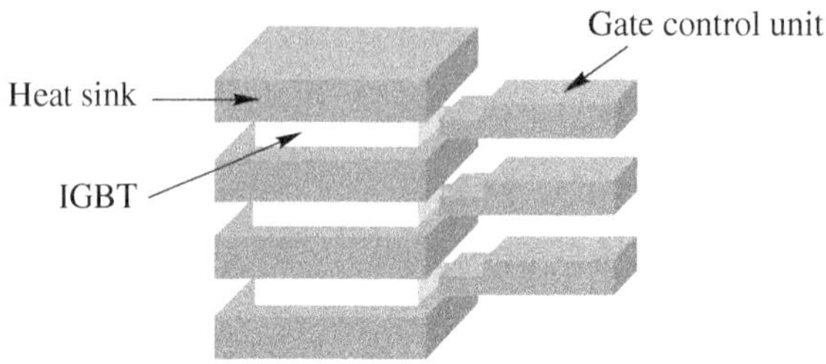

Figure 4.9 Possible arrangement of an IGBT stack assembly. *Source:* Andersen et al. 2005. [6]. Reproduced with permission of CIGRE.

converter structures can be obtained by an imbrication process (Figures 4.10a and 4.10b).

The topology in Figure 4.10a is the modular multilevel using many intermediate floating voltage sources, such as capacitors or batteries.

Cascade converter structures present the following characteristics:

- low power losses and low commutation frequency slightly above to the fundamental frequency;
- voltage scalability due to the simple cascade arrangement of the identical cells;
- almost total elimination of filters on the AC side due to the sine wave form of the voltage;
- mechanical simplicity.

With the new multilevel approach, individual module capacitors are uniformly distributed throughout the topology, and each level is individually controlled to generate a small voltage step. In this way, each module within the multilevel converter is a discrete voltage source in itself, with a local capacitor to define its voltage step without creating ripple voltage distortion across the converter's other phases. By the incrementally controlling each step, a nearly sinusoidal voltage is generated at the AC

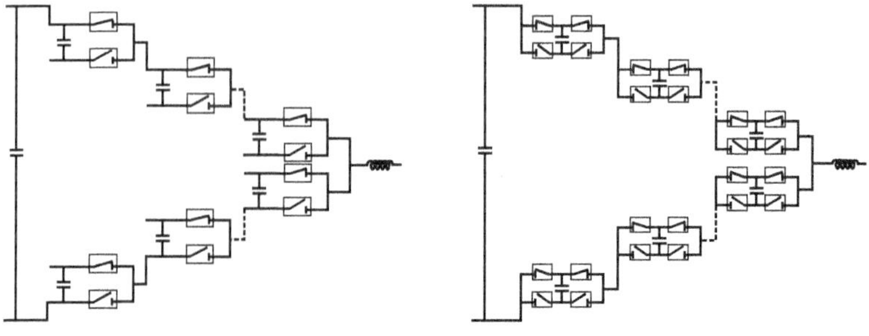

Figure 4.10 Principle of cascade converter structures: (a) imbrication of switching cells; (b) imbrication of double switching cells.

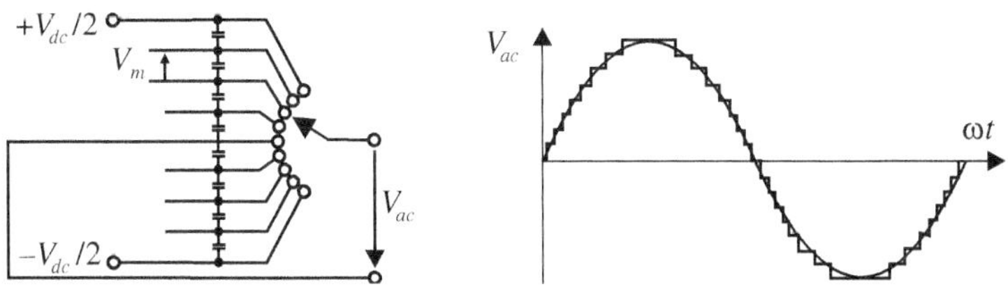

Figure 4.11 Principle of operation of the multilevel converter.

outputs of the "multivalves." The principle of operation of the multilevel converter is shown in Figure 4.11 [1].

Conceptually, the multilevel VSC configurations can be divided into [4]: the H-bridge-based multilevel VSC, the capacitor-clamped multilevel VSC, and the diode-clamped multilevel VSC.

(i) The *H-bridge-based multilevel VSC,* also known as the cascaded H-bridge (CHB) multilevel VSC, is constructed through series connection of the H-bridge modules (Figure 4.2b). Figure 4.12 shows a scheme of a three-phase, wye-connected, H-bridge-based, multilevel VSC.

One salient feature of the configuration of Figure 4.12 is that it permits independent control of the three legs of the converter. If a (grounded) neutral conductor is provided, the converter can also provide independent control over the zero-sequence components of the three-phase current, in addition to the positive-sequence and negative-sequence components.

For this configuration, especially if real power exchange is involved, the DC-bus voltage of each H-bridge module must be independently supplied and regulated by an auxiliary converter system. This renders the H-bridge-based VSC of Figure 4.12 practically unattractive for general-purpose applications; rather, the H bridge-based converter is more suitable for specific applications,

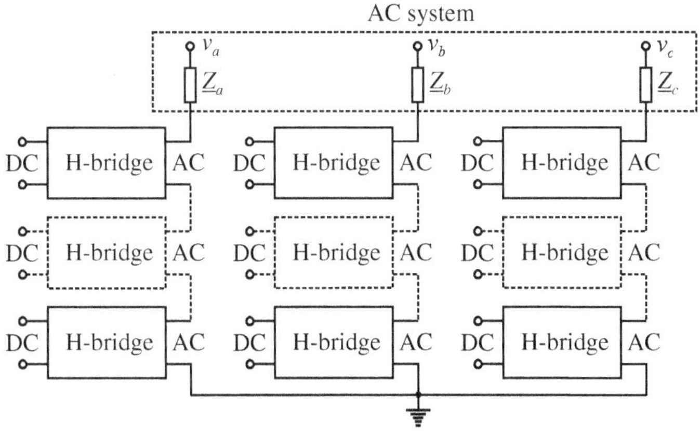

Figure 4.12 Schematic diagram of a wye-connected, H-bridge-based, multilevel VSC. *Source:* Yazdani and Iravani 2010 [5]. Reproduced with permission of IEEE.

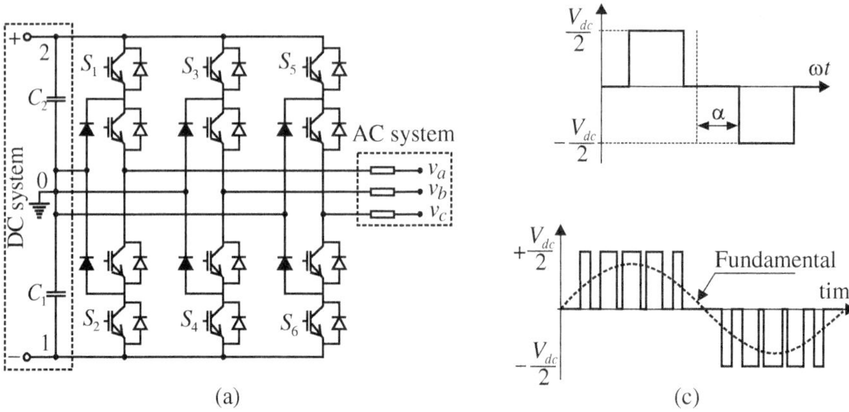

Figure 4.13 Scheme of a VSC (a) three-level three-phase NPC; (b) AC voltage waveform for one phase; (c) PWM line-to-neutral voltage waveform. *Source:* Yazdani and Iravani 2010 [5]. Adapted with permission of IEEE.

for example, the STATCOM, where only reactive power exchange is the objective [9].

(ii) *Multilevel Diode-Clamped Converter* (DCC), also known as NPC converter, is a generalization of the two-level VSC of Figure 4.3a. By using a number of DC capacitors in series and additional diodes, a multilevel DCC can be formed. Figure 4.13a shows the three-level three-phase DCC and output three voltage levels, that is, $+V_{dc}$, 0, $-V_{dc}$ (Figure 4.13b), or PWM line-to-neutral voltage waveform (Figure 4.13c; see also section 4.2.1). For a three-phase unit, the DC capacitors are usually shared by the phases.

The NPC topology can be extended to higher power rates and more output voltage levels by adding additional power switches and clamping diodes to be able to block higher voltages. Figures 4.14a and 4.14b show a three-phase five-level NPC converter and the corresponding single-phase output voltage waveform. Here, the name diode clamped makes more sense since there are more voltage-level clamping nodes than only the neutral M. Note that the number of clamping diodes needed to share the voltage increases dramatically [7].

However, it should be remembered that in practice one degree of freedom would be needed for control of the amplitude of the fundamental frequency. Therefore, in practice, in this example only one of the fifth or the seventh harmonic can be canceled [9].

From Figure 4.14a, it is clear that the circuit complexity rapidly increases with the number of voltage levels. However, compared to the three-level NPC converter, the individual valves are switched approximately half the number of times in the five-level topology and the harmonic performance at the output terminals is still superior.

The advantages of the DCC are reasonable small DC capacitors, lower switch blocking voltages. Compared to a two-level VSC of the same rating,

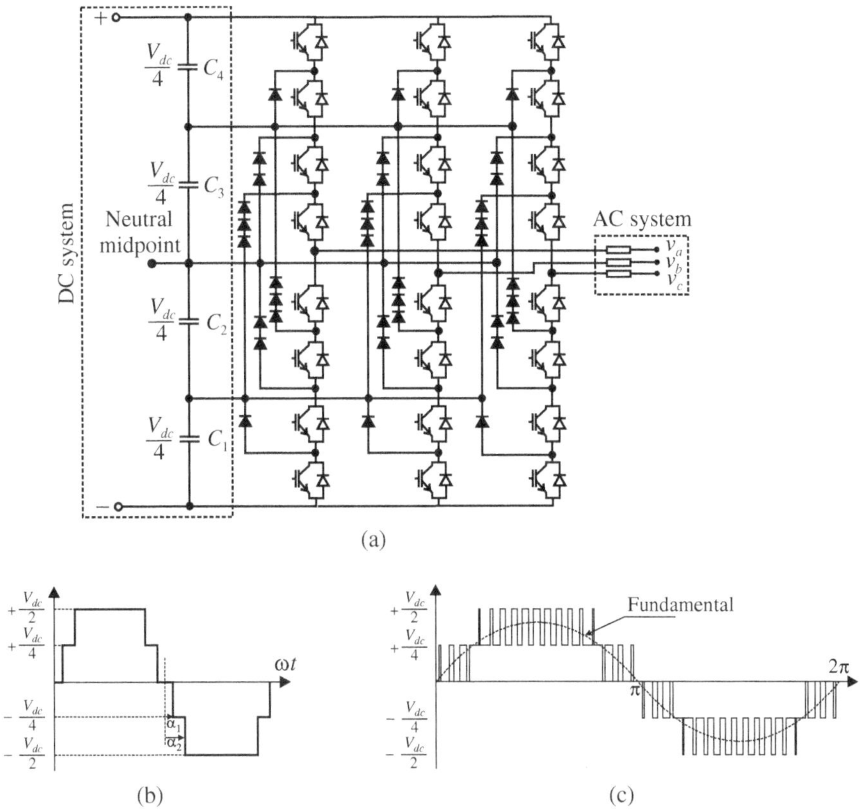

Figure 4.14 (a) Scheme of a three-phase using a five-level NPC converter; (b) AC voltage waveform for one phase; (c) PWM line-to-neutral voltage waveform. *Source:* Rodriguez et al. 2009 [7]. Adapted with permission of IEEE.

the three-level NPC can offer a less distorted synthesized AC voltage, lower switching losses, and reduced switch stress levels.

It should be pointed out that the number of levels of the multilevel DCC could be higher than five. The reader can find more details in [5].

However, with the NPC converter it is difficult to meet some grid code transient requirements, for example, that the converter must remain in operation during AC grid faults in order to provide reactive power to support the grid. This is an issue because the NPC converter DC capacitor voltage balancing at $^1/_2 V_{dc}$ is challenging under asymmetrical AC faults, such as a single-phase open circuit fault, single-phase-to-ground, and phase-to-phase faults. Consequently, HVDC manufacturers eventually abandoned the NPC converters in favor of improved two-level converters.

There are some VSC–HVDC transmission systems based on the NPC converter: Eagle Pass (the United States)—back-to-back system; Murray link (Australia) [10].

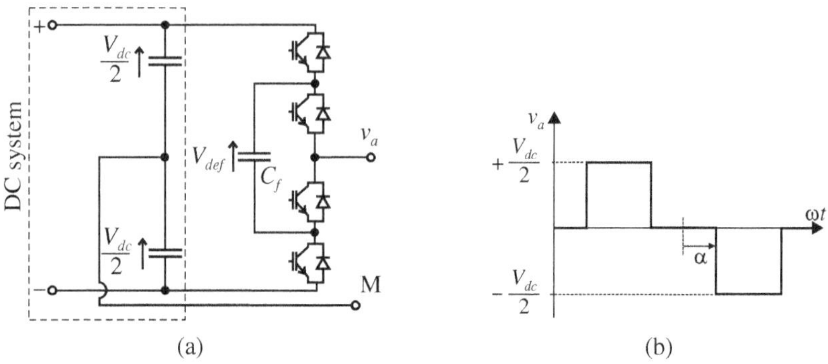

Figure 4.15 Three-level floating capacitor topology (a) and associated AC voltage waveform for one phase (b).

(iii) *Multilevel floating/flying capacitor converter* produces the same AC waveform as the multilevel DCC. The main difference is that the clamping diodes are replaced by floating capacitors. One phase of a three-level floating capacitor converter is shown in Figure 4.15.

Here the load cannot be directly connected to the neutral of the converter to generate the zero voltage level. Instead, the zero level is obtained by connecting the load to the positive or negative bar through the floating capacitor with opposite polarity with respect to the DC link. Like with the NPC, only two gating signals are necessary per phase to avoid DC-link and floating capacitor short circuit. However, in a floating capacitor, the inverted gating signals are related to different switching devices [7].

For a three-phase unit, the main DC capacitors are shared by the three phases but the floating capacitors, marked C_f, are not.

Note: For the output voltage waveform shown, $V_{def} = V_{dc}/2$.

The advantages of the multilevel floating capacitor converters are semiconductor switches having the same duty, lower switch blocking voltage, good basic AC waveform, and low converter switching loss. With the volume of capacitors largely proportional to the square of their nominal voltages, the disadvantage of this topology is the large footprint incurred by the floating capacitors.

4.1.7 Modular Multilevel Converter

The MMC is a novel VSC topology with great potential for HVDC transmission applications. The modular multilevel HVDC converter using half-bridge cells was invented by Marquardt and colleagues [11, 12]. The topology of MMC with "half-bridge" cells is shown in Figure 4.16. The submodules (SMs), composed of IGBTs and capacitors, are connected in series to form a converter valve of MMC.

Comparing with VSC based on two-level or three-level topology, MMC has more advantages. MMC is extendable to any number of levels without adding significant complexity to its control system and the capacitors voltage balancing method.

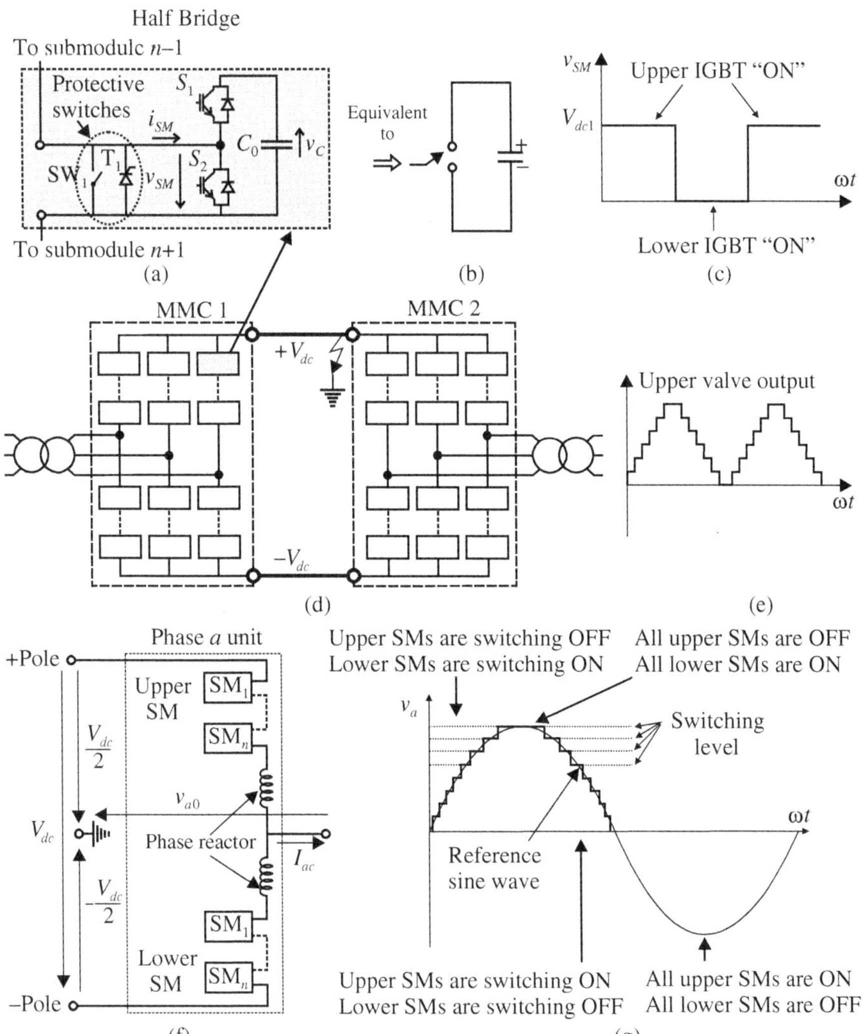

Figure 4.16 "Half-bridge" modular multilevel converter. Adapted from Marquardt and Lesnicar 2004 [4], Peralta et al. 2012 [14], and Saeedifard and Iravani 2010 [15].

With an increased number of levels, it generates high quality output voltage that eliminates the need for high frequency output filters. Furthermore, MMC can run at a lower switching frequency, which can reduce losses. MMC addresses the reliability and fault ride through capabilities.

Following this change of technology, there has been a convergence of VSC solutions from the three major European HVDC manufacturers (Siemens, ABB, and GE/Alstom). This technology, which can have different names according to its manufacturer, is described next as a half-bridge converter (Figure 4.16). One form of this technology is the "full-bridge" converter (Figure 4.18), which has additional functionalities but with higher costs and power losses in operation. The full-bridge converter is better adapted for applications in overhead lines sections or for limited regional multiterminal systems [13].

4.1.7.1 Half-Bridge Modular Multilevel Converter

A VSC-HVDC modular multilevel converter based on "half-bridge SMs" is illustrated in Figure 4.16 [4, 14, 15]. Every converter leg consists of n SMs connected in series with one converter reactor in each arm.

Configuration of a Submodule. A submodule SM_n is mainly composed of two IGBT switches S_1 and S_2 and a local DC storage capacitor C_0 (Figure 4.16a).

When the IGBT S_2 is switched on, the voltage $v_{SM} = 0$. To apply the voltage v_C to the terminals, the IGBT S_1 has to be switched on. In case of switching off both IGBTs, the impressed voltage to the power devices is limited by the capacitor voltage v_C (Figure 4.16b) [16].

In Figure 4.16a, "SW_1" is a highly reliable high-speed bypass switch, that is, a special vacuum switch. It has two following function. For redundant SMs, they must deliver zero voltage at their output during normal operation, that is to say, the inclusion of the redundant SMs number merely results in an increase of the SMs number in a converter arm during normal operation. Due to installing SW_1, it can be realized by closing the switches SW_1 of redundant SMs during normal operation.

In addition, in the event of a SM failure during operation, this fault is detected and the defective SMs can be shorted out by closing its switch SW_1. This provides fail safe functionality, as the current of the failed module can continue to flow. Simultaneously, a corresponding switch SW_1 of a redundant SM is opened, so the defective SM can be replaced by this redundant SM. Therefore, the continuous operation of the VSC without any interruption can be implemented. This results in an increased safety and availability.

"T_1" is a *press-pack thyristor*. In the event of a short circuit between the DC terminals of the VSC or along the transmission route, after the IGBTs switched off within a few microseconds, the current will flow from the three-phase line through the free-wheeling diodes to the short circuit.

The free-wheeling diodes used in VSC have a low capacity for withstanding surge current events related to their silicon surface, that is only a very limited ability to withstand a surge in current without damage. In actual event, the diodes would have to withstand a surge fault current without damage until the circuit breaker opens, that is, in most cases for at least three power frequency periods. So, a protection measure should be adopted. This can be realized by firing T_1 in the event of a fault. As a result, most of the fault current flows through the thyristor and not through the diode. Press-pack thyristors are known for their high capacity to withstand surge currents. This characteristic is also useful in conventional, line-commutated HVDC transmission technology. This fact makes HVDC transmission based on modular multilevel VSC suitable even for overhead transmission lines, an application previously reserved entirely for line-commutated converters with thyristors.

Figure 4.16f illustrates a converter leg consisting of n SMs connected in series with one converter reactor in each arm. We consider:

- each SM is only capable of generating two voltage levels, zero voltage or positive module voltage ($+V_{dc1} = V_{dc}/2n$) (Figure 4.16c);
- converter switching can be shifted to produce stair-way output waveform (Figure 4.16d);
- a reference sine wave is compared with fixed levels every time it crosses a level a SM is turned "on" or "off" (Figure 4.16e);
- equal number of SMs or cells are connected between positive and negative DC poles and the AC terminals;
- SM capacitor voltages are kept almost equal;
- the total capacitor voltages in each half phase module can reach a DC voltage larger than twice the peak phase-to-ground voltage;
- each SM can be either "off" or bypassed (lower IGBT triggered, zero voltage at terminals) or "on" (upper IGBT triggered, capacitor voltage at terminals).

By switching a number of the n SMs in the upper and lower arm, the voltages in each arm can be synthesized, so the voltage V_{dc} and voltages of each phase output terminal with respect to the imaginary DC side neutral point "0"—v_{a0}, $v_{b0,}$ and v_{c0} can be adjusted independently. The interface of the SM is composed solely of two electrical terminals and one bi-directional fiber-optic interface. This reduces the costs for manufacturing and maintenance, too [16].

The voltage of any arm and SM can be freely controlled by software. The individual voltage of the SMs may even be chosen unequally. This can be used to increase the number of resulting voltage steps.

The *phase reactors* shown in Figure 4.16f are a contribution to the interface inductance and are essential for the current control within the phase arms. Furthermore, they also limit the peak current and current gradients in case of severe faults, such as short circuit between DC terminals.

Also, under fault conditions the presence of the antiparallel diodes in each switch cell means that the converter cannot prevent, or block, conduction between the AC terminals of the converter into a fault in the DC system (Figure 4.17). The fault current path can only be blocked by disconnecting the AC feed and then isolating the fault using off-load isolators (Figure 4.17a). This arrangement has been widely used for two-terminal schemes and may be suitable for schemes up to a few terminals. For large multilevel system or HVDC grid, DC breakers are needed to isolate faulty parts of the grid during faults (Figure 4.17b) [17].

Technology of controlled SM *capacitor voltage balancing*. In the above section, it was described how the converter can impress the output voltages, based on the assumption that all the SMs can be regarded as ideal two-level unipolar voltage sources.

In order to keep the capacitors on the same voltage level and to ensure equal stress for the power devices, the algorithm is applied for each arm.

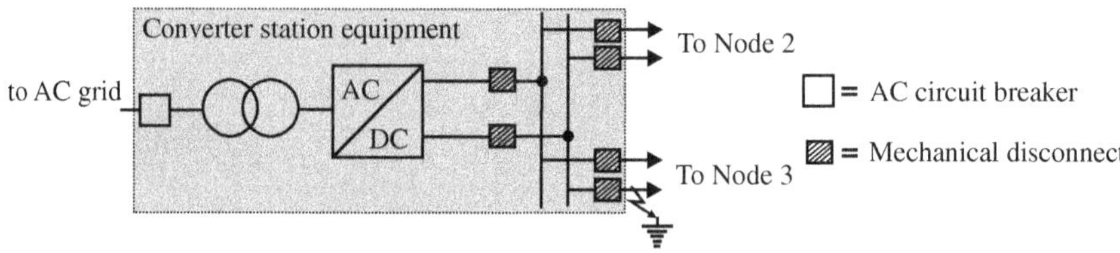

(a)

- DC cable fault cannot be cleared by "half-bridge" AC/DC converter;
- AC circuit breakers at all stations open to clear the fault;
- Appropriate disconnects opened to isolate faulted cable section;
- Complete multi-terminal scheme is re-started.

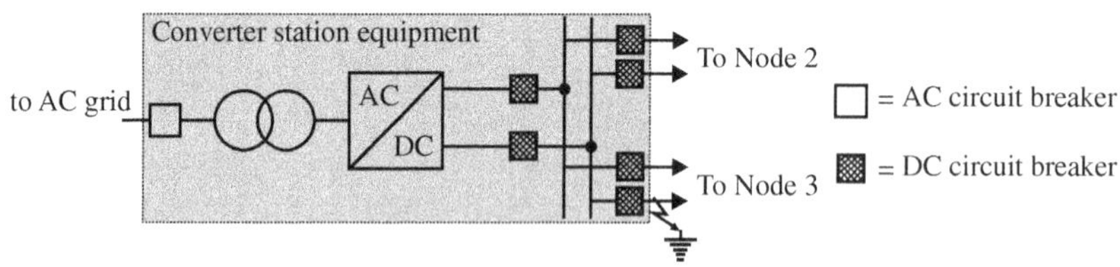

(b)

- DC cable fault can be cleared by the appropriate DC circuit breaker;
- No AC/DC converter action is required;
- Minimal interruption of power flow in the DC grid, except faulted section.

Figure 4.17 DC cable fault in case of half-bridge converters: (a) mechanical disconnect; (b) DC circuit breaker. *Source:* MacLeod 2011 [17]. Reproduced with permission of CIGRE.[1]

The voltages of the capacitors are periodically measured with a typical sampling rate in the millisecond range. According to their voltages, the capacitors are sorted by software [16].

- In case of positive current in the arm, that is, when the converter arm absorbs power, in order to impress the desired arm voltage, the required SMs with the lowest voltages are switched on, determined by output state controller.
- When the current in the corresponding arm is negative, that is, when the converter arm supplies power, in order to impress the desired arm voltage, the demanded number of SMs with the highest voltages is selected. Instead of measuring the current in the arm, the information about the sign may be derived from the voltage difference of two subsequent voltage samples. By this method, continuous balancing of the capacitor voltages is achieved.

Inherently, this technology supports an optimized utilization of the stored energy and evenly distributed power losses for the installed electrical devices.

[1]Material reprinted with permission from CIGRE "MacLeod, N. *DC Grids*, CIGRE Study Committee B4-HVDC and Power Electronics, Tutorials on Introduction to HVDC & HVAC and HVDC Grids, Brisbane, Australia", © Copyright 2011.

Additionally, the power losses can be kept low by switching the SMs solely when a change of the output state is requested.

The entire system has a modular structure, and it can be flexibly configured, what simplifies its standardization. The converter modules are connected on the secondary side of a high-voltage coupling transformer to build the HVDC. Due to the MMC configuration, there is almost no—or in the worst case, very small – need for AC voltage filtering to achieve a clean voltage. The system configuration is very compact and normally occupies 50% less space than a "classic HVDC system."

4.1.7.2 Full-Bridge Modular Multilevel Converter

A VSC HVDC multilevel converter based on full-bridge modules is shown in Figure 4.18.

At first glance, the full bridge module requires more switching devices than the basic half-bridge module, and, at any point in time, two switching devices will be in conduction in each module. Thereby, two situations are identified [18]:

- For capacitor bypassed or "off" state, the $IGBT_1$ and $IGBT_3$ are "on" and $IGBT_2$ and $IGBT_4$ are "off," or the $IGBT_1$ and $IGBT_3$ are "off" and $IGBT_2$ and $IGBT_4$ are "on";
- For capacitor charging or "on" state, the $IGBT_1$ and $IGBT_4$ are "on" and the $IGBT_3$ and $IGBT_2$ are "off," or the $IGBT_1$ and $IGBT_4$ are "off" and the $IGBT_3$ and $IGBT_2$ are "on."

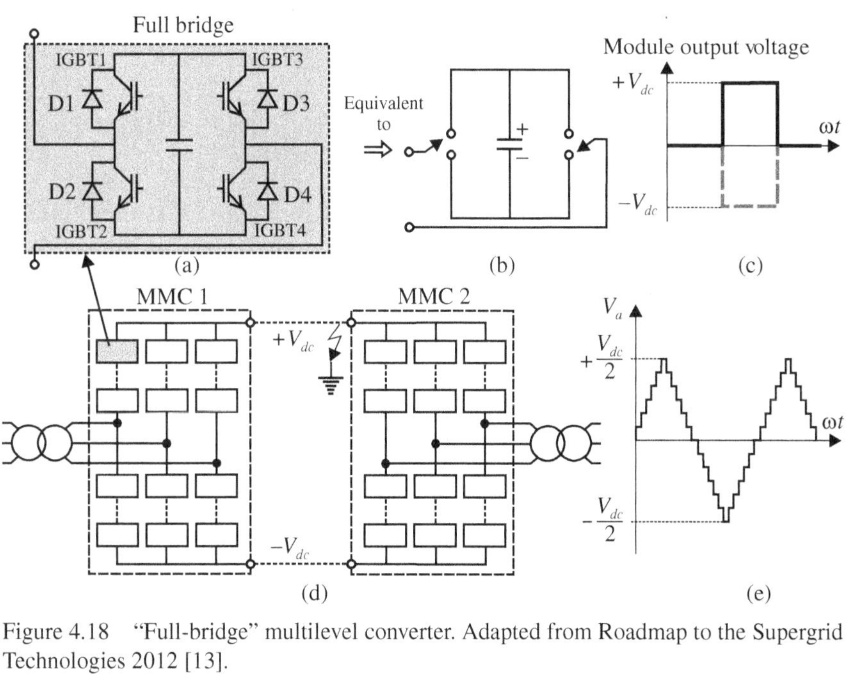

Figure 4.18 "Full-bridge" multilevel converter. Adapted from Roadmap to the Supergrid Technologies 2012 [13].

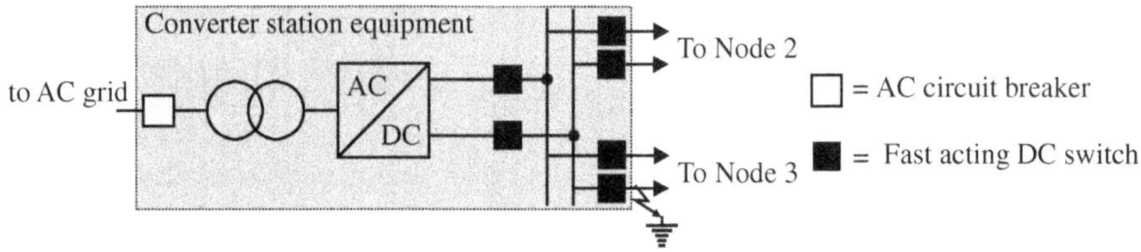

- DC cable fault can be cleared by "Full bridge" AC/DC converter;
- Fast acting (30–40 ms) switches open to clear the faulted cable/line section;
- Complete multiterminal scheme is re-started.

Figure 4.19 DC cable fault in case of full-bridge converters. *Source:* MacLeod 2011 [17]. Reproduced with permission of CIGRE.

Consequently, the conduction losses for this topology are twice that of the half-bridge topology; for the full-bridge converter are probably 30%–50% higher than those of a converter based on a half-bridge module. Each module is capable of generating three different voltages at its terminals: zero voltage, positive module voltage, and negative module voltage.

This allows the converter designer considerably greater freedom over choice of AC terminal voltage and hence AC current flow through the semiconductors.

The availability of the module to generate a negative module voltage gives the full-bridge converter the ability to "oppose" the AC terminal voltage driving a fault current into a short circuit in the DC system and therefore stop the fault current.

The full-bridge converter integrates the converter and DC breaker functionality so that off load isolators can be used to provide for isolating the faulted feeder. These isolators can be fast acting (30–40 ms) devices allowing rapid reconfiguration of the grid and restart of power flow (Figure 4.19).

For HVDC grids to be developed in the future, additional DC breakers are most likely needed [13].

4.1.7.3 The MMC–HVDC INELFE Project

One of the newest HVDC links that uses MMC is the INELFE project designed to increase the transmission capacity on the France–Spain interconnection. The VSC technology was selected because of the dynamic power flow control requirements and the low AC short-circuit ratio between the France and Spain power systems. The technical characteristics of the project are presented in Figure 4.20. A delta

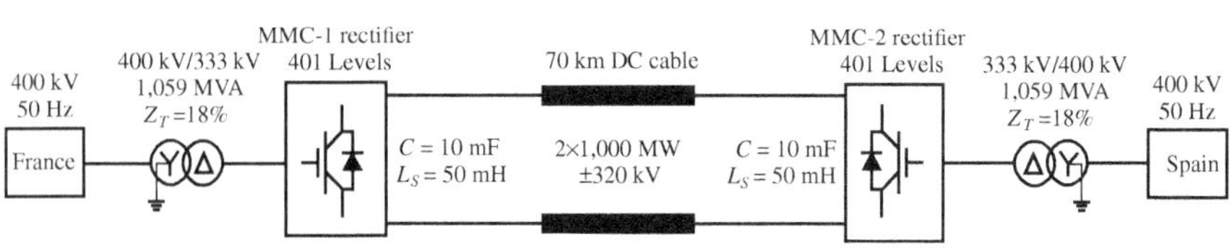

Figure 4.20 MMC–HVDC transmission system. *Source:* Peralta et al. 2012 [14]. Reproduced with permission of IEEE.

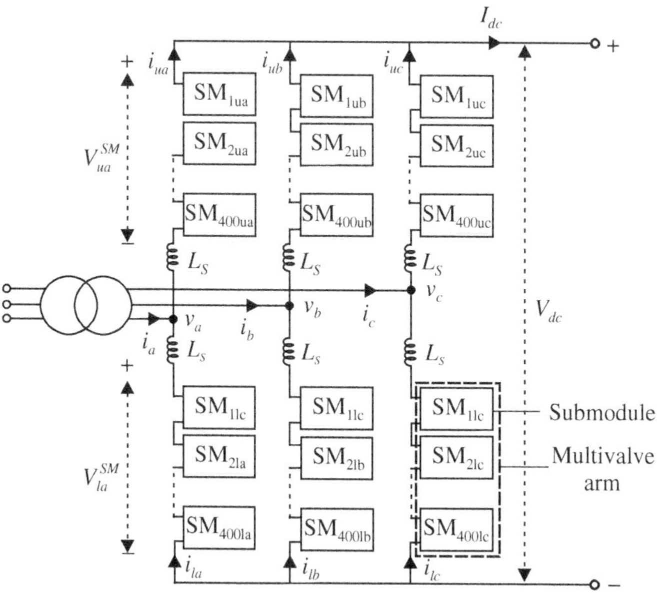

Figure 4.21 Detailed MMC topology. *Source:* Peralta et al. 2012 [14]. Reproduced with permission of IEEE.

connection is used on the converter side of the transformer to block the zero-sequence voltages generated by the MMC. A simulation of this link is presented in [14], where the 320 kV single-core cables are modeled using a frequency-dependent model.

TOPOLOGY. Figure 4.21 shows the MMC topology where every VSC converter includes 800 SMs per phase, that is 400 SMs per multivalve arm. One converter has six arms, thereby the detailed model of the VSC–HVDC that has to be analyzed consists of 4800 ideal switches and 9600 nonideal diodes per MMC [14, 19].

Each SM contains a capacitor C and two IGBT switches, S_1 and S_2 (Figure 4.22a), acting thus as a controllable voltage source. The output voltage is either equal to its capacitor voltage v_C or zero, depending on the switching state (Figure 4.22c). The model considers neither the pack thyristor T_1, which are used for protection during DC faults only, nor the ideal switches SW_1.

As explained in section 4.3.1, at any instant of time, only one of the two switches S_1 and S_2 conducts, the other one being blocked. Therefore, when S_1 is "on," S_2 is "off," and the voltage of the *i*th SM is $v_x = v_C$, whereas when S_2 is "on," S_1 is "off," and the SM voltage is $v_x = 0$. If the two switches S_1 and S_2 conduct, we have invalid operation because the capacitor is shorted.

The *IGBT switches*, shown in Figure 4.22a, can be modeled by an ideal controlled switch, two nonideal diodes (that can be represented by resistances; see Figure 4.22b) and a snubber circuit. The switch SW_1 in Figure 4.22a is a high-speed bypass switch used to increase safety and reliability of the MMC in case of SM failure.

The *arm reactors* L_S helps controlling and balance the currents in the phase arms, provides low-pass filtering of the AC fundamental frequency voltage, and limits

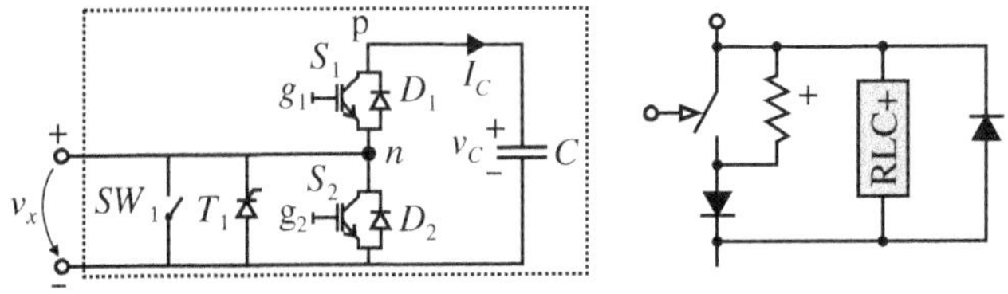

(a) (b)

State		Output voltage, v_x	Module current, I_C	Capacitor state
S_1	S_2			
ON	OFF	v_C	$+v_e$	Charging
ON	OFF	v_C	$-v_e$	Discharging
OFF	ON	0	$+v_e$ or $-v_e$	Bypassed
ON	ON	Invalid operation, capacitor shorted		
OFF	OFF	Open circuit		

(c)

Figure 4.22 (a) MMC submodule; (b) model of an IGBT valve; (c) switching states of a SM. *Source:* Peralta et al. 2012 [14]. Reproduced with permission of IEEE.

the fault currents. Its value is about 15% of the system impedance base. Additional AC filters are used if necessary.

The *DC side capacitor* is designed to provide a stiff voltage source and a low impedance path for the turned-off switching currents, to provide energy storage, as well as to keep the ripple in the DC voltage within a range of $\pm 10\%$. On the other hand, the MMC capacitance is a voltage source that compensates the DC voltage variations caused by faults in the AC system, thus the capacitors should be sized accordingly. The filtering on the DC side of the VSC–HVDC system is provided also by the DC line smoothing reactor. Therefore, The SM capacitance C is estimated as follows [14]:

$$C = \frac{2S\,E_{MMC}}{6N_{arm}v_C^2}$$

where E_{MMC} is the energy per megavolt-ampere (MVA) stored on each MMC, S is the nominal capacity of the MMC (equal to the transformer capacity), N_{arm} is the number of SMs per multivalve arm (400), and the nominal value for v_C is 1.6 kV. For instance, for a stored energy of 30 kJ/MVA, the resulting capacitor value is 10 mF [14].

MMC Control. The control technique employed in this example is the vector control strategy, by means of which four parameters are controlled at the two converters, that is the AC voltage, the DC voltage, the active power, and/or reactive power.

MMC Modulation Technique. Traditional modulation techniques proposed to date for MMCs include phase-disposition modulation [15], phase-shift modulation (PS-PWM) [20], space-vector modulation, and the improved selective harmonic elimination method (SHE) [16, 21, 22].

As the number of levels increased in MMCs, PWM and SHE techniques become cumbersome for electromagnetic transients (EMT)-type simulations. Therefore, more efficient staircase-type method, such as the nearest level control (NLC) technique, has been proposed in [23] (for more details, please see section 4.2.2).

SM CAPACITOR BALANCING CONTROL. In order to achieve appropriate voltage generated by the VSC during normal operation, the capacitor voltages (v_C) at all SMs must be maintained equal. In this regard, the voltage v_C (Figure 4.22a) must be monitored, and the capacitors should be switched on/off based on a balancing control algorithm (BCA). The BCA proposed in [15] determines the number of SMs that have to be switched on per each arm from a two time-dependent switching functions N_{up} and N_{low}. The combination ($N_{up} + N_{low}$) determines the total number of SMs in operation at a certain instant per multivalve arm (400). In the BCA proposed in [20], one SM is switched on and off each time the reference signal crosses one of the triangular signals from the PS-PWM, whereas the BCA proposed in [24] can be directly implemented to each SM by means of proportional-integral (PI) controller. In order to improve the efficiency of the algorithm, the detailed MMC model includes a trigger control that activates the BCA only when N_{up} and N_{low} change [15].

CIRCULATING CURRENT SUPPRESSION. The inherent voltage unbalance between the arm phases of the MMC give rise to a double frequency circulating current, which may require increased rating for the SM capacitors and switches. An active control over the modulated voltage is proposed in [25] to suppress the circulating currents, the solution of adding a parallel capacitor between the midpoints of the upper and lower arm inductances on each phase is proposed in [26].

4.1.8 Cascaded Two-Level Converters[2]

A cascaded two-level (CTL) converter consisting of several smaller two-level building blocks, also called cells, enables the creation of a nearly sinusoidal output voltage from the converter (Figure 4.23).

The technology is scalable up to the highest transmission voltages. Losses are reduced to roughly 1% per converter through a combination of methods.

In principle, the CTL converter has a topology similar to that of the MMC [26]. A different name has been selected to highlight the fact that series-connected press-pack IGBTs are used in the valves, thus extending a technology that has successfully been used for high-voltage two-level VSCs to multilevel VSCs through cascade connection.

Each phase leg of the CTL converter is divided in two arms (Figure 4.23): positive and negative, which, respectively, connect the positive and negative poles of the DC bus.

Each arm is built as a cascade of *N* two-level-converter cells. For instance, for a DC-bus voltage level of ±320 kV, $N = 38$ cells per arm are typically required.

[2] Material reprinted with permission from CIGRE "*VSC-HVDC Transmission with Cascaded Two-Level Converters*, Jacobson et al. 2010 [26].

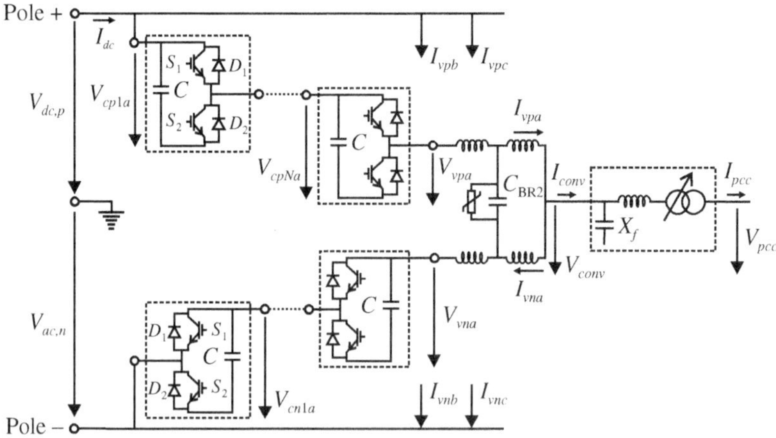

Figure 4.23 Single-line diagram of a cascaded two-level converter. *Source:* Jacobson et al. 2010 [26]. Reproduced with permission of CIGRE.

The cascaded two-level cells in each arm are controlled to provide fundamental-frequency output voltage, related to the desired active- and reactive-power output, through switching of the individual cells.

Each cell consists of a half-bridge two-level converter with two valves, which in turn consist of IGBTs S_1 and S_2 and diodes D_1 and D_2. The valves can be switched in three different ways (Figure 4.24) [9, 18]:

- By turning on S_2 and turning off S_1, the cell is said to be *inserted* and the cell output voltage equals the capacitor voltage. The capacitor then charges when the arm current is positive and discharges otherwise.
- By switching the valves vice versa, that is, S_1 is turned on and S_2 is turned off, the cell will give zero output voltage. The capacitor is then *bypassed*, and its voltage remains constant.
- If both valves are turned off, the cell becomes *blocked*, and current is conducted only through the diodes. The capacitor will then charge only when the arm current is positive. It will (ideally) not discharge.

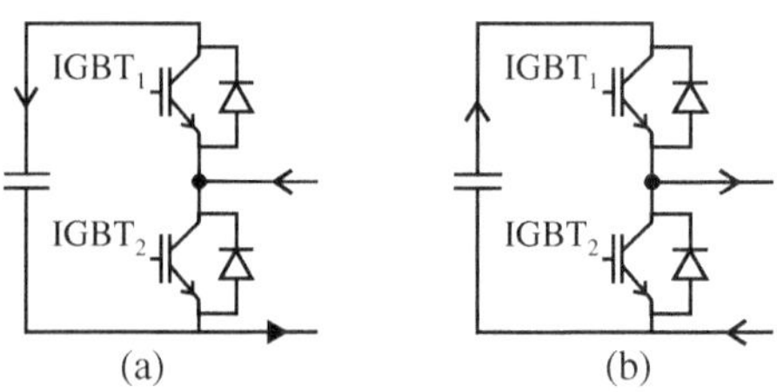

Figure 4.24 Valve switching in a half-bridge: (a) $I_{AC} > 0$, capacitor is charged; (b) $I_{AC} < 0$, capacitor is discharged. *Source:* Mohaddes 2011 [18]. Reproduced with permission of CIGRE.

Each cell is switched at a low switching frequency, typically about f_{swc} = 150 Hz, that is, a pulse number about 3 for a fundamental frequency of 50 Hz. During normal operation, the cells are switched in a staggered fashion [26]. Consequently, the *effective switching frequency per phase leg* becomes $f_{sw} = 2Nf_{swc} = 11.4$ kHz for $N = 38$ cells per arm. This is in the range of 10 times the aforementioned switching frequency of a two-level VSC, which indicates that the dynamic response of a CTL converter is excellent.

To keep capacitor voltage equal, these rules are followed:

- SM turning on and:
  - $I_{ac} > 0$, then select the SM with lowest capacitor voltage;
  - $I_{ac} < 0$, then select the SM with highest capacitor voltage.
- If SM is turning off do to opposite.

The *cell capacitor* is integrated in the valve design, with the main purpose *of serving as energy storage*. The sum cell voltages of the two arms should each have a mean value of twice the pole-to-neutral DC-bus voltage, this in order to allow converter output voltages with maximum amplitude. Since the cell capacitor is subjected to a fundamental-frequency current, there will be ripple superimposed on each cell voltage. The cell capacitance is selected as a trade-off between valve voltage requirements and capacitor size. Typically, the total cell capacitance corresponds to a stored energy of 30–40 kJ/MVA (where MVA refers to the converter rating), giving ripple in the range of 10%. The ripples are predominantly of the fundamental and second-order harmonic frequencies [26].

Each arm includes a *reactor* for the purposes of limiting parasitic currents and fault currents (Figure 4.23). The reactors are of standard air-core type. In the CTL converter, the second harmonic component is suppressed using a *parallel-resonant filter*, realized through a near-center tap of the valve reactors together with a capacitor (C_{BR2}).

The grid current of one phase will split evenly between the positive and negative arms. In addition, each arm current will contain a DC component corresponding to one third of the total DC-side current. Through the sharing of the grid current between the two arms, the corresponding reactive voltage drop will be half compared to a case with a common phase reactor at the point of common coupling. On the other hand, for a DC-side fault, the reactors are effectively placed in series and will effectively limit the fault current together with the transformer and grid impedances [26].

Through the low harmonic content of the output voltage, AC filters are normally not required to meet the typical restrictions on harmonic injection into the grid. Only a small capacitor for high-frequency attenuation is included on the converter bus. Moreover, due to the low harmonic content, a standard transformer with Y/Y configuration can be used, where the converter-side star point is grounded through an arrester. This configuration will allow reduced voltage rating of the converter as well as efficient overvoltage protection for internal faults.

In consequence of the usage of a low switching frequency per cell, the IGBT switching losses are significantly reduced compared to a two-level converter, as are the harmonic losses in the reactors. Furthermore, the conduction losses are reduced

(a) (b)

Figure 4.25 (a) ABB cell module with two valves, each comprising eight series-connected press-pack IGBTs [26]. Reproduced with permission of ABB. ; (b) GE/Alstom MaxSine module. *Source:* Mohaddes 2011 [18]. Reproduced with permission of GE.

through use of state-of-the-art 4.5 kV SPT+ IGBT technology. The total converter station losses are in the range of 1% [26].

The *valve* is the major piece of equipment in the CTL converter in terms of complexity, space requirement, cost, and—still—losses. Multilevel converters can be realized in several different ways, with different types of cells with alternative mechanical and electrical solutions. In [26] some aspects of two principal variants of half-bridge cells are presented, namely, with series connection and without. An example is shown in Figure 4.25a.

Typically, the highest failure rate of components in any converter valve is found in the power electronics part, including the IGBT and its corresponding driving and communication electronics. Passive components have orders of magnitude better reliability, partly because of lower complexity and partly because higher margins of safety can be applied without adversely affecting the cost of the converter.

Two major fault scenarios that could cause immediate problems for the continued service of the converter, unless well cared for in the design, are cell short circuit and DC pole-to-pole short circuit. A cell short circuit can happen for a number of reasons, but the general sequence of events is as follows: One of the switches in the cell is conducting current, whereas the other switch blocks the cell capacitor voltage. The insulation of the blocking switch breaks down causing a short circuit of the cell capacitor through the active switches. Typically, the breakdown occurs in one of the semiconductor chips of the IGBT module or on the edge of such a chip. In any case, it generally happens inside the IGBT module. The current derivative of this event is very high, since the cell is designed to have low inductance, in order to minimize switching stresses during normal operation for the IGBTs.

A stable short circuit is reached in the case with press-pack IGBTs designed for electric power-transmission applications. Open circuit in a transmission-class converter must be avoided, as otherwise the resulting arc and its associated energy dissipation will necessitate a trip of the converter every time an IGBT fails.

In order to safely and securely avoid explosive discharge of the cell capacitor, series connection of devices is introduced also in the multilevel converter. With series connection and press-pack IGBTs, single devices can fail without leading to short circuit or any limitation of operation. When one IGBT fails, the healthy ones in the same switch simply accept the slight voltage increase and transmission continues. Press-pack IGBTs go to a short-circuit failure mode, which means that no extra equipment is necessary to prevent arcing in the valve. During the next scheduled maintenance, the faulty IGBT can be replaced.

4.2 MODULATION TECHNIQUES

4.2.1 PWM Techniques

4.2.1.1 PWM Principle

Because an inverter contains electronic switches, it is possible to control the output voltage as well as optimize the harmonics by performing multiple switching within the inverter with the constant DC input voltage V_{dc}.

In order to improve the quality of the output voltage, the PWM technique can be used. This results in a quasi-sinus waveform with a dominant fundamental component, and with the characteristic harmonics shifted to higher orders [6]. PWM allows the variation of the r.m.s. of the voltage's fundamental and shifts the harmonics of the toward higher frequencies domain, which explains the need of smaller filtering.

With the PWM technique, the output of each converter pole is switched several times during a fundamental cycle between the positive and negative terminals of the DC source and the time intervals between consecutive switching is controlled so that the average of the positive and negative volt-second segments of the output waveform generated follows the wanted sine wave.

By increasing the switching frequency, harmonics in the lower frequency range (typically below the switching frequency) can be reduced or nearly eliminated. As a consequence, harmonics in the higher frequency range increase but the filtering can be achieved more easily. However, increasing the switching frequency also increases the power loss. Therefore, an optimization has to take place to balance the harmonic level, on the one hand, and the capital cost, power loss, footprint, etc., on the other hand [6].

How PWM works? A simple comparator with a sawtooth-shaped carrier can turn a sinusoidal command into a pulse-width modulated output (Figure 4.26a). In general, the larger the command signal, the wider the pulse. Output stays high as long the command is greater than the carrier.

The valves' command signal is an alternating square-shape signal, having unequal width pulses, that is, a width-modulated pulses (Figure 4.26c). PWM is obtained by comparing a triangle carrier wave, called carrier wave ($v_p(t)$), with the f_p frequency and $\hat{V}_p$ magnitude, and a sinusoidal reference wave ($v_{ref}(t)$) with $\hat{V}_{ref}$ magnitude and f_{ref} frequency (Figure 4.26b). A comparator compares the two signals. The frequency f_{ref} of the reference wave must be equal to the desired frequency for the fundamental of the output voltage wave. The carrier wave (triangular wave) repeats itself with a period time T_p. Thus, the switching frequency f_p is equal to $1/T_p$.

The comparator output voltage, processed, is the voltage that commands the inverter's semiconductor devices. This voltage, v_c, is positive if $v_{ref} > v_p$, or negative if $v_{ref} < v_p$. The f_p frequency is giving the commutation frequency for the semiconductor devices. The sine-wave modulation is the most used one, providing a reduced harmonics content for the output voltage and a high magnitude for the fundamental.

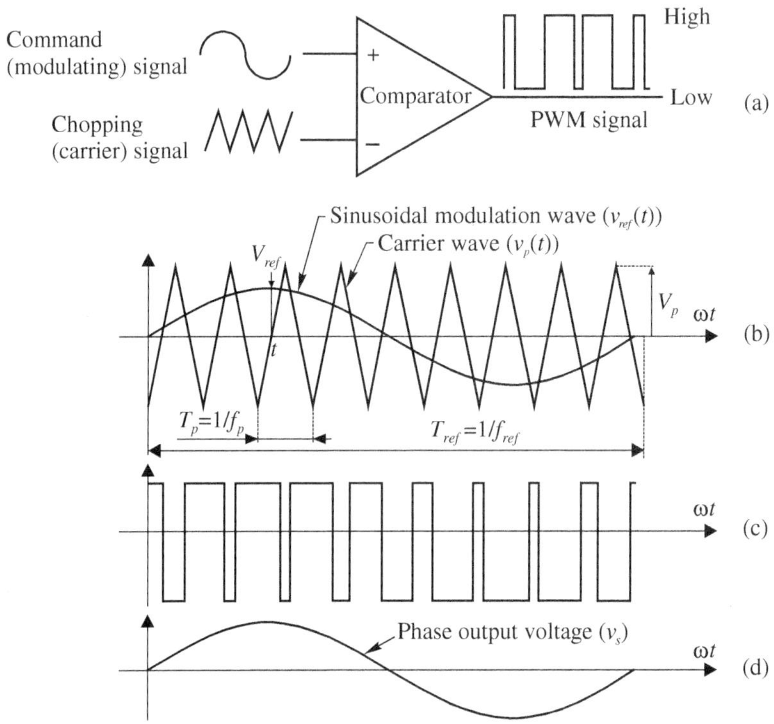

Figure 4.26 Sinusoidal pulse width modulator: (a) the comparator and PWM signal; (b) reference and the carrier waves; (c) resulting PWM pattern; (d) extracted fundamental component of the PWM pattern. Adapted from Sannino 2004 [27].

The PWM strategies can be

- *singular*, when the semiconductor devices receive a single command impulse during each period of the output voltage. The conduction time of the semiconductor device can be modified;
- *multiple*, when the semiconductor devices receive more than one command impulse during each period of the output voltage. These command impulses can have an adjustable length and can be equal (when the reference voltage has a constant value) or unequal (when the reference voltage is alternating) during each period.

A good quality output voltage wave from the inverter is obtained by applying a multiple PWM with a sinusoidal reference wave, also called sinusoidal modulation.

The PWM strategy is characterized by two parameters [27, 28]:

- The *modulation index*, m_a, is defined as the ratio between the amplitude of the two signals:

$$m_a = \frac{\hat{V}_{ref}}{\hat{V}_p} \tag{4.1}$$

where $\hat{V}_{ref}$ is the peak value of the modulating wave and the $\hat{V}_p$ is the peak value of the carrier wave.

When the amplitude of the reference voltage is less or equal to the carrier wave, that is no overmodulation, the value of the modulation index m_a is between zero and one.

- The *frequency modulation ratio* m_f is defined as a ratio between frequencies of the two signals:

$$m_f = \frac{f_p}{f_{ref}} = \frac{T_{ref}}{T_p} \tag{4.2}$$

In Figure 4.26, the frequency modulation ratio is $m_f = 9$.

The voltage's fundamental from the inverter's output depends on the modulation index m_a; hence there is a possibility to regulate the effective value of the output voltage, respectively, the frequency, by modifying the frequency modulation ratio m_f.

The modulation can be synchronous, in which case the fundamental is periodical, with period $T_{ref} = 1/f_{ref}$, and the harmonic content being dependent on the value of m_f. In the case of asynchronous modulation, an f_p = constant and a variable f_{ref} frequency are used.

If synchronous modulation is used, the output voltage wave, v_s, is symmetrical with regard to its center if the modulation is centered from the position of the reference wave. The centering is optimal, and the v_s voltage is symmetrical with regard to its extreme points if v_p passes through one extreme (maximum or minimum point) at the same moment of time as v_s.

(i) *Natural-sampled PWM or sinusoidal PWM* is based on a well-defined modulation process, involving a direct comparison of a sinusoidal modulating wave and a triangular carrier (sampling) wave.

The switching edge of the width-modulated pulse is determined by the instantaneous intersection of the two waves. Natural sampling can therefore be defined as a process of intrinsic natural selection of the sampling points; that is, the time of sampling coincides with the time of appearance of the width-modulated pulse. The natural-sampled pulse width modulator for a single-phase converter is shown in Figure 4.27. The reference voltage signal into the single-phase modulator is written as [27]:

$$u^*(t) = \hat{u}^* \cos(\omega^* t + \phi^*) \tag{4.3}$$

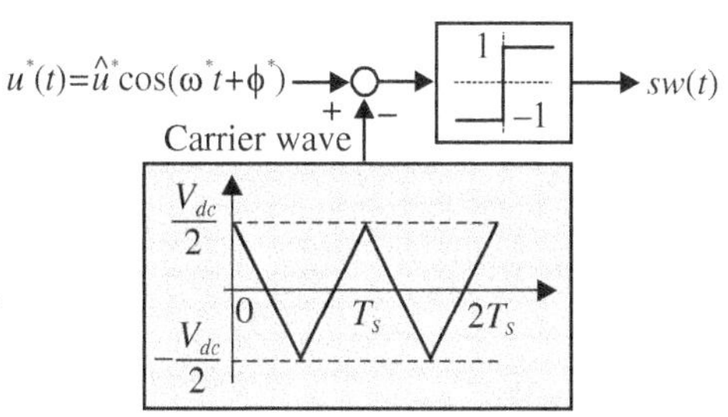

Figure 4.27 Natural-sampled pulse width modulator for a single-phase converter [27].

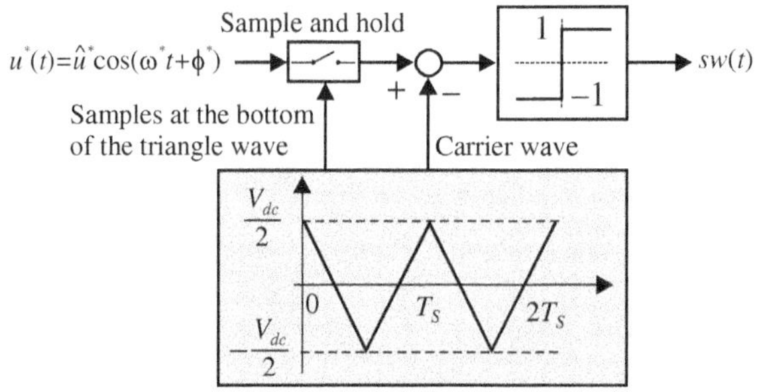

Figure 4.28 Block-scheme of the regular-sampled PWM [27].

where $\hat{u}^*$ and ω^* are the reference values of the voltage amplitude, and the reference angular frequency, respectively. The reference angular frequency is equal to $2\pi f^*$, where f^* is the reference frequency. Furthermore, the reference phases shift is denoted by ϕ^*.

In Figure 4.26a, the implementation of the natural-sampled pulse width modulator is shown. In Figure 4.26b, the reference and the carrier waves are shown during one switching period and the resulting PWM pattern is shown in Figure 4.26c. Furthermore, the extracted fundamental component of the PWM pattern is shown in Figure 4.26d and, as displayed the extracted fundamental voltage is equal to the reference voltage.

(ii) *Regular-sampled PWM.* The block-scheme of the regular-sampled PWM is shown in Figure 4.28.

A typical implementation of this method, which illustrates the general features of double-edge two-level PWM is shown in Figure 4.29. In this case, a triangular carrier wave is used to sample the sinusoidal modulating wave once every carrier cycle at regularly spaced intervals, which correspond to the bottom peaks of the triangular wave (Figure 4.29) [27].

Comparison of the sampled modulating wave with the carrier wave defines the intersection points used to determine the switching instants of the width-modulated pulses. As a result of this process, the sampled modulating signal in Figure 4.29 has a

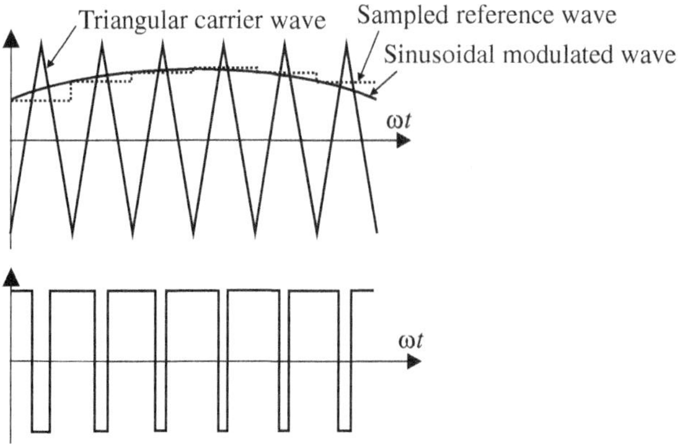

Figure 4.29 (a) Regular-sampled symmetric PWM; (b) width-modulated pulses [27].

constant amplitude during each sampling interval and, consequently, the widths of the pulses are proportional to the amplitude of the modulating signal at regularly spaced sampling times. Hence, the terminology "regular" sampling is used.

An important characteristic of regular sampling is that the sampling positions are sampled values that can be defined unambiguously, such that the pulses produced are predictable both in width and position. The regular-sampling process may therefore be viewed conceptionally as a combination of two modulation processes. The first transforms or maps the sinusoidal modulating wave into a pulse-amplitude modulated wave (PAM) and the second transforms (or maps) the PAM wave into a PWM wave [27].

As illustrated in Figure 4.28, the same "sample" value is used to modulate both edges of the pulse equally with respect to regularly spaced pulse positions; the resulting modulation is referred as "symmetric" PWM.

Further improvements in the harmonic frequency spectrum of the PWM waveform can be achieved if each edge of the pulse is modulated by a different amount. In this case, the sinusoidal modulating wave is sampled at twice the carrier frequency to produce a sampled modulating wave with twice the number of amplitude modulated levels.

Because more information about the modulating wave is contained in the asymmetric-modulated PWM waveform, its harmonic spectrum is superior to that produced using symmetric modulation. Also, since the PWM switching edges can be defined using a simple algebraic equation, the potential for real-time generation of asymmetric PWM using a microprocessor software-based algorithm exists, although it should be noted that the number of calculations required to generate asymmetric PWM is double than for symmetric PWM [27].

4.2.1.2 PWM Strategy Control of a Half-Bridge Converter

Let us consider a converter with one pole and two levels or a half-bridge converter (Figure 4.30a) controlled by PWM strategy. The DC-link voltage is denoted V_{dc}, and the potential of the phase-leg is denoted by $v_a(t)$, whereas the voltage over the load is denoted by $v_L(t)$. The half-bridge converter operates based on the alternate switching of S_1 and S_4. The turn-on/off commands of S_1 and S_4 are issued through a PWM strategy.

The most common PWM strategy compares a high-frequency periodic triangular waveform, the carrier signal, with a slow-varying waveform known as modulating signal (Figures 4.26b and 4.30b). The carrier signal has a periodic waveform with the period T_p and the swings between −1 and +1 (Figure 4.30c).

The intersections of the carrier and modulating signals determine the switching instants of S_1 and S_4. The PWM process is illustrated in Figure 4.30b, where the switching function sw of a switch is defined as [5]

$$sw(t) = \begin{cases} 1, & \text{if the switch is commanded to conduct} \\ 0, & \text{if the switch is turned off.} \end{cases}$$

The switch conducts only if the turn-on command is provided and the current flow through IGBT is from the collector to the emitter.

Thus, as Figure 4.30d shows, when the modulating signal (v_{ref}) is larger than the carrier signal (v_p), a turn-on command is issued for S_1, and the turn-on command of

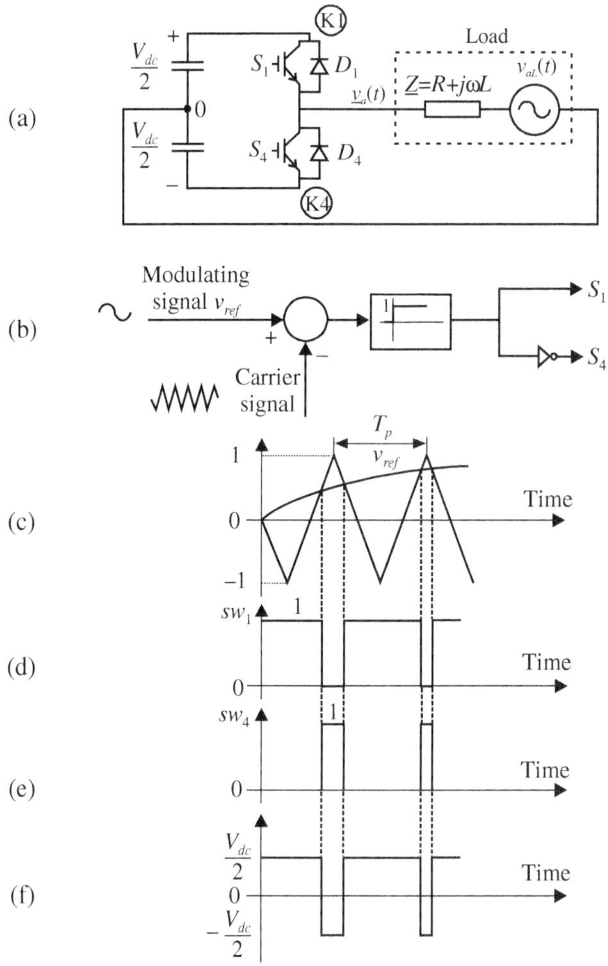

Figure 4.30 PWM: (a) the scheme of one pole and two voltage levels (half-bridge converter VSC); (b) the mechanism to generating PWM pulses for S_1 and S_4 transistors; (c) comparison of the carrier and modulating signals; (d) switching function sw_1 of the S_1 switch; (e) switching function sw_4 of the S_4 switch; (f) phase a output voltage v_{a0}.

S_4 is canceled; a voltage $v_{a0} = +V_{dc}/2$ will result at the output. Once the modulating signal is smaller than the carrier signal, the turn-on command for S_1 is blocked while a turn-on command is issued for S_4; at the output the voltage $v_{a0} = -V_{dc}/2$ results.

Figures 4.30d and 4.30e illustrate the waveforms of the switching functions $sw_1(t)$ and $sw_4(t)$ of the S_1 and S_4 transistors, based on the PWM strategy; it can be observed that $sw_1(t) + sw_4(t) = 1$. Figure 4.30f presents the waveform of the generated voltage with regard to the reference point 0—when the carrier signal has a frequency nine times higher than the reference signal. The two valves, K_1 and K_4, will never be turned on simultaneously; hence the terminal voltage (v_s) will fluctuate between $+V_{dc}/2$ and $-V_{dc}/2$, and the mean value of this voltage will have a sinusoidal form.

Choosing the frequencies (f_p and f_{ref}) and modulation parameters (m_a and m_f). If the frequencies of the two waves verify the $f_{ref} \ll f_p$ expression, then it can be considered that in a short-time interval v_{ref} = const. (see Figure 4.29 top).

It has also been stated that the output wave is closer to a sinusoidal one if the reference wave is also a sinusoidal one (Figures 4.26c and 4.26d). In order to justify this statement, the harmonics contained in the output wave v_s must be studied. The conduction angles θ_i for each switch are dependent only on the value of the modulation index, m_a, and the frequency modulation ratio, m_f. The choosing of the modulation index is not done at random. In order to establish an optimal value, it is necessary to analyze the harmonics of the output wave, $v_s(t)$.

The general Fourier series of the wave can be given as [29]:

$$v_s(t) = \sum_{n=1}^{\infty} [a_n \cos(n\omega t) + b_n \sin(n\omega t)]$$

where $a_n = \frac{1}{\pi} \int_0^{2\pi} v(t) \cos(n\omega t)\, d(\omega t)$

$$b_n = \frac{1}{\pi} \int_0^{2\pi} v(t) \sin(n\omega t)\, d(\omega t)$$

For a waveform with quarter-cycle symmetry, *only odd order harmonics with sine components will be present*. Therefore, $a_n = 0$, and

$$v_s(t) = \sum_{n=1}^{\infty} b_n \sin(n\omega t)$$

where $b_n = \frac{4}{\pi} \int_0^{\pi/2} v_s(t) \sin(n\omega t)\, d(\omega t)$

Assuming that the wave has unit amplitude, that is, $v_s(t) = +1$, b_n can be expanded into a sum of integrals as

$$\begin{aligned} b_n = \frac{4}{\pi} \Bigg[& \int_0^{\alpha_1} (+1) \sin(n\omega t)\, d(\omega t) + \int_{\alpha_1}^{\alpha_2} (-1) \sin(n\omega t)\, d(\omega t) \\ & + \int_{\alpha_2}^{\alpha 3} (+1) \sin(n\omega t)\, d(\omega t) + \cdots + \int_{\alpha_{k-1}}^{\alpha_k} (-1)^{k-1} \sin(n\omega t)\, d(\omega t) \\ & + \int_{\alpha_k}^{\pi/2} (+1) \sin(n\omega t)\, d(\omega t) \Bigg] \end{aligned} \tag{4.4}$$

where $\alpha_1, \alpha_2, \ldots, \alpha_{k-1}, \alpha_k$ are the conduction angles of each static switch.

Using the general relation

$$\int_{\theta_1}^{\theta_2} \sin(n\omega t)\, d(\omega t) = \frac{1}{n}(\cos n\theta_1 - \cos n\theta_2)$$

the first and last terms from (4.4) are

$$\int_0^{\alpha_1} (+1) \sin(n\omega t)\, d(\omega t) = \frac{1}{n}(1 - \cos n\alpha_1) \tag{4.5a}$$

$$\int_{\alpha_k}^{\pi/2} (+1) \sin(n\omega t)\, d(\omega t) = \frac{1}{n} \cos n\alpha_k \tag{4.5b}$$

Integrating the other components of Equation (4.4) and substituting (4.5a) and (4.5b) in it yields:

$$b_n = \frac{4}{n\pi}[1 + 2(-\cos n\alpha_1 + \cos n\alpha_2 - \cdots + \cos n\alpha_k)]$$

$$= \frac{4}{n\pi}\left[1 + 2\sum_{k=1}^{k}(-1)^k \cos n\alpha_k\right] \tag{4.6}$$

Note that Equation (4.6) contains k number of variables (i.e., $\alpha_1, \alpha_2, \ldots, \alpha_{k-1}, \alpha_k$), thus k number of simultaneous equations are required to solve their value. With k number of α angles, the fundamental voltage can be controlled and $k - 1$ harmonics can be eliminated.

Consider, for example, that the fifth and seventh harmonics (lowest significant harmonics) are to be eliminated and the fundamental voltage is to be controlled. The third and other triplen harmonics can be ignored if the machine has an isolated neutral. In this case, $k = 3$ and the simultaneous equations can be written from Equation (4.6) as

- Fundamental: $b_1 = \frac{4}{\pi}(1 - 2\cos\alpha_1 + 2\cos\alpha_2 - 2\cos\alpha_3)$
- Fifth harmonic: $b_5 = \frac{4}{5\pi}(1 - 2\cos 5\alpha_1 + 2\cos 5\alpha_2 - 2\cos 5\alpha_3) = 0$
- Seventh harmonic: $b_7 = \frac{4}{7\pi}(1 - 2\cos 7\alpha_1 + 2\cos 7\alpha_2 - 2\cos 7\alpha_3) = 0$

These nonlinear, transcendental equations can be solved numerically by a computer program for the specified fundamental amplitude and α_1, α_2, and α_3 can be determined.

The r.m.s. value of each harmonics is evaluated with the expression:

$$V_{sn} = \frac{|b_n|}{\sqrt{2}} \tag{4.7}$$

When $m_f \geq 6$, with a good approximation, the value of the fundamental component of the output voltage can be calculated using [28]:

$$v_{s1} = m_a \frac{V_{dc}}{2} \sin\omega t, \text{ for } m_a \leq 1 \tag{4.8a}$$

and

$$\hat{V}_{s1} = m_a \frac{V_{dc}}{2} \tag{4.8b}$$

The input current has

- a continuous component, corresponding to the active power at the inverter's output;
- a term that varies with the frequency $2f_{ref}$ and whose value is proportional with the fundamental of the output voltage v_s;
- other terms with frequencies of $4f_{ref}$, $6f_{ref}$, etc.

For any value of $m_a \in (0, 1)$ and for every $m_f \geq 6$, the output voltage harmonics are grouped into families, centered around the commutation frequency and its multiples (m_f, $2m_f$, $3m_f$,...).

The frequencies of different harmonics inside a family are given by [28]:

$$f_n = (j \cdot m_f \pm k)f_{ref} = f_j \pm k \cdot f_{ref}, \text{ with } j = 1, 2, 3, \ldots$$

where $f_j = j\, m_f f_{ref}$. The magnitudes of the harmonics inside a family, symmetrical with regard to f_j are equal.

Generally, odd values of the modulation index, m_f, are preferred due to the lower harmonics content (only the odd harmonics). In these conditions, from the last expression:

$$j = \text{even} \Rightarrow k = \text{odd}$$
$$j = \text{odd} \Rightarrow k = \text{even}$$

Frequently, inverters with PWM with $m_f \geq 9$ (except for the case where high power is involved) are used.

4.2.1.3 Three-Phase Bridge Inverter with Sinusoidal PWM

The VSC and the grid are modeled as two three-phase voltage sources. The grid filter, often called line filter, is placed between the two three-phase voltage sources (Figure 4.31).

The phase voltages and currents of the grid are denoted by $e_a(t)$, $e_b(t)$, $e_c(t)$, $i_a(t)$, $i_b(t)$, and $i_c(t)$, respectively. The resistance and the inductance of the series filter are denoted by R and L, respectively. The angular frequency and the phase-to-phase rms voltage of the grid are denoted by ω and E. This grid is modeled with three voltage sources connected in Y, and the Y-point is grounded (Figure 4.31a). The instantaneous phase voltages of VSC are denoted by $u_a(t)$, $u_b(t)$, and $u_c(t)$, respectively. The phase voltages of the grid are written as

$$e_a(t) = \sqrt{\frac{2}{3}} E \cos \omega t \tag{4.9a}$$

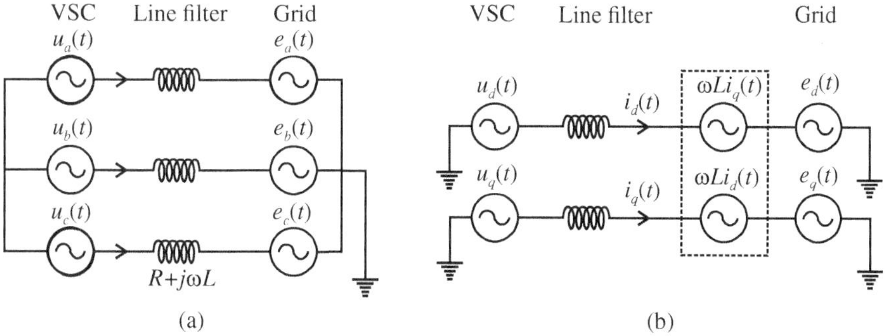

Figure 4.31 (a) Simplified circuit of a grid-connected VSC; (b) schematic circuit in *dq* coordinate system of grid connected VSC [27].

$$e_b(t) = \sqrt{\frac{2}{3}} E\cos(\omega t - 2\pi/3) \tag{4.9b}$$

$$e_c(t) = \sqrt{\frac{2}{3}} E\cos(\omega t - 4\pi/3) \tag{4.9c}$$

The relation between current and voltages of the VSC connected to the grid can be written as differential equations as

$$u_a(t) - e_a(t) - Ri_a(t) - L\frac{di_a(t)}{dt} = 0 \tag{4.10a}$$

$$u_b(t) - e_b(t) - Ri_b(t) - L\frac{di_b(t)}{dt} = 0 \tag{4.10b}$$

$$u_c(t) - e_c(t) - Ri_c(t) - L\frac{di_c(t)}{dt} = 0 \tag{4.10c}$$

The VSC is connected to the grid with three wires, and only the Y-point of the grid is grounded. Therefore, no zero sequence currents will exists and can, thus, be removed from now on. Under these circumstances, if the grid model is represented in the dq frame, as shown in Figure 4.31b, where the cross-coupling voltages are inserted, the voltages in the d- and q-axis can be written as [27]

$$\begin{aligned} u_d(t) - e_d(t) - Ri_d(t) - L\frac{di_a(t)}{dt} + \omega Li_q(t) = 0 \\ u_q(t) - e_q(t) - Ri_q(t) - L\frac{di_q(t)}{dt} + \omega Li_d(t) = 0 \end{aligned} \tag{4.11}$$

The circuit of a three-phase bridge inverter, consisting of three half-bridges, is shown in Figure 4.32. The DC-link voltage is denoted by $V_{dc}(t)$. The phase potentials, the phase voltages, and the potential of the floating-star load are denoted by $v_a(t)$, $v_b(t)$, $v_c(t)$, $u_a(t)$, $u_b(t)$, $u_c(t)$, and $v_0(t)$, respectively.

If the three-phase load is symmetric, the potential of the floating-star load $v_0(t)$ can be written as

$$v_0(t) = \frac{1}{3}(v_a(t) + v_b(t) + v_c(t)) \tag{4.12}$$

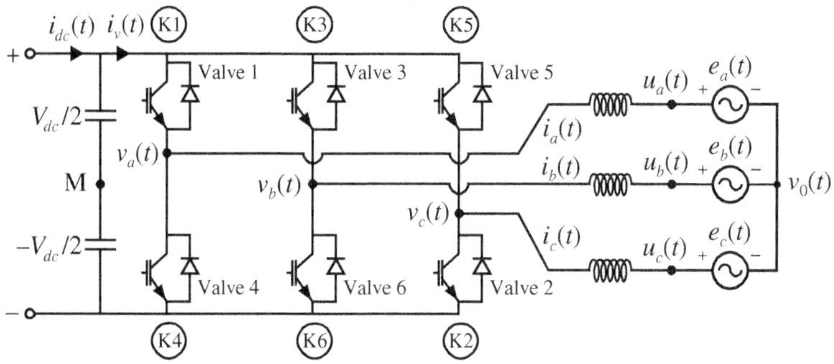

Figure 4.32 Main-circuit scheme of the VSC.

The phase voltages can now be written as

$$u_a(t) = v_a(t) - v_0(t) \tag{4.13a}$$

$$u_b(t) = v_b(t) - v_0(t) \tag{4.13b}$$

$$u_c(t) = v_c(t) - v_0(t) \tag{4.13c}$$

The valves of the VSC are controlled by the digital switching signals $sw_a(t)$, $sw_b(t)$, and $sw_c(t)$. If the $sw_a(t)$ is equal to 1 the phase potential of the phase-leg a is equal to $+V_{dc}/2$, and if the $sw_a(t)$ is equal to −1 the phase potential of the phase-leg a is equal to $-V_{dc}/2$. The reference voltage signals to the modulator consists of three phase voltages and are written as [27]:

$$u_a^*(t) = \sqrt{2/3}\, U^* \cos(\omega^* t + \phi^*) \tag{4.14a}$$

$$u_b^*(t) = \sqrt{2/3}\, U^* \cos(\omega^* t + \phi^* - 2\pi/3) \tag{4.14b}$$

$$u_c^*(t) = \sqrt{2/3}\, U^* \cos(\omega^* t + \phi^* - 4\pi/3) \tag{4.14c}$$

where U^* and ω^* are the reference value of the phase-to-phase rms voltage and the reference angular frequency, respectively. The reference phase shift is denoted by ϕ^*.

This time, the carrier wave, also triangular, variable between $+V_{dc}(t)$ and $-V_{dc}(t)$ should be compared with three reference phase voltages, which forms a symmetrical three-phase system. Voltages on each phase are provided by the potential difference between the middle point of each half-bridge and the middle point of the source. This voltage and the current on each phase are determined in terms of the switches in conduction at each time instant.

Figure 4.33a shows the carriers and the three voltage references as sine wave ($v_{ra}(t)$, $v_{rb}(t)$, $v_{rc}(t)$) for a PWM VSC [30].

In this example, the frequency of the carrier—triangular wave signal—is nine times the fundamental frequency. Figure 4.33b shows the resulting voltage v_{a0} at the AC terminal a, with respect to a hypothetical midpoint M of the DC capacitor.

In comparison with two-level where are only one turn-on, turn-off per device per cycle (Figure 4.1), that is, with two square pulses per cycle, the waveform of Figure 4.33b is made up of nine square pulse cycles of varying width per main frequency cycle. The pulses are wider in the middle of each half sine wave compared to the ends of the half sine wave [30].

For a large modulation index m_f, the sinusoidal PWM three-phase inverter generates at its output a better voltage in terms of the harmonic content. The dominant harmonics are the high frequency ones, which are grouped around the carrier frequency and its multiples. Thus, the unwanted low frequency harmonics at the terminals of a six pulse converter disappear when using the sinusoidal PWM technique.

The modulation index m_f determines the predominant harmonic order in the modulated waveforms of pole voltages. These look like lateral strikes around the carrier frequency and its multiples. For the general case, the harmonic order can be [31]

- In the case of an even j, even stripes of harmonics appear in the pole voltage waveform. If j and k are both odd, superior order harmonics do not appear.

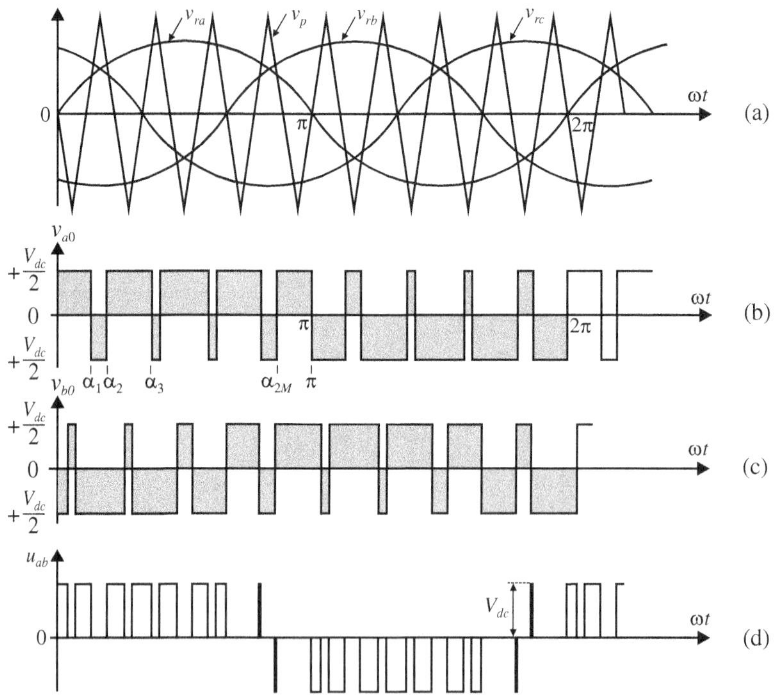

Figure 4.33 Operating of a PWM converter with the switching frequency of nine times the fundamental. *Source:* Hingorani and Gyugyi 2000 [30]. Adapted with permission of IEEE.

Thus, for $j = 2$, harmonics of order $2m_f \pm 1$, $2m_f \pm 3$, $2m_f \pm 5$, etc. appear and their magnitude drops very fast as the lateral stripe order increases.

- For odd j values, lateral stripes of odd order appear in the harmonic spectrum. If j and k are both odd, harmonics of superior order do not appear. Thus, for $j = 1$, harmonics of order $m_f \pm 2$, $m_f \pm 4$, etc. appear.
- For odd multiples of m_f, other harmonics can be seen in the pole voltage waveform, as those of $2m_f \pm 3$ order, known as triplen harmonics. If m_f is a multiple of three, these harmonics do not apply on three-phase load.

When using sinusoidal reference waveforms, the maximum output voltage amplitude at the VSC terminals occurs when the reference voltage in the modulator has the same amplitude as the triangular wave (carrier wave). The maximum output voltage amplitude (rms value) of the fundamental phase-to-phase voltage is [27]

$$U_{\sin(1)} = \frac{V_{dc}}{2}\sqrt{\frac{3}{2}} \approx 0.61 V_{dc} \tag{4.15}$$

In the example of Figure 4.33b the angles 0, α_1, α_2, α_3, …, α_{2M}, π define the switching time. The amplitude of harmonics present in the waveform is given by

$$a_n = \frac{2V_{dc}}{n\pi}\left[1 + \sum_{i=1}^{i=2M} (-1)^i \cos n\alpha_i\right]$$

If the triangular waveform (the carrier) frequency is an odd integer multiple of the fundamental frequency, the waveform shown in Figure 4.33b does not contain even harmonics.

In a three-phase bridge circuit, all triplen harmonics, that is, third, ninth, ... are canceled from the phase-to-phase voltages. Also, if the triangular waveform frequency is a multiple of three, even the harmonics of the order similar to that of the triangular waveform frequency are canceled in the phase-to-phase and phase-to-floating neutral voltages (three-phase converter considered) [30].

In order to minimize the harmonic level, a synchronized carrier wave should be used, where m_f is an odd integer. Furthermore, in order to eliminate zero-sequence harmonic voltages, m_f should be a multiple of three. As shown, the switching frequency voltage component disappears when the three-phase modulator uses the same carrier-wave for all three comparators.

4.2.2 Modulation Techniques for Multilevel Converters

The multilevel power electronic converters have been continuously developed to achieve higher ratings suitable for high-power applications, whereas the number of power electronic devices has inherently increased. The multilevel converters achieved extra degrees of freedom becoming more flexible in operation, at the same time with increased complexity of the modulation algorithm. Various modulation algorithms/methods have been adapted or developed to fit the converter topology, each one having unique advantages and drawbacks. A classification of the most common modulation methods for multilevel inverters was made in [7]. The modulation algorithms shown in Figure 4.34 are classified in terms of the average switching frequency. Here, the border separation between low and high switching frequency is 1 kHz.

A classification of the modulation algorithms can also be made in terms of the strategy applied, as presented below.

4.2.2.1 PWM Algorithms for Multilevel Converters

The PWM technique is the most preferred way of driving modern semiconductor switching devices.

(i) *Phase-shifted carrier pulse width modulation* (PS-PWM) is a derivative of the traditional PWM technique adapted for FC and cascaded H-bridge (CHB) converters. Since each FC device is a two-level converter, and each CHB device is a three-level inverter, the traditional bipolar and unipolar PWM techniques can be used [7]. The PS-PWM technique is mainly applied to multidevice topologies since each carrier can be customized for each switching device. This technique consists of introducing a phase-shift between the carriers applied to contiguous switching devices obtaining thereby a phase-shifted switching pattern. This helps reducing the harmonic level. The greatest reduction is achieved for phase shifts of 180° or 360°/N, where N is the number of switching devices.

(ii) *Level-shifted carrier pulse width modulation* is a method based on amplitude shifts between carriers and is an extension of the bipolar PWM to the multilevel

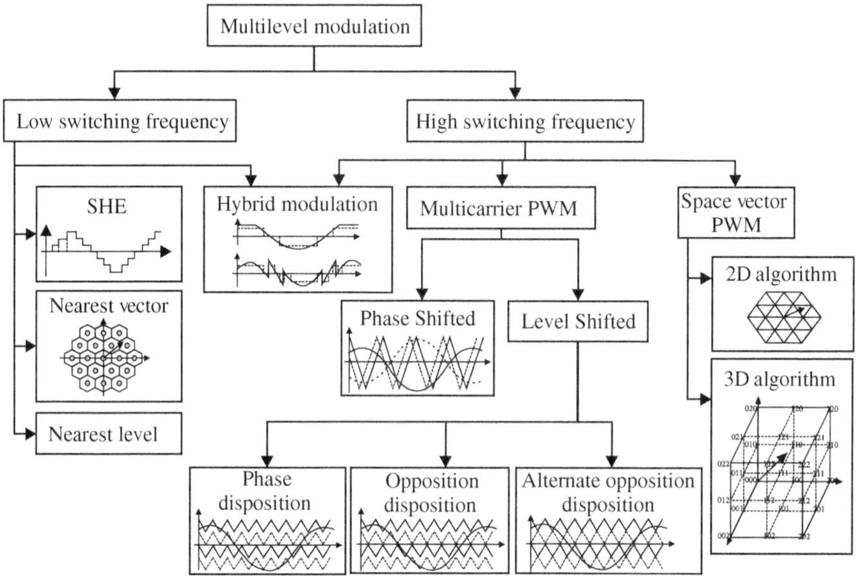

Figure 4.34 Classification of multilevel converter modulation techniques. *Source:* Rodriguez et al. 2009 [7]. Adapted with permission of IEEE.

inverters [32]. The carrier signal, which is associated with a specific voltage, is compared with the reference. When the reference is over one carrier, the corresponding voltage is generated. In a multilevel inverter, $m - 1$ carrier signals are used for the m levels. Each carrier is set between two voltage levels, hence the name "level shifted." The switching devices will be used only when the corresponding level is generated, producing an uneven power distortion and switching conditions between switching devices. This will avoid the current harmonic cancelation at the input and increase the input current distortion. For these reasons, this method is not appropriate for FC and CHB converters. The carriers can be arranged in vertical shifts: (a) with all the signals in phase with each other, called *phase disposition* (PD-PWD); (b) with all the positive carriers in phase with each other and in opposite phase with the negative carriers, known as *phase opposition disposition* (POD-PWM); (c) in *alternate phase opposition disposition* (APOD-PWM), which is obtained by alternating the phase between adjacent carriers [7].

(iii) *Hybrid PWM Modulation* (H-PWM) was developed mainly for CHB converters with unequal DC sources and combines the advantages of unipolar (three-level) and bipolar (two-level) switching schemes. Higher frequency commutation causes higher power losses. Thereby, instead of using high-frequency carrier-based PWM methods in all the switching devices, the power devices with higher DC sources are operated with square waveform patens, switched at low frequency, whereas the power devices with smaller DC sources are operated with unipolar PWM (at higher frequency).

4.2.2.2 Space Vector Modulation Algorithms

The space vector modulation (SVM), initially proposed in [33], is one of the most popular modulation approached for two-level converters and is increasingly applied to multilevel converters. It offers significant flexibility to optimize switching waveforms, and it is well suited for digital implementation. The control of switching devices in the SVM strategy is performed using the PWM technique, but the switching times and switching sequence are determined based on a three-phase vector representation of the reference variables and the inverter switching state. The vectorial representation of the three-phase systems originates in the transformations initially proposed by Park [34] and Clarke.

Depending on the desired converter output voltage, reference voltage vectors are first generated, then mapped into the $\alpha\beta\gamma$ state space, instead of using the *abc* coordinates. The number of states depends on the number of switching devices in the converter. In multilevel converters, it is possible to generate as many output levels of voltage as many levels the converter has. For the NPC, for instance, which has three phases and three output levels or switching states, $3^3 = 27$ possible combinations are achieved, of which only 19 are different and eight are redundant. For a three-phase N-level converter, N^3 states are obtained [7].

The SVM algorithms are classified into two categories, that is, for balanced and unbalanced systems.

- **(i)** *SVM for multilevel balanced converters.* The $\alpha\beta\gamma$ representation offers an interesting information about the zero-sequence component of both currents and voltages (proportional to the γ coordinate). In balanced converters topologies, no zero-sequence component exists and thus the γ-axis is not necessary. Thereby, the reference vector is represented into the bidimensional state space [35]. The closest three state-space vectors to the reference vector are combined in such a way that their time average equals the reference vector.
- **(ii)** *SVM for multilevel unbalanced converters.* In power converters with neutral connection (usually called also unbalanced systems), a zero-sequence component appears and therefore the $\alpha\beta\gamma$ representation, which is a three-dimensional representation, is used [36, 37].

4.2.2.3 Other Modulation and Control Algorithms for Multilevel Converters

Various other topologies and applications of power converters have been developed. A classification of the main algorithms is as follows:

- **(i)** The SHE technique was developed to fit the harmonics requirements for very high power applications controlled with low frequency switching algorithms [21, 22]. In the case of a carrier- based PWM technique, reduction in the carrier frequency results in low-order harmonics in the output voltage. In order to eliminate the undesired low-order harmonics, a few (generally from three to seven) switching angles per quarter fundamental cycle are predefined and precalculated via Fourier analysis.

(ii) *Nearest vector control* (NVC), also known as space vector control, is a low switching frequency control employed in multilevel converters. The method takes advantage of the high number of voltage vectors that can be generated by multilevel converters and approximate the reference vector with the nearest voltage vector that can be generated in the αβ plane [7, 38]. In this way, the number of commutations is reduced, and thus the switching losses are reduced as well. This method does not eliminate the low-order harmonics, but the larger the number of levels the smaller the low-order harmonics level.

(iii) *Nearest level control* (NLC), also known as the *round method*, is an improved version of the NVC method and is efficiently employed to multilevel converters [7]. It has the same principle as for the NVC, except that the nearest voltage level is used instead of using the space level. In other words, the reference vector is approximated with the nearest voltage level that can be generated by the inverter. This is an advantage as against the NVC because it is much easy to approximate a single existing value rather than calculating an average value obtained by a modulation sequence of three vectors. Thereby, NLC is not a modulation method.

4.3 DC/AC CONVERTER ANALYSIS

The detailed models of power-electronics components and modules (where the switching of all diodes and transistors is taken into account) may be readily implemented using a number of commercially available digital programs and/or simulators (EMTP, PSCAD/EMTPDC, EUROSTAG, Simulink Dynamic System Simulation Software, etc.).

The use of detailed switching models often leads to significant increase of the required computing time, which in turn often limits the size of the system that can be practically simulated. At the same time, the switching models are also discontinuous and therefore difficult to use for extracting the small-signal characteristics of various modules for the system-level analysis.

The above challenges have led to development of the so-called *dynamic average-value models* (AVMs), which approximate the original system by "neglecting" or "averaging" the effect of fast switching within a prototypical switching interval [39].

4.3.1 Operation Modes of the Switched-Inductor Cell

The instantaneous "on"/"off" conducting state of the active and passive switches, that is, transistors, thyristors, and diodes, determines the topology of each switching cell in any converter. When the converter operates in steady state, a sequence of topologies will become repetitive within each switching interval defining a certain switching pattern. This repetitive pattern of topologies in turn defines the operating mode of a given switching cell. The prototypical switching interval varies for different converters and is the basis for the averaging window when developing AVMs [39].

For instance, for the switched-inductor cell shown in Figure 4.35a, three topological states can be defined: (i) the transistor is "on" and the diode is "off"; (ii) the transistor is "off" and the diode is "on"; (iii) the transistor is "off" and the diode is "off."

The typical inductor current waveforms for this switching cell are as follows:

- In *continuous conduction mode* (CCM), as shown in Figure 4.35b, each switching interval T_S is divided into two subintervals d_1T_S and d_2T_S, corresponding to the topological states (i) and (ii), where d_1 and d_2 are the so-called relative duty cycles, which are defined such that $T_S = d_1T_S + d_2T_S$, with $d_1 + d_2 = 1$.
- In *discontinuous conduction mode* (DCM), as shown in Figure 4.35c, an additional subinterval d_3T_S is defined, in which both thyristor and diode are off and the current becomes zero. Thereby, the switching interval is calculated by $T_S = d_1T_S + d_2T_S + d_3T_S$, with $d_1 + d_2 + d_3 = 1$.

The overall prototypical switching interval T_S is determined by the PWM strategy and the switching frequency.

In general, an operational mode is a function of loading conditions. Changes in load conditions might lead to a change in the topologies and hence the mode of operation. In more complicated configurations such as 12-pulse or 18-pulse converters, the number and complexity of operational modes significantly increase making it more challenging to develop the AVMs.

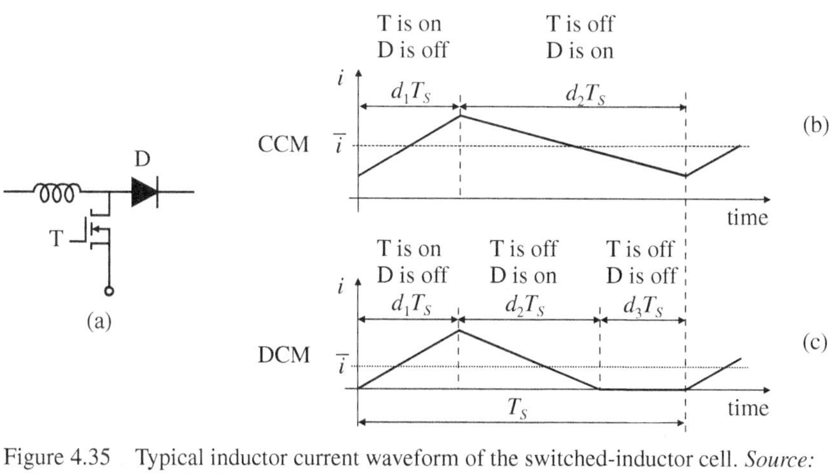

Figure 4.35 Typical inductor current waveform of the switched-inductor cell. *Source:* Chiniforoush et al. 2010 [39]. Reproduced with permission of IEEE.

4.3.2 Ideal DC/AC Half-Bridge Converter

SIMPLIFIED POWER CIRCUIT DIAGRAM. The single-phase half-bridge VSC is the simplest and basic steady-state converter. As it is shown in the simplified scheme from Figure 4.36a, it consists of two switch cells K_1 and K_4, each one containing one switching device S and one diode D.

Nodes M and N in Figure 4.36a denote the DC-side terminals of the half-bridge converter, and the node t specifies the AC-side terminal. On the DC side, the half-bridge converter is connected to two identical capacitors, each one supplying the voltage $V_{dc}/2$. The capacitors are assumed sufficiency large to ensure a stiff DC source. A reference point, denoted by 0, is created at the midpoint between the two capacitors to provide a return path (the negative side) for the current. In case of a three-phase converter, the midpoint is removed. The AC side is represented by a voltage source $\underline{V}_s$ and an impedance $\underline{Z}$ that includes also the smoothing reactor of the converter [5].

In a real case, the converter may operate either as an inverter or as a rectifier and is capable of exchanging leading or lagging reactive power with the AC system. The ideal case is discussed in this section, and the converter is considered to operate as the inverter only.

As for any inverter, the positive direction of active power, P_{dc}, is considered from DC to AC. Although power losses in the converter are neglected, let us denote by P_t the power delivered by the converter at the AC side, and by P_s the power delivered to the AC-side voltage source. In each transistor, the current is positive when it flows from the collector to the emitter, whereas the positive current in a diode flows from the anode to the cathode. Thus, the current through the upper switch cell is positive and is given by $i_M = i_{S1} - i_{D1}$, whereas the current through the lower switch cell is positive and is given by $i_N = -(i_{S4} - i_{D4})$.

CONVERTER OPERATION AND WAVEFORMS. In order to examine the switching waveforms of the half-bridge, we consider the following simplifying assumptions [5]

- Each transistor or diode acts as a short circuit in its conduction state; each transistor or diode acts as an open circuit in its blocking state;
- The transistors have no turn-off tailing current, and the diodes have no turn-off reverse recovery current;
- At any instant of time, during normal operation, only one transistor is in conduction, the other one being blocked; transition of a transistor from a conduction state to a blocking state, and vice versa, takes place instantly; thus, the signal controls for switching on or off the devices S_1 and S_4 are complementary in order to avoid destruction of the bridge;
- Positive current flows from the negative terminal to the positive terminal of the capacitors, and negative current flows from the positive terminal to the negative terminal of the capacitors.
- Since the converter operation is different for positive and negative AC-side currents, we study each case separately.

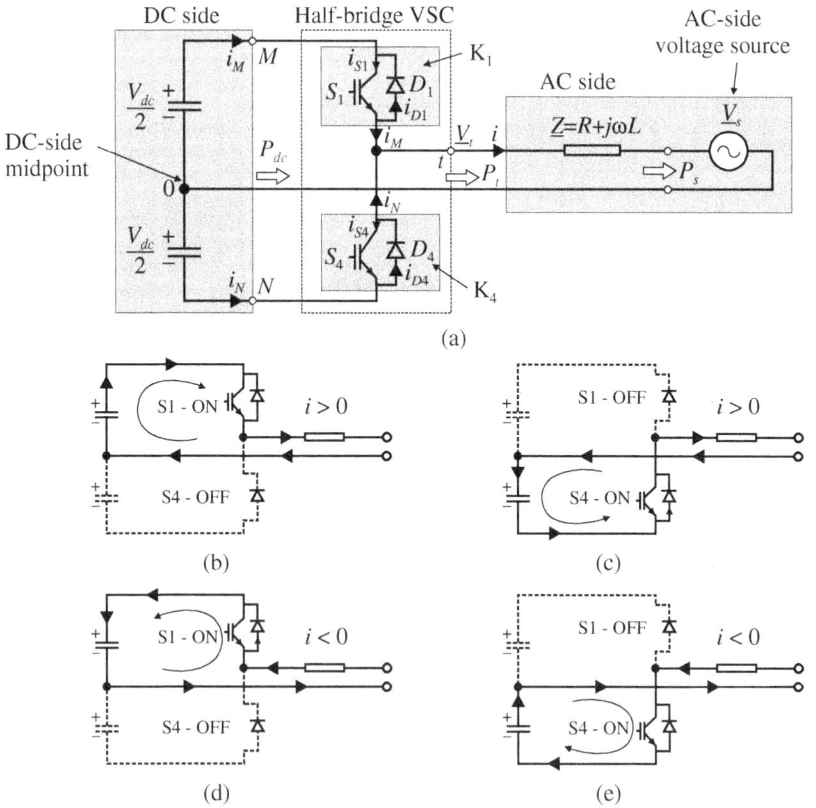

Figure 4.36 An ideal half-bridge converter with ideal switches: (a) simplified scheme; (b)–(e) states and current paths [5].

(i) *Converter waveforms for positive AC-side current.* Let us consider the scheme of an ideal half-bridge converter from Figure 4.36a that handle positive current i on the AC side, and take into account the converter operation characteristics presented above.

Assuming a time at which S_1 conducts (switching function is $sw_1 = 1$) and S_4 is blocked ($sw_4 = 0$), the current follows the path indicated in Figure 4.36b. A voltage $V_t = V_M = V_{dc}/2$ is applied at the converter terminals. Diode D_4 is not conducting because it is reverse biased. Therefore, the current i flows through S_1 only, and the diodes play no role in the converter operation (Figure 4.37).

Considering a time at which S_1 is blocked ($sw_1 = 0$) and S_4 conducts ($sw_4 = 1$), the current follows the path indicated in Figure 4.36c. The converter voltage is now $V_t = V_N = -V_{dc}/2$. Diode D_1 is not conducting because the current through it cannot be negative. In this case, the current i flows through D_4 only (Figure 4.37).

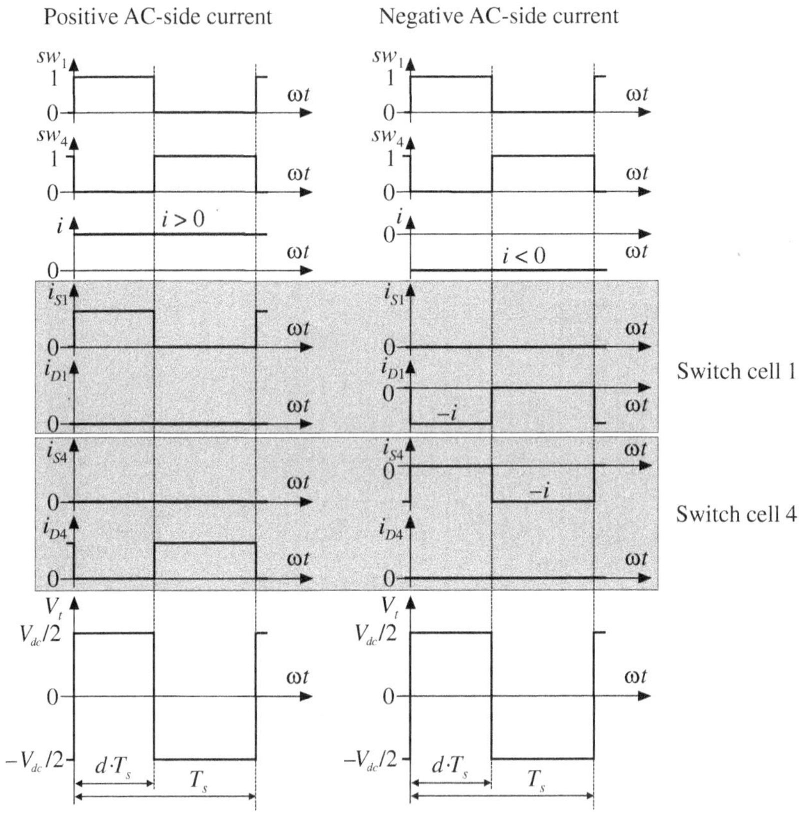

Figure 4.37 Half-bridge converter waveforms for positive and negative AC-side current [5].

(ii) *Converter waveforms for negative AC-side current.* When the AC-side current becomes negative, two cases are identified for the operation of the ideal half-bridge inverter, similar to the AC-side current case.

While for the AC-side current case, the elements D_1 and S_4 played no role in the current flow; in this case the roles are changed. When S_1 conducts ($sw_1 = 1$) and S_4 is blocked ($sw_4 = 0$), while the voltage $V_t = V_M = V_{dc}/2$ is applied to the converter circuit, the current flows through D_1 only (Figure 4.36d). When S_1 is blocked ($sw_1 = 0$) and S_4 conducts ($sw_4 = 1$), while the voltage $V_t = V_N = -V_{dc}/2$ is applied, the current flows through S_4 only (Figure 4.36e).

The half-bridge converter waveforms for both positive AC-side current and negative AC-side current are illustrated in Figure 4.37. The fraction of the switching period T_s during which S_1 conducts is called the *duty ratio* and is denoted by d. The duty ratio takes any value between 0 and 1. When S_1 conducts, the voltage applied to the converter circuit is $V_t = V_M = V_{dc}/2$.

Converter Switched Model. As explained above, the current flowing through the switch cell depends on the direction of the AC-side current. The polarity of the terminal voltage V_t is determined by the switching strategy and is independent

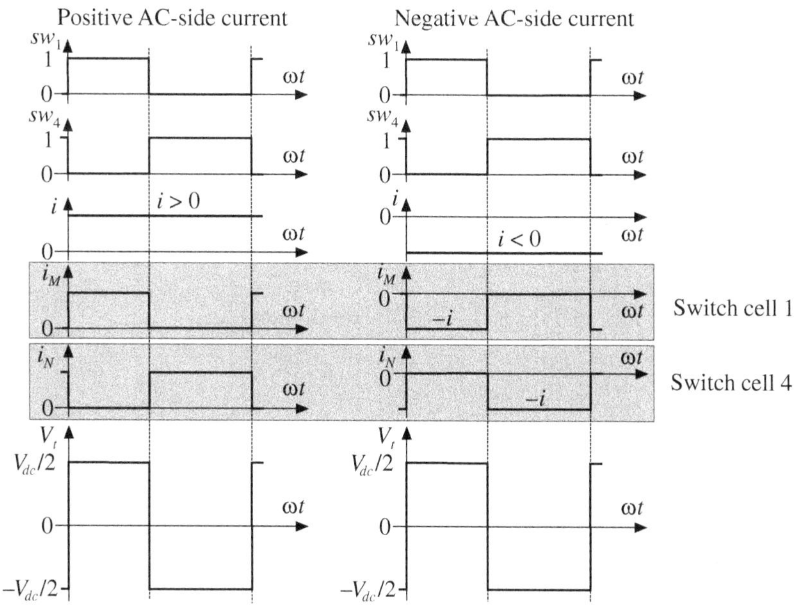

Figure 4.38 Half-bridge converter switching cases [5].

of the polarity of the current i. However, since $i_M = i_{S1} - i_{D1}$ and $i_N = -\, i_{S4} + i_{D4}$, the waveform of the switch cell current is independent of the polarity of i (Figure 4.38). Furthermore, the waveform of the AC-side terminal voltage V_t is independent of the polarity of i and is uniquely determined by the switching functions [5].

Since, at any time, the switching position of the cells 1 and 4 is complementary, we can write

$$sw_1(t) + sw_4(t) = 1 \tag{4.16}$$

The terminal voltage is then given by

$$V_t(t) = (V_{dc}/2)sw_1(t) - (V_{dc}/2)sw_4(t) \tag{4.17}$$

and the currents through the converter circuits are

$$i_M(t) = i(t) \cdot sw_1(t) \tag{4.18a}$$

$$i_N(t) = i(t) \cdot sw_4(t) \tag{4.18b}$$

The active power produced on the DC side is

$$P_{dc}(t) = V_M(t)i_M(t) + V_N(t)i_N(t) = \frac{V_{dc}}{2}[sw_1(t) - sw_4(t)] \cdot i \tag{4.19}$$

The active power transferred by the converter to the AC side is

$$P_t(t) = V_t(t)i(t) = P_{dc}(t) \tag{4.20}$$

then at the source we have

$$P_s(t) = V_s i \tag{4.21}$$

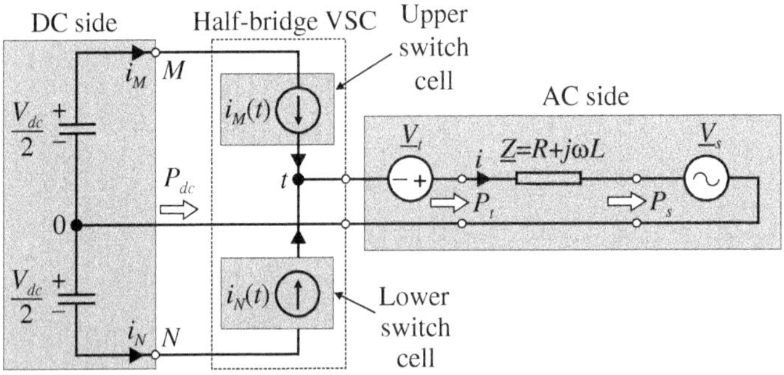

Figure 4.39 Switched equivalent circuit of the ideal half-bridge converter of Figure 4.36.

The power losses in the converter are simply given by

$$P_{loss} = P_{dc} - P_t \tag{4.22}$$

which is zero since we have considered an ideal converter.

Based on equations (4.16)–(4.18), a switched equivalent circuit can be built for the half-bridge converter (Figure 4.39). This model helps computing the instantaneous values of the current and voltage variables in steady-state and dynamic behavior of the converter.

Converter Averaged Model [5]. Starting from the conclusions withdrawn in the above sections, an average model can also be designed to calculate the instantaneous values of the current and voltages. In practice, however, the dynamics of the average values are of greater interest than that of the instantaneous values. In the average model, a modulating signal value/signal, m_a, is used to describe the converter dynamics.

The dynamics of the AC-side current, i, in a half-bridge converter may be described by [5]

$$L\frac{d}{dt}i(t) + R \cdot i(t) = V_t(t) - V_s \tag{4.23}$$

Since $V_t(t)$ is a periodic function with period T_s, it can be developed by Fourier series expansion [29]:

$$V_t(t) = \frac{1}{T_s}\int_0^{T_s} V_t(\tau)d\tau + \sum_{n=1}^{n=+\infty} [a_n \cos(n\omega_s t) + b_n \sin(n\omega_s t)] \tag{4.24}$$

where n is the harmonic order,

$$\omega_s = 2\pi/T_s$$

$$a_n = \frac{2}{T_s}\int_0^{T_s} V_t(\tau)\cos(n\omega_s\tau)d\tau \tag{4.25a}$$

$$b_n = \frac{2}{T_s}\int_0^{T_s} V_t(\tau)\sin(n\omega_s\tau)d\tau \tag{4.25b}$$

Substituting for $V_t(t)$ from (4.24) in (4.23), it results

$$L\frac{di}{dt} + R \cdot i = \underbrace{\left(\frac{1}{T_s}\int_0^{T_s} V_t(\tau)d\tau - V_s\right)}_{Constant\ \text{(DC) component}} + \underbrace{\sum_{n=1}^{n=+\infty} [a_n \cos(n\omega_s t) + b_n \sin(n\omega_s t)]}_{Periodic\ \text{component}} \tag{4.26}$$

Equation (4.26) describes a series RL circuit supplied from a voltage source that produces the output current i. The supply voltage consists of a constant (DC) component and a periodic component. The response of the circuit to the constant component is

$$\bar{i}(t) \equiv L\frac{d\bar{i}}{dt} + R \cdot \bar{i} = \frac{1}{T_s}\int_0^{T_s} V_t(\tau)d\tau - V_s \tag{4.27}$$

and the response to the periodic component is

$$\tilde{i}(t) \equiv L\frac{d\tilde{i}}{dt} + R \cdot \tilde{i} = \sum_{n=1}^{n=+\infty} [a_n \cos(n\omega_s t) + b_n \sin(n\omega_s t)] \tag{4.28}$$

Using the superposition principle, we achieve

$$i(t) = \bar{i}(t) + \tilde{i}(t) \tag{4.29}$$

The periodic component can be seen as a ripple in the DC component, which depends on the inductance L and the rate of change of the current. For high frequencies of the periodic component, this one becomes negligible.

For the purpose of system-level analysis, instead of looking at the instantaneous values of currents and voltages that contain ripple due to the switching, it is useful to consider the *dynamic average value* that is defined over the length of a switching interval. The averaging operator is defined as [40, 41]

$$\bar{x}(t) = \frac{1}{T_s}\int_{t-T_s}^{t} x(\tau)d\tau \tag{4.30}$$

where $x(t)$ is a variable (voltage $v(t)$ or current $i(t)$) and the overbar denotes its average (also known as the moving average). Thus, (4.27) can also be derived by applying the averaging operator (4.30) to both sides of (4.23).

Since the duty ratio changes in time, the modulating waveforms of sw_1 and sw_4 will thereby adequately change in time. Moreover, the pattern of these waveforms may change from one cycle to the other and so is their average. The equation of averaging from (4.30) may be applied to the modulating waveform, and an initial condition for validity of this equation is that the frequency of the carrier waveform should be sufficiently larger than that of the modulating waveform, for instance 10 times larger. Applying equation (4.30) to $sw_1(t)$ and $sw_4(t)$ gives [5]

$$\overline{sw_1}(t) = d \tag{4.31a}$$

$$\overline{sw_4}(t) = 1 - d \tag{4.31b}$$

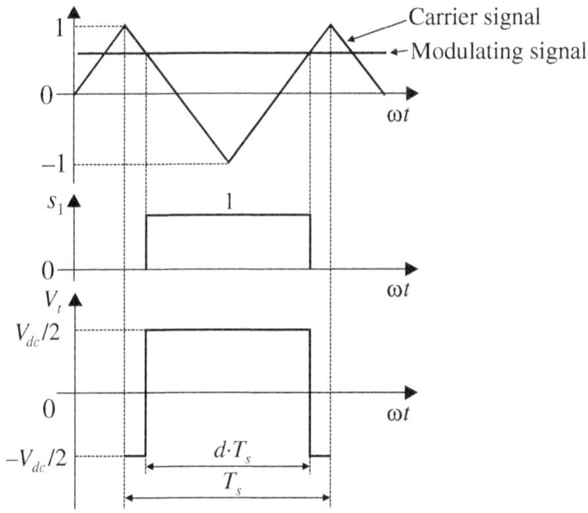

Figure 4.40 Example of generation of the switching signal with a desired duty ratio [5].

The variables $\bar{i}$ and $\bar{V}_{dc}$ can be assumed constant over one switching cycle if the modulating signal does not chance, that is the carrier frequency is higher than the modulating signal frequency (Figure 4.40).

Under these conditions, substituting for $\overline{sw}_1(t)$ and $\overline{sw}_4(t)$ from (equations 4.31a and 4.31b) in equations (4.17)–(4.22) we achieve

$$\bar{V}_t = \frac{V_{dc}}{2}(2d - 1) \tag{4.32}$$

$$\bar{i}_M = d \cdot i \tag{4.33a}$$

$$\bar{i}_N = (1 - d) \cdot i \tag{4.33b}$$

$$\bar{P}_{dc} = \frac{V_{dc}}{2}(2d - 1) \cdot i \tag{4.34a}$$

$$\bar{P}_t = \frac{V_{dc}}{2}(2d - 1) \cdot i \tag{4.34b}$$

$$\bar{P}_s = V_s \cdot i \tag{4.35}$$

$$\bar{P}_{loss} = \bar{P}_{dc} - \bar{P}_t \equiv 0 \tag{4.36}$$

When the PWM strategy is applied, the relationship between the modulating ratio and the duty ratio, that is $m_a = (2d - 1)$, or $d = (m_a + 1)/2$. As it results from Figure 4.40, when m_a changes from −1 to 1 and maintains constant the switching period, d changes linearly from 0 to 1. Substituting for d in (4.32) to (4.34a and 4.34b), it results

$$\bar{V}_t = \frac{V_{dc}}{2} m_a \tag{4.37}$$

$$\bar{i}_M = \left(\frac{1 + m_a}{2}\right) \cdot i \tag{4.38a}$$

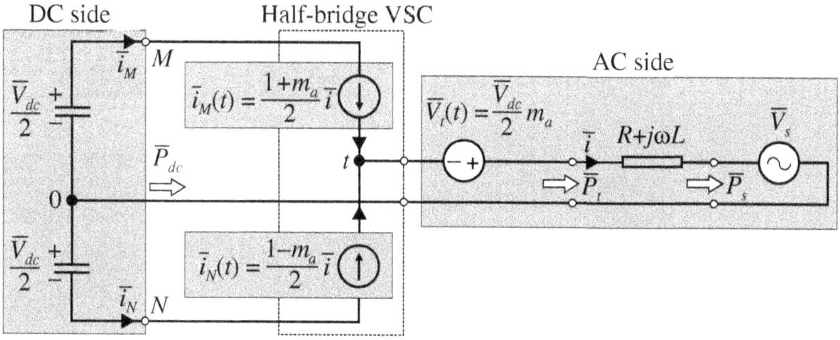

Figure 4.41 Averaged equivalent circuit of the half-bridge converter.

$$\bar{i}_N = \left(\frac{1 - m_a}{2}\right) \cdot i \tag{4.38b}$$

$$\bar{P}_{dc} = m_a \frac{V_{dc}}{2} \cdot i \tag{4.39a}$$

$$\bar{P}_t = m_a \frac{V_{dc}}{2} \cdot i \tag{4.39b}$$

The averaged AC-side terminal voltage $\bar{V}_t$ from equation (4.37) is more easy to analyze. When m_a changes from −1 to +1, $\bar{V}_t$ changes linearly from $-V_{dc}/2$ to $+V_{dc}/2$; zero average voltage is achieved for $m_a = 0$.

The averaged equivalent circuit of the half-bridge converter (Figure 4.41) can be developed starting from the switched equivalent circuit from Figure 4.39, to which the new relationships are applied [5].

An extended average model of the half-bridge converter can also be developed when nonideal characteristics of the switches need to be taken into account (see section 2.6 from [5]).

4.3.3 Averaging Models[3]

Analysis of the average inductor current in DC–DC converters, such as that shown in Figure 4.36, reveals that this is constant in steady state, while slow dynamics is observed during the transients. The idea represented by equation (4.30) can be extended such that the resultant averaged model captures also higher order dynamics, such as those of harmonics; this is referred to as the *extended* or *generalized averaging* [42, 43]. The concept of averaging can also be extended to DC–AC converters. However, simple averaging of the AC variables using (4.30) over the switching interval will not yield the desired result. Instead, the AC-side variables first have to be transformed using an appropriate synchronously rotating reference frame [44].

Let us consider the PWM-controlled two- and three-level voltage source inverter supplying an inductive load. The phase-to-ground voltage v_{ag} in a two-level converter takes two values (levels) only, that is 0 and V_{dc} (Figure 4.42a), whereas in

[3] Material in this section is reproduced by kind permission of IEEE from ref. [39]. (See Chiniforoush et al. 2010 [39]).

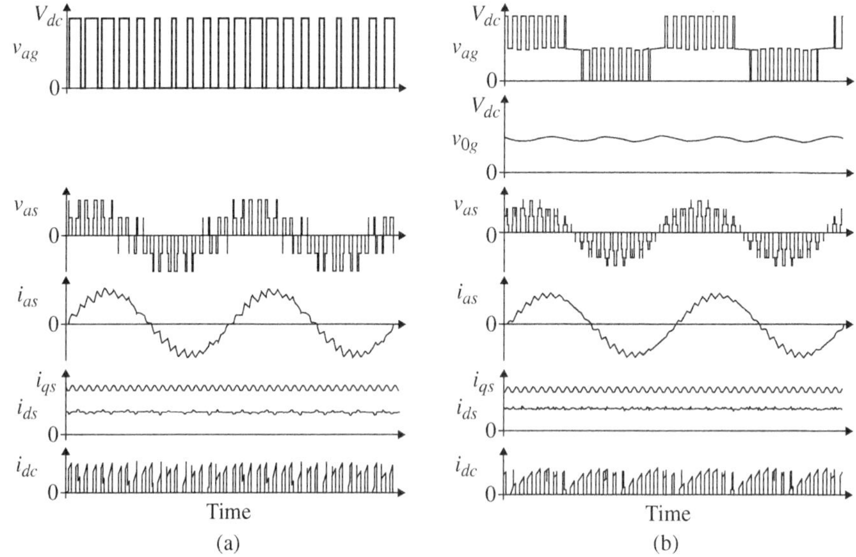

Figure 4.42 Typical voltage and current waveforms of a voltage source inverter (VSI) converter feeding an inductive load: (a) two level; (b) three level. *Source:* Chiniforoush et al. 2010 [39]. Reproduced with permission of IEEE.

a three-level converter it takes three values (levels): that is, 0, $v_{0g}(\approx V_{dc})/2$ and V_{dc} (Figure 4.42b). In both cases, v_{as} represents the phase voltage on the AC load side.

The AC load current, i_{as}, for both converters, consists of the fundamental sinusoidal component with the superimposed high frequency switching ripples (relatively low content of low-frequency harmonics, which depends on many factors, including the PWM strategy and the switching frequency). When transformed into *dq* frame, the i_{qs} and i_{ds} components are DC (constant) terms, but also with superimposed high-frequency ripple with the same switching interval T_S. The constant value in steady state of the *qd* variables allows for using the averaging given by (4.30) in the same way as the variables in the DC–DC converters. The objective of AVM is to replace the discontinuous switching cells with continuous blocks that represent the averaged behavior of the switching cell within a prototypical switching interval (Figure 4.35). However, obtaining the AVMs generally requires detailed analysis of the switching cells and accurate averaging of the converter waveforms.

The variables on the DC side, for example, i_{dc}, will also contain some ripple due to switching, which often necessitates the use of large capacitors on the DC link, and the averaging concept defined by (4.30) may be applied here directly [39].

4.3.3.1 Circuit/Switch Averaging of DC–DC Converters

AVM, also referred to as averaged-switch modeling, of PWM DC–DC converters, can be done by directly averaging the switching cells, such as the switched-inductor cell depicted in Figure 4.43a [39, 45]. The circuit averaging technique allows for equivalencing the active/passive switch pair, of the switched-inductor cell represented in

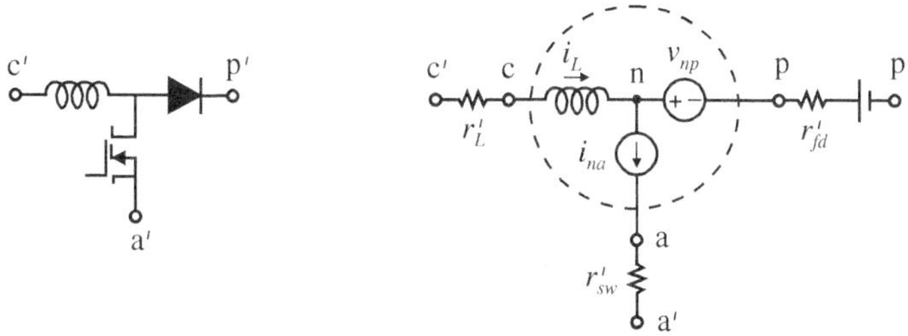

(a) (b)

Figure 4.43 Switched-inductor cell and its averaged equivalent circuit model for nonideal cell that includes dependent sources and parasitics. *Source:* Chiniforoush et al. 2010 [39]. Reproduced with permission of IEEE.

Figure 4.43a, by dependent sources that are functions of the control duty cycle d_1 and the averaged values of the cell's terminal variables, as shown in Figure 4.43b. Such models can be easily obtained analytically for CCM if parasitics are neglected [39, 45]. For operation in DCM, the duty ratio constraint d_2 is also derived as a function of the terminal variables and the averaged inductor current.

Based on the accuracy of portraying the high-frequency dynamics, the reduced-order and the full-order [46] models have been developed in the literature for ideal switching cells. The encircled part (*cna*) shown in Figure 4.43b is the general full-order equivalent cell for the ideal switched-inductor cell [46]. Significant effort has also been done to include the effects of conduction losses. In general, the modified parasitics are placed around the ideal components resulting in the averaged-switch model for the nonideal cell [47]. The values of the parasitics defined in Figure 4.43b are modified to account for the mode of operation and are generally different from the actual values of the parasitics.

4.3.3.2 State-Space Averaging of DC–DC Converters

The state-space averaging method is a small-signal analysis of the PWM DC–DC converters, in which the state equations are obtained for each topology within one switching interval T_S [39, 45] (see Figure 4.35). For a specified simulation time, the average model is achieved through weighted sum of the state-space equations for all subintervals. A corrected full-order state-space averaged model for the ideal converter in DCM is described in [46] under the form

$$\dot{\bar{\mathbf{x}}} = [d_1\mathbf{A}_1 + d_2\mathbf{A}_2 + (1 - d_1 - d_2)\mathbf{A}_3]\mathbf{M}\bar{\mathbf{x}} + [d_1\mathbf{B}_1 + d_2\mathbf{B}_2 + (1 - d_1 - d_2)\mathbf{B}_3]\mathbf{u} \quad (4.40)$$

where $\bar{\mathbf{x}}$ denotes the vector of the average values of the inductor currents (as well as the capacitor voltages), **A** and **B** are the state matrices and input vectors of the circuit, and **u** is an input voltage vector. The so-called correction matrix **M**, which is a diagonal matrix, was introduced in order to make the model work properly in the DCM operation. For the CCM operation, the matrix **M** is simply set to identity, whereas for the DCM operation the elements of **M** depend on the duty cycle d_1, which is determined externally, and the duty ratio constraint d_2, which is algebraically dependent on other system variables. However, the model described by equation (4.40) does not hold and the peak value of inductor current also changes if parasitics are included [45,46]. For a nonideal case, it is then very difficult to evaluate the correction matrix analytically. Thereby, the averaged state-space model was written in a more general form, as given in [48]:

$$\dot{\bar{\mathbf{x}}} = \left(\sum_{k=1}^{3} d_k \mathbf{A}_k \mathbf{W}_k \right) \bar{\mathbf{x}} + \left(\sum_{k=1}^{3} d_k \mathbf{B}_k \right) \mathbf{u} \tag{4.41}$$

to account for arbitrary parasitics and possible nonlinearity of the current/voltage waveforms. The formulation (4.41) does not rely on small- and/or linear-ripple approximation and is therefore more general than (4.40). However, analytical derivation of the weighing-correction matrices $\mathbf{W}_k(d_1, \overline{i_L})$ may be challenging, and as an alternative methodology these functions may be calculated numerically for a desired range of operating conditions [48].

4.3.3.3 *AVM* of DC–AC Converters

To demonstrate dynamic average modeling of DC–AC converters, let us first consider the three-phase full-bridge converter [39].

In many power applications, the efficiency of converters is quite high and therefore the conduction losses are often neglected in analysis of transients.

As described earlier, the AC variables must be expressed in an appropriate reference frame. Typically, the so-called converter reference frame [44], in which the d-axis component of the voltage v_{ds}^c is identically zero, is chosen to facilitate the analysis. The relationship between the converter and arbitrary reference frames may be deduced according to Figure 4.44 as

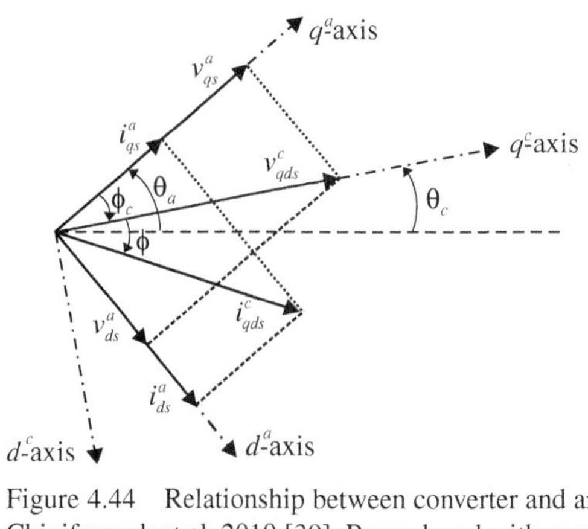

Figure 4.44 Relationship between converter and arbitrary reference frames. *Source:* Chiniforoush et al. 2010 [39]. Reproduced with permission of IEEE.

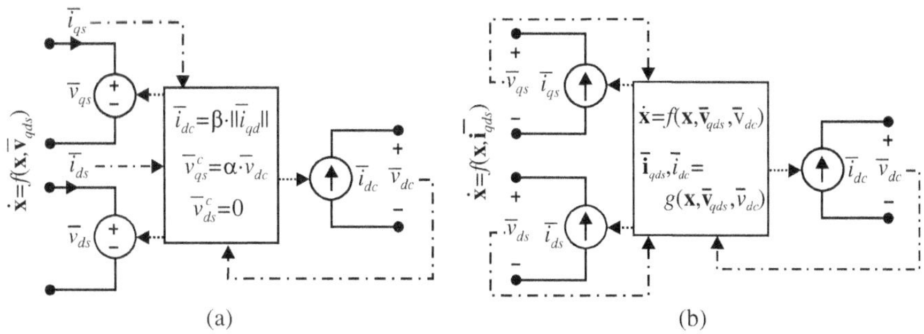

Figure 4.45 AVM for the voltage source inverter: (a) using algebraic–parametric functions relating AC and DC variables; (b) using a state model for representing effective commutating and/or DC – filter inductances. *Source:* Chiniforoush et al. 2010 [39]. Reproduced with permission of IEEE.

$$\begin{bmatrix} v_{qs}^{c} \\ 0 \end{bmatrix} = \begin{bmatrix} \cos(\phi_c) & \sin(\phi_c) \\ -\sin(\phi_c) & \cos(\phi_c) \end{bmatrix} \begin{bmatrix} v_{qs}^{a} \\ v_{ds}^{a} \end{bmatrix} \tag{4.42}$$

Depending on the converter topology and application of the required model, the dynamic AVM can be structured in the form of equivalent circuits shown in Figure 4.45.

Since the switching cell of Figure 4.35a does not contain any energy storage components, the voltages and currents on the AC side can be related to the DC-side variables through functions that are purely algebraic [49]. This approach lands itself on implementation depicted in Figure 4.45a.

In particular, the voltages on the AC side and DC link are related as follows:

$$\|v_{qds}\| = \alpha(\cdot)\bar{v}_{dc} \tag{4.43}$$

where $\alpha(\cdot)$ is an algebraic function. The DC bus current may also be expressed in the following form:

$$\bar{i}_{dc} = \beta(\cdot)\|\bar{\mathbf{i}}_{qds}\| \tag{4.44}$$

where $\beta(\cdot)$ is another algebraic function. Both $\alpha(\cdot)$ and $\beta(\cdot)$ depend on the type of inverter and its operating/loading conditions.

Functions (4.43) and (4.44) can be established by applying the energy conservation principle to the converter cell. In particular, looking at the AC side, the three-phase power can be written as

$$P = \frac{3}{2}\|\bar{v}_{qds}\|\,\|\bar{\mathbf{i}}_{qds}\|\cos\phi \tag{4.45}$$

where ϕ is the power factor angle. Assuming an ideal (lossless) converter, the power calculated using (4.45) is equal to the power on the DC link. Therefore, the DC bus current can be written as follows:

$$\bar{i}_{dc} = \frac{P}{\bar{v}_{dc}} = \left(\frac{3}{2}\alpha(\cdot)\cos\phi\right)\|\bar{\mathbf{i}}_{qds}\| \tag{4.46}$$

Finally, comparing (4.46) and (4.44), $\beta(\cdot)$is obtained as

$$\beta(\cdot) = \frac{3}{2}\alpha(\cdot)\cos\phi \tag{4.47}$$

Since the angle ϕ depends on the load, the value of $\beta(\cdot)$ also depends on loading conditions. The values of parametric functions $\alpha(\cdot)$ and $\beta(\cdot)$, for two modulation strategies, are [44] as follows:

- for a six-step Inverter, $\alpha = \frac{2}{\pi}$, $\beta = \frac{3}{\pi}\cos\phi$
- for a six-step PWM, $\alpha = \frac{2}{\pi}d$, $\beta = \frac{3}{\pi}d\cos\phi$

The corresponding AVM is shown in Figure 4.45a, which assumes that the DC-link voltage is available (which is typically the case due to a large capacitor in the DC link) and that the AC side is connected to an inductive network (e.g., electric machine).

If the state-variable approach is used, the inductive network typically requires input voltages $\bar{\mathbf{v}}_{qds}$. Therefore, the AC-side voltages become the output of the converter AVM and are calculated in terms of the DC bus voltage using (4.49) and the function $\alpha(\cdot)$. A coordinate transformation may be used to recalculate these voltages from the converter reference frame to the physical variables for interfacing with the external inductive network [44]. The AC-side currents $\bar{\mathbf{i}}_{qds}$ are calculated by the external network and become the input for the converter AVM. Finally, the current $\bar{i}_{dc}$ that is injected into the DC bus is calculated using (4.44) and the function $\beta(\cdot)$.

4.3.4 Detailed and Averaged Models for MMC–HVDC Systems

A multilevel converter for HVDC systems includes MMC topology that connects two-level converter modules in cascade to achieve the desired AC voltage. MMC topologies enable using a smaller switching frequency to help reduce converter losses. In addition, filter requirements are eliminated by using a significant number of levels per phase. Scalability to higher voltages is easily achieved, and reliability is improved by increasing the number of SMs.

Detailed models (DMs) for MMC–HVDC systems include the representation of thousands of semiconductor switches and must normally use small numerical integration time steps to accurately represent fast switching events. The computational burden introduced by these models highlights the need to develop *simplified models* that provide similar behavior and dynamic response.

These simplified models are known as *mean-* or *average-value models,* and their purpose is to replicate the average response of switching devices, converters, and controls by using simplified functions and controlled sources [39, 42].

There are several advantages offered by the availability of the DM increased accuracy of SM models (nonlinear, switching losses), capability to account for special switching states direct representation of unbalanced conditions in the converters, and establishment of a reference model.

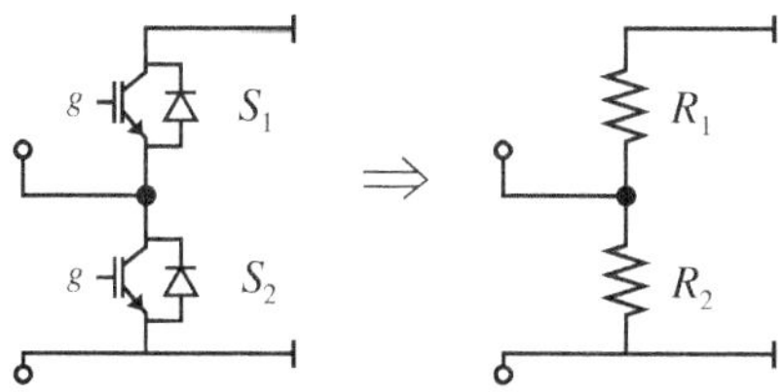

Figure 4.46 The half-bridge equivalent circuit.

4.3.4.1 Detailed Equivalent Models

Equivalent models are preferred for implementation in most EMT software dedicated for simulation of the MMC-type-based VSC converters, for the purpose of reducing the computation time, while achieving satisfactory accuracy [50,51].

In a detailed equivalent model, a switching cell S, consisting of an IGBT and a diode, can be represented by a resistance. Figure 4.46 shows the equivalencing of a half-bridge circuit by resistances R_1 and R_2.

When considering the capacitor in the half-bridge to form a SM (Figure 4.47a), the capacitor can be modeled by a voltage source in series with a resistance (Figure 4.47b), of which voltage can be calculated using the trapezoidal integration method, commonly used in EMT programs to solve differential equations, that is,

$$v_c(t) = R_c i_c(t) + v_{c,eq}(t - \Delta T) \tag{4.48}$$

with

$$R_c = \frac{\Delta T}{2C} \tag{4.49a}$$

$$v_{c,eq}(t - \Delta T) = \frac{\Delta T}{2C} i_c(t - \Delta T) + v_c(t - \Delta T) \tag{4.49b}$$

where ΔT is the integration step.

The resistors R_1 and R_2 are calculated as functions of the gate signals and the direction of current. A more detailed equivalent model proposed in [50] takes into account the voltage across the diode to help choosing the appropriate values of the

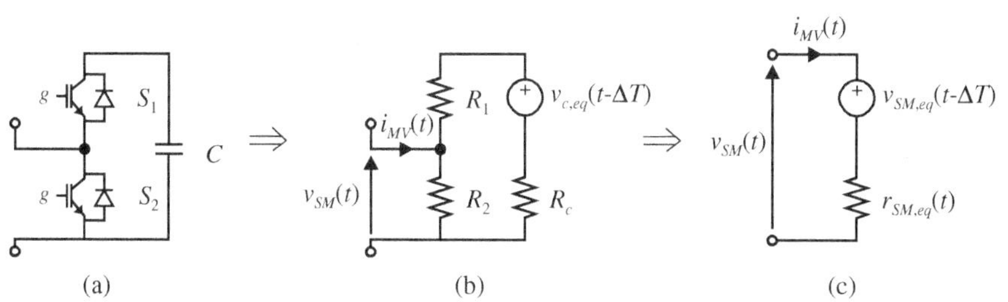

Figure 4.47 SM modeling for a detailed equivalent model: (a) SM circuit; (b) SM equivalent circuit; (c) SM Thevenin equivalent circuit [51].

resistors. Iterative calculus is required to determine the correct values of resistors at each time step.

The blocked state (S_1 and S_2 are off) is considered in the simulation initialization and for the dead band switching patterns. The equivalent detailed model and the network equations are solved sequentially because at each time step the resistors R_1 and R_2 are calculated in terms of the current [51].

A more simple equivalent model of a SM is the Thevenin circuit illustrated in Figure 4.47c, which can be built from the circuit shown in Figure 4.47b. The voltage at the circuit terminals is calculated with

$$v_{SM}(t) = r_{SM,eq}(t) i_{MV}(t) + v_{SM,eq}(t - \Delta T) \tag{4.50}$$

where

$$r_{SM,eq}(t) = \frac{R_2(t) \cdot (R_1(t) + R_c)}{R_2(t) + R_1(t) + R_c} \tag{4.51a}$$

$$v_{SM,eq}(t - \Delta T) = v_{c,\mathrm{eq}}(t - \Delta T) \frac{R_2(t)}{R_2(t) + R_1(t) + R_c} \tag{4.51b}$$

The SMs are connected in series to create a multivalve converter. The multivalve current $i_{MV}(t)$ is the same in all SMs. The equivalent voltage at the multivalve terminals, consisting of nSM submodules, is

$$v_{MV}(t) = r_{eq}(t) i_{MV}(t) + v_{eq}(t - \Delta T) \tag{4.52}$$

where

$$r_{eq}(t) = \sum_{i=1}^{nSM} r_{SM,eq\_i}(t) \tag{4.53a}$$

$$v_{eq}(t - \Delta T) = \sum_{i=1}^{nSM} v_{SM,eq\_i}(t - \Delta T) \tag{4.53b}$$

This approach helps representing a multimodular converter through a simple circuit as shown in Figure 4.48, where an arm consisting of hundreds of SMs connected

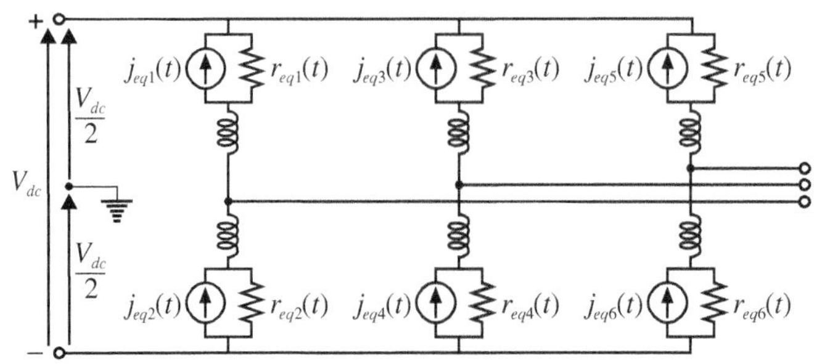

Figure 4.48 MMC equivalent circuit based on the detailed equivalent modeling. *Source:* Wachal et al. 2014 [51]. Reproduced with permission of CIGRE.

in series can be represented through an equivalent current injector in parallel with an equivalent resistor. An equivalent current injector $j_{eq}(t)$ can be calculated as the ratio of the equivalent voltage $v_{eq}(t)$ and the equivalent resistance $r_{eq}(t)$. Therefore, the number of nodes in the circuit is reduced to 11 per converter, and if the arm inductances are included in the Thevenin circuit it can be reduced to five, no matter of the number of SMs.

Detailed equivalent models are well suited for simulation, with good accuracy, of HVDC systems by both real-time simulators and off-line simulation software [50,51].

4.3.4.2 AVM of MMC–HVDC Using Voltage- and Current-Controlled Sources

Detailed representation of the switching devices and small integration time steps to simulate the operation of a VSC require unacceptable calculation efforts. For this reason, simplified models are sought, such as the so-called AVMs, in which the MMC behavior is modeled by controlled voltage and current sources.

(i) *AC-side representation of the AVM.* To describe the operation of the MMC of Figure 4.21, a functional diagram is shown by Figure 4.49 where SM switches are replaced by equivalent two-pole switches: S_{uji} for the SM*s* in the upper arm of phase *j* and S_{lji} for the SM*s* in lower arm. If S_{uji} (S_{lji}) is equal to unity, the output of the *i*th,

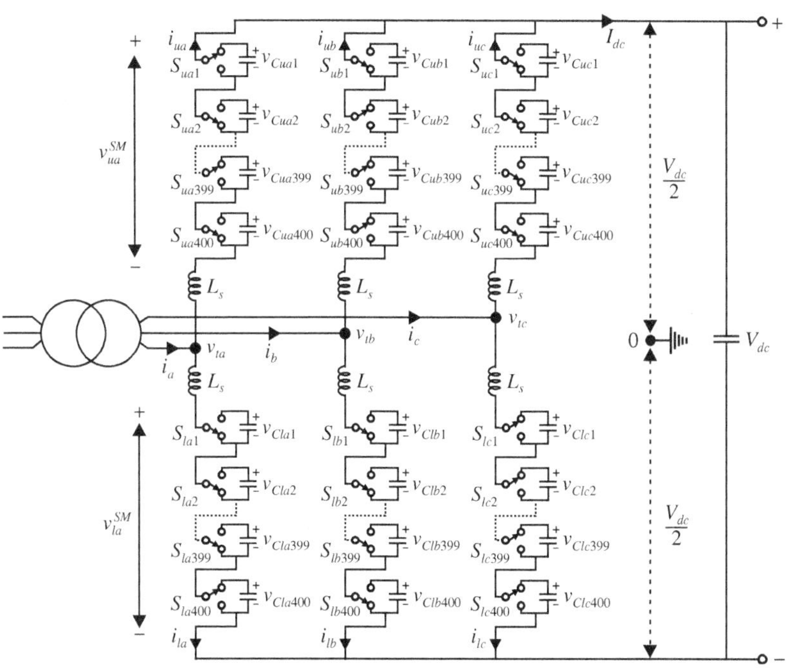

Figure 4.49 Circuit diagram of MMC unit of Figure 4.21. *Source:* Saeedifard and Iravani 2010 [15]. Adapted with permission of IEEE.

$i = 1, \ldots, 400$, SM of upper(lower) arm of phase $j = a, b, c$ is equal to the corresponding SM capacitor voltage; otherwise, it is zero.

In principle, each MMC arm represents a controllable voltage source, which is described by [15]:

$$v_{uj} = v_{uj}^{SM} - L_s \frac{di_{uj}}{dt} = \sum_{k=1}^{N_{arm}} (S_{ujk} \cdot v_{c,ujk}) - L_s \frac{di_{uj}}{dt}, j = a, b, c \tag{4.54a}$$

$$v_{lj} = v_{lj}^{SM} - L_s \frac{di_{lj}}{dt} = \sum_{k=1}^{N_{arm}} (S_{ljk} \cdot v_{c,ljk}) - L_s \frac{di_{lj}}{dt}, j = a, b, c \tag{4.54b}$$

where

$$v_{uj}^{SM} = \sum_{k=1}^{N_{arm}} (S_{ujk} \cdot v_{c,ujk}) \tag{4.55a}$$

$$v_{lj}^{SM} = \sum_{k=1}^{N_{arm}} (S_{ljk} \cdot v_{c,ljk}) \tag{4.55b}$$

where v_{uj} and v_{lj} denote the total voltage of the upper and lower arms of each phase j; the voltage v_{uj}^{SM} is the total voltage of all upper SM, and it is a function of the number of capacitors turned "on" as given by (4.55a). In equation (4.55a), the binary function S_{ujk} gives the state of each capacitor.

Similar definitions are applicable to the lower arm identified with the subscript l.

Considering the point "0" as the fictitious DC-side midpoint (Figure 4.49), the AC-side-phase voltages v_{tj0}, with $j = a, b, c$, are described by [15]

$$v_{tj0} = \frac{V_{dc}}{2} - v_{uj} = -\frac{V_{dc}}{2} + v_{lj} \tag{4.56}$$

It should be noted that the magnitudes and angles of voltages v_{tj0} determine the AC currents going into or out of the MMC.

Based on (4.54) and (4.56), by controlling the switching functions of the SMs in the upper and lower arm of each phase, the arm voltages and, as a result, the AC-side voltages are controlled in discrete steps.

We have also the equations:

$$v_{uj}^{SM} = -v_{tj0} + L_s \frac{di_{uj}}{dt} + \frac{V_{dc}}{2} \tag{4.57a}$$

$$v_{lj}^{SM} = v_{tj0} + L_s \frac{di_{lj}}{dt} + \frac{V_{dc}}{2} \tag{4.57b}$$

The arm currents in each phase j are describe by

$$i_{uj} = \frac{i_j}{2} + \frac{I_{dc}}{3} + i_{zj} \quad (4.58a)$$

$$i_{lj} = -\frac{i_j}{2} + \frac{I_{dc}}{3} + i_{zj} \quad (4.58b)$$

where j is any phase a, b, or c, and i_{zj} is a current composed of one third of the DC current ($I_{dc}/3$) and of the AC second harmonic circulating current within the three phases of the MMC. Its expression for any phase given by

$$i_{zj} = \frac{i_{uj} + i_{lj}}{2} - \frac{I_{dc}}{3} \quad (4.59)$$

subject to $i_{za} + i_{zb} + i_{zc} = 0$.

Since the AVM assumes that voltages $v_{c,uji}$ and $v_{c,lji}$ are perfectly balanced on all capacitors at any time, the second harmonic i_{zj} circulating currents are zero [14].

By subtracting (4.57a) from (4.57b)

$$v_{tj0} = \frac{L_s}{2}\frac{di_j}{dt} + e_j \quad (4.60)$$

where a new voltage is defined as

$$e_j = \frac{v_{lj}^{SM} - v_{uj}^{SM}}{2} \quad (4.61)$$

Replacing (4.58a) and (4.61) into (4.57a) gives

$$v_{uj}^{SM} = -\left(v_{tj0} - \frac{L_s}{2}\frac{di_j}{dt}\right) + \frac{V_{dc}}{2} = -e_j + \frac{V_{dc}}{2} \quad (4.62a)$$

By using the same approach for lower arm equation,

$$v_{lj}^{SM} = v_{tj0} - \frac{L_s}{2}\frac{di_j}{dt} + \frac{V_{dc}}{2} = e_j + \frac{V_{dc}}{2} \quad (4.62b)$$

The AC-side representation of the AVM is presented in Figure 4.50, and only phase-a control blocs are shown.

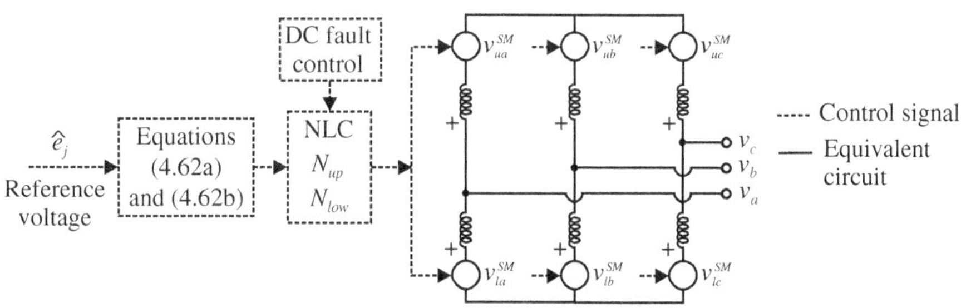

Figure 4.50 MMC AVM AC-side representation. *Source:* Peralta et al. 2012 [14] and Saad et al. 2013 [52]. Reproduced with permission of IEEE.

The new reference voltage $\hat{e}_j$ ($j = a$, b, or c) is generated by adding the inductance $L_s/2$ to the transformer inductance L_R using the inner current controller described in Figures 4.57 and 4.58. The signals out of the PWM modulator (Figure 4.57) are then impressed on the upper and lower voltage-controlled sources of the average MMC model. The AVM can support any modulation technique, but an efficient method is the nearest level control (NLC) technique [23].

The DC voltage of the MMC arm, in each phase, can be expressed by

$$V_{dc} = v_{uj} + v_{lj} = \sum_{i=1}^{400} (S_{uji} \cdot v_{cuji}) + \sum_{i=1}^{400} (S_{lji} \cdot v_{clji}) + L_s \left(\frac{di_{uj}}{dt} + \frac{di_{lj}}{dt} \right), \quad j = a,\, b,\, c \tag{4.63}$$

In the MMC of Figure 4.49, the switching functions S_{uji} and S_{lji} are controlled so that any instant $\sum_{i=1}^{400} (S_{uji} + S_{lji}) = 400$. This implies that at each instant only 400 SMs of the 800 SMs of phase j are "on" (i.e., n_{uj} SMs in the upper arm and $n_{lj} = 400 - n_{uj}$ in the lower arm). The values of n_{uj} and n_{lj} are determined by the desired voltage level of phase j from the modulation technique [14].

(ii) *DC-side representation of the AVM.* The DC-side model is derived using the principle of power balance, meaning that the power on the AC side must be equal to the power on the DC side plus converter losses [14, 51, 52]

$$P_{ac} = P_{dc} + P_{loss} \tag{4.64}$$

$$\sum_{j=a,b,c} e_j i_j = V_{dc} I_{dc} + P_{loss} \tag{4.65}$$

From the definition of the modulation index m_j,

$$m_j = 2\frac{e_j}{V_{dc}}, \quad j = a,\, b,\, c \tag{4.66}$$

the power balance equation (4.64) can be rewritten as

$$\frac{P_{ac}}{V_{dc}} = \frac{1}{2} \sum_{j=a,b,c} m_j i_j = I_{dc} + \frac{P_{loss}}{V_{dc}} \tag{4.67}$$

The converter current loss function I_{loss} is defined as

$$I_{loss} = \frac{P_{loss}}{V_{dc}} = R\frac{I_c^2}{V_{dc}} \tag{4.68}$$

with

$$I_c = \frac{1}{2} \sum_{j=a,b,c} m_j i_j \tag{4.69}$$

where R is the equivalent resistance of MMC, representing both switching and resistive losses, and I_c is the equivalent DC current including converter losses.

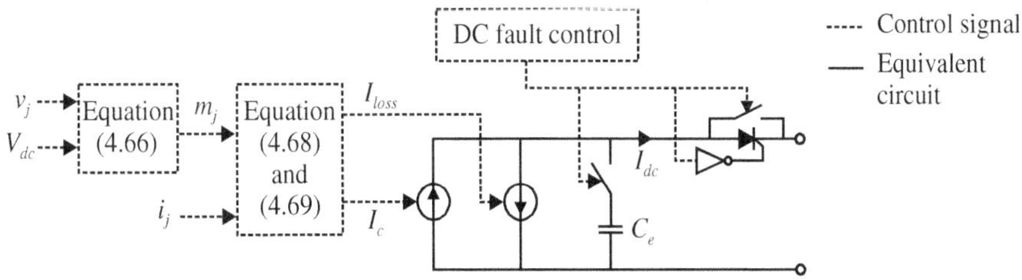

Figure 4.51 AVM DC-side representation of the MMC. *Source:* Peralta et al. 2012 [14] and Saad et al. 2013 [52] Reproduced with permission of IEEE.

The DC current is then derived from (4.67) and (4.69):

$$I_{dc} = I_c - I_{loss}$$

Since converter losses are dependent on the modulation technique and switching frequency, they should be taken into the account when estimating the equivalent resistance *R* of the MMC. The value of *R* is selected using the MMC losses from the full detailed model , which are close to 1% per converter.

A parallel resistance on the DC side of the MMC does not allow modeling current-dependent losses, as the latter will depend on the DC voltage only. Therefore, the loss function of (4.68) is selected to model the MMC losses.

The DC side of the MMC AVM is then represented by two current-controlled current sources, which depend on the current I_c as shown in Figure 4.51.

The equivalent capacitor C_e in Figure 4.51 represents a capacitance equivalent to the detailed model. It is derived using the energy conservation principle [14]:

$$E_{MMC} = 6 \sum_{k=1}^{N_{arm}} \frac{1}{2} C v_{C_k}^2 = \frac{1}{2} C_e V_{dc}^2 \tag{4.70}$$

Assuming all SMs have the same voltage v_C, the equivalent capacitor C_e can be derived from (4.70)

$$C_e = \frac{6C}{N_{arm}}$$

Note that the control parameters remain unchanged for the AVM in order to ensure fast and accurate dynamic response.

During DC faults, all SMs in the detailed MMC are shorted by the thyristor T_1 (see Figure 4.22a) transforming the MMC into a six-pulse bridge diode converter. Therefore, the voltage-controlled sources in the AVM (see Figure 4.50) are shorted (bypassed) and the DC capacitor C_e is opened in order to mimic the effect of T_1 in the DM. A series thyristor is added in the DC-side representation of the AVM (Figure 4.51) to force the DC current flow direction from the AC to DC side.

Finally, we present some conclusions from [14]:

- The detailed model includes all switching devices and, consequently, imposes a higher computation burden,

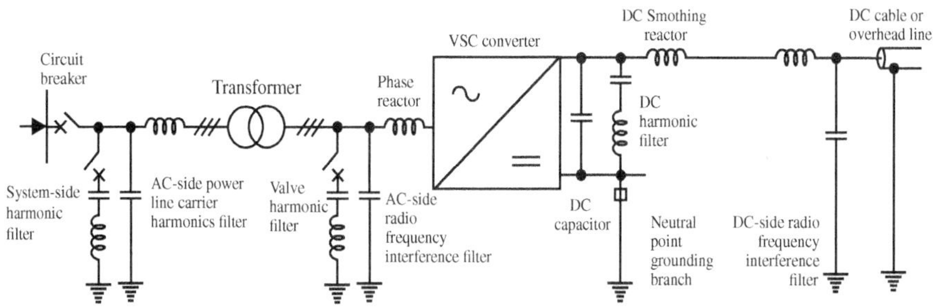

Figure 4.52 Major components of a VSC substation. *Source:* Andersen et al. 2005 [6]. Reproduced with permission of CIGRE.

- the proposed AVM has been shown to accurately replicate the dynamic performance of the detailed model,
- the proposed AVM performs significantly better in terms of computer speed. It is at least 370 times more efficient then the detailed model for the same simulation time step. It is robust and easily scalable from smaller to larger systems.

The detailed model remains useful for simulating higher frequency transients, for studying detailed performance conditions in converters and for calibrating the AVM.

4.4 VSC TRANSMISSION SCHEME AND OPERATION

The CIGRE B4.37 Working Group was formed to approach this important topic as the VSC-HVDC systems are essential for developing the future power systems.[4]

4.4.1 Power Equipment

Figure 4.52 shows the basic structure of a VSC substation and the location of the major power equipment, that is [6],

- The *VSC converter* that converts the direct voltage into alternating voltages or vice versa;
- the *inductive elements*, consisting of the phase reactors, the transformers, or a combination thereof, which are located on the AC side;
- the *DC capacitor(s)*, which is/are used as *voltage source(s)*.

There are also some other similarities between the substation equipment at both the CSC–HVDC and VSC–HVDC technologies: harmonic filters, converter transformer(s), switching equipment, protection equipment, control equipment, and currents and voltages metering equipment.

[4] Material reprinted with permission from CIGRE, Anderson et al. 2005 [6].

Special characteristics related to the harmonic content, associated with the VSC technology, may define peculiar design principles [6]:

- VSC technology can eliminate almost totally the low order harmonics, and thereby high order harmonics tuned filters only are employed, which are cheaper and smaller.
- Higher order harmonics can result in a VSC from the fast commutation process and the associated rapid changes of high voltage (*dv/dt*) and current (*di/dt*) and cause additional power losses, insulation stress, and electromagnetic interference (over the telecommunication systems).
- The harmonic level depends on the technology used.

Figure 4.53 illustrates a general example of a VSC converter together with graphics of voltage curves in various nodes of the converter.

A VSC converter mainly consists of IGBTs. The IGBTs can switch on/off with a frequency up to 20 kHz, eventually independently of the voltage applied, as compared to the conventional thyristors that switch on/off with a frequency of 50 Hz. This is a major advantage that allows a VSC to significantly reduce the harmonic level and control the reactive power. Thereby, a VSC–HVDC does not require capacitor banks for reactive power compensation and harmonic filters are very much reduced.

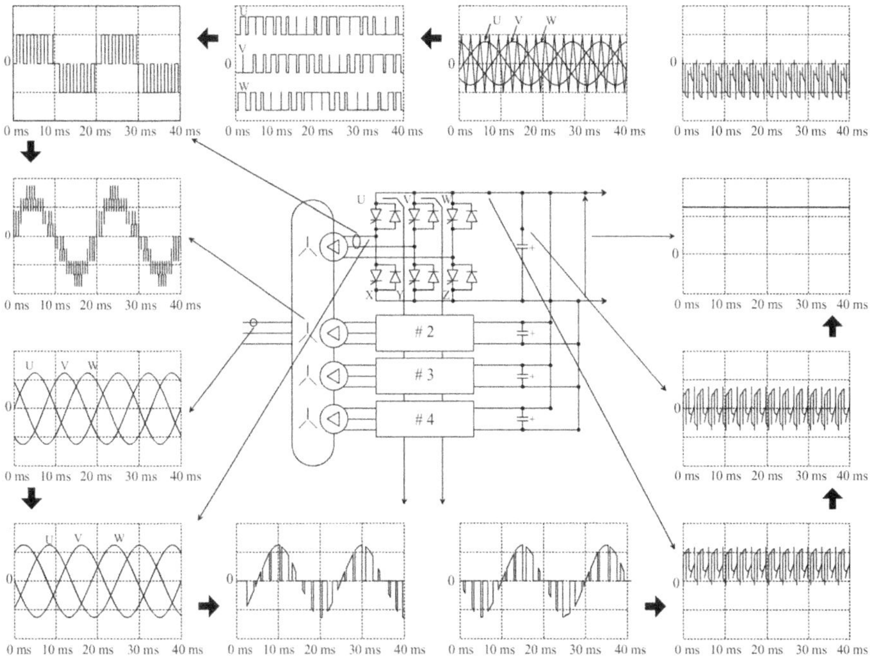

Figure 4.53 Operational characteristics of a voltage-source HVDC system. *Source:* Kim et al. 2009 [53]. Reproduced with permission of John Wiley and Sons, Inc.

The operating mode of a VSC is similar to a CSC. It operates as a rectifier when the active power flows from the AC system to the DC link or as an inverter when the active power flows from the DC link to the AC system. Reactive power is a quantity that can be defined on the AC side only, and it can be controlled independently of the active power.

For reliability purposes, the valves include redundant switching devices. When the integrity of an IGBT is affected, this one enters into a safe short-circuit mode in such a way that the valve continues to operate properly. With this purpose, specially developed IGBT press-pack modules have been developed.

Unlike a CSC-based HVDC link, a VSC-based HVDC system has no inherent capability to clear DC line faults [6]. When a fault occurs, the DC capacitors are sources for the arc created, which is maintained until the VSC substation AC circuit breakers trip the scheme. In case of a temporary fault, normal operation can be restored after a startup of about 10 s. The AC circuit breakers are used to disconnect both converters, and the DC line until a DC circuit breaker will be available for commercial use.

The DC Capacitor is an important component of the VSC station as it is the source for voltage generation. At the rectifier, the DC capacitor will be charged with electric energy to provide DC voltage, and the control system will adjust the PWM controller of the converter to absorb electrical energy from the AC power system. At the inverter, the VSC converts the DC voltage provided by the DC capacitor, which stores electrical energy, into AC voltage. The output voltage of the VSC is approximated with a sinusoidal voltage curve using the PWM strategy (Figure 4.53). The DC capacitor is connected directly in parallel with the converter valve. When designing the DC capacitor, the following aspects should be considered [6]:

- Keep the stray inductance within the commutation circuit, formed by the valves, the DC capacitor and the busbars to low values to minimize the voltage stress on the power electronic devices. The DC capacitor should be located very close to the valves to minimize the circuit inductance. The number of series-connected IGBTs is chosen so that the arm is capable of withstanding the increased voltage stress.
- Appropriately tune the DC capacitor to limit the ripple in the DC voltage. DC harmonics are caused by unbalances in the AC system or in the converter operation, harmonics in the AC network, VSC operation characteristics, or rating of the DC capacitor.
- Limit the harmonic content in the DC voltage to levels that allow avoiding interaction with other VSCs, thus not influencing the DC voltage provided by these;
- Ensure the VSC the capability of keeping the DC voltage within predefined limits in order to allow the active power transfer and AC voltage control by appropriate reactive power support in stable conditions.

THE INTERFACE TRANSFORMERS are included in the VSC substation to [6]

- allow synchronization between the AC system and the VSC, especially when these have different voltages. The transformer is provided with a tap changer that helps optimizing the VSC operation;
- provide a reactance between the AC system and VSC unit. The transformer also allows connecting a reactor in series with the VSC;
- allow connecting two six-pulse valves to achieve a 12-pulse configuration, using a Yyd vector group;
- prevent zero-sequence currents flowing between the AC system and the VSC unit.

AC PHASE REACTORS may be necessary, depending on the converter characteristics, between the interface transformer and VSC to block high frequency harmonics coming from the converter that cause additional voltage stress on the transformer.

AC HARMONIC FILTERS are required depending on the VSC configuration and operation strategy to prevent VSC-generated harmonics to propagate into the AC system. As the filters include capacitors, the reactive power generated by these should be appropriately considered in the voltage control strategy of the VSC. Low frequency harmonics are, in general, not generated by VSC because appropriate methods can be employed, such as the PWM, higher pulse numbers, or multilevel VSC topologies. Thereby, the AC filters have to be tuned for high frequency harmonics only. Such filters are smaller and thus cheaper. When low frequency harmonics are present, appropriately tuned filters can be connected in parallel with the high frequency harmonic filters.

A DC REACTOR may be required in case of long transmission lines, connected in series, to prevent harmonic currents from propagating into the DC line and detune critical resonance with the DC circuit. But, rapid changes in the DC current that are part of the VSC control strategy must be allowed.

DC HARMONIC FILTERS can also be connected in parallel with the DC capacitor, tuned at its frequency, to prevent harmonics from propagating into the DC line. When DC harmonic filters are used, the DC capacitors may be smaller.

RADIO FREQUENCY INTERFERENCE FILTERS, of special design, are sometimes necessary to reduce at acceptable limits the high frequency harmonics that propagate into the AC system.

THE NEUTRAL POINT GROUNDING provides a connection path between the point of zero potential on the DC circuit to the ground, either by direct rigid connection or through an impedance. In a bipolar HVDC system, this point can be the midpoint of the VSC DC capacitors, whereas in the case of a monopolar system this point can be on either of the DC terminals of the VSC unit.

The DC LINE can be either cable or overhead line depending on the environmental constraints and the optimal costs. However, other aspects may be considered when choosing the type of the DC line [6]:

(i) Cables have less impact on the environment, regarding the right-of-way, water or land, electromagnetic influence, protection of woods, visual impact, etc.;

(ii) When VSC is used, cross-linked polyethylene (XLPE)-based cable can be employed, eventually for long distances, because there is no need to change the DC voltage polarity [54];

(iii) Cables are not exposed to lightning strikes, storms, and pollution. Most overhead line faults are temporary, whereas any cable fault is permanent.

When a back-to-back link is involved, some of the above equipments are not required, such as the blocking reactor and DC-side radio frequency interference filter because they are mainly used to reduce the harmonic current flowing in the DC line.

4.4.2 Principles of Active and Reactive Power Control

The VSC has the capability of individually controlling active and reactive power. The exchange of active and reactive power between a VSC and the AC grid is controlled by the phase angle and amplitude of the VSC output voltage in relation to the voltage of the AC grid [6]. Figure 4.54a shows the simplified connection circuits of a VSC on both the DC and AC sides.

The active and reactive power can be controlled simultaneously and independently of each other. In order to understand the way in which the active and reactive

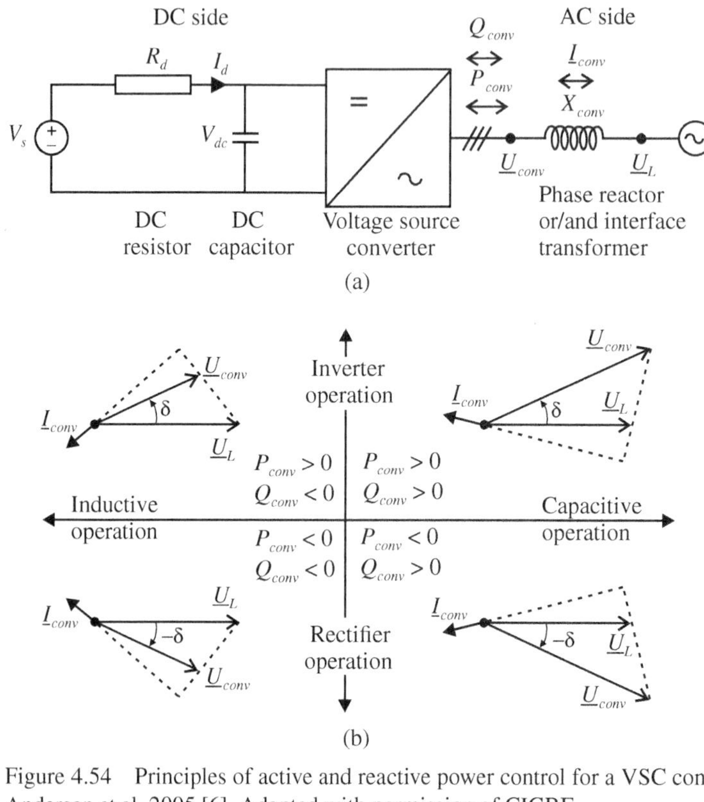

Figure 4.54 Principles of active and reactive power control for a VSC converter. *Source:* Andersen et al. 2005 [6]. Adapted with permission of CIGRE.

powers are exchanged by the VSC with the AC system, let us analyze the basic circuit of the AC side from Figure 4.54a.

For simplicity, the AC filters have been neglected so that I_{conv} equals I_L. The reactance $X_{conv} = \omega L_{conv}$ represents both the interface transformer and phase reactor. Let us take the AC voltage $\underline{U}_L$ as the phase angle reference, that is $\underline{U}_L = U_L\angle 0°$. The fundamental frequency component of the phase-to-phase voltage at the converter terminals is the $\underline{U}_{conv} = U_{conv}\angle\delta$, where δ is the phase angle between the converter AC voltage and the AC system voltage. The rms of the fundamental component of the phase-to-neutral voltage is $V_{conv} = U_{conv}/\sqrt{3}$. In terms of the DC voltage, the fundamental component of the converter AC voltage is [55]

$$\underline{V}_{conv} = \frac{\sqrt{2}}{\pi} k_\lambda V_d e^{-j\delta} \tag{4.71}$$

where $k_\lambda = m_a$, which is the modulation index or voltage ratio factor. For square wave operation, k_λ equals 1. For a PWM converter, k_λ takes values between 0 and 1.

The phase-to-phase voltage at the converter terminals is

$$\underline{U}_{conv} = \frac{\sqrt{6}}{\pi} k_\lambda V_d e^{-j\delta} \tag{4.72}$$

If the resistance is neglected, current on the phase a can be expressed as

$$\underline{I}_{conv,a} = \frac{\underline{V}_{conv} - \underline{V}_L}{\underline{Z}_{conv}} \cong \frac{(\sqrt{2}/\pi) k_\lambda V_d e^{-j\delta} - V_L e^{-j0}}{jX_{conv}} \tag{4.73}$$

THE VSC ACTIVE AND REACTIVE POWER. The active and reactive powers, P_{conv} and Q_{conv}, exchanged between the VSC converter and the AC system are

$$P_{conv} = \frac{U_{conv} \cdot U_L}{X_{conv}} \sin\delta \tag{4.74a}$$

$$Q_{conv} = \frac{U_L^2}{X_{conv}} - \frac{U_L U_{conv}}{X_{conv}} \cos\delta \tag{4.74b}$$

From equation (4.74b), it can be concluded that when U_L is greater than U_{conv}, then the AC system supplies reactive power to the VSC, that is the converter operates in inductive mode. When U_L is lower than U_{conv}, then the VSC supplies reactive power to the AC system, that is the converter operates in capacitive mode. Thus, reactive power can be controlled mainly by controlling the two voltages (U_{conv} and U_L). Furthermore, the reactive power exchanged is independent of DC link power magnitude and direction. On the other hand, if the two voltages are equal in magnitude, there is no reactive power exchanged between the VSC and the AC system.

From equation (4.74a), it is obvious that the active power can be controlled by controlling the phase angle δ between voltages $\underline{U}_{conv}$ and $\underline{U}_L$; the change in the magnitudes of the two voltages will have small influence on the active power flow because they have to be maintained within the tight secure limits. When the phase angle δ is zero, no active power is exchanged between the VSC and the AC system. Changing the sign of δ will result in reversing the power flow direction on the AC side.

The direct current can be derived from the equality of powers on the AC and DC sides, that is [55]

$$I_d = \frac{\sqrt{6}}{\pi}\frac{U_L}{X}\sin\delta \tag{4.75}$$

On the DC side of a VSC–HVDC link, the power magnitude and direction can be changed by proportionally changing the DC current magnitude and direction. Furthermore, under normal operating conditions, the DC voltage is maintained approximately constant.

In terms of the phase angle δ, the converter may operate either as a rectifier or as an inverter [6] (see Figure 4.54b):

- If the voltage $\underline{U}_{conv}$ lags the voltage $\underline{U}_L$, which means that the VSC absorbs active power from the AC system, then the converter operates as a rectifier. On the DC side, an equivalent current will be injected in the DC source and the voltage V_{dc} will increase in accordance with Ohm's law, that is, $V_{dc} = V_s + R_d I_d$.
- If the voltage $\underline{U}_{conv}$ leads the voltage $\underline{U}_L$, which means that the VSC injects active power into the AC system, then the converter operates as an inverter. On the DC side, an equivalent current will be drawn from the DC source and the voltage V_{dc} will decrease, that is, $V_{dc} = V_s - R_d I_d$.

In order to understand the power transfer on a point-to-point VSC link, let us consider the scheme shown in Figure 4.55.

In a VSC–HVDC link, the DC voltage polarity is always the same, and thereby the direction of the power flow on the DC line is determined by the direction of the DC current. According to Ohm's law, the current flowing from converter 1 to converter 2 is determined by

$$I_d = \frac{V_{dc1} - V_{dc2}}{R_d}$$

Thereby, the direction of a DC current is always from the converter of higher DC voltage to the converter of lower DC voltage. The DC voltage at the rectifier must therefore be higher than the DC voltage at the inverter. In Figure 4.55, the current flows from converter 1 (sending end) to converter 2 (receiving end).

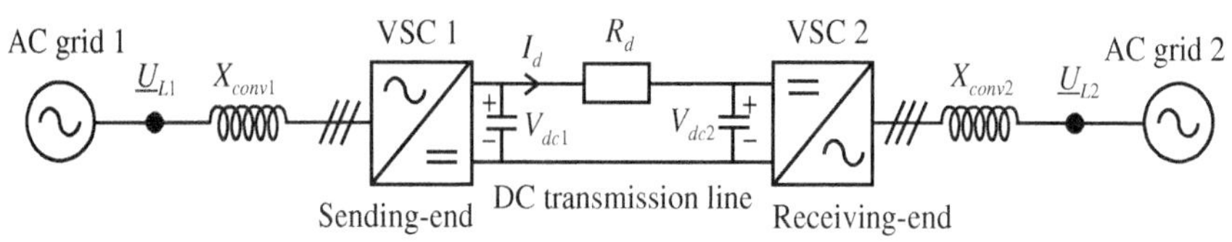

Figure 4.55 A point-to-point VSC transmission scheme. [6].

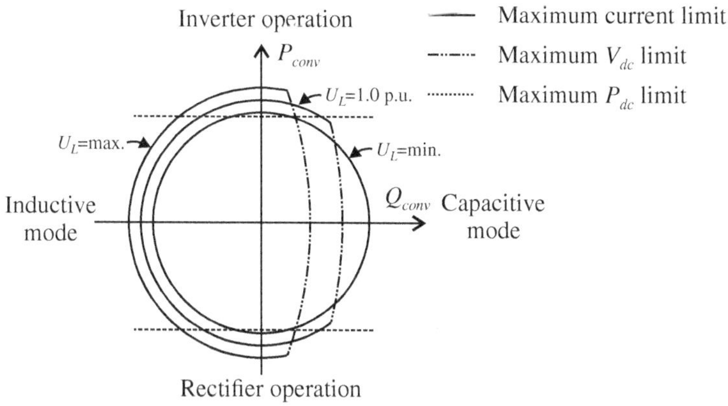

Figure 4.56 A basic simplified *P–Q* diagram of a VSC. *Source:* Johansson et al. 2004 [56]. Adapted with permission of CIGRE.

The control of active power on a CSC–HVDC link is different by that on a VSC–HVDC link. In a CSC–HVDC, the current direction is always the same and thereby the power flow direction can be changed by changing the DC voltage polarity. Furthermore, one converter (the inverter) maintains the voltage at a reference value whereas the other converter (the rectifier) is controlling the DC current.

PQ CAPABILITY DIAGRAM. Figure 4.56 illustrates the *P–Q* capability diagram of a VSC converter seen from stability point of view. As also shown in Figure 4.56, a VSC can operate within four quadrants of the *P–Q* plan, which defines the operation at the interface point with the AC system.

An inverter injects active power into the AC system and thus $P_{conv} > 0$, whereas a rectifier absorbs active power from the AC system and $P_{conv} < 0$. The two converters can operate either in capacitive mode, with $Q_{conv} > 0$, or in inductive mode, with $Q_{conv} < 0$.

There are mainly three factors that limit the active and reactive powers produced/absorbed by a VSC converter [56]:

- *Maximum current through IGBTs*. As this current is constant, the *P–Q* capability limits depend on the AC voltage. The larger is the AC voltage, the larger *P–Q* capability is achieved, but it is limited by the minimum (U_{min}) and maximum (U_{max}) security limits of the AC voltage. Full reactive power that can be produced is achieved at minimum AC voltage. At high grid voltage (U_{max}), however, the reactive power is limited by the maximum converter AC voltage U_{conv}, which in turn is limited due to the DC voltage limitation.
- *Maximum DC voltage level*. The reactive power is mainly dependent on the voltage difference between the AC voltage at the VSC terminals, which is a function of the DC voltage, and the grid AC voltage. In capacitive mode, if the grid AC voltage is high, since the VSC AC voltage must be higher, which is turn is limited by the DC voltage, the difference between the maximum DC voltage and the AC voltage will be low. Thus, in capacitive mode, the

reactive power capability increases with decreasing AC voltage. This makes sense from a stability point of view. The on-load tap-changing transformer can optimize the AC converter voltage so that to achieve maximum steady-state power capability. The DC voltage is also given by the maximum allowed limit of the storage capacitor.

– *Maximum active power limit*. This limit is determined by the rating of the VSC, in terms of current, and the rating P_{dc} of the transmission cable.

4.4.3 VSC Transmission Control

All configurations for VSCs possess a series of inductive interface separating the switching valves from the AC system. The switching valves generate a fundamental frequency AC voltage from a DC voltage. The magnitude and the phase of the fundamental frequency component of this AC voltage at the valve side of the series inductive interface can be controlled.

A VSC therefore has the capability of acting as a rectifier or as an inverter, and/or as a generator or an absorber of reactive power. It is the control of the modulation index m_a and the phase angle δ that dictates the strategies for controlling voltage-sourced converters.

The means of controlling the magnitude of the fundamental frequency AC voltage is dependent upon the configuration of the VSC. The AC voltage generated by a multipulse VSC (24-pulse or 48-pulse) is directly proportional to the DC-side capacitor voltage. Therefore, AC-side voltage control can be achieved by controlling the DC-side capacitor voltage. In turn, the DC-side capacitor voltage depends on the charging level. The more energy is stored into the capacitor the higher is its DC voltage. As the capacitor is discharged, voltage will decrease. The more usual and preferred case is to maintain the DC-side voltage constant. This is readily achieved with multilevel converters and/or, if PWM is used, with two-level or multilevel converters.

Various control strategies can be applied to control the modulation index m_a and phase angle δ for a VSC transmission link. In this book, the vector control only is described.

4.4.3.1 VSC Converter Control Using the Vector Control Strategy

Vector control is a current control strategy that allows a decoupled control of the active and reactive powers, whereas the modulation index m_a is adjusted independent from the phase angle δ. The advantage is that the current can be directly limited to avoid overloading of the valves, but this is an additional control loop that may slow down the controller's response.

Figure 4.57 shows the basic VSC control scheme using the vector control strategy. In normal operation mode, each VSC controls its reactive power flow independently from the other station. The reactive power is controlled in terms of the requirements for the AC voltage control. However, active power flow into the converter must equal the active power out the converter, neglecting the converter losses. Any active power unbalance results in immediate change in the DC voltage.

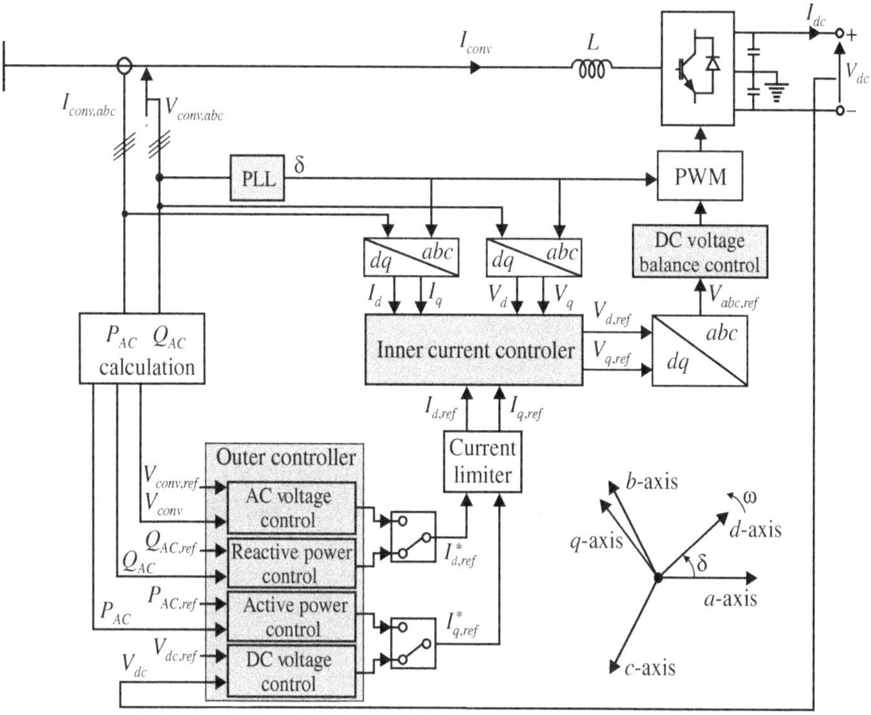

Figure 4.57 The control scheme of a VSC converter using the vector control strategy.

In order to facilitate the VSC control, the *abc* three-phase stationary coordinates system is transformed into the *dq* rotating coordinates. This requires transformation from *abc* to *dq*. In the *dq* coordinates the subscript *conv* that denotes the converter will be omitted. For implementation and design purposes, it is more efficient to use time-dependent variables instead of using amplitude and phase of a state space. Moreover, the number of control loops required reduces from three to two. Several control loops are designed in the VSC control scheme for these purposes.

THE OUTER CONTROLLER. Using sensing devices for voltages and currents, the controlled variables, that is the active and reactive powers, are determined using a calculation block. There are two controllers employed to control each variable, as shown Figure 4.57. Only one of them is used at a given time, depending on the system specifications. The active power can be controlled by either direct control, at the reference value $P_{AC,ref}$, or indirectly by controlling the DC voltage, at the predefined value $V_{dc,ref}$. On the other hand, the reactive power can be controlled either directly, at the reference value $Q_{AC,ref}$, or indirectly by controlling the AC voltage, at the predefined value $V_{AC,ref}$. All controllers include a PI regulator to achieve zero steady-state error.

The outer controller provides the reference currents ($I_{d,ref}$, $I_{q,ref}$) to the inner controller. In order to avoid converter overloading, a current limiter is used to maintain the total reference current within the converter current limit.

Active and reactive power control. In terms of *dq* quantities, and assuming that the *abc* system is balanced, the instantaneous active and reactive powers are calculated with

$$P = \frac{3}{2}(V_d I_d + V_q I_q) \tag{4.76a}$$

$$Q = \frac{3}{2}(V_q I_d - V_d I_q) \tag{4.76b}$$

In the vector control strategy, the q-axis of the dq-frame is aligned with the AC network voltage phasor, and the d-axis voltage becomes zero, that is $V_d = 0$. Hence, analyzing equations (4.76), it can be said that the active and reactive powers can be independently controlled. The active power can be controlled by controlling the q-axis current, I_q, and the reactive power can be controlled by controlling the d-axis current, I_d.

The active power into one converter of a HVDC link is equal to the active power out of the other converter, to which the power losses should be added. For this reason, the active power control at the two converters must be coordinated. The active power control on the AC side is achieved by controlling the phase angle δ of the fundamental frequency component of the AC voltage generated by the VSC [57]. Power is withdrawn or feed into the AC system depending on whether δ lags or leads the phase angle of AC bus voltage.

The VSC converters can provide an important voltage control in the AC power system. When other voltage control equipment are located in close proximity, they should be integrated into a coordinated voltage control system through a carefully designed droop characteristic. Droop control characteristic may be difficult to define in many situations, and thereby reactive power control might be preferred.

AC voltage control. In case of multilevel converters, at which the switching strategy is achieved by the PWM technique, the magnitude of the AC voltage generated by the VSC is controlled by appropriate control of the modulation index m_a. The DC capacitor voltage is usually held constant. The phase shift between voltage and current is thereby controlled, and more or less reactive power flows to/from the converter depending on the need for voltage control. The AC voltage controller is used to override the reactive power controller in order to maintain the AC voltage within acceptable limits.

When the VSC supplies power to an isolated AC system, in which there is no other active power source, that VSC will control the power and the VSC at the other end of the HVDC link will independently control the DC side voltage.

DC voltage control. A DC voltage regulator is used to override the active power controller in order to maintain the DC voltage within secure range, especially during disturbances occurring in the AC system. A proportional controller, and sometimes a PI controller, is used (Figure 4.57).

THE INNER CONTROLLER. The inner controller presented in Figure 4.58 allows controlling the modulated voltages, V_d and V_q, that will be used to generate the required PWM-switching pattern. In order to decouple the d- and q-axis, a feed-forward technique is used to compensate cross-coupling terms. Thus, the active and

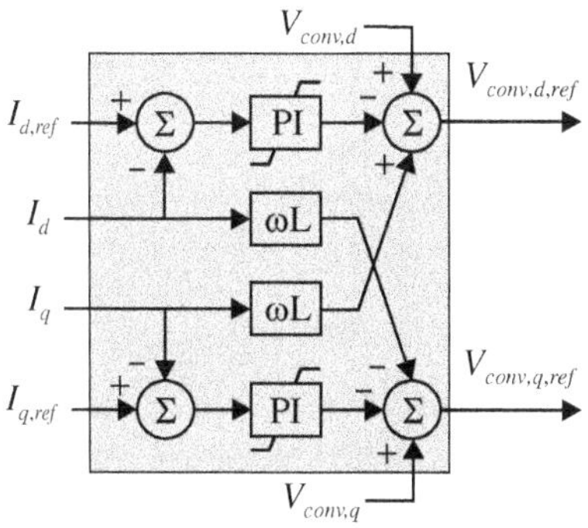

Figure 4.58 Inner current control loop.

reactive currents through the inductance L in the dq frame can be independently controlled via a PI control [58,59]. These currents can be controlled by varying the magnitude and phase angle of the converter voltage $\underline{V}_{conv}$.

THE PHASE-LOCKED LOOP CONTROL. The main function of the phase-locked loop (PLL) is to synchronize the converter output voltage with the AC grid voltage. The PLL is employed to generate the current and voltage references. PLL provides guidance in the vector control strategy to align the q-axis of the dq-frame with the AC network voltage phasor, and also in the modulation process to help determining the commutation sequence.

DC-VOLTAGE BALANCE CONTROL. A DC-voltage balance control is added on the three-level converters to balance the DC voltages on the positive and negative poles. Unbalances between the DC voltages may occur due to lack of precision in the modulation process, due to inherent unbalances in the converter circuits, etc.

FREQUENCY CONTROL. This is performed by appropriate design of the phase oscillator that determines the valve pulse firing sequence.

When integrated within an AC power system where other frequency controls are active, besides setting the number of cycles per second, the PLL oscillator has to synchronize the fundamental frequency AC voltage generated by the VSC with the AC system voltage. In this case the oscillator is phase-locked.

When supplying an isolated AC power system, only the number of cycles per second has to be controlled by the oscillator. In this case, the oscillator is called unlocked.

The ability of a VSC to control frequency and AC voltage, and absorb and deliver real power, makes it useful for assisting in black start conditions.

4.4.3.2 Levels of Control

There are three levels of controls of a VSC, as shown in Figure 4.59 [6].

(i) *Firing control.* The firing strategy depends on the configuration of the VSC and type of power electronic devices, which in turn depends on the compromise between the level of harmonics accepted and the size of the AC filters. If PWM modulation is used, the valve-firing sequence is regulated by the type

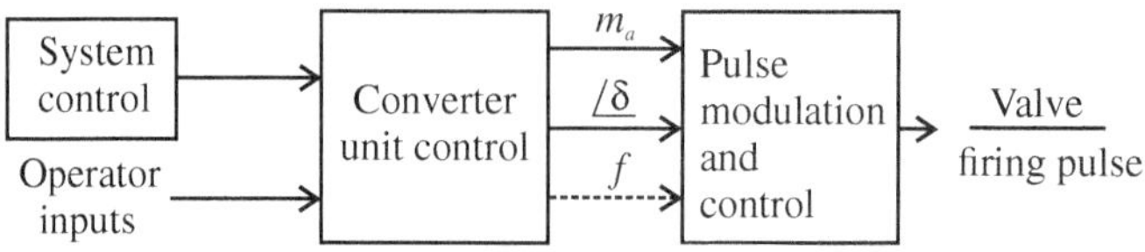

Figure 4.59 Levels of controls for a VSC [6].

of PWM applied. In order to obtain the harmonic cancelation possible for the least number of valve switchings each cycle, a SHE method (SHEM) may be used [60]. The level of harmonic distortion should meet certain standards [61]. The correlation of firing to the switching devices depends on an oscillator used, which can be either phase locked, when synchronized to an active AC system is required, or independently locked, when the HVDC link supplies an isolated power system with no other form of frequency control.

(ii) *The converter unit control* determines the DC voltage across the capacitor if the VSC is configured as a multipulse two-level converter. This is done by adjusting the phase of the phase-locked oscillator so that the phase angle difference created across the interface reactor causes the necessary power to flow into or out of the converter. The DC side capacitor is charged accordingly to the level of DC voltage required. If the converter is multilevel or operates with PWM, the converter control acts to change both the phase angle of the generated fundamental frequency component of the AC voltage at the converter side of the interface reactor, and the magnitude of that voltage.

The converter control system may perform the following functions [6]:

- protects series converters;
- balances currents between parallel converters;
- generates and controls the voltage phasor for the shunt or series component;
- performs negative phase sequence control and minimizes unbalance effects that result in negative sequence currents;
- reduces DC current offsets in the AC-side currents;
- controls and limits the DC-side capacitor voltage;
- limits converter current;
- controls the DC voltage balance for multilevel converters.

(iii) The *system control* performs the control of active and reactive powers injected/withdrawn into/from the AC power system, maintains the voltage magnitude, angle and impedance within security limits, and when necessary contributes to transient stability enhancement, oscillation damping, etc. The system control depends on the appropriate reference signals.

4.4.3.3 Coordination of Controls

The flexibility in control of a VSC is a significant advantage for a VSC–HVDC technology as compared to a CSC–HVDC technology or even AC equipments. There are two operation scenarios for which coordination of the controllers of a VSC have to be coordinated [6]:

(i) *Supplying an isolated load.* When supplying an AC load that has no other source of generation, the inverter may be set to control the frequency (which becomes the frequency of the load) and the AC voltage. The VSC may be an excellent source that supplies the load at appropriate quality voltage and frequency, irrespective of the load change. On the sending end side, the rectifier connected to the main grid or generation may be set to control the DC voltage and the AC voltage.

When integrated within an AC system that includes synchronous generators or is interconnected with the main AC power system through an AC transmission line, the frequency control may be disabled and the firing pulses may be switched from an independent clock to phase-locked onto the AC voltage. Alternatively, a droop characteristic for the frequency control and the AC voltage control may be required so that the VSC–HVDC link can operate synchronously with the AC system.

(ii) *Interconnection of two or more AC power systems.* When a VSC–HVDC link interconnects two or more AC systems or is simply integrated into a bulk power system, there are some choices in setting the control modes. All converters can control the AC voltage, but if the AC system has a very high short circuit ratio, it may be preferable to control the reactive power, if possible at zero. When all converters control the AC voltage, one of the rectifiers may include the DC voltage control V_{dc} (DC voltage control), whereas the other converters may control the power P_d (power control). The DC voltage control and power control between converters are interchangeable. However, if electromechanical damping in the AC system is required, the power control should be located at the converters along with the damping control [6].

An example of *coordinated control for a two-terminal VSC–HVDC link* is shown in Figure 4.60, where the terminal A converter controls the DC voltage, V_{dc}, and the terminal B converter controls the active power, P_d.

The DC voltage for the two-terminal DC link can be controlled using the voltage margin method [62]. Figure 4.61 illustrates the V_{dc}–P_d characteristic when the DC voltage control is set as a fundamental control at terminal A.

The V_{dc} controller tries to keep the DC voltage to the predefined value $V_{dc,ref}$ by adjusting P_A within its lower and upper limits. If the DC voltage V_{dc} is lower than the reference value, $V_{d,ref}$, the controller increases P_A until it reaches P_B. When P_A is equal to P_B, then V_{dc} is maintained constant. If V_{dc} is higher than $V_{dc,ref}$, the controller reduces P_A until it reaches P_B. On the other hand, when P_B exceeds the upper limit, the DC voltage will decrease, whereas if P_B is smaller than the lower limit, the DC voltage will increase.

In order to avoid unfavorable interference between DC voltage controllers at the two terminals, a voltage margin ΔV_{dc} is introduced In the DC voltage regulator, as shown in Figure 4.62. The voltage margin is defined as the difference between the DC reference voltages of the two terminals. At terminal B, $V_{dc,ref}$ is subtracted from the voltage margin, and the power controller

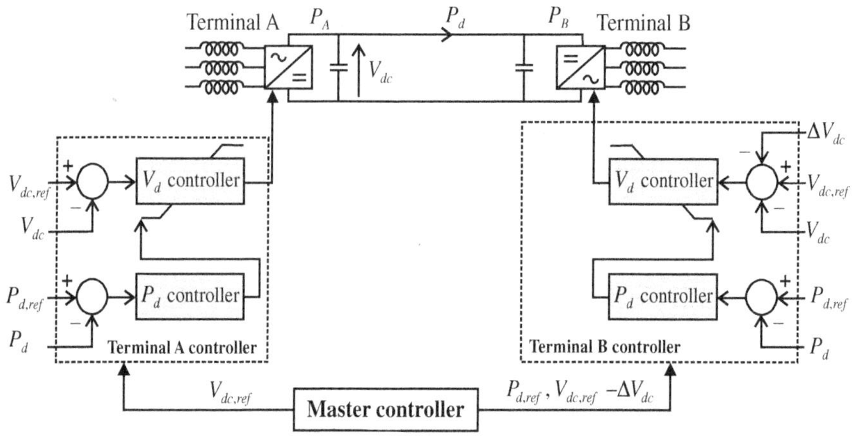

Figure 4.60 Example of coordinated control for a two-terminal VSC–HVDC link. *Source:* Andersen et al. 2005 [6]. Adapted with permission of CIGRE.

with $P_{d,ref}$ controls the lower limit of the DC voltage controller. The operating point is located at the intersection of the V_{dc}–P_d characteristics of each terminal. Terminal A controls the DC voltage, and terminal B controls the active power. The active power can be controlled by changing the reference power $P_{d,ref}$ [62].

(iii) *Telecommunication between converter stations.* Coordinated control between the VSC converters of a HVDC link that requires fast communications between the converters may be applied for conditions such as

– coordinated damping of electromechanical oscillations;

– reconfiguration of the control modes between converters, etc.

(iv) *Supply from a wind farm.* When DC transmission is required to bring power from a wind farm to a substation, VSC transmission can be integrated into the wind turbine design for maximum performance and economy. This technology may be particularly applicable for offshore wind farms. At the sending end, AC voltage control, power control, and frequency control can be coordinated with the generators. Best overall performance is achieved when the controls of the

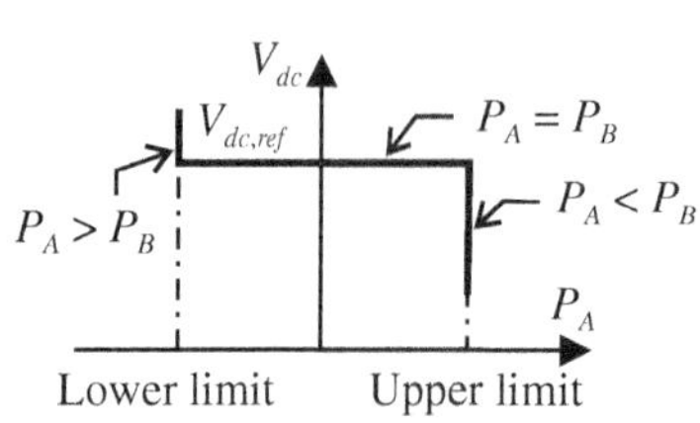

Figure 4.61 The V_{dc}–P_d characteristic for DC voltage control at terminal A. *Source:* Nakajima and Irokawa 1999 [62]. Reproduced with permission of IEEE. *Source:* Asplund et al. 2013 [63]. Reproduced with permission of CIGRE.

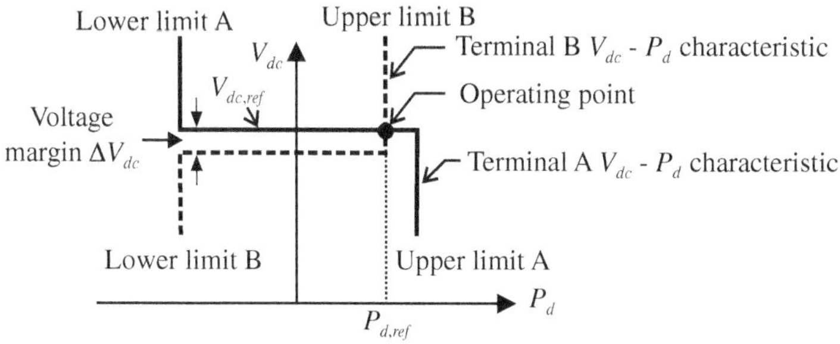

Figure 4.62 Operating point in the voltage margin method. *Source:* Nakajima and Irokawa 1999 [62]. Reproduced with permission of IEEE.

VSC sending end converter are coordinated with the turbine pitch controller (if used), the generator type, and the wind velocity. The receiving end converter may incorporate AC voltage control and DC voltage control [6].

4.5 MULTITERMINAL VSC–HVDC SYSTEMS AND HVDC GRIDS

4.5.1 On the Conventional Multiterminal HVDC Configurations

In the countries with older transmission grids, new challenges are coming up such as integration of renewable energy sources (wind and solar, etc.) and new load flow patterns due to increasing energy trades. This poses new requirements on the grid. AC lines could be built, but the permitting process for any new overhead line is often extremely time consuming if not impossible. The problem might be solved technically by converting AC lines to DC, which could increase the transmission capability in the same right of way, or by using underground cables. However, AC cables at 400 kV or 500 kV are not viable since they can only be used for very short distances (up to around 50 km). For longer distances, DC cables are the only solution if going underground.

In a CSC–HVDC multiterminal system, reversing the power flow direction on a certain branch requires the power flow direction reversal on the other branches including the reversal of polarity of the converters. The complexity that this introduces into multiterminal operation has limited the widespread use of multiterminal systems (MTDC).

As also presented in Chapter 3, only two multiterminal schemes, with a tapping connection on the DC system, have been in operation until now. Most notable are the links from Hydro-Quebec to New England (1990) and the connection between Italy, Corsica, and Sardinia (1989), which are in operation, both with three terminals. After more than 20 years, a new multiterminal HVDC system called NEA (±800 kV, 6000 MW, 1728 km) between Biswanath—Alipurduar–Agra, is under construction in India and is expected to be fully commissioned by 2016. In 2015, only one pole

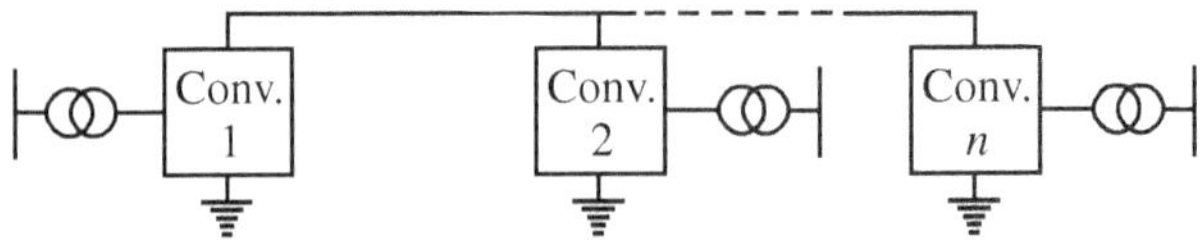

Figure 4.63 A monopole HVDC grid with earth return. *Source:* Asplund et al. 2013 [63]. Reproduced with permission of CIGRE.

was in operation. The system comprises four terminals located at three converter stations with a 33% continuous overload rating, thus with the possibility of transferring 7000 MW, which is the largest HVDC transmission ever built.

4.5.2 Multiterminal HVDC Grid Configurations[5]

The HVDC systems can be developed in monopolar or bipolar configurations or a combination of the two as shown in the following [63].

(i) *Asymmetric monopole system with earth return.* In its most simple form, a long distance HVDC system can consist of two converters each connected between earth and a high voltage conductor linking the two converters. Current flows between the two terminals through the high voltage conductor and returns through the earth is shown in Figure 4.63. This arrangement is called *asymmetric monopole with earth return.* The earth electrodes at the stations must be designed for continuous full current. The monopole system with earth return can be expanded to a multiterminal system and eventually to a complete HVDC grid.

Monopole systems with earth return are most attractive for the cable systems. In a cable HVDC project, usually the cost of the high voltage cable is a large portion of the total project cost and therefore avoiding extra cables brings considerable savings. However, continuous flow of current through earth is not acceptable in many parts of the world on the basis of environmental concerns.

(ii) *Asymmetric monopole system with metallic return.* To avoid continuous earth current, most new monopole systems use a similar arrangement but with a dedicated low voltage conductor for the return path. This configuration is called *asymmetric monopole with metallic return.* The neutral points of all converters should be connected by a low voltage conductor, which can be directly earthed at one terminal only (Figure 4.64). In this case, the neutral cable or metallic return conductor is insulated for low voltage, enough to withstand the voltage drop along the conductor and the neutral point voltage rises during abnormal (fault) operating conditions. Compared to the system with earth return, the metallic return system has higher transmission losses. This is due to the higher resistance of the metallic return path compared to the earth return.

(iii) *Symmetrical monopole grid* named "symmetrical monopole configuration" is mainly used for VSC-based HVDC systems, although CSC converters can also

[5] Reprinted with permission from CIGRE, Asplund et al. 2012 [63].

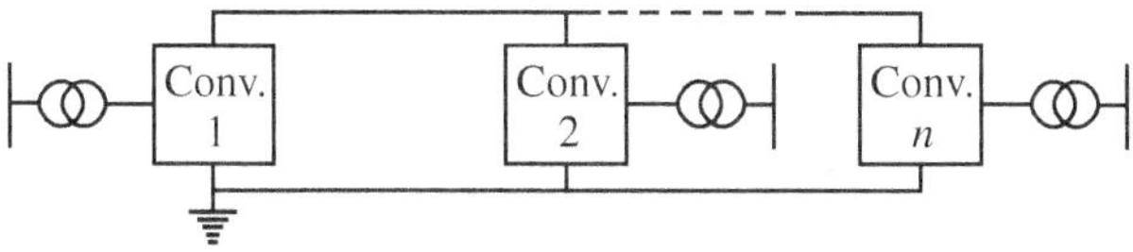

Figure 4.64 A monopole HVDC grid with metallic return. *Source:* Asplund et al. 2013 [63]. Reproduced with permission of CIGRE.

be designed with this configuration. In this configuration, the DC-side converters are connected between two high voltage conductors of the same magnitude but of opposite polarity (Figure 4.65).

The earth reference can be provided with various methods including the connection of the DC capacitors' midpoint as shown in Figure 4.66, high impedance reactors on the AC side of the converters, and high resistances from both DC buses to earth. In either case, no current will flow through the earth under normal operating conditions. Although in this configuration two conductors of opposite polarity carry power, they cannot operate independently, that is, if one conductor is out of service the healthy conductor cannot continue transmitting power using earth as return path. This is the main difference between symmetrical monopole and full bipole configurations. The majority of VSC-based HVDC systems to date utilize the symmetrical monopole configuration.

(iv) *Bipolar hvdc grid.* If an HVDC grid is configured as a monopole system, an outage of a converter will reduce the power flow through the corresponding terminal to zero. Depending on the grid structure, the outage of a conductor can be tolerating by rerouting the power. Under extreme situations, a conductor outage can cause the grid to be divided into separate islands. A bipolar arrangement can greatly improve system reliability by using two converters at each HVDC terminal, one connected between earth and the positive pole and the other between earth and the negative pole (Figure 4.66).

In this configuration if any of the converters or high voltage conductors is out of service, power transmission can continue (perhaps at a lower level) through the healthy pole using earth or a dedicated low voltage conductor as the return path. In case of a single pole converter outage in a two-terminal HVDC system, the healthy pole conductor can be used as the return path. In an HVDC grid application, this is only possible if the converter is connected radially to the rest of the grid. Otherwise, it is necessary to use a low voltage dedicated conductor to operate as the return path in case of a single pole outage.

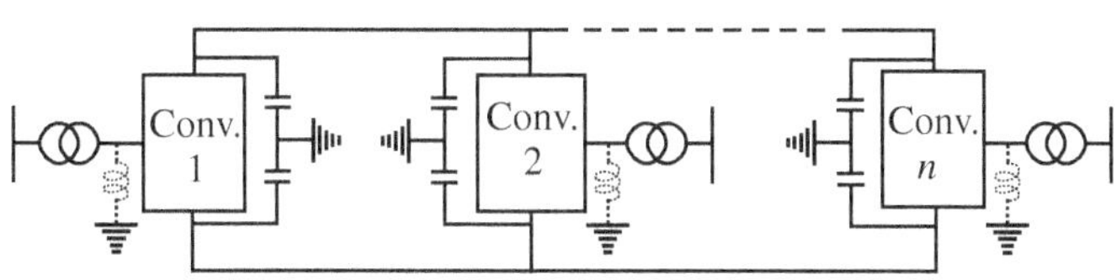

Figure 4.65 Symmetrical monopole configuration. *Source:* Asplund et al. 2013 [63]. Reproduced with permission of CIGRE.

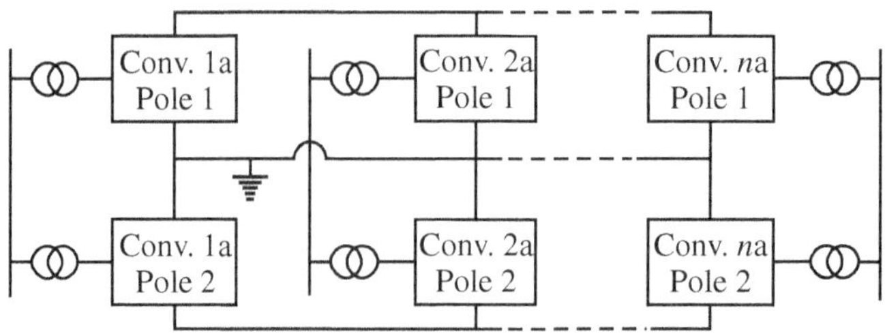

Figure 4.66 A bipolar HVDC grid with a dedicated metallic return conductor. *Source:* Asplund et al. 2013 [63]. Reproduced with permission of CIGRE.

In a normal bipolar system, the two poles are symmetrical with equal DC voltage and current ratings. During normal operation, the DC current in the two poles is controlled to be equal so that the neutral current remains near zero. If required, the metallic return conductor can be solidly earthed only at one point. This is necessary to avoid the DC current flowing through earth due to its lower resistance compared to the metallic return path.

(v) *Connecting monopolar converters to a bipolar HVDC grid.* Bipolar configuration with a dedicated metallic conductor is a promising configuration for future HVDC grids. Voltage-sourced converters built as symmetrical or asymmetrical monopoles can be connected to a bipolar grid as shown in Figure 4.67.

Under normal operating conditions, the symmetrical monopole converter carries balanced DC currents. If there is any imbalance in the DC current in the two poles due to a monopole or a monopolar outage in another terminal, the symmetrical monopole converter cannot adjust itself to absorb (or supply) unbalanced DC currents. The behavior under the pole to earth fault depends on the protection and earthing methods of the grid and the converter.

(vi) *System earthing.* The earthing principles of an electric transmission system affect the electrical equipment rating. The earthing principles define the personnel safety and the system protection principles that have to be applied. A high impedance earth system will result in significant voltage stresses during the grid contingencies, meanwhile in the case of a low impedance earth the system will experience significant current stresses.

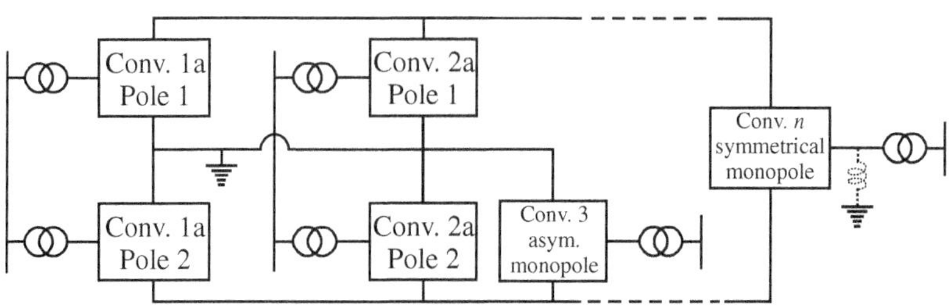

Figure 4.67 Connection of an asymmetrical monopole converter (three) and a symmetrical monopole converter (*n*) to a bipolar HVDC grid. *Source:* Asplund et al. 2013 [63]. Reproduced with permission of CIGRE.

The earthing strategy of HVDC grids may also be affected by the chosen grid configuration. The system can in principle be developed with monopolar converter stations, bipolar converter stations with two independently controlled poles, or symmetric monopole converter with one control system for two fully insulated poles with symmetric potential related to earth, as described earlier.

At present it seems to be increasingly difficult to get permission for operation with earth return, and several transmissions are now using isolated return conductors. Therefore, only the neutral metallic return alternative is considered in this section. Considering this fact a symmetric monopole arrangement with two fully insulated conductors/cables, instead of a neutral conductor/cable, offers the following advantages:

- two poles of opposite polarity but with the same voltage amplitude simplify the design of the converter transformer (combined AC and DC stresses can be avoided) and opens for possible configurations without converter transformers;
- higher transmission voltage between the positive and the negative poles reduces the DC current and thereby the thermal rating and the losses;
- the investment cost difference between a fully operational low voltage metallic return conductor and a fully insulated operational one is not so large, once the laying costs are also considered.

On the one hand, a bipolar system will have the advantage of the second bullet point together with an increased reliability. On the other hand, assuming that the use of ground electrodes is excluded, this solution will have to carry the cost of neutral return cables.

Based on the above aspects together with the capability of a higher transmission capacity, a configuration with two fully insulated poles is chosen as a starting point for the evaluation of an HVDC grid earthing strategy. For the bipolar configuration, the grid is assumed to include a low voltage neutral conductor. A bipolar grid may then during its development contain sections with both monopole and symmetric monopole converter stations.

- *High impedance earth.* During earth faults in high impedance earthed HVDC grids, the fault currents are effectively limited. However, the equipment in the grid will be subjected to large overvoltages, up to twice the normal operating voltage (Figure 4.68). This fact, together with the difficulties to implement an effective and selective protection system for earth faults are strong arguments against high impedance earthed HVDC grid.
- *Low impedance earth—Directly earthed systems.* During earth faults in effectively earthed bipolar HVDC grids, the overvoltages are limited meanwhile the earth fault currents may reach extreme values. With effective and fast HVDC breakers available, this grid configuration seems attractive (Figure 4.69).

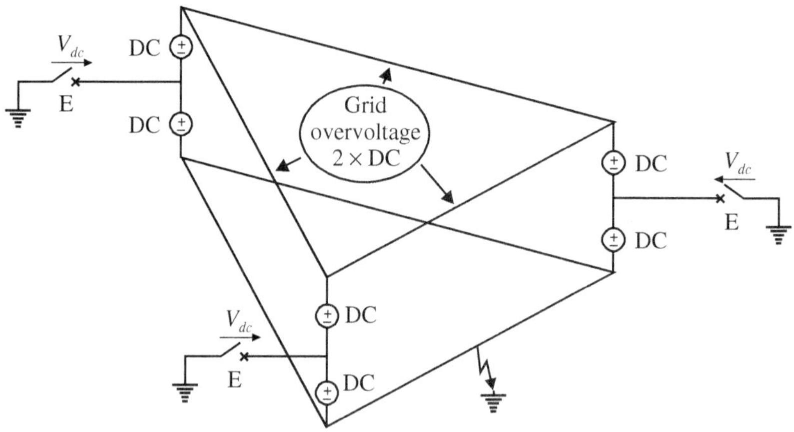

Figure 4.68 High impedance earthed DC grid. *Source:* Asplund et al. 2013 [63]. Reproduced with permission of CIGRE.

The earthing strategy with regard to the number of earth connections in a grid with low impedance earth will depend on the overall transmission configuration and the control and protection principles, for example,

- the HVDC grid is divided into areas with only one node earthed. Control supervision and switched earth connections may assure that one node is always earthed;
- several nodes are directly earthed. Special control loop regulates neutral current to acceptable levels (HVDC grids with symmetrical monopoles);
- a combination of directly earthed and impedance earthed converter stations together with supervision (and control) of the actual earth connections.

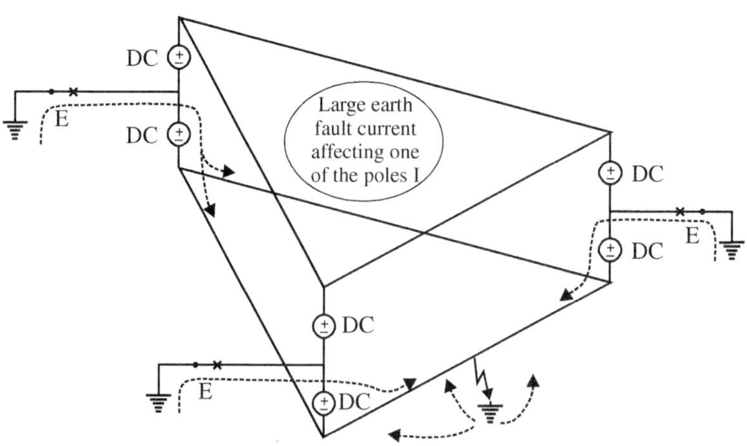

Figure 4.69 Directed earthed DC grid. *Source:* Asplund et al. 2013 [63]. Reproduced with permission of CIGRE.

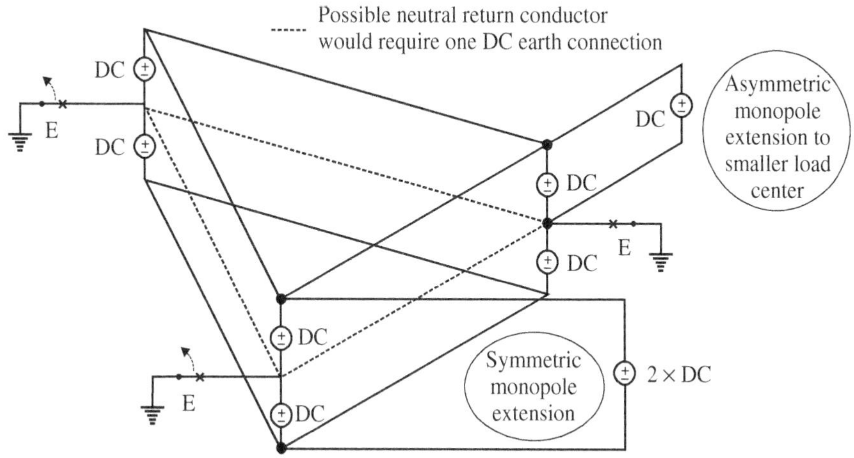

Figure 4.70 Single Pole extension connected to DC grid. *Source:* Asplund et al. 2013 [63]. Reproduced with permission of CIGRE.

- *Combination of bipolar and monopolar stations.* During the development of an HVDC grid, it may at some point be of interest to include asymmetric and symmetric monopoles in the network. Accordingly, it is of interest to evaluate how this may affect the chosen earthing strategy.

In the high impedance earthed DC grid, the poles of the converters will have to be rated for high voltages occurring in connection with an earth fault. In case of an asymmetric monopole also the neutral cable would have to be designed with increased insulation, and it may often be natural to design it with the same capability as the pole cable in order to cater for possible future upgrades. In these alternatives, the cost savings may be counteracted by the required higher insulation levels.

In the directly earthed DC grid, a monopolar converter station may be added without significant increased insulation levels. For an asymmetric monopole, a metallic neutral return conductor is required with a moderate increase of the neutral insulation level. The protections of a network with different types of converter stations will be strongly dependent on the overall strategy and the location of DC breakers. An example of a bipolar HVDC grid that was extended with monopoles (symmetric and asymmetric) is presented in Figure 4.70.

4.5.3 Meshed HVDC Grid Configurations

The advent of the DC breaker for higher voltages enables designing the HVDC systems from the simple point-to-point configuration to multilink, multiterminal (MTDC) or multistation configurations, eventually including technologies from different vendors (Figure 4.71).

Most of the HVDC systems in operation, either CSC- or VSC-based technology, are of point-to-point type (Figure 4.71a). Theoretically, the AC nodes can be connected by several point-to-point links in a so-called multiple-pair configuration

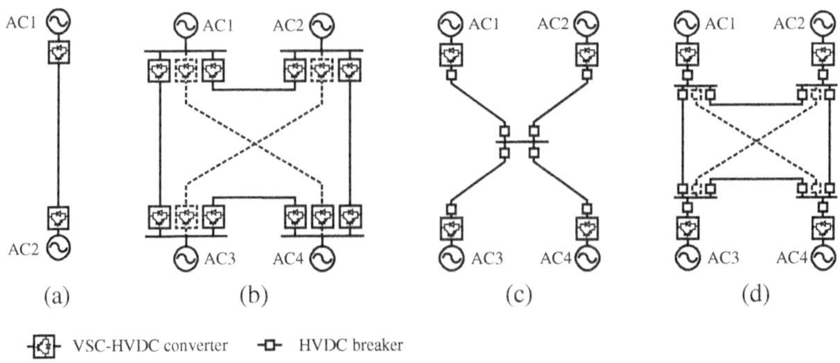

Figure 4.71 HVDC configurations: (a) point-to-point; (b) HVDC grid based on point-to-point links; (c) radial MTDC grid; (d) meshed MTDC grid; (Converter terminal; HVDC circuit breaker).

(Figure 4.71b). However, for technical and economical reasons, it is more effective to connect the lines on the DC side, forming an HVDC network (Figures 4.71c and 4.71d). The advantage is the reduced number of converters which results in reduced costs and power losses.

The VSC–HVDC technology is totally flexible when developing HVDC networks. Each VSC station can send/receive power or reverse the power flow direction while providing similar control only at the station involved in the transaction. With VSC technology, power reversal is achieved by reversing the direction of current flow, but requires no change to the polarity of the converter terminals. This opens up the possibility of designing multiterminal systems and hence full meshed DC grids [13].

The advantage of an HVDC grid increases with increasing amount of power, long distance, and increasing number of stations. If only a few reinforcements are needed, normal point-to-point HVDC transmissions will certainly do the job. However, when the number of point-to-point transmissions increases and the load flow will change frequently or is not known at the planning stage of these schemes then it could be advisable to consider a HVDC grid solution.

With more and more HVDC systems developed, it will soon be interesting to consider joining these links into an HVDC grid or "*supergrid*" (Figure 4.72) [63]. The schematic pictures show two possible situations where the increasing need for transmission of power has been solved by building only point-to-point transmission

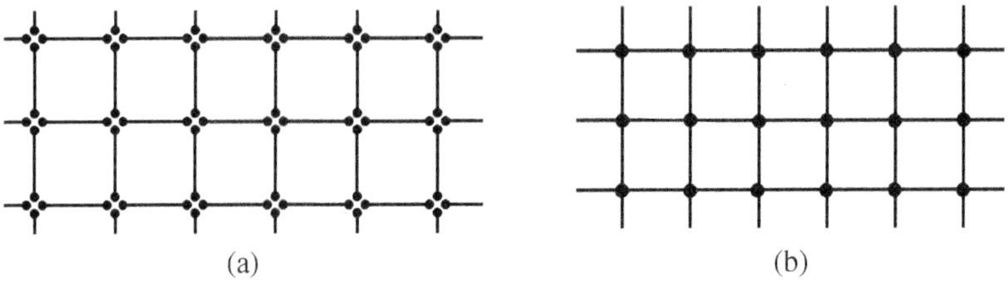

Figure 4.72 Overlay HVDC with multiple end-to-end transmission HVDC (a); overlay HVDC grid (b) [63].

HVDC or in the other case an HVDC grid. In the pictures, only the DC lines and DC nodes are shown, where the DC nodes are drawn as dots. It is assumed that a single converter is connected at each DC node.

The functional difference between the two solutions is very small; however, in the case with multiple point-to-point transmission links, all the transmitted power will be converted, whereas in the HVDC grid a smaller amount will be converted compared to the sum of the power to the node, that is the power will transmitted through the node without any conversion. The function of the nodes in the HVDC grid is to act as buses with protection and switching facilities. Power can in both cases be transmitted freely within the capacity of the transmission lines/cables. In the case of the HVDC grid, there is no power limitation in the nodes by the converter stations. If we assume that both converter stations and HVDC breakers have the capacity needed the performance regarding transmitting power should be similar.

An economic benefit of meshed HVDC grids might result from increased redundancy of the entire system including the underlying interconnected AC system. The conventional way to provide redundancy in a meshed AC system is to back up a branch device by means of a one-by-one redundancy. This is the well-known "$N - 1$" principle. In case of an overlay DC grid, commutating a fraction of the AC power flow into the DC grid and thereby deloading certain AC corridors can compensate the loss of one AC branch. This principle also works in the opposite direction, i.e. one-by-one redundancy is not entirely needed in an HVDC grid because of the existence of a meshed AC grid underneath.

4.5.4 Need for Fast and Low Loss HVDC Breakers

4.5.4.1 Preconditions

The relatively low impedance in HVDC grids is a challenge when a short circuit fault occurs, because the fault penetration is much faster and deeper. Consequently, fast and reliable HVDC circuit breakers (DCCB) are needed to isolate faults and avoid collapse of the common HVDC grid voltage. Furthermore, maintaining a reasonable level of HVDC voltage is a prerequisite for the converter station to operate normally. In order to minimize disturbance on converter operation, particularly the operation of stations not connected to the faulty line or cable, it is necessary to clear the fault within a few milliseconds [64].

Existing mechanical HVDC breakers which are capable of interrupting DC currents within several tens of milliseconds are too slow to fulfill the requirement of a reliable DC grid. Semiconductor-based DC breakers easily overcome the limitations in operation speed but generate large transfer losses typically in the range of 30% of the losses of a VSC station. To overcome these obstacles, a hybrid HVDC breaker was proposed in [65]. The alternative design has negligible conduction losses while preserving fast circuit interrupting capability.

At present, DC fault currents are still best and most swiftly interrupted by the converters themselves, which is applicable to CSC technology only. Existing HVDC switches have been used for more than 30 years in the neutral switchyard of bipolar HVDC installations. Here they perform various functions, such as rerouting HVDC current during reconfiguration of the main circuit, or helping to extinguish fault currents. For example, the metallic return transfer breaker (MRTB) is used to commutate the current from the ground path to a metal conductor, when there are restrictions on how long an HVDC current can be routed through the ground (see also section 3.1.2). Other examples include the ground return transfer switch, neutral bus switch, and neutral bus grounding switch.

The main difficulty in realizing a HVDC circuit breaker resides in the absence of natural current zero crossings in DC circuits. Thereby, the breakers have to fulfill the following basic requirements [66]:

(i) create a current zero crossing to interrupt the current;

(ii) dissipate the energy stored in the parallel circuit inductance;

(iii) withstand the voltage response of the network after current interruption.

Depending on the breaker application, additional requirements may be needed, such as

- as stated before, for VSC topology, the HVDC breaker has to be able to interrupt quickly;
- the maximum voltage generated by the breaker must not exceed a value given by the insulation coordination of the DC system. This is particularly important for switching of load currents, where the network is at nominal voltage.

The fault clearing time varies depending on the system configuration (radial or meshed network), design parameters of the VSC, transmission capacity, the presence of a series-connected DC reactor, and impedance of line/cable. However, DCCBs are required to clear the fault with a shorter time as compared with AC circuit breakers [67].

4.5.4.2 *Schemes for the Current Zero Formation*

The DC current can be brought to zero by generating a countervoltage of similar or larger amplitude than the system voltage. When the circuit breaker have to be tripped a commutation circuit is introduced in parallel with it thus creating a countervoltage and artificial current zeroes across the breaker contacts. The parallel circuit must have the capability of dissipating the energy stored during this period. The larger the countervoltage, the smaller the time needed to interrupt, but the larger is the energy that have to be dissipated. Figure 4.73 shows various schemes for the current zero formation [67].

Current limiting switch scheme is more suited to low voltage class DC no-fuse breaker (Figure 4.73a). An example of such scheme is the 2-kV air-blast type high-speed switch used for railway power systems. This scheme is not suitable for HVDC circuit breakers because the arc voltage induced by air or SF6 across the contacts of the breaker (several kilovolts at maximum) limits the DC current and creates a current zero.

Passive resonant current zero formation scheme (Figure 4.73b) is often employed to MRTB, which clears currents up to a few thousands of amperes. The parallel commutation circuit is a passive reactor/capacitor (L/C) resonant circuit that

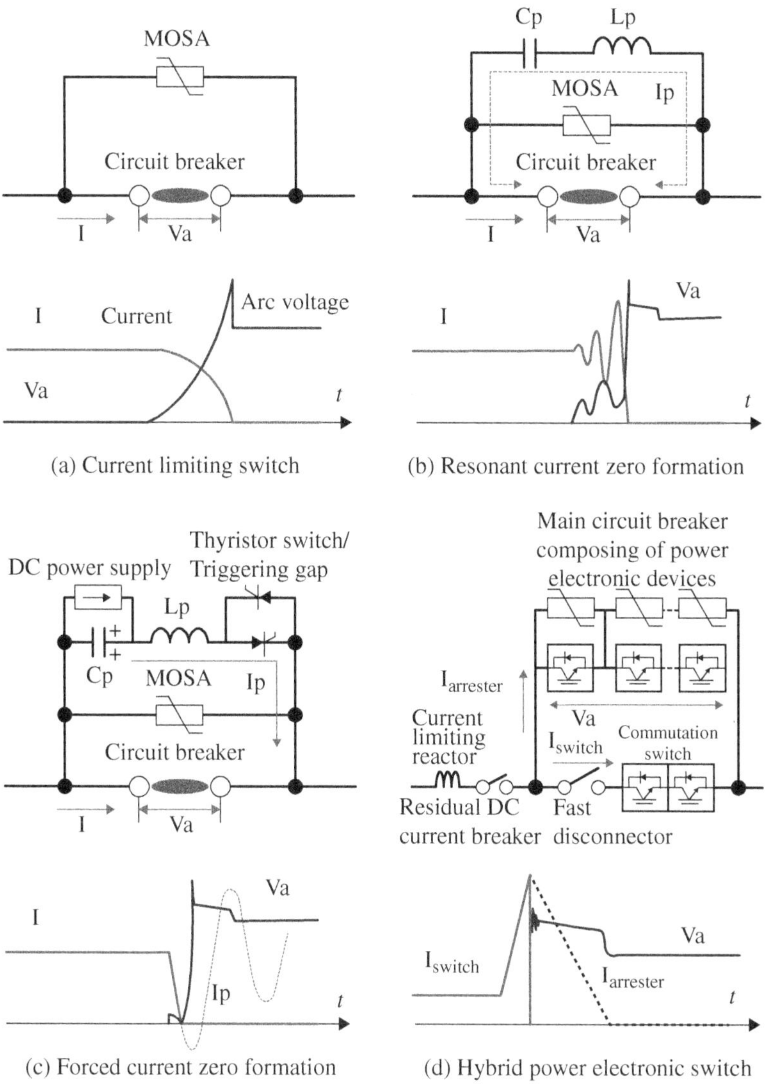

Figure 4.73 DC circuit breakers with different interrupting schemes. *Source:* Tahata et al. 2014 [67]). Reproduced with permission of CIGRE.

causes the current to oscillate around the initial value, with a frequency in the range of 1–3 kHz, until current zeroes are created where the arc can be extinguished. When the arc is extinguished, the capacitor is rapidly charged until the nonlinear resistor takes over the current and limits the voltage to a suitable value. This scheme can interrupt the fault generally within 20–40 ms.

Active resonant (forced) current zero formation scheme (Figure 4.73c) is also used for electromechanical circuit breakers. With the help of a thyristor switch/triggering gap in the active resonant circuit, a precharged large-capacity capacitor in series with a reactor generates a high frequency (several kilohertzs) inverse current that cause the current to instantly decrease to zero. This allows the DC fault to be eliminated within 8–10 ms. This scheme may be suitable for high voltage applications.

Hybrid power electronic switch scheme (Figure 4.73d) is composed of mechanical switches and semiconductor devices. When a DC fault occurs, the current is diverted toward the series and parallel connected power electronic devices which can block the current within a few milliseconds.

4.5.4.3 Types of DC Circuit Breakers

THE ELECTROMECHANICAL CIRCUIT BREAKER [68, 69]. Most of the practically realized HVDC circuit breakers include also separate commutation and energy-absorbing paths (Figure 4.74a):

- The *normal path* comprises only the circuit breaker, with low ohmic losses in closed position, which is so far, only possible with movable metallic contacts. Upon opening of these contacts, an arc is established and its arcing voltage is used to commutate the current to the resistive path where the energy of the system is then dissipated.
- The *commutation path* is a series resonance circuit consisting of a capacitance C_c and inductance L_c that cause the current to oscillate with the natural frequency $\omega_0^2 = 1/L_cC_c$, as well the switch S_c;
- the *dissipation path* that consists either of energy absorbing linear or nonlinear resistors in series with a switch S_e or nonlinear ZnO varistors; this path becomes partly conductive only above a certain applied voltage.

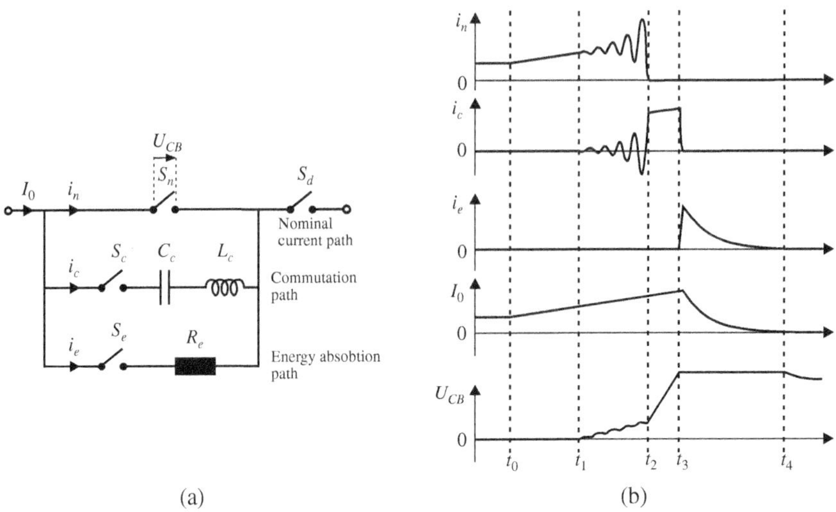

Figure 4.74 (a) Electromechanical circuit breaker structure [68]; (b) basic current and voltage development during interruption. Adapted from Vithayathil 1983 [69].

Operation in time of this circuit breaker scheme is summarized as follows (Figure 4.74b):

- *Period* t_0 *to* t_1: At instant t_0, a DC fault occurs and the DC current starts to increase. The interrupter contacts of the nominal current path separate at t_1;
- *Period* t_1 *to* t_2: at instant t_1, the commutation path is introduced in the circuit, the current is oscillating with increasing amplitude started by the natural fluctuation in the arc voltage;
- *Period* t_2 *to* t_3: when at instant t_2, the amplitude of the oscillating current i_n becomes larger than the system DC current, I_0, a current zero crossing occurs in the nominal path, and the interrupter S_n interrupts the current. Current I_0 continues to flow in the commutation path, charging the capacitor C_c.
- *Period* t_3 *to* t_4: at instant t_3, when the capacitor voltage exceeds a given value, which is chosen to be the voltage capability of the circuit breaker or the insulation coordination of the HVDC system, the energy absorption path acts causing the system current I_0 to decrease. The voltage U_{CB} is limited by the elements of the dissipation path, current only flows through the energy-absorbing path, and the current I_0 of the system ceases.

In this scheme, optimization (better use) of the costly components (varistor, capacitor) is required to reduce the cost. Also, the optimum capacitance value would minimize the breaker's interruption time and improve the whole interruption performance. At the same time, a large C/L ratio can help maximize the break's interruption performance. The cost of components required for an electromechanical DC circuit breaker would not be significantly higher than that of an AC circuit breaker.

Electromechanical HVDC circuit breakers were available up to 500 kV, 5 kA and have a fault-clearing time of the order of 30–50 ms. Recently, a new MRTB DC commutation breaker with a rated current of 5000 A has been tested on a HVDC transmission system (± 800 kV, 8000 MW), in China, by Siemens.

The two main differences between these transfer breakers and the hybrid HVDC breaker are that (i) the transfer breakers operate considerably slower that the hybrid breaker and (ii) that part of the current is transferred, rather than interrupted. These high-voltage switches or breaker systems use an AC-type of high voltage breaker; the zero crossing of the HVDC fault current is imposed by discharge of a capacitor bank to generate a current opposite to the fault current, in order to extinguish the arc (see section 4.5.4.2).

Solid-State Circuit Breakers are capable of interrupting faults in the medium voltage grids, eventually supplied from multiple sources. They are based on IGCT, which compared to IGBT (bipolar transistors) have lower on-static losses (Figure 4.75) [70]. Also, since no mechanical parts are included, they provide interruption time of a few milliseconds, much faster than the electromechanical circuit breakers.

Under normal operating conditions, the current flows through the gate commutated thyristors (GCTs). Once a fault is detected, the semiconductors are turned-off, the voltage across the varistor (that is parallel to the thyristor) increases rapidly until it starts to conduct. Any voltage higher than the grid voltage is blocked due to the

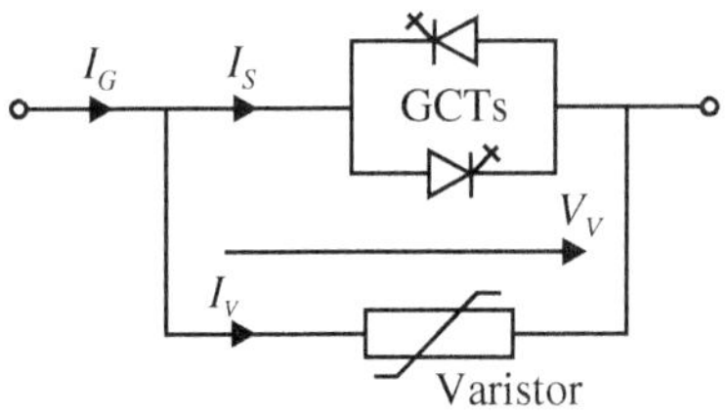

Figure 4.75 Solid-state circuit breaker.

design of the varistor. This in turn, leads to the demagnetization of the line inductance. The higher the voltage across the inductor the higher the overvoltage which allows the demagnetization process to be performed faster [71].

There are still some problems that have to be solved to achieve static and dynamic stability, that is (1) synchronization of firing signal for each thyristors. Asynchronous firing of the electronic devices causes some of them to burn out; (2) problems with balancing the voltage and current in each electronic device, which may damage the DC breaker; (3) increased switching loss: the larger the number of electronic devices the greater the switching losses. For these reasons, DC breakers are difficult to realize.

Typical ratings of solid-state circuit breakers in operation are 4 kV, 2 kA, although in ratings of up to 150 kV, 2 kA were considered.

Hybrid DC Circuit Breaker consists of a main breaker and a bypass breaker (Figure 4.76). The bypass breaker is formed by a semiconductor-based load commutation switch that matches lower voltage and energy capability in series with a fast mechanical disconnector. The main breaker has several modules each comprising a string of IGBTs connected back-to-back with antiparallel diodes. Individual arrester banks, dimensioned for full voltage and current breaking capability, are connected in parallel with these modules to limit the rapid increase of voltage after the main breaker is switched off.

Fault clearance by a hybrid DC circuit breaker requires the following steps [64, 65, 72]:

(1) Under normal operating conditions, the normal current flows through the bypass breaker, that ensures very low losses, while the current in the main breaker is zero.

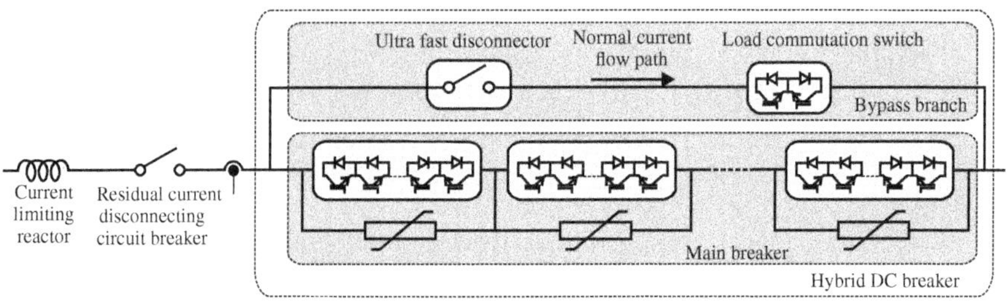

Figure 4.76 Hybrid HVDC breaker. *Source:* Callavik et al. 2012 [64]). Reproduced with permission of ABB.

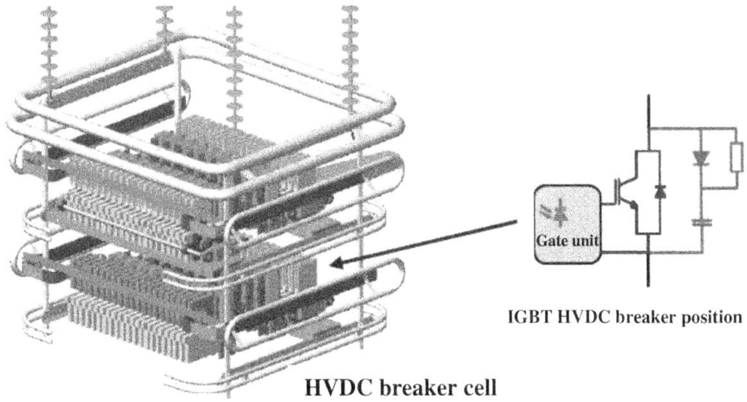

Figure 4.77 Design of the 80 kV main HVDC breaker cell. Reproduced with permission of ABB.

(2) When a DC fault occurs the current start to rise. When the current exceed the preset overcurrent value, the load commutation switch immediately commutates the current to the main DC breaker, then the mechanical disconnector opens at zero current and very low voltage stress, thus without an arc. It isolates the load commutation switch from the primary voltage across the main breaker during current breaking.

(3) The main breaker modules open and divert the fault current into the corresponding arrester banks.

(4) After fault clearance, the disconnecting circuit breaker interrupts the residual current and isolates the faulty line from the HVDC grid to protect the arrester banks of the hybrid HVDC breaker from thermal overload [65].

The Swiss company ABB has announced in 2012 the successful tests performed on a hybrid HVDC breaker having a current breaking capability of 9 kA in an HVDC grid with rated voltage of 320 kV and rated HVDC transmission current of 2 kA [64].

The main HVDC breaker consists of several 80 kV HVDC breaker cells. Each cell contains four stacks as shown in Figure 4.77, two stacks being used to break the current in either current direction. Arrester banks are in parallel with each cell to limit the maximum voltage across them to a specific level during current breaking. The stacks are composed of up to 20 series-connected IGBT HVDC breaker positions. Due to the large *di*/*dt* stress during current breaking, a mechanical design with low stray inductance is required. Application of press pack IGBTs with 4.5 kV voltage rating enables a compact stack design and ensures a stable short-circuit failure mode in case of individual component failure.

Resistor–capacitor–diode snubbers across each IGBT position ensure equal voltage distribution during current breaking. Optically powered gate units enable operation of the IGBT HVDC breaker independent of current and voltage conditions in the HVDC grid. For the design of the load commutation switch, one IGBT HVDC breaker position for each current direction is sufficient to fulfill the requirements of the voltage rating. In order to improve the reliability of the switch, redundant IGBT modules are used. Parallel connection of IGBT modules increases the rated current of the hybrid HVDC breaker. The load commutation switch requires a cooling system since it is continuously exposed to the line current [64].

4.5.5 HVDC Grid Protection

Protection strategy of an MTDC against faults differs from that of an AC grid due to the different nature of the short-circuit phenomenon [73]. As explained earlier, the DC fault must be cleared faster than the AC faults and the challenge is to create current zero crossings. The HVDC converters together with the AC grid breakers are designed to provide faults clearing in point-to-point configurations. In HVDC grids, however, the HVDC breakers are essential to meet the protection requirements.

Similar to an AC protection system, a DC protection system has to meet the following requirements: (*i*) speed – fast fault clearing is mandatory in order to prevent the fault spreading to other parts of the system; (*ii*) sensitivity – all faults must be detected, whereas transients in the normal operation should not trigger a protection action; (*iii*) selectivity – a correct fault location has to be provided; (*iv*) robustness – the backup protection should work when the main protection is not functioning; (*v*) seamless – after fault clearing, the rest part of the power system should continue to operate in secure conditions.

The most stringent requirement is the speed of operation. The total time necessary to clear the fault is

$$t_c = t_d + t_i + t_b$$

where t_d and t_i are the time intervals necessary for fault detection and identification (localization), which depend on the protection algorithms and relay operation, and t_b is the time necessary for the effective fault clearing (after receiving a tripping signal), which depends on the method of fault current interruption.

The protections of an MTDC can be coordinated either at the grid level or at the converter level. The protections coordinated at the grid level are DC line voltage unbalance protection, DC line overvoltage protection, and DC undervoltage protection. At the converter level, the main protections are DC overcurrent protection, DC current differential protection, busbar current differential protection, and asymmetry protection.

Detection and identification. The main characteristics that are considered in the fault detection algorithm are the fault resistance and grounding. The fault resistance can vary from zero to dozens of ohms. There are two types of grounding solutions employed: (*i*) nongrounded systems – this solution provides better reliability of power supply in case of pole-to-ground faults and (*ii*) grounded systems (including high resistance grounding).

The protective relaying algorithms must detect and identify the fault within a very short amount of time. As the timescale of the HVDC grid protection is 10–100 times shorter than for AC grid protection, customized algorithms are needed. Similar to AC grid protection, these relaying algorithms can be divided into nonunit and unit protection methods [73].

Nonunit protection makes use of open protection zones, which are only at one side clearly determined. They make use only of measurements local at the relay, which enables them to be fast.

Unit protection is based on closed protection zones and requires communication between the relays at both ends of the protection zone. Unit protection methods are inherently selective, but suffer from the time delays caused by communication. In AC grid protection, communication delays are typically in the order of 10 ms.

METHODS OF FAULT CURRENT INTERRUPTION. The DC faults can be cleared by intervention of one of the following components [73]:

(i) *Converter AC breakers*. This method requires several cycles of the fundamental AC frequency to operate, thereby current breaking is done within tens of milliseconds. Furthermore, this method results in nonselective fault current interruption at the DC side.

(ii) *Converters with fault-blocking capability*. The HVDC converters can be blocked within microseconds upon fault detection, leading to faster fault current interruption than using the AC breakers. The same strategy has to be applied to all rectifier terminals in order to prevent the DC arc from being supplied, thus requiring to shut down the entire HVDC grid.

(iii) *HVDC breakers*. Effective HVDC grid protection is possible by using HVDC breakers. By placing the breakers at both ends of all HVDC lines, only the line on which the faults occurs is disconnected thus minimizing the affected parts of the HVDC grid. As the HVDC breaker is available for greater voltages and currents, VSC–HVDC grids can be designed at larger scale.

COORDINATION. The HVDC grid converters must be coordinated in order to maintain the power balance within the HVDC grid when a DC fault leads to disconnection of a terminal station. A fast coordination between stations is thereby required in order to ensure the seamless capability at the HVDC grid level. Depending on the operation mode of the disconnected station, the other HVDC grid station must increase or decrease the power injected or absorbed.

FAULT CLEARING STRATEGIES FOR MTDC SYSTEMS [73]. Depending on the type and location of the fault, five protection strategies can defined, as shown in Figure 4.78.

STRATEGY (A): LINE BREAKERS PROTECTION. This strategy assumes the use of the breakers at the end of the faulted line to clear the fault within a time that allows preventing the fault from affecting other parts of the grid, a philosophy similar to that employed in the AC grids (Figure 4.78a). Thereby, fast and selective fault clearing, also with the use of nonunit protection, is required. Fault current limiters should be used when extension of the clearing time is intended.

STRATEGY (B): LINE BREAKERS AND CONVERTERS PROTECTION. If the converts adjacent to the ends of the faulted line are allowed to operate in blocking mode, then the current that the DC breakers have to break is significantly reduced (Figure 4.78b). The required time for fault clearing is greater than for strategy (a), but should be limited to few tens of milliseconds in order to prevent the HVDC grid from collapsing.

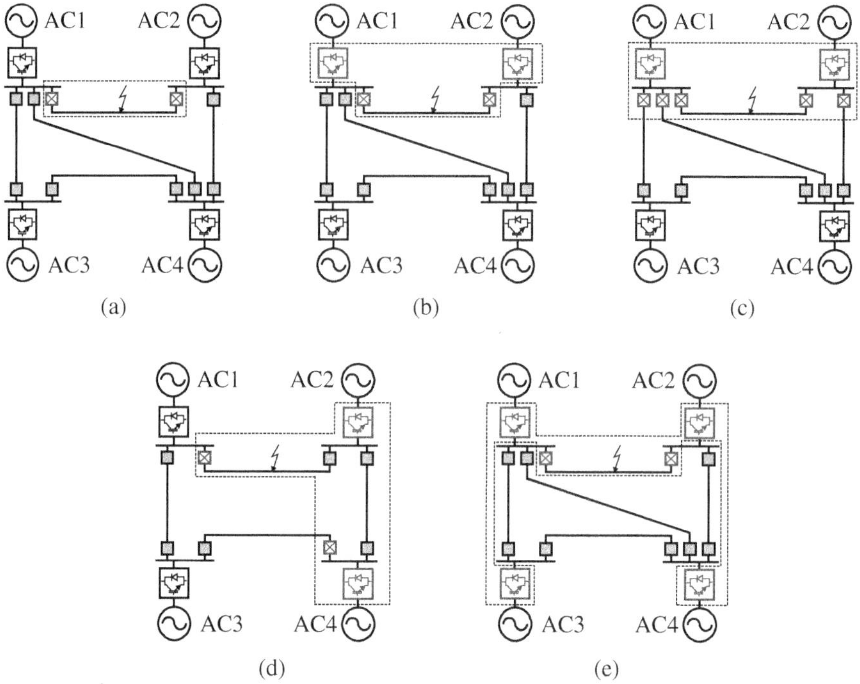

Figure 4.78 Fault clearing strategies in a sample MTDC system. The dashed line indicates the HVDC grid elements involved in fault clearing [73].

Strategy (c): Open Grid Protection. In this strategy, the breakers and the converters adjacent to the two ends of the faulted line cooperate to clear the fault (Figure 4.78c). After fault clearing, the breakers of the healthy lines reclose to restore normal operation of the grid. Since more lines are now disconnected, stability of the HVDC grid becomes the determinant in designing a reliable relaying algorithm that will also clear the fault within the required timeframe.

Strategy (d): Grid-Splitting Protection. The HVDC grid, in this case, is split into different protection zones which are separated by fast breakers (Figure 4.78d). The fault is removed in two stages. In the first stage, the zone within which the fault occurred is isolated by the action of breakers and converters, within a timeframe imposed by the converter IGBTs of the healthy part of the grid. In the second stage, the faulted line is safely disconnected, within a timeframe that is either imposed by the converter diode surge withstand capability or AC system stability, which allows slower fault clearing.

Strategy (e): Low-Speed HVDC Grid Protection. This protection philosophy may be employed when the entire HVDC grid is affected in case of faults (Figure 4.78e). One option would be to clear the fault by actions at all converters in the network. This strategy may have a great impact on the AC system, and thereby it should be limited to small HVDC grids. Another option is to limit the fault current infeed of the converters, which enables the use of slower protective algorithms and breakers.

For this option, the HVDC grid is not completely switched off, which enables faster fault recovery. The time constraint for this option is determined by the AC network constraints.

The decision for choosing the protection strategy is taken depending on the HVDC grid size (in terms of number of terminals and transmission capacity, the existence of breakers, characteristics of converters, and others. Strategies (a)–(c) are suitable when fast fault clearing is required, whereas strategies (d) and (e) can be chosen when the impact of the DC grid on the AC grid is limited.

4.6 LOAD FLOW AND STABILITY ANALYSIS[6]

The establishment of DC grids is one of the most recent technological challenges in power systems and energy transmission. This new technique offers the integration of the many different renewable and nonrenewable energy sources such as wind, photovoltaic, biomass, hydroelectric, thermal, etc.

The use of VSC technology in HVDC systems opens the possibility to

- The development of a new super DC grid over the already existing AC network and
- Multiterminal connections for long line transmissions.

Recent developments in the VSC technology for HVDC systems suggests that in the future it will be possible to create highly controllable multiterminal DC grids operating along with the existing AC networks. The perspective is a much more stable and reliable power system. Therefore, it is necessary to develop methods and improve the existing tools to analyze possible configurations for adequate operation and planning of the new integrated power system. With the expansion of the existing AC network and the development of a new DC grid, the complexity of the system increases. Predicting its possible behavior during abnormal conditions will demand the renewal of the existing tools to incorporate the effects of this new technology. The available programs currently used do not present the means to foresee the static or dynamic behavior of DC grids and its impact over AC networks.

4.6.1 Load Flow in Meshed AC/DC Grids

4.6.1.1 Generalities

The advent of a more flexible technology for integrating AC with DC networks demands a follow-up of analysis tools. Since a high level of independence between both networks can be achieved through active control, a decoupled method to study the load flow problem in meshed AC/DC grids is herein proposed. Because the DC converters are set to control DC voltage or active power, it is necessary to calculate first the DC grid in order to determine all the boundaries conditions for the AC network. The AC network is then solved.

[6] The section was contributed by Marco Antonio Barbosa Horita and Marcos Tiago Bassini, PhD students at University of Sao Paulo, Brazil.

The main focus of this chapter is to analyze the distribution of power within the DC grid, its converters, and the AC network. It may include the independent effects over positive and negative poles and the results related to the neutral grid due to an unbalance between poles. The power flow may also be predicted during abnormal conditions such as the loss of a pole, an entire DC line, or a converter.

Algorithms for solving the DC grid are proposed including the interaction between positive, negative, and neutral grids. An example showing the solution of a particular DC grid through the proposed method is presented in section 4.6.1.3.

One of the greatest advantages of the VSC over the traditional CSC–HVDC systems is the possibility to independently control active and reactive power, or the magnitude of the AC and DC voltages. This greater flexibility tends to separate the AC and DC networks into subsystems as they operate independently.

With the proper control of the switches in the VSC, it is possible to generate an AC voltage whose magnitude and phase angle can be controlled as desired. This ability enables another level of control; it is possible to handle variables such as active and reactive power flowing through the converter independently. This is one of the most important advantages of this new kind of the AC/DC converter over the traditional thyristor-based converter. These new possibilities of control, stability management, and integration of different AC networks, along with the capacity of using VSC for black start of an AC network, make this new technology very promising. More interesting is the recent trend in the VSC technology based on MMC. This kind of VSC is capable of constructing an AC voltage with a waveform very close to a sine wave, avoiding the need of low harmonics filters, thus reducing the overall complexity of the system. The operation principles and control of a VSC-based HVDC system is illustrated in Figure 4.79.

As depicted in Figure 4.79, it is possible to control directly the magnitude and phase angle of the generated AC voltage. It can also be used another strategy and/or control variables such as (i) active and reactive power, (ii) active power and AC voltage magnitude, (iii) DC voltage and reactive power, or (iv) DC voltage and AC voltage magnitude.

The VSC converter in the AC network can be represented as type 2 bus (generator bus) in which P and V are the known variables, or a type 3 bus (load bus) where P and Q are known variables. In a conventional load flow program, there may be a

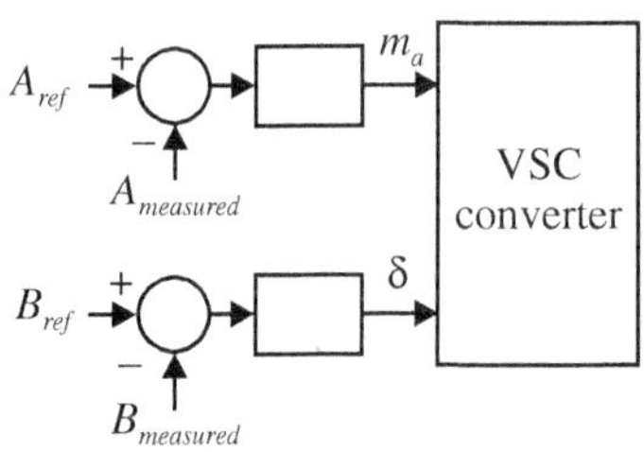

Figure 4.79 Direct control of modulation index m_a and phase angle δ by parameters A and B.

minor difference from normal usage. The converter can have negative P for type 2 bus and positive P for type 3 bus, this situation is rather uncommon in conventional AC load flow, though it, should not introduce any constraints.

4.6.1.2 Load Flow Calculation in a DC Grid

Basically ,there are two configurations of VSC HVDC systems [74]: the symmetrical monopole and the bipole configuration (section 4.5.2.). The difference between these two systems is that in the bipole configuration each pole can be controlled and operated independently whereas in the symmetrical monopole this is not possible. The loss of a line or a converter in the symmetrical monopole means the loss of that entire segment, thus in the symmetrical monopole only a balanced operation is viable. In a bipole configuration, the loss of a pole, or a converter, means that only half of the link is lost.

In a bipolar VSC HVDC system, there may be a total of three subnetworks: the positive, the negative, and the neutral networks. The operating condition will affect all of them, and the load flow calculation will be a result of an interactive process involving these three networks. The difference between currents in the positive and negative pole will cause a current injection in the neutral grid, leading to an electrical potential difference between the local neutral and the ground (or reference). The fact that the local neutral voltage for each converter may be different from zero affects the load flow in the positive and negative poles. Thus, finding the load flow solution for an unbalanced operation is essential.

Methods like Newton–Raphson and Gauss–Seidel may be applied to solve the problem, and the only adaptation is that there will be the interaction between the neutral grid and the poles affecting the converters' local references. Some adjustments need to be made in the algorithms to include this interaction during unbalanced operation. The current injection approach should be accounted for this purpose.

$$[I_P] = [Y_{bus,P}][V_P]; [I_N] = [Y_{busN}][V_N] \rightarrow [I_n] = [I_P] + [I_N] \tag{4.77}$$

$$[V_n] = [Z_{bus,n}][I_n]; [V'_P] = [V_p] - [V_n]; [V'_N] = [V_N] - [V_n] \tag{4.78}$$

The subscripts P, N, and n account for the positive pole, negative pole, and neutral system, respectively. V is the voltage measured to ground reference, and V' is the voltage measured to local neutral reference. The Y_{bus} matrix in each network may be different; the injected power and slack busses may also be different in the positive and negative grids. It is important to consider that the injected power in each bus is referenced to the local neutral and not to the ground reference; this can generate a different injected power from the initial set point when calculated to ground reference. With the adjustments described above, it is possible to calculate the load flow during contingencies and unbalanced operation.

The nodal analysis must be done separately for the three subnetworks (+/–/n).

$$[I] = [G][V] \tag{4.79a}$$

where

I is the injected currents;

V – the voltage between pole and neutral reference;

G – the real part of the Y_{bus} matrix.

Equation (4.79a) can be rewritten as

$$I_j = \sum_n G_{jk} V_k \tag{4.79b}$$

Assuming that from n buses m are not controlled (type u) and that there is at least one controlled bus (type c). I_j can be expressed as

$$I_j = \sum_{l\ldots m} G_{jk} V_{u_k} + \sum_{m+l\ldots n} G_{jw} V_{C_w} \tag{4.79c}$$

$$I_{S_J} = I_j - \sum_{m+l\ldots n} G_{jw} V_{C_w} = \sum_{l\ldots m} G_{jk} V_{u_k} \tag{4.80}$$

In matrix form ($m \times n$ equation):

$$[I_S] = [G][V_u] \tag{4.81a}$$

$$[V_u] = [R][I_s] \tag{4.81b}$$

The system is solved through iterations. The z iteration is calculated by:

$$I_{S_J}^{(z)} = \frac{P_j}{V_{u_j}^{(z-1)}} - \sum_{m+l\ldots n} G_{jw} V_{C_w} \tag{4.82a}$$

$$V_u^{(z)} = R I_S^{(z)} \tag{4.82b}$$

This procedure must be applied to the positive and negative networks separately. The following equations apply for the neutral grid:

$$[I_r] = [G_r][V_r] \tag{4.83a}$$

where I_{r_j} is the neutral injected current, which is the sum of the injected currents in the positive and negative poles

$$I_{r_j} = -\left(\frac{P_{j+}}{V_{u_{j+}}} + \frac{P_{j-}}{V_{u_{j-}}} \right) \tag{4.83b}$$

where $[G_r]$ is the neutral grid conductance matrix. In the $[G_r]$ matrix, there must be a common reference, which is a grounded bus where all voltages are referenced.

The positive and negative grids are solved by similar equations. For the positive grid, we have

$$I_{j+} = \sum_n G_{jk} V_{k+} \tag{4.84a}$$

$$P_{C_{j+}} = (V_{j+} - V_{jr}) \sum_n G_{jk} V_{k+} = (V_{j+} - V_{jr}) \left\{ G_{jj} V_{j+} + \sum_{n \neq j} G_{jk} V_{k+} \right\} \tag{4.84b}$$

where

$P_{C_{j+}}$ is the injected active power in the positive pole at the converter terminal;

V_{j+} – the voltage between the positive pole and the local neutral reference;

V_{jr} – the voltage between the local neutral and the global reference.

Defining the matrix J as

$$J_{jj} = \frac{\partial P_{C_{j+}}}{\partial V_{j+}} = \frac{\partial}{\partial V_{j+}}(V_{j+} - V_{jr})\left\{ G_{jj}V_{j+} + \sum_{n \neq j} G_{jk}V_{k+} \right\} =$$

$$= 2G_{jj}V_{j+} - V_{jr}G_{jj} + \sum_{n \neq j} G_{jk}V_{k+} \tag{4.85a}$$

$$J_{jk} = \frac{\partial P_{C_{j+}}}{\partial V_{k+}} = G_{jk}V_{k+} \tag{4.85b}$$

It is important to note that $\frac{\partial P_{C_{j+}}}{\partial V_{k+}}$ is equal zero if k is a bus with controlled voltage (swing or generator).

In matrix form, we have

$$[\Delta P_{C_{j+}}] = [J][\Delta V] = \left(P_{C_{j+}} - P_j^{sp}\right) \tag{4.86}$$

where P_j^{sp} is the target injected power determined by the user. The objective is to get as closes as possible to $\Delta P_{C_{j+}} = 0$ (or, equivalently, $\Delta V = 0$).

By solving equation (4.86) with respect to ΔV, it is possible to determine the voltage for the next iteration:

$$V^{(z)} = \Delta V + V^{(z-1)} \tag{4.87}$$

There are many methods and algorithms to obtain the condition $V^{(z)} - V^{(z-1)} \cong 0$ (or $\Delta P \cong 0$). The most used one for solving load flow problems is the Newton–Raphson thus its application in DC grids will be explained and developed here.

THE NEWTON–RAPHSON METHOD. The Newton–Raphson algorithm is a numerical iterative method that has the objective of estimating the roots of a function. So, for a function $F(x) = 0$ we have

$$x_{k+1} = x_k - \frac{F(x_n)}{F'(x_n)}$$

For solving a DC power flow, we can use the following F function:

$$H(V_i) = P \tag{4.88a}$$

$$H(V_i) - P^{sp} = 0 = F(V_i) \tag{4.88b}$$

where P^{sp} is the target injected power (the condition defined by the user and that defines the boundaries of the problem) in the DC buses.

$$H(V_i) = [V_i] \cdot ([G][V_i])$$

where $[V_i]$ is a vector with the voltages between the DC buses and the local reference.

Applying the Newton–Raphson method:

$$[V_i]^{k+1} = [V_i]^k - \frac{F(V_i)}{F'(V_i)} \tag{4.89a}$$

$$H(V_i) = \begin{bmatrix} V_1 \\ \vdots \\ V_n \end{bmatrix} \cdot \begin{bmatrix} G_{11} & \cdots & G_{1n} \\ \vdots & \ddots & \vdots \\ G_{n1} & \cdots & G_{nn} \end{bmatrix} \begin{bmatrix} V_1 \\ \vdots \\ V_n \end{bmatrix} = \begin{bmatrix} \sum_{n=1}^{n} G_{1i} V_i V_1 \\ \vdots \\ \sum_{i=1}^{n} G_{ni} V_i V_n \end{bmatrix} \tag{4.89b}$$

$$F(V_i) = \begin{bmatrix} f_1(V_1) \\ \vdots \\ f_n(V_n) \end{bmatrix} = \begin{bmatrix} \sum_{n=1}^{n} G_{1i} V_i V_1 \\ \vdots \\ \sum_{i=1}^{n} G_{ni} V_i V_n \end{bmatrix} - \begin{bmatrix} P_1 \\ \vdots \\ P_n \end{bmatrix} = 0 \tag{4.89c}$$

where $f_j(V_i) = G_{jj} V_j^2 + V_j \sum_{\substack{i=1 \\ i \neq j}}^{n} G_{ji} V_i - P_j^{sp}$.

The Jacobian matrix is

$$F'(V_i) = \begin{bmatrix} \frac{\partial f_1(V_i)}{\partial V_1} & \cdots & \frac{\partial f_1(V_i)}{\partial V_n} \\ \vdots & \ddots & \vdots \\ \frac{\partial f_n(V_i)}{\partial V_1} & \cdots & \frac{\partial f_n(V_i)}{\partial V_n} \end{bmatrix} \tag{4.90}$$

where

$$\frac{\partial f(V_i)}{\partial V_k} = \frac{\partial}{\partial V_k}\left(G_{jj} V_j^2\right) + \frac{\partial}{\partial V_k}\left(V_j \sum_{\substack{i=1 \\ i \neq j}}^{n} G_{ji} V_i \right) - \frac{\partial}{\partial V_k}\left(P_j^{sp}\right) \tag{4.91a}$$

$$\frac{\partial f(V_i)}{\partial V_k} = \begin{cases} 2G_{kk} V_k + \sum_{\substack{i=1 \\ i \neq k}}^{n} G_{ki} V_i, & k = j \\ G_{jk} V_j, & k \neq j \end{cases} \tag{4.91b}$$

For the calculation of the algorithm presented above, first it is necessary to eliminate the buses in which the voltage is known. We consider the matrix equation:

$$\begin{bmatrix} I_1 \\ \vdots \\ I_m \\ \vdots \\ I_n \end{bmatrix} = \begin{bmatrix} G_{11} & \cdots & G_{1n} \\ \vdots & \ddots & \vdots \\ G_{n1} & \cdots & G_{nn} \end{bmatrix} \begin{bmatrix} V_1 \\ \vdots \\ V_m \\ \vdots \\ V_n \end{bmatrix} \tag{4.92a}$$

Between and including V_{m+1} and V_n are the known voltage buses. It is possible to reduce the system order from n to m.

$$\begin{bmatrix} I_1 - \sum_{i=m+1}^{n} G_{1i}V_i \\ \vdots \\ I_m - \sum_{i=m+1}^{n} G_{mi}V_i \end{bmatrix} = \begin{bmatrix} G_{11} & \cdots & G_{1m} \\ \vdots & \ddots & \vdots \\ G_{m1} & \cdots & G_{mm} \end{bmatrix} \begin{bmatrix} V_1 \\ \vdots \\ V_m \end{bmatrix} \tag{4.92b}$$

THE GAUSS–SEIDEL METHOD is an iterative technique for solving a square system of n linear equations with unknown x:

$$Ax = b$$

In DC power-flow , the above equation translates to

$$[G][V] = [I]$$

The Gauss-Seidel method defines the following iteration:

$$LV^{(k+1)} = I - UV^{(k)}$$

The matrix G is decomposed into its lower triangular component L, and a strictly upper triangular component U: $R = L + U$.

The Gauss–Seidel method now solves the left-hand side of this expression for V, using a previous value of V on the right-hand side. Analytically, this may be written as

$$V^{(k+1)} = L^{-1}(I - UV^{(k)}) \tag{4.93}$$

To solve the linear system stated by (4.93), it is necessary to set initial starting points for the injected currents and voltages. It is usually correct to assume that the system bus voltages are around 1 p.u., and the injected powers at each converter are known. It is important to remember that before applying the algorithm it is necessary to eliminate the buses where the voltage is known. This process is explained by equations (4.92a) and (4.92b).

A flowchart illustrating the calculation of the DC grid is shown in Figure 4.80.

4.6.1.3 Application

In order to test the proposed procedure for calculating the power flow, let us consider a network with three converters (Figure 4.81).

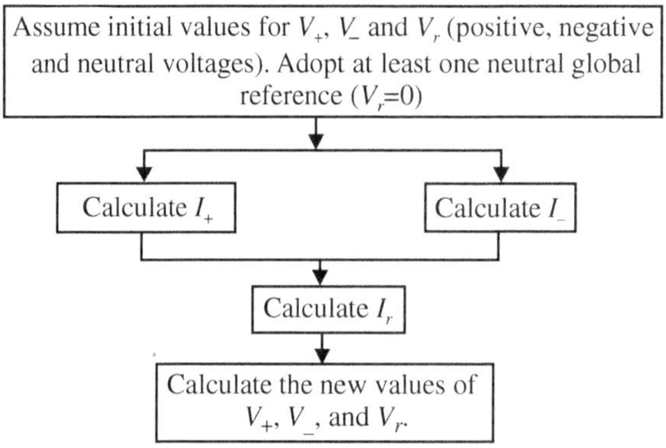

Figure 4.80 Flowchart illustrating the steps for load flow calculation in DC grids.

In a DC system, variables are P and V, and therefore there are two types of bar: type u (uncontrolled voltage; where P is given, and V is calculated) and type C (controlled voltage; V is given and P is calculated).

The three DC subnetworks types, positive, negative, and neutral (or return) are represented. The subnetworks must be solved separately and combined at each iteration. As there may contingencies with the loss of a cable or converter, all the subnetworks may become different. The neutral grid does not necessarily have to be similar to the positive and negative ones.

Note: The voltages + / – / r will be referred to the reference bar in the neutral network that is grounded (global reference). The calculations are performed using absolute units. The inputted data are presented in Table 4.1.

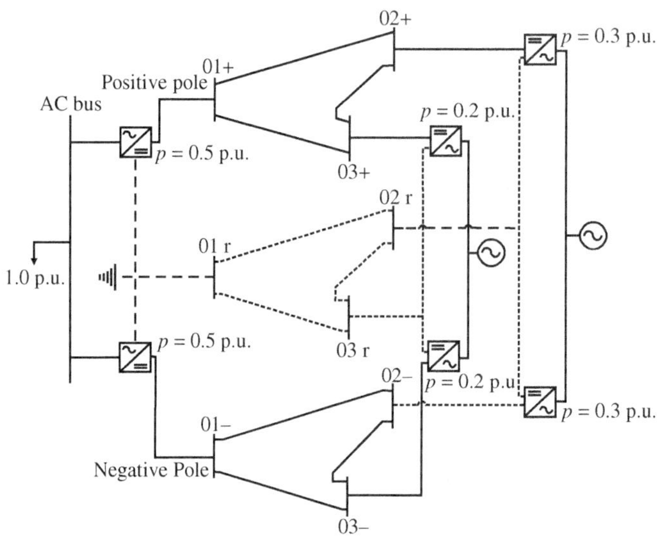

Figure 4.81 DC grid with three converters. 1 p.u.= 1500 MW, ± 400 kV.

TABLE 4.1 Cable and branch characteristics

P1+	P1−	P2+	P2−
−750 MW	−750	450	250
	mm^2	A	r (ohm/km)
Cable (+/−)	400	850	0.02288
Neutral cable	200	750	0.04575
Branch	km	Ohm +/−	Ohm r
01-02	400	9.15	18.30
01-03	300	6.86	13.73
02-03	100	2.29	4.58

The positive and negative load flow calculations and the current flow in the neutral grid are detailed next.

Negative and positive poles

$$\begin{bmatrix} I_1 \\ I_2 \\ I_3 \end{bmatrix} = \begin{bmatrix} 0.255009 & -0.109290 & -0.145719 \\ -0.109290 & 0.546448 & -0.437158 \\ -0.437158 & -0.437158 & 0.582878 \end{bmatrix} \begin{bmatrix} V_1 \\ V_2 \\ V_3 = \pm 400\,\text{kV} \end{bmatrix}$$

After the elimination of the bus 3 (voltage controlled bus), it is obtained:

$$\begin{bmatrix} I_1 - G_{13}V_3 \\ I_2 - G_{23}V_3 \end{bmatrix} = \begin{bmatrix} 0.255009 & -0.109290 \\ -0.109290 & 0.546448 \end{bmatrix} \begin{bmatrix} V_1 \\ V_2 \end{bmatrix}$$

with:

$$G_{13}V_3 = -58.2878 \text{ and } G_{23}V_3 = -174.863$$

The inverse of $G(R)$ is

$$\begin{bmatrix} V_1 \\ V_2 \end{bmatrix} = \overset{R_{jk}\pm}{\begin{bmatrix} 4.289063 & 0.857813 \\ 0.857813 & 2.001563 \end{bmatrix}} \begin{bmatrix} I_1 - G_{13}V_3 \\ I_2 - G_{23}V_3 \end{bmatrix}$$

The neutral conductance matrix and its inverse are

$$\begin{bmatrix} I_{1r} \\ I_{2r} \end{bmatrix} = \overset{G_{jkr}}{\begin{bmatrix} 0.127505 & -0.054640 \\ -0.054640 & 0.273224 \end{bmatrix}} \begin{bmatrix} V_{1r} \\ V_{2r} \end{bmatrix}$$

$$\Rightarrow \begin{bmatrix} V_{1r} \\ V_{2r} \end{bmatrix} = \overset{R_{jkr}}{\begin{bmatrix} 8.578125 & 1.7415625 \\ 1.715625 & 4.003125 \end{bmatrix}} \begin{bmatrix} I_{1r} \\ I_{2r} \end{bmatrix}$$

TABLE 4.2 Load flow calculation using the Gauss–Seidel method

			V3+	V3−		
			400	−400		
Variable	V_{1+}	V_{1-}	V_{1r}	V_{2+}	V_{2-}	V_{2r}
Iteration 1	400.000	−400.000	0.000	400.000	−400.000	0.000
Iteration 2	392.923	−392.494	−0.858	400.643	−399.643	−2.002
Iteration 4	392.790	−392.326	−0.928	400.603	−399.616	−1.975
Iteration 5	392.789	−392.321	−0.936	400.603	−399.615	−1.977
Iteration 6	392.789	−392.320	−0.937	400.603	−399.615	−1.977
Iteration 7	392.789	−392.320	−0.937	400.603	−399.615	−1.977
Iteration 8	392.789	−392.320	−0.937	400.603	−399.615	−1.977
Iteration 9	392.789	−392.320	−0.937	400.603	−399.615	−1.977

Variable	$I_{1+}-G_{13}V_{3+}$	$I_{1-}-G_{13}V_{3-}$	$I_{1r}=-(I_{1+}+I_{1-})$	$I_{2+}-G_{23}V_{3+}$	$I_{2-}-G_{23}V_{3-}$	$I_{2r}=-(I_{2+}+I_{2-})$
iteration 1	56.41280	−56.41280	0.00000	175.98839	−175.48839	−0.50000
iteration 2	56.38318	−56.37275	−0.01043	175.98100	−175.49210	−0.48890
iteration 4	56.38288	−56.37158	−0.01129	175.98118	−175.49210	−0.48909
iteration 5	56.38291	−56.37152	−0.01139	175.98118	−175.49210	−0.48908
iteration 6	56.38292	−56.37152	−0.01140	175.98118	−175.49210	−0.48907
iteration 7	56.38292	−56.37152	−0.01140	175.98118	−175.49210	−0.48907
iteration 8	56.38292	−56.37152	−0.01140	175.98118	−175.49210	−0.48907
iteration 9	56.38292	−56.37152	−0.01140	175.98118	−175.49210	−0.48907

Power Flow [MW]						
	Positive		Negative		Neutral	
Cable	From	To	From	To	From	To
..01–02	−335.463	−342.137	−312.755	−318.570	−0.053	−0.112
..01–03	−412.753	−420.331	−439.040	−447.634	0.064	0.000
..02–03	105.653	105.494	−67.327	−67.392	0.855	0.000

Current Flow [kA]		
	Negative	Neutral
..01-02	0.797	0.057
..01-03	1.119	−0.068
..02-03	0.168	−0.432

Each step of the iterative process using the Gauss–Seidel method is presented in Table 4.2.

The results of the calculation (power flow in megawatts, current in kiloamperes, and voltage in kilovolts) are presented in Figure 4.82.

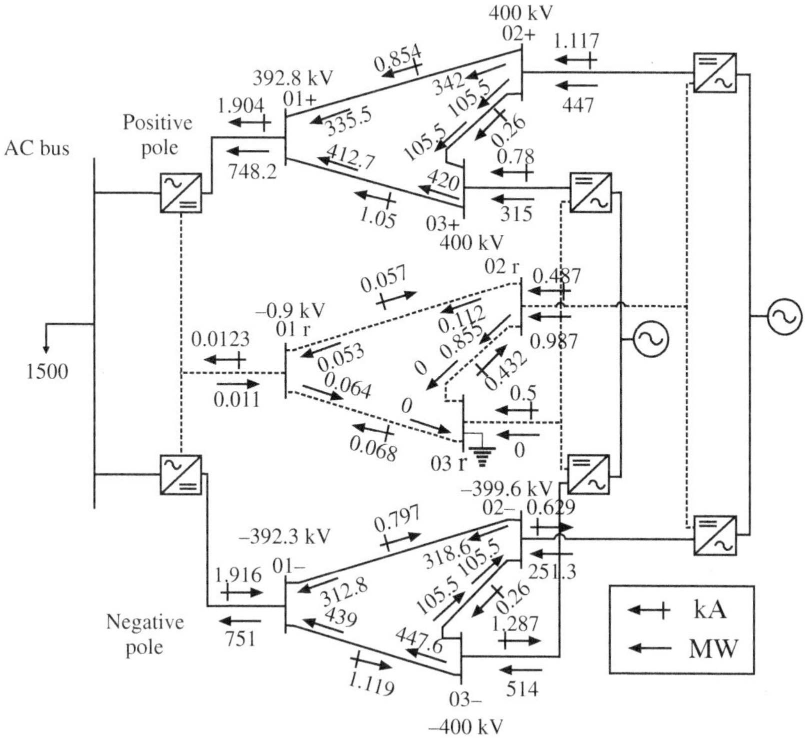

Figure 4.82 The load flow calculation.

4.6.2 Dynamic Stability in Meshed AC/DC Grids

4.6.2.1 Generalities

Meshed DC networks are the latest technological challenge in the area of electric power transmission.

The VSC transmission enables reliable and controllable power transfer between networks. In principle, the converters operation (rectifiers and inverters of the VSC–HVDC system) does not depend on the robustness of the connected AC systems. Moreover, the VSC system provides reactive power control at all its terminals independently of the active power transferred by it.

Recent refinements in the technology used in this equipment, such as the increase in the electronic switches (that is IGBT) rating allowed the VSC to reach higher power capacity, rendering the VSC transmission a competitive alternative. There are VSC–HVDC systems already employed at levels around 1000 MW and 500 kV. This work focuses on the topology of the latest VSC technology, called MMC (see section 4.1.7). Unlike previous settings, the MMC simplifies the construction of the VSC because of its modular structure, making it possible to increase the number of levels in the design without increasing the complexity of wiring. The MMC also has some advantages over the classical VSC types, because of its low frequency modulation, lower losses, and lower harmonic distortion. Figure 4.83 shows a

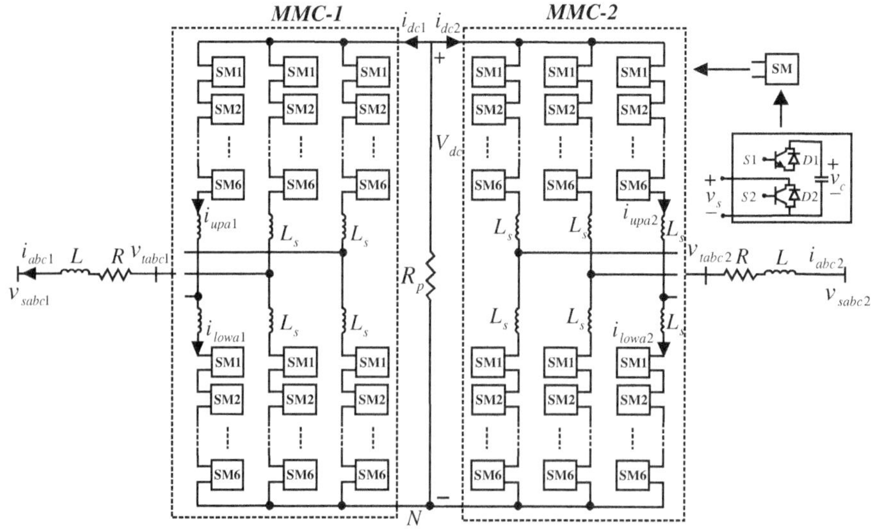

Figure 4.83 General topology of a VSC—MMC. *Source:* Saad et al. 2013 [75]. Reproduced with permission of IEEE.

general diagram for a back-to-back VSC–MMC. The voltage waveform is built by the adequate switching of each SM [15, 75].

For electromechanical stability calculations, a symmetrical three-phase or phasor representation of the network is generally used, as it presents an advantage in terms of computational complexity. To analyze the behavior of a VSC within a stability study, it is necessary to accurately model the power transfer between AC and DC sides and voltage swings. It is also important to model the converter controllers, since the fast acting power electronics plays an important role on the dynamic behavior of the system.

The control strategy employed on VSC equipment may vary depending on the manufacturer, but it generally takes the vector control approach, which is heavily based on instantaneous measurements and transformations [75].

In the next section, it is presented a phasor approach to model the VSC at the fundamental frequency, which is more suitable for implementation of stability programs.

A VSC model has to meet some requirements in order to be useful for an AC/DC system analysis, modeling, planning, etc. The same requirements imposed to the CSC models generally used for dynamic studies are employed: it has to provide the correct interaction with other network elements, but it does not need to reproduce the detailed internal states of the converters (such as unbalanced currents between converter arms); it must be consistent with other fast acting power electronics based system elements; it has to be a synchronous representation on the power frequency on the AC side and a mean value representation on the DC side, with harmonics and filters neglected; it has to allow the evaluation of the system during load changes, capacitors and reactors switching, load rejection, recovery from faults and enable

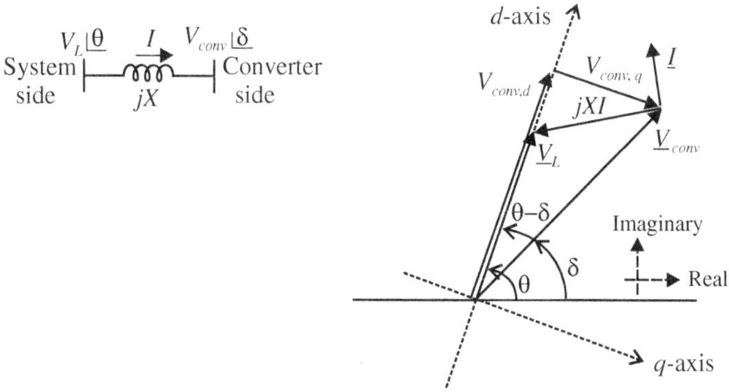

Figure 4.84 Phasor diagram of the VSC.

the investigation of control performance and enhancement; also, it should be useful to study the damping of active power and frequency oscillations through control measures.

4.6.2.2 Description of the VSC Model for Stability Analysis

The VSC is able to generate an AC voltage with controlled magnitude and phase angle, thus being modeled as a symmetrical controlled voltage source on the AC side. This characteristic allows a high level of controllability, making possible for a converter station to control: (i) active and reactive power, (ii) active power and AC voltage magnitude, (iii) DC voltage and reactive power, or (iv) DC and AC voltage magnitude [6].

The root mean square (RMS) stability model proposed is validated by comparing the results obtained with a point-to-point system based on an electromagnetic transient AVM (EMT–AVM) [75]. The AVM was implemented in an EMT-type software, whereas the RMS model was created in the linear system simulation software MatLab Simulink®.

A phasor strategy is used for the RMS and is presented below. It is analogous to the *d-q* vector control strategy usually adopted in the VSC. The phase-to-neutral voltage $\underline{V}_{conv}$ built internally by the VSC is separated from the AC system bus by the converter reactance X. Figure 4.84 presents a phasor diagram of the connection between the VSC and the AC bus.

From Figure 4.84, the voltage drop equation is

$$\underline{V}_L - \underline{V}_{conv} = jX\underline{I} \tag{4.94}$$

The *d–q* synchronous frame is set with the d axis upon the system voltage $\underline{V}_L$, making it the reference voltage for the converter. Then, the projection of $\underline{V}_{conv}$ on the reference frame is

$$\underline{V}_{conv} = V_{conv}[\cos(\theta - \delta) + j\sin(\theta - \delta)] \tag{4.95a}$$

and

$$V_{conv,d} = V_{conv}\cos(\theta - \delta) \tag{4.95b}$$

$$V_{conv,q} = V_{conv}\sin(\theta - \delta) \tag{4.95c}$$

Replacing (4.95a) in (4.94) yields

$$\underline{V}_L - \underline{V}_{conv} = \underline{V}_L - V_{conv}\cos(\theta - \delta) - jV_{conv}\sin(\theta - \delta)$$

After rearranging the terms, from (4.94) it results

$$\underline{I} = \frac{\underline{V}_L - \underline{V}_{conv}}{jX} = \frac{(\underline{V}_L - V_{conv}\cos(\theta - \delta)) - jV_{conv}\sin(\theta - \delta)}{jX}$$
$$= \frac{V_{conv}\sin(\theta - \delta)}{X} + j\frac{V_L - V_{conv}\cos(\theta - \delta)}{X} \tag{4.96}$$

from where we achieve

$$I_d = \frac{V_{conv}\sin(\theta - \delta)}{X} \tag{4.97a}$$

$$I_q = \frac{V_L - V_{conv}\cos(\theta - \delta)}{X} \tag{4.97b}$$

where the index d or q indicates a projection on the d-axis or q-axis.

The active power flowing from the AC system to the converter is

$$P = V_L I_d = \frac{V_L V_{conv}\sin(\theta - \delta)}{X} = \frac{V_L V_{conv,q}}{X} \tag{4.98a}$$

and the reactive power is

$$Q = V_L I_q = \frac{V_L(V_L - V_{conv}\cos(\theta - \delta))}{X} = \frac{V_L^2 - V_L V_{conv,d}}{X} \tag{4.98b}$$

Both the active and reactive power are dependent on the AC system voltage magnitude, V_L, but are otherwise independent.

At any instant, the lossless converter has AC power input equal to the DC power output

$$P_{AC} = P_{DC} = V_{DC} I_{DC} \tag{4.99}$$

Combining Equations (4.98a) and (4.99), one obtains

$$I_d = I_{DC}\frac{V_{DC}}{V_L} \tag{4.100}$$

Defining the q-axis projection of the converter voltage as a function of the DC voltage

$$V_{conv,q} = V'_{conv,q} V_{DC} \tag{4.101}$$

Equation (4.101) links the AC and DC sides of the converter in terms of switching operation, since the converter is able to build an AC voltage only when there is a compatible DC voltage level on its terminals.

Rearranging the equations, it can be shown that

$$I_{DC} = \frac{V'_{conv,q} V_L}{X} \tag{4.102}$$

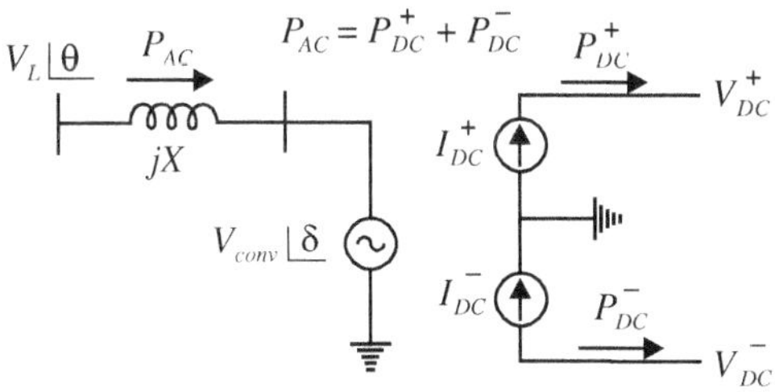

Figure 4.85 VSC model for stability studies, with one converter per pole [51].

Equation (4.102) suggests that the DC side of the converter may be represented by a current source that varies linearly with $\underline{V}'_{conv,q}$ Fluctuations on the system voltages will cause the power to oscillate as well.

The model of Figure 4.85 is proposed to represent the AC–DC VSC for stability studies, with a controlled synchronous voltage source on the AC side and a DC current source with the reference current as a function of active power flowing into the converter and the DC voltage at its terminals.

This model describes the power transfer through the lossless converter. Losses may be represented by a combination of shunt and series resistances [76].

4.6.2.3 Control Models

The controller will act upon the state variables of the converter, and thus may have different topologies. Here, three control models are presented, even though there is only one converter model, being that they may be easily associated with the converter depending on the level of detail necessary to the study. The three control types are (i) direct control on the variables V_L–δ; (ii) one stage vector control on the variables $V_{conv,d}$–$V_{conv,q}$; (*iii*) two-stage vector control on the variables I_d –I_q and $V_{conv,d}$ –$V_{conv,q}$.

- **(i)** *Control on the variables* V_L–δ. Figure 4.86 shows the feedback loop to control active power via the q-axis component of the voltage. The disadvantage of this control is the significant coupling between controllers (active power/DC voltage and reactive power/AC voltage).
- **(ii)** *Control on the variables* $V_{conv,d}$–$V_{conv,q}$. Figure 4.87 shows the feedback loop to control active power via the q-axis component of the voltage. It provides good decoupling between the active power/DC voltage and reactive power/AC voltage controllers for a symmetrical power system (positive sequence).

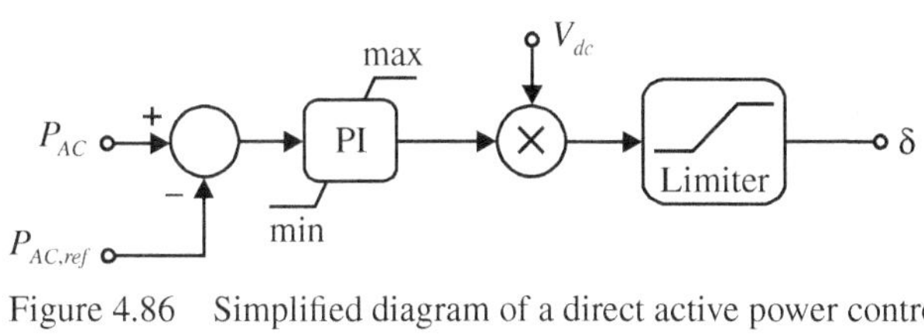

Figure 4.86 Simplified diagram of a direct active power control.

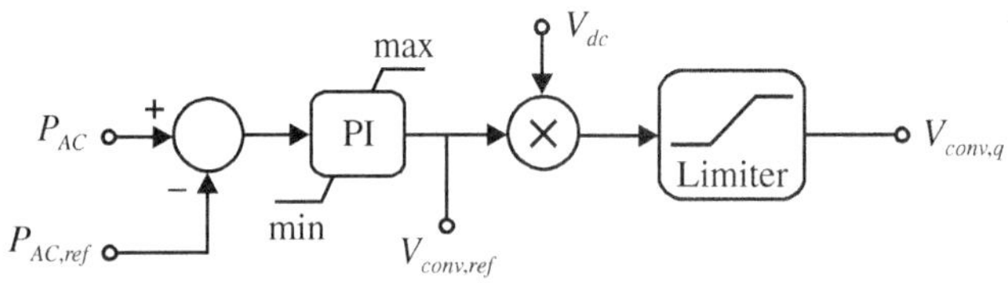

Figure 4.87 Simplified diagram of a one stage active power control.

(iii) *Control on the variables $V_{conv,d}$–$V_{conv,q}$ and I_d–I_q.* Based on the diagram of Figure 4.85 and equations of the previous section, a two-stage control [75, 76] model can be derived and realized for RMS stability studies. Figure 4.88 shows the feedback loop to control active power via the d-axis component of the current and the q-axis component of the voltage. It provides excellent decoupling between controllers for a symmetrical power system (positive sequence) [51, 77].

Assuming that the converter is connected to AC and DC stiff busses, the transfer function for the block diagram of Figure 4.88 can be shown to be

$$\mathrm{TF}_{P\rightarrow V_{conv,q}} = k_{gain}\frac{\left(s+\dfrac{k_{I_{\mathrm{Id}}}}{k_{P_{Id}}}\right)\left(s+\dfrac{k_{I_{V_{conv,q}}}}{k_{P_{V_{conv,q}}}}\right)}{\left(s+\dfrac{k_{I_{Id}}}{k_{P_{Id}}+1}\right)\left(s+\dfrac{k_{I_{Id}}}{k_{P_{Id}}+X}\right)} \tag{4.103}$$

where $\mathrm{TF}_{P\rightarrow V_{conv,q}}$ means the transfer function of $V_{conv,q}$ controlling the active power, k_{gain} is a global gain that depends on the PI controller parameters $k_{I_{Id}}$, $k_{P_{Id}}$, $k_{I_{V_{conv,q}}}$ and $k_{P_{V_{conv,q}}}$, which are the integral and proportional gains for each stage of control. $\mathrm{TF}_{pv\rightarrow conv,q}$ is also highly dependent on the converter model reactance X. Similar transfer functions may be derived for other types of operation, such as the reactive power control through the direct axis controlled voltage $V_{conv,d}$.

All controllers can be easily modified to include the interference of variations of AC and DC system voltages for a noninfinite bus. Also, it can be shown that this configuration has a high disturbance rejection.

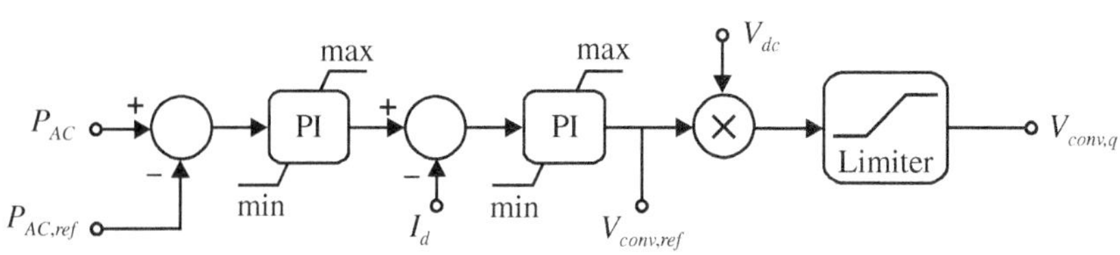

Figure 4.88 Simplified diagram of a two-stage active power control.

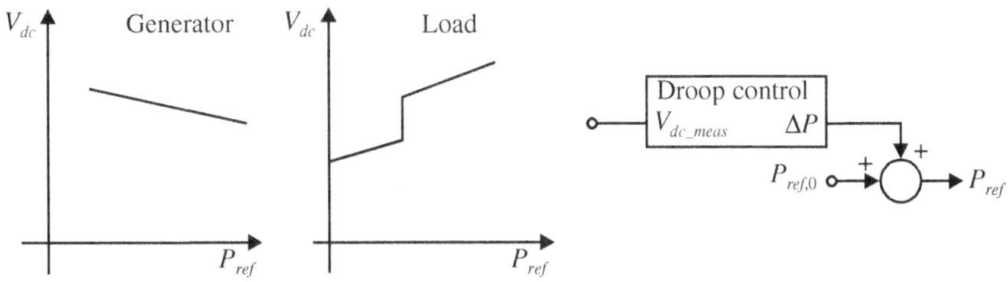

Figure 4.89 Converter droop controller representation.

4.6.2.4 *P–V Droop Control*

Since the action of the droop control in the DC systems is faster than in the AC system (due to a lower inertia of electronic actuators), it has to be included in the model. Another difference with respect to the AC system is that the droop on the VSC has an important protective function. Since the converters must operate independently in a DC network, the detection of a fault activates the droop control, which affects the performance of each converter.

Because of this rapid control in networks with VSC, contrary to what happens in conventional CSC systems, telecommunication between terminals and a centralized control is necessary in the long-term operation only. The transient response must be very fast so the actions are localized at each converter.

The droop controller in VSC will dictate the dynamics during severe events on the network, being highly responsible for knocking and restoring the converter and the system during and after a fault. The droop characteristic is different for load converters and for generator converters, as shown in Figure 4.89.

4.6.2.5 *Current and Voltage Limits*

As any power system element, the VSC has both maximum current and maximum voltage capability. The former is more of a protection to the converter and its components, whereas the latter is the maximum possible voltage the converter may build on the AC side and depends on the V_{DC}/V_{AC} relation. The implementation of such limits is different for each control model.

- In the control model I, the voltage and current have one degree of freedom each:

$$V_{AC} \le V_{L\max} \tag{4.104a}$$

$$I_{AC} \le I_{\max} \tag{4.104b}$$

- In the control model II, the controlled variables (*d–q* voltages) have two degrees of freedom, while the current is dependent on the voltage and its limits are implicitly defined by (4.105b).

$$V_{conv,d}^2 + V_{conv,q}^2 \le V_{L,\max}^2 \tag{4.105a}$$

$$V_{conv,q}^2 + (V_L - V_{conv,d})^2 \le (X \cdot I_{\max})^2 \tag{4.105b}$$

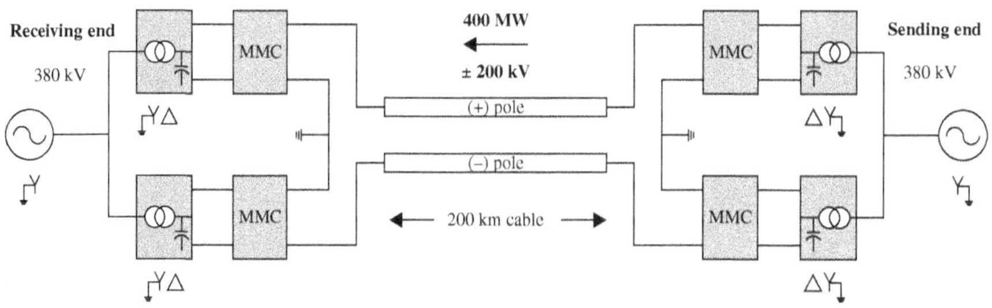

Figure 4.90 VSC point-to-point transmission configuration.

- In the control model III, both the *d–q* voltage and the current have two degrees of freedom, and may be defined as

$$V_{conv,d}^2 + V_{conv,q}^2 \leq V_{L\max}^2 \tag{4.106a}$$

$$I_d^2 + I_q^2 \leq I_{\max}^2 \tag{4.106b}$$

The projected variables approach allows the controller to favor one axis over the other, giving preference to the control of active parameters (active power or DC voltage) over reactive parameters (reactive power or AC voltage magnitude).

4.6.2.6 *RMS Model Testing*

To evaluate the accordance of the phasor model, a 200-km, 400 MW, ±200 kV point-to-point system shown in Figure 4.90 was used. The transmission cable was described as a full frequency-dependent parameters model on the EMT software. For the VSC, a three-phase AVM was used. The RMS proposed model was implemented in Matlab-Simulink®, where the cable was represented using its positive sequence equivalent π section and implemented with an input/output transfer function in the frequency domain, given in the matrix form:

$$\begin{bmatrix} V_{dc,\,sending} \\ V_{dc,\,receiving} \end{bmatrix} = \begin{bmatrix} Z_{11} & Z_{12} \\ Z_{21} & Z_{22} \end{bmatrix} \begin{bmatrix} I_{dc,\,sending} \\ I_{dc,\,receiving} \end{bmatrix} \tag{4.107}$$

The RMS model control parameters (the controller gains) were set the same as in the EMT model.

The converters were set to control active and reactive power on the sending end, and DC voltage and reactive power on the receiving end. Each converter is connected to an equivalent AC source with a short-circuit level of 30 GVA and the reactance between the AC network, and the VSC was set to 0.255 p.u. on both converters.

The integration step in the EMT software was set to 40 μs, whereas in the MatLab®Simulink® it was set a fixed time step of 1000 μs, which implied a much shorter simulation time. The simulation results are compared through two test cases.

Case I ***Step on the active power on the sending end.*** To assess the dynamic response of the point-to-point VSC transmission, the step response was evaluated on both models through a change of 25% on the active power set point during 500 ms. Figure 4.91 shows the power transient on both terminals.

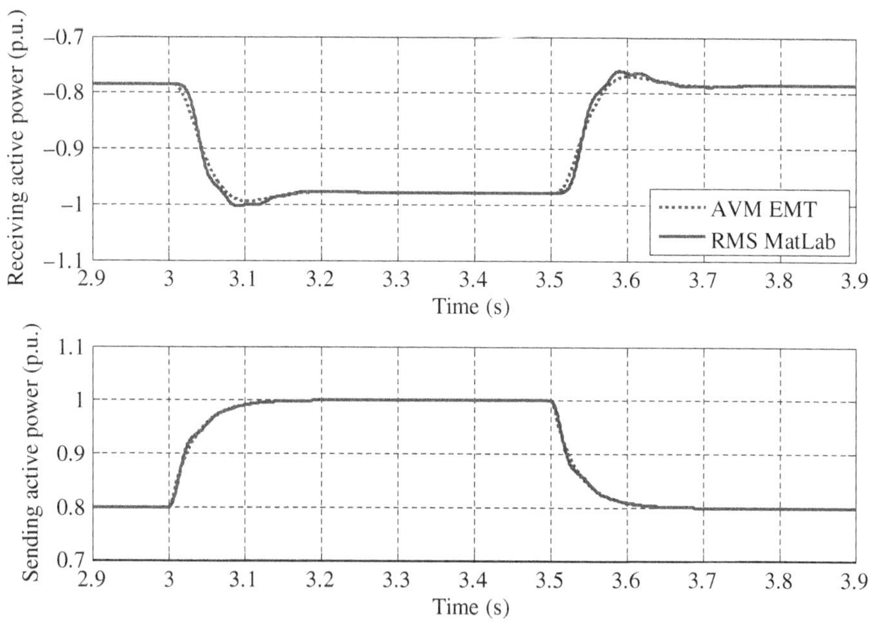

Figure 4.91 Active power (p.u.) during a step on the sending converter active power set point.

During this transient, there is more power being injected on the sending end and less being drawn from the receiving end, which results in a net power absorbed by the transmission lines and a corresponding voltage swell. The converter controlling DC voltage then acts to bring the voltage to the preset value, as seen in Figure 4.92.

Case II ***Three-phase AC short circuit on the sending end.*** To evaluate the accuracy of the RMS model during severe faults near the converter, a three-phase solid short circuit with duration of 200 ms is simulated (Figures 4.93 and 4.94). Since the RMS model is simplified, it cannot precisely represent the power flow, voltages, and fast transient effects, but it successfully mimics the input/output characteristics (min/max values and transient time) of the VSC as a power system *black box* element. This event was chosen to show the strength of the model even during severe faults close to the converter.

From the results presented in Figures 4.93 and 4.94, it can be concluded that the models developed are suitable for stability studies, since the behavior of the transmission system plus converter dynamics is equivalent on both cases presented.

4.6.2.7 Simulations on an AC/DC Meshed Grid

To show the robustness and applicability of the phasor model, a 13 converter DC grid test system was simulated (Figure 4.95) [78].

It has 11 AC/DC voltage sourced converters where five of them are connected to islanded AC asynchronous wind-based power plants or off-shore loads. Also, there are two DC/DC converters in the grid, which were modeled as inverse back-to-back

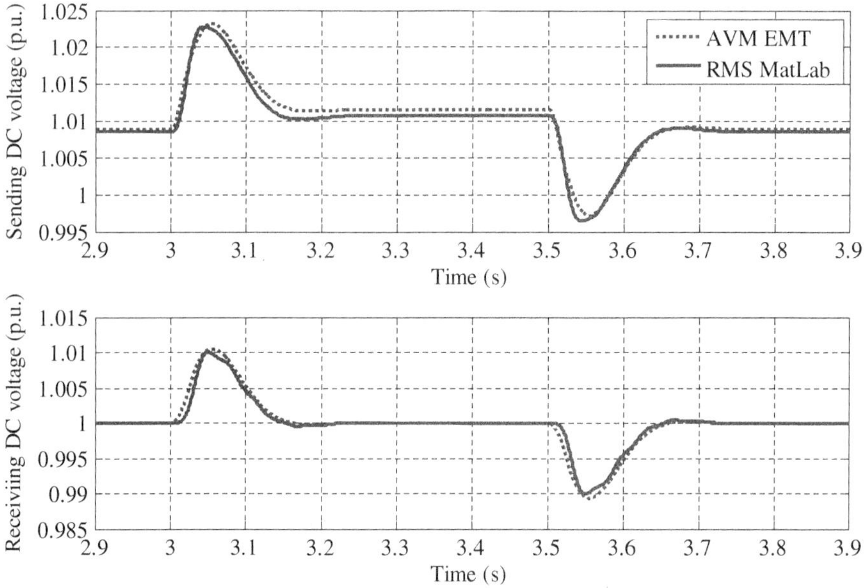

Figure 4.92 DC voltages (p.u.) during a step on the sending converter active power setpoint.

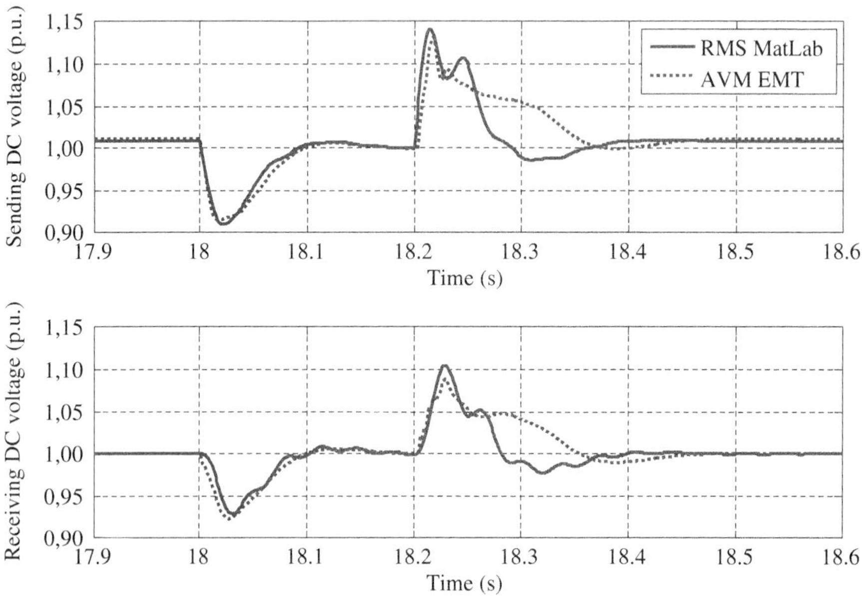

Figure 4.93 DC voltage (p.u.) during a three-phase short circuit on the sending end.

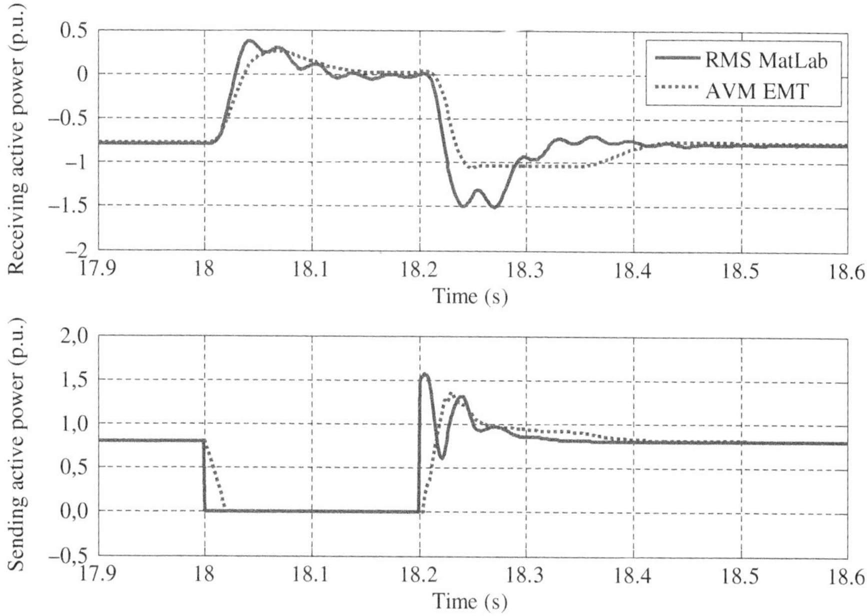

Figure 4.94 Active power (p.u.) during a three-phase short circuit on the sending end.

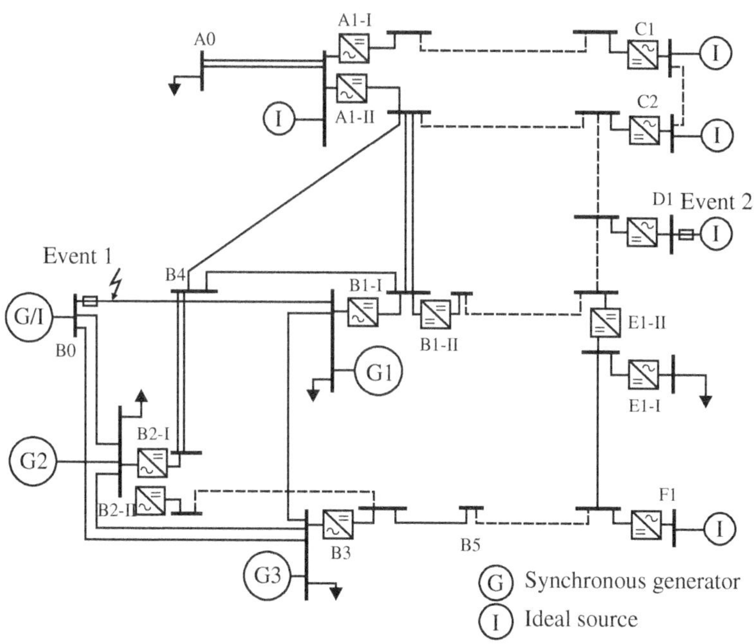

Figure 4.95 Location of the ideal sources, generators and loads. *Source:* Vrana et al. 2013 [78]). Reproduced with permission of CIGRE.

VSCs. One acts like a DC version of a unified power flow controller, adjusting the transmitted power in a circuit; the other one acts like a DC transformer, allowing the connection of two systems with different operating voltages.

The DC grid has transmission lines and submarine cables, with rated voltages of ±200 kV or ±400 kV [78]. Figure 4.95 shows the AC/DC system used, where four machines were included to evaluate the dynamic behavior of the asynchronous connections as well as the power exchange between networks and its influence on their electromechanical stability [79]. Two events were analyzed: (1) a three-phase fault in the AC system close to B0 with disconnection of the faulted AC line; (2) disconnection of an entire power plant (D1).

The 380-kV AC subsystem has a total load of 6400 MW and a local AC system generation of approximately 3000 MW, demanding 3500 MW from the DC interconnection. On the ideal sources (I), the voltage values, with zero internal reactance, no mechanical equations and no governor or exciter, are specified. The synchronous generators (G) are modeled using the complete Park and mechanical equations, governor and exciter with typical hydraulic power plant parameters; all with 1800 MVA rated power connected through a 2000 MVA 13.8 kV/380 kV transformer.

The main objectives of the simulations are to verify and present: the synchronous generators power and angle oscillations; the VSC control actions; the transfer of oscillations across the DC and AC grid.

Event 1 ***Three-phase fault with line opening.*** A solid three-phase fault was simulated with duration of 300 ms with consequent opening of a 200-km AC line from B0 to B1. The impact on the converters power dispatch and DC voltages are those shown in Figures 4.96 and 4.97.

The fault led to a power oscillation in the system lines, with the machines and converters acting to dynamically restore the prefault condition. The angle oscillations after the elimination of the fault are damped, as seen in Figure 4.98, indicating that the system is overall stable.

Event 2 ***Loss of 1,000 MW power plant.*** The outage of an entire power plant at bus D1 causes the loss of 1000 MW of power supply to the system. The deficit of generation on the ±400 kV system had to be supplied by the slack converters A1-I, A1-II, and B2-II (Figure 4.99a), as the rest of the converters are controlling its active power.

It can be seen that most of the power was supplied by converter A1-II, which draws power from the radial AC subsystem. This is because the power flow distribution on the meshed AC subsystem is highly determined by the injection of power from the converters through B1, B2-I, and B3. The slack converter at A1-I showed great independence from the rest of the grid, since it is connected by a radial DC link to C1 transmitting a fixed power.

Figure 4.100 shows that, shortly after the outage, there is a drop in the DC voltages over the whole system, because there is less power injection to support the operating voltage. The slack converters guarantee the return to the prefault condition, indicating the need of fast acting controllers on these converters.

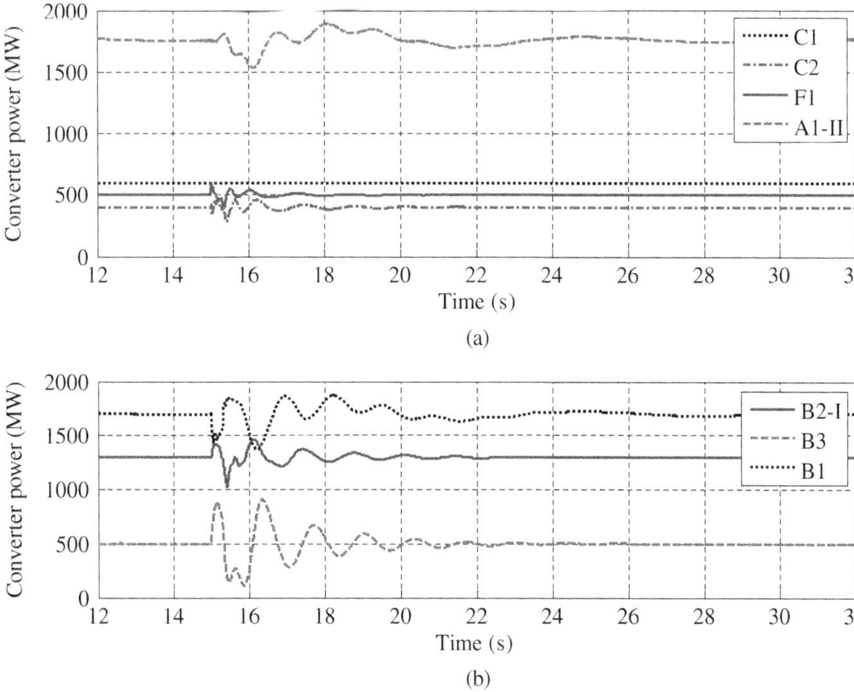

Figure 4.96 Converter power: (a) impact on DC grid; (b) AC/DC interchange converters.

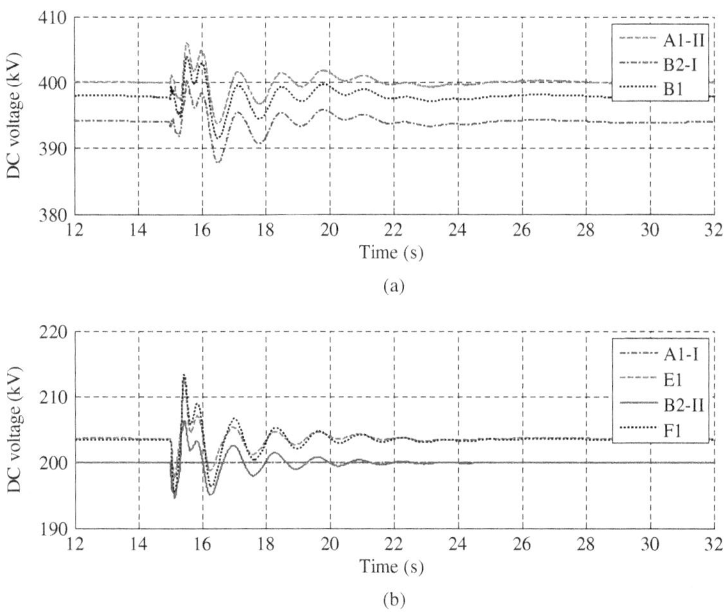

Figure 4.97 DC voltages: (a) ±400 kV network; (b) ±200 kV network.

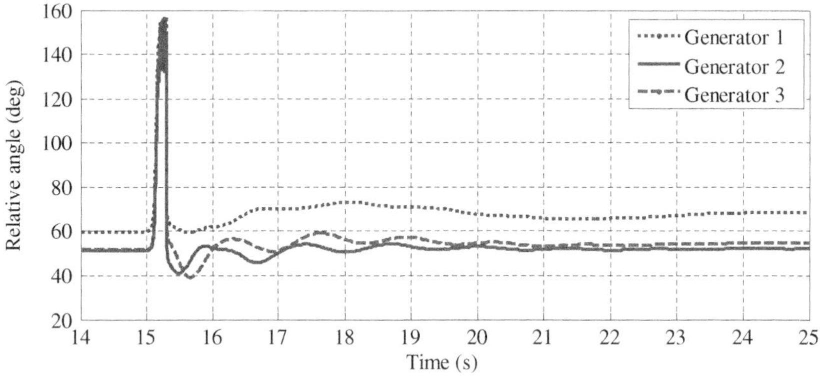

Figure 4.98 Phase angles (degrees) relative to the 380 kV AC slack bus B0.

The effect of this outage on the machines of the AC subsystem are very small compared to the total generation lost, showing that the DC grid significantly absorbs the power transient, as shown in Figure 4.101.

The concept of a DC-meshed networks is new within power systems, and there is a lack of necessary tools to adequately analyze them today. This chapter provided

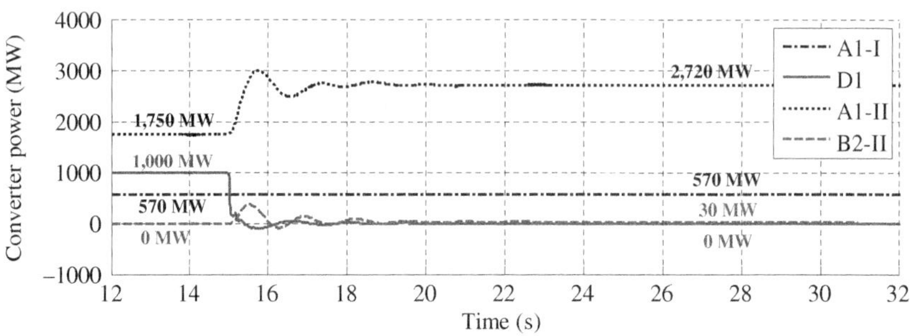

(a)

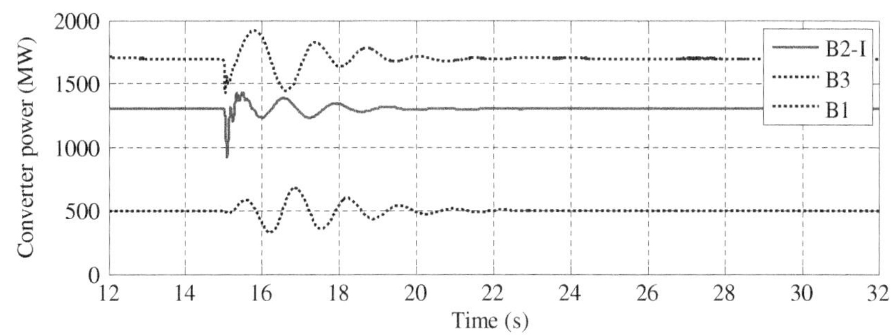

(b)

Figure 4.99 Converter power: (a) *slack* converters redispatch; (b) AC/DC exchange converters.

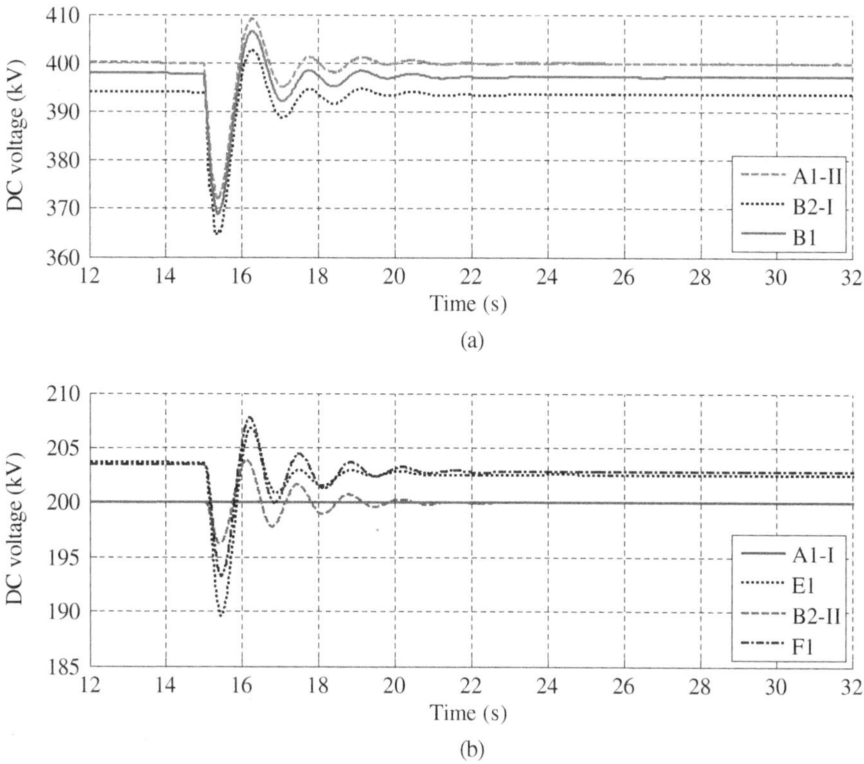

Figure 4.100 DC voltages: (a) ± 400 kV network; (b) ± 200 kV network.

a method for modeling and analyzing the AC/DC converters for stability studies. All the models are phasor based, i.e., positive sequence on the fundamental frequency only, thus making it easy to be included in standard power system stability programs available today. It is an invaluable tool to assess future connections, limits of power transmission, AC/DC interchange, operation, and planning of AC/DC meshed systems.

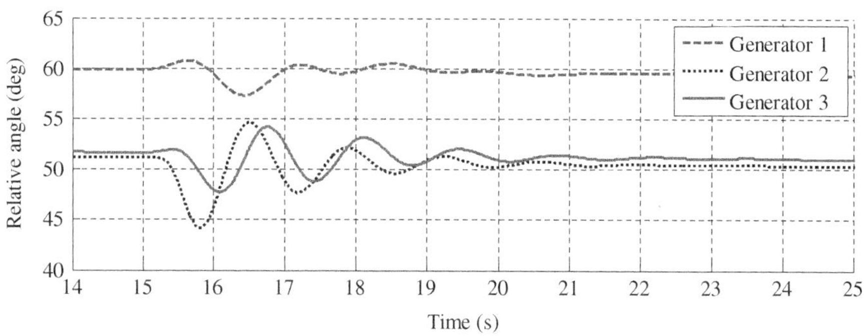

Figure 4.101 Phase angles (degrees) relative to the 380 kV AC slack bus B0.

4.7 COMPARISON OF CSC–HVDC VERSUS VSC–HVDC TRANSMISSION[7]

4.7.1 Differences Resulting from the Commutation Principle

The differences between CSC–HVDC and VSC–HVDC transmission technologies are originated from the power electronic devices used: CSC (called also LCC, line commutated converter) because the DC voltage is formed from the AC line voltage) based on thyristors, which conduct at the firing angle every half-cycle and are blocked naturally when the current reaches zero, whereas a VSC based on IGBTs is suited with turn-on and turn-off capability that enables self-commutation.

(i) *Dependence on a AC voltage source.*

The voltage source of a CSC–HVDC converter for the commutation process is the AC system voltage. The DC voltage is formed by successive conduction of the thyristor valves during the peak values of the AC phase voltages supplied by the AC system. If there is no generator in the AC system, a synchronous compensator or a STATCOM is required to provide the necessary voltage source.

In a VSC–HVDC link, the DC voltage is formed from the DC capacitors that are part of the converter, irrespective of the AC system voltage or the current carried, by a forced commutation of the converter valves.

(ii) *Reactive power compensation or generation.*

In a CSC–HVDC converter. the current lags behind the voltage by an angle given by the thyristor valve firing/extinction angle resulting thus in reactive power consumptions. For security reasons and for operation purposes, there is always a delay between voltage and current, with a minimum around 12° under normal conditions. On the inverter side, due to the turn-off time of the thyristor, the extinction angle should not be below 15° in order to prevent a commutation failure. Reactive power is lost also in the leakage reactance of the converter transformer. Under normal operating conditions of the CSC–HVDC link, the reactive power consumption is about 50%–60% of the active power. Thereby, capacitor banks are necessary on the AC side to compensate the reactive power.

The current absorbed by the CSC–HVDC converter from the AC system is not sinusoidal, and it varies in terms of the power transferred. Thereby, harmonic filters are required on the AC side, which also consist of capacitors. Since capacitors provide also reactive power, their size have to be taken into account when designing the capacitor banks. A special CSC configuration, called capacitor commutated converter HVDC or CCC-HVDC (see section 2.3.2), includes a capacitor in series with the converter transformer that can also provide reactive power compensation.

In a VSC converter ,the delay between current and voltage can be controlled as strategy to generate or absorb reactive power, when required, using

[7] Reprinted with permission from CIGRE, Andersen et al. 2005 [6].

the converter capacitors as energy source. This control is independent of the active power control, except for the overall design limits, is very rapid and versatile so that a VSC–HVDC converter can operate as a STATCOM even when the DC line is disconnected.

(iii) *Short-circuit level requirement for stable operation.*

As explained above, the CSC–HVDC is dependent on the AC system voltage. The AC voltage stability and the AC system capacity of providing reactive power depend on how strong or weak is it. The strength of the AC system is evaluated by the short circuit ratio (SCR) at the AC connection point (see section 3.6.2), which is defined by the expression SCR = SCL/P_{dc}, where SCL is the short-circuit level at the connection point and P_{dc} is the rated power of the HVDC transmission system. At the connection point, the voltage may fluctuate either because the reactive power demanded by the HVDC converter may change as the active power change or because of the changes in the voltage drops on the AC system that are caused by changes in the AC system load. Thereby, a certain minimum SCR is required to ensure the stable operation of a CSC–HVDC configuration, typically greater than 2 [80].

Since the operation of a VSC–HVDC link does not depend on the AC system voltage, no conditions related to the SCR are required. On the contrary, such a technology can significantly contribute to the AC system stability.

(iv) *Harmonics and filter requirements.*

A CSC converter station includes usually two six-pulse converter, one being supplied by a Y–Y connected transformer and the other by a Y–Δ connected transformer, that form a 12-pulse converter. Since a thyristor performs 50/60 switches per seconds, nonsinusoidal waves result on the AC side from the commutation process of the thyristor valves. Some of the harmonic currents are canceled in the transformer connections, and only the order of ($12n \pm 1$) remains. Thereby, harmonic filters are necessary on either side of the CSC–HVDC link, of which rating would be 20%–30% of the converter rating.

With the advent of power electronics and creation of VSC converters, higher switching frequency is achieved, up to few kilohertzs, thus reducing the harmonic content. Therefore, filters may not be required or maybe only just a small one may be installed to cancel the higher order harmonics. On the other hand, the higher the switching frequency is, the higher the active power losses are. For this reason, in IGBT-based converters, the switching frequency is an optimum between the harmonic content and the active power losses. Optionally, the filters may be designed to compensate for the AC system harmonics and suppress the flicker generated by industrial process.

(v) *Overvoltages in the AC system.*

In case of sudden disconnection of the CSC–HVDC link, when a large amount of active power is transferred, the sudden interruption of reactive power absorption by the converters results in AC fundamental frequency overvoltage due also to the surplus of reactive power from the capacitor banks and filters. If the system impedance is relatively, resonance-type overvoltages may also appear [6].

In a VSC–HVDC system, since there is no capacitor bank for reactive power compensation and the filters rating is low, overvoltages may not occur. Furthermore, the converter itself has the capability of rapidly handling the reactive power for voltage control.

(vi) *Robustness against system faults.*

Operation of a CSC–HVDC link requires communication between the converter stations. In case of a fault in the AC system, the communication between converters may be affected resulting even in the disconnection of the HVDC link.

On the contrary, operation of the VSC–HVDC system converters is independent from each other. This helps such a system to continue operating, even at limited active power transfer, in case of disturbances in the AC system.

4.7.2 Differences Resulting from the Converter Type

Differences between CSC–HVDC and VSC–HVDC technologies may be identified when the converter capabilities are analyzed [6].

(i) *Protection against DC system faults.*

In the case of a two-terminal link, when a transient fault occurs on the DC side, the overcurrent can be limited and the fault can be cleared by means of the thyristor valve control and protection systems, thereby no breaker is required.

In the case of a two-terminal VSC–HVDC link, since the free-wheeling diodes allow the DC current to supply the fault, even if the IGBTs are blocked, the entire HVDC link must be disconnected by opening the AC circuit breakers at both ends. Multiterminal VSC systems were not built so far also because there is no commercial circuit breaker available. Multiterminal VSC systems are planned to be built in soon, but at lower voltage and power ratings.

(ii) *Power flow reversal in the Multiterminal HVDC system.*

In a CSC–HVDC system, since the current flow is unidirectional, the power flow can be reversed by changing the DC voltage polarity, which is not a problem in a two-terminal link. But, low cost extruded polymeric cables cannot be used because they can be destroyed when reversing the voltage polarity. In a multiterminal HVDC link, however, reversal of the voltage polarity at one converter involves the change in the voltage polarity at all converter, thus the rectifiers will become inverters and the inverters will become inverters. In this case, reversal of power direction is on a terminal basis, and only one or two terminal stations are suite with such capability. High speed transfer switches are installed in a configuration that allows reversing the position of the converter instead.

In a VSC–HVDC system, changing the power flow direction no longer needs changing the voltage polarity but simply is achieved by changing the current direction. In a multiterminal configuration, on the other hand, the power flow direction can be changed very fast by each converter through appropriate commutation strategy, without any communication with the other converters.

(iii) *Cost, losses and reliability of large-scale HVDC systems.*

Important advantages of the VSC technology versus the CSC technology were identified above, but higher price must be currently paid for these. On the other hand, the power ratings and the lengths of the CSC–HVDC links are greater than those available to the VSC–HVDC, which is a developing technology. Improvements have been done in VSC converters, and their active power losses are about to be similar with those of the CSC converters. The many years of operation of the CSC–HVDC systems have shown that these perform well in terms of reliability, which cannot be said yet about VSC–HVDC because this is a much younger technology.

4.8 FORWARD TO SUPERGRID

4.8.1 Challenges and Solutions for Developing Supergrid

Fossil and nuclear power are dominating sources of electricity today. The corresponding power plants were built at location providing best conditions for fuel supply and operation. The existing transmission system was built to accommodate for the power flows supplying the loads from these power plants. The increased use of renewable energy, creates significant challenges for the electric power transmission system, mainly due to the location of the planned large-scale renewable energy sources, whether offshore wind in the north or large solar power in the south of the continent. A as consequence [13]:

- The load centers in central Europe need to be supplied more and more from renewable energy sources over long distance, resulting in a demand for increased transmission capacity. Long distances require efficient power transmission solutions keeping transmission losses low.
- The shutdown of existing power plant means that at the same time their voltage and reactive power control capability "is lost" to the power system. This is even more significant as the demand for voltage control increases with longer transmission distances.
- The disconnection of large generators reduces inertia of the system and consequently the reducing dynamic system stability.
- The fluctuating nature of renewable energy sources requires new methods for providing adequate frequency control reserve.

Because the load centers are situated in central Europe—the region roughly comprising of the Benelux, the north of France (including Paris), western Germany, and north of Italy and England, it will be inevitable to build new north–south and west–east connections within Europe to secure electricity supply to the demand center. Furthermore, the load centers are getting more concentrated as there in a noticeable shift of the population and industry toward the city.

Rather than extending the current highly loaded 400 kV AC system a fundamental upgrade is required with new energy corridors or an "electricity highways" with a transmission capacity that is one order of magnitude higher than that of the system which is now use.

There are many visions of pan-European overlay grids proposing different grid topologies and transmission technologies. Most visions propose VSC–HVDC transmission as the preferred technology for such transmission grids [81].

4.8.1.1 Connecting Renewable Energy Sources and Increased Transmission System Capacity

The first large-scale offshore wind parks in the power range up to about 500 MW have been connected in Denmark and England with transmission distances that allow connection by HVAC cables. Today, wind parks up to 900 MW with transmission distances to shore of more than 100 km are built. HVDC systems based on VSC technology are used to connect the wind parks by separate point-to-point connections.

MMCs in the so-called halfbridge design are applied today providing the conversion from AC to DC and vice versa.

Using today's XLPE cable technology, about 500 MW can be transmitted per cable at transmission voltages up to 320 kV; and for MI cable technology these figures are 1000 MW and 500 kV. Cables for higher voltages and power greater than 1500 MW per cable are under development.

Some offshore wind park projects include considerable transmission distances on-shore in addition to offshore. Compared to offshore conditions, where long cable sections can be laid by ship, physical limitations (e.g., drum capacity) reduce the maximum section length on land making more joints necessary.

As XLPE is lighter and requires less time to complete a joint (approximately one day instead of several), it delivers significant advantages for onshore installation cost and reliability. VSC–HVDC transmission via XLPE land cables provides an attractive alternative where environmental or other constrains prevent overhead lines from being used. The Spain–France interconnection INELFE (France–Spain Interconnector) having a power rating of 2×1,000 MW is one example of using VSC transmission with XLPE land cables.

The utilization of renewable energy leads to an increased power flow north–south in Europe. New links will be required to strengthen the connection between central Europe with northern and southern Europe. In northern Europe, subsea cable connections are needed whereas in the central and southern part overhead lines and land cables can be used.

A map of VSC–HVDC projects commissioned and under delivery around the world is presented in Figure 4.102.

4.8.1.2 Compensating Reactive Power

Long distance power transmission, which becomes necessary due to replacement of existing power plants, changes the reactive power flow within the HVAC system and can have unwanted side effects. Reactive power compensation can therefore be an effective measure to increase the utilization of existing systems. FACTS devices or mechanically switched branches can be used to compensate reactive power where needed.

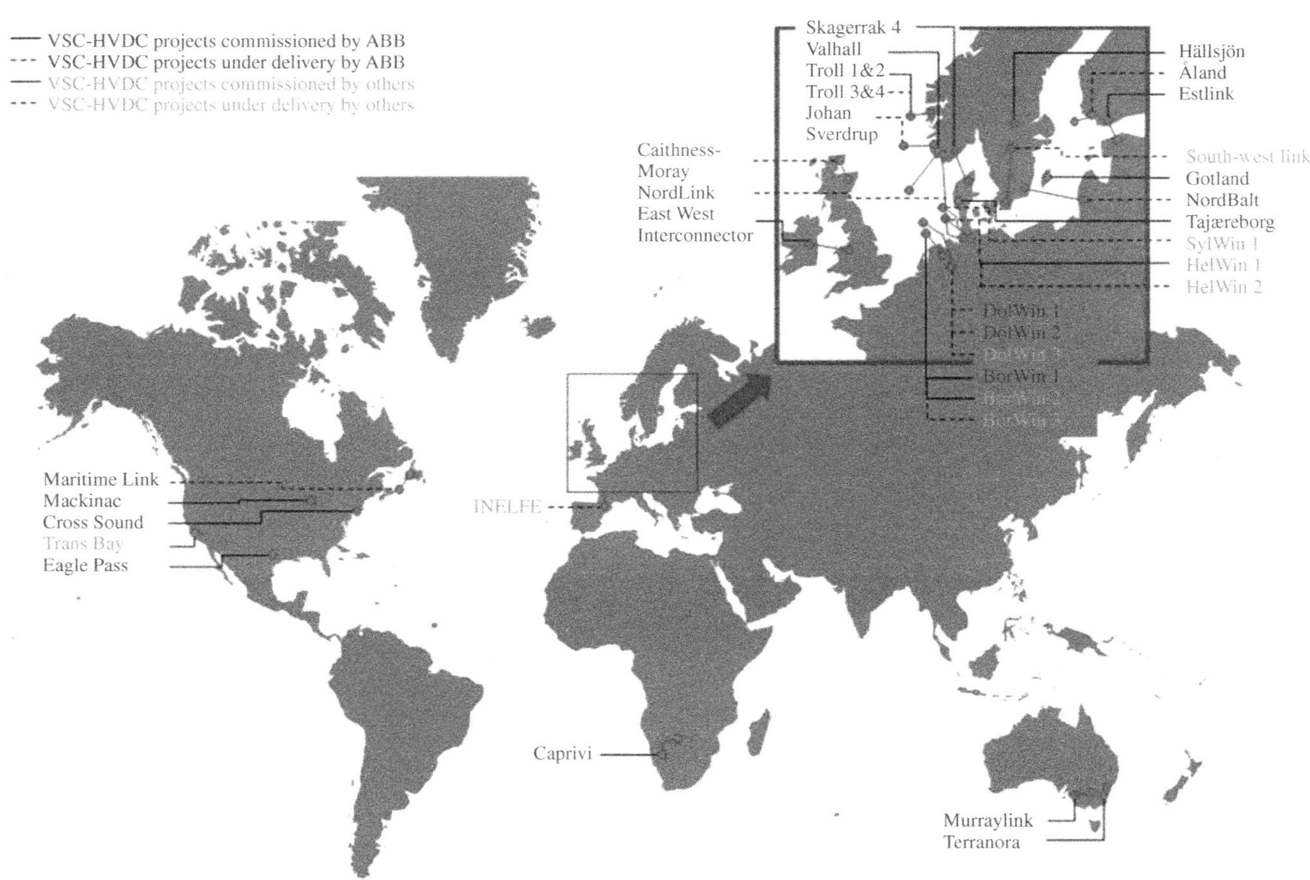

Figure 4.102 VSC–HVDC projects commissioned and under delivery. Reproduced with permission of ABB.

Wind parks connected by HVAC cables are often equipped with FACTS devices to meet the reactive power requirements at their point of connection.

State-of-the-art HVDC transmission based on VSC combines both the transmission of the power with control of reactive power at the converter station. Moreover, a VSC station, designed appropriately, can be used to energize AC networks, which means that they have black-start capability.

4.8.1.3 Maintaining System Stability

The stability of an AC system can be precisely studied today using appropriate computer modeling including the controls of the generators and large loads. Thus possible weaknesses and effective countermeasures can be identified at an early stage of system planning.

With its precise power flow control capability, the HVDC system can provide damping to power oscillations. This feature has been successfully implemented in various HVDC projects in the world.

Another important aspect of the AC system stability is frequency control. The fluctuating nature of generation from renewable energy sources requires new methods of balancing load and generation. A strong European overlay network will help to level out differences in the local generation and provide connection to large-scale energy storages like pumped hydro power plants. In order to achieve that, various high power links should be integrated into one supergrid.

The supergrid, a high-voltage electricity transmission system mainly based on DC transmission, will help connecting European countries to an enormous sustainable energy resource, bringing national energy markets together, and altering the way Europe produces, transmits, and consumes electricity. The supergrid as an energy network, allows large-scale power generation to be transmitted over a large distance to consumers.

4.8.2 Hybrid AC and DC Systems

The two-terminal CSC–HVDC links has shown impressive evolution in the last decade, breaking barriers that are not possible with the AC technology. Such links allow direct transfer of large amounts of active power from generation areas to load remote areas, or interconnection of AC systems when the AC technology is inapplicable. The CSC–HVDC multiterminal systems are today limited to just a few installations due to some technical constraints. However, it is expected that with the advent of VSC technology and the availability of a DC circuit breaker, the multiterminal HVDC networks will be a solution for strengthening the AC transmission to allow further integration of renewable energy sources.

Figure 4.103 shows a possible configuration of the future AC–DC hybrid power grid, with a multiterminal HVDC grid developed in parallel with the actual AC grid. Since large costs are involved in an electric grid, the concerns would be related to the rating of the DC grid that includes also the power electronic converters. In fact, this depend on the location and size of the generation and load areas between which DC transmission is used, as well as on the configuration and rating of the existing AC grid.

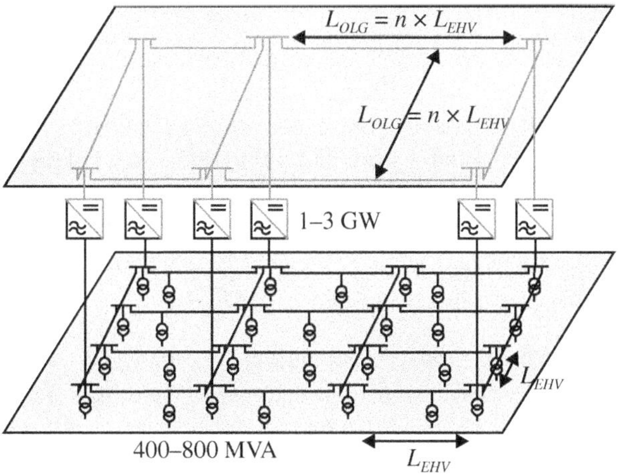

Figure 4.103 Example of layered configuration of the hybrid AC–DC grid. *Source:* Ergun et al. 2012 [81]. Adapted with permission of IEEE.

A DC transmission line should have a rating somewhere in the range of 4–8 or even up to 10 GW per circuit. The converters rating will be correlated with the AC grid rating, maybe in the range of 1–3 GW for each. In the AC grid, the power is transferred from the extra high voltage (EHV) level to lower voltage levels through 400–800 MVA transformers (possibly with several transformers in parallel) [81].

In order to handle large amounts of energy transferred from the HVDC grid, the AC grid would require strengthening by more EHV lines, more parallel transformers, substations with splited bus-bars and special configurations to limit the short-circuit currents, and eventually in a more meshed grid. Such upgrades are customized for each point of connection between the DC grid and the AC grid and may differ from country to country.

The idea of a supergrid, in Europe for instance, was born as a need to exploit the wind energy potential from offshore locations. There will therefore be changes of the power flows, sometimes with predominant flows from shore to inland.

The DC–AC converters can be seen as controllable generators in the AC grid. They have to be included in the coordinated voltage and frequency controls since they may handle large amounts of power. The active power injections and withdrawals into the same synchronized power system, through the multiple converters, have to be coordinated in such a way to avoid loop flows. The N-1 security criterion have to be considered when designing the AC–DC hybrid power grid so that, in case of a converter outage, the impact on the grid operation has to be minimum [80].

These integrated hybrid AC–DC systems provide significant advantages in terms of technology, economics as well as system security. They reduce transmission costs and help bypass heavily loaded AC systems. The DC grid can be seen as a highway for the AC grid, fully suitable for a secure and sustainable access to huge renewable energy resources. This approach is an important step in the direction of environmental sustainability of power supply [82, 83].

4.8.3 Supernodes

The Supernode is a hybrid system, which uses an islanded AC network to provide collection and routing of power on the supergrid. Connecting the HVDC systems via AC combines the advantages of AC systems with those of long distance HVDC transmission while providing effective solutions for connecting offshore wind parks or oil and gas platforms.

The supernode concept is largely based on technology existing today. The development needed to build supernodes is mainly in the field of control and protection for the islanded AC network, which includes frequency control, fault detection, and fault clearing strategies.

The preferred DC transmission technology for building supernodes is VSC. This is because a VSC transmission system can generate and maintain the AC voltage at the node with respect to amplitude and frequency, a feature also referred to as black start capability. As long as there are VSC systems providing sufficient short-circuit power available at the AC node, CSC-based DC transmission can also be connected.

A DC grid requires significant investment and will likely be completed in steps over many years. A possible scenario is to first build point-to-point HVDC and regional multiterminal HVDC systems which later may be connected and integrated to form a large DC grid. As technology progresses, the planned HVDC links will be operating at different DC voltages from the existing ones [84, 85].

A LCL DC/DC converter proposed in [86] has the abilities of interconnecting two DC transmission lines with different DC voltages, provide power flow control and inherently prevent DC fault propagation from one zone to another. In order to provide single connection point for multiple DC lines and to overcome the production challenge in DC grid, Jovcic and Lin [87] have proposed a *multiport DC hub concept*. The aim is to facilitate power exchange between numerous DC lines and to restrict the DC fault to only the faulted DC transmission line.

An LCL DC hub plays the role of a DC substation in a DC grid. The hub is capable of connecting multiple DC transmission lines with different DC voltage. Each DC transmission line can be added to or isolated from the hub without affecting the operation of the other DC transmission.

Figure 4.104 shows a proposed topology of DC grid and the position of the hub. The DC grid has four terminals (AC/DC stations), and the hub has four ports that are connected to DC lines. The term “ports” is used with hubs to avoid the confusion with DC grid “terminals.”

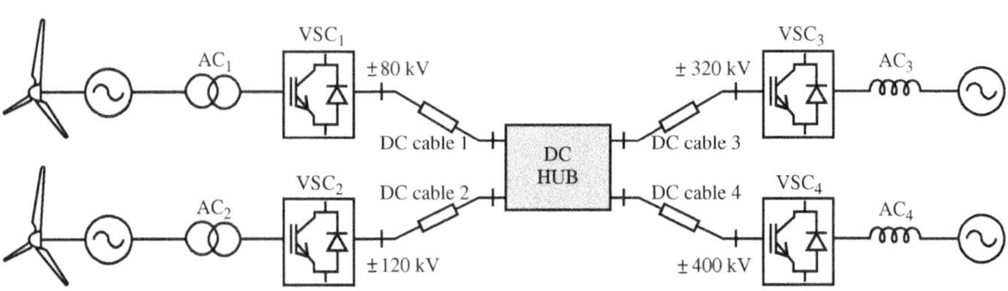

Figure 4.104 Four terminal DC grid with four port DC hub. *Source:* Jovcic and Lin 2013 [87]. Adapted with permission of IEEE.

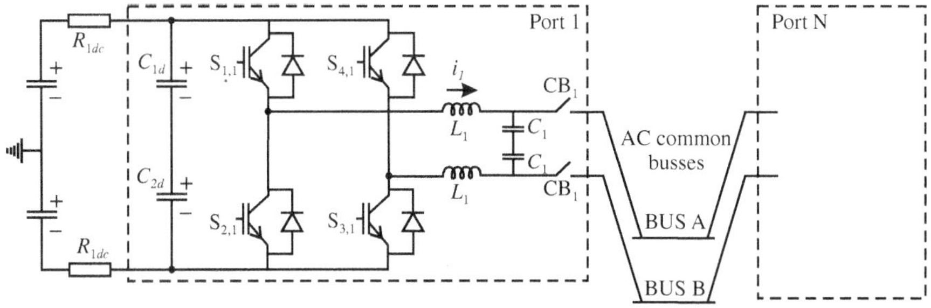

Figure 4.105 Two-phase N port DC hub topology. Adapted from Jovcic and Zhan 2012 [86].

The hub is based on multiple AC/DC IGBT-based converters and an internal passive LCL circuit without internal AC transformers. Each AC/DC bridge controls its active power independently. The designed hub has the ability to ride through the DC faults by keeping the fault current within the order its rated value.

As the DC hub is located centrally in the DC grid, it will have critical role in the security of DC grid power transfer.

A DC grid with DC hubs has concentric layered structure—passive circuit, converter, passive circuit, etc.—where the inner passive circuit of DC hub is in center.

The power transfer values have the most relevance at the outer layer, i.e., at terminals, which are the trading points for AC network operators. It is assumed that each terminal can set independently and arbitrarily the local desired power. The desired power at each terminal will be moderated by local droop feedback which ensures power balance within the whole DC grid.

The topology of an LCL hub is shown in Figure 4.105, where a two-phase hub is illustrated [86].

The hub comprises 1..*N* ports and two common AC bus (A, B). Each of the *N* ports is composed of two inductors L_i, two capacitors C_i, and one DC/AC bridge with switches ($S_{1,i}$–$S_{4,i}$). Each port is connected to the common AC buses through AC circuit breaker CB_i. Each inductor L_i is designed to enable transmitting of rated power, whereas each capacitor C_i is designed to locally compensate the reactive current generated by L_i. A circuit breaker is placed between each of the capacitors and the common AC bus such that on tripping a phase or tripping a port, the inductor together with its associated capacitor will be tripped from the hub. The remaining inner LCL circuit still maintains reactive power balance and operates without interruption.

4.8.4 Stepwise Development of the European Supergrid

The supergrid is planed to be implemented in various steps initially connecting the current crop of offshore wind generators to existing grids.

In Phase I, the so-called supernodes will be constructed on marine platforms, beginning in 2020, and will serve as collecting hubs for the energy produced by

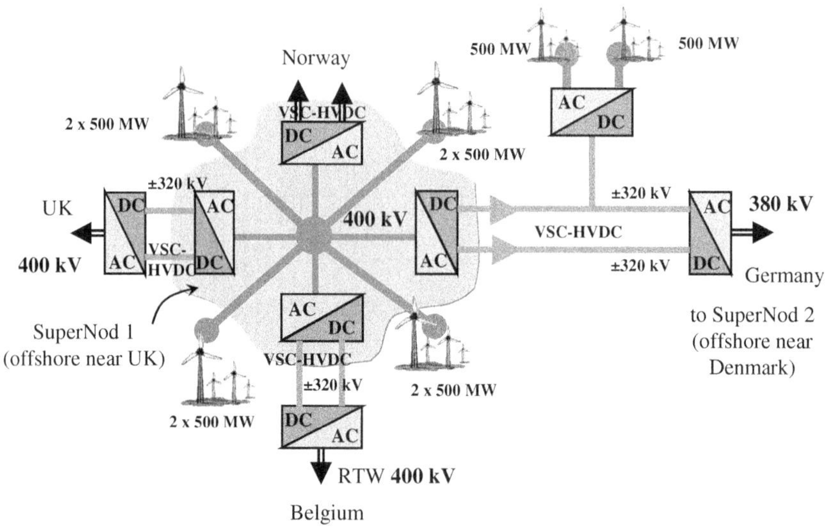

Figure 4.106 Details of a supernode near United Kingdom in the vision of Friends of SuperGrids. *Source:* Roadmap to the Supergrid technologies [13]. Public domain.

the offshore wind generation clusters, energy that will be then delivered in large quantities using VSC-HVDC links to the existing grids from UK, Belgium, Germany and Norway [13]. The vision of the Friends of the Supergrids regarding the development of the supernodes in North Sea is illustrated in Figure 4.106.

Energy from wind generation clusters of the United Kingdom east coast is collected at the SuperNodes at Firth of Forth, Dogger Bank/Horsea, and Norfolk Bank, which are connected together and interconnected at DC with the German and Belgium North Sea clusters, as well as with the Norway power system. The grid then delivers this power to the existing grids at the terminals in Glasgow, Hull, and Zeebrugge, as well as to the SuperNodes at London and Munich.

ISLES is a similar project planned for development, stretching from the west coast of the Isle of Lewis down to the north coast of Northern Ireland and the North Channel, then further south into the Irish Sea along the east coast of Ireland and across to the west coast of England. It is estimated that the project will reach 12 GW power installed in offshore wind generation and 4 GW in wave and tidal systems. The resulting links can be identified in Figure 4.108.

Phase II assumes creation of a ring, until 2030, by constructing a node in the Netherlands. Subsequently, construction of HVDC lines to link this node with France, Germany and another node from the North Sea is planned. Interconnection of Denmark to the ring is also planned at this phase. A proposal for interconnection of the supernodes in North Sea envisaged for 2030 is shown in Figure 4.107.

The Phase III is projected for 2050 and aims to expand the supergrid toward the south and the east of the European continent (Figure 4.108).

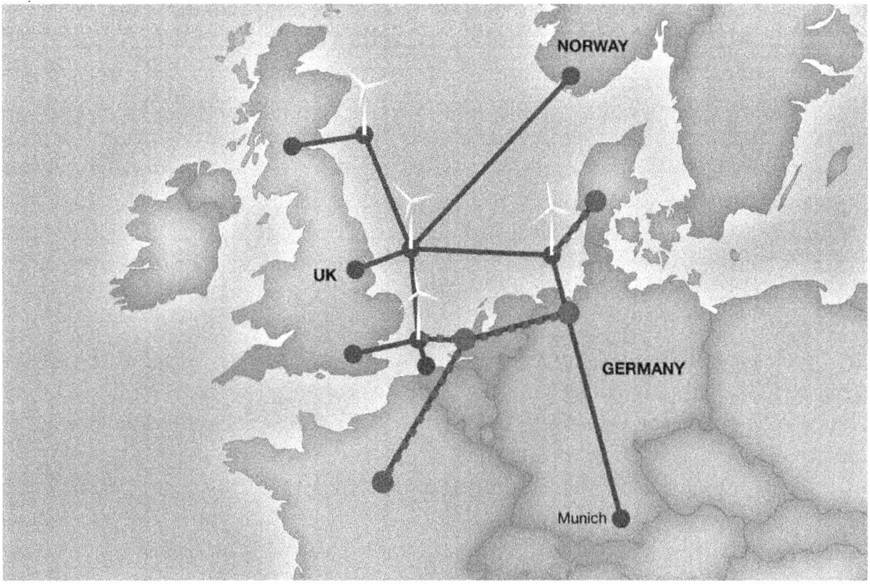

Figure 4.107 Interconnection of SuperNodes in the vision of friends of Supergrids. *Source:* Roadmap to the Supergrid technologies [13]. Public domain.

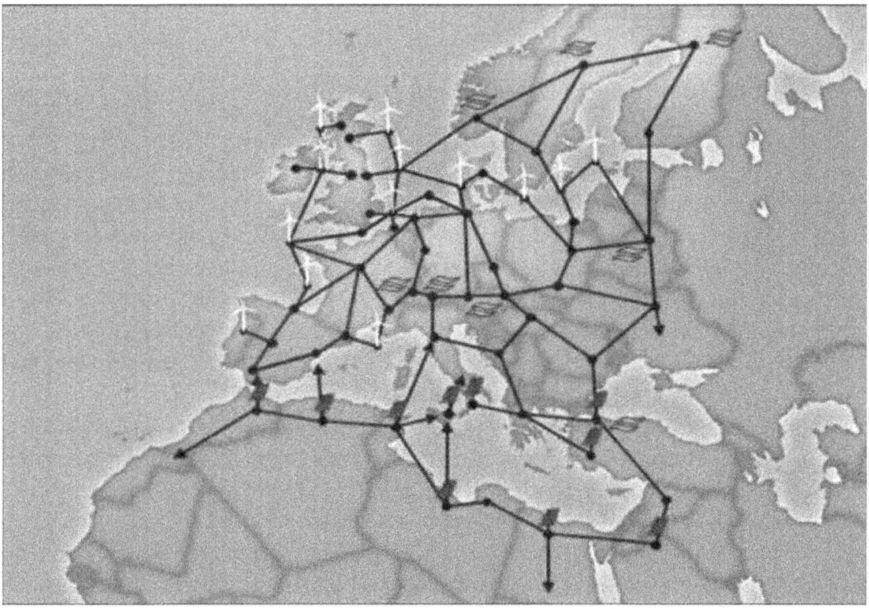

Figure 4.108 Developing the DC interconnection in Europe in the vision of Friends of the Supergrid. *Source:* Roadmap to the Supergrid technologies [13]. Public domain.

Some other projects were initiated in Europe focusing on large-scale interconnection to deliver power from remote generation locations [88]. For instance, the Medgrid project analysis the older idea of creating an interconnection ring around the Mediterranean Sea that includes the North Africa power systems, in order to allow the transfer of large amount of power from solar resources from Africa to Europe. The Medgrid project is also based on the more recent initiative called Desertec, started by a consortium of 12 companies and banks in 2009, which aim to develop large solar-based power installations. The Dii initiative was started, also in 2009, in order to find the way for developing a local market in North Africa and Middle East for solar and wind power generation, and eventually for exporting to Europe.

However, all the ideas for transferring electricity from Africa to Europe support the need for developing HVDC interconnections between the North and the South of Europe. The vision for a possible HVDC layer in Europe comes from the Friends of the Supergrid for year 2050, as shown in Figure 4.108.

There are also some disadvantages associated with the concept of SuperNode including the number of AC/DC and DC/AC conversions necessary. The power converters needed are costly; require relatively expensive space on offshore platforms, and cause extra power losses. Eliminating some of the power converters requires HVDC links to be interconnected on the DC-side forming HVDC multiterminal systems or grids. There are only very few HVDC systems having three CSC terminals today which are operated under specific conditions. In general, larger multiterminal systems have to be considered as a new field of technology.

4.8.5 Steps Toward a Planetary Supergrid

The power systems have been interconnected with the purpose of sharing the primary energy resources, including coal, natural gas, hydro, nuclear. Aiming to achieve clean energy and protect the environment, in the recent years huge amounts of financial resources have been used around the world to develop wind and photovoltaic power plants, which shifts the generation centers and thus change the power flow directions. As the power plants are constructed mainly near the location of the primary sources, large amounts of power are, in many cases, transmitted over long distances.

Border-to-border lines were constructed to synchronously interconnect national power systems and develop continental networks. The stability problems, the different frequencies used, differences in the technical codes, and other aspects have limited the AC systems to expand over longer distances. Paul Dimo's vision on a planetary power system [89] is today possible as the HVDC transmission lines can transmit 8000 MW or more over distances of up to 2700 km.

Europe is doing big steps to quit using the classical power plants based on unfriendly environment fuels and develops wind generation in North. Impressive projects have been initiated around the North Sea, as well as Baltic Sea, and challenging targets to increase the installed power in offshore wind power plants were set. As explained before, and also shown in Figure 4.109, several initiatives, like

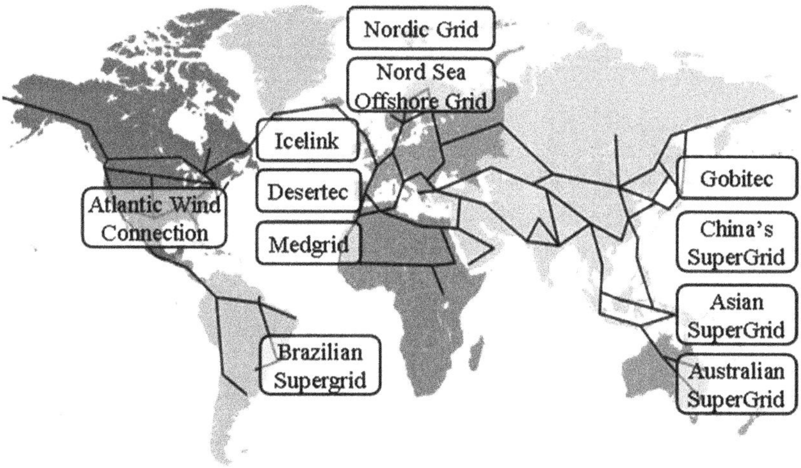

Figure 4.109 Vision on the global supergrid [90].

Desertec and Medgrid, emerged in Europe aiming to harvest the solar power in the Mediterranean Sea. While the European power systems are already synchronously interconnected, HVDC links are needed in order to transport the planned/expected power generated in wind and solar units to the populated centers.

Similar projects were initiated in Asia and Australia. The Gobitec project is intended to develop wind and photovoltaic systems in and around the Gobi Desert, which is a reason to interconnect Irkutsk with Shanghai, Seoul and Tokyo through an HVDC supergrid. Also, in order to exploit the abundant solar potential from Australia, it is expected to expand the HVDC grids through Philippine, Malaysia, and Indochina.

China's strategy to satisfy its energy needs aimed to develop hydro, wind, and photovoltaic power plants. The transmission of large amounts of power from the new generation facilities to the industrial centers in the east and south is today possible with the HVDC technology. China became a leader in development and deployment of new technologies with the most extensive network of high-voltage AC and HVDC in the world, and its plans to build another 13–20 ultra HVDC lines (±800 kV and ±1100 kV range) will allow creating a real hybrid supergrid.

After the two ±600 kV HVDC links, that is the 785 + 805 km long Itaipu link and the 2375 km long Rio Madeira link, Brazil makes a new step to develop its supergrid, by planning the 2092 + 2518 km long ±800 kV Belo Monte HVDC project. The HVDC technology is the only solution for power transmission from the large hydro power plants to the most populated areas around Sao Paulo, Rio de Janeiro, and Brasilia.

The Atlantic Wind Connection is an offshore transmission line that will span the mid-Atlantic region of the United States and will be developed as a backbone transmission system from north (New Jersey) to south (Virginia). This project will

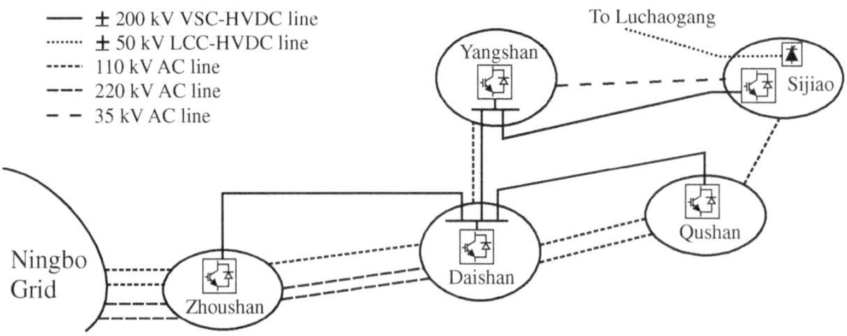

Figure 4.110 Zhoushan five terminal VSC–HVDC transmission.

allow advancing the offshore wind generation projects but, on the other hand, will be a step toward a supergrid.

4.8.6 VSC Multiterminal in China

Zhoushan city located in the Zhejiang province of China comprises a series of small islands. Zhoushan electrical grid has been suffering from poor power quality for long time due to weak system, inadequate power flow and lack of reactive power compensation. Also, the power capacity installed in wind generators has increased considerably in the past years.

After evaluating several possible alternative solutions, it was decided to construct a five-terminal VSC–HVDC transmission system to interconnect and supply power to the islands as a science and technology demonstration project, as shown in Figure 4.110 [91].

A multiterminal DC transmission system consists of three or more converter stations, some of them can be rectifier stations to send power from multiple sources and some of them can be inverter stations to receive and supply power to multiple load areas.

The construction of the Zhoushan ±200 kV VSC five-terminal project has began in 2013 and was commissioned in operation on July 4, 2014. It features the highest level of voltage, largest number of terminals, and highest capacity for single terminal among world's multiterminal flexible DC transmission projects.

Its five converter stations at Dinghai, Daishan, Qushan, Sijiao, and Yangshan were designed for a capacity of 400/300/100/100/100 MW and were connected by newly built ±200 kV DC transmission lines, having a total length of 141.5 km, of which 129 km are undersea cables.

APPENDIX 4.1 VSC–HVDC PROJECTS AROUND THE WORLD

Name	Country	Cable (km)	Overhead line	Converter topology	Voltage (kV)	Rated power (MW)	Year	Remarks
Tjuja—Ashu–Susamyr	Kirgiziya SSR	3.5	22	Two-level, with thyristors ETT	14	3.2	1984	Out of service in 1995
Hellsjön – Grängesberg	Sweden		10	Single two-level, phase unit	±10	3 MW/ 3 MVAr	1997	Experimental (ABB)
Gottland (Näs–Backs)	Sweden	2×70		Two-level, IGBT	± 80	50 MW / ± 30 MVAr	1999	Land cable for connecting of Näs wind farms to the city of Visby (ABB)
Eagle Pass	USA -Mexico			Three-level NPC, IGBT	± 15.9	36 MW/ ± 36 MVAr	2000	BTB (ABB)
Direct Link (Mullumbimby-Bungalora)	Australia	6×59		Two-level, IGBT	3× ± 80	180 MW/ ± 75 MVAr	2000	Land cable (ABB)
Murraylink	Australia	2×180		Three-level NPC, IGBT	± 150	200 MW/ +140/ -150	2002	The longest underground HVDC cable in the world (ABB)
Cross Sound Cable	USA	2×40		Three-level NPC, IGBT	± 150	330	2002	Sub-marine cable (ABB)
Troll A1-2	Norway	4×70		Two-level, IGBT	± 60	88	2005	Sub-marine cable to a gas platform for directly control of motor speed (ABB)
Estlink	Estonia to Finland	2×31 2×74		Two-level, IGBT	± 150	350	2006	Land + submarine cable (ABB)
Nord E.ON 1	Germany	200				400	2009	Interconnect the world largest off-shore wind park in Germany
Caprivi Link	Namibia		950		± 350	600 Ultimate	2010	Connecting weak AC networks
Transbay Cable	USA	88		MMC	± 200	400	2010	First multilevel VSC converter (Siemens)
Valhall	Norway	292			± 150	78	2011	Sub-marine cable supplying off-shore platform

(continued)

Name	Country	Cable (km)	Overhead line	Converter topology	Voltage (kV)	Rated power (MW)	Year	Remarks
East–West interconnector	UK –Ireland	186	75	Two-level	± 200	500	2012	Interconnector between Ireland and Wales (ABB)
BorWin1	Germany	200		IGBT	± 150	400	2012	Land + submarine cable (ABB)
HelWin1	Germany	85			± 259	576	2013	Submarine cable to a platform (Siemens)
BorWin2	Germany	125			± 300	800	2014	Submarine cable to a platform (Siemens)
INELFE	France – Spain	64		MMC	± 320	2×1000	2014	Underground XLPE cable (Siemens)
SylWin1	Germany	204		MMC	± 320	864	2014	Offshore wind farm connection
SK4	Denmark - Norway	240		Cascaded converter, part of a bipole	500	700 (± 80 MVAr)	2014	Submarine cable
Tres Amigas Substation	USA	0		IGBT	± 345	750	2014	BTB (Alstom)
Zhoushan Islands	China	129		IGBT	± 200	1000	2014	Five-terminal HVDC undersea/OHL cables
DolWin1	Germany	165		Cascaded, Two-level converters	± 320	800 (±260 MVAr)	2015	Land and submarine cable connected off-shore wind generation (ABB)
NordBalt	Sweden – Lithua-nia	450			± 300	700 (± 350 MVAr)	2015	400 km submarine + 50 km land cable, Longest HVDC light
DolWin2	Germany	135			± 320	900 (– 300/ +380)	2015	Largest off-shore VSC – HVDC
Troll A3-4	Norway	70			±60	2 × 50 MVA ± 24 MVAr	2015	Directly controls motor speed loaded with compressor
Aland	Finland	158			±80	100 MW ± 30 MVAr	2015	Interconnecting grids, length of sea crossing.
NordLink	Norway–Germany	516+54	53		±525	1400 MW	2020	Longest interconnector link. First full bipole VSC–HVDC.

APPENDIX 4.2 EXAMPLES OF VSC–HVDC ONE-LINE DIAGRAMS

- The Helljön Project (1997): the first VSC–HVDC transmission project in the world

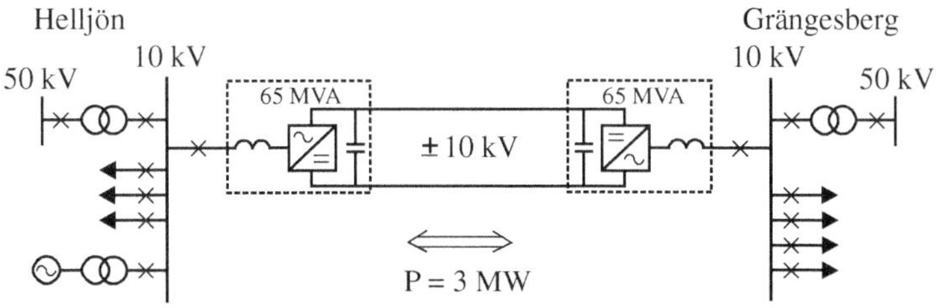

- The Terranora Interconnector (2000): Australia.

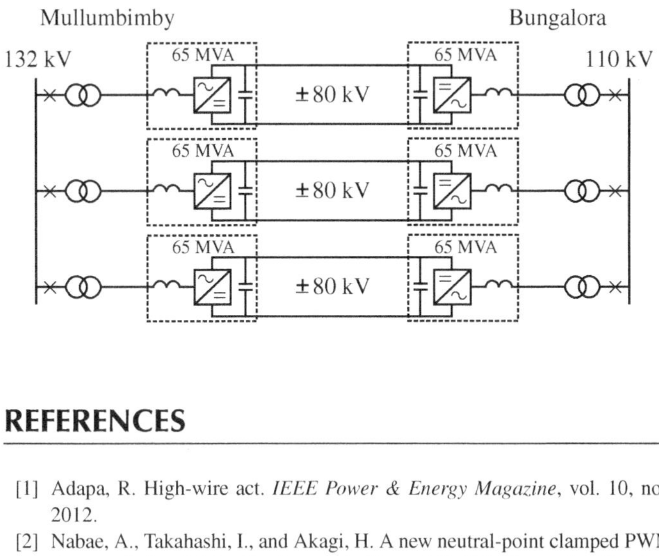

REFERENCES

[1] Adapa, R. High-wire act. *IEEE Power & Energy Magazine*, vol. 10, no. 6, November/December 2012.

[2] Nabae, A., Takahashi, I., and Akagi, H. A new neutral-point clamped PWM inverter. *IEEE Transactions on Industrial Applications*, vol. 17, pp. 518–523, September/October 1981.

[3] Rodriguez, J., Lai, J. S., and Peng, F. Z. Multilevel inverters: a survey of topologies, controls, and applications. *IEEE Transactions on Industrial Electronics*, vol. 49, no. 4, pp. 724–738, August 2002.

[4] Marquardt, R., and Lesnicar, A. New concept for high voltage – Modular Multilevel Converter. Proceedings of Power Electronic Specialists Conference PESC '04, Aachen, Germany, June 20–25, 2004.

[5] Yazdani, A., and Iravani, R. *Voltage-Sourced Converters in Power Systems. Modeling, Control, and Applications*. IEEE, John Wiley & Sons, Inc., Hoboken, NJ, 2010.

[6] Andersen, B. (Convenor), Baerd, H., Borgen, H., Dhaliwal, N. S., Edris, A. A., Hyttinen, M., Irokawa, S., Joncquel, É., Jyrinsalo, J., Litzenberger, W., Pincella, C., Schettler, F., Søbrink, K., Takasaki, M., Vovos, N. A., and Woodford, D. VSC transmission. CIGRE B4.37 Working Group, Technical Brochure no. 269, April 2005.

[7] Rodriguez, J., Franquelo, L. G., Kouro, S., Léon, I. J., Portillo, R., Prats, M. M., and Pérez, M. A. Multilevel converters: an enabling technology for high-power applications. *Proceedings of the IEEE*, vol. 97, no. 11, pp. 1786–1817, November 2009.
[8] Dijkhuizen, F. Multilevel converters: review, form, function and motivation, EVER 2012 Monaco, pp. 10–12 October 2012.
[9] Erinmez, I. A. (chairman), Foss, A. M., Bergmann, K., Cease, T. W., Ferry, G. A., Gyugyi, L., Irokawa, S., Pincella, C., Poumarede, C., Reichert, K., Renz., K., Takeda, M., and Weinhold, M. Static synchronous compensator (STATCOM). CIGRE WG 14-19, Technical Brochure no.144, 1999.
[10] Jovcic, D., and Ahmed, K. *High Voltage Direct Current Transmission. Converters, Systems, and DC Grids*. John Wiley & Sons., Ltd., 2015.
[11] Marquardt, R., Lesnicar, A., and Hildinger, J. *Modulares stromrichterkonzept für netzkupplungsanwendung bei hohen spannungen*. ETG-Fachtagung, Bad Nauheim, Germany, 2002.
[12] Lesnicar, A., and Marquardt, R. A new modular voltage source inverter topology. Proceedings of European Conference on Power Electronics and Applications, EPE'03, Toulouse, France, 2003.
[13] Friends of the Supergrid *Roadmap to the Supergrid Technologies*, WG 2 Technological, Final Report, March 2012.
[14] Peralta J., Saad, H., Dennetiere, S., Mahseredjian, J., and Nguefeu, S. Detailed and averaged models for a 401-level MMC-HVDC system. *IEEE Transactions on Power Delivery*, vol. 27, no. 3, pp. 1501–1508, July 2012.
[15] Saeedifard, M., and Iravani, R. Dynamic performance of a modular multilevel back-to-back HVDC system. *IEEE Transactions on Power Delivery*. vol. 25, no. 4, pp. 2903–2912, October 2010.
[16] Ding, G., Tang, G., He, Z., and Ding, M. New technologies of voltage source converter (VSC) for HVDC transmission system based on VSC. Proceedings of IEEE Power Engineering Society General Meeting, Pittsburgh, PA, June 20–24, 2008.
[17] MacLeod, N. DC grids. CIGRE Study Committee B4-HVDC and Power Electronics, Tutorials on Introduction to HVDC & HVAC and HVDC Grids, Brisbane, Australia, October 19, 2011.
[18] Mohaddes, M. Introduction to VSC HVDC. CIGRE Study Committee B4- HVDC and Power Electronics, Tutorials on Introduction to HVDC & HVAC and HVDC Grids, Brisbane, Australia, October 19, 2011.
[19] Retzmann, D. HVDC station layout, equipment, LCC & VSC and integration of renewables using HVDC, CIGRE Study Committee B4 – HVDC and Power Electronics, Tutorials on Introduction to HVDC & HVAC and HVDC Grids, Brisbane, Australia, October 19, 2011.
[20] Solas, E., Abad, G., Barrena, J. A., Carcar, A., and Aurtenetxea, S. Modulation of modular multilevel converter for HVDC application. Proceedings of 14th Power Electronics and Motion Control Conference, Mondragon, Spain, pp. 84–89, September 2010.
[21] Patel, H. S., and Hoft, R. G. Generalized technique of harmonic elimination and voltage control in thyristor inverters—Part I: harmonic elimination. *IEEE Transactions on Industry Applications*, vol. IA-9, pp. 310–317, May/June 1973.
[22] Li, L., Czarkowski, D., Liu, Y., and Pillay, P. Multilevel selective harmonic elimination PWM technique in series-connected voltage inverters. *IEEE Transactions on Industry Applications*. vol. 36, no. 1, pp. 160–170, January/February 2000.
[23] Tu, Q., and Xu, Z. Impact of sampling frequency on harmonic distortion for modular multilevel converter. *IEEE Transactions on Power Delivery*. vol. 26, no. 1, pp. 298–306, January 2011.
[24] Hagiwara, M., and Akagi, H. PWM control and experiment of modular multilevel converter. Proceedings of IEEE Power Electronics Specialist Conference, Tokyo, Japan, pp. 154–161, June 2008.
[25] Tu, Q., Xu, Z., and Xu, L. Reduced switching frequency modulation and circulating current supression for modular multilevel converters. *IEEE Transactions on Power Delivery*, vol. 23, no. 3, pp. 2009–2017, July 2011.
[26] Jacobson, B., Karlsson, P., Asplund, G., Harnefors, L., and Jonsson, T. VSC-HVDC transmission with cascaded two-level converters. B4-110, CIGRE, Paris, 2010.
[27] Sannino, A. *Power Electronics 2*, EEK180, Chalmers University of Technology, January 2004.
[28] Ionescu, F., Six, J. P., Delarue, P., and Floricău, D. *Power Electronics. Static Converters*, Technical Publisher, Bucharest, 1998. (in Romanian)
[29] Bose, B. K. *Modern Power Electronics and AC Drives*. Prentice Hall, 2001.

[30] Hingorani, N. G., and Gyugyi, L. *Understanding FACTS. Concepts and technology of Flexible AC Transmission Systems*, IEEE Press, Inc., New York, 2000.

[31] Floricău, D., and Hapiot, J. C. *Power Static Converters. Structures and command.* Printech Publisher, Bucharest, 2000. (in Romanian)

[32] Agulo, M., Lezana, P., Kouro, S., Rodriguez, J., and Wu, B., Level shifter PWM for cascaded multilevel inverters with even power distribution. Power Electronics Specialists Conference, PESC, Orlando, Florida, June 17–21, 2007.

[33] Huber, L., and Borojevic, D., Space vector modulator for forced commutated cycloconverters. *IEEE-IAS Annual Meeting*, vol. 1, pp. 871–876, 1989.

[34] Park, R.H. Two-reaction theory of synchronous machines. Generalized method of analysis. Part I. *AIEE Transactions*, vol. 48, pp. 716–730, July 1929.

[35] Choi, S., and Saeedifard, M. Capacitor voltage balancing of flying capacitor multilevel converters by space vector PWM. *Transactions on Power Delivery*, vol. 27, no 3, pp. 1154–1161, July 2012.

[36] Prats, Mª A. N., Franquelo, L. P., Portillo, R., León, J. I., Galván, E., and Carrasco, J. M. A 3-D space vector modulation generalized algorithm for multilevel converters. *IEEE Power Electronics Letters*, vol. 1, no. 4, pp. 110–114, December 2003.

[37] Franquelo, L. P., Prats, Mª A. N., Portillo, R. C., Galvan, J. I. L., Perales, M. A., Carrasco, J. M., Díez, E. G., and Jiménez, J. L. M. Three-dimensional space-vector modulation algorithm for four-leg multilevel converters using abc coordinates. *IEEE Transactions on Industrial Electronics*, vol. 53, no. 2, pp. 458-466, April 2006.

[38] Mekhilef, S., and Kadir, M. N. A., Novel vector control method for three-stage hybrid cascaded multilevel inverter. *IEEE Transactions on Industrial Electronics*, vol. 58, no. 4, pp. 1339–1349, April 2011.

[39] Chiniforoush, S., Jatskevich, J., Yazdani, A., Sood, V., Dinavahi, V., Martinez, J. A., and Ramirez, A. Definitions and applications on dynamic average models for analysis of power systems. *IEEE Transactions on Power Delivery*, vol. 25, no. 4, pp. 2655–2669, October 2010.

[40] Sanders, J. A., and Verhulst, F. *Averaging Methods in Nonlinear Dynamic Systems*. Springer-Verlag, 1985.

[41] Krein, P. T., Bentsman, J., Bass, R. M., and Lesieutre, B. L. On the use of averaging for the analysis of power electronic systems. *IEEE Transactions on Power Electronics*, vol. 5, pp. 182–190, April 1990.

[42] Sanders, S., Noworolski, J. M., Liu, X. Z., Verghese, G. C. Generalized averaging method for power conversion circuits. *IEEE Transactions Power Electronics*, vol. 6, no. 2, pp. 251–259, April 1991.

[43] Sanders, S., and Verghese, G. C. Synthesis of average circuit models for switched converters. *IEEE Transactions Circuits Systems*, vol. 38, no. 8, pp. 905–915, August 1991.

[44] Krause, P. C., Wasynczuk, O., and Sudhoff, S. D. *Analysis of Electric Machinery and Drive Systems*, 2nd edition. IEEE Press/Wiley, Piscataway, NJ, 2002.

[45] Erickson, R. W., and Maksimovic, D. *Fundamentals of Power Electronics*, 2nd edition. Kluwer, Norwell, MA, 2001.

[46] Sun, J., Mitchell, D. M., Greuel, M. F., Krein, P.T., and Bass, R. M. Average modelling of PWM converters operating in discontinuous conduction mode. *IEEE Transactions Power Electronics*, vol. 16, no. 4, pp. 482–492, July 2001.

[47] Davoudi, A., Jatskevich, J., and Chapman, P. L. Averaged modeling of switched-inductor cells considering conduction losses in discontinuous mode. *Institute of Engineering Technology Electric Power Applications*, vol. 1, no. 3, pp. 402–406, May 2007.

[48] Davoudi, A., Jatskevich, J., and De Rybel, T. Numerical state-space average-value modeling of PWM DC–DC converters operating in DCM and CCM. *IEEE Transactions Power Electronics*, vol. 21, no. 4, pp. 1002–1012, July 2006.

[49] Jatskevich, J., Perakek, S. D., and Davoudi, A. Parametric average-value model of synchronous machine-rectifier systems. *IEEE Transactions on Energy Conversion*, vol. 21, no. 1, pp. 9–18, March 2006.

[50] Saad, H., Dufour, C., Mahseredjian, J., Dennetière, S., and Nguefeu, S. Real time simulation of MMCs using the state-space nodal approach. International Conference on Power System Transients, Vancouver, BC, Canada, July 18–20, 2013.

[51] Wachal, R. (convenor), Jindal, A., Dennetiere, S., Saad., H., Rui, O., Cole, S., Barnes, M., Zhang., L., Song., Z., Jardini, J., Garcia, J. C., Mosallat, F., Suriyaarachich, H., Le-Huy, P., Totterdell, A., Zeni, L., Kodsi, S., Deepak, T., Thepparat, P., Beddard, T., Velasquez, J., D'Arco, S., Morales, A., Kono, Y., Vrana, T. K., and Yang, Y. Guide for the development of models for HVDC converter in a HVDC grid. CIGRE WG B4-57, Technical Brochure no. 604, December 2014.

[52] Saad, H., Peralta, J., Dennetière, S., Mahseredjian, J., Jatskevich, J., Martinez, J. A., Davoudi, A., Saeedifard, M., Sood, V., Wang, X., Cano, J., and Mehrizi-Sani, A. Dynamic averaged and simplified models for MMC-Based HVDC transmission systems. *IEEE Transactions on Power Delivery*, vol. 28, no. 3, July 2013.

[53] Kim, C. K., Sood, V. K., Jang, G. S., Lim, S. J., and Lee, S. J. *HVDC Transmission*. John Wiley & Sons (Asia) Pte. Ltd, 2009.

[54] Karlstrand, J., Bergman, G., and Jönsson, H. A. Cost-efficient XLPE cable system solutions. *IEE 7th International Conference on AC-DC Power Transmission*, Conference Publication No. 485, pp. 33–38, November 2001.

[55] Ekstrom, A. *High Power Electronics HVDC and SVC*. The Royal Institute of Technology, Stockholm, June 1990.

[56] Johansson, S. G., Asplund, G., Jansson, E., and Rudervall, R. Power system stability benefits with VSC DC-transmission systems. Proceedings of CIGRE General Meeting, Paris, August 2004.

[57] Kontos, E. Control and protection of VSC-based Multi-terminal DC Networks. Master thesis, Technical University of Delft, 2013.

[58] Song, R., Zheng, C., Li, R., and Zhou, X. VSC-based HVDC and its control strategy, IEEE Transmission and Distribution Conference and Exhibition: Asia and Pacific, IEEE/PES, pp. 1-6, Dalian, December 2005.

[59] Lindberg, A. PWM and Control of two and three level high power voltage source converters, Licentiate degree, Royal Institute of Technology, Stockholm, Sweden, 1995.

[60] Boost, M. A., and Ziogas, P. D. State of the art carrier PWM techniques: a critical evaluation. *IEEE Transaction on Industry Applications*, vol. 24, no. 2, pp. 271–280, March/April 1988.

[61] IEEE Standard 519-1992, Revision of IEEE Standard 519-1981. IEEE Recommended Practices and Requirements for Harmonic Control in Electrical Power Systems.

[62] Nakajima, T., and Irokawa, S. A control system for HVDC transmission by voltage sourced converters. IEEE Power Engineering Society Summer Meeting 1999 Proceedings, Edmonton, vol. 2, pp. 1113–1119, 1999.

[63] Aplund, G. (convener), Lindén, K., Barker, C., Baur, U., Christensen, P., Cole, S., Van Hertem, D., Jialiang, W., Jovcic, D., Labra, P., Marzin, A., Pahalawaththa, N., Rashwan, M., Rittiger, J., Søgaard, K., Westermann, D., Wilkening, E.-D., and Yue, C. HVDC Grid feasibility study. CIGRE B4-52 Technical Brochure no. 533, April 2013.

[64] Callavik, M., Blomberg, A., Häfner, J., and Jacobson, B. The hybrid HVDC breaker. An innovation breakthrough enabling reliable HVDC grids. ABB Grid Systems, Technical Paper, November 2012.

[65] Häfner, J., and Jacobson, B. - Proactive hybrid HVDC breakers - A key innovation for reliable HVDC grids. CIGRE Symposium Bologna 2011.

[66] Pucher, W. Fundamentals of HVDC interruption. CIGRE Electra, no.5, June 1986.

[67] Tahata, K., Ka, S., Tokoyoda, S., Kamei, K., Kikuchi, K., Yoshida, D., Kono, Y., Yamamoto, R., and Ito, H. HVDC circuit breakers for HVDC grid applications. CIGRE AORC Technical Meeting, 2014, Paper B4-1120, Paris, 2014.

[68] Franck, C. M. HVDC circuit breakers: A review identifying future research needs. *IEEE Transactions on Power Delivery*, vol. 26, pp. 998–1007, April 2011.

[69] Vithayathil, J. HVDC breaker and its application. Symposium on HVDC Technology "Sharing the Brazilian experience," Paper 2-4, Rio de Janeiro, 1983.

[70] Kostoulas, N., Sitokonstantinou, V., Idris, M., Sterling, P., Sayed, S., and Karaiskos, P. DC circuit breakers and their use in HVDC grids, Report, Imperial College, London, UK.

[71] Meyer, C., Kowal, M., and De Doncker, R. W. Circuit breaker concepts for future high-power DC-applications. IEEE IAS, 2005.

[72] Rosenqvist, R. HVDC breakers and multiple terminal configurations. ABB PSG Raleigh, October 18, 2012.

[73] Leterme, W., and Van Hertem, D. Classification of fault clearing strategies for HVDC grids. CIGRE Symposium Lund, Sweden, 2015.

[74] Horita, M. A. B., Jardini, J., and Saiki, G. Y. Load flow calculation in a hybrid AC and DC grid. CIGRE Colloquium on HVDC and Power Electronics to Boost Network Performance, Brasilia, Brazil, October 2013.

[75] Saad, H., Dennetière, S., Mahseredjian, J., Delarue, P., Guillaud, X., Peralta, J., and Nguefeu, S. Modular multilevel converter models for electromagnetic transients. *IEEE Transactions Power Delivery*, vol. 29, no. 3, pp. 1481–1489, November 2013.

[76] Cole, S., Beerten, J., and Belmans, R. Generalized dynamic VSC MTDC model for power system stability studies. *IEEE Transactions Power Systems*, vol. 25, no. 3, pp. 1655–1662, August 2010.

[77] Wang, W., Beddard, A., Barnes, M., and Marjanovic, O. Analysis of active power control for VSC–HVDC. *IEEE Transactions on Power Delivery*, vol. 29, no. 4, pp. 1978–1988, August 2014.

[78] Vrana, T. K., Dennetière, S., Yang, Y., Jardini, J., Jovcic, D., and Saad, H. The Cigré B4 DC grid test system. *Cigré Electra*, No. 270, pp. 10–19, October 2013.

[79] Bassini, M., Horita, M., Jardini, J., Casolari, R., and Cavalheiro, M. Modelling hybrid AC/DC network for dynamic stability studies. CIGRE Colloquium on HVDC and Power Electronics to Boost Network Performance, Brasilia, Brazil, October 2013.

[80] CIGRÉ WG 14.07 Guide for planning DC lines terminating at AC locations having low short-circuit capacities, Part I: AC/DC interaction phenomena, CIGRÉ Brochure No. 68, 1992.

[81] Ergun, H., Beerten, J., and Van Hertem, D. Building a new overlay grid for Europe. Proceedings of IEEE-PES General Meeting, San Diego, July 22–26, 2012.

[82] Breuer, W., Retzmann, D., and Uecker, K. Highly efficient solution for smart and bulk power transmission of green energy, 21th World Energy Congress, Montreal, September 2010.

[83] Friends of the Supergrid. Position paper on the EC Communication for a European Infrastructure Package, Brussels, December 2012.

[84] Van Hertem, D., and Ghandhari, M. Multi-terminal VSC HVDC for European supergrid: obstacles. *Renewable & Sustainable Energy Reviews*, vol. 14, no. 9, pp. 3156–3163, December 2010.

[85] Jovcic, D., Linden, K., Van Hertem, D., and Taisne, J. P. Feasibility of DC transmission networks. ISGT Europe, Manchester U.K., December 2011.

[86] Jovcic, D., and Zhan, J. High power IGBT-based DC/DC converter with DC faults tolerance, Proceedings of the 15th IPEC, Novi Sad, Serbia, September 2012, pp. 1–6.

[87] Jovcic, D., and Lin, W. Multiport high-power LCL DC Hub for use in DC Transmission grids. *IEEE Transactions on Power Delivery*, vol. 29, no. 2, pp. 760–768, September 2013.

[88] European Climate Foundation (ECE) Roadmap 2050, a practical guide to prosperous, low-carbon Europe. Technical Report, 2010.

[89] Dimo, P. *The Planetary Power System*. Publishing House of the Romanian Academy, Bucharest, 1988 (in Romanian)

[90] Gellings, C. W. A globe-spanning upper grid. *IEEE Spectrum*, vol. 52, no. 8, pp. 48–54, August 2015.

[91] Zhejiang Zhoushan Islands multiterminal HVDC flexible projects. State Grid Corporation of China, 2012. Available at http://www.sgcc.com.cn/ztzl/newzndw/sdsf/04/271599.shtml (in Chinese)

PART II

FACTS TECHNOLOGIES

(Courtesy of ABB)

Advanced Solutions in Power Systems: HVDC, FACTS, and Artificial Intelligence, First Edition.
Edited by Mircea Eremia, Chen-Ching Liu, and Abdel-Aty Edris.

5. Static VAr Compensator (SVC)
Mircea Eremia, Aniruddha Gole, and Lucian Toma

6. Series Capacitive Compensation
Mircea Eremia and Stig Nilsson

7. Phase Shifting Transformer: Mechanical and Static Devices
Mylavarapu Ramamoorty and Lucian Toma

8. Static Synchronous Compensator – STATCOM
Rafael Mihalic, Mircea Eremia, and Bostjan Blazic

9. Static Synchronous Series Compensator (SSSC)
Laszlo Gyugyi, Abded-Aty Edris, and Mircea Eremia

10. Unified Power Flow Controller (UPFC)
Laszlo Gyugyi

11. Interline Power Flow Controller (IPFC)
Laszlo Gyugyi

12. Sen Transformer: A Power Regulating Transformer
Kalyan K. Sen

13. Medium Voltage Power Electronics Devices for Distribution Grids
Ion Etxeberria-Otadui, David Frey, Seddik Bacha, and Bertrand Raison

CHAPTER 5

STATIC VAr COMPENSATOR (SVC)

Mircea Eremia, Aniruddha Gole, and Lucian Toma

5.1 GENERALITIES

A static VAr system (SVS) is defined as a combination of discretely and continuously switched VAr sources that are operating in a coordinated fashion by an automated control system. This includes the static VAr compensator (SVC) and the static synchronous compensator (STATCOM).

A static VAr compensator (SVC) is, according to IEEE and CIGRE, "a static VAr generator whose output is varied to exchange capacitive or inductive current so as to maintain or control specific parameters of the electric power system, typically bus voltages" [1].

The primary purpose of the SVC is usually the rapid control of voltage at weak points in a network. Installation may be at the midpoint of interconnection lines or in load areas.

In comparison to mechanically switched capacitors (MSCs) or mechanically switched reactors (MSRs), usually connected at a high voltage bus, the SVCs have a very rapid reaction and a good reliability. By using power thyristors, the static compensators present many advantages from several points of view, such as high reaction speed, insignificant contribution to the short-circuit power, and low maintenance. The static compensator overcame mechanical commutation problems (sudden voltage variations and transient perturbations) and this is the reason why it gradually became the most used Flexible Alternating Current Transmission System (FACTS) device.

The SVCs can improve the performances of the power system and consumers in several ways as follows.

- Transmission system application:
  - stabilizes the voltages in weak power systems;
  - reduces power losses in transmission lines;
  - increases the transmission capacity to reduce, defer, or eliminate the need for new lines;
  - contributes to power and voltage oscillations damping;
  - enhances voltage and stability control.

Advanced Solutions in Power Systems: HVDC, FACTS, and Artificial Intelligence, First Edition.
Edited by Mircea Eremia, Chen-Ching Liu, and Abdel-Aty Edris.

- Distribution system application:
  - stabilizes the voltage at the receiving end of long distribution lines;
  - reduces the reactive power absorbed from the main grid, which leads to lower losses and improved tariffs;
  - balances the asymmetrical loads;
  - reduces the voltage fluctuations and light flicker.

The SVC device has been used for over four decades, and it was employed as practical application long before the concept of flexible systems existed, having their place in the great family of FACTS devices.

The first static compensators were used in industrial distribution systems, while various solutions were used in transmission systems to increase the stability limits. With the development of power electronic devices, the number of SVC applications has drastically increased, being preferred in many power system applications (see Section 5.9).

Figure 5.1 represents the simplified scheme of a static VAr compensator.

The following components can be distinguished:

- Thyristor-controlled reactor (TCR), which is a continuously controlled inductance, between $L = 0$ (thyristors are blocked) and the maximum value $L = \max$ (thyristors are in full conduction).
- Thyristor-switched capacitors (TSCs), which are switched on/off by static devices. By appropriate control of the TCR, a continuous control domain of the SVC between the maximum inductive and maximum capacitive is achieved. This combination allows the SVC to absorb/produce the exact necessary amount of reactive power.
- Fixed filters, used especially for low order harmonics filtering, produced by the TCR. The capacitors included in the filters produce reactive power also.

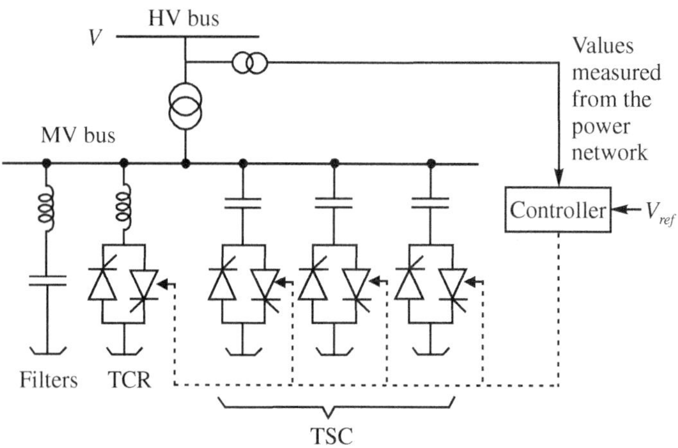

Figure 5.1 Simplified scheme of a static VAr compensator.

Pairs of anti-parallel connected thyristors are used in order to ensure conduction on both voltage waveform cycles. Both thyristors of the pair have to be maintained in conduction state, or a command impulse has to be ensured at the beginning of each semi-period at the thyristor that controls the respective cycle, in order to maintain a constant current through the capacitor and/or the reactor. If the command impulse is interrupted, the current through the thyristor will not be immediately interrupted and the thyristor will be blocked only at the natural zero crossing of the voltage. This is the instant at which the other thyristor of the valve can be switched on.

5.2 THYRISTOR-CONTROLLED REACTOR

An elementary TCR consists of a fixed reactor of inductance L and a bidirectional thyristor valve (Figure 5.2a).

Currently available large thyristors can block voltages up to 4–9 kV and conduct currents up to 3–6 kA. Therefore, in practical applications, the valve consists of many thyristors connected in series to meet the required blocking voltage levels at a given power rating.

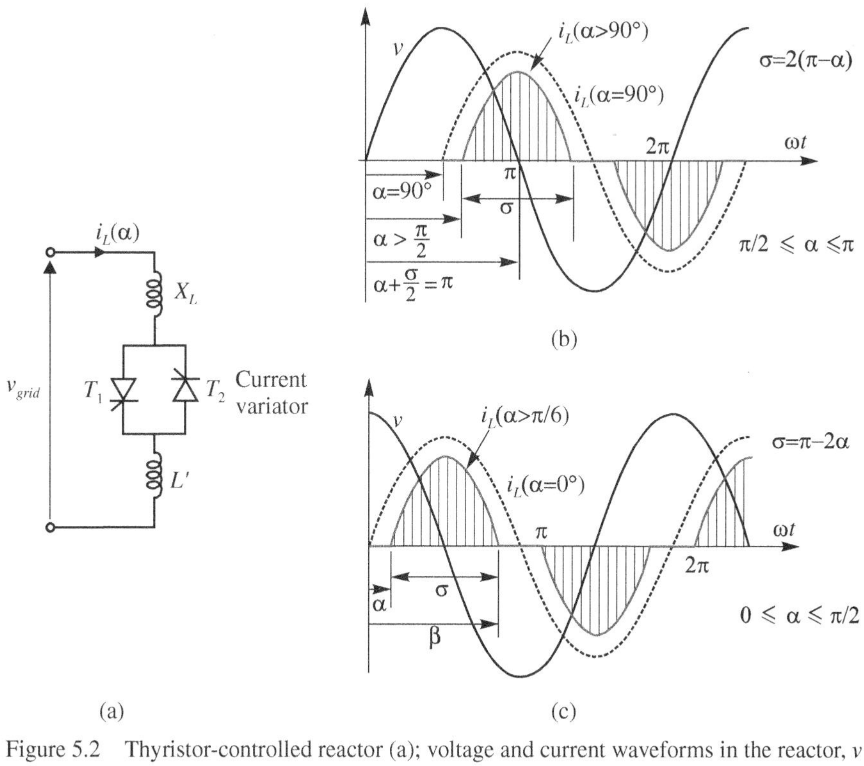

Figure 5.2 Thyristor-controlled reactor (a); voltage and current waveforms in the reactor, v and $i_L(\alpha)$ (b, c).

The current in the reactor can be continuously modified from maximum conduction to zero by firing delay angle control. Variation of the reactor current $i_L(\alpha)$ is achieved by controlling the instant of ignition of the thyristors and thus by controlling the conduction period for each half-cycle. If the conduction is initiated at maximum input voltage, then full thyristor conduction period and thus maximum current in the reactor is achieved.

As regards the operating principles of the bidirectional module, the command of thyristors T_1 and T_2 is sequential, that is when one thyristor is in conduction mode, the other one is blocked. In the literature, there are two main hypotheses concerning the point at which the delay angle α is referred to:

(i) the zero crossing instant of the voltage waveform positive half-cycle (Figure 5.2b);

(ii) the instant at which the maximum voltage value occurs (Figure 5.2c).

If the delay angle is measured with respect to the zero crossing instant of the voltage (Figure 5.2b) a full conduction is achieved for $\alpha = 90°$ and a minimum conduction for $\alpha \approx 180°$. Thereby, the TCR can be used for $\pi/2 \leq \alpha \leq \pi$. These constraints emerge due to the fact that one thyristor has to be blocked before the other enters in conduction.

- If the thyristor firing is initiated at $\alpha > \pi/2$, then the conduction period will be $\sigma = 2\pi - 2\alpha$, and the current will be blocked during the negative semi-period of the voltage. Therefore, the other thyristor, having a negative input voltage, can be fired at the angle $\alpha + \pi$.
- If the thyristor firing is initiated at $\alpha < \pi/2$, then the current will not be interrupted before the instant at which the other thyristor is fired (i.e., $\alpha + \pi$), and the current will flow in one direction only. Thus, the condition for normal operation is that the interruption angle β must be lower than the angle at which the other thyristor is fired, that is, $\beta < \alpha + \pi$, where $\beta = \sigma + \alpha = 2\pi - \alpha$ and therefore $2\pi - \alpha < \alpha + \pi$, concluding that $\alpha \geq \pi/2$.

For the second hypothesis, two cases emerge (Figure 5.2c):

- At the delay angle $\alpha = 0°$, the thyristor valve closes when the voltage reaches the maximum; the current in the reactor will, in this case, have a maximum value, equal to the steady-state value when the valve is permanently closed (full conduction).
- When the valve gating is delayed by an angle α $(0 \leq \alpha \leq \pi/2)$ with respect to the instant of maximum voltage, partial conductions are obtained (between full conduction and minimum conduction).

Calculation of Current in TCR. Considering the electric circuit from Figure 5.2a, where the instantaneous phase voltage

$$v(t) = \sqrt{2}V \cos \omega t \tag{5.1}$$

and the angle α are metered with respect to the instant at which the supply voltage reaches the peak value (Figure 5.2c), the current in the circuit is given by:

$$L\frac{di}{dt} = \sqrt{2}V\cos\omega t \tag{5.2}$$

By integration, we achieve:

$$i(t) = \frac{\sqrt{2}V}{\omega L}\sin\omega t + C \tag{5.3}$$

where L is the TCR's inductance, and C is an integration constant.

The integration constant C can be determined at the limit, at the exact firing instance when the thyristor is blocked. Let the delay angle α be metered with respect to the instant at which the maximum voltage is achieved (Figure 5.2c). Since before the instant $\omega t = \alpha$, the current $i_L(\omega t = \alpha) = 0$, the value of integration constant from equation (5.3) is:

$$C = -\frac{\sqrt{2}V}{\omega L}\sin\alpha \tag{5.4}$$

and the expression (5.3) of the instantaneous current becomes:

$$i_L(\omega t) = \frac{\sqrt{2}V}{X_L}(\sin\omega t - \sin\alpha),\ \alpha < \omega t < \beta(= \alpha + \sigma) \tag{5.5}$$

where:

V is the amplitude of the applied AC voltage at the TCR.

X_L – the fundamental frequency inductive reactance of the reactor.

At $\omega t = \beta$, the current in the reactor, $i(\beta)$, crosses zero again, thereby equation (5.5) becomes:

$$0 = \frac{\sqrt{2}V}{X_L}(\sin\beta - \sin\alpha)$$

respectively,

$$\sin\beta = \sin\alpha$$

As $\alpha < \beta\ (= \alpha + \sigma)$, the acceptable solution for the turn-off angle β is:

$$\beta = \pi - \alpha$$

The thyristors are in conduction between α and $\pi - \alpha$, which means that the conduction period σ is:

$$\sigma = \pi - \alpha - \alpha = \pi - 2\alpha \tag{5.6}$$

Calculation of Fundamental and Harmonic Components in TCR. The current $i_L(t)$ consists of sinusoidal half-cycles, containing only odd harmonics, depending on the delay angle α. The fundamental component of the current (I_{L1}), which corresponds to a variable reactive power absorption, depending on the delay angle, can be computed using Fourier analysis.

The current $i_L(t)$ in the TCR can be written as a Fourier series [2, 3]:

$$i_L(t) = a_0 + \sum_{n=1}^{\infty} a_n \sin(n\omega t) + \sum_{n=1}^{\infty} b_n \cos(n\omega t) \tag{5.7}$$

where:

$a_0 = 0$, due to the symmetry with respect to the x axis;

$b_n = 0$, due to the symmetry of odd waveforms $f(x) = -f(-x)$;

$a_n = 0$, for n = 2, 4, 6, 8, …, due to the symmetry within each half-wave $f(x + T/2) = -f(x)$.

The nonzero terms become:

$$a_n = \frac{2}{\pi} \int_0^{\pi} i_L(t) \sin(n\omega t) d(\omega t) \tag{5.8}$$

The peak value of the fundamental component of the current in the TCR can be computed using equations (5.5) and (5.8) [4]:

$$\begin{aligned}
\hat{i}_{L1} &= \frac{2}{\pi} \int_0^{\pi} i(t) \cdot \sin \omega t \cdot d(\omega t) = \frac{2}{\pi} \int_{\alpha}^{\beta} \frac{\sqrt{2}V}{X_L} (\sin \omega t - \sin \alpha) \cdot \sin \omega t \cdot d(\omega t) = \\
&= \frac{\sqrt{2}V}{X_L} \frac{2}{\pi} \left[\int_{\alpha}^{\pi-\alpha} \sin^2 \omega t \cdot d(\omega t) - \sin \alpha \int_{\alpha}^{\pi-\alpha} \sin \omega t \cdot d(\omega t) \right] = \\
&= \frac{2\sqrt{2}V}{\pi X_L} \left[\int_{\alpha}^{\pi-\alpha} \frac{1 - \cos 2\omega t}{2} \cdot d(\omega t) + \sin \alpha \cos \omega t|_{\alpha}^{\pi-\alpha} \right] = \\
&= \frac{2\sqrt{2}V}{\pi X_L} \left[\frac{\pi - 2\alpha}{2} - \frac{1}{4} \sin 2\omega t|_{\alpha}^{\pi-\alpha} + \sin \alpha(-\cos \alpha - \cos \alpha) \right] = \\
&= \frac{2\sqrt{2}V}{\pi X_L} \left\{ \frac{\pi - 2\alpha}{2} - \frac{1}{4} [\sin(2\pi - 2\alpha) - \sin 2\alpha] - 2 \sin \alpha \cos \alpha \right\} = \\
&= \frac{2\sqrt{2}V}{\pi X_L} \left(\frac{\pi - 2\alpha}{2} + \frac{1}{2} \sin 2\alpha - \sin 2\alpha \right) = \frac{\sqrt{2}V}{X_L} \left(1 - \frac{2\alpha}{\pi} - \frac{\sin 2\alpha}{\pi} \right)
\end{aligned}$$

Thus, the rms value of the reactor fundamental current $I_{L1}(\alpha)$ can be expressed as a function of angle α:

$$I_{L1}(\alpha) = \frac{V}{\omega L} \left(1 - \frac{2\alpha}{\pi} - \frac{\sin 2\alpha}{\pi} \right) \tag{5.9}$$

where V is the rms value of the phase voltage $v(t)$.

In Figure 5.3a, the normalized fundamental current is plotted in p.u. of the maximum sinusoidal current as a function of angle α. TCR can continuously control the fundamental component of the current from zero (open valve) to maximum value (valve closed) in a manner similar to a variable reactive admittance.

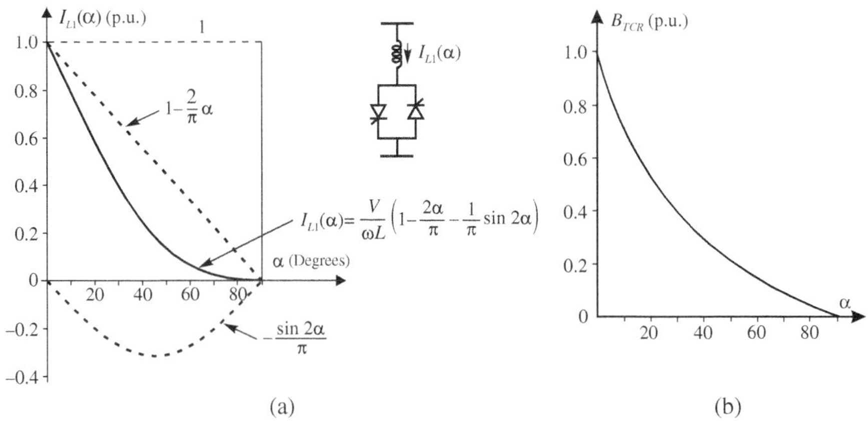

Figure 5.3 Amplitude variation of the fundamental TCR current with the delay angle α.

An effective reactive admittance $B_{TCR}(\alpha)$ for the TCR can be expressed from equation (5.9) as a function of angle α:

$$B_{TCR}(\alpha) = B_{\max}\left(1 - \frac{2}{\pi}\alpha - \frac{1}{\pi}\sin 2\alpha\right) \tag{5.10}$$

where $B_{\max} = 1/\omega L$.

Obviously, the admittance $B_{TCR}(\alpha)$ varies with angle α in the same manner as the fundamental current $I_{L1}(\alpha)$ (Figure 5.3b): the per-unit value of B_{TCR} is obtained with respect to its maximum value $B_{\max}$ as the base quantity.

In practice, the TCR can operate anywhere within a defined *V–I* area, whose limits are defined by the maximum admittance, voltage, and current (Figure 5.4).

The rms value of the fundamental component of current can also be expressed, from equation (5.9), as a function of the conduction angle $\sigma = \pi - 2\alpha$:

$$I_{L1}(\sigma) = B_{\max} V \frac{(\sigma - \sin\sigma)}{\pi} = B_{LCT}(\sigma) \cdot V \tag{5.9$'$}$$

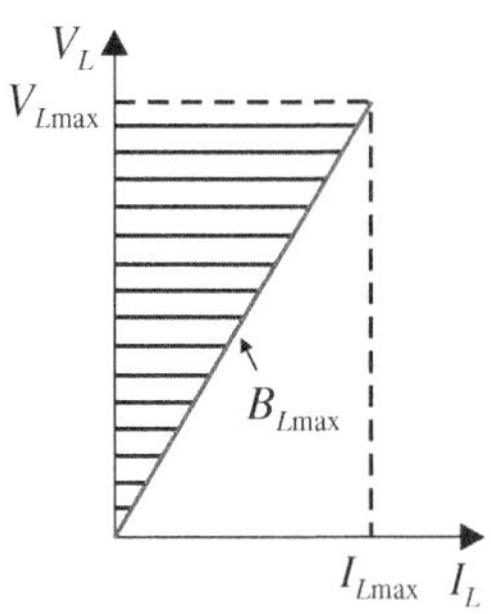

Figure 5.4 Operating *V–I* area of the TCR.

The reactive power absorbed by the reactor is:

$$Q_L = \frac{\sigma - \sin\sigma}{\pi X_L} V^2 \tag{5.11}$$

If *angle* α *is measured with respect to the zero-crossing instant of the supply voltage* $v(t)$, the same expression (5.9) results, but with $\alpha' = \alpha + \pi/2$ and thus $\alpha = \alpha' - \pi/2$ [2].

Thereby, equation (5.9) can be transformed into

$$I_{L1}(\alpha') = \frac{V}{X_L}\left\{1 - \frac{2}{\pi}(\alpha' - \pi/2) - \frac{\sin[2(\alpha' - \pi/2)]}{\pi}\right\}$$

which is equal to

$$I_{L1}(\alpha') = \frac{V}{\pi X_L}\left[2\pi - 2\alpha' + \sin(2\alpha')\right] \tag{5.9''}$$

The rms value of the fundamental component of the current in terms of the angle α measured from the zero-crossing instant of the supply voltage ($\pi/2 < \alpha < \pi$) becomes

$$I_{L1}(\alpha) = \frac{V}{\pi X_L}[2(\pi - \alpha) + \sin 2\alpha]$$

Also, the conduction period is obtained as

$$\sigma = 2\pi - 2\alpha = 2(\pi - \alpha)$$

and thus the same expression as in (5.9′).

From equation (5.10), it can be observed that the maximum value for B_{TCR} is $B_{max} = 1/\omega L$, obtained at $\sigma = \pi$ ($\alpha = 90°$), meaning full conduction (Figure 5.5).

The minimum value for B_{TCR} is zero, obtained at $\sigma = 0°$ ($\alpha = 180°$). This control principle relies on the zero crossing of the voltage waveform positive half-cycle and it is an asymmetric control on both cycles.

A first effect of the increase in angle α is the decrease in magnitude of the fundamental component I_{L1} of the current. This decrease is equivalent to an increase

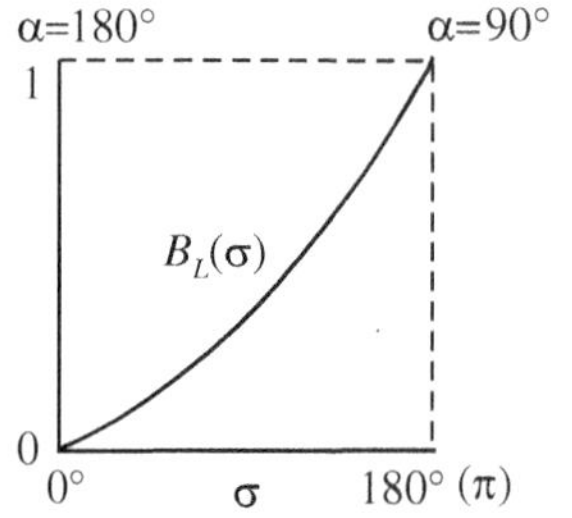

Figure 5.5 The asymmetric control principle at zero crossing of the voltage waveform positive half-cycle.

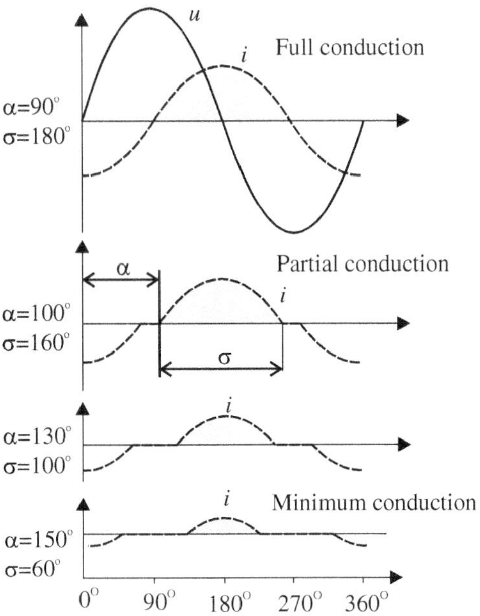

Figure 5.6 Modifying the fundamental component of the current using the delay angle α.

in the reactor's inductive reactance, reducing the reactive power absorbed and the current intensity in its circuit (Figure 5.6).

Another effect of increasing the firing angle α is an increased distortion coefficient of the current waveform, thus the current becomes nonsinusoidal and harmonics are generated.

Analysis of Figures 5.2b, 5.2c, and 5.6, respectively, shows that the conduction angle control, characterizing the operation of TCR, results in a nonsinusoidal current waveform in the reactor. In other words, the TCR, in addition to the wanted fundamental current, also generates *harmonics of superior order*.

If the firing angles are equal for the two thyristors, only odd-order harmonics are generated and the magnitude for the nth order harmonic component of the current, $\hat{i}_{Ln}$, is equal to component a_n of the Fourier series [1–3]:

$$\hat{i}_{Ln} = \frac{2}{\pi} \int_{\alpha}^{\pi-\alpha} \frac{\hat{v}}{\omega L} (\sin \omega t - \sin \alpha) \sin (n\omega t)\, d(\omega t) =$$

$$= \frac{2\hat{v}}{\pi X_L} \left[\int_{\alpha}^{\pi-\alpha} \sin (\omega t) \sin (n\omega t)\, d(\omega t) - \int_{\alpha}^{\pi-\alpha} \sin \alpha \sin (n\omega t)\, d(\omega t) \right]$$

$$\hat{i}_{Ln} = \frac{\hat{v}}{X_L} \frac{4}{\pi} \left[\frac{\sin \alpha \cos (n\alpha) - n \sin (n\alpha) \cos \alpha}{n \left(n^2 - 1\right)} \right] \tag{5.12}$$

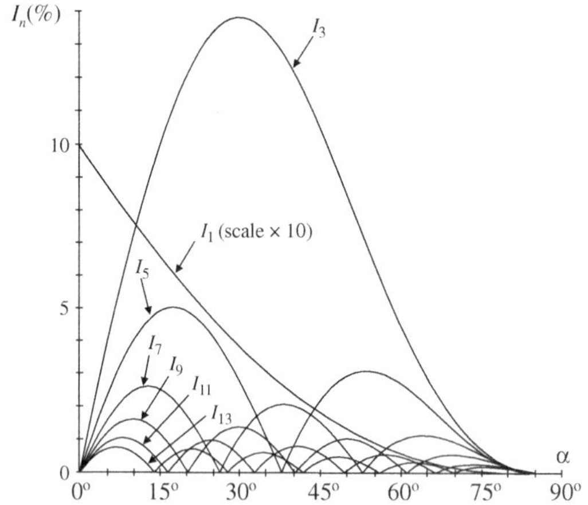

Figure 5.7 Amplitudes of the harmonic current of the TCR as a function of the delay angle α. *Source*: Hingorani & Gyugyi 1999 [1]. Reproduced with permission of IEEE.

or

$$I_{Ln}(\alpha) = \frac{4}{\pi}\frac{V}{X_L}\left[\frac{\sin\alpha\cos(n\alpha) - n\sin(n\alpha)\cos\alpha}{n\left(n^2-1\right)}\right] \tag{5.12'}$$

where the angle α is measured with respect to the peak voltage value and therefore $0 < \alpha < \pi/2$, and the order n is equal to 3, 5, 7 …, that is, only the odd-order harmonics are generated. However, this observation is true only if the TCR is controlled with the same angle α on both positive and negative half-cycle.

Figure 5.7 presents the amplitude of harmonic components as a function of angle α, expressed as percent of the maximum sinusoidal current. It can be seen that the maximum value of the third harmonic component is obtained at $\alpha = 30°$, of the fifth harmonic component at $\alpha = 20°$, and the one for the seventh component is obtained at $\alpha = 10°$. Therefore, only these components (3, 5, and 7) are taken into consideration in technical calculus, as higher order harmonic components have negligible magnitude values.

All the above statements refer to the TCR as a single-phase device.

A *three-phase TCR* consists of three identical single-phase TCRs, delta connected, as shown in Figure 5.8a. Each TCR branch contains two equally sized reactors, one on each side of the thyristor valve, to prevent the full AC voltage appearing across the valves in the case of a short circuit occurring at the branch terminals. With this arrangement, if the three grid voltages are balanced, the three branches are identical, the thyristors are controlled with the same firing delay angle α, and the triplen order harmonics (3, 9, 15, 21, 27, etc.) only circulate within the delta connection and do not enter the line.

Let i_A, i_B, and i_C be the three line currents and i_{An}, i_{Bn}, and i_{Cn} their n-order harmonic components. Let i_{AB}, i_{BC}, and i_{CA} be the three currents in the delta-configured

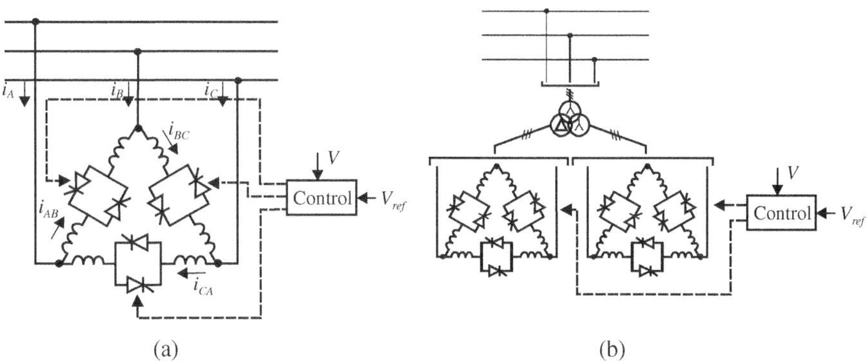

Figure 5.8 Three-phase TCR connected in delta: (a) six-pulse arrangement and (b) 12-pulse arrangement. *Source*: Erinmez et al. 1986 [7]. Reproduced with permission of CIGRE.

TCR, and i_{ABn}, i_{BCn}, and i_{CAn} their n-order harmonic components. Then, for the third harmonic current components, we have [2, 3]:

$$i_{AB3} = \hat{i}_3 \cos(3\omega t + \phi_3) \tag{5.13a}$$

$$i_{BC3} = \hat{i}_3 \cos\left(3\omega t + \phi_3 - 3\frac{2\pi}{3}\right) = \hat{i}_3 \cos\left(3\omega t + \phi_3 - 2\pi\right) = \hat{i}_3 \cos\left(3\omega t + \phi_3\right) \tag{5.13b}$$

$$i_{CA3} = \hat{i}_3 \cos\left(3\omega t + \phi_3 - 3\frac{4\pi}{3}\right) = \hat{i}_3 \cos\left(3\omega t + \phi_3 - 4\pi\right) = \hat{i}_3 \cos\left(3\omega t + \phi_3\right) \tag{5.13c}$$

Thus,

$$i_{AB3} = i_{BC3} = i_{CA3}$$

All three currents are in phase and circulate in the delta configuration, forming a zero sequence system. It follows that the third harmonic line currents reduce to zero, as follows:

$$i_{A3} = i_{AB3} - i_{CA3} = 0 \tag{5.14}$$

and likewise for i_{B3} and i_{C3}.

A similar analysis can be done for the fifth and seventh harmonic current. The fifth harmonic currents do not cancel out in the line and form a negative sequence system of currents. This also holds for all harmonics of the order n with $n = 6k + 5$, $k = 0, 1, 2, 3, \ldots$ (5, 11, 17), etc. The seventh harmonic currents do not cancel out in the line and form a positive sequence system of currents. This also holds out for all harmonics of order n with $n = 6k + 1$, $k = 0, 1, 2, 3, \ldots$ (7, 13, 19), etc. In a three-phase installation, fifth and seventh harmonic filters are generally needed [2].

The reduction of fifth- and seventh-order harmonic currents can be obtained by using two six-pulse arrangements, one operated from wye-connected windings and the other from delta-connected windings of the secondary of a coupling transformer (Figure 5.8b). Because of the 30° phase shift between the related voltages of the

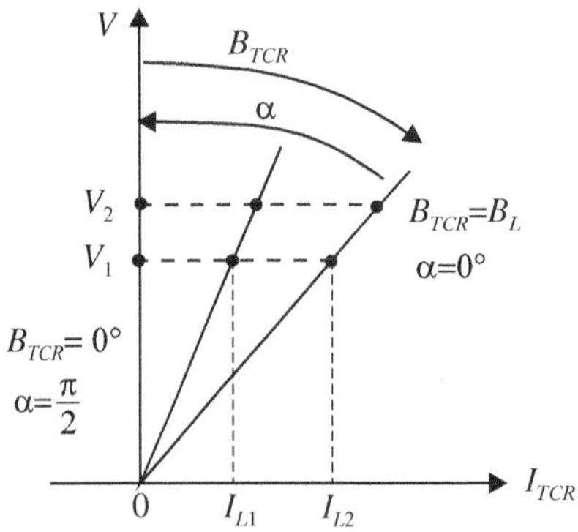

Figure 5.9 The voltage–current characteristic for an open loop controlled TCR.

two transformer windings, the fifth- and seventh -order harmonic components of the output current cancel out. This connection is known as the 12-pulse arrangement, as there are 12 thyristors in conduction state on each cycle of the three-phase voltage waveform. With this arrangement, the highest orders of harmonics are 11 and 13, which can further be reduced by employing harmonic filters.

Operating Characteristics of a TCR. By varying the firing delay angle of a TCR, a variable inductive reactance in a continuous range between zero and the value of the natural reactance of the inductor can be obtained. Each value of the reactance thus obtained corresponds to a characteristic in the *V–I* plane. All these characteristics are straight lines that cross the origin of the axes and have a slope equal to the corresponding value of inductive reactance. Changing the firing delay angle α corresponds in the *V–I* plane to switching from one characteristic to another in the whole control range, as shown in Figure 5.9.

For a given value of the applied voltage, the value of the current is given by the abscissa of the corresponding point on the characteristic. For different applied voltages, V_1 and V_2, for example, different values of the current are absorbed [2].

In order to use the TCR for controlling the voltage at the bus where it is connected, the TCR has to be accompanied by a voltage measurement circuit and a voltage controller (Figure 5.10).

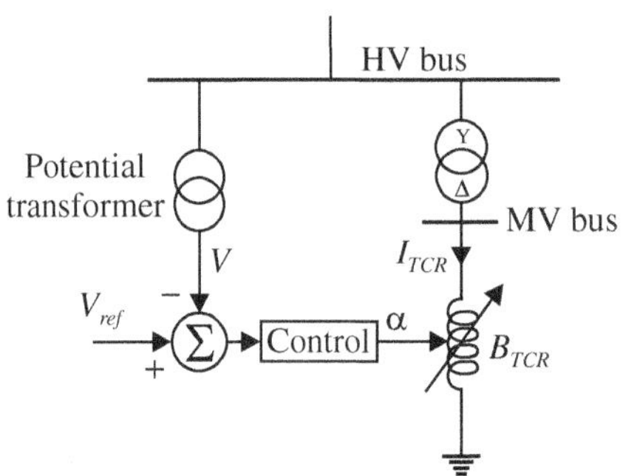

Figure 5.10 Single-line diagram of a three-phase TCR with firing delay angle controller [2].

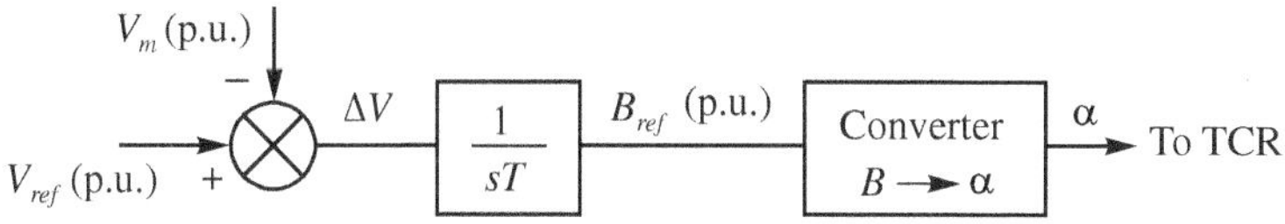

Figure 5.11 Closed-loop voltage controller for TCR.

The task of the controller is to set the firing delay angle applied to the thyristors so that the TCR absorbs a current, which maintains the bus voltage at a constant level, independent of the system voltage changes.

The simplest closed-loop controller used for this purpose, shown in Figure 5.11, compares the measured bus voltage with the reference or desired voltage and integrates the computed voltage error in order to obtain the reference current that has to be absorbed by the TCR (or, equivalently, the reference susceptance that has to be realized by the TCR) [2].

However, computing the firing delay angle α from the susceptance B_{ref} is cumbersome because of the nonlinearity between the delay angle and the TCR fundamental current, given by equation (5.9), so this is usually done by using a look-up table.

The characteristic of a TCR with a voltage controller is shown in Figure 5.12. The voltage is controlled at the reference value of 1 p.u. within the control range of the TCR.

Thus:

- If the system voltage has the value $E^{(1)}$, the controller will detect a difference between $E^{(1)}$ and the reference voltage $E^{(3)}$ and automatically calculate the corresponding delay angle α_1 to be applied so that the bus voltage becomes equal to $E^{(3)}$. If the voltage further increases up to $E^{(2)}$, the controller will compute an angle $\alpha = 0°$ so that the bus voltage to be maintained at the reference level corresponds to a completely inserted reactor.

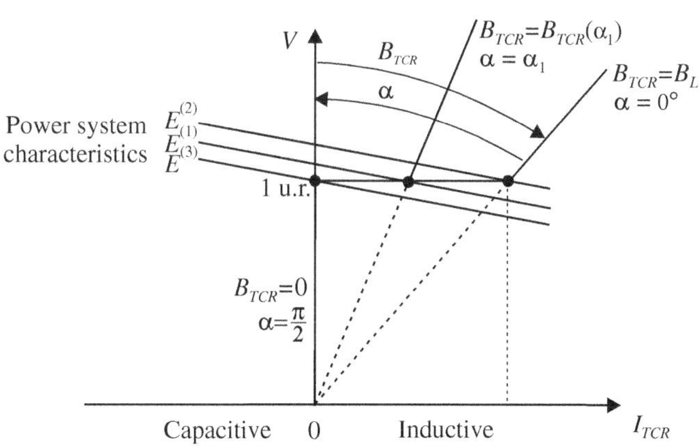

Figure 5.12 V–I characteristic of a TCR with closed-loop voltage controller.

- If the voltage decreases toward the reference $E^{(3)}$, the controller will compute an angle equal to $\pi/2$, meaning that the TCR having $B_{TCR} = 0$ should not absorb any current. It has to be stated that the control range for the example at hand is $E^{(3)} < E < E^{(2)}$.
- Voltages lower than $E^{(3)}$or higher than $E^{(2)}$ are out of the TCR control range, and the TCR will behave as an open circuit or a fixed reactor, respectively.

It has also to be mentioned that the lower boundary of the control range is equal to the reference voltage, while the higher boundary is determined both by the size of the reactor and the slope of the system load characteristic, in other terms, the system reactance.

5.3 THYRISTOR-SWITCHED CAPACITOR

The single-phase TSC, shown in Figure 5.13a, consists of a capacitor, a bidirectional thyristor valve, and a relatively small reactor L', which is primarily used to limit the surge current in the thyristor valve (during the commutation process) and to avoid resonance with the AC system at specific frequencies. The circuit is completed by an RC series connected in parallel with each thyristor bank; their function is to limit dangerous surge currents that may appear at thyristor blockings and to balance the voltage levels on the thyristor banks.

The two thyristor modules are connected anti-parallel and are controlled by a control and command device. Depending on the bus voltage, the control and command device connects a lower or higher number of elements.

Switching Transients in the TSC. The main problem relates to choosing the suitable instant of commutation so as the voltage on the thyristor to be minimum, ideally equal to zero. The presence in the same circuit of both inductivities and capacities makes the transient state to be oscillatory. The natural frequency of the oscillations

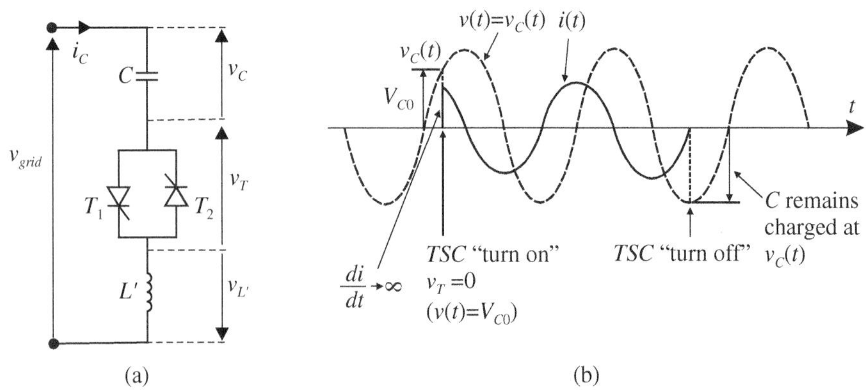

Figure 5.13 Thyristor-switched capacitor at a voltage source: (a) circuit diagram and (b) associated current and voltage waveforms. *Source*: Mathur & Varma 2002 [3]. Reproduced with permission of IEEE.

is an important factor for the voltage and current values after the commutation takes place.

As a first step, consider that there is *no inductance or resistance in the circuit.* The transient state analysis of the current after closing the switch implies two cases (Figure 5.13):

(i) The voltage at the capacitor gate terminals is different from the system voltage when the thyristors are fired (enter conduction), $v_C \leq v$. Immediately after closing the switch, an infinite magnitude current appears in the circuit and the capacitor charges at the system voltage in a very short time. The thyristors in the switch get damaged, not being able to face this stress.

(ii) The voltage at the capacitor gate terminals is equal to the system voltage when the thyristors enter conduction. In this case, the current immediately reaches the steady-state current value. The steady-state conditions are reached in an infinite small time. Even though the current magnitude does not outrun the steady-state value, the thyristors have a superior boundary for the *di/dt* values, which in this case is infinite, and the thyristors would be destroyed.

In conclusion, this simple circuit of the TSC is not suitable and it needs an inductance to be connected in series with the capacitor.

Consider that an *inductance* L' is introduced in the electric circuit of a TSC (Figure 5.13a).

If $v(t)$ is the voltage applied to the TSC and $i(t)$ is the current in the TSC branch, the equation of the circuit is

$$v(t) = L\frac{di(t)}{dt} + v_C(t) \tag{5.15}$$

where

$$i(t) = C\frac{dv_C(t)}{dt}$$

and therefore

$$v(t) = LC\frac{d^2v_C(t)}{dt^2} + v_C(t) \tag{5.15'}$$

Note that the losses are not considered. The solution of this equation is given by the solution of the homogenous equation $v_{C0}(t)$, that is, the response of the circuit when the voltage source is zero, and the steady-state solution is $v_{cp}(t)$. The complete solution is [2]:

$$v_C(t) = v_{C0}(t) + v_{cp}(t) \tag{5.16}$$

Under steady-state conditions, when the thyristor valve is closed and the TSC branch is connected to a sinusoidal AC voltage source, the expression of current is:

$$i(t) = \hat{I}_{AC}\cos(\omega_0 t + \alpha) - nB_C\left(V_{C0} - \frac{n^2}{n^2-1}\hat{V}\sin\alpha\right)\sin\omega_n t - \hat{I}_{AC}\cos\alpha\cos\omega_n t \tag{5.17}$$

where:

$$\hat{I}_{AC} = \hat{V}\frac{B_C B_L}{B_C + B_L} \tag{5.18}$$

is the fundamental frequency component peak value of the current in the capacitor, and

$B_C = \omega_0 C$ is the capacitive susceptance of the capacitor at system frequency.
$\omega_n = n\omega_0 = 1/\sqrt{L/C}$ is the natural pulsation of the circuit.
$n = \sqrt{X_C/X_L}$ is the natural pulsation in p.u.
V_{C0} is the initial capacitor voltage at $t = 0$.

The first term of equation (5.17), containing the fundamental pulsation ω_0, represents the steady-state solution. Current I_{AC} leads the supply voltage by 90° and its magnitude is given by equation (5.18). It can also be written as:

$$\hat{I}_{AC} = \hat{V} \cdot B_C \frac{n^2}{n^2 - 1} \tag{5.18$'$}$$

The term $n^2/(n^2-1)$ is a gain factor that represents the degree of tuning of the LC circuit; for high values of L', n can reach 2.5 and therefore a gain factor of 1.2. In practical applications, n is generally greater than 3 (between the fourth and fifth harmonics).

The last two terms of equation (5.17) that depend on the natural pulsation constitute the oscillatory transients upon TSC switching.

A major design challenge for this circuit is related to minimizing the transients at TSC switching. An ideal case, with transient-free switching, is obtained by canceling the last two terms of equation (5.17), the oscillatory transients resulting in two conditions that have to be satisfied simultaneously:

$$\begin{aligned} &\cos\alpha = 0, \text{ respectively, } \sin\alpha = \pm 1, \\ &V_{c0} = \pm\hat{V}\tfrac{n^2}{n^2-1} = \pm X_c \hat{I}_{AC} \end{aligned} \tag{5.19}$$

The first equation in (5.19) expresses the fact that the switching should take place exactly when the voltage reaches its peak, whereas the second equation reveals that the capacitor should be precharged at predetermined voltage $\pm\hat{V}n^2/(n^2 - 1)$ of the same polarity. For the particular case where $L' = 0$ and $R = 0$, presented in Figure 5.13b, the switching instant is chosen when the bus voltage (V) reaches its peak value on the same polarity as the capacitor voltage polarity ($V_{c0} = \pm\hat{V}$). This ensures, theoretically, that the capacitor is connected transient-free. The capacitor is disconnected by suppressing the firing impulses sent to the thyristors. They are blocked as soon as the current becomes zero.

The capacitor remains charged at maximum positive or negative voltage value, being ready for the next switch. Of the two conditions in (5.19), the second one cannot be controlled by thyristor gating command circuits, because V_{C0}, n, and $\hat{V}$ can

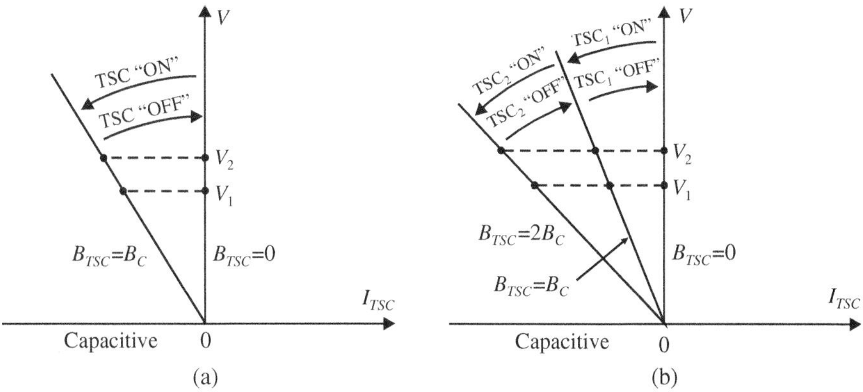

Figure 5.14 *V–I* characteristic for the TSC: (a) one TSC and (b) two TSCs in parallel.

vary during nonconduction periods, before the thyristors switching; therefore, it is impossible to have ideal transient-free switching in practice.

The current maximum transient state will occur if firing takes place when the system AC voltage crosses zero or at maximum thyristor gate terminal voltage.

Operating Characteristics of the TSC. As known, the characteristic of a TSC is a simple straight line and the maximum possible delay in switching is one full cycle of the applied voltage, which is the time between two successive peaks of the same polarity [2].

Consider a TSC with a reactance that varies in discrete steps; for each voltage value, there are only two operating points (Figure 5.14a): one in which the TSC is ON and absorbs maximum capacitive current for that voltage, and one in which the TSC is OFF and does not absorb any current.

When the TSC is ON, the capacitor absorbs a higher negative current for a higher applied voltage.

The controllability can be improved, but for higher costs, by splitting the total capacitance into two capacitive branches that can be controlled separately by two anti-parallel thyristor switches. Three possible operating points for each value of the applied voltage are thereby obtained: one with both TSCs OFF, corresponding to a zero current, one with TSC_1 ON, corresponding to a certain capacitive current depending on the applied voltage, and one with both TSCs ON, corresponding to a higher capacitive current for the same applied voltage (Figure 5.14b).

5.4 CONFIGURATIONS OF SVC

5.4.1 Fixed Capacitor and Thyristor-Controlled Reactor

The two elements, fixed capacitor (FC) and TCR, are connected in parallel and thus the total current in the device is the sum of the capacitive current, which is negative and constant for a certain applied voltage, and the current in the TCR, which is positive and can be continuously controlled. The TCR can thus be controlled so that to

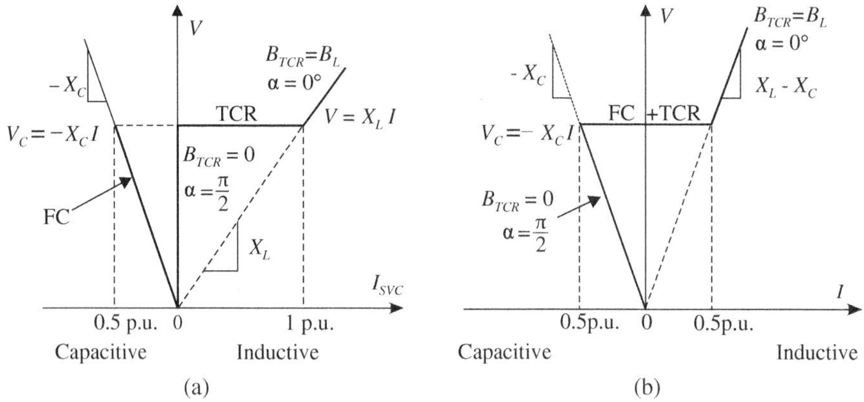

Figure 5.15 (a) Individual *V–I* characteristics of an FC and a TCR and (b) combined *V–I* characteristic for the FC–TCR [2].

"cancel" a part of the capacitive current produced by the FC. If the reactor in the TCR has the same reactance as the capacitor, the capacitive current can be continuously controlled within the whole range, between zero and the maximum capacitive value, by a firing angle control of the TCR.

The individual characteristics of an FC and a TCR of double reactance are shown in Figure 5.15a. The FC–TCR characteristic, shown in Figure 5.15b, can be obtained by summing the two currents for the corresponding voltage values.

The TCR is switched off at the maximum capacitive output by setting the firing delay angle α to 90°. The capacitive output power is reduced by decreasing the firing delay angle; if the current absorbed by the reactor is equal to the current produced by the capacitor, then the net output power is zero.

If the TCR inductive reactance is higher than the capacitor reactance, a further decrease in the firing delay angle will produce a positive reactive power output. At zero delay angle, the TCR is in full conduction over a half-cycle and the net inductive power output is the difference between the total VAr generated by the capacitor and the power absorbed by the reactor in full conduction.

The total losses in the FC–TCR device are due to the capacitor, the reactor, and the thyristors. The capacitor losses are small but constant, whereas the reactor losses increase with the square of the current and the thyristor losses increase with the current in a linear manner. Overall, losses increase with increasing TCR current and therefore decrease with increasing capacitive VAr output. A major disadvantage of this configuration is that even at zero reactive power generation, there are high losses in the device because two equal currents circulate through the reactor and capacitor. This is the reason why such configuration is not suitable for applications, the average degree of compensation being very low (e.g., compensation of transmission systems). This configuration is better suited for applications where the average degree of compensation is close to the maximum capacitive compensation, but still the necessity of providing full inductive compensation for short periods exists (e.g., industrial applications requiring power factor correction) [2].

5.4.2 The SVC Device (TSC–TCR)

5.4.2.1 V–I Characteristics

As previously shown, in order to have a control range of 1 p.u. for an FC–TCR, the reactor rating should be double as compared to the capacitor. Moreover, the losses at zero load operation are very high. A more flexible VAr generator can be obtained by connecting in parallel a TSC and a TCR, controlled by the firing delay angle α. The controlled characteristic of a TCR and the characteristic of a TSC are shown in Figure 5.16a. Similar to the case of an FC–TCR, the two elements are connected in parallel and therefore the total current is the algebraic sum of the current absorbed by the TCR and the current generated by the TSC, when the latter is ON.

If the TSC is OFF, the total current will be equal to the inductive current in the reactor only. Thus, by switching off the capacitor, the resulting characteristic of the SVC (Figure 5.16b) is the same as the individual characteristic of the TCR, and the current can be controlled from maximum inductive to zero by varying the delay angle α from 0 to π/2. [2].

The same control range as that shown in Figure 5.15b can be obtained, but the size of the reactor must be equal to that of the capacitor instead of a double one, as previously needed. On the other hand, the cost of valves and the cost of special equipment used for capacitor switching must be taken into account with extra care. This solution is acceptable if the same control range is needed for both the capacitor and reactor.

Depending on the power system characteristics, a wider range of control in the capacitive region might be necessary. In such situations, one TCR is combined with more than one TSCs. For instance, if one TCR of 0.5 p.u. is combined with two TSCs of 0.5 p.u. each, a total reactive power control range between 1 p.u. (capacitive) and 0.5 p.u. (inductive) is achieved. Their individual characteristics are shown in Figure 5.17, and the combined characteristic is shown in Figure 5.18.

In a general case, if n TSC braches are sized for a total rated reactive power of $Q_{C\max}$, one TSC branch may produce $Q_{C\max}/n$. Usually, the rating of the TCR is equal to the rating of one TSC. Thereby, the total reactive power control lies in the range

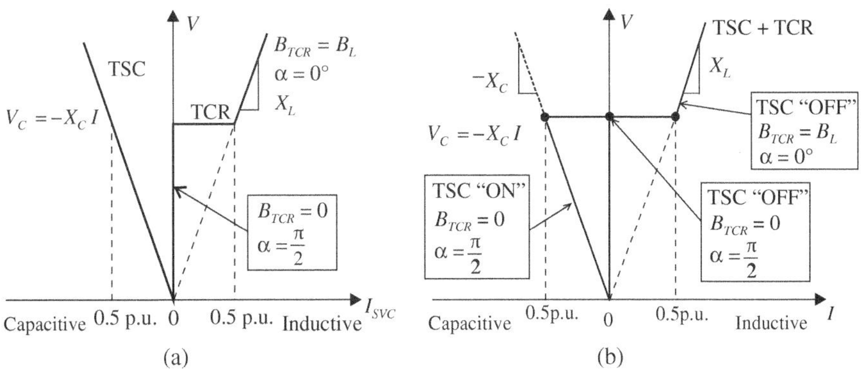

Figure 5.16 *V–I* characteristics: (a) TCR and TSC operating separately and (b) TCR and TSC connected in parallel [2].

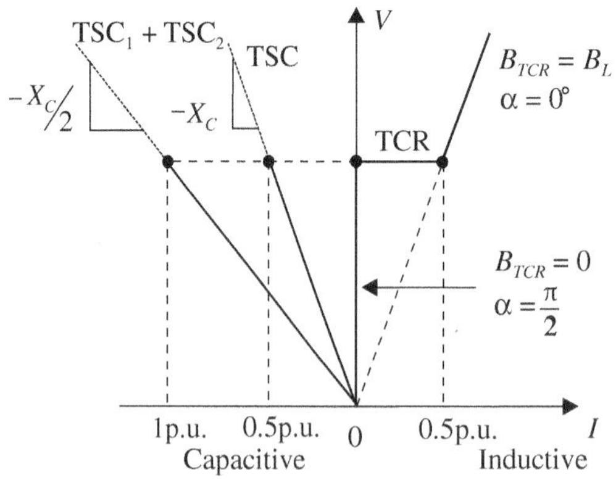

Figure 5.17 *V–I* characteristics for two TSCs and a TCR [2].

[$Q_{C\max}$ and $Q_{L\max}$]. The capacitors provide step changes only, and the continuous control is achieved with the contribution of the TCR.

Depending on the reactive power demand for voltage control at the HV bus, the SVC's controller determines the number of TSC branches to be in operation, of which combined output reactive power is at least equal to the demanded reactive power. At the same time, the controller calculates the firing delay angle α for the TCR that corresponds to the reactive power to be absorbed, which is in surplus from TSCs.

For instance, when only two TSCs are required, the reactive power that is produced by these is $2Q_{C\max}/n$. If the resulted voltage is too high, the TCR is used to absorb the surplus of reactive power by varying its inductive current.

5.4.2.2 Operating Domain

The *V–I* operating domain of the SVC (FC–TCR) is defined by the maximum attainable capacitive and inductive admittances and by the voltage and current ratings of the main power components (capacitor, reactor, and thyristor valve) (Figure 5.19).

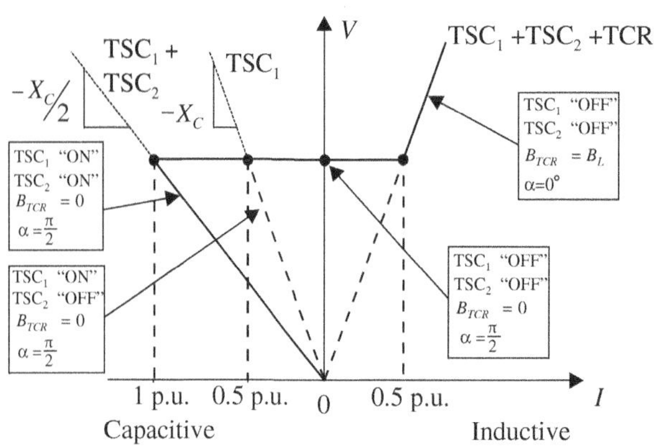

Figure 5.18 Resulting *V–I* characteristic for an SVC composed of two TSCs and a TCR [2].

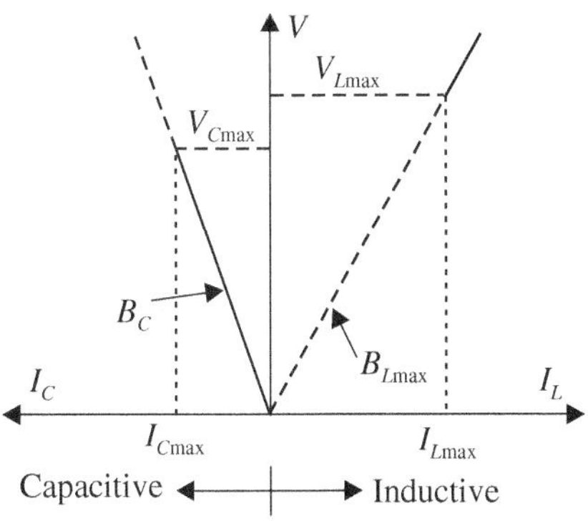

Figure 5.19 Operating V–I domain of the FC–TCR-type VAr generator.

The variables from Figure 5.19 are as follows:

V_{Cmax} and V_{Lmax} are the voltage limits for capacitor and TCR.
I_{Cmax} and I_{Lmax} are the capacitive and inductive current limits.
B_{Lmax} is the maximum inductive admittance.
B_C is the capacitor admittance.

The operating characteristics of an SVC resulting from the combination between the characteristics of a TCR and a capacitor (TSC) are shown in Figures 5.20a and 5.20b.

The steady-state operating domain of an SVC is split into three subdomains (Figure 5.20):

(a) The linear control domain, in which the control system is provided with the required power for bus voltage control, so that to allow setting the operating point on the AB characteristic. This is bounded by the reactive power Q_{Cmax}

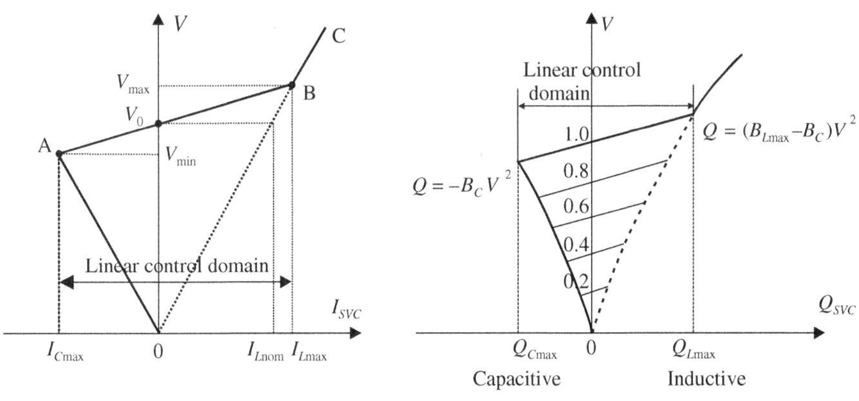

Figure 5.20 The operating characteristics (V–I and V–Q) and domains of the SVC.

supplied by the equivalent capacitive susceptance B_C of the capacitors, with the reactor blocked, and by the reactive power $Q_{L\max}$ absorbed by the equivalent inductive susceptance ($B_{L\max}$–B_C) corresponding to a full conduction state of the thyristors, respectively.

$$Q_{C\max} \le Q \le Q_{L\max} \tag{5.20}$$

where

$$Q_{C\max} = -B_C V^2 \tag{5.20a}$$

$$Q_{L\max} = (B_{L\max} - B_C)V^2 \tag{5.20b}$$

(b) The high voltages domain (BC), where $Q_{SVC} > Q_{L\max}$; the static compensator is, in this case, out of the control area and it behaves like a fixed inductive susceptance, whose value can be determined by equation (5.20b).

(c) The low voltages domain (0A), where the system voltage is so low that the static compensator reaches its maximum capacitive limit; in this case, the SVC behaves like a fixed capacitive susceptance, whose value can be computed by using equation (5.20a).

When operating within the control domain, the SVC characteristic can be expressed as:

$$V = V_0 \mp X_{SL} I_S \tag{5.21}$$

where:

V_0 is the reference voltage (V_{ref}) for $I_S = I_{SVC} = 0$.

SL is the characteristic slope within the control domain:

$$SL = \frac{\Delta V \text{ (at nominal current)}}{V_{nom}} \tag{5.22}$$

X_{SL} is the reactance corresponding to the characteristic slope within the control domain:

$$X_{SL} = \frac{SL \cdot V_{nom}}{\sqrt{3} I_{nom}} \tag{5.22$'$}$$

$X_{SL} = SL$ [p.u.].

The power system in which the static compensator is connected can be modeled as an equivalent Thévenin circuit of electromotive force E_{Th} and source reactance X_{Th}, as "seen" from the SVC connection point (Figure 5.21).

Considering Figure 5.21a, the characteristic of the power system can be expressed as:

$$\underline{V} = \underline{E}_{Th} \mp X_{Th} \underline{I}_S \tag{5.23}$$

The combined characteristic of the static compensator and the power system is illustrated in Figure 5.22.

Consider the V–I characteristics of the power system corresponding to three different values for the equivalent supply voltage, $E_{Th}^{(1)} > E_{Th}^{(0)} > E_{Th}^{(2)}$. The middle

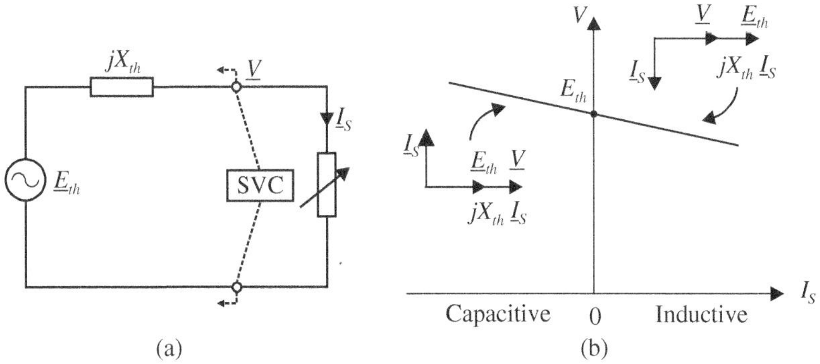

Figure 5.21 Modeling the electric power system and its V–I characteristic: (a) equivalent Thévenin circuit and (b) V–I characteristic [5].

characteristic, $E_{th}^{(0)}$—which corresponds to the power system operating in normal conditions, crosses the static compensator characteristic in a point where $V_A = V^{(0)}$ and $I_s = 0$. Two cases are additionally identified:

- If the system voltage increases, up to $E_{th}^{(1)}$, following the decreasing in the load demand for instance, the voltage at the connection terminal of the equivalent circuit increases to $V^{(1)}$, when the circuit does not have a static compensator connected. If a VAr compensator that absorbs the inductive current $I_s = I_B$ is connected at the connection terminal, the operating point changes to B and the voltage drops to $V_B > V_A$ (reference value).
- If the supply voltage decreases, up to $E_{th}^{(2)}$ following the increasing in the load demand, the voltage at the connection terminal of the equivalent circuit drops to $V^{(2)}$, when a static compensator is not present. If an SVC that generates capacitive current is connected, the operating point moves to C and the voltage becomes $V_C < V_A$, but with $V_C > V^{(2)}$.

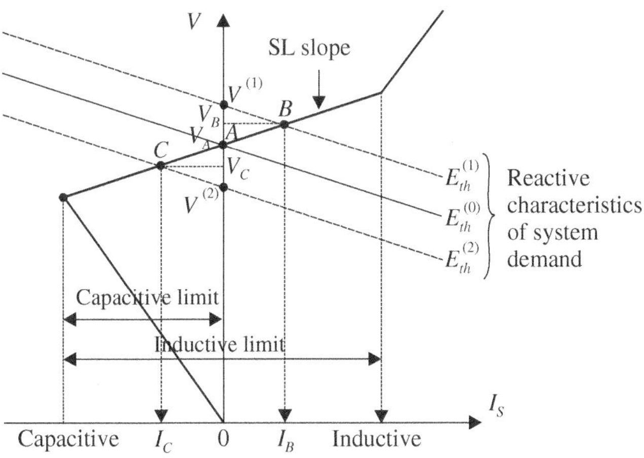

Figure 5.22 The V–I mixed characteristic of the static compensator and the electric system.

When the characteristic slope SL of the static compensator is zero, the voltage is maintained at $V^{(0)}$ in both of the above situations.

Note: The reactance X_{SL}, corresponding to the slope SL, has important contribution to the SVC performances; a high value of X_{SL} leads to great voltage variations at the bus where the SVC is connected.

5.5 CONTROL OF SVC OPERATION

The static compensator control is performed using an automatic voltage controller (Figure 5.1). It determines the demand for capacitive/inductive reactive power by comparing the measured voltage with the reference voltage, and then generates command signals for the thyristors. These signals are converted either into analogue signals for firing delay angle control of the thyristors (depending on the reactive demand in the TCR circuits), or into digital signals through which the thyristors are switched ON or OFF (for the TSCs).

If the bus voltage drops below a critical level, defined by the power system operator for stability reasons, for a longer period, the compensator control is interrupted, and then is resumed upon voltage recovery. However, the voltage can drop to low values for short periods, for instance, during transient contingencies, without causing interruptions of the static compensator operation.

The command has to take place so as to minimize the transients at capacitor switching. If the current in the capacitor is zero, the command for blocking the thyristors is given at voltage zero crossing. Otherwise, the command sets, depending on the current in the capacitor, a delay angle α that minimizes the effect of the transient phenomenon.

The overall control scheme of the SVC is shown in Figure 5.23. The controller can be divided into two parts: the outer controller, which measures the variables to be regulated, usually the voltage, and after comparison with a desired value determines the reference control signal for the SVC, which can be a susceptance reference or a current reference; the inner controller (gate pulse generator [GPG] unit) determines the switching pulses to the thyristors of both the TCR and TSCs, in such a way to achieve the desired total susceptance [2].

5.5.1 The Voltage Regulator

The SVC voltage regulator determines the susceptance B_{ref} that corresponds to the reactive power necessary to achieve the desired voltage V_{ref}. The reference value V_{ref} is always compared with the measured value V_{meas}, as well as with other variables

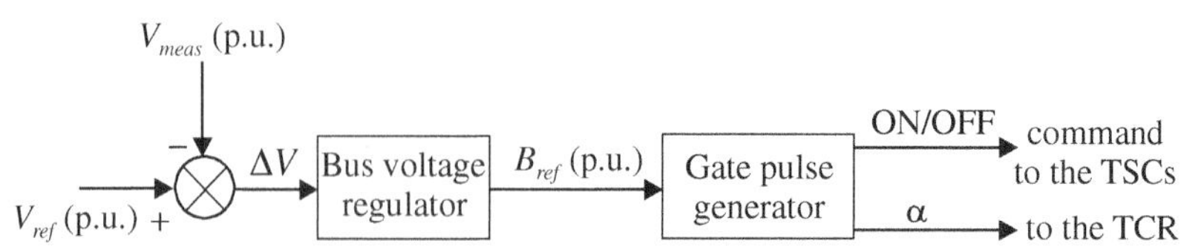

Figure 5.23 Block scheme of SVC controller [2].

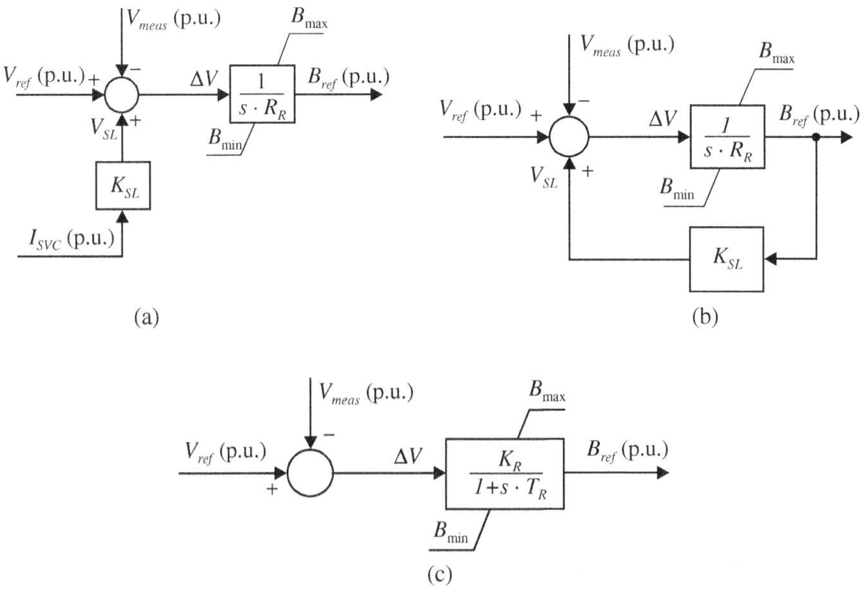

Figure 5.24 Implementation of current droop: (a) an integrator with current droop feedback, (b) an integrator with susceptance droop feedback, and (c) regulator with gain-time constant model. *Source*: Mathur & Varma 2002 [3] and Taylor [6]. Reproduced with permission of IEEE.

depending on the SVC application (Figure 5.24). The result is feed forwarded into a transfer function that provides the reference susceptance B_{ref}, which is subsequently transmitted to the GPG circuit. The capability limits B_{min} and B_{max} are applied when generating the reference susceptance value [3, 6].

When the SVC current is explicitly measured, a scheme with current droop feedback is used (Figure 5.24a). The measured current I_{SVC} is multiplied by a factor K_{SL}, which represents the slope factor (or droop) of the steady-state characteristic, to produce an additional voltage signal V_{SL} that is fed into the summing junction. In this way, for inductive current (positive), the reference voltage will increase, whereas for capacitive current (negative) the reference voltage will decrease. The regulator is of integral type, where the response rate R_R represents the time needed for going from fully inductive to fully capacitive response, in response to 1 p.u. voltage error [3]. The current is difficult to measure when the SVC is in floating mode, that is, it does not exchange reactive power with the power system. The solution would be to measure the reactive power.

Another solution is to create a feed-backward loop to adjust the susceptance (Figure 5.24b). In per units, the voltage is around 1 p.u., thereby the SVC current is about equal to the susceptance B_{ref} since $I_{SVC} = B_{ref} \cdot V_{SVC}$. The regulator is the same as in Figure 5.24a.

When the voltage only is used to determine the reference susceptance, a first-order control system, as that in Figure 5.24c, can be used. The static gain is the inverse of the current droop, $K_R = 1/K_{SL}$, and the time constant of the integrator is equal to $T_R = R_R/K_{SL}$.

The ratio between K_R and T_R is termed "transient gain" K_T and is representative of the dynamic performance of the regulator. The gain-time constant model has been very much used in SVC modeling for transient studies, but in reality the physical realization of most installed SVCs is the current-droop model of Figure 5.24a. The advantage of this model is that steady-state characteristics—described by K_R—, and dynamic performance—described by R_R—can be specified independently. The readers can find more details in References [3] and [6].

5.5.2 Gate Pulse Generator

The reference susceptance B_{ref} determined by the voltage regulator is transmitted to the GPG unit, which produces the appropriate firing pulses at the thyristors (Figure 5.23). The GPG unit has to perform the following functions [1,2]:

(i) Determine the number of TSC branches n_c to be switched ON in order to achieve at least the required susceptance, but not more than one step higher. This is done by dividing the reference susceptance to the susceptance B_C of one capacitor bank and rounding off the result to the next integer.

$$n_c = round(B_{ref}/B_C)$$

(ii) Determine the value of susceptance B_{TCR} to be inserted by the TCR in order to cancel the excess of capacitive susceptance, that is the difference between the capacitive susceptance of all TSC branches ($n_c B_C$) and the reference B_{ref}:

$$B_{TCR} = n_c B_C - B_{ref}$$

(iii) Determine the sequence in which the TSC branches should be switched, depending on the polarity and magnitude of the residual voltage across them, in order to meet the criteria for minimizing the effect of the transient phenomenon.

(iv) Calculate the firing angle α for the TCR to achieve the desired susceptance B_{TCR}. For this purpose, a conversion from B to α is required, which can be done in a microprocessor by a look-up table.

5.6 SVC MODELING

Modeling the SVC for steady state and dynamic simulation has been a preoccupation for power system specialists since the beginning of the modern power electronic. One reference work is the technical brochure "*Static VAr compensators*" elaborated by the CIGRE Working Group 38-01, Task Force No. 2 on SVC [7], which has also been the guidance for this section of the book.

5.6.1 Steady-State SVC Modeling

The static characteristic presented in Figure 5.20a provides the support for representing an SVC device in the load flow mathematical problem. It expresses the *V–I* relationship of (5.21) between the absorbed or injected current and the voltage of the controlled bus, under the assumption of balanced three-phase operation, at fundamental frequency and covering the normal and abnormal voltage bands [6–10].

5.6.1.1 Modeling of an SVC That Operates Within or Outside the Linear Control Domain

Consider the ideal case, when the slope (SL) of the steady-state characteristic is zero. The network bus in which SVC is installed will thus be treated during the iterative process of the load-flow computation as a generator bus (PV) with $P = 0$ and $V = V_{ref}$ (Figure 5.25a).

The *operation within the linear control domain* (AB from Figure 5.20a) is drawn for $I_{\min} < I_{SVC} < I_{\max}$ and $V_{\min} < V < V_{\max}$.

As the connection terminal voltage linearly depends on the SVC injected/absorbed current, it is seen from the high voltage network bus where SVC is installed as a constant supply voltage (E) behind a reactance X_{SL} that models the V–I characteristic slope (Figure 5.25b). Representation of the SVC slope is essential in the case of weak systems [6].

The reactance X_{SL} can be computed as:

$$X_{SL} = \frac{\Delta V}{I_{nom}}$$

where:

ΔV is the voltage drop in the reactance (1÷10%).

I_{nom} – the nominal current when the SVC operates with $B_{SVC} = B_{L\max}$.

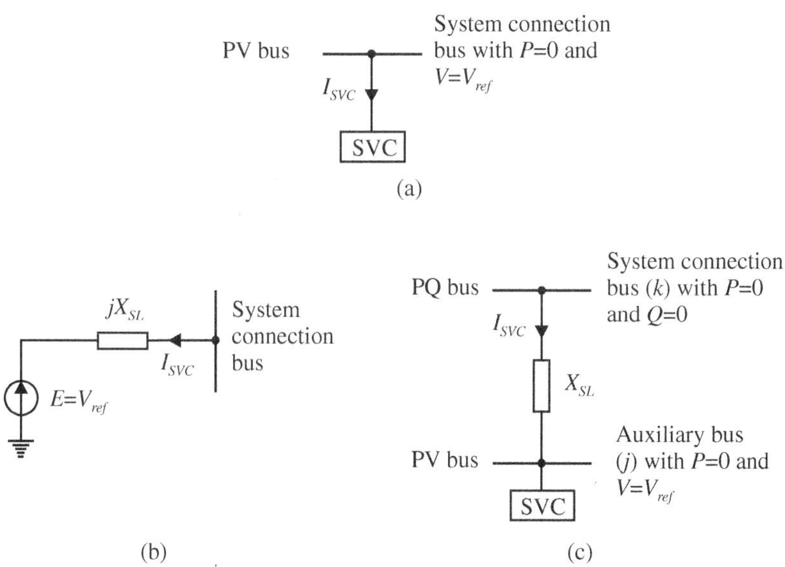

Figure 5.25 Models representing the SVC operation within the linear control domain: (a) ideal model ($SL = 0$), (b) simplified model with slope ($SL \neq 0$), and (c) basic model with slope [7].

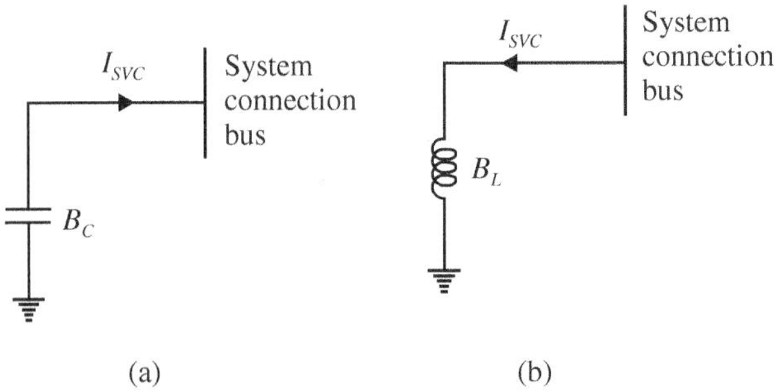

Figure 5.26 Equivalent SVC configurations, operating outside the linear control domain: (a) at capacitive limit and (b) at inductive limit.

In the *real case*, when the characteristic slope SL is nonzero, the SVC is modeled in the load flow problem by introducing an auxiliary bus (j) that will be treated as a generator bus PV, with $P = 0$, $V = V_{ref}$, and the network connection bus (i) will be treated as a PQ bus with $P = 0$, $Q = 0$ (Figure 5.25c). In normal operating conditions, in order to ensure the voltage control at the connection bus, the SVC operates within the linear control domain and therefore this model constitutes the basic model for the steady-state computation.

Operation outside the linear control domain implies two cases [7,9, 11]:

(a) Capacitive operation, for $V < V_{\min}$, when the SVC is seen from the connection bus as a capacitive susceptance jB_C (Figure 5.26a).

(b) Inductive operation, for $V > V_{\max}$ or $I_{SVC} > I_{L\max}$, when the SVC is seen from the connection bus as an inductive susceptance jB_L (Figure 5.26b).

In order to develop compatible models for steady-state computation, some sign conventions regarding the absorbed/injected reactive power meaning and the equivalent susceptance (B_{SVC}) are adopted:

- $Q_{SVC} > 0$ or $B_{SVC} < 0$ means that the SVC has an inductive behavior, that is, it absorbs reactive power from the power system.
- $Q_{SVC} < 0$ or $B_{SVC} > 0$ means that the SVC has a capacitive behavior, that is, it injects reactive power into the power system.
- If the SVC required current is $I_{SVC} > I_{\max}$, the inductive limit is violated (Figure 5.27a) and the susceptance value is calculated as:

$$B = B_{\min} = -\frac{Q_{\max}}{V_{\max}^2}$$

where $Q_{\max}(= Q_{L\max})$ is the maximum inductive reactive power at the voltage $V_{SVC} = V_{\max}$; $Q_{\max}$ has positive value ($Q_{\max} > 0$).

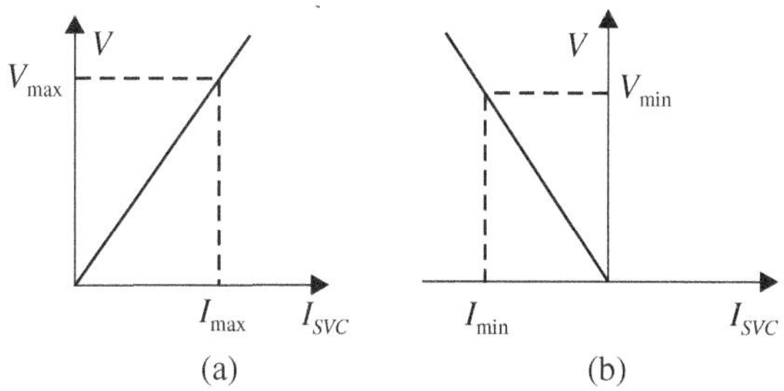

Figure 5.27 SVC operating limits.

- If the bus voltage is $V < V_{\min}$, the capacitive limit is violated (Figure 5.27b) and the susceptance value is calculated as:

$$B = B_{\max} = \frac{Q_{\min}}{V_{\min}^2}$$

where $Q_{\min}(= Q_{C\max})$ is the maximum capacitive reactive power at the voltage $V_{SVC} = V_{\min}$; $Q_{\min}$ has negative value ($Q_{\min} < 0$).

5.6.1.2 Improved Models for SVC Representation

In certain overload conditions—caused by low voltage levels, or at low loads—which cause too high voltage levels, the SVC could start operating outside the control domain. Treating the SVC as a classical generator bus with fixed reactive limits $Q_{\min}$ and $Q_{\max}$ is not suitable anymore because the compensator behaves as a capacitive or inductive susceptance and therefore the reactive power supplied or consumed is not constant, but varies with the square of the voltage magnitude. Thereby, some changes are required in the base model in order to provide a suitable modeling of the SVC for various operating conditions.

(i) SVC modeling by a variable susceptance

The susceptance B_{SVC} depends on the voltage level V_k at the network connection bus. Depending on the voltage error ΔV, computed as the difference between the reference voltage (V_{ref}) and the bus voltage (V_k^{meas}), the controller (Figure 5.28) orders step-by-step changes in the susceptance B_{SVC} until one of the limits, $B_{\min}$ or

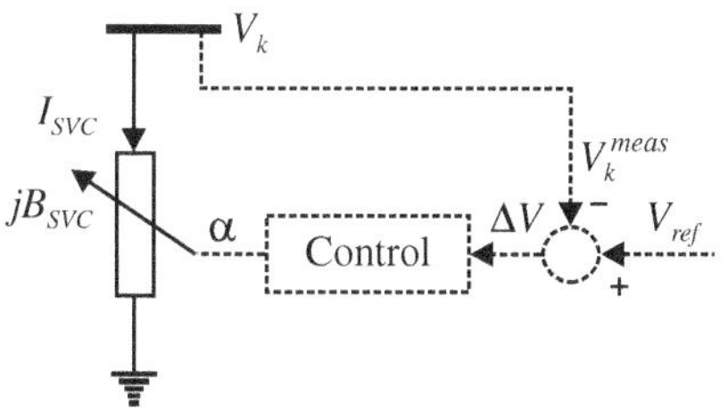

Figure 5.28 SVC modeling by a voltage controlled susceptance.

B_{max}, is reached. The accuracy of this model is given by the number of susceptance control steps.

Let X_L stand for the TCR reactance, and X_C for the TSC reactance. Using equation (5.10), the equivalent variable susceptance of the SVC can be written as a function of the firing delay angle $\alpha = \alpha_{SVC}$ as:

$$B_{SVC} = B_C - B_{TCR} = \frac{1}{X_C X_L}\left[X_L - \frac{X_C}{\pi}(\pi - 2\alpha - \sin 2\alpha)\right] \tag{5.24}$$

where the firing delay angle α is referenced to the maximum voltage value (Figure 5.2c).

Taking into account Figure 5.28, the current in the variable susceptance B_{SVC} is given by:

$$I_{SVC} = jB_{SVC}V_k$$

and the reactive power generated by the SVC at the connection bus k becomes:

$$Q_{SVC} = Q_k = -B_{SVC}V_k^2 \tag{5.25}$$

Consider the equivalent susceptance B_{SVC} as state variable. Using equation (5.25), the linear matrix equation for load-flow computation by Newton–Raphson method is obtained:

$$\begin{bmatrix} \Delta P_k \\ \Delta Q_k \end{bmatrix}^{(i)} = \begin{bmatrix} 0 & 0 \\ 0 & Q^{(i)} \end{bmatrix}^{(i)} \cdot \begin{bmatrix} \Delta\theta_k \\ \Delta B_{SVC}/B_{SVC} \end{bmatrix}^{(i)} \tag{5.26}$$

Thus, each Newton–Raphson iteration consists of the following steps:

(1) Evaluate the reactive power in terms of the connection bus voltage and the operating state of the SVC:

- within the linear control domain:

$$Q_k^{(i)} = \left(V_k^{(i-1)} - V_{ref}\right)\frac{V_k^{(i-1)}}{X_{SL}} \tag{5.27}$$

- outside the linear control domain, for $B_{SVC} = B_{min}$ or $B_{SVC} = B_{max}$:

$$Q_k^{(i)} = -B_{SVC}\left[V_k^{(i-1)}\right]^2 \tag{5.27'}$$

(2) Adjust the corresponding diagonal element from the Jacobian matrix and solve the system (5.26), and then extract the term $(\Delta B_{SVC}/B_{SVC})^{(i)}$.

(3) The variable shunt susceptance B_{SVC} is updated using the expression:

$$B_{SVC}^{(i)} = B_{SVC}^{(i-1)} + \left(\frac{\Delta B_{SVC}}{B_{SVC}}\right)^{(i)} B_{SVC}^{(i-1)} \tag{5.28}$$

(ii) SVC modeling by firing delay angle α_{SVC}

An SVC can also be modeled in terms of the firing delay angle α_{SVC} rather than the susceptance B_{SVC}. Replacing equation (5.24) in equation (5.25), the SVC reactive power is obtained as a function of the firing delay angle:

$$Q_k = -\frac{V_k^2}{X_C X_L}\left[X_L - \frac{X_C}{\pi}\left(\pi - 2\alpha_{SVC} - \sin 2\alpha_{SVC}\right)\right] \tag{5.29}$$

Since α_{SVC} is now a state variable, using the expression of the reactive power from (5.29), the linear matrix equation for load-flow computation becomes:

$$\begin{bmatrix} \Delta P_k^{(i)} \\ \Delta Q_k^{(i)} \end{bmatrix} = \begin{bmatrix} 0 & 0 \\ 0 & -(2V_k^2/\pi X_L)[\cos 2\alpha_{SVC} + 1] \end{bmatrix}^{(i)} \cdot \begin{bmatrix} \Delta\theta_k^{(i)} \\ \Delta\alpha_{SVC}^{(i)} \end{bmatrix} \tag{5.30}$$

Then, at the end of iteration i, the angle α_{SVC} is updated with the expression:

$$\alpha_{SVC}^{(i)} = \alpha_{SVC}^{(i-1)} + \Delta\alpha_{SVC}^{(i)} \tag{5.31}$$

(iii) SVC modeling by controlled PV and PQ buses

When modeling the SVC by controlled buses, a PV-type auxiliary (or fictitious) bus is created and a reactance equal (on the SVC base) to the per unit slope (X_{SL}) is added between the network bus (point of coupling to the system) and the auxiliary bus (Figure 5.29). Then, a control logic is included in the basic model for load-flow calculation to supervise the load nature of the SVC in the auxiliary bus.

By evaluating the SVC current I_{SVC} that flows through the reactance X_{SL}, the control logic will decide whether the SVC will be treated as an equivalent PV-type bus or as a constant capacitive or inductive susceptance [7]:

- If the SVC operates within the admissible limits $I_{\max}$ and $V_{\min}$, then the auxiliary bus will be of PV-type, with $P = 0$ and $V = V_{ref}$.

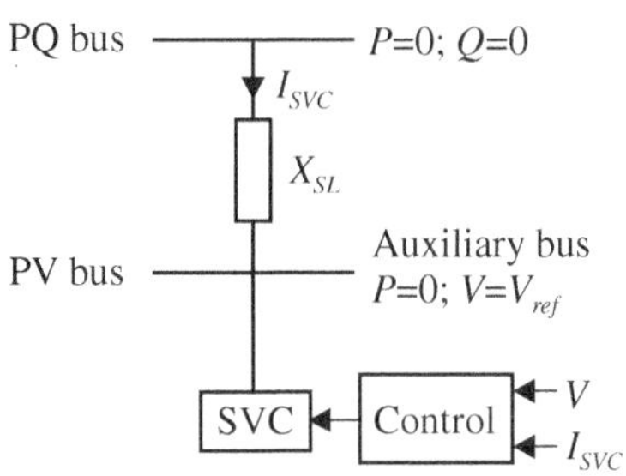

Figure 5.29 SVC base model with control procedure [7].

- If the SVC operates outside the limits I_{max} and V_{min}, then the auxiliary bus will be treated as a PQ-type bus, with $P = 0$ and voltage calculated at the previous iteration for PQ-type buses; the SVC reactive power is set at:
  - $Q = B_{min} V^2$ if $I_{SVC} > I_{max}$;
  - $Q = -B_{max} V^2$ if $V < V_{min}$.

When the SVC is back to its normal operation (within the linear control domain), the auxiliary bus becomes again of PQ-type.

(iv) Improving the basic model by considering the coupling transformer

The coupling transformer may be represented in the SVC model, with good accuracy, by its leakage reactance X_T, in series with the SVC slope reactance X_{SL} (Figure 5.30). A more exact representation of the power system is thus achieved when considering the coupling transformer, and the SVC required reactive power is more correctly determined when the point of coupling to the system is the medium voltage bus.

This arrangement consists of the medium voltage bus where the SVC is connected (V_2) and an auxiliary bus alongside the high voltage system bus (V_1). The auxiliary bus is of PV-type with voltage equal to V_{ref} when the SVC operates within the control domain, and of PQ-type with uncontrolled voltage when the SVC operates outside the linear voltage domain. The point of connection to the system and the medium voltage bus where the SVC is physically installed are treated as PQ buses, and the voltages V_1 and V_2 are computed at each iteration.

The equivalent reactance between the system connection bus and the medium voltage bus is equal to the transformer leakage reactance, whatever the nature of the auxiliary bus. In order to conserve this value and consider the SVC slope when operating within the linear control domain, the reactance X_{SL} is added and subtracted on both sides of the auxiliary bus.

When operating outside the control domain, the SVC is treated as a fixed shunt device. Thereby, capacitive and inductive susceptance limits are defined rather than the reactive power limits. Depending on whether the SVC transformer is represented or not, the voltage is controlled at the medium voltage or at the high voltage system bus.

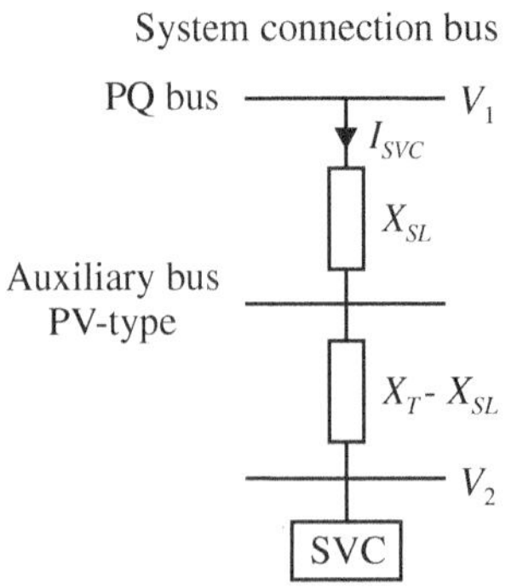

Figure 5.30 The SVC improved model by considering the coupling transformer.

If the SVC transformer is represented as an external element of the SVC model, it is necessary to convert the SVC operating range given on the high voltage side of the SVC transformer into the medium voltage side where the SVC is modeled. The effect of the reactance of the SVC transformer must be accounted for so that the correct range of reactive power is delivered to the high voltage bus.
Based on the parameters given on the high voltage bus side, the susceptance limits, referred to the medium voltage bus, in per unit, are [6]:

$$B_{\max,cap} = \frac{1}{\left|\frac{S_n}{Q_{cap}}\right| + |X_{t,p.u}|}; \; B_{\min,ind} = -\frac{1}{\left|\frac{S_n}{Q_{ind}}\right| - |X_{t,p.u}|}$$

where

S_n is the MVA base.
$X_{t,p.u}$ – the reactance of SVC transformer in per unit on S_n base.
Q_{cap} – the maximum capacitive reactive power on high voltage side bus.
Q_{ind} – the maximum inductive reactive power on high voltage side bus.

When the SVC transformer is included in the SVC model, *the parameters for the voltage regulator gain*, which are given for the high voltage side, will also need to be adjusted for the medium voltage side where the SVC is connected.

Example [6]: Let us consider the reactive power range for an SVC connected to a 230 kV bus between $Q_{cap} = -350$ MVA and $Q_{ind} = 300$ MVA, given for the nominal voltage, and the reactance of the 230 kV/20 kV SVC transformer of 12% on a 350 MVA rating. Taking the base power as $S_b = 300$ MVA, the per unit transformer reactance is:

$$X_{t,p.u.} = \frac{12\%}{100} \cdot \frac{350}{300} = 0.14 \text{ p.u.}$$

and the susceptance ranges are:

$$B_{\max,cap} = \frac{1}{\left|\frac{350}{-350}\right| + |0.14|} = 0.8772 \text{ p.u.}$$

$$B_{\min,ind} = -\frac{1}{\left|\frac{350}{300}\right| - |0.14|} = -0.9740 \text{ p.u.}$$

The susceptance values correspond to 20 kV power ratings of 307 MVAr capacitive and 341 MVAr inductive.

(*v*) *Improving the SVC model using the delay angle* α_{SVC} *by introducing the network coupling transformer*

If the device is modeled by its equivalent impedance $\underline{Z}_{SVC} \approx jX_{SVC}(\alpha_{SVC})$ in series with the coupling transformer impedance $\underline{Z}_T = R_T + jX_T$ (Figure 5.31a), then the two impedances can be summed up to obtain the reduced model as in Figure 5.31b; this model allows controlling the voltage on the power system side [10].

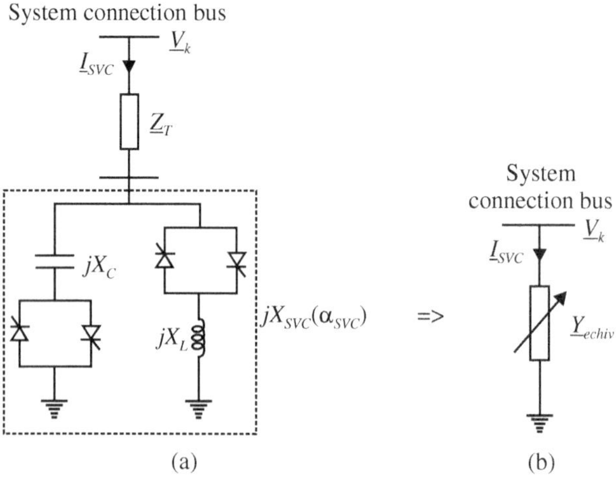

Figure 5.31 Transformer—SVC mixed model: (a) detailed diagram and (b) equivalent model.

The equivalent admittance $\underline{Y}_{eq}$ is:

$$\underline{Y}_{eq} = \frac{\underline{Y}_T \underline{Y}_{SVC}}{\underline{Y}_T + \underline{Y}_{SVC}} = G_{eq} + jB_{eq} \tag{5.32}$$

or

$$\underline{Y}_{eq} = \frac{\dfrac{1}{R_T + jX_T} \cdot \dfrac{1}{jX_{SVC}}}{\dfrac{1}{R_T + jX_T} + \dfrac{1}{jX_{SVC}}} = \frac{1}{R_T + j(X_{SVC} + X_T)} = \frac{R_T - jX_{eq}}{R_T^2 + X_{eq}^2} \tag{5.32$'$}$$

then

$$G_{eq} = \frac{R_T}{R_T^2 + X_{eq}^2}; \tag{5.33a}$$

$$B_{eq} = -\frac{X_{eq}}{R_T^2 + X_{eq}^2} \tag{5.33b}$$

Since the TCR reactance is a function of the angle α_{SVC}, we get:

$$X_{SVC}(\alpha_{SVC}) = \frac{X_C X_{TCR}(\alpha_{SVC})}{X_C - X_{TCR}(\alpha_{SVC})} \tag{5.34}$$

where

$$X_{TCR}(\alpha_{SVC}) = \frac{\pi X_L}{\pi - 2\alpha_{SVC} - \sin(2\alpha_{SVC})} \tag{5.35}$$

Considering Figure 5.31b, the active and reactive powers can be written as:

$$\begin{aligned} P_k &= G_{eq} V_k^2 \\ Q_k &= B_{eq} V_k^2 \end{aligned} \tag{5.36}$$

The linearized matrix equation for the steady-state computation using the Newton–Raphson method is:

$$\begin{bmatrix} \Delta P_k^{(i)} \\ \Delta Q_k^{(i)} \end{bmatrix} = \begin{bmatrix} 0 & V_k^2 \dfrac{\partial G_{eq}}{\partial \alpha_{SVC}} \\ 0 & -V_k^2 \dfrac{\partial B_{eq}}{\partial \alpha_{SVC}} \end{bmatrix}^{(i)} \begin{bmatrix} \Delta \theta_k^{(i)} \\ \Delta \alpha_{SVC}^{(i)} \end{bmatrix} \tag{5.37}$$

Given the notation $D = R_T^2 + X_{eq}^2$, the partial derivatives of the parameters from equations (5.33a and 5.33b) are obtained as:

$$\frac{\partial G_{eq}}{\partial \alpha_{SVC}} = -\frac{R_T}{D^2} \frac{\partial D}{\partial \alpha_{SVC}} \tag{5.38'}$$

$$\frac{\partial B_{eq}}{\partial \alpha_{SVC}} = \frac{1}{D^2} \left(-D \frac{\partial X_{SVC}}{\partial \alpha_{SVC}} + X_{eq} \frac{\partial D}{\partial \alpha_{SVC}} \right) \tag{5.38''}$$

where

$$\frac{\partial D}{\partial \alpha_{SVC}} = 2X_{eq} \frac{\partial X_{SVC}}{\partial \alpha_{SVC}} \tag{5.39}$$

$$\frac{\partial X_{SVC}}{\partial \alpha_{SVC}} = \frac{2X_{SVC}}{\pi X_L} (1 - \cos 2\alpha_{SVC}) \tag{5.40}$$

At the end of the *i*th iteration, the firing delay angle is adjusted by:

$$\alpha_{SVC}^{(i)} = \alpha_{SVC}^{(i-1)} + \Delta \alpha_{SVC}^{(i)} \tag{5.41}$$

5.6.1.3 Newton–Raphson Modified Algorithm to Include the SVCs

When an SVC device is connected to bus k, the Newton–Raphson method for steady-state computation is modified as follows.

Provide the input data of the SVC parameters:

- the reactive power limits, Q_{cap} and Q_{ind}, in MVAr;
- the reactance X_{SL}, corresponding to the slope of the V–I characteristic, p.u.;
- the inductive and capacitive reactance (X_L and X_C) of the SVC components;
- the minimum and maximum values of the firing delay angle, α_{min} and α_{max};
- the reference voltage at bus k, $V_{k,ref}$.

- ***Changes at each iteration (p)***

  1. *Check if the voltage at the controlled bus k, $V_k^{(p)}$, is within the static characteristic control domain*

  1.1. If $(1 - X_{SL})V_{k,ref} < V_k^{(p)} < (1 + X_{SL})V_{k,ref}$ then the auxiliary (controlling) bus j (Figure 5.25) keeps working as PV-type with $P = 0$ and $V = V_{k,ref}$, and go to step 2;

1.2. If $V_k \geq (1 + X_{SL})V_{k,ref}$
then set $\alpha = \alpha_{\min}$, switch bus j from PV-type to PQ-type, and go to step 3.1;

1.3. If $V_k \leq (1 - X_{SL})V_{k,ref}$
then set $\alpha = \alpha_{\max}$, switch bus j from PV-type to PQ-type, and go to step 3.2;

2. *Determine the SVC reactive power*

2.1. Determine the reactive power contribution from SVC. Determine first the bus reactive power Q_k in the absence of SVC, then determine the reactive power, $Q_{k,sp}$, required at bus k to achieve a voltage $V_k = V_{k,ref}$. Then, calculate the required reactive power for compensation as:

$$\Delta Q_k = Q_k - Q_{k,sp}$$

2.2. The reactive power injected or absorbed by SVC is thus:

$$Q_{SVC} = \Delta Q_k$$

where:

$$Q_{SVC} = V_k^{(p)2}\left[\frac{1}{X_C} - \frac{1}{\pi X_L}(\pi - 2\alpha - \sin 2\alpha)\right]$$

2.3. Using this last equation, determine the value of angle α through successive attempts until

$$\left|\frac{Q_{SVC} - \Delta Q}{Q_{SVC}}\right| \leq \varepsilon$$

where ε can be considered 1%.

2.4. Check for the value of angle α

2.4.1. If $\alpha \in (\alpha_{\min}, \alpha_{\max})$, then the static compensator operates within the control domain and the auxiliary bus j is of PV-type; the Jacobian matrix is then modified by inserting the elements $\partial P_k/\partial\theta_j$, $\partial Q_k/\partial\theta_j$, and go to step 4.

2.4.2. If $\alpha \leq \alpha_{\min}$, then set $\alpha = \alpha_{\min}$ and go to step 3.1.

2.4.3. If $\alpha \geq \alpha_{\max}$, then set $\alpha = \alpha_{\max}$ and go to step 3.2.

3. *Treat the static compensator when α reaches limit values*

3.1. When $\alpha = \alpha_{\min}$, the compensator will be treated as a fixed inductive reactance, and the auxiliary bus j is of PQ-type with:

$$P = 0;\ Q = Q_{SVC}\big|_{\alpha_{\min}} = V_k^{(p)2}\left[\frac{1}{X_C} - \frac{\pi - 2\alpha_{\min} - \sin 2\alpha_{\min}}{\pi X_L}\right]$$

In this case, the terms $\partial P_k/\partial\theta_j$, $\partial P_k/\partial V_j$, $\partial Q_k/\partial\theta_j$, and $\partial Q_k/\partial V_j$ are added to the Jacobian.

3.2. When $\alpha = \alpha_{max}$, the compensator will be treated as a fixed inductive reactance, and the auxiliary bus j is of PQ-type, with:

$$P = 0;\ Q = Q_{SVC}\big|_{\alpha_{max}} = V_k^{(p)2}\left[\frac{1}{X_C} - \frac{\pi - 2\alpha_{max} - \sin 2\alpha_{max}}{\pi X_L}\right]$$

In this case, the terms $\partial P_k/\partial\theta_j$, $\partial P_k/\partial V_j$, $\partial Q_k/\partial\theta_j$, and $\partial Q_k/\partial V_j$ are added to the Jacobian.

4. *Continue with the classical Newton–Raphson algorithm (Jacobian factorizing, ΔV, $\Delta\theta$ deviations computing, convergence test, etc.).*

5.6.2 SVC Dynamic Modeling

For the dynamic modeling—transient and medium term—the SVC mathematical models for time domain simulations should represent its operation only in the positive sequence system. This is due to the fact that the negative sequence reactances of the SVC, of the reactor and capacitor more precisely, are equal to the positive sequence reactances. The zero sequence reactances do not affect the power system under the short-circuit (single- and three-phase) conditions because the SVC connection to the transformer tertiary (delta connection for a six-pulse SVC and wye connection no grounding for a 12-pulse SVC) makes the zero sequence impedances arrangement to be open.

Most of the dynamic models proposed in the literature focus on the steady-state model and consider different control loops [6, 10, 11, 12].

5.6.2.1 The Basic Dynamic Model

The basic structure of an SVC operating within the linear control domain is shown in Figure 5.32a. Its dynamical model that considers a balanced power system operating at fundamental frequency is shown in Figure 5.32b [11].

This model is described by the following set of algebraic-differential equations, expressed in p.u. [11]:

$$\begin{bmatrix} \dot{x}_c \\ \dot{\alpha} \end{bmatrix} = f(x_c, \alpha, V, V_{ref})$$

$$0 = \begin{bmatrix} B_{SVC} - \frac{1}{X_C X_L}\left\{X_L - \frac{X_C}{\pi}\left[\pi - 2\alpha_{SVC} - \sin(2\alpha_{SVC})\right]\right\} \\ I_{SVC} - V_i B_{SVC} \\ Q - V_i^2 B_{SVC} \end{bmatrix} \tag{5.42}$$

Most of the variables from (5.42) are defined in Figure 5.35b, where x_c is the vector of control variables, and *f*() is their equations. These equations allow representing not only the firing delay angle α, but also those of the current I_{SVC} and voltages V, V_i; they also allow modeling the different control modes, such as the control mode of the reactive power injected by the SVC.

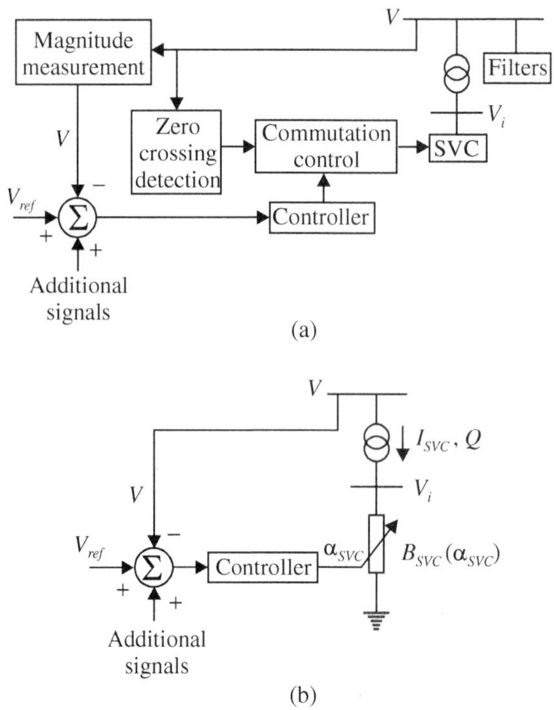

Figure 5.32 SVC base model: (a) control structure and (b) dynamic model [11].

The differential equations from (5.42) depend on the type of controller considered in the study.

5.6.2.2 *First-Order Dynamic Model*

This model determines the susceptance B_{SVC} as a function of the connection bus voltage V_k of the SVC to the system. Figure 5.33 shows the block diagram of the simplest controller, thus obtaining the first-order dynamic model.

In this simple model, the parameter $K_{SVC} = 1/X_{SL}$ models the SVC characteristic slope in the linear control area, T_{SVC} is the time constant of the system usually associated with the firing delay of the thyristors, POD states for power oscillations damping, and s is the Laplace operator.

From Figure 5.33, we derive

$$(1 + sT_{SVC})x_1 = K_{SVC}(V_{POD} + V_{ref} - V) \tag{5.43}$$

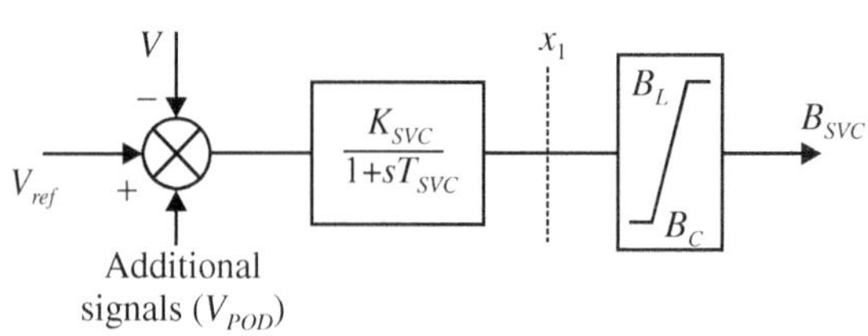

Figure 5.33 First-order dynamic model of the SVC.

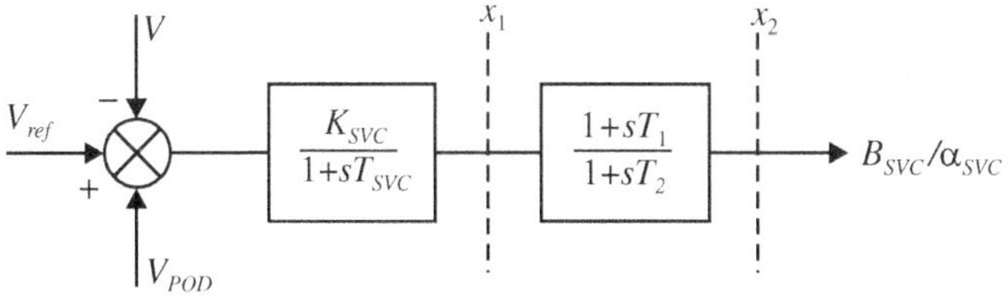

Figure 5.34 The SVC basic dynamic model.

or expressing the above equation in time-domain

$$\dot{x}_1 = \frac{1}{T_{SVC}}[-x_1 + K_{SVC}(V_{POD} + V_{ref} - V)] \tag{5.43'}$$

The differential equation (5.43′) describes the dynamic behavior of the controller. The state variable x_1 can represent either B_{SVC} or α_{SVC} depending on the adopted model (5.42).

In Figure 5.34, the basic controller model is shown [11]; this scheme has the advantage of providing a very good model of the real regulator behavior and, moreover, allows the coordination with the other regulators in the system. We observe that the difference between the model depicted in Figure 5.33 and the one from Figure 5.34 consists of a supplementary lead-lag block in order to ensure a better stability of the voltage loop.

From Figure 5.34, we derive for x_1 the same equation as in (5.43′). Regarding the second state variable, x_2, we have

$$(1 + sT_1)x_1 = (1 + sT_2)x_2$$

Knowing that (5.43) still holds for x_1, we replace sx_1 in the above equation to obtain the time domain differential equation

$$\dot{x}_2 = \frac{1}{T_2}\left[-x_2 + \left(1 - \frac{T_1}{T_{SVC}}\right)x_1 + \frac{T_1 K_{SVC}}{T_{SVC}}(V_{POD} + V_{ref} - V)\right] \tag{5.44}$$

Equations (5.43′) and (5.44) describe the dynamic behavior of the regulator from Figure 5.34. In this case, the state variable x_1 is an intermediate variable, while x_2 can represent either B_{SVC} or α_{SVC} depending on the adopted model.

These models can be easily implemented in a computer program for small-signal, transient, and voltage stability assessment.

5.6.2.3 Complex SVC Dynamic Models

More complex SVC dynamic models have been standardized by CIGRÉ and IEEE [6, 7, 9]. Two basic models for transient stability programs are recommended as standard models. The structures of basic model 1 and basic model 2 are similar except for the method of representing the slope; model 1 gives a voltage to susceptance linear relation, and the model 2 gives a voltage to current linear relation. Both models are suitable for continuously controlled SVCs (TCR–FC, TCR–TSC, and TCR–MSC), but can be modified to represent the TSC/thyristor-switched reactor (TSR) types of compensators.

(i) IEEE/CIGRÉ SVC basic model 1

Figures 5.35a and 5.35b show the IEEE/CIGRÉ basic SVC model 1 and its voltage regulator model. This model corresponds to the gain-time-constant format.

This model controls the current I_{SVC} using the voltage value V measured at the connection point between the network and the SVC. The static characteristic slope is given by the voltage controller block, which controls the susceptance through the thyristors. An interface block computes the SVC current as a function of its admittance.

The measuring circuit module is characterized by a first-order transfer function $1/(1 + sT_m)$, where the measuring system time constant T_m has values of 1÷8 ms. The measuring equipment converts the three-phase voltage into continuous signals depending on the gain of the measured variable.

The voltage regulator (Figure 5.35b) is of the proportional type and the gain, K_R, is the inverse of the slope setting; a gain of 100 p.u. B_{SVC} / p.u. ΔV on the SVC base means of a 1% slope. The time constant, T_R, is usually between 20 and 150 ms and the time constants T_1 and T_2 are zero in most cases [3, 6].

A phase lead can, however, be provided to enhance the damping contribution of SVC, although this may have a slightly detrimental effect on the synchronizing torque. The lead-lag time constants can also be used to provide adequate phase and gain margins when large values of steady-state gain are

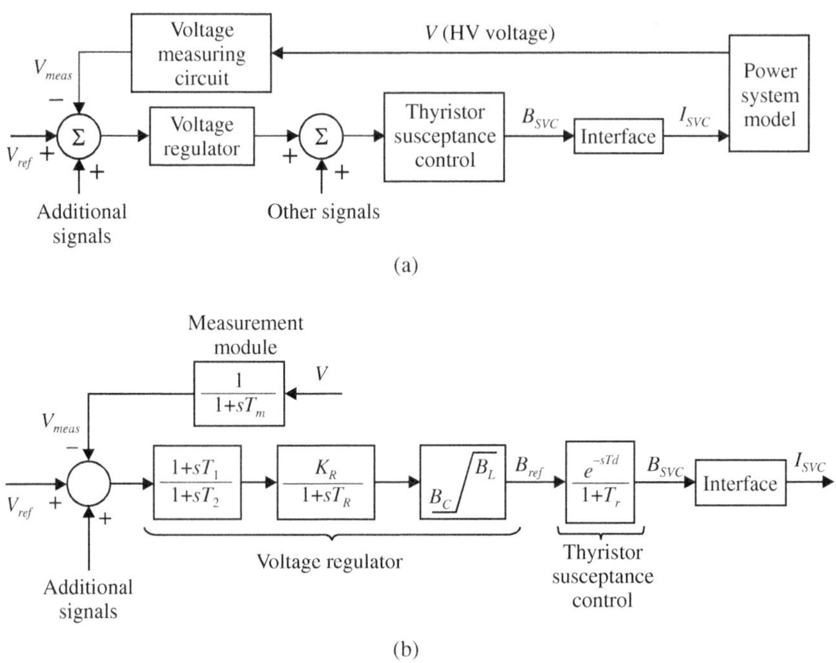

Figure 5.35 (a) The IEEE/CIGRÉ basic model 1 of the SVC and (b) the voltage regulator model. *Source*: Therond et al. 1999 [9]. Reproduced with permission of CIGRE. *Source*: Taylor et al. 1994 [6]. Reproduced with permission of IEEE.

employed. The maximum and minimum limits on the susceptance output B_{ref} are given by B_C and B_L, respectively.

The transfer function of voltage regulator in the basic model 1 is given by [6]:

$$G(s) = \frac{K_R}{1 + sT_R}\left(\frac{1 + sT_1}{1 + sT_2}\right) \tag{5.45}$$

The thyristor susceptance control module has the transfer function:

$$H_B = e^{-sT_d}/(1 + sT_r)$$

where T_d corresponds to the gating delay of the thyristors—around 1 ms, and T_r is a time constant representing the thyristor statistic firing delay time, with a value between 3 and 6 ms [6].

The SVC interface module with the network can be by either of the two methods. The variable susceptance, B, can be used to update the admittance matrix. Alternatively, B is multiplied by voltage to obtain the SVC current. The SVC current is an injection into the network. The choice depends on the network solution method.

(ii) IEEE/CIGRÉ SVC basic model 2

The main difference from the above-presented SVC basic model 1 is the current measuring mode. The V–I characteristic slope in the control area is given by the current loop. The IEEE/CIGRE basic model 2 of an SVC is presented in Figure 5.36.

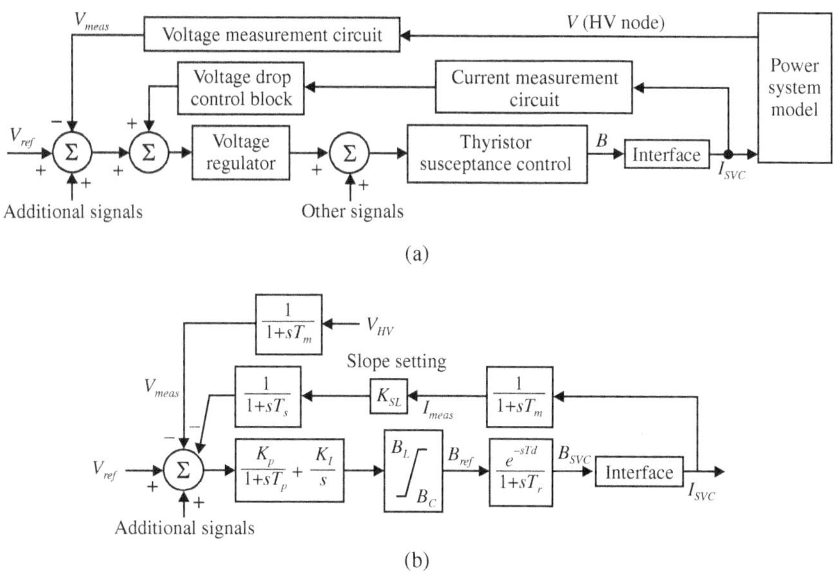

Figure 5.36 IEEE/CIGRÉ basic dynamic model 2 scheme. *Source*: Therond et al. 1999 [9]. Reproduced with permission of CIGRE. *Source*: Taylor et al. 1994 [6]. Reproduced with permission of IEEE.

The voltage regulator uses a proportional-integral (PI) corrector. All other functions of the model are identical with the functions described for the IEEE/CIGRÉ basic model 1.

The transfer function of voltage regulator in model 2 is given by

$$G(s) = \frac{K_I}{s}\left(\frac{1 + sT_Q}{1 + sT_P}\right) \quad (5.46)$$

where $T_Q = T_P + \frac{K_P}{K_I}$.

The proportional gain K_P is employed to increase the speed of response, and T_P is kept to zero, thereby rendering the controller to be of the simple PI type. The integrators in both basic models 1 and 2 are of the nonwindup type [3].

The nonlinear relationship between B_{ref} and TCR firing angle is compensated by a linearizing function in the thyristor susceptance control module. The delay represented by e^{-sT_d} is obtained from a filtered development that only retains the first terms, resulting in the numerator $1- sT_d$.

Note: Modular models for other functions can interface with the basic models (for longer term dynamic simulations, for subsynchronous resonance damping, etc.).

5.7 PLACEMENT OF SVC

Generally, the SVC devices are installed in existing power stations, that is, either in a network bus or in the middle of an electric line [13].

SVC Installed in a Network Bus. In this case, the SVCs are installed in buses where sensitive loads or highly power fluctuating loads are connected. SVCs can also be installed in generator buses that cannot supply/absorb sufficient reactive power in order to maintain the desired voltage level.

If the SVC is installed at bus i (Figure 5.37), then only the diagonal term $\underline{Y}_{ii}$ of the bus admittance matrix is changed:

$$\underline{Y}'_{ii} = \underline{Y}_{ii} + \underline{y}_{SVC}$$

where $\underline{y}_{SVC}$ is the corresponding admittance of the SVC.

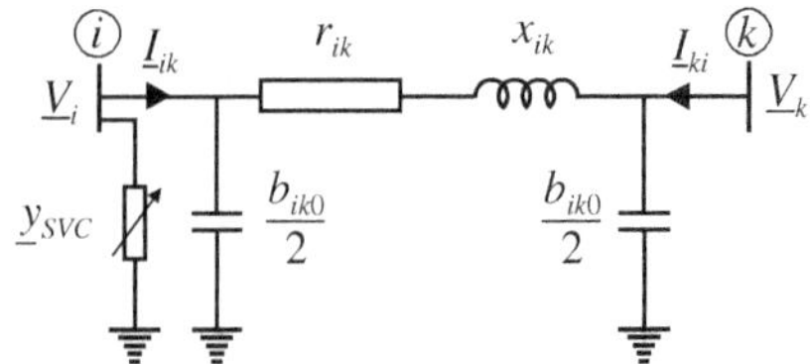

Figure 5.37 SVC installed in a network bus.

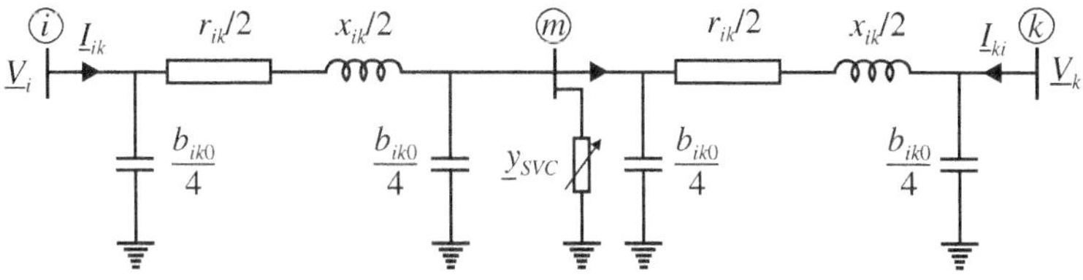

Figure 5.38 SVC installed in the midpoint of a transmission line.

In this case, the bus admittance matrix becomes:

$$\left[\underline{Y}_{nn}\right] = \begin{bmatrix} \underline{y}_{ik} + \dfrac{\underline{y}_{ik0}}{2} + \underline{y}_{SVC} & -\underline{y}_{ik} \\ -\underline{y}_{ik} & \underline{y}_{ik} + \dfrac{\underline{y}_{ik0}}{2} \end{bmatrix} \tag{5.47a}$$

SVC Installed in the Midpoint of a Transmission Line. If the static compensator is installed in the middle of an electric line at bus m, the configuration can be represented as in Figure 5.38.

By adding a new bus m on the transmission line, a new row and a new column have to be added in the bus admittance matrix. This implies changing the matrix dimensions, but it can be avoided by using a wye-delta transfiguration and computing the equivalent line parameters (Figure 5.39).

Following this transformation, all the bus admittance matrix terms are modified:

$$\left[\underline{Y}_{nn}^{\,\mathrm{mod}}\right] = \begin{bmatrix} \underline{y}'_{ik} + \underline{y}'_{ik0}/2 & -\underline{y}'_{ik} \\ -\underline{y}'_{ik} & \underline{y}'_{ik} + \underline{y}'_{ik0}/2 \end{bmatrix} \tag{5.47b}$$

The shunt parameter at the intermediary point m is given by:

$$\underline{y}_{m0} = \underline{y}_{ik0}/2 + \underline{y}_{SVC} \tag{5.48}$$

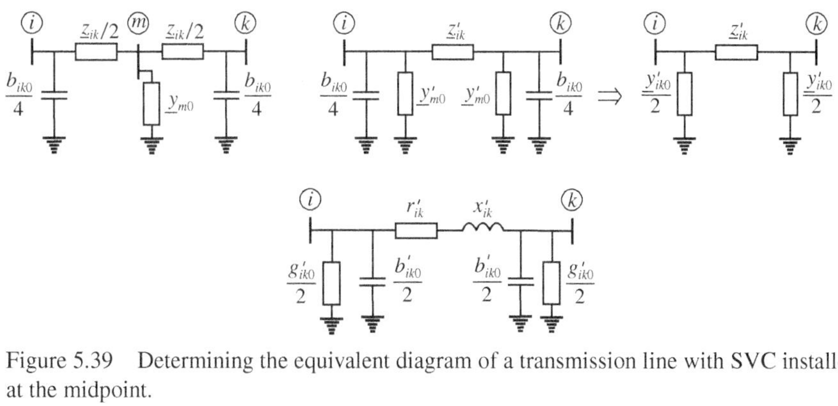

Figure 5.39 Determining the equivalent diagram of a transmission line with SVC installed at the midpoint.

The series impedance of the reduced equivalent diagram is:

$$\underline{z}'_{ik} = \frac{4\underline{y}_{ik} + \underline{y}_{m0}}{4\underline{y}^2_{ik}} = \frac{1}{\underline{y}_{ik}} + \frac{\underline{y}_{m0}}{4\underline{y}^2_{ik}} = \underline{z}_{ik} + \frac{1}{4}\underline{z}^2_{ik}\left(\frac{\underline{y}_{ik0}}{2} + \underline{y}_{SVC}\right) \tag{5.49}$$

where

$$r'_{ik} = \underline{r}_{ik} - \frac{1}{2}r_{ik}x_{ik}\left(\frac{b_{ik0}}{2} + b_{SVC}\right) \tag{5.50a}$$

$$x'_{ik} = x_{ik} + \frac{1}{4}\left(r^2_{ik} - x^2_{ik}\right)\left(\frac{b_{ik0}}{2} + b_{SVC}\right) \tag{5.50b}$$

The shunt admittance of the reduced equivalent diagram is:

$$\frac{\underline{y}'_{ik0}}{2} = \frac{\underline{y}_{ik0}}{4} + \frac{2\underline{y}_{ik}\underline{y}_{m0}}{4\underline{y}_{ik} + \underline{y}_{m0}} = \frac{\underline{y}_{ik0}}{4} + \left(\frac{\underline{y}_{ik0}}{2} + \underline{y}_{SVC}\right)\Big/\left[2 + \frac{1}{2}\underline{z}_{ik}\left(\frac{\underline{y}_{ik0}}{2} + \underline{y}_{SVC}\right)\right] \tag{5.51}$$

where

$$\frac{g'_{ik0}}{2} = \frac{\frac{1}{2}r_{ik}\left(\frac{b_{ik0}}{2} + b_{SVC}\right)^2}{4 - 2x_{ik}\left(\frac{b_{ik0}}{2} + b_{SVC}\right) + \frac{1}{4}\left(r^2_{ik} + x^2_{ik}\right)\left(\frac{b_{ik0}}{2} + b_{SVC}\right)^2} \tag{5.52a}$$

$$\frac{b'_{ik0}}{2} = \frac{b_{ik0}}{4} + \frac{2\left(\frac{b_{ik0}}{2} + b_{SVC}\right) - \frac{1}{2}x_{ik}\left(\frac{b_{ik0}}{2} + b_{SVC}\right)^2}{4 - 2x_{ik}\left(\frac{b_{ik0}}{2} + b_{SVC}\right) + \frac{1}{4}\left(r^2_{ik} + x^2_{ik}\right)\left(\frac{b_{ik0}}{2} + b_{SVC}\right)^2} \tag{5.52b}$$

Note: The capacitance g'_{ik0} can be neglected since it takes low values as compared to the susceptance.

5.8 APPLICATIONS OF SVC

The main purpose of the SVC is to increase the transmission capacity of the power system. This can be achieved by providing voltage support and increasing the stability margins. In order to provide voltage support at the receiving-end of a transmission line and to contribute to transient (first swing) stability improvement, the SVC is operated essentially as a voltage regulator, the VAr output being varied so as to reduce and rapidly damp the bus voltage variations during and, in particular, following major power disturbances [14].

5.8.1 Maintaining the Voltage Level of a Bus or into an Area

An SVC device is designed to control the voltage level of a bus by injecting or absorbing reactive power. The voltage within a certain area of a power system can also be influenced by changing the voltage set-point of the SVC.

In systems of small short-circuit power, which incorporate long transmission lines, the voltage is highly affected by load variations or network components disconnection. At significant loads, the voltage can drop drastically and low voltage levels can lead to system voltage instability; SVCs are suitable to correct these situations.

5.8.2 Increasing the Transmission Capacity

Consider a transmission line linking two power systems (Figure 5.40a).

When the voltages at both ends of the line are kept equal, $V_A = V_B = V$, the active and reactive power transfer is calculated as [14]:

$$P_A = \frac{V^2}{X_L} \sin \delta \tag{5.53a}$$

$$Q_A = \frac{V^2 - V^2 \cos \delta}{X_L} = \frac{V^2}{X_L} (1 - \cos \delta) \tag{5.53b}$$

where $\delta = \delta_A - \delta_B$.

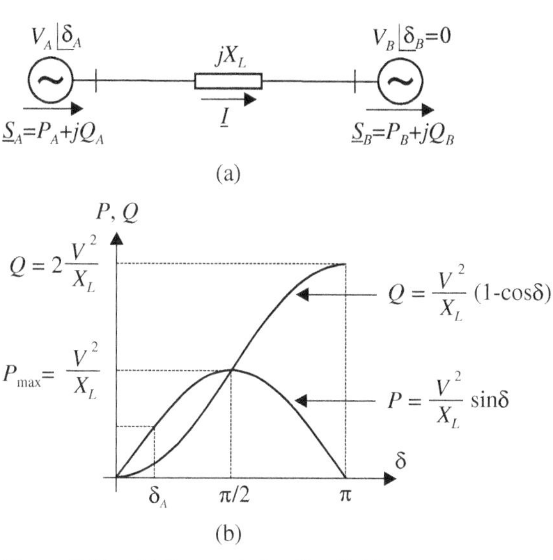

Figure 5.40 (a) Transmission line linking two power systems and (b) the variation of real and reactive power versus the phase angle difference δ.

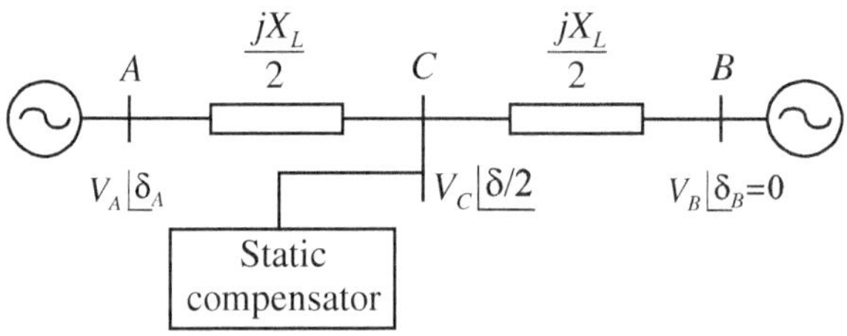

Figure 5.41 Simple two-machine power system with ideal midpoint compensator.

Figure 5.40b illustrates the variation of the active and reactive power with the phase angle difference δ. The theoretical maximum transmissible power defining the steady-state stability limit on the line is obtained for $\delta = \pi/2$:

$$P_{A\max} = \frac{V^2}{X_L} \tag{5.54}$$

Transmission lines cannot be operated too close to their steady-state stability limit because a sufficient margin in the power transfer is needed to recover from dynamic disturbances during which the transmitted power and transmission angle can significantly change from, or oscillate around the steady-state value [14].

Consider that in the midpoint of the line, on bus C, a shunt static compensator is installed, which aims maintaining the voltage at value V, the same as that at the sending and receiving ends. Equation (5.53a) can then be applied for each half of the line (Figure 5.41).

The transmissible power in the case with compensation, expressed by (5.55a), is illustrated in Figure 5.42a, where P is plotted against angle δ.

$$P_{sh} = \frac{V^2}{X_L/2}\sin\frac{\delta}{2} = \frac{2V^2}{X_L}\sin\frac{\delta}{2} \tag{5.55a}$$

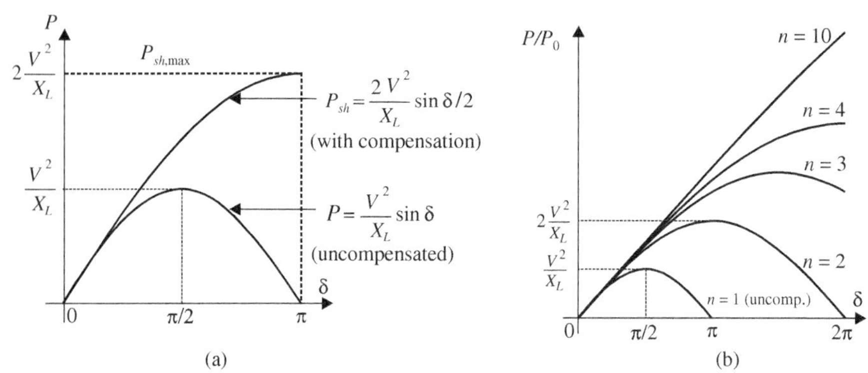

Figure 5.42 Real power variation with angle δ for shunt compensation in the midpoint of the line (a), respectively, for n sections (b) [5].

It is obvious that the maximum transmissible active power is obtained at angle $\delta/2 = \pi/2$, and the steady-state limit is twice than that achieved for the uncompensated case:

$$P_{sh,\max} = \frac{2V^2}{X_L}\sin\frac{\pi}{2} = \frac{2V^2}{X_L} = 2P_{\max} \tag{5.55b}$$

In general, the transmission reactance X_L can be divided into n equal sections with a perfect synchronous compensator in the joint points of the sections (Figure 5.42b). In this case, the transmissible power is characterized theoretically by the equation [14]

$$P = \frac{V^2}{\frac{X_L}{n}}\sin\frac{\delta}{n}$$

which gives a maximum transmissible power of nV^2/X_L, which is n times the steady-state limit of the uncompensated case.

In practical applications, this solution with multiple sections is limited because of high costs or thermal rating of the conductor.

5.8.3 Static and Transient Stability Reserve Improvement

In order to improve the transient stability of the power system, the line impedance can be varied with a control strategy. In this way, the network can operate, while maintaining the stability, at power levels higher than the ones it was originally designed for.

In order to analyze the influence of shunt compensators upon the static stability reserve, consider the power characteristics illustrated in Figure 5.43 [15].

As illustrated in Figure 5.43, the intersection of the mechanical power characteristic $P = P_m$ with different characteristics generates the following operating points: $F_1 \rightarrow \delta_1$ (uncompensated case); $F_2 \rightarrow \delta_2$(with shunt compensation in the midpoint

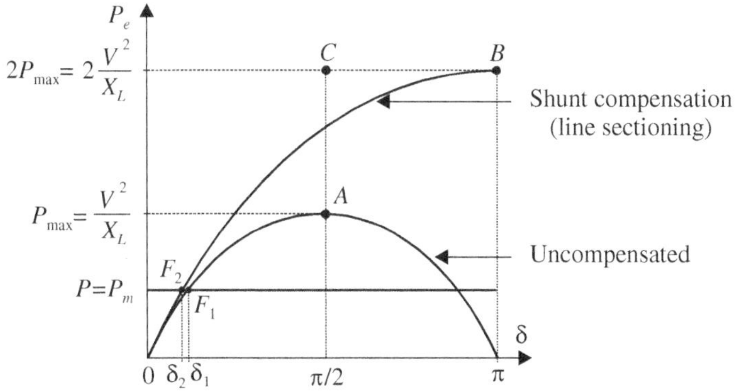

Figure 5.43 Transmissible power versus angle δ, for different compensation methods.

of the line). From the steady-state stability point of view, the limit operating points are A and B. The static stability reserve, substantially improved by using compensation devices, is given by the segments:

- AF_1 $(\pi/2 - \delta_1)$, without compensation;
- BF_2 $(\pi - \delta_2)$, with shunt compensation.

As regards the transmissible power, the static stability reserve is: $(P_{max} - P)$ in the arrangement without compensation, and $(2P_{max} - P)$ in the case with shunt compensation.

When no compensation is provided, the maximum power that can be transmitted between two source buses linked by an electrical line is achieved for $\delta = \pi/2$, that is, $P_{max} = V^2/X_L$. In practice, for economical and technical reasons, the transmission line is operated below this maximum power. From technical point of view, it is necessary to maintain an available transmissible power reserve that would allow efficient generator breaking after an incident (e.g., an insulation fault or a sudden load variation). Shortly after the incident—which leads to load reduction, following the resulted unbalance between the mechanical power and the generated electrical power, the generators accelerate and, in case that the automatic systems are not properly tuned, the synchronism can be lost.

From *transient stability* point of view, the necessary time for the protection and switching devices to isolate the fault must be smaller than the critical clearing time, to which it corresponds δ_{crit} (the limit time to eliminate a fault beyond which one or more generators loss the synchronism). This duration is a mean of measuring the robustness of an electric power system relative to the perturbations that occur. If the fault is not cleared in due time, then a generator or group of generators are separated from the rest of the system, leading to transient instability phenomenon or loss of synchronism.

In order to determine the transient stability reserve resulted by shunt compensation, let us consider the scheme of a double-circuit transmission line with midpoint SVC connection (Figure 5.44a). Figure 5.44b shows the P–δ characteristics in a scheme without any compensation. In this case, for a transferred power P_1, the operating point in normal operating conditions is A, at the intersection with the normal characteristic (I).

In case of a short-circuit K_1^3 occurring on one of the two circuits of the line, close to the generator connected at bus 1, G1, the transmitted power becomes zero, while the mechanical power of the generator remains constant (P_m = const.). The operating point crosses from A to B, on the fault characteristic II (situated on the abscissa axis), and the electrical power becomes zero ($P_e \approx 0$). Since the mechanical power is greater than the electrical power, generator 1 is accelerated until the fault is cleared, in point C, and the internal angle increases from δ_0 to δ_{disc}. The kinetic energy stored by the generator during this time is represented by the acceleration surface S_1. At this instant, the normal scheme is restored and the operating point crosses back on the normal characteristic, but in the point D (characteristic III). The angle δ keeps increasing because of the rotating mass inertia. The electric power is now higher than the mechanical power provided by the rotating shaft, and the generator starts to

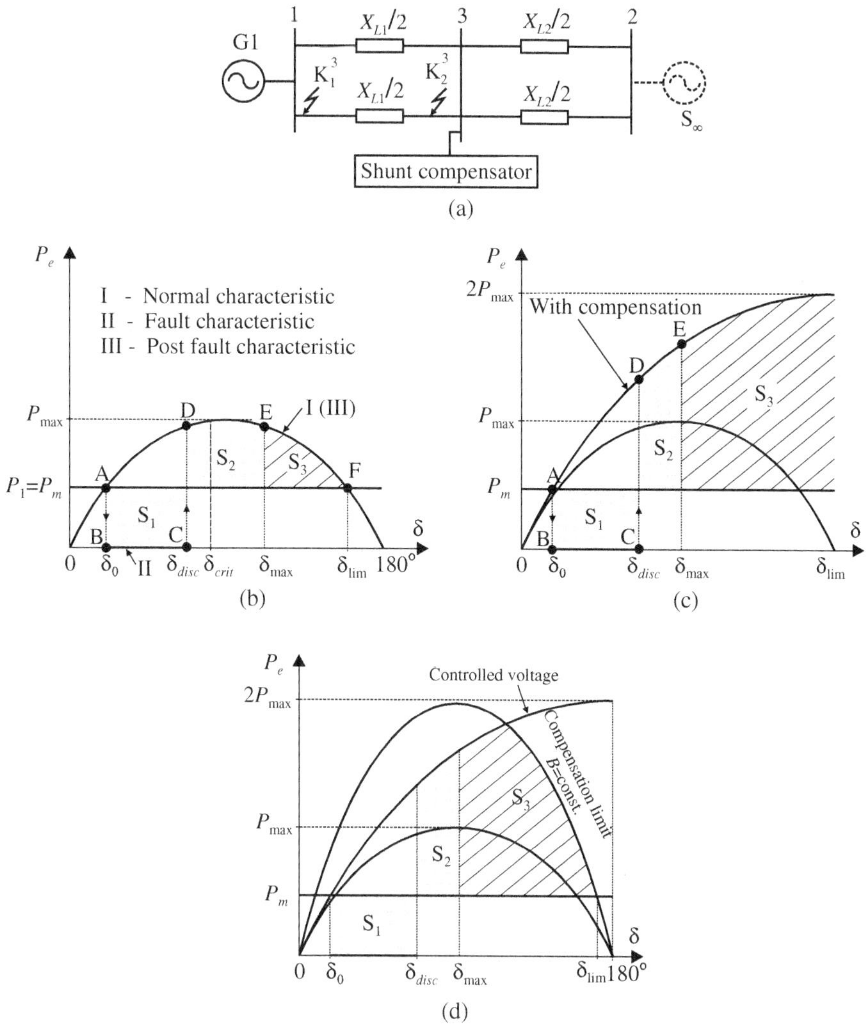

Figure 5.44 Equal area criterion in transient stability: (a) one-line diagram, (b) uncompensated case, (c) with ideal compensator in the midpoint of the line, and (d) with SVC in the midpoint of the line operating at compensation limit (B_{max}).

decelerate. The energy lost by the generator during this process is represented by the decelerating surface S_2.

In accordance with the equal area criterion, the generator finds its stability only if $S_1 < S_2$. The maximum angle reached during the first swing, δ_{max}, is obtained when the kinetic energy stored during the fault was "lost," which happens at $S_1 = S_2$ (Figure 5.44b).

The critical clearing angle δ_{crit}, associated with the critical fault clearing time t_{crit}, is that angle value at which the fault is cleared and ensures that the maximum angle δ_{max} does not exceed the boundary value δ_{lim} at the first swing. When the fault is cleared at $\delta_{disc} < \delta_{crit}$, a stability reserve represented by the surface S_3 is ensured. The maximum power that can be produced by the generator 1 and transferred to the power system, while maintaining its transient stability, is achieved for $S_1 = S_2 + S_3$.

If line circuit disconnection occurs at $\delta_{disc} > \delta_{crit}$, the angle δ rises over $\delta_{\lim}$ (unstable operating point) at the first swing, the accelerating area S_1 increases, and the decelerating area S_2 is no longer sufficient to ensure the stored kinetic energy being released, and the stability reserve surface S_3 becomes zero. Under these conditions, the generator synchronism is lost.

A device designed to improve the transient stability should try to "help" the generator to release the kinetic energy stored within a shorter period ($\delta_{\max} - \delta_{disc}$). Therefore, the device should increase the difference between $P_e(\delta)$ and P_m, either by diminishing the mechanical power P_m (by using the speed regulator) or by increasing $P_e(\delta)$ (by using the voltage regulator or the static compensator). Any of the improvements will reduce the maximum angle $\delta_{\max}$ and will increase the stability reserve S_3. Under these conditions, the power system would be capable of supporting a higher disconnection angle δ_{disc} and therefore a longer fault duration.

If an ideal reactive power compensator is considered at the midpoint of the line, at bus 3 (Figure 5.44a), which maintains the voltage constant at a desired value, the power transmitted on the double-circuit transmission line is given by equation (5.55a).

For the same fault K_1^3, Figure 5.44c shows the way in which the stability reserve (the area S_3) is substantially improved. Note that in order to obtain this improvement, the compensator must be capable of generating a reactive power given by:

$$Q = \frac{4V^2}{X_L}\left(1 - \cos\frac{\delta}{2}\right) \tag{5.56}$$

In this case, the equilibrium point may be reached for $\delta_{\lim} = 180°$. The reactive power demand at constant midpoint voltage increases rapidly with increasing power transmission, reaching a maximum value equal to $4V^2/X_L$ at the maximum steady-state power transmission limit of $2P_{\max}$ ($P_{\max}$ is the maximum transmissible power of the uncompensated system).

There are two arguments to consider this ideal case as being unrealistic [16]:

- The compensating device requires a larger power in comparison with the transferred power; thus, any economical advantage of using this kind of system is canceled.
- The capacity necessary to generate the reactive power for angle $\delta = 180°$ would create a resonance phenomenon that would lead to a high risk of overvoltages. This is the cause for which the calculated values will always be smaller. Taking into account the capability of the SVC to continuously control its reactive power, it can be seen as an ideal compensator that maintains the voltage level in the controlled bus. This is possible only if the susceptance B is:

$$B = \frac{4}{X_L}\left(1 - \cos\frac{\delta}{2}\right)$$

Usually, the susceptance B has a maximum value $B_{max} < 4/X_L$, B_{max} representing the SVC compensation limit. When this limit is reached, the SVC acts like a fixed capacitive susceptance and the transmitted power is given by:

$$P'_e = \frac{V^2}{X_L \left(1 - \frac{B_{max} \cdot X_L}{4}\right)} \sin\delta \tag{5.57}$$

where $B_{max} = \omega C_{max}$, and C_{max} is the maximum capacitance of the SVC.

If this last case (Figure 5.44d) is compared with the ideal compensation case, presented in Figure 5.44c, a substantial downsize of the area S_3, corresponding to the stability reserve, and therefore a reduction of the critical clearing time (t_{crit}), can be observed.

From the above stated, it can be seen that the stability reserve and the critical clearing time depend, mainly, on two parameters:

(i) The first parameter is the mechanical power P_m that delimits the frontiers from the left and bottom areas of S_3. In order to generalize, P_m will be expressed as percent of the maximum transmissible power on the uncompensated line, P_{max}. The ratio P_m/P_{max} is highly related to the angle δ_0 and characterizes the initial conditions of the generator, whatever its size.

(ii) The second parameter, the curve $P_e(\delta)$, delimits the top frontier of S_3. This characterizes the way in which the generator releases the kinetic energy stored during the fault. When an SVC is used, this parameter is determined through the size of the compensator, defined as the ratio Q_{max}/P_{max}, where Q_{max} is the maximum reactive power that the SVC is able to generate at the voltage V, $Q_{max} = B_{max}V^2$. The critical clearing time is determined by iterative computation until the decelerating area S_2 embeds the reserve area S_3.

Figure 5.45 presents the influence of the static compensator size upon the transmissible power (a) and upon the critical clearing time (b) for different values of the ratio P_{mec}/P_{max}. These curves are general, as they were obtained independently of the voltage and line impedance values [16].

These kinds of diagrams built for actual cases permit to compute the reactive power necessary for the SVC to enhance the robustness of an electric power system.

For example, for a given network configuration, let t_{crit} be 90 ms, and the transfer capacity be 80%P_{max} (Figure 5.45b). By installing an SVC—having the power, in MVAr, equal to P_{max}, in MW—the critical clearing time rises to 140 ms (the dash line) [16].

As regards the improvement of the transmissible power and maintaining the stability reserve, for the same configuration, it can be observed that without compensation, for $t_{crit} = 90$ ms no more than $0.8P_{max}$ can be transferred. By using an SVC of $Q_{max} = 0.7P_{max}$, the transmissible power rises to $0.9P_{max}$.

The above mentioned statements can also be generalized for a short circuit in K_2^3, anywhere on the transmission line (Figure 5.46a). Figures 5.46a and 5.46b present the characteristics $P_e = f(\delta)$ in the normal state (I), fault (II), and postfault (III), without and with compensation (I′, II′, and III′) [16].

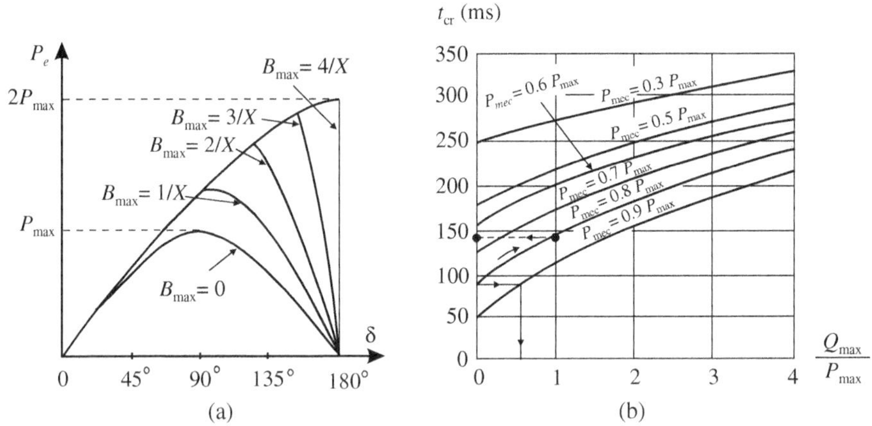

Figure 5.45 SVC power influence upon: (a) the transmitted power and (b) the critical disconnection time [16].

It can be observed that, in the case of shunt compensation with SVC, for the same transmitted power, the maximum fault clearing time increases considerably. If the same stability reserve as in the uncompensated case is considered, using an SVC would allow increasing the transmissible power ($P_2 > P_1$).

In order to improve the line power transfer capability in transient state, as well as in steady state, a FACTS device should be able to act very fast—faster than the electromechanical oscillation speed of a generator (e.g., 0.5 … 2 s) upon a perturbation in the electrical network.

5.8.4 Oscillations Damping

When two power systems are interconnected through a "weak" (large impedance) transmission line, low frequency (reaching 1 Hz) power oscillations may result from transient states (faults and their removal). Taking into account several factors, the oscillations can be damped slower or faster and they can sometimes be augmented, leading to generators falling out of synchronism.

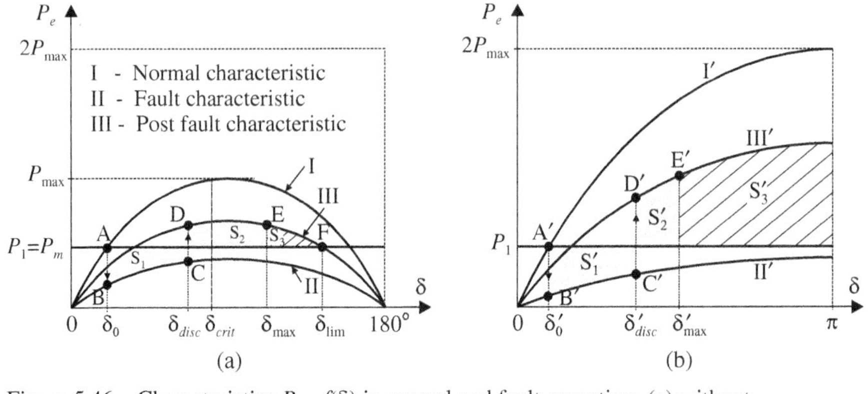

Figure 5.46 Characteristics $P = f(\delta)$ in normal and fault operation: (a) without compensation and (b) with static compensator in the midpoint of the line [16].

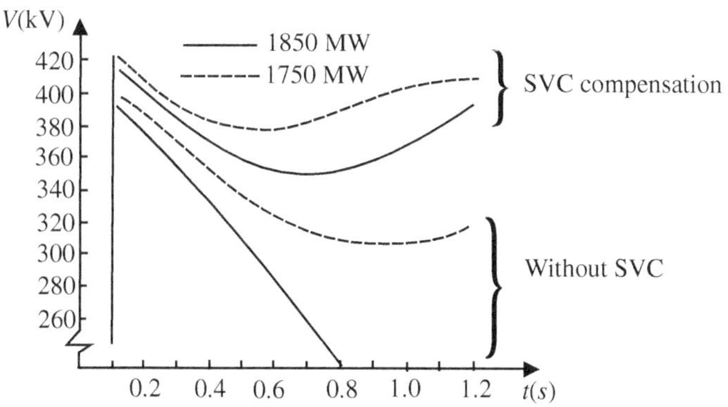

Figure 5.47 SVC stabilizing effect upon voltage-level changing [17].

In practice, in order to damp these kinds of oscillations, special power system stabilizers (PSS) are used as additional signals of the automatic voltage generators. There are cases when the oscillations frequency reaches very low values and the PSSs become inefficient, and therefore different solutions must be used, including FACTS devices. When these devices are designed to damp power oscillations, they have to be equipped with special command/control elements to monitor the oscillations and provide fast control of the device parameters.

One example of FACTS device that can provide power oscillation damping is the SVC. The stabilizing effect of voltage oscillations by contribution of SVC is illustrated in Figure 5.47. The SVCs) can rapidly and optimally damp line oscillations by continuous reactive power and voltage control, respectively. The compensator controls the power flow by varying the reactive current injected in the connection bus on the transmission line. Capacitor banks are connected or disconnected depending on the power derivate sign—connected for positive derivatives and disconnected otherwise. The resulting effect is oscillations damping; these are considered to be damped when the difference between the most recent maximum and minimum values is lower than a predefined level. At the same time, the system voltage is monitored and if it drops below certain values capacitor, disconnections are halted in order to avoid a voltage collapse.

Figure 5.48 concerns the main constraints upon the transfer capability of an electric line: thermal charge limit, transient state limit, and power oscillations charge limit. As it results, for certain network configurations, SVC oscillation damping can lead to transfer capability enhancement without building new facilities. As an approximate, installing an SVC of 1 MVAr can result in an improvement of 1–2 MW in transfer capability.

5.8.5 Reducing the Transient Overvoltages

Short-term overvoltages are produced as a result of maneuvers undertaken in order to connect or disconnect elements in the electrical network. An SVC can absorb reactive power and therefore can contribute to overvoltages diminishing.

Moreover, besides the above-mentioned applications, the SVC holds an important advantage, which is its rapid response. As an SVC does not comprise rotating

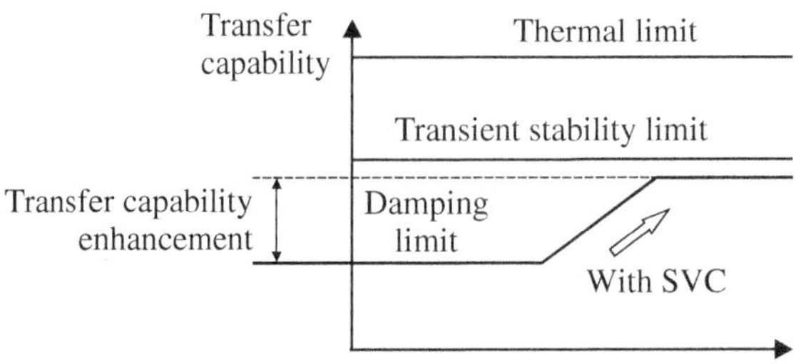

Figure 5.48 The possibility of power lines transfer capability enhancement through oscillations damping. *Source*: Lindström et al. 1984 [18]. Reproduced with permission of ABB.

elements, and therefore does not have inertia, its response constant is better than the time response of synchronous machines. A rapid action upon the voltage control avoids raising the reactive power generated; the reaching of their supply/absorption of reactive power limit is avoided.

For a certain charge, a rapid SVC response helps in avoiding for the dynamic charges (induction motors) to increase the total system load.

5.9 SVC INSTALLATIONS WORLDWIDE

At global level, there is an experience of over three decades in using FACTS devices. There are about 1500 SVC devices installed throughout the world, with a total installed power of about 100,000 MVAr. The main manufacturers are ABB, Alstom/Areva, GE, Siemens, and Mitsubishi/TMEIC.

Some of the applications of SVC devices in the world are:

- The first shunt compensation solution, consisting of a saturable reactor together with fixed capacitors, was used for overvoltages control on a transmission line of 420 km in length, of 330 kV rated voltage, in Zambia (1969).
- A combination of a saturable reactor and a mechanical switched reactor was also used on an 132 kV transmission line, and 785 km long, in the intermediary points and at the end of the transmission line, in 1978 in Nigeria for both dynamic stability and system voltage stability.
- A TSC shunt compensator combined with a TCR was installed in 1979 by ASEA in the Hydro-Quebec (Canada) network, at 735 kV. The total power of the installation was 450 MVAr. Afterward, four SVC devices were installed in the 735 kV network that links the power plants from James Bay and the Montreal area (around 1000 km). The SVCs were used in this case in order to stabilize the voltage under normal operating conditions and to improve the system stability during perturbations.
- In order to control the voltage and improve the stability of a transmission line rated at 220 kV, of 650 km in length situated in western Australia, three SVC devices were installed and another one afterward.

- Two of the greatest static compensators, rated at 580 MVAr each, were installed in Venezuela on a line of 800 kV that links the Guri hydroelectric plant and the load areas from Caracas and Valencia.
- Two static compensators were installed in the 385 kV electric network in Saudi Arabia, having the role of rapid voltage control at different loads.
- An SVC device of ±160 MVAr is installed in the 420 kV Sylling electrical network, near Oslo, to maintain the voltage during and after a short circuit in the power grid.
- Two identical SVC devices of 100 MVAr inductive/ 150 MVAr capacitive are installed at 330 kV in the Kemps Creek substation, where three electric power systems of Australia are interconnected; these were designed to ensure a transmission of 500 MW. The installation was put into commercial operation in 1990 and contributes to transient stability enhancement.
- Since 1995, three SVCs are operating in the Natal-Kwazulu system from South Africa. Their effect covers voltage control with superior performances and avoided building new lines. Figure 5.49 presents the scheme of the static compensator in Natal, a device of 2 × 150 MVAr inductive and 100 MVAr capacitive (from the 3, 5, and 7 harmonics filters).
- A combined compensating scheme comprising of four series capacitor banks and two reactive power static compensators is operating in Argentina. Thus, on the double-circuit line of 500 kV between El-Chocon and Buenos Aires, reaching 1000 km, in two different substations, series capacitors are used, each rated at 2 × 182 MVAr. The choice of series compensation with capacitors took into consideration an increase in the transfer capability. The static compensators, each rated at 160 MVAr inductive and 160 MVAr capacitive, help stabilizing the voltage in the Buenos Aires area for different load levels.
- In 1995, the first mobile SVCs were installed, rated at 0/+60 MVAr, in England. These compensators were connected to the tertiary winding of the

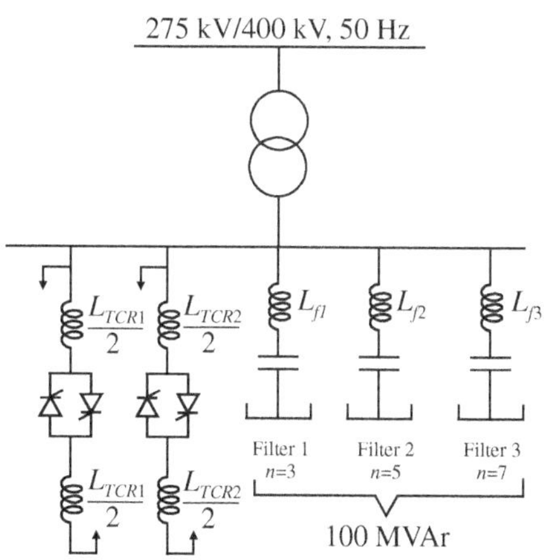

Figure 5.49 Scheme of static compensator in Natal, South Africa.

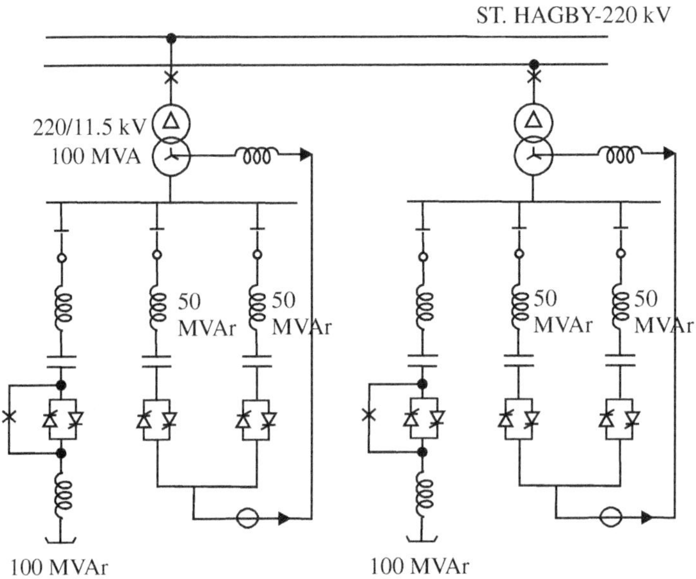

Figure 5.50 Hagby compensator scheme of 200 MVAr [18].

autotransformers. The advantage is that they can be moved to different substations, depending on necessities, in maximum 3 months. This breakthrough was followed by a demand of four mobile SVCs in 1997.

5.9.1 SVC at Hagby, in Sweden[1]

The static compensator from the Hagby 220 kV substation in Sweden comprises of 100 MVAr thyristor-controlled capacitors on each bus and 50 MVAr thyristor-controlled inductances (Figure 5.50). The command can be done manually or computer-aided. The monitored parameters are obtained through measuring transducers and are transmitted to a computer using analog and digital converters. Depending on the parameters, the computer determines the necessary steps in the following cycle. The computer outputs represent commands of connecting or disconnecting steps of compensation.

The VAr compensator can be controlled automatically by a microcomputer or manually with the microcomputer disconnected. The supervised parameters are obtained through measuring transducers and transmitted to a process computer by analog and numeric converters. Depending on the supervised parameters, the computer stabilizes the steps number necessary in the next cycle. The outputs of the microcomputer represent commands for connecting or disconnecting the compensation steps.

The system is supervised in three different levels:

[1] Reproduced with permission of CIGRE "*The 200 MVAr static compensator in Hagby, Sweden*, Lindström, C.O. et al., Report 38.02, CIGRE 1984", © Copyright 1984 [18].

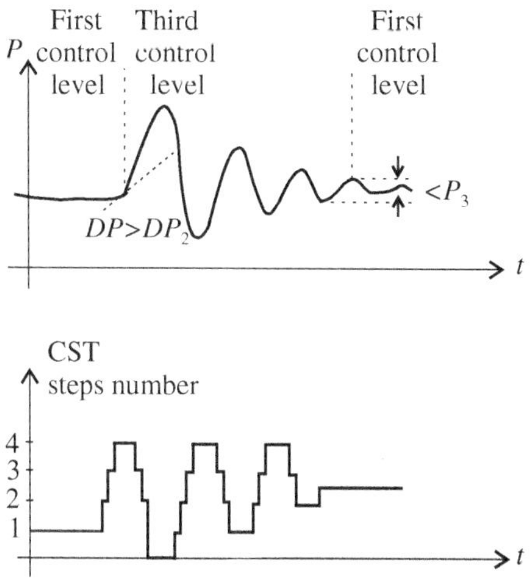

Figure 5.51 Power oscillations damping. *Source*: Lindström et al. 1984 [18]. Reproduced with permission of ABB.

(i) Voltage control (200 kV), which is the maximal control level. The voltage busbars are monitored individually and an appropriate number of TSC steps are switched in. This switching is associated with a certain hysteresis and time delay in order to prevent the too frequent at small voltage variations. Voltage control can also be done manually by the operator from the central control room.

(ii) Extreme voltage control: if the voltage of one of the 220 kV busbars exceeds the maximum permissible level, then all the TSC steps on the busbar will be switched off after a very short time delay. Similarly, if the voltage drops below a minimum permissible level, a corresponding number of TSC steps will be switched on.

(iii) Damping of power oscillations. Depending on the sign of the power derivative, the computer will order in (if positive derivative) or out (if negative derivative) all the TSC steps. When the power flow has reached its maximum or minimum value and the derivative changes sign, a number of TSC steps will be switched in or out depending on the magnitude of the power swing. The procedure is repeated at each extreme value. The power oscillation is considered to be damped out if the difference between the latest maximum and minimum value is smaller than a preset level P_3 (Figure 5.51).

5.9.2 SVC at Forbes, in United States

Northern States Power Co. (NSP) of Minnesota is operating an SVC in its 500 kV power transmission network between Winnipeg and Minnesota. This device is located at Forbes substation, in the state of Minnesota, and is shown in Figure 5.52. The purpose is to increase the power interchange capability on existing transmission lines

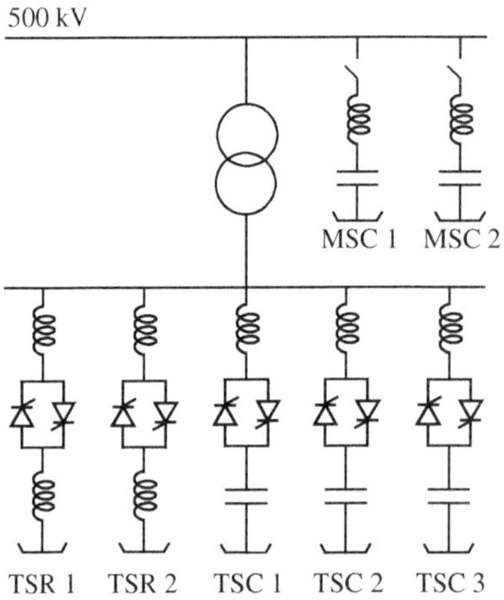

Figure 5.52 SVC at Forbes, United States. *Source*: [19, 20]. Reproduced with permission of ABB.

[19, 20]. This solution was chosen instead of building a new line as it was found superior with respect to increased advantage utilization as well as reduced environmental impact. With the SVC in operation, the power transmission capability was increased in about 200 MW.

The system has a dynamic range of 450 MVAr inductive to 1000 MVAr capacitive at 500 kV, making it one of the largest of its kind in the world. It consists of an SVC and two 500 kV, 300 MVAr, MSC banks. The large inductive capability of the SVC is required to control the overvoltage during loss of power from the incoming HVDC at the northern end of the 500 kV line.

The SVC consists of two TSRs, 190 MVAr + 260 MVAr, and three TSCs, 55 MVAr + 110 MVAr + 235 MVAr. Additionally, the SVC has been designed to withstand brief (<200 ms) overvoltages up to 150% of rated voltage.

Without the SVC, power transmission capacity of the NSP network would be severely limited, either due to excessive voltage fluctuations following certain fault situations in the underlying 345 kV system, or to severe overvoltages at loss of feeding power from HVDC lines coming from Manitoba.

5.9.3 SVC in Temascal, Mexico

An SVC compensator with a range of 600 MVAr is operating since 1982 in the Temascal 400 kV substation in Mexico. The compensator was installed in order to increase the transmission capacity (with 200 MW) and ensure secure power supply in the transmission system between the hydropower generating stations in the south area of the country and the large consumer district of Mexico City [20, 21].

The SVC consists of four TSC groups and four TCRs, each rated at 75 MVAr (Figure 5.53). This compensator permits continuous regulation of reactive power from 300 MVAr inductive to 300 MVAr capacitive. The operation range is further extended by the control of nine MSRs with a total rating of 490 MVAr.

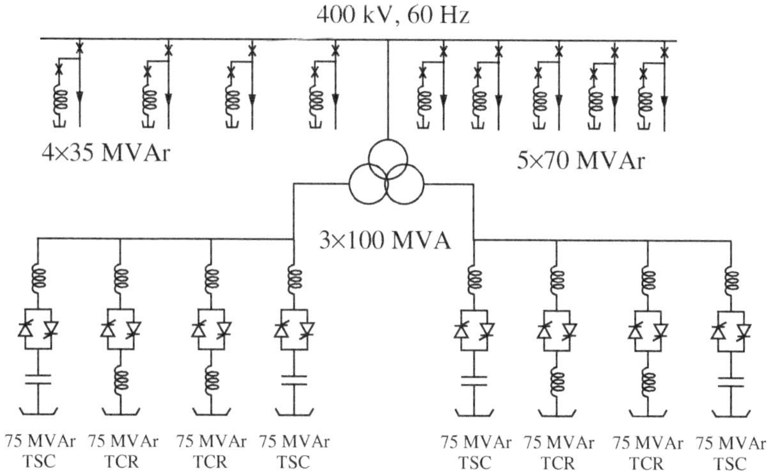

Figure 5.53 SVC compensator scheme in Temascal. *Source*: [20, 21]. Reproduced with permission of ABB.

The TCRs eliminate the voltage rise that can occur during operation with light load and during abnormal conditions and in addition reduce surges due to switching operations. The TSCs stabilize the system during heavy load or peak load periods.

5.9.4 Complex Compensation Scheme in Argentina

As the first utility in Argentina, Servicios Electricos del Grande Buenos Aires has installed two thyristor-controlled SVC s in its 500/220 kV substation at Rodriguez, situated approximately 60 km from down-town Buenos Aires (Figure 5.54a) [20, 22].

The national Argentinian system has its main supply sources in El Chocon and the Salto Grande districts, considerably distant from the principal load areas in and around Buenos Aires. The power is transmitted via 500 kV lines to a semi-circular 500 kV ring of substations around Buenos Aires, of which the Rodriguez substation is one.

The compensators have been incorporated to stabilize the 500 kV supply in its important power-receiving end (Figure 5.54b). They are parallel-operating, connected to the 500 kV bus via 132 kV tertiary windings of the station's main 800 MVA transformers. An interconnecting bus also enables the two units to be operated over one of the two existing three-phase groups of transformers when so required, thus permitting considerable flexibility in operation.

The installed dynamic range is 320 MVAr for each compensator. By means of two breaker-switched regulators, the total controllable range will be 852 MVAr.

5.9.5 SVC in the 735 kV Transmission System in Canada

The 735 kV network of Hydro-Quèbec transmits a total of 10,000 MVA of hydro power over five lines from the generating stations along La Grande Rivière at James Bay down to the Montreal district some 1000 km further south [20, 23].

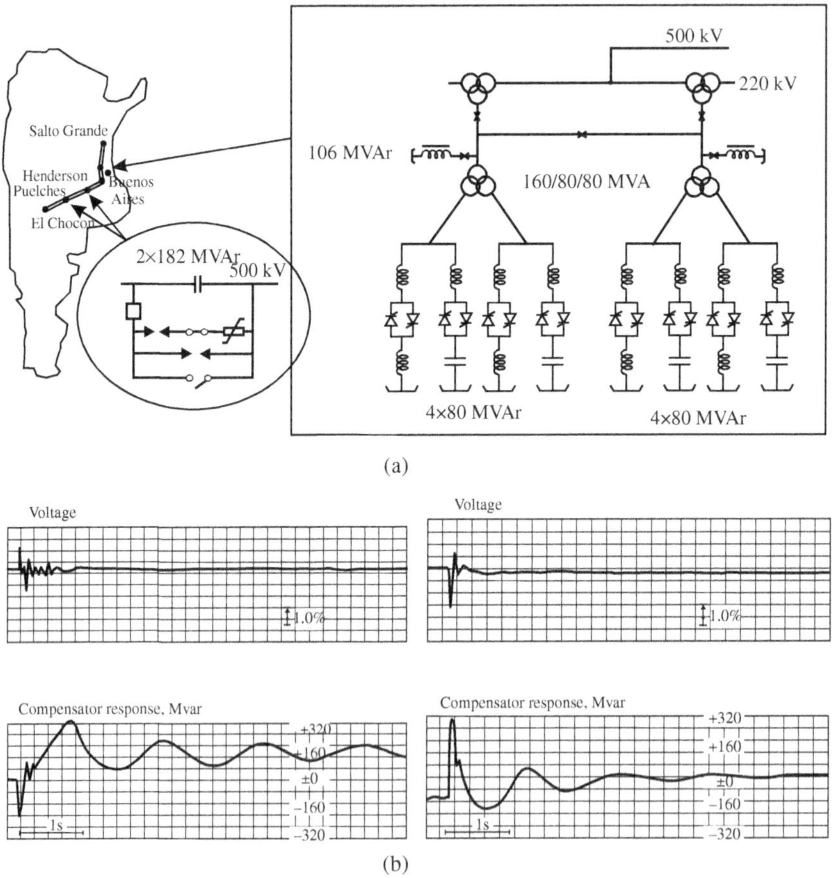

Figure 5.54 Compensation scheme with SVC and TCSC in Argentina. *Source*: [20, 22]. Reproduced with permission of ABB.

In the 735 kV transmission system of Hydro-Quèbec, Canada, six SVCs are operating. The compensators, each having a dynamic rating of 110 MVAr inductive to 330 MVAr capacitive at 735 kV, form a turn-key delivery commitment comprising two units at La Vérendrye substation (since 1984), two at Chibougamau (since 1985), and two at Chamouchouane (since 1990) (Figures 5.55a and 5.55b). The SVCs at Chamouchouane have extended to yield a total of 330 MVAr inductive to 330 MVAr capacitive.

The purpose of the compensator is twofold: to stabilize the voltage during normal operating conditions and to maintain the system stability during network perturbations.

5.9.6 SVC at Auas, in Namibia

Namibia is located at South-West of Africa, between Angola, Botswana, South Africa, and the Atlantic Ocean. The load in Namibia is concentrated in Windhoek

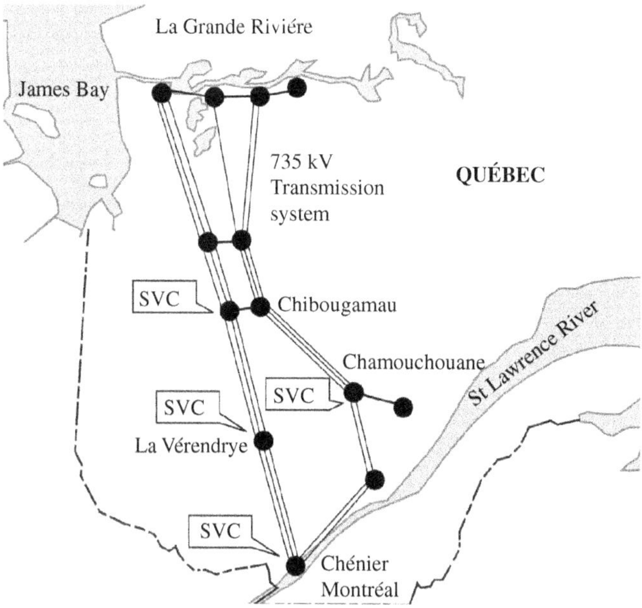

(a)

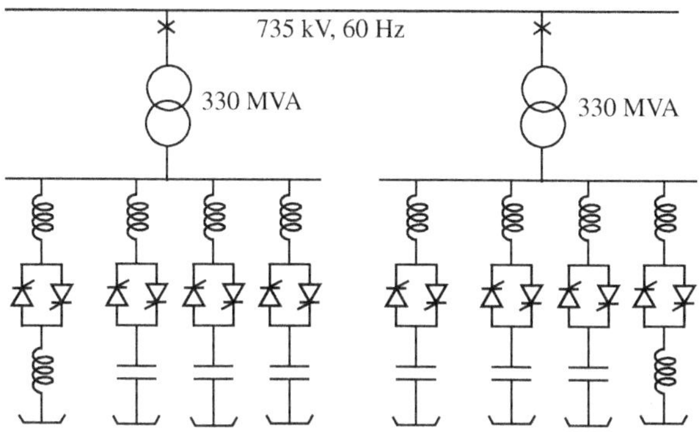

(b)

Figure 5.55 (a) The North Canadian 735 kV transmission system and (b) an SVC single-line diagram. *Source*: [20, 23]. Reproduced with permission of ABB.

and the northern region, where the most mine exploitations and mineral industries are localized. Since not long ago, the NamPower electric system consisted of a radial network with a hydro power plant at Ruacana and a circuit line rated at 330 kV, of 520 km in length, interconnected with the ESKOM system (South Africa) through a 400 kV, 890 km line (Figure 5.56) [24, 25].

This network was often loaded to its stability limit during Ruacana hydro power plant outages. Furthermore, the great lengths of the 220 and 330 kV lines and the

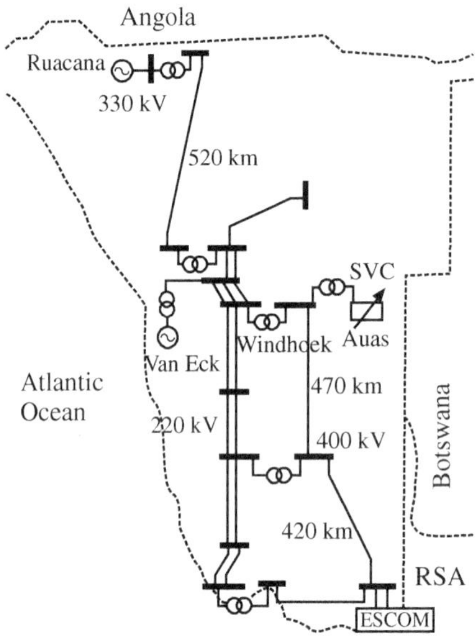

Figure 5.56 Namibia's power system. *Source*: Grünbaum et al. 2003 [25]. Reproduced with permission of ABB.

low power loads in comparison with the generators' capacity are two aspects that aggravate the stability problems in reduced load conditions.

In order to solve these problems, a 400 kV network between Auas and Kokerboom was put in service in 2000 [24, 25]. This single circuit line of 400 kV strengthened the NamPower system by connecting it to the ESKOM system. On the other hand, having a total length of 890 km, being the longest 400 kV line in the world, and still being supplied from a plant through a radial network lead to important capacitive operating states. Their effect is a modification in the existing parallel resonance near 50 Hz, making the network more sensitive to voltages during transient states, for example, the case when the 400 kV line is put under voltage or during its reconnection after a fault. Each of these phenomena leads to important overvoltages.

The NamPower network has a natural parallel resonance frequency of 55–70 Hz. After adding the Aries-Kokerboom-Auas new line section and the four shunt compensator reactors of 100 MVA, the first resonance frequency was modified toward 60–75 Hz.

Under these conditions, because of extremely high overvoltages at Auas substation, in some regimes even to peak values of 1.7 p.u., voltage control was decided by installing an SVC of 250 MVAr inductive and 80 MVAr capacitive in the Auas substation (Figure 5.57a) [24, 25].

The SVC device consists of three TCRs (TCR1, TCR2, and TCR3), a fourth, continuously energized TCR (TCR4), and two identical double-tuned filters, each rated at 40 MVAr. The filters take care of harmonics and supply capacitive reactive power during steady-state operation.

If, for any reason, it should have to be taken out of service, the 400 kV transmission system could not be operated without risking dangerous overvoltages. As a

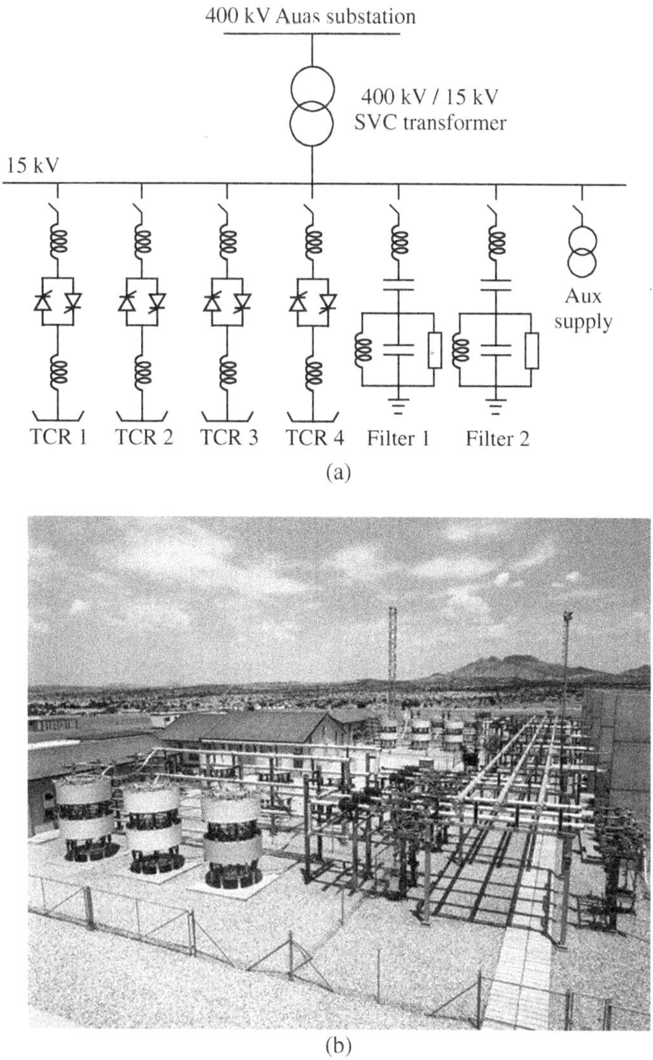

Figure 5.57 SVC at Auas substation: (a) one-line diagram and (b) view of the SVC. *Source*: Gene 2014 [24] and Grünbaum et al. 2003 [25]. Reproduced with permission of ABB.

result, an availability figure of 99.7% was specified, and this led to high investment costs. Figure 5.57b shows a view of the SVC installed at Auas [24, 25].

5.9.7 SVC at the Channel Tunnel Rail Link

Today, it is possible to travel between London and Paris in just over 2 h, at a maximum speed of 300 km/h. The railway power system is designed for power loads in the range

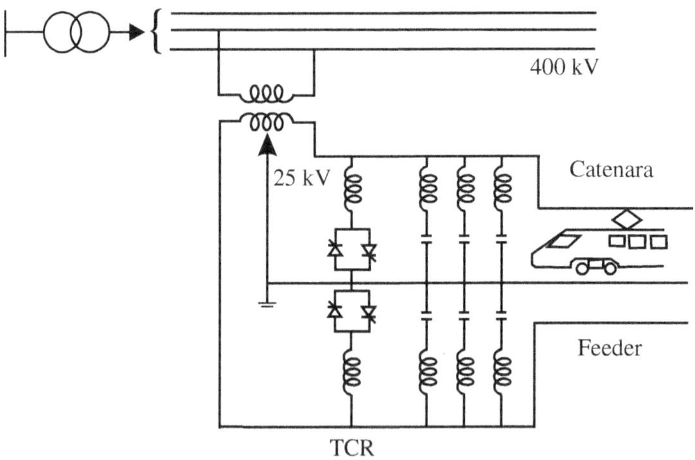

Figure 5.58 VAr compensating system for the Channel Tunnel rail link. *Source*: Grünbaum et al. 2003 [26]. Reproduced with permission of ABB.

of 10 MW. The traction feeding system is a modern 50 Hz, 2 × 25 kV supply incorporating an autotransformer scheme to keep the voltage drop along the traction lines low. Power step-down from the grid is direct, via transformers connected between two phases. A major feature of this power system, shown in Figure 5.58, is the SVC support. The primary purpose of VAr is to balance the unsymmetrical load and to support the railway voltage in the case of a feeder station trip—when two sections have to be fed from one station. The second purpose of the SVCs is to ensure a low tariff for the active power by maintaining unity power factor during normal operation. Thirdly, the SVCs alleviate harmonic pollution by filtering the harmonics from the traction load [26].

Harmonic compensation is important because strict limits apply to the traction system's contribution to the harmonic level at the supergrid connection points. The SVCs for voltage support only are connected on the traction side of the interconnecting power transformers. The supergrid transformers for the traction supply have two series-connected medium-voltage windings, each with its midpoint grounded. This results in two voltages, 180° apart, between the winding terminals and ground. The SVCs are connected across these windings; consequently, there are identical single-phase SVCs connected feeder to ground and catenary to ground. The traction load up to 120 MW is connected between two phases. Without compensation, this would result in an approximately 2% negative phase sequence voltage. To counteract the unbalanced load, a load balancer (an asymmetrically controlled SVC) has been installed in the Sellindge substation. This has a three-phase connection to the grid.

5.9.8 SVC at Harker, in United Kingdom

Two SVC s with a nominal rating of –75 to +150 MVAr each have been installed at Harker substation in the north of UK grid near the Scottish border in 1992 [27]. The main task of these SVCs is not only voltage control, but also a stability controller for damping of system power oscillations is implemented.

The National Grid Company (NGC) owns and operates the 400 and 275 kV transmission system in England and Wales. The NGC transmission system is interconnected with the power system in Scotland by means of two double-circuit lines.

The SVCs at Harker were specifically installed for the purpose of increasing the transient and dynamic stability margins to meet the appropriate planning standards when designing for an increase in the power transfer from Scotland to England.

The basic control requirement for the Harker SVCs was for constant voltage control with the SVCs to be operated at zero output, so as to have full capability of the SVCs available for transient stability enhancement.

The real power flow across the tie-lines between Scotland and England was chosen for the input signal for the POD controller, based on the experience that there are light damped power oscillations with a frequency of about 0.5 Hz, which is the lowest electro-mechanical frequency seen in the system. This mode of oscillation is strongly present in the tie-line power flow.

The curves in Figure 5.59 show the calculated system response following a three-phase fault east of Harker cleaned by line tripping. The power flow in the tie-line as well as the output of one SCV is shown for the case that no SVC is in operation, that the SVCs operate in voltage control mode only, and the POD controller is active additionally. While improving the stability performance also by controlling the voltage, a distinctly more rapid damping can be achieved with the POD control activated.

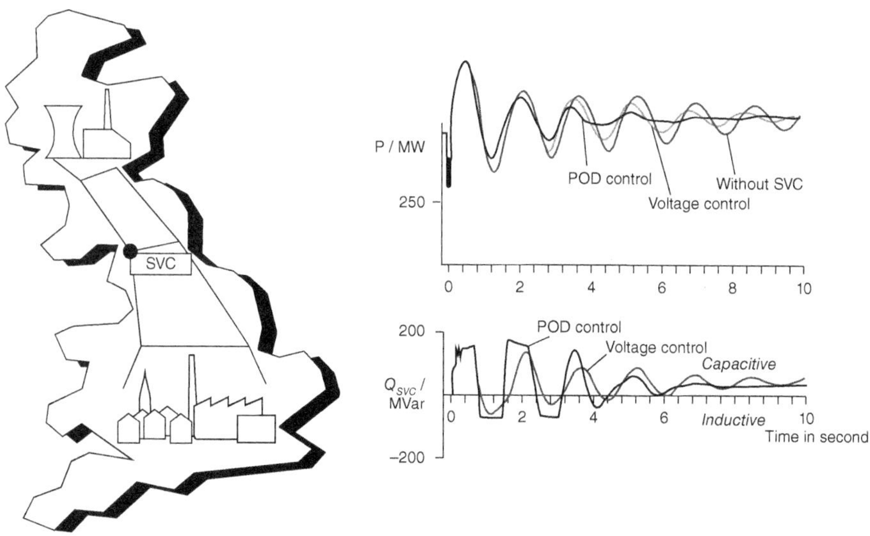

Figure 5.59 UK grid and system response to a three-phase fault with and without SVC–POD control. *Source*: Dwek 2001 [27]. Reproduced with permission of CIGRE.

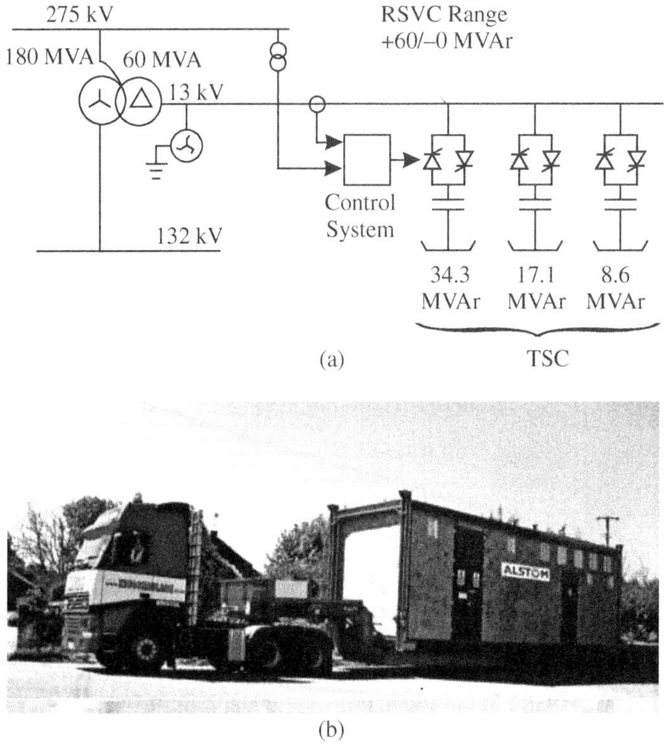

Figure 5.60 RSVC: (a) single-line diagram and (b) cabin. Courtesy of GE.

5.9.9 Relocatable SVCs

It can be very difficult for system planners to forecast future load growth, and the corresponding future development and expansion of their supply network, for more than a few years ahead. The installation of new generating plant or high voltage overlays may greatly strengthen weak points and render some SVCs redundant after only a few years of service. Other developments may result in the closure of some generating plants and a consequent weakening of the network at that point. The SVCs may then not be well placed to help the network in its hour of need [27].

In general, SVCs have been installed by utilities as fixed substation plant, to meet a perceived long-term need, with a typical anticipated lifetime of 25–30 years or more. Although it is always possible to move SVCs to another site, as has been done in South Africa, their design and layout are usually not arranged so that this relocation is a simple task.

In order to simplify relocatability, SVC installations need to be compact, avoiding permanent buildings; outdoor equipment needs to be arranged in-groups of components that are capable of being carried, by road or rail, with a minimum of dismantling. Packed substations have been used for many years and the use of transportable cabins to provide a weatherproof housing for sensitive components is a similar concept [27]. Thus, the thyristor valves, controls, protection, auxiliary power sources, etc., need to be inside a cabin, whereas reactors, capacitors, switchgear, auxiliary/earthing transformers, etc., can be mounted outdoors on easily transportable skids of frameworks, complete with many of their interconnecting busbars. NGC developed a specification for standardized relocatable SVCs, which can easily and quickly be moved from one substation to another, as system needs dictate, without requiring dedicated transformers and without requiring any design changes to match the system parameters at their new location. This system parameter can vary over a very wide range.

REFERENCES

[1] Hingorani, N. G., and Gyugyi, L. *Understanding FACTS. Concepts and Technology of Flexible AC Transmission Systems*. IEEE Press, New York, 2000.

[2] Sannino, A., and Svensson, J. *Part 3: Reactive Power Compensation – Power Electronics 2*. EEK, Chalmers University of Technology, January 2004.

[3] Mathur, R. M., and Varma, R. K. *Thyristor – Based FACTS Controllers for Electrical Transmission Systems*. IEEE Press, Wiley Interscience, New York, 2002.

[4] Eremia, M., Trecat, J., and Germond, A. *Réseaux electriques. Aspects actuels*. Editura Tehnică, Bucureşti, 2000.

[5] Kundur, P. *Power System Stability and Control*. McGraw-Hill, NY, 1994.

[6] Taylor, C. W. (chairman), Scott, G., Hammad, A., Wong, W., Osborn, D., Fkmos, A. J. P., Johnson, B., Mc Nabb, D., Arabi, S., Martin, D., Thanawala, H. L., Luini, J., Gonzalez, R., Concordia, C. Static VAr compensator models for power flow and dynamic performance simulation. *IEEE Special Stability Controls Working Group, IEEE Transactions on Power Systems*, vol. 9, no. 1, pp. 229–239, February 1994.

[7] Erinmez, I. A. (chairman), Jervis, W. B., Gavrilovic, M., Gyugyi, L., Pasternack, B., McGranaghan, F., Lasseter, R. H., Norman, H. B., Suzuki, H., Christensen, J. F., Leost, J. Y., Compagnion, C., Reichert, K., Meringdal, P., Olsen, K., Frank, H., Engberg, K., Testi, G., Nicola, G., Thanawala, H. L., and Petry, L. Static VAr compensators. CIGRE Working Group 38-01, Task Force No. 2 on SVC, Paris, 1986.

[8] Song, Y. H., and Johns, A. T. (Eds.) *Flexible AC Transmission Systems (FACTS)*. IEE Power and Energy Series, London, 1999.

[9] Therond, P. G. (Convenor), Adielson, T., Bortoni, P., Byrne, D., Ceresoli, Daniel, D., Foss, A. M., Fosso, O. B., Iyoda, I., Mihalic, R., Miller, N. W., Modlitba, P., Nelson, R., Thanawala, H. L., Uhlen, K., and Witzmann, R. Modeling of power electronics equipment (FACTS) in load flow and stability programs. CIGRE Task Force 38.01.08, Brochure 145, August 1999.

[10] Fuerte-Esquivel, C. R., Acha, E., and Ambriz-Perez, H. Integrated SVC and step-down transformer model for Newton Raphson load flow studies. *IEEE Power Engineering Review*, pp. 45-46, February, 2000.

[11] Canizares, C. A. Power flow and transient stability models of FACTS controllers for voltage and angle stability studies. *Proceedings of 2000 IEEE/PES Winter Meeting*, Singapore, January, 2000.

[12] Pourbeik, P., Sullivan, D. J., Boström, A., Sanchez-Gasca, J., Kazachkov, Y., Kowalski, J., Salazar, A., Meyer, A., Lau, R., Davies, D., and Allen, E. Generic model structure for simulating static VAr systems in power system studies. *A WECC Task Force Effort, IEEE Transactions on Power Systems*, vol. 27, no. 3, pp. 1618–1627, August 2012.

[13] Gerbex, S. *Metaheuristiques appliquées au placement optimal de dispositifs FACTS dans un réseau électrique – Thèse de doctorat*. Ecole Polytechnique Fédérale de Lausanne, Lausanne, 2003.

[14] Gyugyi, L. Power electronics in electric utilities: static VAr compensators. *Proceedings of the IEEE*, vol. 76, no. 4, pp. 483–494, April 1988.

[15] Le Du, A. Réflexion sur l'insertion des FACTS dans un réseau d'interconnexion. CIGRE SC37, Melbourne 1991.

[16] Cholley, P., Lacoste, J., and Trotignon, M. Dynamic shunt compensation performance for transient stability enhancement. EDF, DER, 95NR00104.

[17] Aboytes, F., Arroyo, G., and Villa, G. Application of static VAr compensators in longitudinal power systems. *IEEE Transactions on Power Apparatus and Systems*, vol. PAS-102, no. 10, pp. 340–3466, October 1983.

[18] Lindström, C. O., Walve, K., Waglund, G., Frank, H., and Ivner, S. The 200 MVAr static compensator in Hagby, Sweden. Report 38.02, CIGRE, SC 38, 1984.

[19] ABB. SVC for increased power interchange capability between Canada and USA. ABB Technical Brochure, A02-0147 E, 2011-03.

[20] ABB. Reactive power compensation: SVC and series capacitors for transmission. ABB Power Systems, 1995.

[21] ABB. SVC to increase the transmission capacity and stability of the Mexican 400 kV system. ABB Technical Brochure, A02-0105 E, 2011-01.

[22] ABB. The first SVC installed in the Argentinean 500/220 kV transmission system. ABB Technical Brochure, A02-0119 E, 2011-01.

[23] ABB. SVCs to stabilize a large 735 kV transmission system in Canada. ABB Technical Brochure, A02-0106 E, 2012-02.

[24] Gene, W. Grid FACTS: building blocks for the grid. ABB AB, Transmission & Distribution World, pp. 14-20, July 2014. Available at www.tdworld.com

[25] Grünbaum, R., Halonen, M., and Rudin, S. Power factor, ABB static VAr compensator stabilizes Namibian grid voltage. ABB Review (Special report on Power Technologies), pp. 19–24, 2003.

[26] Grünbaum, R., Petersson, A., and Thorvaldsson, B. FACTS improving the performance of electrical grids. ABB Review (Special report on Power Technologies), pp. 13-18, 2003.

[27] Dwek, M.G. (convenor), Cleobury, E. G., Eremia, M., Loughran, J., Pramayon, P., Witzmann, R., Weber, T., Stoa, P., Halvarsson, P., Kula, M., Gama, C., and Taylor, C. W. FACTS technology for open access. CIGRE Joint Working Group 14/37/38/39.24, Technical Brochure, April 2001.

CHAPTER 6

SERIES CAPACITIVE COMPENSATION

Mircea Eremia and Stig Nilsson

6.1 GENERALITIES

Series capacitive compensation method is very well known and it has been widely applied on transmission grids; the basic principle is capacitive compensation of portion of the inductive reactance of the electrical transmission, which will result in increased power transfer capability of the compensated transmissible line. Series capacitive compensation is very widely used because it is very inexpensive. Inserting a capacitive reactance in series with the long—typically more than 200 km—transmission line reduces both the voltage drop and the angular deviation between supplying generator and the receiving end. Series compensation can provide increased transmission capacity, improved voltage profile of the grid, enhanced angular stability of power corridor, damping of power oscillations, and optimizing power sharing between parallel lines [1].

Thyristor-controlled series capacitors (TCSCs) introduce a number of important benefits in the application of series compensation which are as follows: elimination of subsynchronous resonance (SSR) risk, damping of active power oscillations, postcontingency stability improvement, and dynamic power flow control.

6.2 MECHANICAL COMMUTATION-BASED SERIES DEVICES

The origin of the flexible system idea is the classical methods, for example, *insertion of a series capacitor (SC) into the electrical line* (Figure 6.1), for improving the steady-state stability (by increasing the synchronizing power P_e) and/or the reduction of overvoltages. The first series-compensated transmission line was put into operation in Sweden in 1952.

Insertion of one capacitor C in series with a transmission line results in an apparent reduction of the line reactance, that is, $X_{equiv.} = X_L - X_C$ (Figure 6.1c), and

Advanced Solutions in Power Systems: HVDC, FACTS, and Artificial Intelligence, First Edition.
Edited by Mircea Eremia, Chen-Ching Liu, and Abdel-Aty Edris.

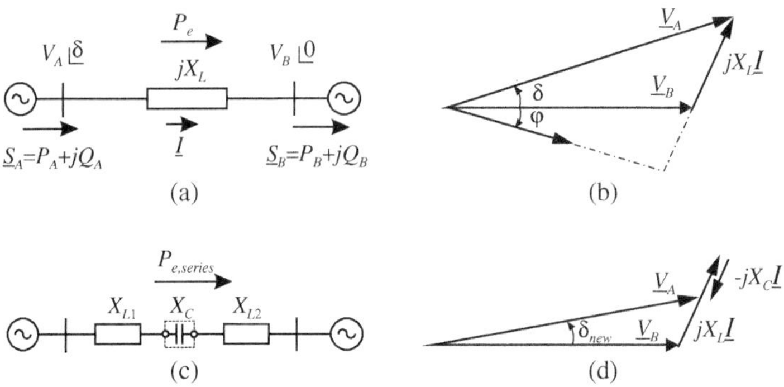

Figure 6.1 Modification of the line reactance by insertion of a capacitor bank: (a) one-line diagram, (b) phasor diagram, (c) one-line diagram with an inserted capacitor, and (d) phasor diagram with contribution of the capacitor.

increase in the transmitted power, respectively [2]:

$$P_{e,series} = \frac{V_A V_B}{X_L - X_C} \sin\delta > P_e \tag{6.1}$$

A detailed presentation of a series capacitive compensation scheme is shown in Figure 6.2 [1].

One or more SC banks can be installed in each phase. The capacitor banks themselves consist of parallel- and series-connected capacitors. In series with each group is mounted a current transformer that feeds signals to an unbalance protection system to detect faulted capacitors. These components together with other auxiliary components are placed on a platform insulated from ground. Metal oxide varistor (MOV)

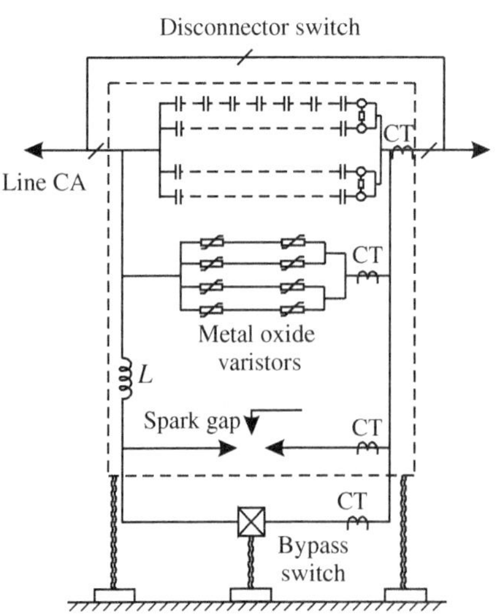

Figure 6.2 The scheme of the series capacitor compensation installation [1].

arresters, current transformers, spark gap, and a bypass breaker are installed in parallel with the SC bank. An inductor L, which limits the current magnitude through the capacitor when the spark gap is ignited, is mounted in the bypass circuit. In normal operating conditions, the current flows through the SC. Only a small amount of current flows through the varistors, but the spark gap is not conducting and the bypass switch is open. If a fault occurs, the bypass switch will be closed.

The main spark gap is triggered at a certain voltage level to protect the capacitors from severe overvoltage during a fault on the compensated transmission line. The voltage values for triggering the gap are between 2.5 and 4 times the operation voltage. A typical command sequence is the following: the current transformer from the spark gap circuit detects the current in the circuit and sends a signal to close the bypass switch. When the bypass switch is closed, the current is redirected from the spark gap circuit. At that moment when the line current, measured by the current transformer from the bypass switch circuit, falls back to its normal value, a signal is given for automatic opening of the bypass switch, which accomplishes the reinsertion of the SC.

Overvoltage protection is designed to protect the capacitor bank in the case of a sustained overload, but below the level of operation of the spark gap. When needed, a command is given to close the bypass switch, which will prevent overstressing of the SCs. The bypass switch is automatically opened to reinsert the SC, once the current falls back to its normal level.

The MOVs are typically used for limitation of overvoltage on the SCs.

The major disadvantage of this scheme is the lack of flexibility because the capacitor banks are, in general, fixed and with no adjusting possibilities.

A *scheme with adjusting possibility* is presented in Figure 6.3. In this scheme, a set of capacitor bank modules are connected in series. The adjustment of the series capacitive compensation is achieved through the bypassing or switching in respective modules using mechanical switches. The adjustment is, therefore, slow sometimes manually initiated. Such bypassing and switching operations result in electromagnetic transients that may result in overvoltage and resonance situations.

Furthermore, with compensation degrees $k_s = X_C/X_L = 25\% \ldots 70\%$, series resonance frequency f_{series} of the series circuits (transmission line reactance and the SC) is below the power frequency:

$$f_{series} = f\sqrt{X_C/X_L}$$

Consequently, the oscillatory mode of the compensated line will then be in the subsynchronous oscillatory region, which at certain condition may result in SSR situations.

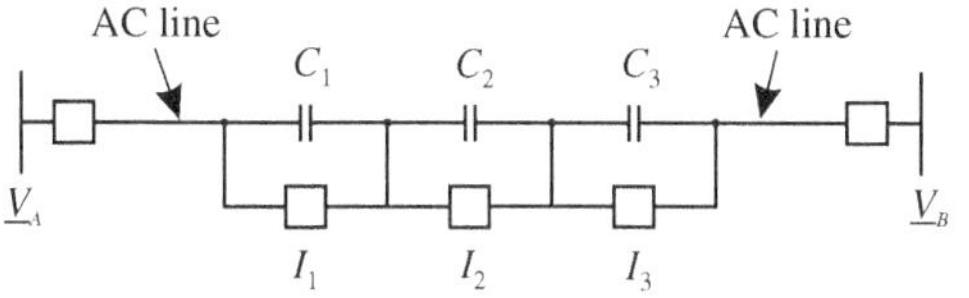

Figure 6.3 Step regulation with capacitor banks controlled by mechanical switches.

Series compensation of long transmission lines may cause the following types of oscillations [2]:

(i) subsynchronous oscillations, caused by the interaction between the electrical network and the generator-shaft torsional system;

(ii) low frequency, about 5–10 Hz, oscillations, caused by interaction between the SCs and the shunt inductors, especially during line switching and faults. These oscillations have large amplitudes and last for long periods because of the high shunt-reactor $\underline{Q}$-factors;

(iii) switching oscillations, caused by the switching of lines.

6.3 STATIC-CONTROLLED SERIES CAPACITIVE COMPENSATION

This series compensator is a reciprocal of the shunt compensator. The shunt compensator is functionally a controlled reactive current source shunt connected on the transmission line to control its voltage. The series compensator is functionally a controlled voltage source connected in series with the transmission line to control its current. This reciprocity suggests that both the admittance and voltage source-type shunt compensators have a corresponding series compensator [2].

The series compensator can be implemented either as a variable reactive impedance or as a controlled voltage source in series with the line.

Just as in reactive shunt compensation, variable impedance-type series compensators are composed of thyristor—switched/controlled capacitors or thyristor-controlled reactors (TCRs) with fixed capacitors.

6.3.1 GTO-Controlled Series Capacitor

The GTO-controlled series capacitor (GCSC) device (GTO TCSC), proposed by a team lead by George Karady in 1992, consists of a fixed capacitor and a pair of GTO switches connected in anti-parallel. To insert the capacitor in the circuit, the GTOs are turned off. To remove the capacitor from the circuit, the GTOs are turned on just before the capacitor voltage becomes zero. The result is that the capacitor is in the circuit for a part of the cycle. When the switch is turned on, the compensation is zero. When the switch is turned off, the capacitor compensates for a part or all of the circuit inductance. The switch-off angle (insertion angle) controls the level of the compensation [3].

The controlled SC-circuit shown in Figure 6.4a is the dual of the TCR, which is operated from a voltage source (see Chapter 5). The proposed circuit is supplied by a current source. The TCR has an inductor in series with the control switch, and controls the current drawn from a constant voltage source by a fixed inductor.

The GCSC device has a capacitor in shunt with the control switch, and controls the voltage, v_C, developed by a constant current (CC) source across a fixed capacitor.

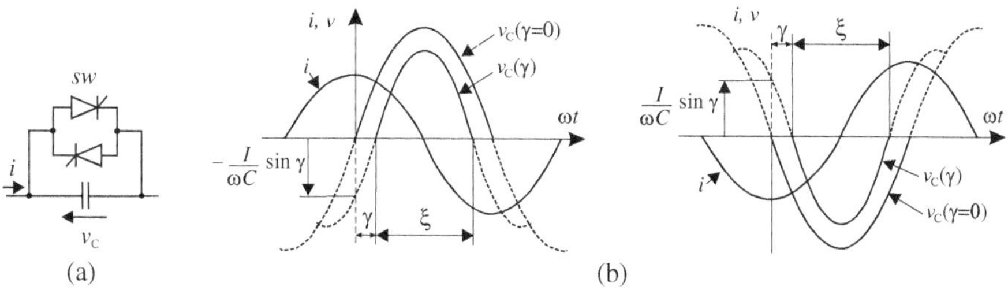

Figure 6.4 GTO thyristor-controlled series capacitor: (a) simplified scheme and (b) operating principle of GTO with firing angle γ in a positive and negative half-cycle. *Source*: Hingorani & Gyugyi 2000 [2]. Reproduced with permission of IEEE.

The main objective of GCSC device is to control the AC voltage, v_c, which is across the capacitor, at a certain line current *i*. When the GTO thyristor is closed, the voltage at capacitor terminals is zero, and when it is opened the voltage reaches the maximum value.

To control the v_c voltage, the switching *on* and *off* of the GTO thyristor is produced in every half-cycle in synchronism with the power system frequency. The GTO valve is stipulated to close automatically whenever the capacitor voltage crosses zero. The closing instant of GTO thyristor is defined by a firing angle γ ($0 \leq \gamma \leq \pi/2$) measured from the instant when the current *i* reached the maximum value.

In Figure 6.4b, the line current, *i*, and the capacitor voltage, $v_c(\gamma)$, are shown at $\gamma = 0$ (valve open) and at an arbitrary turn-off delay angle γ, for a positive and a negative half-cycle. If the valve sw (switch) is opened at the crest of the line current (when $\gamma = 0$), the capacitor voltage v_c will be the same as that obtained in steady state when the valve is permanently open.

Delaying the firing angle γ with respect to the crest of the line current $i(t) = I\cos\omega t$, the capacitor voltage can be expressed by the equation [2, 3]:

$$v_c(t) = \frac{1}{C}\int_{\gamma}^{\omega t} i(t)\cdot dt = \frac{I}{\omega C}(\sin\omega t - \sin\gamma) \tag{6.2}$$

If the valve opens at γ angle and closes at the first voltage zero, then equation (6.2) is true for the interval $\gamma \leq \omega t \leq \pi - \gamma$. For the subsequent positive half-cycle, the same expression remains valid, but for the subsequent negative half-cycle intervals, the sign of the terms in equation (6.2) becomes opposite.

In equation (6.2), the term $(I/\omega C)\sin\gamma$ is simply a γ-dependent constant by which the sinusoidal voltage obtained at $\gamma = 0$ is offset, shifted down for positive and up for negative voltage half-cycles (Figure 6.4b), and up for negative voltage half-cycles. The magnitude of the capacitor voltage can be varied continuously by this method of turn-off delay angle control from maximum ($\gamma = 0$) to zero ($\gamma = \pi/2$).

Comparison of Figure 6.4b to Figure 5.2c shows that the waveshape obtained for the current of the TCR is identical to that derived above for the voltage GCSC and confirms the duality between the GCSC and the TCR [2].

This duality makes it possible to use analytical results given in (5.9), (5.10), and (5.12′), and to extend at the GCSC and related series capacitive compensators.

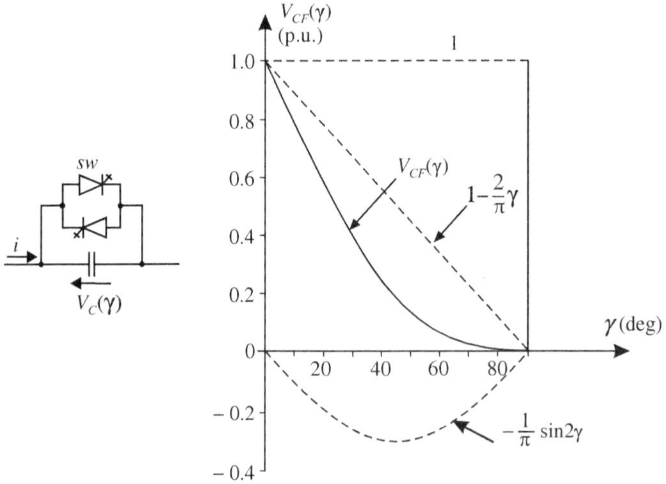

Figure 6.5 Fundamental component of the series capacitor voltage versus the turn-off delay angle γ. *Source*: Hingorani & Gyugyi 2000 [2]. Reproduced with permission of IEEE.

The amplitude $V_{CF}(\gamma)$ of the fundamental capacitor voltage $v_{CF}(\gamma)$ can, therefore, be expressed as a function of angle γ by the following equation:

$$V_{CF}(\gamma) = \frac{I}{\omega C}\left(1 - \frac{2}{\pi}\gamma - \frac{1}{\pi}\sin 2\gamma\right) \tag{6.3}$$

where

I is the amplitude of the line current;

C – the capacitance of the GTO thyristor-controlled capacitor;

ω – the angular frequency of the AC system.

The variation of the amplitude $V_{CF}(\gamma)$, in p.u. (normalized to the maximum voltage $V_{CF\,\max} = I/\omega C$), is shown in Figure 6.5 versus the delay angle γ.

Figure 6.5 reveals that varying the fundamental capacitor voltage at a fixed line current is equivalent to a variable capacitive compensation. In this way, the effective capacitive reactance, X_C, as a function of angle γ, can be expressed as follows:

$$X_C(\gamma) = \frac{1}{\omega C}\left(1 - \frac{2}{\pi}\gamma - \frac{1}{\pi}\sin 2\gamma\right) \tag{6.4}$$

The reactance $X_C(\gamma)$ varies with the angle γ in the same manner as the fundamental capacitor voltage $V_{CF}(\gamma)$.

Practically, the application of the GCSC can be operated in two different control modes [2]:

(i) *Voltage compensation mode*, $V_{CF}(\gamma)$. In this case, the GCSC device has to maintain the rated compensating voltage $V_{CF}(\gamma)$, while the line current is changing over a defined interval $I_{\min} \leq I \leq I_{\max}$. For this operating mode, the compensating voltage—line current characteristic is shown in Figure 6.6a.

The capacitive reactance X_C is selected so as to produce the rated compensating voltage with $I = I_{\min}$, that is, $V_{C\max} = X_C I_{\min}$. As current $I_{\min}$ is

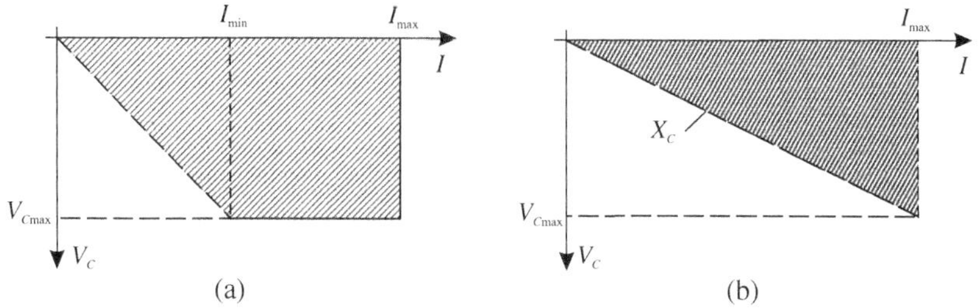

Figure 6.6 Compensating voltage—line current characteristics of the GCSC device when operated: (a) in compensating voltage control mode and (b) in compensating reactance control mode. *Source*: Hingorani & Gyugyi 2000 [2]. Reproduced with permission of IEEE.

increased toward I_{max}, the turn-off delay angle γ is increased to reduce the duration of the capacitor injection and thereby to maintain the compensating voltage with increasing line current.

(ii) *Impedance compensation mode*, $X_C(\gamma)$. The GCSC has to maintain the maximum rated compensating reactance at any line current up to the rated maximum. The compensating voltage—line current characteristic for this operating mode is shown in Figure 6.6b.

In this compensation mode, the capacitive reactance is chosen so as to provide the maximum series compensation at rated current, $X_C = V_{C\max}/I_{\max}$, so that the GCSC can vary in the $0 \leq X_C(\gamma) \leq X_C$ range by controlling the effective capacitor voltage $V_{CF}(\gamma)$, that is, $X_C(\gamma) = V_{CF}(\gamma)/I$.

The turn-off delay angle control of the GCSC, just like the turn-off delay angle control of the TCR, generates harmonics. For identical positive and negative voltage half-cycles, only odd harmonics are generated. The amplitudes of these harmonics are a function of angle γ and can be expressed by the following expression [2, 3]:

$$V_{Cn}(\gamma) = \frac{4}{\pi} \cdot \frac{I}{\omega C} \left[\frac{\sin\gamma \cos(n\gamma) - n\cos\gamma \sin(n\gamma)}{n(n^2 - 1)} \right] \tag{6.5}$$

where $n = 2k + 1$, $k = 1, 2, 3 \ldots$

The magnitude of odd harmonics, expressed as a percentage of the maximum fundamental capacitor voltage, is shown plotted versus angle γ in Figure 6.7; the magnitude decreases as the harmonics order increases.

Note: The drawback of the scheme is that the GTO cannot handle short-circuit currents up to the level of a regular thyristor.

6.3.2 Thyristor-Switched Series Capacitor

The simplified scheme of the thyristor-switched series capacitor (TSSC) device consists of a number of capacitors, inserted in series with the compensated transmission line, each capacitor is shunted by an appropriately rated bypass valve composed of a string of anti-parallel connected thyristors (Figure 6.8) [1–4].

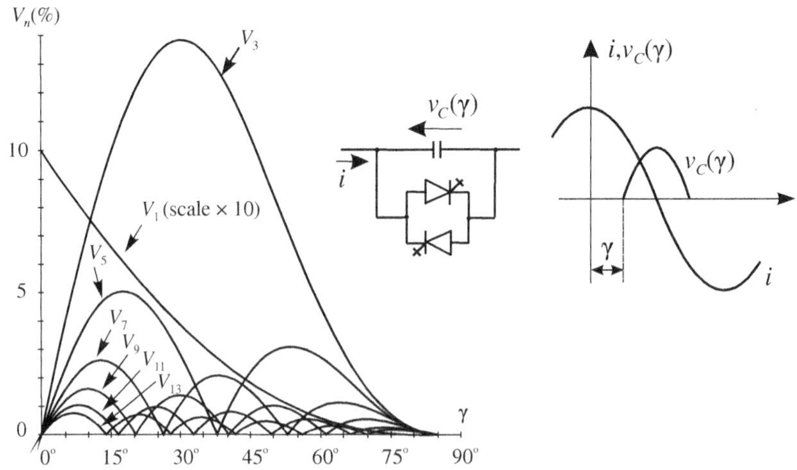

Figure 6.7 The amplitudes of the harmonic voltages, expressed as percents of the maximum fundamental capacitor voltage versus the turn-off delay angle γ. *Source*: Hingorani & Gyugyi 2000 [2]. Reproduced with permission of IEEE.

As seen, TSSC is similar to the circuit structure of the sequentially operated GCSC, but its operation is different due to the imposed switching restrictions of the conventional thyristor valve.

The operating principle of the TSSC is straightforward: the degree of series compensation is controlled in a stepwise manner by increasing or decreasing the number of the inserted capacitors. A capacitor is inserted by blocking the gating signals to the anti-paralleled thyristor valves, and it is bypassed by fully conducting the thyristor valves.

A thyristor valve commutates "naturally", that is, it turns off when the current crosses zero. Thus, a capacitor can be inserted in series with the transmission line by the thyristor valve only at the zero crossings of the line current. Since the insertion takes place at line current crossing zero, a full half-cycle of the line current will charge the capacitor from zero to maximum and the successive opposite polarity half-cycle of the line current will discharge it from this maximum to $v_c = 0$ (Figure 6.9) [2]. The same offset will arise if a capacitor module is switched by a mechanical switch. However, a thyristor can be turned on to short circuit the capacitor momentarily to drain the offset, which will reduce the time the capacitors are exposed to a 2 p.u. voltage.

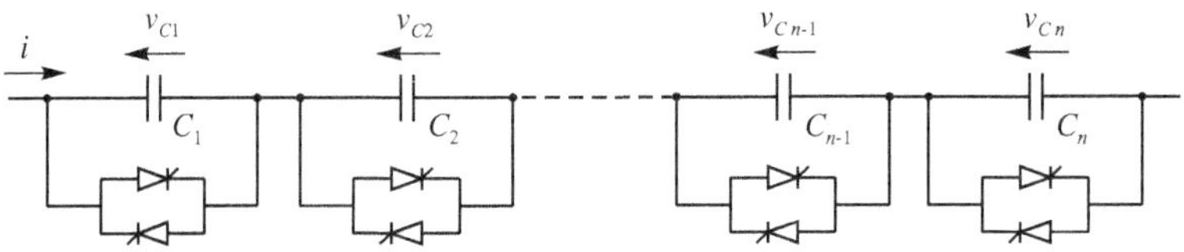

Figure 6.8 Basic TSSC scheme.

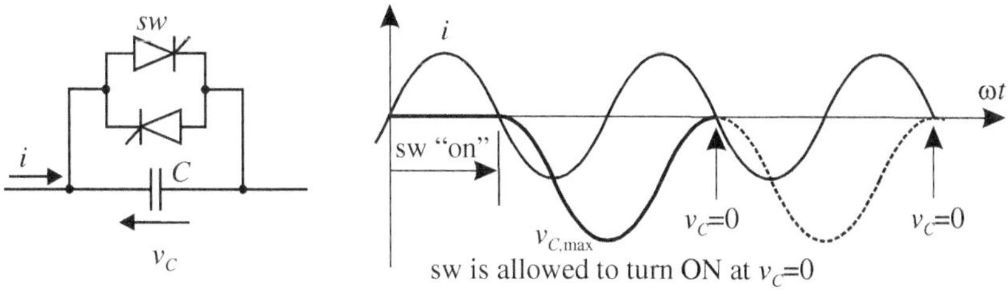

Figure 6.9 Illustration of capacitor offset voltage resulting from the restriction of inserting at zero line current. *Source*: Hingorani & Gyugyi 2000 [2]. Reproduced with permission of IEEE.

The TSSC can control the degree of series capacitive compensation by either inserting or bypassing SCs, but it does not change the natural characteristic of the classical SC-compensated line.

More compensating steps can be performed if different capacitive reactances are employed. One such scheme can have n-1 capacitors, each having a reactance of $0.5X_C/n$ and one capacitor with $0.5X_C$ reactance. Through an appropriate switching, different combinations between these reactances can be obtained [1]:

$$X_{ef} = \frac{0,5X_C}{n}p;\; p = 0, 1,\; 2, \ldots,\; 2n$$

The TSSC device has, similar to the GCSC, two operating modes [2]:

(i) ***Compensating voltage control mode***. In this mode, the total reactance of the n capacitor banks (n is the total number of the "capacitor-thyristor valve" modules) is chosen to produce the maximum compensating voltage ($V_{C\max} = nX_C I_{\min}$) when the line current varies between $I_{\min}$ and $I_{\max}$. The compensating voltage versus line current characteristic for this operating mode, for example, with $n = 4$, is shown in Figure 6.10a. As the current $I_{\min}$ is increasing toward $I_{\max}$, the capacitor banks are progressively bypassed by the respective valves to reduce the overall capacitive reactances in a stepwise manner and thereby maintains the compensating voltage with increasing line current.

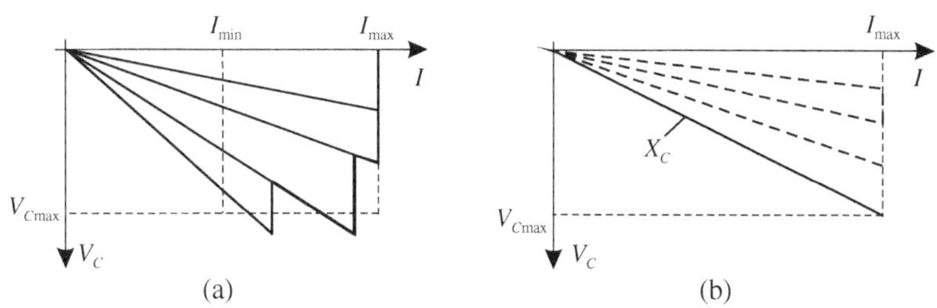

Figure 6.10 The compensating voltage versus line current characteristic for TSSC device with four modules: (a) compensating voltage control mode and (b) compensating reactance control mode [2, 5].

(ii) ***Compensating reactance control mode.*** The total capacitive reactance is chosen to obtain a maximum compensating voltage when the line current I varies between 0 and I_{max}. The compensating voltage versus line current characteristic for this operating mode (for the particular case with $n = 4$ modules) is shown in Figure 6.10b. The capacitive reactance is chosen such that the maximum series compensation at rated current, $4X_C = V_{C\,max}/I_{max}$, is provided, that is, TSSC can vary in a stepwise manner by bypassing one or more capacitor banks.

TSSC installation offers the following advantages compared with the classical compensation mode (fixed capacitors with mechanical switches):

- The commutation with thyristors allows unlimited number of switching operations. This capability allows utilization of the TSSC installation at any moment for increasing the degree of the line compensation and for having an efficient control regarding the steady-state operation.
- The utilization of thyristors allows choosing precisely the commutation instant (the point on the voltage waveform), thus minimizing the switching transients, in contrast to the classical commutation (based on mechanical switches) which is, normally, not synchronized.
- A very fast response, for example, the interval between the command impulse and the insertion of one capacitor in the line is less than a half-cycle (10 ms for 50 Hz). So that, in case of losing an important line, the power will be transferred to the TCSC-compensated line.
- It does not generate harmonics.
- The minimization of power losses in the conduction period for a TSSC installation is possible if it is used together with a very fast switch.
- If the TSSC system is configured as a series of small capacitor modules, then modules can be quickly inserted or bypassed if SSR frequencies were to be detected. This will detune the AC system and move it away from the critical turbine-generator resonance frequencies.

The TSSC compensation scheme provides stepwise series capacitive compensation.

6.3.3 Thyristor-Controlled Series Capacitor

The TCSC is essentially a series-connected static VAr compensation (SVC) system. TCSC consists of the SC shunted by a small TCR. TCSC provides a continuous control of series capacitive compensation degrees. In practical TCSC implementation, several such basic compensating modules may be connected in series to obtain the desired voltage rating and operating characteristics. This was the idea behind the Slatt system installed at Bonneville Power Administration (BPA) (see Section 6.7.3). This arrangement is similar in structure to the TSSC and, if the impedance of the reactor (X_L) is sufficiently smaller than that of the capacitor (X_C), it can be operated in an on/off manner like the TSSC [2].

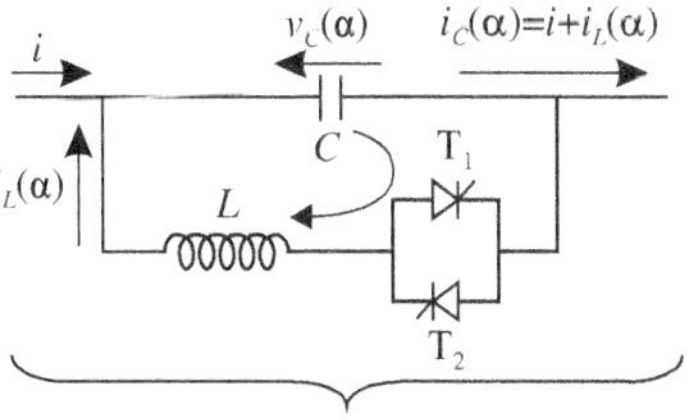

Figure 6.11 Basic structure of a TCSC module.

6.3.3.1 Basic Structure

The basic structure of the TCSC scheme was, first, proposed in 1986 by a research team lead by J.J. Vithayathil as a method of "rapid adjustment of network impedance (RANI)", which improved transient stability and controlled loop flows [6,7]. General information on the TCSC scheme has been presented in References [7] and [8].

The basic model consists of a SC connected in parallel with a TCR (Figure 6.11).

Proper selection of the thyristor firing angle increases the equivalent series compensation by developing a "loop flow" current through the SC. The thyristors are switched on at the zero current crossings to minimize the generation of harmonics. A detailed analysis on the TCR /fixed reactor scheme (rather than the SC) was presented by Vithayathil in paper [7][1], including the derivation of the equivalent impedance as a function of the thyristor firing angle:

$$X = \frac{\pi X_L}{(\sigma - \sin\sigma) + \pi \dfrac{X_L}{X_T}} \tag{6.6}$$

where $\sigma = 2\pi - 2\alpha$, X_T is the fixed impedance (including filters), and X_L is the controlled inductance.

In addition to the above equation, Reference [8] presented the equation:

$$X(\alpha) = X_C \frac{V_C(\alpha)}{I_{line}} \tag{6.7}$$

where X_C is the nominal capacitor reactance of the capacitor, $V_C(\alpha)$ is the capacitor voltage in p.u. of the nominal voltage, and I_{line} is the line current in p.u. of the nominal current.

Equation (6.6), which was derived for the case of a TCR connected in parallel with a fixed reactor, is valid only for approximately one cycle when the TCR is connected in parallel with a fixed capacitor due to the ability of the capacitor to store charge on the capacitor. This "stored charge" will increase the steady-state voltage on the capacitor, and thus the overall impedance of the TCSC element. Therefore, the equation for the overall impedance that understates the steady-state

[1] Material reprinted with permission from CIGRE "*Case studies of conventional and novel methods of reactive power control on an AC transmission system*, Vithayathil, J.J., Taylor, C.W., Klinger, M., Mittelstadt, W.A., CIGRE Paper 38-02, 1986", © CIGRE 1986.

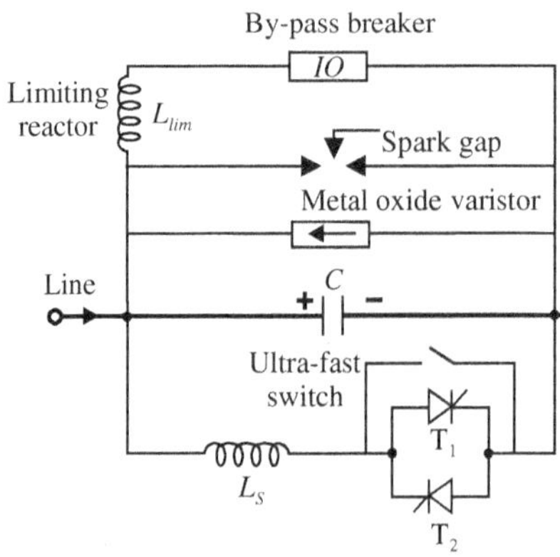

Figure 6.12 Protection equipment for the series capacitor.

impedance is introduced in [9]. This paper has derived a set of equations that enables a detailed study of the system operation of an advanced scheme for series compensation (TCSC). Using an iterative process to solve the equations developed, the circuit's steady-state condition can be calculated on a cycle-by-cycle basis.

The module of TCSC is also equipped with protection equipment for the series capacitor (SC) (Figure 6.12).

Thus, a variable nonlinear resistor (arrester) with MOV is connected in parallel with the SC for limiting overvoltages on the capacitor terminals. The MOV varistor (arrester) does not only limit the voltage at capacitor terminals, but also allows fast reinsertion of the capacitor banks in the circuit in case of disturbances, which will result in improving transient stability of the system by providing maximum synchronizing torque for the generators in the AC system after the AC system short circuit is cleared.

The SC is also shunted a bypass breaker (IO), which has the function to control the series connection of the capacitor. At the same time, IO has the function of bypassing the capacitor in the case of major fault. A current limiting reactor, L_{lim}, is connected in the circuit to limit the magnitude and frequency of the current that passes through the SC when operated in "bypass through thyristor" regime.

If the TCSCs have to operate in "full conduction" mode for a long period, the power losses are minimized through the installation of a breaker in parallel with the thyristor. This metallic breaker could be used to minimize the power losses. The breaker is closed a short time after the thyristor valve enters its full conduction mode. When the breaker is switched open, the thyristors will be turned on, which reduces the arcing time in the breaker upon opening. The breaker is normally open when the thyristor valves are blocked and the SC fully inserted. In case of overloaded thyristor valves, the breaker is closed for reduction of valve stresses. However, during AC system faults, the breaker should only be closed to prevent thyristor failures since if the TCSC is bypassed during the fault, it may not be able to operate quickly to help stabilize the AC system.

In general, a typical TCSC device is composed of several TCSC modules connected in series with a fixed capacitor C_F (Figure 6.13) [10, 11]. The principal

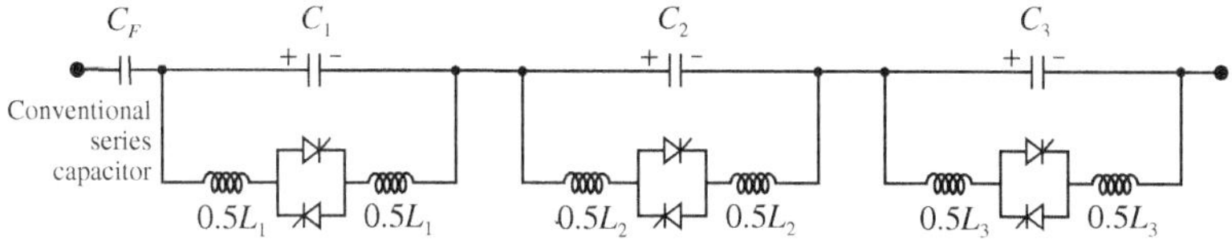

Figure 6.13 Series compensation scheme including three modules of TCSC.

role of the fixed capacitor is to reduce the overall cost of series capacitive compensation equipment. Each TCSC module is composed of a SC C_i in parallel with a TCR L_i.

The capacitors C_1, C_2, and C_3 used to realize different TCSC modules may have different capacitances so that TCSC has the possibility to control the value of capacitive reactance in a wide range of values. The reactor, which is in series with the thyristor valve, is split into two for protection in case of faults at the reactor level.

6.3.3.2 *Operating Principles of TCSC. Steady-State Approach and Synchronous Voltage Reversal*

The TCSC operation is similar to that of a parallel reactor-capacitor circuit (LC). The difference is that the current and the voltage are sinusoidal in the LC circuit, whereas in the TCSC circuit the voltage and the current through the fixed capacitor are not sinusoidal because of the thyristor commutations.

The idea of the TCSC basic scheme is to achieve an effective variable capacitive compensation. This is realized by partial compensation of capacitive reactance of the capacitor through the control of inductive reactance of the reactor connected in series with the thyristors.

The theoretical description of the TCSC naturally will form the base for the design of the control system. Basically, two approaches could be followed, one originating from a description of steady-state conditions, the other one from a description of transient conditions [12, 13].

STEADY-STATE APPROACH. Traditionally, thyristor converters have been described using the firing/trigger angle, α, as the control variable.

It is known that a TCR provides a continuous variable inductive reactance at the fundamental frequency of electrical network through firing angle, α control of the series thyristor valves.

The equivalent reactance of the parallel circuit is

$$X_{echiv}(\alpha) = \frac{X_C \cdot X_L(\alpha)}{X_L(\alpha) - X_C} = -j\frac{1}{\omega C - \dfrac{1}{\omega L}}$$

Thus, we can have the following three situations:

a. If $\omega C - (1/\omega L) > 0$, or $\omega L > 1/\omega C$, then the fixed capacitor impedance is smaller than the parallel variable reactor reactance and this combination provides a variable-capacitive impedance. Moreover, this reactor increases the equivalent capacitive reactance of the LC combination.

b. If $\omega C - (1/\omega L) = 0$, a parallel resonance situation is produced, which results in an infinite equivalent capacitive reactance, this should be avoided.

c. If $\omega C - (1/\omega L) < 0$, then the LC circuit results in an equivalent inductive reactance.

In real circuits, $X_L(\alpha)$ from the last equation is the inductive reactance of the TCR and is a function of the angle α in the equation [2]:

$$X_L(\alpha) = X_L \frac{\pi}{\pi - 2\alpha - \sin 2\alpha}, X_L \leq X_L(\alpha) \leq \infty \tag{6.6}$$

where $X_L = \omega L$, and the angle α is measured from the crest of the capacitor voltage (or, equivalently, the zero crossing of the line current).

Thus, the TCSC device behaves like a parallel tuned LC circuit:

- If the controlled reactor impedance $X_L(\alpha)$ varies from the maximum value (infinite) to the minimum value (ωL), then the TCSC device increases its capacitive reactance from the minimum capacitive value $X_{TCSC,\min} = X_C = 1/\omega C$ until the parallel resonance is established, that is for the $X_C = X_L(\alpha)$, so consequently the equivalent reactance $X_{TCSC,\max}$ becomes infinite.
- If $X_L(\alpha)$ starting to decrease, then the impedance $X_{TCSC}(\alpha)$ becomes inductive and reaches the minimum value $(X_L X_C)/(X_L - X_C)$ for $\alpha = 0$, when the fixed capacitor is in effect bypassed and the current circulated only through the TCR.

Therefore, for the usual TCSC arrangement in which the reactance of the TCR reactor, X_L, is smaller than the reactance of the capacitor, X_C, the TCSC has two operating ranges around the value α_r of the firing angle for which the internal circuit resonance is obtained: the first domain in which $X_{TCSC}(\alpha)$ is capacitive, for $\alpha_{C\,\mathrm{lim}} \leq \alpha < \pi/2$ and the second in which the $X_{TCSC}(\alpha)$ is inductive, for the $0 \leq \alpha \leq \alpha_{L\,\mathrm{lim}}$ (Figure 6.14).

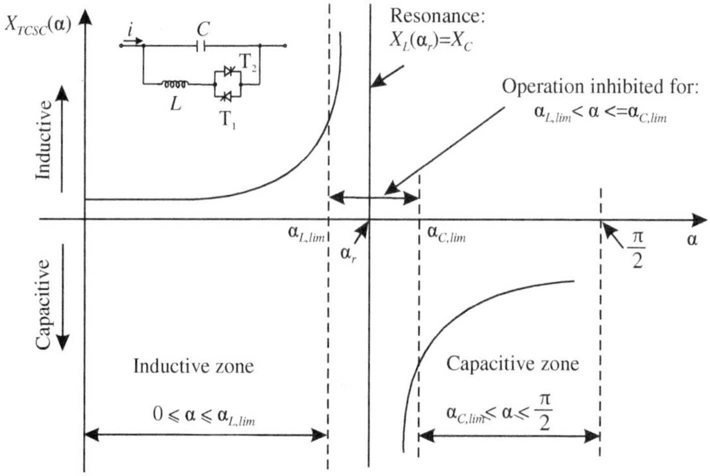

Figure 6.14 The impedance versus firing angle α characteristic of the TCSC.

The steady-state approach of the TCSC described above is based on the characteristics of the TCR element established for the SVC device, where the TCR is supplied from a constant voltage source. This model is useful to attain a basic understanding of the functional behavior of the TCSC. However, in the TCSC scheme, the TCR element is connected in shunt with a capacitor, instead of a fixed voltage source. The dynamic interaction between the capacitor and reactor changes the operating voltage curve from that of the basic sine wave established by the constant line current [2].

One way of describing the TCSC characteristics is to specify its *apparent impedance* X_{app}, that is, the quotient between the phasors representing the voltage across the TCSC ($\underline{V}_{C1}$) and the injected line current ($\underline{I}_1$) at rated frequency versus firing angle (see Section 6.3.3.3) [14, 15].

The steady-state characteristic is strongly nonlinear and exhibits an asymptotic behavior for at least one control angle. This asymptote corresponds to the firing angle for which the apparent reactance of the thyristor-controlled inductive equals the reactance of the capacitor. At this point, the apparent reactance of the TCSC becomes infinite. The control angle close to the asymptote must be avoided by the controller [12].

The steady-state description of the TCSC only involves the factor $\lambda(= \omega_n/\omega)$, which describes the ratio between the power frequency and resonance frequency of the combination of the thyristor branch inductor and the capacitor in the TCSC.

However, in transient conditions, the capacitor voltage (v_c) of the TCSC depends on the disturbance, the phase of line current change, the boost level, etc. [12]. Therefore, it appears that a generally applicable theory, valid in a wide frequency range, cannot be derived using the firing angle as the control variable.

The transient approach. Synchronous Voltage Reversal [12, 13]. An alternative description of the TCSC can be derived starting from the transient occurring when a thyristor is fired in the TCSC mode. Using a new control variable, it is possible to describe the TCSC in transient conditions in a way that is much easier to conceptually understand. It becomes possible to describe the TCSC, also transiently, decoupled from the external parameters. In this respect, the method used is the synchronous voltage reversal (SVR), which will be described below.

Figure 6.15 shows the waveforms associated with a TCSC (see Figure 6.11) when operating in steady state. The thyristor conduction intervals are substantially

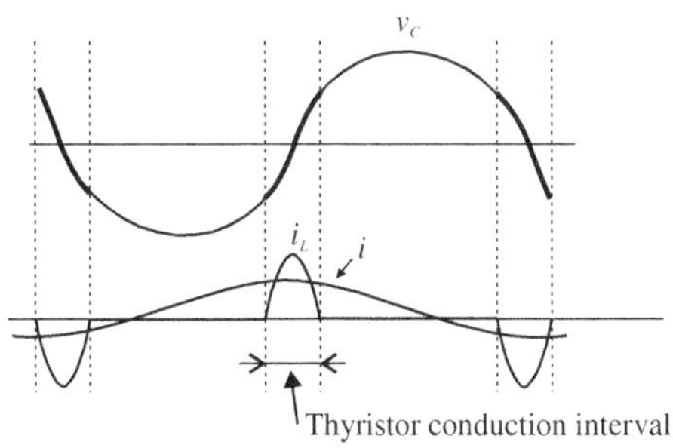

Figure 6.15 Waveforms associated with a TCSC when operating in steady state [12, 13].

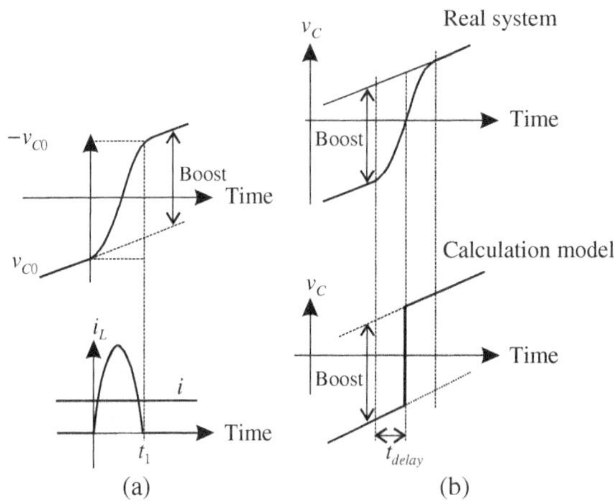

Figure 6.16 Capacitor voltage reversal: (a) real transient of capacitor voltage v_c, and line current (i) and thyristor current (i_L); (b) modeling of equivalent voltage reversal approximation [12, 13].

shorter than the network frequency half-cycle time, and they occur around the peak values of the line current.

Accordingly, the variation of the line current (i) during the thyristor conduction interval is rather small and the line current can be approximated by a constant current. With this approximation, the capacitor voltage (v_C) varies linearly before and after the conduction interval as shown in Figure 6.16.

Whenever the thyristor is triggered in the TCSC mode, a capacitor voltage reversal process is initiated. When the thyristor is fired, a current pulse passes through the thyristor and the capacitor voltage is repolarized. Ideally, with no losses, the thyristor ceases to conduct exactly when the capacitor voltage magnitude is equal to the initial magnitude at turn-on.

In other words, a capacitor voltage reversal occurs during the thyristor conduction interval. The charge passing through the thyristor branch brings about a parallel displacement of the lines representing the capacitor voltage before and after the conduction interval. This displacement represents the boost that is achieved by the thyristor action.

From the power system point of view, the amount of boost (i.e., the parallel displacement) is of particular interest, but the exact waveform of the transient is not. Figure 6.16b shows that the correct amount of boost can be obtained by an equivalent, instantaneous reversal of the capacitor voltage in the middle of the thyristor conduction interval.

It can be concluded that the capacitor voltage can be described adequately if the instants of these equivalent, instantaneous reversals of the capacitor voltage are known. It will be noted that the length of the thyristor conduction interval varies depending on the initial conditions, so the delay between thyristor turn-on and the equivalent reversal also varies.

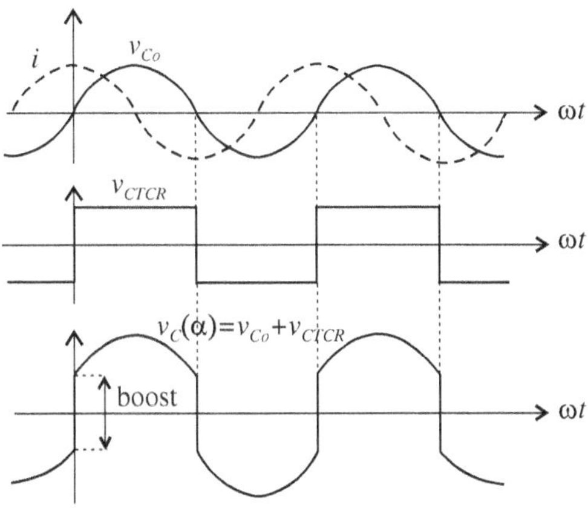

Figure 6.17 Composition of the idealized TCSC compensating voltage waveform from the line current produced (sinusoidal) capacitor voltage and the square wave voltage generated by capacitor voltage reversal. *Source*: Hingorani & Gyugyi 2000 [2]. Adapted with permission of IEEE.

The reversal of the capacitor voltage is clearly the key to the control of the TCSC. The time duration of the voltage reversal is dependent primarily on X_L/X_C ratio, as well as on the magnitude of the line current [2]:

- If $X_L \ll X_C$, then the reversal is almost instantaneous, and the periodic voltage reversal produces a square wave across the capacitor that is added to the sine wave produced by the line current. Thus, the steady-state compensating voltage across the SC comprises an uncontrolled and a controlled component. The uncontrolled component is v_{C0}, a sine wave whose amplitude is directly proportional to the amplitude of the prevailing line current, and the controlled component is v_{CTCR}, substantially a square wave whose magnitude is controlled through charge reversal by the TCR (Figure 6.17).
- For a finite, but still relatively small, X_L, the time duration of the charge reversal is not instantaneous, but is quite well defined by the natural resonant frequency, $f = 1/(2\pi\sqrt{LC})$, of the TCSC circuit, since the TCR conduction time is approximately equal to the half-period corresponding to this frequency: $T/2 = 1/2f = \pi\sqrt{LC}$. However, as X_L is increased relative to X_C, the conduction period of TCR increases and the zero crossings of the capacitor voltage become increasingly dependent on the prevailing line current.

Generally, the X_L/X_C ratio for practical TCSCs would likely be in 0.1–0.3 range, depending on the application requirements and constraints. It is important that the natural resonance frequency of the TCSC does not coincide with, or is close to, two and three times the fundamental [2].

The mechanism of controlling the DC offset by charge reversal is illustrated for the increase and decrease of the capacitor voltage in Figures 6.18b, 6.18c, 6.18d, and

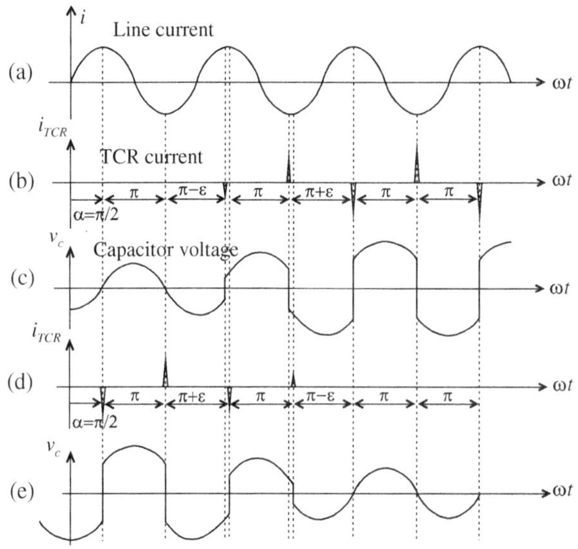

Figure 6.18 Mechanism of controlling the DC offset by charge reversal: (a) line current; (b,c) increase in the capacitor voltage by advancing the voltage reversal from $\alpha = \pi$ to $\alpha = \pi - \varepsilon$; (d,e) decrease in the capacitor voltage by retarding the voltage reversal from $\alpha = \pi$ to $\alpha = \pi + \varepsilon$. *Source*: Hingorani & Gyugyi 2000 [2]. Adapted with permission of IEEE.

6.18e, respectively; we consider the theoretically ideal case of instantaneous voltage reversal (with an infinite small X_L).

Initially, the TCR is gated on an $\alpha = \pi/2$, at which the TCR current is zero and the capacitor voltage is entirely due to the line current. To produce a DC offset, the periodically repeated gating in the second cycle is advanced by a small angle ε, that is, the prevailing half-period is reduced by ε to π-ε (Figures 6.18b and 6.18c).

This action produces a phase advance for the capacitor voltage with respect to the line current and, as a result, the capacitor absorbs energy from the line, charging it to a higher voltage. With the phase advance, the $\int idt$ is greater for the positive current segment than for the negative one in the half-period considered, and the resulting difference between these two integrals produces a net positive offset charge for the capacitor [2]. Should this phase advance be maintained, then the offset charge of the capacitor would keep increasing its charge at every half-cycle (Figure 6.18c). If the ε phase advance is negated, when the sufficient offset level of the capacitor voltage is reached, then the capacitor voltage at the desired magnitude can be maintained by continuing periodic gating at line current zeros ($\alpha = 0$).

In Figures 6.18d and 6.18e, the opposite process is illustrated, that is, when magnitude of the capacitor voltage is decreased by retarding the periodic gating from the current zeros until the desired offset voltage level is reached.

Thus, the TCSC boost level can be described from a control point of view as an integrating system that increases or decreases the boost proportionally to the time displacement of the zero-crossings of the capacitor voltage from their positions relative to the line current.

6.3.3.3 Operation Modes and the Characteristics of the TCSC

The characteristic of the TCSC main circuit depends on the relative reactances of the capacitor bank ($X_C = -1/\omega_n C$), and the thyristor branch $X_L = \omega_n L$, where ω_n is the fundamental angular speed, C is the capacitance of the capacitor bank, and L is the inductance of the parallel reactor.

The TCSC can operate in several different modes with varying values of apparent reactance, X_{app}. In this context, X_{app} is defined simply as the imaginary part of the quotient [14, 15]:

$$X_{app} = \text{Im}\left\{\frac{\underline{V}_{C1}}{\underline{I}_1}\right\}$$

where the phasors represent the fundamental value of the capacitor voltage $\underline{V}_{C1}$, and the line current $\underline{I}_1$, at rated frequency.

It is also practical to define a boost factor, K_B, as the quotient of the apparent and physical reactance, X_C, of the TCSC (Figure 6.19)

$$K_B = \frac{X_{app}}{X_C}$$

In Figure 6.19, the conduction angle is $\beta = \pi - \alpha$.

Also, a parameter λ is defined as the quotient between the resonant frequency ($\omega_r = \omega_n\sqrt{-X_C/X_L}$) and the network frequency (f_n):

$$\lambda = \frac{\omega_r}{\omega_n} = \sqrt{\frac{-X_C}{X_L}}$$

with $2 \leq \lambda \leq 4$ (the reactance of the inductors is much smaller than that of the capacitor bank at the rated frequency).

The possibility to control the apparent reactance of the TCSC at network frequency can be utilized to improve the power system performance: power flow control, additional damping of electromechanical (0.5–2 Hz) power oscillations, and temporarily increase the degree of compensation of one line in order to alleviate the voltage drops in nodes that do not have sufficient reactive power support in cases of contingencies.

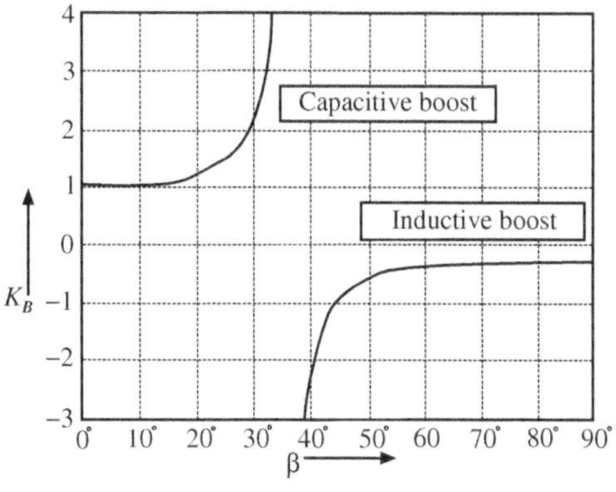

Figure 6.19 Boost factor, K_B, versus conduction angle β, for a TCSC [14].

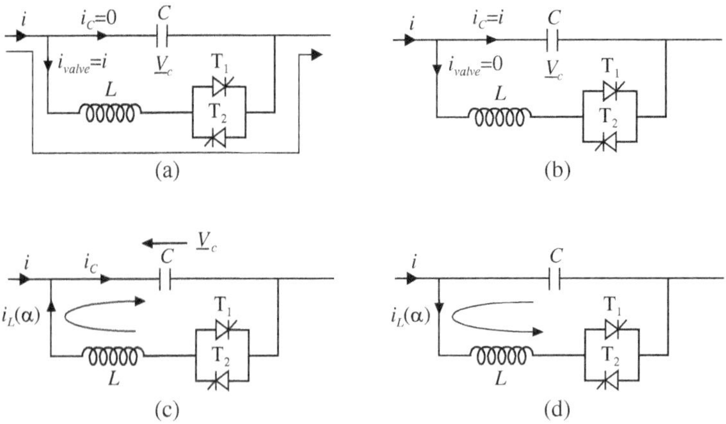

Figure 6.20 Operating modes of TCSC: (a) bypass thyristor (full thyristor conduction), (b) blocked thyristors (no thyristor current), (c) capacitive vernier operation, and (d) inductive vernier operation [11].

THE OPERATION MODES. TCSC has three principal operating modes (Figure 6.20) [14, 16].

(i) *Bypassed mode.* If the thyristor valve is triggered continuously, the valve stays conducting all time and the TCSC behaves like a parallel connection of the SC bank with the inductor in the thyristor valve. The resulting voltage in the steady-state across the TCSC is given by [14]:

$$\underline{V}_C = (-jX_C/(\lambda^2 - 1))\underline{I},\ X_C < 0$$

The voltage is inductive and the boost factor is negative (Figure 6.19).

$$K_B = -\frac{1}{\lambda^2 - 1}$$

In this case, the valve current is somewhat larger than the line current due to the current generation in the capacitor bank. When λ is considerable larger than unity, the capacitor voltage at a given line current is much smaller in bypass than in the blocking mode.

The operating mode when the thyristors are bypassed, which is when the bypassed breaker in parallel with the SC is closed, is used for capacitor and TCSC module protection against internal fault.

This operating mode is used for the control and protection of the SC bank. Each time a TCSC module is bypassed to prevent current flows through the thyristors larger than the limit current, some time must pass before the TCSC module can be reintegrated in the system after the line current decreases under specified limit.

(ii) *Blocking mode* (Figure 6.20b). When the thyristor valves are not triggered, that is the thyristor remains nonconducting, the TCSC operates in blocking mode.

In that case, the line current passes through the capacitor bank only. The capacitor voltage phasor, $\underline{V}_{C1}$, is given in terms of the line current, $\underline{I}_1$, according to equation [14]

$$\underline{V}_C = jX_C\underline{I},\ X_C < 0$$

In this mode, the TCSC performs in the same way as a fixed SC with a boost factor equal to one.

The compensation voltages across the capacitors are monitored and the capacitors can be quickly discharged using the thyristor control system to avoid damaging overvoltages across the capacitors. Thus, the thyristor switch can act as an overvoltage protection in lieu of arresters.

(iii) *The thyristors are in partial conduction operating mode or fine adjustment operating* (Figures 6.20c and 6.20d). This mode allows a TCSC module to have a behavior similar with either an inductive reactance or capacitive reactance, where both of them are continuously adjustable. Nevertheless, direct transition from the capacitive mode to the inductive mode is not allowed because of the parallel resonance zone that exists between the two operating modes.

A variant of this operating mode is the *capacitive boost* or *fine capacitive adjustment operating mode*, in which the thyristors are in conduction when the voltage at capacitor terminals and the current through it have opposite polarity. In this operating condition, the current through the TCR is opposite to the current through the capacitor. This results in current loop inside the TCSC module (Figure 6.20c). This current causes voltage increase across the fixed capacitor terminals and an increase in the equivalent capacitive reactance, thus resulting larger capacitive compensation. To avoid parallel resonance, the firing angle α, measured from the zero crossing of the positive voltage, is restricted to the $\alpha_{min} \leq \alpha \leq 180°$ interval. This restriction ensures a fine control of the TCSC module impedance. The current in this loop grows when α decreases from 180° to α_{min}. The maximum reactance allowed for the TCSC at $\alpha = \alpha_{min}$ is typically 2.5–3 times larger than the reactance of the capacitor at the nominal power system frequency.

Another variant of this operating mode is the *inductive boost* or *continuous inductive adjustment operating mode*, in which the TCSC can operate with a high level of conduction. This time, the current circulation is changed in such a way that the TCSC has net inductive reactance (Figure 6.20d).

Figure 6.21 illustrates the capacitor voltage and current waveforms, together with the TCR current and voltage waveforms, when TCSC operates under vernier control.

THE CHARACTERISTICS OF TCSC. The compensation voltage versus line current (V–I) characteristic of a basic TCSC is shown in Figure 6.22a [2, 17]. As can be seen, in the capacitive domain, the minimum firing angle, α_{Clim}, sets the limit for the maximum compensation voltage up to a value of the line current (I_{min}) at which the maximum rated voltage, V_{Cmax}, constrains the operation until the rated maximum current I_{max} is reached.

In the inductive mode, the maximum firing angle, α_{Llim}, limits the voltage at low line currents and the maximum rated thyristor current at high line currents.

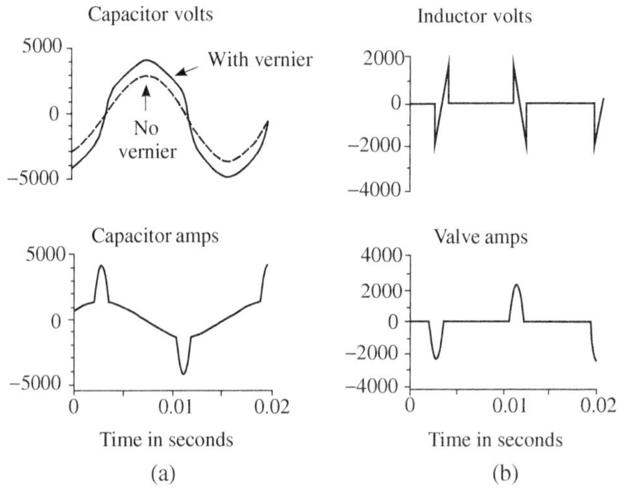

Figure 6.21 An example of the voltage and currents of the individual components of the TCSC under vernier control.

In the reactance compensating mode, the TCSC is operated to maintain the maximum rated compensation reactance at any line current up to the rated maximum. For this operating mode, the TCSC capacitor and the TCR are chosen so that at $\alpha_{C\lim}$, the maximum capacitive reactance can be maintained at and under the maximum rated line current as is shown in Figure 6.22b. The minimum capacitive compensating reactance that TCSC can provide is the impedance of the capacitor itself, theoretically obtained $\alpha = 90°$ (with nonconducting thyristor valve).

It should be realized that the voltage and reactance compensating modes are interchangeable by means of control actions; however, limitations imposed by component ratings will constrain the attainable operating ranges. For example, the compensation voltage versus line current characteristic shown in Figure 6.22a can be

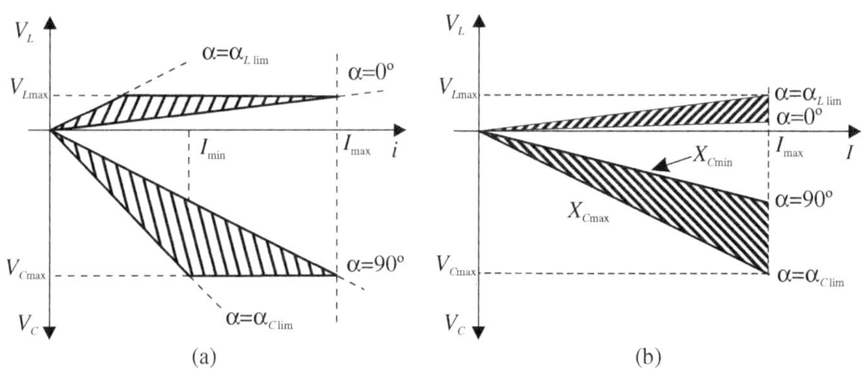

Figure 6.22 The *V–I* characteristics of TCSC when operated in: (a) voltage control mode and (b) reactance control mode. *Source*: Hingorani & Gyugyi 2000 [2]. Reproduced with permission of IEEE. *Source*: Mathur & Varma 2002 [16]. Reproduced with permission of IEEE.

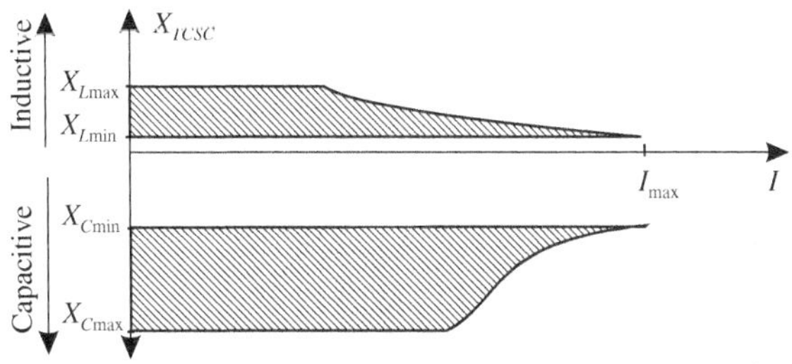

Figure 6.23 The compensation reactance versus line current characteristic of the TCSC, corresponding to the voltage compensation mode V–I characteristic (Figure 6.22a).

transformed into the compensation reactance versus line current characteristic shown in Figure 6.23.

From Figures 6.22a and 6.23, it can be observed that constant compensation voltage necessarily results in varying compensation reactance, and vice versa, constant reactance produces varying compensation voltage with changing line current.

The maximum limits for voltage and currents are design parameters for which the thyristor valve, the reactor, and capacitor banks are rated to meet specific application requirements. The TCSC, like its switched counterpart the TSSC, typically has transient voltage and current ratings defined for specific time durations.

The TCSC design is complicated by the fact that the harmonics generated aggravate the limit conditions. The harmonic currents cause additional losses and result in temperature increase in both, the thyristor valve and the reactor. However, the harmonic voltage generated by the TCSC system is relatively small. Therefore, in applications where the TCSC system is placed in a relatively long line, the line reactance limits the harmonic current injection into the AC system, which may eliminate the need for a harmonic filter connected across the TCSC. The harmonic voltages produced across the capacitor increase the crest voltage and the stress on the TCSC power components. The effects of the harmonics must be taken into account under the worst-case scenario of the operating conditions to determinate the necessary maximum voltage and current ratings of the TCSC components to satisfy specified operating conditions. Outside these defined operating limits, TCSC is protected against excessive voltage and current surges either by a shunt-connected external protection (e.g., MOV arresters, triggered spark gap, and bypass breaker) or by the TCR itself (with a backup breaker) in bypass operation.

The TCSC, with partial conduction of TCR, injects harmonic voltages into the line. These harmonic voltages are caused by the TCR harmonic currents, which circulate through the series compensation capacitor. The TCR, as established in Section 5.2, generates all odd harmonics, the magnitudes of which are a function of the firing angle α. The harmonic voltages corresponding to these currents in a TCSC circuit are clearly dependent on the impedance ratio of the TCR reactor to the SC, X_L/X_C. For a ratio $X_L/X_C = 0.133$ used in the existing TSCS installations, the most important harmonic voltages are the third, fifth, and seventh generated in the capacitive operating region; in Figure 6.24, these harmonics are plotted against the line current I and as a percentage of the fundamental capacitor V_{C0}, with the TCR off at rated current.

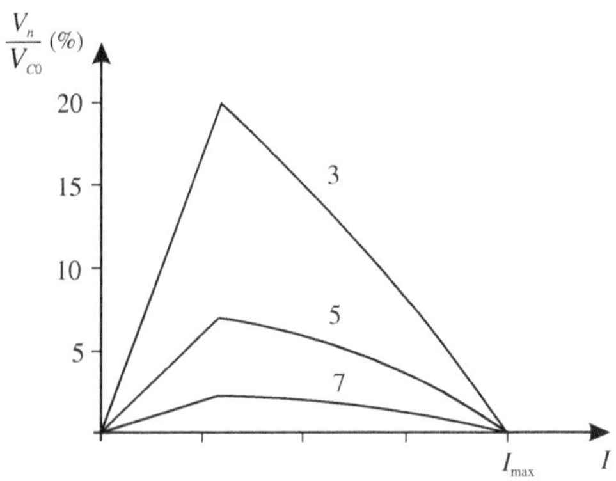

Figure 6.24 The dominant harmonic voltages generated by the TCSC at a X_L/X_C (= 0.133) ratio. *Source*: Hingorani & Gyugyi 2000 [22]. Reproduced with permission of IEEE.

6.3.3.4 *Capability Characteristics of the TCSC*

Although the TCSC is designed to operate in specific conditions, the operating limits are determined by the characteristics of the selected components. The major limits are:

(i) *Voltage limit* is determined by the insulation level of each component of the TCSC installation. This limit can vary in terms of the duration in which the voltage is applied. The application of overvoltages for short period (usually, less than 10 s) is more stressful for the MOV arresters than for the capacitor.

(ii) *Current limit* is imposed by the current that passes through the thyristor valve, fixed capacitor, and the limitation reactor to avoid the overheating of these components. The harmonics produced, also, increase the thermal stress and limit the TCSC operating modes; the limit of the firing angle for the thyristors must be carefully chosen so that the TCSC will not enter in the parallel resonance zone (even for a short period).

SINGLE MODULE TCSC. Figure 6.25 shows a single module TCSC capability characteristic in terms of module voltage versus line current. For operation in the capacitive region, the maximum firing advance (β) limits the capability up to a value of line current where the maximum voltage constraints operation.

The maximum voltage constraint will typically be given for three durations: continuous, for short time operating regime (30 min) and for the 1–10 s range. In both the capacitive and inductive regimes, the TCSC operating is, in general, restricted between a minimum and maximum value of its impedance. To these restrictions are added limitations for the voltage, current, and for the presence of harmonics. In this way, the operating region for TCSC is reduced [11].

The maximum value of the reactance X_{TCSC} is about 2–3 p.u. of the X_C. This restriction is imposed by the superior limit of the angle β. The minimum value of the capacitive reactance of TCSC is obtained when the thyristors are blocked and the current circulation through the thyristor is absent—corresponding to a reactance X_{TCSC} = 1 p.u. As the line current grows, the voltage at the TCSC terminals grows

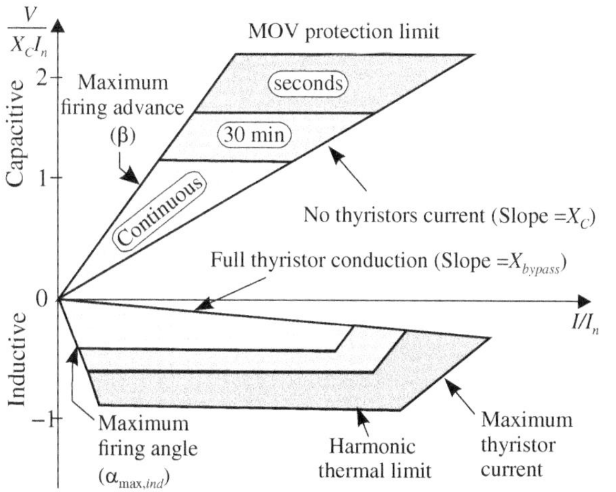

Figure 6.25 The *V–I* capability characteristics of single TCSC module [11, 17].

until the maximum value is reached. This limit depends on the duration of the applied voltage.

Inductive operation is limited by the maximum firing angle (α) at the low line currents and maximum thyristor current at high line currents. Between these constraints is an additional limiting characteristic related to harmonics. The maximum value of the inductive impedance is, usually, 2 p.u.; the minimum value of inductive reactance is reached when the thyristors are continuously conducting. Increasing line current harmonics are produced that leads to thermal stresses in the reactor and the thyristors, and the peak voltage is increased close to the admissible limit for the capacitor and the MOV arresters as can be seen in the capability characteristics of the TCSC represented in Figure 6.25. In case of a very high line current, the maximum value of the thyristor current in the fine inductive adjustment operating mode limits the operating domain of the TCSC.

Also, the operating characteristics of TCSC can be defined in terms of *impedance versus line current* as shown in Figure 6.26. The impedance value of TCSC in dynamic domain is reduced by increasing the line current. As mentioned above, direct transition from the inductive domain to the capacitive domain is not possible. In the normal operation mode, the TCSC operates in the first quadrant for both axes systems *V–I* and *X–I*.

Multiple Modules TCSC. In some power system applications, such as the power flow control or power oscillations damping, a continuous control of the line impedance can be useful. This can be accomplished by dividing a single TCSC into multiple modules and use them separately in the inductive and capacitive operating modes. For example, by dividing a TCSC into two modules, each of these submodules having an equivalent power equal with a half MVA rating of the overall system, the result is a *V–I* capability characteristic for these modules that combines the operating characteristics for each module operating on its own or with both modules operating together (Figure 6.27).

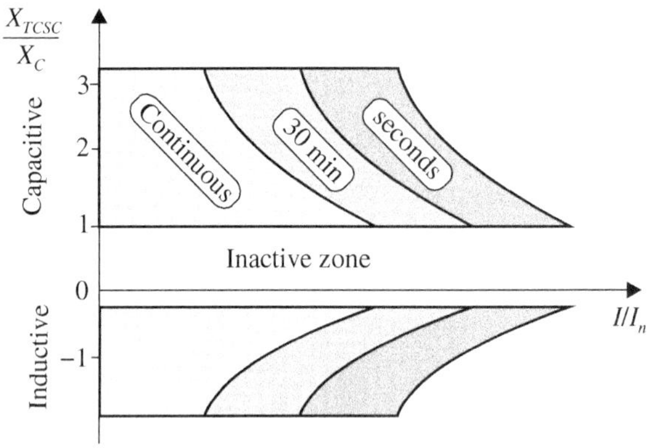

Figure 6.26 The *X–I* capability characteristics for a single TCSC module [11, 17].

When the two modules operate together in inductive/capacitive region, they produce a curve identical to that produced by a single TCSC module as expected. When the two modules operate separately, they produce intermediate characteristics. From Figure 6.27, we can see that the limits for the minimum capacitive reactance and for minimum inductive reactance disappear and a continuous transition from the inductive domain to the capacitive domain can be realized.

The capability characteristics, *V–I* and *X–I*, respectively, for a TCSC with more than one module are presented in Figures 6.28a and 6.28b. It is obvious that as the number of TCSC modules is increased, the compensation domain of TCSC for reactive power is also increased; but this is obtained with higher costs because of the larger number of cross connections on the platform and the number of reactors.

For a TCSC with more than one module, these modules are connected/disconnected if the desired impedance for TCSC is smaller than the nominal capacitive reactance of a fixed capacitor. The fine adjustment operating mode is characteristic to the situation in which the TCSC module impedance is bigger than the nominal capacitive reactance of a fixed capacitor. The system can have one or a few TCSC modules, and the others can be TSSC modules.

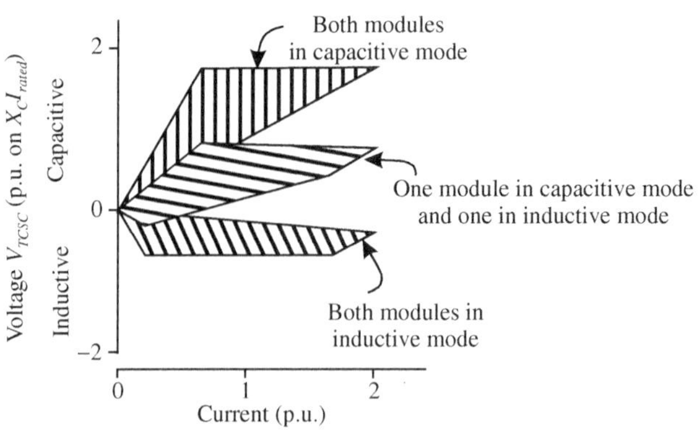

Figure 6.27 The *V–I* capability characteristics for a two module TCSC [11, 17].

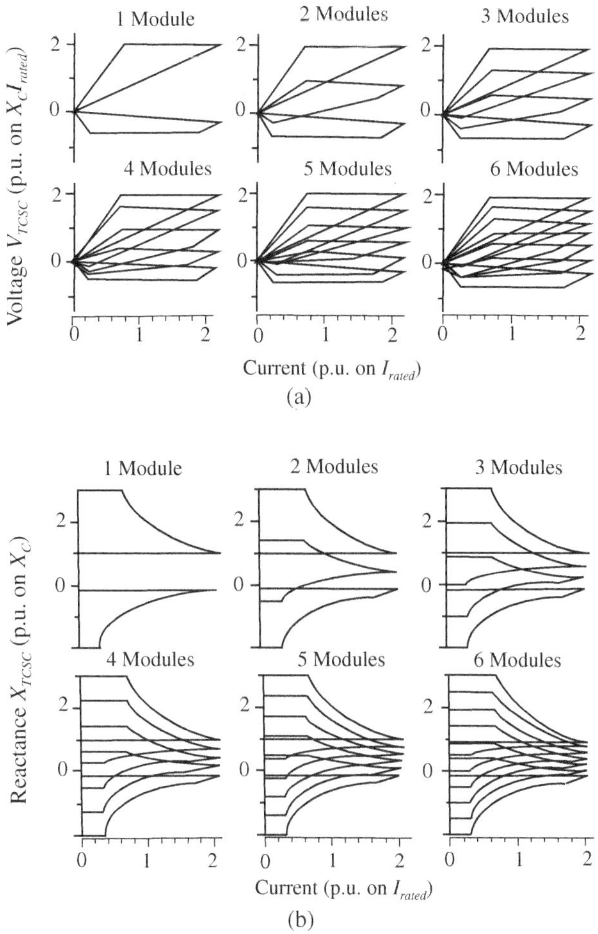

Figure 6.28 The capability characteristics for increasing number of equally sized TCSC modules: (a) *V–I* characteristics and (b) *X–I* characteristics [11, 17].

6.4 CONTROL SCHEMES FOR THE TCSC[2]

6.4.1 Open Loop Impedance Control

The impedance control represents the base strategy for TCSC control and is used mainly for power-flow control. The block scheme of such control system is presented in Figure 6.29.

The desired level of the series compensation in the electrical line, in normal regime, is expressed by a reference impedance (X_{ref}), which is applied at the control device input. This control scheme contains a *delay block* that represents the TCSC

[2] This section was inspired from the book "Thyristor-based FACTS controllers for electrical transmission systems" Mohan Mathur and Rajiv Varma, IEEE Press, 2002 [16].

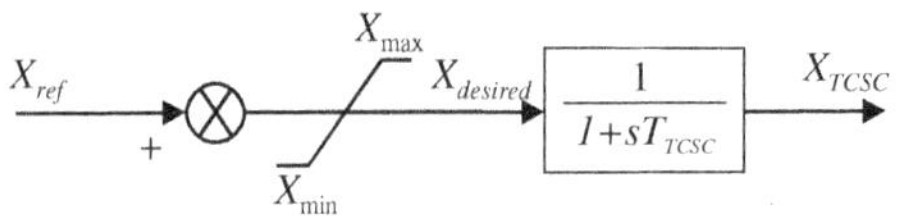

Figure 6.29 The block scheme of the TCSC constant-impedance controller [16].

action; normally, the delay is chosen about 15 ms. The output of the control device is a reactance order signal (X_{TCSC}), which is linearized to obtain the desired firing angle; from here, the signal "angle command" is transmitted to the control pulse generator that produces the gate command signals for the thyristors and the desired impedance is obtained.

6.4.2 Closed Loop Control

(i) Constant current control strategy of TCSC

The role of the constant current (CC) control device is to make a reference signal that has to maintain the line current amplitude at a desired value. Figure 6.30 shows the block diagram of a basic control structure of a TCSC device [15, 17, 19].

The TCSC controller is based on a proportional-integral (PI) regulator. A derivative line current feedback was introduced in the current controller to increase the damping of the SSR modes.

The three phase currents are measured and rectified in the "measurement block." The rectified signal is first passed through a filter block and is normalized ($I_{p.u.}$) so that it can be compared, in p.u., with the reference current signal (I_{ref}). After that, the signal goes to a proportional-integral (PI) regulator, the output of this block is the desired susceptance signal (Y) within the preset limits. The linearization block transforms the susceptance signal into the firing-angle signal.

An operation mode selector block is used for protection of the TCSC during faults. The "thyristor switched reactor (TSR) mode" is activated

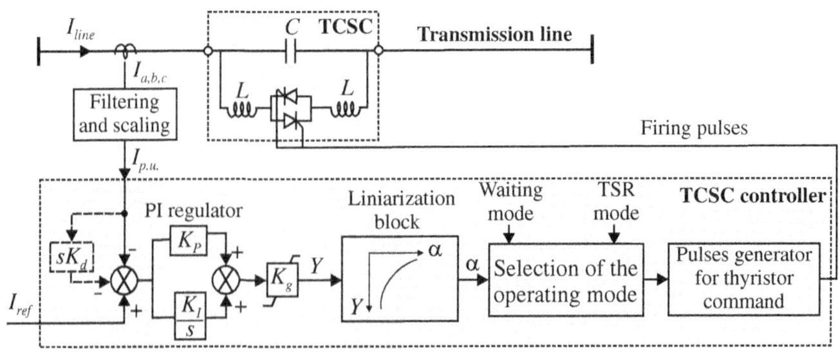

Figure 6.30 The constant-current (CC) control device model of TCSC. *Source*: Pilotto et al. 2003 [19]. Reproduced with permission of IEEE.

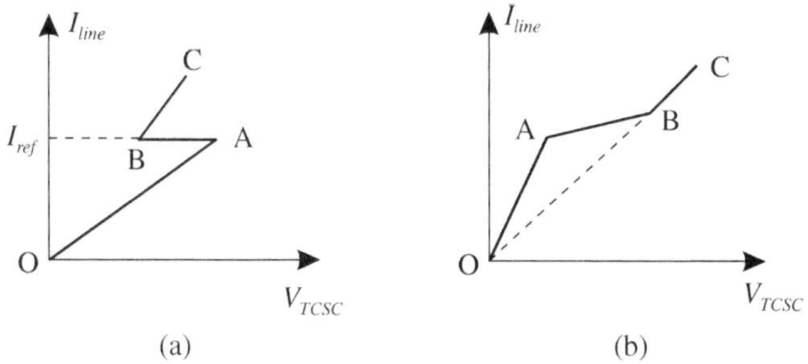

Figure 6.31 The control characteristics of the TCSC: (a) CC control and (b) CA control [16].

during short-circuit situations. Whenever the current in the MOV exceeds a certain limit, the TSR mode is activated. This mode basically fires all valves at 90° and therefore sets the TSR in full conduction. This decreases the energy requirements of the arrester and makes it possible to have fewer MOV columns installed in parallel with the TCSC equipment. The "waiting mode" operates during fault removal and is responsible for discharging the DC offset voltage that may build-up on a series capacitor during the reinsertion process [19].

The control characteristics in the steady-state operation of the CC control device on V_{TCSC}–I_{line} axes system are presented in Figure 6.31a. The convention used in the figure is to consider the voltage at the capacitor terminals positive (opposite to the convention used for load flow).

The CC control characteristic is formed by three segments: OA, AB, and BC. The OA and BC segments represent the maximum and minimum limits of TCSC impedance. The AB segment represents the control domain in which the TCSC impedance can be modified by controlling the firing angle to maintain the line current to an imposed value (I_{ref}) [16].

For improving the damping of certain oscillatory modes, such as subsynchronous oscillations, an optimized derivative line-current feedback is embedded in the TCSC controller with dashed line in Figure 6.30 [19].

(ii) *Constant angle control strategy*

The constant angle (CA) control strategy is utilized when the transmission line compensated with TCSC operates in parallel with other transmission lines. The control objective during the transient time period or during contingences is to maintain unchanged the power flow through the parallel transmission lines, and pulling the power variation through the line compensated with TCSC.

To maintain constant power flow through the parallel transmission lines, the angular difference between voltages at the line terminals needs to be maintained constant, which is CA control strategy.

If the voltage magnitudes at the two line terminals are regulated, then to maintain a constant angular difference a constant voltage drop (ΔV_{line}) across the line needs to be maintained constant. If the line resistance R_L is neglected, then to realize the control objective is starting from the expression [16]:

$$\Delta V_{line} = I_{line} X_{Lnet} - V_{TCSC} = \Delta V_{Lref} \tag{6.8}$$

and

$$I_{line} = \frac{1}{X_{Lnet}} (\Delta V_{Lref} + V_{TCSC}) \tag{6.9}$$

or

$$I_{ref} - \left(I_{line} - \frac{V_{TCSC}}{X_{Lnet}} \right) = 0 \tag{6.10}$$

where:

ΔV_{Lref}	is	constant;
I_{line}	–	the magnitude of the current in the TCSC compensated line;
X_L	–	the total line inductive reactance;
X_{FC}	–	the reactance of the series fixed capacitor (if exists);
$X_{Lnet} = X_L - X_{FC}$	–	the net line impedance, including the effect of the fixed capacitor;
V_{TCSC}	–	the voltage across the TCSC: positive for capacitive voltage; negative for inductive voltage;
$I_{ref} = \frac{\Delta V_{Lref}}{X_{Lnet}}$	–	the reference line current.

The control characteristic, in V_{TCSC}–I_{line} coordinates, for CA control, is shown in Figure 6.31b. The AB segment represents the control domain, with a slope X_{Lnet}. The OA and BC segments represent the maximum limit and minimum limit, respectively, for the TCSC impedance. Generally, the CA control is very efficient in the power oscillation damping.

In Figure 6.32 is shown the control block diagram of TCSC that incorporates the capability of both control devices CC and CA.

In Figure 6.32a, T_m represents the time constant associated with the measurement circuit, which is typically a first order, low pass filter. The measurement circuits are used for voltage measurement at the TCSC terminals and for the line current measurement. In case of CC control, the multiplier block S is set to zero, and in case of CA control the S block is set to the value $1/X_{Lnet}$.

The regulator is typically a PI controller, which might be connected in cascade with one phase-lead circuit (Figure 6.32b). If a pure integrator is used, then K_P is set to zero.

For the CC control scheme, the integral gain factor K_I is considered positive. In this control scheme, a positive current-error signal means that the capacitive reactance of TCSC must be increased to enhance the line current and to reduce the error signal. For the CA control scheme, the integral gain factor K_I is considered negative. If the current-error signal is positive in equations (6.8)

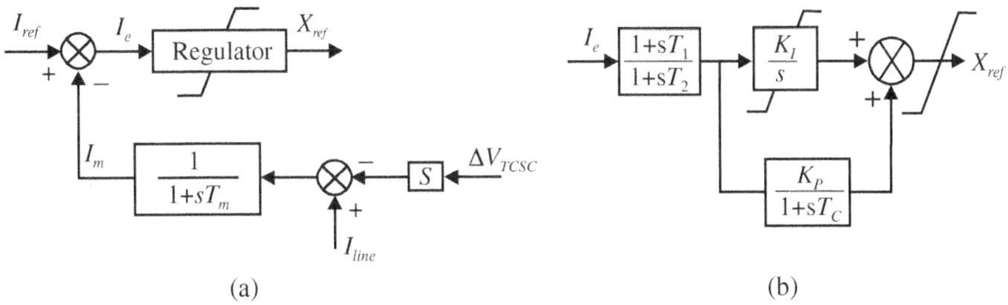

Figure 6.32 The block diagram of TCSC control system: (a) CC or CA control device and (b) the block scheme of the regulator. *Source*: Grünbaum et al. 1999 [15]. Reproduced with permission of ABB. *Source*: Paserba et al. 1995 [17]. Reproduced with permission of IEEE.

through (6.10), the voltage drop ΔV_{line} on the electrical line is less than ΔV_{Lref}, so that a decrease in the TCSC voltage V_{TCSC} and consequently a decrease in the TCSC impedance X_{TCSC} (or X_{ref}) are necessary. For this reason, K_I has a negative sign in CA control.

Although the TCSC firing angle delays are modeled through a single time constant of 15 ms, they may be ignored in the electromechanical stability studies as their effect is insignificant [16].

(iii) *Constant power control strategy*

The block diagram for a conventional TCSC power control device is presented in Figure 6.33 [19]. The power flow through the line is computed from voltage and current signals measured at local level after the *abc* → αβ0 transformation.

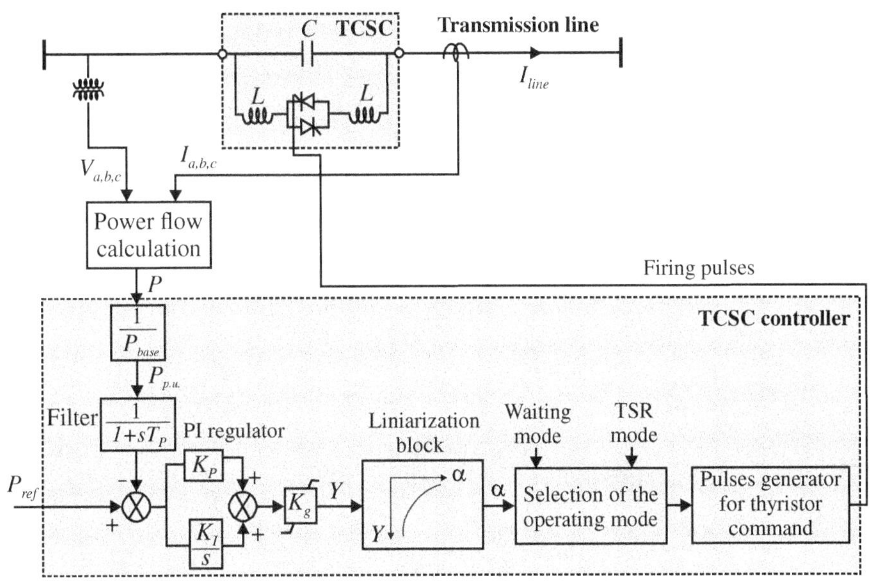

Figure 6.33 The conventional structure for power control device. *Source*: Pilotto et al. 2003 [19]. Reproduced with permission of IEEE.

The calculated power signal (P) is first transformed into a per-unit quantity ($P_{u.r.}$) and filtered, then transmitted to the summation block in the power control system. The reference signal, P_{ref}, represents the desired level of active power flow in the line compensated with TCSC, and the power controller device in based on a PI structure. After the limiter, the control signal is linearized to generate a firing angle that provides a linear relationship between the TCSC susceptance and the control signal. This block finally excites the firing pulse generation scheme.

The power controller device of TCSC is efficient if it is used as a slowly operating control device for damping power oscillations or subsynchronous power oscillations. An attempt to increase the response speed of the controller device by reducing the power-controller time constant T_P renders the response oscillatory.

Remark. For improving the performance of the power controller device in a TCSC installation, the advantages of the power control strategy can be combined with the advantages of the current control strategy [19]. One of these two loops can be fast acting through the current loop and the other slowly operating through an external power control loop [16]. Such a controller can provide fast response to system faults and provide a slow response to electromechanical oscillations.

6.5 TCSC MODELING

6.5.1 Steady-State Modeling of TCSC

6.5.1.1 TCSC Modeling Through Series Variable Impedance

This model is based on the simple concept of a variable series reactance with a value that is changing automatically so that the active power that transits the line will be held at a constant value, equal with the impose value (Figure 6.34).

The introduction of a TCSC device on the *i-k* line assumes creation of an additional bus *m*. Thereby, two series elements are created on the line *i-k*, that is, TCSC on the element *i-m* and the initial line on the element *m-k* (Figure 6.34b). In the bus admittance matrix $[\underline{Y}_{nn}]$, the transversal element *i-m* is given by $G_{im} = 0$ and $B_{im} = 0$. The value of this reactance X_{TCSC} can be determined easily with Newton–Raphson method. So, the reactance from Figure 6.34b represents the equivalent reactance of all TCSC series modules, operating in both domains, inductive and capacitive [18, 20].

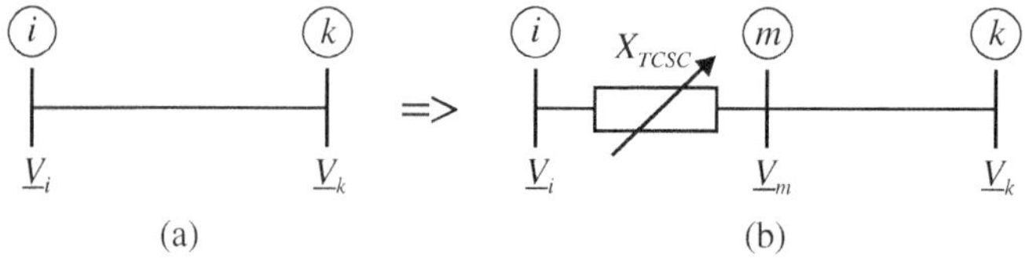

Figure 6.34 TCSC variable series impedance model.

If the Kirchhoff laws are applied for this scheme, the nodal equations for the circuit are obtained:

$$\left.\begin{aligned}\underline{I}_i &= (\underline{V}_i - \underline{V}_m)\frac{1}{jX_{TCSC(1)}}\\ \underline{I}_m &= (\underline{V}_m - \underline{V}_i)\frac{1}{jX_{TCSC(1)}}\end{aligned}\right\} \Rightarrow \begin{bmatrix}\underline{I}_i\\ \underline{I}_m\end{bmatrix} = \begin{bmatrix}-j\dfrac{1}{X_{TCSC(1)}} & j\dfrac{1}{X_{TCSC(1)}}\\ j\dfrac{1}{X_{TCSC(1)}} & -j\dfrac{1}{X_{TCSC(1)}}\end{bmatrix}\begin{bmatrix}V_i\\ \underline{V}_m\end{bmatrix}$$

where $X_{TCSC(1)}$ is the fundamental frequency equivalent reactance of TCSC, when operating in either the inductive or the capacitive regions.

The transfer admittance matrix of the variable series compensator is given by:

$$\begin{bmatrix}\underline{I}_i\\ \underline{I}_m\end{bmatrix} = \begin{bmatrix}jB_{ii} & jB_{im}\\ jB_{mi} & jB_{mm}\end{bmatrix}\begin{bmatrix}\underline{V}_i\\ \underline{V}_m\end{bmatrix} \tag{6.11}$$

- For inductive operation

$$B_{ii} = B_{mm} = -\frac{1}{X_{TCSC}} \tag{6.12a}$$

$$B_{im} = B_{mi} = \frac{1}{X_{TCSC}} \tag{6.12b}$$

- For capacitive operation

$$B_{ii} = B_{mm} = \frac{1}{X_{TCSC}} \tag{6.12c}$$

$$B_{im} = B_{mi} = -\frac{1}{X_{TCSC}} \tag{6.12d}$$

If we take into account the particular form of the matrix $[\underline{Y}_{nn}]$, given by equation (6.11), the expression for nodal powers will be given by the next equation:

$$P_i = V_i V_m B_{im} \sin(\theta_i - \theta_m) \tag{6.13a}$$

$$Q_i = -V_i^2 B_{ii} - V_i V_m B_{im} \cos(\theta_i - \theta_m) \tag{6.13b}$$

$$P_m = V_m V_i B_{mi} \sin(\theta_m - \theta_i) \tag{6.13c}$$

$$Q_m = -V_m^2 B_{mm} - V_m V_i B_{mi} \cos(\theta_m - \theta_i) \tag{6.13d}$$

The active and reactive power "forced" to flow from bus i toward bus m $(P_{im}^{TCSC}, Q_{im}^{TCSC})$ is equal to the active and reactive power flowing from m to i, but taken with a negative sign, that is:

$$P_{im}^{TCSC} = V_i V_m B_{im}^{TCSC} \sin(\theta_i - \theta_m) \tag{6.14a}$$

$$P_{mi}^{TCSC} = V_m V_i B_{im}^{TCSC} \sin(\theta_m - \theta_i) \tag{6.14b}$$

$$Q_{im}^{TCSC} = -V_i^2 B_{ii} - V_i V_m B_{im}^{TCSC} \cos(\theta_i - \theta_m) \tag{6.14c}$$

$$Q_{mi}^{TCSC} = -V_m^2 B_{ii} - V_m V_i B_{im}^{TCSC} \cos(\theta_m - \theta_i) \tag{6.14d}$$

Thereby, in order to consider the device installed between buses i and m, the power flow through this device is added to the nodal powers at buses i and m, that is:

$$P_i^{new} = P_i + P_{im}^{TCSC} \tag{6.15a}$$

$$P_m^{new} = P_m + P_{mi}^{TCSC} \tag{6.15b}$$

$$Q_i^{new} = Q_i + Q_{im}^{TCSC} \tag{6.15c}$$

$$Q_m^{new} = Q_m + Q_{mi}^{TCSC} \tag{6.15d}$$

In order to represent the power control in the Newton–Raphson steady-state calculation, a new balance equation is added. This represents the active power flow through the TCSC device, which must be equal to the specified value, that is:

$$\Delta P_{im}^{TCSC} = P_{im}^{sp} - P_{im}^{TCSC,calc} \tag{6.16}$$

where P_{im}^{sp} is the specified active power on the *i-m* element.

The last expression for ΔP_{im}^{TCSC} will be added to the vector of power mismatches ΔP and ΔQ. Similarly, the TCSC reactance mismatch is added in the vector of mismatches of unknown variables, which will allow determining the TCSC reactance required to achieve the specified power flow on the element *i-m*.

The set of linearized power flow equation is:

$$\begin{bmatrix} \Delta P_i \\ \Delta P_m \\ \Delta Q_i \\ \Delta Q_m \\ \Delta P_{im}^{TCSC} \end{bmatrix} = \left[\begin{array}{ccccc} \frac{\partial P_i}{\partial \theta_i} & \frac{\partial P_i}{\partial \theta_m} & \frac{\partial P_i}{\partial V_i}V_i & \frac{\partial P_i}{\partial V_m}V_m & \frac{\partial P_i}{\partial X_{TCSC}}X_{TCSC} \\ \frac{\partial P_m}{\partial \theta_i} & \frac{\partial P_m}{\partial \theta_m} & \frac{\partial P_m}{\partial V_i}V_i & \frac{\partial P_m}{\partial V_m}V_m & \frac{\partial P_m}{\partial X_{TCSC}}X_{TCSC} \\ \frac{\partial Q_i}{\partial \theta_i} & \frac{\partial Q_i}{\partial \theta_m} & \frac{\partial Q_i}{\partial V_i}V_i & \frac{\partial Q_i}{\partial V_m}V_m & \frac{\partial Q_i}{\partial X_{TCSC}}X_{TCSC} \\ \frac{\partial Q_m}{\partial \theta_i} & \frac{\partial Q_m}{\partial \theta_m} & \frac{\partial Q_m}{\partial V_i}V_i & \frac{\partial Q_m}{\partial V_m}V_m & \frac{\partial Q_m}{\partial X_{TCSC}}X_{TCSC} \\ \hline \frac{\partial P_{im}^{TCSC}}{\partial \theta_i} & \frac{\partial P_{im}^{TCSC}}{\partial \theta_m} & \frac{\partial P_{im}^{TCSC}}{\partial V_i}V_i & \frac{\partial P_{im}^{TCSC}}{\partial V_m}V_m & \frac{\partial P_{im}^{TCSC}}{\partial X_{TCSC}}X_{TCSC} \end{array}\right] \begin{bmatrix} \Delta\theta_i \\ \Delta\theta_m \\ \frac{\Delta V_i}{V_i} \\ \frac{\Delta V_m}{V_m} \\ \frac{\Delta X_{TCSC}}{X_{TCSC}} \end{bmatrix} \tag{6.17}$$

In general case, when the TCSC is connected in series on a line *i-k*, the simplified form of the matrix Jacobian becomes:

$$JAC^{new} = \begin{bmatrix} H^{new} & M^{new} & J_{13} \\ K^{new} & L^{new} & J_{23} \\ J_{31} & J_{32} & J_{33} \end{bmatrix} \tag{6.18}$$

The terms corresponding to the TCSC device added into the new forms of the submatrices *H*, *M*, *K*, and *L* are:

$$H_{ii}^{TCSC} = H_{mm}^{TCSC} = \frac{\partial P_{im}^{TCSC}}{\partial \theta_i} = \frac{\partial P_{mi}^{TCSC}}{\partial \theta_m} = V_i V_m B_{im}^{TCSC} \cos(\theta_i - \theta_m)$$

$$H_{im}^{TCSC} = H_{mi}^{TCSC} = \frac{\partial P_{im}^{TCSC}}{\partial \theta_m} = \frac{\partial P_{mi}^{TCSC}}{\partial \theta_i} = -V_i V_m B_{im}^{TCSC} \cos(\theta_i - \theta_m)$$

$$M_{ii}^{TCSC} = M_{mm}^{TCSC} = \frac{\partial P_{im}^{TCSC}}{\partial V_i} V_i = \frac{\partial P_{mi}^{TCSC}}{\partial V_m} V_m = V_i V_m B_{im}^{TCSC} \sin(\theta_i - \theta_m)$$

$$M_{im}^{TCSC} = M_{mi}^{TCSC} = \frac{\partial P_{im}^{TCSC}}{\partial V_m} V_m = \frac{\partial P_{mi}^{TCSC}}{\partial V_i} V_i = V_i V_m B_{im}^{TCSC} \sin(\theta_i - \theta_m)$$

$$K_{ii}^{TCSC} = K_{mm}^{TCSC} = \frac{\partial Q_{im}^{TCSC}}{\partial \theta_i} = \frac{\partial Q_{mi}^{TCSC}}{\partial \theta_m} = V_i V_m B_{im}^{TCSC} \sin(\theta_i - \theta_m)$$

$$K_{im}^{TCSC} = K_{mi}^{TCSC} = \frac{\partial Q_{im}^{TCSC}}{\partial \theta_m} = \frac{\partial Q_{mi}^{TCSC}}{\partial \theta_i} = -V_i V_m B_{im}^{TCSC} \sin(\theta_i - \theta_m)$$

$$L_{ii}^{TCSC} = \frac{\partial Q_{im}^{TCSC}}{\partial V_i} V_i = -2V_i^2 B_{im}^{TCSC} - V_i V_m B_{im}^{TCSC} \cos(\theta_i - \theta_m)$$

$$L_{mm}^{TCSC} = \frac{\partial Q_{mi}}{\partial V_m} V_m = -2V_m^2 B_{im}^{TCSC} - V_i V_m B_{im}^{TCSC} \cos(\theta_m - \theta_i)$$

$$L_{im}^{TCSC} = L_{mi}^{TCSC} = \frac{\partial Q_{im}^{TCSC}}{\partial V_m} V_m = \frac{\partial Q_{mi}^{TCSC}}{\partial V_i} V_i = -V_i V_m B_{im}^{TCSC} \cos(\theta_i - \theta_m)$$

The terms of the J_{31} submatrix are:

$$\frac{\partial P_{im}^{TCSC}}{\partial \theta_i} = V_i V_m B_{im}^{TCSC} \cos(\theta_i - \theta_m)$$

$$\frac{\partial P_{im}^{TCSC}}{\partial \theta_m} = -V_i V_m B_{im}^{TCSC} \cos(\theta_i - \theta_m)$$

The terms of the J_{32} submatrix are:

$$\frac{\partial P_{im}^{TCSC}}{\partial V_i} V_i = \frac{\partial P_{im}^{TCSC}}{\partial V_m} V_m = V_i V_m B_{im}^{TCSC} \sin(\theta_i - \theta_m)$$

The terms of the J_{33} submatrix are:

$$\frac{\partial P_{im}^{TCSC}}{\partial X_{TCSC}} X_{TCSC} = -V_i V_m B_{im}^{TCSC} \sin(\theta_i - \theta_m)$$

The terms of the J_{13} submatrix are:

$$\frac{\partial P_{im}^{TCSC}}{\partial X_{TCSC}} X_{TCSC} = -V_i V_m B_{im}^{TCSC} \sin(\theta_i - \theta_m)$$

$$\frac{\partial P_{mi}^{TCSC}}{\partial X_{TCSC}} X_{TCSC} = -V_i V_m B_{im}^{TCSC} \sin(\theta_m - \theta_i)$$

The terms of the J_{23} submatrix are:

$$\frac{\partial Q_{im}^{TCSC}}{\partial X_{TCSC}} X_{TCSC} = V_i^2 B_{ii} + V_i V_m B_{im}^{TCSC} \cos(\theta_i - \theta_m)$$

$$\frac{\partial Q_{mi}^{TCSC}}{\partial X_{TCSC}} X_{TCSC} = V_m^2 B_{ii} + V_i V_m B_{im}^{TCSC} \cos(\theta_m - \theta_i)$$

The other elements of the Jacobian matrix are assumed to be known from the classical Newton–Raphson method for a steady-state calculation.

At the end of iteration *iter*, the value of the series reactance is corrected as follows:

$$X_{TCSC}^{(iter)} = X_{TCSC}^{(iter-1)} + \left(\frac{\Delta X_{TCSC}}{X_{TCSC}}\right)^{(iter)} X_{TCSC}^{(iter-1)}$$

After calculation of the equivalent reactance of the TCSC, X_{TCSC}, the corresponding firing angle α_{TCSC} can be determined. It must be considered that this calculation has meanings only when all the TCSC modules, which make up the compensation device, have identical parameters and are set to operate with the same angle α_{TCSC}. In that case, the calculation of the α_{TCSC} angle is an iterative process because the reactance X_{TCSC} and α_{TCSC} angle are not linearly dependent. A solution to eliminate these disadvantages is to model each TCSC module from the compensation device structure through the α_{TCSC} angle.

6.5.1.2 TCSC Impedance Modeling as a Function of the Firing Angle

Let us consider the circuit of one TCSC module and firing angle α_{TCSC} (Figure 6.35).

This model takes into account again the expression of the fundamental frequency TCSC equivalent reactance as a function of TCSC firing angle α_{TCSC} [20,21]:

$$\begin{aligned} X_{TCSC(1)}(\alpha_{TCSC}) = & -X_C + C_1\{2(\pi - \alpha_{TCSC}) + \sin[2(\pi - \alpha_{TCSC})]\} - \\ & - C_2 \cos^2(\pi - \alpha_{TCSC})\{\overline{\omega} \tan[\overline{\omega}(\pi - \alpha_{TCSC})] - \tan(\pi - \alpha_{TCSC})\} \end{aligned} \quad (6.19)$$

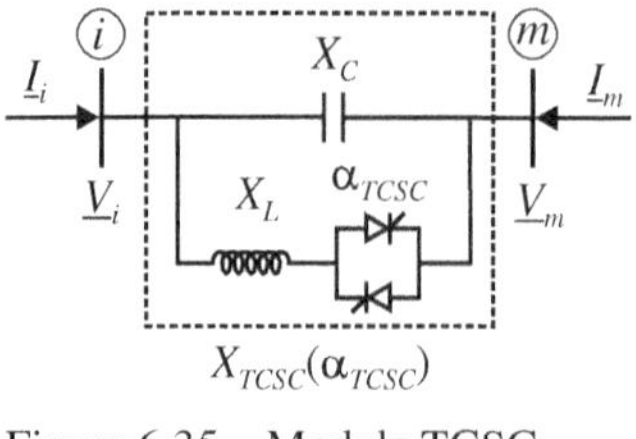

Figure 6.35 Module TCSC.

where:

$$C_1 = \frac{X_C + X_{LC}}{\pi}; C_2 = \frac{4X_{LC}^2}{\pi X_L}$$

$$X_{LC} = \frac{X_C X_L}{X_C - X_L}$$

$$\overline{\omega} = \sqrt{\frac{X_C}{X_L}}$$

The equivalent reactance $X_{TCSC(1)}$in equation (6.19) replaces X_{TCSC} in equations (6.12a), (6.12b), and (6.11), and the TCSC active and reactive power equations at busses i and m are

$$P_i^{TCSC} = V_i V_m B_{im(1)}^{TCSC} \sin(\theta_i - \theta_m) \tag{6.20a}$$

$$Q_i^{TCSC} = -V_i^2 B_{ii(1)}^{TCSC} - V_i V_m B_{im(1)}^{TCSC} \cos(\theta_i - \theta_m) \tag{6.20b}$$

$$P_m^{TCSC} = V_m V_i B_{mi(1)}^{TCSC} \sin(\theta_m - \theta_k) \tag{6.20c}$$

$$Q_m^{TCSC} = -V_m^2 B_{mm(1)}^{TCSC} - V_m V_i B_{mi(1)}^{TCSC} \cos(\theta_m - \theta_i) \tag{6.20d}$$

where:

$$B_{ii(1)} = -B_{im(1)} = B_{TCSC(1)}$$

For the case when the TCSC controls active power flowing from bus i to bus m, at a specified value, the set of linearized power flow equations is

$$\begin{bmatrix} \Delta P_i \\ \Delta P_m \\ \Delta Q_i \\ \Delta Q_m \\ \Delta P_{im}^{\alpha_{TCSC}} \end{bmatrix} = \left[\begin{array}{cccc|c} \frac{\partial P_i}{\partial \theta_i} & \frac{\partial P_i}{\partial \theta_m} & \frac{\partial P_i}{\partial V_i} V_i & \frac{\partial P_i}{\partial V_m} V_m & \frac{\partial P_i}{\partial \alpha} \\ \frac{\partial P_m}{\partial \theta_i} & \frac{\partial P_m}{\partial \theta_m} & \frac{\partial P_m}{\partial V_i} V_i & \frac{\partial P_m}{\partial V_m} V_m & \frac{\partial P_m}{\partial \alpha} \\ \frac{\partial Q_i}{\partial \theta_i} & \frac{\partial Q_i}{\partial \theta_m} & \frac{\partial Q_i}{\partial V_i} V_i & \frac{\partial Q_i}{\partial V_m} V_m & \frac{\partial Q_i}{\partial \alpha} \\ \frac{\partial Q_m}{\partial \theta_i} & \frac{\partial Q_m}{\partial \theta_m} & \frac{\partial Q_m}{\partial V_i} V_i & \frac{\partial Q_m}{\partial V_m} V_m & \frac{\partial Q_m}{\partial \alpha} \\ \hline \frac{\partial P_{im}^{\alpha_{TCSC}}}{\partial \theta_i} & \frac{\partial P_{im}^{\alpha_{TCSC}}}{\partial \theta_m} & \frac{\partial P_{im}^{\alpha_{TCSC}}}{\partial V_i} V_i & \frac{\partial P_{im}^{\alpha_{TCSC}}}{\partial V_m} V_m & \frac{\partial P_{im}^{\alpha_{TCSC}}}{\partial \alpha_{TCSC}} \end{array}\right] \begin{bmatrix} \Delta \theta_i \\ \Delta \theta_m \\ \frac{\Delta V_i}{V_i} \\ \frac{\Delta V_m}{V_m} \\ \Delta \alpha_{TCSC} \end{bmatrix} \tag{6.21}$$

where the active power flow mismatch for the TCSC model is

$$\Delta P_{im}^{\alpha_{TCSC}} = P_{im}^{sp} - P_{im}^{\alpha_{TCSC},calc} \tag{6.22}$$

and the $P_{im}^{\alpha_{TCSC},calc} (= P_i^{TCSC})$ is the calculated power as given by (6.20a).

The elements of the Jacobian matrix are identical with the ones from the earlier model (6.18), with the following exceptions:

$$\frac{\partial P_{im}}{\partial \alpha_{TCSC}} = P_{im} B_{TCSC(1)} \frac{\partial X_{TCSC(1)}}{\partial \alpha_{TCSC}}; \quad \frac{\partial Q_{im}}{\partial \alpha_{TCSC}} = Q_i B_{TCSC(1)} \frac{\partial X_{TCSC(1)}}{\partial \alpha_{TCSC}} \quad (6.23a,b)$$

$$\frac{\partial P_{m}}{\partial \alpha_{TCSC}} = P_{m} B_{TCSC(1)} \frac{\partial X_{TCSC(1)}}{\partial \alpha_{TCSC}}; \quad \frac{\partial Q_{m}}{\partial \alpha_{TCSC}} = Q_m B_{TCSC(1)} \frac{\partial X_{TCSC(1)}}{\partial \alpha_{TCSC}} \quad (6.23c,d)$$

where

$$\frac{\partial B_{TCSC(1)}}{\partial \alpha_{TCSC}} = B^2_{TCSC(1)} \frac{\partial X_{TCSC(1)}}{\partial \alpha_{TCSC}}$$

$$\begin{aligned}\frac{\partial X_{TCSC(1)}}{\partial \alpha_{TCSC}} &= -2C_1[1+\cos(2\alpha_{TCSC})] + C_2 \sin(2\alpha_{TCSC}) \cdot \\ &\cdot \{\overline{\omega} \tan[\overline{\omega}(\pi - \alpha_{TCSC})] - \tan \alpha_{TCSC}\} + \\ &+ C_2 \left\{ \overline{\omega}^{-2} \frac{\cos^2(\pi - \alpha_{TCSC})}{\cos^2[\omega_n(\pi - \alpha_{TCSC})} - 1 \right\} \end{aligned} \quad (6.24)$$

At the end of iteration *i*, the firing angle is corrected according to the following expression:

$$\alpha^{(i)}_{TCSC} = \alpha^{(i-1)}_{TCSC} + \Delta\alpha^{(i)}_{TCSC} \quad (6.25)$$

Note: The performance of the TCSC model is affected by the number of internal resonant points exhibited by the TCSC in the range 0–90°. The resonant points (poles) of equation (6.24) are determined by the following expression [22]:

$$\alpha_{TCSC} = \pi \left[1 - \frac{(2n-1)\omega\sqrt{LC}}{2} \right], \quad n = 1, 2, 3 \ldots \quad (6.25)$$

In practice, a well-designed TCSC scheme will only have one resonant peak within its range of operation.

6.5.2 TCSC Dynamic Models

6.5.2.1 *Transient Stability Model*

For transient—and oscillatory—stability studies, the TCSC can be represented by the variable-reactance, which is a quasi-static approximation model. Consequently, the TCSC dynamics during power-swing frequencies are modeled by a variable reactance at fundamental frequency (Figure 6.36) [16, 17].

The model has provisions for an open loop auxiliary signal (X_{aux}), which could be, for example, the input from an external power flow controller and also for a small-signal modulation input (X_{mod}). The reference X_{ref} represents the initial operating point of the TCSC impedance [18].

The sum of these signals is performed in the *control block;* the transfer function corresponding to this block depends on the regulator type and the output of this is the desired value for the TCSC reactance ($X^{desired}_{TCSC}$). This value is passed through

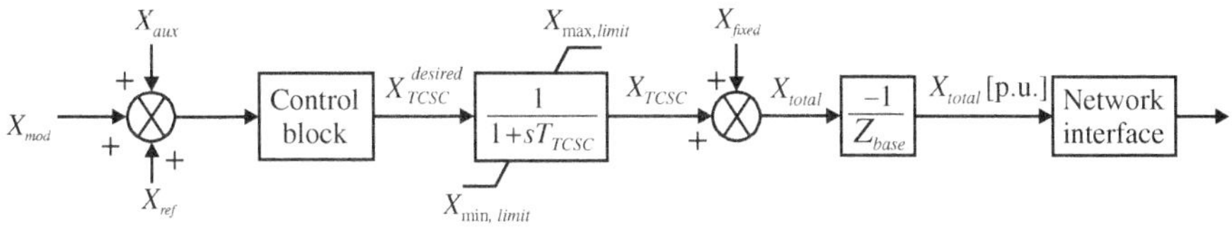

Figure 6.36 Dynamic model for a variable TCSC reactance type [16, 17].

a *delay block* (first-order lag); the transfer function corresponding to this block is $1/(1 + sT_{TCSC})$. This lag is associated with the firing controls and the natural response of the TCSC and is represented by a single time constant (T_{TCSC}). For a TCSC similar to that installed at the Slatt substation (see Section 6.7.3), this time constant is on the order of 15 ms [17].

The output of the lag block is called X_{TCSC} and has nonwindup limits associated with the integration. These limits are variable limits based on the TCSC reactance capability curve and equations as shown in Figure 6.36 and described later. The value for the TCSC reactance X_{TCSC} is added to the value of the fixed capacitor X_{fixed} (if there is any), to obtain a total compensation value for the reactance X_{total} [18].

Since most stability programs use a per unit system for the network equations, this total reactance X_{total} is converted from ohms to per unit (X_{total}, p.u), where $Z_{base} = (\mathrm{kV_{TCSC}})^2/\mathrm{MVA_{sys}}$ is calculated from the TCSC rated rms line-to-line voltage and the study-system MVA base. Here, the convention is a positive per unit value for net capacitive compensation and a negative per unit value for net inductive compensation. This reactance value is used to modify the line impedance of the series-compensated branch during the calculation of the network solution [17].

When creating a "load flow database" for study of a SC application including a TCSC, it is best to designate a "separate branch" for the transmission line compensation. Since the value of the compensation in the load flow (X_{total}, p.u) could include both the fixed and TCSC reactances, the initial operating point of the TCSC (X_{ref}) in the stability model is defined as:

$$X_{ref} = X_{total} - X_{fixed} \quad \text{(for initialization)}$$

As previously described, the capability curve for a single TCSC module (Figure 6.26) presents a discontinuity between the capacitive and inductive regime. However, this discontinuity can be eliminated if a TCSC with more than one module is used (Figure 6.28). The variable reactance type for TCSC modules presumes a continuous variation domain for reactance and therefore is applicable for multi-module configurations. This model is, in general, used for the analysis of *the interzone oscillations modes* and this analysis is realized with a high precision when the amplification factor for the reactance X_{TCSC}/X_C is less than 1.5 [16].

As we can see in the block diagram from Figure 6.36, the value of the X_{TCSC} reactance is limited by superior and inferior values. These limits are included because of the technological and stability restrictions of the device.

The capability curve for the reactance of multi-module TCSC presented in Figure 6.28 can be approximated by the equivalent capability curve presented in Figure 6.37 [16, 17]. This figure can be used for the variable reactance-type TCSC

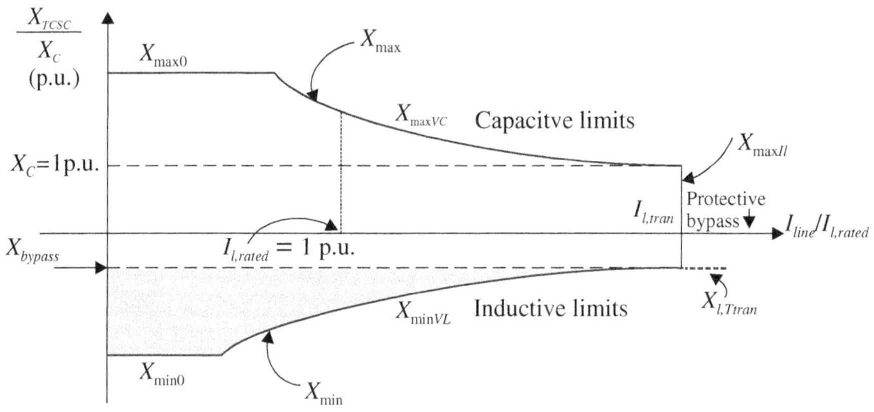

Figure 6.37 Transient reactance limits. The simplified reactance-capability curve for a multi-module TCSC [17, 18].

model; the capability curve from the picture includes the effect of the TCSC transient overload levels. It must be noted that the reactance limit for the high currents is represented in Figure 6.37 as a group of discrete points for different modules.

During periods of overcurrent, only some TCSC modules move into the bypassed mode, for the bypassing of a module causes the line current to decrease and thus reduces the need for the remaining TCSC modules to go into the bypass mode. However, for the case of modeling, only one continuous reactance limit—denoted by a vertical line in Figure 6.37 is considered for all TCSC modules [17, 18].

Referring to the limits as shown in Figure 6.37, the TCSC model permits operation anywhere within the enclosed region. These boundaries are due to a number of constraints, as subsequently described. All reactances are in per unit on X_C except as noted, all voltages in per unit on $(I_{l,rated}X_C)$, and all currents are in, or converted to amperes.

In the *capacitive zone,* the different TCSC reactance constraints are caused by the following limitations [17, 18]:

(i) Limit on the firing angle α_{TCSC}, this limit is represented by the constant reactance $X_{\max 0}$.

(ii) Limit on the voltage across the TCSC ($V_{C,tran}$). The restriction for the corresponding reactance is given by the following expression:

$$X_{\max VC} = V_{Ctran}\frac{I_{l,rated}}{I_l} \tag{6.26}$$

where I_l is the value of the line current, and $I_{l,rated}$ is the rated rms line current.

(iii) Limit on the line current ($I_{l,tran}$), at which point the TCSC will go into the "thyristor bypassed" protective mode:

$$X_{\max,I_l} = \begin{cases} \infty, & \text{for } I_l < I_{l,tran} \cdot I_{l,rated} \\ X_{bypass}, & \text{for } I_l \geq I_{l,tran} \cdot I_{l,rated} \end{cases} \tag{6.27}$$

where X_{bypass} is the value of the TCSC for bypass mode.

Once the TCSC is bypassed on this overcurrent constraint, it is subject to a time delay (T_{delay}) on reinsertion after line current falls back below $I_{l,tran}$. In a multi-module TCSC, it is possible that only some of the modules will bypass, since once one module bypasses the line current will drop, which in turn may allow the remaining modules to stay in capacitive mode.

Note: For simplicity in typical stability studies, it is suggested that this nuance be neglected [18].

Finally, the capacitive limit for the X_{TCSC} reactance is given by the minimum between the three limits described above:

$$X_{\max limit} = \min(X_{\max 0}, X_{\max VC}, X_{\max,I_l}) \cdot X_C[\Omega] \tag{6.28}$$

In the *inductive zone,* the TCSC operating is restricted by the following limits:

(i) Limit on the firing angle α_{TCSC}, expressed as a constant reactance limit $X_{\min 0}$.

(ii) The limit imposed by the harmonics, approximated as a constant voltage across the TCSC ($V_{l,tran}$). In this condition, the effective reactance constraint is:

$$X_{\min VL} = V_{l,tran} \frac{I_{l,rated}}{I_l} \tag{6.29}$$

(iii) The limit on thyristor current. As an approximation, the fundamental component of thyristor current is limited to that at which the TCSC can operate in thyristor bypass for the duration of a transient. This is expressed as input data in terms of the corresponding line current ($I_{l,Ttran}$):

$$X_{\min IlT} = 1 - \frac{I_{l,Ttran} \cdot I_{l,rated}(1 - X_{bypass})}{I_l} \tag{6.30}$$

Finally, the limit for the inductive reactance in the vernier operating mode is given by the maximum between the three limits described above:

$$X_{\min limit} = \max(X_{\min 0}, X_{\min VL}, X_{\min IlT}) \cdot X_C[\Omega] \tag{6.31}$$

If a TCSC is not expected to operate in the inductive-vernier mode, the minimum reactance limit is X_{bypass} irrespective of the line-current magnitude [16].

The typical data used for stability studies in TCSC analysis are presented in Tables 6.1 and 6.2; in these tables, the reactances are given in p.u. from X_C, all voltages are given in p.u. on ($I_{l,rated} X_C$), and all currents are given in amperes.

6.5.2.2 *Long-Term Stability Model*

The TCSC model for typical stability studies described in section 6.5.2.1 considers only the transient overload capability. The time-limited overload capability must also be considered for long-term dynamic stability studies. Figure 6.38 illustrates the

TABLE 6.1 Typical TCSC data for stability studies. Capacitive vernier region [17]

Notation	Name	Units	Specific value
U_{rated}	Rated voltage (rms) phase-to-phase	kV	*
$I_{l,rated}$	Rated line current (rms)	A	*
X_C	Nominal TCSC reactance	Ω	*
X_{fixed}	Fixed reactance compensation	Ω	*
T_{TCSC}	Time constant associated with the TCSC firing controls	Seconds	0.015–0.030
X_{bypass}	TCSC reactance for bypass mode	p.u. of X_C	−0.15
$X_{\max 0}$	Upper limit of TCSC reactance	p.u. of X_C	3.0
V_{Ctran}	Maximum voltage for capacitive vernier region for transient overload	p.u. of $(I_{l,rated}X_C)$	2.0
$I_{l,tran}$	Maximum transient line current	p.u. of $I_{l,rated}$	2.0
T_{delay}	Protective bypass recovery delay	Seconds	0.025

* Based on a specific TCSC device rating for a given application.

capability curve for a multi-module TCSC, including the time-overload limits for both capacitive and inductive vernier operation [17, 23].

For consideration of time-overload, in [18] are introduced two time frames: temporary and rated, in addition to the transient time frame:

- The transient time frame used for all the preceding aspects is the least restrictive, but is valid only for a short time, typically 3–10 s.
- The temporary time frame is more restrictive and is valid for several minutes (usually 30 min).

The rated time frame is most restrictive and is valid for continuous duty.

The relationship between the time frames is illustrated with typical values in Table 6.3.

The purpose of this function is to monitor line current and the TCSC voltage and then to transition the voltage and current limits from the transient level to the temporary level to the continuous level within the appropriate time constraints for each overload regions of operation. The transient overload period is typically 3–10 s

TABLE 6.2 Typical TCSC data for stability studies. Inductive vernier region [17]

Notation	Name	Units	Specific value
$X_{\min 0}$	Lower limit of TCSC reactance	p.u. of X_C	−2.0
V_{ltran}	Maximum voltage for inductive vernier region for transient overload	p.u. of $(I_{l,rated}X_C)$	−0.6
I_{lTtran}	Maximum line current for full conduction operation for inductive region transient overload	p.u. of $I_{l,rated}$	2.0

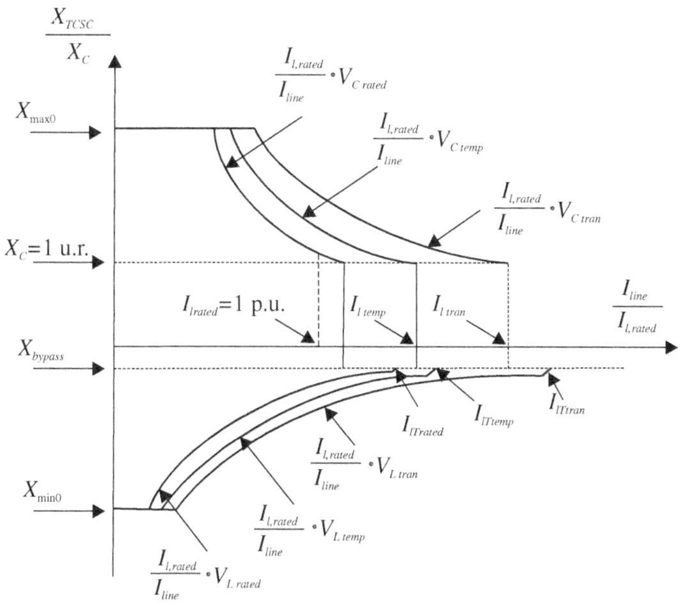

Figure 6.38 The *X–I* capability characteristic of a multi-module TCSC, including time-overload capability [11, 16, 17].

and the temporary level is typically 30 min. By transitioning the voltage and current limits, the overload management function, in effect, modifies $X_{\max limit}$ and $X_{\min limit}$ as a function of time [11]. Once the TCSC is outside of an overload region for a specified amount of time, the voltage and current limits transition back to the original capability level [17].

TABLE 6.3 Typical TCSC data for stability studies. The case of a TCSC having only the capability for capacitive vernier region [17]

Specific value Time frame	Transient overload	Temporary overload	Rated
Duration	3–10 s	30 min	Continuous
Maximum voltage for capacitive vernier region (V_C)	2 p.u.	1.5 p.u.	1.15 p.u.
Maximum voltage for inductive vernier region (V_L)	−0.6 p.u.	−0.45 p.u.	−0.3 p.u.
Maximum current through the line when the thyristors are in full conduction in capacitive region (I_{lTtemp})	2 p.u.	1.5 p.u.	1.15 p.u.
Maximum current through the line when the thyristors are in full conduction operation in inductive region (I_{lT})	2 p.u.	1.35 p.u.	1.35 p.u.

TABLE 6.4 Additional data for the TCSC model for long-term stability studies [17]

Input data	Description	Units	Typical value
V_{Crated}	Maximum capacitor voltage for continuous operation	pu of $(I_{lrated})(X_C)$	1.15
V_{Ctemp}	Maximum capacitor voltage for temporary overload	pu of $(I_{lrated})(X_C)$	1.5
I_{ltemp}	Maximum temporary line current	pu of (I_{lrated})	1.35
V_{Lrated}	Maximum voltage for inductive vernier region for continuous operation	pu of $(I_{lrated})(X_C)$	−0.3
V_{Ltemp}	Maximum voltage for inductive vernier region for temporary overload	pu of $(I_{lrated})(X_C)$	−0.45
$I_{lTrated}$	Rated line current for full-conduction operation in the inductive region	pu of (I_{lrated})	1.35
I_{lTtemp}	Maximum line current for full-conduction operation for inductive region temporary overload	pu of (I_{lrated})	1.35
T_{tran}	Maximum time for transient overload condition	Seconds	3–10
T_{temp}	Maximum time for temporary overload condition	Seconds	1800
$Reset_{tran}$	Reset for the transient overload capability	Multiple of T_{tran}	2
$Reset_{temp}$	Reset for the temporary overload capability	Multiple of T_{tran}	2

Table 6.4 describes the additional information for the overload management function required for long-term dynamic simulations.

6.6 APPLICATIONS OF TSSC/TCSC INSTALLATIONS

Series compensation is a known technology for increasing the transmission capacity of electrical lines and for controlling the power flow on the parallel transmission lines.

SCs are installed in electrical transmission networks to solve the following problems:

- SCs will increase the power flows on the transmission lines.
- Dynamic stability problems following a major disturbance after a loss of the synchronism.
- Dynamic instability characterized by low frequency undamped oscillations. This situation appears when an important quantity of energy is transmitted through a weak transmission line.
- Voltage stability problems characterized by a voltage decrease when the transmission lines loading is increased. The SCs compensate a part of the reactive power generated in the transmission network.

The *conventional SCs* (with fixed capacity) are economical means that can be used to increase the transfer capacity of the transmission lines. Despite the encouraging economical factor, still the number of such installations is limited especially because of the possibility that a SSR can arise. This is demonstrated through a slow shift of energy between the electric system and the shafts of nearby generators, which can lead to fatigue of the shaft and even to shaft failures.

Other critical factors in the application of the series compensation are the admissible thermal load of the transmission lines and the total electrical losses in the electric system.

The *SCs commanded with thyristors (TCSC)* offer much more benefits to the power systems than the series fixed capacitors. Some of these benefits are:

- In the meshed networks, the TCSC devices enable the operators to control the value of the series capacitive reactance and consequently they can easily control the current and the power flow on the compensated line.
- The capacitive and inductive reactance of the TCSC can be adjusted very fast enabling dynamic control of the power flow between two zones in the power system. Thus, the TCSC devices can be used for damping the low frequency oscillations that appear in the system in the interval 0.2–2 Hz.
- The control mechanisms of the TCSC can be designed to eliminate the risk of SSR even for the high levels of compensation.
- During the faults on the transmission line compensated with TCSC, the thyristors can bypass the SCs so that the duration of the overvoltage, the energy of the MOV arresters, and the fault current (decreasing the solicitation of the power-switch) are reduced.
- The controlled blocking of the thyristors, followed by the reinsertion of the capacitors in the circuit, enables the compensation device to quickly be reinserted into the systems after a fault.

The placement of FACTS controllers at appropriate locations is a critical issue. An optimally placed FACTS device requires a lower rating to achieve the same control objective than if it were located elsewhere [16, 24]. At times, the FACTS controllers may need to be placed at nonoptimal locations to minimize costs, especially when land prices and environmental concerns become important.

In Figure 6.39 are presented two possible TCSC locations.

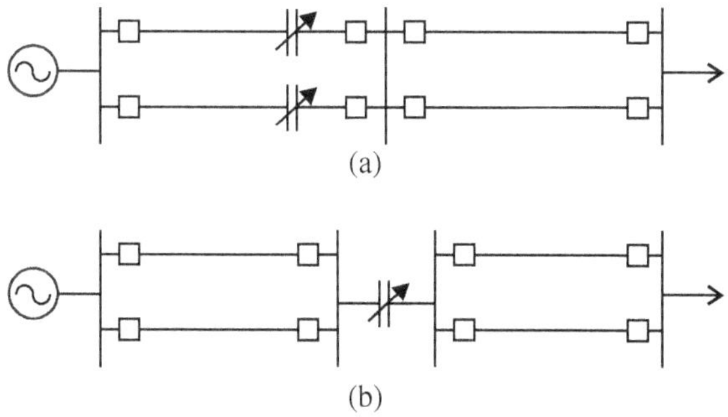

Figure 6.39 Possible TCSC locations: (a) two TCSCs located at substation entrance with one equipment in series with each of two line sections and (b) one TCSC located within substation with one equipment in series with both lines.

The following conditions generally apply when considering the placement of TCSC:

(i) The TCSC should be located in lines that experience limiting power oscillations.

(ii) The swing of voltages on each side of the TCSC must be within acceptable limits; otherwise, multiple sites may be necessary.

(iii) The control action of the TCSC in one transmission path should not cause undue power swings in a parallel path. If it does, then variable series compensation may become necessary in the parallel path.

(iv) Sometimes, it may be advisable to distribute the control action among multiple TCSCs rather than confining the control action to one large-rating TCSC. Doing so, ensures some system reliability if one of the TCSCs should fail.

Improvement of the Power Transfer Capability. The power transfer between two different zones from a strong power system can take two forms:

- The radial power transfer, when exists only one way for power transferring. This situation characterized the systems recently interconnected, where two zones are linked through a single transmission system.
- The power transfer in a loop when more than one corridor for power transfer exists.

For radial power transfer, the principal problem is that the transport capacity of the interconnection line between the two zones is limited. This requires planning studies to establish if the existing transmission system can transfer the desired power without perturbing the operation of the electrical power systems. The series compensation (the fixed SCs and TCSC) can be an efficient method from the economical point of view to increase the transport capacity of the system and to improve the voltage profile along the transmission line. Figure 6.40 shows a conceptual scheme for the interconnection between two zones with a high voltage level. The SVC or the on-load tap changing transformers can have the same effect in this application.

Regarding the power transfers in the loop, some problems such as power losses in the electrical network, power flow, overloaded transmission lines, and transport capacity must be carefully considered.

TCSC offers flexibility for modification of the series compensation level with the aim to increase the transport capacity of the compensated lines. This permits the operators to adjust the power transfers on those high voltage circuits, which can lead

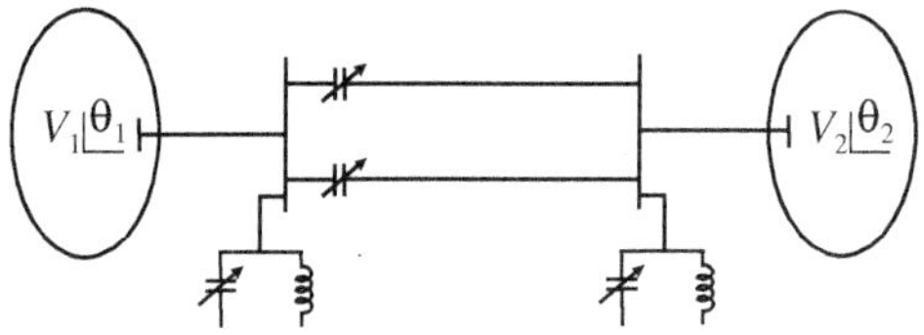

Figure 6.40 Control of power flow between regions [25, 26].

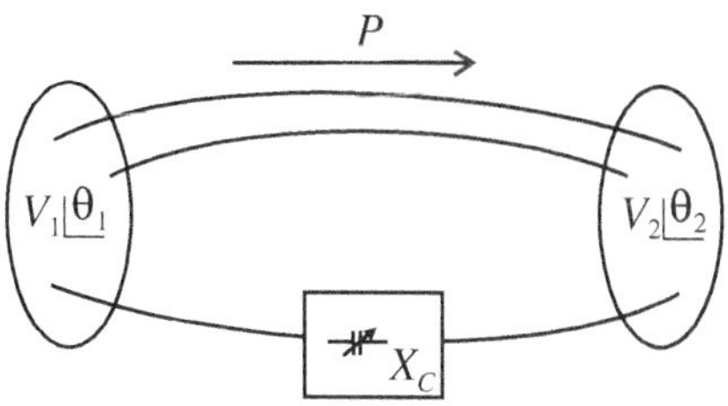

Figure 6.41 Interconnection between two zones [25].

to the reduced power losses and elimination of overloading in other lower capacity circuits. Plus, TCSC can reduce loop power flows, automatically or manually.

Redirecting the Power Flow into the Loop. In interconnected power system, the power flow from one region to another may take unintended routes depending on the impedances of the transmission lines connecting the areas. A conceptual diagram showing the use of TCSC for controlling the power flow on the transmission lines is illustrated in Figure 6.41.

Other technologies such as phase-shifting transformers, unified power flow control devices, static VAr compensators and the HVDC transmission lines also exist for controlling the power flow or to increase the transfer capacity of a transmission system.

In some applications, a phase-shifting transformer can be a cost-effective approach in particular where the angle across a line is small. In that case, the controllability of the power flow is greater than for a TCSC to increase or reduce the power flow through the transmission line.

In a meshed network, an SVC installation can also be more expensive than a TCSC installation for increasing the transport capacity of the transmission lines. In general, to establish which is the better technology to apply a very detailed analysis of the power system is needed.

Improving the Stability Reserve. The compensation with SC can improve the transient stability of a power system if this compensation is done in a transmission line where the power can be effectively controlled. The series compensation has the effect of making the compensated lines to appear shorter from the electric point of view. This will lead to an increase in the synchronization torque. Thus, the dynamic stability limit of the power system is increased. However, a detailed analysis including a contingency analysis must be performed considering all parallel, uncompensated transmission for the entire power system in case the compensated transmission lines are interrupted.

TCSC can increase the dynamic stability limit of the power system over the level obtained by using fixed SCs. Using TCSC, the short duration for nominal voltage application at the capacitor terminals can be used to assure a high compensation level for the period following to a fault. In addition, through the reduction of the line impedance, the synchronizing torque produced through the compensated line is improved; $P_e = EV \sin\delta/(X_L - X_C)$.

Improving the System Damping. Interarea power oscillations (0.2–0.7 Hz) very often appear in the power systems when the transported energy between two areas overtakes the limit threshold (e.g., during a system fault). The limit threshold depends, among others, on how strong the link between the interconnected systems is. The TSSC and TCSC installations offer possibilities to improve the damping of the power oscillations and to increase the power transmission capacity between two areas.

The local measurements of the line current and voltage are essential for detecting the oscillations and to control the TCSC for amplifying the power oscillations damping.

Subsynchronous Resonance Damping. The phenomenon of SSR has caused concern in the past in situations where the risk for occurrence of SSR has acted as an impediment to the use of series compensation systems in cases where otherwise the technology would have offered the best and the most economical solution. These concerns are alleviated if correctly designed TCSC systems are used.

The SSR risk used to be linked to the utilization of series compensation of transmission lines fed by thermal generation, particularly in cases with high degrees of compensation, where analysis showed that the complementary series resonance frequency of the compensated line(s) coincided with some poorly damped torsional vibration frequency of the turbo-generator shaft, and could hence induce increased mechanical stresses in the shafts. The TCSC can be used to eliminate this risk for coinciding resonance frequencies by making the SC(s) act inductive in the subsynchronous frequencies band, thereby rendering the occurrence of series resonance in the transmission system for subsynchronous frequencies altogether impossible.

This inductive character of the TCSC is made possible by the use of a thyristor-controlled inductor in parallel with the SC. The system is governed by an advanced, patented control algorithm called Synchronous Voltage Reversal (SVR), by means of which the TCSC exhibits an inductive virtual impedance in the complete frequency range of concern for SSR (Figure 6.42). As a consequence of this property of the series capacitor, SSR is efficiently ruled out.

Note: The principle of SVR mitigation of TCSCs has been obtained from the pioneering work done by Dr. N. G. Hingorani, for whom the NGH scheme of damping

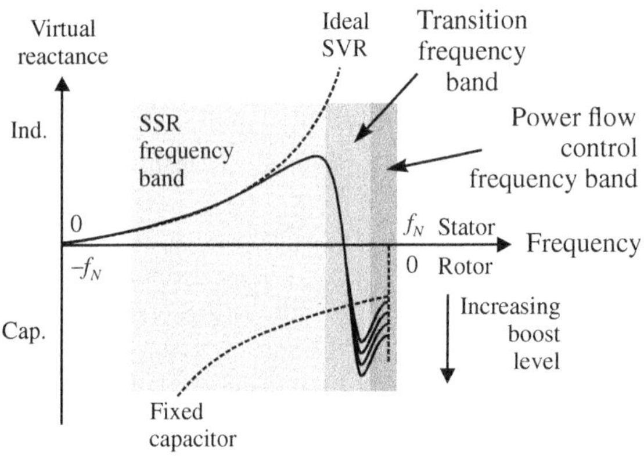

Figure 6.42 Apparent reactance characteristic of TCSC [27].

SSR [2, 27] was named. This scheme involves a thyristor-controlled discharge resistor connected in shunt with the SC and is installed in practical systems.

6.7 SERIES CAPACITORS WORLDWIDE

6.7.1 Kanawha River Mechanically Switched Series Capacitor in United States

Kanawha River mechanically switched series Capacitor, commissioned in 1991, is used in order to adopt the compensation level of a backbone 345 kV line in order to have sufficient stability margins during an outage of the parallel 765 kV system (Figure 6.43) [28, 29].

The outage of either the Baker-Broadford or Broadford-Jackson's Ferry 765 kV circuits can result in overloading of the Kanawha River-Funk 345 kV line and the parallel 138 kV transmission facilities. In order to alleviate this overload, the decision to install a 788 MVAr, 2500 A, 42 ohms SC bank on the Kanawha River-Funk 345 kV, which is 174 km long, was taken. Each phase of installation consists of two platforms, one with both a 10% (7 Ω) and a 20% (14 Ω) segment, and the other with the remaining 30% (21 Ω) segment (Figure 6.44).

Each segment is made up of a number of parallel/SC units. Each unit is rated 17.5 kV, 60 Hz, and 405 kVAr. Each capacitor segment is split into two parallel groups. The current in each group is monitored using optical current transformers, and capacitor failures, detected by a difference in the currents, result in alarm or trip. MOVs are connected across each capacitor segment. These highly nonlinear

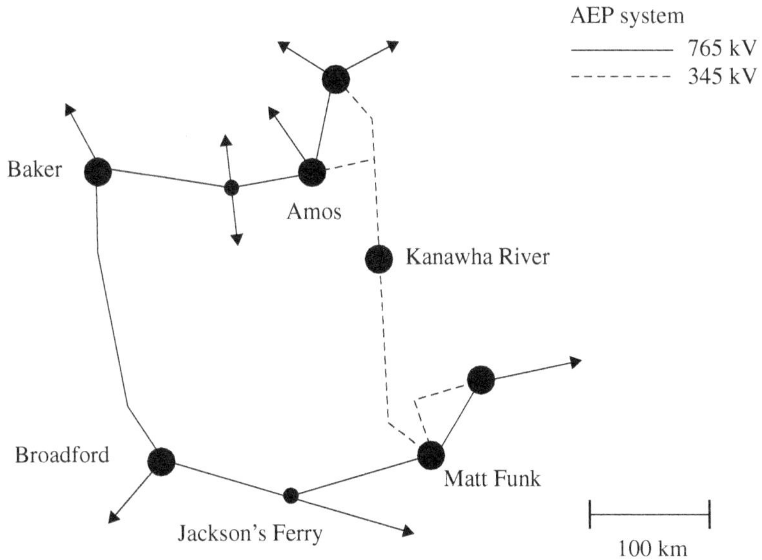

Figure 6.43 Southeastern AEP EHV system [28].

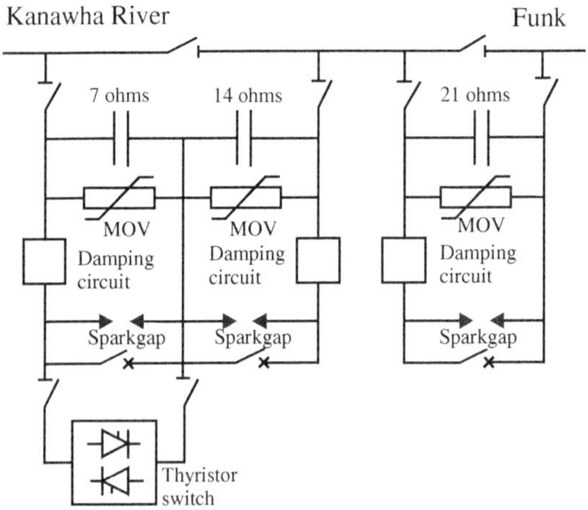

Figure 6.44 Simplified one line of Kanawha River substation [28, 29].

varistors protect the capacitors in case of excessive line currents during fault conditions. In the event of an internal fault, that is, a fault located along the compensated line, the energy ratings of the MOV could be exceeded. Therefore, a triggered spark gap is installed across each segment to bypass the capacitor for these conditions. A damping circuit—an air core reactor in parallel with a resistor—is connected in the capacitor bypass circuit in order to limit the capacitor discharge current and damp the oscillations resulting from the operation of the spark gap, the bypass breaker, or the thyristor switch.

Each capacitor segment is switched by a bypass breaker, which is a traditional SF_6 but with operating mechanism modified for fast closing operation [26][3].

In the last stage of the project was installed a light triggered thyristor switch across the middle phase of the 10% (137 MVAr, 7 Ω) segment. The middle phase of the untransposed Kanawha River-Funk 345 kV line has higher magnitudes of current (114%) and voltage (102%) than the other two phases [26, 28, 29]. It is implemented as a self-standing unit, and connected to the platform by disconnect switches. The disconnect switches provide the flexibility to inspect the thyristor without interfering with the normal operation of the conventional SC segment.

The thyristors can withstand the maximum system fault current for a four-circle period, until the line breaker or the bypass breaker operates. The switch is assembled from two series-connected modules of thyristor strings. The thyristor switch is automatically controlled by load level on the line or by remote dispatch. The switch protection system includes thermal overload, thyristor failure, and unsymmetrical triggering.

[3] Material reprinted with permission from CIGRE "*Improving transmission system performance using controlled series capacitors*, Keri, A.J.F., Ware B.J., Byron, R.A., Mehraban, A.S., Chamia, M., Halvarsson, P., Ängquist, L., CIGRE, Paper 14/37/38-07, Paris, 30 August - 5 September 1992", © Copyright 1992.

The thyristor switch at Kanawha River operates in switching mode. The damping inductor accordingly was designed with low inductance. It has a 60 Hz reactance of 2% of the reactance corresponding to the capacitor segment. Therefore, the requirements of the surge current capability of the thyristors are very demanding for a turn-on from the MOV protection voltage. This would produce a substantial discharge current from the SC segment. If a fault occurs close to the SC installation, the thyristor current could peak at values above 70 kA.

The introduction of SCs affects electrical losses. The topology of the system resulted in higher system losses with SCs. Further system analysis revealed that segmented capacitors, resulting in 0–60% compensation, would be justified. This provides the flexibility to use only the minimum amount of compensation required to meet the system conditions, and thus minimize system losses.

Thyristor-switched SCs can be utilized for power flow control and damping of power oscillations. The advantages of thyristor application, compared to the traditional circuit breaker technology, are the speed of operation, number of possible operation in a short time, ease of maintenance, and possibility of point-on-wave switching for lower transients.

A consideration in the application of SCs is the phenomenon of SSR, which may evolve as a result of energy exchanges between the series-compensated system and an electrically close turbine-generator shaft. This condition can cause shaft fatigue and damage. The result of some studies [30] shows that SSR may occur at a nearby generator unit (Amos Unit No. 2) if the compensation level is 50%, while the system is weakened by a specific set of outages. Relay protection is being provided to trip the generator in case SSR oscillations occur [25].

6.7.2 Kayenta TCSC in United States

The Kayenta TCSC, commissioned in 1993 by Siemens at the Kayenta substation in Arizona, United States, is the world's first three-phase TCSC. The TCSC was installed in the Western States Coordination Council (WSCC) regional system, on a 230 kV line, 320 km, between Glen Canyon and Shiprock substations (Figure 6.45) in order to increase the transmission capacity from 300 to 400 MW. The WSCC system includes a large number of power plants both hydro and thermal, and many SC-compensated lines.

As shown in Figure 6.46, the Kayenta TCSC is part of an Advanced Series Compensation ([ASC], name proposed by Siemens) system, which consists of two conventional SC banks, each rated at 165 MVAr, with a 55 Ω single-phase impedance, and 1000 A. One of the two capacitor banks is split into a 40 and a 15 Ω segment. A TCR was introduced in parallel with the 15 Ω segment thereby creating a TCSC, which is capable of providing a control of the transmission line impedance. By varying the firing angle at the thyristors, the impedance of the parallel reactor is changed so is the equivalent impedance of the TCSC, seen by the transmission line. In this way, the ASC can smoothly control the capacitive impedance from 15 to 60 Ohms [31–34].

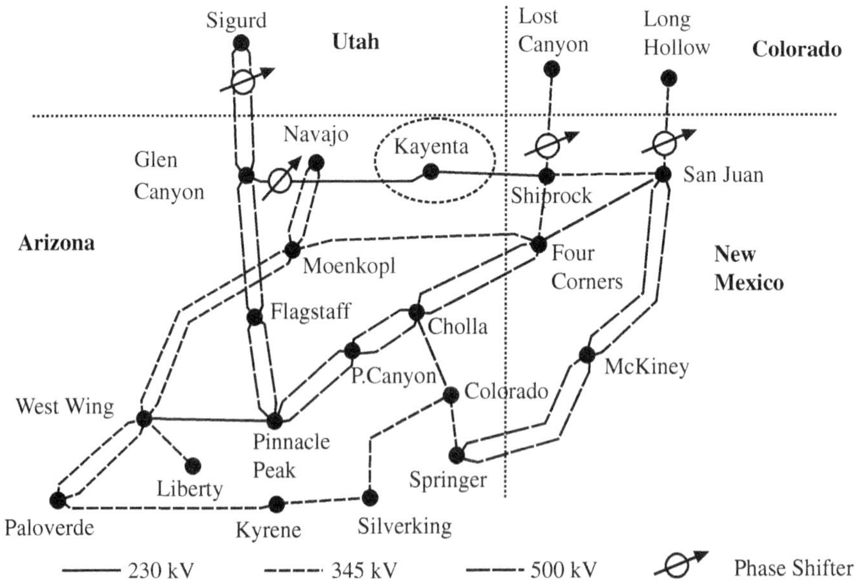

Figure 6.45 System diagram of transmission system near Kayenta substation [5].

The TCSC is able to withstand an 1100 A steady-state overloading, 1330 A for 30 min, and 1660 A for 1 min, respectively.

The TCSC segment provides the following advantages [2]: flexible, continuous control of SC compensation; direct and dynamic control of power flow; short-circuit current reduction by rapidly changing from a controlled capacitive to inductive impedance; SSR mitigation; improved protection and rapid reinsertion of the SCs during system faults; and reduces the DC offset of the capacitor voltage within a few cycles.

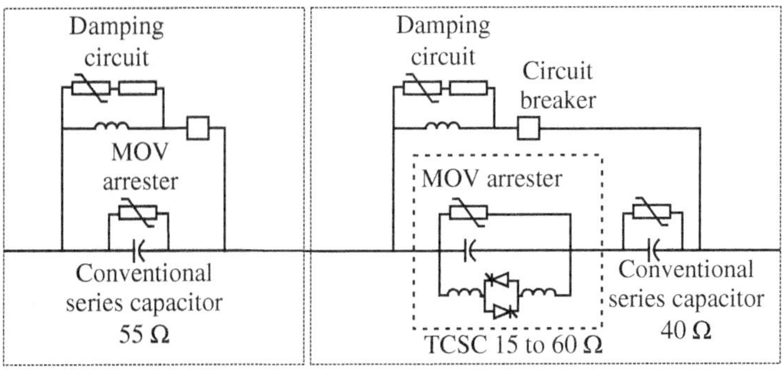

Figure 6.46 The one-line diagram of the Kayenta ASC [2, 5, 32].

Figure 6.47 ASC at Kayenta substation [32, 33].

Figure 6.47 shows two of the three platforms, one for each phase. Each platform has one 15 Ω (45/3 MVAr) TCSC and one 40 Ω (120/3 MVAr) conventional SC. The 15 Ω TCSC portion occupies the front part of the entire installation including the 40 Ω conventional capacitor located at the back. A reactor of approximately 3.0 Ω is divided into two reactors of approximately 1.5 Ω each, located on the two sides of the valve module.

Each of the three thyristor valves has 11 bidirectional parallel-connected thyristor pairs forming individual levels. Of the 11 series-connected thyristor levels, one level is redundant, allowing the valves to provide normal service in the event of failure of any one thyristor level. The thyristors are 100 mm cell diameter, rated for 3.5 kA and 5.5 kV.

Figure 6.48 shows the steady-state impedance characteristics of the TCSC in both the capacitive and the inductive range. The inductive range can be obtained by varying the thyristor turn-on angle from 90° up toward 180°, and the capacitive range is obtained by varying the angle from 90° down toward 0°. Dynamically, the reactance in either inductive or capacitive range can be varied up to 60 Ω. However, the control for the variable inductive side is not installed and therefore the available inductive side operation is only at fixed impedance of 3.0 Ω. Capacitive side operation is available from −15 to −60 Ω; however, this range is subject to the maximum capacitor voltage of 2.0 p.u. Therefore, its continuous operation at −60 Ω will be permissible line current up to 250 A [2].

Since the TCSC is a single module, it cannot cover the variable reactance range from inductive 3.0 Ω to capacitive 15 Ω. Thus, change in reactance from capacitive to inductive involves a jump from −15 to +3.0 Ω, and vice versa; however, this reactance band is not needed for this line.

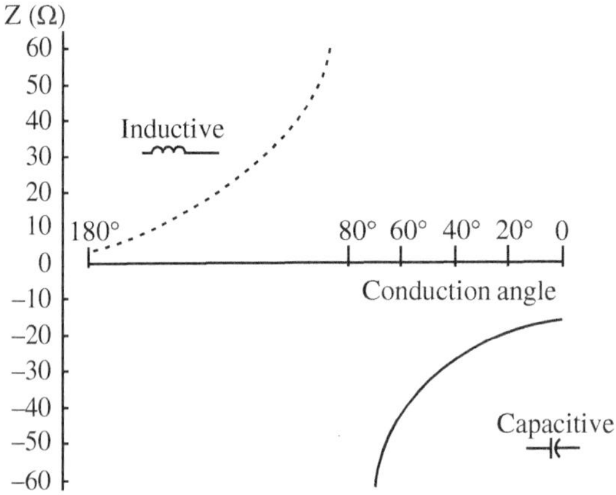

Figure 6.48 Kayenta TCSC impedance characteristic. *Source*: Hingorani & Gyugyi 2000 [2]. Reproduced with permission of IEEE.

6.7.3 Slatt TCSC in United States

The world's first and the multi-module TCSC system has been installed and is in operation from 1995 on the Slatt-Buckley 500 kV line in Northwestern Oregon (Figure 6.49). Transmission lines from four generation sites; two dams, one coal-fired steam power plant, and one nuclear plant, terminate at Slatt substation (BPA). The surplus generated power in this area is sent to western load centers and to California along the Third AC Inertie. Slatt was chosen for the TCSC site because it offered the best opportunity to test the TCSC capabilities, specifically SSR mitigation and also offered some system benefits such as loss savings. The Slatt TCSC was a research and development project designed to test the operational and dynamic capabilities of a TCSC [35, 36].

The objectives of the project were to produce a commercial multi-module TCSC, to test it thoroughly under severe conditions, and to operate continuously as an integral part of BPA transmission system. Apart from being the highest AC voltage (500 kV), in the WSCC grid system, this location has high short-circuit current (20.3 kA), and it also provides an opportunity for evaluating TCSC interaction with a thermal power plant.

Nevertheless, this TCSC has provided benefits of transmitting more power from Northwest to Southwest and alleviated SSR concerns for the nearby Boardman thermal power plant.

Figure 6.50 shows a one-line diagram of the Slatt TCSC installation. Each phase consists of six identical TCSC modules connected in series. Each module includes a capacitor, MOV, reactor, and bidirectional thyristor valve. A bypass breaker with its associated reactor is connected across each platform for use in operational and functions protection. Also, three motor operate disconnect switches are used to bypass and isolate the TCSC from the 500 kV Slatt-Buckley transmission line [36].

The nominal rating of the Slatt TCSC is 8 Ω at 2900 A, which is equivalent to 202 MVAr. The basic ratings are summarized in Table 6.5.

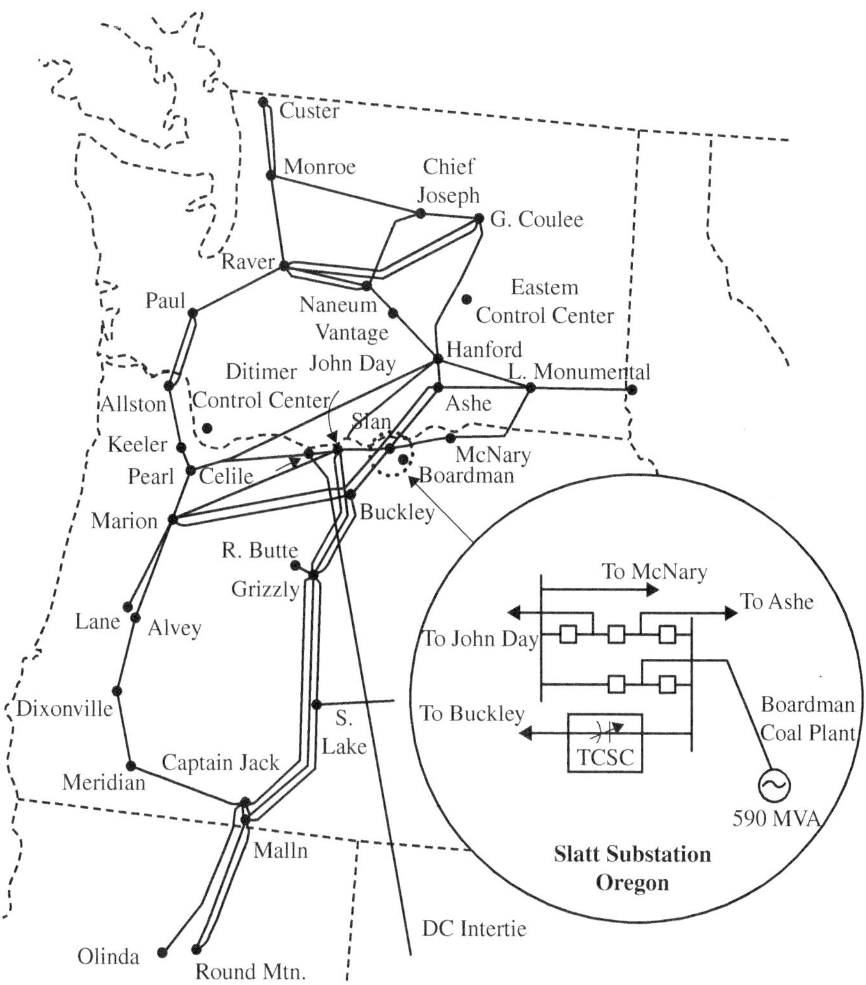

Figure 6.49 BPA's Slatt TCSC location in Pacific Northwest 500 kV transmission system. *Source*: Urbanek et al. 1993 [35]. Reproduced with permission of IEEE.

Each module is independently controllable and can be operated with thyristors fully blocked, in which case its impedance is 1.33 Ω capacitive. Each module can also be fully bypassed with its thyristor switch (thyristors conducting with zero delay angle), in which case its impedance is approximately 0.2 Ω inductive. In this case, the module impedance is a parallel combination of 0.18 Ω inductor and 1.33 Ω capacitor. For an input current of I_{ac}, a current of 1.15 I_{ac} flows through the inductor and – 0.15 I_{ac} through the capacitor; that is, a current of 0.15 I_{ac} circulates in the inductor-capacitor loop. Being inductive in this bypass mode, it can also serve as a current limiter, as required.

Each module can also be operated with its thyristor switch turn-on angle controlled from 180° (1.33 Ω capacitive) to lower than 180° up to a range where its

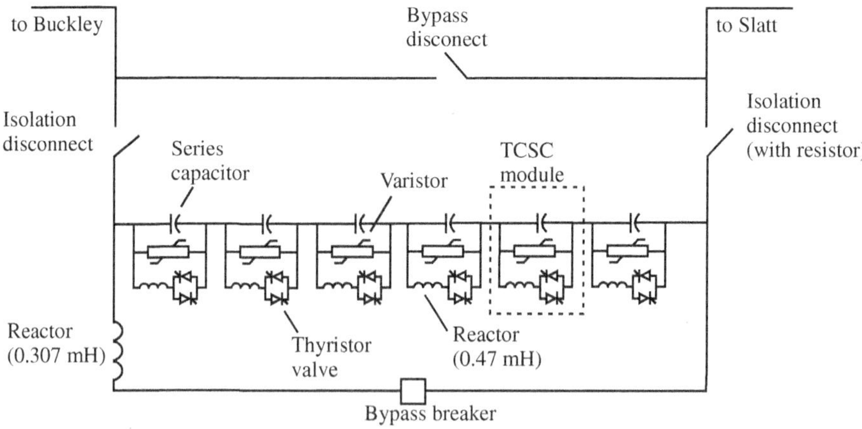

Figure 6.50 Multi-module TCSC system installed at the 500 kV Slatt substation [5, 35, 36].

reactance is 4.0 Ω capacitive. In this case, the module is an effective combination of a 2.0 Ω inductor in parallel with a 1.33 Ω capacitor at fundamental frequency (Figure 6.51) [2, 35].

Impedance of the inductor with partial conduction has effectively decreased from infinity (at 180°) to 2.0 Ω at some angle of advance from 180°. For a fundamental input current I_{ac}, a fundamental current of 2.0 I_{ac} flows in the inductor (and the thyristor switch) and 3.0 I_{ac} flows in the capacitor, that is, 2.0 I_{ac} circulates in the inductor-capacitor loop.

At 4.0 Ω, the fundamental voltage across the capacitor would be $4I_{ac}$. Given the limitation of capacitor voltage capability, the continuous current rating at 4.0 Ω

TABLE 6.5 Slatt TCSC parameters (six modules) [35]

Parameter	Value
– Maximum system voltage (line-to-line)	550 kV
– Nominal capacitive reactance (29% series compensation with all six modules)	8.0 Ω
– Continuous effective capacitive reactance (with normal minimum thyristor control)	9.2 Ω
– Net inductive reactance (all six modules bypassed)	1.2 Ω
– Maximum dynamic range of capacitive reactance	24.0 Ω
– Nominal three-phase MVAr compensation	202 MVAr
– Rated line current (1.0 p.u.)	2900 A rms
– 30 min overload current (1.5 p.u.)	4350 A rms
– 10 s overload current (2.0 p.u.)	5800 A rms
– 10 s overvoltage (2.0 p.u.)	53.3 kV
– Maximum fault current through TCSC	20.3 kA rms
– Maximum fault current in the thyristor valve	60 kA peak

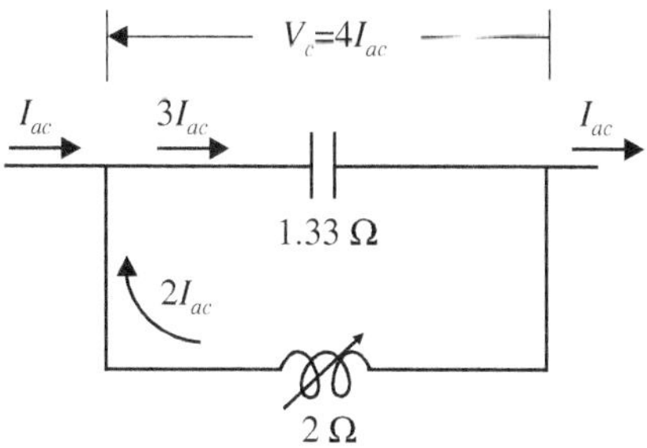

Figure 6.51 Current flow with TCSC in operation at 4 Ω (3 p.u.) [2, 35].

module-reactance is limited to a lower limit than the rated line current of 2900A (Table 6.5).

The range of the permissible operation is a function of the line current and its duration. Figure 6.52 illustrates this reactance capability.

Figure 6.52a shows the reactance-current capability curve for one TCSC module. It shows that it can continuously operate at high ohmic value up to 4.0 Ω capacitive, but with declining current capability. Also shown are the areas of short time overload capabilities, which show that the TCSC's dynamic operating range is twice its continuous rating. It can operate at 16 Ω for 10 s and 12 Ω for 30 min. The temporary overload limit is determined by the IEEE standard 824, for SCs. A module can be operated at any point in the shaded areas and also on the line of –0.2 Ω inductive. An individual module cannot operate between –0.2 and 1.33 Ω.

However, since the individual modules can be operated independently at any point of their characteristic, including bypass with their thyristor switches, it is easy to realize that a *combination of six modules* would cover the entire range of operation from –1.2 to 26.0 Ω, as shown in Figure 6.52b. If all six modules were to be combined into one large unit, the operating characteristic would be the same as for one module, Figure 6.52a with ohmic scale multiplied by 6. There would therefore be a prohibited

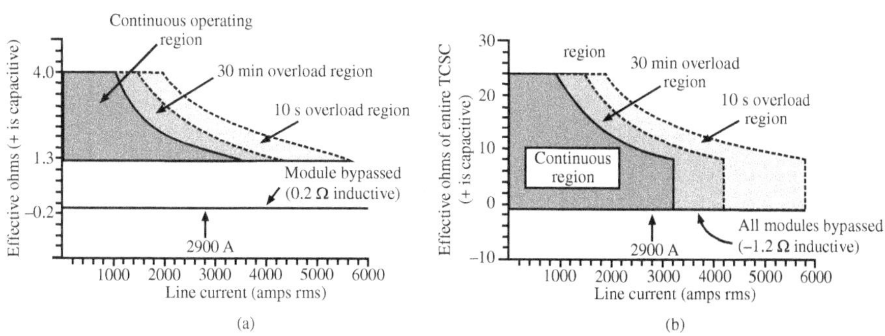

Figure 6.52 Capability curves for Slatt TCSC: (a) capability curves for one TCSC module and (b) capability curves for the entire six-module TCSC. *Source*: Hingorani & Gyugyi 2000 [2]. Reproduced with permission of IEEE. *Source*: Urbanek et al. 1993 [35]. Reproduced with permission of IEEE.

impedance zone from –1.2 to 8.0 Ω. The continuous ohmic control range would only be from 8.0 to 26.0 Ω and along the bypass mode line of –1.2 Ω, as against the entire range of –1.2 to 26.0 Ω. This is a clear advantage of the modular approach to the design of a TCSC, apart from the fact that the modular approach allows continued partial operation of the TCSC in the event of problems with individual modules.

The modules are not generally operated with thyristors in blocking mode, but rather at an angle somewhat less than 180°. It is operated at 1.53 Ω instead of 1.33 Ω (9.2 Ω for all six modules) in steady-state minimum capacitive mode, to provide damping for SSR modes [2]. This TCSC is not designed to operate as a thyristor-controlled series reactor.

Based on the application studies and the demonstrated capabilities of the equipment at Slatt, the TCSC offers the following power system performance benefits [36]:

- The TCSC does not participate in SSR and does not interact with machine torsional oscillations. The TCSC can be used to avoid and mitigate SSR problems and to achieve higher series compensation and power transfer levels.
- Can modulate its impedance to damp power system oscillations, thus increasing system stability and power transfer levels.
- Can utilize its short-term overload capability and temporarily increase its impedance to increase transient stability and power transfer levels.
- Can effectively regulate power flow on a transmission line by adjusting the line impedance. This can be useful in balancing power flow among transmission paths.

6.7.4 Stöde TCSC in Sweden

The 420 kV transmission system between Northern and Middle Sweden comprises eight lines with eight series capacitors (SCs), having a total rating of 4800 MVAr. The Stöde SCs are located in the middle of Sweden about 400 km North of Stockholm and 300 km North of Forsmark nuclear power station [37].

The Stöde conventional SCs, which were in service for many years, were designed to increase the power flow capability of the 400 kV transmission system from the hydro resources in the northern part of Sweden to the load center in the middle and southern parts of Sweden (Figure 6.53). It was recognized in the 1970s that the introduction of series compensation in lines that are closely connected to thermal generating stations might introduce a negative damping on the mechanical tensional model in the turbine-generator shaft system. In the electrical system, these oscillations occur in the subsynchronous frequency range. Therefore, these phenomena are generally referred to as SSR effect [37]. A solution for eliminating the problem was to split the existing SC of the Stöde unit into two segments. Of the original SC, 70% (51 Ω) was kept intact as a fixed component and the balance, representing 30% (22 Ω) segment has been converted into a TCSC. The 30% balance was selected so that if any failure of TCSC equipment occurs, the line could still operate with a reduced value of series compensation, while the TCSC could be bypassed [16].

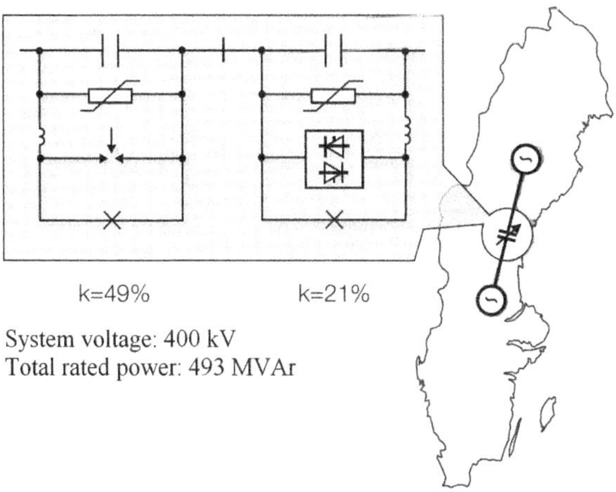

Figure 6.53 One-line diagram Stöde TCSC [38].

TABLE 6.6 Stöde basic design parameters [37]

Parameter	Value
• Physical capacitor reactance	18.25 Ω/ph
• Apparent TCSC reactance in the steady state	21.9 Ω/ph
• Rated current	1500 A, rms continuous
• Short-time current	2025 A, rms 30 min 2250 A, rms 10 min

The design parameters of the Stöde TCSC installation, which was commissioned in 1998, are given in Table 6.6.

With the TCSC in operation, the SSR risk will be eliminated for all possible system operating conditions. Unwanted bypass or tripping operations will be avoided, resulting in the benefits of the increased availability of active power from the Forsmark generating station, as well as a high level of power transmission being preserved throughout the system.

A significant aspect of the Stöde TCSC is that it demonstrates how an existing series-compensation installation can be reorganized partly by employing existing equipment and partly by employing some newly acquired apparatuses. If the TCSC component is set up on independent platforms, the line is also not required to be out of service for a long period. The Stöde TCSC was a precursor for the part conversion of existing series-compensated lines into TCSC schemes [16].

6.7.5 Imperatriz-Serra da Mesa TCSC in Brazil

The North-South interconnection in Brazil, connecting the north-northeast (NNE) system and the south-southeast (SSE) system, was developed in three steps. The first

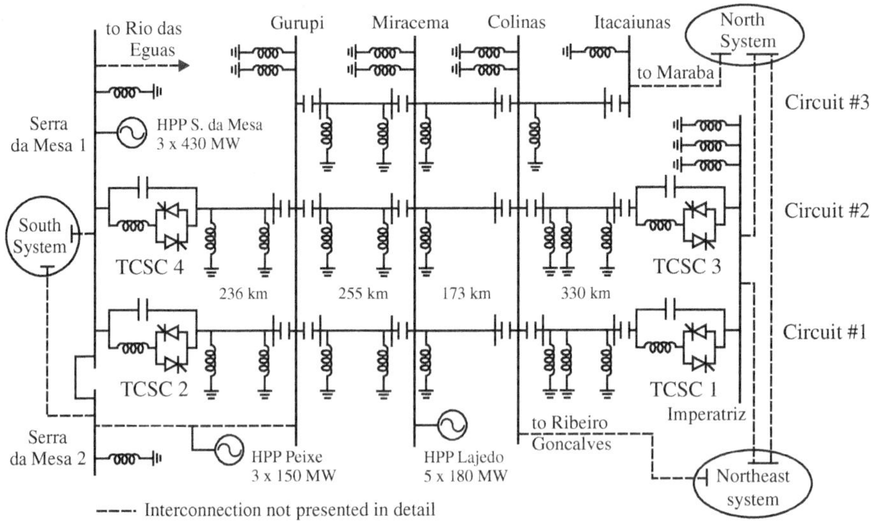

Figure 6.54 One-line diagram of the Imperatriz -Serra da Mesa transmission line. Adapted from [39–41].

500 kV compact transmission line (4 × 954 MCM bundle), 1020 km long, linking the Imperatriz substation in the North system to the Serra da Mesa substation in the South system, was energized in 1999 (Circuit #1 in Figure 6.54). At that time, the NNE system included 14 GW installed power, while the SSE system included 48 GW, primarily in hydroelectric power plants. It was dimensioned to transmit power up to 1300 MW, with suitable operation being required from no load up to maximum flow, in both directions [39–41].

Following the increased load demand, new power plants were commissioned in the Brazil transmission system. In order to increase the transmission capacity between the North and South systems, a second circuit (Circuit #2 in Figure 6.54) was commissioned in 2004 between Serra da Mesa and Imperatriz. A third circuit (Circuit #3 in Figure 6.54) was constructed in 2008 between Gurupi and Colinas, while other interconnection lines were constructed between various stations of the NS interconnection and other parts of the Brazil power system.

Circuits #1 and #2 are provided with 100% shunt compensation (eight line reactor banks and other shared bus reactor banks, as shown in Figure 6.54, each rated for 136 MVAr, except for one bus reactor bank in Imperatriz which is 165 MVAr) and 54% fixed-series compensation distributed on six series-capacitor banks (161 MVAr or −23.8 Ω each) along the transmission trunk. The rated capacity of the SCs was defined based on 1500 A (1300 MW). Circuit 3 is identical with the other circuits between Gurupi and Colinas.

The SCs installed in the NS interconnection have the task of raising the steady state and dynamic stability of the intertie. The transient stability is not an issue because of the fixed SCs. However, due to the interconnection length, low frequency poorly damped interarea oscillation modes emerge [15]. For this reason, in order to

TABLE 6.7 The TCSC controller data [39]

Parameter	Value	
	TCSC 1	TCSC 2
– Maximum system voltage (phase-to-phase)	550 kV	
– Nominal reactive power	108 MVAr	107.5 MVAr
– Nominal capacitive reactance (X_C = 1 p.u.)	13.27 Ω/ph	
– Nominal capacitive boost factor (X_{app}/X_C)	1.2 p.u.	
– Continuous effective capacitive reactance (X_{ef} = 1.2 p.u.)	15.92 Ω/ph	
– Maximum dynamic range of capacitive reactance ($3X_C$)	39.81 Ω/ph	
– Reactor inductance (TCR)	5.61 mH	5.5 mH
– Inductive reactance in bypass	2.46 Ω/ph	2.52 Ω/ph
– Rated line current (1 p.u.)	1500 A	
– 30 min overload current (1.5 p.u.)	2025 A	
– 10 s overload current (2 p.u.)	3000 A	
– Rated continuous TCSC voltage (V_{max}=1 p.u.)	23.88 kV	
– 30 min overload voltage (1.5 p.u.)	35.82 kV	
– 10 s overload voltage (2.0 p.u.)	47.76 kV	
– Maximum thermal short-circuit current	23 kA	

make the interconnection feasible, four TCSCs were installed at the ends of Circuits #1 and #2.

The TCSC bank of Circuit #1 installed at Serra da Mesa, supplied by Siemens for Furnas, consists of a 13.25 Ω SC in parallel with a 2.07 TCR, whereas the TCSC bank of the same circuit installed at Imperatriz, supplied by ABB for Eletronorte, consists of a 13.27 Ω SC in parallel with a 2.11 Ω TCR. [39]. In steady-state conditions, the TCSCs act as a fixed capacitor and their reactance is maintained at –15.92 Ω (6% of the total line reactance). This corresponds to a continuous capacitive voltage boost factor of 1.2, which means that the apparent reactance at fundamental frequency is 1.2 times the physical reactance of the SC. The vernier control of both TCSCs ranges from 6% (13.25 Ω, with blocked thyristor valve) up to 15% (39.81 Ω, corresponding to a capacitive voltage boost factor of 3.0) of the transmission line reactance, providing controllable reactance to counteract power oscillations (interarea mode) [40]. The basic parameters of TCSCs 1 and 2 are listed in Table 6.7.

The impedance-current capability characteristic of the TCSC installed in the Imperatriz substation of Brazil's NS interconnection is presented in Figure 6.55 [39], where X_{ef} is the nominal boost level, X_{bypass} is the boost level at TCSC bypass, and X_C is the unity boost level.

The TCSC banks installed on the Circuit #2 were supplied by GE for Novotrans, and they have been designed with characteristics similar to those installed on the Circuit #1.

During the commissioning works in 1999 of the first 500 kV interconnection circuit, the TCSCs were tuned for damping 0.2 Hz power oscillations, that is for magnitudes of oscillations up to ±300 MW. With the integration of Circuit #2 in 2004, then Circuit #3 in 2008, on the NS interconnection, as well as other transmission lines

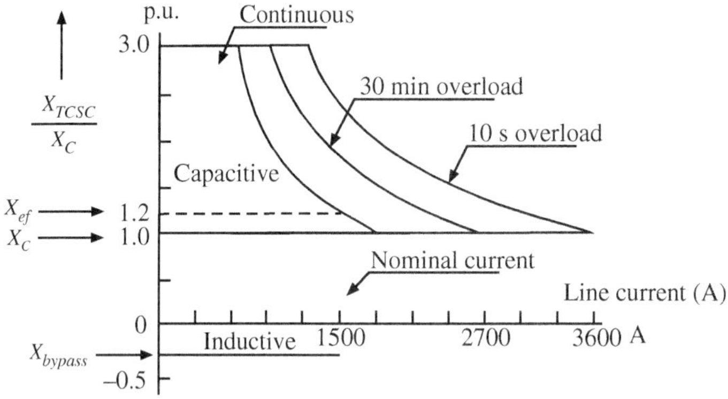

Figure 6.55 Impedance-current capability characteristic of the TCSC installed in the Imperatriz substation [39].

in the system, the interarea mode has changed so that today all TCSCs are tuned to damp power oscillations of 0.35 Hz.

The challenge is that the four TCSCs are under the responsibility of different transmission agents and use unlike input signals of active power flow. By the contribution of the TCSCs, the NS interconnection lines are operated at different loadings even if they are operated in parallel. This aspect is a great challenge for the power system operation, especially during the various disturbances occurring in the power system.

6.7.6 Purnea and Gorakhpur TCSC/FSC in India

The world's biggest series compensation installations consisting of fixed series compensation (FSC) and TCSC were put into commercial operation in 2006 at Purnea and Gorakhpur substations, at the ends of a double circuit 420 kV, 475 km, transmission line in the northern part of India. The transmission line was built to carry power from the Tala hydroelectric power plant, located in Buthan, toward the industrialized region around Delhi. The series compensation installations were required in order to improve the system stability. The TCSCs were designed to provide appropriate damping of power oscillations that may occur during or after perturbations.

Figure 6.56 shows the single-line diagram of either series compensation installation. A fixed SC segment is an assembly of capacitor banks in parallel with MOVs. The aim of the MOVs is to limit the voltage across the capacitors during a fault on the system. Additionally, a triggered spark gap is used to spark over in the event of excess varistor energy. The bypass-switch, which is connected in parallel with the spark gap, closes automatically in the case of prolonged gap conduction, and the current capability of the MOV or the spark gap is exceeded during AC faults. The damping reactor circuit is introduced in series with the triggered spark gap and the

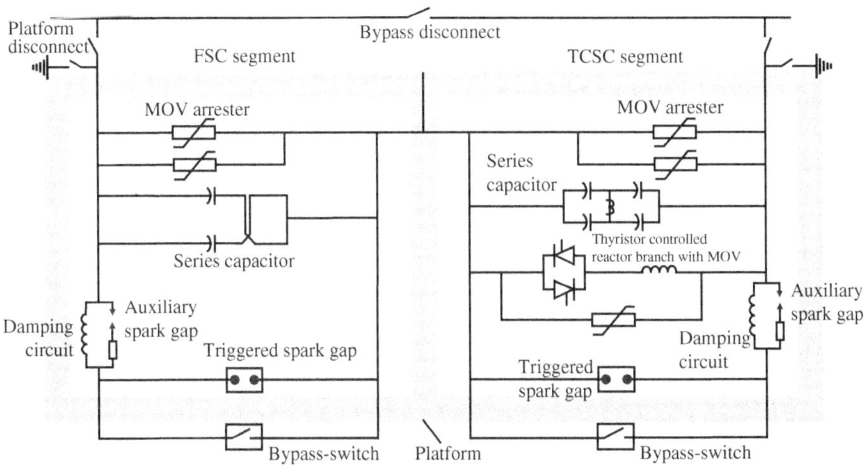

Figure 6.56 Single-line diagram of Purnea/Gorakhpur substation for FSC and TCSC [42].

bypass-switch in order to reduce the capacitor discharge resulting from gap sparkover or bypass breaker closure.

The technical data of the worldwide biggest FSC/TCSC project are presented in Table 6.8.

The main components of the TCSC segment are also shown in Figure 6.56. Typical protection elements, similar to the FSC segment, are also used at the TCSC segment. However, a TCR branch is connected in parallel with the capacitor to provide variable impedance. The impedance of the parallel L-C circuit can be changed by varying the firing angle of the thyristor valves. In case of AC line faults, the thyristor valve can be blocked in order to reduce the capacitive reactance and consequently the fault current [42].

Basically, the thyristor valve could have been used to provide the bypass function alone, without the spark gap. For this project, however, it was decided to use a separate triggered spark gap as a fast protecting device, because of very high thermal stress for the thyristors caused by extremely high fault currents. This enables immediate operation of the thyristor valve in the full dynamic range, for example, for the power oscillations damping which is normally required after fault

TABLE 6.8 Technical data for Purnea-Gorakhpur TCSC/FCS [42]

Substation parameter	Purnea	Gorakhpur
Rated nominal voltage	420 kV	420 kV
Rated nominal power/ compensation degree	743 MVAr FSC / 40% 112 MVAr TCSC / 5–15%	716 MVAr FSC / 40% 108 MVAr TCSC / 5–15%

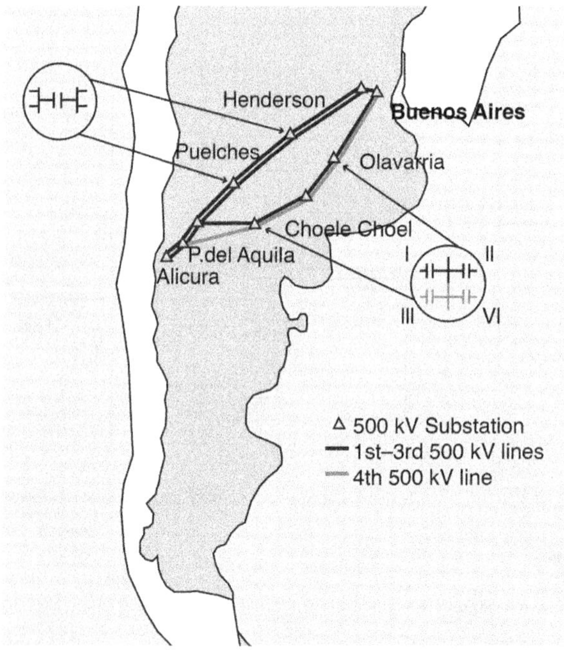

Figure 6.57 Transmission corridors in Argentina [43].

clearing in order to stabilize the system. The capacitor discharge current is limited by means of the bypass damping circuit in the same way as in the case of the FSC [42].

6.7.7 Series-Compensated 500 kV Power Transmission Corridors in Argentina

Transener's 500 kV AC power transmission corridor, enabling vast amounts of hydro power to be transmitted from the Comahue region in south-western Argentina to the large consumer areas around Buenos Aires, a distance of more than 1000 km. Transmission is performed by means of four parallel 500 kV AC lines, all series compensated (Figure 6.57).

In addition to series compensation, shunt reactor compensation at intervals along the corridor has been implemented, as well, in order to maintain satisfactory system conditions over the full range of power transfer.

Ratings of individual SCs range from less than 100 MVAr to close to 700 MVAr (Table 6.9) [43].

The evolution/expansion procedure over the years can be summarized as follows:

MID-1970s: first and second lines in operation, including four SCs (Puelches and Henderson); total power transmission capacity: $\Sigma P = 1650$ MW.

MID-1980s: third line in operation.

TABLE 6.9 Technical data, series capacitors

			Choele Choel				Olavarria			
	Puelches	Henderson	I	II	III	IV	I	II	III	IV
Rated power, MVAr (3-phase)	681	596	215	170	156	104	245	135	98	72
Rated current, kA	2.80	2.62	1.49	1.32	1.35	1.32	1.59	1.18	1.20	1.00
Varistor rating, MJ (3-phase)	107	128	125	12	71	19	29	96	41	64
Degree of comp, %	40	40	40	40	30	30	40	40	30	30

Type of protection: ZnO with bypass gap.

MID-1990S: four SCs added to the third line (Olavarria and Choele Choel). Power transmission capacity increased to Σ P = 2900 MW.

1996: first and second lines SC upgrade (Puelches and Henderson). Power transmission capacity further increased to Σ P = 3300 MW.

1999: fourth line in operation, including four SCs (Olavarria and Choele Choel); overall power transmission capacity Σ P = 4600 MW.

Thus, part of ABB undertaking with the fourth line was also the supply and commissioning of a number of switched shunt reactors with ratings between 150 and 250 MVAr (three phase) at the intermediate transformer stations to limit the voltage rise during low load conditions. Furthermore, two SVCs each rated at −160/+160 MVAr have been in operation at the receiving end of the corridor since 1983 for dynamic control of the voltage and power factor in the system.

Totally, there are 10 SCs in operation in the power corridor, altogether rated at close to 2500 MVAr at 500 kV. The purpose of the series compensation of the four lines is to enable an increase in the active power transmission capability of the power corridor, a task which is fulfilled in several ways at the same time:

- by raising the transient stability limit of the lines;
- by improving the reactive power balance and voltage regulation of the lines;
- by improving the active power sharing between the lines.

The 10 SCs are all protected by means of zinc-oxide varistors, protecting the capacitors in the event of faults in the transmission system. In case of external faults, that is faults not located in the same line section as the SC, the SCs are reinserted into the circuit instantly upon clearing of the fault. This safeguards the highest possible dynamic stability of the system. In case of internal faults, that is faults located inside a series-compensated line segment, the SC is allowed to be bypassed, and is reinserted in due course after reinsertion of the faulted line.

Without the SCs, several additional 500 kV lines would have to be built to enable stable transmission of the same amount of power through the corridor.

The TCSCs commissioned in various power systems so far are shown in Appendix 6.1 [23].

APPENDIX 6.1 TCSC SYSTEMS AROUND THE WORLD. REPRODUCED WITH PERMISSION OF CIGRE [23]

Installed (year)	Location	Configuration	Rating	Purpose
1991	Kanawha River AEP, United States	3 × FC	345 kV, 788 MVAr, and 2500 A	
1992	Kayenta, WAPA, United States	FC plus TCSC	230 kV, 2 × 165 MVAr (TCSC module: 45 MVAr); 1000 A	Increase power transfer, SSR (TCSC system converted to FC)
1993	Slatt System, BPA, United States	6 TCSC modules in series	500 kV, 208 MVAr, and 2900 A	System to demonstrate power flow control, SSR damping, stability control, and fault performance
1997	Stöde, Svenska Kraftnät, Sweden	FC plus TCSC	400 kV, 493 MVAr total (TCSC 148 MVAr); 1500 A	SSR control
1999	Imperatriz, Eletronorte, Brazil		500 kV, 108 MVAr, and 1,500 A	System damping
1999	Serra da Mesa, FURNAS Centrais Elétricas S.A Brazil		500 kV, 107 MVAr, and 1500 A	System damping
2002	Pingguo, State Power South Company, Guangzhou, China	2 × FC plus 2 × TCSC	500 kV, TCSC 55 MVAr; FSC 350 MVAr, 2000 A	System damping
2004	Serra da Mesa, Novatrans, Brazil		500 kV, 107.5 MVAr, and 1500 A	System damping
2004	Imperatriz, Novatrans, Brasil		500 kV, 107.5 MVAr, and 1500 A	System damping
2004	Chengxian, Gansu Electric Power Co. China	TCSC	220 kV, 86.7 MVAr, and 1000 A	Damping and improving power transmission capability
2004	Raipur end of Rourkela - Raipur 400 kV DC line; Power Grid Corporation of India, Ltd.	2 × FC plus 2 × TCSC	400 kV, 394 MVAr FC and 71 MVAr TCSC, 1550 A	System damping

Installed (year)	Location	Configuration	Rating	Purpose
2006	Gorakhpur end of Muzaffarpur—Gorakhpur 400 kV DC line; Power Grid Corporation of India, Ltd.	2 × FC plus 2 × TCSC	420 kV, 716 MVAr FC, and 107 MVAr TCSC	System damping
2006	Purnea end of Muzaffarpur—Purnea 400 kV DC line; Power Grid Corporation of India, Ltd.	2 × FC plus 2 × TCSC	420 kV, 743 MVAr FC, and 112 MVAr TCSC	System damping
2009	Fengtun, Northeast China Grid Co., Ltd.	TCSC	500 kV, 326 MVAr, and 2330 A	System damping

REFERENCES

[1] Anderson, P. M., and Farmer, R. G. *Series Compensation of Power Systems*. PBLSH!Inc., Encinitas, California, USA, 1996.

[2] Hingorani, N. G., and Gyugyi, L. *Understanding FACTS. Concepts and Technology of Flexible AC Transmission Systems*. IEEE Press, NY, 2000.

[3] Karady, G. G., Ortmeyer, T. H., Pilvelait, B. R., and Maratukulam, D. Continuously regulated series capacitor. *IEEE/PES Summer Meeting*, Paper No. 92 SM 492-9 PWRD, Seattle, WA, July 1992.

[4] Nilsson, S. Apparatus for controlling the reactive impedance of a transmission line. US Patent No. 4,999,565, issued March 12, 1991.

[5] Torgerson, D. R. (convener) Allaire, J. F., Chakravorty, S., Clark, K., Edris, A., Gama, C., George, T., Kinney, S., Othman, H.A., Ping, W. W., Renz, K., Thanawala, H. L., Wegner, C., Weiss, S. Thyristor-controlled series capacitor. Working Group 14.18, CIGRE Technical Brochure No. 123, 1997.

[6] Vithayathil, J. J. Scheme for rapid adjustment of network impedance. US Patent No. 5,032,738, January 22, 1986.

[7] Vithayathil, J. J., Taylor, C. W., Klinger, M., and Mittelstadt, W. A. Case studies of conventional and novel methods of reactive power control on an AC transmission system. CIGRE Paper 38-02, Paris, 1986.

[8] Christl, N., Lutzelberger, P., Pereira, M., and Sadek, K., Krause, E., Montoya, A. H., Torgerson, D. R., and Vossler, B. A. Advanced series compensation with variable impedance. *Presented at EPRI Workshop, The Future in High-Voltage Transmission Flexible AC Transmission Systems (FACTS)*, Cincinnati, Ohio, November 14–16, 1990.

[9] Helbing, S. G., and Karady, G. G. Investigation of an advanced form of series compensation. *IEEE Transactions on Power Delivery*, vol. 9, no. 2, pp. 939–947, April 1994.

[10] Larsen, E. V., Bowler, C. E. J., Damsky, B. L., and Nilsson, S. L. Benefits of thyristor-controlled series compensation. CIGRE Paper 14/37/38-04, Paris, 1992.

[11] Larsen, E. V., Clark, K., Miske Jr., S. A., and Urbanek, J. Characteristics and rating considerations of thyristor controlled series compensation. *IEEE Transactions on Power Delivery*, vol. 9, no. 2, pp. 992–1001, April 1994.

[12] Ängquist, L., Ingeström, G., and Othman, H. Synchronous voltage reversal (SVR) scheme – a new control method for thyristor controlled series capacitors. *Conference on Flexible AC Transmission Systems (FACTS 3): The Future of High Voltage Transmission*, Baltimore, MD, October 5–7, 1994.

[13] Ängquist, L., Ingestrom, G., and Jonsson, H. A. Dynamical performance of TCSC schemes. CIGRE Paper 14-302, Paris, 1996.

[14] Song, Y. H., and Johns, A. T., Eds. *Flexible AC Transmission System (FACTS)*. IEE Power and Energy Series, London, 1999.

[15] Grünbaum, R., Noorozian, M., and Thorvaldsson, B. *FACTS – Powerful Systems for Flexible Power Transmission*. ABB Review, pp. 4–17, May 1999.

[16] Mathur, R. M, and Varma, R. K. *Thyristor – Based FACTS Controllers for Electrical Transmission Systems*. IEEE Press, Willey Interscience, New York, 2002.

[17] Paserba, J. J., Miller, N. W., Larsen, E. V., and Piwko, R. J. A thyristor-controlled series compensation model for power system stability analysis. *IEEE Transactions on Power Delivery*, vol. 10, no. 3, pp. 1471–1478, July 1995.

[18] Therond, G. (convener), Adielson, T., Bortoni, P., Byrne, D., Ceresoli, Daniel, D., Foss, A. M., Fosso, O. B., Iyoda, I., Mihalic, R., Miller, N. W., Modlitba, P., Nelson, R., Thanawala, H. L., Uhlen, K., Witzmann, R. Modeling of power electronics equipment (FACTS) in load flow and stability programs. CIGRE Task Force 38.01.08, Brochure No. 145, August 1999.

[19] Pilotto, L. A. S., Bianco, A., Long, W. F., and Edris, A. A. Impact of TCSC control methodologies on subsynchronous oscillations, *IEEE Transactions on Power Delivery*, vol. 18, no. 1, pp. 243–252, January 2003.

[20] Fuerte-Esquivel, C. R., Acha, E., and Ambriz-Perez, H. A thyristor controlled series compensator model for the power flow solution of practical power networks. *IEEE Transactions on Power Systems*, vol. 15, no. 1, pp. 58–64, February 2000.

[21] Acha, E., Fuerte-Esquivel, C. R., Ambriz-Perez, H., and Angeles-Camacho, C. *FACTS: Modelling and Simulation in Power Networks*. John Wiley & Sons, 2004.

[22] Christl, N., Hedin, R., Sadek, K., Lutzelberger, P., Krause, P. E., McKenna, S. M., Montoya, A. H., and Torgerson, D. R. Advanced series compensation (ASC) with thyristor controlled impedance. *International Conference on Large High Voltage Electric Systems (CIGRE)*, Paris, September 1992, Paper 14/37/38-05.

[23] Nilsson, S. (convener), Aho, J., Chandrasena, W., Gama, C., Shen, H., Powers, K., Reynolds, M. A., Rudin, S., Santo, S. E., Sen, S., Tenório, R., Qin, X. Performance evaluation and applications. Reviews of existing thyristor control series capacitor devices - TCSC. Technical Brochure No. 554, CIGRE Working Group B4-49, 2013.

[24] Wang, H. F., Swift, F. J., and Li, M. Selection of installing locations and feedback signals of FACTS-based stabilizers in multimachine power systems by reduced-order modal analysis. *IEE Proceedings - Generation, Transmission and Distribution*, vol. 144, no. 3, pp. 263–270, May 1997.

[25] Ängquist, L., Halvarsson, P., and Othman, H. Adapting controllable series capacitors to different transmission purposes. *Presented at the Canadian Electric Association Meeting*, Toronto, March 1994.

[26] Keri, A. J. F., Ware, B. J., Byron, R. A., Mehraban, A. S., Chamia, M., Halvarsson, P., and Ängquist, L. Improving transmission system performance using controlled series capacitors. CIGRE, Paper 14/37/38-07, Paris, August 30–September 35, 1992.

[27] Hingorani, N. G. A new scheme for subsynchronous resonance damping of torsional oscillations and transient torque. *Part I, IEEE Transactions on Power Apparatus and Systems*, vol. PAS-100, no. 04, pp. 1852–1855, April 1981.

[28] Maliszewski, R. M., Pasternack, B. M., Scheren, H. N. Jr., Chamia, M., Frank, H., and Paulsson, L. Power flow control in a highly integrated transmission network. CIGRE 1990, Paper 37-303, Paris, August 26–September 1, 1990.

[29] Ängquist, L. New concepts in series compensation. *Canadian Electrical Association, SVC & HVDC Controls Committee*, 1992 Spring Meeting, Vancouver, 1992.

[30] Rama, R. D., and Hulf, S. W., Hayes, R.M., Schulz, R.P., AEP's Kanawha River 345 kV series capacitor installation – subsynchronous resonance studies and torsional measurements. *American Power Conference*, April 29–May 1, 1991.

[31] Christl, N., Hedin, R., Johnson, R., Krause, P., and Montoya, A. Power system studies and modeling for the Kayenta 230 kV substation advanced series compensation. *IEEE Fifth International Conference on AC and DC Power Transmission*, London, September 1991.

[32] Renz, K., Thumm, G., and Weiß, St. Advanced series compensator enhances flexibility for AC transmission systems. *IEEE 2nd International Conference on Advances in Power System Control, Operation and Management*, Hong Kong, December 1993.

[33] Juette, G., and Renz, K. FACTS controllers and HVDC enhance power transmission (a manufacturer's perspective). *IEEE Technical Applications Conference and Workshops Northcon95*, Portland, Oregon, pp. 294–304, 1995.

[34] Siemens. FACTS – Flexible AC Transmission Systems. Series compensation. Technical Brochure, Siemens AG, Energy Sector, 2010.

[35] Urbanek, J., Piwko, R. J., Larsen, E. V., Damsky, B. L., Furumasu, B. C., Mittlestadt, W., Eden, J. D. Thyristor-controlled series compensation prototype installation at the Slatt 500 kV. *IEEE Transactions on Power Delivery*, vol. 8, no. 3, pp. 1460–1469, July 1993.

[36] Miske, S. A. Jr., Lang, R. J., Bilodeau, P., and Granger, M. Recent series capacitor applications in North America. *Presented at CEA Electricity '95 Vancouver Conference*, March 25–31, 1995.

[37] Holmberg, D., Danielsson, M., Halvarsson, P., and Angquist, L. The Stöde thyristor-controlled series capacitor. CIGRE Paper 14-105, Paris, 1998.

[38] ABB. TCSC thyristor controlled series compensation. Keeping grids together. Technical Brochure, ABB, A02-0158 E, 2010-03.

[39] Gama, C., Leoni, R. L., Gribel, J., Fraga, R., Eiras, M. J., Ping, W., Ricardo, A., Cavalcanti, J., and Tenório, R. Brazilian North-South Interconnection: application of thyristor controlled series compensation (TCSC) to damp inter-area oscillation mode. CIGRE Paper 14-101, Paris, 1998.

[40] Gama, C., Ängquist, L., Ingeström, G., and Noroozian, M. Commissioning and operative experience of TCSC for damping power oscillation in the Brazilian North-South interconnection. CIGRE Paper 14-104, Paris, 2000.

[41] Noroozian, M., and Gama, C. Application of thyristor controlled series capacitor for damping of electromechanical oscillations in Brazilian system. *CIGRE Symposium on Power System Issues in the Rapidly Industrialising Countries*, Kuala Lumpur, Malaysia, Paper 240-05, September 20–23, 1999.

[42] Braun, K., Krummholz, A., Retzmann, D., Rohr, U., and Thumm, G. The world's biggest FACTS project with series compensation. *Energize*, pp. 36–39, April 2007.

[43] ABB. Series compensated 500 kV power transmission corridor in Argentina. ABB, Power Systems, Technical Brochure, Vasteras, Sweden, 1995.

CHAPTER 7

PHASE SHIFTING TRANSFORMER: MECHANICAL AND STATIC DEVICES

Mylavarapu Ramamoorty and Lucian Toma

7.1 INTRODUCTION

The electrical power systems experience increased fluctuations in the power flows due to the scheduled power exchanges under the liberalized electricity market and the increased penetration of wind and solar energy. Congestion in the transmission grid is a phenomenon that is encountered more often than before, and therefore power flow control is an issue that becomes increasingly important in meshed systems. The proof is that throughout Europe, voltage and phase angle regulating transformers, which are examples of power flow controllers, are installed at numerous locations. Steady-state power flow regulation by means of conventional phase-shifters has been a common practice by the utility industry for a long time.

The phase and voltage regulating transformers, also known as *phase-shifting transformer* (PST), *phase angle regulating transformer*, *phase angle regulator* (PAR, American usage), *phase shifter* (West coast American usage), or *quadrature booster* (QB) (quad booster, British usage), are a specialized form of transformer used to control the active power flow in three-phase electricity transmission networks.

The term *phase-shifter* is more generally used to indicate a device that can inject a voltage with a controllable phase angle and/or magnitude under no-load (off-load) and load (on-load) conditions.

A comprehensive description of conventional phase-angle regulating transformers is available in [1]. Operation of a conventional phase-shifter is characterized by [2]:

(i) high response time as a result of inertia of moving parts;

(ii) high level of maintenance due to mechanical contacts and oil deterioration.

Conceptually, the above mentioned drawbacks of a conventional phase-shifter transformer are overcome if mechanical switches are replaced with semiconductor (static) switches.

Advanced Solutions in Power Systems: HVDC, FACTS, and Artificial Intelligence, First Edition.
Edited by Mircea Eremia, Chen-Ching Liu, and Abdel-Aty Edris.

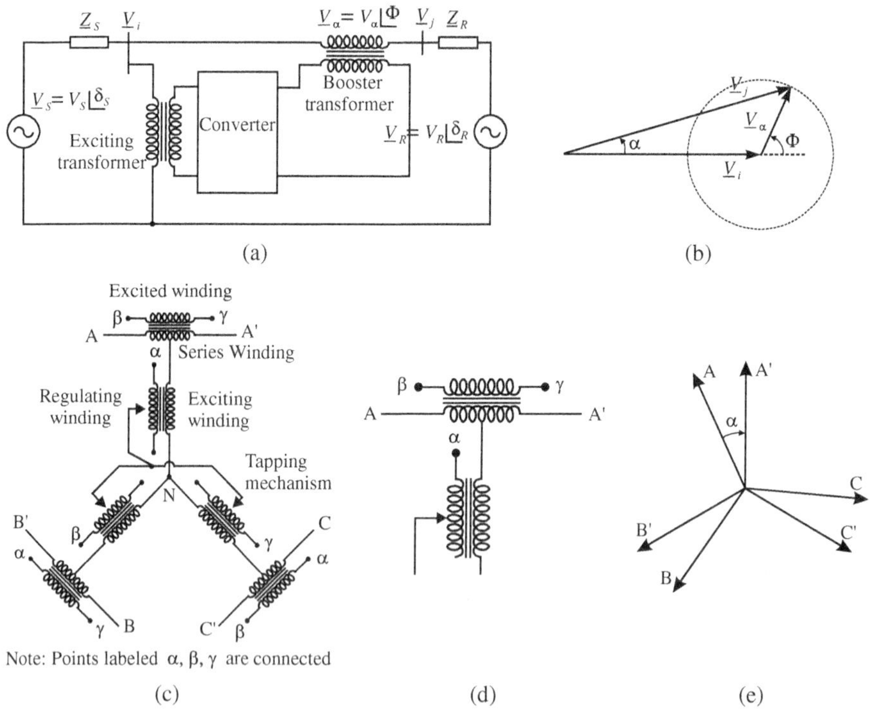

Figure 7.1 Diagrams of a phase-shifting transformer: (a) schematic diagram, (b) phasor diagram, (c) winding diagram, (d) simple schematic of quadrature booster, and (e) phase shift for advance [4].

The term *static phase shifter* (SPS) is used to distinguish semiconductor-controlled phase shifters from conventional (mechanical) phase shifters. The IEEE FACTS Terms & Definitions Task Force [3] defines "a Thyristor Controlled Phase Shifting Transformer (TCPST) as a phase-shifting transformer adjusted by thyristor switches to provide rapidly variable phase angle." It should be noticed that TCPST defines a subgroup of apparatus under more general category introduced by the term SPS.

7.2 MECHANICAL PHASE SHIFTING TRANSFORMER

7.2.1 Principle of Operation of the PST

The PST is a device that is used for power flow control in order to relieve congestions and minimize power losses in electrical grids. Figure 7.1a shows a schematic diagram of a phase shifter installed on a transmission line between buses *i* and *j*. The sending and receiving ends of the transmission line are represented by voltage phasors $\underline{V}_S$ and $\underline{V}_R$, and corresponding impedances $\underline{Z}_S$ and $\underline{Z}_R$, respectively [5].

The main power circuitry of the PST consists of [4] (Figure 7.1a):

- the exciting transformer, which provides input voltage to the phase shifter;
- the series transformer, which injects a controllable voltage $\underline{V}_\alpha$ in the transmission line;
- the converter or tap changer, which controls the magnitude and/or phase angle of the injected voltage.

A conventional PST is normally constructed as a three-phase device, with a total of 12 windings on six magnetic cores. Figure 7.1c shows the windings arrangement of a typical device. The exciting and the series windings, which have direct connections to the transmission line, do not have tapping devices (Figure 7.1d). All switching is accomplished by the regulating winding, which operates at a lower voltage. As the turns ratio of the regulating/exciting winding pair is increased by the tapping mechanism, a voltage is induced in the series winding, which is in quadrature to the phase voltage. This effectively introduces a phase shift in the voltage between the two terminals of the transformer. Figure 7.1e shows a phasor diagram for angular advance between nodes A and A′. Angular retard is achieved by the use of a reversing switch between the regulating and exciting winding.

Depending on the magnitude and phase angle of the injected voltage $\underline{V}_\alpha$, magnitude and/or phase-angle of the system voltage $\underline{V}_j$ is varied (Figure 7.1b). With a flexible PST, the control range achieved is a circle with center in the tip of the phasor $\underline{V}_i$ and radius equal to the amplitude of $\underline{V}_\alpha$. The output voltage of the PST is controlled by varying the amplitude and angle of the phasor $\underline{V}_\alpha$, that is V_α and Φ.

The active power flow on the transmission line that incorporates a PST is given by:

$$P = \frac{V_S V_R}{X_{eq}} \sin\left(\delta_S - \delta_R \pm \alpha\right) \tag{7.1}$$

where X_{eq} is the net equivalent reactance of the line and sources, and δ_S and δ_R are phase angles of phasors $\underline{V}_S$ and $\underline{V}_R$, respectively. Based on equation (7.1), the angle α is the dominant variable for power flow control. For a given system operating condition, the converter characteristics define the ranges that phase angle and amplitude of the injected voltage are controllable.

The converter section of a conventional phase shifter comprises of mechanical switches, which are usually embedded within the exciting transformer. A conventional phase shifter can vary angle α approximately within $\pm 40°$ in discreet steps about 1–2°.

Rapid phase angle control could be accomplished by replacing the mechanical tap changer by a thyristor switching network.

A particular class of similar applications is the Sen transformer, developed in detail in Chapter 12 of this book, as well as in the book "Introduction to FACTS controllers. Theory, Modeling and Applications" by Sen K. Kalyan and Mey Ling Sen, published by Wiley and IEEE Press in 2009 [5].

7.2.2 PST Topology

The PSTs are designed in various topologies, that is [6, 7]:

- **Direct-type PSTs** that provide a phase shift between voltages at its ends by connecting the windings in an appropriate manner to each other; they have one three-phase core design.
- **Indirect-type PSTs** are based on a construction using two separate transformers: one variable tap exciter to regulate the amplitude and one series transformer to inject the quadrature voltage in the line.
- **Symmetrical PSTs** have the capability to inject a voltage that alters the line voltage angle compared to the input voltage with no change in the line voltage magnitude.
- **Asymmetrical PSTs** have the capability to inject a voltage directly in quadrature with the line voltage resulting in an alteration in the voltage magnitude.

Depending on the number of cores, the PSTs can be:

- **Single-core PST**, which has all windings mounted on a single core.
- **Two-core PST**, which consists of a series unit and an exciting unit. The series and the exciting unit can be either in one tank or in separate tank.

7.2.2.1 Direct-Type Asymmetrical PSTs

Figure 7.2a shows the configuration of a direct-type asymmetrical PST. The input terminals are represented by A, B, and C, whereas the output terminals are represented by A′, B′, and C′. This PST type consists of a delta-connected exciting unit and regulating windings wound on the same phase core limb as the exciting winding.

Each regulating winding is fitted with a tap changer (T_A, T_B, and T_C) and a selector switch. This configuration allows adding a variable quadrature voltage, $\Delta\underline{V}_{PST}$, to the input voltage thereby achieving a phase shift α between the input and the output terminal voltages (Figure 7.2b). The direction of the phase shift can be

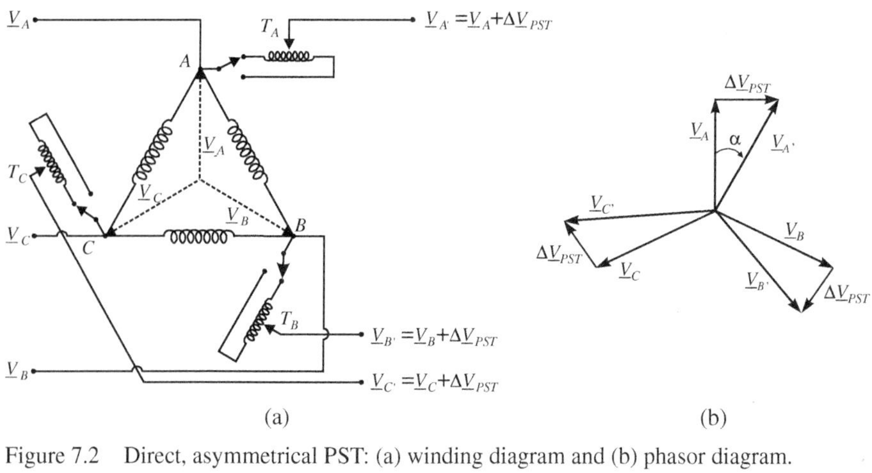

Figure 7.2 Direct, asymmetrical PST: (a) winding diagram and (b) phasor diagram.

changed by using switches. In this way, the power flow in the line can be increased or decreased [6].

The phase shift angle α is a nonlinear function of the tap position, and can be derived from the phasor diagram of Figure 7.2b, that is [7]:

$$\alpha = \text{atan}\frac{\Delta V_{PST}}{V_A} \tag{7.2}$$

With the phase shift angle thereby known, the magnitude of the secondary voltage, $V_{A'}$, can also be determined in terms of the injected quadrature voltage, ΔV_{PST}, that is:

$$V_{A'} = \frac{\Delta V_{PST}}{\sin\alpha} \tag{7.3}$$

Substituting for α from (7.2) into (7.3), we achieve:

$$V_{A'} = \frac{\Delta V_{PST}}{\sin\left(\text{atan}\dfrac{\Delta V_{PST}}{V_A}\right)} \tag{7.4}$$

The active power flow on a lossless transmission line ($R_L = 0$), when adding the angle α to the natural phase angle difference δ ($= \delta_S - \delta_R$), between voltages at the two ends of the line, is given by:

$$P = \frac{V_{A'}V_R}{X_L + X_{PST}}\sin(\delta + \alpha) \tag{7.5}$$

where X_L is the line reactance, and X_{PST} is the series reactance of the PST.

The greater the phase shift, the greater the secondary voltage $V_{A'}$ with respect to the input voltage V_A. Changes in either of the two terms ($V_{A'}$ and α) will result in changes in the transmitted power. It will increase by increasing the shift angle α. However, the phase shift is controllable within certain capability limits. On the other hand, if the change in the reactance is analyzed, the maximum transmissible power decreases by a factor $X_L/(X_L + X_{PST})$ when using a PST.

Using (7.2) and (7.4), equation (7.5) can be rewritten as:

$$P = \frac{V_R}{X_L + X_{PST}}\frac{\Delta V_{PST}}{\sin\left(\text{atan}\frac{\Delta V_{PST}}{V_A}\right)}\sin\left(\delta + \text{atan}\frac{\Delta V_{PST}}{V_A}\right) \tag{7.6a}$$

$$P = \frac{V_R}{X_L + X_{PST}}\frac{\Delta V_{PST}\left(\sin\delta\cos\left(\text{atan}\dfrac{\Delta V_{PST}}{V_A}\right) + \cos\delta\sin\left(\text{atan}\dfrac{\Delta V_{PST}}{V_A}\right)\right)}{\sin\left(\text{atan}\dfrac{\Delta V_{PST}}{V_A}\right)} \tag{7.6b}$$

or

$$P = \frac{V_R}{X_L + X_{PST}}\left(V_A\sin\delta + \Delta V_{PST}\cos\delta\right) \tag{7.6c}$$

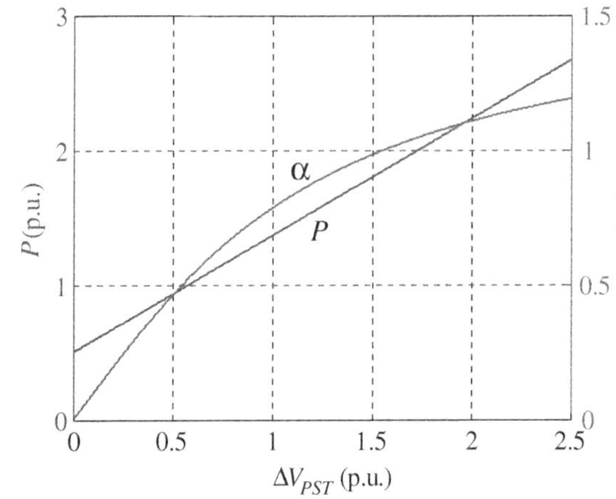

Figure 7.3 Relation between P, α, and ΔV_{PST} for a direct and asymmetrical PST with $\delta = 30°$.

For constant δ, the active power flow on the line varies linearly with the quadrature voltage. Equations (7.2) and (7.6c) are plotted in Figure 7.3, for $\delta = 30°$, $V_A = 1$ p.u., and $V_R/(X_L + X_{PST}) = 1$ p.u. The curve of α is relatively linear up to a value of about $\alpha = 0.5$ rad $\approx 29°$ [7].

7.2.2.2 Direct-Type Symmetrical PSTs

A direct asymmetrical PST can be transformed into a symmetrical one by designing a symmetrical regulating winding (Figure 7.4a). Each of the two half regulating windings is provided with a tap changer.

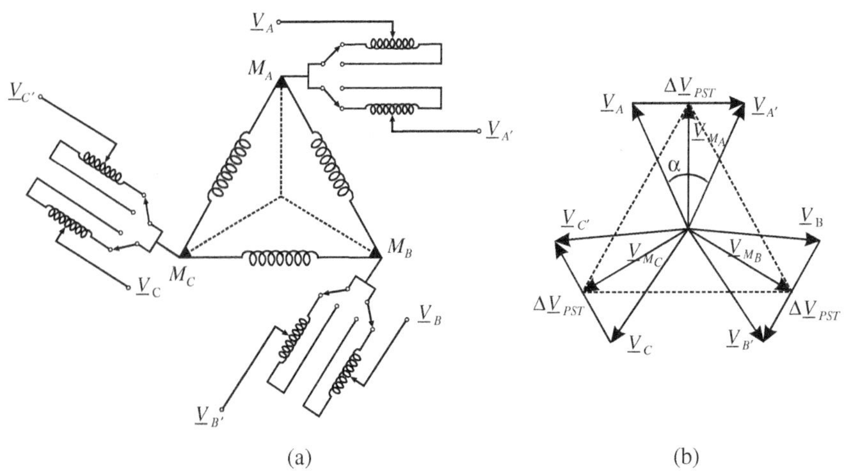

(a) (b)

Figure 7.4 Direct and symmetrical PSTs: (a) winding diagram and (b) phasor diagram.

In order to change the direction of phase angle shift, to either increase or decrease the total phase angle difference between voltages at the line ends, a commutation between the two half regulating windings using a selector switch is done. In this configuration, the same voltage magnitude for both ends of the phase shifter is achieved no matter of the phase shift direction, thus the power flow is influenced by the phase shift angle only.

The phase angle shift is again a nonlinear function of the quadrature voltage, as it results from Figure 7.4b [6]:

$$\alpha = 2\,\mathrm{asin}\frac{\Delta V_{PST}}{2V_A} \tag{7.7}$$

For the phase angle shift α given in (7.7), the active power transferred between the line ends is:

$$P = \frac{V_A V_R}{X_L + X_{PST}} \sin\left(\delta + 2\,\mathrm{asin}\frac{\Delta V_{PST}}{2V_A}\right) \tag{7.8}$$

The phase angle shift and the transmitted power are plotted against the quadrature voltage in Figure 7.5, for $V_A = 1$ p.u. and $\frac{V_R}{X_L + X_{PST}} = 1$ p.u. The transmitted power reaches a maximum for $\Delta V_{PST} = 1$ p.u. ($\alpha = 2.5$ rad), thus any increase in either of the two variables causes the transmitted power to decrease toward zero.

A solution to reduce the number of taps per phase from two to one is the delta-hexagonal configuration depicted in Figure 7.6a. A controllable voltage is injected in every phase, which is proportional to the voltage between the primary and secondary terminal of the two other phases. The resulting phasor diagram is shown in Figure 7.6b.

A disadvantage of this arrangement is the need for additional impedances to protect the tap changers when the phase shift is set to zero, because in that case, short-circuit currents occur [6].

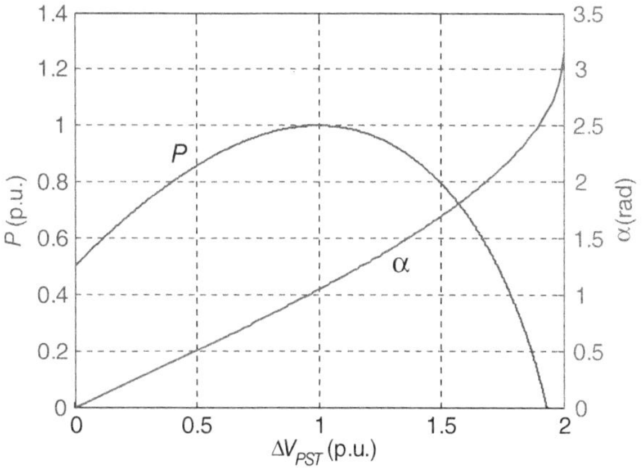

Figure 7.5 Relation between P, α, and the quadrature voltage for a direct symmetrical PST with $\delta = 30°$.

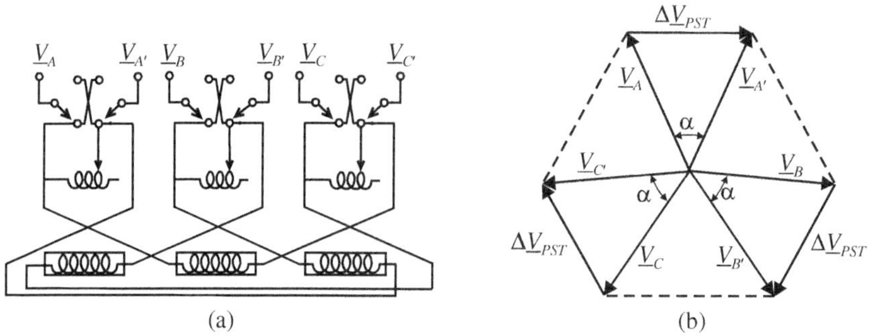

Figure 7.6 Direct-type symmetrical PST with hexagonal winding connection (a) and phasor diagram (b). Adapted from Verboomen et al. 2005 [6].

7.2.2.3 *Indirect-Type Asymmetrical and Symmetrical PSTs*

The indirect-type PST can be either asymmetrical or symmetrical. They both consist of an exciting unit and a series unit. Depending on the PST rating, the two units are housed either in separate tanks or in a single tank. The two-tank design has the advantage of an easier transport.

Figure 7.7a shows the configuration of an asymmetrical PST.

- The exciting transformer unit provides the phase voltages. Both primary and secondary windings are star connected. The secondary of the exciting unit is connected to the primary of the series-connected unit.

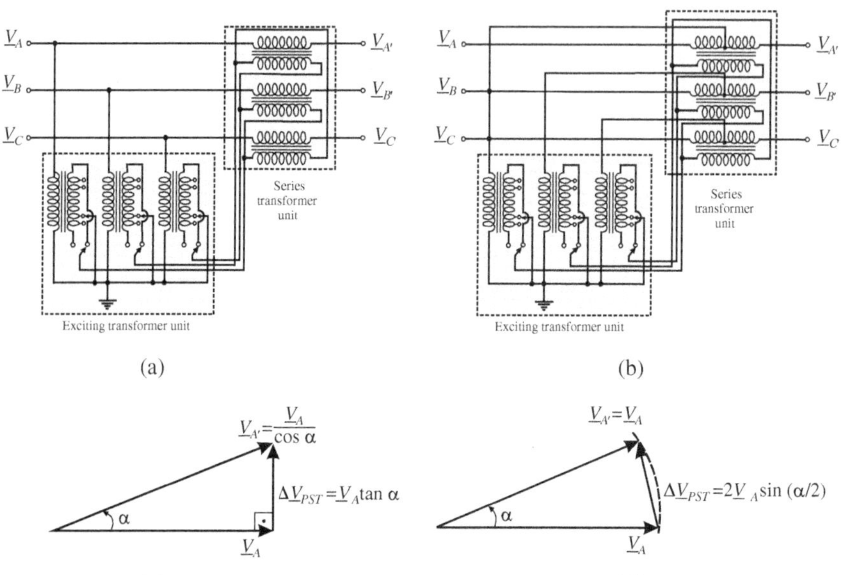

Figure 7.7 Windings connection of the indirect-type asymmetrical (a), and the indirect-type symmetrical (b) topologies of PSTs; phasor diagram of phase *A* voltages for an asymmetric PST (c), and a symmetric PST (d).

- The series-connected transformer unit is a booster and provides a phasorial summation of voltages. For this reason, this type of PST is also called "Quadrature booster" or "Quadbooster." Its secondary winding is series connected with the transmission line. The primary winding is delta connected. With appropriate connection of phases, a voltage proportional with ($\underline{V}_B - \underline{V}_C$), by a factor equal to the product of the transformation ratios of the exciting and series transformers, is achieved at the secondary winding terminals, which is series connected with the phase A.

The phasor diagram of phase A voltages for an indirect-type asymmetrical PST is depicted in Figure 7.7c. Since $\Delta\underline{V}_{PST}$ is in quadrature with the input voltage $\underline{V}_A$, for any phase shift angle greater than zero the output voltage $\underline{V}_A$ is greater than the input voltage.

The indirect-type asymmetrical PST can be made symmetrical by splitting the secondary winding of the series-connected transformer unit into two and tapping the voltage for the exciter from the middle (Figure 7.7b). The phasor diagram of phase A voltages for an indirect-type symmetrical PST is depicted in Figure 7.7d. Note that the symmetric design keeps constant the voltage amplitude on both sides of the PST, while it provides phase shift only.

If phase shifting is intended, the change in the quadrature voltage amplitude must be also possible. This can be done by changing the tap position of the exciting transformer fitted with off-load or on-load tap changer. The voltages phase diagram from Figure 7.7c shows that the voltage amplitude is not constant; if the transformer impedance is neglected, the ratio of the two voltages is the cosine of the phase shifting angle. One way of canceling this effect is to connect the primary of the exciting transformer in the middle point of the secondary winding of the series transformer (Figure 7.7b). By this special construction, the voltage amplitude can thereby be maintained constant.

It is also possible to fit the exciting transformer with a second secondary winding equipped with an on-load tap changer. The primary winding is supplied with a voltage ($kV_B + k'V_C$), which allows controlling the series and quadrature components of the additional voltage by acting on the terms k and k', thus achieving a control in amplitude and phase of the output voltage with respect to the input voltage.

7.2.2.4 Comparison of the Topologies

As it regards the configuration, asymmetric PSTs are less complex and thus less expensive. However, the symmetrical topologies are more popular because they do not affect the voltage amplitude and can attain a larger angle than their asymmetric counterparts.

A direct-type configuration is easier to construct and hence cheaper compared to the indirect-type one since no exciter is needed. A major drawback is the fact that the tap changer and the regulating winding are directly exposed to system disturbances, making them very vulnerable. Also, the indirect topology allows more flexibility in the design phase, as the regulator circuit can be sized independently, which is a major asset when selecting the tap changer [7].

7.2.3 Steady-State Model of a Mechanical Phase Shifter

A possible conventional PST construction consists of a parallel and series transformer coupled with a mechanical tap-changer (Figure 7.8a).

The PAR is a type of PST, which from the system point of view, has the ability to introduce a phase shift between terminal voltage phasors more or less independent of throughput current. The controllable PAR parameter may be assumed to be the terminal voltage phasor separation, i.e. the angle α.

An equivalent model is presented in Figures 7.8b as a combination of a series injected voltage source $\underline{V}_T$ and a parallel connected current source $\underline{I}_T$. The corresponding phasor diagrams for the phase angle regulator (PAR) and the quadrature boosting transformer (QBT) are presented in Figures 7.8c and 7.8d, respectively.

Let us consider the particular case of a PST inserted on a transmission line, as shown in Figure 7.9.

Using basic relations we can write [8]:

$$\underline{V}_{PST1} = \underline{V}_1 - j\underline{I}_1 X_{L1} \tag{7.9a}$$

$$\underline{V}_{PST2} = \underline{V}_{PST1} e^{j\alpha} \tag{7.9b}$$

$$\underline{I}_2 = \underline{I}_1 e^{j\alpha} = \frac{\underline{V}_{PST2} - \underline{V}_2}{jX_{L2}} \tag{7.9c}$$

If losses (active and reactive) are neglected then the PST does not produce nor consume active and reactive power, i.e. it is balanced, the PAR input and output powers are equal: $\underline{V}_{PST1}\underline{I}_1^* = \underline{V}_{PST2}\underline{I}_2^*$. From this relation we can infer that the phasors $\underline{I}_1$ and $\underline{I}_2$ are shifted by the same angle α similar to the PAR terminal voltages, their magnitudes being equal as it results from equation (7.9c).

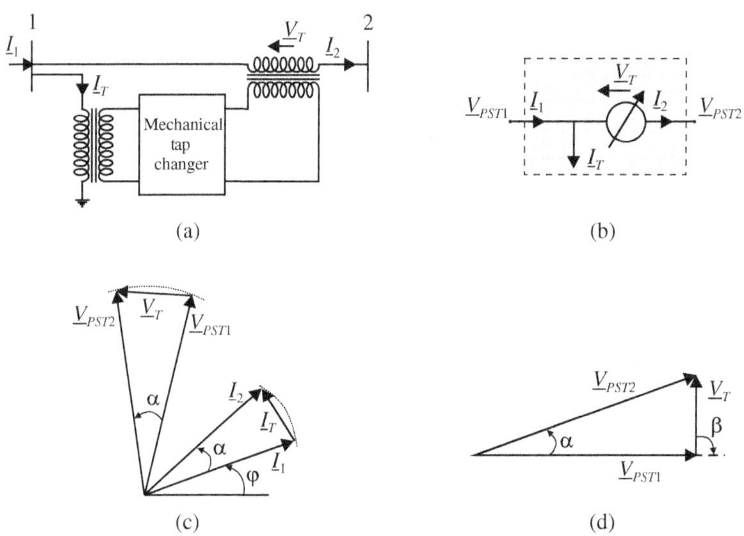

Figure 7.8 Basic PST modeling [8]: (a) conventional PST structure; (b) equivalent model; (c) operation of a PAR; (d) phasor diagram of QBT.

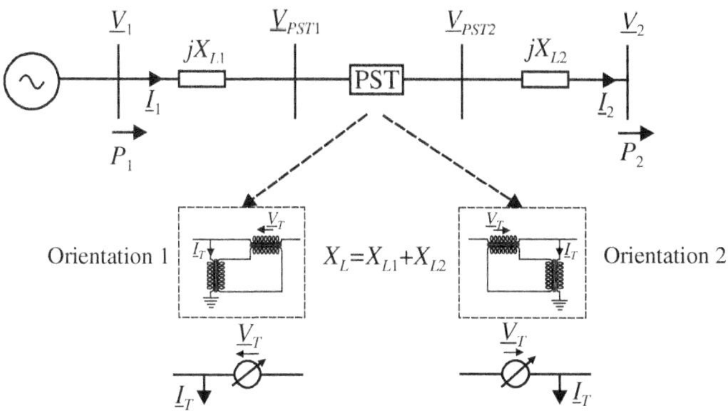

Figure 7.9 PST in a longitudinal transmission system [8].

From (7.9c), considering equations (7.9a) and (7.9b), we can calculate [4]:

$$\underline{I}_1 = \frac{-j}{X_{L1} + X_{L2}} \left(\underline{V}_1 - \underline{V}_2 e^{-j\alpha}\right) \tag{7.10a}$$

$$\underline{I}_2 = \frac{-j}{X_{L1} + X_{L2}} \left(\underline{V}_1 e^{j\alpha} - \underline{V}_2\right) \tag{7.10b}$$

Considering $\underline{V}_2$ as a reference phasor ($\underline{V}_2 = V_2$), the transmitted power can be determined:

$$P_{PAR} = P_1 = P_2 = V_2 \mathrm{Re}\{\underline{I}_2^*\} = \frac{V_1 V_2}{X_{L1} + X_{L2}} \sin(\delta + \alpha) \tag{7.11}$$

The quadrature boosting transformer (QBT) can also separate its terminal voltage phasors but the injected voltage $\underline{V}_T$ phase is fixed with regard to the input voltage phasor $\underline{V}_{PST1}$, i.e. $\underline{V}_T$ is perpendicular on $\underline{V}_{PST1}$ (Figure 7.8d). The controllable parameter here may be assumed to be the terminal voltage phasor separation, i.e. angle α. The system is described by the equations [4]:

$$\underline{V}_{PST1} = \underline{V}_1 - \underline{I}_1 jX_{L1} \tag{7.12a}$$

$$\underline{V}_{PST2} = \underline{V}_{PST1} e^{j\alpha} \frac{1}{\cos\alpha} \tag{7.12b}$$

$$\underline{I}_2 = \underline{I}_1 e^{j\alpha} \cos\alpha = \frac{\underline{V}_{PST2} - \underline{V}_2}{jX_{L2}} \tag{7.12c}$$

Equation (7.12c) is, as in the PAR case, derived from balanced PST power conditions, i.e. $\underline{V}_{PST1}\underline{I}_1^* = \underline{V}_{PST2}\underline{I}_2^*$. After calculating the currents $\underline{I}_1$ and $\underline{I}_2$, we can obtain:

$$P_{QBT} = P_1 = P_2 = V_2 \mathrm{Re}\{\underline{I}_2^*\} = \frac{V_1 V_2}{\frac{X_{L1}}{\cos\alpha} + X_{L2}\cos\alpha} \sin(\delta + \alpha) \tag{7.13}$$

From expressions (7.11) and (7.13), it is evident that a QBT is a non-symmetrical device. This is because the transferred power is dependent on both the orientation and the location of QBT; these parameters do not influence the transmitted power in the case of a PAR.

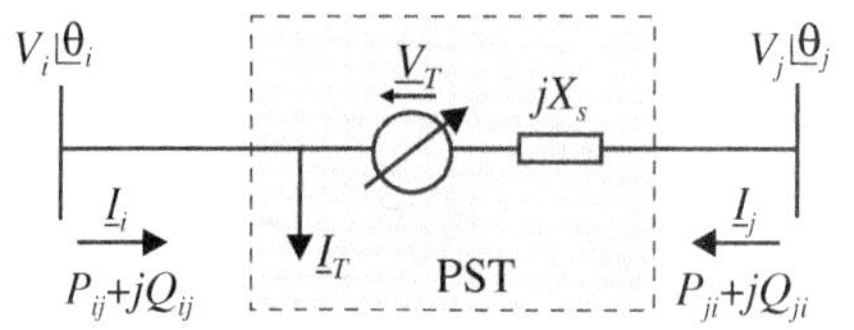

Figure 7.10 PST model with a transformer reactance.

Consider now a PST with its series-transformer reactance inserted between nodes i and j in Figure 7.10.

Omitting the reactance X_{L1} in Figure 7.9 and in Equations (7.11) and (7.13), we can give the following mathematical expressions for the reactive and real power flow that are valid for Figure 7.10, which shows the PST in orientation 1 [8]:

- PAR case

$$P_{ij} = -P_{ji} = \frac{V_i V_j}{X_s} \sin(\theta_{ij} + \alpha) \tag{7.14}$$

$$Q_{ij} = \frac{V_i}{X_s} \left[V_i - V_j \cos(\theta_{ij} + \alpha) \right] \tag{7.15}$$

$$Q_{ji} = \frac{V_j}{X_s} \left[V_j - V_i \cos(\theta_{ij} + \alpha) \right] \tag{7.16}$$

- QBT case

$$P_{ij} = -P_{ji} = \frac{V_i V_j}{X_s \cos\alpha} \sin(\theta_{ij} + \alpha) \tag{7.17}$$

$$Q_{ij} = \frac{V_i}{X_s \cos\alpha} \left[\frac{V_i}{\cos\alpha} - V_j \cos(\theta_{ij} + \alpha) \right] \tag{7.18}$$

$$Q_{ji} = \frac{V_j}{X_s \cos\alpha} \left[V_j - \frac{V_i}{\cos\alpha} \cos(\theta_{ij} + \alpha) \right] \tag{7.19}$$

where X_s represents the PST's series-transformer reactance, while $\theta_{ij} = \theta_i - \theta_j$.

If we consider an alternative orientation of the QBT, it is necessary to make the following replacements $V_i \leftrightarrow V_j$, $\alpha \leftrightarrow -\alpha$ in $\theta_{ij} \leftrightarrow -\theta_{ji}$ in formulae (7.14)–(7.16).

7.2.4 Equivalent Series Reactance as a Function of the Phase Shift Angle[1]

7.2.4.1 Symmetrical Phase Shifter

The simplified one-line diagram and the voltage phasorial diagram of a symmetrical PST are illustrated in Figure 7.11 [9].

[1] This section was written based on ENTSO-E documentation "ENTSO-E: *Phase shift transformers modeling*, Version 1.0.0, CGMES V2.4.14, 28 May 2014" [9].

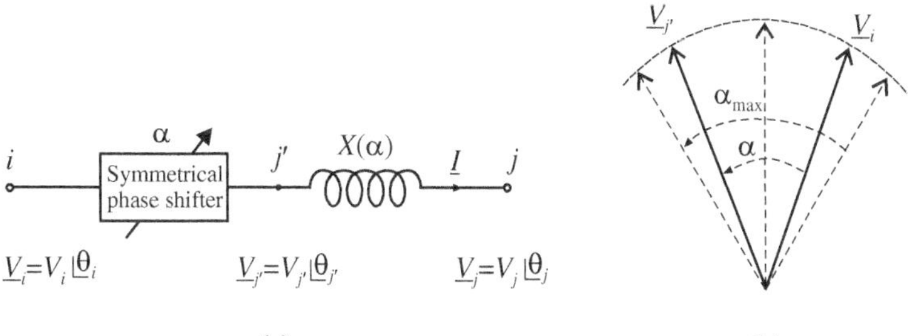

Figure 7.11 (a) One-line diagram of symmetrical PST and (b) voltage phasorial diagram. *Source*: ENTSO-E 2014 [9]. Public domain.

From Figure 7.11a, we infer the voltages at the two ends of the ideal PST [9]:

$$\underline{V}_j = \underline{V}_i e^{j\alpha} - jX(\alpha)\underline{I} \tag{7.20a}$$

$$\underline{V}_{j'} = \underline{V}_i \left(1 + 2j \sin\frac{\alpha}{2} \cdot e^{j\frac{\alpha}{2}}\right) = e^{j\alpha}\underline{V}_i \tag{7.20b}$$

The expressions of the angle and ratio per tap are:

$$\alpha = (n - n_0) \cdot \delta\alpha \quad \text{or} \quad \alpha = 2\text{atan}\left(\frac{(n - n_0) \cdot \delta u}{2}\right) \tag{7.21}$$

$$r = 1$$

where:

- α is the phase shift angle.
- n – the phase tap changer step.
- n_0 – the tap changer neutral step.
- $\delta\alpha$ – the phase shifter increment.
- δu – the voltage step increment.
- r – the ratio.

In order to develop the expression of $X(\alpha)$, let us consider the detailed three-phase diagram of the symmetrical phase shifters with two cores (Figure 7.12).

When considering an ideal phase shifter, the electric power through it is conserved, that is:

$$\underline{S}_i = 3\underline{V}_i\underline{I}_i^* = 3\underline{V}_o\underline{I}_o^* \tag{7.22}$$

where $\underline{V}_i$ and $\underline{I}_i$ are the input variables, and $\underline{V}_o$ and $\underline{I}_o$ are the output variables of PST.

As the output voltage is shifted from the input voltage by α:

$$\underline{V}_o = e^{j\alpha}\underline{V}_i, \tag{7.23a}$$

the output current is also shifted by α from the input current:

$$\underline{I}_o = e^{j\alpha}\underline{I}_i \tag{7.23b}$$

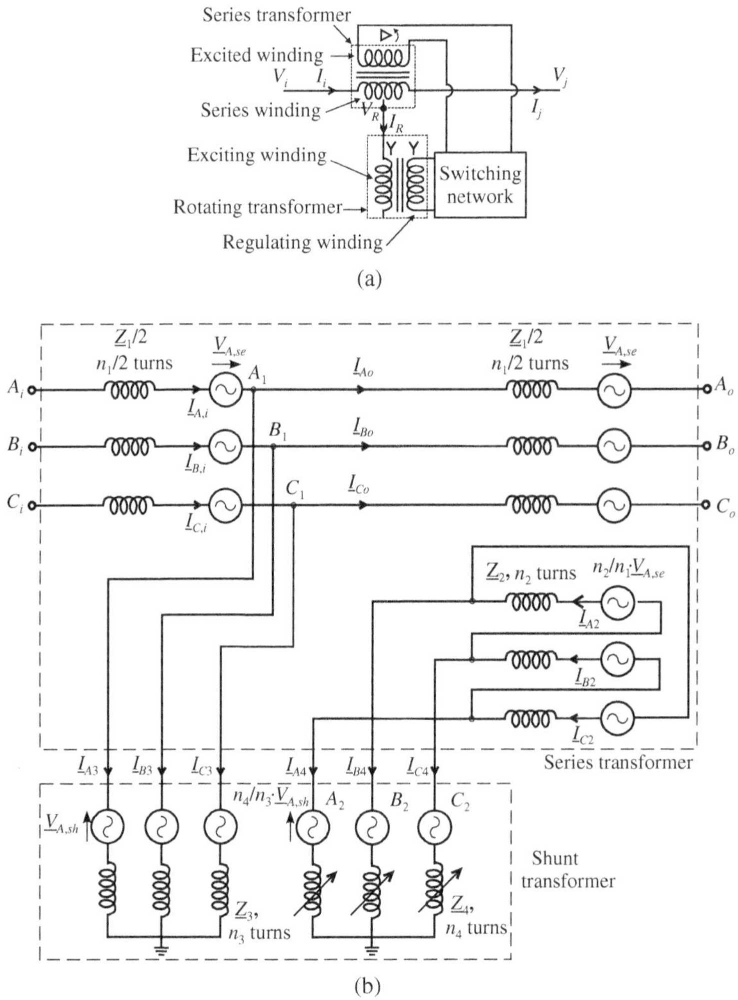

Figure 7.12 Symmetrical phase shifters with two cores [9].

The current flowing to the shunt transformer unit is calculated as:

$$\underline{I}_3 = \underline{I}_i - \underline{I}_o = \left(1 - e^{j\alpha}\right) \underline{I}_i, \tag{7.24}$$

THE SHUNT REACTANCE $X_{shunt}(\alpha)$ is defined as "the equivalent reactance which crossed by the series input current I_i would produce the reactive losses of the transformer" [9]:

$$Q_{shunt} = 3X_{shunt}(\alpha) \cdot I_i^2 \tag{7.25}$$

with

$$Q_{shunt} = 3X_3 I_3^2 + 3X_4 I_4^2 \tag{7.26}$$

and

$$I_3^2 = I_i^2 \cdot \left|1 - e^{j\alpha}\right|^2 = 2\left(1 - \cos\alpha\right) I_i^2 = 4\left(\sin\frac{\alpha}{2}\right)^2 I_i^2$$

$$I_4^2 = \frac{n_3^2}{n_4^2} I_3^2$$

Then, from equations (7.25) and (7.26), we achieve

$$X_{shunt}(\alpha) = 4\left(\sin\frac{\alpha}{2}\right)^2 \left(X_3 + X_4\left(\frac{n_3}{n_4}\right)^2\right)$$

Assuming that $X_4 = X_{4\max}\left(\frac{n_4}{n_{4\max}}\right)^2$, the expression of the shunt reactance becomes:

$$X_{shunt}(\alpha) = 4\left(\sin\frac{\alpha}{2}\right)^2 \left(X_3 + X_{4\max}\left(\frac{n_3^2}{n_{4\max}^2}\right)\right) \quad (7.27)$$

THE SERIES REACTANCE $X_{series}(\alpha)$ is defined as "the equivalent reactance which crossed by the series input current I_i would produce the reactive losses of the series transformer" [9]:

$$Q_{series} = 3X_{series}(\alpha) \cdot I_i^2 \quad (7.28)$$

with

$$Q_{series} = 3\frac{X_1}{2} I_i^2 + 3\frac{X_1}{2} I_o^2 + 3X_2 I_2^2 \quad (7.29)$$

The relationship between currents in the series transformer is

$$\frac{n_1}{2}\underline{I}_i + \frac{n_1}{2}\underline{I}_o = +n_2\underline{I}_2$$

and knowing from equation (7.23b) that $\underline{I}_o = e^{j\alpha}\underline{I}_i$, we achieve

$$\underline{I}_2 = \frac{n_1}{2n_2}\left(1 + e^{j\alpha}\right)\underline{I}_i$$

or

$$I_2^2 = \left(\frac{n_1}{2n_2}\right)^2 I_i^2 \left|1 + e^{j\alpha}\right|^2 = \frac{1}{2}\frac{n_1^2}{n_2^2}\left(1 + \cos\alpha\right) I_i^2$$

Then,

$$X_{series}(\alpha) = X_1 + \frac{1}{2}\left(\frac{n_1}{n_2}\right)^2 \left(1 + \cos\alpha\right) X_2 = X_1 + \left(\frac{n_1}{n_2}\right)^2 \left(1 - \left(\sin\frac{\alpha}{2}\right)^2\right) X_2$$

$$\boxed{\cos(2\alpha) = 1 - 2(\sin\alpha)^2}$$

or

$$X_{series}(\alpha) = \left(X_1 + \left(\frac{n_1}{n_2}\right)^2 X_2\right) - \left(\frac{n_1}{n_2}\right)^2 X_2 \left(\sin\frac{\alpha}{2}\right)^2 \quad (7.30)$$

Assuming the reactance of the regulating winding varies with the square of the number of turns, the expression of the total equivalent reactance in terms of the phase shift angle can be written as:

$$X(\alpha) = \left[X_1 + \left(\frac{n_1}{n_2}\right)^2 X_2\right] + 4\left(\sin\frac{\alpha}{2}\right)^2 \left[X_3 + X_{4\,\max}\left(\frac{n_3^2}{n_{4\,\max}^2}\right) - \left(\frac{n_1}{2n_2}\right)^2 X_2\right]$$

or

$$X(\alpha) = X(0) + \left[X\left(\alpha_{\max}\right) - X(0)\right]\left(\frac{\sin\frac{\alpha}{2}}{\sin\frac{\alpha_{\max}}{2}}\right)^2 \tag{7.31}$$

with

$$X(0) = X_1 + \left(\frac{n_1}{n_2}\right)^2 X_2 \tag{7.32a}$$

$$X\left(\alpha_{\max}\right) - X(0) = 4\left(\sin\frac{\alpha_{\max}}{2}\right)^2 \left[X_3 + X_{4\,\max}\left(\frac{n_3^2}{n_{4\,\max}^2}\right) - \left(\frac{n_1}{2n_2}\right)^2 X_2\right] \tag{7.32b}$$

with three parameters:

- $\alpha_{\max}$ is the maximal phase shift.
- $X(0)$ is the equivalent series reactance at zero phase shift $\alpha = 0$.
- $X\left(\alpha_{\max}\right)$ is the equivalent series reactance at maximum phase shift $\alpha = \alpha_{\max}$,

and one variable, which is the current phase shift, α.

Equation (7.31) is valid for single or double core symmetrical phase shifters except for the hexagonal technology. For single core symmetrical phase shifters, $X(0) = 0$.

7.2.4.2 Quadrature Booster

The simplified one-line diagram and the voltage phasorial diagram of a QB are illustrated in Figure 7.13 [9].

The voltage equations are:

$$\underline{V}_{j'} = \underline{V}_i(1 + j\tan\alpha) = \rho e^{j\alpha}\underline{V}_i \tag{7.33a}$$

$$\underline{V}_j = \underline{V}_i\rho e^{j\alpha} - jX(\alpha)\underline{I} \tag{7.33b}$$

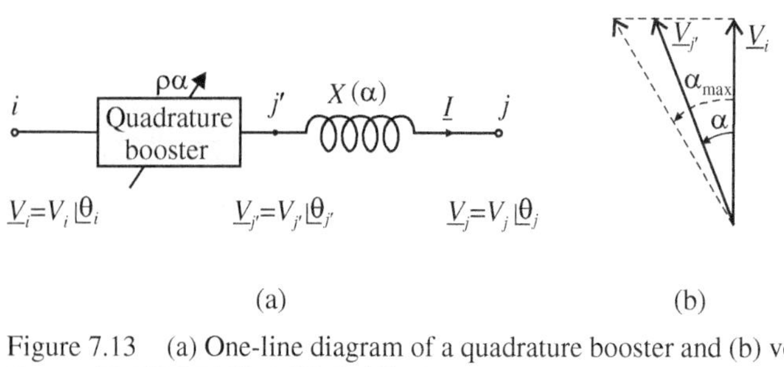

Figure 7.13 (a) One-line diagram of a quadrature booster and (b) voltage phasorial diagram. *Source*: ENTSO-E 2014 [9]. Public domain.

ρ and α vary with $\rho\cos\alpha$ = constant.

Expression of the angle and ratio per tap:

$$\alpha = \text{atan}\left(\left(n - n_0\right)\delta u\right) \tag{7.34a}$$

$$r = \frac{1}{\sqrt{\left(\left(n - n_0\right)\delta u\right)^2 + 1}} \tag{7.34b}$$

Assuming the reactance of the regulating winding varies with the square of the number of turns, the equivalent reactance of the QB can be written as [9]:

$$X(\alpha) = X(0) + \left(X\left(\alpha_{\max}\right) - X(0)\right)\left(\frac{\tan\alpha}{\tan\alpha_{\max}}\right)^2 \tag{7.35}$$

with three parameters

- $\alpha_{\max}$ is the maximal phase shift.
- $X(0)$ is the equivalent series reactance at zero phase shift.
- $X\left(\alpha_{\max}\right)$ is the equivalent series reactance at maximal phase shift

and one variable α, which is the current phase shift.

For QBs with a single core, $X(0) = 0$.

7.2.4.3 Asymmetrical Phase Shifter

The simplified one-line diagram and the voltage phasorial diagram of an asymmetrical phase shifter are illustrated in Figure 7.14 [9].

The voltage equations are [9]:

$$\underline{V}_{j'} = \underline{V}_i\left(1 + e^{j\theta}\frac{\tan\alpha}{\sin\theta - \tan\alpha\cos\theta}\right) = \rho e^{j\alpha}\underline{V}_i \tag{7.36a}$$

$$\underline{V}_j = \underline{V}_i\rho e^{j\alpha} - jX(\alpha)\underline{I} \tag{7.36b}$$

For no load conditions, $\left(\underline{I} = 0\right)$:

$$\Delta\underline{V} = -e^{j\theta}\cdot V_j\frac{\tan\alpha}{\sin\theta - \tan\alpha\cos\theta}$$

The boost voltage angle θ is fixed, but ρ and α vary.

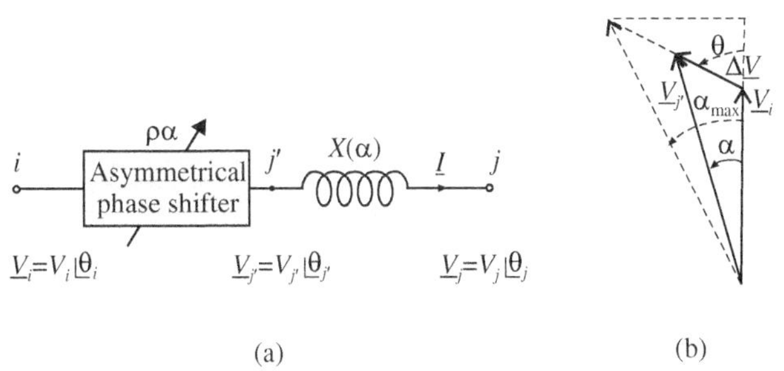

Figure 7.14 (a) One-line diagram of an asymmetrical shifter and (b) voltage phasorial diagram. *Source*: ENTSO-E 2014 [9]. Public domain.

Expressions of the angle and ratio per tap (Figure 7.14b) are written as:

$$\alpha = \text{atan}\left(\frac{(n-n_0)\,\delta u \cdot \sin\theta}{1+(n-n_0)\,\delta u \cdot \cos\theta}\right) \tag{7.37a}$$

$$r = \frac{1}{\sqrt{\left((n-n_0)\,\delta u \cdot \sin\theta\right)^2 + \left(1+(n-n_0)\,\delta u \cdot \cos\theta\right)^2}} \tag{7.37b}$$

Assuming the reactance of the regulating winding as the square of the number of turns, the equivalent series reactance can be written [9] as:

$$X(\alpha) = X(0) + \left(X(\alpha_{max}) - X(0)\right)\left(\frac{\tan\alpha}{\tan\alpha_{max}}\,\frac{\sin\theta - \tan\alpha_{max}\cos\theta}{\sin\theta - \tan\alpha\cos\theta}\right)^2 \tag{7.38}$$

with four parameters

- α_{max} is the maximal phase shift.
- $X(0)$ is the equivalent series reactance at zero phase shift.
- $X(\alpha_{max})$ is the equivalent series reactance at maximal phase shift.
- θ is the boost voltage angle,

and a variable

- α which is the current phase shift.

7.2.4.4 In-Phase Transformer and Symmetrical/Asymmetrical Phase Shifter

(i) *In-phase transformer and symmetrical phase shifter* (Figure 7.15) [9].

Assuming the reactance of the regulating winding varies as the square of the number of turns and the equivalent reactance is the sum of the reactance of the in-phase transformer X_r and the reactance of the phase shifter part X_α, the phase shifting angle α does not depend on the in-phase ratio r and the expression of the equivalent series reactance can be written as:

$$X(r,\alpha) = X_r(r_{nom})\left(\frac{r}{r_{nom}}\right)^2 + X_\alpha(0) + \left(X_\alpha(\alpha_{max}) - X_\alpha(0)\right)\left(\frac{\sin\frac{\alpha}{2}}{\sin\frac{\alpha_{max}}{2}}\right)^2 \tag{7.39}$$

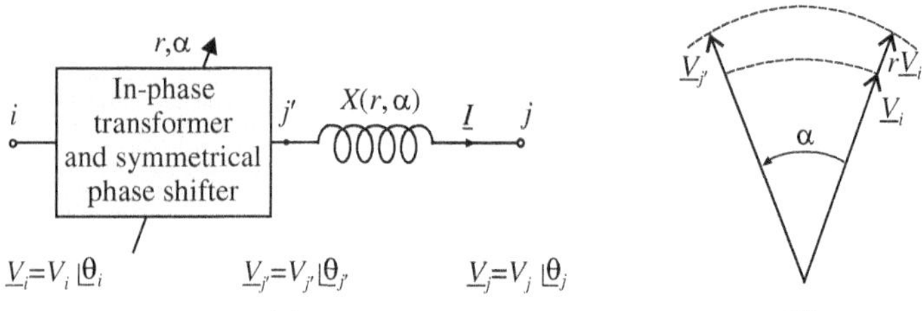

Figure 7.15 Voltage phasorial diagram of the in-phase transformer and symmetrical phase shifter. *Source*: ENTSO-E 2014 [9]. Public domain.

with six parameters:

- r_{nom} is the nominal ratio of the in-phase transformer;
- $X_r(r_{nom})$ is the equivalent series reactance of the in-phase transformer at nominal in-phase ratio;
- α_{max} is the maximal phase shift angle;
- $X_\alpha(0)$ is the equivalent series reactance of the phase shifter part zero phase shift;
- $X_\alpha(\alpha_{max})$ is the equivalent series reactance of the phase shifter part at maximal phase shift at nominal in-phase ratio (r_{nom}),

and two variables:

- r is the current ratio of the in-phase transformer;
- α is the current phase shift angle.

(ii) *In-phase transformer and asymmetrical phase shifter* (Figure 7.16) [9].

Assuming the reactance of the regulating winding varies as the square of the number of turns and the equivalent reactance is the sum of reactance of the in-phase transformer X_r and the reactance of the phase shifter part X_α, the equivalent series reactance given the angle and the in-phase transformer ratio can be written as:

$$X(r,\alpha) = X_r(r_{nom})\left(\frac{r}{r_{nom}}\right)^2 + X_\alpha(0) + \left(X_\alpha\left(\alpha^0_{max}\right) - X_\alpha(0)\right) \times \left(\frac{\tan\alpha}{\tan\alpha_{max}(r)}\frac{\sin\theta - \tan\alpha_{max}(r)\cos\theta}{\sin\theta - \tan\alpha\cos\theta}\right)^2 \quad (7.40)$$

with

$$\alpha_{max}(r) = \operatorname{atan}\left(\frac{\sin\theta}{\frac{r}{r_{nom}\tan\alpha^0_{max}}\left(\sin\theta - \tan\alpha^0_{max}\cdot\cos\theta\right) + \cos\theta}\right) \quad (7.41)$$

and

$$\alpha^0_{max} = \alpha_{max}(r_{nom}).$$

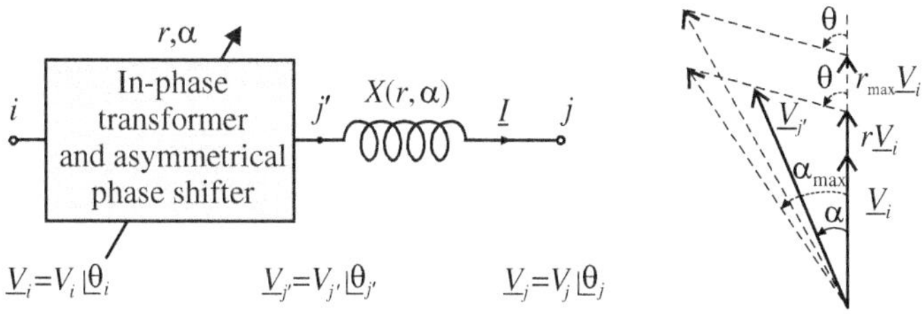

Figure 7.16 Voltage phasorial diagram of the in-phase transformer and asymmetrical phase shifter. *Source*: ENTSO-E 2014 [9]. Public domain.

This time, there are six parameters:

- r_{nom} is the nominal ratio of the in-phase transformer;
- $X_r(r_{nom})$ is the equivalent series reactance of the in-phase transformer at nominal in-phase ratio r_{nom};
- θ is the fix boost voltage angle;
- $\alpha^0_{\max} = \alpha_{\max}(r_{nom})$ is the maximal phase shift for nominal in-phase ratio r_{nom};
- $X_\alpha(0)$ is the equivalent series reactance of the phase shifter part at zero phase shift;
- $X_\alpha\left(\alpha^0_{\max}\right)$ is the equivalent series reactance of the phase shifter part at maximal phase shift at nominal in-phase ratio (r_{nom}),

and two variables:

- r is the current ratio of the in-phase transformer;
- α is the current phase shift.

For $\theta = \pi/2$, that is, in case of the QB:

$$\alpha_{\max}(r) = \operatorname{atan}\left(\frac{r_{nom}}{r}\tan\alpha^0_{\max}\right)$$

7.3 THYRISTOR-CONTROLLED PHASE SHIFTING TRANSFORMER

7.3.1 Configurations of the Static Phase Shifter

As previously shown, a PST can be designed either as a single unit or as two distinct units (a series and a shunt transformer). In the second case, when a distinct boosting (shunt) transformer is used (Figure 7.17), the primary winding of the boosting

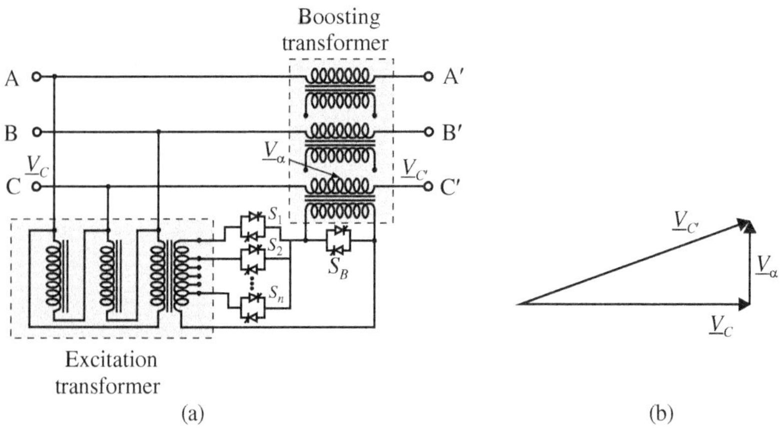

Figure 7.17 (a) Schematic diagram of an SPS based on substitution of mechanical tap changer by electronic switches and (b) voltage phasor diagram.

transformer is connected in series with respect to line. Therefore, the terminals of the secondary winding of the transformer must be either shorted or must be connected through a small impedance to provide a path for current flow under all system conditions [4].

The primary winding of the exciting transformer is shunt connected to the transmission line. Therefore, a short circuit of its secondary winding is equivalent to a fault for the system.

Since the power flow at the side where a phase shifter is located can change many times during one day, the number of operations that a PST tap changer has to perform is 10–15 times higher than for the other transformers in the network. A PST mechanical tap changer requires more attention when designing.

PSTs are usually installed on sensitive corridors of an electrical network, where the power flow must be tightly controlled, in many cases to avoid network overloading. The time reaction of an on-load tap changer (OLTC) is of the order of several seconds. For this reason, studies are done to find solutions for replacing the mechanical switched OLTC with a static switched one at acceptable costs.

Several SPS configurations can be identified depending on the semiconductor switches and converter topologies used [4, 10–14].

7.3.1.1 Substitution of Mechanical Tap Changer by Electronic Switches

Figure 7.17a shows the configuration of a SPS where the tap changer of the exciting transformer consists of electronic switches (classical thyristor). The figure illustrates the tap changer for one phase only, while for the other phases the structure is the same.

The tap switches S_i and the switch S_B consist of anti-parallel connected thyristors. Only one of the switches $S_1 \ldots S_n$ can be turned on at a time, all other switches being turned off. A control logic of the thyristors avoids turning on two switches simultaneously in order to avoid short circuiting the excitation transformer. When a switch S_i is turned on, the switch S_B is turned off. When all switches $S_1, \ldots, S_n$ are turned off, S_B must be turned on in order to short circuit the primary of the series transformer. This is necessary to prevent introducing the transformer exciting reactance in series with the transmission line. Switches S_i are not phase controlled, but they are turned on at the corresponding current zero crossing.

Similar to a mechanical tap changer, the ratio of the shunt transformer is determined by the activated switch $S_1 \ldots S_n$. When S_1 is conducting, the maximum voltage is injected in the system, whereas the minimum voltage is injected when S_n is conducting. Magnitude of the injected voltage is adjusted in either equal or unequal discrete steps between the maximum and the minimum values. Depending on either delta or star connection of the primary windings of exciting transformer, the injected voltage is either in quadrature-phase or in-phase with respect to the corresponding system phase voltage. In-phase voltage injection can be used for rapid compensation of voltage sag in subtransmission and distribution systems [4].

7.3.1.2 Thyristor-Controlled Quadrature Voltage Injection

Figure 7.18 shows one phase of a thyristor-controlled quadrature voltage transformer used as an interface converter between the exciting and the boosting units [12].

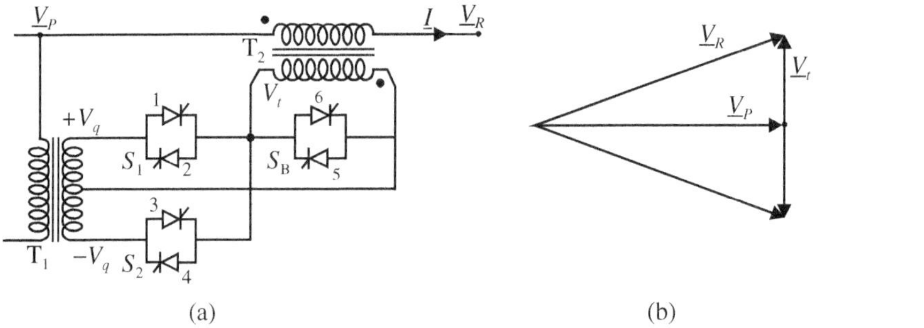

Figure 7.18 (a) One phase diagram of the quadrature voltage injection transformer and (b) voltage phasor diagram. *Source*: Arnold et al. 1981 [12]. Adapted with permission of IEEE.

As compared to the diagram from Figure 7.17a, a midpoint connection is created on the secondary winding of the exciting transformer to achieve a symmetry of voltage phase angle shift. The secondary winding of the boosting transformer (T_2) and the secondary winding of the exciting transformer (T_1) are wound in opposition. This allows achieving both leading ($+V_q$, when S_1 is on and S_2 is off) and lagging ($-V_q$, when S_2 is on and S_1 is off) quadrature voltage. S_B is a short circuiting switch, which is necessary to prevent an open-circuit condition of the series transformer during the nonboosting periods.

The firing angles α_1, α_2, and α_B for thyristor switches S_1, S_2, and S_B, respectively, are all measured with respect to the zero crossing of the appropriate quadrature voltage.

When the current $\underline{I}$ lags behind the voltage $\underline{V}_R$, by angles $0° < \phi < 90°$, two operating modes can be used, that is, boosting and bucking.

- In the *quadrature boosting control,* the thyristor switches S_1 and S_B are conducting whereas the switch S_2 is open, so that a quadrature voltage ($+V_q$) leading the primary voltage by 90° is achieved. Theoretical waveforms of voltages and currents are shown in Figure 7.19a, where the range of variation for α_1 is $(90 + \phi)° < \alpha_1 < 180°$, and for α_B is $0° < \alpha_B < (90 + \phi)°$ [10].

During the positive half-cycle of the line current I, when V_P is positive, the voltage V_t across the secondary winding of the series transformer is negative. This voltage forward biases thyristor 5 and reverse biases thyristor 6. The provision of a gate pulse to thyristor 5 will therefore turn it on, thus short circuiting the secondary winding of T_2. V_t is now equal to $-V_f$, where V_f is the forward voltage drop of thyristor 5.

When V_q is positive, thyristor 1 is forward-biased. Therefore, a gate pulse applied to thyristor 1 will turn it on, and V_t will become positive. Because thyristor 5 is now reverse-biased, it will turn off.

When V_q is negative, thyristor 5 is again forward-biased. Hence, the firing of thyristor 5 will turn it on, and a commutation from thyristor 1 to thyristor 5 takes place, thus turning thyristor 1 off.

The operation during the second half-cycle is similar to the first, but with all voltage polarities reversed and with the alternate thyristor of each back-to-back pair

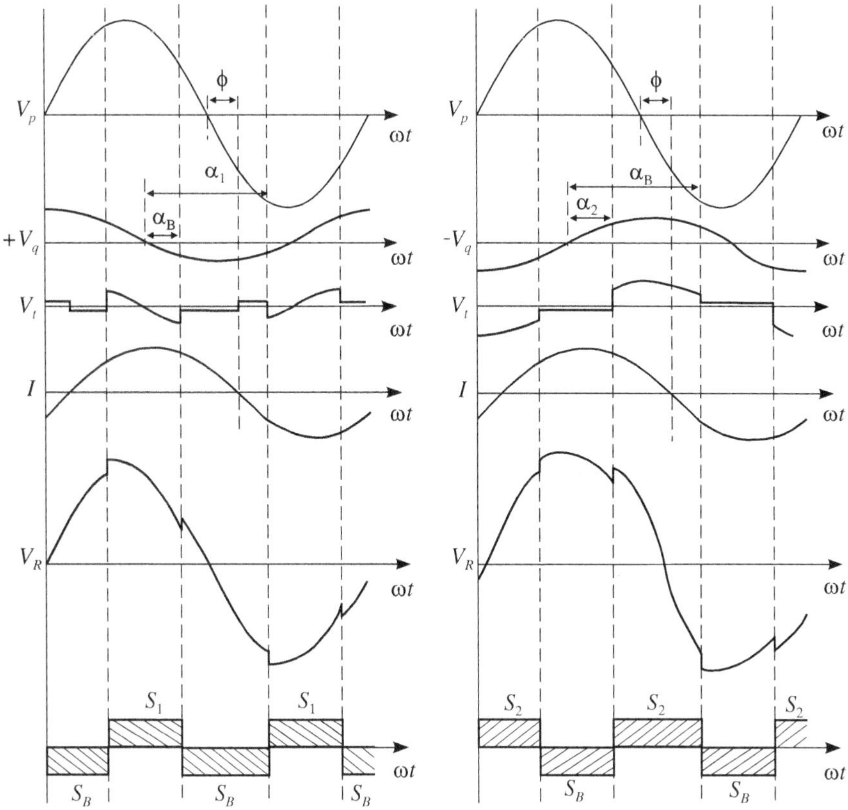

Figure 7.19 Theoretical waveforms for the cases of: (a) quadrature boosting and (b) quadrature bucking. *Source*: Arrillaga & Duke 1979 [10]. Adapted with permission of IEEE.

conducting. Thus, for operation in quadrature boosting mode, with lagging power factor, the effective range of the firing angle α for S_1 is $(90 + \phi)° < \alpha < 180°$ and the range of firing angle α_B for S_B is $0° < \alpha_B < (90 + \phi)°$. The firing angles α and α_B are both measured with respect to the zero crossings of $+V_q$.

- *Quadrature bucking control* is achieved by means of thyristor switches S_2 and S_B, together with a lagging quadrature voltage $(-V_q)$. The operation of quadrature bucking is described with reference to the theoretical waveforms of Figure 7.19b [12].

At the beginning of the positive half-cycle of line current I, with thyristor 3 conducting, $-V_q$ is negative and the voltage V_t across the secondary winding of the series transformer is negative. This voltage forward biases thyristor 5 and reverse biases thyristor 6. The provision of a gate pulse to thyristor 5 will therefore turn it on, thus short circuiting the secondary winding of T_2. V_t is now equal to $-V_f$, where V_f is the forward voltage drop of thyristor 6, and since $-V_q$ is more negative than V_t

thyristor 3 is reverse-biased and turns off. The current flow in thyristor 5 can be turned off as $-V_q$ changes polarity because thyristor 3 is again forward-biased. Providing a gate pulse to thyristor 3 will turn it on and a commutation from thyristor 5 to thyristor 3 will take place, turning thyristor 5 off.

The operation during the second half-cycle of line current is similar to the first, but with all voltage polarities reversed and with the alternate thyristor of switches S_2 and S_B conducting. The range of variation of the firing angle α_2 is $0° < \alpha_2 < (90 + \phi)°$, and for α_B is $(90 + \phi)° < \alpha_B < 180°$.

In order to avoid short circuiting the thyristors, a control logic can be implemented to ensure that the switch S_B will be gated only when S_1 and S_2 are off.

If the SPS of Figure 7.18 is a three-phase scheme, it can be used for in-phase/180° out-of-phase voltage injection, if the primary side windings of exciting transformer are star connected [11]. A scheme for four-quadrant phase-angle control is presented in Reference [15].

The main drawbacks of delay-angle control of Figure 7.18 are [4]:

- Firing-angle control of thyristor switches S_1, S_2, and S_B generates harmonic voltage components in addition to the fundamental frequency voltage component. Injection of harmonic voltage components can be of concern with respect to the power quality of the system.
- Necessary conditions to turn-on each thyristor switch depend on the system operating condition, that is, phase-angle between corresponding currents and voltages. There exist operating scenarios for which the necessary conditions to turn-on the required thyristor switch are not satisfied. Therefore, the SPS may not be able to provide voltage control under all possible system operating conditions.

7.3.1.3 Pulse-Width Modulation AC Controller

The limitations of the phase shifter configuration of Figure 7.18 are conceptually eliminated by employing a three-phase pulse-width modulation (PWM) AC controller (Figure 7.20a). Each of the switches S_1, S_2, and S_3 consists of a diode and a semiconductor switch with on-off control capability (forced commutation, for example, a gate turn-off [GTO] thyristor), connected in anti-parallel. A PWM switching pattern is used to turn the switches on and off so that to control the magnitude of the quadrature voltage $\underline{V}_\alpha$. When the forced-commutated switch of S_1 (S_2 and S_3) is on, the corresponding secondary windings of exciting transformer and boosting transformer are electrically connected. Current flow is either through the forced-commutated switch or its anti-parallel diode, depending on the current direction. When S_1 (S_2 and S_3) is off, forced-commutated switch S_B through a three-phase diode-rectifier provides a free-wheeling path for the secondary side current of the boosting transformer [4].

The magnitude of the injected voltage is controlled by varying the conduction interval (duty-cycle control) of the forced-commutated switches. The exciting transformer provides in-phase and quadrature-phase voltage if its primary windings are either star- or delta-connected, respectively. The advantages of an SPS based on a PWM AC controller are [4]:

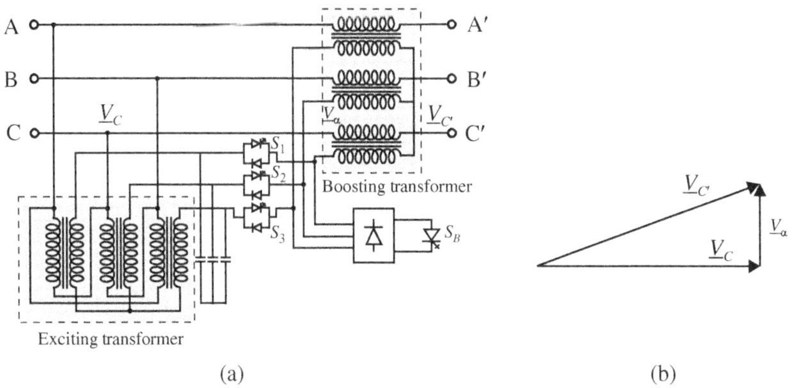

Figure 7.20 (a) Schematic diagram of a PWM AC controller-based SPS and (b) voltage phasor diagram. *Source*: Iravani 1999 [4]. Adapted with permission of IEEE.

- High frequency voltage harmonics only are generated, which do not propagate widely and are easily filtered.
- The PWM AC converter operation does not depend on the power system operating conditions.
- A PWM-controlled converter provides higher speed of response, but the side effect is the larger power losses, which increase with the switching frequency.

7.3.1.4 Delay-Angle Controlled AC-AC Bridge Converter

As shown in Figure 7.21, the interface between the exciter and the boosting transformers can be formed by uses of single-phase AC-AC bridge converters. Each leg of the converters consists of a bidirectional switch. The magnitude of the inserted quadrature voltage is determined by the delay-angle at the thyristor switches. The main drawback of this solution is the significant voltage harmonic content [4].

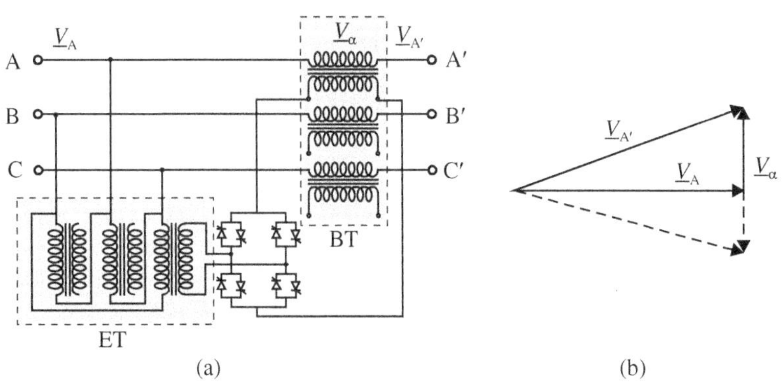

Figure 7.21 (a) Schematic diagram of an SPS based on firing-angle controlled AC-AC converter bridge and (b) voltage phasor diagram [4]. *Source*: Iravani 1999 [4]. Adapted with permission of IEEE.

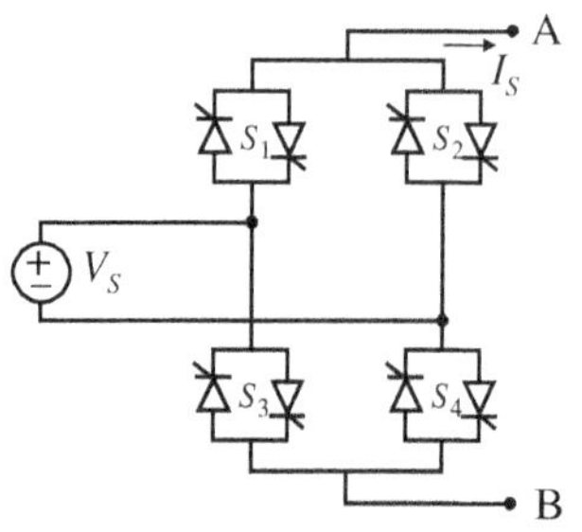

(a)

S_1	S_2	S_3	S_4	V_{AB}
on	off	off	on	$+V_s$
off	on	on	off	$-V_s$
on	off	on	off	0
off	on	off	on	0

(b)

Figure 7.22 (a) Single-phase AC-AC bridge converter and (b) output voltage of bridge converter based on discrete-step control [4].

7.3.1.5 Discrete-Step Controlled AC-AC Bridge Converter

In order to avoid harmonic generation, an AC-AC bridge converter with discrete-step delay-angle control can be employed. Each thyristor switch is turned on only at the zero crossing instant of the corresponding voltage.

The single-line diagram of a discrete-step controlled AC-AC bridge converter is shown in Figure 7.22. Its output voltage, V_{AB}, can take three distinct values ($+V_s$, 0, and $-V_s$) in terms of the operating state of the four switches. This voltage is independent of the direction of current i.

Simultaneous on-state of the switches S_1 (S_3) and S_2 (S_4) must be prevented in order to avoid a short circuit across the input voltage source V_S. Simultaneous off-state of all switches must also be prevented in order to avoid the load current i from being interrupted [4].

Figure 7.23 shows the schematic diagram of an SPS equipped with a *multiple discrete-step controlled* AC-AC converter bridge [14, 16]. By appropriate operation of the converter bridge, a voltage is injected in quadrature with respect to the line voltage, of which amplitude can be adjusted in steps. The exciting transformer of the configuration shown in Figure 7.23 includes three secondary windings, each one connected to the series transformer through a bridge converter. The turns ratios of the windings differ by a factor of three. This arrangement allows achieving $3^N = 27$ discrete-step of quadrature voltage amplitude: 13 steps of boost, 13 steps of buck, and zero. The maximum booster voltage V_α is 25% of the input voltage [12, 16, 17].

7.3.1.6 PWM Voltage Source Converter

An AC-DC-AC converter is used in the arrangement from Figure 7.24 to connect the exciting transformer with the boosting transformer [2, 18]. The converter is composed of two PWM-controlled voltage source converters (VSC) that share a DC link capacitor. The converters are three-phase full-bridge, where each arm consists of a bidirectional switch. The switches consist of a diode and a switch with turn on/off capability, for example, GTO thyristor (Figure 7.24b), which are anti-parallel connected.

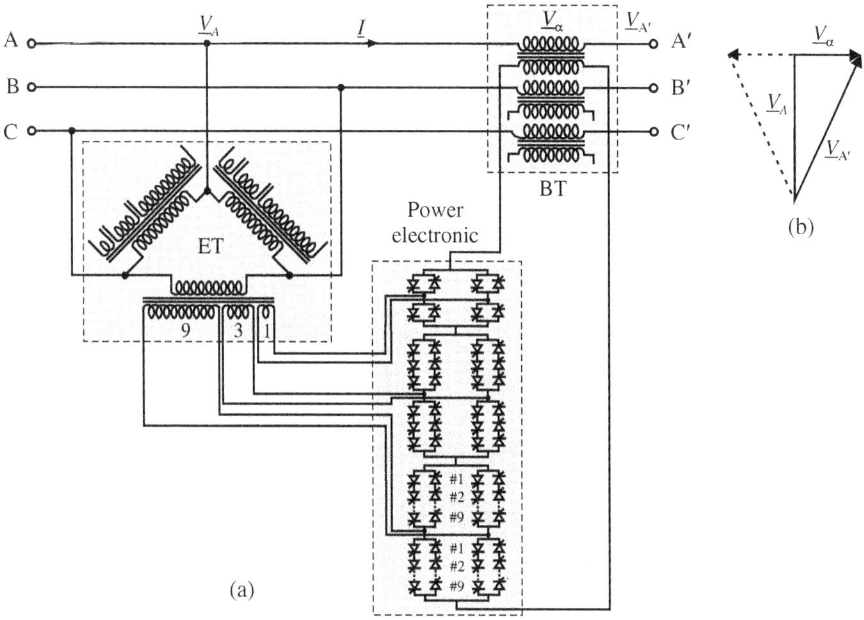

Figure 7.23 (a) Schematic diagram of a static thyristor-controlled phase shifting transformer and (b) voltage phasor diagram [16].

The booster side converter, VSC_{bo}, provides independent control of magnitude and phase angle of the injected voltage ($\underline{V}_\alpha$), whereas the exciting side converter, VSC_{ex}, regulates the DC-link capacitor voltage. The VSC converters can provide reactive power compensation so that the shunt-connected transformer, which is linked through the VSC_{ex} unit, can control the magnitude of $\underline{V}_C$. Thereby, a VSC-based phase shifter can provide independent and continuous control of both the active and reactive power flow on the line. Using these capabilities, the injected voltage phasor $\underline{V}_\alpha$ can be continuously varied within the circled control area as shown in Figure 7.24c. The two parts of the VSC can operate independent of each other so that this phase shifter configuration (Figure 7.24a) can operate as a shunt compensator, series compensator, or shunt/series active power filter [4]. This configuration is more widely known as unified power flow controller (see Chapter 10 of this book).

Note that a PWM current source converter can be used instead of PWM VSC in the AC-DC-AC converter-based phase shifter of Figure 7.24.

In modern times, insulated-gate bipolar transistors (IGBTs) for power applications are available. These are high frequency self-commutating switches. If the GTOs are replaced by the IGBTs in the two converters of Figure 7.24 and sinusoidal pulse width modulation technique is used for voltage and phase angle control (refer to Chapter 4), then the distortion in the output voltages will be very much reduced and LC filters can be replaced by a simple capacitor. For more information on VSC-controlled static PAR, the following papers may be referred [19, 20].

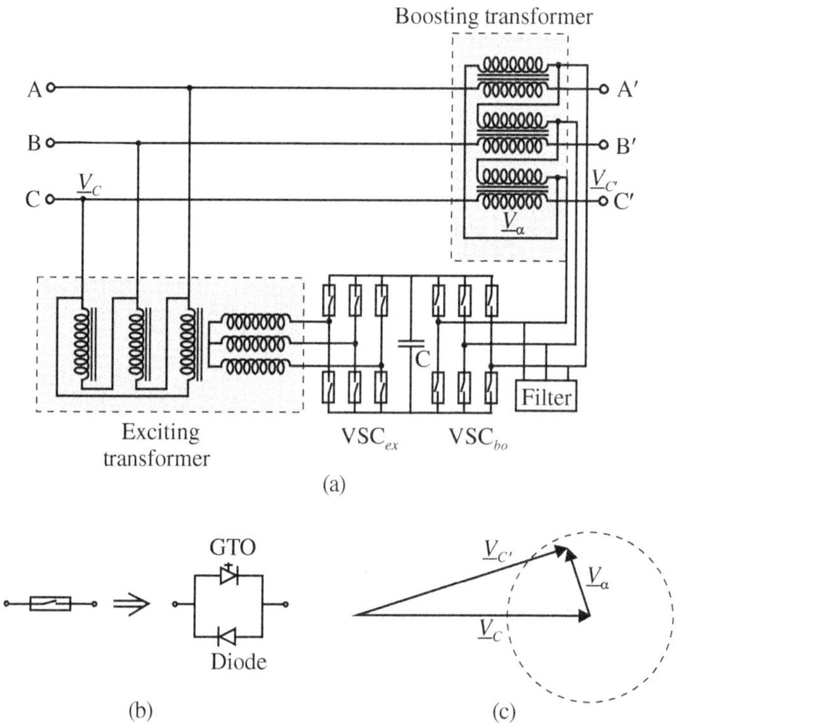

Figure 7.24 (a) Schematic diagram of an SPS based on PWM VSC converter, (b) switch structure, and (c) voltage phasor diagram.

7.3.2 Modeling of TCPST

7.3.2.1 Model of a Transmission System with a TCPST

There are many possible realizations of PSTs. While in Europe (especially the United Kingdom) QBTs are widely used, in the United States it is the PAR that is the most common device. If losses are neglected, the PAR terminal voltage phasors are separated without changing their magnitude. The QBT also separates terminal voltage phasors but the injected voltage $\underline{V}_T$ phase is fixed with regard to the input voltage phasor $\underline{V}_{PST1}$ ($\beta = \pm\, 90°$), (Figure 7.25) [21–23].

In both cases, the angle α of the terminal voltage phasors separation may be assumed to be a controllable parameter.

Although at first sight they appear to be similar devices, a PAR and a QBT differ in terms of their impact on power flow. While the PAR is asymmetrical device (no impact of orientation—the way its terminals are connected to the system—on power flow), the QBT is not and its orientation has an impact on the power flow. In addition, the PAR's location in the corridor does not influence power flow, while the QBT's location is important [4, 22].

The model of transmission system with a TCPST, and phasor diagrams for both a PAR and QBT are presented in Figures 7.25 [22, 23].

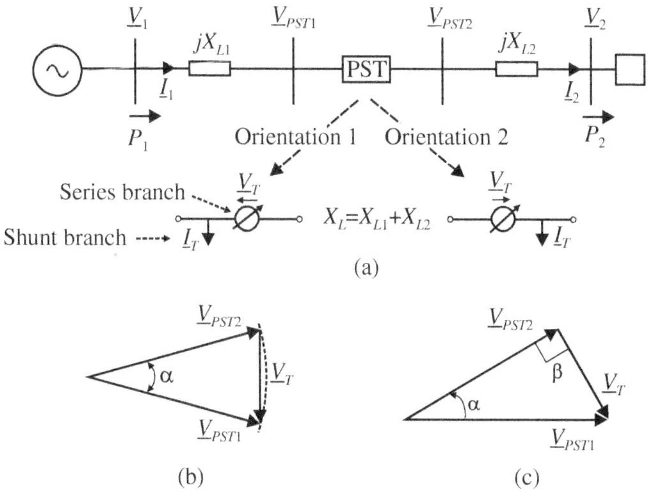

Figure 7.25 (a) Network scheme, (b) phasor diagram of a PAR, and (c) phasor diagram of a QBT [22].

The PAR equation for real power flow is:

$$P_{PAR} = P_2 = P_1 = \frac{V_1\,V_2}{X_L}\sin(\delta+\alpha) \tag{7.42}$$

The QBT equation for real power flow in the case of orientation 1 results in

$$P_{QBT} = P_2 = P_1 = \frac{V_1\,V_2}{X_{L_1}/\cos\alpha + X_{L_2}\cos\alpha}\sin(\delta+\alpha) \tag{7.43}$$

7.3.2.2 Line Model with Thyristor-Controlled Phase Angle Regulator

The real power flow, P_{ij}, and reactive power flow, Q_{ij}, in a transmission line connected between bus i and bus j without any thyristor-controlled phase angle regulator (TCPAR) can be written as:

$$P_{ij} = V_i^2 g_{ij} - V_i V_j[g_{ij}\cos(\delta_i-\delta_j) + b_{ij}\sin(\delta_i-\delta_j)] \tag{7.44a}$$

$$Q_{ij} = V_i^2(b_{ij}-b_c) - V_i V_j[g_{ij}\sin(\delta_i-\delta_j) + b_{ij}\cos(\delta_i-\delta_j)] \tag{7.44b}$$

where V_i and δ_i are the voltage magnitude and angle at bus i, $g_{ij}+jb_{ij}$ is the series admittance of the line connected between bus i and bus j, and b_c is the half-line shunt admittance.

Figure 7.26 shows the static model of a TCPAR. The effect of TCPAR can be modeled by a series inserted voltage source $\underline{V}_T$ and a tapped current $\underline{I}_T$.

The voltage at the output terminals of a TCPAR, $\underline{V}_{i'}$, is achieved by adding or subtracting a variable amplitude voltage component, $\underline{V}_T$, in quadrature with the phase voltage at the input terminals, $\underline{V}_i$. In this way, the output voltage $\underline{V}_{i'}$ is shifted with respect to the input voltage $\underline{V}_i$ by an angle α, and thereby:

$$\underline{V}_{i'} = t\cdot e^{j\alpha}\underline{V}_i \tag{7.45}$$

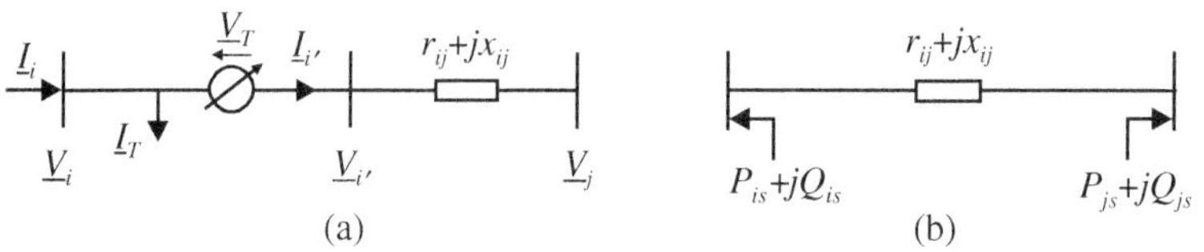

Figure 7.26 (a) Static model representation of TCPAR and (b) injection model of TCPAR.

where $t = 1/\cos\alpha$ is a transformation ratio of the voltage magnitudes.

Equation (7.45) can be written under the form:

$$\underline{V}_{i'} = \frac{\cos\alpha + j\sin\alpha}{\cos\alpha}\underline{V}_i = (1 + j\tan\alpha)\underline{V}_i \tag{7.46}$$

thereby, the quadrature voltage is:

$$\underline{V}_T = \underline{V}_i \tan\alpha \tag{7.47}$$

Neglecting losses in the TCPAR and using the principle of conservation of energy, the power flowing into a PST is equal to the power flowing out of the transformer, thereby:

$$\underline{V}_i\underline{I}_i^* = \underline{V}_{i'}\underline{I}_{i'}^* \tag{7.48}$$

The relationship between voltages and currents is thus given by

$$\frac{\underline{V}_{i'}}{\underline{V}_i} = \frac{\underline{I}_i^*}{\underline{I}_{i'}^*} = t \cdot e^{j\alpha} \tag{7.49}$$

The quadrature current is the phasorial difference between the input and the output currents:

$$\underline{I}_T = \underline{I}_i - \underline{I}_{i'} = \underline{I}_{i'}\left(\frac{e^{-j\alpha}}{\cos\alpha} - 1\right) = -j\underline{I}_i' \tan\alpha \tag{7.50}$$

The power flow equation from bus i to bus j can be written as:

$$\underline{S}_{ij} = P_{ij} + jQ_{ij} = \underline{V}_i\underline{I}_{ij}^* = \underline{V}_i(\underline{I}_T + \underline{I}_i')^* \tag{7.51}$$

The active and reactive power flow equations are thereby achieved [23]:

$$P_{ij} = t^2V_i^2g_{ij} - tV_iV_j[g_{ij}\cos(\delta_i - \delta_j + \alpha) + b_{ij}\sin(\delta_i - \delta_j + \alpha)] \tag{7.52a}$$

$$Q_{ij} = t^2V_i^2(b_{ij} + b_c) - tV_iV_j[g_{ij}\sin(\delta_i - \delta_j + \alpha) - b_{ij}\cos(\delta_i - \delta_j + \alpha)] \tag{7.52b}$$

The injection model of TCPAR is shown in Figure 7.26b. The injected active power P_{is} and P_{js}, and reactive power Q_{is} and Q_{js} of a line having phase shifter are:

$$P_{is} = -T^2V_i^2g_{ij} - TV_iV_j(g_{ij}\sin\delta_{ij} - b_{ij}\cos\delta_{ij}) \tag{7.53a}$$

$$Q_{is} = T^2V_i^2b_{ij} + TV_iV_j(g_{ij}\cos\delta_{ij} + b_{ij}\sin\delta_{ij}) \tag{7.53b}$$

$$P_{js} = -TV_iV_j(g_{ij}\sin\delta_{ij} + b_{ij}\cos\delta_{ij}) \tag{7.53c}$$

$$Q_{js} = -TV_iV_j(g_{ij}\cos\delta_{ij} - b_{ij}\sin\delta_{ij}) \tag{7.53d}$$

where $T = \tan\alpha$.

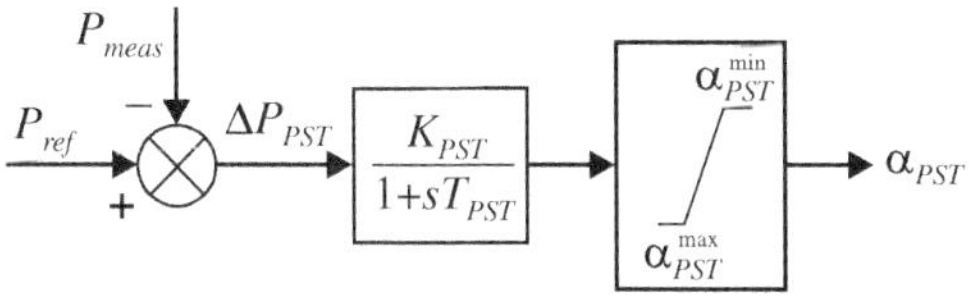

Figure 7.27 The dynamic model of the phase shifter.

7.3.2.3 The Dynamic Model of the Phase Shifter

Figure 7.27 shows the dynamic model of the PST. The measured power flow P_{meas} is compared with the scheduled value P_{ref} to provide a power error ΔP_{PST}, which is then feed forward into a first-order transfer function, where K_{PST} and T_{PST} are the gain and time constant of the phase shifter, and s is the Laplace operator. The output of the dynamic model is the phase angle α_{PST}, which is limited by physical capability limits (α_{PST}^{min}, α_{PST}^{max}).

7.4 APPLICATIONS OF THE PHASE SHIFTING TRANSFORMERS

The on-load tap changer has been invented in the early period of creation of the AC transmission systems as a need to regulate the voltage on the load side. The change in the tap position leads to a change in the reactive power flow direction through the transformer. Thus, voltage boosting or bucking also involves exchange of reactive power between the two sides of the transformer. Since transmission line impedances are predominantly reactive, an "in-phase voltage" component introduced into the transmission circuit causes a substantially quadrature (reactive) current flow that, with appropriate polarity and magnitude control, can be used to improve prevailing reactive power flow [24].

Although reactive compensation and voltage regulation by on-load tap changers appear to provide the same transmission control function, there is an important operating difference to note between them. While a reactive compensator supplies reactive power to, or absorbs that from the AC system to change the prevailing reactive power flow and thereby indirectly control the transmission line voltage, the tap changer-based voltage regulator cannot supply or absorb reactive power. It directly manages the transmission voltage on one side and leaves it to the power system to provide the necessary reactive power to maintain that voltage. Should the power system be unable to provide the reactive power demand, overall voltage collapse in the system could occur. On-load tap changers contribution to voltage collapse under certain conditions is well recognized [25, 26].

Chapter 6 presented the control of transmitted power by series capacitive compensation, including mainly compensation by TCSC, which can be a highly effective means to control power flow in the line as well as improving the dynamic behavior of the power system. However, their costs are higher than those of the PSTs, and for this reason the PSTs are preferred when very fast power flow control is not desired [24].

Apart from steady-state voltage and power flow control, the role of modern voltage PARs with fast electronic control can also be extended to handle dynamic system events. Potential areas of application include transient stability improvement, power oscillation damping, and the minimization of postdisturbance overloads and the corresponding voltage dips [5, 27].

7.4.1 Power Flow Control by Phase Angle Regulators

In order to present the basic concept of power flow control by angle regulation, let us consider the simple case of two equivalent machines interconnected through a transmission line (Figure 7.28a). A PAR is inserted between the sending-end generator and the transmission line [24].

A PAR can be designed to insert in the transmission line a controllable amplitude and phase angle voltage, $\underline{V}_\alpha$. The resulted voltage of the sending-end side of the system from Figure 7.28 is

$$\underline{V}'_A = \underline{V}_A + \underline{V}_\alpha$$

Depending on the phase angle of the inserted voltage, the phase angle difference between voltages of the two machines can increase or decrease, as shown in Figure 7.28b. By appropriate design of the PAR, in the ideal case, the phasor of

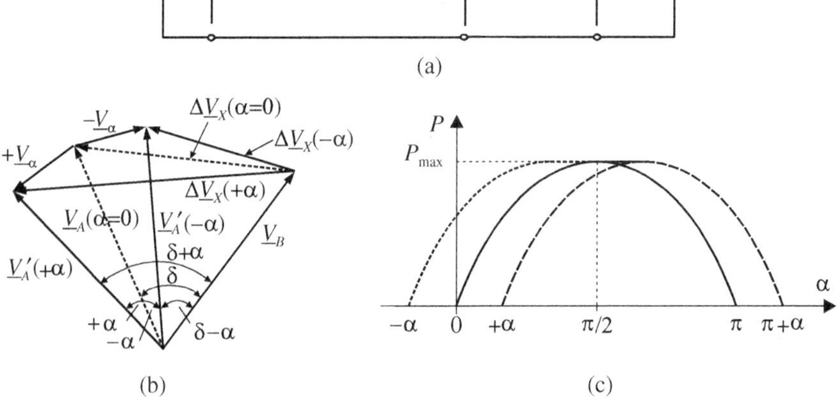

Figure 7.28 (a) Two-machine power system with a PAR, (b) phasor voltage diagram, and (c) transmitted power versus angle characteristics. *Source*: Hingorani & Gyugyi 2000 [24]. Reproduced with permission of IEEE.

voltage $\underline{V}_\alpha$ can be shifter with respect to phasor $\underline{V}_A$ by angle α, whereas their magnitudes remain unchanged, that is

$$V_A' = V_A = V \tag{7.54}$$

The main purpose of the PAR is to maintain the power flow on the transmission line at the desired value, independent of the prevailing phase angle difference, δ, between the voltages at the two ends of the transmission line.

Under the control characteristic stipulated by (7.54) and the effective phase angle (δ–α) between the sending- and receiving-end voltages, the active and reactive power flows are as follows:

$$P = \frac{V^2}{X_L} \sin(\delta \pm \alpha) \tag{7.55a}$$

$$Q = \frac{V^2}{X_L} [1 - \cos(\delta \pm \alpha)] \tag{7.55b}$$

The steady-state active power characteristic in terms of the angles δ and α resulted from equation (7.55a) is plotted in Figure 7.28c. Naturally, for the uncompensated line, from stability point of view, the maximum transmissible power is achieved for a phase angle difference δ=π/2. When using a PAR, the active power flow can theoretically be kept at its peak value even for angles δ that exceed π/2, eventually in the range $\pi/2 < \delta < (\pi/2 + \alpha)$. This can be possible by controlling the amplitude of the inserted quadrature voltage $\underline{V}_\alpha$ so that the effective phase angle difference (δ–α) between the sending-end and receiving-end voltages does not exceed π/2.

The power characteristic can be displaced to the left or to the right depending on the direction/polarity of the inserted voltage. A PAR equipment does not increase the maximum transmissible power, but it can increase or decrease the actual transmitted power for the same phase angle difference δ.

Quadrature Booster Case. If the phasor $\underline{V}_\alpha$ is shifted relative to the phasor $\underline{V}_A$ by a fixed angle ±90°, the PAR becomes a QB, and the phasorial and algebraic relationship between voltages on the sides on the PAR is given by:

$$\underline{V}_A' = \underline{V}_A + \underline{V}_\alpha$$

$$V_A' = \sqrt{V_A^2 + V_\alpha^2}$$

When the PAR is a QB, the transmitted power P can be expressed by:

$$P = \frac{V^2}{X_L} \left(\sin\delta + \frac{V_\alpha}{V} \cos\delta \right) \tag{7.56}$$

Figure 7.29a illustrates the phasor diagram of the system voltages when PAR behaves like a QB. The active power characteristic given by (7.56) is plotted in Figure 7.29b in terms of angle δ for various values of the injected quadrature voltage $\underline{V}_\alpha$.

The maximum transmissible power increases in this case because, by insertion of the quadrature voltage $\underline{V}_\alpha$, the magnitude of the sending-end voltage increases.

In contrast to the previous investigated reactive shunt (Chapter 5) and series compensator schemes (Chapter 6), the PARs generally have to handle both real and

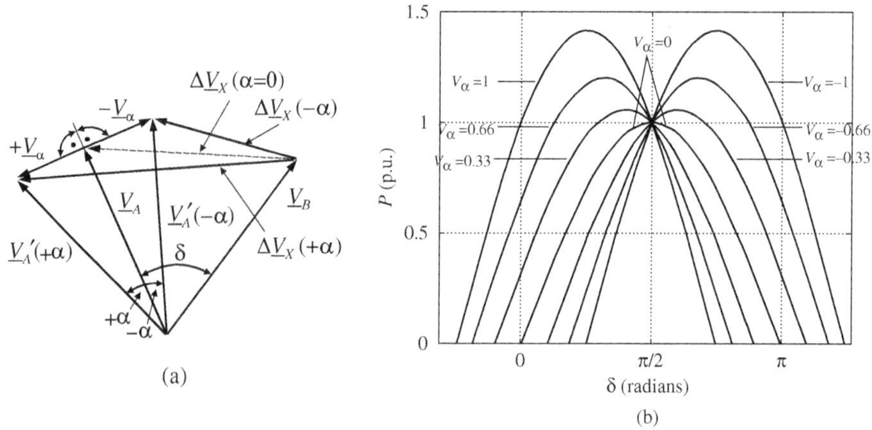

Figure 7.29 Phasor diagram and transmission power versus angle characteristics of a quadrature booster. *Source*: Hingorani & Gyugyi 2000 [24]. Adapted with permission of IEEE.

reactive powers. The total VA (apparent power) throughput power of the angle regulator (viewed as voltage source) is

$$VA = \left|\underline{V}_A' - \underline{V}_A\right| \cdot |\underline{I}| = \left|\underline{V}_\alpha\right| |\underline{I}| = V_\alpha I \tag{7.57}$$

The rating of the angle regulator is thus determined by the product of the maximum injected voltage and the maximum continuous line current.

7.4.2 Real and Reactive Loop Power Flow Control

The power flow on an AC transmission line is given by the difference in the phase angles of the voltages at the two line ends. When two parallel corridors of different parameters between two network points are involved, manipulation of the phase angles allows controlled division of the power flows between corridors, thus preventing overloads. A PAR can control the power flow on a certain corridor to a desired value.

When a power system experiences, on a regular basis, unintended power flows caused by a neighbor interchange partner, an appropriate solution is to install a PST at a frontier station or on a tie line. As the wind generation has been developed, fluctuating power flows are experienced between neighbor power systems. This is the case of Europe, where several power system operators have installed or are planning to install PSTs in order to tightly control the power flow on their frontier transmission lines.

Consider two power systems A and B connected by a single transmission line as shown in Figure 7.30.

The operating conditions of the two systems and the transmission of active power P between them result in a difference in magnitude and phase angle between the terminal voltages, $\underline{V}_A$ and $\underline{V}_B$ (Figure 7.30b). Phasorial voltage difference $\Delta\underline{V}_L = \underline{V}_A - \underline{V}_B$ that appears across the transmission line impedance $\underline{Z}_L (= R_L + jX_L)$

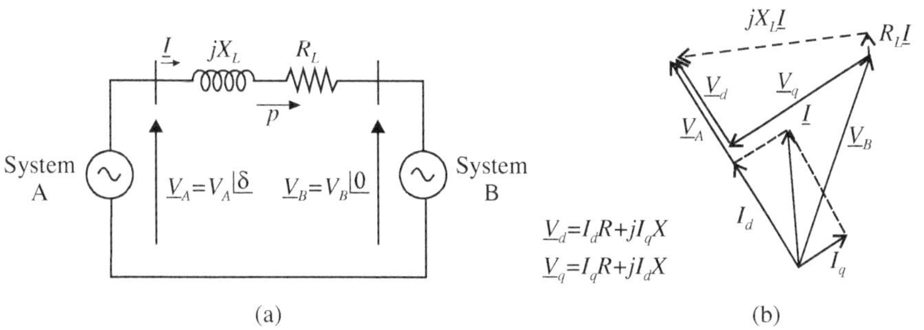

Figure 7.30 (a) Two systems with a single line intertie and (b) phasor voltage diagram [24].

results in the line current $\underline{I}$. Phasor $\Delta\underline{V}_L$ is normally considered to be composed of the resistive and inductive voltage drops $\underline{I}R$ and $j\underline{I}X$, respectively.

For the case of loop power flow, it is more meaningful to decompose $\Delta\underline{V}_L$ into two components, one in phase (V_d) and the other in quadrature (V_q) with the sending-end voltage phasor $\underline{V}_A$ (Figure 7.30b). These voltage components determine the reactive and active power supplied by the sending-end system [24].

In practice, power systems are normally connected by two or more parallel transmission paths, resulting in one or more circuit loops with the potential for circulating current flow. Consider the above system with two parallel transmission lines (Figure 7.31).

Basic circuit considerations indicate that if the *X/R* ratios for the two lines are not equal, that is, if $X_{L1}/R_{L1} \neq X_{L2}/R_{L2}$, then a circulating current $\underline{I}_C$ will flow through the two lines. Assuming such an inequality and decomposing both line currents, $\underline{I}_1$ and $\underline{I}_2$ into an in-phase and a quadrature component with respect to the voltage $\underline{V}_A$ (see Figure 7.30b), then the corresponding in-phase and quadrature voltage components for the lines, V_{1d}, V_{1q} and V_{2d}, V_{2q} can be expressed as follows [24]:

$$\begin{aligned} V_{1d} &= \left(I_{1d} + I_{cd}\right) R_{L1} + j\left(I_{1q} + I_{cq}\right) X_{L1} \\ V_{1q} &= \left(I_{1q} + I_{cq}\right) R_{L1} + j\left(I_{1d} + I_{cd}\right) X_{L1} \\ V_{2d} &= \left(I_{2d} - I_{cd}\right) R_{L2} + j\left(I_{2q} - I_{cq}\right) X_{L2} \\ V_{2q} &= \left(I_{2q} - I_{cq}\right) R_{L2} + j\left(I_{2d} - I_{cd}\right) X_{L2} \end{aligned} \qquad (7.58)$$

Figure 7.32a illustrates the case when there is a difference in the quadrature voltage components, V_{1q}–V_{2q}. Considering the practical assumption that $R_{L1} \ll X_{L1}$

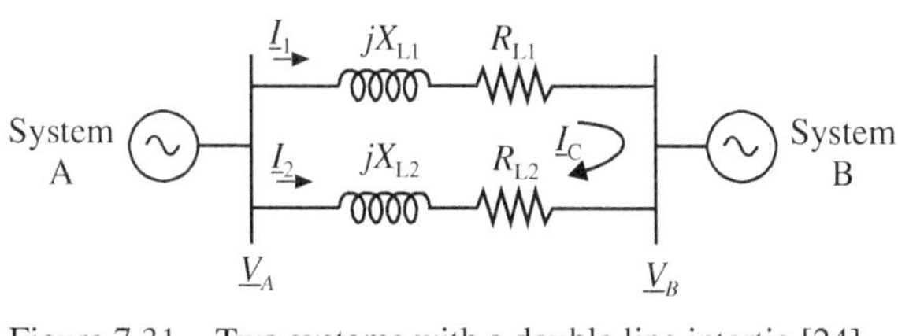

Figure 7.31 Two systems with a double line intertie [24].

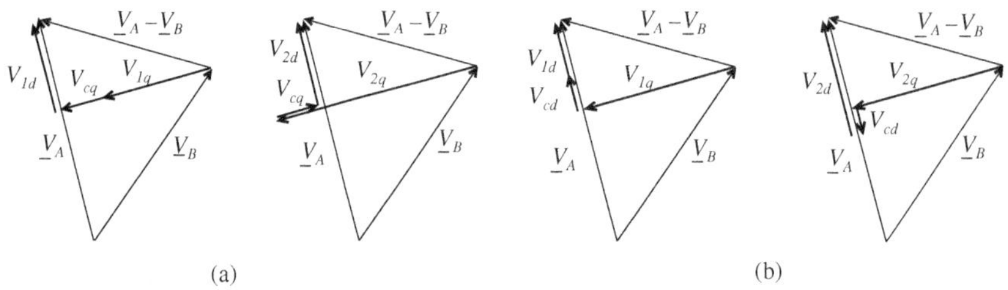

Figure 7.32 Phasor diagrams illustrating voltage unbalances resulting in active (a) and reactive (b) loop power flows [24].

and $R_{L2} \ll X_{L2}$, this difference will primarily maintain the in-phase circulating current component I_{cd}, thus increasing the active power in the line 1 and decreasing it in the line 2.

Figure 7.32b illustrates the case when a difference in the in-phase voltage components, V_{1d}–V_{2d}, exists. This difference, under the above assumption, will maintain primarily the quadrature circulating current component, I_{cq}, and thus change the reactive power flow balance between the lines. In a general case, differences may exist between both the in-phase and quadrature voltage components, which will maintain both the in-phase and quadrature components of the circulating current, changing both the active and reactive power flow balance between the lines.

Generally, the distribution of active power flow over interconnections forming loop circuits can be controlled by PARs. The flow of reactive power can be controlled by voltage regulators. These statements follow from the fact that the transmission circuit impedances are predominantly reactive. The PAR injects a quadrature voltage in series with the circuit loop resulting in the flow of in-phase circulating current. The voltage regulator introduces a series in-phase voltage into the loop, and quadrature current is circulated through the loop since the impedances are substantially reactive.

Thus, the insertion of a PAR in either tie-line 1 or tie-line 2 in the system of Figure 7.31 can correct the difference in quadrature voltage drops and can generally control the active power distribution between the lines. An added voltage regulator, separate or embedded in the PAR, can cancel the in-phase voltage difference and generally control the reactive power flow balance.

In conclusion, both voltage and angle regulators can clearly provide major benefits for multiline and meshed systems: the full utilization of transmission assets by decrease of reactive power flow, control and balance of active power flow, and reduction of overall system losses through the elimination of circulating loop currents.

7.4.3 Improvement of Transient Stability with PST

When a synchronous generator is connected at the sending-end of a transmission line, a QB reduced the effective angle between the generator rotor angle and the voltage at the receiving-end of the line. Power-system busbar is reduced and the power-angle curve is thus shifted to the right. The thyristor-controlled quadrature voltage injection (QVI) transformer provides any intermediate power-angle characteristic between the no-boosting and full-boosting curves.

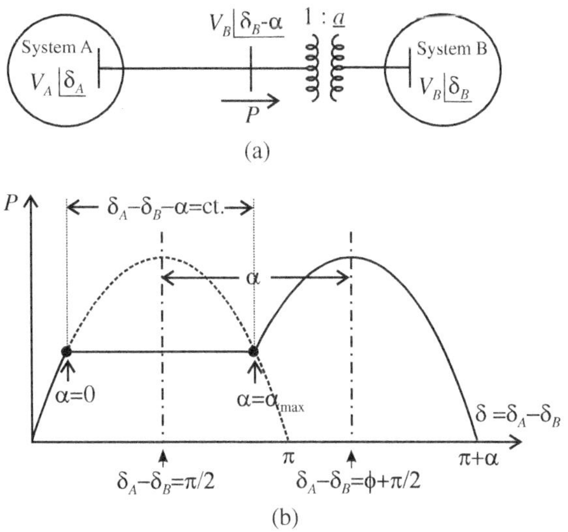

Figure 7.33 Interconnected power system with phase shifter: (a) interconnected systems and (b) power-angle diagram.

Without boosting, the maximum power transfer occurs at approximately 90°, while with full boosting it occurs at approximately (90+α)°, where α is the phase shift introduced by the QVI transformer (Figure 7.33b).

Figure 7.33a shows the interconnecting line between the system buses A and B, with a PST that provides a phase shift of α [27].

The complex transformation ratio $\underline{a} = 1 \cdot e^{j\alpha}$ means that the secondary voltage is equal in magnitude to the primary voltage but has a phase lead of α. If X_L is the line reactance and X_T is the leakage reactance of the transformer referred to the primary, the power flow from the bus A to bus B is given by:

$$P = \frac{V_A V_B}{X_L + X_T} \sin(\delta_A - \delta_B - \alpha)$$

If the system B is swinging with respect to system A, then the power flow through the interconnecting line can be controlled by adjusting the shift angle α such that $(\delta_A - \delta_B + \alpha)$ remains constant over the available range of α. If there is no artificial phase shift, then the steady-state system stability is lost when the angle $(\delta_A - \delta_B)$ becomes 90°. With the introduction of the PST and making α negative, the angle $(\delta_A - \delta_B)$ can become greater than 90° as shown in Figures 7.33b and 7.34. In Figure 7.34, δ_1 represents the angle of fault initiation, δ_2 is the angle of fault removal, and δ_3 is the first swing maximum angle.

By contribution of a phase shifter, the maximum power transfer can theoretically be achieved for angles up to (90+α)°, when a positive quadrature voltage (boosting) is injected, as shown in Figure 7.34b. The first swing stability can therefore be improved because the area on the power-angle curve between transmitted and generated power is increased and the maximum angle that ensures generator stability is increased from δ_{lim1} to δ_{lim2} [12].

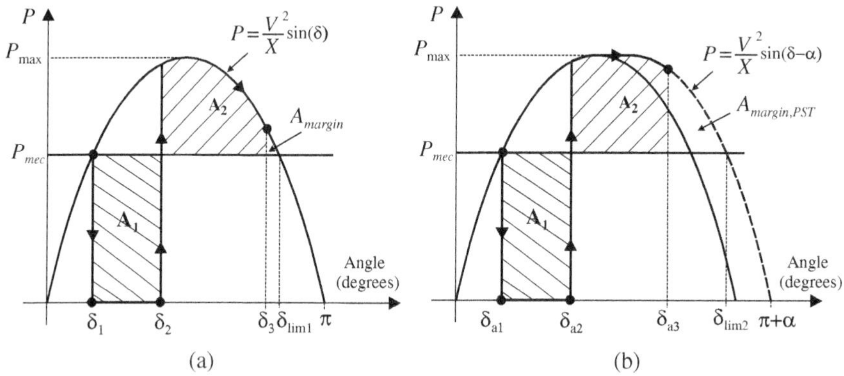

Figure 7.34 Equal area criterion illustrating transient stability margin for a simple two-machine system: (a) without phase-angle control and (b) improvement in first swing stability with quadrature boosting [12].

Figure 7.34 has been simplified for clarity by assuming that the network prior to and after the fault is unchanged, that there is zero power transfer during the fault, and that fault clearance occurs before the power-angle reaches 90°.

Using an SPS, the accelerating area A_1 can be limited if bucking control is initiated on fault inception, which means that a negative quadrature voltage (bucking) is injected. The full-bucking is achieved immediately after the fault is removed. The maximum possible effect of both boosting and bucking is shown in Figure 7.35.

This improvement is smaller in practice than would appear from the diagram, as during the period in which the system is faulted, bucking will have little or no effect.

7.4.4 Power Oscillation Damping with PST

Since a PST is capable of controlling the power flow on certain network corridors, it can also be employed to damp power oscillations.

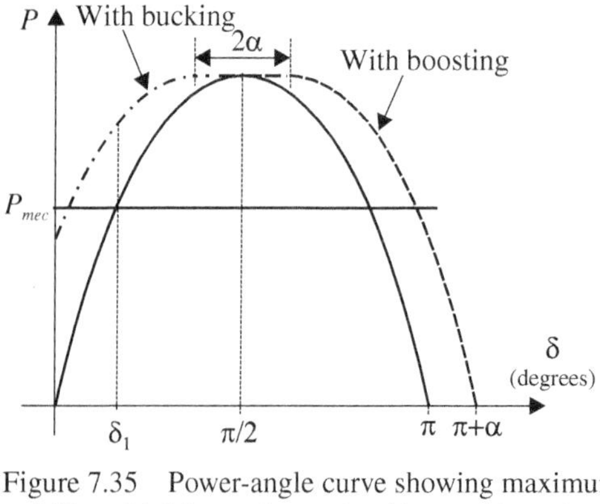

Figure 7.35 Power-angle curve showing maximum possible improvement in first swing stability with 20° quadrature bucking and boosting [12].

Let us analyze the case of a synchronous generator located at one end of the transmission corridor. Considering a control angle of the phase shifter within the range of $\alpha_{min} < \alpha < \alpha_{max}$, and knowing that stability can theoretically be maintained for $0 \leq \delta \leq \pi/2$, two cases are identified [24]:

- When the rotationally oscillating generator accelerates and angle δ increases, that is $d\delta/dt > 0$, the electric power transmitted must be increased to compensate for the excess mechanical input power. In order to increase the allowed angle between the generator and the opposite end of the transmission corridor and to increase the transmitted power, a negative angle α has to be introduced, thus making the P-δ curve shift to the left.
- When the generator decelerates and angle δ decreases, that is $d\delta/dt < 0$, the electric power must be decreased to balance the insufficient mechanical input power. This can be done by introducing a positive angle α, which shifts the P-δ curve to the right, thus decreasing the overall transmission angle and real power transmitted.

In the absence of system damping, and disregarding the quadrature voltage effects, the system would oscillate about the new steady-state operating point such that the areas between the power generation and transfer curves would be equal. Instead, after the first swing maximum angle excursion, it is desirable to reduce the decelerating area. This can be achieved by reversing the controls so that for power-angles in excess of 90°, there is quadrature bucking and for power-angles less than 90°, there is quadrature boosting.

In theory, it would be possible to change from full boosting to full bucking immediately after the maximum swing has been reached. However, the power-angle cannot change instantly, and the power transfer would be reduced considerably as a result, as shown in Figure 7.36a. There is therefore a risk of instability when the angle of maximum swing is close to the stability limit.

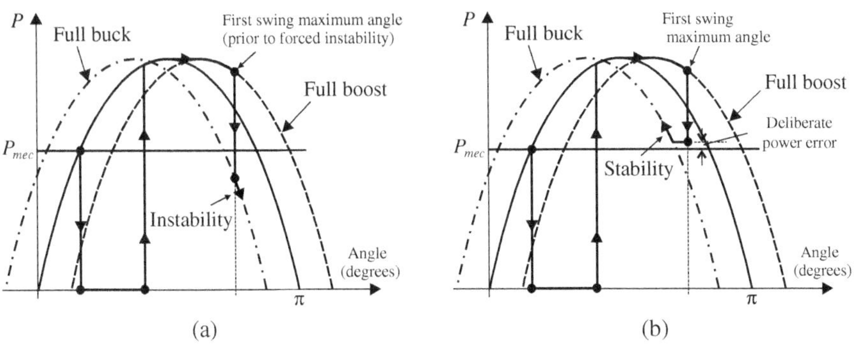

Figure 7.36 Power angle curves showing the effects of control strategies: (a) danger of causing instability at first swing maximum by immediate change from full boost to full buck and (b) instability avoided at first swing maximum by controlled injection. *Source*: Arnold et al. 1981 [12]. Reproduced with permission of IEEE.

With this scheme, there is a risk of instability if the angle of maximum swing is close to the stability limit. Thus, injection must be modified in a controlled manner to ensure that this does not occur.

Different means of achieving are described in paper [12].

- *Damping mode A* (Figure 7.36b). At maximum swing (greater than 90°), the injection is immediately modified so that the power transfer is just above the input power (P_{mec}). This ensures that moderate deceleration does occur and will probably result in the PST operating in a controlled mode, that is as the angle reduces, the injection is modified to full bucking. For angles less than 90°, the control is on full boost. In the event of the maximum swing being less than 90°, changeover can take place immediately. In the unlikely event of this causing acceleration instead of deceleration, the acceleration will be small and there will be plenty of safety margin.

  Alternatively, provided that the maximum swing is greater than 90°, the PST can be operated in a controlled manner to keep the decelerating power equal to the value at maximum swing as long as possible. Otherwise, this is the same as "mode A."
- *Damping mode B*. Full boost is maintained over the whole of the second swing period. This mode is the easiest to implement.

  Similar schemes can be adopted at maximum backswing to further reduce the next accelerating period. In this way, each subsequent swing can be reduced. At maximum backswing, the operation can be changed from full boost to full buck instantaneously. It is not important whether the new power transfer causes a subsequent period of acceleration or deceleration.
- *Damping mode C*. In the ideal solution, the amount of injection at maximum backswing should be changed to force the power transfer to a value as close as possible to the input power (P_{mec}). This reduces the amplitude of the next swing. Subsequent swings can be similarly reduced and at the same time, if each new QVI setting is regarded as a maximum value from then on, the injection will slowly be reduced. Because of the difficulty of calculating the correct amount of injection to reduce the transients to zero, there will be an error and small swings will continue. This swing will be sufficient to remove almost all the injection.

As an alternative method of removing the voltage injection after a predetermined number of swings in mode C, the injection may be reduced in a controlled manner with no further attempts to control damping [12].

7.4.4.1 Application to Damp Power Oscillations

Let us consider the simple case of a single machine connected to an infinite bus power system (SMIB) by a transmission line and a PST (Figure 7.37). The generator is modeled by an electromotive voltage $\underline{E}_g$ behind reactance X_g, the line is represented by its reactance X_L, and the phase shifter by the reactance X_T.

In order to damp out the power oscillations in a simple SMIB system, the phase angle of the output voltage from the machine can be changed, to reduce or increase

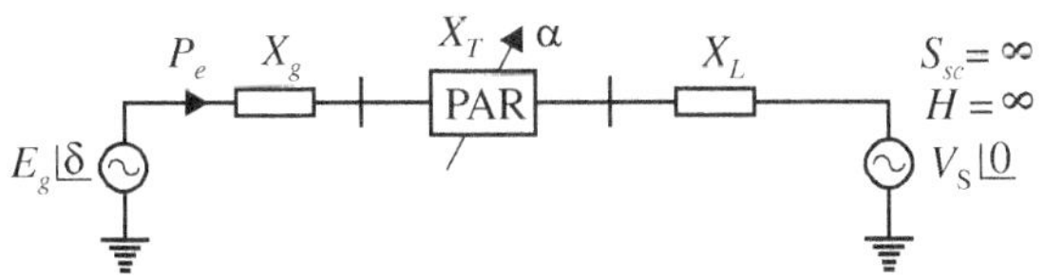

Figure 7.37 The one-line diagram of an SMIB.

the power angle δ between the generator bus and the infinite bus depending upon whether the machine speed is lower or higher than the rated speed. Accordingly, the power flow from the machine gets controlled. Therefore, the change in the power angle Δα must be proportional to the negative of the change in speed of the machine Δω [26, 28].

The swing equation of the generator is expressed, in per unit, as:

$$\begin{cases} \dfrac{d\omega}{dt} = \dfrac{1}{T_a}(P_m - P_e) \\ \dfrac{d\delta}{dt} = \omega_0 \omega \end{cases}$$

where:

P_m is the mechanical power.

P_e – the generated active power.

T_a – the mechanical start-up time.

Linearization about the initial operating point gives

$$\begin{cases} \dfrac{d\Delta\omega}{dt} = \dfrac{1}{T_a}(\Delta P_m - \Delta P_e) = \dfrac{1}{T_a}\Delta P_a \\ \dfrac{d\Delta\delta}{dt} = \omega_0 \Delta\omega \end{cases} \tag{7.59}$$

where ΔP_a is the change in the accelerating power.

The active power transferred from the single machine to the power system, in the presence of a PAR, is given by:

$$P_g = \frac{E_g V_S}{X_{eq}} \sin(\delta - \alpha) \tag{7.60}$$

where X_{eq} is the total reactance between the machine and the infinite bus.

The phase angle provided by PAR is related to the mismatch of the transferred power with respect to the reference value, P_{ref}, that is [28]:

$$\alpha = \frac{1}{sT_p}(P_{ref} - P_e) \tag{7.61}$$

where T_p is the time constant of the PAR controller, and s is the Laplace operator.

Linearizing equations (7.60) and (7.61), about the actual operating point, assuming that the machine voltage is kept constant by fast acting automatic voltage

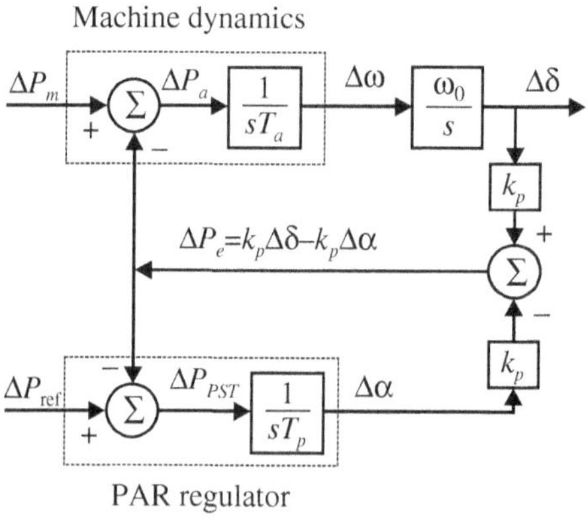

Figure 7.38 Block of electromechanical loop in the presence of PAR [28].

regulator (AVR) $\Delta E_g = 0$, and that there is no change in the reference active power transfer for the PAR, that is $\Delta P_{ref} = 0$, we achieve:

$$\begin{cases} \Delta P_e = k_p \Delta\delta - k_p \Delta\alpha \\ \Delta\alpha = -\dfrac{1}{sT_p}\Delta P_e \end{cases} \tag{7.62}$$

where k_p is determined for the initial conditions, that is

$$k_p = \left.\frac{\partial P_e}{\partial \delta}\right|_{\delta=\delta_0} = \frac{E_g V_S}{X_{eq}} \cos\delta_0$$

Based on equations (7.59) and (7.62), the combined block diagram of the generator and the PAR, for damping small perturbation, is built (Figure 7.38).

7.5 PHASE SHIFTING TRANSFORMER PROJECTS AROUND THE WORLD

The electricity market, the cross-border energy trades, and the intermittent energy sources have been the greatest challenges for the operators in ensuring the power system security and reliability. In large and strongly interconnected power systems, parallel flows and sometimes network overloadings are experienced by many countries. The mechanical PSTs have been the first choice, and also the cheapest, for the power system operators around the world to control the power flows.

BENELUX MAKES A STEP AHEAD TO DEAL WITH INCREASED WIND GENERATION. The central area of European Network of Transmission System Operators for Electricity (ENTSO-E), which includes the power systems from Benelux (The Netherlands, Belgium, and Luxembourg), Germany, and France, is subjected to power flows originated from the wind generation produced in north of Germany or from the nuclear generation located in north of France, which may cause network overloading.

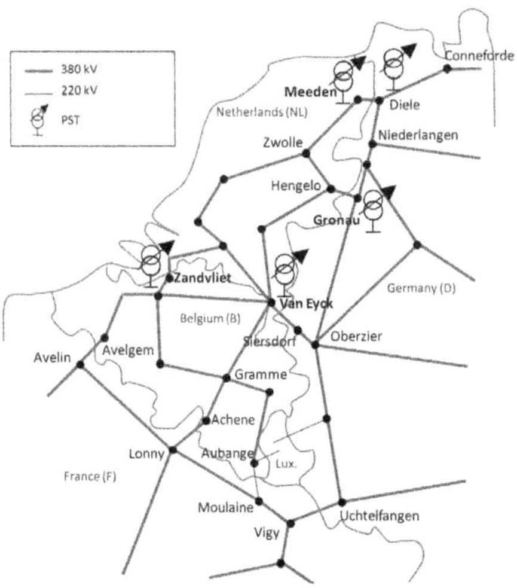

Figure 7.39 Location of PSTs in Benelux [32].

The increased trading between Germany and France creates loop flows through the Belgium and Dutch grids. In order to control these power flows, the transmission and system operators from The Netherlands and Belgium have decided to install PSTs on the interconnection lines, at the border substations. As regards the Dutch grid, all tie lines are controlled by phase shifters [29–31].

Figure 7.39 shows the location of phase shifters on the borders of the Dutch grid, and their characteristics are provided in Table 7.1.

Two additional PSTs are installed in the 150 kV network of the Dutch power system, at Delft and Leiden.

As wind generation capacity is continuously increasing in the North Sea, new power flow control equipment may be required in the near future. A PST would be placed on the existing 225 kV line between Luxembourg and Belgium by 2016.

TABLE 7.1 PSTs installed on the Benelux's interconnections [32]

Transmission line	Voltage	Throughput rating	Regulation	Taps	Year
Hengelo (Netherlands)—Gronau* (Germany)	380 kV	1250 MVA	±12°	−17/17	—
Meeden* (Netherlands)—Diele (Germany)	380 kV	2 × 1000 MVA	±30°	−30/30	2003
Diele* (Germany)	380 kV	2 × 1450 MVA	±24°	−31/31	2007
Zandvliet* (Belgium)	380 kV	1400 MVA	±25°	−17/17	2008
Van Eyck* 1 and 2 (Belgium)	380 kV	2 × 1400 MVA	±25°	−17/17	2009

*The PST location.

TABLE 7.2 Phase shifting transformers installed on the Italian border

Transmission line	System voltage	Regulation	Taps	Throughput rating	Year
Rondissone* (Italy)—Albertville (France)	400 kV	+18°	33	1630 MVA	2003
Padriciano* (Italy)—Divaca (Slovenia)	220 kV	±31°	33	370 MVA	2008
Redipuglia (Italy)—Divaca* (Slovenia)	400 kV	±40°	±32	2 × 600 MVA	2010
Trinité Victor / Menton (France)—Camporosso* (Italy)	220 kV	—	—	450 MVA	—
Lienz* (Austria)—Soverzene (Italy)	220 kV				2012

*The PST location.

THE 2003 ITALIAN BLACKOUT—A LESSON LEARNED. The northern Italian grid is interconnected with the neighboring countries by 17 line circuits (eight circuits at 380 kV, eight circuits at 220 kV, and one circuit at 132 kV). These interconnections can be ideally grouped into three corridors connecting Italy to France, Switzerland, and Austria-Slovenia. Italy is a net importer of electrical energy, mainly from France. After the 2003 Italian blackout, TERNA—the Italian transmission and system operator (TSO)—has decided to make some investments in order to increase the transmission capacity on its interconnection corridors and to increase the reliability, among which installation of PSTs (Table 7.2).

The two PSTs installed by TERNA—the Italian TSO—on the 380 kV Rondissone (Italy)—Albertville (France) double circuit line are among the largest in the world (Figure 7.40). Located on the Italian side in the Rondissone substation, the two PSTs have enabled increased transmission capacity from France to Italy [33].

A new 400 kV double circuit transmission line is planned for construction in 2016 between Okroglo (Slovenia) and Udine (Italy), as well as installation of a PST in the Okroglo substation.

FRANCE LOOKS FOR FLEXIBLE POWER FLOW CONTROL. In France, there are six PSTs, operated by RTE, aiming to optimize the use of existing power lines. They are only used to control the active power flow, and offer a temporary solution to postpone the construction of new power lines. However, the choice for PSTs provides both environmental—they reduce the impact of the network development—and economical—they postpone the investment in a new power line. The PSTs installed in France are presented in Table 7.3.

France and Spain are planning to increase the transmission capacity between their power systems. After installation of the only underground VSC–HVDC link, a new PST is planned for installation in the Arkale 220 kV substation to control the flows on the 220 kV interconnection line between Arkale (ES) and Argia (FR).

GREAT BRITAIN. The GB transmission system is composed of a highly interconnected series of relatively short (10–100 km) transmission circuits. Power flow is

Figure 7.40 Picture of the PST installed at Rondissone. Reproduced with permission of ABB.

predominantly from the north, where there is an excess of generation, to the south where there is an excess of demand. Thereby, there is a particular need to manage the power flows on certain circuits, following the loss of one or more of the several transmission lines that connect a region of high net generation to one of high net demand. The ability to control the power flow in one or more of the lowest capacity lines linking two parts of the network allows an increase in the overall secured capacity (the capacity following a fault or faults) of that part of the transmission system [34].

Currently, there are 15 QBTs installed in Great Britain's high voltage transmission network. Nine of them are connected to 400 kV circuits, five having nominal ratings of 2000 MVA, and four rated at 2750 MVA. At the same time, six are connected to 275 kV circuits having nominal rating of 750 MVA. They are owned and operated by the National Grid plc that is also the Great Britain System Operator [35].

Germany and its neighbors. Germany's shift to green power and the electricity market in Europe may lead to additional investments in the power grid infrastructure. The installed wind generation in Germany has already reached

TABLE 7.3 PSTs in France

Substation	System voltage	Regulation	Throughput rating	Type	Year
Pragnères	225 kV	$\pm 2 \times 21.4°$	312 MVA	2 dual core PST in series	1998
Rance	225 kV	$\pm$ 10°	413 MVA	Dual core	2001
La Praz	400 kV	$\pm$ 10°	1181 MVA	Dual core	2002
Sainte-Cécile	63 kV	$\pm$ 23°	61 MVA	Single core (delta-hexagonal)	2005
Guarbecque	225 kV	$\pm$ 11°	438 MVA	Dual core	2006
Niort	225 kV	$\pm$ 11°	438 MVA	Dual core	2006

40 GW in 2015, and is going to increase as Germany has decided to shut down nuclear power plants after the Fukushima disaster. Since the largest wind generation capacity is located in north, undesired power flows may be observed through France, Poland, and Czech power systems.

A phase shifter was installed in 2005 in the 400 kV substation of Nosovice (Czech Republic) to control the active power flows between Czech and Poland power system. Installation of phase shifters is planned in the coming years on all four connections between Polish power system and 50 Hertz (one of the four German TSOs) that is Krajnik (PL)—Vierraden (DE), Mikulowa (PL)—Hagenwerder (DE), Krajnik (PL)—Vierraden (DE), and Mikułowa (PL), and Hagenwerder (DE) transmission lines.

A PST at the CEPS (the Czech TSOs)—50 Hertz cross-border connection is envisaged by 2016. These transformers only deflect the flows, which means that the Czech Republic and Poland could end up moving even more of the excess on to the Austrian grid.

The Austrian power system takes part in the power transfers between Germany to Italy, as Italy is a net energy importer. Besides that, the greatest challenge for the Austrian TSO is to ensure the power transfer from north-east, where there is a surplus of installed power, to south, where there is an excess of demand, through the three 220 kV transmission lines. In 2006, three 600 MVA PSTs were installed on these lines, in the Tauern, Ernsthofen, and Ternitz substations, in order to avoid power system braking down.

Figure 7.41 shows the main cross-border congestions in Europe identified by ENTSO-E. Daily major decisions are taken by the European TSOs in order to avoid a

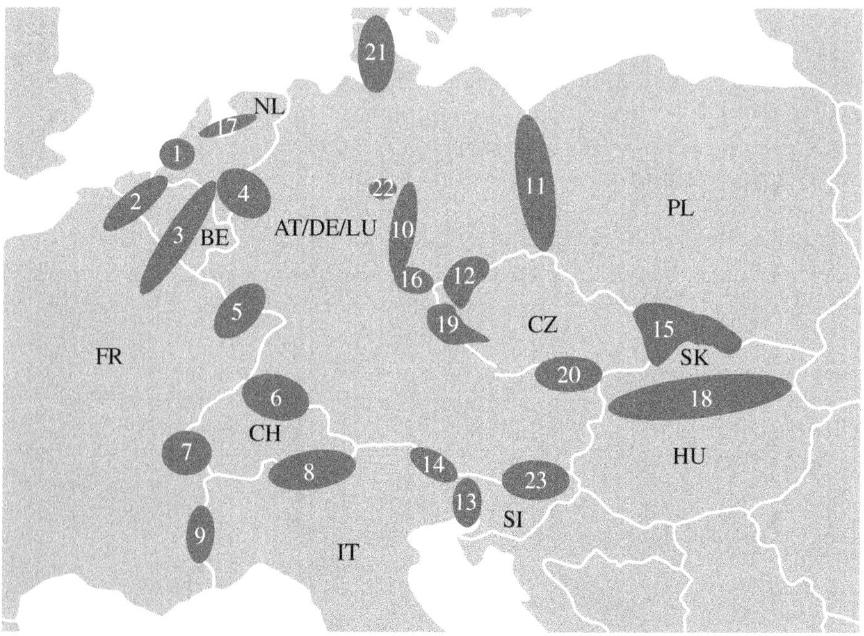

Figure 7.41 Network congestions in the planning phase in Europe.

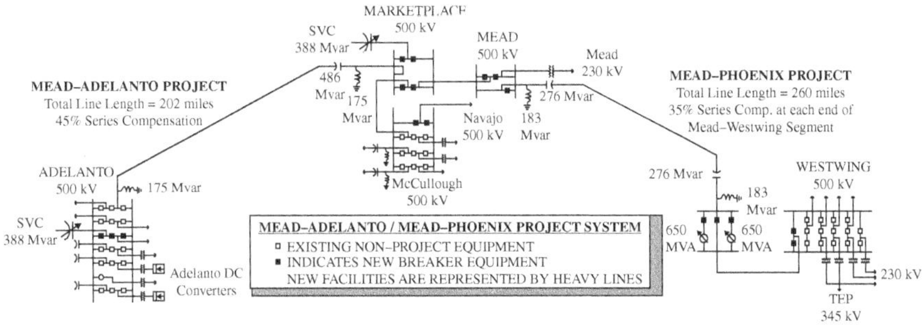

Figure 7.42 Schematic of Mead-Adelanto and Mead-Phoenix Project System. *Source*: Lee et al. 1994 [37]. Reproduced with permission of IEEE.

new incident like the one that occurred in November 2006, which affected 15 million households in Germany, France, Italy, Belgium, Spain, Portugal, Austria, and Croatia, after splitting the former Union for the Co-ordination of Transmission of Electricity power system into three synchronous zones.

THE MEAD-PHOENIX PROJECT—THE HIGHEST VOLTAGE. The highest voltage at which a PST is installed is 500 kV. Two PSTs are installed at Perkins, an extension of the Westwing substation, in Phoenix-Arizona, on the 500 kV Mead-Phoenix transmission line (Figure 7.42). The line is also 70% series compensated. With 650 MVA rated power each, the two PSTs are capable of providing up to ±24° phase angle shift between the input and output terminals. The on-load variable phase angle is provided by a tap changer in the secondary winding of the exciting transformer. Due to the size and phase angle requirements, the series and exciting units are built into

Figure 7.43 The 845 MVA Ontario Hydro PST. Reproduced with permission of ABB.

two separate tanks [36]. The PSTs enabled a 1300 MW transmission capacity on the Mead-Phoenix 500 kV transmission line. Without PSTs, the capability of the line is limited to about 800 MW.

Figure 7.42 shows an extended one-line diagram of the Mead-Phoenix project and the Mead-Adelanto project, where various shunt and series compensation equipment were designed, besides the Mead-Phoenix PST intended for power flow control.

THE LARGEST PHASE SHIFTING TRANSFORMERS IN CANADA. Two 240 kV/845 MVA PSTs were installed by Ontario Hydro in 2000/2001, one on the interconnection circuit L4D and one on interconnection circuit L51D (Figure 7.43). The PSTs, with an effective phase angle control range of ±40° under full load, provide the capability of controlling Lake Erie Circulation by approximately 500–600 MW. They also ensure a power flow control between the Ontario and Michigan power systems.

REFERENCES

[1] Dobsa, J. Transformer for in-phase, phase-angle and quadrature phase regulation. *Brown Boveri Review*, vol. 8, pp. 376–383, 1972.

[2] Iravani, M. R., and Maratukulam, D. Review of semiconductor-controlled (static) phase shifters for power system applications. *IEEE Transactions on Power Systems*, vol. 9, no. 4,pp. 1833–1839, November 1994.

[3] IEEE FACTS Working Group Terms & Definitions. Proposed terms and definitions for flexible AC transmission systems (FACTS). *IEEE Transactions on Power Delivery*, vol. 12, no. 4, pp. 1848–1853, 1997.

[4] Iravani, M. R. Chapter 6. Phase shifter. pp. 243–267. In: *Flexible AC Transmission Systems (FACTS)*, Y. H. Song and A. T. Johns, Eds., IEE, Power and Energy Series, London, UK, 1999.

[5] Sen, K. K., and Sen, M. L. *Introduction to FACTS Controllers. Theory, Modeling and Applications*. Wiley and IEEE Press, Hoboken, New Jersey, 2009.

[6] Verboomen, J., Van Hertem, D., Schavemaker, P. H., Kling, W. L., and Belmans, R. Phase shifting transformers: principles and applications. *Proceedings of the Future Power Systems Conference 2005*, Amsterdam, the Netherlands, November 2005.

[7] Verboomen, J. Optimization of transmission systems by use of phase shifting transformers. Ph.D. Thesis, Delft University of Technology, the Netherlands, 2008.

[8] Gabrijel, A., and Mihalič, R. Phase-shifting transformers in a structure-preserving energy function. *Electric Power Systems Research*, vol. 74, pp. 323–330, 2005.

[9] ENTSO-E, Phase shift transformers modeling. Version 1.0.0, CGMES V2.4.14, May 28, 2014.

[10] Arrillaga, J., and Duke, R. M. Thyristor-controlled quadrature boosting. *Proceedings of IEE*, vol. 126, no. 6, pp. 493–498, 1979.

[11] Arrillaga, J., Wood, G., and Duke, R. M. Thyristor-controlled in phase boosting for HVDC converters. *Proceedings of IEE*, vol. 127, no. 4, pp. 221–227, 1980.

[12] Arnold, C. P., Duke, R. M., and Arrillaga, J. Transient stability improvement using thyristor controlled quadrature voltage injection. *IEEE Transactions on Power Apparatus and Systems*, vol. PAS-100, no. 3, pp. 1382–1388, March 1981.

[13] Mathur, R. M., and Basati, R. S. A thyristor-controlled static phase-shifter for AC power transmission. *IEEE Transactions on Power Apparatus and Systems*, vol. PAS-100, no. 5, pp. 2650–2655, May 1981.

[14] Güth, G., Baker, R., and Eglin, P. Static thyristor controlled regulating transformer for AC transmission. *IEE International Conference on Thyristor and Variable Static Equipment for AC and DC Transmission*, London, November 30–December 3, 1981.

[15] Arrillaga, J., Duke, R. M., and Arnold, C. P. Thyristor-controlled 4-quadrant voltage injection. *IEE AC and DC Conference Record*, London, pp. 65–68, 1985.

[16] Baker, R., Güth, G., Egli, W., and Eglin, P. Control algorithm for a static phase shifting transformer to enhance transient and dynamic stability of large systems. *IEEE Transactions on Power Apparatus and Systems*, vol. PAS-101, no. 9, pp. 3532–3542, September 1982.

[17] Stemmler, H., and Güth, G. The thyristor-controlled static phase shifter. A new tool for power flow control in AC transmission systems. *Brown Boveri Review*, vol. 69, no. 3, pp. 73–79, March 1982.

[18] Gyugyi, L. Unified power flow control concept for flexible AC transmission systems. *IEE Proceedings-C*, vol. 139, no. 4, pp. 323–331, 1992.

[19] Moore, P., and Ashmore, P. Flexible AC transmission systems. Part 4: Advanced FACTS controllers. *Proceedings of Power Engineering Journal*, vol. 12, no. 2, pp. 95–100, April 1998.

[20] Johnson, B. K., and Venkataramanan, G. A hybrid solid state phase shifter using PWM AC converters, IEEE Transactions on Power Delivery, vol. 13, no. 4, pp. 1316–1321, October 1998.

[21] Mihalič, R., and Žunko, P. Phase-shifting transformer with fixed phase between terminal voltage and voltage boost:tool for transient stability margin enhancement. *IEE Proceedings Generation, Transmission and Distribution*, vol. 142, no. 3, pp. 257–262, May 1995.

[22] Mihalič, R., and Gabrijel, U. Transient stability assessment of systems comprising phase-shifting FACTS devices by direct methods. *Elsevier - Electric Power and Energy Systems*, vol. 26, pp. 445–453, 2004.

[23] Kumar, A., Singh, S. N., and Lai, L. L. Impact of TCPAR on congestion clusters and congestion management used mixed integer nonlinear programming approach. *IEEE PES Meeting*, Montreal, Quebec, 2006.

[24] Hingorani, N. G., and Gyugyi, L. *Understanding FACTS. Concepts and Technology of Flexible AC Transmission Systems*. IEEE Press, New York, 2000.

[25] Van Cutsem, T., and Vournas, C. D. *Voltage Stability of Electric Power Systems*. Kluwer Academic Publishers, Boston, 1998.

[26] Eremia, M., and Shahidehpour, M., Eds. *Handbook of Electrical Power System Dynamics: Modeling, Stability, and Control*. Wiley-IEEE Press, Hoboken, New Jersey, 2013.

[27] Rammamooty, M. *Introduction to Thyristors and Their Applications*. The MacMillan Press, London, England, 1978.

[28] Marconato, R. *Electric Power Systems. Volume 3. Dynamic Behaviour, Stability, and Emergency Controls*. CEI – Italian Electrotechnical Committee, Milano, 2008.

[29] Spoorenberg, C. J. G., Van Hulst, B. F., and Reijnders, H. F. Specific aspects of design and testing of a phase shifting transformer. *Proceedings of the 13th International Symposium on High Voltage Engineering*, Delft, the Netherlands, 2003.

[30] Kling, W. L., Klaar, D. A. M., Schuld, J. H., Kanters, A. J. L. M., Koreman, C. G. A., Reijnders, H. F., and Spoorenberg, C. J. G. Phase shifting transformers installed in the Netherlands in order to increase available international transmission capacity. *CIGRE Session 2004*, C2-207, Paris, France, 2004.

[31] Rimez, J., Van Der Planker, R., Wiot, D., Claessens, G., Jottrand, E., and Declercq, J. Grid implementation of a 400 MVA 220/150 kV -15°/+3° phase shifting transformer for power flow control in the Belgian network: specification and operational considerations. *CIGRE Session 2006*, A2-202, Paris, France, 2006.

[32] Häger, U., Schwippe, J., Görner, K. Improving network controllability by coordinated control of HVDC and FACTS devices, Delivery D1.2.2 REALISEGRID Project, 2010.

[33] Carlini, E. M., Manduzio, G., and Bonmann, D. Power flow control on the Italian network by means of phase-shifting transformers. *CIGRE General Meeting*, Paris, August 2006.

[34] Jarman, P., Hynes, P., Bickley, T., Darwin, A., Hayward, H., and Thomas, N. The specification and application of large quadrature boosters to restrict post-fault power flows. *Proceedings of CIGRE General Meeting*, Paris, August 2006.

[35] Belivanis, M., and Bell, K. Use of phase-shifting transformers on the transmission network in Great Britain. *45th International Universities Power Engineering Conference UPEC 2010*, Cardiff, United Kingdom, August 31–September 3, 2010.

[36] Brochu, J., Beauregard, F., Cloutier, R., Bergeron, A., Garant, L., Sirois, F., and Henderson, M. I. Innovative applications of phase-shifting transformers supplemented with series reactive elements. *CIGRE General Meeting*, Paris, August 2006.

[37] Lee, R. L., Beshir, M. J., Finley, A. T., Hayes, D. R., Hsu, J. C., Peterson, H. R., DeShazo, G. L., and Gerlach, T. W. Application of static VAr compensators for the dynamic performance of the Mead-Adelanto and Mead-Phoenix transmission projects. *Proceedings of the 1994 IEEE PES Transmission and Distribution Conference*, pp. 444–453, April 10–15, 1994.

CHAPTER 8

STATIC SYNCHRONOUS COMPENSATOR – STATCOM

Rafael Mihalic, Mircea Eremia, and Bostjan Blazic

8.1 PRINCIPLES AND TOPOLOGIES OF VOLTAGE SOURCE CONVERTER

8.1.1 Basic Considerations[1]

STATic synchronous COMpensator (STATCOM), sometimes denoted as an "Advanced Static VAr Compensator," is a flexible alternating current transmission system (FACTS) device connected in parallel with a power system that is capable to exchange reactive power with the system in both directions (by absorbing or by generating). The terms "compensator" and "synchronous" indicate that the device is equivalent to an ideal synchronous generator, which produces a set of three-phase fundamental frequency sinusoidal voltages. "Static," further, denotes the fact that it is realized without moving parts (i.e., by power electronic elements). The device allows a very fast control of the three-phase network voltages at the connection point, enabling amplitude and phase shift control of the generated voltage. With a voltage source converter (VSC), one can in general control the magnitude, the phase angle, and frequency of the output voltage. Therefore, the STATCOM essentially injects an almost sinusoidal reactive current of variable magnitude at the point of compensation. This reactive current, in turn, regulates the transmission line voltage.

The STATCOM device is mainly used for dynamic compensation in power systems, it can provide fast (dynamic) voltage support (e.g., for flicker mitigation) and might also be used for transient stability margin enhancement or oscillation damping improvement.

The STATCOM consists of a VSC, a magnetic circuit (MC), a shunt coupling transformer, a shunt breaker, and a control and protection unit (Figure 8.1) [1,2].

In a VSC, a number of square wave voltages are generated at fundamental frequency by operating the controllable semiconductor switches once per cycle of the

[1] Note: In this chapter for phase-to-earth and DC voltages, the letter "*V*" is used (capital for phasors and small letters for momentary values). In other cases (general explanation, phase-to-phase quantities, etc.), "*U*" is used.

Advanced Solutions in Power Systems: HVDC, FACTS, and Artificial Intelligence, First Edition.
Edited by Mircea Eremia, Chen-Ching Liu, and Abdel-Aty Edris.

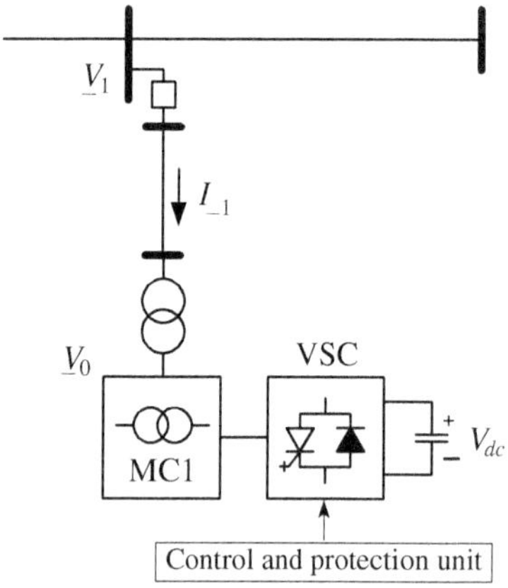

Figure 8.1 General scheme of STATCOM.

fundamental frequency (see Section 4.1.3, Figure 4.3). These square wave voltages are combined using a MC in order to produce a high quality sinusoidal voltage with acceptable harmonic content. It should be noted that a shunt-connected VSC always injects some harmonic current components into the power system at the point of compensation.

As direct current flows in both directions through the VSC, the converter valves must be bidirectional; also, because the direct current does not change its polarity, the gate turn-off (GTO) thyristors do not have to be provided with the ability to function at a voltage with reversed polarity. Therefore, a VSC valve consists of a "turn-off asymmetric"-type device, containing a GTO and an anti-parallel connected diode (Figure 8.2a) [3].

In practical high voltage applications, units consisting of several GTO-type thyristor structures are used—diodes mounted in series.

Figure 8.2b illustrates the basic scheme of a VSC operation. On the DC side, the voltage is unipolar (+ and –) and is maintained by a capacitor (C); the direct current (i_d) may circulate in both directions, and so the converter may change power continuously with the DC system. It can be noticed that the AC area of the converter is being connected with the AC system via a reactor or a transformer.

Being an AC voltage source with low internal impedance, a series inductive interface with the AC system (a series inductor and/or a transformer) is essential to ensure that the DC capacitor is not short-circuited and discharged rapidly into a capacitive load such as a transmission line. Also, an AC filter may be necessary (not

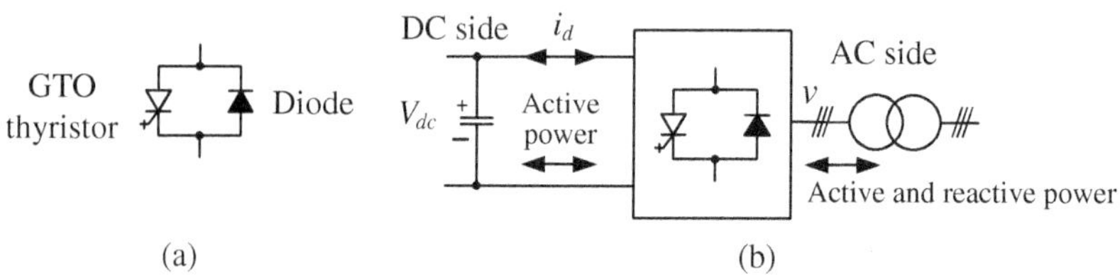

Figure 8.2 The concept of the voltage source converter based on the valve of GTO-diodes structure.

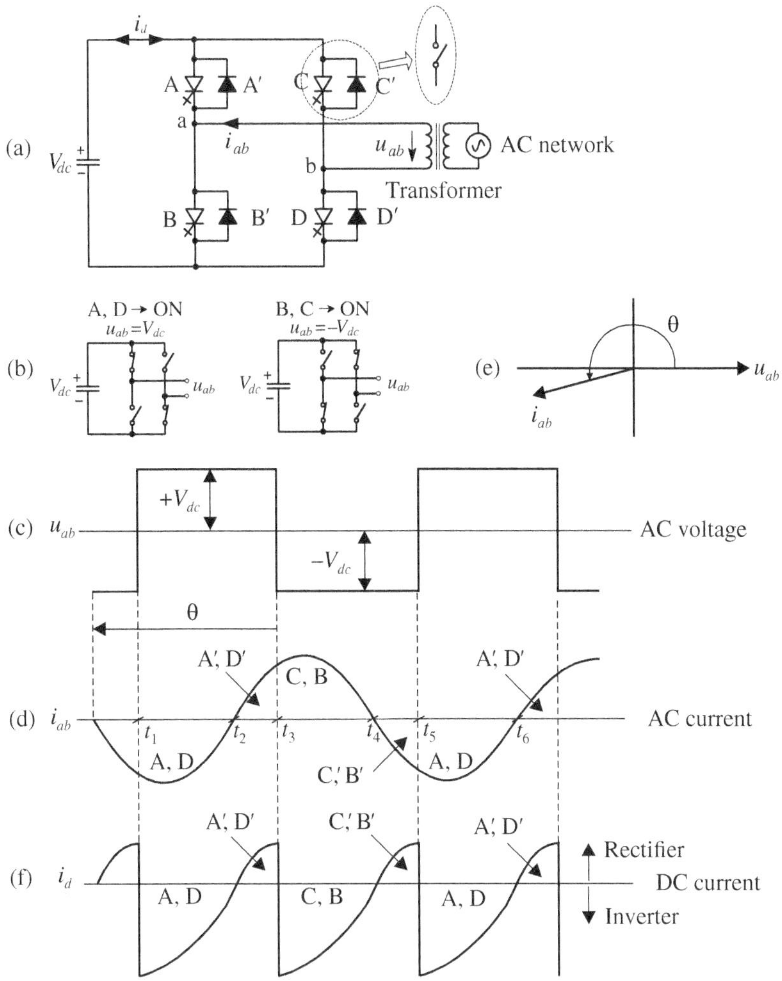

Figure 8.3 Operation of single-phase two-level full bridge converter. *Source*: Hingorani & Gyugyi 2000 [4]. Adapted with permission of IEEE.

shown) following the series inductive interface to limit the consequent current harmonics entering the system side [3].

Although FACTS devices generally use three-phase converters, a *single-phase two-level H bridge* converter can be used in certain projects (Figure 8.3a). This scheme of converter consisting of four valves (A, B, C, and D), a capacitor as DC voltage source, and two nodes, a and b, connected to the AC network via a transformer.

The V_{dc} voltage is converted into AC voltage through an adequate sequence of switching of the valves (Figure 8.3b). It can be noticed that when valves A and D are switched ON, voltage u_{ab} becomes $+V_d$ for one half-cycle, and when valves B and C are switched ON (and valves A and D are turned OFF), the u_{ab} voltage becomes $-V_d$ for the other half-cycle (Figure 8.3c).

The AC current is the result of interaction of the converter-generated AC voltage with the AC system voltage and impedance. For example, suppose that the current flow from the AC system is a sinusoidal waveform i_{ab}, leading the square wave voltage by angle θ (Figures 8.3c and 8.3d). The relationship between the AC voltage and the phase current is illustrated in Figure 8.3e, showing power flow from AC to DC with a lagging power factor.

Starting from instant t_1, it is seen from the circuit and waveform that [3]:

(i) From instant t_1 to t_2, with turn-off devices A and D ON, and B and C OFF, AC voltage u_{ab} is positive and i_{ab} is negative. The current flows through device A into AC phase a, and then out of AC phase b through device D, with power flow from DC to AC (inverter action).

(ii) From instant t_2 to t_3, the current reverses, that is, it becomes positive and flows through diodes A' and D' with power flow from AC to DC (rectifier action). Note that during interval t_2–t_3, although devices A and D are still ON and voltage u_{ab} is $+V_{dc}$, devices A and D cannot conduct in a reverse direction. In reality, devices A and D are ready to turn ON by turn-on pulses when required by the direction of actual current flow.

(iii) From instant t_3 to t_4, devices A and D are turned OFF and devices C and B are turned ON, thereby u_{ab} becomes negative, while i_{ab} is still positive. The current now flows through devices C and B with power flow from DC to AC (inverter action).

(iv) From instant t_4 to t_5, with devices C and B still ON, and A and C OFF, and u_{ab} negative, current i_{ab} reverses and flows through diodes C' and B' with power flow from AC to DC (rectifier action).

(v) From instant t_5, the cycle starts again similar to instant t_1 with devices A and D turned ON and C and B turned OFF.

Figure 8.3f shows the waveform of current flow i_d in the DC bus with the positive side flowing from AC to DC (rectifier action), and the negative side flowing from DC to AC (inverter action). Clearly, the average DC current is negative. The current I_d contains the DC current and the harmonics. The DC current must flow into the DC system and for a large DC capacitor, virtually all of the harmonic current will flow through the capacitor. Being a single-phase, full-wave bridge, the DC harmonics have an order of $2k$, where k is an integer, that is, 2nd, 4th, 6th, …, all of the even harmonics.

The square wave, shown in Figure 8.3c as the AC voltage u_{ab}, has substantial harmonics in addition to the fundamental. These harmonics are of the order $n = 2k \pm 1$, where k is an integer, that is, 3rd, 5th, 7th,…

Integration of the waveform in Figure 8.3c gives the rms value of the square wave AC voltage with a peak of V_{dc} [3]:

$$U_{ab} = \sqrt{\frac{1}{\pi}\int_{-\pi/2}^{+\pi/2} V_{dc}^2\, d(\omega t)} = V_{dc}$$

which includes the fundamental and the harmonics.

The fundamental and individual harmonics are given by

$$v = \frac{4}{\pi} V_{dc} \left(\cos \omega t - \frac{1}{3} \cos 3\omega t + \frac{1}{5} \cos 5\omega t - \frac{1}{7} \cos 7\omega t + \ldots \right) \tag{8.1}$$

which gives

$$v_n = \frac{4}{\pi} V_{dc} \left(\frac{1}{n} \cos n\omega t \right), \quad n = 1, 3, 5, 7, \ldots \tag{8.2}$$

The rms value of v_n is

$$V_n = \frac{1}{n} \frac{2\sqrt{2}}{\pi} V_{dc} \tag{8.2'}$$

Thus, the rms fundamental component of a square wave AC voltage u_{ab} is

$$V_1 = \frac{2\sqrt{2}}{\pi} V_{dc} = 0.9 V_{dc}$$

The magnitude of each voltage harmonic is $1/n$ of the fundamental.

A multilevel converter may be constructed using different combinations of the GTO thyristors and diodes. An example of such combinations, which form a multilevel converter, is the chain-circuit, where a few converter bridges are connected as shown in Figure 8.4a.

Each connection consists of a capacitor and four GTO devices with diodes connected in anti-parallel where through the voltage of the capacitor can be injected

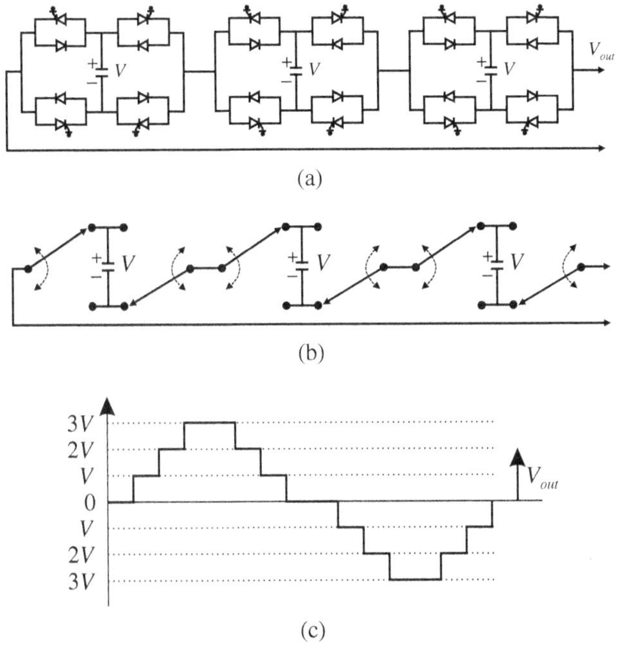

Figure 8.4 "Chain link" circuit with multiple series connections (Analogy with the mechanic switch, (b)).

in the AC circuit in any direction, or reversed, or it may be omitted. To obtain the required voltage, a series of extendable connections are made (Figure 8.4a). The converter produces an independent control of each phase, and, if a significant number of connections in each valve are used, an approximately sinusoidal voltage is obtained (Figure 8.4c).

8.1.2 Converter Topologies

8.1.2.1 Two-Level Topologies

SIX-PULSE TWO-LEVEL VSC. A basic three-phase two-level VSC consists of six semiconductor valves (GTO+Diode or IGBT+Diode) and a common DC capacitor (Figures 8.5a, 8.5b).

In basic VSC operation, the semiconductor switches are switched once per cycle, generating a square wave voltage with amplitude of $V_{dc}/2$ and $-V_{dc}/2$ (measured to the fictitious neutral point M). The generated voltages are shown in Figure 8.6 where the numbers of the conducting valves are also indicated. As the generated voltage has two levels ($V_{dc}/2$ and $-V_{dc}/2$), this circuit is referred to as a two-level circuit.

As it is clear from the waveform of the generated voltages, the three-phase two-level converter also generates harmonics. The generated phase voltage can be expressed in a Fourier series [1, 3]:

$$v_a = \frac{4}{\pi}\frac{V_{dc}}{2}\left(\sin\omega t + \frac{1}{3}\sin 3\omega t + \frac{1}{5}\sin 5\omega t + \frac{1}{7}\sin 7\omega t + \ldots\right) \tag{8.3}$$

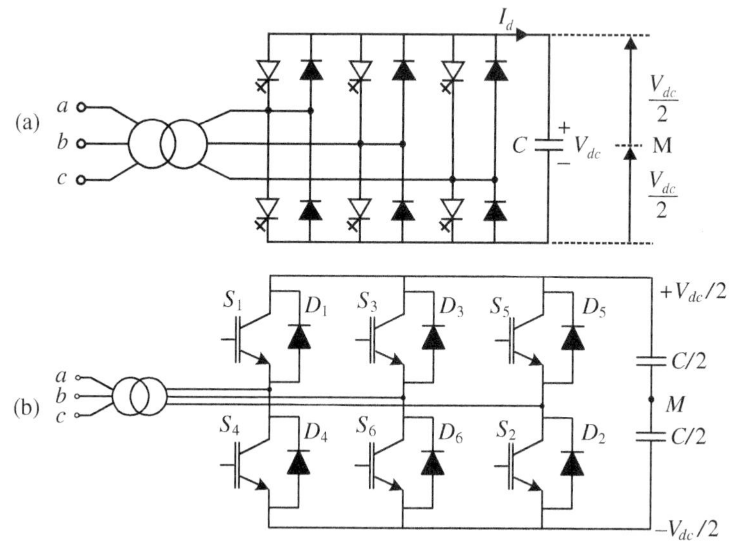

Figure 8.5 Three-phase two-level six-pulse voltage source converter: (a) with GTO thyristors and diodes and (b) with insulated-gate bipolar transistors (IGBTs).

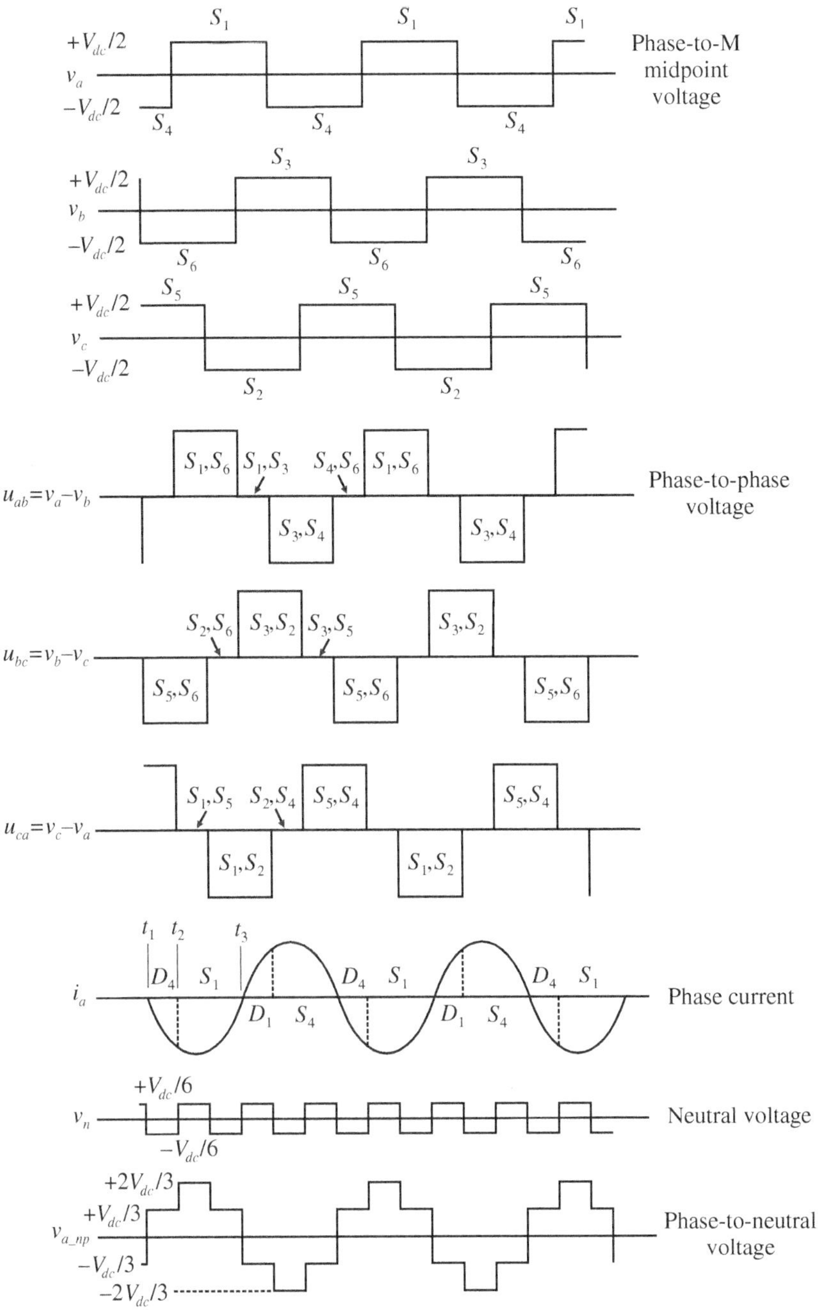

Figure 8.6 AC waveform for a three-phase two-level VSC [3].

The constant $2/\pi$ represents the ratio between the AC-voltage amplitude and the DC-side voltage. As already mentioned, a three-phase three-wire converter has no return path for zero-sequence currents, therefore zero-sequence components of phase voltages appear between the fictitious neutral point M of the DC capacitor and the network neutral point (NP). As a result, phase voltages, measured to the network neutral point, contain no triplen (zero sequence) harmonics:

$$v_{a\_np} = \frac{4}{\pi}\frac{V_{dc}}{2}\left(\sin\omega t + \frac{1}{5}\sin 5\omega t + \frac{1}{7}\sin 7\omega t + \frac{1}{11}\sin 11\omega t + \frac{1}{13}\sin 13\omega t + ...\right) \tag{8.3'}$$

Voltage v_{a_np} exhibits a six-pulse waveform, and thereby such VSC is referred to as a six-pulse VSC. The converter generates harmonics $n = 6k \pm 1$, where k is an integer. The amplitudes of harmonic components n (v_{a_n}), when compared to the fundamental component (v_{a_1}), are:

$$\frac{v_{a\_n}}{v_{a\_1}} = \frac{1}{n}$$

Multi-Pulse Two-Level VSC. Multi-pulse converters consist of multiple six-pulse converters connected in parallel through an MC (transformers). By increasing the number of six-pulse units, harmonic cancelation (or harmonic reduction) can be achieved. As shown before, the voltage generated by a six-pulse converter contains harmonics $n = 6\mathrm{k} \pm 1$, where k is an integer. A $6m$-pulse converter, where m is the number of six-pulse converters, generates only harmonics $n = 6mk \pm 1$. The displacement angle between two consecutive converters has to be $2\pi/6m$. Multi-pulse converters use the same DC capacitor.

A 12-*pulse converter* is composed of two six-pulse converters connected in parallel using two standard transformers with star and delta primary (Figure 8.7a) [1, 3]. The configuration of the MC is such that if a VSC pole voltage is time-shifted by an equivalent fundamental angle of –30°, the fundamental component of the pole voltage gets phase-shifted by an angle of +30°. The simplest way to give the fundamental components in the D-E-F pole voltages a +30° phase shift is to apply these voltages to the terminals of a delta winding.

The square wave voltages from A-B-C VSC are fed to the primary windings of a Y-Y transformer and those from the D-E-F VSC are fed to the primary windings of a Δ-Y transformer. In order to keep the same volts per turn in both Y and Δ windings, the delta-connected primary windings have $\sqrt{3}$ times the turns (n_1) of the Y-connected primary windings. Both sets of secondary Y windings are connected in series in the respective phases and the final output voltage is connected to a three-phase load (a, b, and c) (Figure 8.7a) [1].

The fundamental pole voltage phasors (V_{A1}, V_{B1}, and V_{C1}), shown in Figure 8.7c (bottom), are at 0° (reference), –120°, and –240°, respectively. The A-B-C set of square wave voltages is said to be placed at 0°. The fundamental pole voltage phases (V_{D1}, V_{E1}, and V_{F1}), shown in Figure 8.7c (upper), are at –30°, –150°, and 90°, respectively. The D-E-F set of square wave voltages is said to be placed at –30°.

Figure 8.7b shows a symbolic representation of two six-pulse VSCs (A-B-C) and (D-E-F), which are connected across a shared DC link with a pair of capacitors

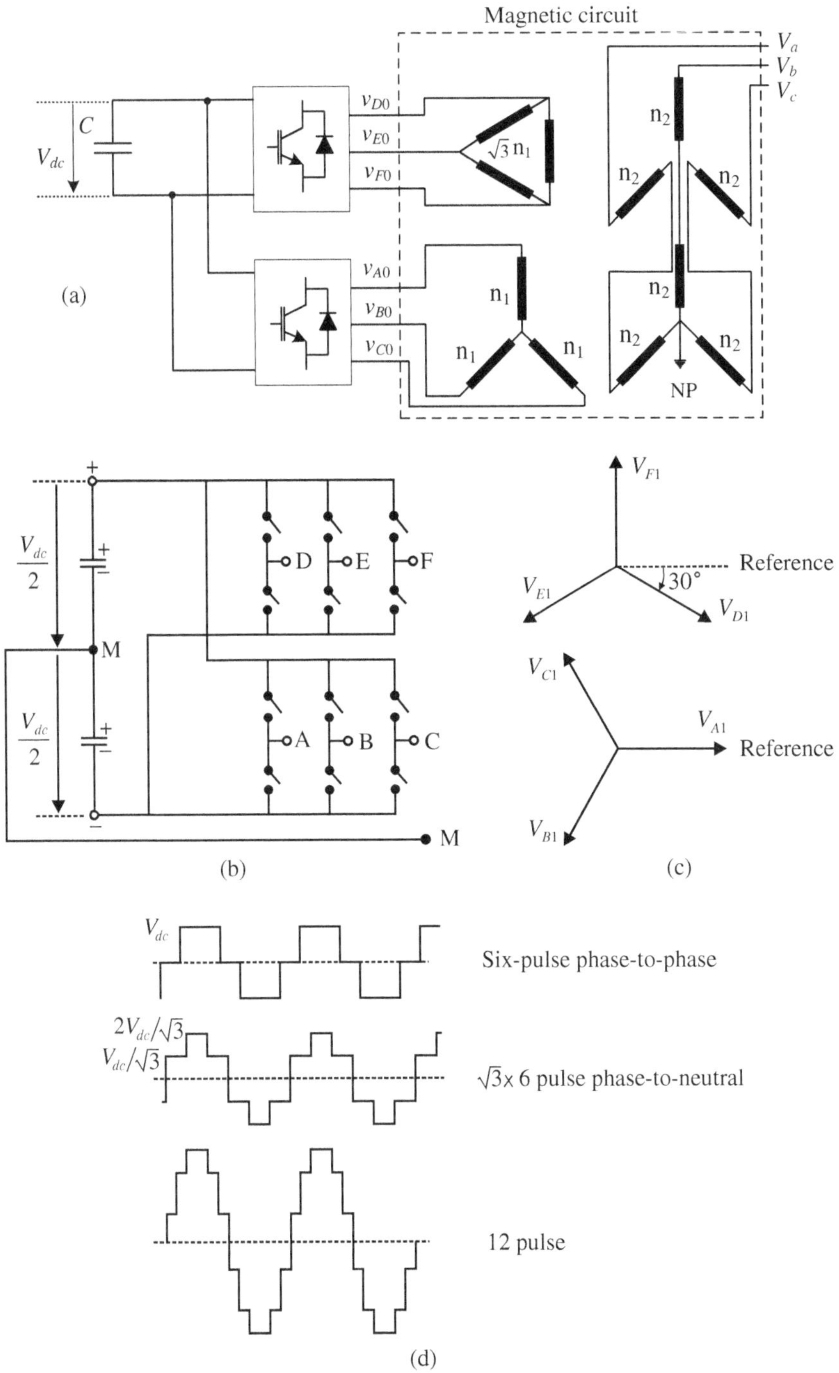

Figure 8.7 Twelve-pulse VSC configuration: (a) 12-pulse converter with wye and delta secondaries, (b) 12-pulse converter simplified representation of the valves, (c) phasor diagram of fundamental voltages, and (d) 12-pulse waveform from two six-pulse waveforms.

and the related fundamental voltage phasors. The pole voltages are combined by an MC to form a three-phase (a-b-c) voltages system. The poles A, B, C, D, E, and F are operated in such a way that the pole voltages (v_{A0}, v_{B0}, and v_{C0}) of the A-B-C VSC are time-shifted from one another by one-third of the time period (T) of the pole voltage.

The pole voltages (v_{D0}, v_{E0}, and v_{F0}) of the D-E-F VSC are time-shifted with respect to the corresponding A, B, and C pole voltages by one-twelfth of the time period, equivalent to a fundamental angle of –30°. Therefore, the fundamental phasors in a group (V_{A1}, V_{B1}, and V_{C1} of the A-B-C VSC, and V_{D1}, V_{E1}, and V_{F1} of the D-E-F VSC) are 120° apart and the fundamental voltage phasors set of the D-E-F VSC lags the fundamental voltage phasors set of the A-B-C VSC by 30° (Fig. 8.7c). Note that the fundamental components in the square wave voltages ($V_{D0,1}$, $V_{E0,1}$, and $V_{F0,1}$) of the D-E-F VSC with respect to the midpoint of the DC link capacitor are referred to as V_{D1}, V_{E1}, and V_{F1} [1].

With the configuration from Figure 8.7a, the 5th, 7th, 17th, and 19th harmonic voltages cancel out, and the two fundamental voltages add up, as shown in Figure 8.7d, and the combined unit becomes a true 12-pulse converter [3].

The generated voltage contains harmonics $n = 12k\pm1$, where k is an integer, and n_1 and n_2 are the numbers of the transformers primary and secondary winding turns [1]:

$$v_{a\_np} = 2\frac{n_2}{n_1}\frac{4}{\pi}\frac{V_{dc}}{2}\left(\sin\omega t + \frac{1}{11}\sin 11\omega t + \frac{1}{13}\sin 13\omega t + \frac{1}{23}\sin 23\omega t + \frac{1}{25}\sin 25\omega t \ldots\right) \tag{8.4}$$

The amplitudes of harmonic components n, when compared to the fundamental component, are:

$$\frac{v_{a\_n}}{v_{a\_1}} = \frac{1}{n} \tag{8.5}$$

For *24-pulse and 48-pulse converters*, which are usually used for high power STATCOMs, complex phase shift transformers have to be used. The phase shift between adjacent transformers in a 24-pulse design should be 15° and in a 48-pulse design should be 7.5°.

The generated voltage of a 24-pulse converter contains harmonics $n = 24k\pm1$, where k is an integer:

$$v_{a\_np} = 4\frac{n_2}{n_1}\frac{4}{\pi}\frac{V_{dc}}{2}(\sin\omega t + \frac{1}{23}\sin 23\omega t + \frac{1}{25}\sin 25\omega t + \frac{1}{47}\sin 47\omega t + \frac{1}{49}\sin 49\omega t + \ldots) \tag{8.6}$$

The amplitudes of harmonic components n (v_{a_n}), when compared to the fundamental component (v_{a_1}), are:

$$\frac{v_{a\_n}}{v_{a\_1}} = \frac{1}{n} \tag{8.7}$$

TABLE 8.1 The relative values of harmonics for different converter topologies

VSC \ n	5	7	11	13	17	19	23	25	29	30	35	37
6-p	0.200	0.142	0.091	0.077	0.059	0.053	0.043	0.040	0.034	0.032	0.029	0.027
12-p			0.091	0.077			0.043	0.040			0.029	0.027
24-p							0.043	0.040				
quasi 24-p			0.012	0.010			0.043	0.040			0.0038	0.0036

An alternative solution to the complex and expensive MC is the use of quasi 24-pulse and quasi 48-pulse designs.

A *quasi 24-pulse converter* consists of two 12-pulse units connected in parallel where the firing angles of individual converters are set 15° apart. In such a design, the harmonic components $n = 12k\pm1$ (where $k = 1, 3, 5\ldots$) are not completely canceled. However, their values are much lower than the values of the 12-pulse design:

$$\begin{aligned} v_{a\_np} = 4\frac{n_2}{n_1}\frac{4}{\pi}\frac{V_{dc}}{2}\Big[&\cos\left(\frac{\pi}{24}\right)\sin\omega t + \frac{1}{11}\cos\left(\frac{11\pi}{24}\right)\sin 11\omega t \\ &+\frac{1}{13}\cos\left(\frac{13\pi}{24}\right)\sin 13\omega t + \frac{1}{23}\cos\left(\frac{23\pi}{24}\right)\sin 23\omega t \\ &+\frac{1}{25}\cos\left(\frac{25\pi}{24}\right)\sin 25\omega t + \ldots\Big] \end{aligned} \tag{8.8}$$

The amplitudes of harmonic components n (v_{a_n}), when compared to the fundamental component (v_{a_1}), are:

$$\frac{v_{a\_n}}{v_{a\_1}} = \frac{\cos(n\pi/24)}{n\cos(\pi/24)} \tag{8.9}$$

A *quasi 48-pulse converter* can be built by connecting two quasi 24-pulse converters with a phase displacement between converters of 7.5°.

Table 8.1 shows the relative values of harmonics (compared to the fundamental component value) for different converter topologies [1].

8.1.2.2 Multilevel Topologies

Multilevel converter topologies offer the possibility to decrease the harmonic content of the generated voltage or to regulate the fundamental-frequency voltage amplitude at constant DC-side voltage.

One phase leg of a three-level converter is shown in Figure 8.8a. The other two phase legs (not shown) would be connected across the same DC busbars clamping diodes (D_1' and D_2') connected to the same midpoint M of the DC capacitor [3].

Figures 8.8b, 8.8c, and 8.8d show the output voltage corresponding to one three-level phase leg. The first waveform shown is a full 180° square wave obtained by closing the devices S_1 and S_2 to give $+V_{dc}/2$ for 180°, and by closing the devices S_4 and S_3 for 180° to give $-V_{dc}/2$ for 180° (Figure 8.8b) [3].

Now consider the voltage waveform in Figure 8.8c, in which the upper device S_1 is turned OFF and device S_3 is turned ON an angle α earlier than they were in the 180°

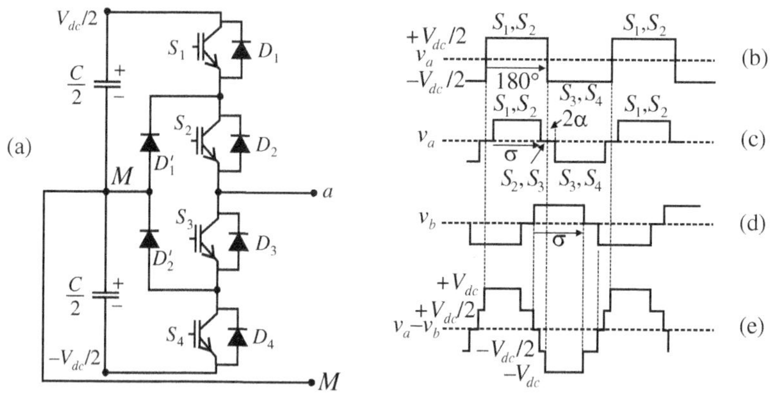

Figure 8.8 Operation of a three-level converter: (a) one phase leg of a three-level converter and (b, c, d, and e) output AC voltage.

square operation. This leaves only devices S_2 and S_3 ON, which in combination with diodes D_1' and D_2', clamp the phase voltage v_a to zero with respect to the DC midpoint M regardless of which way the current is flowing. This continues for a period 2α until the device S_2 is turned OFF, and device S_4 is turned ON, and the voltage jumps to $-V_{dc}/2$ with both the lower devices S_4 and S_3 turned ON and both the upper devices S_1 and S_2 turned OFF, and so on. Of course, angle α is variable, and the output voltage v_a is made up to $\sigma = 180° - 2\alpha°$, square waves. This variable period σ per half-cycle potentially allows the voltage v_a to be independently variable with potentially a fast response. It can be seen that devices S_2 and S_3 are turned ON for 180° during each cycle, devices S_1 and S_4 are turned ON for $\sigma = 180° - 2\alpha°$ during each cycle, while diodes D_1' and D_2' conduct for $2\alpha° = 180° - \sigma$ during each cycle.

Figures 8.8d and 8.8e show the output voltage of the b phase v_b and the phase-to-phase voltage u_{ab} for a three-level converter. The converter is referred to as three-level because the DC voltage has three levels, that is, $-V_{dc}/2$, 0, and $+V_{dc}/2$.

The generated phase voltage can be expressed in terms of a Fourier series:

$$v_{a\_np} = \frac{4}{\pi}\frac{V_{dc}}{2}\left(\cos\sigma\sin\omega t + \frac{1}{5}\cos(5\sigma)\sin 5\omega t + \frac{1}{7}\cos(7\sigma)\sin 7\omega t + \ldots\right) \tag{8.10}$$

which contains harmonics $n = 6k\pm1$, where k is an integer.

The amplitudes of harmonic components n (v_{a_n}), when compared to the fundamental component (v_{a_1}), are:

$$\frac{v_{a\_n}}{v_{a\_1}} = \frac{\cos(n\sigma)}{n\cos\sigma} \tag{8.11}$$

The angle σ represents the width between the pulses as shown in Figure 8.8c. As follows from equation (8.11), with proper selection of σ, a selected harmonic can be eliminated or two adjacent harmonics can be decreased.

For $\sigma = 0°$, the three-level converter behaves like a two-level converter and offers maximum DC-voltage utilization. For $\sigma > 0°$, the fundamental voltage decreases and the voltage waveform improves. As follows from equation (8.11), the amplitudes of harmonics vary sinusoidally. At $\sigma = 15°$, the fundamental component is decreased by 3.41% and the fifth and seventh harmonic components are reduced by 74.1% when compared to the values of a two-level converter.

8.1.2.3 PWM Converter

As was shown in Section 4.2, PWM can be illustrated on the basis of operation of a single-pole of a three-phase two-level converter (Figure 8.9a).

As shown in Figure 8.9b, a fundamental frequency signal (reference, v_{a_ref}) is compared with a triangular signal (carrier). When the reference signal is higher than the carrier signal, a firing pulse for switch S_1 is generated. In the opposite case, a pulse for S_4 is generated. Figure 8.9c shows the generated voltage waveform (to the reference point M) when the carrier signal has a frequency of 9 times the reference signal frequency. The output voltage is:

$$v_{aM} = \begin{cases} \dfrac{V_{dc}}{2}, & \text{if } v_{ref} > v_{carrier\,signal}\ (S_1 \text{ in conduction}) \\ -\dfrac{V_{dc}}{2}, & \text{if } v_{ref} < v_{carrier\,signal}\ (S_4 \text{ in conduction}) \end{cases}$$

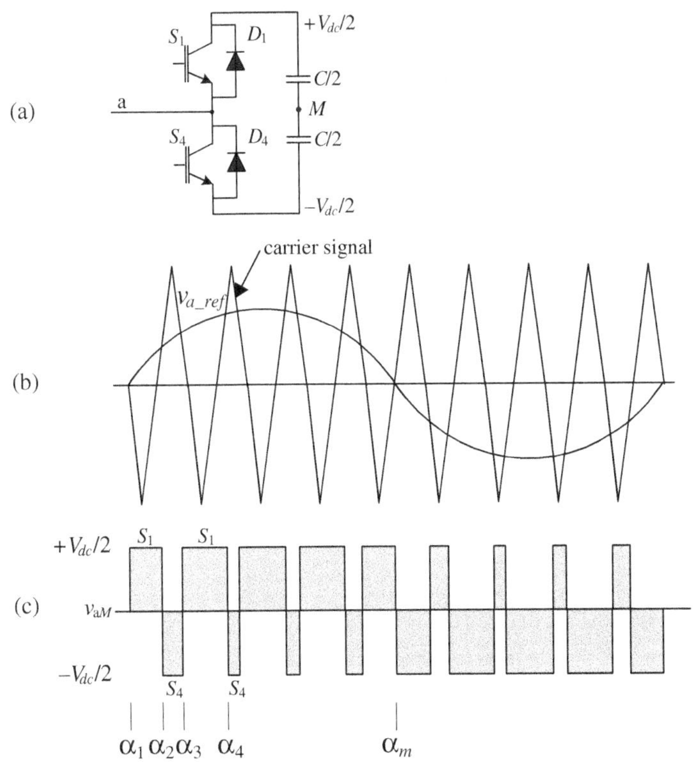

Figure 8.9 (a) Single-pole of a three-phase two-level converter, (b) carrier signal and reference signal, and (c) generated voltage.

8.1.3 Switching Function

The *switching function* provides a general description of AC-voltage generation from DC voltage and therefore gives a mathematical description of the connection between the converters' AC and DC side.

If we consider one phase leg of a three-phase two-level converter with once-per-cycle switching, each switching element carries the current for half a cycle.

The generated AC voltage can be expressed as follows:

$$v_a = k_p S_a(t) V_{dc} \tag{8.12}$$

where k_p is a constant connecting the DC voltage and AC voltage amplitude and is dependent upon the converter type, and S_a is the switching function, which is shifted for a phase angle δ from the reference voltage (Figure 8.10). The Fourier series for S_a is:

$$S_a(t) = m_p \sum_{n=1,3,5\ldots}^{\infty} (-1)^{(n+3)/2} \frac{\cos n(\omega_0 t + \delta)}{n} \tag{8.13}$$

where m_p ($\equiv m_a$) represents the amplitude modulation ratio that can take any value between zero and one.

The generation of three-phase voltages can be written as:

$$\begin{bmatrix} v_a \\ v_b \\ v_c \end{bmatrix} = k_p \begin{bmatrix} S_a(t) \\ S_b(t) \\ S_c(t) \end{bmatrix} V_{dc} \tag{8.14}$$

If we neglect the harmonics generated by the converter, the switching function simplifies to:

$$\begin{bmatrix} S_a(t) \\ S_b(t) \\ S_c(t) \end{bmatrix} = m_p \begin{bmatrix} \cos(\omega_0 t + \delta) \\ \cos(\omega_0 t + \delta - 2\pi/3) \\ \cos(\omega_0 t + \delta + 2\pi/3) \end{bmatrix} \tag{8.15}$$

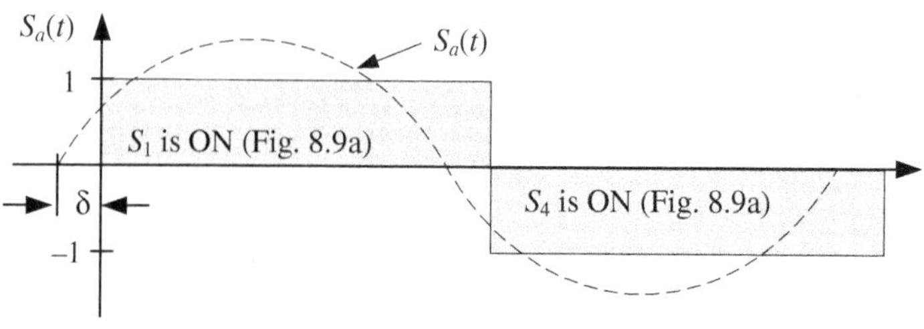

Figure 8.10 Switching function.

8.2 STATCOM OPERATION

A STATCOM consists mainly of a VSC, supplied from a capacitor operating at DC, and which is connected to the power grid via a transformer (Figure 8.11a).

The operating principle consists of maintaining a capacitor charged to a DC voltage V_{dc} and to generate via the conversion bridge, using the energy stored in the capacitor, an AC voltage $\underline{V}_0$ in phase with the network voltage $\underline{V}$. The difference between the amplitudes of these voltages across the inductive reactance X_T causes a reactive current flow. The amplitude of this voltage difference determines the amplitude of the reactive current, and its polarity determines the direction of the phase angle (+ or – 90°) of the reactive current with respect to the network voltage. The STATCOM will be "seen" by the electrical network as a reactive element of variable reactive current. The reactive current produced by a STATCOM depends only on the amplitude of the AC voltage V_0, assuming that the capacitor is charged. The value of voltage V_0 is proportional to the voltage V_{dc} of the DC-side capacitor.

As it can be seen in Figure 8.11b, when the current $\underline{I}$ lags behind the voltage $\underline{V}$ by $\pi/2$, the device acts like a reactor, absorbing reactive power. When $\underline{I}$ leads the voltage $\underline{V}$ by $\pi/2$, then it acts like a capacitor, supplying reactive energy (Figure 8.11c). If $V = V_0$, the current through the reactance X_T is zero and there is no energy exchange.

Figure 8.12 shows a typical 48-pulse output voltage waveform generated by the combined outputs of either eight two-level six-pulse, or four three-level 12-pulse converters.

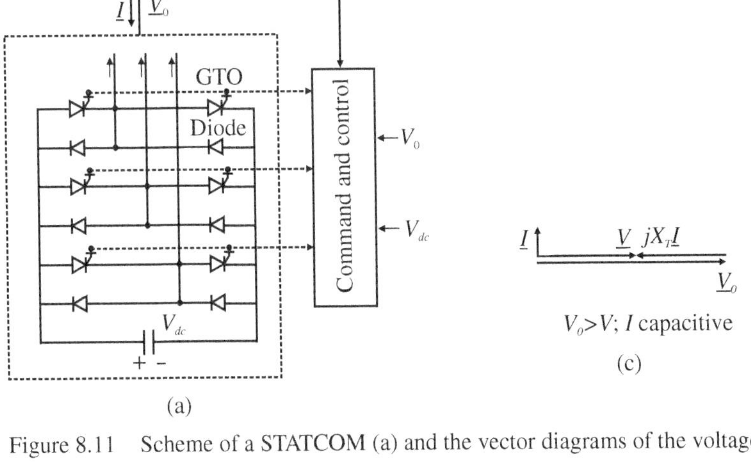

Figure 8.11 Scheme of a STATCOM (a) and the vector diagrams of the voltages (b and c).

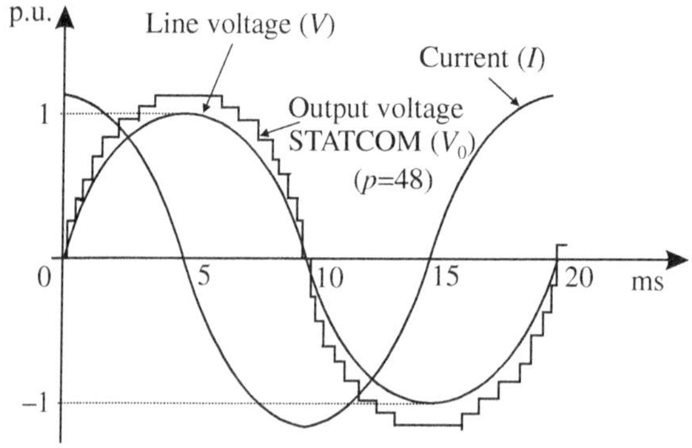

Figure 8.12 The output voltage (V_0) and current waveforms (I) for a converter with 48-pulse.

The operation of the VSC, used as a controllable static VAr generator, can be explained by the physical fact that, similar to all switching power converters, the net instantaneous power at the AC output terminals must always be equal to the net instantaneous power at the DC input terminals neglecting the losses in the semiconductor switches. Since the converter supplies only reactive power, its output voltages are generated in phase with the AC system voltages. The real input power provided by the DC source (charged capacitor) must be zero, as the total instantaneous power on the AC side is also zero. Furthermore, since reactive power at zero frequency—at the DC capacitor—by definition is zero, the DC capacitor plays no role in the reactive power generation. In other words, the converter simply interconnects the three AC terminals in such a way that the reactive output currents can flow freely between them. Seen from the terminals of the AC system, one could say that the converter establishes a circulating current among the phases with zero net instantaneous power exchange [3].

The need for the DC storage capacitor is theoretically due to the above stipulated equality of the instantaneous output and input powers. The output voltage waveform of the DC to AC converter is not a perfect sine wave. For this reason, the net instantaneous output power (VA) has a fluctuating component even if the converter output currents would be pure sine waves. Thus, in order not to violate the equality of the instantaneous output and input power, the converter must draw a fluctuating "ripple" current from the DC storage capacitor that provides a constant voltage at the input.

The presence of the input ripple current components is thus entirely due to the ripple components of the output voltage, which depends on the output waveform fabrication method used. In a practical VAr generator, the elementary two- or three-level converters do not meet the harmonic requirements either for the output voltage or for the input (DC capacitor) current. However, by combining a number of these basic converters into a multi-pulse structure, and/or using appropriate pulse-width modulation (PWM), the output voltage distortion and capacitor ripple current can be theoretically reduced to any desired degree. The readers can find more details in the book [3].

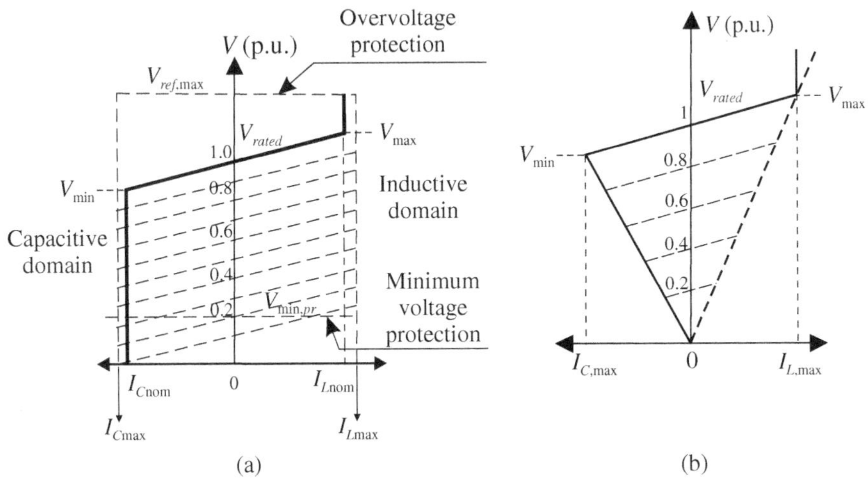

Figure 8.13 The *V–I* characteristic of (a) a STATCOM and (b) an SVC.

Figure 8.13a illustrates the *V–I* characteristic of a STATCOM. It can be noticed that STATCOM can generate either inductive or capacitive currents, independent of the network voltage. The device can generate maximum capacitive current at any AC voltage, practically to a zero value; it represents a significant difference as compared to an SVC of which current drops linearly with the network voltage drop (because it is determined in conformity with the maximum equivalent admittance). The characteristic of an SVC is depicted in Figure 8.13b.

When comparing a STATCOM with an SVC, the following conclusions may be drawn:

- STATCOM is faster, as the switching frequency is higher, especially when IGBTs and PWM are used.
- An SVC is, as seen from the network, a controllable reactance, and its VAr output is a quadratic function of the terminal voltage. On the other hand, a STATCOM is capable to provide constant reactive current in the majority of operating area. Its VAr output is linearly dependent on terminal voltage.
- In case of an SVC, the reactive power is actually produced by VAr elements (reactors, capacitors, and filter). On the contrary, the rating of the capacitor on the DC side of a STATCOM is much smaller than for the SVC. As a container for temporary energy storage, needed for reactive power generation/consumption, an electrical power system (EPS) itself is used by suitable phase connections by electronic valves. STATCOM on the other hand is inherently capable to operate in both regimes up to the same current.
- In contrast to the SVC, a STATCOM has no capacitive elements connected directly to the AC system and thus it cannot produce resonance phenomenon with the network.

- Another relevant aspect is related to the transient behavior of STATCOM. It can boost, for limited periods, the value of the current, either capacitive or inductive, depending on the thermal conditions (temperature) of the semiconductors' junctions (see $I_{L\max}$ and $I_{C\max}$ in Figure 8.13a). In other words, it can be overloaded temporarily. In the case of an SVC, the behavior is different: its maximum current is determined strictly by the value of the admittance and the AC voltage of the network. An SVC device is not capable of providing transient characteristics higher than the steady-state one.

Some of the differences between a STATCOM and an SVC are also reflected from their characteristics. It is shown that the maximum STATCOM reactive current is constant, whereas for an SVC it depends linearly on voltage.

8.3 STATCOM MODELING

8.3.1 STATCOM Model for Steady-State Analysis

The static behavior of a STATCOM device is described by its *V–I* characteristic, presented in Figure 8.13a. Analyzing these characteristics, the following areas of potential steady-state operation can be identified:

- The linear control area, which is limited by the nominal currents ($I_{C\text{nom}}, I_{L\text{nom}}$) in steady-state operation, or by the maximum currents ($I_{C\max}, I_{L\max}$) for short-term operation, respectively.
- The capacitive area, where a STATCOM injects constant reactive current in the node of connection ($I_{C\text{nom}}$ for long-term operation or $I_{C\max}$ for short-term operation).
- The inductive area, where a STATCOM absorbs constant reactive current from the node of connection ($I_{L\text{nom}}$ for long-term operation or $I_{L\max}$ for short-term operation).

Note that the short-term operation is possible until the thermal limit of the semiconductor valves is reached.

From operational point of view, besides the current limitations, there are also voltage limitations, such as:

- the maximum limit of the reference voltage ($V_{ref,\max}$);
- the protection limit at minimum voltage ($V_{\min,pr}$).

Thereby, from a static point of view, the STATCOM's behavior is somehow similar to the one of an SVC (Figure 8.13b). The major difference consists of the fact that outside the linear area of control (i.e., $I_{STATCOM} = I_{L\max}$ when $V > V_{\max}$ and $I_{STATCOM} = I_{C\max}$ when $V < V_{\min}$), a STATCOM behaves as a constant reactive current source, whereas SVC is in fact a reactance.

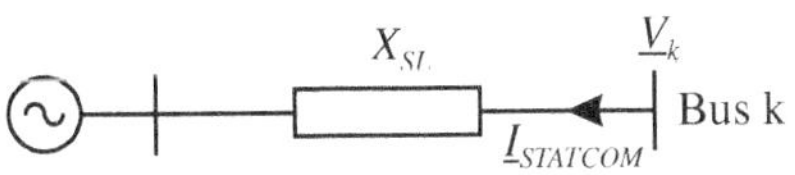

Figure 8.14 STATCOM representation in the linear area of control.

In the linear area of control, the voltage of the connection node (k) is kept inside the interval [$V_{\min}$, $V_{\max}$] and, correspondingly, the STATCOM may be equivalated by an electromotive voltage behind an X_{SL} reactance, which defines the slope of the adjustment characteristic (Fig. 8.14).

The controlled bus voltage is given by

$$\underline{V}_k = V_{ref} + jX_{SL}\underline{I}_{STATCOM}$$

The network connection node (k) is modeled as a PV node. In case the limits of the reactive power of the STATCOM are exceeded, the node k becomes a PQ-type node with its reactive current given by the exceeded limit of the device. Because such a node is not a "standard" element for load-flow calculations, it should be modeled as explained further on.

To deduce the mathematical model for steady-state calculation using the Newton–Raphson method, the equivalent circuit from Figure 8.15 is considered. Two possibilities for STATCOM modeling will be presented, that is, the so-called voltage-based model and the current-based model.

In the single-phase representation (Figure 8.15a), a STATCOM consists of a converter, supplied a DC capacitor, which is connected to a network bus via a one-phase transformer. The transformer primary is earthed.

In the three-phase representation, the converter is connected to the secondary of a three-phase transformer (or three single-phase transformers), which provides the

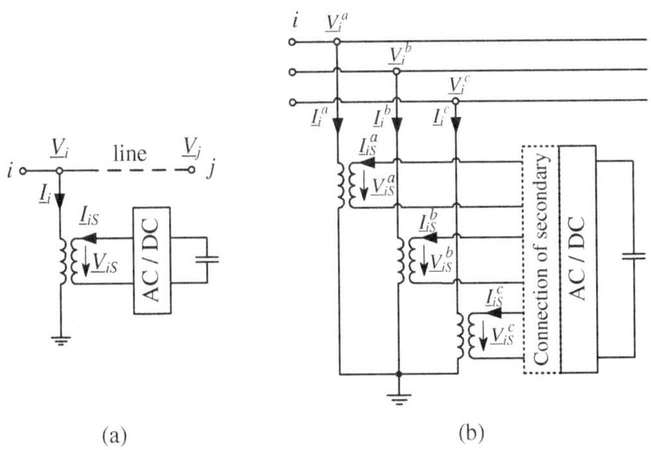

Figure 8.15 The STATCOM scheme: (a) single-phase representation and (b) three-phase representation.

link with the network bus. The transformer primary is earthed, while the secondary may be connected in various ways (e.g., star and delta).

8.3.1.1 Basic Load Flow Equations

Although it is considered that the readers are familiar with the Newton–Raphson-based power flow method, basic equations are given for convenience [3, 5].

SINGLE-PHASE REPRESENTATION. The single-phase representation considers only one phase and is not suitable for description of an arbitrary three-phase power system. When, however, a power system is presumed symmetrical, a single-phase model is also applicable.

From strictly mathematical point of view, the single-phase model considers the phase-to-ground voltages. There are actually two possibilities for calculation of the three-phase powers in a symmetrical three-phase system by applying the single-phase procedures, that is:

- to multiply the resulted powers by 3;
- instead of phase voltages (denoted by "*V*"), simply use the phase-to-phase voltages (denoted by "*U*"), which in a symmetrical system have the same magnitude.

The power of node i has to be balanced, that is, the sum of node production, consumption, and flows toward n neighboring nodes has to be zero.

$$P_i = -P_{Gi} + P_{Bi} + \sum_{k=1}^{n} V_i V_k (G_{ik}\cos(\delta_i - \delta_k) + B_{ik}\sin(\delta_i - \delta_k)) = 0 \tag{8.16}$$

$$Q_i = -Q_{Gi} + Q_{Bi} + \sum_{k=1}^{n} V_i V_k (G_{ik}\sin(\delta_i - \delta_k) - B_{ik}\cos(\delta_i - \delta_k)) = 0 \tag{8.17}$$

where

P_{Gi} and Q_{Gi}	are	the active and reactive powers generated at bus i.
P_{Bi} and Q_{Bi}	–	the active and reactive power demands at bus i.
V_i and δ_i	–	the voltage magnitude and the angle at bus i.
V_k and δ_k	–	the voltage magnitude and the angle at bus k.
G_{ik} and B_{ik}	–	the conductance and the susceptance of the transmission line between buses i and k.
n	is	the number of neighboring nodes.
k	–	the counter denoting neighboring nodes.

THREE-PHASE REPRESENTATION. This is, like in the case of a single-phase load flow, based on balance equations that describe the balance of the active, P, and the reactive, Q, powers in each bus of the EPS. The main difference is that in the three-phase representation, load-flow calculations of the power-balancing equations have to be completed for each of the phases:

$$P_i^a = -P_{Gi}^a + P_{Bi}^a + \sum_{k=1}^{n} V_i^a \sum_{r=a,b,c} V_k^r \left(G_{ik}^{ar}\cos\left(\delta_i^a - \delta_k^r\right) + B_{ik}^{ar}\sin\left(\delta_i^a - \delta_k^r\right)\right) = 0$$

$$P_i^b = -P_{Gi}^b + P_{Bi}^b + \sum_{k=1}^{n} V_i^b \sum_{r=a,b,c} V_k^r \left(G_{ik}^{br}\cos\left(\delta_i^b - \delta_k^r\right) + B_{ik}^{br}\sin\left(\delta_i^b - \delta_k^r\right)\right) = 0$$

$$P_i^c = -P_{Gi}^c + P_{Bi}^c + \sum_{k=1}^{n} V_i^c \sum_{r=a,b,c} V_k^r \left(G_{ik}^{cr}\cos\left(\delta_i^c - \delta_k^r\right) + B_{ik}^{cr}\sin\left(\delta_i^c - \delta_k^r\right)\right) = 0 \tag{8.18}$$

$$Q_i^a = -Q_{Gi}^a + Q_{Bi}^a + \sum_{k=1}^{n} V_i^a \sum_{r=a,b,c} V_k^r \left(G_{ik}^{ar}\sin\left(\delta_i^a - \delta_k^r\right) - B_{ik}^{ar}\cos\left(\delta_i^a - \delta_k^r\right)\right) = 0$$

$$Q_i^b = -Q_{Gi}^b + Q_{Bi}^b + \sum_{k=1}^{n} V_i^b \sum_{r=a,b,c} V_k^r \left(G_{ik}^{br}\sin\left(\delta_i^b - \delta_k^r\right) - B_{ik}^{br}\cos\left(\delta_i^b - \delta_k^r\right)\right) = 0$$

$$Q_i^c = -Q_{Gi}^c + Q_{Bi}^c + \sum_{k=1}^{n} V_i^c \sum_{r=a,b,c} V_k^r \left(G_{ik}^{cr}\sin\left(\delta_i^c - \delta_k^r\right) - B_{ik}^{cr}\cos\left(\delta_i^c - \delta_k^r\right)\right) = 0 \tag{8.19}$$

where

P_{Gi}^p, Q_{Gi}^p	are	the production of active and reactive powers at bus i, phase p.
P_{Bi}^p, Q_{Bi}^p	–	the consumptions of active and reactive power at bus i, phase p.
U_i^p, δ_i^p	–	the voltage magnitude and the angle at bus i, phase p.
U_k^p, δ_k^p	–	the voltage magnitude and the angle at bus k, phase p.
G_{ik}^{pr}, B_{ik}^{pr}	–	the conductance and the susceptance between buses i and k and between phases p and r.
p, r	–	the phases a, b, and c.
n	is	the number of neighboring nodes.
k	–	the counter denoting neighboring nodes.

Compared to the single-phase calculation, the admittance matrix of the three-phase calculation extends by a factor of three in both dimensions, as the interactions among the phases have to be taken into consideration. The admittance matrix of the line between buses i and k, for instance, is presented in (8.20):

$$\left[\underline{Y}_{ik}^{abc}\right] = \begin{bmatrix} \underline{Y}_{ik}^{aa} & \underline{Y}_{ik}^{ab} & \underline{Y}_{ik}^{ac} \\ \underline{Y}_{ik}^{ba} & \underline{Y}_{ik}^{bb} & \underline{Y}_{ik}^{bc} \\ \underline{Y}_{ik}^{ca} & \underline{Y}_{ik}^{cb} & \underline{Y}_{ik}^{cc} \end{bmatrix} = \begin{bmatrix} G_{ik}^{aa} + jB_{ik}^{aa} & G_{ik}^{ab} + jB_{ik}^{ab} & G_{ik}^{ac} + jB_{ik}^{ac} \\ G_{ik}^{ba} + jB_{ik}^{ba} & G_{ik}^{bb} + jB_{ik}^{bb} & G_{ik}^{bc} + jB_{ik}^{bc} \\ G_{ik}^{ca} + jB_{ik}^{ca} & G_{ik}^{cb} + jB_{ik}^{cb} & G_{ik}^{cc} + jB_{ik}^{cc} \end{bmatrix} \tag{8.20}$$

With the previous expressions, the linear model for the calculus of the steady-state power flow is obtained using the Newton–Raphson method for the STATCOM device.

Description of STATCOM's Basic Feature. As stated earlier, a STATCOM does not generate nor consume active power. Its losses are taken into consideration in the impedance connecting STATCOM to the network or in an additional resistive element on the DC side. Therefore, the real power absorption of this idealized representation (P_0) equals zero.

For single-phase representation, the following is valid:

$$P_0 = 0 \tag{8.21}$$

For the three-phase representation, there are two possibilities, that is, a STATCOM may have a common DC link for all the phases or each phase may have its own DC link. In the first case, the sum of the active powers of all the phases (p) has to be equal to zero (equation 8.22). In the second case, however, the sum of the active powers of each single phase (p), separately, has to be equal to zero (equation 8.23). In the case of the common DC link, one equation is enough, but in the case of separated DC links, three equations have to be written:

$$P_0 = \sum_{p=a,b,c} P_0^p = 0 \tag{8.22}$$

$$P_0^a = 0, \quad P_0^b = 0, \quad P_0^c = 0 \tag{8.23}$$

Definition of the Controlled Parameters. For controlling the system quantities (reference quantities—*REF*), by employing a STATCOM, the mathematical description for each controlled variable is required. In the following text, the typical controlled parameters are mathematically described. In general, for inclusion of any quantity as control parameter, it has to be mathematically described and equalized to its reference value.

The most common controlled parameters of STATCOM are:

- the bus voltage magnitude of the i-th node;
- the magnitude of the internal voltage of the parallel branch;
- the reactive power into a node in which STATCOM is connected;
- the magnitude of the device current.

8.3.1.2 The Single-Phase Voltage-Based Model

A STATCOM may, for single-phase modeling, be represented by a parallel branch, connected to a system. It consists of a series connection of an impedance $\underline{Z}_{sh}$ and a controllable voltage source $\underline{V}_{sh}$, as seen in Figure 8.16. The impedance $\underline{Z}_{sh}$ represents the STATCOM transformer impedance and the converter losses.

As it results from Figure 8.16, the following equations for power-flow can be written:

$$P_{i\text{-}sh} = V_i^2 G_{sh} - V_i V_{sh}(G_{sh}\cos(\delta_i - \delta_{sh}) + B_{sh}\sin(\delta_i - \delta_{sh})) \tag{8.24}$$

$$Q_{i\text{-}sh} = -V_i^2 B_{sh} - V_i V_{sh}(G_{sh}\sin(\delta_i - \delta_{sh}) - B_{sh}\cos(\delta_i - \delta_{sh})) \tag{8.25}$$

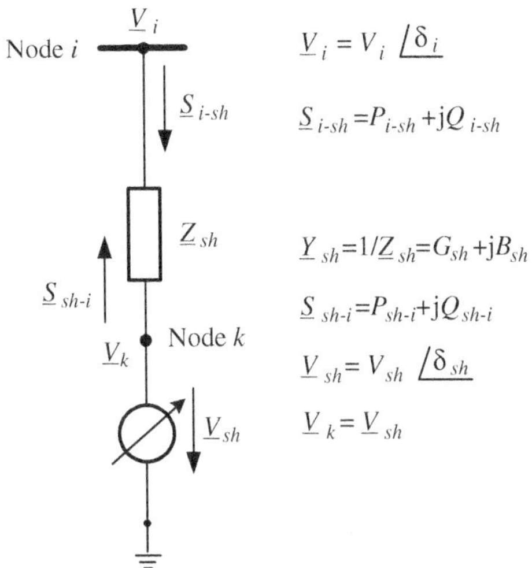

Figure 8.16 STATCOM single-phase voltage-based model.

The active power flowing from the voltage source toward the network is:

$$P_{sh-i} = V_{sh}^2 G_{sh} - V_i V_{sh}(G_{sh} \cos(\delta_{sh} - \delta_i) + B_{sh} \sin(\delta_{sh} - \delta_i)) \tag{8.26}$$

As already mentioned, STATCOM is assumed as a reactive source, so its active power equals zero. In this way, the device is described by the equation

$$P_0 = P_{sh-i} = 0 \tag{8.27}$$

The simplified matrix equation of the Newton–Raphson method for power flow calculation, when considering the STATCOM device, is:

$$\begin{bmatrix} \dfrac{\partial P}{\partial \delta} & \dfrac{\partial P}{\partial V} & \dfrac{\partial P_{sh-i}}{\partial \delta_{sh}} & \dfrac{\partial P_{sh-i}}{\partial V_{sh}} \\ \dfrac{\partial Q}{\partial \delta} & \dfrac{\partial Q}{\partial V} & \dfrac{\partial Q_{sh-i}}{\partial \delta_{sh}} & \dfrac{\partial Q_{sh-i}}{\partial V_{sh}} \\ \dfrac{\partial REF}{\partial \delta} & \dfrac{\partial REF}{\partial V} & \dfrac{\partial REF}{\partial \delta_{sh}} & \dfrac{\partial REF}{\partial V_{sh}} \\ \dfrac{\partial P_0}{\partial \delta} & \dfrac{\partial P_0}{\partial V} & \dfrac{\partial P_0}{\partial \delta_{sh}} & \dfrac{\partial P_0}{\partial V_{sh}} \end{bmatrix} \cdot \begin{bmatrix} \Delta\delta \\ \Delta V \\ \Delta\delta_{sh} \\ \Delta V_{sh} \end{bmatrix} = \begin{bmatrix} \Delta P \\ \Delta Q \\ \Delta REF \\ \Delta P_0 \end{bmatrix} \tag{8.28}$$

where the quantities P, Q, V, and δ stand for nodal active power, reactive power, voltage, and phase angle, respectively. As it results from equation (8.28), the classical Jacobean is completed with two rows and two columns corresponding to the control variables of the STATCOM device.

The equations for parameters control in the case of a single-phase voltage-based STATCOM are:

– the bus voltage magnitude of the ith node (V_i):

$$V_i - V_{REF} = 0 \tag{8.29}$$

– the magnitude of the internal parallel branch voltage (V_{sh}):

$$V_{sh} - V_{sh,REF} = 0 \tag{8.30}$$

– the reactive ($Q_{i\text{-}sh}$) power flowing into a parallel branch (see Figure 8.16):

$$\begin{aligned} Q_{i\text{-}sh} &= -V_i^2 B_{sh} - V_i V_{sh}(G_{sh}\sin(\delta_i - \delta_{sh}) - B_{sh}\cos(\delta_i - \delta_{sh})) \\ &= Q_{P,REF} \end{aligned} \tag{8.31}$$

– the reactive (Q_{ij}) power flowing on a line between nodes i and j (see Figure 8.15a):

$$Q_{ij} = -V_i^2 B_i - V_i V_j(G_{ij}\sin(\delta_i - \delta_j) - B_{ij}\cos(\delta_i - \delta_j)) = Q_{ij\text{-}REF} \tag{8.32}$$

where

Q_{ij}	is the reactive power flowing from node i to node j.
B_i	– the shunt admittance of bus i.
$\underline{Y}_{ij} = G_{ij} + jB_{ij}$	– the admittance between buses i and j.
V_i, δ_i	– the voltage magnitude and angle at bus i.
V_j, δ_j	– the voltage magnitude and angle at bus j.

– the magnitude of the device's capacitive current $I_{QC\text{-}REF}$:

$$I_{QC\text{-}REF}\cos(\delta_{sh} + 90°) = \mathrm{Re}\left\{\frac{\underline{V}_i - \underline{V}_{sh}}{\underline{Z}_{sh}}\right\} \tag{8.33a}$$

$$I_{QC\text{-}REF}\sin\left(\delta_{sh} + 90°\right) = \mathrm{Im}\left\{\frac{\underline{V}_i - \underline{V}_{sh}}{\underline{Z}_{sh}}\right\} \tag{8.33b}$$

– the magnitude of the device's inductive current $I_{QL\text{-}REF}$:

$$I_{QL\text{-}REF}\cos\left(\delta_{sh} - 90°\right) = \mathrm{Re}\left\{\frac{\underline{V}_i - \underline{V}_{sh}}{\underline{Z}_{sh}}\right\} \tag{8.34a}$$

$$I_{QC\text{-}REF}\sin\left(\delta_{sh} - 90°\right) = \mathrm{Im}\left\{\frac{\underline{V}_i - \underline{V}_{sh}}{\underline{Z}_{sh}}\right\} \tag{8.34b}$$

8.3.1.3 The Single-Phase Current-Based Model

In the single-phase current-based model of a STATCOM, a current $\underline{I}_{sh}$ flowing through the branch is represented. Impedance $\underline{Z}_{sh}$ is separated from the voltage source $\underline{V}_{sh}$ by a node k. The model is shown in Figure 8.17.

The current through the branch is represented by the real $\mathrm{Re}\{\underline{I}_{sh}\}$ and the imaginary $\mathrm{Im}\{\underline{I}_{sh}\}$ components. In balanced power systems, the following equations can be written:

$$P_{i\text{-}sh} = V_i^2 \cdot G_{sh} - V_i \cdot V_{sh}(G_{sh}\cdot\cos(\delta_i - \delta_{sh}) + B_{sh}\cdot\sin(\delta_i - \delta_{sh})) \tag{8.24}$$

$$Q_{i\text{-}sh} = -V_i^2 \cdot B_{sh} - V_i \cdot V_{sh}(G_{sh}\cdot\sin(\delta_i - \delta_{sh}) - B_{sh}\cdot\cos(\delta_i - \delta_{sh})) \tag{8.25}$$

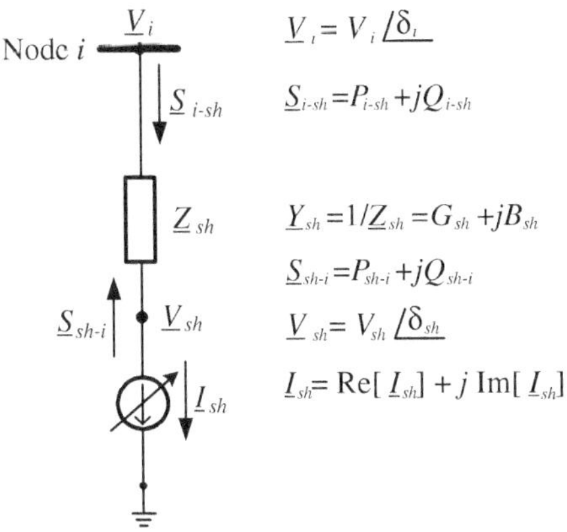

Figure 8.17 The STATCOM single-phase current-based model.

$$P_{sh} = V_{sh}^2 \cdot G_{sh} - V_{sh} \cdot V_i \cdot (G_{sh} \cdot \cos(\delta_{sh} - \delta_i) + B_{sh} \cdot \sin(\delta_{sh} - \delta_i)) \\ -(-V_{sh} \cdot (\mathrm{Re}[\underline{I}_{sh}] \cdot \cos(\delta_{sh}) + \mathrm{Im}[\underline{I}_{sh}] \cdot \sin(\delta_{sh}))) \tag{8.35}$$

$$Q_{sh} = -V_{sh}^2 \cdot B_{sh} - V_{sh} \cdot V_i \cdot (G_{sh} \cdot \sin(\delta_{sh} - \delta_i) - B_{sh} \cdot \cos(\delta_{sh} - \delta_i)) \\ -(-V_{sh} \cdot (\mathrm{Re}[\underline{I}_{sh}] \cdot \sin(\delta_{sh}) - \mathrm{Im}[\underline{I}_{sh}] \cdot \cos(\delta_{sh}))) \tag{8.36}$$

Since the STATCOM device does not exchange real power, we have:

$$P_0 = -V_{sh} \cdot (\mathrm{Re}[\underline{I}_{sh}] \cdot \cos(\delta_{sh}) + \mathrm{Im}[\underline{I}_{sh}] \cdot \sin(\delta_{sh})) = 0 \tag{8.37}$$

When the current-based model is used, the basic matrix equation for load flow calculation becomes:

$$\begin{bmatrix} \frac{\partial P}{\partial \delta} & \frac{\partial P}{\partial V} & \frac{\partial P}{\partial \mathrm{Re}[\underline{I}_{sh}]} & \frac{\partial P}{\partial \mathrm{Im}[\underline{I}_{sh}]} \\ \frac{\partial Q}{\partial \delta} & \frac{\partial Q}{\partial V} & \frac{\partial Q}{\partial \mathrm{Re}[\underline{I}_{sh}]} & \frac{\partial Q}{\partial \mathrm{Im}[\underline{I}_{sh}]} \\ \frac{\partial REF}{\partial \delta} & \frac{\partial REF}{\partial V} & \frac{\partial REF}{\partial \mathrm{Re}[\underline{I}_{sh}]} & \frac{\partial REF}{\partial \mathrm{Im}[\underline{I}_{sh}]} \\ \frac{\partial P_0}{\partial \delta} & \frac{\partial P_0}{\partial V} & \frac{\partial P_0}{\partial \mathrm{Re}[\underline{I}_{sh}]} & \frac{\partial P_0}{\partial \mathrm{Im}[\underline{I}_{sh}]} \end{bmatrix} \cdot \begin{bmatrix} \Delta\delta \\ \Delta V \\ \Delta \mathrm{Re}[\underline{I}_{sh}] \\ \Delta \mathrm{Im}[\underline{I}_{sh}] \end{bmatrix} = \begin{bmatrix} \Delta P \\ \Delta Q \\ \Delta REF \\ \Delta P_0 \end{bmatrix} \tag{8.38}$$

The controlled parameters, indicated by *REF*, for the single-phase current-based STATCOM model, are considered by the following equations:

– the bus voltage magnitude of the i-th node (V_i):

$$V_i - V_{REF} = 0 \tag{8.39}$$

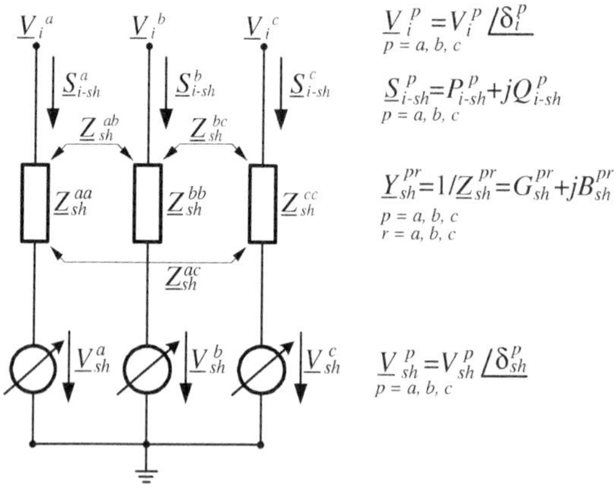

Figure 8.18 The STATCOM three-phase voltage-based model.

– the magnitude of the internal parallel branch voltage (V_{sh}):

$$V_{sh} - V_{sh,REF} = 0 \tag{8.40}$$

– the reactive power exchanged by STATCOM with the i-th node:

$$Q_{i\text{-}sh} = -V_i^2 \cdot B_{sh} - V_i \cdot V_{sh}(\cdot G_{sh} \cdot \sin(\delta_i - \delta_{sh}) - B_{sh} \cdot \cos(\delta_i - \delta_{sh})) \tag{8.41}$$

– the shunt current magnitude I_{sh}:

$$I_{sh} = \sqrt{\mathrm{Re}[\underline{I}_{sh}]^2 + \mathrm{Im}[\underline{I}_{sh}]^2} \tag{8.42}$$

Note: The current-based single-phase model increases the number of nodes in the system by 1.

8.3.1.4 Three-Phase Voltage-Based Model

The three-phase voltage-based model of the STATCOM is shown in Figure 8.18.

The balance equations for phase active and reactive powers in a three-phase model are:

$$\begin{aligned} P_{i\text{-}sh}^a = V_i^a \cdot & \sum_{r=a,b,c} V_i^r \cdot \left(G_{sh}^{ar} \cdot \cos\left(\delta_i^a - \delta_i^r\right) + B_{sh}^{ar} \cdot \sin\left(\delta_i^a - \delta_i^r\right)\right) \\ & - V_i^a \cdot \sum_{r=a,b,c} V_{sh}^r \cdot \left(G_{sh}^{ar} \cdot \cos\left(\delta_i^a - \delta_{sh}^r\right) + B_{sh}^{ar} \cdot \sin\left(\delta_i^a - \delta_{sh}^r\right)\right) \end{aligned} \tag{8.43a}$$

$$\begin{aligned} P_{i\text{-}sh}^b = V_i^b \cdot & \sum_{r=a,b,c} V_i^r \cdot \left(G_{sh}^{br} \cdot \cos\left(\delta_i^b - \delta_i^r\right) + B_{sh}^{br} \cdot \sin\left(\delta_i^b - \delta_i^r\right)\right) \\ & - V_i^b \cdot \sum_{r=a,b,c} V_{sh}^r \cdot \left(G_{sh}^{br} \cdot \cos\left(\delta_i^b - \delta_{sh}^r\right) + B_{sh}^{br} \cdot \sin\left(\delta_i^b - \delta_{sh}^r\right)\right) \end{aligned} \tag{8.43b}$$

$$P_{i\text{-}sh}^{c} = V_i^c \cdot \sum_{r=a,b,c} V_i^r \cdot \left(G_{sh}^{cr} \cdot \cos\left(\delta_i^c - \delta_i^r\right) + B_{sh}^{cr} \cdot \sin\left(\delta_i^c - \delta_i^r\right)\right)$$
$$-V_i^c \cdot \sum_{r=a,b,c} V_{sh}^r \cdot \left(G_{sh}^{cr} \cdot \cos\left(\delta_i^c - \delta_{sh}^r\right) + B_{sh}^{cr} \cdot \sin\left(\delta_i^c - \delta_{sh}^r\right)\right) \quad (8.43c)$$

$$Q_{i\text{-}sh}^{a} = V_i^a \cdot \sum_{r=a,b,c} V_i^r \cdot \left(G_{sh}^{ar} \cdot \sin\left(\delta_i^a - \delta_i^r\right) - B_{sh}^{ar} \cdot \cos\left(\delta_i^a - \delta_i^r\right)\right)$$
$$-V_i^a \cdot \sum_{r=a,b,c} V_{sh}^r \cdot \left(G_{sh}^{ar} \cdot \sin\left(\delta_i^a - \delta_{sh}^r\right) - B_{sh}^{ar} \cdot \cos\left(\delta_i^a - \delta_{sh}^r\right)\right) \quad (8.44a)$$

$$Q_{i\text{-}sh}^{b} = V_i^b \cdot \sum_{r=a,b,c} V_i^r \cdot \left(G_{sh}^{br} \cdot \sin\left(\delta_i^b - \delta_i^r\right) - B_{sh}^{br} \cdot \cos\left(\delta_i^b - \delta_i^r\right)\right)$$
$$-V_i^b \cdot \sum_{r=a,b,c} V_{sh}^r \cdot \left(G_{sh}^{br} \cdot \sin\left(\delta_i^b - \delta_{sh}^r\right) - B_{sh}^{br} \cdot \cos\left(\delta_i^b - \delta_{sh}^r\right)\right) \quad (8.44b)$$

$$Q_{i\text{-}sh}^{c} = V_i^c \cdot \sum_{r=a,b,c} V_i^r \cdot \left(G_{sh}^{cr} \cdot \sin\left(\delta_i^c - \delta_i^r\right) - B_{sh}^{cr} \cdot \cos\left(\delta_i^c - \delta_i^r\right)\right)$$
$$-V_i^c \cdot \sum_{r=a,b,c} V_{sh}^r \cdot \left(G_{sh}^{cr} \cdot \sin\left(\delta_i^c - \delta_{sh}^r\right) - B_{sh}^{cr} \cdot \cos\left(\delta_i^c - \delta_{sh}^r\right)\right) \quad (8.44c)$$

The DC link can be represented in two ways: common DC link or separated DC link for each phase.

For three-phase voltage-based model, equations for each phase are presented by:

$$P_0^a = V_{sh}^a \cdot \sum_{r=a,b,c} V_{sh}^r \cdot \left(G_{sh}^{ar} \cdot \cos\left(\delta_{sh}^a - \delta_{sh}^r\right) + B_{sh}^{ar} \cdot \sin\left(\delta_{sh}^a - \delta_{sh}^r\right)\right)$$
$$-V_{sh}^a \cdot \sum_{r=a,b,c} V_i^r \cdot \left(G_{sh}^{ar} \cdot \cos\left(\delta_{sh}^a - \delta_i^r\right) + B_{sh}^{ar} \cdot \sin\left(\delta_{sh}^a - \delta_i^r\right)\right) \quad (8.45a)$$

$$P_0^b = V_{sh}^b \cdot \sum_{r=a,b,c} V_{sh}^r \cdot \left(G_{sh}^{br} \cdot \cos\left(\delta_{sh}^b - \delta_{sh}^r\right) + B_{sh}^{br} \cdot \sin\left(\delta_{sh}^b - \delta_{sh}^r\right)\right)$$
$$-V_{sh}^b \cdot \sum_{r=a,b,c} V_i^r \cdot \left(G_{sh}^{br} \cdot \cos\left(\delta_{sh}^b - \delta_i^r\right) + B_{sh}^{br} \cdot \sin\left(\delta_{sh}^b - \delta_i^r\right)\right) \quad (8.45b)$$

$$P_0^c = V_{sh}^c \cdot \sum_{r=a,b,c} V_{sh}^r \cdot \left(G_{sh}^{cr} \cdot \cos\left(\delta_{sh}^c - \delta_{sh}^r\right) + B_{sh}^{cr} \cdot \sin\left(\delta_{sh}^c - \delta_{sh}^r\right)\right)$$
$$-V_{sh}^c \cdot \sum_{r=a,b,c} V_i^r \cdot \left(G_{sh}^{cr} \cdot \cos\left(\delta_{sh}^c - \delta_i^r\right) + B_{sh}^{cr} \cdot \sin\left(\delta_{sh}^c - \delta_i^r\right)\right) \quad (8.45c)$$

– when common DC link is preferred, the sum of all three equations has to be equal to zero:

$$P_0 = \sum_{p=a,b,c} P_0^p = 0 \quad (8.46)$$

– when three separated DC links are represented, all three equations have to be equal to zero:

$$P_0^a = P_0^b = P_0^c = 0 \tag{8.47}$$

The matrix form of the system equations, with extended Jacobian, is:

$$\begin{bmatrix}
\frac{\partial P^p}{\partial \delta^p} & \frac{\partial P^p}{\partial V^p} & \frac{\partial P^p}{\partial \delta_{sh}^a} & \frac{\partial P^p}{\partial \delta_{sh}^b} & \frac{\partial P^p}{\partial \delta_{sh}^c} & \frac{\partial P^p}{\partial V_{sh}^a} & \frac{\partial P^p}{\partial V_{sh}^b} & \frac{\partial P^p}{\partial V_{sh}^c} \\
\frac{\partial Q^p}{\partial \delta^p} & \frac{\partial Q^p}{\partial V^p} & \frac{\partial Q^p}{\partial \delta_{sh}^a} & \frac{\partial Q^p}{\partial \delta_{sh}^b} & \frac{\partial Q^p}{\partial \delta_{sh}^c} & \frac{\partial Q^p}{\partial V_{sh}^a} & \frac{\partial Q^p}{\partial V_{sh}^b} & \frac{\partial Q^p}{\partial V_{sh}^c} \\
\frac{\partial REF^a}{\partial \delta^p} & \frac{\partial REF^a}{\partial V^p} & \frac{\partial REF^a}{\partial \delta_{sh}^a} & \frac{\partial REF^a}{\partial \delta_{sh}^b} & \frac{\partial REF^a}{\partial \delta_{sh}^c} & \frac{\partial REF^a}{\partial V_{sh}^a} & \frac{\partial REF^a}{\partial V_{sh}^b} & \frac{\partial REF^a}{\partial V_{sh}^c} \\
\frac{\partial REF^b}{\partial \delta^p} & \frac{\partial REF^b}{\partial V^p} & \frac{\partial REF^b}{\partial \delta_{sh}^a} & \frac{\partial REF^b}{\partial \delta_{sh}^b} & \frac{\partial REF^b}{\partial \delta_{sh}^c} & \frac{\partial REF^b}{\partial V_{sh}^a} & \frac{\partial REF^b}{\partial V_{sh}^b} & \frac{\partial REF^b}{\partial V_{sh}^c} \\
\frac{\partial REF^c}{\partial \delta^p} & \frac{\partial REF^c}{\partial V^p} & \frac{\partial REF^c}{\partial \delta_{sh}^a} & \frac{\partial REF^c}{\partial \delta_{sh}^b} & \frac{\partial REF^c}{\partial \delta_{sh}^c} & \frac{\partial REF^c}{\partial \delta_{sh}^a} & \frac{\partial REF^c}{\partial V_{sh}^b} & \frac{\partial REF^c}{\partial V_{sh}^c} \\
\frac{\partial P_0^a}{\partial \delta^p} & \frac{\partial P_0^a}{\partial V^p} & \frac{\partial P_0^a}{\partial \delta_{sh}^a} & \frac{\partial P_0^a}{\partial \delta_{sh}^b} & \frac{\partial P_0^a}{\partial \delta_{sh}^c} & \frac{\partial P_0^a}{\partial V_{sh}^a} & \frac{\partial P_0^a}{\partial V_{sh}^b} & \frac{\partial P_0^a}{\partial V_{sh}^c} \\
\frac{\partial P_0^b}{\partial \delta^p} & \frac{\partial P_0^b}{\partial V^p} & \frac{\partial P_0^b}{\partial \delta_{sh}^a} & \frac{\partial P_0^b}{\partial \delta_{sh}^b} & \frac{\partial P_0^b}{\partial \delta_{sh}^c} & \frac{\partial P_0^b}{\partial V_{sh}^a} & \frac{\partial P_0^b}{\partial V_{sh}^b} & \frac{\partial P_0^b}{\partial V_{sh}^c} \\
\frac{\partial P_0^c}{\partial \delta^p} & \frac{\partial P_0^c}{\partial \delta^p} & \frac{\partial P_0^c}{\partial \delta^p} & \frac{\partial P_0^c}{\partial \delta^p} & \frac{\partial P_0^c}{\partial \delta^p} & \frac{\partial P_0^c}{\partial \delta^p} & \frac{\partial P_0^c}{\partial \delta^p} & \frac{\partial P_0^c}{\partial \delta^p}
\end{bmatrix}
\cdot
\begin{bmatrix}
\Delta\delta^p \\ \Delta V^p \\ \Delta\delta_{sh}^a \\ \Delta\delta_{sh}^b \\ \Delta\delta_{sh}^c \\ \Delta V_{sh}^a \\ \Delta V_{sh}^b \\ \Delta V_{sh}^c
\end{bmatrix}
=
\begin{bmatrix}
\Delta P^p \\ \Delta Q^p \\ \Delta REF^a \\ \Delta REF^b \\ \Delta REF^c \\ \Delta P_0^a \\ \Delta P_0^b \\ \Delta P_0^c
\end{bmatrix} \tag{8.48}$$

where $p = a, b, c$.

The controlled parameters, denoted by *REF*, for three-phase voltage-based model of STATCOM are:

– the bus voltage magnitude on the p (a, b, c) phase at the ith node (V_i):

$$V_i^a - V_{REF}^a = 0 \tag{8.49a}$$

$$V_i^b - V_{REF}^b = 0 \tag{8.49b}$$

$$V_i^c - V_{REF}^c = 0 \tag{8.49c}$$

– the magnitude of the parallel branch internal voltage (V_{sh}) on the phase p (= a, b, c):

$$V_{sh}^{a} - V_{sh,REF}^{a} = 0 \tag{8.50a}$$

$$V_{sh}^{b} - V_{sh,REF}^{b} = 0 \tag{8.50b}$$

$$V_{sh}^{c} - V_{sh,REF}^{c} = 0 \tag{8.50c}$$

– the reactive power on the phase p (a, b, c) of an i node:

$$\begin{aligned} Q_{i\text{-}sh}^{a} = {} & V_{i}^{a} \cdot \sum_{r=a,b,c} V_{i}^{r} \cdot \left(G_{sh}^{ar} \cdot \sin\left(\delta_{i}^{a} - \delta_{i}^{r}\right) - B_{sh}^{ar} \cdot \cos\left(\delta_{i}^{a} - \delta_{i}^{r}\right)\right) \\ & - V_{i}^{a} \cdot \sum_{r=a,b,c} V_{sh}^{r} \cdot \left(G_{sh}^{ar} \cdot \sin\left(\delta_{i}^{a} - \delta_{sh}^{r}\right) - B_{sh}^{ar} \cdot \cos\left(\delta_{i}^{a} - \delta_{sh}^{r}\right)\right) \end{aligned} \tag{8.51a}$$

$$\begin{aligned} Q_{i\text{-}sh}^{b} = {} & V_{i}^{b} \cdot \sum_{r=a,b,c} V_{i}^{r} \cdot \left(G_{sh}^{br} \cdot \sin\left(\delta_{i}^{b} - \delta_{i}^{r}\right) - B_{sh}^{br} \cdot \cos\left(\delta_{i}^{b} - \delta_{i}^{r}\right)\right) \\ & - V_{i}^{b} \cdot \sum_{r=a,b,c} V_{sh}^{r} \cdot \left(G_{sh}^{br} \cdot \sin\left(\delta_{i}^{b} - \delta_{sh}^{r}\right) - B_{sh}^{br} \cdot \cos\left(\delta_{i}^{b} - \delta_{sh}^{r}\right)\right) \end{aligned} \tag{8.51b}$$

$$\begin{aligned} Q_{i\text{-}sh}^{c} = {} & V_{i}^{c} \cdot \sum_{r=a,b,c} V_{i}^{r} \cdot \left(G_{sh}^{cr} \cdot \sin\left(\delta_{i}^{c} - \delta_{i}^{r}\right) - B_{sh}^{ar} \cdot \cos\left(\delta_{i}^{c} - \delta_{i}^{r}\right)\right) \\ & - V_{i}^{c} \cdot \sum_{r=a,b,c} V_{sh}^{r} \cdot \left(G_{sh}^{cr} \cdot \sin\left(\delta_{i}^{c} - \delta_{sh}^{r}\right) - B_{sh}^{ar} \cdot \cos\left(\delta_{i}^{c} - \delta_{sh}^{r}\right)\right) \end{aligned} \tag{8.51c}$$

8.3.1.5 Three-Phase Current-Based Model

The equivalent diagram of a three-phase current-based STATCOM model is shown in Figure 8.19.

The three-phase current-based model is represented by the following equations:

$$\begin{aligned} P_{i\text{-}sh}^{a} = {} & V_{i}^{a} \cdot \sum_{r=a,b,c} V_{i}^{r} \cdot \left(G_{sh}^{ar} \cdot \cos\left(\delta_{i}^{a} - \delta_{i}^{r}\right) + B_{sh}^{ar} \cdot \sin\left(\delta_{i}^{a} - \delta_{i}^{r}\right)\right) \\ & - V_{i}^{a} \cdot \sum_{r=a,b,c} V_{sh}^{r} \cdot \left(G_{sh}^{ar} \cdot \cos\left(\delta_{i}^{a} - \delta_{sh}^{r}\right) + B_{sh}^{ar} \cdot \sin\left(\delta_{i}^{a} - \delta_{sh}^{r}\right)\right) \end{aligned} \tag{8.52a}$$

$$\begin{aligned} P_{i\text{-}sh}^{b} = {} & V_{i}^{b} \cdot \sum_{r=a,b,c} V_{i}^{r} \cdot \left(G_{sh}^{br} \cdot \cos\left(\delta_{i}^{b} - \delta_{i}^{r}\right) + B_{sh}^{br} \cdot \sin\left(\delta_{i}^{b} - \delta_{i}^{r}\right)\right) \\ & - V_{i}^{b} \cdot \sum_{r=a,b,c} V_{sh}^{r} \cdot \left(G_{sh}^{br} \cdot \cos\left(\delta_{i}^{b} - \delta_{sh}^{r}\right) + B_{sh}^{br} \cdot \sin\left(\delta_{i}^{b} - \delta_{sh}^{r}\right)\right) \end{aligned} \tag{8.52b}$$

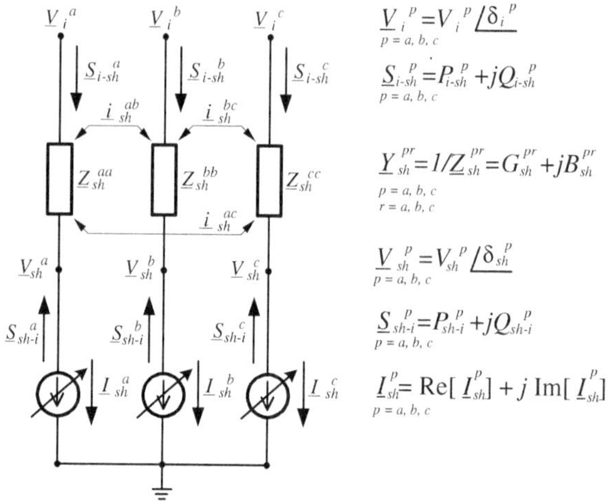

Figure 8.19 STATCOM three-phase current-based model.

$$P^c_{i\text{-}sh} = V^c_i \cdot \sum_{r=a,b,c} V^r_i \cdot \left(G^{cr}_{sh} \cdot \cos\left(\delta^c_i - \delta^r_i\right) + B^{cr}_{sh} \cdot \sin\left(\delta^c_i - \delta^r_i\right)\right)$$

$$-V^c_i \cdot \sum_{r=a,b,c} V^r_{sh} \cdot \left(G^{cr}_{sh} \cdot \cos\left(\delta^c_i - \delta^r_{sh}\right) + B^{cr}_{sh} \cdot \sin\left(\delta^c_i - \delta^r_{sh}\right)\right) \quad (8.52c)$$

$$Q^a_{i\text{-}sh} = V^a_i \cdot \sum_{r=a,b,c} V^r_i \cdot \left(G^{ar}_{sh} \cdot \sin\left(\delta^a_i - \delta^r_i\right) - B^{ar}_{sh} \cdot \cos\left(\delta^a_i - \delta^r_i\right)\right)$$

$$-V^a_i \cdot \sum_{r=a,b,c} V^r_{sh} \cdot \left(G^{ar}_{sh} \cdot \sin\left(\delta^a_i - \delta^r_{sh}\right) - B^{ar}_{sh} \cdot \cos\left(\delta^a_i - \delta^r_{sh}\right)\right) \quad (8.53a)$$

$$Q^b_{i\text{-}sh} = V^b_i \cdot \sum_{r=a,b,c} V^r_i \cdot \left(G^{br}_{sh} \cdot \sin\left(\delta^b_i - \delta^r_i\right) - B^{br}_{sh} \cdot \cos\left(\delta^b_i - \delta^r_i\right)\right)$$

$$-V^b_i \cdot \sum_{r=a,b,c} V^r_{sh} \cdot \left(G^{br}_{sh} \cdot \sin\left(\delta^b_i - \delta^r_{sh}\right) - B^{br}_{sh} \cdot \cos\left(\delta^b_i - \delta^r_{sh}\right)\right) \quad (8.53b)$$

$$Q^c_{i\text{-}sh} = V^c_i \cdot \sum_{r=a,b,c} V^r_i \cdot \left(G^{cr}_{sh} \cdot \sin\left(\delta^c_i - \delta^r_i\right) - B^{cr}_{sh} \cdot \cos\left(\delta^c_i - \delta^r_i\right)\right)$$

$$-V^c_i \cdot \sum_{r=a,b,c} V^r_{sh} \cdot \left(G^{cr}_{sh} \cdot \sin\left(\delta^c_i - \delta^r_{sh}\right) - B^{cr}_{sh} \cdot \cos\left(\delta^c_i - \delta^r_{sh}\right)\right) \quad (8.53c)$$

$$P^a_{sh-i} = V^a_{sh} \cdot \sum_{r=a,b,c} V^r_{sh} \cdot \left(G^{ar}_{sh} \cdot \cos\left(\delta^a_{sh} - \delta^r_{sh}\right) + B^{ar}_{sh} \cdot \sin\left(\delta^a_{sh} - \delta^r_{sh}\right)\right)$$

$$-V^a_{sh} \cdot \sum_{r=a,b,c} V^r_i \cdot \left(G^{ar}_{sh} \cdot \cos\left(\delta^a_{sh} - \delta^r_i\right) + B^{ar}_{sh} \cdot \sin\left(\delta^a_{sh} - \delta^r_i\right)\right)$$

$$-\left(-V^a_{sh} \cdot \left(\mathrm{Re}\left[\underline{I}^a_{sh}\right] \cdot \cos\delta^a_{sh} + \mathrm{Im}\left[\underline{I}^a_{sh}\right] \cdot \sin\delta^a_{sh}\right)\right) \quad (8.54a)$$

$$P^b_{sh-i} = V^b_{sh} \cdot \sum_{r=a,b,c} V^r_{sh} \cdot \left(G^{br}_{sh} \cdot \cos\left(\delta^b_{sh} - \delta^r_{sh}\right) + B^{br}_{sh} \cdot \sin\left(\delta^b_{sh} - \delta^r_{sh}\right)\right)$$
$$-V^b_{sh} \cdot \sum_{r=a,b,c} V^r_i \cdot \left(G^{br}_{sh} \cdot \cos\left(\delta^b_{sh} - \delta^r_i\right) + B^{br}_{sh} \cdot \sin\left(\delta^b_{sh} - \delta^r_i\right)\right)$$
$$-\left(-V^b_{sh} \cdot \left(\mathrm{Re}\left[\underline{I}^b_{sh}\right] \cdot \cos\delta^b_{sh} + \mathrm{Im}\left[\underline{I}^b_{sh}\right] \cdot \sin\delta^b_{sh}\right)\right) \quad (8.54b)$$

$$P^c_{sh-i} = V^c_{sh} \cdot \sum_{r=a,b,c} V^r_{sh} \cdot \left(G^{cr}_{sh} \cdot \cos\left(\delta^c_{sh} - \delta^r_{sh}\right) + B^{cr}_{sh} \cdot \sin\left(\delta^c_{sh} - \delta^r_{sh}\right)\right)$$
$$-V^c_{sh} \cdot \sum_{r=a,b,c} V^r_i \cdot \left(G^{cr}_{sh} \cdot \cos\left(\delta^c_{sh} - \delta^r_i\right) + B^{cr}_{sh} \cdot \sin\left(\delta^c_{sh} - \delta^r_i\right)\right)$$
$$-\left(-V^c_{sh} \cdot \left(\mathrm{Re}\left[\underline{I}^c_{sh}\right] \cdot \cos\delta^c_{sh} + \mathrm{Im}\left[\underline{I}^c_{sh}\right] \cdot \sin\delta^c_{sh}\right)\right) \quad (8.54c)$$

$$Q^a_{sh-i} = V^a_{sh} \cdot \sum_{r=a,b,c} V^r_{sh} \cdot \left(G^{ar}_{sh} \cdot \sin\left(\delta^a_{sh} - \delta^r_P\right) - B^{ar}_{sh} \cdot \cos\left(\delta^a_{sh} - \delta^r_P\right)\right)$$
$$-V^a_{sh} \cdot \sum_{r=a,b,c} V^a_i \cdot \left(G^{ar}_{sh} \cdot \sin\left(\delta^a_{sh} - \delta^a_i\right) - B^{ar}_{sh} \cdot \cos\left(\delta^a_{sh} - \delta^a_i\right)\right)$$
$$-\left(-V^a_{sh} \cdot \left(\mathrm{Re}\left[\underline{I}^a_{sh}\right] \cdot \sin\delta^a_{sh} - \mathrm{Im}\left[\underline{I}^a_{sh}\right] \cdot \cos\delta^a_{sh}\right)\right) \quad (8.55a)$$

$$Q^b_{sh-i} = V^b_{sh} \cdot \sum_{r=a,b,c} V^r_{sh} \cdot \left(G^{br}_{sh} \cdot \sin\left(\delta^b_{sh} - \delta^r_P\right) - B^{br}_{sh} \cdot \cos\left(\delta^b_{sh} - \delta^r_P\right)\right)$$
$$-V^b_{sh} \cdot \sum_{r=a,b,c} V^b_i \cdot \left(G^{br}_{sh} \cdot \sin\left(\delta^b_{sh} - \delta^b_i\right) - B^{br}_{sh} \cdot \cos\left(\delta^b_{sh} - \delta^b_i\right)\right)$$
$$-\left(-V^b_{sh} \cdot \left(\mathrm{Re}\left[\underline{I}^b_{sh}\right] \cdot \sin\delta^b_{sh} - \mathrm{Im}\left[\underline{I}^b_{sh}\right] \cdot \cos\delta^b_{sh}\right)\right) \quad (8.55b)$$

$$Q^c_{sh-i} = V^c_{sh} \cdot \sum_{r=a,b,c} V^r_{sh} \cdot \left(G^{cr}_{sh} \cdot \sin\left(\delta^c_{sh} - \delta^r_P\right) - B^{cr}_{sh} \cdot \cos\left(\delta^c_{sh} - \delta^r_P\right)\right)$$
$$-V^c_{sh} \cdot \sum_{r=a,b,c} V^c_i \cdot \left(G^{cr}_{sh} \cdot \sin\left(\delta^c_{sh} - \delta^a_i\right) - B^{cr}_{sh} \cdot \cos\left(\delta^c_{sh} - \delta^c_i\right)\right)$$
$$-\left(-V^c_{sh} \cdot \left(\mathrm{Re}\left[\underline{I}^c_{sh}\right] \cdot \sin\delta^c_{sh} - \mathrm{Im}\left[\underline{I}^c_{sh}\right] \cdot \cos\delta^c_{sh}\right)\right) \quad (8.55c)$$

The phase real powers to ensure the basic feature of STATCOM are expressed by the following equations:

$$P^a_0 = -V^a_{sh} \cdot \left(\mathrm{Re}\left[\underline{I}^a_{sh}\right] \cdot \cos\delta^a_{sh} + \mathrm{Im}\left[\underline{I}^a_{sh}\right] \cdot \sin\delta^a_{sh}\right) \quad (8.56a)$$

$$P^b_0 = -V^b_{sh} \cdot \left(\mathrm{Re}\left[\underline{I}^b_{sh}\right] \cdot \cos\delta^b_{sh} + \mathrm{Im}\left[\underline{I}^b_{sh}\right] \cdot \sin\delta^b_{sh}\right) \quad (8.56b)$$

$$P^c_0 = -V^c_{sh} \cdot \left(\mathrm{Re}\left[\underline{I}^c_{sh}\right] \cdot \cos\delta^c_{sh} + \mathrm{Im}\left[\underline{I}^c_{sh}\right] \cdot \sin\delta^c_{sh}\right) \quad (8.56c)$$

– for common DC link, the sum of all three equations has to be equal to zero:

$$P_0 = \sum_{p=a,b,c} P^p_0 = 0 \quad (8.57)$$

– when three separated DC links are considered, all three equations have to be equal to 0:

$$P_0^a = P_0^b = P_0^c = 0 \tag{8.58}$$

The basic matrix equation for load flow calculation, extended to consider the current-based model of STATCOM, is:

$$\begin{bmatrix}
\frac{\partial P^p}{\partial \delta^p} & \frac{\partial P^p}{\partial V^p} & \frac{\partial P^p}{\partial \delta_{sh}^a} & \frac{\partial P^p}{\partial \delta_{sh}^b} & \frac{\partial P^p}{\partial \delta_{sh}^c} & \frac{\partial P^p}{\partial V_{sh}^a} & \frac{\partial P^p}{\partial V_{sh}^b} & \frac{\partial P^p}{\partial V_{sh}^c} \\
\frac{\partial Q^p}{\partial \delta^p} & \frac{\partial Q^p}{\partial V^p} & \frac{\partial Q^p}{\partial \delta_{sh}^a} & \frac{\partial Q^p}{\partial \delta_{sh}^b} & \frac{\partial Q^p}{\partial \delta_{sh}^c} & \frac{\partial Q^p}{\partial V_{sh}^a} & \frac{\partial Q^p}{\partial V_{sh}^b} & \frac{\partial Q^p}{\partial V_{sh}^c} \\
\frac{\partial REF^a}{\partial \delta^p} & \frac{\partial REF^a}{\partial V^p} & \frac{\partial REF^a}{\partial \mathrm{Re}\left[\underline{I}_{sh}^a\right]} & \frac{\partial REF^a}{\partial \mathrm{Re}\left[\underline{I}_{sh}^b\right]} & \frac{\partial REF^a}{\partial \mathrm{Re}\left[\underline{I}_{sh}^c\right]} & \frac{\partial REF^a}{\partial \mathrm{Im}\left[\underline{I}_{sh}^a\right]} & \frac{\partial REF^a}{\partial \mathrm{Im}\left[\underline{I}_{sh}^b\right]} & \frac{\partial REF^a}{\partial \mathrm{Im}\left[\underline{I}_{sh}^c\right]} \\
\frac{\partial REF^b}{\partial \delta^p} & \frac{\partial REF^b}{\partial V^p} & \frac{\partial REF^b}{\partial \mathrm{Re}\left[\underline{I}_{sh}^a\right]} & \frac{\partial REF^b}{\partial \mathrm{Re}\left[\underline{I}_{sh}^b\right]} & \frac{\partial REF^b}{\partial \mathrm{Re}\left[\underline{I}_{sh}^c\right]} & \frac{\partial REF^b}{\partial \mathrm{Im}\left[\underline{I}_{sh}^a\right]} & \frac{\partial REF^b}{\partial \mathrm{Im}\left[\underline{I}_{sh}^b\right]} & \frac{\partial REF^b}{\partial \mathrm{Im}\left[\underline{I}_{sh}^c\right]} \\
\frac{\partial REF^c}{\partial \delta^p} & \frac{\partial REF^c}{\partial V^p} & \frac{\partial REF^c}{\partial \mathrm{Re}\left[\underline{I}_{sh}^a\right]} & \frac{\partial REF^c}{\partial \mathrm{Re}\left[\underline{I}_{sh}^b\right]} & \frac{\partial REF^c}{\partial \mathrm{Re}\left[\underline{I}_{sh}^c\right]} & \frac{\partial REF^c}{\partial \mathrm{Im}\left[\underline{I}_{sh}^a\right]} & \frac{\partial REF^c}{\partial \mathrm{Im}\left[\underline{I}_{sh}^b\right]} & \frac{\partial REF^c}{\partial \mathrm{Im}\left[\underline{I}_{sh}^c\right]} \\
\frac{\partial P_0^a}{\partial \delta^p} & \frac{\partial P_0^a}{\partial V^p} & \frac{\partial P_0^a}{\partial \mathrm{Re}\left[\underline{I}_{sh}^a\right]} & \frac{\partial P_0^a}{\partial \mathrm{Re}\left[\underline{I}_{sh}^b\right]} & \frac{\partial P_0^a}{\partial \mathrm{Re}\left[\underline{I}_{sh}^c\right]} & \frac{\partial P_0^a}{\partial \mathrm{Im}\left[\underline{I}_{sh}^a\right]} & \frac{\partial P_0^a}{\partial \mathrm{Im}\left[\underline{I}_{sh}^b\right]} & \frac{\partial P_0^a}{\partial \mathrm{Im}\left[\underline{I}_{sh}^c\right]} \\
\frac{\partial P_0^b}{\partial \delta^p} & \frac{\partial P_0^b}{\partial V^p} & \frac{\partial P_0^b}{\partial \mathrm{Re}\left[\underline{I}_{sh}^a\right]} & \frac{\partial P_0^b}{\partial \mathrm{Re}\left[\underline{I}_{sh}^b\right]} & \frac{\partial P_0^b}{\partial \mathrm{Re}\left[\underline{I}_{sh}^c\right]} & \frac{\partial P_0^b}{\partial \mathrm{Im}\left[\underline{I}_{sh}^a\right]} & \frac{\partial P_0^b}{\partial \mathrm{Im}\left[\underline{I}_{sh}^b\right]} & \frac{\partial P_0^b}{\partial \mathrm{Im}\left[\underline{I}_{sh}^c\right]} \\
\frac{\partial P_0^c}{\partial \delta^p} & \frac{\partial P_0^c}{\partial \delta^p} & \frac{\partial P_0^c}{\partial \mathrm{Re}\left[\underline{I}_{sh}^a\right]} & \frac{\partial P_0^c}{\partial \mathrm{Re}\left[\underline{I}_{sh}^b\right]} & \frac{\partial P_0^c}{\partial \mathrm{Re}\left[\underline{I}_{sh}^c\right]} & \frac{\partial P_0^c}{\partial \mathrm{Im}\left[\underline{I}_{sh}^a\right]} & \frac{\partial P_0^c}{\partial \mathrm{Im}\left[\underline{I}_{sh}^b\right]} & \frac{\partial P_0^c}{\partial \mathrm{Im}\left[\underline{I}_{sh}^c\right]}
\end{bmatrix}
\cdot
\begin{bmatrix}
\Delta \delta^p \\
\Delta V^p \\
\Delta \mathrm{Re}\left[\underline{I}_{sh}^a\right] \\
\Delta \mathrm{Re}\left[\underline{I}_{sh}^b\right] \\
\Delta \mathrm{Re}\left[\underline{I}_{sh}^c\right] \\
\Delta \mathrm{Im}\left[\underline{I}_{sh}^a\right] \\
\Delta \mathrm{Im}\left[\underline{I}_{sh}^b\right] \\
\Delta \mathrm{Im}\left[\underline{I}_{sh}^c\right]
\end{bmatrix}
=
\begin{bmatrix}
\Delta P^p \\
\Delta Q^p \\
\Delta REF^a \\
\Delta REF^b \\
\Delta REF^c \\
\Delta P_0^a \\
\Delta P_0^b \\
\Delta P_0^c
\end{bmatrix}
\tag{8.59}$$

The controlled parameters, denoted by *REF*, for the three-phase current-based model of STATCOM, are presented by the following equations:

- the bus voltage magnitude on the p (a, b, c) phase of the ith node (V_i):

$$V_i^a - V_{REF}^a = 0 \tag{8.60a}$$

$$V_i^b - V_{REF}^b = 0 \tag{8.60b}$$

$$V_i^c - V_{REF}^c = 0 \tag{8.60c}$$

- the magnitude of the parallel branch internal voltage (V_{sh}) on the phase p (= a, b, c):

$$V_{sh}^a - V_{sh,REF}^a = 0 \tag{8.61a}$$

$$V_{sh}^b - V_{sh,REF}^b = 0 \tag{8.61b}$$

$$V_{sh}^c - V_{sh,REF}^c = 0 \tag{8.61c}$$

- the reactive power of the p (a, b, c) phase to an i node:

$$\begin{aligned} Q_{i\text{-}sh}^a = {} & U_i^a \cdot \sum_{r=a,b,c} U_i^r \cdot \left(G_{sh}^{ar} \cdot \sin\left(\delta_i^a - \delta_i^r\right) - B_{sh}^{ar} \cdot \cos\left(\delta_i^a - \delta_i^r\right)\right) \\ & - U_i^a \cdot \sum_{r=a,b,c} U_{sh}^r \cdot \left(G_{sh}^{ar} \cdot \sin\left(\delta_i^a - \delta_{sh}^r\right) - B_{sh}^{ar} \cdot \cos\left(\delta_i^a - \delta_{sh}^r\right)\right) \end{aligned} \tag{8.62a}$$

$$\begin{aligned} Q_{i\text{-}sh}^b = {} & U_i^b \cdot \sum_{r=a,b,c} U_i^r \cdot \left(G_{sh}^{br} \cdot \sin\left(\delta_i^b - \delta_i^r\right) - B_{sh}^{br} \cdot \cos\left(\delta_i^b - \delta_i^r\right)\right) \\ & - U_i^b \cdot \sum_{r=a,b,c} U_{sh}^r \cdot \left(G_{sh}^{br} \cdot \sin\left(\delta_i^b - \delta_{sh}^r\right) - B_{sh}^{br} \cdot \cos\left(\delta_i^b - \delta_{sh}^r\right)\right) \end{aligned} \tag{8.62b}$$

$$\begin{aligned} Q_{i\text{-}sh}^c = {} & U_i^c \cdot \sum_{r=a,b,c} U_i^r \cdot \left(G_{sh}^{cr} \cdot \sin\left(\delta_i^c - \delta_i^r\right) - B_{sh}^{cr} \cdot \cos\left(\delta_i^c - \delta_i^r\right)\right) \\ & - U_i^c \cdot \sum_{r=a,b,c} U_{sh}^r \cdot \left(G_{sh}^{cr} \cdot \sin\left(\delta_i^c - \delta_{sh}^r\right) - B_{sh}^{cr} \cdot \cos\left(\delta_i^c - \delta_{sh}^r\right)\right) \end{aligned} \tag{8.62c}$$

- the magnitude of the current I_{sh} on the phases p (a, b, c):

$$I_{sh}^a = \sqrt{\operatorname{Re}\left[\underline{I}_{sh}^a\right]^2 + \operatorname{Im}\left[\underline{I}_{sh}^a\right]^2} \tag{8.63a}$$

$$I_{sh}^b = \sqrt{\operatorname{Re}\left[\underline{I}_{sh}^b\right]^2 + \operatorname{Im}\left[\underline{I}_{sh}^b\right]^2} \tag{8.63b}$$

$$I_{sh}^c = \sqrt{\operatorname{Re}\left[\underline{I}_{sh}^c\right]^2 + \operatorname{Im}\left[\underline{I}_{sh}^c\right]^2} \tag{8.63c}$$

Note: The current-based three-phase model increases the number of nodes in the system by 3.

INITIAL CONDITIONS. A successful convergence of the Newton–Raphson method depends on the initial conditions. For power systems without FACTS devices, it is usual to initialize the bus voltage magnitudes with 1 p.u. and the bus-voltage angles with 0° (a flat start). The same logic may apply for voltage source-based model ($V_{sh} = 1$, $\delta_{sh} = 0$).

When current source models are used, the initial values for the real and the imaginary components of the branch currents (Re[$\underline{I}_{sh}$] and Im[$\underline{I}_{sh}$], respectively) have to be set. References [6] and [7] are recommended to find the initial conditions. The tests performed showed that the current method is almost insensitive to the initial conditions. The following values have proven to be useful. However, the "flat start" (both currents being zero) is not possible. The currents have to be initialized with value different from zero, such as:

$$\text{Re}[I_{sh}] = 0 \text{ p.u.}, \text{Im}[I_{sh}] = 0.1 \text{ p.u.} \tag{8.64}$$

HANDLING CONSTRAINTS. When setting the reference quantities *REF* (e.g., the power flows or the bus voltages), the FACTS parameters that are needed in the load flow model are, in general, unknown. They may fall, for example, beyond the device rating. Therefore, it is important to consider the constraints of the device parameters, for example, the magnitude of the injected voltage and/or the magnitude of a branch current and/or the injected active power of a branch. In the case when one or more of the FACTS parameters violate the limit (*LIM*), the original control parameter *REF* is replaced by the corresponding limit parameter *LIM*.

In the case of "multi-branch" devices and/or several devices in the system, the main issue is to identify which *REF* parameter (in the case that there are several) should be replaced in order to maintain the constraints. Normally, it should be the one that affects the device parameter, the constraint of which is violated, that is the highest value. If, for example, the series branch reactive power is chosen as an *REF* parameter and the injected current of the STATCOM branch goes beyond its rating, it is clear that the rated STATCOM current should be taken as the new *REF* parameter. Nevertheless, there are situations imaginable in which the situation might not be so clear.

8.3.2 Dynamic Models of STATCOM

The dynamic models are based upon SVC's dynamic models, the distinction between the two being their limitations. STATCOM is confined in current (between $I_{C\max}$ and $I_{L\max}$), while SVC is confined in susceptance (B_C and B_L).

8.3.2.1 Simplified Dynamic Model

The modeling of a STATCOM in the dynamic state may be accomplished through an injection of current of which value is corrected by a command and control block (Figure 8.20).

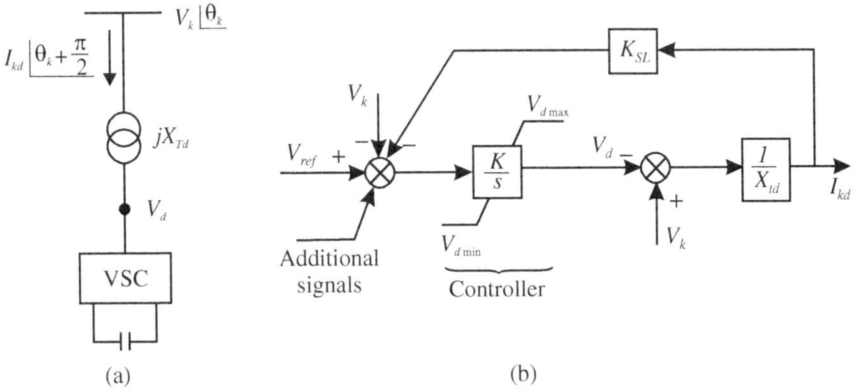

Figure 8.20 The simplified dynamic model of the STATCOM.

With the controlling device, the voltage V_d, generated by the VSC, is calculated, and, based upon its results, the current $\underline{I}_{kd}$ injected into the electric network is computed as:

$$\underline{I}_{kd} = (\underline{V}_k - \underline{V}_d)/jX_{Td} \tag{8.65}$$

where X_{Td} is the leakage reactance of the coupling transformer.

The simplified model consists of the following models:

- the voltage regulator, modeled through the transfer function *K/s*, where *K* is the integral gain factor;
- the block that models the gradient of the *V–I* characteristic of the device through the K_{SL} transfer function, K_{SL} being the inverse of the gradient;
- a lead-in for the additional signals introduced, for example, to improve oscillation damping;
- the voltage limits corresponding to the limit currents of the STATCOM (see the *V–I* characteristic illustrated in Figure 8.13):

$$\begin{aligned} V_{d\min} &= V_k - X_{Td} I_{C\min} \\ V_{d\max} &= V_k + X_{Td} I_{L\max} \end{aligned} \tag{8.66}$$

where $I_{C\min}$ and $I_{L\max}$ are the maximum capacitive and inductive currents of the STATCOM device.

If the STATCOM device has the possibility to increase the injection/absorption capacity during the transient state, then these limit currents depend upon the duration of the transient state (Figure 8.21).

These time-dependent limits can be used in case that the transient characteristics of the equipment are known. Generally, to study the small disturbance stability, these limits are considered constant.

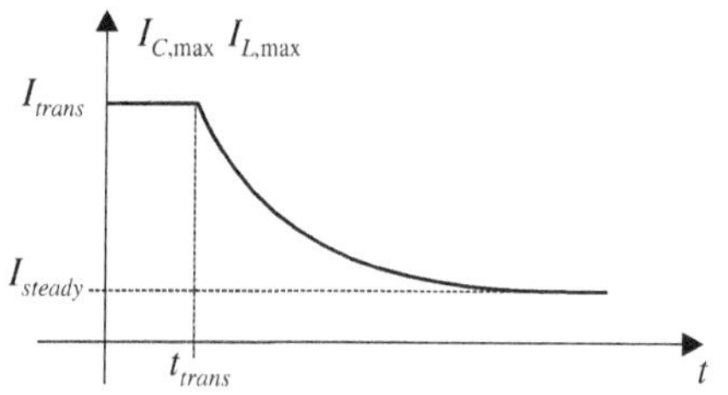

Figure 8.21 The time dependence of the current limits through the STATCOM.

It should be mentioned that the simplified dynamic model presents some limitations, as it does not include the DC voltage source, or the AC voltage generated by the STATCOM. Therefore, the simplified model is used especially to study the short-term stability phenomenon.

Figure 8.22 illustrates the dynamic model of the STATCOM device implemented in the EUROSTAG software. This model is based upon the generic model of the SVC proposed by CIGRE.

The current $I_{STATCOM}$ is calculated based upon the susceptance B and the voltage V. The adjusted measured voltage ($V{-}X_{SL}I_{STATCOM}$) is compared with the reference value V_{ref} to provide a voltage error to the proportional-integral controller.

The open loop STATCOM is modeled by a static boost factor and by a transfer function, which takes account of the transit track and the statistic delay in the breaker's command. The output current of the STATCOM is limited between the $I_{C\max}$ and $I_{L\max}$. The susceptance B is calculated starting from the current $I_{STATCOM}$ and the voltage V_{meas}.

8.3.2.2 *Detailed Dynamic Model*

A detailed mathematical model of STATCOM is presented as follows [9, 10], which is defined in the synchronously rotating *d-q* coordinate system. An important issue of the *d-q* transformation is its frequency dependence. For proper synchronization with the network, the fundamental frequency signals are transformed into DC quantities

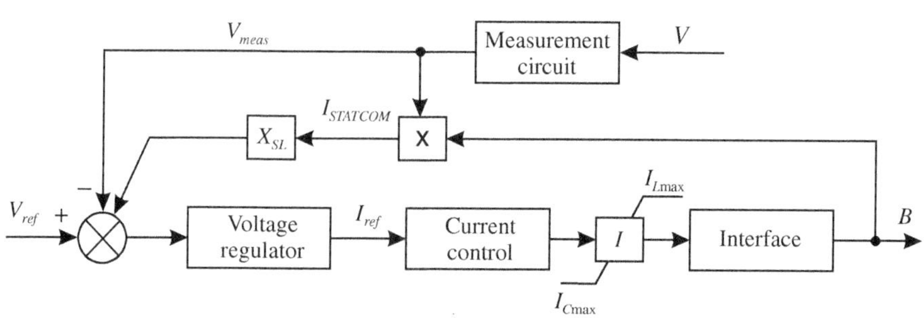

Figure 8.22 The STATCOM model implemented in the EUROSTAG software [8].

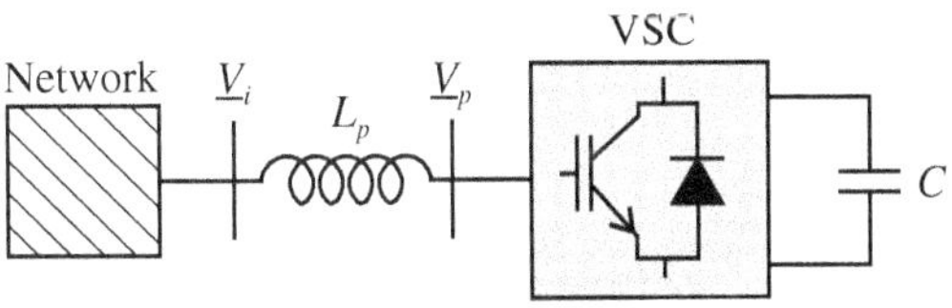

Figure 8.23 Basic circuit of a STATCOM.

(in steady state) that can be used as inputs of a control circuit. The mathematical model forms the base for the development of the control algorithm.

In all subsequent derivations, a per-unit (p.u.) system is used where i_B and u_B are the base values and ω_B is the angular speed at the system nominal frequency:

$$\begin{aligned}
i'_{pa} &= \frac{i_{pa}}{i_B} \quad i'_{pb} = \frac{i_{pb}}{i_B} \quad i'_{pc} = \frac{i_{pc}}{i_B} \\
v'_{pa} &= \frac{v_{pa}}{v_B} \quad v'_{pb} = \frac{v_{pb}}{v_B} \quad v'_{pc} = \frac{v_{pc}}{v_B} \\
v'_{ia} &= \frac{v_{ia}}{v_B} \quad v'_{ib} = \frac{v_{ib}}{v_B} \quad v'_{ic} = \frac{v_{ic}}{v_B} \\
z_B &= \frac{v_B}{i_B} \quad L'_p = \frac{\omega_B L_p}{z_B} \quad R'_p = \frac{R_p}{z_B} \\
v'_{dc} &= \frac{v_{dc}}{v_B} \quad i'_{dc} = \frac{i_{dc}}{i_B} \quad C' = \frac{1}{\omega_B C z_B} \quad R'_c = \frac{R_c}{z_B}
\end{aligned} \tag{8.67}$$

Figure 8.23 shows the basic circuit of a STATCOM.

The equivalent circuit of a STATCOM is presented in Figure 8.24. The device is modeled as a voltage source ($v'_{pa}, v'_{pb}, v'_{pc}$) that is connected to the network through the inductance L_p' and the resistance R_p', which models the losses of the coupling circuit. The STATCOM current is denoted as i'_p. The DC circuit is represented by a current source (i'_{dc}) connected to the capacitor C'. The connection of resistance R'_c enables representation of semiconductor and DC-circuit losses. The link between the AC and the DC circuit is described by the switching function approximating the VSC valves switching operation.

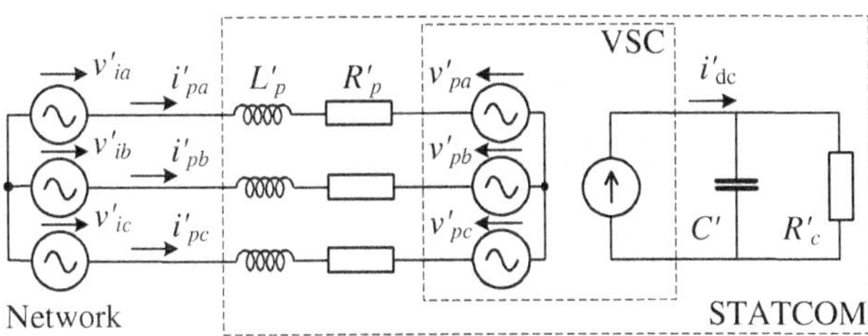

Figure 8.24 Equivalent three-phase circuit of a STATCOM.

Based on the three-phase equivalent circuit, we can write the differential equations for all three STATCOM currents:

$$\frac{d}{dt}\begin{bmatrix} i'_{pa} \\ i'_{pb} \\ i'_{pc} \end{bmatrix} = \frac{-R'_p\omega_B}{L'_p}\cdot\begin{bmatrix} i'_{pa} \\ i'_{pb} \\ i'_{pc} \end{bmatrix} + \frac{\omega_B}{L'_p}\begin{bmatrix} \left(v'_{ia} - v'_{pa}\right) \\ \left(v'_{ib} - v'_{pb}\right) \\ \left(v'_{ic} - v'_{pc}\right) \end{bmatrix} \tag{8.68}$$

For accurate modeling of the STATCOM operation in transient conditions, we must take into account also the DC-side circuit. Namely, the DC-side capacitor is usually relatively small and the DC voltage is therefore not constant. The DC-side current can be expressed as:

$$i'_{dc} = \frac{1}{\omega_B C'}\frac{dv'_{dc}}{dt} + \frac{v'_{dc}}{R'_c} \tag{8.69}$$

The connection between the VSC's AC and DC side can be described (see (8.14)) with the already described switching function:

$$\begin{bmatrix} v'_{pa} \\ v'_{pb} \\ v'_{pc} \end{bmatrix} = k_p\begin{bmatrix} S_a \\ S_b \\ S_c \end{bmatrix} v'_{dc} \tag{8.70}$$

To describe the connection between the AC and DC side, we consider the instantaneous power. Namely, the instantaneous power on the DC side of the converter is always equal to the power on the AC side:

$$v'_{dc}i'_{dc} = v'_{pa}i'_{pa} + v'_{pb}i'_{pb} + v'_{pc}i'_{pc} \tag{8.71}$$

Taking into account (8.70), we get:

$$i'_{dc} = k_p S_a i'_{pa} + k_p S_b i'_{pb} + k_p S_c i'_{pc} \tag{8.72}$$

The complete set of equations describing STATCOM operation and including the DC-side circuit is [9, 10]:

$$\frac{d}{dt}\begin{bmatrix} i'_{pa} \\ i'_{pb} \\ i'_{pc} \\ v'_{dc} \end{bmatrix} = \begin{bmatrix} \frac{-R'_p\omega_B}{L'_p} & 0 & 0 & \frac{-k_p\omega_B}{L'_p}S_a \\ 0 & \frac{-R'_p\omega_B}{L'_p} & 0 & \frac{-k_p\omega_B}{L'_p}S_b \\ 0 & 0 & \frac{-R'_p\omega_B}{L'_p} & \frac{-k_p\omega_B}{L'_p}S_c \\ k_p\omega_B C' S_a & k_p\omega_B C' S_b & k_p\omega_B C' S_c & -\frac{\omega_B C'}{R'_c} \end{bmatrix}\begin{bmatrix} i'_{pa} \\ i'_{pb} \\ i'_{pc} \\ v'_{dc} \end{bmatrix} + \frac{\omega_B}{L'_p}\begin{bmatrix} v'_{ia} \\ v'_{ib} \\ v'_{ic} \\ 0 \end{bmatrix} \tag{8.73}$$

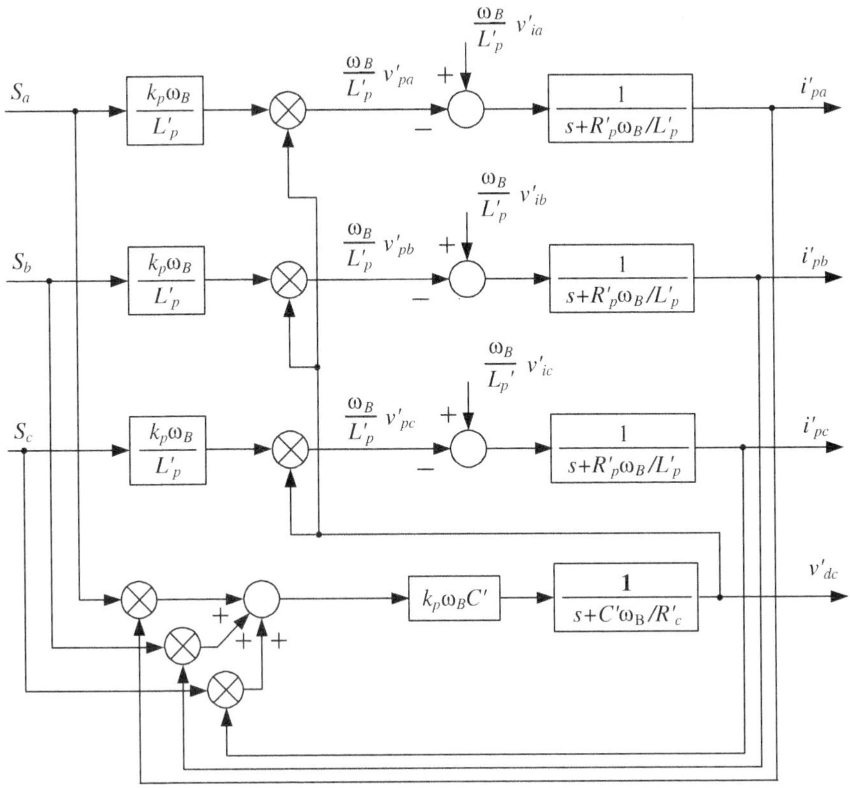

Figure 8.25 Equivalent block diagram of the three-phase STATCOM mathematical model.

Figure 8.25 shows the equivalent block diagram of the STATCOM model [9].

Figure 8.26 shows the block diagram of the STATCOM model in the rotating *d-q* frame. In the next step, the mathematical model, given in equation (8.73), is transformed into the *d-q* synchronous frame [9]:

$$\frac{d}{dt}\begin{bmatrix} i'_{pd} \\ i'_{pq} \end{bmatrix} = \begin{bmatrix} \dfrac{-R'_p\omega_B}{L'_p} & \omega \\ -\omega & \dfrac{-R'_p\omega_B}{L'_p} \end{bmatrix} \begin{bmatrix} i'_{pd} \\ i'_{pq} \end{bmatrix} + \frac{\omega_B}{L'_p}\begin{bmatrix} v'_{id} - v'_{pd} \\ v'_{iq} - v'_{pq} \end{bmatrix} \tag{8.74}$$

Connection between the AC- and DC-side is given with the following expression in the *d-q* frame:

$$\begin{bmatrix} v'_{pd} \\ v'_{pq} \end{bmatrix} = k_p \begin{bmatrix} S_d \\ S_q \end{bmatrix} v'_{dc} \tag{8.75}$$

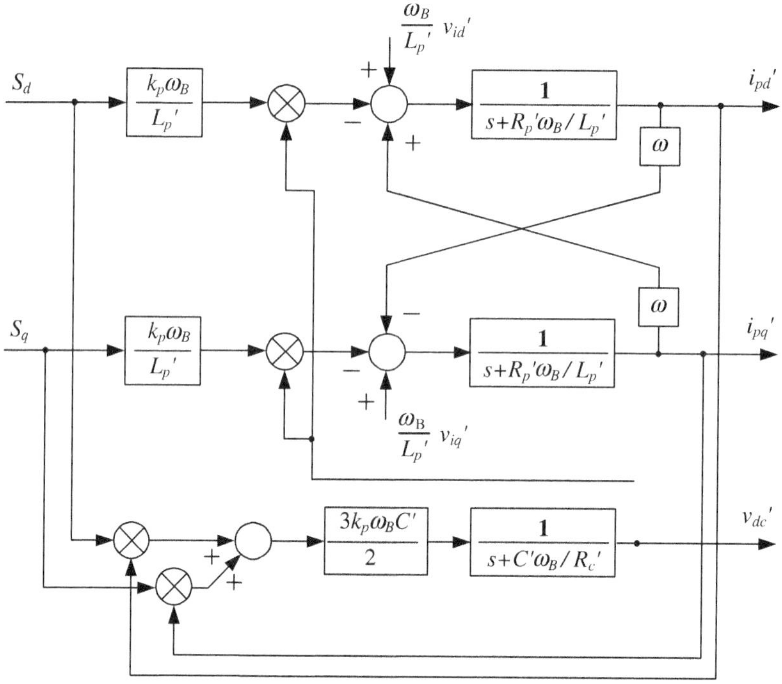

Figure 8.26 Equivalent block diagram of the STATCOM mathematical model in the rotating *d-q* frame [9].

The average switching function in *d-q* coordinates, where the effect of switching harmonics is neglected, can be written as follows:

$$\begin{bmatrix} S_d \\ S_q \end{bmatrix} = m_p \begin{bmatrix} \cos\delta \\ \sin\delta \end{bmatrix} \tag{8.76}$$

where angle δ represents the phase shift of the converter output voltage from the reference position, and m_p represents the control factor (modulation ratio) that can take any value between zero and one. The connection between the AC and the DC voltages is described by expression (8.75), where factor k_p is dependent upon the converter type and the coupling transformer ratio. The connection between AC and DC currents is given in the following equation, derived from the power balance equation written in the *d-q* coordinates:

$$i'_{dc} = \frac{3}{2}\left(k_p S_d i'_{pd} + k_p S_q i'_{pq}\right) \tag{8.77}$$

On the basis of the equivalent circuit in Figure 8.24, and taking into account (8.74) and (8.77), the mathematical model in the *d-q* coordinate system can be derived and is given in (8.78). The constant ω is the angular speed of the fundamental frequency voltage component.

Because the d-q transformation is frequency dependent, the derived mathematical model is valid only for the fundamental frequency component. For the evaluation of any other harmonic or negative-sequence component, an additional set of equations should be included into (8.78).

$$\frac{d}{dt}\begin{bmatrix} i'_{pd} \\ i'_{pq} \\ v'_{dc} \end{bmatrix} = \begin{bmatrix} \frac{-R'_p\omega_B}{L'_p} & \omega & \frac{-k_p\omega_B}{L'_p}S_d \\ -\omega & \frac{-R'_p\omega_B}{L'_p} & \frac{-k_p\omega_B}{L'_p}S_q \\ \frac{3k_p\omega_B C'}{2}S_d & \frac{3k_p\omega_B C'}{2}S_q & -\frac{\omega_B C'}{R'_c} \end{bmatrix} \begin{bmatrix} i'_{pd} \\ i'_{pq} \\ v'_{dc} \end{bmatrix} + \begin{bmatrix} \frac{\omega_B}{L'_p}v'_{id} \\ \frac{\omega_B}{L'_p}v'_{iq} \\ 0 \end{bmatrix} \tag{8.78}$$

8.3.3 Control Algorithm

The control algorithm is based on the STATCOM mathematical model in the rotating d-q frame. The control algorithm is decoupled, enabling independent control of d-axis and q-axis currents in steady-state conditions [11]. Basic working algorithm can be assessed already from the mathematical model: we have two controllable parameters (S_d and S_q) and three dependent variables (i'_{pd}, i'_{pq}, and v'_{dc}). Therefore, two variables can be controlled independently, while the third one is a dependent variable. The usual choice is to use i'_{pd} to regulate the DC-side voltage v'_{dc}, while i'_{pq} can be used according to STATCOM purpose.

We introduce first the variables w'_{pd}, w'_{pq}, and w'_{dc}:

$$\begin{bmatrix} w'_{pd} \\ w'_{pq} \\ w'_{dc} \end{bmatrix} = \begin{bmatrix} 0 & 0 & \frac{-k_p\omega_B}{L'_p}S_d \\ 0 & 0 & \frac{-k_p\omega_B}{L'_p}S_q \\ \frac{3k_p\omega_B C'}{2}S_d & \frac{3k_p\omega_B C'}{2}S_q & 0 \end{bmatrix} \begin{bmatrix} i'_{pd} \\ i'_{pq} \\ v'_{dc} \end{bmatrix} + \begin{bmatrix} \frac{\omega_B}{L'_p}v'_{id} \\ \frac{\omega_B}{L'_p}v'_{iq} \\ 0 \end{bmatrix} \tag{8.79}$$

so that the STATCOM mathematical model equation is transformed to

$$\frac{d}{dt}\begin{bmatrix} i'_{pd} \\ i'_{pq} \\ v'_{dc} \end{bmatrix} = \begin{bmatrix} \frac{-R'_p\omega_B}{L'_p} & \omega & 0 \\ -\omega & \frac{-R'_p\omega_B}{L'_p} & 0 \\ 0 & 0 & -\frac{\omega_B C'}{R'_c} \end{bmatrix} \begin{bmatrix} i'_{pd} \\ i'_{pq} \\ v'_{dc} \end{bmatrix} + \begin{bmatrix} w'_{pd} \\ w'_{pq} \\ w'_{dc} \end{bmatrix} \tag{8.80}$$

As it can be seen from (8.79), w'_{pd} and w'_{pq} represent the voltage drop on the coupling reactor and w'_{dc} is proportional to the DC-side current. Variables w'_{pd}, w'_{pq},

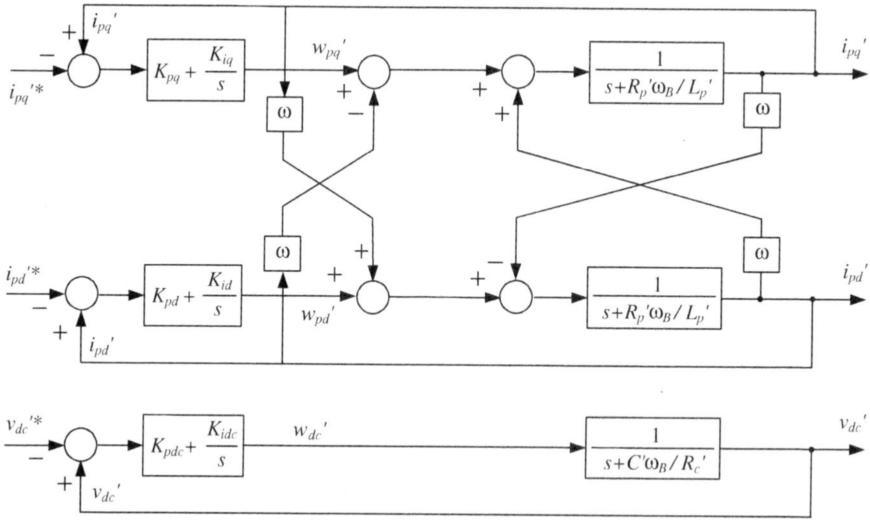

Figure 8.27 Block diagram of the STATCOM control algorithm in the rotating *d-q* frame [11].

and w'_{dc} form the output of current and voltage controllers, and decoupling between *q* and *d* axes is also made by adding/subtracting the terms $\omega \cdot i'_{pd}$ and $\omega \cdot i'_{pq}$, as shown in Figure 8.27. The *d*-axis reference current $i'_{pd}{}^{*}$ is calculated from w'_{dc} by using equation (8.80).

As can be seen in Figure 8.27, the cross-coupling terms $\omega \cdot i'_{pd}$ and $\omega \cdot i'_{pq}$ cancel out, so that the block diagram of control loop simplifies as shown in Figure 8.28.

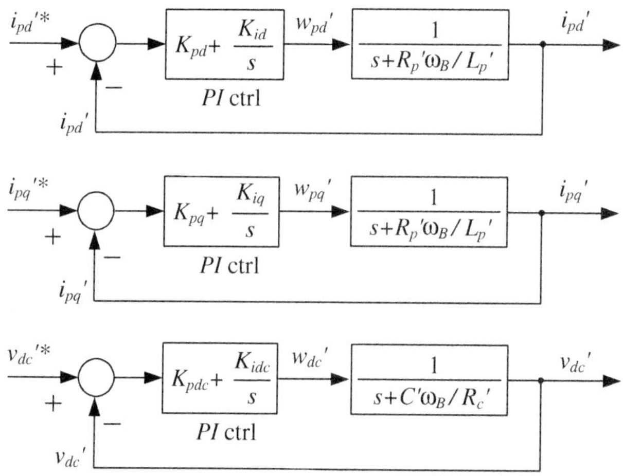

Figure 8.28 The simplified block diagram of the STATCOM control algorithm in the rotating *d-q* frame [11].

The outputs of proportional-integral (PI) controllers can be defined as:

$$w'_{pd} = \left(K_{pd} + \frac{K_{id}}{s}\right)\left(i'^{*}_{pd} - i'_{pd}\right) \tag{8.81a}$$

$$w'_{pq} = \left(K_{pq} + \frac{K_{iq}}{s}\right)\left(i'^{*}_{pq} - i'_{pq}\right) \tag{8.81b}$$

$$w'_{dc} = \left(K_{pdc} + \frac{K_{idc}}{s}\right)\left(v'^{*}_{dc} - v'_{dc}\right) \tag{8.81c}$$

and the transfer functions of the control loops are:

$$F_d(s) = \frac{K_{id} + sK_{pd}}{K_{id} + s(R'_p\omega_B/L'_p + K_{pd}) + s^2} \tag{8.82a}$$

$$F_q(s) = \frac{K_{iq} + sK_{pq}}{K_{iq} + s(R'_p\omega_B/L'_p + K_{pq}) + s^2} \tag{8.82b}$$

$$F_{vdc}(s) = \frac{K_{idc} + sK_{pdc}}{K_{idc} + s(C'\omega_B/R'_c + K_{pdc}) + s^2} \tag{8.82c}$$

If a simple relationship between the integral and proportional constants of PI controllers is used, the transfer function reduces to a first-order function, resulting in fast and stable operation:

$$K_{id} = \frac{R'_p\omega_B}{L'_p}K_{pd} \Rightarrow F_d(s) = \frac{K_{pd}}{K_{pd} + s} \tag{8.83a}$$

$$K_{iq} = \frac{R'_p\omega_B}{L'_p}K_{pq} \Rightarrow F_q(s) = \frac{K_{pq}}{K_{pq} + s} \tag{8.83b}$$

$$K_{idc} = \frac{C'\omega_B}{R'_c}K_{pdc} \Rightarrow F_{vdc}(s) = \frac{K_{pdc}}{K_{pdc} + s} \tag{8.83c}$$

8.3.4 STATCOM Model for Unbalanced Operation

The sources of unbalanced conditions can be roughly divided into two cases. The first case concerns voltage unbalances caused by unbalanced loads or by unbalanced power system faults. Consequently, large negative-sequence currents may flow between the compensator and the network. The second case concerns a compensator that is used for compensation of unbalanced power flows. This section discusses the operation of a STATCOM under unbalanced conditions. A mathematical model for unbalanced conditions is presented and a control algorithm is introduced, enabling operation STATCOM with independent control of the positive- and negative-sequence currents. The mathematical model is developed in the *d-q* reference frame and is based on the model for balanced operation [12].

Positive- and negative-sequence components of the measured currents and voltages must be computed in the *d-q* frame. An unbalanced system can be written as a sum of positive- and negative-sequence parts (phasors $\underline{X}_p$ and $\underline{X}_n$, respectively) rotating in opposite directions:

$$\underline{X}_{\alpha\beta}(t) = \underline{X}_p e^{j\omega t} + \underline{X}_n e^{-j\omega t} \tag{8.84}$$

Multiplying equation (8.84) by $e^{-j\omega t}$ and by $e^{j\omega t}$, we obtain the expressions in the positive- and negative-sequence *d-q* reference frame:

$$\underline{X}_{dq\_p}(t) = \underline{X}_p + \underline{X}_n e^{-j2\omega t} \tag{8.85}$$

$$\underline{X}_{dq\_n}(t) = \underline{X}_p e^{j2\omega t} + \underline{X}_n \tag{8.86}$$

The positive-sequence component appears as DC in (8.85), while the negative-sequence component appears as a second harmonic. The opposite holds for (8.86), where the positive-sequence component appears as a second harmonic and the negative-sequence component appears as DC. The DC components of (8.85) and (8.86) are extracted with the use of a low pass filter. The STATCOM model for the positive sequence is derived in the positive-sequence *d-q* reference frame, while the negative-sequence part is derived in the negative-sequence *d-q* reference frame.

For the control of unbalanced currents, the compensator has to generate unbalanced voltages. This is achieved with the use of an unbalanced switching function that consists of a positive- and a negative-sequence component:

$$\begin{bmatrix} S_d \\ S_q \end{bmatrix} = \begin{bmatrix} S_{d\_p} \\ S_{q\_p} \end{bmatrix} + \begin{bmatrix} S_{d\_n} \\ S_{q\_n} \end{bmatrix} \tag{8.87}$$

A set of equations is added to (8.80) describing the compensator operation with negative-sequence components. *The complete mathematical model for unbalanced conditions* is given in (8.88) and (8.89).

$$\frac{d}{dt}\begin{bmatrix} i'_{pd\_p} \\ i'_{pq\_p} \\ i'_{pd\_n} \\ i'_{pq\_n} \\ v'_{dc} \end{bmatrix} = A_{dq}\begin{bmatrix} i'_{pd\_p} \\ i'_{pq\_p} \\ i'_{pd\_n} \\ i'_{pq\_n} \\ v'_{dc} \end{bmatrix} + \begin{bmatrix} \frac{\omega_B}{L'_p} v'_{id\_p} \\ \frac{\omega_B}{L'_p} v'_{iq\_p} \\ \frac{\omega_B}{L'_p} v'_{id\_n} \\ \frac{\omega_B}{L'_p} v'_{iq\_n} \\ 0 \end{bmatrix} \tag{8.88}$$

where

$$A_{dq} = \begin{bmatrix} \frac{-R'_p\omega_B}{L'_p} & \omega & 0 & 0 & \frac{-k_p\omega_B}{L'_p}S_{d\_p} \\ -\omega & \frac{-R'_p\omega_B}{L'_p} & 0 & 0 & \frac{-k_p\omega_B}{L'_p}S_{q\_p} \\ 0 & 0 & \frac{-R'_p\omega_B}{L'_p} & \omega & \frac{-k_p\omega_B}{L'_p}S_{d\_n} \\ 0 & 0 & -\omega & \frac{-R'_p\omega_B}{L'_p} & \frac{-k_p\omega_B}{L'_p}S_{q\_n} \\ \frac{3k_p\omega_B C'}{2}S_{d\_p} & \frac{3k_p\omega_B C'}{2}S_{q\_p} & \frac{3k_p\omega_B C'}{2}S_{d\_n} & \frac{3k_p\omega_B C'}{2}S_{q\_n} & \frac{-C'\omega_B}{R'_c} \end{bmatrix} \tag{8.89}$$

The STATCOM mathematical model for unbalanced conditions includes the positive-sequence part, the negative-sequence part, and a common DC circuit. The positive- and negative-sequence parts have the same form as they are derived in the positive- and negative-sequence d-q reference frames, respectively. Positive-sequence d-axis and q-axis currents are coupled with the term ω. The same holds for the negative-sequence d- and q-axis currents. The DC component of the DC-side voltage is affected by both, the positive- and negative-sequence currents. Therefore, the DC component of the DC-side voltage should also be maintained fairly constant in transient conditions to enable good compensator dynamics. The currents and the DC-side voltage are connected with the switching function. The harmonics generated on the DC side are not included in the model. The block diagram of the presented mathematical model is shown in Figure 8.29. On the basis of the presented mathematical model, a control algorithm is developed next.

The purpose of the *control algorithm* is to enable separate control of positive- and negative-sequence currents and also decoupled control of d- and q-axis current components. The decoupling of the d- and q-axis currents (in both, the positive- and negative-sequence) enables a faster dynamic response of the converter [12].

Firstly, five new variables are introduced (w'_{pd_p}, w'_{pq_p}, w'_{dc}, w'_{pd_n}, and w'_{pq_n}):

$$\begin{bmatrix} w'_{pd\_p} \\ w'_{pq\_p} \\ w'_{pd\_n} \\ w'_{pq\_n} \\ w'_{dc} \end{bmatrix} = B_{dqR} \begin{bmatrix} i'_{pd\_p} \\ i'_{pq\_p} \\ i'_{pd\_n} \\ i'_{pq\_n} \\ v'_{dc} \end{bmatrix} + \begin{bmatrix} \frac{\omega_B}{L'_p}v'_{id\_p} \\ \frac{\omega_B}{L'_p}v'_{iq\_p} \\ \frac{\omega_B}{L'_p}v'_{id\_n} \\ \frac{\omega_B}{L'_p}v'_{iq\_n} \\ 0 \end{bmatrix} \tag{8.90}$$

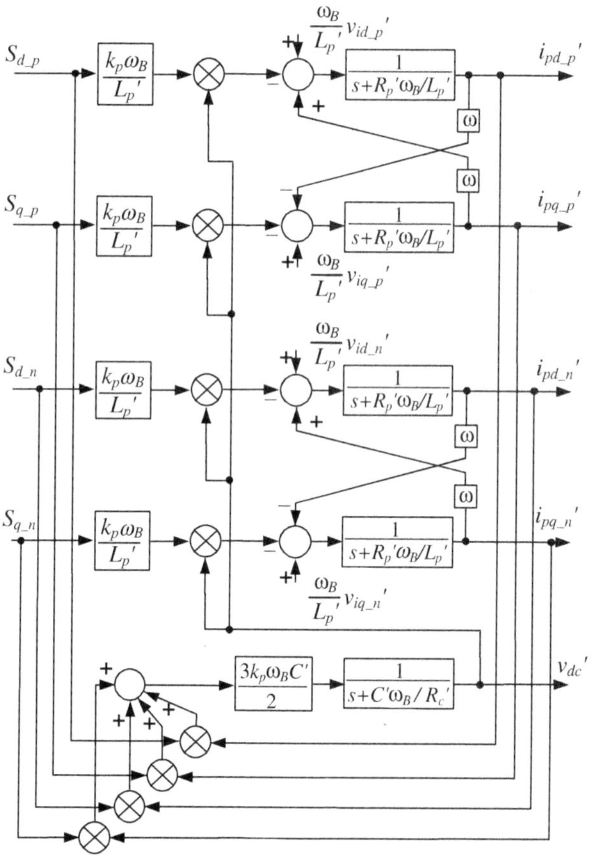

Figure 8.29 Block diagram of the D-STATCOM (distribution-STATCOM) mathematical model for unbalanced conditions [12].

$$B_{dqR} = \begin{bmatrix} 0 & 0 & 0 & 0 & \frac{-k_p\omega_B}{L_p'}S_{d\_p} \\ 0 & 0 & 0 & 0 & \frac{-k_p\omega_B}{L_p'}S_{q\_p} \\ 0 & 0 & 0 & 0 & \frac{-k_p\omega_B}{L_p'}S_{d\_n} \\ 0 & 0 & 0 & 0 & \frac{-k_p\omega_B}{L_p'}S_{q\_n} \\ \frac{3k_p\omega_B C'}{2}S_{d\_p} & \frac{3k_p\omega_B C'}{2}S_{q\_p} & \frac{3k_p\omega_B C'}{2}S_{d\_n} & \frac{3k_p\omega_B C'}{2}S_{q\_n} & 0 \end{bmatrix} \tag{8.91}$$

so that the STATCOM mathematical model equation is transformed to

$$\frac{d}{dt}\begin{bmatrix} i'_{pd\_p} \\ i'_{pq\_p} \\ i'_{pd\_n} \\ i'_{pq\_n} \\ v'_{dc} \end{bmatrix} = \begin{bmatrix} \frac{-R'_p\omega_B}{L'_p} & \omega & 0 & 0 & 0 \\ -\omega & \frac{-R'_p\omega_B}{L'_p} & 0 & 0 & 0 \\ 0 & 0 & \frac{-R'_p\omega_B}{L'_p} & \omega & 0 \\ 0 & 0 & -\omega & \frac{-R'_p\omega_B}{L'_p} & 0 \\ 0 & 0 & 0 & 0 & \frac{-C'\omega_B}{R'_c} \end{bmatrix} \begin{bmatrix} i'_{pd\_p} \\ i'_{pq\_p} \\ i'_{pd\_n} \\ i'_{pq\_n} \\ v'_{dc} \end{bmatrix} + \begin{bmatrix} w'_{pd\_p} \\ w'_{pq\_p} \\ w'_{pd\_n} \\ w'_{pq\_n} \\ w'_{dc} \end{bmatrix} \quad (8.92)$$

The newly introduced variables represent the output of a general PI controller. With the known values of w'_{pd_p}, w'_{pq_p}, w'_{dc}, w'_{pd_n}, and w'_{pq_n}, all four adjustable parameters (S_{d_p}, S_{q_p}, S_{d_n}, and S_{q_n}) can be computed from (8.90). With the introduction of new variables, decoupling is also achieved. The STATCOM mathematical model is reduced to five first-order functions, improving the converter performance.

The developed control scheme is shown in Figure 8.30. The controlled signals (positive- and negative-sequence currents, DC voltage) are first compared to the reference ones (marked with an asterisk). The difference is then applied to a PI controller. The output of each controller is a corresponding variable v'_{pac}. With the solution of (8.90), the values of adjustable parameters (S_{d_p}, S_{q_p}, S_{d_n}, and S_{q_n}) can be calculated. The two adjustable parameters for the positive sequence (S_{d_p} and S_{q_p}) are calculated with the known values of w'_{pd_p} and w'_{pq_p}. The first parameter is used to control positive-sequence active current (i'_{pd_p}) and the second one is used to control the positive-sequence reactive current (i'_{pq_p}). Furthermore, with the known values of w'_{pd_n} and w'_{pq_n}, the adjustable parameters for the negative sequence (S_{d_n} and S_{q_n}) can be calculated. One is used for negative-sequence active current (i'_{pd_n}) control and the other is used for negative-sequence reactive current (i'_{pq_n}) control. From the

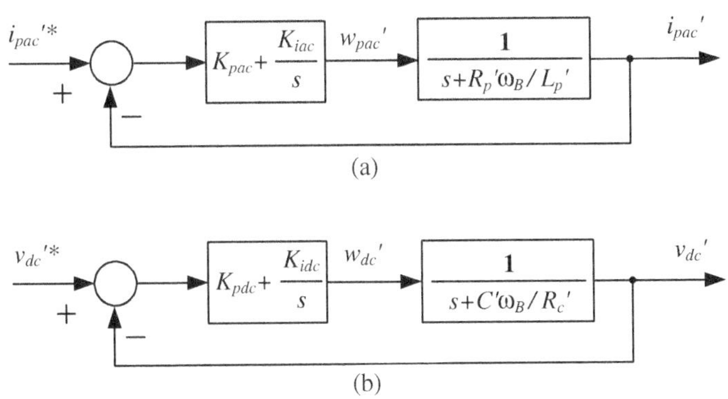

Figure 8.30 (a) Current control loop and (b) DC-voltage control loop [12].

output of the DC-voltage controller (w'_{dc}), the reference positive-sequence active current ($i'_{pd\_p}*$) is calculated so that the DC component of the capacitor voltage is maintained at constant value. To sum up, with four adjustable parameters, all four currents can be controlled. With the DC control loop variable (w'_{dc}), the reference value of the positive-sequence active current is calculated.

With the described control algorithm, all four currents (*d*- and *q*-axis currents of the positive- and negative-sequence) are decoupled. However, as all the currents share a common DC bus, a change in the DC component of the DC-side voltage affects all the currents. Therefore, the positive-sequence active current control loop has to be made fast enough to keep the DC voltage at a constant level.

8.4 STATCOM APPLICATIONS

STATCOMs may be used in power systems to solve a broad range of problems.

8.4.1 Fast Voltage Control and Maintaining Voltage Levels of a Bus or an Area

As already described, STATCOM's advantage, as compared to fixed parallel compensation or SVC, are its response speed and its ability to maintain a constant reactive current in a wide range of system voltages. These make it a preferred candidate for fast voltage changes suppression (flickers, arc furnace compensator, etc.). Of course, it is also an effective tool for system voltage control in systems subjected to frequent voltage variations. Typical examples are weak systems (small surge power) with load and/or generation variations (distributed generation) or variations of other network components. In such a situation, the voltage can drop drastically and low voltage levels can lead to system voltage instability.

8.4.2 Flicker Compensation

Flicker is defined as a visual impression of fluctuating light intensity or spectrum. It is caused by variations of voltage amplitude, where changes with frequencies within 0.5 and 25 Hz are of special concern. Voltage variations are the consequence of strongly fluctuating power consumption.

One of the main sources of flicker in power networks is arc furnaces used for scrap melting. They consume stochastically variable active and especially reactive power. Figure 8.31 shows the measured steelwork voltages and measured currents of a 36 MVA arc furnace [13]. The following signals are displayed:

- phase *c* voltage at the 110 kV bus (U_110_L3);
- phase *c* voltage at the 20 kV bus (U_20_L3);
- phase *c* arc furnace current (I_EAF_L3).

As it can be seen from the measured voltages and currents, the arc furnace operation causes voltage fluctuations, which are the cause of flicker.

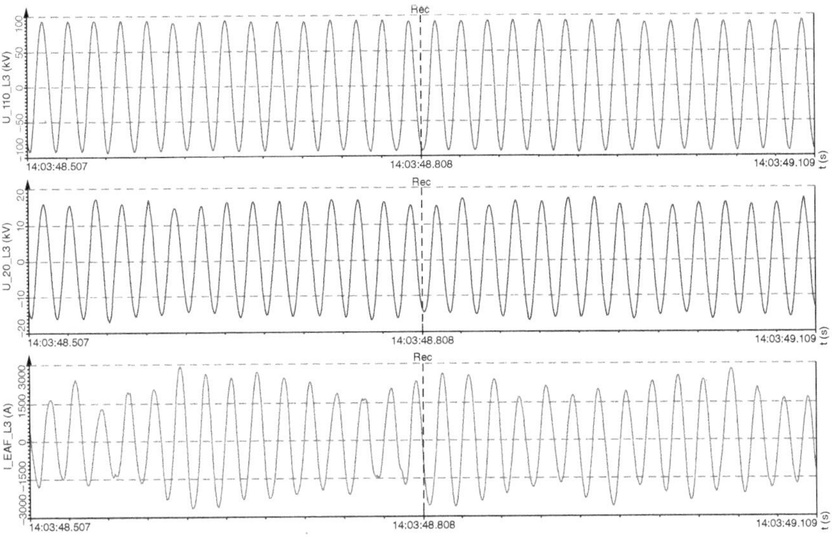

Figure 8.31 Measured arc furnace currents and steelwork voltages [13].

The installation of a STATCOM in parallel with the arc furnace is a straightforward solution of the flicker problem. The compensator can reduce reactive power fluctuations between the network and the arc furnace. A more stable voltage at the point of common coupling also enables a more efficient arc furnace operation.

The STATCOM circuit for flicker compensation is shown in Figure 8.32.

The STATCOM may supply either inductive or capacitive reactive power. Fixed capacitors are connected in parallel with STATCOM and provide constant reactive power. They are tuned for different harmonic frequencies and must prevent harmonic amplification due to resonance in the power system. The compensator is connected in parallel with the arc furnace at the MV level. The STATCOM was rated at 22.5 MVA and fixed capacitors are tuned at 5100, 100, and 141 Hz, and have a total rated power 22.5 MVAr. The STATCOM has therefore a control range of 0–45 MVAr.

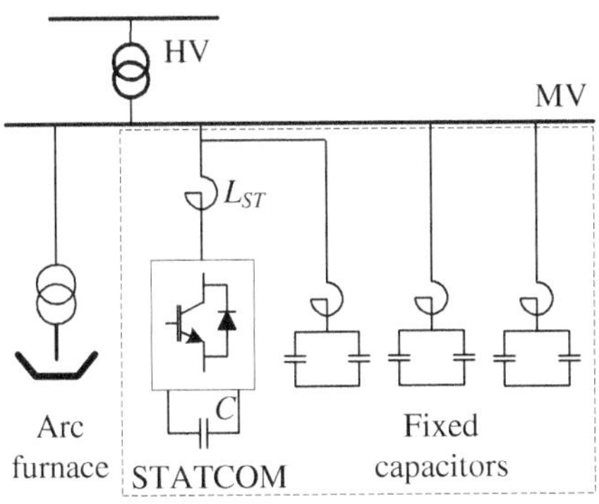

Figure 8.32 Single-line diagram of STATCOM connection [13].

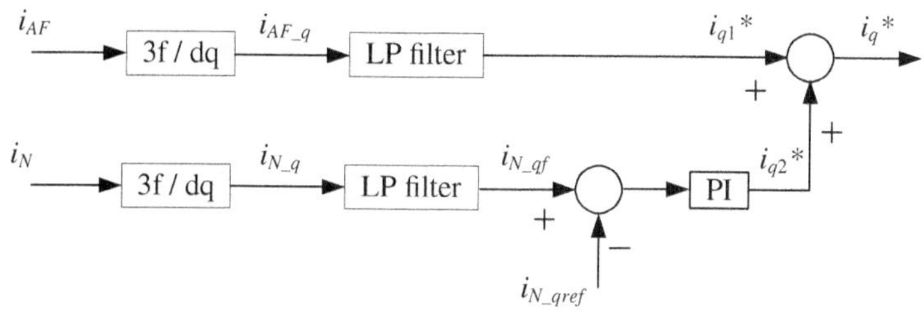

Figure 8.33 Calculation of reference currents [13].

Figure 8.33 shows a simplified diagram for the definition of STATCOM reference current. The controller is composed of an open loop, where the arc furnace current (i_{AF}) is measured. The reactive component of the current (i_{AF_q}) is filtered through a low pass filter and the output represents one part of the reference current. The second part of the reference current is derived from a closed loop controller, where network currents (i_N) are measured. The reactive component of the currents (i_{N_q}) is compared to a reference value and the difference forms the input of a PI controller.

STATCOM simulation results are shown in Figure 8.34. The graphs show the active and reactive powers absorbed from the network, the load active and reactive power, and the compensator active and reactive power. At the instant of STATCOM operation initiation, the reactive power from the network is almost completely compensated also resulting in less voltage fluctuation. Based on simulation results and field experiences, it can be concluded that a STATCOM enables maximum flicker

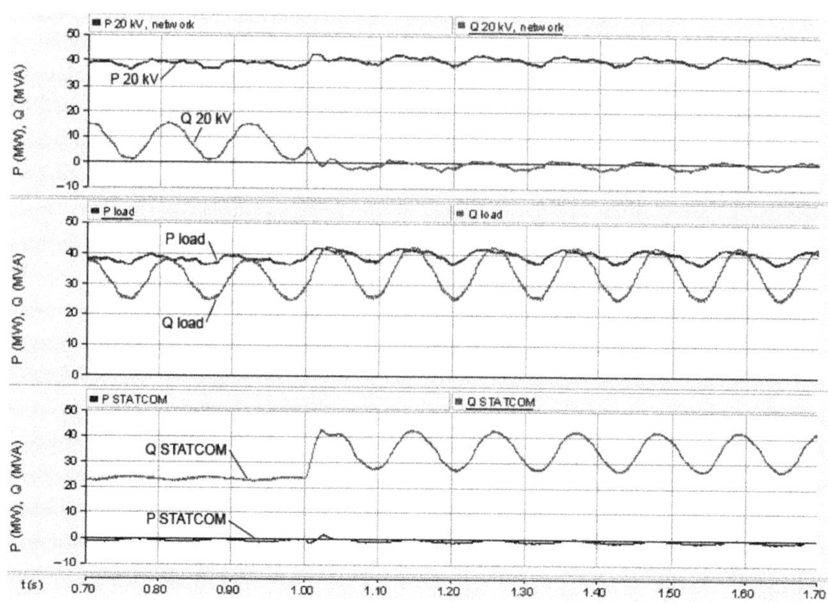

Figure 8.34 Simulation results for flicker compensation with STATCOM [13].

reduction of approximately 75%. In the presented case, the flicker P_{lt} at 110 kV was reduced from 3.0 to approximately 0.8.

8.4.3 Improvement of the Network Transmission Capability

From a system point of view, STATCOM may be described as a parallel-connected current source because in the major part of its operating area its current is independent of the terminal voltage magnitude $V_{STATCOM}$ (Figure 8.35). The STATCOM controllable parameter may therefore be its current magnitude I_Q. When losses are neglected, $\underline{I}_Q$ represents the STATCOM reactive current phasor, which is perpendicular on the terminal voltage phasor $\underline{V}_{STATCOM}$ [2].

According to Kirchhoff's laws, the following three equations can be written based on Figure 8.35a:

$$\underline{V}_{STATCOM} = \underline{V}_1 - j\underline{I}_1 X_1 \tag{8.93}$$

$$\underline{I}_2 = \frac{\underline{V}_{STATCOM} - \underline{V}_2}{jX_2} \tag{8.94a}$$

$$\underline{I}_2 = \underline{I}_1 - \underline{I}_Q \tag{8.94b}$$

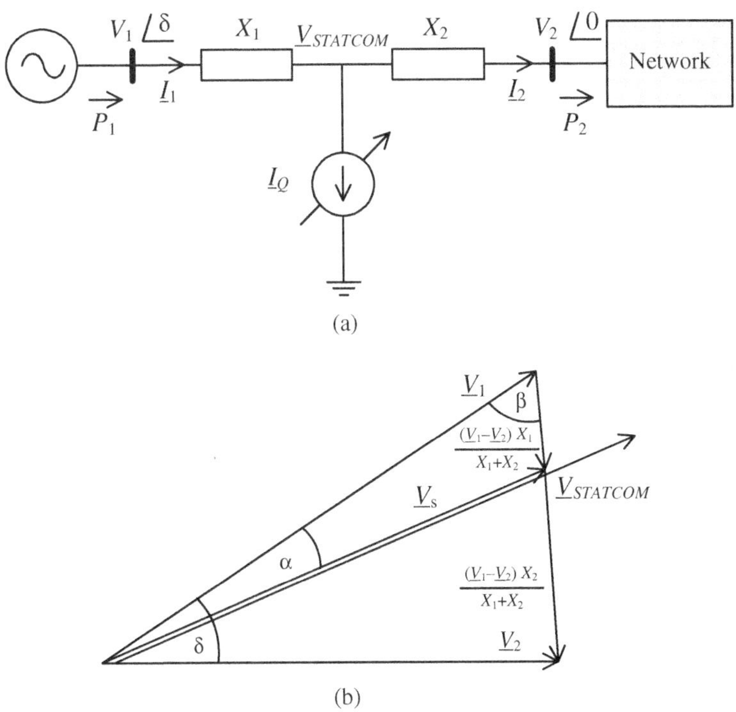

Figure 8.35 Model of the transmission system with STATCOM: (a) network scheme and (b) corresponding phasor diagram.

By equating right-hand terms of the (8.94a) and (8.94b) and by considering (8.93), the current $\underline{I}_1$ can be easily calculated as:

$$\underline{I}_1 = \frac{\underline{V}_1 - \underline{V}_2}{j(X_1 + X_2)} + \underline{I}_Q \frac{X_2}{(X_1 + X_2)} \tag{8.95}$$

Replacing (8.95) in (8.93) and inserting $\underline{V}_S$, we obtain:

$$\underline{V}_{STATCOM} = \underline{V}_1 - \frac{(\underline{V}_1 - \underline{V}_2)X_1}{(X_1 + X_2)} - j\underline{I}_Q \frac{X_1 X_2}{(X_1 + X_2)} = \underline{V}_S - j\underline{I}_Q \frac{X_1 X_2}{(X_1 + X_2)} \tag{8.96}$$

As shown in equation (8.96), the new variable $\underline{V}_S$ has been introduced. It is in fact the STATCOM terminal voltage if STATCOM is out of operation (is not connected to the system, i.e., $\underline{I}_Q = 0$).

Taking into consideration that:

$$\underline{I}_Q = j\,I_Q \frac{\underline{V}_S}{V_S} \tag{8.97}$$

equation (8.96) can be rewritten as follows:

$$\underline{V}_{STATCOM} = \underline{V}_S + I_Q \frac{\underline{V}_S}{V_S} \frac{X_1 X_2}{(X_1 + X_2)} = \underline{V}_S \left(1 + \frac{I_Q}{V_S} \frac{X_1 X_2}{(X_1 + X_2)}\right) \tag{8.98}$$

Equation (8.97) describes the fact that $\underline{I}_Q$ is shifted by 90° with respect to $\underline{V}_S$ ("$\underline{V}_S$ / V_S" being unity phasor in the $\underline{V}_S$ direction). This is proved as follows. Let us consider that STATCOM is not represented by reactive current source but with its operating point reactance $X_{STATCOM}$. Then, in equation (8.96), $\underline{I}_Q$ should be substituted by the term $\underline{V}_{STATCOM}/(jX_{STATCOM})$. Now (8.96) can be transformed into the form $\underline{V}_{STATCOM}\, K = \underline{V}_S$. Thus, K is a scalar quantity, $\underline{V}_{\mathrm{STATCOM}}$ and $\underline{V}_S$ have the same phase angle, and $\underline{I}_Q$ is perpendicular on both the terminal voltage phasor $\underline{V}_{STATCOM}$ and $\underline{V}_S$.

The corresponding voltage phasor diagram is presented in Figure 8.35b. According to this diagram, applying the sine law, the following two equations can be written [2]:

$$\frac{\sin(\beta)}{V_2} = \frac{\sin(\delta)}{V_1 - V_2} \tag{8.99a}$$

$$\frac{\sin(\alpha)}{(V_1 - V_2)\,\frac{X_1}{(X_1+X_2)}} = \frac{\sin(\beta)}{V_S} \tag{8.99b}$$

from which the sine of α can be calculated:

$$\sin(\alpha) = \frac{V_2 \sin(\delta) X_1}{V_S(X_1 + X_2)} \tag{8.100}$$

The transmitted power P_1 can (considering the equations for transmitted power via reactance X_1 (8.101) and (8.100)) be calculated from:

$$P_1 = P_2 = P = \frac{V_{STATCOM} V_1}{X_1} \sin(\alpha) = \frac{V_1 V_2 \sin(\delta)}{(X_1 + X_2)} \frac{V_{STATCOM}}{V_S} \tag{8.101}$$

Considering (8.98), equation (8.101) can be rewritten as:

$$P = \frac{V_1 V_2 \sin(\delta)}{(X_1 + X_2)} \frac{\left|\underline{V}_S \left(1 + \frac{I_Q}{V_S} \frac{X_1 X_2}{(X_1+X_2)}\right)\right|}{V_S} = \frac{V_1 V_2 \sin(\delta)}{(X_1 + X_2)} \left(1 + \frac{I_Q}{V_S} \frac{X_1 X_2}{(X_1 + X_2)}\right) \tag{8.102}$$

If $\underline{V}_2$ is the reference phasor, therefore $\underline{V}_2 = V_2$ and $\underline{V}_1 = V_1 \, e^{j\delta}$, the expression of the amplitude of the variable $\underline{V}_S$ is:

$$V_S = |\underline{V}_S| = \left|\frac{\underline{V}_1 X_2 + V_2 X_1}{(X_1 + X_2)}\right| = \frac{\sqrt{V_1^2 X_2^2 + V_2^2 X_1^2 + 2\,V_1\,V_2 X_1\,X_2 \cos(\delta)}}{(X_1 + X_2)} \tag{8.103}$$

Finally, transmission characteristic is described by the following equation:

$$P_1 = P_2 = P = \frac{V_1 V_2 \sin(\delta)}{(X_1 + X_2)} \left(1 + \frac{I_Q X_1 X_2}{\sqrt{V_1^2 X_2^2 + V_2^2 X_1^2 + 2\,V_1\,V_2 X_1\,X_2 \cos(\delta)}}\right) \tag{8.104}$$

From equation (8.104), it is obvious that enhancement of transmission capacity is in direct proportion to STATCOM current.

The question regarding the location for achieving the maximum STATCOM effect in the system can be answered by deriving the power transfer characteristic by the ratio of X_1 to X_2, (X_{RATIO}). Then, the derivative of this expression is equaled to 0 and finally X_{RATIO} is calculated. The expressions are quite comprehensive. It should only be noted that X_{RATIO} is dependent on system terminal voltages V_1 and V_2 as well as on transmission angle δ. In case the terminal voltages are equal, $V_1 = V_2$, X_{RATIO} equals to 1 and thus in this case, *the most effective STATCOM location is the electrical middle of the system* [2].

The set of transmission characteristics achieved when STATCOM is located in the middle of the system is shown in Figure 8.36a. Note that some kind of dualism with static synchronous series compensator (SSSC) appears. The form of transmission characteristics is similar to the SSSC case. However, the STATCOM

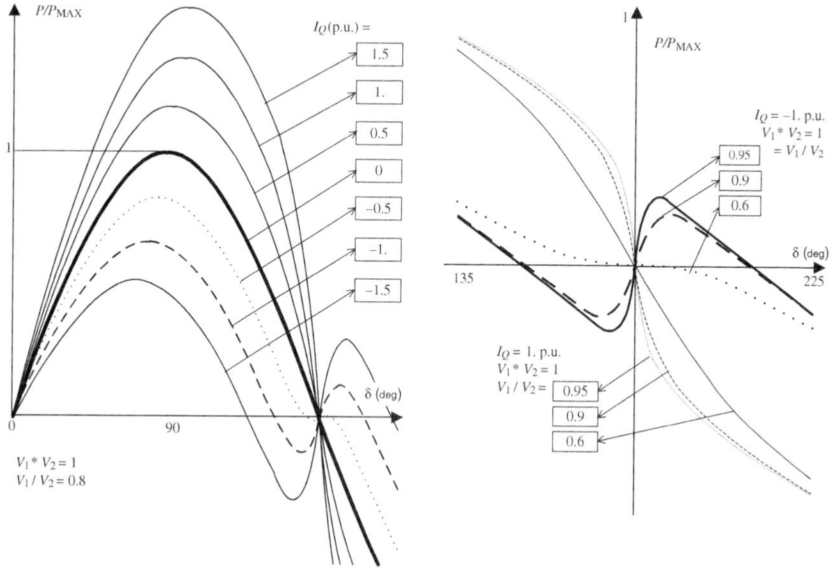

(a) (b)

Figure 8.36 Model of the transmission system characteristics with STATCOM: (a) transmission characteristics and (b) impact of the system terminal voltage ratio on transmission characteristics [2].

characteristics are "mirror" picture of those in the SSSC case. Also here, the ratio between the system terminal voltage magnitudes impacts the transmission characteristics (especially in the area of transmission angles around 180°), although their product remains constant. Let us assume $V_1 / V_2 = V_{ratio}$, $V_1 \times V_2 = 1$, $V_1 = \sqrt{V_{ratio}}$, and $V_2 = \sqrt{V_{ratio}}$. Variation of the transmission characteristic in terms of V_{ratio} is illustrated in Figure 8.36b.

8.4.4 Improvement of Static and Transient Stability Reserve

Although basic stability considerations are out of scope of this book, for reader's convenience let us summarize shortly basic features.

Transient stability (also referred to as a *first swing stability*—although multi-machine system may become transient instable in second or even later swing) may be defined as an ability of the EPS to remain in synchronism after being subjected to a mayor system disturbance such as a short circuit on transmission facilities, loss of generating unit, or large load. The reaction of an EPS after such kind of a disturbance is characterized by large deviations in machine rotor relative angular positions, power flows, system voltage profile, etc. A generator or group of generators are transient unstable when angular separation between machines occurs. If they remain in certain bounds, an EPS keeps the synchronism and is consequently stable. It should be noted that the rotor angular speed or velocity cannot be a criterion for stability (instability), thus angular speed or velocity of system rotating masses may change more or less

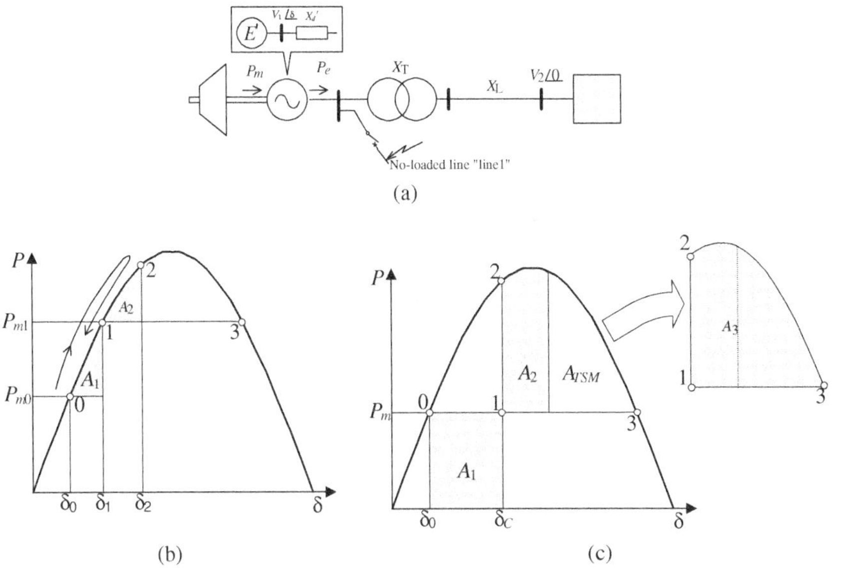

Figure 8.37 Generator-infinite bus system: (a) network configuration, (b) related transmission characteristic, and (c) illustration of the transient stability.

simultaneously (system frequency deviations). In this case, system may in fact be stable, although such operating conditions may not be tolerated.

In numerous references, the basic machine swing equations are derived that lead to the so-called *equal area criterion* [14]. Let us assume that a generator is connected to the infinite bus via a transformer and a transmission line, as shown in Figure 8.37a. Let the infinite bus voltage $\underline{V}_2$ be the reference phasor. The angle difference between the generator transient voltage $\underline{E}'$ and the reference phasor is referred to as δ and the sum of system reactances $(X_d' + X_T + X_L)$ to as X. The transmission characteristic is presented in Figure 8.37b.

Let us examine the behavior of the system if the mechanical power is stepwise changed from P_{m0} to P_{m1}. In this case, the system operating point follows the transient transmission characteristic from the old equilibrium point (point 0 on the characteristic) to the new equilibrium point (point 1). However, a rotor cannot be pushed to a new point instantaneously and thus to move from 0 to 1. The mechanical power accelerates the rotating masses because there is a surplus of mechanical power over the electrical one. This excess results in enlarged kinetic energy of rotating masses. Therefore, the angle δ advances until the excess of kinetic energy is no longer transformed into potential energy in the point 2.

From point 1 to 2, generator is decelerating because there is an excess of electric power over the mechanic power P_{m1}. In point 2, the relative rotor movement direction is turned over and rotor swings back. Without any damping, rotor would swing around the point 1 between the points 0 and 2. It is obvious that the point 3 of a transmission characteristic is the point of no return. By passing over the point 3, the electrical power becomes higher than the mechanical power and the rotating masses are accelerated again and the system becomes unstable. The mathematical basis for equal area criterion is [14]:

$$\int_{\delta_0}^{\delta_1} (P_m - P_e)d\delta = \int_{\delta_1}^{\delta_2} (P_m - P_e)d\delta \qquad (8.105)$$

Now, let us assume a three-phase fault occurring on the transmission line near the generator bus (see Figure 8.37a) at the instant t_0 ($\delta = \delta_0$), which is cleared after a certain time t_C ($\delta = \delta_C$). Let us consider that during the fault, the generator electrical power is zero. The transmission characteristics for such operation are presented in Figure 8.37c. It is obvious that according to previous considerations, the system is stable as long as the area A_1 is smaller than the area between points 1-2-3 (area A_3). The difference between the areas A_3 and A_2 can be denoted as the transient stability margin A_{TSM}. As soon as the stability margin becomes negative, the system becomes unstable.

In other words, enhancing the transmission characteristic as high as possible after fault clearing (in this way, maximizing area A_3 in Figure 8.37c) is optimal behavior. Therefore, it can be concluded that after fault clearing, a STATCOM should be set to its capacitive limit.

Simulations, carried out on the longitudinal system, similar to the one presented in Figure 8.37, have shown that the rating of the STATCOM device necessary to achieve stability is lower than it would be necessary when employing an SVC. The effectiveness of a STATCOM depends, among others, on the fault duration. The STATCOM's required rating is 25% lower than the SVC's rating for slightly higher fault durations, decreasing down to 2% for larger fault durations. However, the larger the fault duration, the greater (unrealistically) the FACTS device rating is required.

The specific quantities of the simulation and the dynamics of the transmitted power in terms of the transmission angle are presented in Figure 8.38.

8.4.5 Oscillations Damping

In large systems, where large amounts of power are transmitted over long distances, the power transfer limiting factor may be a problem relating to dynamic phenomena, especially when the outages are taken into consideration. Poor damping of the power oscillations represents a problem for such systems. If this is the limiting factor to power transfer, it can be shifted higher by enhancement of the system damping. The power system stabilizers (PSSs) at generators can only partly increase the inter-area oscillation damping mode; therefore, fast devices are required, which have the capability of fast line power flow control. Thus, the FACTS devices offer effective solutions.

In general, the modulating power, in order to ensure effective damping, is relatively small as compared to the oscillating power. Therefore, besides the series controllable devices, the parallel ones (SVC and STATCOM) may also be effective tools for improving the power oscillations damping. In the actual conditions, for damping purposes, the existing devices may be used. Of course, appropriate signals and power

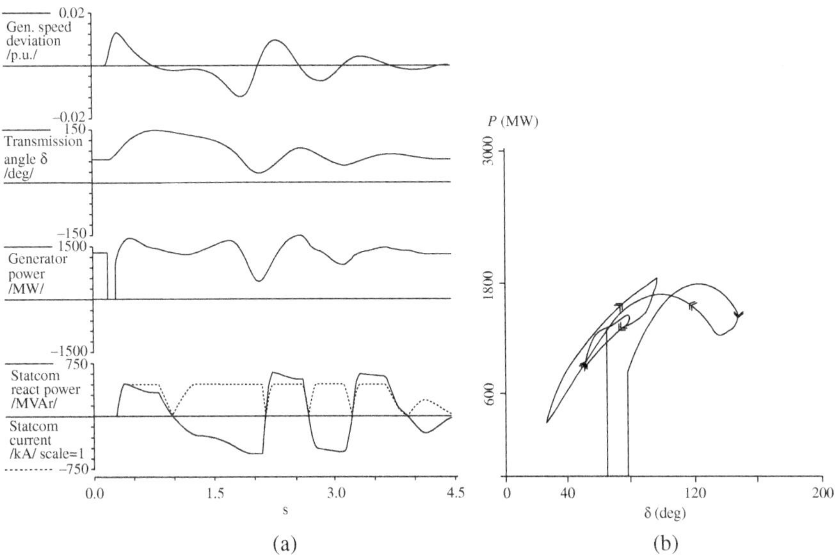

Figure 8.38 Simulation of transient stability margin enhancement applying STATCOM: (a) oscillograms and (b) dynamic P–δ behavior.

oscillation mode controls have to be included into device's controllers. In fact, the implementation of a STATCOM as a power oscillation damping device resembles PSS action. A PSS represents a separate channel in the generator's voltage control and modulates reactive power output. The same does a STATCOM.

Simulations, carried out on the longitudinal system, have shown that similar damping effect of both a STATCOM and an SVC is achieved, but with for a rating of the STATCOM ca. 10% lower than the SVC's rating. The reason for better results of a STATCOM versus an SVC may reside in the fact that STATCOM behaves better at low voltages, thus being capable of providing more reactive power. As the power oscillations issues arise in general in heavy loaded systems, low voltages may occur in dynamic conditions, case for which the STATCOM device is preferred.

8.5 STATCOM INSTALLATIONS IN OPERATION

8.5.1 ± 80 MVAr STATCOM in Japan

A prototype STATCOM installation rated ±80 MVAr was installed for service in Japan in 1991. The main converter circuit configuration is given in Figure 8.39. This STATCOM uses eight VSCs, each of 10 MVA rating, connected to a main STATCOM transformer via eight converter transformers producing 7.5° phase angle displacement from each other, resulting in 48-pulse operation.

The control system incorporates power system voltage control, power oscillation damping, and constant reactive power output control. The control system varies

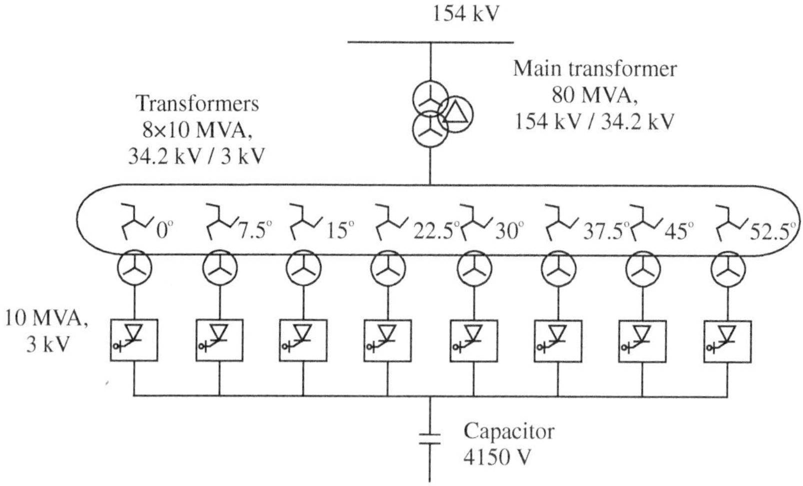

Figure 8.39 ±80 MVAr STATCOM in Japan [15, 16].

the width of the rectangular output of each converter to achieve voltage magnitude control and to ensure low losses.

A gapped-core design is used for the eight phase-displacement transformers to reduce the effects of DC magnetization, decrease magnetic impedance, and improve the uniformity of voltage sharing between windings. This is especially important when the STATCOM is energized from the power system during start-up sequences or following system faults. Initially, the converter start-up system used a relatively large, separate, "start-up converter" to supply DC voltage to the STATCOM main converter. This method was found to be slow and a new system now allows the STATCOM converter to start immediately after the energization of the STATCOM transformer and the eight converter transformers from the power system.

8.5.2 ± 100 MVAr STATCOM at Sullivan, in United States

A ±100 MVAr STATCOM designed by EPRI and Westinghouse was put to use in 1995 at the 500 kV/161 kV Sullivan substation, which belongs to the Tennessee Valley Authority (TVA), to control the voltage and damp the power oscillations. The basic elements of the STATCOM device from Sullivan substation are presented in Figure 8.40 [14, 5]. This station energizes seven electric power distributors and a large industrial consumer.

Eight VSCs of 12.5 MVA each are connected to the 161 kV line through a main transformer and through "magnetic interfaces," which include phase-shift transformers to obtain a 48-pulse functioning.

The STATCOM's adjustment domain of the reactive power and the 84 MVAr capacitor banks with mechanic command are used to regulate the 161 kV node's voltage and to minimize the maneuver of the 500 kV/161 kV transformer's contact plates. Separately from the voltage control in the steady state at the 161 kV node,

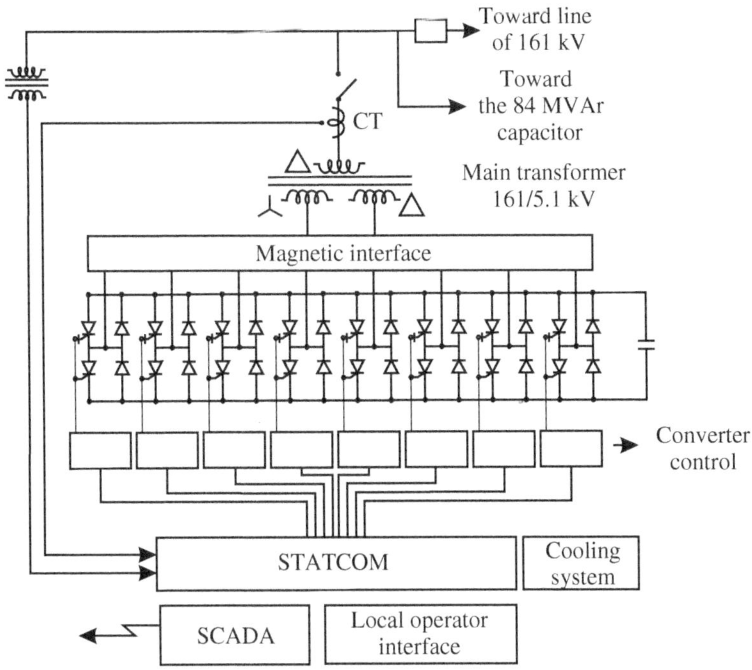

Figure 8.40 The ± 100 MVAr STATCOM structure at Sullivan [4].

the STATCOM can help to prevent the voltage failures in contingency conditions, for example, the loss of a 500 kV line.

The current operations of the STATCOM consist of the voltage regulation at the 161 kV line depending on the daily demands, thereby the contact plates of the transformers are used as little as possible. Most of the breakdowns at the transformers in the TVA's stations were at their contact plates. Thus, using the STATCOM to regulate the voltage and not the transformers' contact plates, the ensured life span of the Sullivan substation transformers will be significantly raised.

Along with the STATCOM, a switch for the 84 MVAr capacitor bank, which is also controlled through the STATCOM's control system, was installed at the Sullivan substation. This allows a fast control of the voltage on a large scale of action. If a malfunction at the main transformer at the Sullivan substation occurs, for example in the winter time, during the peak-load the voltage on the 161 kV line may drop from 10% to 15%. It is expected that STATCOM will provide a fast response and will maintain the voltage at an acceptable level until the capacitor banks at the other stations are being linked to the line.

The necessary area needed for the location of a STATCOM is considerably smaller than the area needed for a conventional SVC because the reactive power generated by the electronic components is very substantial than the one generated by the capacitors or the reactors.

Figure 8.41 shows the image of the STATCOM valves. Each valve has an electronic control system and gate circuits situated in the proximity, which communicate

Figure 8.41 An image of the electronic valves at Sullivan station. *Source*: Song et al. 1999 [3]. Reproduced with permission of IEEE.

with the central control system through a connection system made from fiber optics. The central control system receives feedback signals from the voltage transformers connected to the 161 kV line and from the current transformers linked to the main transformer.

The basic parameters of the converter in the STATCOM's composition are presented in Table 8.2 [3].

The local operator uses a graphic display terminal, but the STATCOM is generally controlled with the (SCADA) Supervisory Control and Data Acquisition system. The power losses in the electronic valves generate heat that is removed with a cooling

TABLE 8.2 The parameters of the STATCOM at Sullivan [3]

Device rating	± 100 MVAr
Transmission line voltage	161 kV
Converter rated output voltage	5.1 kV
Rated DC voltage	6.6 kV
Number of basic six-pulses three-phase converter	8
Type of converter poles	Two levels (two valves)
Total number of converter poles	24
Number of converter pulses	48
Number of series GTO in each valve	5 (one is supplemental)
Total number of GTOs	240
GTO rating	4.5 kV/4 kA (peak)

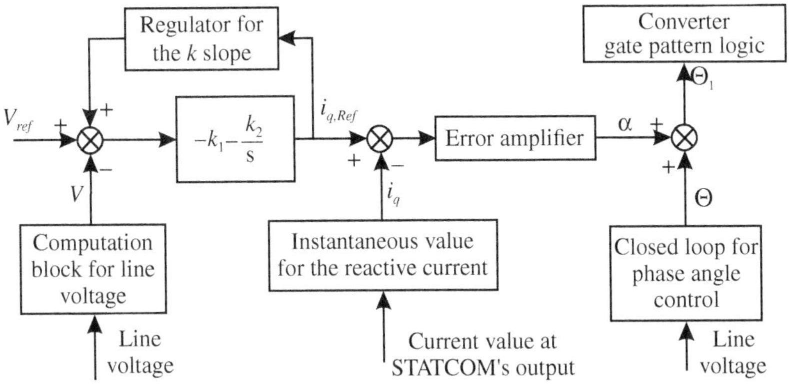

Figure 8.42 The operating block diagram of the control system of the STATCOM [3].

fluid (deionized water and a glycol mixture), which circulates through the valves and then through a heat exchanger outside the building.

The STATCOM device used in the Sullivan substation utilizes a converter that neutralizes the harmonics. This converter was designed to respond to the demands regarding the quality of energy from the harmonics point of view without using filters.

The control system of the STATCOM device is situated in a separate chamber inside the main building and consists of two distinct subsystems, which carry out the analysis of the installation and, respectively, the real time control. The first subsystem is responsible for collection and analysis of data from all the STATCOM devices and the display of the system's status; also, this subsystem accomplishes the tasks regarding the starting up, the switching-off, or the connection to the line of the STATCOM.

The real-time control system is responsible for the command of the valves, thus the STATCOM would maintain a required level of the reactive power to regulate the voltage on the 161 kV line at a reference value. The block diagram of the utilized algorithm by the control subsystem is displayed in Figure 8.42.

An interior loop with reversed reaction is used to regulate the instantaneous value of the reactive current of STATCOM. The real-time control of the system is obtained through the modification of the base phase angle α of the relative output voltage of the converter at the value of the phase angle of the transport line's voltage. This indirect control of the voltage makes possible the maintaining of a constant maximum report between the output voltage of the converter and the DC voltage at the capacitor's hubs. The reference value for the control loop of the reactive current is generated by another loop responsible for the control of the system's voltage. The control loop is similar to the one used for SVC and contains a regulation loop for the gradient k. An inevitable delay in the voltage control reversed reaction loop appears due to the necessary amount of time needed to calculate the positive sequence of the node's voltage. A very quick response can be obtained for the control of the reactive current; the time response of a voltage regulator is approximately a period of the line voltage.

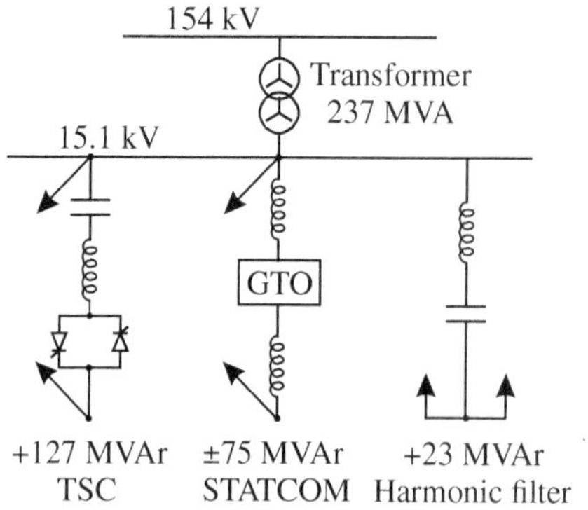

Figure 8.43 The +225/–52 MVAr SVC and the ± 75 MVAr STATCOM in the United Kingdom.

8.5.3 +225/–52 MVAr TSC and STATCOM Mixed System at East Claydon, in Great Britain

The East Claydon station is approximately 70 km north-west to London and was selected to locate within it a VAr compensation system for the south of England.

The design consists of a mixed system of ±75 MVAr STATCOM in parallel with a +127 MVAr TSC (thyristor-switched capacitors and a 23 MVAr harmonics filter); it results that the adjustment domain is between +225 and –52 MVAr (Figure 8.43). This STATCOM uses multilevel converters in a chained circuit configuration.

All the power electronics equipment and the control devices are localized in plastic fiber cabins and the exterior components are grouped together on metal frames thereby, when it is necessary they can be easily relocated to another station.

8.5.4 +133/–41 MVAr STATCOM at Essex, in United States

The +133/– 41 MVAr STATCOM device started its activity in 2001, in the 115 kV Essex Station of the Vermont Electric Power Company (VELCO). In Figure 8.44, single-wire scheme of the STATCOM in the Essex station is presented, and the main technical characteristics are provided in Table 8.3.

TABLE 8.3 The main characteristics of the STATCOM at Essex station

	Characteristics	Observations
Device rating	+133/–41 MVA	
Rated AC voltage	115 kV/3.2 kV	Two capacitor banks(24 MVAr)
Structure	Two transformers (43 MVA)	
Voltage source converters	• 3 modules of 12.5 MVA each • 5 MVA harmonics filter • 2 × 43 MVA, 3.2 kV/115 kV transformer	Three-level inverter composition and five pulse with MLI
GTO thyristors	6 kV and 6 kA	

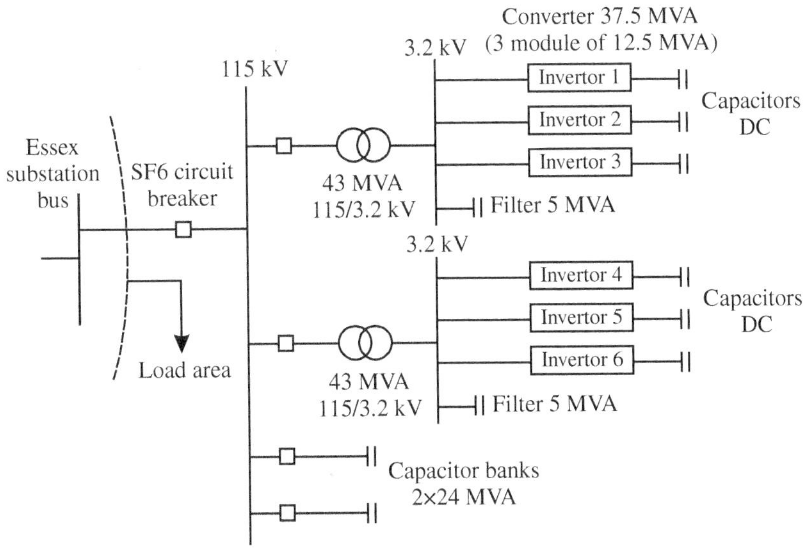

Figure 8.44 The +133/–41 MVA STATCOM diagram at Essex station.

8.5.5 STATCOM (+80/–110 MVAr) and Mechanic-Switched Capacitor (–93 MVAr) Mixed System, at Holly, in United States

Since 2005, a STATCOM with +80 MVAr inductive-generated reactive power and –110 MVAr capacitive-generated reactive power is functioning at Holly, for Austin Energy in Texas.

This device has replaced an old station that was based on gas and oil, in the near vicinity, which was put out of function after 45 years of activity and because of its reduced usage. As a result of the station's retirement and the existence of some sensible consumers in the area, there would have been stability and voltage problems in the transport system. Figure 8.45 illustrates the single-wire scheme of the IGBT-based STATCOM device installed at Holly-Texas [17].

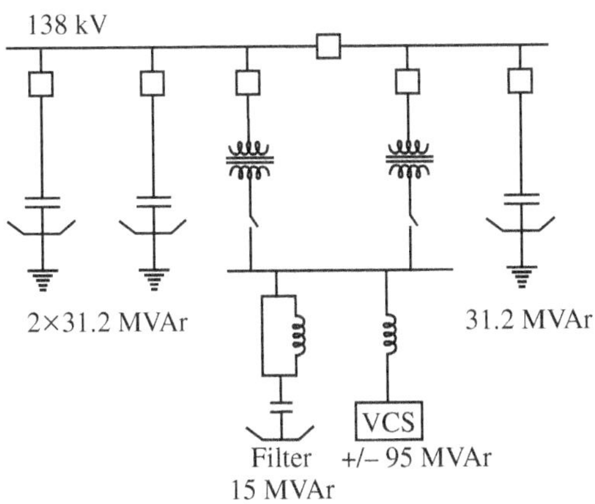

Figure 8.45 The single-wire diagram of the mixed STATCOM and MSC system at Holly.

TABLE 8.4 The main technical characteristics of the STATCOM at Holly

Rated voltage	138 kV
Rated power in dynamic state	80 MVAr inductive
	110 MVAr capacitive
Voltage source converter	Three-level scheme
	IGBT utilization, with MLI
	21 pulses
Rated power of the SVC	95 MVA
The control system	Digital-controlled voltage device
	The control includes the switching over the three MSCs (31 MVAr each), connected at the 138 kV line, for the voltage control in steady state

The scheme consists of two 110 MVA power transformers, one of them being in reserve, which powers the STATCOM device and three mechanically-switched capacitor (MSC) banks 31 MVAr each, connected to the 138 kV line.

The STATCOM device uses a VSC, which operates in symmetric state. It is manufactured for ± 95 MVAr and has two-by-two 15 MVAr harmonic filters each. As a result, it has an operating range between 80 MVAr inductive reactive power and 110 MVAr capacitive reactive power. The STATCOM is connected to the network coupling transformer via a reactor. If we consider the three 31.2 MVAr capacitor banks, provided to ensure the support of the reactive power in the electric network, then the adjustment area of the system at Holly is between 80 MVAr inductive through approximately 200 MVAr capacitive. The dynamic capacity is 190 MVAr, namely, double than the 95 MVAr of the SVC.

In Table 8.4, the main technical characteristics of the STATCOM at Holly are specified.

It is mentioned that the control system at Holly contains the contact plate adjustment-under-charge switch controller, of the three MSCs. In Figure 8.46, an outside view of the Holly station is presented.

8.5.6 ±100 MVAr STATCOM at Talega, in United States

In October 2002, a ±100 MVAr STATCOM was commissioned at the Talega power station (138 kV), near San Clemente, California. The device, manufactured by the Mitsubishi Electric Power Products, belongs to San Diego Gas & Electric and is employed for the dynamic control of the reactive power during the high consumption periods. The STATCOM consists of two groups of VSC converters (50 MVA each). Each group has four sets of 12.5 MVA modules and one 5 MVA harmonics filter (plus one spare filter switchable to either group), with a nominal phase-to-phase AC voltage of 3.2 kV and a DC-link voltage of 6 kV. The two 50 MVA converters are connected to the 138 kV transmission system by three 138 kV/3.2 kV transformers (one is for reserve), each rated at 55 MVA (Figure 8.47). The converters incorporate gate commutated turn-off thyristors, rated at 6 kA and 6 kV, arranged in modules for practical purposes. The system also contains breakers, a cooling system, and

Figure 8.46 An outside view of the STATCOM at Holly station. Reproduced with permission of ABB.

an automatic control and protection system. There are also three 69 MVA shunt capacitors that are connected directly at the 230 kV system.

The main advantages of introducing a STATCOM in the Talega transmission system are [18]:

- fast response to various disturbances that may occur in the system;
- provides smooth voltage control over a wide range of operating conditions;

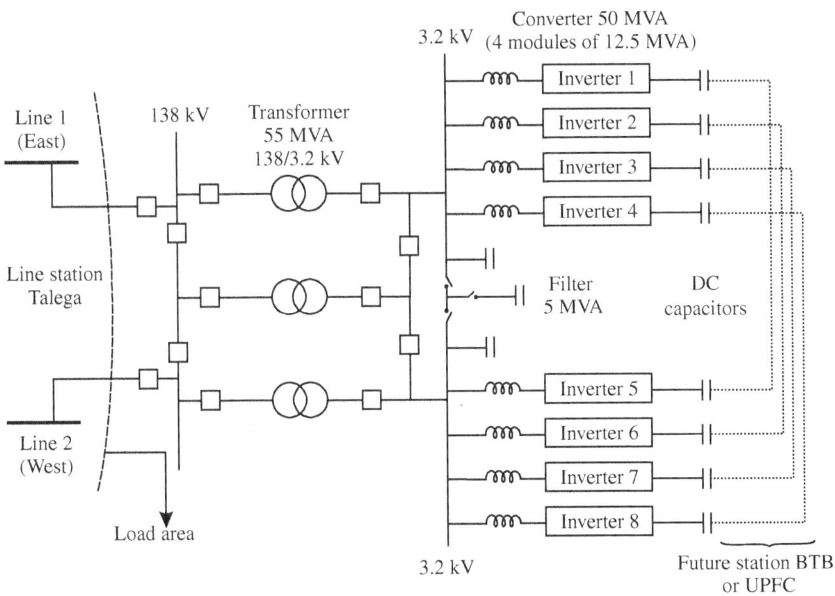

Figure 8.47 ± 100 MVAr STATCOM at Talega [18].

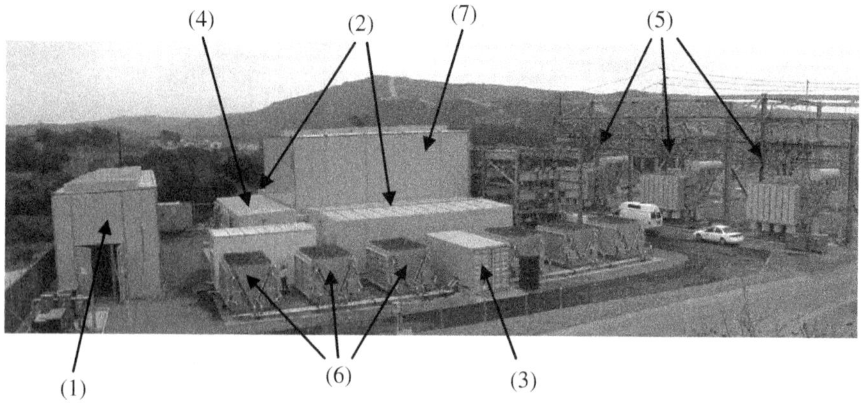

Figure 8.48 The outside view of the STATCOM at Talega station. *Source*: Reed et al. 2003 [19]. Reproduced with permission of IEEE.

- the device has a high built-in redundancy due to its modularity;
- can be automatically reconfigured to handle same equipment failure (i.e., a transformer or a filter) avoiding thereby device shut-down.

The system was designed for future operation as a back-to-back station between the Talega East and West buses, and it would be able to deliver power bidirectionally (Figure 8.47). Also, the DC voltage connection may be used to create a unified power flow controller (UPFC) type device in the future.

Figure 8.48 illustrates an outside view of the STATCOM station at Talega, where various components can be identified, such as LV filters (1), converter houses (2), cooling house (3), control house (4), the supply transformers (5), heat exchangers (6), and the series reactors (7).

REFERENCES

[1] Sen, K. K., and Sen, M. L. *Introduction to FACTS Controllers: Theory, Modeling, and Applications*. John Wiley & Sons, New Jersey, 2009.

[2] Holtz, J. Pulse-width modulation for electronic power conversion. *Proceedings of the IEEE*, vol. 82, no. 8, pp. 1194–1214, August 1994.

[3] Song, Y. H., and Johns, A. T., Eds. *Flexible AC Transmission Systems (FACTS)*, vol. 30. IEE Power and Energy Series, London, 1999.

[4] Hingorani, N. G., and Gyugyi, L. *Understanding FACTS. Concepts and Technology of Flexible AC Transmission Systems*. IEEE Press, New York, 2000.

[5] Zhang, X.-P., Rehtanz, C., and Pal, B. *Flexible AC Transmission Systems: Modelling and Control*. Springer, Verlag, 2006.

[6] Vinkovic, A., and Mihalic, R. A current-based model of an IPFC for Newton-Raphson power flow. *Electric Power Systems Research*, vol. 79, no. 8, pp. 1247–1254, August 2009.

[7] Vinkovic, A., and Mihalic, R. A universal method for the modeling of the 2nd generation FACTS devices in Newton-Raphson power flow. *International Journal of Electrical Power & Energy Systems*, vol. 33, no. 10, pp. 1631–1637, December 2011.

[8] EUROSTAG package documentation. Tractebel-EDF, May 1999.

[9] Schauder, C., and Mehta, H. Vector analysis and control of advanced static VAr compensators. *IEE Proceedings Generation, Transmission and Distribution*, vol. 140, no. 4, pp. 299–306, July 1993.

[10] Papič, I. Mathematical analysis of FACTS devices based on a voltage source converter – part I. *Electric Power System Research*, vol. 56, no. 2, pp. 139–148, November 2000.

[11] Papič, I., and Žunko, P. UPFC converter-level control system using internally calculated system quantities for decoupling. *Electrical Power and Energy Systems*, vol. 25, no. 8, pp. 667–675, October 2003.

[12] Blažič, B., and Papič, I. Improved D-StatCom control for operation with unbalanced currents and voltages. *IEEE Transactions Power Delivery*, vol. 21, no. 1, pp. 225–233, January 2006.

[13] Blažič, B., Matvoz, D., and Papič, I. Evaluation of solutions for flicker mitigation in the Slovenian transmission network. *CIRED 2007 Conference Proceedings*, Vienna, pp. 1–4, May 21–24, 2007.

[14] Eremia, M., and Shahidehpour, M., Eds. *Electric Power Systems Dynamics. Modeling, Stability and Control*. IEEE Press & Wiley, Hoboken, 2013.

[15] Schauder, C. D., Gernhardt, M., Stacey, E., Lemak, T., Gyugyi, L., Cease, T. W., Edris, A. Development of a ±100 MVAr static condenser for voltage control of transmission systems. *IEEE Transactions on Power Delivery*, vol. 10, no. 3, pp. 1486–1496, July 1995.

[16] Schauder, C. D., Gernhardt, M., Stacey, E., Lemak, T., Gyugyi, L., Cease, T. W., Edris, A., Wilhelm, M. TVA STATCON project: design, installation and commissioning. CIGRE paper 14–106, 1996.

[17] ABB. STATCOM replaces critical urban generation for voltage control and stability, ABB. Application Note A02-0200E, Västeräs, 2007.

[18] Reed, G., Paserba, J., Croasdaile, T., Westover, R., Jochi, S., Morishima, N., Takeda, M., Sugiyama, T., Hamasaki, Y., Show, T., and Abed, A. SDG&E Talega STATCOM project – system analysis, design, and configuration. *The Asia Pacific IEEE/PES Transmission and Distribution Conference and Exhibition*, Yokohama, Japan, vol. 2, October 2002.

[19] Reed, G., Paserba, J., and Salavantis, P. The FACTS on resolving transmission gridlock. The case for implementing power electronics control technologies to enhance reliability and upgrade capacity. *IEEE Power & Energy Magazine*, vol. 1, no. 5, pp. 41-46, September/October 2003.

CHAPTER 9

STATIC SYNCHRONOUS SERIES COMPENSATOR (SSSC)

Laszlo Gyugyi, Abded-Aty Edris, and Mircea Eremia

9.1 INTRODUCTION

Within[1] the general concept of using static converter technology for series compensation, parallel compensation, and phase angle control of transmission lines, Dr. Laszlo Gyugyi proposed in 1989 the FACTS controller named Static Synchronous Series Compensator (SSSC), also often called series STATCOM [1].

In accordance with the definition of IEEE–CIGRE, the SSSC represents "the outline of series compensation which uses a switching power converter as a synchronous voltage source, to produce a controllable voltage in quadrature with the line current." [2]

Similarly to the STATCOM, the SSSC employs a voltage-sourced converter (VSC), but it is connected in series with the transmission line by a coupling transformer. The converter generates a controllable AC voltage source $v = V \sin(\omega t - \psi)$ that is applied to the series coupling transformer to affect the active and reactive power flow in the transmission line. The SSSC represents an alternative to the series capacitors compensation of the transmission lines.

A VSC, with suitable internal control, can be assimilated to a synchronous voltage source, that is, an ideal synchronous generator, which can produce a set of three alternating voltages, at the desired fundamental frequency, of which magnitude and the phase angle are controllable. This synchronous voltage source can generate or absorb reactive power similar to a synchronous compensator. Moreover, the synchronous voltage source can exchange active power with the AC power system, when the DC terminal of the synchronous voltage source is connected to an adequate energy source, for example, a suitably rated storage battery. Thus, the transmitted power becomes a function of the controllable injected voltage.

[1] This chapter contains excerpts from the CIGRE Working Group B4-40 Technical Brochure *Static Synchronous Series Compensator (SSSC)*, Edris, A.A. (Convener), with permission from CIGRE, © Copyright 2008 [1].

Advanced Solutions in Power Systems: HVDC, FACTS, and Artificial Intelligence, First Edition.
Edited by Mircea Eremia, Chen-Ching Liu, and Abdel-Aty Edris.

In terms of practical applications, such as power flow control and stability improvement, the SSSC provides a more versatile and effective power transmission control at the same rated power than the series capacitors do. However, series capacitor compensation may presently have a cost advantage, particularly at high transmission voltage levels. Up to this date, no transmission SSSC has been placed in service although several studies are being done for this purpose.

9.2 ARCHITECTURE AND OPERATING PRINCIPLES

9.2.1 The Basic Structure and Principles of Operation

The SSSC is a reactive power series compensator employing a voltage source converter in series with the transmission line, as shown in Figure 9.1 [1–5].

This operating mode emulates a controlled series reactive compensation (similar with that obtained by the thyristor-controlled series capacitor [TCSC]), but provides a wider control range since it can operate equally in capacitive or inductive operating domains. Thyristor-based FACTS controllers, like TCSC, are based on impedance control, so they are dependent on the line current. The SSSC can operate as a voltage source, in which case the series compensation becomes independent of the line current.

To explain the operating principle of the SSSC, we will compare it to the classical series capacitor compensation with the use of phasor diagram (Figure 9.2).

In the case of series compensation, the capacitor produces a voltage ($\Delta \underline{V}_C$) at the operating frequency of 50 (or 60) Hz, which is of opposite polarity to the inductive voltage across the line reactance [2, 4]:

$$\Delta \underline{V}_C = -jk_S X_L \underline{I}; \quad j = \sqrt{-1}$$

where

$\Delta \underline{V}_C$ is the injected compensating voltage phasor;

$\underline{I}$ – the line current;

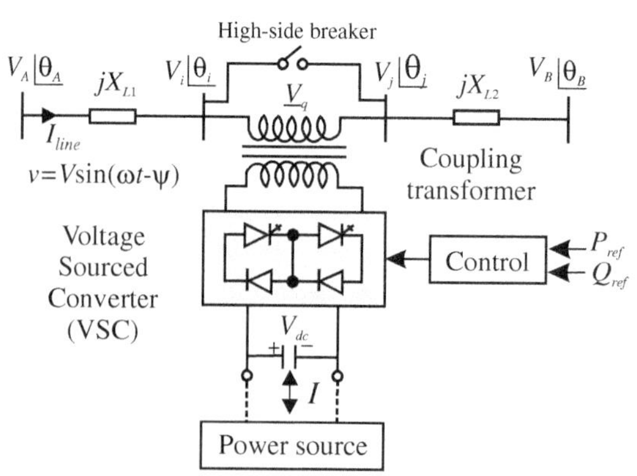

Figure 9.1 Schematic diagram of the SSSC [2].

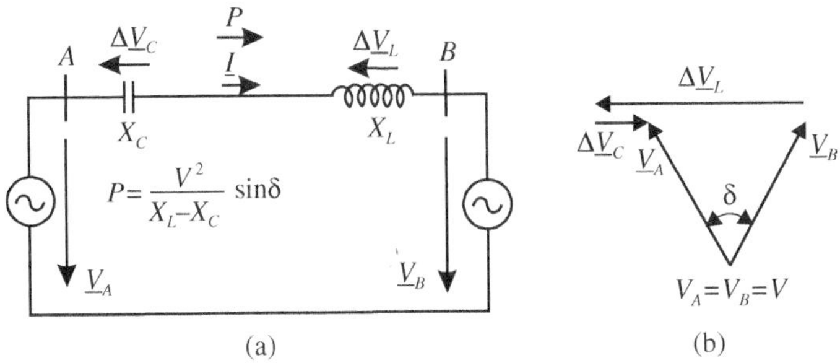

Figure 9.2 Classical compensation of a transmission line using series capacitors: (a) equivalent scheme and (b) phasor diagram [2, 3].

k_S – the degree of series compensation, $k_S = X_C/X_L$;

X_L – the inductive line reactance;

X_C – the reactance of the series capacitor, which produces the same effect as a suitably controlled VSC.

As a result of the series capacitive compensation, a decrease in the equivalent line impedance is achieved and, consequently, a corresponding increase in the electric power transmitted on the line. The same transmitted power on the line could be obtained by using a VSC for series compensation, which is controlled to produce the same voltage as the one across the series capacitor (Figure 9.3).

The VSC within the SSSC is operated in synchronism with the transmission line current. The voltage generated by the VSC is kept in quadrature with the line current, lagging or leading it by 90°. The injection of a lagging voltage with respect to the line current emulates a series capacitor, whereas a leading voltage emulates a reactor in series with the line. Thus, the SSSC can provide either series capacitive or series inductive compensation, without any rating increase or additional reactive components, by its inherent capability to reverse the polarity of the output voltage it generates.

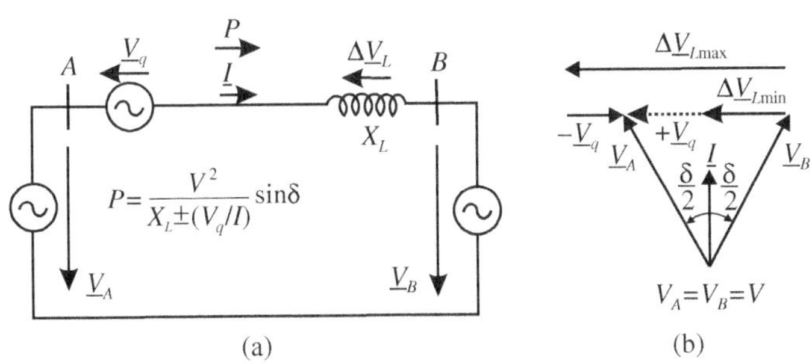

Figure 9.3 Equivalent compensation with an SSSC: (a) equivalent scheme and (b) phasor diagram illustrating capacitive and inductive operation modes [2, 3].

A generalized expression for the injected voltage ($\underline{V}_q$) can be written as [2]:

$$\underline{V}_q = \pm jV_q(\xi)\frac{\underline{I}}{I} \tag{9.1}$$

where $V_q(\xi)$ is the magnitude of the injected compensating voltage, with $0 \leq V_q(\xi) \leq V_{q\,\max}$, and ξ is a chosen control parameter.

The quadrature relationship between the output voltage of the SSSC and the line current ensures substantially zero real power exchange between the SSSC and the AC system, except for the small amount (about 1% at full output) required to replenish the internal losses of the converter. This power is drawn from the line by the converter, by a small (typically less than 1°) deviation from the ideal 90°, to keep the DC capacitor charged without an external DC power supply [1].

In capacitive compensation, a voltage is inserted in series with the line that is in phase opposition to the voltage produced by the line current across the series line reactance. As a result, the voltage across the series line reactance is forced to increase. At the same time, the apparent line reactance seen from the system is reduced which will cause a proportional increase in the current and thus in the transmitted power.

Similarly, inductive compensation (when the SSSC's output voltage leads the line current) injects a voltage in phase with the voltage across the line reactance. As a result, the voltage across the line reactance decreases, as if its inductance was increased, causing the line current and the corresponding transmitted power to proportionally decrease. The line current and the corresponding power increase or decrease is proportional to the magnitude of the series compensating voltage relative to the voltage across the series line reactance. Traditionally, this is expressed by the *degree of series compensation,* which is defined as the impedance ratio of the series line reactance to the effective series capacitor, or, equivalently, the amplitude ratio of the voltage across the series line reactor to the series compensating voltage.

The adjustment of the converter voltage amplitude can be accomplished by the control of the DC voltage through the momentary charge or discharge of the DC capacitor. Alternatively, the voltage amplitude can be adjusted internally in the converter by pulse-width-modulation, or other techniques, from a fixed (regulated) DC voltage (see Section 4.2).

The SSSC can control the effective transmission line impedance by injecting a controllable compensating voltage in quadrature with the line current to emulate a series capacitor or reactor. Emulated capacitive compensation will decrease the effective series line impedance, whereas emulated inductive compensation will increase it. Thus, the SSSC can increase or decrease steady-state power flow or change it dynamically to counteract dynamic disturbances in order to increase transient stability and damp power oscillation.

9.2.2 Operating Modes of SSSC

Due to the almost ideal voltage source characteristic of the VSC, the SSSC can provide capacitive or inductive compensating voltage independent of the line current, up to its specified voltage rating. Thus, the SSSC can maintain the rated maximum capacitive or inductive compensating voltage in the face of changing line current,

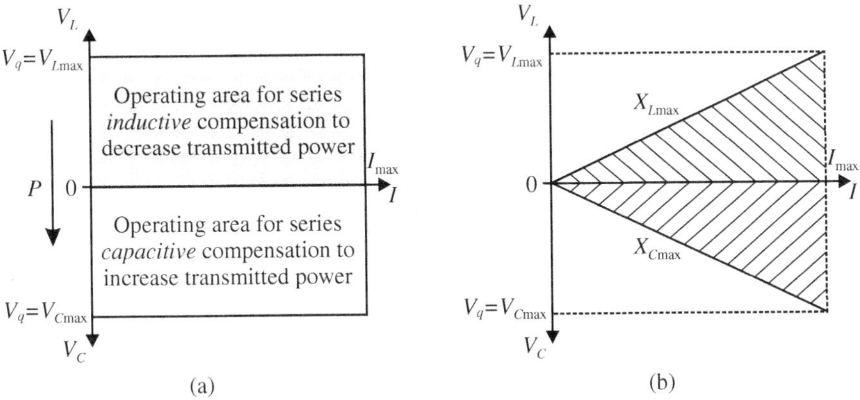

Figure 9.4 Operating range characteristics of SSSC [1,2].

theoretically in the total operating range of 0 to I_{max} (Figure 9.4). The practical minimum is the line current value at which the SSSC is still able to absorb enough power to replenish its operating losses [2].

The SSSC has two operating modes:

- *Control of voltage compensation*, where the SSSC operates so as to maintain the capacitive or inductive compensation voltage at a maximal value, regardless of the current variations in the transmission line between 0 and I_{max} (Figure 9.4a).
- *Control of reactance compensation*, where the SSSC operates so as to maintain the capacitive or inductive compensation reactance at a maximal value, when the current in the line varies between 0 and I_{max} (Figure 9.4b).

In principle, an SSSC is able to exchange active power and reactive power with the transmission grid. If considering only reactive power compensation, then the size of the power source can be relatively small. In this case, it will only control the magnitude of the voltage phasor, which is kept perpendicular on the line current phasor, leading or lagging it by 90°. This means that, in the case of the SSSC, the magnitude of the compensating voltage, at fixed +90° or –90° angle, can be controlled continuously in the operating domain of the VSC.

If the power source has real power control capability, then the injected voltage could be controlled both in magnitude and phase. The behavior of an SSSC is the same as that of a series capacitor and a series reactor, both controllable. The main difference is that the voltage generated by the SSSC can be controlled independently of the line current because it does not depend on it. Consequently, an SSSC can operate both in the cases of increased and decreased line loads.

Since the SSSC emulates the line compensation of the series capacitor by the direct injection of the required compensating voltage at the fundamental system frequency, without reproducing the impedance versus frequency characteristic of a physical capacitor, it is, in contrast to the series capacitor, unable to form a classical series resonant circuit with the inductive line impedance to initiate subsynchronous oscillations. In other words, independent of whether the SSSC is operated in the capacitive or inductive compensation domain, it is seen by the system as a (zero-impedance) voltage source in series with a small inductive impedance (the leakage impedance of the coupling transformer). Because of its fast response, the SSSC could theoretically provide effective damping of subsynchronous oscillation by suitable control, should the condition for subsynchronous oscillation be established by (existing) series capacitors.

9.2.3 The P_q-δ Characteristic of SSSC

The transmitted power versus transmission angle, that is, the P_q-δ characteristic of the SSSC at various per unit values of the compensating voltage, V_q, (which may be capacitive or inductive) is shown in Figure 9.5. This figure shows that the SSSC is able to decrease the transmitted power, subject to a suitable rating, and also to reverse the direction of power flow.

The well-known expression, in which the electric power transmitted on the line is dependent on the angle δ, becomes a function of the injected voltage $V_q(\xi)$. Thus, the transmitted power, P, of a system with two synchronous generators connected through the AB transmission line that is controlled at its midpoint by an SSSC, can be expressed, at $V_A = V_B = V$, as:

$$P = \frac{V^2}{X_L}\sin\delta \pm \frac{V}{X_L}V_q\cos\frac{\delta}{2} \tag{9.2}$$

We can observe that, in contrast to a series capacitor which can increase the electric power transmitted on the line with a fixed percentage of the transmitted power obtained on a uncompensated line at a given angle δ, the SSSC can increase the electric power, transmitted on an uncompensated line by a fraction of the maximum transmissible power independent of the angle δ, for values between 0 and $\pi/2$.

Also, the SSSC can be used for the control of power flow by increasing or decreasing the power $P_q[= \pm(V/X_L)V_q\cos(\delta/2)]$, by means of changing the polarity

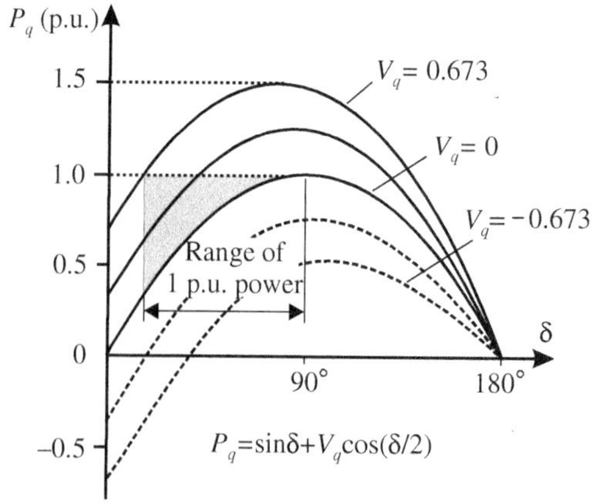

Figure 9.5 P_q-δ characteristic with SSSC [1].

of the injected voltage. Thus, from Figure 9.5, for $\delta = 90°$, $V_q = +0.673$ results in an increase in the transmitted power by 50%, and for $V_q = -0.673$ the transmitted power is decreased by 50%.

A comparison of the basic operating characteristic of the SSSC to that of the STATCOM reveals their duality. In a given operating range, determined by the rating of the VSC employed, the series-connected SSSC can maintain reactive compensating *voltage* independent of the line current, whereas the shunt-connected STATCOM can maintain reactive compensating *current* independent of the transmission line voltage.

9.3 COMPARISON OF SSSC WITH OTHER TECHNOLOGIES

This section presents a brief comparison of SSSC and other technologies such as fixed series capacitor, fixed series reactor, phase-angle regulator (PAR), TCSC, and gate-controlled series capacitor (GCSC). The comparison shown below is for the case of conventional long transmission lines with equivalent length smaller than 1/4 of wavelength (1250 and 1500 km, respectively, for 60 and 50 Hz system) [1].

9.3.1 Comparison with Fixed Series Capacitor

Figure 9.6a shows a transmission system with fixed series capacitive compensation, which naturally has no controllability and, therefore, has no flexibility. Figure 9.6b shows the compensation characteristics for three different levels of fixed capacitive compensation. The operating area of an SSSC is a rectangle in the V–I plane as shown in Figure 9.6b, and for the fixed capacitive compensation we have only a fixed line for each value of the compensation capacitance. In this compensation, the larger the capacitive reactance, the larger will be the transmitted power.

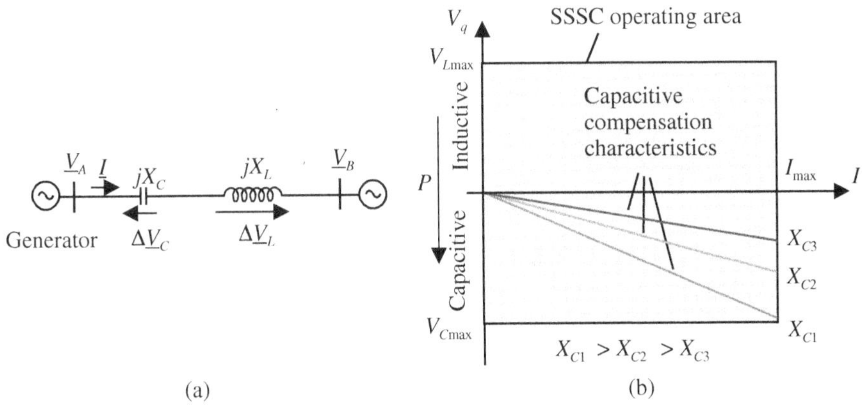

Figure 9.6 (a) Transmission system with fixed capacitor compensation and (b) fixed capacitive compensation characteristics [1].

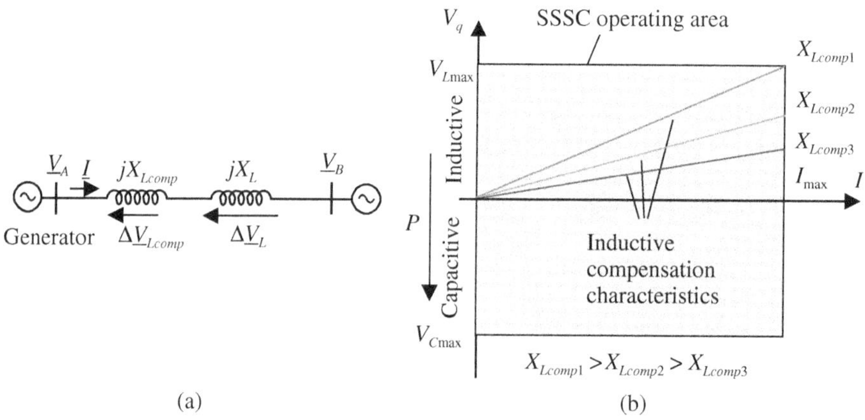

Figure 9.7 (a) Transmission system with fixed inductive compensation and (b) fixed inductive compensation characteristics [1].

9.3.2 Comparison with Fixed Series Reactor

Figure 9.7a shows a transmission line with fixed inductive compensation, which is not very common as it works in a way to decrease the transmitted power as shown in Figure 9.7b. As in the case of fixed capacitive compensation, the power transmission characteristic is given just by a line for each compensation level and the larger the reactance, the smaller the transmitted power.

9.3.3 Comparison with Phase Angle Regulator

Figure 9.8a shows a transmission system with a thyristor-controlled phase angle regulator (TCPAR). The TCPAR shown in this figure is composed of a transformer with primary winding connected in shunt between two phases and three secondary windings connected in series with the other phase (see Chapter 7). A TCPAR scheme is shown in Figure 9.8a for the 1:3:9 transformer winding arrangement to provide discrete-level control.

The voltage injected in series with the line is the sum of the voltages in the three secondary windings controlled by thyristor valves. The winding that is fully connected adds a sinusoidal voltage in series with the line. The winding that is partially connected has its voltage controlled by the firing angle of the thyristors.

Figure 9.8b shows the phasor diagram of the sending-end voltage $\underline{V}_A$ and the compensating voltage $\underline{V}_C$. This compensation voltage can have +90° or –90° phase difference with respect to the sending-end voltage $\underline{V}_A$. Different from the SSSC, the phase angle of the compensating voltage of a TCPAR can lead the current phasor or lag it by an angle that does not need to be in quadrature. Therefore, depending on this phase difference, the series transformer may have to inject or absorb active power, which is supplied or absorbed by the shunt winding.

If we compare the operating area of the TCPAR, in terms of voltage and current magnitudes, with the case of SSSC we will find that they are basically equal,

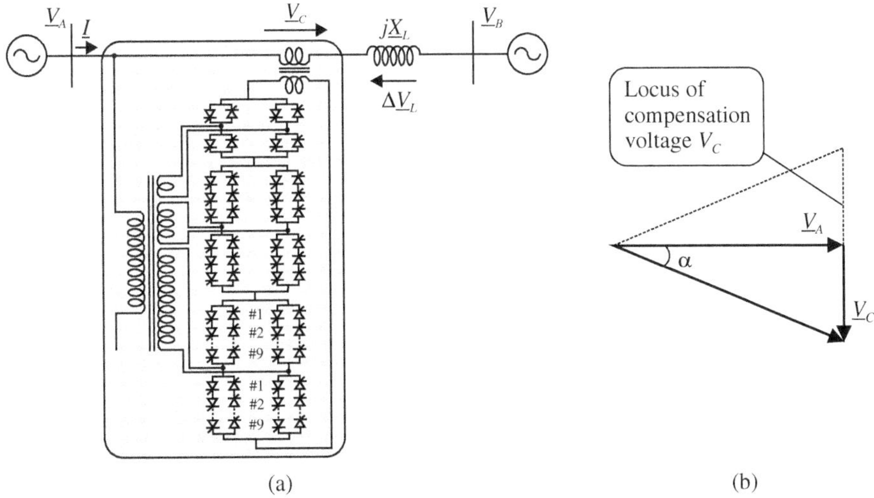

Figure 9.8 (a) Transmission line with phase angle regulator and (b) phasor diagram for the phase angle compensation [2].

that is, the voltage can vary from zero to its maximum value, so happens for the current. Therefore, the operating area in the *V–I* plane is a rectangle like in the case of SSSC. The difference is that in the case of SSSC, the compensation voltage is always in quadrature with the line current while in the TCPAR, this angle between compensation voltage and current can have any value. However, this phase difference is not controllable. It appears as a function of the current phase with respect to the sending-end voltage. Depending on the number of secondary windings and the switching control of the thyristor, the TCPAR may present higher level of low order harmonic components injected in series with the line. However, the number of secondary windings is chosen in a way that harmonics be kept in an acceptable range.

9.3.4 Comparison with Thyristor-Controlled Series Capacitor

The operation of the SSSC is based on VSCs and, therefore, it can operate in a rectangular area in the *V–I* plane, that is, it has a wide controllability range as shown in section 9.2.2. The operation of the TCSC, on the other hand as explained by Hingorani and Gyugyi, is a compensator based on the concept of impedance control [2]. The control is executed by thyristor switches or *valves,* constructed of individual thyristors. The thyristor is a semiconductor switch, at least one generation older than the semiconductor switches used in the SSSC (see Chapter 2). However, due to the higher rating and lower cost of thyristors, there are relatively large number of various actual TCSC applications around the world for power oscillation damping and/or power flow control (see Chapter 6).

Figure 9.9a shows a transmission system with a line compensated by a TCSC. As shown, the TCSC is composed of a fixed series capacitor in parallel with a thyristor-controlled reactor (TCR). If the thyristors are not conducting, only the

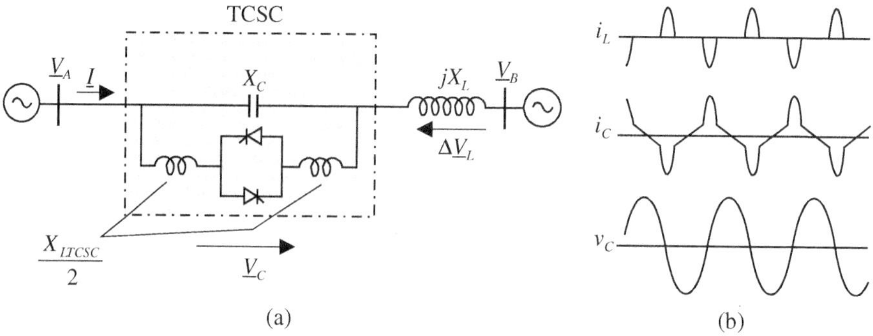

Figure 9.9 (a) Transmission line with thyristor-controlled series capacitor (TCSC) and (b) current and voltage waveforms in a TCSC [1].

capacitor is in series with the line. If the thyristors are switched on repeatedly at a given firing angle, then the current and voltage waveforms become as shown in Figure 9.9b. The reactor current i_L has a pulse-like waveform. This current flows through the capacitor, and together with the line current, produces a current waveform i_C as shown in Figure 9.9b; it also shows the resultant capacitor voltage waveform, v_C.

In general, harmonics generated by the thyristor switching is not a serious problem if the reactor and capacitor size, as well as firing angle, are correctly chosen [1].

The equivalent impedance of the fixed capacitor in parallel with the TCR can produce an equivalent impedance that may be capacitive or inductive (Figure 9.10).

Figure 9.10 shows the equivalent impedance of a TCSC for firing angle varying from 90° to 180°; for firing angle close to 90°, the equivalent impedance is inductive and for firing angle close to 180°, it is capacitive. At about the middle of the range of 90–180°, the fixed capacitor and the controlled reactor present a resonance and the device cannot be operated in this region because too large current may flow in the LC circuit. When the TCSC operates with capacitive characteristics, normally the firing angle has a minimum limit that produces the maximum compensating capacitive impedance Z_{max}; in Figure 9.10, this limit is slightly below 150°.

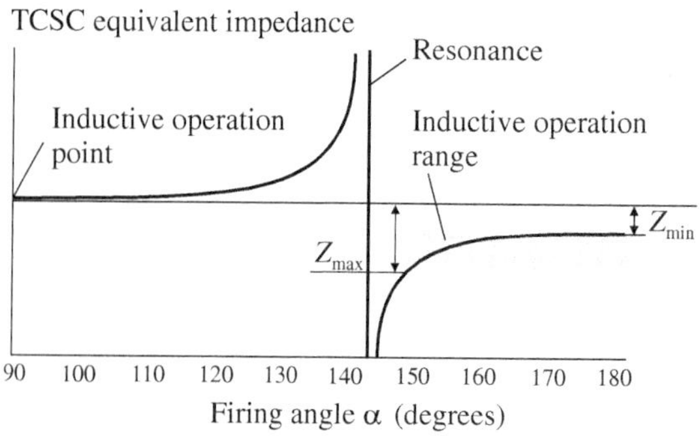

Figure 9.10 TCSC equivalent impedance characteristics as a function of the thyristor firing angle.

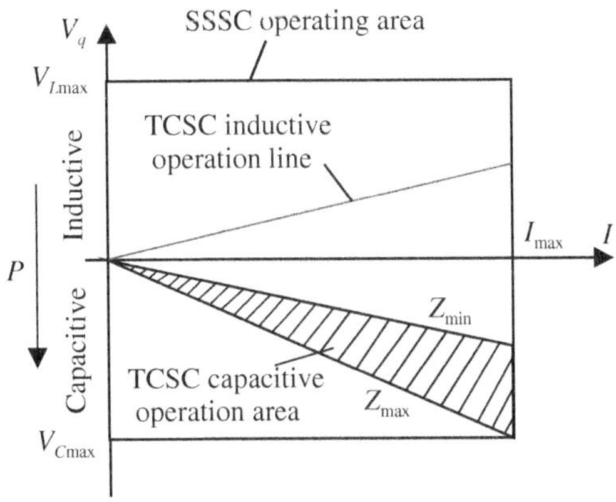

Figure 9.11 TCSC operating area [1].

This device has minimum compensating impedance $Z_{\min}$ that is given by the capacitor alone when the firing angle of the thyristors is equal to 180°, that is, the thyristors are off. Then, the capacitive continuous impedance control range goes from $Z_{\min}$ to $Z_{\max}$.

In some cases, when it is necessary to decrease the transmitted power in the line, the TCSC can be operated with firing angle equal to 90° and it will present inductive characteristic. In this operating condition, the firing angle is not varied, and therefore there is no continuous control in the inductive side.

The TCSC operating area in the V–I plane is shown in Figure 9.11. In the capacitive region, it can be controlled continuously from $Z_{\min}$ to $Z_{\max}$, which is represented by the area defined between the lines for $Z_{\min}$ and $Z_{\max}$. If the TCSC is designed for power flow control, it can operate in any point of this area continuously without problem of overheating the capacitor, reactor, or the thyristors. On the other hand, if the TCSC is designed only to damp power oscillations, it may operate continuously on the line defined by $Z_{\min}$ and operate for short time (few seconds) in the area between $Z_{\min}$ and $Z_{\max}$. In this last case, a capacitor with smaller VA rating may be used as compared with the case of power flow control.

Comparing the TCSC with SSSC with the information given above allows us to conclude that [1]:

- The operating area of the TCSC is much smaller than the SSSC in the capacitive region.
- In the inductive region, the TCSC operating characteristic is reduced to a line only as compared to a rectangle in the case of SSSC.
- TCSC may present problem of internal resonance that has to be avoided.

Despite the advantages of the SSSC as compared with the TCSC, this thyristor-based device is still used as a practical option when the wide range of controllability of the SSSC is not necessary. When the application allows, the TCSC is usually the preferred option due to its lower cost.

9.3.5 Comparison with Gate-Controlled Series Capacitor

Figure 9.12a shows a transmission line with a GCSC connected in series. This is a device also based on the concept of variable impedance and was originally presented by Karady et al. [6] and they called it "continuously regulated series capacitor." Later, De Souza et al. [7] introduced the name "GTO-controlled series capacitor," which was adopted by Hingorani and Gyugyi [2]. However, due to the fact that different switching device can be used, provided that it can be turned off, recently Edris [8] has changed "GTO-controlled" to "Gate-controlled." In all cases, the acronym GCSC was kept to make a contrast with TCSC.

Figure 9.12a shows that the GCSC has a very simple structure with basically one capacitor and in parallel with bidirectional gate-controlled switches. This switch arrangement in parallel with the capacitor is exactly the dual circuit of a TCR, where a reactor is in series with a bidirectional thyristor valve. In the case of the TCR, the control is done by the variation of the firing angle. However, in the case of GCSC, the switches are turned on automatically always when the capacitor voltage crosses zero and they have to be turned off to insert the capacitor in series with the line. Therefore, instead of firing angle, in the case of the GCSC, the turn-off angle is controlled.

One interesting point here is that the TCR is theoretically well matched for shunt connection, whereas the GCSC is well matched for series connection. One advantage of the GCSC over the TCSC is the fact that it does not need a reactor and, therefore, it may result in a more compact equipment. Reference [7] presents a detailed comparison between GCSC and TCSC. If a perfect dual circuit of the TCSC is to be produced, then a reactor has to be connected in series with the capacitor, which is in parallel with the switches. However, this option may not be necessary as the line is normally inductive and putting more series inductance has no meaning except to increase the cost of the compensation equipment.

Figure 9.12b shows the current and voltage waveforms in the transmission line and in the capacitor of the GCSC. The line current is assumed sinusoidal. As explained above, when the voltage on the capacitor crosses zero, the switches are turned on and this voltage is kept zero until the switch is turned off at a given turn-off angle.

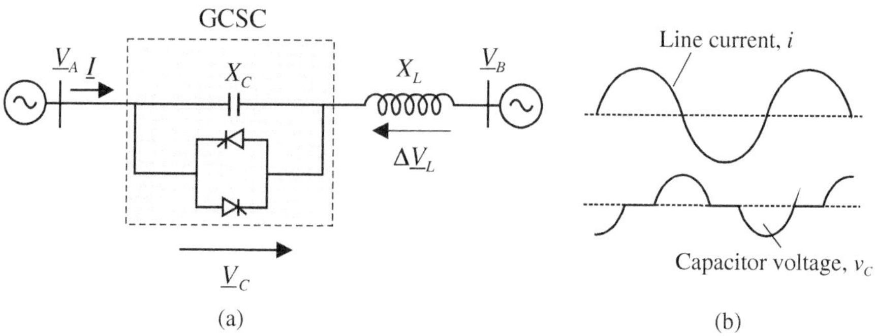

Figure 9.12 (a) Transmission line with a gate-controlled series capacitor (GCSC) and (b) current and voltage waveform in the GCSC.

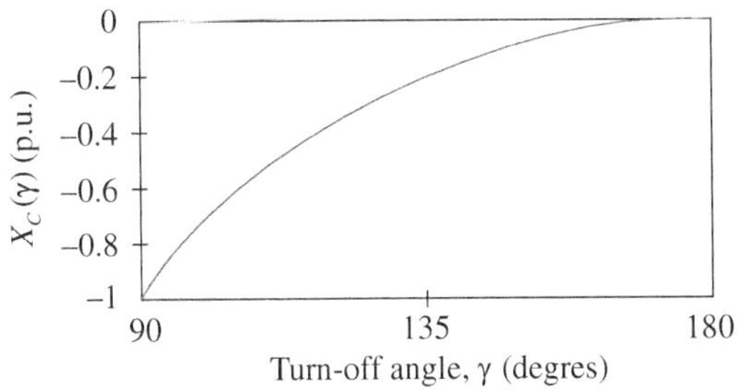

Figure 9.13 Equivalent capacitive reactance of the GCSC [1,9].

If voltage and current waveforms of the GCSC and the TCR are compared, it becomes evident that they are dual of each other, that is, the current waveform in a shunt-connected TCR has the same shape as the voltage waveform in the series-connected GCSC, and the voltage waveform (line voltage) on the TCR has the same shape as the current in the GCSC (line current).

Figure 9.13 shows the equivalent capacitive reactance of the GCSC as a function of the turn-off angle; this reactance is nonlinear and varies smoothly from its maximum value to zero. No resonance problem is present.

Figure 9.14a shows the GCSC operating area in the *V–I* plane for the case where the maximum capacitor voltage occurs for the maximum line current. In this case, the operating area of the GCSC is similar to that of the TCSC with two differences: (i) the minimum capacitive reactance is zero and (ii) there is no inductive reactance (because no reactor was inserted in series). This device has a smaller control range when compared with the SSSC, although it is wider than in the case of the TCSC. The GCSC has fewer components than the TCSC and is much less complex than the SSSC.

Figure 9.14b shows the GCSC operating characteristics for the case where the capacitor maximum voltage occurs for a line current smaller than its maximum value. In this case, the capacitor voltage has to be controlled by the turn-off angle to avoid

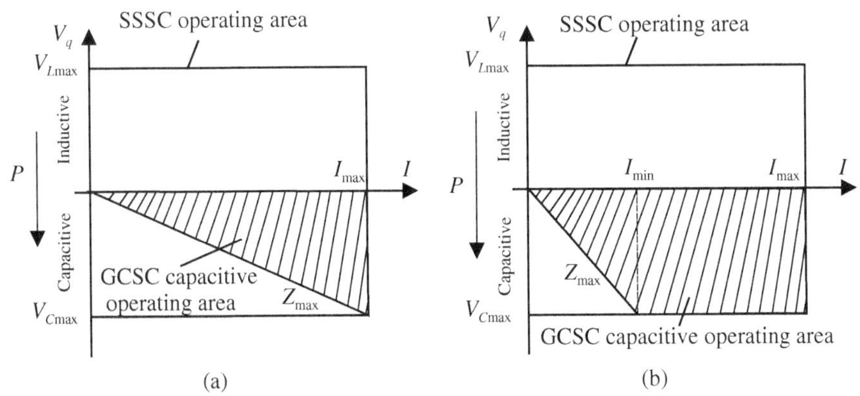

Figure 9.14 GCSC operating area.

steady-state overvoltage, except if a high compensation voltage is necessary for a short time, for example, for power oscillation damping.

The SSSC is more flexible and operates in a wider range than the GCSC. The GCSC may be a good option in situation where there is no need for inductive compensation and the compensation can be done by a controlled impedance instead of a controlled voltage source as in SSSC.

9.3.6 Dynamic Flow Controller

By combining the TCSC/reactor or the thyristor-switched series capacitor/reactor and a phase-shifting transformer, a new FACTS system is suggested. The main advantages of this system are that [1]:

- controllability even at low transfer levels;
- dynamically controllable;
- rating is split between several modules and future upgrades, while changes are possible;
- the weight of each of the building blocks is limited;
- power oscillation damping and transient control.

The dynamic flow controller (Dynaflow) concept so far is presented in several conference papers, and it is believed that more optimization of networks and interconnections increases the market for controllable devices such as Dynaflow.

9.4 COMPONENTS OF AN SSSC

9.4.1 Overview of the Functional SSSC Components

The main functional components of the SSSC are the DC–AC converter, the series coupling transformer, and the by-pass protection switch. In a practical SSSC, the converter itself can be broken down into a number of major constituents, such as the semiconductor switch assembly, the DC capacitor, and the AC harmonic filters, if needed. Figure 9.15 gives an overview of the SSSC main components (highlighted in boxes), with a particular converter design.

(i) *Converters*. The converters of the existing stations are made up by GTO devices and are based on three-level arrangements. Generally, converters can be made up by thyristors (forced-commutated configurations), GTOs, integrated gate commutated thyristors (IGCTs), and IGBTs. The principal arrangement can vary from two-level configurations using pulse width modulation (PWM) modulation up to multilevel configurations using fundamental switching technology.

(ii) *Series transformer*. SSSC is connected in series with the line. To couple the voltages that are generated in the converter circuits at lower voltage levels

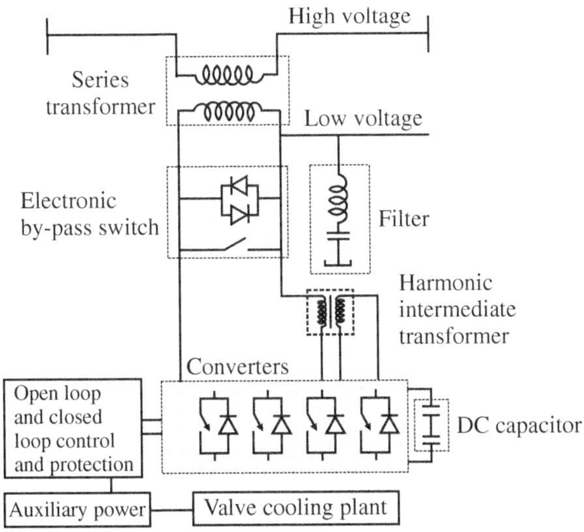

Figure 9.15 Overview of an SSSC single-line diagram [1].

toward the system, a transformer is connected in series with the line at high voltage side.

(iii) *Electronic by-pass switch*. In the case of system short circuits, where the system short-circuit current can also flow through the SSSC system-side transformer winding, the respective current will also arise on the low voltage side transformed by the transformer ratio. This current will also flow through the converter power electronic devices. Therefore, electronic by-pass switches, which can react immediately, are typically installed in parallel to the converter terminals. They are designed to take over the short circuit for a short time and are released from these stresses by the operation of a primary conventional by-pass breaker.

(iv) *DC capacitor*. The DC capacitor across the DC terminals of the converter provides energy storage for the VSC. The size of the DC capacitors is determined from the allowable DC ripple contents and how fast voltage changes are required on the DC side. Depending on the converter design, the DC capacitor is connected across the common DC terminals or across the multiple levels of the converter structure.

(v) *Harmonic neutralizer intermediate transformer/series reactor*. This is used in existing SSSC installations employing converters with GTO semiconductors. The converter topology is built up in such a way that the cancelation of selected harmonics is achieved by specific magnetic interconnections of individual converters operated with appropriate phase-shifts with respect to each other. With this scheme, the converters are operated at the fundamental frequency of the AC system, and still the generated output harmonic content can be reduced to the desired level, and the switching losses of the converters can also be kept at a minimum.

If several VSCs are operating in parallel, then they have to be decoupled by reactors. Each of the parallel VSC is equipped with its own series reactor. The series reactor is designed to limit the d*i*/d*t* stresses of the converter components. Paralleling is used in newer installations based on IGBT devices.

(vi) *Harmonic elimination requirements.* Future installations based on IGBT technology may employ several converters operating in parallel, using PWM control, with appropriate phase-shift, to suppress major harmonics. Typically, they will use common filtering of higher frequency harmonics at the bus where they are connected in parallel.

(vii) *Valve cooling plant.* A cooling system for the semiconductor devices is necessary to limit their thermal stresses. The size of the cooling system depends on the continuous and transient operation of the SSSC, as well as consideration to the internal and system fault conditions. Typically, ionized water-cooling technology can be used, configured as a single-circuit or double-circuit cooling system depending on the environmental conditions. The cooling system purification is located inside the building with the pump skids. Water to air heat exchangers are located outside. The cooling system is the major load for the auxiliary power supply.

(viii) *Auxiliary power.* The major auxiliary load for the operation of an SSSC is the cooling system. DC power is also provided to the converter poles for GTO/IGBT gating and local pole control. Station service power is typically backed up by battery equipment (lead acid batteries and charging equipment).

(ix) *Control and protection equipment.* The control includes the open-loop (sequence control) and closed-loop control (CLC). The sequence control acts on start-up and shutdown, and communicates with all major components collecting status information of the equipment. The CLC controls the SSSC according to its requirements from the system. Depending on the converter configuration, a different kind of control is used. Each converter pole is typically made up of submodules, which include associated gate drive controls for each semiconductor level. Each pole has additional electronics that are associated with the particular switching of the pole. The control signals from the CLC are sent to the converter pole controls where the device firing and blocking are coordinated with the other series-connected devices. Communication is provided by fiber-optic cables. The control is connected by serial communication to all relevant parts of the SSSC including the cooling system. The control monitors continuously the operation of all subsystems, collecting and analyzing the relevant status information.

9.4.2 Control

The basic operating principle of the SSSC is to inject the voltage V_{inj} in quadrature (i.e., at ±90°) with respect to the transmission line current as illustrated in Figure 9.16. Therefore, the SSSC exchanges only reactive power with the transmission line. The SSSC has the effect of changing the equivalent line impedance and therefore only the magnitude of the line current. As shown in Figure 9.16a, with $\underline{V}_{inj} = 0$,

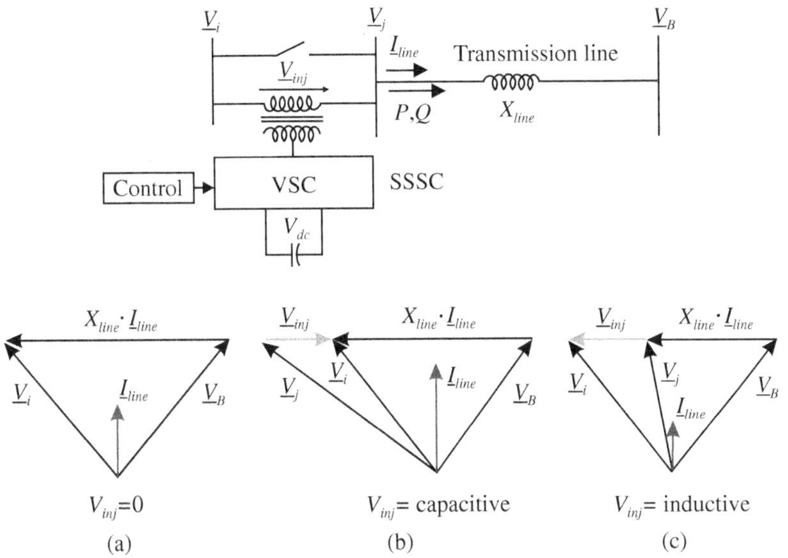

Figure 9.16 SSSC inserted voltage with respect to the line current: (a) neutral, (b) capacitive and (c) inductive [1].

the magnitude of the line current is unchanged; with capacitive $\underline{V}_{inj}$ the line current magnitude is increased (as shown in Figure 9.16b), and with inductive $\underline{V}_{inj}$ the line current magnitude is decreased (as shown in Figure 9.16c).

The control of the magnitude of the line current, in effect, controls the real power flow P (MW) in the transmission line. That is, the real power flow can be increased or decreased from the nominal (uncompensated) line power flow by the polarity of the injected voltage. In fact, the real power flow in the line can be reversed depending on the achievable change in the line current magnitude (or the rating of the SSSC) with respect to the nominal (uncompensated) line power flow value.

The SSSC is implemented with a VSC connected in series with the transmission line. The basic operation of the VSC is essentially to convert the DC voltage at their DC terminals into an AC voltage output, in accordance with a vector reference that defines:

- the ratio between the DC-side and AC-side fundamental voltage;
- the phase angle of the AC output voltage.

The basic controller structure for the SSSC is shown in Figure 9.17. For SSSC control (as well as for STATCOM), the VSC can be operated with a constant ratio between the DC voltage and the correspondingly defined AC output voltage, as one control option. In this control method, the phase angle of the VSC output voltage is thus the only controlled variable. Changes in the magnitude of the output (series injected) voltage are made by linearly charging and discharging the DC bus capacitors by momentarily changing the phase angle of the VSC output voltage, or by changing V_{dc} voltage with external means (e.g., a DC to AC converter). The inverter voltage

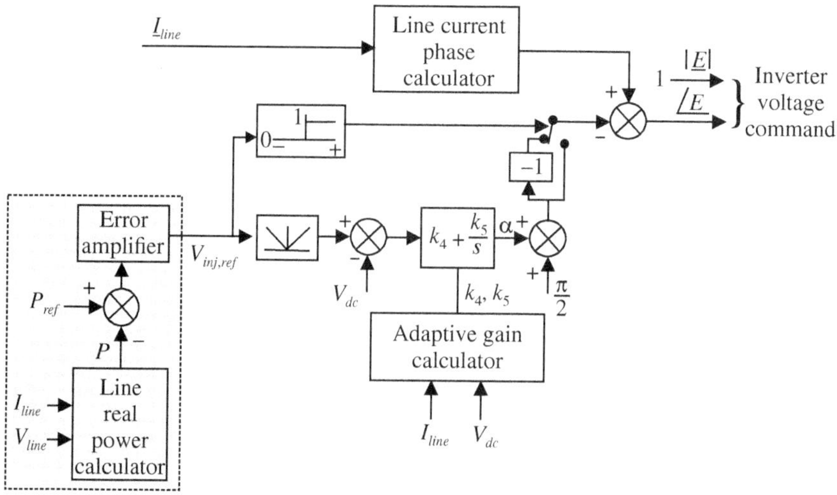

Figure 9.17 SSSC controller structure [1].

$\underline{V}_{inj}$ is kept substantially in quadrature with the line current $\underline{I}_{line}$. This can be done with leading or lagging polarity so that the injected voltage $\underline{V}_{inj}$ appears in the line as inductive or capacitive, respectively.

Charging of the DC bus capacitors (or V_{dc} voltage) is controlled by allowing the injected voltage to deviate slightly from a quadrature relationship with the line current, thereby exchanging real power between the VSC (i.e., the DC capacitors) and the line. Clearly, this mechanism will only work when there is a sufficient level of line current to allow power exchange—-typically, 0.05–0.1 p.u. line current. The control scheme thus has a singular region around zero line current, which must be avoided. In addition, the DC voltage control loop gain varies with the DC voltage and line current. Adaptively changing the loop gains can compensate for this problem as shown in Figure 9.17.

As a second control option, the VSC can also be vector controlled where both the magnitude and phase angle of the output voltage are independently controlled. In this case, the DC bus voltage V_{DC} is regulated to a constant value, as compared to the SSSC controller in Figure 9.17, where the DC bus voltage is linearly proportional to the magnitude of the output voltage.

The SSSC injected voltage (V_{inj}) control method shown in Figure 9.17 is an *open-loop control*—-that is, the operator "dials" in an injected voltage reference (as a percentage of the system bus voltage) and depending on the line impedance, the magnitude of the line current changes. Because the injected voltage range is small (typically 5% to maximum of 10% of the line/bus voltage), the line current magnitude change is linear. This is an important result because the operator can a priori predict the voltage injection required for a given line current magnitude change. This even allows the calibration of the real power changes to required voltage injection V_{inj} for a given transmission line in p.u. This "linear" control aspect of SSSC is attractive to the operator for real power dispatch, even with open-loop voltage injection V_{inj}.

If a vector control structure and a closed loop power flow controller are used (the shaded block from Figure 9.17), then the line real power P_{line} (MW) can be regulated to a desired P_{ref}. The real power on the line is calculated based on the line/bus voltage and current at the point of SSSC voltage injection bus. The power error is used to generate an injected voltage reference $V_{inj,ref}$, which is required to increase or decrease the line current magnitude. Note that in this case, the line real power flow can be increased or decreased from the nominal or uncompensated line power flow value, and can even be reversed.

For the SSSC, the reactive voltage injection is chosen to influence the power flow on the line. Capacitive injection tends to increase the magnitude of the line current, and inductive insertion tends to decrease it. With the SSSC, real and reactive power on the line cannot be independently influenced. In principle, the choice of the reactive voltage reference for a particular change in line power can be made automatically by a control loop.

It is important to note that the reactive power flow Q_{line} (MVAr) on the transmission line with the SSSC in operation is uncontrolled, that is, the reactive power flow on the line can increase or decrease depending on the line impedance and voltage injection.

9.4.3 Protection

The protection includes conventional protection devices for the conventional equipment and fast action protection functions in the control for the electronic devices of the VSC. The protection must ensure that all components are operated within their design limits. Conventional overcurrent and overvoltage protection is used for the conventional equipment. Special control-embedded overcurrent and overvoltage protections limit the voltage and current stresses of the semiconductors.

The protective action for the SSSC is to bypass the series insertion transformer as shown in Figure 9.18 [1].

This can be done mechanically by means of the bypass breaker, and/or electronically by means of the secondary side fast-acting thyristor bypass switches (TBSs). Protective bypass can be initiated at any time and will also be initiated by the SSSC under irrecoverable internal failures and line fault conditions. Because the SSSC inverter carries the line current, but does not dictate its value, the principal reason for bypass is inverter overcurrent protection. The method by which the SSSC comes out of bypass after an overcurrent needs to be defined. Also the possibility of initiating only an electronic bypass and not closing the mechanical bypass breaker can

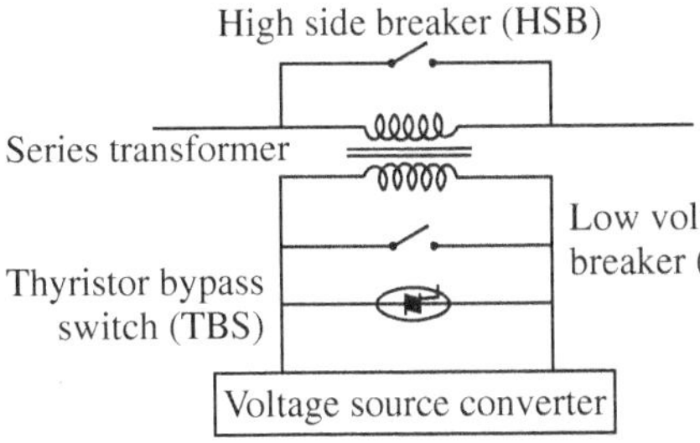

Figure 9.18 Thyristor bypass system (TBS) for protection of series-connected VSC-based SSSC.

also be investigated. Several options exist—-including reinsertion of the SSSC after the system line fault condition. Consider the case of a six-cycle line fault. Due to the line fault, the overcurrent limit of the VSC is reached and the VSC is bypassed by the fast-acting thyristor bypass circuit. The low voltage breaker (LVB) and the high side breaker are both remain open and after the fault, the SSSC is reinserted into the line with the same voltage injection as the prefault condition. If the system line fault recurs within a certain predetermined period or persists, both the TBS and the LVB can be closed to isolate the SSSC from the line, forcing a restart of the SSSC after the fault condition.

9.5 SSSC MODELING

9.5.1 Steady-State SSSC Model

9.5.1.1 VSC Controller Load Flow Models

In load flow and system stability studies, series VSC controllers introduce additional complexities. To accommodate a series VSC controller installed on a transmission line, the decoupled FACTS controller load flow model [10] uses a generator and a load to represent the series VSC controller, and the power injection model [11] adds additional equivalent power injections at the from-bus and the to-bus of the series VSC controller in the conventional load flow. However, both of these two bus injection models require additional computation effort in the regulation mode, and are not applicable to the fixed voltage injection mode or the rated capacity operation mode of the series VSCs.

On the other hand, the voltage source model (VSM) [12, 13], where the VSCs are modeled directly as voltage sources, is intuitive and efficient. Directly using the voltage sources, the Newton–Raphson algorithm shows good convergence properties by simultaneously adjusting all voltage variables, without needing an outer loop for changing the equivalent injected bus power or current. Another advantage is that the line current and power are readily computed, allowing direct enforcement of device limits. The VSM is capable of modeling different operating modes of various VSC controllers.

BALANCED POSITIVE-SEQUENCE VOLTAGE SOURCE MODEL. The balanced positive-sequence steady-state operation of an SSSC can be modeled as an injected voltage source ($\underline{V}_q$) in series with the series transformer reactance (X_t), as shown in Figure 9.19.

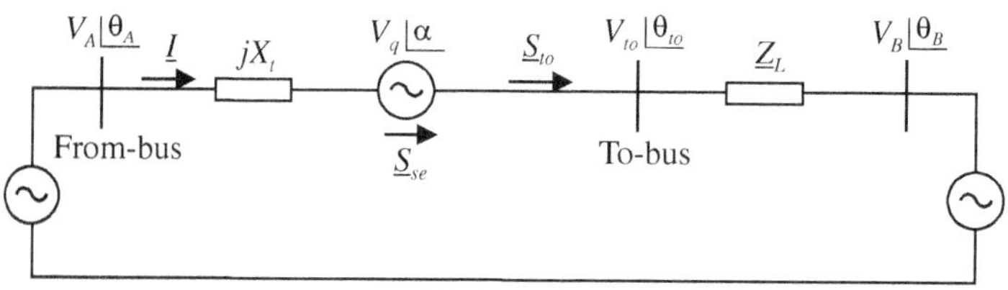

Figure 9.19 Injected voltage source model of the SSSC.

The line current flowing through the series VSC is given by

$$\underline{I} = \frac{\underline{V}_A - (\underline{V}_q + \underline{V}_{to})}{jX_t} \tag{9.3}$$

such that the power injected by the series VSC is

$$\begin{aligned}\underline{S}_{se} &= (\underline{V}_A - \underline{V}_{to})\underline{I}^* = (\underline{V}_A - \underline{V}_{to})\frac{(\underline{V}_A^* - (\underline{V}_q^* + \underline{V}_{to}^*))}{-jX_t} \\ &= \frac{|\underline{V}_A - \underline{V}_{to}|^2 - (\underline{V}_A - \underline{V}_{to})\underline{V}_q^*}{-jX_t}\end{aligned} \tag{9.4}$$

and the power injected into the "to-bus" is

$$\underline{S}_{to} = \underline{V}_{to}\underline{I}^* = \underline{V}_{to}\frac{\underline{V}_A^* - (\underline{V}_q^* + \underline{V}_{to}^*)}{-jX_t} = \frac{\underline{V}_{to}(\underline{V}_A^* - \underline{V}_q^*) - V_{to}^2}{-jX_t} \tag{9.5}$$

Control of active power with SSSC. As shown in Figure 9.5, the SSSC affects the active power transfer on the transmission line. It can be controlled in two different ways [1]:

(i) The SSSC controls the line active power flow P_{to} to a desired value P^{sp}:

$$P_{to} = V_{to}[V_q \sin(\theta_{to} - \alpha) - V_A \sin(\theta_{to} - \theta_A)]/X_t = P^{sp} \tag{9.6}$$

where V_A, θ_A, V_{to}, and θ_{to} are the series branch "from-bus" and "to-bus" voltage magnitudes and angles, respectively, V_q and α are the series voltage injection magnitude and angle, respectively, and X_t is the equivalent series transformer winding reactance.

(ii) The injected voltage is kept at a fixed voltage magnitude, in either the quadrature leading or lagging direction with respect to the transmission line current:

$$V_q = V_q^{sp}$$

In both control modes, the active power injected into the transmission line by the SSSC is zero

$$P_{se} = 0 \tag{9.7}$$

which can also be expressed as

$$V_A \sin(\theta_A - \alpha) - V_{to} \sin(\theta_{to} - \alpha) = 0 \tag{9.8}$$

In (9.7), the VSC losses are neglected. Active power can be injected into the transmission line if the series VSC has an energy storage system connected to the DC capacitor or it is connected to the DC bus of another VSC.

9.5.1.2 Newton–Raphson Load Flow Solution

In an N-bus power system with N_g generators and without any VSC, the load flow equations can be formulated as N–1 equations for the active power bus injections/loads P and $(N - N_g - 1)$ equations of reactive power bus injections/loads Q [1]:

$$\begin{cases} f_P([v]) = P \\ f_Q([v]) = Q \end{cases} \tag{9.9}$$

where $[v] = \begin{bmatrix} V_1 & V_2 & \dots & V_N & \theta_1 & \theta_2 & \dots & \theta_N \end{bmatrix}^T$ is a $2N$ vector of bus voltage magnitudes and angles.

The Jacobian used in the Newton–Raphson solution is:

$$J = \begin{bmatrix} \partial f_P/\partial V & \partial f_P/\partial \theta \\ \partial f_Q/\partial V & \partial f_Q/\partial \theta \end{bmatrix} \tag{9.10}$$

When the N-bus power network also includes M VSCs, then the load flow equation (9.9) will expand by $2M$ equations, resulting in:

$$\begin{aligned} f'_P\left(\left[v'\right]\right) &= P \\ f'_Q\left(\left[v'\right]\right) &= Q \\ f'_{VSC}\left(\left[v'\right]\right) &= R \end{aligned} \tag{9.11}$$

where [1]

$$[v'] = V_1 \quad \dots \quad V_N \quad \theta_1 \quad \dots \quad \theta_N \quad V_{q1} \quad \dots \quad V_{qM} \quad \alpha_1 \quad \dots \quad \alpha_M]^T \tag{9.12}$$

is a $2N+2M$ variables vector of bus voltage magnitudes and angles for both the network busses and for the VSCs equivalent sources, and the last equation in (9.11) is determined by the VSC operating modes, where R is a vector of the FACTS controller set points or reference values. In this approach, the P and Q expressions in (9.11) will remain unchanged for all operating modes.

To apply the Newton–Raphson algorithm to the augmented system (9.11), the Jacobian matrix $[J]$ from (9.10) is expanded to:

$$[J] = \begin{bmatrix} \partial f'_P/\partial V & \partial f'_P/\partial \theta & \partial f'_P/\partial V_q & \partial f'_P/\partial \alpha \\ \partial f'_Q/\partial V & \partial f'_Q/\partial \theta & \partial f'_Q/\partial V_q & \partial f'_Q/\partial \alpha \\ \partial f'_{VSC}/\partial V & \partial f'_{VSC}/\partial \theta & \partial f'_{VSC}/\partial V_q & \partial f'_{VSC}/\partial \alpha \end{bmatrix} \tag{9.13}$$

Note that the first 2 × 2 blocks of $[J]$ (namely, the (1,1), (1,2), (2,1), and (2,2) entries) are identical to the Jacobian J in (9.10), except for the additional terms due to the shunt and series VSC transformer reactances and injection terms. Thus, an attractive feature of this Newton–Raphson algorithm for solving load flow is that the formulation (9.11)–(9.13) can be readily built into a conventional Newton–Raphson algorithm. For large data sets, sparse factorization techniques can be used to achieve an efficient solution.

The operating limits are imposed on the load flow computation to ensure that the VSCs are not overloaded, and that the voltages of the adjacent buses are within an acceptable range. For the SSSC, the operating limits include:

- the maximum series injected voltage;
- the maximum series line current;

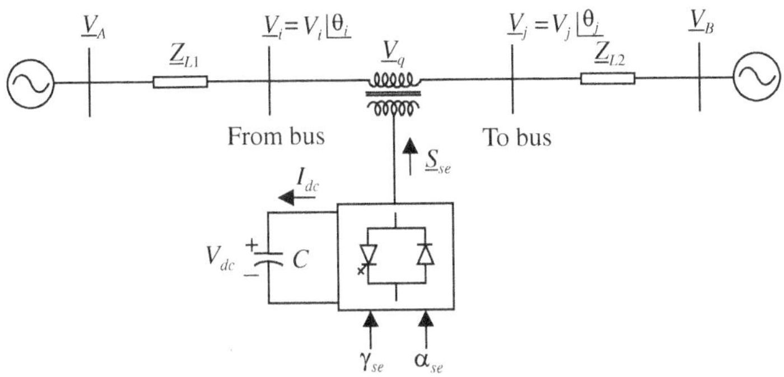

Figure 9.20 Series voltage-sourced converters showing DC capacitor [1].

- the series VSC MVA rating;
- the maximum and minimum voltage magnitudes at adjacent buses;
- the maximum power transfer between the converter and the energy storage system, if the series VSC is coupled with an energy storage system.

If any of these limits are violated, the voltage or flow setpoints of the SSSC can no longer be enforced.

9.5.2 SSSC Dynamic Model

Figure 9.20 shows the schematic diagram of the SSSC, where γ_{se} is the conversion ratio signal to control the series converter voltage magnitude, and α_{se} is the phase angle of the series VSC, $\underline{V}_q$. Note that α_{se} is measured with respect to the angle of the "from-bus" voltage phasor $\underline{V}_i$.

In the time scale of transient stability analysis, in which the VSC switching dynamics are neglected, the voltage insertion model of the series VSC can be represented as:

$$\underline{V}_q = kV_{dc}e^{j(\alpha_{se}+\theta_i)} = V_q e^{j\alpha} \tag{9.14}$$

where k is a factor that relates the inverter DC-side voltage to its AC-side terminal voltage, and θ_i is the angle of $\underline{V}_i$; k is dependent on the ratio γ_{se}.

The dynamic balanced positive-sequence model of the SSSC is shown in Figure 9.21.

The series VSC is modeled as a controllable injected voltage source $\underline{V}_q$ behind an equivalent transformer reactance X_t. The line current $\underline{I}$, the power $\underline{S}_{se}$ injected by

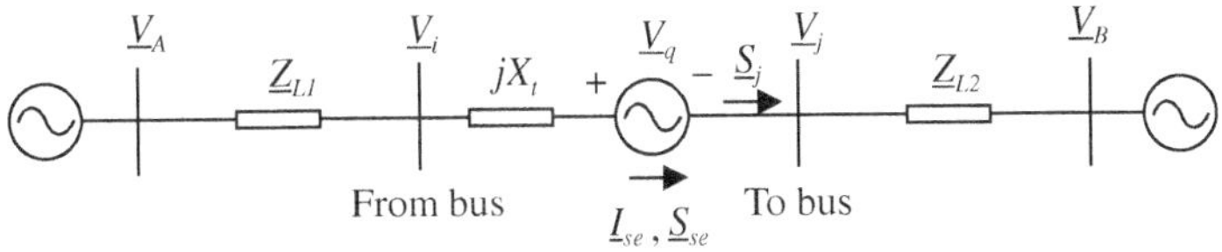

Figure 9.21 SSSC balanced positive-sequence model

the series VSC, and the power $\underline{S}_j$ injected into the to-bus (bus j) are the same as given in (9.3,...,9.5) for the steady-state series VSC model, respectively.

During transient stability studies, the DC link capacitor of the SSSC will exchange energy with the AC system and consequently its voltage varies. The variation of the DC capacitor voltage is dependent on its current inflow, which can be modeled as:

$$C\frac{dV_{dc}}{dt} = I_{dc}, \tag{9.15}$$

where I_{dc} is the current flowing into the DC capacitor from the VSC, and C is the capacitance of the DC capacitor. In steady-state operation, the power transfer is balanced, and hence, $I_{dc} = 0$ and $dV_{dc}/dt = 0$.

The SSSC dynamic model will be interfaced with the other dynamic components in a power system, such as synchronous generators and excitation systems, through the transmission network. In using injected voltage source for the VSC in the load flow formulation, this transition to dynamic model will be seamless because $\underline{V}_q$ is a variable in the load flow model.

The AC instantaneous active power injected into the power system for an SSSC is given by (9.7). Assuming that the VSC model is ideal, the AC instantaneous active power in the AC-side is equal to the DC-side active power, that is,

$$P_{se} = V_{dc}I_{dc} \tag{9.16}$$

From (9.15) and (9.16), we have

$$\frac{dV_{dc}}{dt} = \frac{1}{CV_{dc}P_{se}} \tag{9.17}$$

Equation (9.17) is in general not per-unitized.

The block diagram of the DC link dynamics is shown in Figure 9.22.

The SSSC can be either in the line active power control mode or the inverter voltage magnitude control mode. The block diagrams for these two modes are shown in Figures 9.23a and 9.23b, respectively. The magnitude V_q and the angle α of the inverter voltage are generated by the control systems.

The SSSC is operated with a constant conversion factor k between its AC-side voltage V_q and DC-side voltage V_{dc}, and hence the changes in the magnitude of the inverter output voltage are achieved by charging or discharging the DC bus capacitor to a different voltage.

A proportional-integral (PI) controller K_p+K_i/s and a low pass (LP) filter $1/(1 + Ts)$ are applied in the angle control loop of the SSSC. The input signal for the line active power control mode is the difference of the line active power setpoint P_{ref} and its measured value P, while the input signal for the inverter voltage magnitude control mode is the difference of the inverter voltage magnitude setpoint $|V_{qref}|$ and kV_{dc}.

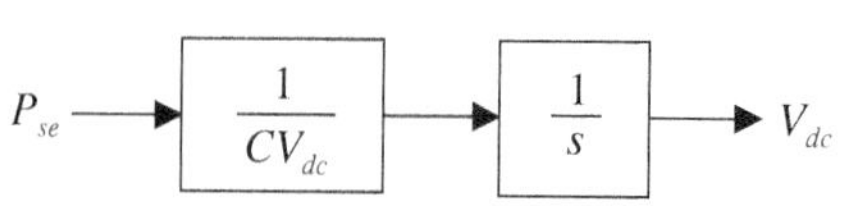

Figure 9.22 DC link dynamics.

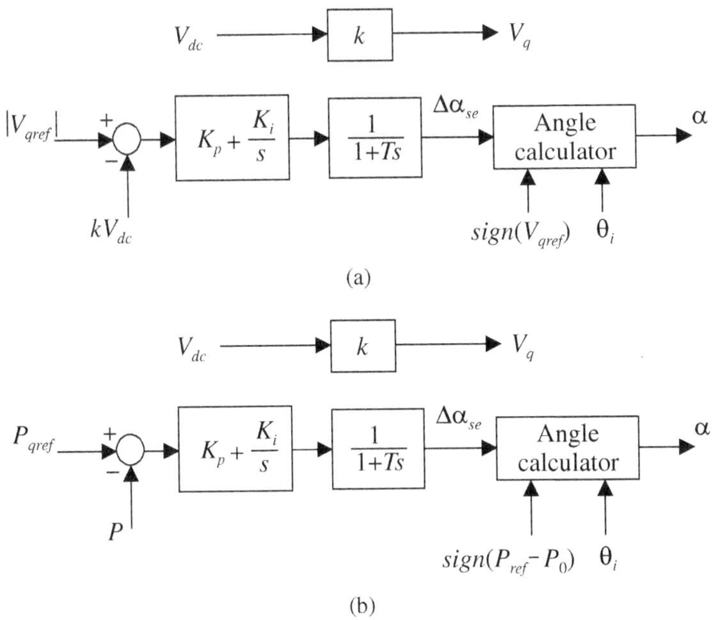

Figure 9.23 SSSC set point control diagrams: (a) inverter voltage magnitude control mode and (b) line P control mode.

In each operating mode, the output signal from the LP filter is the angle deviation $\Delta\alpha_{se}$. In steady state, $\Delta\alpha_{se}$ is zero, implying that the inverter output voltage is kept essentially in quadrature with the current of the compensated line. Small transient positive or negative deviations in the phase angle of the inverter voltage will cause nonzero active power to go through the DC capacitor, and thus result in changing the DC bus voltage.

9.6 APPLICATIONS

The SSSC is one of the most powerful FACTS controllers for power flow control. Although it can provide transmission line voltage regulation by the control of the effective line impedance, particularly for the end-voltage of the radial line, the primary purpose of the SSSC is usually the direct, and dynamic, control of the transmitted power in a given line. The main applications within the broad area of adjustable or dynamic power flow control are next presented [1].

(i) Compensation of long transmission lines

Because of the relatively large series inductance of long transmission lines and the correspondingly large transmission angle required, the voltage profile substantially drops toward the midpoint with increasing line current, causing heavy flow of reactive power and an uneconomically low limit on the transmittable real power. Series capacitive compensation is the most effective means to increase the transmittable power by reducing the overall transmission impedance and angle. To handle changing load conditions and line contingencies, the series compensation may need to be variable. In addition, concerns of subsynchronous resonance may limit safe degree of series compensation by conventional capacitors. The SSSC can provide an economical solution in combination with conventional series capacitor banks to provide a Vernier control by adding to, or subtracting from, the fixed compensation provided by the capacitors, and also to increase the immunity against subsynchronous oscillation.

(ii) **Equalization of power flow in lines and prevention of loop-flows of real power**

The inherent capability of the SSSC to decrease as well as to increase (real) power flow in the line makes it eminently suitable to equalize line power flows and prevent loop-flows of power. Both of these allow higher power transmission, and thus higher utilization of the power system, without thermal limit problems.

(iii) **Receiving-end voltage regulation of a radial line**

The receiving-end voltage of a radial line varies with the load and, in particular, with the power factor of the load due to the reactive line impedance. The SSSC can be used to regulate the end-voltage of a radial line by controlling the degree of series compensation so as to keep the end-voltage constant in face of changing load and load power factor. Note that, although the end-voltage can be regulated, the load power factor cannot be corrected by series compensation.

(iv) **Improvement of transient stability and dynamic stability (power oscillations damping)**

Transient stability improvement and power oscillations damping by shunt-connected STATCOM are achieved through the increase and subsequent decrease in the transmission line (midpoint) voltage during the accelerating and decelerating swings of the disturbed machine(s). The powerful capability of the series-connected SSSC to control the transmitted power can be utilized more effectively to increase the transient stability limit and to provide power oscillations damping. In these applications, the SSSC is controlled to force the line current and power flow to increase, by rapid injection of the necessary compensating voltage, when the relevant machines in the system accelerate, and, conversely, force the line current and transmitted power to decrease when the machines decelerate. The function of rotational stability improvement can be, of course, combined with power flow objectives relevant to the dynamically stable power system (e.g., power flow optimization and line current equalization).

9.7 SSSC INSTALLATION

9.7.1 SSSC in Operation

There are three SSSCs installed on transmission grids, SSSC portions included in the interline power flow controller installed on New York Power Authority (NYPA) transmission system (see Chapter 11), SSSC included in the unified power flow controller (UPFC) installed at Inez substation of American Electric Power (AEP) transmission system, and the UPFC installed on the Korean transmission system at Kangjin.

9.7.2 SSSC for Power Flow Control: A Project in Spain

9.7.2.1 *Project Overview*

With the installation of an SSSC, the risk of overload conditions in a 220 kV line that has a high variability in its flow in certain conditions of demand and generation will be minimized [14].

The SSSC operates by injecting an AC voltage at the fundamental frequency in series with the transmission power line. The amplitude of the injected voltage is controllable, and the phase angle is set to lead or lag the transmission line current by 90° in order to increase or decrease the line current and thereby the transmitted power.

There are numerous other reasons to consider the applications of new technologies to achieve a better use of the existing infrastructure, such as:

- Some new technologies are reaching levels of sufficient reliability and more affordable prices.
- Massive integration of delocalized new sources of electricity generation has substantially modified the energy flows through the network, even more in the case of renewable generation, dependent on primary resources at any time. As a result, some facilities may operate with high loads in some situations and with low loads in others.
- The environmental factors and public opinion are making it more and more difficult to install new infrastructures. From the environmental point of view, the installation of a new FACTS equipment results in less impact than the construction of a new transmission or distribution line.

This physical reality gives little opportunities for controlling flows through lines other than altering the amount of power that must be transmitted from A to B by redispatching generation or loads or modifying the topology of the grid. A voltage reduction at the line ends V_1 and V_2 in order to reduce the power flow implies a voltage change also in the parallel paths, so the angle difference would increase in order to transmit the same amount of power. The higher angle difference would then increase as well the flow through the line trying to be controlled. The impedance X_{line} of a line depends on the physical configuration of the line and is therefore given.

A solution that is normally used nowadays to control power flows across a line is by changing the topology of the system to modify the impedance of the parallel paths. These topology changes are generally obtained by opening a line or a coupling switch between bus-bars in a double bus-bar substation and are only done if in the final scenario the N–1 security criterion is met. These maneuvers in general reduce the redundancy of the grid as they make it less meshed and, although the N–1 criterion is fulfilled, the probability of a cascading event is higher in case of multiple simultaneous failures not considered in the security analysis. Reducing the meshing of the system is usually not desirable unless the network was planned and developed taking into account that some lines or switches would normally be operated open, which happens only rarely.

In this context, the use of load-flow control equipment that is able to alleviate congestions in specific areas is desirable as a temporary measure until the final transmission system enforcements are put into service. In a few lines of the Spanish system, which are located in areas where it is no longer possible to reinforce the grid, load-flow equipment (phase shifting transformers) are being installed permanently to prevent overloading of the line.

The function of the SSSC is to introduce in series with the transmission line (where the power flow needs to be regulated) a voltage with a controlled magnitude and phase. The voltage will lead or lag the line current by 90°.

9.7.2.2 *Components of the SSSC*

The voltage source will be implemented with the use of a coupling transformer whose primary winding is connected in series with the line and the secondary feeds a power electronic converter (Figure 9.24).

The converter used is of the VSC type that converts DC voltage into AC with controllable phase and magnitude. The VSC is composed of several three-phase neutral-point clamped (NPC) diode clamped three-level inverters. These power electronic units are used in series and/or in parallel in order to achieve the required power rating. The series/parallel combination of the unit will be accomplished by appropriate intermediate magnetic components.

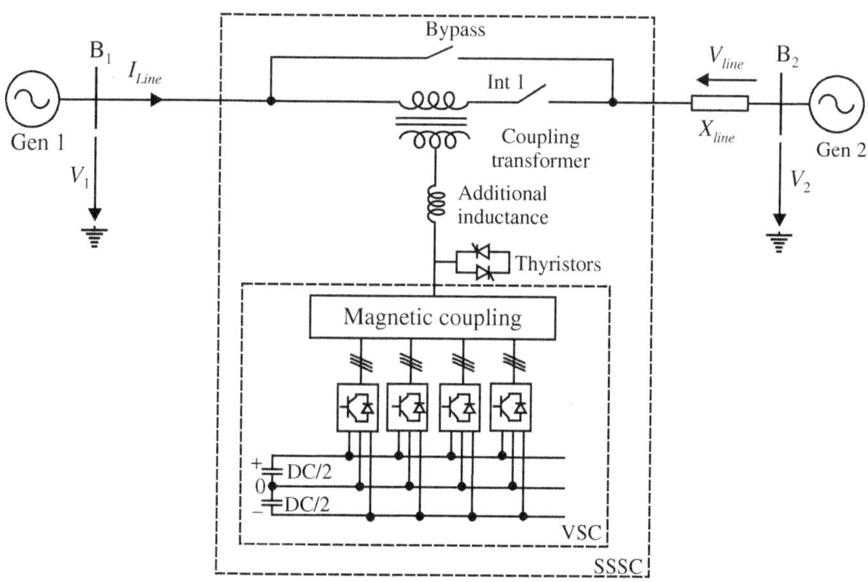

Figure 9.24 Schematic diagram of SSSC in a transmission line and main components [14].

The inverters will be implemented with the use of state-of-the-art IGCT semiconductor technology. These high power-medium voltage devices, which belong to the thyristor family, have an optimized gate with an integrated driver, allowing switching frequencies in the range of 500 Hz. The initial studies will consider the use of two different modulation strategies to obtain the converter output voltage: PWM and synchronous full-wave multipulse modulation. The former obtains a very fast output response, whereas the latter reduces the semiconductor loss.

To reduce the total power rating of the VSC, the addition of an external series inductance has been considered that reinforces the predominant inductive response of the SSSC's output power. This external inductance is so that the total static impedance (external inductance plus leakage inductance of the transformer) of the SSSC is 3 Ω.

Taking into account the possible large short-circuit current that may be encountered in the 220 kV transmission grid, corresponding to about 50 times that present during the normal power transmission of the line, it is evident that it is necessary to protect the VSC semiconductors in case of a short circuit. This can be done with a dedicated bypass installed in the secondary of the coupling transformer. The connection and disconnection of the SSSC will be done using the bypass, which is placed in the primary of the coupling transformer. The state of the primary and secondary switches will define the next functional states for the SSSC:

- Nonoperative when it is bypassed by the breaker placed in the primary of the transformer.
- Operative when the breaker of the primary is opened. Under this situation, two cases are to be distinguished:
  - Dynamic compensation: When the line current is less than 1250 A and the thyristor bypass of the secondary is open.
  - Static compensation: When the current across the line is greater than 1250 A and the bypass of the secondary of the coupling transformer is closed. In this case, the reactance that is imposed by the SSSC is the additional external reactance of the secondary plus the leakage inductance of the coupling transformer (more or less 3 Ω).

9.7.2.3 Location Selection for Prototype Installation

The SSSC is going to be used to avoid overloads in 220 kV lines of the region of Navarra, especially in high wind situation. Nowadays, in the location chosen, during some situations of generation and demand, the only available solutions to this problem are to unmesh the grid (and thereby reduce the security margin) or to reduce wind power generation (and thereby its economic benefits). With the installation of the SSSC in series with one of the 220 kV lines of the chosen region, it is expected that the number of problems requiring the above solutions will be appreciably reduced.

The objective of the SSSC equipment is, as stated above, to avoid the overloading of one (or several) 220 kV line(s) by deviating part of the flow that would naturally pass through that line to another alternative path.

In order to achieve these objectives, the following *functional*, *physical*, and *system security* factors have to be taken into account during location selection.

(i) From *functional* considerations, the following premises have to be fulfilled by the location selected:

a. There must exist an alternative path to the line that becomes overloaded.

b. The desired performance has to be achievable with the equipment rating and design characteristics.

c. Defined regulation range (equivalent range from 4 capacitive ohms to 10 inductive ohms), for the scenarios considered.

d. The conditions in which the equipment will actuate have to be frequent enough for assessing its performance.

(ii) For the *physical implementation* of the equipment, a minimum surface of $20 \times 20\ m^2$ is needed, plus the space for the high voltage (HV) switching and measuring equipment. The connection of the equipment to the grid will be done through a series transformer.

Three breakers are foreseen, one of them to configure the by-pass that will allow the disconnection of the equipment without disconnecting the line, and the other two to facilitate the protection of the equipment. The already existing breaker for switching the line will be used. Disconnectors and current and voltage transformers are also to be installed. Interconnection of the different equipment will be done with Lapwing-type aluminum/steel cable.

(iii) From *system security* point of view it is, on the one hand, important to minimize the impact of a possible malfunctioning of the equipment and, on the other hand, it is difficult to obtain permission to put out of service critical lines to do the works, and this may delay the connection of the equipment to the grid.

The following are some of the advantages that have been identified if the SSSC is installed as compared with more conventional solutions, for example, with the installation of a new line:

- It occupies less real estate. This leads to:
  - less environmental impact and thus less social rejection and, consequently, easier to obtain all necessary permits;
  - less terrain costs.
- The time that is needed to put in service the SSSC is shorter (partially, because it is easier to obtain permits).
- The total cost of the necessary equipment is less.
- It is foreseen that in the future, the SSSC dynamic capabilities could be exploited, even if today is not one of the main application concerns.

REFERENCES

[1] Edris, A. A. (Convener), Chow, J., Watanabe, E., Barbosa, P., Halvarsson, P., Angquist, L., Fardenesh, B., Uzunovic, E., Huang, A., Tyll, H., Batacharia, S., Enslin, J., Ivakin, V., and Wiik, J. Static

synchronous series compensator (SSSC). CIGRE Working Group B4-40, Technical Brochure no. 371, January 2008.

[2] Hingorani, N., and Gyugyi, L. *Understanding FACTS: Concepts and Technology of Flexible AC Transmission Systems*. IEEE Press, New York, 2000.

[3] Gyugyi, L., Schauder, C. D., and Sen, K. K. Static synchronous series compensator: a solid-state approach to the series compensation of transmission lines. *IEEE Transactions on Power Delivery*, vol. 12, no. 1, pp. 406–417, January 1997.

[4] Song, Y. H., and Johns, A. T. (Eds.) *Flexible AC Transmission Systems (FACTS)*. IEEE Power and Energy Series, IEEE London, 1999.

[5] Sen, K. K. SSSC static synchronous series compensator: theory, modeling and application. *IEEE Transactions on Power Delivery*, vol. 13, no. 1, pp. 241–246, January 1998.

[6] Karady, G. G., Ortmeyer, T. H., Pilvelait, B. R., and Maratukulam, D. Continuously regulated series capacitor. *IEEE Transactions on Power Delivery*, vol. 8, no. 3, pp. 1348–1355, July 1993.

[7] De Souza, L. F. W., Watanabe, E. H., and Aredes, M. GTO controlled series capacitors: multimodule and multipulse arrangements. *IEEE Transactions on Power Delivery*, vol. 15, no. 2, pp. 725–731, April 2000.

[8] Edris, A. A., Gyugyi, L. Power electronics-based T&D controllers technology developments, 2003 *IEEE PES Transmission and Distribution Conference and Exposition*, vol.3, pp. 1131–1137, 2003.

[9] De Souza, L. F. W., Watanabe, E. H., Alves, J. E. R., and Da Pilatto, L. A. S. Thyristor and gate controlled series capacitors: comparison of components rating. *Proceedings of IEEE Power Engineering Society General Meeting*, vol. 4, Toronto, Canada, pp. 2542–2547, July 2003.

[10] Gotham, D. I., and Heydt, G. T. Power flow control and power flow studies for systems with FACTS devices. *IEEE Transactions on Power Systems*, vol. 13, no. 1, pp. 60–65, February 1998.

[11] Noroozian, M., Angquist, L., Ghandhari, M., and Anderson, G. Use of UPFC for optimal power flow control in electrical power networks. *IEEE Transactions on Power Delivery*, vol. 12, pp. 1629–1634, 1997.

[12] Fuerte-Esquivel, C. R., and Acha, E. A Newton-type algorithm for the control of power flow in electrical power networks. *IEEE Transactions on Power Systems*, vol. 12, pp. 1474–1480, 1997.

[13] Therond, P.G. (Convenor), Adielson, T., Bortoni, P., Byrne, D., Ceresoli, Daniel, D., Foss, A. M., Fosso, O. B., Iyoda, I., Mihalic, R., Miller, N. W., Modlitba, P., Nelson, R., Thanawala, H. L., Uhlen, K., Witzmann, R. Modeling of power electronics equipment (FACTS) in load flow and stability programs. CIGRE TF 38.01.08, Brochure 145, August 1999.

[14] Alvira, D., Torre, M., Bola, J., Burdalo, U., Marquez, M., Rodriguez, M. A., Chivite, J., Hernandez, A., and Alvarez, S. The use of a static synchronous series compensator (SSSC) for power flow control in the 220 kV Spanish transmission network. CIGRE 2010, paper B4-107, Paris, 2010.

CHAPTER 10

UNIFIED POWER FLOW CONTROLLER (UPFC)

Laszlo Gyugyi

10.1 INTRODUCTION

The unified power flow controller (UPFC) is a generalized AC transmission controller that has multiple functional capabilities and thus, it is capable of performing simultaneously or selectively the compensation and control functions of different individual line compensators and controllers.

There are several different ways to introduce and explain the concept of the UPFC. These include mathematical approach, vector or phasor diagrams, or various plots showing graphically relationships among the main transmission parameters, power (active and reactive), voltage, line impedance, and transmission angle. In a similar manner, the understanding of the UPFC, that is, the visual concept one's mind ultimately develops, is heavily dependent upon the individual's established way of looking the operation and control of power systems based upon his or her theoretical understanding and background experience. Since the broad practical applications of sophisticated power electronics-based transmission controllers depend so much on how utility planners, engineers, and operators can relate them to power system requirements and problems, the introduction of the UPFC concept here, in contrast to the customary theoretical approach, is carried out from well-established power transmission operating and control principles.

10.1.1 UPFC as the Functional Combination of Conventional Transmission Controllers

Consider the simple case of transmitting power between bus 1 and bus 2, as shown in Figure 10.1. Each of the two busses may be represented with a generator and source impedance. The two generators are in steady-state synchronism and at zero power transmission, the magnitudes of the two bus voltages are the same.

It is well understood from basic power transmission theory that the transmitted power P between bus 1 and bus 2 and the reactive power Q supplied from the busses are dependent on the magnitude V of the bus voltage $\underline{V}$ (considered as a phasor or

Advanced Solutions in Power Systems: HVDC, FACTS, and Artificial Intelligence, First Edition.
Edited by Mircea Eremia, Chen-Ching Liu, and Abdel-Aty Edris.

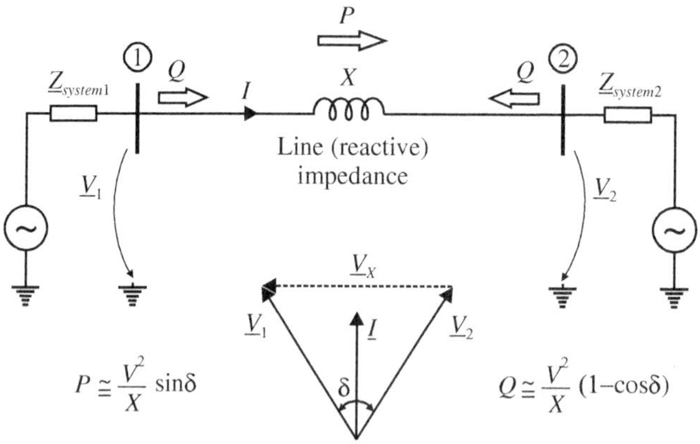

Figure 10.1 Two-machine, single-line power system with corresponding phasor diagram and transmission equations.

vector), the prevailing angle δ, between the voltages at the two busses, and the (reactive) line impedance X, as expressed by the simple equations:

$$P \cong \frac{V^2}{X} \sin \delta \tag{10.1}$$

$$Q \cong -\frac{V^2}{X}(1 - \cos \delta) \tag{10.2}$$

If we assume for this example that the utilization of the conduction capacity of the line and corresponding desired power transmission are not achievable with the existing line impedance and prevailing transmission angle (due to long distance and other interconnections in the overall system), then existing practice would suggest the well-established solutions of applying *series capacitive compensation* to obtain the desired overall transmission impedance and, if the required angle still remains insufficient, a *phase angle shifting transformer* as well. Thus, it can be concluded that in a general case to ensure complete control of power flow in the line, both the line impedance and transmission angle may have to be controlled.

However, further examining the possible transmission problems in the present example, let us also assume that the source impedance behind bus 1 is relatively large and the corresponding voltage drop constrains the desired power transmission (and also causes voltage-related operating problems in the system). The existing practice would suggest the obvious solution of using a *shunt compensator* to provide the necessary voltage support for bus 1.

Thus, as illustrated in Figure 10.2, the full control of a transmission line may require three independent controllers. It should be noted here that, although each of the three controllers is independent with regard to controlling its "own" parameter (i.e., impedance, angle, and voltage), they may have an interaction on each other. For example, the action of changing the effective transmission impedance by series compensation may also require an appropriate change in the overall transmission angle to get the desired power flow and vice versa.

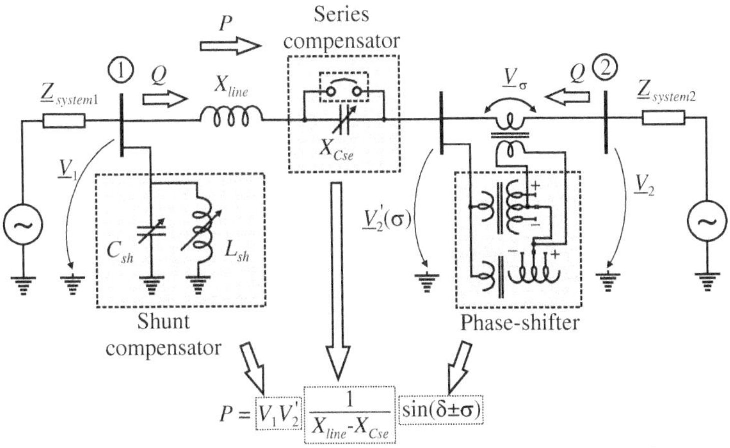

Figure 10.2 Full transmission control: reactive shunt compensation for voltage, series compensation for impedance, and phase-shifting for angle control.

The example so far considered assumed a steady-state condition for the power transmission. Under this condition, the desired transmission can be achieved with some fixed capacitance/reactance for series and shunt compensation, as well as with some fixed angular adjustment. However, it is evident that under different transmission conditions, these values would require readjustment, and, in particular, under the conditions of dynamic disturbances, the stabilization of the system could necessitate the rapid controllability of these parameters.

In the above example, on the basis of classical transmission control practice, we employed conventional reactive series and shunt compensation, as well as phase-shifting to achieve the required transmission control. Although the name of each action seems to imply the ways the required result is achieved, it is still worthwhile to examine the actual physical phenomena taking place with these to prepare for the understanding of the UPFC concept.

Considering first the series capacitive compensation (phase shifter is assumed bypassed), the voltage between bus 1 and bus 2 is applied across the series-connected compensating capacitor C_{se} and reactance of the line X. The line current produces a voltage across X that leads, and across the capacitor that lags, this current by 90°. The sum of these two voltages must equal the prevailing voltage between the two busses, as illustrated by the simple phasor diagram in Figure 10.3. It is to be observed that, in a physical sense, the series capacitor in effect increased the voltage across the existing physical line reactance, and thereby increased the line current, and the transmitted power, in the same way, as if the line reactance had been reduced. This observation leads to the conclusion that series capacitive compensation in the actual physical sense achieves power flow control by changing, that is, increasing the *voltage* across the existing line reactance, and *not* the reactance itself. This means that the same end objective of power flow control can be achieved by any practical means that can increase (or decrease) the magnitude of the voltage across the existing (reactive) impedance of the transmission line. An additional observation also needs to be noted,

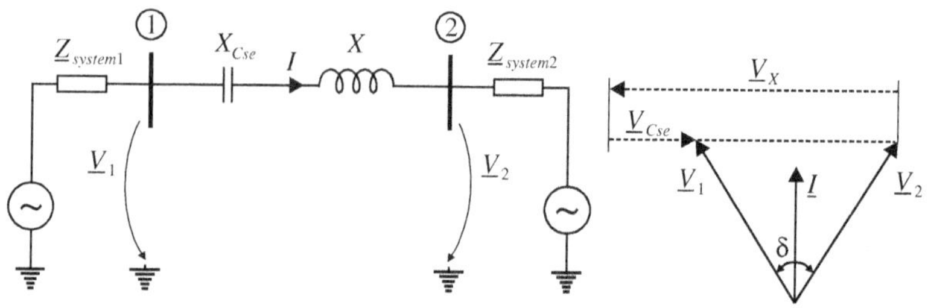

Figure 10.3 Series capacitive line compensation and corresponding phasor diagram.

that is, the *voltage across the line reactance is in quadrature with the line current*, and therefore the adjustment of this voltage, whether achieved by a series capacitor or other means, does involve only an exchange of *reactive* power, but no active power.

Considering now the operation of the phase-shifter (bypassed series compensation), the voltage across the line is the (vectorial) sum of three voltages. To obtain the desired power transmission, the voltage at bus 1 is modified to that at bus 1′, to which the line is actually connected (see Figure 10.4).

Thus, the voltage across the line is the voltage between bus 1′ and bus 2. The phase shifter shown in Figure 10.4 is a conventional transformer type that can inject, via the insertion transformer, a voltage composed of two components, one in phase, the other in quadrature with the voltage at bus 1. This is necessary to maintain the magnitude of the phase-shifted voltage at the required value independent of the amount of phase-shift applied. (It is to be noted that this arrangement would also facilitate the control of the bus voltage while maintaining the phase shift at a fixed value.) The physical operation of the phase shifter leads again to the same conclusion as reached for series line compensation, that is, the phase shifting needed to obtain the transmission angle necessary to achieve the desired power flow is done by *injecting a specific voltage* between one of the two busses and the line, to establish the needed voltage *across* the line. It is evident that, from the standpoint of this objective, the means by which the voltage injection is accomplished is unimportant.

Besides the conclusion that power flow control using series reactive compensation or phase shifting is essentially achieved by the injection of an appropriate

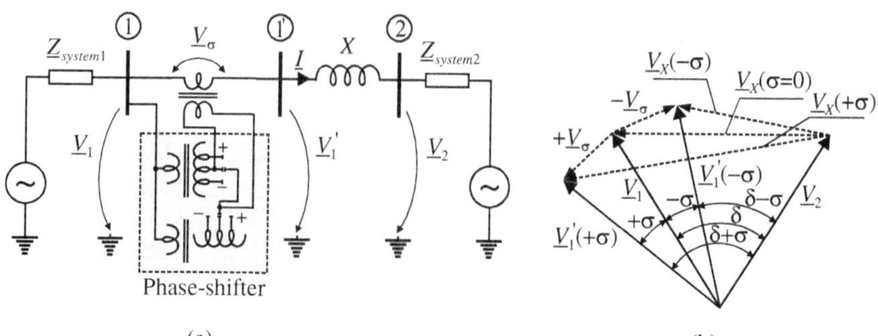

Figure 10.4 Phase-shifting for transmission angle control (a) and corresponding phasor diagram (b).

voltage between the line and the "sending-" and/or "receiving-end" busses of the power system, there are some important differences between the two approaches. One important difference is that series (capacitive) compensation by definition is *reactive*, and the injected voltage is always in quadrature with the line current. By contrast, phase shifting injects a voltage to establish and maintain the desired transmission angle, and therefore this voltage has no predefined relationship with the line current. Consequently, the phase shifter *does,* in general, it exchanges *both* active and reactive power with the system. Thus, while the series capacitive compensation is self-supporting (the capacitor supplies the reactive power that the line reactance absorbs), the transformer-type phase shifter has no capability to this and both the reactive and active power exchanged in the process of phase shifting must be supplied externally, usually by the system itself. (In practice, shunt capacitive compensation is often applied to supply the reactive power.)

In comparing further the phase shifter to the series reactive compensator, the deduction can readily be made that the generalized phase shifter with combined in-phase and quadrature voltage injection capability could perform the specific function of series compensation by keeping the injected voltage in quadrature with the line current (of course, without supplying the necessary reactive power itself). Moreover, because of the two-dimensional voltage injection capability of the generalized phase shifter, the more general deduction that the such phase shifter could actually provide *both* functions of series compensation and angle control by injecting the resultant voltage, that is, the sum of that the two controllers would separately provide to satisfy the specific power transmission, as illustrated in Figures 10.5a and 10.5b. (Figure 10.5a shows an example for combined series compensation and phase shift, and Figure 10.5b illustrates the case when the same transmission line condition is accomplished by the phase shifter alone.) Note, however, that in the (b) case, the reactive power, previously supplied by the series capacitor, is supplied now by the system from bus 1.

Despite the attractive functional flexibility of the generalized phase shifter for power flow control, it has very rarely been applied in practice. (Even when it was, the in-phase injection capability was used separately for bus voltage control and the quadrature one for angle regulation.) There are several reasons for this: the multi-transformer arrangement with taps is complex and difficult to realize for high voltage levels; the tap-changer gear is slow and requires regular maintenance; since it has no internal capability to supply reactive power, the total VA the arrangement exchanges through the series voltage injection with the line must be provided at the primary side by the system itself, which, due to the possible high reactive power content, can make the overall transformer rating and reactive power flow in the primary-side supply line high, usually necessitating a separate reactive power compensator.

These considerations led to the conclusion that a practically viable generalized power flow controller capable of controlling all the parameters determining power transmission (voltage, impedance, and angle) selectively or in combination must have an implementation with inherent reactive power generation/absorption capability.

A conceptual model for the generalized power flow controller with inherent reactive power generation capability can be derived once again with a conventional

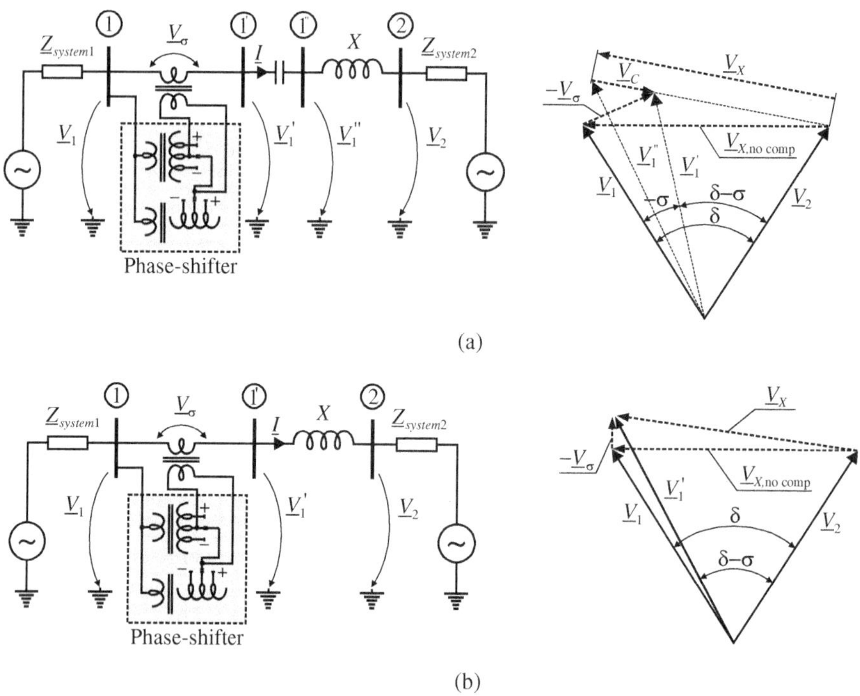

Figure 10.5 (a) Series capacitive compensation plus phase-shifting and (b) identical transmission end-result with the phase shifter providing independent two-dimensional (in-phase and quadrature) voltage injection.

power system component, the rotating synchronous compensator or, in a more general form, the synchronous machine. Such a model is illustrated in Figure 10.6.

To understand the operation of this model, one has to recall that a synchronous machine can be operated both as a generator and as a motor. In the former case, the machine receives mechanical input power and it converts into electrical output power; in the latter case, its input is electrical power that it converts into mechanical output power. Independently whether it is operated as a generator or as a motor, it can generate or absorb reactive power (VAr) by excitation control (VAr generation by overexcitation; VAr absorption by underexcitation). If two synchronous machines are mechanically coupled to run at a common synchronous speed, as shown in Figure 10.6, then controllable active electrical power can flow from one electrical terminal to the other in either directions. At the same time, regardless of the direction of active power flow, reactive power can be generated or absorbed at the two terminals independently of each other.

The operation of the two-machine model is illustrated by the phasor diagram in Figure 10.6b. Machine 2 inserts a voltage $\underline{V}_{pq}$ between the bus 1 and the line at bus 1′. The voltage across the line, $\underline{V}_X$, the line current I, and the transmitted power is thus determined by the voltage between bus 1′ and bus 2. The magnitude of the inserted voltage $\underline{V}_{pq}$ is fully controllable from zero and a maximum, and its phase between

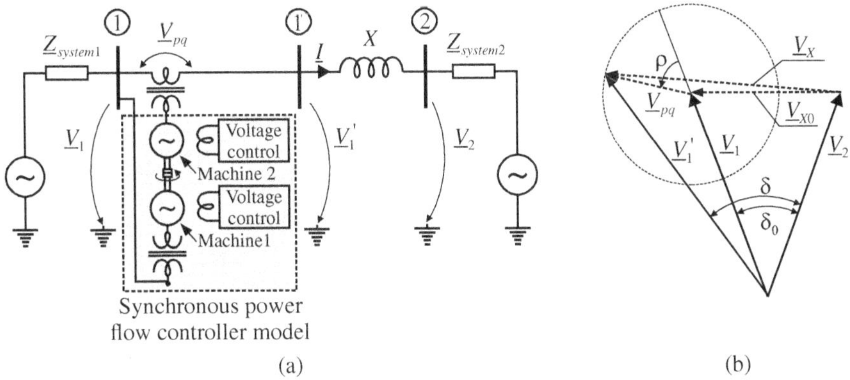

Figure 10.6 (a) Two-machine synchronous power flow controller model with internal reactive power generation for self-sufficient series and shunt reactive compensation and (b) corresponding phasor diagram.

0° and 360°, relative to the voltage at bus 1. This inserted voltage, in general, results in an exchange of both active and reactive power with machine 1 and the line. The reactive power exchanged is generated by machine 2 itself. However, the active power is supplied by the system at bus 1 via machine 1 that, as a motor, drives machine 2. In addition, machine 1 can generate (or absorb) reactive power, if needed, to regulate the voltage at bus 1.

From the previous discussion, it should be evident that the angularly unconstrained voltage injection can not only emulate the operation any one of the traditional transmission compensators/controllers used for voltage regulation, series compensation, and angle control (e.g., see Figure 10.5), but also provide an equivalent control action as if all three of those would be separately operated by synthesizing the resultant voltage of the three components they would separately provide.

It should be observed that if the generalized power flow controller formed from two synchronous machines is mechanically decoupled, then its operation is restricted to that of two independent, synchronous, reactive compensators, one connected in shunt and the other in series with the line.

The synchronous power flow controller model, shown in Figure 10.6, has not been realized in practice due probably to cost, possible operating difficulties under dynamic conditions, maintenance requirements, and also to relatively slow response to handle disturbances. However, this model can be eminently realized with fast response and unprecedented control flexibility by static, high power switching converters using semiconductor switches. Such a realization is shown schematically in Figure 10.7, and called *UPFC*.

The operation of this arrangement is analogous to that of the rotating synchronous machines: converter 2 is used to inject the compensating voltage, $\underline{V}_{pq}$, in series with the line, which is in synchronism with the AC system, but its magnitude and phase angle with respect to bus voltage $\underline{V}_1$ are freely variable between zero and a maximum, and between 0° and 360°, respectively. Converter 2 is able to internally provide the reactive power exchanged as a result of the voltage injection, but the active

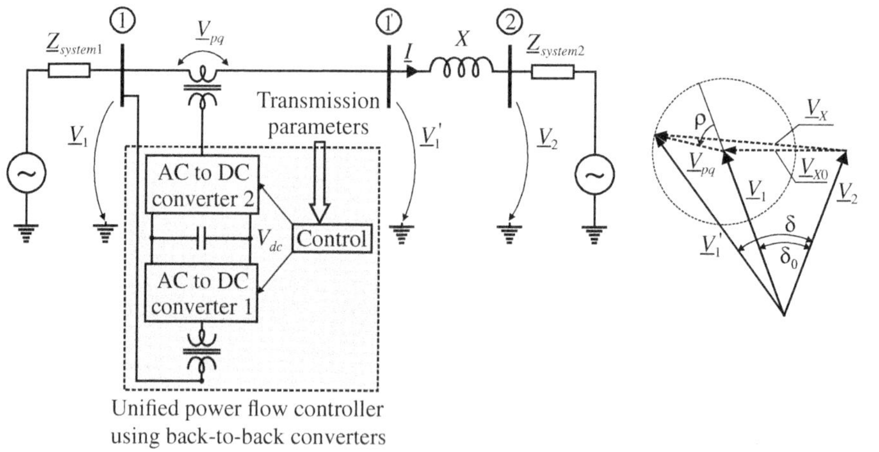

(a) (b)

Figure 10.7 Realization of the unified power flow controller (UPFC) with two, back-to-back connected voltage-sourced converters providing identical characteristics to that of the synchronous power flow controller model with fast response time.

power must be provided to it, or absorbed from it. This function is provided by converter 1, which absorbs/supplies active power from the system at bus 1, and delivers it to converter 2 via the common DC link. Converter 1 can also provide reactive shunt compensation by appropriately controlling both the magnitude and phase angle of its voltage. (Recall the operation of the voltage-sourced AC to DC converters discussed in Chapters 4 and 8.)

10.1.2 UPFC Directly Providing Line Current Forcing Function

Section 10.1.1 introduced the functional concept of the UPFC by the methodology of conventional transmission control. However, with a more direct approach, one could simply state that, ultimately, the function of a generalized transmission controller is to optimize or control the power transmission over a transmission line. In this sense, one could argue that the conventional controllers, such as phase shifters and series capacitors, really aim to vary the line current in order to control the transmitted power by means of changing a particular transmission parameter, such as the transmission angle or line impedance, to accomplish this objective indirectly. By contrast, the UPFC, with its versatile voltage injection and reactive power generation capability, can directly *force* the line current, via the use of an appropriate closed-loop control of the injected voltage phasor (vector) to assume an angle and magnitude with respect to a given bus voltage, to deliver the desired active power and, simultaneously, minimize the reactive power flow in the line. In other words, the UPFC has, with sufficient rating, the functional capability of directly, and independently, controlling the active and reactive power flow in the line, as illustrated schematically in Figure 10.8, establishing automatically—within allowed limits—the corresponding transmission parameters.

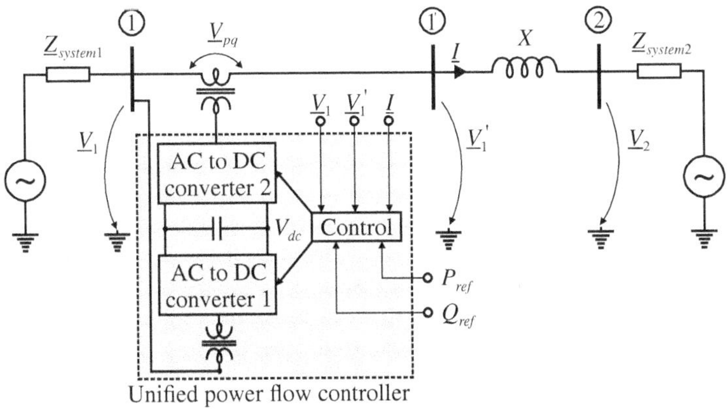

Figure 10.8 Unified power flow controller providing closed-loop control of the reference defined active and reactive power flow.

10.2 BASIC CHARACTERISTICS OF THE UPFC

In a conventional uncompensated transmission line connecting two busses with equal voltages, the relationship between the active power P, reactive line power Q, and the transmission parameters bus voltage V, line impedance X, and transmission angle δ can be described by two simple equations, as given in (10.1) and (10.2). These equations result in simple curves when P and Q plotted against the usually controllable variable, the transmission angle δ, showing graphically the variation of the active and reactive line power flow with the transmission angle. Similarly, a single curve graphical relationship can also be established between the active and reactive line power.

When a transmission line is controlled by a UPFC, the mathematical and corresponding graphical relationships become somewhat more complex, because the UPFC can vary the conventionally fixed transmission parameters, the magnitude and angle of the transmission voltage, as well as the line impedance, in addition to the effective transmission angle. Thus, the simple curves of the uncompensated transmission become operating areas or regions, the boundaries of which are defined by the rating or control imposed restrictions of the UPFC. These characteristic curves will be presented here in a manner analogous to those of conventional power transmission for easy understanding, and also for the visual illustration of the hitherto unattainable transmission control capability of the UPFC.

The mathematical expressions for the active and reactive power for the UPFC-controlled line can be expressed in an analogous form of those given by equations (10.1) and (10.2), that is:

$$P = \frac{V^2}{X}\sin\delta + \frac{VV_{pq}}{X}\sin\left(\frac{\delta}{2}+\rho\right) \tag{10.3}$$

and

$$Q = -\frac{V^2}{X}(1-\cos\delta) + \frac{VV_{pq}}{X}\sin\left(\frac{\delta}{2}+\rho\right) \tag{10.4}$$

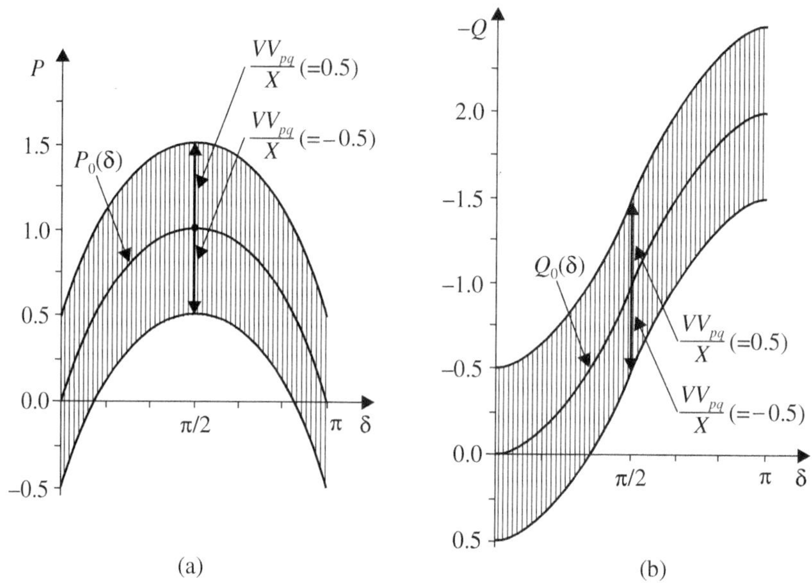

Figure 10.9 Range of transmittable active power P (a) and receiving-end reactive power demand Q, (b) versus the transmission angle δ.

where V_{pq} is the amplitude, and ρ is the angle (with respect to the bus 1 voltage, $\underline{V}_1$) of the $\underline{V}_{pq}$ voltage injected by the UPFC in series with the line (refer to Figure 10.8). It can be seen that equations (10.3) and (10.4) contain the customary terms of the uncompensated line given by equations (10.1) and (10.2), plus an additional term that is a function of both the magnitude and angle of the injected voltage $\underline{V}_{pq}$. Recalling that angle ρ is freely variable between 0° and 360° (and is independent of the MVA rating of the equipment), a plot showing the expansion of the single P versus δ curve into an operating band, the upper and lower boundary of which is determined by the magnitude V_{pq} (i.e., the MVA rating of the UPFC), can be constructed for both P and Q, as shown in Figure 10.9. Inspection of Figures 10.9a and 10.9b shows that at *any* transmission angle δ, the active power P or the reactive power Q can be increased or decreased by amount VV_{pq}/X, which in the illustration is equal to 0.5 p.u. MW or MVAr, since $V_{pq}/V = 0.5$. (Note that this ratio in practical applications would be rather large, requiring a UPFC with high MVA rating.) It is important to realize that the possible maximum increase in power is a fixed amount, which is independent of the operating point defined by the transmission angle. This means that as the transmission angle is decreased from 90° (the theoretical operating point for maximum power transmission), the percentage capability of the UPFC to increase the transmitted power is continuously increasing. Thus, at small transmission angles, a UPFC with relatively low MVA rating can provide highly effective power control. This particular characteristic of the UPFC (similarly to the static synchronous series compensator (SSSC) discussed elsewhere) is due to its inherent capability to inject and maintain a fixed series compensating voltage in face of a variable line current (in contrast to, e.g., the series capacitor, which provides a fixed compensating impedance).

Although equations (10.3) and (10.4) indicate that both the active and reactive line power can be controlled over the same range, and can be increased or decreased by the same amount, they do not imply that the maximum values are simultaneously attainable. Simple physical reasoning suggests this, since the control of active line power flow requires quadrature voltage injection with respect to the line current, whereas that of the reactive line power is accomplished by in-phase voltage injection. The overall control capability of the UPFC with regard to simultaneous active and reactive line power flow control can best be illustrated by a P versus Q plot, which shows the corresponding active and reactive power values (in per unit) over the transmission angle range of interest. The starting point for the derivation of such a plot is again the two simple equations (10.1) and (10.2), characterizing the uncompensated transmission line. Normalizing these equations by making $V^2/X = 1$, and by summing the squares of them, the following expression can be obtained:

$$(Q_0 + 1)^2 + P_0^2 = \sin^2 \delta + \cos^2 \delta = 1 \tag{10.5}$$

where the subscript zero indicates that P_0 and Q_0 represent the uncompensated transmission at angle δ. Equation (10.5) describes a circle with a radius of 1.0 around the center defined by coordinates $P_0 = 0$ and $Q_0 = -1$ in a $\{Q_0,P_0\}$ plane, as illustrated for positive values of P_0 in Figure 10.10a.

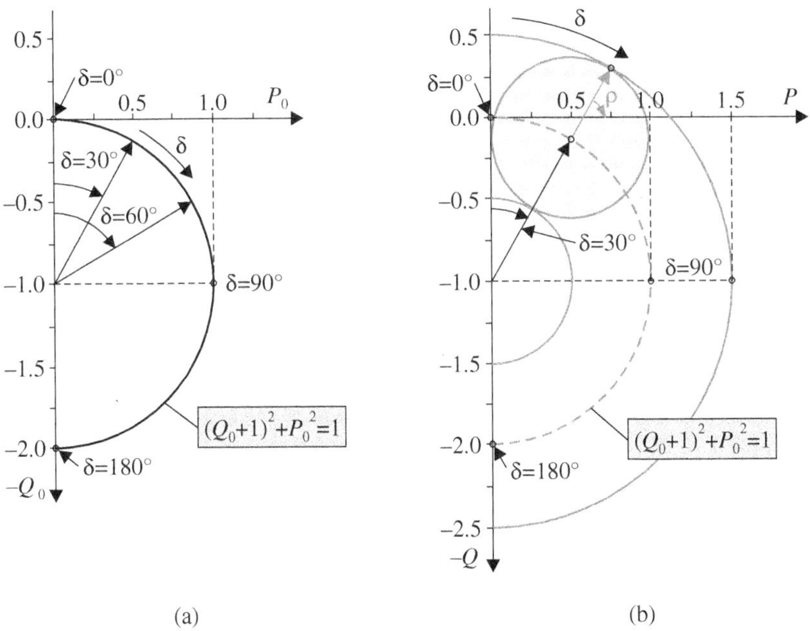

Figure 10.10 (a) Reactive versus active power of the uncompensated line in the $\{Q,P\}$ plane at transmission angle range of $0 \leq \delta \leq 180$ and (b) control range of the UPFC in the corresponding $\{Q,P\}$ plane.

Each point of this circle gives the corresponding P_o and Q_o values of the uncompensated system at a specific value of transmission angle δ. For example, at $\delta = 0$, $P_o = 0$, and $Q_o = 0$; at $\delta = 30°$, $P_o = 0.5$, and $Q_o = -0.134$; and at $\delta = 90°$, $P_o = 1.0$, and $Q_o = -1.0$.

Inspection of Figures 10.9a and 10.9b suggests that with the maximum positive value of V_{pq}, the radius of this P_o versus Q_o semi-circle would increase from 1 to $(1+V_{pq}/V)$, and the maximum negative value would decrease it by that amount. (In the present example used in Figure 10.9, $V_{pq}/V = 0.5$ and thus $(1+V_{pq}/V)$ is equal to $1+0.5 = 1.5$.) Of course, these values are obtained at particular values of ρ, the angle of voltage $\underline{V}_{pq}$, which is freely variable from 0° to 360°. The operation of the UPFC could be visualized in the $\{Q,P\}$ plane by correlating the voltage phasor diagram shown in Figure 10.7b to the P versus Q plots shown in Figure 10.10. The transmission angle δ defines a related P_o and Q_o value pair for the uncompensated line with given transmission parameters. These P_o and Q_o value pairs define a semi-circular curve in the $\{Q,P\}$ plane as δ varied from 0° to 180°. When compensating voltage $\underline{V}_{pq}$ is injected between bus 1 and the line, the uncompensated P_o and Q_o are pushed off the defined semi-circle curve in the $\{Q,P\}$ plane into a location that is determined by the magnitude and angle of $\underline{V}_{pq}$. It is evident that as $\underline{V}_{pq}$ is rotated over the full range of 0–360°, the P and Q value pair will move along a full circle around the point in the $\{Q,P\}$ plane defined by the original P_o and Q_o value pair of the uncompensated line. It is further evident that, as the magnitude of the compensating voltage varied between zero and its maximum value, a circular area around the original P_o and Q_o value pair defined point will be established, in which any P and Q can be established by setting the amplitude and angle of $\underline{V}_{pq}$ to corresponding values. Changing the transmission angle δ will, of course, move the original P_o and Q_o values along the semi-circular curve, and around this new operating point the UPFC will establish its own circular operating area, expanding it into an eventual operating band around the uncompensated semi-circle curve, as illustrated in Figure 10.10b.

The operating circular area of the UPFC with a center defined by coordinates P_o and Q_o and a radius of VV_{pq}/X in the $\{Q,P\}$ plane can be expressed mathematically from equations (10.3) and (10.4):

$$[P(\rho) - P_o]^2 + [Q(\rho) - Q_o]^2 = \left(\frac{VV_{pq\,\max}}{X}\right)^2 \tag{10.6}$$

This circular control region defined by equation (10.6) is shown in Figure 10.10b for $V = 1.0$, $V_{pq\max} = 0.5$, and $X = 1.0$ (per unit or p.u. values) with their centers on the semi-circular arc characterizing the uncompensated system (equation 10.5) at transmission angles $\delta = 30°$. It can be observed in the illustrated example that the UPFC would be able to increase the transmitted active power P almost by a factor of two (or reduce it to zero) and, at the same time, maintain zero reactive power flow in the line. It is mentioned again that to accomplish this, the UPFC would need a significant MVA rating (equal to half of the power transmitted by the uncompensated line), but considerably smaller rating could still provide significant control for the active and reactive power flow in the line.

10.3 UPFC VERSUS CONVENTIONAL POWER FLOW CONTROLLERS

10.3.1 UPFC versus Series Reactive Compensators

Controlled series reactive compensators provide a series compensating voltage that, by definition, is in quadrature with the line current. Consequently, *they can affect only the magnitude of the current flowing through the transmission line.* As discussed in Chapter 3, there are various types of series compensators using different solid-state circuit configurations to emulate variable capacitors or provide quadrature voltage injection directly. However, whatever technique is used, at any given setting of the effective capacitive/inductive impedance, an overall transmission impedance is defined at which the transmitted power is strictly determined by the transmission angle (assuming a constant amplitude for the end voltages). Therefore, the reactive power demand at the end-points of the line is determined by the transmitted active power in the same way as if the line was uncompensated but had lower impedance. Consequently, the relationship between the transmitted power P and the reactive-power demand of the line can be represented by a single Q–P circular locus, similar to that shown for the uncompensated system in Figure 10.10a. This means that for a continuously controllable compensator, an infinite number of Q–P circular loci can be established by using the basic transmission relationships:

$$P \cong \frac{V^2}{X - X_q} \sin \delta \tag{10.7}$$

$$Q \cong -\frac{V^2}{X - X_q}(1 - \cos \delta) \tag{10.8}$$

where the effective compensating reactance X_q is varying between 0 and $X_{q\max}$ ($X_q = X_C$ for series compensators emulating a variable capacitor, and $X_q = |\underline{V}_q/\underline{I}| = V_q/I$ for compensators executing quadrature voltage injection.) Evidently, a given transmission angle defines a single point on each Q–P locus obtained with a specific value of X_q. Thus, the progressive increase in X_q from zero to $X_{q\max}$ could be viewed as if the point defining the corresponding P and Q values at the given transmission angle on the first Q–P locus (uncompensated line) moves through an infinite number of Q–P loci representing progressively increasing series compensation, until it finally reaches the last Q–P locus that represents the system with maximum series compensation. The first Q–P locus, representing the uncompensated power transmission, is the lower boundary curve for the type of series compensators that emulate a variable series capacitor and identified by $(Q\text{–}P)_{Xq=0}$. The last Q–P locus, representing the power transmission with maximum capacitive impedance compensation, is identified by $(Q\text{–}P)_{X_{C\max}}$. The lower and upper boundary curves for the type of series compensators that directly inject quadrature compensating voltage, such as the SSSC, are different and therefore they are identified by $(Q\text{–}P)_{+V_{q\max}}$ and $(Q\text{–}P)_{-V_{q\max}}$. The difference is partially due to their capability to inject the compensating voltage both with 90° lagging (capacitive) or 90° leading (inductive) relationship with respect to the line current. Thus, they can both increase and decrease the transmitted power. Also, they can usually maintain maximum compensating voltage with decreasing line current

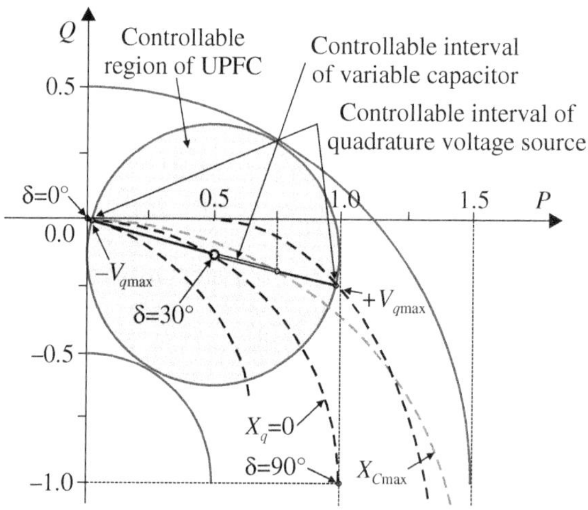

Figure 10.11 Attainable Q and P values with controlled capacitive compensation (points on the heavy filled/unfilled straight line inside the circle) and those with the UPFC (any point inside the circle) at $\delta = 30°$.

(theoretically, even with zero line current). For these reasons, the series compensators with quadrature voltage injection capability have a considerably wider control range at low transmission angles than those emulating variable series capacitors. Note that all Q–P circular curves are considered only for the normal operating range of the transmission angle ($0 \leq \delta \leq 90°$).

The plots in Figure 10.11 correspond to the plots shown for the UPFC in Figure 10.10 at $\delta = 30°$ transmission angle. Both types of series reactive compensators have the same 0.5 p.u. maximum voltage rating stipulated for the UPFC. The previously derived circular control region of the UPFC is also shown in Figure 10.11 for comparison. In each figure, the upper and lower boundary curves identified above for the variable capacitor-type and voltage injection-type compensators.

Inspection of Figure 10.11 indicates that the relationship between Q and P, for variable capacitor-type series compensator, is a straight line connecting the two related points on the lower and upper boundary curves, obtained at $X_q = 0$ and $X_{C\max}$, which represent the power transmission at $\delta = 30°$ with zero and, respectively, maximum series compensation. Series compensators with direct voltage injection capability can provide maximum compensating voltage, theoretically down to zero line current. They can also reverse the polarity of the compensating voltage to reduce the power. For these reasons, this type of compensators can provide compensation over a wider range, between the boundaries determined by $+V_{q\max}$ and $-V_{q\max}$, than can the variable capacitor types. However, although they lengthen the attainable control interval, their basic control characteristic, fundamental to all series reactive compensators, remains unchanged: *active power increase or decrease in the line inevitably associated with a proportional change of reactive line power*, independent active and reactive power flow control is *not* possible.

Figure 10.11 vividly illustrated the superior power flow control characteristic of the UPFC in that providing full control for the transmitted active power and, at the same time, also able to control reactive line power flow, and thereby facilitating the maximum utilization of transmission assets and minimizing losses.

10.3.2 UPFC versus Phase Shifters

As discussed in Chapter 4, the function of the phase shifter (also called phase angle regulator) is to change the angle of a given bus voltage to make it compatible with prevailing transmission requirements. Ideal phase shifters provide voltage injection in series with the line so that the voltage applied at their input terminals appears with the same magnitude but with $\pm\sigma$ phase difference at their output terminals. For simplicity, mechanical and thyristor-controlled phase shifters usually provide controllable quadrature voltage injection ("Quadrature Boosters") in series with the line and thus the magnitude of their phase-shifted output voltage increases with increasing angular change. These phase shifters employ a shunt-connected regulation (excitation) transformer with appropriate taps on the secondary windings, a mechanical or thyristor-based tap changer, and a series insertion transformer. The excitation transformer has wye to delta (or delta to wye) windings and thus the phase-to-phase secondary voltages are in quadrature with respect to the corresponding primary phase to neutral voltages.

Apart from their different power flow control characteristics, and despite their similarity in employing both a shunt-connected and series-connected transformer for coupling their inputs and outputs to the transmission system, there is a fundamental, and from the application standpoint important, difference between the above (mechanical and thyristor-controlled) phase shifters and the UPFC. In the former case, the total *VA* (i.e., both the active power and the VArs), exchanged by the series insertion transformer, appears at the primary of the regulation transformer as a load demand on the power system. However, in the latter case, only the active power exchanged is supplied by the shunt transformer from the power system, because the reactive power exchanged is generated by the UPFC series converter itself.

The purpose of the present investigation is to compare the power flow control capability of the UPFC to that of the phase shifter. For this comparison, the practical implementation of the phase shifter and its ability to generate reactive power is unimportant. For simplicity, phase shifters are considered as "ideal," that is, their "quadrature boosting" is ignored, and thus they are assumed to be able to vary the phase angle between the voltages at the two ends of the insertion transformer in the control range of $-\sigma_{max} \leq \sigma \leq \sigma_{max}$, without changing the magnitude of the phase-shifted voltage from that of the original line voltage.

As previously established, the transmitted power and the reactive power demands at the sending-end and receiving-end can be described by relationships analogous to those characterizing the uncompensated system:

$$P \cong \frac{V^2}{X}\sin(\delta - \sigma) \tag{10.9}$$

$$Q \cong -\frac{V^2}{X}[1 - \cos(\delta - \sigma)] \tag{10.10}$$

It follows from these equations that phase shifters *cannot* increase the maximum transmittable power, $P = V^2/X$, or change Q at a fixed P. Consequently, the Q–P relationship, with transmission angle $\delta' = \delta\text{-}\sigma$ $(0 \leq \delta' \leq 90°)$ controlling the actual power transmission, is identical to that of the uncompensated system. The function of the phase shifter is simply to establish the effective transmission angle, δ', required for the transmission of the desired power P, by appropriately adjusting the phase-shift angle σ. In other words, the phase shifter can vary the transmitted power at a fixed δ, or maintain the actual transmission angle δ' constant in the face of a varying δ, but it cannot increase the maximum transmittable power or control the reactive power flow independently of the active power.

The plots in Figure 10.12 show the Q–P relationship for the ideal phase shifter in comparison to that of the UPFC at $\delta = 30°$, the transmission angle used in the previous examples of transmission control characteristic. It can be observed that the control range centers around the point defined by the given angle between the two ends of the line ($\delta = 30°$) on the Q–P locus characterizing the uncompensated power transmission ($V_\sigma = 0$ for the phase shifter and $V_{pq} = 0$ for the UPFC). As the phase-shift angle σ is varied in the range of $-30° \leq \sigma \leq 30°$ (which corresponds to the maximum inserted voltage of $V_\sigma = 0.5$ p.u.), this point moves on the uncompensated Q–P locus in the same way as if the fixed angle δ was varied in the range of $(\delta - 30°) \leq \delta \leq (\delta + 30°)$.

The Q versus P plots in Figure 10.12 clearly show the superiority of the UPFC over the phase shifter for power flow control: The UPFC has a wider range for active power control and facilitates the independent control of the reactive power demand of the line over a broad range. For example, it is seen that the UPFC can facilitate up to almost 1.0 p.u. active power transmission with unity power factor at the receiving-end ($Q = 0$), whereas with phase-shift the reactive power demand at the receiving-end

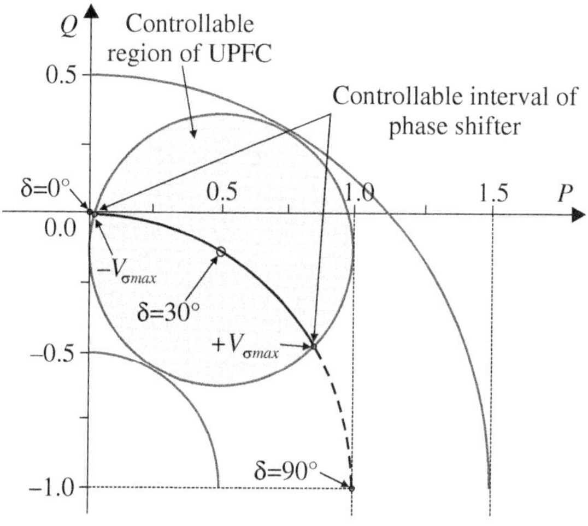

Figure 10.12 Attainable Q and P values with phase-shifting (points on the heavy arc inside the circle) and those with the UPFC (any point inside the circle) at $\delta = 30°$.

would increase with increasing active power (Q would reach 0.5. p.u. at the end of the phase shifter's control range, at about $P = 0.9$ p.u.).

10.4 UPFC CONTROL SYSTEM

The superior operating characteristics of the UPFC, described in the previous sections, are due to its unique ability to inject an AC compensating voltage with variable magnitude and angle in series with the line to independently control the active and reactive components of the line current, and thereby the transmitted active power and the reactive power demand of the line. Apart from establishing and maintaining the desired power flow, the inherent characteristic features of the UPFC offer a great potential to effectively counteract dynamic system disturbances if its operation is controlled by a suitable electronic control system. With this, the UPFC can cause the series-injected voltage to vary rapidly and in magnitude and angle as desired to increase transient stability and provide highly effective oscillation damping.

The control of the UPFC is based upon the so-called "vector-control," which was originally developed for controlling converter-supplied AC motor drives, but today it is generally used for transmission compensators and controllers employing the converter-based approach. The term *vector*, instead of phasor, is used in this type of systems to represent a set of three instantaneous phase variables, voltages, or currents, which sum to zero. The symbols $\tilde{\boldsymbol{v}}$ and $\tilde{\boldsymbol{i}}$ are often used for voltage and current vectors. Since these vectors represent instantaneous phase variables, they are not stationary, but move around a fixed point in the plane as the values of the phase variables change, describing various trajectories. *It is important to note that these trajectories become circles when the phase variables represent a balanced, steady-state condition,* which means that under these conditions the *vectors become phasors* that are well known in traditional power engineering. Thus, utility (and other) engineers who may not be familiar with "vector- control" may consider vectors as phasors (mentally changing the low case letters into capital letters, like $\tilde{\boldsymbol{v}}$ to $\underline{V}$, $\tilde{\boldsymbol{i}}$ to $\underline{I}$, p to P and q to Q, without affecting the understanding of the basic control principles and structures). (This simplifying replacement is used in the illustration of the basic control system structure in Figure 10.13.)

For the purpose of power control, it is useful to view these vectors in an orthogonal coordinate system with p and q axes such that the p axis is always coincident with the instantaneous voltage vector $\tilde{\boldsymbol{v}}$ and the q axis is in quadrature with it. In this coordinate system, the p-axis current component, i_p, accounts for the instantaneous active (real) power and the q-axis current component, i_q, for the reactive power. Under balanced steady-state conditions, the p-axis and q-axis components of the voltage and current vector are constant quantities. This characteristic of the described vector representation makes it highly suitable for the control of the UPFC by facilitating the decoupled control of the active and reactive current components.

The UPFC control system may be divided functionally into *internal* (or *converter*) *control* and *functional* (or *system*) *operation* control, as illustrated in Figure 10.13. The internal controls operate the two converters so as to produce the

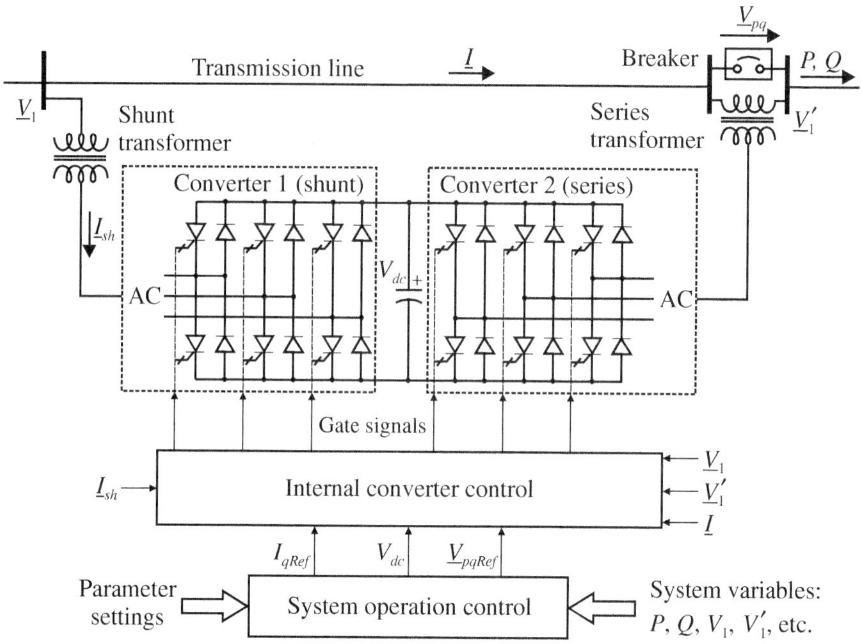

Figure 10.13 Overall UPFC control structure with separated internal converter control and system operation control functions.

commanded series-injected voltage for power flow control and, simultaneously, draw the desired shunt reactive current for bus voltage regulation. The internal controls provide gating signals to the converter valves so that the converter output voltages will properly follow the internal reference variables, i_{pRef}, i_{qRef}, and $\tilde{v}_{pqRef}$.

The basic structure for *internal* (or *converter*) *control* is shown in Figure 10.14. The functions of the two converters are different, and thus they are controlled according to their functional roles. The role of the series converter is to execute the voltage injection as commanded. Accordingly, the series converter, being supplied from the shunt converter-regulated DC source, can respond directly and independently to the demand for series voltage vector injection. Changes in series voltage vector, $\tilde{v}_{pq}$, can therefore be affected virtually instantaneously. The role of the shunt converter is to maintain the DC voltage level to support the series voltage injection (by providing the necessary active power) and reactive shunt compensation demand. Thus, in contrast to the series converter, the shunt converter operates under a closed-loop current control structure whereby the active power component, needed for the regulation of the common DC link voltage, and the reactive power component, used for the AC bus voltage regulation, are independently controlled. The shunt converter therefore fulfills two functions with two independent control loops: It provides shunt reactive power in response to reactive shunt current reference, and it maintains the necessary voltage level on the DC link, and thereby provides the active power supply or sink

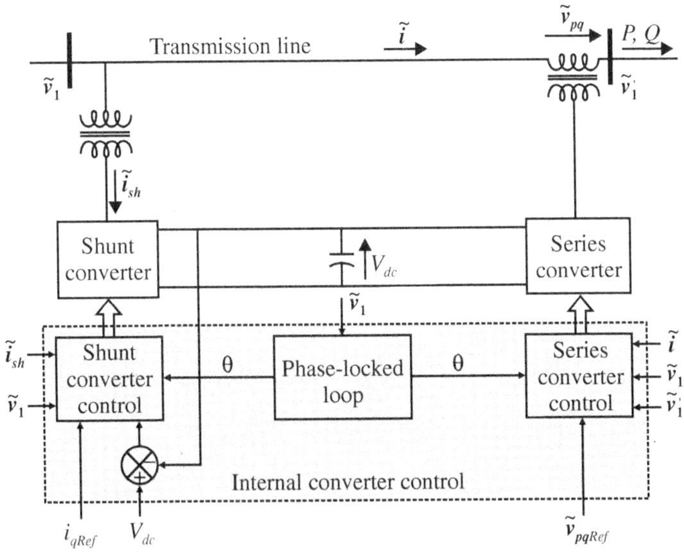

Figure 10.14 UPFC internal control structure.

needed for the support of the series voltage injection. In other words, the control loop for the active power ensures the required active power balance between the output of the two converters (i.e., $P_{shunt} + P_{series} = 0$). As emphasized previously, the converters do not (and could not) exchange reactive power through the DC link.

The *internal converter control* is thus composed of two separate control blocks, one for the series and the other for the shunt converter, as shown in Figure 10.14. They are operated in synchronism with the AC system, but their functional roles are executed independently. The synchronization is provided by a common *phase-locked loop,* which ensures that all variables run in a common, synchronous time frame.

The *external* (or *functional*) *operation control* defines the operating mode of the UPFC and is responsible for generating the *internal* references, $\tilde{v}_{pqRef}$ for the series converter, and i_{qRef} for the shunt converter to meet the prevailing demands of the transmission system. As explained in Section 10.3, the UPFC has, in addition to providing an unconstrained two-dimensional voltage injection (in a defined region) to independently control active and reactive power flow, capability to emulate the operation of line impedance compensation and transmission angle regulation. The functional operating modes, such as defined $\tilde{v}_{pq}$ voltage injection (determined by an independent power system optimization control), active and reactive power regulation to given P_{Ref} and Q_{Ref} references, line impedance compensation to a Z_{Ref} or X_{Ref} reference, or transmission angle regulation to a σ_{Ref} reference, can be set manually (via a computer keyboard) by the operator or dictated by an automatic *system optimization control* to meet specific operating and contingency requirements. An overall control structure, showing the *internal,* the *system operation*, and *system optimization* controls with the internal and external references, is shown in Figure 10.15.

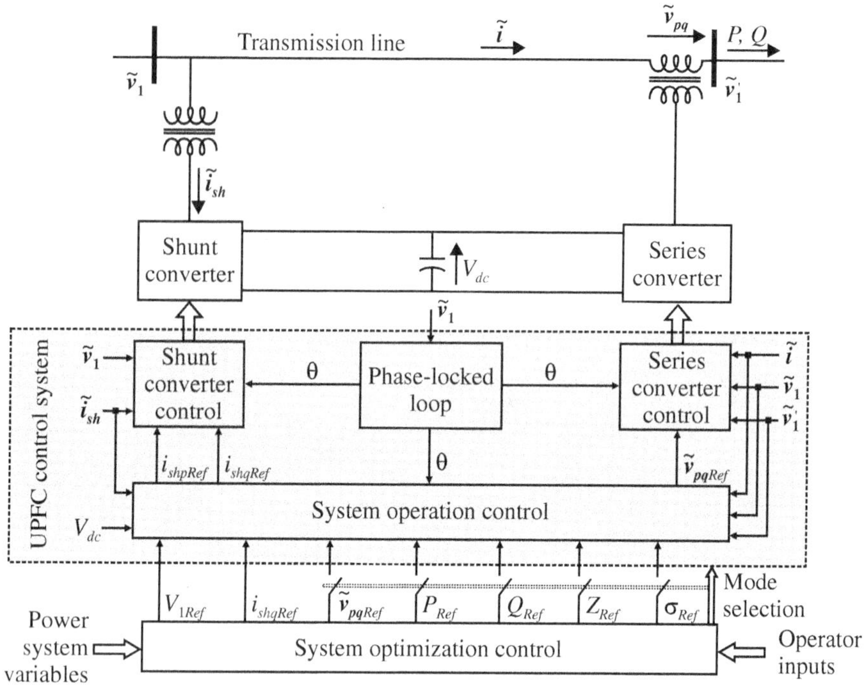

Figure 10.15 UPFC control operated by the inputs of a system optimization control managing the area transmission.

The unique functional capability of the UPFC, based on unrestricted series voltage injection together with independently controllable reactive shunt compensation, is facilitated by the circuit structure of two back-to-back converters. This flexible structure not only facilitates several operating and control modes in normal operation, but it also allows the total decoupling of the two converters (i.e., separating the DC terminals of the two converters) to provide independent reactive shunt compensation (STATCOM) and reactive series compensation (SSSC) without any active power exchange if network conditions or contingency situations so require.

The possible functional operating modes of the UPFC with the available control options for the shunt and series converters are detailed next.

10.4.1 Functional Control of the Shunt Converter

The shunt converter is operated so as to draw a controlled current, i_{sh}, from the line. One component of this current, i_{shp}, is automatically determined by the requirement to balance the active power of the series converter by maintaining the common DC link voltage. The other current component, i_{shq}, is reactive and can be set to any desired reference level (inductive or capacitive) within the capability of the converter. The reactive compensation control modes of the shunt converter are, of course, very similar to those commonly employed for the STATCOM and conventional static VAr compensator (SVC).

- *Reactive power (VAr) control mode.* In *reactive power control mode,* the reference input is a capacitive or inductive VAr request. The shunt converter control translates the VAr reference into a corresponding shunt current request and adjusts the gating of the converter to establish the desired compensating current. The control in a closed-loop arrangement uses current feedback signals obtained from the output current of the shunt converter to enforce the current reference. A feedback signal representing the DC bus voltage, V_{dc}, is also used to ensure the necessary DC link voltage.
- *Automatic voltage control mode.* In voltage control mode (which is normally used in practical applications), the shunt converter reactive current is automatically regulated to maintain the transmission line voltage to a reference value at the point of connection, with a defined droop characteristic. The droop factor defines the per unit voltage error per unit of converter reactive current within the current range of the converter. The automatic voltage control uses voltage feedback signals, usually representing the magnitude, V_1, of the positive sequence component of bus voltage $\widetilde{v}_1$.

10.4.2 Functional Control of the Series Converter

The series converter controls the magnitude and angle of the voltage vector $\widetilde{\boldsymbol{v}}_{pq}$ injected in series with the line. This voltage injection is, directly or indirectly, always intended to influence the flow of power on the line. However, $\widetilde{\boldsymbol{v}}_{pq}$ is dependent on the operating mode selected for the UPFC to control power flow. The possible operating modes include:

- *Direct voltage injection mode.* The series converter simply generates the voltage vector, $\widetilde{\boldsymbol{v}}_{pq}$, with the magnitude and phase angle requested by the reference input. This operating mode may be advantageous when a separate system optimization control coordinates the operation of the UPFC and other transmission controllers employed in the transmission system. Special functional cases of direct voltage injection include those having dedicated control objectives, for example, when the injected voltage vector, $\widetilde{\boldsymbol{v}}_{pq}$, is kept in phase with the system voltage for voltage magnitude control, or in quadrature with it for controlled "quadrature boosting," or in quadrature with the line current vector, $\widetilde{\boldsymbol{i}}$, to provide controllable reactive series compensation.
- *Bus voltage regulation and control mode.* The injected voltage vector, $\widetilde{\boldsymbol{v}}_{pq}$, is kept in phase with the "input" bus voltage vector $\widetilde{\boldsymbol{v}}_1$ and its magnitude is controlled to maintain the magnitude "output" bus voltage vector $\widetilde{\boldsymbol{v}}'_1$ at the given reference value.
- *Line impedance compensation mode.* The magnitude of the injected voltage vector, $\widetilde{\boldsymbol{v}}_{pq}$, is controlled in proportion to the magnitude of the line current, $\widetilde{\boldsymbol{i}}$, so that the series insertion emulates an impedance when viewed from the line. The desired impedance is specified by reference input and, in general, it may be a

complex impedance with resistive and reactive components of either polarity. A special case of impedance compensation occurs when the injected voltage is kept in quadrature with respect to the line current to emulate purely reactive (capacitive or inductive) compensation. This operating mode may be selected to match the existing series capacitive line compensation in the system.

- *Phase angle regulation mode*. The injected voltage vector $\widetilde{v}_{pq}$ is controlled with respect to the "input" bus voltage vector $\widetilde{v}_1$ so that the "output" bus voltage vector $\widetilde{v}_1'$ is phase shifted, without any magnitude change, relative to $\widetilde{v}_1$ by an angle σ in response to the reference input. One special case of phase shifting occurs when $\widetilde{v}_{pq}$ is kept in quadrature with $\widetilde{v}_1$ to emulate the "quadrature booster."
- *Automatic power flow control mode*. The magnitude and angle of the injected voltage vector, $\widetilde{v}_{pq}$, are controlled so as to force such a line current vector, $\widetilde{i}$, to assume a magnitude and angle that result in the desired active and reactive power flow in the line. In automatic power flow control mode, the series-injected voltage is determined automatically and continuously by a closed-loop control system to ensure that the desired P and Q are maintained despite system changes. The transmission line containing the UPFC in this operating mode appears to the rest of the power system as a high impedance power source or sink. This operating mode, which is not achievable with conventional line compensating equipment, provides great potential for power flow scheduling and management, as well as asset utilization. It can also be applied effectively to handle dynamic system disturbances (e.g., to damp power oscillations).

10.4.3 Stand-Alone Shunt and Series Compensation

The UPFC circuit structure offers the possibility of operating the shunt and series converters independently of each other by disconnecting their common DC terminals and splitting the capacitor bank. In this case, the shunt converter operates as a stand-alone STATCOM, and the series converter as a stand-alone SSSC. This feature may be included in the UPFC structure in order to handle contingencies (e.g., one converter failure) and be more adaptable to future system changes (e.g., the use of both converters for shunt only or series only compensation). In the stand-alone mode, of course, neither converter is capable of absorbing or generating active power so that their operation is restricted to the reactive power domain, and their control modes to those of the reactive shunt and series compensators.

10.4.4 Basic Control Structure for the Series and Shunt Converters

Although the UPFC has many possible operating modes, the basic control structures for the series and shunt converter within the overall *internal control* remain substantially the same. The block diagram-based explanations here focus on the automatic power flow control mode providing independent control for active and reactive power flow in the line. This control mode utilizes most of the unique capabilities of the UPFC and it is expected to be the basic mode of series compensation in the majority

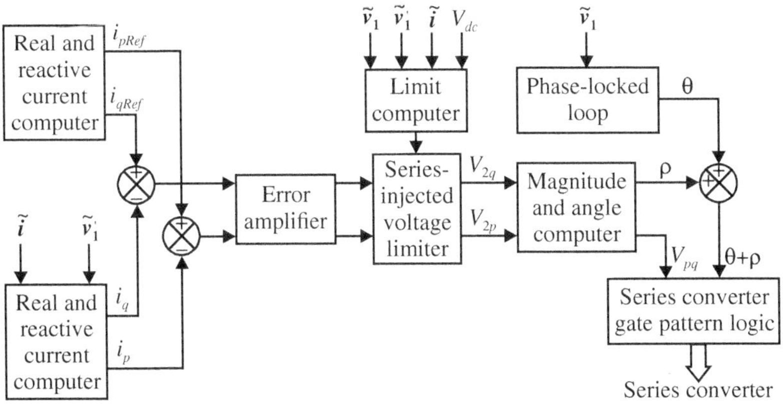

Figure 10.16 Functional block diagram of the series converter control.

of practical applications. Similarly, the UPFC shunt compensation capability is normally used for automatic bus voltage regulation. These control modes could easily altered for different series and shunt compensation objectives by deriving appropriate references for other system variables to be controlled.

The block diagram for the control scheme of the series converter is shown in Figure 10.16. (Note that for clarity only the most significant features are shown in this and related figures ignoring some of signal processing and limiting functions that may be used in practical circuits.) It is assumed that the magnitude and phase angle of the output voltage generated by both the series and shunt converter are independently controllable at a fixed DC link voltage.

As shown in Figure 10.16, the automatic power flow control for the series converter is achieved by means of a vector control scheme that regulates the transmission line current using a synchronous reference frame (established with an appropriate phase-locked loop producing reference angle θ) in which the control quantities appear as DC signals in the steady state. (Recall that the synchronous reference frame with coordinates q and p rotates synchronously with the vector/phasor variables at the power system frequency. Consequently, a steady-state vector—that is, phasor—with some angle with respect to the p axis will have a constant, but generally different, projection to both the p and the q axes, defining the in-phase and quadrature or "active" and "reactive" components of the vector.) The appropriate references for active and reactive line current components, i_{pRef} and i_{qRef}, are determined for the desired active and reactive line power, P_{Ref} and Q_{Ref}. These are compared with the measured line currents, i_p and i_q, and, within a closed control-loop, used to derive the magnitude and angle of the series converter voltage, V_{pq} and ρ, respectively. Note that a voltage limiter in the forward path is employed to enforce practical limits on the series voltage injected. These limits may result from system restrictions (e.g., voltage and current limits) or equipment and component ratings.

The block diagram for the control scheme of the shunt converter is shown in Figure 10.17. This is also a vector control scheme in which the controlled quantity is the current vector $\tilde{i}_{sh}$ drawn from the line by the shunt converter. The active and

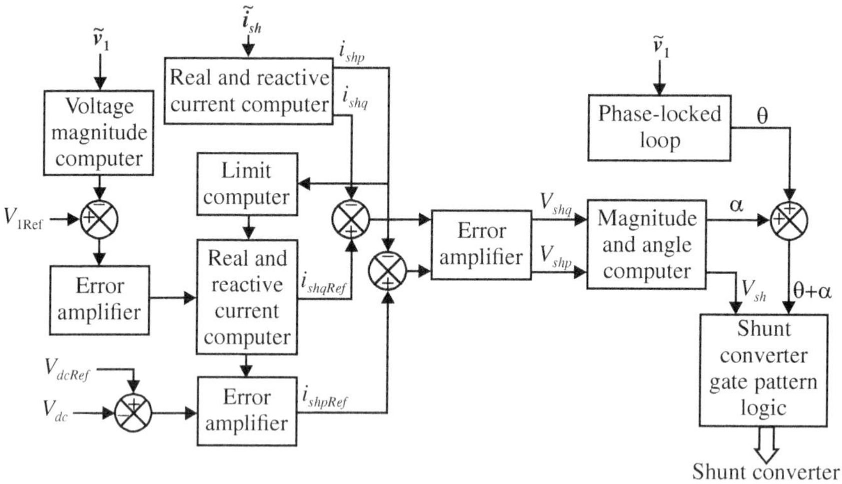

Figure 10.17 Functional block diagram of the shunt converter control.

reactive components of this current, however, have a different significance. The reference for the reactive current component of the shunt converter, i_{shqRef}, is generated by an outer voltage control loop that is responsible for regulating the AC bus voltage. The reference for the active current component, i_{shpRef}, on the other hand, is generated by a second voltage control loop that regulates the DC bus voltage. This second control loop thus achieves the automatic balancing of the active power need of the series converter in executing the series voltage injection for the desired power flow control. The DC voltage reference, V_{dcRef}, may be kept substantially constant.

It is pointed out here that shunt reactive compensation may also be controlled by varying the DC voltage and thereby changing the magnitude of the shunt converter's output voltage. In this case, the outer voltage loop would regulate the AC bus voltage and would also control the DC capacitor voltage by changing the angle α of the converter voltage with respect to the AC bus voltage until the DC capacitor voltage reaches the value necessary to achieve the reactive compensation demanded. With this operation of the shunt converter, the closed-loop controlling the output of the series converter would be responsible for maintaining the magnitude of the injected voltage, v_{pq}, in face of the variable DC voltage. This method may be used to minimize harmonics and losses due to direct voltage magnitude control (e.g., executed by pulse-width modulation). However, disadvantage of this "indirect" control technique is that the DC voltage is allowed to vary (typically up to $\pm 12\%$) according to the prevailing shunt compensation demand, which would inevitably reduce the attainable magnitude of the injected series voltage when the shunt converter is operated with high reactive power absorption. However, this, in many applications, may be an acceptable trade-off.

It can be observed that in the control scheme of Figure 10.16, a limit is imposed on the reactive current component of the shunt converter. This shows that operating priority is given for power flow control and, for this reason, the shunt reactive compensating current may be limited in order to use the maximal current capability of the shunt converter to satisfy the active power demand of the series converter. Of course, priority could be given to bus voltage regulation, if system operation would prefer that. Also, current limit could be imposed without priority, affecting both series and shunt compensation equally. As indicated earlier, the control block diagrams shown in Figures 10.16 and 10.17 represent only a selected operating mode to illustrate the basic control structure. Other control algorithms needed for additional operating modes of the UPFC could similarly be implemented. It should also be noted that the block diagrams here omit control functions related to converter protection, as well as sequencing routines during operating mode changes and start-up and shut-down procedures. These are similar to those used in other converter-based compensators and controllers, such as the STATCOM, SSSC, and others.

10.4.5 Practical Control Considerations

Apart from the basic active time control functions discussed above in the internal and external controls of the UPFC, the total control system also has many other elements to manage the proper and safe operation of the equipment with high reliability and availability, as well as to accommodate proper interface with local and remote operators.

The main elements of a modern UPFC control system, shown schematically in Figure 10.18, include:

1. Interface between a high power, high voltage semiconductor valves of the UPFC switching converters, and a highly sophisticated real-time control system, incorporating the internal and external control functions. This interface, transmitting gating commands from the control to the valves and status information from the valves to the control, is usually implemented by optical links.
2. Signal measuring and processing circuits for system and equipment variables. The real-time control and protection relays (and operator displays) need certain

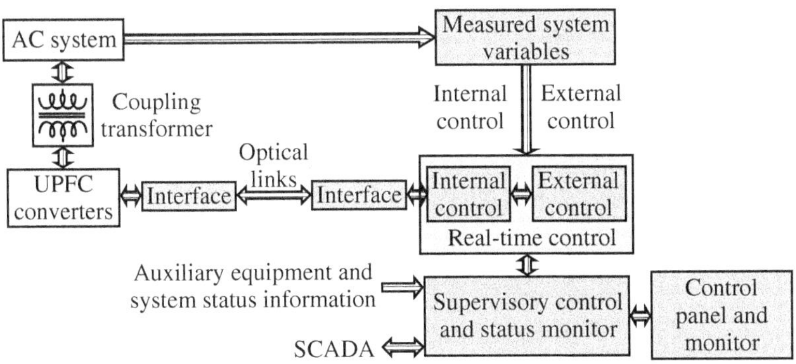

Figure 10.18 Main elements of modern, practical UPFC control system.

system variables as inputs, such as appropriate bus voltages and line currents, as well as corresponding internal voltages and currents of the equipment, from which magnitude, phase, frequency, and other relevant information characterizing the prevailing system conditions, to provide the desired transmission control and monitoring of equipment operation and performance.

3. Supervisory control and status monitor that interfaces with all parts of the UPFC, including all essential components of the converters and their support equipment (e.g., cooling system, power supplies, breakers, switches, and interlocks) It collects status information from every part of the system, usually via serial communication links, organizes, and interprets the status data to determine the operational integrity of the UPFC and to provide diagnostics for possible malfunctions and failures. It also carries out the start-up and shutdown sequencing and other operating routines of the compensator and provides appropriate communication links for the local and remote operators.
4. User interface with graphical displays is usually provided by a stand-alone computer with an appropriate monitor, key board, and pointing device for data entry. This computer usually has a serial link to the status processor and runs a graphical display and control software. Through the interface, a large amount of information is available for the operation, diagnostic, and maintenance purposes in graphical and numerical form. The information includes: Status information from the valves, identifying failed power semiconductors and other components and associated circuits; selected operating modes of the compensator and associated control and operational parameter settings; control operating and redundancy; and, status of support equipment such as cooling system, auxiliary power supplies, breakers, switches, etc., and building climate status (temperature, humidity, etc.).

10.5 EQUIPMENT STRUCTURAL AND RATING CONSIDERATIONS

This section is intended for providing some guidance in defining the electrical capabilities of the equipment at its *terminals where it is connected to the electrical system*, and it does not deal with the ratings of the individual components used in the equipment. Clearly, the ratings of the internal components used in the converter are the responsibility of the manufacturer and it involves detailed design knowledge of the electrical components and their applications, design philosophies and margins the designer applies; protection strategies and other, often proprietary operating considerations, which would generally vary from one manufacturer to the other. The typical user would usually not have sufficient knowledge and expertise in these highly specialized issues to judge them and, also, the manufacturer should have a relatively free hand in these to be able to provide the required guaranty on the equipment. On the other hand, the terminal capabilities would have to be clearly defined by the user with the appreciation of the special nature of an electronic equipment in order for the manufacturer to understand the application requirements and be able to properly rate the internal components for the application duties and, at the same time, minimize the overall cost of the equipment.

It is usually the normal practice for the user to carry out necessary studies, such as load flow, transient and/or voltage stability, power oscillation damping, and other studies, which would provide the required data in terms of voltage and current magnitudes and angles (when applicable). The aim is to select the appropriate values, which would define the electrical capability of the equipment in rms and peak values both for steady-state and short-term transient conditions. It is also an aim to give guidance for the manufacturer regarding the desired functional capabilities and corresponding operating regions to enable the manufacturer to consider different structural options for cost minimization.

In order to appreciate the rating issues and its effect on the cost of the equipment, it is necessary to understand the significant difference between conventional and power electronics-based transmission equipment in that the conventional electrical components are quite tolerant against short-term overvoltage and overcurrent excursions, whereas the electronic ones are generally not. Thus, transient rating requirements set without due consideration of their impact on the equipment rating can, on the one hand, unnecessarily increase the cost or, on the other hand, impact the reliability of the power electronics-based transmission controller. For this reason, in this section, the nature of, and the basic considerations for, power electronics-based equipment will briefly be summarized.

The technological base of the transmission controllers considered here, such as the UPFC, is the voltage-sourced converter (VSC). As explained elsewhere in this book, the VSC is a power electronics-based *switching converter* using high power semiconductors to generate from a DC voltage source the necessary, usually three-phase alternating (ac) voltage for compensation. The required rating of the converter in transmission applications is at least several tens of MVA, and often falls in the range of 100–300 MVA or more. The maximum rating of suitable power semiconductors is presently in the range of 1–3 kA and 4–6 kV; therefore, a considerable number of individual semiconductor devices are required to provide the desired output rating. Depending on the design, the desired converter rating may be achieved with high voltage, solid-state *valves* using an appropriate number of individual semiconductor devices in series connection. An alternative approach is to use a number of basic, single device converters in parallel or series connection to get the necessary rating by summing their voltage or current at the output. A frequently used approach, particularly for large ratings, is the combination of the above two, using series-connected semiconductor valves to get a basic converter of appreciable rating, and applying a number of such converters in parallel to get the final output.

Various types of power semiconductors (gate turn-off [GTO], gate-commutated thyristors, or insulated gate bipolar transistors [IGBTs], etc.) may be used in the power converter. These devices have different advantages and disadvantages related to switching capabilities and operating losses, which are still debated by technical experts and representatives of various manufacturers. From the standpoint of equipment rating and optimized circuit configurations, the issues related to the merits of the individual semiconductor device are not important; those are to be resolved by the equipment manufacturer in meeting the overall requirements of the user.

All power semiconductors used today are made of thin wafers of large silicon crystals (up to 150 mm, or 6″, in diameter) with rigorously controlled impurities and must be operated within a defined temperature range not exceeding 125°C or 150°C). From this physical reality, it follows that their capability in handling transient ratings (excess voltage and/or current) could be rather limited. Indeed, the inherent limitation of semiconductors is *absolute* for voltage rating, strictly defined for maximum magnitude of instantaneous current in terms of milliseconds, and short (in terms of seconds) by conventional standards for sustained overcurrent.

In the actual converter, there are usually a sophisticated control and protection circuit arrangements to ensure that the semiconductor voltage and current limits are not exceeded even under extreme system conditions. However, the activation of these arrangements usually means a limitation or interruption in the normal function of the equipment. Therefore, it is extremely important that the user provides credible information regarding voltage and current magnitudes and necessary *time durations* under worst-case steady state and contingency transient system conditions, during which the substantially uninhibited functional operation of the equipment is expected. It is then the manufacturer's responsibility to establish the ratings and internal safety margins for the semiconductors and related components used in the converter, taking into account statistical and temperature-dependent parameter variations, internal voltage transients resulting from the turn-off actions of the semiconductors, and all other factors that could unduly increase device voltage stresses, unmanageable rate of current increases, and dangerously high internal (junction) temperatures. It is also evident that an overly generous specification requiring operation without limitation under high overvoltage and overcurrent conditions for long time intervals characterizing conventional equipment could lead to considerable increase in the cost, since these requirements would generally necessitate substantial increase in the steady-state rating of the power electronics-based equipment. Thus, judicious trade-off between cost and functional operation during system contingencies and fault conditions is required to arrive at an optimal system rating meeting essential system requirements at minimum capital cost.

10.5.1 Circuit Structural Considerations

The UPFC is composed of two, back-to-back connected, VSCs. Thus, the basic and only fundamental restriction of the constituent converters is that their DC terminals must be freely available and of the same voltage. This restriction allows the application of practically all types of converters, with the possible exception of some of the multilevel types, in which multiple and isolated DC capacitors may make the connection complex and difficult to operate.

The main converter types with different advantages and disadvantages are the harmonic neutralized, multilevel, and pulse-width-modulated (PWM) converters, and their combinations. Their operations are not detailed here, but their main operating characteristics, with advantages and disadvantages, are summarized for convenience.

The *harmonic neutralized converter* is composed of n basic so-called six-pulse elementary converters that are operated with some phase displacement and their individual outputs are combined to yield a high, $6 \times n$ pulse number that approximate a sine-wave sufficiently well to avoid filtering. The harmonic neutralized converter has the following advantages:

- valve switching at the fundamental 60 Hz—*low operating losses;*
- multi-pulse output voltage—*little or no output filtering;*
- high MVA output by parallel operation—*relatively low DC voltage (insulation).*

The main disadvantage of the harmonic neutralized converter is a relatively complex and expensive magnetic structure.

The *multilevel converter* can produce multi-pulse-type waveforms by using an increasing number of DC capacitors in different configurations and, in addition to retaining the same main advantages (low switching frequency and no or little output filtering) as the harmonic neutralized converter, it has no auxiliary magnetic components and requires only a single "standard" coupling transformer. However, its DC terminal becomes complex often with additional semiconductors, and the total rating of its DC capacitor, and the cost thereof, is many times that of the harmonic neutralized converter.

The *PWM* converter can have a simple six-pulse structure with a single coupling transformer. However, its valves are switching at several or even many times of the fundamental frequency, which requires semiconductor switches capable of high frequency operation. Their output voltage contains relatively large unwanted components ("harmonics") generated as the side-bands of the basic high frequency switching frequency. Therefore, the operating losses of the PWM converter are significantly higher than those of the other two types, and it does require considerable high frequency filtering at the output. It is mentioned that two or more PWM converters can be combined at the expense of some circuit complexity to achieve carrier cancelation and thereby considerable reduction in filtering requirement.

It can be concluded that there is no basic converter circuit approach that would provide optimum solution for output terminal simplicity (magnetics), DC terminal simplicity (DC capacitors), and converter operation and performance (switching frequency, losses, and filtering). With present semiconductor technology, the best approach may be the combination of all three basic approaches, each only with a minimum level complexity (e.g., 12-pulse and three-level converter with relatively low switching, like 180 or 300 Hz).

Generally, it appears that it can presently *not* be specified the type of converter to be used in all UPFC applications. Particular operating and performance characteristics and features desired for the specific application may determine the best converter alternative. Naturally, critically important requirements (e.g., operating losses), application experience, the manufacturer's preference, and equipment cost may ultimately be the decisive factors for the type of converter eventually used.

10.5.2 Rating Considerations for Series and Shunt Converters

Since the UPFC consists of two converters in back-to-back connection, one connected in series and the other in parallel with the transmission line, it is convenient to separate initially the application requirements for the series and shunt compensation. Indeed, the *reactive* compensations provided by the series and shunt converters *are* operationally independent; there is no direct power exchange or necessary functional rating coordination between the two converters.

The initial rating requirements for the UPFC can easily be derived from load flow studies and, depending on the application, this base rating can be subsequently modified by superimposed additional requirements obtained from appropriate dynamic (transient stability, power oscillation damping, etc.) studies.

10.5.2.1 Series Converter Rating to Meet Line Compensation Requirements

The primary objectives of the load flow studies are:

1. Establish sustainable limits for power transmission (transmitted active power and allowed reactive line power) under normal as well as contingency system conditions that the UPFC has to maintain or, in other words, to define the boundary limits for the operating domain(s) in the $\{Q,P\}$ plane.
2. Derive the corresponding compensating voltage phasors (magnitudes and angles) that need to be injected in series with line.
3. Ensure that the electrical parameters of the compensated line stay within established transmission operating constraints (e.g., transmission voltage limits).
4. Establish the required maximum values of the *compensating voltage*, *active power*, and *line current* that the UPFC needs to produce to meet all compensation requirements; in other words, define the rating of the series converter that has to have at the transmission line voltage insertion terminals. (Note that the maximum values of the three parameters, i.e., the magnitude of compensating voltage, of the line current, and injected active power, which determine the rating of the series converter of the UPFC, may *not* occur simultaneously, each may be related to a different operating condition).

An operating region for the UPFC, representing all possible values of active and reactive line power, would generally be derived around an initial operating point in the $\{Q,P\}$ plane obtained with a given system set-up (or contingency) at a given transmission angle with no series compensation (the UPFC is nonoperational). Several such regions, covering all the operating conditions of interest, would need to be established and evaluated for maximum compensation and rating requirements. The worst-case requirements, in terms of the injected voltage, active power, and line current, would then be selected *individually* from these regions (i.e., the maximum value of each variable could come from a different region representing nonconcurrent operating conditions), and checked against prevailing transmission parameter constraints to finally define the system-side, series injection capability of the UPFC.

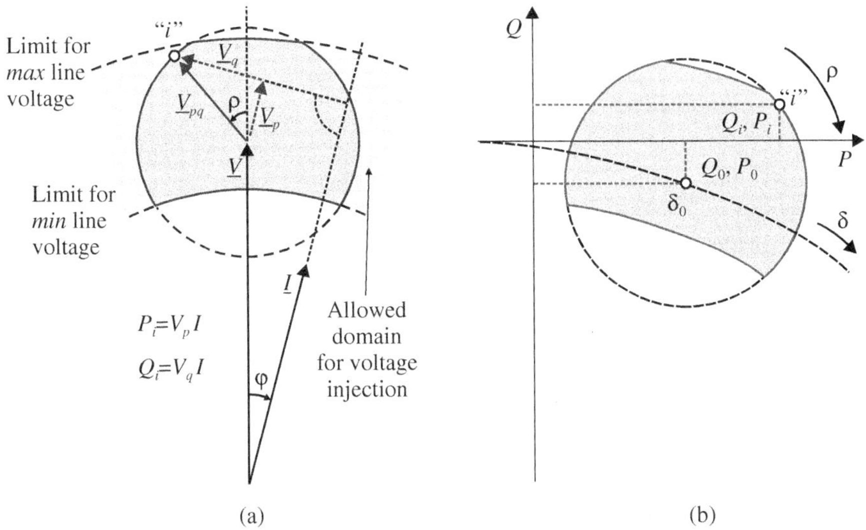

Figure 10.19 Illustration for determining the operating region for injected voltage from power flow control requirements and practical system and equipment constraints.

The above process is illustrated in Figure 10.19, where at (b) a circular operating region under a particular system condition is shown around an initial operating point, marked P_o, Q_o, representing the power transmission at the prevailing transmission angle δ_o without any series compensation (the UPFC is nonoperational). The operating point "i" defined by coordinates Q_i, P_i at the boundary of this region represents an extreme transmission condition that requires the UPFC to inject the maximum series compensating voltage $V_{pq\max}$ at angle ρ.

The phasor diagram shown at (a) in Figure 10.19 illustrates the relationship between the system bus voltage $\underline{V}$, the injected voltage $\underline{V}_{pq}$ and its corresponding circular operating area, and the line current $\underline{I}$, under the system operation defined by point "i" in the $\{Q,P\}$ plane. (Note that these phasors, $\underline{V}_{pq} = V_{pq}e^{j\rho}$ and line current $\underline{I} = Ie^{j\varphi}$, corresponding to operating point "i," define the active and reactive power, $P_i = IV_{pq}\cos(\rho + \varphi)$ and $Q_i = IV_{pq}\sin(\rho + \varphi)$, which the series converter of the UPFC has to supply for the line to sustain the operation under the system conditions "i.")

Note that in Figure 10.19a limits are established for the maximum and minimum allowed transmission voltage. These limits constrain the boundary for the injected voltage phasor at the upper and lower part of the original circular region as shown in the figure. The restriction on the allowed voltage injection results in a corresponding reduction in the related original circular operating region in the $\{Q,P\}$ plane, as illustrated in Figure 10.19b.

It follows from the above simple illustration that the rating requirements for the series converter can be simply derived by establishing quantitative relationships between the worst-case P and Q values (defining the transmission requirements at the boundary of the operating region in the $\{Q,P\}$ plane), and the corresponding compensating voltage and active power that the UPFC has to supply to maintain the resulting line current.

The steps of the process can be summarized with the help of the illustration in Figure 10.19, as follows:

1. Derive the circular locus of the largest compensating voltage phasor $\underline{V}_{pq\max}$ around the end point of bus voltage phasor $\underline{V}$ from the boundary of the operating area in the $\{Q,P\}$ plane representing the worst-case transmission requirements. This automatically defines the maximum magnitude of the compensating voltage.
2. Check if the injection of $\underline{V}_{pq\max}$ at any value of angle ρ will violate the allowed maximum or minimum value of the line voltage. If so, change the operating area of $\underline{V}_{pq\max}$ with the given voltage limits.
3. Modify accordingly the corresponding operating area of power transmission in the $\{Q,P\}$ plane.
4. Check if the worst-case operating point yielding the maximum value for $\underline{V}_{pq\max}$ is still within the modified area in the $\{Q,P\}$ plane. If not, determine the new value for $\underline{V}_{pq\max}$ (and make a decision if the attainable transmitted active power and reactive power load with the defined voltage limit are still acceptable or additional measures are needed).
5. Once the locus corresponding to the worst-case $\underline{V}_{pq\max}$ is determined, the locus of the resultant line current phasor $\underline{I}$ can be derived and, with these, the active and reactive power, P_{UPFC} and Q_{UPFC}, the series converter exchanges with the AC system at the boundary points of the P, Q operating domain, can be calculated.
6. The maximum magnitudes of the three variables $\underline{V}_{pq}$, $\underline{I}$, and P_{UPFC}, obtained from the sets of values defining the boundary of the operating area, will define the rating of the UPFC to meet the *series* compensation requirements for this operating region at the terminals of voltage injection. (Note that the computational process for generation of the necessary data, illustrated above in a simplified manner, would, of course, be carried out automatically by appropriate computational routines of the power flow program and, possibly, only the selection and evaluation of the data would require special human attention.)

In the case of having a number of operating regions around a different initial P_o, Q_o, values in the $\{Q,P\}$ plane due to changes in the transmission angle δ or other system parameters, it would become necessary to determine the maximum magnitudes of the three variables of interest, $\underline{V}_{pq}$, $\underline{I}$, and P_{UPFC}, for each of the regions. The largest *individual* value found for voltage magnitude V_{pq}, line current magnitude I, and active power P_{UPFC} among these, *regardless* whether they belong to the same or different operating regions, would define the ultimate rating requirements for the series converter of the UPFC.

If the application would involve power oscillation damping, or some other temporary compensation functions, then it would be necessary to assess by additional studies, or appropriate risk analysis, whether the incorporation of the additional function would require an increase in the rating of the series converter of the UPFC.

Issues of Transient Ratings. Considering possible transient rating requirements for the series converter of the UPFC, it should be remembered that practically any required increase over the specified maximum steady-state compensating voltage will likely become the design value for the actual steady-state rating of the converter, since, as mentioned before, power semiconductors do not have repetitive overvoltage capability. However, it may be possible to obtain some short-term overvoltage operation at the expense of reducing the voltage margin of the converter (that the equipment is likely to have by design) for these infrequent and short time durations. (This approach implies certain operating risks and thus its feasibility should be explored with the equipment manufacturer.) It should be made sure, too, that the coupling transformer of the series converter has sufficient voltage rating to accommodate the increased compensating voltage without saturation. In considering the need for overvoltage capability, it should also be kept in mind that the series converter of the UPFC, in contrast to conventional series capacitive compensation, is able to maintain (or even to decrease) the compensating voltage with increasing line current and, therefore, it can continue the operation if it has sufficient overcurrent rating to accommodate this system condition.

The series converter, depending on the design, is likely to have some (perhaps 20%) overcurrent capability for a short time, about 1 s, until the junction temperature of the power semiconductors reaches the maximum operating limit, typically 125° or 150°C. The maximum overcurrent with the time duration during which operation is required has to be specified for the manufacturer and, for economic reasons, needs to be kept at minimum acceptable. Operation with higher than about 20% overcurrent and/or time durations exceeding about one second will likely necessitate the increase in the actual steady-state rating of the equipment by means of modifying the converter valves or structure so that, under normal operating conditions, the current through the individual power semiconductors will be reduced (together with the corresponding junction temperature) relative to their rated value, in order to provide sufficient margin accommodating the desired overcurrent conditions. Clearly, the overcurrent requirement can affect the equipment cost considerably. However, in contrast to transient overvoltage rating, short-term operation with line currents significantly higher than the steady-state maximum will not require an increase in the rating of the coupling transformer.

If the line current, under fault or some contingency conditions, exceeds the overcurrent (magnitude and/or time) limit, then the UPFC protection (to be discussed later) would be activated to provide an appropriate by-pass for the line current.

1. The maximum rating of the series converter in the UPFC is defined by the maximum line current, the maximum injected voltage, and the maximum active power provided for the line for compensation under the worst-case operating conditions when sustained operation is required.
2. The maximum line current, injected voltage, and active power may *not* occur simultaneously under a single operating condition. Still, the converter *must be rated individually for each of these variables.*

3. The series converter of the UPFC is composed of power semiconductors. As such, in contrast to conventional power equipment, it has no inherent overvoltage and only very short duration overcurrent capability. The manufacturer usually includes judicious safety margins for voltage and current ratings for increased reliability. If system operation does require equipment functioning under overvoltage and/or overcurrent conditions, this should be precisely specified for the manufacturer with the understanding of probable cost consequences. Overvoltage requirements usually considered as steady state, almost independent of practical time durations, thus it will likely increase equipment rating and cost. Moderate overcurrent capability may be provided for short time, up to about 1 s, after which it would also become a steady-state requirement with likely cost increase.

10.5.2.2 Shunt Converter Rating to Meet UPFC Operation Requirements

In considering the rating requirements for the shunt converter of the UPFC, the following application options and asset investment issues should be considered:

1. The sole purpose of the UPFC is to provide the prescribed two-dimensional series compensation for active and reactive power flow control in the line. Support of transmission voltage at the bus of connection is *not* necessary (now or ever).
2. In addition to the required two-dimensional power flow control, the transmission voltage *does need* support *now* by reactive compensation at the bus of connection.
3. In spite of not needing transmission voltage support now in addition to the required two-dimensional power flow control, there is a *concern* that unpredictable system, load, or operational changes will necessitate the support of the transmission voltage to at the UPFC bus in the *future*.

The rating considerations for shunt converter of the UPFC under the above three operational scenarios would be distinctly different.

Under option no. 1, the rating of the shunt converter is determined simply by the active power, P_{UPFC}, which the shunt converter needs to supply at its DC terminal from the AC bus for the series converter to execute the prescribed two-dimensional compensation. (The shunt converter thus will have of minimum MVA rating, but no reactive compensation capability.)

Under option no. 2, the reactive power rating of the shunt converter is to be determined by normal load flow studies under prescribed system conditions to maintain the bus voltage within defined limits. Once the maximum reactive rating of the shunt converter is established, the total shunt converter MVA can be determined by the simple formula:

$$\mathrm{MVA}_{shunt} = \sqrt{Q^2_{shunt} + P^2_{UPFC}}$$

where Q_{shunt} is the maximum reactive power that the shunt converter has to supply to, or absorb from the AC bus to keep the bus voltage within defined limits, and P_{UPFC} is the maximum active (real) power that it has to provide at the DC bus of the series converter to sustain the two-dimensional series line compensation in the total P–Q operating domain specified. (Note that in practice, P_{UPFC} would also include the losses of the two converters—about 1.5–2.0% of their total MVA rating at maximum output—in order to maintain the required DC link voltage.)

Under *option no.* 3, there is no definite requirement for reactive shunt compensation. However, from the standpoint of asset investment, the formula established for *option no.* 2 provides a useful indication of the high economic incentive for incorporating reserve capacity into the shunt converter for reactive compensation. This is because, as seen in the simple expression above, the active and reactive power sums as two phasors in quadrature and, thus, for a moderate overall VA rating increase, much more than proportional dynamic VAr compensation capacity can be realized. Consider, for example, that the required active power rating of the shunt converter to support the series compensation is 100 MW. Then, in order to also provide a maximum of 100 MVAr controllable capacitive (and inductive) reactive compensation capability, the rating of the shunt converter would have to be increased only by 41% ($\sqrt{2}$) to 141 MVA. Considering further that (a) the continuous variable reactive output from 100 MVAr inductive to 100 MVAr capacitive represents 200 MVAr dynamic compensation range; (b) this dynamic range can simply be converted into a zero to 200 MVAr compensation range with the addition of a fixed 100 MVAr capacitor bank, should the requirement for voltage support increase in the future; and (c) since most of the control, auxiliary equipment, and the operational infrastructure hardware, as well as much of the installation labor, are already included in the cost of the basic UPFC (with no shunt compensation capability), it is quite evident that the inclusion of reactive shunt compensation capability with the UPFC is, in general, an economically savvy decision for transmission asset investment.

Issues of Transient Rating. The main issue to consider is the maximum system overvoltage at which the shunt converter must be kept operating. Under overvoltage condition, the converter would need to be absorbing reactive power by drawing inductive current, the magnitude of which must be limited to the maximum rated converter current. In order to do that, the converter control must increase the DC link voltage in order to enable the converter to produce a large enough output voltage so that the corresponding output current determined by the difference between the system and converter voltage will not exceed the maximum allowed value, that is, $(V_{line} - V_{converter})/X_{transformer} \leq I_{converter\ \max}$. It is evident that the system overvoltage condition must be carefully evaluated from the standpoint of operation benefit versus cost to specify the level at which the converter operation is required since, above a nominal design margin, it does directly affect the converter rating and cost. It is worth mentioning that at relatively high overvoltage (which typically would be caused by load rejection), not only the coupling transformer of the shunt converter, but also all other system transformers in the affected part of the system would saturate, resulting in a very large overall VAr absorption, albeit with very distorted current.

Thus, under this condition, the converter's VAr absorption may be irrelevant, and its operation could anyway be interrupted by large harmonic currents resulting in current peaks exceeding the turn-off capability of the converter valves.

The overcurrent considerations within the normal operating voltage range are similar to those characterizing the series converter. With increasing overcurrent requirement (magnitude and time duration), the cost of the converter equipment will increase in proportion. It should be noted that if the operation of the system would require higher than maximum rated current from the shunt converter, the control could be programmed to give either a predefined or condition-based priority to supply first the active current need of the series converter for the desired power flow, or to use the capacity primarily for reactive compensation to regulate the bus voltage.

10.5.3 UPFC Rating Optimization by Combined Compensation

As previously explained and illustrated (e.g., in Figure 10.19), the unconstrained voltage injection of the UPFC is in a circular region around the end of the transmission bus voltage phasor. This means that the UPFC voltage injection in general results in an advance or a retard of the prevailing transmission angle. Thus, the UPFC could be considered to have two equal half-circle segments as operating regions, one characterized by the advancement, the other by the retardation of the effective transmission angle. Although this inherent capability of the UPFC to operate symmetrically in both regions may well be utilized in many practical applications (e.g., power oscillation damping), there are other applications in which voltage injection resulting in unequal operating regions for advancement and retardation, or in unidirectional change, either advancement or retardation, of the transmission angle is quite satisfactory. For example, if the attainable transmission angle is too small for the desired power transmission, the UPFC would always have to provide first a leading and essentially fixed phase angle advancement to establish the correct steady-state operating point around which the control of active and reactive power would be executed under the prevailing system conditions. In this type of applications, the series converter of the UPFC would not be utilized effectively because with a given MVA rating it could control, for example, the transmission angle over the range of $-\sigma_{max} \leq \sigma \leq +\sigma_{max}$, but it would actually be used to control the transmission angle over only either the positive range of $0 \leq \sigma_{max}$ or the negative range of $-\sigma_{max} \leq 0$. In other words, the rating capacity of the converter is utilized only 50%. Another consideration in this type of application is that a significant portion of the UPFC rating may be used up for steady-state angle control, which could be provided by more economical means.

The operating requirements for unequal operating range and steady-state angular shift can be satisfied if the UPFC is combined with a phase shifting transformer providing fixed or selectable angle of advancement or retardation. A possible circuit arrangement to implement this hybrid arrangement is shown schematically in Figure 10.20a. As seen, the overall circuit arrangement includes the usual UPFC configuration with two converters, one is in shunt- and the other is in series-connection with the transmission line. This hybrid arrangement, in its simplest form, however, also includes an additional winding on the secondary (converter) side of each phase of the shunt-connected coupling transformer. (In other arrangements, the phase shifting

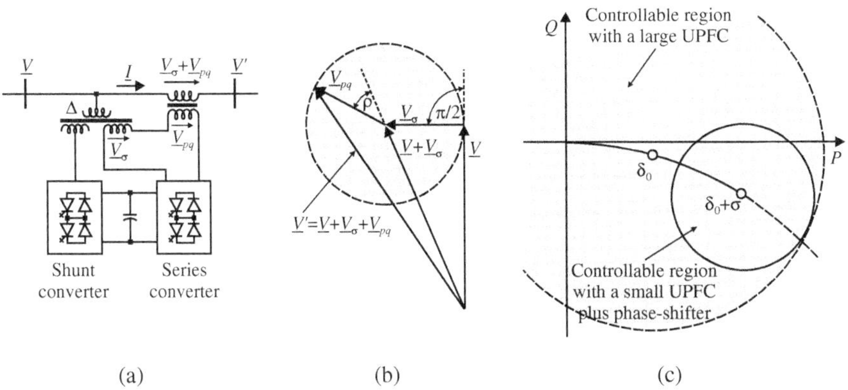

Figure 10.20 UPFC combined with (fixed) phase-shifting to reduce its rating to that necessary only for handling frequent angle variations and dynamic compensation needs.

could also be accomplished by a separate transformer, and in both arrangements a tap changer could be applied to adjust the steady-state phase shift.) The transformer connection is such (e.g., delta-connected primary) that the voltages obtained at the phases *a, b,* and *c* of the secondary windings are in quadrature with the phases *a, b,* and *c* phase-to-neutral primary (line) voltages. Considering one phase, for example, phase *a*, it is seen that this winding is connected in series with the secondary winding of the series-connected coupling transformer and the phase *a* output of the series converter. As a result, the voltage injected in series with the line to control the flow of power through it is the vectorial sum of the fixed voltage provided by the shunt-connected coupling transformer and the controllable output voltage of the series converter.

The operation of the hybrid power flow controller is summarized with reference to the phasor diagram shown Figure 10.20b. As explained above, the total injected voltage, used to control the power flow in the line, is made up of two components: Voltage component $\underline{V}_\sigma$, which is the fixed quadrature voltage provided by the shunt-connected transformer to advance (or retard) the existing transmission angle by a fixed angle σ, and voltage component $\underline{V}_{pq}$, which is the controllable component provided by the series converter of the UPFC. The magnitude of $\underline{V}_{pq}$ is variable in the range of $0 \leq V_{pq} \leq V_{pq\max}$ ($<V_\sigma$) and its angle ρ in the range of $0 \leq \rho \leq 2\pi$ with respect to the fixed $\pi/2$ angle of V_σ. The magnitude and angle of the effective transmission voltage, $\underline{V}'$, at the bus of the transmission line, are obtained by vectorially adding the total injected voltage $\underline{V}_\sigma + \underline{V}_{pq}$ to the original bus voltage $\underline{V}$. In the hybrid arrangement, the circular operating region of the UPFC is centered around the end of the voltage phasor $\underline{V}_\sigma$, providing in effect a fixed angular shift for voltage phasor $\underline{V}$ and, with the stipulation of $V_{pq\max} \leq V_\sigma$, the total control region allows only unidirectional change (retardation or advancement depending on the sign of angle σ) for the transmission angle. (Note that the figure illustrates the case of angle advancement.)

The economic benefits of the hybrid arrangement for suitable applications, such as the present example of steady-state angle advancement, can be further explained with reference to Figure 10.20c that shows the resulting operating region of the power transmission in the $\{Q,P\}$ plane.

Without any compensation, the "natural" steady-state transmission angle of the line is δ_o. The corresponding power is too small for the adequate utilization of the line. To obtain the desired steady-state power transmission, the existing transmission angle is increased by an additional angle σ to yield the desired overall angle of $\delta_o+\sigma$. This steady-state angular phase shift is simply accomplished by a fixed-angle phase shifting transformer. In this way, the required rating of the UPFC is significantly reduced and used only for optimizing and managing the real-time power transmission demands (minimization of reactive power flow, matching load demands, counteracting dynamic disturbances, etc.) in the desired operating $\{Q,P\}$ domain (shown by the solid-line circle in Figure 10.20c) around the fixed operating point corresponding to the *effective* transmission angle, $\delta_o+\sigma$, obtained by the fixed phase shift.

The stipulated transmission requirements could, of course, be met with a single UPFC of much higher rating than that needed with the phase-shifter (over 2.5 times higher in the illustrative example), as shown by the large circular operating domain inside the dashed line boundary in the $\{Q,P\}$ plane of Figure 10.20c. However, as can be observed, the advancement of the transmission angle to the required operating region would also establish a much larger operating domain (including capability for power flow reversal and reactive power flow control over a wide range from inductive to capacitive) than the application requires, at a cost that could not be justified.

10.6 PROTECTION CONSIDERATIONS

The generalized power flow controllers using combinations of shunt- and series-connected converters with common DC link require different types of protection for their series- and shunt-connected constituents. Thus, the protection issues of the UPFC can generally be broken down to the individual protection of the series-connected and the shunt-connected constituent converter.

10.6.1 Protection of the Series Converter

All types of compensators and power flow controllers *connected in series with the line* require special rating and/or protection considerations to handle surge currents caused by line faults or other system disturbances. One consideration is, of course, whether the series-connected equipment has sufficient current rating to handle the worst-case line surge currents. The other consideration is whether it would have sufficient voltage rating, or suitable voltage limitation characteristic, to support or limit the voltage that would develop between the two connection-points in response to the flow of line surge current. The VSC generally has some limited inherent surge current capability, but for rather short time duration (from a few cycles up to about 60 cycles). On the other hand, the output voltage of the converter (i.e., the injected series compensating voltage) can be controlled independently of the line current. Thus, the rating and protection issues of series-connected converter are very much dependent on the particular application, primarily on the prevalent line impedance ("weak" or "strong" line) and the desired behavior of the controller during surge current conditions. In general, there are three possibilities:

1. The line is "weak" and the possible surge current is within (the economically practical) rating of the converter.
2. The converter during surge current conditions is put in a "bypass" operation to provide a path for the line current at zero voltage injection.
3. The converter limits the injected voltage to a maximum value for line current higher than a given maximum.

The simplest case is the rare (almost nonexistent) very "weak" line case, in which the converter has sufficient current rating to handle the anticipated line fault current while maintaining the (reactive) voltage injection up to the maximum voltage rating of the converter. In this case, there is no other protection than appropriate control algorithm together with the practically compulsory breaker bypass to handle converter or control failures, or thermal problems due to a sustained fault.

The protection approaches based on bypass or voltage limitation can be implemented on the basis of two protection philosophies. One is to provide a fast solid-state bypass and the other is to have some kind of voltage limiter parallel with the voltage injection terminal, which takes over the current conduction, but maintains the injected voltage below a given maximum, when the line current exceeds the normal operating range. Some implementation possibilities of these approaches are illustrated in Figures 10.21 and 10.22.

Due to the large range of possible surge currents at different line voltages and system structures, as well as to the variety of differing application requirements and constraints, it is not possible to devise or select a single protection solution that would uniquely meet in all cases the operation needs and cost objectives. For this reason, in this section, the basic practical possibilities for protection are summarized only for the purpose of giving guidance for selection and evaluation in different applications the designer may encounter.

The first arrangement in Figure 10.21 illustrates the case of handling modest line surge currents by operating the converter switches so as to provide an *internal* bypass via two series-connected valves with zero voltage injection. The idea behind this solution is that the power semiconductors, particularly those in the thyristor family, can conduct significantly higher current, and can be operated at higher junction temperature, when they are kept in full conduction than when it normally operated with repeated turn-on and off. The main feature of the arrangement is that it is "free" (no extra components are needed), and it may be a sufficient practical solution, particularly for relatively low voltage and weak transmission lines. (One actual installed UPFC at AEP's Inez substation employs this method.) In the future, when the anticipated silicon-carbide power semiconductor devices become available, this solution may be more generally applicable due to their anticipated much higher surge current and operating temperature capability.

The second arrangement allows the use of all four valves of the commonly used bridge converter for bypass protection. This is achieved through the rapid disconnection of the DC capacitor by a fast auxiliary valve constructed of suitable

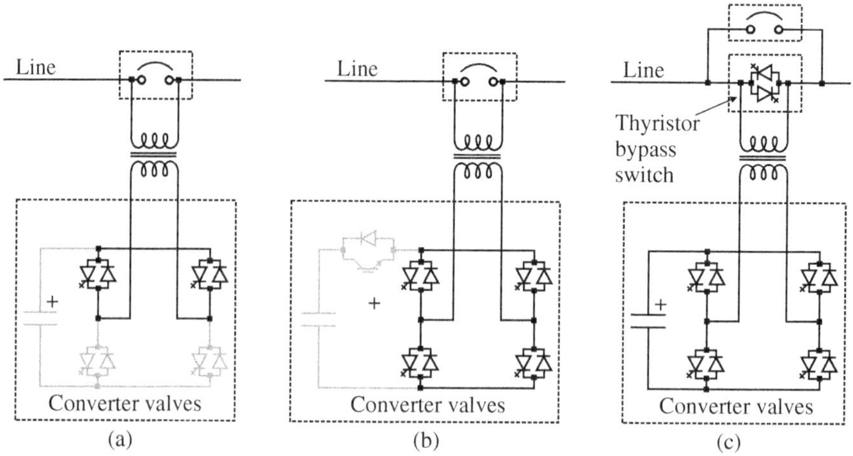

Figure 10.21 Current bypass approaches to the protection of series-connected converter during line faults and major disturbances: (a) bypass protection by *single converter valve* (weak line), (b) bypass protection by *two parallel converter valves,* and (c) bypass protection by *external thyristor switch.*

power semiconductors (e.g., IGBTs), and by the simultaneous gating-on of all converter valves. This arrangement in effect *doubles* the bypass capability of the previous one at a very modest cost and thus it would likely be satisfactory in appreciably increased number of practical cases.

In connection with these two approaches, it should be noted that in many actual converter configurations, several individual converters may in effect operate in parallel. With this type of structures, of course, the role of the single valve shown in both illustrations would be replaced by a number of valves sharing the current in parallel connection.

The third arrangement in Figure 10.21 illustrates the case of handling high surge currents by an external thyristor bypass switch of sufficiently high short-term current rating. This approach is quite universally applicable due to the very high attainable surge current rating of conventional thyristors (i.e., those having no turn turn-off capability). It should be mentioned that this, as well as the previous two solutions, would be backed-up by a fast mechanical breaker to handle sustained fault currents under unsuccessful fault clearing.

The first arrangement in Figure 10.22 is similar to that used for the protection of series capacitors. A (set of) ZnO arrester(s) of appropriate rating provides a bypass for the line current at the break-over voltage of the arrester(s). Since the ZnO break-over voltage needs to be about twice as high as the peak (injected) voltage during normal operation, the converter voltage rating must be correspondingly high. This is because during surge currents, the converter gating would be blocked, and the DC capacitor would get charged to the maximum voltage determined by the limit of ZnO arrester via (the series string of) the anti-parallel diodes in the valves that the (string of) controlled semiconductors (e.g., GTOs) in off-state would have to be able to block.

The second arrangement shown in Figure 10.22 solves the voltage rating problem of the conventional ZnO limitation by applying a thyristor-switched ZnO voltage

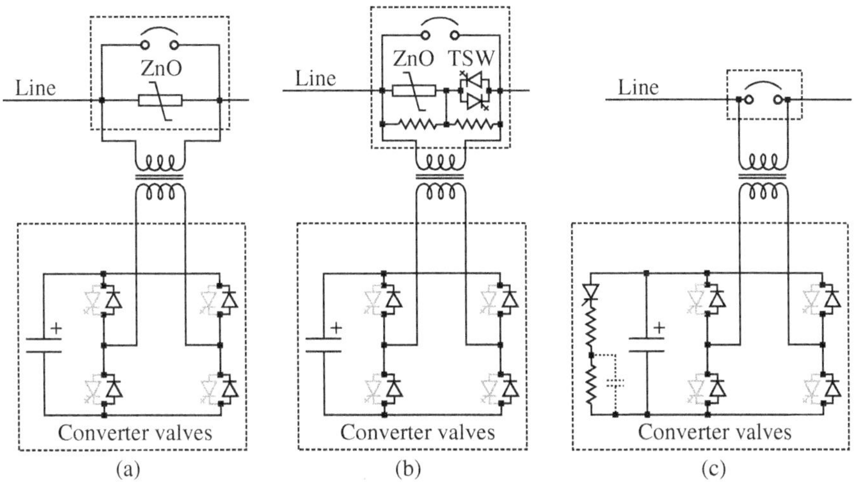

Figure 10.22 Voltage limiting approaches to the protection of series-connected converter during line faults and major disturbances: (a) protection by *ZnO arrester* with converter off (high converter voltage rating), (b) protection by *switched ZnO arrester* with converter off, and (c) protection by *DC voltage limitation* with converter off (high current converter diodes).

limiter. The ZnO break-over voltage is now equal to the maximum voltage rating of the converter required for normal operation (with sufficient safety margin). In normal operation, the thyristor switch is opened (blocked) and a low current voltage divider reduces the prevailing operating voltage across the ZnO arrester to a safe design value (about half of the converter injected voltage). Under surge current conditions, the thyristor switch is gated on, the converter gating is blocked, and thus the ZnO is forced to conduct the line current and limit the converter terminal voltage to the nominal maximum. Note that this can be an economical solution in many applications, since the voltage rating of both the thyristor switch and the ZnO arrester is half of that required in the previous arrangement.

The third and final arrangement is different in that the voltage limitation during surge current conditions is provided by a (controlled) voltage limiter across the DC capacitor. In this case, the controlled semiconductors in the converter valves are again blocked and the line current through the valve diodes is allowed to charge the DC capacitor. However, as soon as a preset DC voltage level is reached, the auxiliary thyristor switch across the DC capacitor turned-on to discharge the capacitor via appropriately rated linear or nonlinear resistors. At the end of the fault conditions, the thyristor switch is commutated off as the converter is brought back to normal operation. This protection arrangement would require the converter diodes to have a sufficient surge current rating, which may not be compatible with their switching requirements under the normal converter operation, particularly if PWM is used for output voltage construction and control.

It is pointed out that the above voltage limiting approaches, just as well as those with the previously discussed current bypass arrangements, are complemented with a suitable mechanical bypass breaker to handle equipment failures and other unforeseen contingences. In addition, it is usually a normal practice to apply appropriate disconnect switches to be able to isolate and remove failed equipment from the line after breaker bypass is applied.

It should also be kept in mind that, as mentioned before, the control system of the series converter does have sophisticated protection functions that monitor internal circuit parameters, component failures, redundancy conditions, and the overall performance of the system and give alarm indication or, in extreme cases initiate shut downs, if parameter limits are exceeded or reliable operation no longer can be ensured due to internal component failures or circuit malfunctions.

10.6.2 Protection of the Shunt Converter

The protection of the shunt-connected converter does, in general, require no external means, except for the normal circuit breaker connecting the coupling transformer to the system bus. However, the internal, control-based protection of the shunt converter has the functional task of ensuring that the maximum instantaneous current does not exceed the turn-off capability of the valves in the converter. Similarly, the converter protection must ensure that operating voltage is kept below the defined maximum. Since the internal protection usually results in temporary suspension of operation or total shut-down, it is important to determine the necessary operating voltage and current limits within the process of defining the specified ratings of the equipment, as discussed in section 10.5 on *Rating considerations*.

Overvoltage exceeding the operation limit results in suspended gating and subsequent shut down. High instantaneous current magnitudes in the converter valves (usually due to harmonics caused by saturated transformers or large system voltage unbalances) result in temporary short duration stoppage (from a fraction of a cycle to one or two cycles, or until the condition sufficiently improves) in the operation of the converter by gate signal suppression, followed by automatic synchronous restart. The instantaneous current magnitude caused operation stoppage can be greatly reduced by sophisticated control techniques, which can keep the converter output current substantially sinusoidal despite external transformer saturation and system voltage unbalance.

10.7 APPLICATION EXAMPLE: UPFC AT AEP'S INEZ STATION

Following the introduction of the UPFC concept [1], three UPFCs of different ratings were designed and two of these were commissioned in the United States and one in South Korea. All three have convertibility for different multi-functional operations. Here, the first UPFC ever built and installed at the Inez substation of the American Electric Power (AEP), and has gone through extensive performance evaluation, is selected for detailed discussion. This UPFC installation, with a total rating of ±320 MVA, was commissioned in the mid-1998 at the Inez station of the AEP in Kentucky, United States, for voltage support and power flow control. The project was jointly sponsored by the Electric Power Research Institute and AEP, and designed and manufactured by the Westinghouse Electric Corporation.

10.7.1 Background and Planning Information at the Time of Installation

System Description. AEP is an investor-owned electric utility company, which operates across seven Midwestern states, from Michigan in the north and west, to Tennessee in the south, and Virginia in the east. These facilities, including the generating plants, 765 kV extra high voltage (EHV), 345 kV high voltage, and 138 kV transmission lines, distribution system, and associated stations, are all interconnected and operated as a major power network within the eastern US power system. AEP provides electric service to approximately 1.7 million customers in an area of about 48,000 square miles (124,313 square km).

AEP produces over 124 billion kilowatt-hours a year, representing a peak system demand of over 26,000 MW. A major portion of the system's generation is located on the Ohio River and its tributaries. The areas south and east of these rivers depend on long transmission lines to meet the electric demand. The load area of concern regarding this project is one of these areas, the Inez Area identified in Figure 10.23. The project involves the Inez Area and the neighboring Tri-State Area. These two areas are quite different in their physical and electrical characteristics. The reinforcement plan adopted to mitigate the problems strongly ties them as one area.

The Inez Area located in eastern Kentucky is rural in nature and has a population of about 670,000 which is spread over an area of about 6300 square miles (16,316 square km). Generating plants and EHV/138 kV substations are located only at the periphery of the area. The Inez Area's winter power demand of approximately

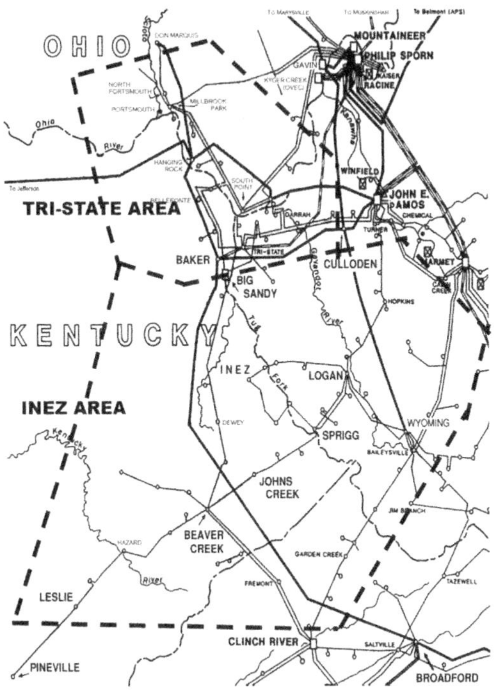

Figure 10.23 Inez and Tri-State areas of the AEP power system.

2000 MW is served by several long 138 kV transmission lines. System voltages and area reactive power requirements are supported by a ±125 MVAr SVC and a large assortment of switched shunt capacitor banks located at several 138 kV and lower voltage subtransmission stations.

The Tri-State Area located adjacent to and north of the Inez Area has a population of about 480,000 people and is spread over an area of approximately 3500 square miles (9064 square km). The Tri-State Area load is summer peaking with a requirement of about 1400 MW. It is served by a strong network of EHV lines, EHV/138 kV substations, and generating plants within and near its boundaries. The area transmission system supports power flows into the Inez Area via 138 kV interarea line connections.

System Problems. The Inez Area depends on long 138 kV transmission lines to support its customers' demand. Even under normal system conditions, many 138 kV transmission lines carry power flows reaching 300 MVA. These power flow levels are well above the surge impedance loading for 138 kV lines and as such are indications of the system stress. Moreover, these high loadings, as compared to the thermal capabilities of the lines, left little margin for system contingencies. In addition, voltage levels in the Inez Area were generally as low as 95% of nominal, which is considered to be the lowest acceptable voltage level for reliable service. Heavy 138 kV line loadings and comparatively long line distances resulted in excessive voltage gradients across the transmission network, further indicating stress on the system.

The Tri-State Area, on the other hand, is served by a strong network of EHV lines and EHV/138 kV station. The load is concentrated in and around a comparatively smaller area. The 138 kV line exits are heavily loaded over relatively short distances. Voltage levels are maintained at all substations within a narrow bandwidth around 98% of nominal.

Several single contingency outages in the Inez Area could have resulted in depressed voltage and/or thermal overload conditions. As an example, a simulation of a 765 kV line outage, during winter peak load periods, indicated increases in system active and reactive power losses by about 85 MW and 700 MVArs, respectively. The need to supply these increased system losses under a contingency condition would have further reduced the area voltages to levels which would certainly have resulted in wide area customer complaints. In addition, underlying transmission system would be loaded to well above their winter emergency capabilities. A second contingency under these circumstances was intolerable.

Single contingency outages in the Tri-State Area would not have caused any thermal overloads or depressed bus voltage conditions. However, double contingency outages involving one or more EHV/138 kV transformers or 138 kV generating units would have resulted in thermal overloads of the EHV/138 kV transformer and/or 138 kV line exits.

System Reinforcement Plan. The analysis of the system performance indicated that additional voltage support and power supply facilities were required to resolve normal, single contingency, and double contingency problems in the Inez Area. In addition, the EHV/138 kV transformer and line exit capacities of the Tri-State Area needed to be increased.

Following extensive analysis, the conclusion was reached that constructing a high capacity 138 kV line having thermal capabilities approaching a 345 kV line would provide an economical means of adding thermal capacity to the area. However, such a high capacity line would not carry its share of line loading according to its capacity. Power flow on such a line would still be governed by its impedance and other AC transmission system parameters. Concurrent with thermal considerations, peak and off-peak voltage performances in the Inez Area also dictated the need for a dynamic voltage support facility in the area.

Series capacitors together with SVCs and converter-based FACTS controllers were then evaluated to enhance the power flow and provide sufficient voltage support. The evaluation showed that a combined converter-based FACTS controller, such as the UPFC, which can provide both the voltage and line flow control capabilities, was a logical choice to be an integral part of the overall reinforcement plan for the Inez and Tri-State areas, which include the following:

- A high capacity 950 MVA, 138 kV line between the Big Sandy and the Inez stations.
- A ±320 MVA UPFC at the Inez station to fully utilize the high capacity of the new 138 kV line, provide dynamic voltage support, and control several mechanically switched capacitors in the area.
- A 345/138 kV transformer at the Big Sandy Station in the Tri-State Area to provide for the power flow requirements of the new high capacity 138 kV line.
- Series reactors to constrain loadings on existing thermally limited facilities.

The reinforcement plan was executed in two phases. The first phase of the plan included the installation of a 345/138 kV transformer bank at the Baker/Big Sandy Station and line power flow limiting series reactors at two stations. The last portion of this first phase involved the installation of the ±160 MVA shunt converter part of the UPFC at the Inez station. The 345/138 kV transformers were designed to have sufficient capacity to provide the Tri-State Area needs and fulfill the future loading requirements of the high capacity Big Sandy-Inez 138 kV line. The series reactors installed to limit the loadings on critical, but low capacity lines. During the first phase, the shunt part of the UPFC functioned as a STATCOM and supported the reactive power and the dynamic voltage needs of the Inez Area. In addition, it provided signals to control the switching operations of several 138 kV shunt capacitor banks in the area.

The second phase of the plan included the construction of a high capacity (950 MVA) 138 kV line between the Big Sandy and Inez substations, installation of the series part of the Inez UPFC and two 138 kV mechanically switched shunt capacitor banks at the Inez station. The series converter was specified to be identical to the shunt converter in order to increase the operating flexibility of UPFC.

The third and the last phase of the project included additional 138 kV line construction, switching configuration changes at several stations, and additional reactive power correction to supplement the existing system reactive power margins.

10.7.2 UPFC Operation Strategy

Voltage Control. The UPFC is required to regulate Inez substation 138 kV bus voltages and control six 138 kV shunt capacitor banks (a total of over 330 MVAr) located at the Inez and three other nearby stations. This way, proper capacitor switching strategy is established to reduce daily and seasonal voltage fluctuations to within acceptable limits. During system disturbances, mechanically switched shunt capacitor banks and associated controls are generally slow to react. Over 60 shunt capacitor banks are connected on the 138 kV and lower voltage transmission system in the Inez Area. Under actual system contingency conditions, all of these banks may not switch-on (hunting concerns) or some may over-correct the voltage and lock-out. To resolve this situation, the UPFC is required to maintain a predetermined reactive power margin to maximize the shunt converter's dynamic reactive power reserve for system contingency conditions. This ensures that the controllable reactive power range of the shunt converter (from –160 to +160 MVAr, i.e., a maximum control range of 320 MVAr) can be made available at all times to compensate for dynamic system disturbances. The shunt capacitor banks will be switched on and off to maintain the reserve UPFC and SVC margins during steady-state load fluctuations.

POWER FLOW CONTROL. With all transmission system facilities in service, the Big Sandy-Inez 138 kV line loading to be maintained at a level which would minimize system losses. For the peak and near peak load conditions, load flow studies indicate that a 300 MW line loading would meet this objective. Actual flow on the line could be higher depending on other prevailing system conditions. The UPFC, therefore, may be required to slightly reduce the line loadings. Line reactive power flow and its direction will be monitored to help maintain the dynamic reactive power margin of the shunt converter.

The series power flow control becomes important during contingency conditions. Major double contingency outages heavily load critical 138 kV facilities (especially the Big Sandy-Beaver Creek line). Controls of the UPFC have to monitor loadings on the three critical lower capacity lines. The control objective is to increase the Big Sandy-Inez 138 kV line loading to decrease loadings on the three lines. This control is to be activated as soon as any one of the three line loadings exceeds 90% of their respective emergency thermal ratings. The UPFC has to increase the Big Sandy-Inez line loading until the critical line loadings are reduced below the defined levels or the UPFC reaches its rating limit. Under severe contingency conditions, the UPFC-controlled Big Sandy-Inez line will be capable of transferring 950 MVA (4000 A).

10.7.3 Description of the UPFC

The UPFC for the Inez station was designed to meet the above defined system requirements. In particular, to provide fast reactive shunt compensation with a total control range of 320 MVAr (–160 to +160 MVAr) and control power flow in the 138 kV high capacity transmission line, forcing the transmitted power, under contingency conditions, up to 950 MVA.

In order to increase the system reliability and provide flexibility for future system changes, the UPFC installation was required to allow self-sufficient operation of the shunt converter as an independent STATCOM and the series converter as an independent SSSC. It is also possible to couple both converters together to provide either shunt only or series only compensation over a doubled control range.

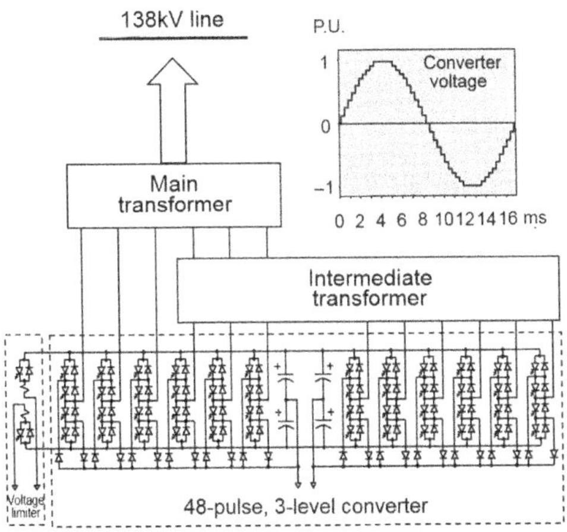

Each converter

Nominal rating:	±160 MVA
Transmission voltage:	138 kV
Converter output voltage: (to coupling transformer)	37 kV (phase-to-phase)
Pole output voltage:	9.3 kV
No. of converter pulses:	48
Type of converter pole:	3-level
No. of poles:	12
No. of 3-phase converters:	4
Type of semiconductor:	GTO
Rating of GTO:	4.5 kV, 4 kA
No. of GTOs per valve:	8 (7+1)
DC voltage:	10.5+10.5 kV
DC capacitor rating:	110 kJ

Note: All ratings are nominal

Figure 10.24 Basic structure of the UPFC converters and the corresponding technical data.

Power Circuit Structure. The UPFC equipment comprises two identical GTO thyristor-based converters, each rated for ±160 MVA. Each converter includes multiple high power, three-level, 48-pulse GTO valve structures feeding an intermediate (low voltage) transformer rated for approximately 80 MVA. The converter and magnetic configuration provides harmonic neutralization and thus the output is a three-phase voltage set of nearly-sinusoidal quality that is coupled to the transmission line via the intermediate transformer by a conventional (three-winding to three-winding) main coupling transformer rated for 160 MVA. The converter-side voltage of the main transformer is 37 kV phase-to-phase (for both shunt and series transformers.) The shunt-connected transformer has a 138 kV delta-connected primary, and the series transformer has three separate primary windings each rated at 16% of the phase voltage. The converter structure is shown schematically with relevant information in Figure 10.24. The installed converter is shown in Figure 10.25.

To maximize the versatility of the installation, two identical main shunt transformers and a single main series transformer have been provided as shown by the power circuit schematic at the left-hand side of Figure 10.26. (At the right-hand side of the figure, the major components of the actual installation are identified.) With this arrangement, a number of power circuit configurations are possible. Converter 1 can operate as a STATCOM with either one of the two main shunt transformers, while converter 2 operates as an independent SSSC. Alternatively, converter 2 can be connected to the spare main shunt transformer and can operate as an additional STATCOM. With the latter configuration, a shunt reactive capability of ±320 MVA becomes available. The power circuit arrangement indicates the priority established for shunt compensation at this location.

Control System. Both converters comprising the UPFC are controlled from a single central control system housed in three cabinets in the control room. Two of the cabinets house the relay interface and signal conditioning, while a single cabinet contains the control electronics. The conceptual structure of the control system is shown in Figure 10.27.

Figure 10.25 The UPFC converter hall.

The actual control algorithms that govern the instantaneous operation of the two converters are performed in the real-time control electronics, which employs multiple digital signal processors. The real-time control communicates with the two-valve pole electronics mounted on each pole of the converter via the valve interface that is linked to the poles by fiber optic cables. The status processor is connected to every part of the system, including the cooling system and all of the poles, by serial communications. During runtime, it continually monitors the operation of all subsystems, collecting and analyzing status information. It is responsible for all startup and shutdown sequences and for organizing and annunciation of all alarm conditions. The

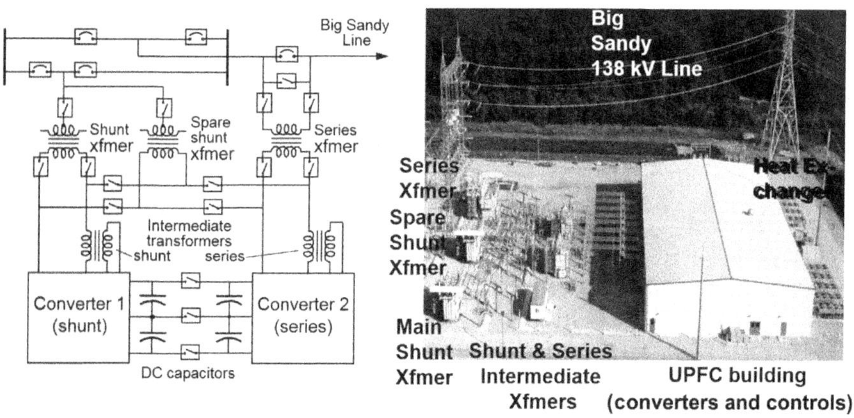

Figure 10.26 Simplified UPFC power circuit in the AEP system at Inez and the corresponding practical installation.

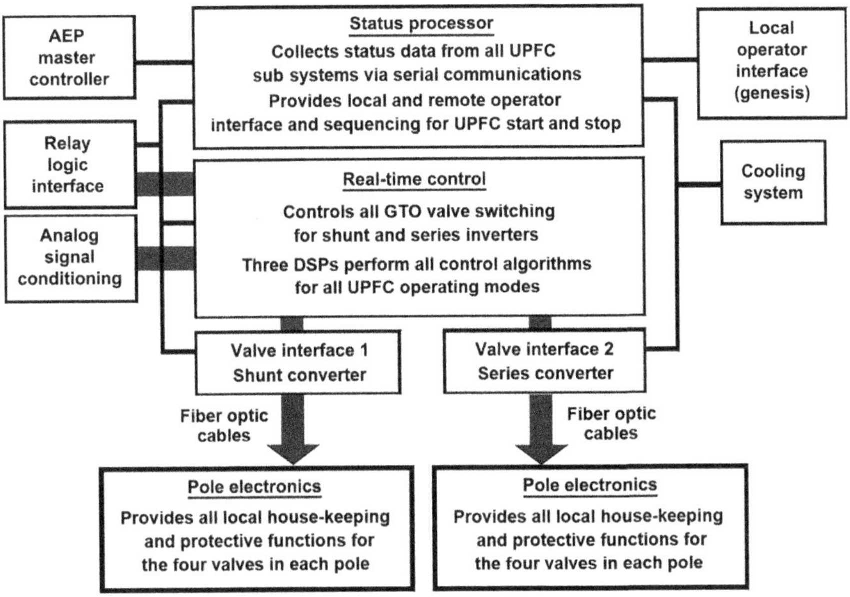

Figure 10.27 Basic functional control structure of the Inez UPFC.

status processor is serially connected to a graphical display terminal, which provides the local operator interface. A hierarchical arrangement of graphical display screens gives the operator access to all system settings and parameters, and provides extensive diagnostic information right down to the individual GTO modules of each valve.

10.7.4 Performance of the UPFC

In the course of commissioning the UPFC, tests were performed to verify its predicted capability. For this purpose, the UPFC was directed to produce large controlled swings of active and reactive power on the Big Sandy line, and sizeable swings of voltage at the Inez station, while measurements were recorded. It is emphasized that these swings do not represent "normal" duty for the UPFC in this application. Arbitrary variations of this kind can be disruptive to the power system because they force active power flows to be redistributed in the network and affect voltage regulation at other stations. AEP system operators defined acceptable boundary limits for the tests and, in addition, slow ramping functions were applied to the control references for the UPFC automatic power flow controller.

Three representative cases have been selected for the purpose of demonstrating the capabilities of the UPFC. The first two cases show the UPFC independently controlling line P, line Q, and Inez bus voltage, respectively. In both cases, the UPFC is operating with the shunt converter in automatic voltage control mode and the series converter in automatic line power flow control mode. In the third case, the series converter of the UPFC is operated as an independent series reactive compensator, SSSC. Each set of results is annotated using the sign convention as defined in Figure 10.28. Note in particular that P and Q for the line are measured at the line-side terminals of the series insertion transformer. This is the actual power at the end of the line, and is

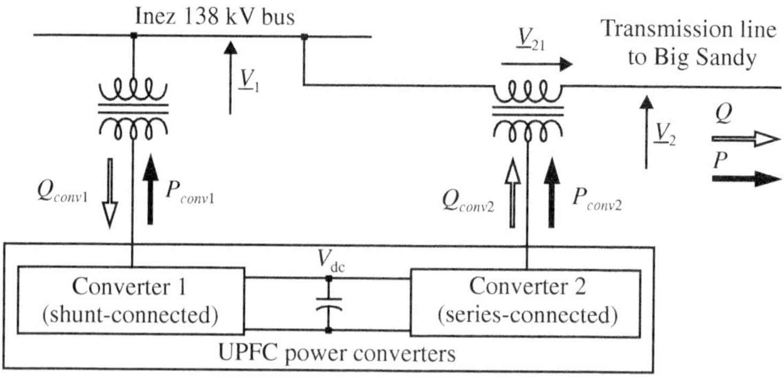

Figure 10.28 Definitions of polarity conventions used for the commissioning tests of the AEP UPFC.

defined as positive toward Big Sandy. The active and reactive power for each of the two converters is also shown. Note that the converters independently generate reactive power but that their active power is substantially equal and opposite. In each case, a few stationary points (between major transitions) have been chosen and a phasor diagram drawn to represent the operating condition at that point. This is considered to be helpful, because it is difficult to interpret each situation from the time plots alone. The interpretation is made even more difficult by the fact that the power network "adjusts" following each major transition, especially with regard to the voltage phase angles at Inez and Big Sandy. The following three cases will take the form of a specific commentary for each of the cases.

Case I ***UPFC Changing Active Power (P)***. Refer to Figure 10.29. This case starts with the UPFC idling near zero-injected voltage and the active power flow on the line near the "natural" level of 150 MW from Big Sandy. The shunt converter is regulating the Inez bus to 1.0 p.u. by generating about 60 MVAr capacitive, and about 36 MVAr is being delivered into the line. The objective for this case is to maintain the Inez bus voltage and the line reactive power flow Q unaltered, while making big step changes in line P.

The UPFC is first commanded to raise the line power to 240 MW. It does this by injecting a voltage of about 0.16 p.u. roughly in quadrature (lagging) with the Inez bus voltage. To satisfy the required conditions, the shunt converter drops its capacitive output to about 20 MVAr, and the series converter delivers about 40 MVAr capacitive to the line. Active power exchange between the converters is about 8 MW.

The second transition commanded is a 170 MW drop in the line power to 70 MW. This is accomplished with little change in the magnitude of the injected voltage, but about 180° phase shift, so that the injected voltage is now still roughly in quadrature with the Inez bus voltage, but leading. The shunt converter produces about 85 MVArs capacitive, the series converter reverses its output to 10 MVAr inductive, and about 8 MW flows between the converters through the DC bus. The final transition returns the system to the initial operating point.

Throughout this case, the Inez bus voltage is tightly regulated at 1.0 p.u. and the reactive line power Q stays constant. Note, however, that the voltage, V_2, applied

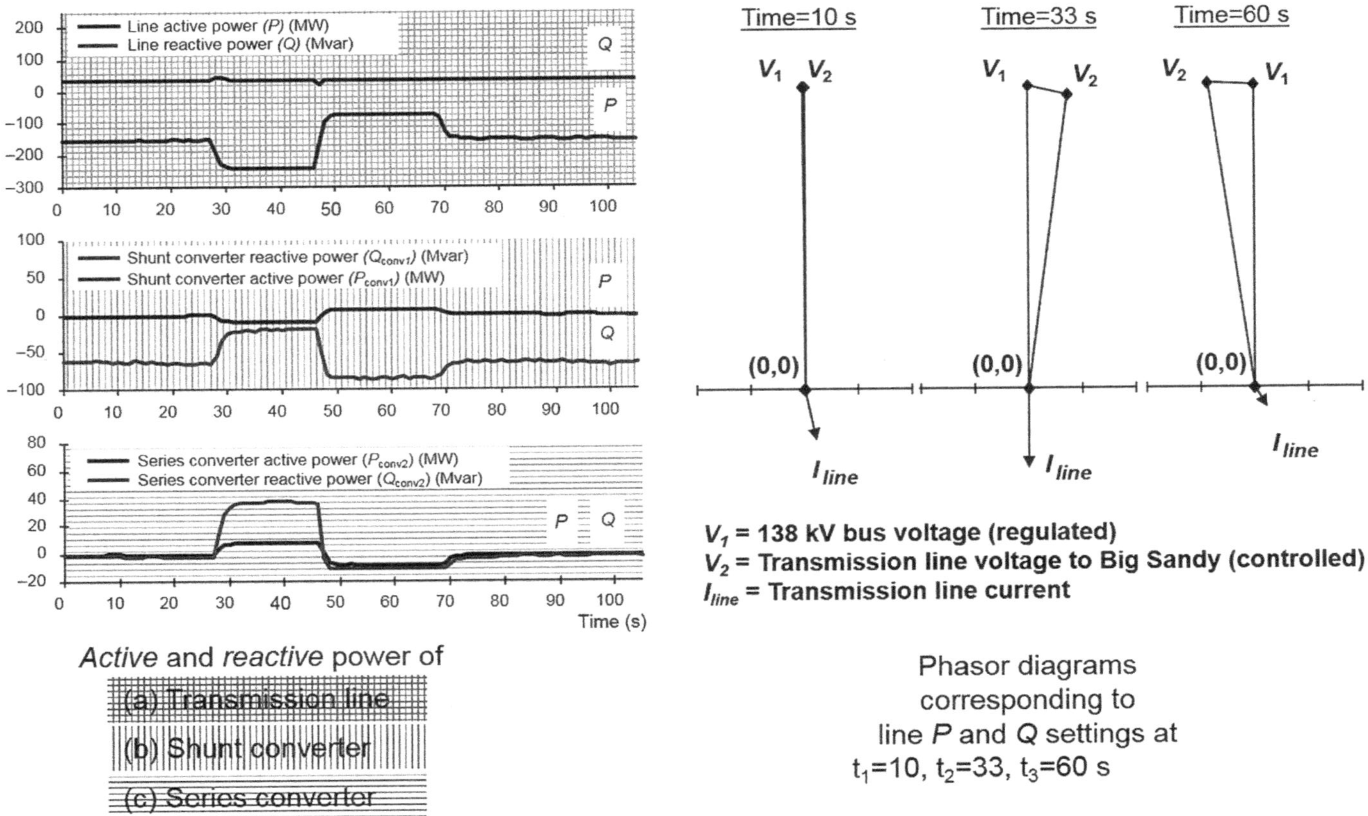

Figure 10.29 Inez UPFC controlling active power *P* while maintaining constant reactive power *Q* in the line.

to the transmission line, is lowered by a few percent relative to V_I to achieve the first swing and raised by a few percent for the second. It should also be noted that the large changes in active power arriving at Inez must, of course, be balanced by an equal and opposite total change in the active power on the other lines leaving the station. For this to happen, a change in the phase angle of the Inez bus voltage, $\underline{V}_1$, is unavoidable.

At the time when these tests were performed, the natural power flow on the line was too high for the UPFC to demonstrate its unique ability to reverse power flow. On other occasions, however, when the line has been lightly loaded, this has been demonstrated successfully and the UPFC has driven active power back toward Big Sandy.

Case II ***UPFC Changing Reactive Power (Q).*** Refer to Figure 10.30. The initial conditions for this test are similar to Case 1, but the objective is to regulate the Inez voltage at 1.0 p.u. and keep the line active power, P, constant while causing large steps in the line reactive power, Q.

For the first swing, the UPFC reference for Q is changed from +30 to −30 MVAr. After the change, 30 MVAr is being received from the line, compared with the 30 MVAr delivered *to* the line initially. The UPFC forces the change by injecting about 0.05 p.u. voltage roughly in anti-phase with $\underline{V}_1$. The line voltage, $\underline{V}_2$, is consequently reduced in magnitude by about 5%. For the second step, the Q reference is taken to +100 MVAr (i.e., 100 MVAr to the line). This time, the injected voltage is in phase with $\underline{V}_1$ so that V_2 is increased by about 5%. The final step reduces the Q reference to zero. The line is now fed at unity power factor with V_2 reduced by about 2.5% relative to V_1.

It is interesting to note that the changes in Q at the line terminals are balanced almost entirely by equal and opposite changes in the reactive output of the shunt converter, which acts to maintain the Inez voltage. This brings to light a fascinating capability of the UPFC. In essence, it can "manufacture" inductive or capacitive MVArs using the shunt converter and "export" this reactive power into a particular transmission line (i.e., the one with the series insertion transformer), without changing the local bus voltage and without changing the reactive power on any of the other lines leaving the substation. It can therefore regulate the station bus voltages at *both* the sending and receiving ends of a transmission line, while still freely controlling the active power flow, P, on the line.

Case III ***The Series Converter Operating in SSSC Mode.*** Refer to Figure 10.31. For this case, the shunt converter is disconnected from the DC terminals of the series converter and is completely out of service. Consequently, the Inez bus voltage is not regulated. The series converter injects voltage into the line essentially in quadrature with the prevailing line current. The injection angle deviates from true quadrature to draw active power from the line for converter losses and to charge and discharge the DC bus capacitor banks as operation requires. By means of the quadrature voltage injection, the SSSC is able to raise or lower the line current, but cannot independently alter P and Q. The SSSC operation is an important subset of full UPFC operation, since it can be used when the shunt converter is not available. Also, a single series-connected converter installation may be the most cost-effective solution for

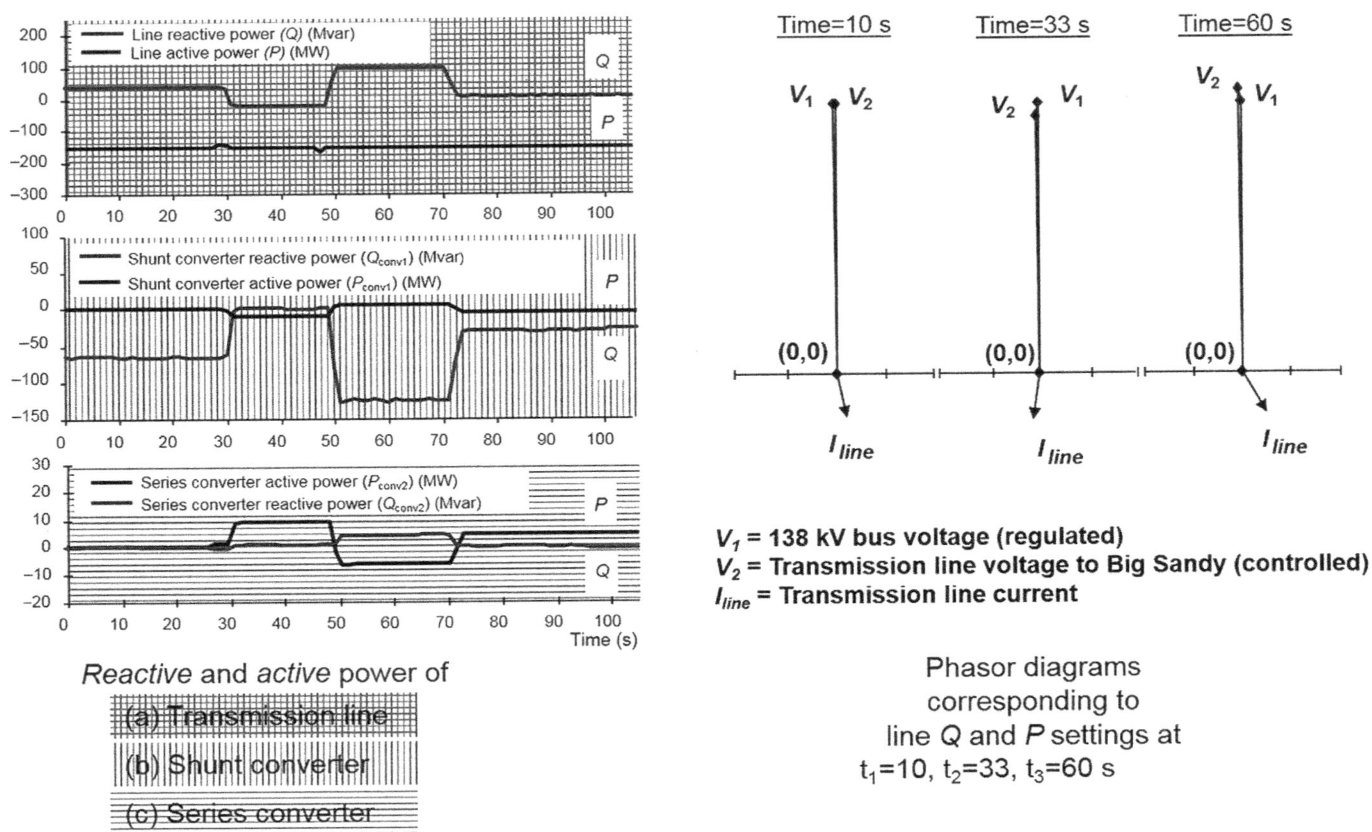

Figure 10.30 Inez UPFC controlling reactive power Q while maintaining constant active power P in the line.

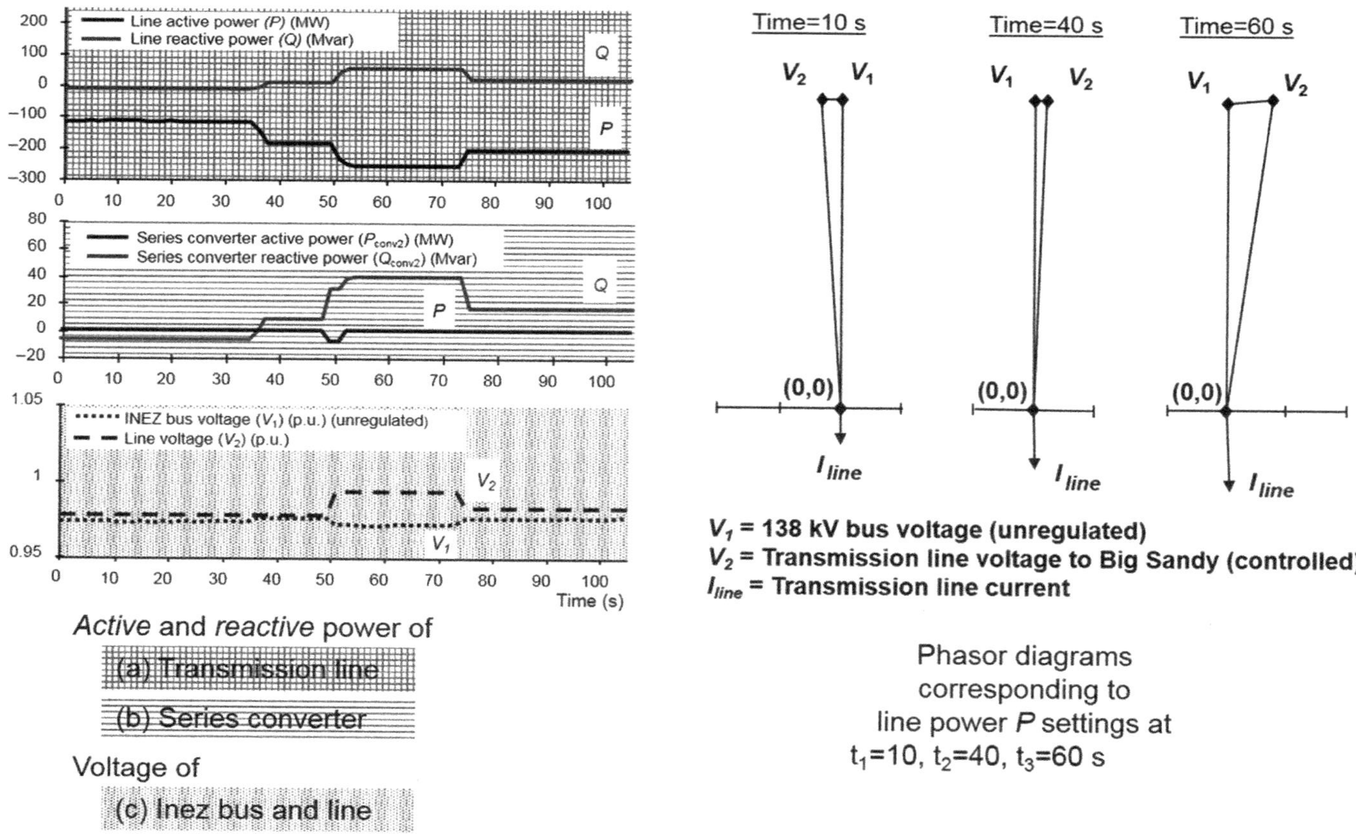

Figure 10.31 Inez UPFC operated as a static synchronous series compensator (SSSC).

applications, particularly for lines of moderate length, where a simpler form of power flow control is sufficient.

The objective of this case is simply to show the SSSC raising and lowering the power on the Big Sandy line. The SSSC is operated to control the magnitude and the polarity of the injected voltage (i.e., the line power is not automatically controlled). Voltage injections are selected to give a sequence of approximately 100, 180, 250, and finally 200 MW on the line. Note the corresponding changes in Q. From the phasor diagrams, it can be clearly seen how $\underline{I}_{line}$ maintains a constant phase relationship to $\underline{V}_1$ and is always in quadrature with $\underline{V}_{21}$. The natural flow on the line is between 100 and 180 MW. Consequently, the polarity of $\underline{V}_{21}$ is reversed from lagging $\underline{I}_{line}$ to leading in this transition. The polarity reversal is accomplished by taking the converter DC bus voltage to zero, then raising it again with 180° phase shift in the converter output voltage.

10.7.5 Importance of Results and Possible Future Trends

This first UPFC installation at Inez has fully proven the conceptual predictions regarding the unique operating and functional characteristics of the UPFC:

- Unique capability to provide independent and concurrent control for the active and reactive line power flow, as well the regulation of the bus voltage.
- Functional adaptability for a variety selectable operating modes and control functions.
- Flexible circuit structure that be reconfigured for independent shunt (STATCOM) and series (SSSC) compensation, as well as for only shunt or only series compensation at double rating.
- Operation in utility environment.

Considering the present technological trends and advancement, the UPFC would benefit most from technological developments in two areas, which could have great impact on its capital and operating cost, and reliability. One is in the device area, which expected to yield more rugged, faster switching power semiconductors (such as the silicon-carbide devices) with higher voltage and current ratings and, in particular, lower operating losses. The other is in power converter circuit area which, as one objective, would come up with a simple converter structure producing low output distortion, and, as a second and—arguably—a more important objective, would create a suitable circuit topology in which the series converter is connected directly, without an insertion transformer, in series with the high voltage line.

10.8 MODELING OF THE UPFC DEVICE

10.8.1 The Steady-State Model of UPFC

In steady-state studies, the UPFC device can be represented by two voltage sources, $\underline{V}_{sh}$ and $\underline{V}_{pq}$, and their corresponding impedances (Figure 10.32b). In this model, the impedance of the shunt transformer and the line impedance (including the impedance

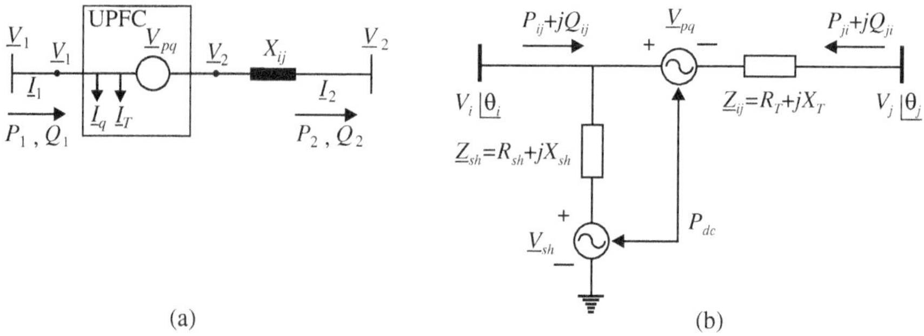

Figure 10.32 UPFC models: (a) the model with a voltage source and two independent current sources and (b) the model with two voltage sources.

of the series transformer) are considered constant. Furthermore, the power losses of the device are neglected.

THE POWER INJECTION MODEL OF UPFC. The two voltage sources are converted into two power injections expressed in rectangular form. The advantage of this representation is that the symmetry character of the admittance matrix is not affected. Expressing the power flow equations in rectangular form leads to a quadratic form that offers some advantages in numerical analysis.

In the first step, by converting the shunt source into a power injection at bus i, the equivalent diagram from Figure 10.33 is obtained.

The power injected at bus i is written as:

$$\underline{S}_{i0} = P_{i0} + jQ_{i0} = \underline{V}_i \left(\frac{V_i - \underline{V}_{sh}}{\underline{Z}_{i0}} \right)^* \tag{10.11}$$

where $\frac{1}{\underline{Z}_{i0}^*} = Y_{i0}^* = G_{i0} - jB_{i0}$

From (10.11), it can be written that:

$$\begin{aligned} \underline{S}_{i0} &= P_{i0} + jQ_{i0} = \underline{V}_i \left(\frac{\underline{V}_i - \underline{V}_{sh}}{\underline{Z}_{i0}} \right)^* = \left(\underline{V}_i \cdot \underline{V}_i^* - \underline{V}_i \cdot \underline{V}_{sh}^* \right) \cdot \frac{1}{\underline{Z}_{i0}^*} \\ &= \left[V_i^2 - \left(V_i' + jV_i'' \right) \cdot \left(V_{sh}' - jV_{sh}'' \right) \right] \cdot \left(G_{io} - jB_{i0} \right) \\ &= \left(V_i^2 - V_i'V_{sh}' - V_i''V_{sh}'' - jV_i''V_{sh}' + jV_i'V_{sh}'' \right) G_{io} - \\ &\quad - jB_{i0} \left(V_i^2 - V_i'V_{sh}' - V_i''V_{sh}'' - jV_i''V_{sh}' + jV_i'V_{sh}'' \right) \end{aligned} \tag{10.11'}$$

Separating the real and imaginary parts yields:

$$\begin{aligned} P_{i0} &= G_{i0} \left(V_i^2 - V_i'V_{sh}' - V_i''V_{sh}'' \right) + B_{i0} \cdot \left(V_i'V_{sh}'' - V_i''V_{sh}' \right) \\ Q_{i0} &= G_{i0} \left(V_i'V_{sh}'' - V_i''V_{sh}' \right) + B_{i0} \cdot \left(V_i'V_{sh}' + V_i''V_{sh}'' - V_i^2 \right) \end{aligned} \tag{10.12}$$

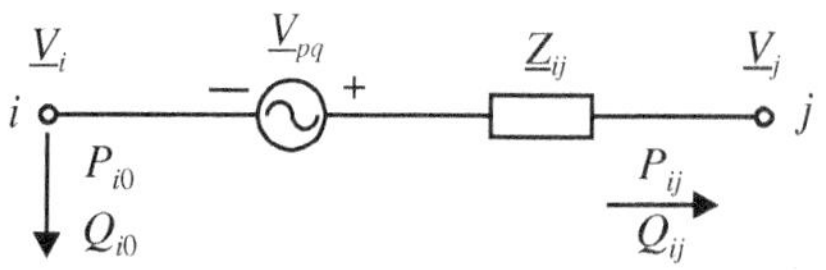

Figure 10.33 Conversion of the shunt part of UPFC to a power injection only at bus i.

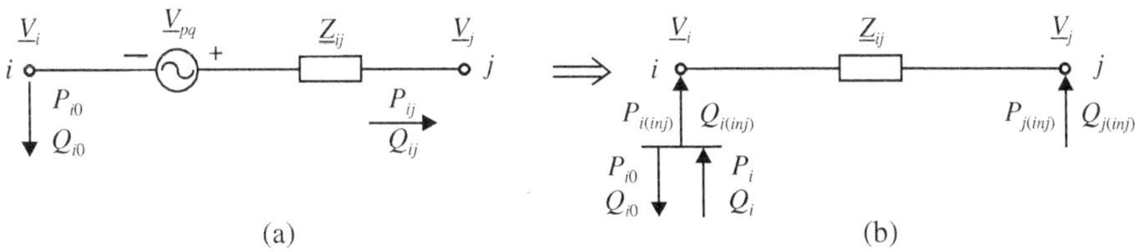

Figure 10.34 Converting the series voltage source of UPFC into two power injections at both i and j busses.

In the second step, the series voltage source of UPFC is converted into two power injections, one at each of the two busses, i and j (Figure 10.34).

By converting the series voltage source, the following power injected at bus i results:

$$\underline{S}_i = P_i + jQ_i = \underline{V}_i\left(-\frac{\underline{V}_{pq}}{\underline{Z}_{ij}}\right)^* = (V_i' + jV_i'')\cdot\left(-V_{pq}' + jV_{pq}''\right)\cdot(G_{ij} - jB_{ij})$$

$$= G_{ij}\left(-V_i'V_{pq}' - V_i''V_{pq}'' + jV_i'V_{pq}'' - jV_i''V_{pq}'\right) +$$

$$+ jB_{ij}\left(V_i'V_{pq}' + V_i''V_{pq}'' - jV_i'V_{pq}'' + jV_i''V_{pq}'\right)$$

Separating the real and imaginary parts yields:

$$\begin{aligned} P_i &= G_{ij}\left(-V_i'V_{pq}' - V_i''V_{pq}''\right) + B_{ij}\left(V_i'V_{pq}'' - V_i''V_{pq}'\right) \\ Q_i &= G_{ij}\left(V_i'V_{pq}'' - V_i''V_{pq}'\right) + B_{ij}\left(V_i'V_{pq}' + V_i''V_{pq}''\right) \end{aligned} \tag{10.13}$$

Similarly, the power injected at bus j can be written as:

$$\underline{S}_j = P_j + jQ_j = \underline{V}_j\left(\frac{\underline{V}_{pq}}{\underline{Z}_{ij}}\right)^*$$

and, by substitutions and appropriate calculations, it yields:

$$\begin{aligned} P_j &= G_{ij}\left(V_j'V_{pq}' + V_j''V_{pq}''\right) + B_{ij}\left(V_j''V_{pq}' - V_j'V_{pq}''\right) \\ Q_j &= G_{ij}\left(V_j''V_{pq}' - V_j'V_{pq}''\right) - B_{ij}\left(V_j'V_{pq}' + V_j''V_{pq}''\right) \end{aligned} \tag{10.14}$$

The active power transferred from the shunt side to the series one (P_{dc}):

$$P_{dc} = \text{Re}\left[\underline{V}_{pq}\left(\frac{\underline{V}_i + \underline{V}_{pq} - \underline{V}_j}{\underline{Z}_{ij}}\right)^*\right]$$

respectively,

$$\begin{aligned} P_{dc} &= G_{ij}\left(V_{pq}^2 + U_i'V_{pq}' - U_j'V_{pq}' + U_i''V_{pq}'' - U_j''V_{pq}''\right) + \\ &+ B_{ij}\left(U_i'V_{pq}'' - U_j'V_{pq}'' - U_i''V_{pq}' + U_j''V_{pq}'\right) \end{aligned} \tag{10.15}$$

If the power losses of the UPFC device are neglected, then:

$$P_{i0} = P_{dc}$$

Based on the power balance at bus i, the injected apparent complex power is given by (Figure 10.34b):

$$\underline{S}_i(inj) = \underline{S}_i - \underline{S}_{i0}$$

The active power injected into bus i is:

$$\begin{aligned}
P_i(inj) = P_i - P_{dc} = {} & G_{ij}\left(-V_i'V_{pq}' - V_i''V_{pq}''\right) + B_{ij}\left(V_i'V_{pq}'' - V_i''V_{pq}'\right) - \\
& - G_{ij}\left(V_{pq}^2 + V_i'V_{pq}' - V_j'V_{pq}' + V_i''V_{pq}'' - V_j''V_{pq}''\right) - \\
& - B_{ij}\left(V_i'V_{pq}'' - V_j'V_{pq}'' - V_i''V_{pq}' + V_j''V_{pq}'\right)
\end{aligned}$$

Reducing the terms, and performing similar operations for the injected reactive power, gives:

$$\begin{aligned}
P_i(inj) = {} & G_{ij}\cdot\left(-V_{pq}^2 - 2V_i'V_{pq}' - 2V_i''V_{pq}'' + V_j'V_{pq}' + V_j''V_{pq}''\right) \\
& + B_{ij}\cdot\left(V_j'V_{pq}'' - V_j''V_{pq}'\right) \\
Q_i(inj) = {} & G_{ij}\left(V_i'V_{pq}'' - V_i''V_{pq}'\right) + B_{ij}\left(V_i'V_{pq}' + V_i''V_{pq}''\right) \\
& -G_{i0}(V_i''V_{sh}' + V_i'V_{sh}'') + B_{i0}(V_i^2 - V_i'V_{sh}' + V_i''V_{sh}'')
\end{aligned} \tag{10.16}$$

As it results from Figure 10.34b, the powers balance at bus j is $\underline{S}_j(inj) = \underline{S}_j$, thereby the injected apparent power remains unchanged:

$$\begin{aligned}
P_j(inj) &= G_{ij}\left(V_j'V_{pq}' + V_j''V_{pq}''\right) + B_{ij}\left(V_j''V_{pq}' - V_j'V_{pq}''\right) \\
Q_j(inj) &= G_{ij}\left(V_j''V_{pq}' - V_j'V_{pq}''\right) - B_{ij}\left(V_j'V_{pq}' + V_j''V_{pq}''\right)
\end{aligned} \tag{10.17}$$

Therefore, the two power injections ($P_i(inj)$, $Q_i(inj)$) and ($P_j(inj)$, $Q_j(inj)$) define the static model of UPFC.

Normally, the control parameters of UPFC are given in polar form because it provides more intuitive information, the state variables being written as magnitude and angle. They can be easily converted from one representation form to the other using the following expressions:

$$\begin{aligned}
V_{pq} &= \sqrt{V_{pq}'^2 + V_{pq}''^2} \quad & V_{sh} &= \sqrt{V_{sh}'^2 + V_{sh}''^2} \\
\alpha_{pq} &= \operatorname{atan}\left(\frac{V_{pq}''}{V_{pq}'}\right) \quad & \theta_{sh} &= \operatorname{atan}\left(\frac{V_{sh}''}{V_{sh}'}\right)
\end{aligned} \tag{10.18}$$

10.8.2 Power Flow and Active Power Balance Restrictions

Let us consider again the equivalent circuit from Figure 10.32b, where $\underline{V}_{sh} = V_{sh}\underline{|\theta_{sh}}$, $\underline{V}_{pq} = V_{pq}\underline{|\alpha_{pq}}$, $\underline{V}_i = V_i\underline{|\theta_i}$, $\underline{V}_j = V_j\underline{|\theta_j}$. The active and reactive powers exchanged by the UPFC device through the shunt branch are [17]:

$$\begin{aligned} P_{sh} &= V_i^2 G_{sh} - V_i V_{sh}[G_{sh}\cos(\theta_i - \theta_{sh}) + B_{sh}\sin(\theta_i - \theta_{sh})] \\ Q_{sh} &= -V_i^2 B_{sh} - V_i V_{sh}[G_{sh}\sin(\theta_i - \theta_{sh}) - B_{sh}\cos(\theta_i - \theta_{sh})] \end{aligned} \tag{10.19}$$

whereas the active and reactive power flows controlled through the series branch are:

$$\begin{aligned} P_{ij}^{se} &= V_i^2 G_{ij} - V_i V_j (G_{ij}\cos\theta_{ij} + B_{ij}\sin\theta_{ij}) - \\ &- V_i V_{pq}[G_{ij}\cos(\theta_i - \alpha_{pq}) + B_{ij}\sin(\theta_i - \alpha_{pq})] \end{aligned} \tag{10.20a}$$

$$\begin{aligned} Q_{ij}^{se} &= -V_i^2 B_{ij} - V_i V_j (G_{ij}\sin\theta_{ij} - B_{ij}\cos\theta_{ij}) - \\ &- V_i V_{pq}\left[G_{ij}\sin(\theta_i - \alpha_{pq}) - B_{ij}\cos(\theta_i - \alpha_{pq})\right] \end{aligned} \tag{10.20b}$$

$$\begin{aligned} P_{ji}^{se} &= V_j^2 G_{ij} - V_i V_j (G_{ij}\cos\theta_{ji} + B_{ij}\sin\theta_{ji}) + \\ &+ V_j V_{pq}\left[G_{ij}\cos(\theta_j - \alpha_{pq}) + B_{ij}\sin(\theta_j - \alpha_{pq})\right] \end{aligned} \tag{10.21a}$$

$$\begin{aligned} Q_{ji}^{se} &= -V_j^2 B_{ij} - V_i V_j (G_{ij}\sin\theta_{ji} - B_{ij}\cos\theta_{ji}) + \\ &+ V_j V_{pq}\left[G_{ij}\sin(\theta_j - \alpha_{pq}) - B_{ij}\cos(\theta_j - \alpha_{pq})\right] \end{aligned} \tag{10.21b}$$

where $G_{sh} + jB_{sh} = 1/\underline{Z}_{sh}$; $G_{ij} + jB_{ij} = 1/\underline{Z}_{ij}$

$$\theta_{ij} = \theta_i - \theta_j; \quad \theta_{ji} = \theta_j - \theta_i$$

THE ACTIVE POWER BALANCE CONSTRAINTS. One operational constraint is written from the power balance through the UPFC's dc link. Since there is no active power storage on the dc link, the power out from one converter equals the power entering the other converter. Thereby:

$$\Delta P_{\Sigma} = PE_{sh} - PE_{pq} = 0 \tag{10.22}$$

where $PE_{sh} = \text{Re}(\underline{V}_{sh}\underline{I}_{sh}^*)$ is the active power exchanged by the shunt converter with the dc link, and $PE_{pq} = \text{Re}(\underline{V}_{pq}\underline{I}_{ji}^*)$ is the active power exchanged by the series converter with the dc link, respectively.

In steady-state conditions, UPFC is capable of controlling the voltage at bus i as well as the active and reactive powers, P_{ij} and Q_{ij} (or P_{ji} şi Q_{ji}), flowing through the transmission line.

In addition to these control functions that are frequently used, the UPFC device has the possibility to independently control the phase angle, the voltage amplitude, the line series impedance, or a combination of these. The equations of these control modes can be written in general form as:

$$\begin{aligned} \Delta G([x], G^{spec}) = 0 \\ \Delta R([x], R^{spec}) = 0 \end{aligned} \tag{10.23}$$

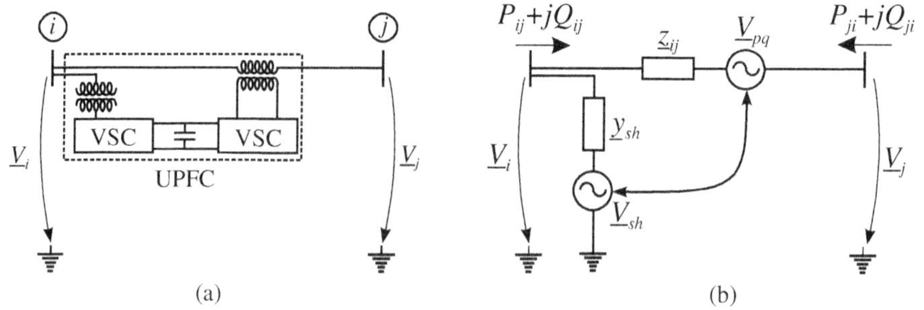

Figure 10.35 The representation of the UPFC device installed on a power line.

where $[x] = [\theta_i, V_i, \theta_j, V_j, \alpha_{pq}, V_{pq}, \theta_{sh}, V_{sh}]^T$, and G^{spec} and R^{spec} are the control references.

10.8.3 Implementing the UPFC Model in the Newton–Raphson Method

The UPFC device consists of two parts, one shunt connected at bus i, and one connected in series with the transmission line between busses i and j. The two components can exchange active power through the dc link (Figure 10.35a).

If each of the two parts operates independently, the shunt one behaves similar to a STATCOM device, while the series one behaves like an SSSC device. Therefore, the equivalent diagram of a UPFC can be built by combining the equivalent diagrams of the STATCOM and SSSC devices and adding the dc link between them (Figure 10.35b).

As previously stated, a UPFC device can have several control functions/modes at the same time. However, the most common control variables are the voltage magnitude, V_i, at the bus where the shunt part is connected, and the active and reactive power flows, P_{ij} and Q_{ij}, on the transmission line where the series part is connected. The control of the three variables implies the control of the magnitude and angle of the two voltage sources, $\underline{V}_{sh}$ and $\underline{V}_{pq}$.

The active and reactive powers transferred from bus i to bus j are given by:

$$\begin{aligned} P_{ij}^{UPFC} &= P_{ij}^{se} + P_{sh} = \\ &= V_i^2(G_{ij} + G_{sh}) - V_iV_j\left[G_{ij}\cos(\theta_i - \theta_j) + B_{ij}\sin(\theta_i - \theta_j)\right] - \\ &- V_iV_{pq}\left[G_{ij}\cos(\theta_i - \alpha_{pq}) + B_{ij}\sin(\theta_i - \alpha_{pq})\right] - \\ &- V_iV_{sh}\left[G_{sh}\cos(\theta_i - \theta_{sh}) + B_{sh}\sin(\theta_i - \theta_{sh})\right] \end{aligned} \tag{10.24a}$$

$$\begin{aligned} Q_{ij}^{UPFC} &= Q_{ij}^{se} + Q_{sh} = \\ &= -V_i^2(B_{ij} + B_{sh}) - V_iV_j\left[G_{ij}\sin(\theta_i - \theta_j) - B_{ij}\cos(\theta_i - \theta_j)\right] - \\ &- V_iV_{pq}\left[G_{ij}\sin(\theta_i - \alpha_{pq}) - B_{ij}\cos(\theta_i - \alpha_{pq})\right] - \\ &- V_iV_{sh}\left[G_{sh}\sin(\theta_i - \theta_{sh}) - B_{sh}\cos(\theta_i - \theta_{sh})\right] \end{aligned} \tag{10.24b}$$

and active and reactive powers transferred from bus j to bus i are given by

$$\begin{aligned} P_{ji}^{UPFC} &= P_{ji}^{se} = \\ &= V_j^2 G_{ij} - V_i V_j \left[G_{ij}\cos(\theta_j - \theta_i) + B_{ij}\sin(\theta_j - \theta_i)\right] + \\ &\quad + V_j V_{pq}\left[G_{ij}\cos(\theta_j - \alpha_{pq}) + B_{ij}\sin(\theta_j - \alpha_{pq})\right] \end{aligned} \tag{10.24c}$$

$$\begin{aligned} Q_{ji}^{UPFC} &= Q_{ji}^{se} = \\ &= -V_j^2 G_{ij} - V_i V_j \left[G_{ij}\sin(\theta_j - \theta_i) - B_{ij}\cos(\theta_j - \theta_i)\right] + \\ &\quad + V_j V_{pq}\left[G_{ij}\sin(\theta_j - \alpha_{pq}) - B_{ij}\cos(\theta_j - \alpha_{pq})\right] \end{aligned} \tag{10.24d}$$

These powers, which are "forced" by the UPFC device to flow between the i and j busses, are added to the nodal powers of the busses i and j such that:

$$\begin{aligned} P_i^{new} &= P_i + P_{ij}^{UPFC} \quad ; \quad P_j^{new} = P_j + P_{ji}^{UPFC} \\ P_j^{new} &= P_j + P_{ji}^{UPFC} \quad ; \quad Q_j^{new} = Q_j + Q_{ji}^{UPFC} \end{aligned} \tag{10.25}$$

The active power exchanged by the two parts of the UPFC device through the dc link is given by the following expression:

$$\begin{aligned} P_\Sigma &= PE_{sh} - PE_{pq} = \\ &= V_{sh}^2 G_{sh} - V_i V_{sh}\left[G_{sh}\cos(\theta_i - \theta_{sh}) - B_{sh}\sin(\theta_i - \theta_{sh})\right] + \\ &\quad + V_i V_{pq}\left[G_{ij}\cos(\theta_i - \alpha_{pq}) - B_{ij}\sin(\theta_i - \alpha_{pq})\right] - \\ &\quad - V_j V_{pq}\left[G_{ij}\cos(\theta_j - \alpha_{pq}) - B_{ij}\sin(\theta_j - \alpha_{pq})\right] \end{aligned} \tag{10.26}$$

As provided in equation (10.22), we have $P_\Sigma = 0$. This identity is modeled in the Newton–Raphson method as:

$$\Delta P_\Sigma = 0 - P_\Sigma$$

which is added to the vector of power mismatches.

As mentioned above, the control variables are the voltage magnitude at bus i, V_i, and the active and reactive power flows on the transmission line, P_{ij} and Q_{ij}. Thereby, three more terms are added in the vector of power mismatches, that is:

$$\begin{aligned} F &= V_i - V_i^{spec} \\ G &= P_{ij}^{UPFC} - P_{ij}^{spec} \\ R &= Q_{ij}^{UPFC} - Q_{ij}^{spec} \end{aligned} \tag{10.27}$$

where *spec* stands for the specified/scheduled value.

By introducing these terms, the amplitude and phase angle of the series and shunt equivalent voltage sources of the UPFC device are determined, that is, V_{pq}, α_{pq}, V_{sh}, and θ_{sh}.

The extended form of the matrix equation that has to be solved within the Newton–Raphson method is given in equation (10.28).

$$
\begin{bmatrix} \cdots \\ \Delta P_i \\ \cdots \\ \Delta P_j \\ \cdots \\ \cdots \\ \Delta Q_i \\ \cdots \\ \Delta Q_j \\ \cdots \\ \Delta G \\ \Delta R \\ \Delta P_{\Sigma} \\ \Delta F \end{bmatrix}
=
\begin{bmatrix}
\cdots & \cdots & \cdots & \cdots & \cdots & \cdots & \cdots & \cdots & \cdots & \cdots & \cdots & \cdots & \cdots \\
\cdots & H_{ii} + \dfrac{\partial P_{ij}^{UPFC}}{\partial \theta_i} & \cdots & H_{ij} + \dfrac{\partial P_{ij}^{UPFC}}{\partial \theta_j} & \cdots & M_{ii} + \dfrac{\partial P_{ij}^{UPFC}}{\partial V_i} V_i & \cdots & M_{ij} + \dfrac{\partial P_{ij}^{UPFC}}{\partial V_j} V_j & \cdots & \dfrac{\partial P_{ij}^{UPFC}}{\partial \alpha_{pq}} & \dfrac{\partial P_{ij}^{UPFC}}{\partial V_{pq}} V_{pq} & \dfrac{\partial P_{ij}^{UPFC}}{\partial \theta_{sh}} & \dfrac{\partial P_{ij}^{UPFC}}{\partial V_{sh}} V_{sh} \\
\cdots & \cdots & \cdots & \cdots & \cdots & \cdots & \cdots & \cdots & \cdots & \cdots & \cdots & \cdots & \cdots \\
\cdots & H_{ji} + \dfrac{\partial P_{ji}^{UPFC}}{\partial \theta_i} & \cdots & H_{jj} + \dfrac{\partial P_{ji}^{UPFC}}{\partial \theta_j} & \cdots & M_{ji} & \cdots & M_{jj} + \dfrac{\partial P_{ji}^{UPFC}}{\partial V_j} V_j & \cdots & \dfrac{\partial P_{ji}^{UPFC}}{\partial \alpha_{pq}} & \dfrac{\partial P_{ji}^{UPFC}}{\partial V_{pq}} V_{pq} & \dfrac{\partial P_{ji}^{UPFC}}{\partial \theta_{sh}} & \dfrac{\partial P_{ji}^{UPFC}}{\partial V_{sh}} V_{sh} \\
\cdots & \cdots & \cdots & \cdots & \cdots & \cdots & \cdots & \cdots & \cdots & \cdots & \cdots & \cdots & \cdots \\
\cdots & \cdots & \cdots & \cdots & \cdots & \cdots & \cdots & \cdots & \cdots & \cdots & \cdots & \cdots & \cdots \\
\cdots & K_{ii} + \dfrac{\partial Q_{ij}^{UPFC}}{\partial \theta_i} & \cdots & K_{ij} + \dfrac{\partial Q_{ij}^{UPFC}}{\partial \theta_j} & \cdots & L_{ii} + \dfrac{\partial Q_{ij}^{UPFC}}{\partial V_i} V_i & \cdots & L_{ij} + \dfrac{\partial Q_{ij}^{UPFC}}{\partial V_j} V_j & \cdots & \dfrac{\partial Q_{ij}^{UPFC}}{\partial \alpha_{pq}} & \dfrac{\partial Q_{ij}^{UPFC}}{\partial V_{pq}} V_{pq} & \dfrac{\partial Q_{ij}^{UPFC}}{\partial \theta_{sh}} & \dfrac{\partial Q_{ij}^{UPFC}}{\partial V_{sh}} V_{sh} \\
\cdots & \cdots & \cdots & \cdots & \cdots & \cdots & \cdots & \cdots & \cdots & \cdots & \cdots & \cdots & \cdots \\
\cdots & K_{ji} + \dfrac{\partial Q_{ji}^{UPFC}}{\partial \theta_i} & \cdots & K_{jj} + \dfrac{\partial Q_{ji}^{UPFC}}{\partial \theta_j} & \cdots & L_{ji} & \cdots & L_{jj} + \dfrac{\partial Q_{ji}^{UPFC}}{\partial V_j} V_j & \cdots & \dfrac{\partial Q_{ji}^{UPFC}}{\partial \alpha_{pq}} & \dfrac{\partial Q_{ji}^{UPFC}}{\partial V_{pq}} V_{pq} & \dfrac{\partial Q_{ji}^{UPFC}}{\partial \theta_{sh}} & \dfrac{\partial Q_{ji}^{UPFC}}{\partial V_{sh}} V_{sh} \\
\cdots & \cdots & \cdots & \cdots & \cdots & \cdots & \cdots & \cdots & \cdots & \cdots & \cdots & \cdots & \cdots \\
\cdots & \dfrac{\partial G}{\partial \theta_i} & \cdots & \dfrac{\partial G}{\partial \theta_j} & \cdots & \dfrac{\partial G}{\partial V_i} V_i & \cdots & \dfrac{\partial G}{\partial V_j} V_j & \cdots & \dfrac{\partial G}{\partial \alpha_{pq}} & \dfrac{\partial G}{\partial V_{pq}} V_{pq} & \dfrac{\partial G}{\partial \theta_{sh}} & \dfrac{\partial G}{\partial V_{sh}} V_{sh} \\
\cdots & \dfrac{\partial R}{\partial \theta_i} & \cdots & \dfrac{\partial R}{\partial \theta_j} & \cdots & \dfrac{\partial R}{\partial V_i} V_i & \cdots & \dfrac{\partial R}{\partial V_j} V_j & \cdots & \dfrac{\partial R}{\partial \alpha_{pq}} & \dfrac{\partial R}{\partial V_{pq}} V_{pq} & \dfrac{\partial R}{\partial \theta_{sh}} & \dfrac{\partial R}{\partial V_{sh}} V_{sh} \\
\cdots & \dfrac{\partial P_{\Sigma}}{\partial \theta_i} & \cdots & \dfrac{\partial P_{\Sigma}}{\partial \theta_j} & \cdots & \dfrac{\partial P_{\Sigma}}{\partial V_i} V_i & \cdots & \dfrac{\partial P_{\Sigma}}{\partial V_j} V_j & \cdots & \dfrac{\partial P_{\Sigma}}{\partial \alpha_{pq}} & \dfrac{\partial P_{\Sigma}}{\partial V_{pq}} V_{pq} & \dfrac{\partial P_{\Sigma}}{\partial \theta_{sh}} & \dfrac{\partial P_{\Sigma}}{\partial V_{sh}} V_{sh} \\
\cdots & \dfrac{\partial F}{\partial \theta_i} & \cdots & \dfrac{\partial F}{\partial \theta_j} & \cdots & \dfrac{\partial F}{\partial V_i} V_i & \cdots & \dfrac{\partial F}{\partial V_j} V_j & \cdots & \dfrac{\partial F}{\partial \alpha_{pq}} & \dfrac{\partial F}{\partial V_{pq}} V_{pq} & \dfrac{\partial F}{\partial \theta_{sh}} & \dfrac{\partial F}{\partial V_{sh}} V_{sh}
\end{bmatrix}
\cdot
\begin{bmatrix} \cdots \\ \Delta \theta_{sh} \\ \cdots \\ \Delta \theta_{sh} \\ \cdots \\ \cdots \\ \Delta V_i / V_i \\ \cdots \\ \Delta V_j / V_j \\ \cdots \\ \Delta \alpha_{pq} \\ \Delta V_{pq} / V_{pq} \\ \Delta \theta_{sh} \\ \Delta V_{sh} / V_{sh} \end{bmatrix}
\tag{10.28}
$$

The Newton–Raphson algorithm for load flow calculation in a network that includes a UPFC device is as follows:

1. Read the bus and network data.

2. Create the bus admittance matrix $[\underline{Y}_{nn}]$.

3. Set the iteration step $p = 0$.

4. Initialize the bus voltage magnitudes and phase angles:

– the magnitudes: $V_i^{(0)}$, with $i = \overline{1,n}$, $i \neq$ slack bus;

– the phase angles: $\theta_i^{(0)} = 0$;

– the magnitudes of the equivalent voltages: $V_{pq}^{(0)} = V_{pq}^{spec}$, $V_{sh}^{(0)} = V_{sh}^{spec}$;

– the phase angles of the equivalent voltages: $\alpha_{pq}^{(0)} = 0$, $\theta_{sh}^{(0)} = 0$.

5. Calculate the bus active and reactive powers

5.1. The bus powers $P_i^{(p)}$ and $Q_i^{(p)}$ ($i = \overline{1,n}$, $i \neq$ slack bus), without considering the *i-j* line.

5.2. Calculate the powers forced by the UPFC device to flow on the *i-j* line: P_{ij}^{UPFC}, P_{ji}^{UPFC}, Q_{ij}^{UPFC}, and Q_{ji}^{UPFC}. Adjust the bus powers:

$$P_i^{new} = P_i + P_{ij}^{UPFC} \quad ; \quad P_j^{new} = P_j + P_{ji}^{UPFC}$$
$$Q_i^{new} = Q_i + Q_{ij}^{UPFC} \quad ; \quad Q_j^{new} = Q_j + Q_{ji}^{UPFC}$$

6. Calculate the mismatches at iteration p:

$$\Delta P_i^{(p)} = P_i^{spec} - P_i^{(p)}, \text{ for PQ and PV busses}$$
$$\Delta Q_i^{(p)} = Q_i^{spec} - Q_i^{(p)}, \text{ for PQ busses}$$
$$\Delta G = 0 - G;\ \Delta R = 0 - R;\ \Delta PE = 0 - PE\ ;\ \Delta F = 0 - F$$

7. Treat the generator busses.

8. Build the column vector of power mismatches at the nodes and the those corresponding to the UPFC functions:

$$\Delta PQ = [\Delta P;\ \Delta Q;\ \Delta G;\ \Delta R;\ \Delta P_{\Sigma};\ \Delta F]^T$$

9. Calculate the Jacobian matrix as:

$$JAC = \begin{bmatrix} H^{new} & M^{new} & J_{13} & J_{14} & J_{15} & J_{16} \\ K^{new} & L^{new} & J_{23} & J_{24} & J_{25} & J_{26} \\ J_{31} & J_{32} & J_{33} & J_{34} & J_{35} & J_{36} \\ J_{41} & J_{42} & J_{43} & J_{44} & J_{45} & J_{46} \\ J_{51} & J_{52} & J_{53} & J_{54} & J_{55} & J_{56} \\ J_{61} & J_{62} & J_{63} & J_{64} & J_{65} & J_{66} \end{bmatrix}$$

Note 1: The row and column corresponding to the slack bus are eliminated because the voltage (in both magnitude and angle) is specified.
Note 2: The component matrices of the Jacobian matrix will have the following size:

H $((n-1)\times(n-1))$	M $((n-1)\times n_l)$	J_{13} $((n-1)\times n_{UPFC})$	J_{14} $((n-1)\times n_{UPFC})$	J_{15} $((n-1)\times n_{UPFC})$	J_{16} $((n-1)\times n_{UPFC})$
K $(n_l\times(n-1))$	L $(n_l\times n_l)$	J_{23} $(n_l\times n_{UPFC})$	J_{24} $(n_l\times n_{UPFC})$	J_{25} $(n_l\times n_{UPFC})$	J_{26} $(n_l\times n_{UPFC})$
J_{31} $(n_{UPFC}\times(n-1))$	J_{32} $(n_{UPFC}\times n_l)$	J_{33} $(n_{UPFC}\times n_{UPFC})$	J_{34} $(n_{UPFC}\times n_{UPFC})$	J_{35} $(n_{UPFC}\times n_{UPFC})$	J_{36} $(n_{UPFC}\times n_{UPFC})$
J_{41} $(n_{UPFC}\times(n-1))$	J_{42} $(n_{UPFC}\times n_l)$	J_{43} $(n_{UPFC}\times n_{UPFC})$	J_{44} $(n_{UPFC}\times n_{UPFC})$	J_{45} $(n_{UPFC}\times n_{UPFC})$	J_{46} $(n_{UPFC}\times n_{UPFC})$
J_{51} $(n_{UPFC}\times(n-1))$	J_{52} $(n_{UPFC}\times n_l)$	J_{53} $(n_{UPFC}\times n_{UPFC})$	J_{54} $(n_{UPFC}\times n_{UPFC})$	J_{55} $(n_{UPFC}\times n_{UPFC})$	J_{56} $(n_{UPFC}\times n_{UPFC})$
J_{61} $(n_{UPFC}\times(n-1))$	J_{62} $(n_{UPFC}\times n_l)$	J_{63} $(n_{UPFC}\times n_{UPFC})$	J_{64} $(n_{UPFC}\times n_{UPFC})$	J_{65} $(n_{UPFC}\times n_{UPFC})$	J_{66} $(n_{UPFC}\times n_{UPFC})$

where n is the number of network busses, n_l is the number of PQ nodes, and n_{UPFC} is the number of UPFC devices connected in the system. Thereby, the size of the Jacobian matrix is:

$$((n-1)+n_l+(4\times n_{UPFC}))\times((n-1)+n_l+(4\times n_{UPFC})).$$

10. Calculate the corrections of the unknown variables as:

$$[\Delta X]=[JAC^{-1}]\cdot[\Delta PQ]$$

where $[\Delta PQ]$ is the vector of mismatches.

11. After determining the corrections $\Delta V_i^{(p)}$, for PQ busses, and $\Delta\theta_i^{(p)}$, for PQ and PV busses, at the current iteration p, the voltage magnitudes and phase angles are updated for the next iteration p+1:

$$V_i^{(p+1)}=V_i^{(p)}+\Delta V_i^{(p)}\quad,\quad i\in C$$

$$\theta_i^{(p+1)}=\theta_i^{(p)}+\Delta\theta_i^{(p)}\quad,\quad i\in C\cup G$$

Furthermore, for the active UPFC devices, the equivalent voltages, in magnitude and phase angle, will also be updated:

$$\begin{matrix}\alpha_{pq}^{(p+1)}=\alpha_{pq}^{(p)}+\Delta\alpha_{pq}^{(p)} & & \theta_{sh}^{(p+1)}=\theta_{sh}^{(p)}+\Delta\theta_{sh}^{(p)}\\ V_{pq}^{(p+1)}=V_{pq}^{(p)}+\Delta V_{pq}^{(p)} & ; & V_{sh}^{(p+1)}=V_{sh}^{(p)}+\Delta V_{sh}^{(p)}\end{matrix}$$

12. Perform the convergence test. This test is usually performed after step 6. It consists of verifying if the following conditions are met:

$$\begin{cases}\left|\Delta P_i^{(p)}\right|\le\varepsilon_{Pi}\quad,\quad i\in\text{PQ and PV buses}\\ \left|\Delta Q_i^{(p)}\right|\le\varepsilon_{Qi}\quad,\quad i\in\text{PQ buses}\end{cases}$$

where ε_{Pi} and ε_{Qi} are the accepted mismatches, with usual values in the range of 0.001 … 0.5 MW/MVAr.

Moreover, for all the active UPFC devices, the following mismatches are also checked:

$$\left|\Delta G^{(p)}\right|\le\varepsilon_G,\left|\Delta R^{(p)}\right|\le\varepsilon_R,\left|\Delta PE^{(p)}\right|\le\varepsilon_{PE},\left|\Delta F^{(p)}\right|\le\varepsilon_F$$

13. Calculate the power flows and the power losses for all the network branches.

MODELING THE UPFC VOLTAGE AND CURRENT CONSTRAINTS

The voltage and current constraints for the shunt side of UPFC are given by the following inequalities:

$$V_{sh}^{\min} \leq V_{sh} \leq V_{sh}^{\max} \text{ and } I_{sh} \leq I_{sh}^{\max}$$

whereas the voltage and current constraints are given by:

$$0 \leq V_{se} \leq V_{se}^{\max} \text{ and } I_{se} \leq I_{se}^{\max}.$$

Initialization of the UPFC variables in the Newton–Raphson-based program for load-flow analysis.

The active and reactive power flows on the transmission line can be controlled by simply specifying the desired values, so that the power mismatches are defined by $P_{ji} - P_{ji}^{spec} = 0$ and $Q_{ji} - Q_{ji}^{spec} = 0$.

Considering that the shunt-side converter is used for voltage amplitude control at the bus i, the equivalent voltage V_{sh} can be determined by:

$$V_{sh} = \left(V_{sh}^{\max} + V_{sh}^{\min}\right)/2 \text{ or } V_{sh} = V^{spec} \tag{10.29}$$

In these conditions, the phase angle, θ_{sh}, can be obtained by solving equation (10.22):

$$\theta_{sh} = -\text{asin}\left[B/\left(V_i V_{sh}\sqrt{(G_{sh}^2 + B_{sh}^2)}\right)\right] + \text{atan}(-G_{sh}/B_{sh}) \tag{10.30}$$

where

$$\begin{aligned} B = {} & V_{sh}^2 G_{sh} + V_{pq}^2 G_{ij} + V_i V_{pq}\left[G_{ij}\cos(\theta_j - \alpha_{pq}) - B_{ij}\sin(\theta_j - \alpha_{pq})\right] - \\ & - V_i V_{pq}\left[G_{ij}\cos(\theta_i - \alpha_{pq}) - B_{ij}\sin(\theta_i - \alpha_{pq})\right] \end{aligned} \tag{10.31}$$

10.8.4 The Dynamic Model of UPFC

The UPFC device combines both series and shunt control equipment. The device can be seen as being built from a STATCOM and SSSC, connected between them through the dc link. This equipment can independently control both the active and reactive power flowing on the transmission line.

In order to build the dynamic model of UPFC, it is necessary that, in the first step, to build its steady-state model. For this, the two voltage sources model, presented in Figure 10.32b, can be used. In this model, the series part is similar to the SSSC model, whereas the shunt part is similar to the STATCOM model. The shunt part is represented by the $R_{sh}+jX_{sh}$ impedance, corresponding to the shunt transformer, connected to a variable voltage source $\underline{V}_{sh} = V_{sh}\angle\theta_{sh}$. The series side is represented by the R_T+jX_T impedance, corresponding to the series transformer of the SSSC, connected to a variable voltage source $\underline{V}_{pq} = V_{pq}\angle\alpha_{pq}$.

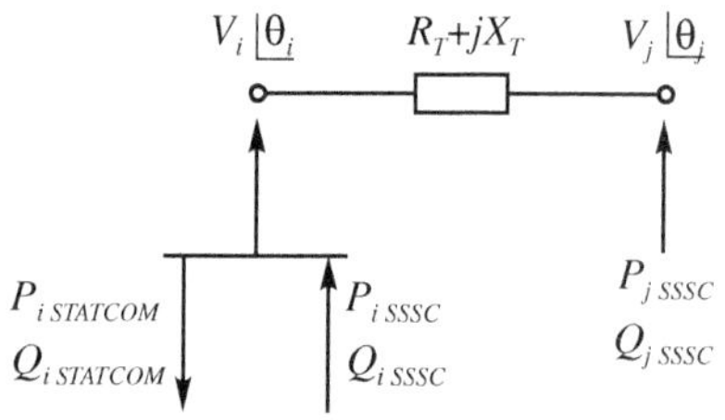

Figure 10.36 The two voltage sources diagram for UPFC modeling.

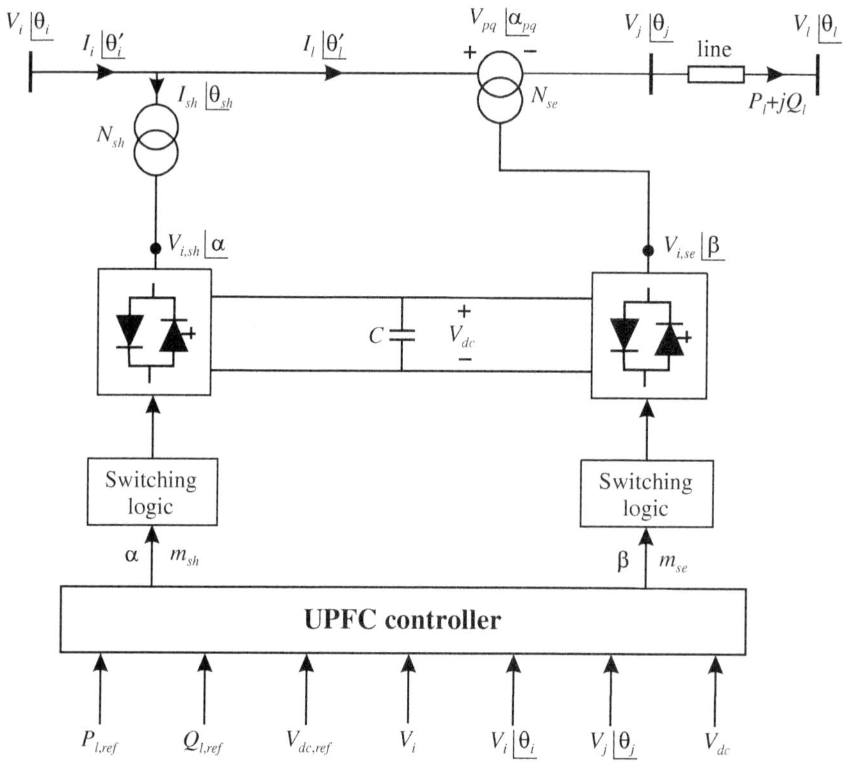

Figure 10.37 Block diagram of UPFC.

The power P_{dc} represents the power transfer between the shunt and series sides of UPFC.

According to Section 10.8.1, the two model of voltage sources (Figure 10.32b) is converted into a model of power injections at busses (Figure 10.36). The steady state of UPFC is composed from the two pairs of power injections ($P_i(inj)$,$Q_i(inj)$) and ($P_j(inj)$,$Q_j(inj)$) given by equations (10.16) and (10.17).

The UPFC model for transient stability studies[1]

In order to develop this model, the structure of a UPFC device presented in Figure 10.37 is considered together with the corresponding model for transient stability studies from Figure 10.39, proposed in [18].

[1] Parts of this subsection have been taken from [18].

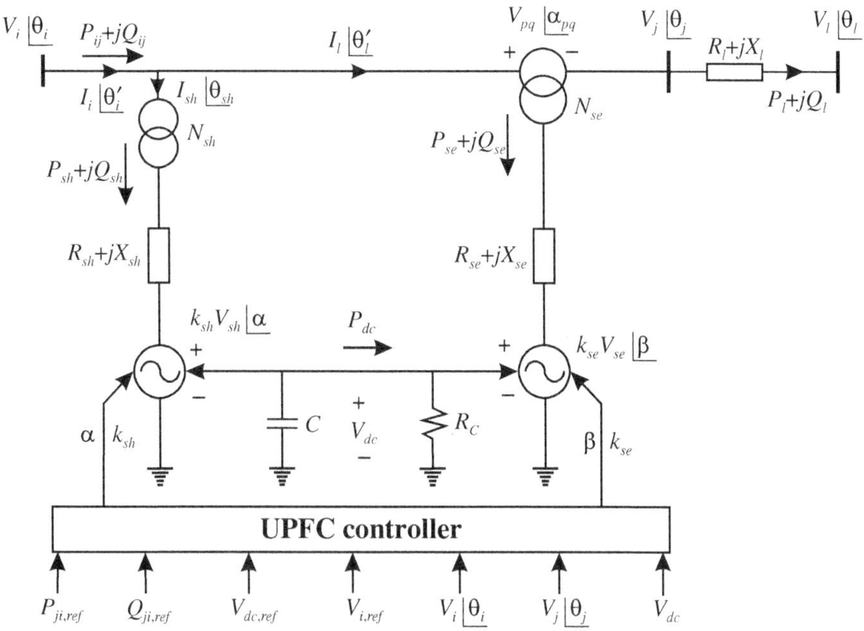

Figure 10.38 The UPFC model for stability studies.

The UPFC model for stability studies (Figure 10.38) is defined by a set of algebraic-differential equations:

- The set of equations for the shunt side and *dc* link is

$$\begin{bmatrix} \dot{x}_c \\ \dot{\alpha} \\ \dot{\beta} \\ \dot{m}_{sh} \\ \dot{m}_{se} \end{bmatrix} = f(x_c, \alpha, \beta, m_{sh}, m_{se}, V_i, V_l, V_{dc}, \theta_i, \theta_l, P_{l,ref}, Q_{l,ref}, V_{i,ref}, V_{dc,ref})$$

$$\dot{V}_{dc} = \frac{V_i}{C}\frac{I_{sh}}{V_{dc}}\cos(\theta_i - \theta_{sh}) + \frac{V_j}{C}\frac{I_l}{V_{dc}}\cos\left(\theta_j - \theta_j'\right) - \frac{1}{R_C C}V_{dc} - \frac{R_{sh}}{C}\frac{I_{sh}^2}{V_{dc}} - \frac{R_{se}}{C}\frac{I_l^2}{V_{dc}}$$

$$g_{sh}\left(\alpha, k_{sh}, V_i, V_{dc}, \theta_i, I_{sh}, \theta_{sh}, P_{sh}, Q_{sh}\right) = \\ = \begin{bmatrix} P_{sh} - V_i I_{sh}\cos(\theta_i - \theta_{sh}) \\ Q_{sh} - V_i I_{sh}\sin(\theta_i - \theta_{sh}) \\ P_{sh} - V_i^2 G_{sh} + k_{sh}V_{dc}V_i G_{sh}\cos(\theta_i - \alpha) + k_{sh}V_{dc}V_i B_{sh}\sin(\theta_i - \alpha) \\ Q_{sh} - V_i^2 B_{sh} + k_{sh}V_{dc}V_i B_{sh}\cos(\theta_i - \alpha) + k_{sh}V_{dc}V_i G_{sh}\sin(\theta_i - \alpha) \end{bmatrix} = 0$$

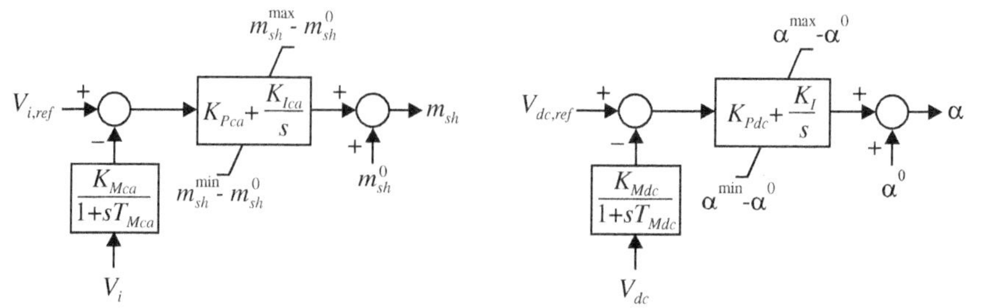

Figure 10.39 The shunt-side control blocks of UPFC.

- The additional set of algebraic equations

$$g_{con}\left(V_i,\theta_i,I_i,I_{sh},I_l,\theta'_i,\theta_{sh},\theta'_l,P_i,Q_i\right)=$$
$$=\begin{bmatrix} I_i\cos(\theta'_i)-I_{sh}\cos(\theta_{sh})-I_l\cos(\theta'_l) \\ I_i\sin(\theta'_i)-I_{sh}\sin(\theta_{sh})-I_l\sin(\theta'_l) \\ P_i-V_iI_i\cos(\theta'_i-\theta_i) \\ Q_i-V_iI_i\sin(\theta'_i-\theta_i) \end{bmatrix}=0$$

- The set of equations for the series side and *dc* link

$$g_{se}(\beta,k_{se},V_{dc},V_i,V_l,V_{pq},\theta_i,\theta_l,\alpha_{pq},I_l,\theta'_l,P_i,P_l,P_{sh},P_{se},Q_i,Q_l,Q_{sh},Q_{se})=$$
$$=\begin{bmatrix} P_i-P_{sh}-V_iI_l\cos(\theta_i-\theta'_l) \\ Q_i-Q_{sh}-V_iI_l\sin(\theta_i-\theta'_l) \\ P_l-V_jI_l\cos(\theta_j-\theta'_l) \\ Q_l-V_jI_l\sin(\theta_j-\theta'_l) \\ P_i-P_l-P_{sh}-P_{se} \\ Q_i-Q_l-Q_{sh}-Q_{se} \\ P_{se}-V_{pq}^2G_{se}+k_{se}V_{dc}V_{pq}G_{se}\cos(\alpha_{pq}-\beta)+k_{se}V_{dc}V_{pq}B_{se}\sin(\alpha_{pq}-\beta) \\ Q_{se}+V_{pq}^2B_{se}-k_{se}V_{dc}V_{pq}B_{se}\cos(\alpha_{pq}-\beta)+k_{se}V_{dc}V_{pq}G_{se}\sin(\alpha_{pq}-\beta) \end{bmatrix}=0$$

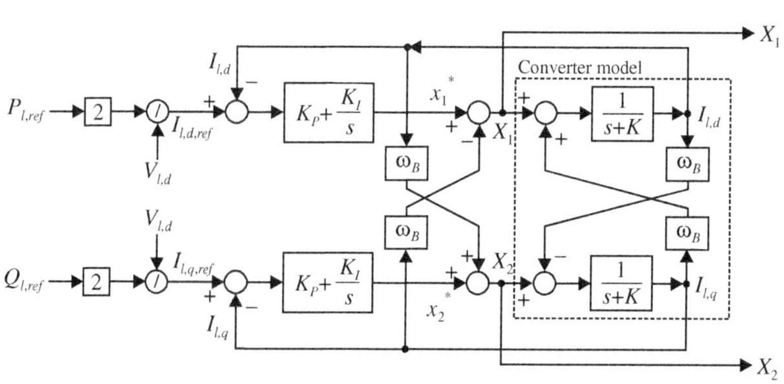

Figure 10.40 The *d–q* control of the series side of UPFC, referred to the voltage $V_j\angle\theta_j$ of the bus.

Note that these equations represent a combination of the STATCOM and SSSC equations. The main difference subsists in the control variables and equations, represented here by x_c and $f(\cdot)$, and in the additional set of restrictions $g_{con}(\cdot) = 0$, which stand for the interconnection between the shunt (STATCOM) and series (SSSC) sides of UPFC.

The diagrams corresponding to the control system for the shunt and series sides are presented in Figures 10.39 and 10.40.

It is reminded that the control part for the shunt side is the same as the one used for STATCOM. In order to decouple the control of active and reactive power of the inverter, the control block for the series side is a PQ controller, based on the two axes, *d* and *q*, decomposition of the control.

For the case of a *d–q* control of the series side of UPFC, the control variables m_{se}, β, and the parameters of the control diagram are given by the following expressions (all the variables are in p.u. except for the fundamental frequency ω_B of the system, which is expressed in rad/s) [18]:

$$K = \frac{R_T \omega_B}{X_T}$$

$$R_T = R_l + R_{se} \ ; \ X_T = X_l + X_{se}$$

$$V_{ld} = \sqrt{2} V_l$$

$$V_{i,d} = \sqrt{2} V_i \cos(\theta_l - \theta_i); V_{i,q} = \sqrt{2} V_i \sin(\theta_l - \theta_i)$$

$$V_{i,se,d} = V_{i,d} - V_{l,d} - \frac{X_T}{\omega_B} x_1; V_{i,se,q} = V_{i,q} - \frac{X_T}{\omega_B} x_2$$

$$V_{i,se} = \frac{1}{\sqrt{2}} \sqrt{V_{i,se,d}^2 + V_{i,se,q}^2}$$

$$m_{se} = \sqrt{\frac{8}{3}} \frac{V_{i,se}}{V_{dc}}$$

$$\beta = \theta_l - \operatorname{atan}\left(\frac{V_{i,se,q}}{V_{i,se,d}}\right)$$

REFERENCES

[1] Gyugyi, L. A unified power flow control concept for flexible AC transmission systems. *IEE Fifth International Conference on AC and DC Power Transmission*, London, Publication No. 345, pp. 19–26, 1991. Reprinted in *IEE Proceedings C - Generation, Transmission and Distribution*, vol. 139, no. 4, July 1992.

[2] Schauder, C. D., and Mehta, H. Vector analysis and control of advanced static VAr compensators. *IEE Proceedings C – Generation, Transmission and Distribution*, vol. 140, no. 4, pp. 299–306, July 1993.

[3] Gyugyi, L. Dynamic compensation of AC transmission lines by solid-state synchronous voltage sources. *IEEE Transactions on Power Delivery*, vol. 9, no. 2, pp. 904–911, April 1994.

[4] Lerch, E., Povh, D., Witzmann, R., Hlebcar, B., Mihalic, R. Simulation and performance analysis of unified power flow controller. CIGRE Paper No. 14-205, 1994.

[5] Gyugyi, L., Schauder, C.D., Williams, S.L., Rietman, T.R., Torgerson, D.R., Edris, A. A. The unified power flow controller: a new approach to power transmission control. *IEEE Transactions on Power Delivery*, vol. 10, no. 2, pp. 1085–1097, April 1995.
[6] Maliszewski, R. M., and Rahman, M. Planning and operating considerations for the world's first UPFC installation. CIGRE Report no. 37-96 (US) 13 (E), Paris, 1996.
[7] Mehraban, A. S. Application of the world's first UPFC on the AEP system. *EPRI Conference - The Future of Power Delivery*, Washington, D.C., April 9–11, 1996.
[8] Mihalic, R., Zunko, P, and Povh, D. Improvement on transient stability using unified power flow controller. *IEEE Transactions on Power Systems*, vol. 11, no. 1, pp. 485–492, January 1997.
[9] Rahman, M., Ahmed, M., Gutman, R., Nelson, R.J., and Bian, J. UPFC application on the AEP system: planning considerations. *IEEE Transactions on Power Systems*, vol. 12, no. 4, pp. 1695–1701, November 1997.
[10] Papie, I., Zunko, P., Povh, D., Weinhold, M. Basic control of unified power flow controller. *IEEE Transactions on Power Systems*, vol. 12, no. 4, pp. 1734–1739, November 1997.
[11] Schauder, C. D., Gyugyi, L., Lund, M.R., Hamai, D.M., Rietman, T.R., Torgerson, D.R., Edris, A. A. Operation of the unified power flow controller (UPFC) under practical constrains. *IEEE Transactions on Power Delivery*, vol. 13, no. 2, pp. 630–639, April 1998.
[12] Sen, K. K., and Stacey, E. J. UPFC – unified power flow controller: theory, modeling, and applications. *IEEE Transactions on Power Delivery*, vol. 13, no. 4, pp. 1453–1460, October 1998.
[13] Edris, A., Mehraban, A.S., Rahman, M., Gyugyi, L., Arabi, S., Reitman, T. Controlling the flow of real and reactive power. *IEEE Computer Applications in Power*, Vol. 11, no. 1, pp. 20–25, January 1998.
[14] Renz, B. A., Keri, A. J. F., Mehraban, A. S., Kessinger, J. P., Schauder, C. D., Gyugyi, L., Kovalsky, L. J., Edris, A. A. World's first unified power flow controller on the AEP system. CIGRE Paper no. 14-107, 1998.
[15] Keri, A., Mehraban, A. S., Schauder, C. AEP unified power flow controller performance. IEEE Transactions on Power Delivery, vol. 14, no. 4, pp. 1374-1381, October 1999.
[16] Hingorani, N. G., and Gyugyi, L. *Understanding FACTS – Concepts and Technology of Flexible AC Transmission Systems*. IEEE Press, and Wiley, New York, 2000.
[17] Zhang, X. P., Rehtanz, C., and Pal, B. *Flexible AC Transmission System: Modelling and Control*. Springer-Verlag, Berlin Heidelberg, 2006.
[18] Canizares, C. A. Power flow and transient stability models for FACTS controllers for voltage and angle stability studies. *Proceedings of IEEE/PES Winter Meeting*, Singapore, January 2000.

CHAPTER 11

INTERLINE POWER FLOW CONTROLLER (IPFC)

Laszlo Gyugyi

11.1 GENERALITIES

It has been shown in Chapter 10 that the unified power flow controller (UPFC) has the unique capability of independently controlling both the active and reactive power flow in the line. The UPFC concept provides a powerful tool for the cost-effective utilization of individual transmission lines by facilitating the independent control of both the active and reactive power flow, and thus the maximization of active power transfer at minimum losses, in the line. The interline power flow controller (IPFC) is essentially an extension of the UPFC concept to the economic control and optimization of *multiline* power transmission systems [1,3].

The IPFC addresses the problem of compensating a number of transmission lines at a given substation. Conventionally, series capacitive compensation (fixed or controlled) is employed to increase the transmittable active power over a given line and also to balance the loading of a normally encountered multiline transmission system. However, series reactive compensators are unable to control the reactive power flow in, and thus the proper load balancing of, the lines. This problem becomes particularly evident in those cases where the ratio of reactive to resistive line impedance (X/R) is relatively low. Series reactive compensation reduces only the effective reactive impedance X and, thus, significantly decreases the effective X/R ratio and thereby increases the reactive power flow and losses in the line. The IPFC scheme, in addition to independently controllable reactive series compensation of each line, provides means to transfer active power between the compensated lines. This capability makes it possible to: equalize both active and reactive power flow between the lines; reduce both the active *and* reactive loop-flow of power; compensate against resistive line voltage drops and the corresponding reactive power demand; and increase the effectiveness of the overall compensating system for dynamic disturbances. In other words, the IPFC can potentially provide a highly effective scheme for power transmission management at a multiline substation.

Advanced Solutions in Power Systems: HVDC, FACTS, and Artificial Intelligence, First Edition.
Edited by Mircea Eremia, Chen-Ching Liu, and Abdel-Aty Edris.

11.2 BASIC OPERATING PRINCIPLES AND CHARACTERISTICS OF THE IPFC

In its general form, the IPFC employs a number of DC to AC converters, each providing series compensation, similarly to that used in the UPFC, for one of the lines of a multiline system. However, within the general concept of the IPFC, the DC terminals of these converters are not supplied from a shunt-connected converter like in the UPFC scheme (although this possibility in some applications can be advantageously utilized), but are linked together at their DC terminals, as illustrated in Figure 11.1. With this arrangement, each converter, in addition to providing independently controllable series reactive compensation, can be controlled to supply (or draw) active power to (or from) the common DC link from (or to) its own transmission line. Thus, an overall surplus power can be made available from the underutilized lines, which then can be used by other lines to reduce reactive power flow by active power compensation. In this way, some of the converters, compensating lines with a heavy burden of reactive power flow, or high priority lines, can be equipped with full two-dimensional, reactive and active power control capability, similar to that offered by the UPFC. Evidently, this arrangement mandates the rigorous maintenance of the overall power balance at the common DC terminal to make the total active power exchange between the IPFC and all the lines equal to zero by proper coordination of the control system of the converters.

Consider an elementary IPFC scheme consisting of two back-to-back DC to AC converters, each compensating a transmission line by series voltage injection. This arrangement is shown functionally in Figure 11.2a, where the outputs of two back-to-back DC to AC converters are connected in series with transmission lines 1 and 2 to inject voltage phasors $\underline{V}_{1pq}$ and $\underline{V}_{2pq}$, respectively. The common DC link of the converters provides a path for active power flow between the two lines. Transmission line 1, represented by reactance X_1, has a sending-end bus with voltage phasor $\underline{V}_{1s}$

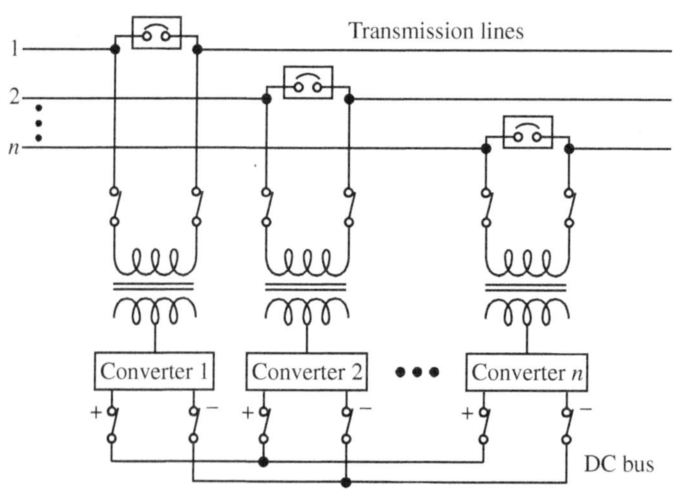

Figure 11.1 Interline power flow controller (IPFC) scheme.

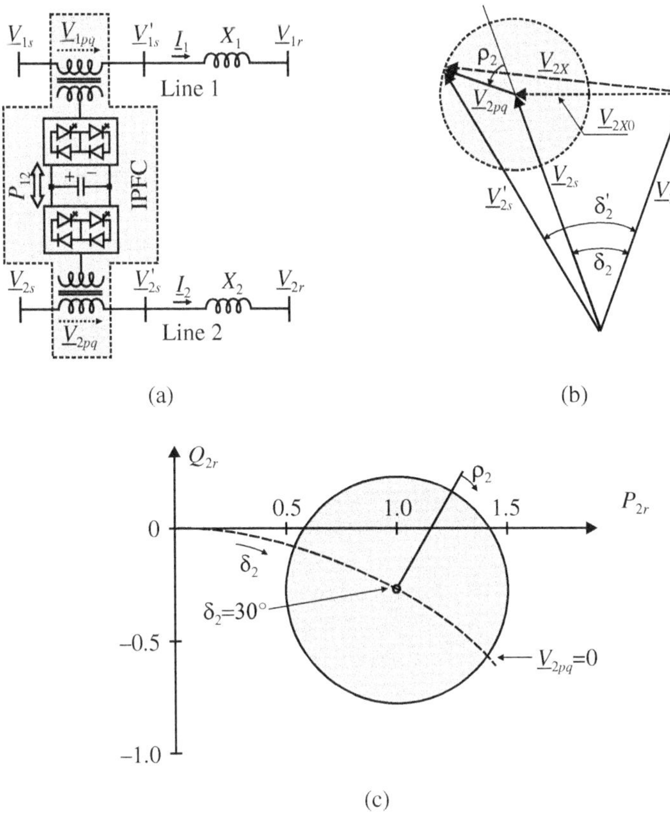

Figure 11.2 (a) Two-line IPFC scheme; (b) phasor diagram for "priority" line 2; and (c) corresponding operating region in the {P,Q} plane at $\delta = 30°$

and a receiving-end bus with voltage phasor $\underline{V}_{1r}$. The sending-end voltage phasor of line 2, represented by reactance X_2, is $\underline{V}_{2s}$ and the receiving-end voltage phasor is $\underline{V}_{2r}$. For clarity of the illustrations, all the sending-end and receiving-end voltages are assumed to be constant with fixed amplitudes, $V_{1s} = V_{1r} = V_{2s} = V_{2r} = 1.0$ p.u. The impedance of the two lines and their transmission angle, as well as the rating of the two compensating voltage sources produced by the converters, are also assumed to be identical, that is, $X_1 = X_2 = 0.5$ p.u., $\delta_1 = \delta_2$ (= 30°), and $V_{1pq\max} = V_{2pq\max} = 0.25$ p.u. Thus, the maximum active or reactive power increase is $VV_{pq}/X = 0.5$ p.u. for both lines. Although lines 1 and 2 could be (and in practice are likely to be) different with different transmission line voltage, impedance, and angle, in order to make the relationships governing the operation of the IPFC clear, the above stipulated operational identity of the two lines is maintained throughout this section.

Note that the operational identity of the two lines does not violate their assumed independence; they could even belong to different systems, having arbitrary phase relationship with each other since the active power transfer between them through the DC link of the converters is asynchronous.

In order to establish the transmission relationships between the two independent lines, line 2 is arbitrarily selected to be the one with *transmission priority* for which free controllability of both active and reactive line power flow is stipulated. The reason for this stipulation is to derive the constraints by which the free controllability of line 2 restricts the power flow control of line 1.

A phasor diagram representing transmission in line 2, which defines the relationship between $\underline{V}_{2s}$, $\underline{V}_{2r}$, $\underline{V}_{2X}$ (the voltage phasor across X_2) and the inserted voltage phasor $\underline{V}_{2pq}$, with controllable magnitude ($0 \leq V_{2pq} \leq V_{2pq\max}$) and angle ($0 \leq \rho_2 \leq 360°$), is shown in Figure 11.2b. It should be noted that this phasor diagram is similar to that characterizing a UPFC compensated line (refer to Figure 10.7).

As illustrated for the UPFC in, for example, Figure 10.11, the effect of the rotating compensating voltage $\underline{V}_{2pq}$ on the power transmission can best be illustrated in the $\{Q_{2r},P_{2r}\}$ plane, where each value of $\underline{V}_{2pq}$ (magnitude plus angle) defines a pair of corresponding P_{2r} and Q_{2r} values that represents the total MVA the transmission line carries. As established for the UPFC, the full (360°) rotation of phasor $\underline{V}_{2pq\max}$ produces a circular locus around the operating point of the uncompensated line defined by transmission angle δ_2, which determines the attainable maximum P_{2r} and Q_{2r}. At the selected transmission angle, $\delta_2 = 30°$, the uncompensated line 2 ($V_{2pq} = 0$) with the stipulated $X = 0.5$ p.u. reactance value transmits $P_{2r,30°} = 1.0$ p.u. active power and absorbs $Q_{2r,30°} = 0.268$ p.u. reactive power at the receiving-end bus. These $P_{2r,30°}$ and $Q_{2r,30°}$ coordinate values define the center of the circular control region. With the maximum amplitude of the compensating voltage, $V_{2pq\max} = 0.25$ p.u., the radius of the circular region in the $\{Q_{2r},P_{2r}\}$ plane is $[P_{2pq}{}^2 + Q_{2pq}{}^2]^{1/2}$ (= 0.5 p.u. VA), as illustrated in Figure 11.2c. This circular locus provides the boundary for the two-dimensional control range within which any corresponding Q_{2r} and P_{2r} values are achievable by the proper setting of the magnitude, V_{2pq}, and angle, ρ_2, of the compensating voltage phasor $\underline{V}_{2pq}$.

The compensation of the priority line 2 described above is identical to that characterizing the operation of the UPFC. However, in the case of the UPFC, the active power exchanged through the series voltage insertion is supplied via the shunt-connected converter from the sending-end bus. This power can be positive or negative, as the prevailing compensation requirements dictate with usually no other limit, but the rating of the shunt converter. In the case of the simple two-line IPFC considered here, the active power is obtained from the other line via the series-connected compensating converter of that line. In order to establish the possible compensation range for line 1, under the constraints imposed by the unrestricted priority compensation of line 2, it is helpful to decompose the total compensating MVA provided for line 2 into reactive power Q_{2pq} (MVAr) and active power P_{2pq} (MW). To this end, the injected voltage phasor $\underline{V}_{2pq}$ is decomposed into two components, one, $\underline{V}_{2q}$, in quadrature with the line current and the other, $\underline{V}_{2p}$, in phase with it. The products of these with the line current define the reactive and active power, Q_{2pq} and P_{2pq}. The component Q_{2pq}, generated internally by converter 2, evidently provides self-sufficient series reactive compensation for line 2, just like a controllable series capacitor. The component P_{2pq}, on the other hand, provides active power compensation for line 2, but this power must be supplied for converter 2 by converter 1 from line 1.

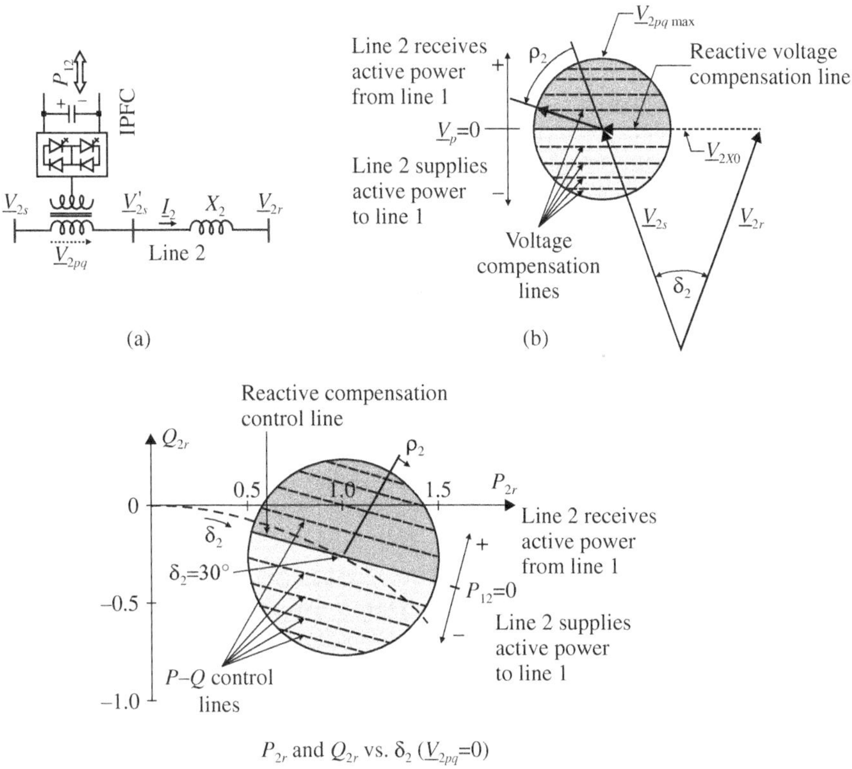

Figure 11.3 *P–Q* control lines in the operating region of the {*P,Q*} plane corresponding to different amount active power compensation of the "priority" line 2 controlled by in-phase voltage injection.

Consider first the special case in which the compensation of line 2 requires no active power. In this case, P_{2pq} is zero, converter 2 operates as a controllable series compensator injecting a compensating voltage that is in quadrature with the line current, that is, $\underline{V}_{2pq} = \underline{V}_{2q}$. This means that this voltage directly adds to or subtract from the voltage across the reactive line reactance X_2, which is, by definition, also in quadrature with the line current. The loci of the compensating voltage are on a straight line, termed here as *reactive compensation line,* passing through the end points of the sending- and receiving-end voltage phasors, as illustrated in Figure 11.3b.

Suppose now that a compensating voltage component, $\underline{V}_{2p}$ that is in phase with the line current, is added with a fixed magnitude to the variable quadrature component $\underline{V}_{2q}$. This will have an effect of shifting the reactive compensation line up or down, depending on the polarity of $\underline{V}_{2p}$ with respect to the line current phasor $\underline{I}$, creating a new *voltage compensation line*. This line can be visualized as loci of the end points of the compensating phasor, $\underline{V}_{2pq} = \underline{V}_{2p}+\underline{V}_{2q}$, when in-phase component $\underline{V}_{2p}$ is kept constant and quadrature component $\underline{V}_{2q}$ is continuously varied over its total range of $-V_{2q\max} \leq V_{2q} \leq + V_{2q\max}$. In a different way, it may also be said that if the magnitude of voltage phasor $\underline{V}_{2pq}$ is controlled over the attainable range of angle ρ so that its end point stays on a straight line trajectory ("voltage compensation line"), parallel to the line connecting the end points of the sending-end and receiving-end phasors ("reactive voltage compensation line"), as illustrated in Figure 11.3b, then the product of $\underline{V}_{2pq}$ with the prevailing line current phasor $\underline{I}_2$ will result in constant active power demand (that the converter must supply or absorb) independent of angle ρ. This means that within the circular operating region of $\underline{V}_{2pq}$, the reactive compensation is freely variable along a "voltage compensation line" at any value of the active power compensation within this operating range. As indicated, the active power demand is, by definition, zero when the trajectory of $\underline{V}_{2pq}$ coincides with the "reactive voltage compensation line" ($\underline{V}_{2p} = 0$), which divides the circular operating region into two equal halves. An increasing amount of active power is to be supplied to line 2 as the voltage compensation line is shifted higher and higher above the "reactive voltage compensation line" in the upper half of the control region. Conversely, increasing active power is to be absorbed from line 2 as the compensation line is shifted lower and lower below the "reactive voltage compensation line" in the lower half of the compensation region.

The "voltage compensation lines," prescribing the trajectory of the phasor $\underline{V}_{2pq}$ for constant active power demand in the phasor diagram characterizing line 2, define a linear relationship between the receiving-end reactive and active power, Q_{2r} and P_{2r}, respectively, within the circular locus representing the boundary of the control region in the $\{Q_{2r}, P_{2r}\}$ plane, as shown in Figure 11.3c. Thus, a "reactive compensation control line," which crosses the center of the boundary circle, defines the Q_{2r} versus P_{2r} relationship for purely reactive variable compensation. An infinite number of parallel Q_{2r} versus P_{2r} control lines of decreasing length, corresponding to the "voltage compensation lines" of the phasor diagram, can be established above and below the reference control line corresponding to the "reactive compensation line," as a function of the active power exchanged. A number of such lines, corresponding to the "voltage compensation lines" of the related vector diagram, are drawn in the same manner (dashed with different spacing), in the Q_{2r} versus P_{2r} control region of the $\{Q_{2r}, P_{2r}\}$ plane shown in the figure. Comparison of Figures 11.3b to 11.3c indicates that the lagging reactive power flow decreases (and can become leading) in proportion with the active power *supplied* to the line and, conversely, it increases with the active power *absorbed* from the line by series compensation.

With respect to the total IPFC scheme shown in Figure 11.1, the following conclusions can be drawn from the above discussion. The operating point of converter 2 compensating the "priority" line 2 can be considered to be a point of a particular "voltage compensation line." Consequently, a corresponding point on the related Q_{2r} versus P_{2r} control line in the $\{Q_{1r}, P_{1r}\}$ plane defines the resultant active and reactive power flow in the transmission line. In general, at a selected operating point, converter 2 has to exchange both active and reactive power with line 2. However, the converter can internally generate only the reactive power and thus it must be supplied with the active power (plus or minus) it exchanges. The active power demand remains constant, while the internally generated reactive power changes, as the operating point of the converter is shifted along a selected "voltage compensation line." With the shifting operating point (i.e., with the resultant variable reactive compensation), the receiving-end active and reactive power moves along the relevant Q_{2r} versus P_{2r} control line in the $\{Q_{1r}, P_{1r}\}$ plane, changing primarily the transmitted active power P_{2r}. Moving the operating point from one "voltage compensating line" to another changes the active power demand of the converter and shifts the resulting receiving-end active and reactive power to a parallel Q_{2r} versus P_{2r} control line in the $\{Q_{2r}, P_{2r}\}$ plane, and thus primarily changing the reactive power Q_{2r} in the line.

It follows therefore that in order to satisfy the active power demand of converter 2 operated along a selected "voltage compensation line," converter 1 must be operated along a complementary "voltage compensation line" so as to precisely supply this demanded active power from line 1 via the common DC link for converter 2. (The term "complementary" means here that if converter 2 is operated on a voltage compensation line *above* the reference "reactive compensation line," then converter 1 must be operated on a corresponding voltage compensation line *below* "reactive compensation line.") In other words, the relationship $P_{1pq} = -P_{2pq}$ must be continuously satisfied. In practice, this condition is usually satisfied by controlling converter 1 so as to maintain the voltage of the common DC link in face of the variable active power demand of converter 2.

The operation of the two-converter IPFC scheme is illustrated with the help of the complementary regions of "voltage compensation" and Q_r versus P_r control lines, characterizing the operations of the "priority" line 2 and the "support" line 1, in Figure 11.4. For clarity, the previously stipulated transmission line and operating parameters are assumed (two independent lines with identical impedances and operating parameters, transmission voltage and angle). In this case, the selection of

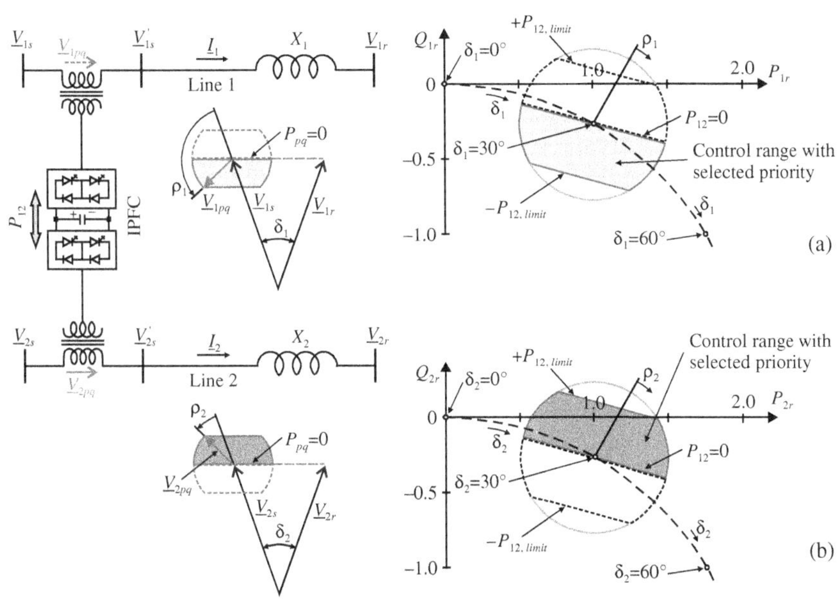

Figure 11.4 Illustration of the basic complementary relationship governing the operation of a two-line IPFC. Line 2 is operated as the selected "priority" line and line 1 is operated as the "support" line.

"priority" and "support" is arbitrary; in practice, the priority line usually would have heavier reactive load.

The right-hand side of Figure 11.4b shows the circular operating area of the "priority" line 2 in the $\{Q_{2r}, P_{2r}\}$ plane. The "priority" designation, however, here means that the desired power transmission necessitates the reduction of the reactive line power flow, while maintaining the full controllability of the active power up to the maximum power of 1.5 p.u. This requirement puts the operation of the IPFC controlled line 2 in the upper region of the circular operating area, above the reference control line representing purely reactive compensation ($P_{2pq} = 0$). (As shown in the figure, this semicircle region may further be limited on the top by the available maximum transferable active power or by the allowed maximum magnitude of corresponding voltage.)

The left-hand side of Figure 11.4b shows the operating area for the IPFC injected voltage phasor $\underline{V}_{2pq}$. As can be seen, this semicircle area is *above* the "reactive compensation line," meaning that converter 2 supplies active compensating power P_{12} for line 2 by series voltage injection. This P_{12} power must, of course, be provided for converter 2 by converter 1 via the common DC link from "support" line 1. To absorb the P_{12} power from line 1, the active component of voltage phasor $\underline{V}_{1pq}$ must be in phase opposition to the line current phasor $\underline{I}_1$, which requirement constrain the operation region of phasor $\underline{V}_{1pq}$ in the semicircle area *below* the "reactive compensation line," as shown on the left-hand side of Figure 11.4a. (Note that the operating regions of both $\underline{V}_{1pq}$ and $\underline{V}_{2pq}$ show a possible limit for the maximum value allowed for the injected active voltage component due to the increase in transmission voltage.)

The operating region of line 1 in the $\{Q_{1r}, P_{1r}\}$ plane is shown on the right-hand side of Figure 11.4a. As could be expected, the restriction of the $\underline{V}_{1pq}$ to the operating region below the "reactive compensation line" constrains the corresponding semicircle operating region of the attainable P_{1r} and Q_{1r} values below the reference control line representing purely reactive compensation. This means that although the active power in line 1 is still well controllable, the reactive power flow in line 1 increases in proportion to the active power it supplies to line 2. It is emphasized again that in practice, the IPFC-type compensation would be typically applied if the two lines reactive loading would be different or if the capacity of one line would be appreciably greater than that of the other, or some other network or transmission reasons in which reactive power flow control or balancing would increase asset utilization or aid network optimization.

11.3 GENERALIZED INTERLINE POWER FLOW CONTROLLER FOR MULTILINE SYSTEMS

Although the detailed explanation of the IPFC has been restricted to a simple two-line, two-converter arrangement, the general principle is applicable to an n-line transmission system in which each line is compensated by a DC to AC converter, which is part of an *n*-converter IPFC compensator with common DC terminals, as illustrated schematically in Figure 11.1. The operation of this is similar to the two-converter

IPFC in that a number of "priority" lines could be designated as long as their active power demand could be satisfied by the remaining "support" converters, since it is a fundamental operating requirement that the total of active power exchanged between the converters and the transmission lines must be zero. However, there can be compensation requirements for particular multiline transmission systems, which would not be compatible with this basic operating constraint of the IPFC. This fundamental requirement could be particularly restrictive under an emergency contingency when those lines that were to support the "priority" lines would also become overloaded or even disconnected due to fault. Such potential problems can be solved, and a more general applicability of the multiline compensator established, by combining the UPFC and IPFC concepts to realize a generalized IPFC arrangement, in which a shunt-connected converter is added to the number of converters providing series compensation, as illustrated in Figure 11.5.

With this scheme, the net power difference at the DC terminal is supplied or absorbed by the shunt converter, and ultimately exchanged with the AC system at a suitable shunt-bus where active power exchange is not subject to rigorous limitation. Apart from removing the restriction for series active power compensation for the lines, this generalized IPFC scheme greatly simplifies the control by using the shunt-connected converter for the exclusive role of maintaining the common DC link voltage and thereby automatically maintain the required power balance. The arrangement can also be economically attractive because the shunt converter has to be rated only for the maximum active power difference anticipated for the whole system. Furthermore, it can also facilitate relatively inexpensive shunt reactive compensation, if this is needed at the particular substation bus, because the shunt converter would need to be rated only for the vectorial sum, that is, the square-root of the squares, of active

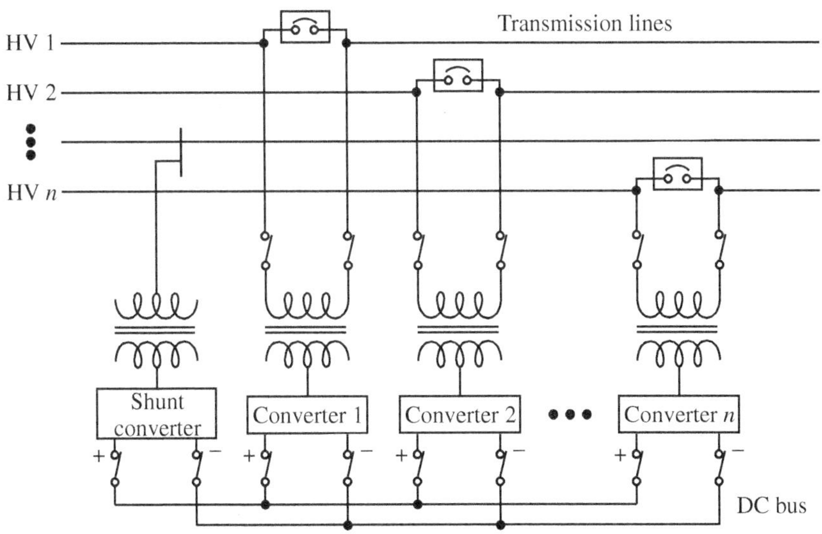

Figure 11.5 Generalized multiline interline power flow controller scheme in which a shunt-connected converter is used to support the common DC link voltage and thereby allow unrestricted reactive power flow control in each line.

power it exchanges with the other converters and the reactive power it provides for shunt compensation.

It is also possible that the compensation and power flow control requirements at a given substation change due to system contingency, maintenance, and system operational changes. Under these conditions and scenarios, complete functional changes in compensation and control requirements may be possible. For example, in a weakened system or under heavy line loading, voltage support of a given bus may be more important than the power flow control of a given line. In this case, a series compensator could be converted into an additional shunt compensator within or outside of generalized IPFC system. Overall, a multiconverter compensating system could provide a hitherto unachievable operational flexibility together with almost unrestricted functional convertibility.

11.4 BASIC CONTROL SYSTEM

The control structure of the IPFC is similar to that of the UPFC shown in Figures 10.15, 10.16, and 10.17 with appropriate changes in the controlled variables and the necessary constraints imposed by the possible limitations of active power transfer. A possible IPFC internal control scheme is shown in the form of a block diagram in Figure 11.6. For this structure, the previous assumption, stipulating system 2 with

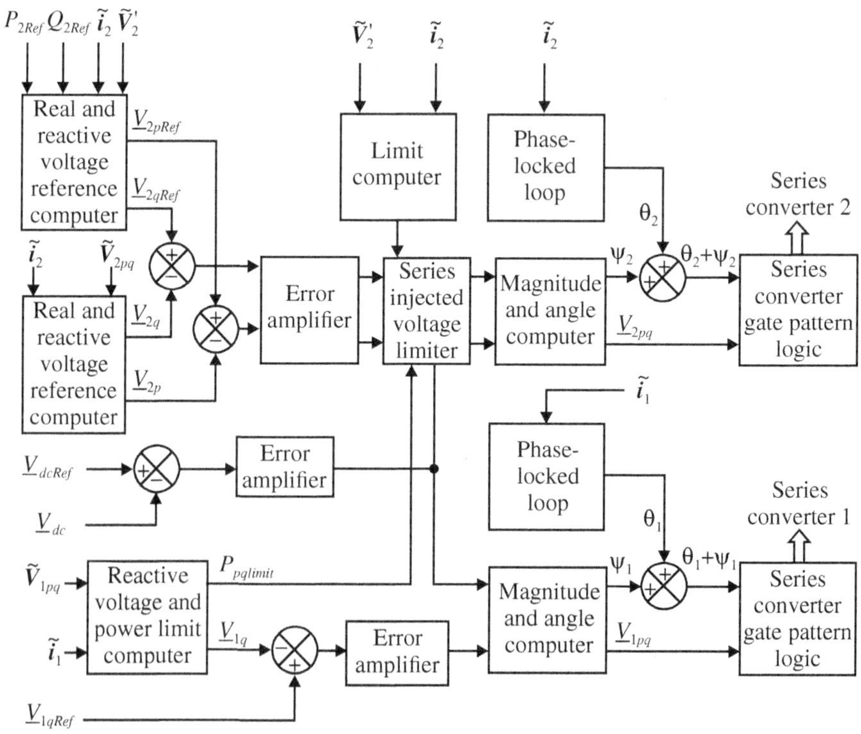

Figure 11.6 Block diagram for the basic control of a two-line IPFC.

IPFC converter 2 as the "priority" system requiring the independent control of both active and reactive power is, for continuity and clarity, retained. This stipulation makes the control of the two converters functionally somewhat different. However, the readers should note that in practice, the two converter controls would be identical with control inputs selecting either converter for the "prime" or "support" operating role.

As shown in Figure 11.6, the operation of converter 1 is synchronized to line current $\underline{I}_1$ and converter 2 to line current $\underline{I}_2$ by two independent phase locked loops. This enables each converter to provide independent series reactive compensation, and to keep operating under contingency conditions when the other line or converter is out of service.

The input to the "priority" control is either the desired active and reactive line power, P_2 and Q_2 (indicated in the figure by P_{2Ref} and Q_{2Ref}) or it could be the desired quadrature and in-phase compensating voltage V_{2q} and V_{2p}, shown in the figure as internal references, V_{2qRef} and V_{2pRef}, derived from P_2 and Q_2. Voltage component V_{2qRef}, being in quadrature with the prevailing line current $\underline{I}_2$, represents series reactive compensation to control the transmitted active power, and component V_{2pRef}, being in phase with that, represents series active compensation to control the reactive power flow in the line.

The internally derived references, V_{2qRef} and V_{2pRef}, are compared to the actual voltage components, V_{2q} and V_{2p}, derived from the measured line current and injected voltage vectors, $\tilde{\boldsymbol{i}}_1$ and $\tilde{\boldsymbol{v}}_{2pq}$. Thus, the obtained error signals after appropriate amplification and possible limitation provide the input for the computation of the magnitude, V_{2pq}, and angle, ψ_2, of the injected voltage vector $\tilde{\boldsymbol{v}}_{2pq}$.

The limiter preceding the computation of V_{2pq} and ψ_2 is an important function of the IPFC control to ensure system operation within predefined constraints. One set of constraints may be provided by the voltage and current limitations of line 2. (This set of constraints would also apply to the series converter of the UPFC.) The second set of constraints is unique to the IPFC and related to the possible limitations of line 1 to supply the active power demand resulting from the "priority" compensation of line 2. As indicated in Figure 11.6, these limitations may result from insufficient current in line 1 to supply active power for the maintenance of the DC bus voltage (which is burdened by the active power demand of line 2) or from limitations imposed for active power transfer by the allowed reactive power flow constraints on line 1.

The control of converter 1 is different from that of converter 2 due to its functional role to support the operation of the "priority" converter 2 by supplying the necessary active power from line 1. This requirement means that, since the in-phase component of the injected compensating voltage is imposed on line 1 by the active power demand of line 2, the control of converter 1 can vary only the transmitted active power in its own line by controlling the quadrature component, V_{1q}, of the injected voltage vector $\tilde{\boldsymbol{v}}_{1pq}$. Thus, the reference input to the control of converter 1 is the desired quadrature compensating voltage, V_{1qRef}. (The readers should note that an equivalent input could be the desired compensating reactance. It would also be possible to provide a closed regulation loop for the desired active power P_1 that would yield V_{1qRef} as an internal reference.) Reference voltage V_{1qRef} is compared to voltage component V_{1q} derived from the measured injected voltage $\tilde{v}_{1pq}$. From this, and from the amplified error representing the difference between the desired and actual DC bus voltage, V_{dcRef} and V_{dc}, the magnitude, V_{1pq}, and angle, ψ_1, of the injected voltage vector $\tilde{v}_{1pq}$ are derived to generate the output voltage of converter 1, as illustrated in Figure 11.6, to provide the required series reactive compensation and maintain the DC link voltage.

The constituents and functional arrangement of the overall practical control system for the IPFC would be similar to that illustrated for the UPFC in Figure 10.18.

In the case of a multiconverter IPFC compensating several lines, without a shunt-support converter, the control system could become quite complex. The fundamental problem is to maintain the balance for the active power exchanged by the IPFC with a multiline system. There is yet no precedent for the practical realization of such a system, although its functional advantages are well established. Qualitatively, it appears that all lines compensated would have to have a defined priority order (or criteria thereof) together with rigorous limits for critical variables (e.g., maximum line current, reactive power, and line voltage increase/decrease), and a computer employed in the control system would determine the attainable decrease and allowed increase in the reactive power flow in the lines according to this priority order and within the limits, and set the compensating voltages for each line accordingly.

However, if a shunt converter is employed, then the control system would be simple and rugged, and the series and shunt converter controls shown in Figures 10.16 and 10.17 for the UPFC could be used practically without change. The reason for this is that in this case, each series converter could be controlled freely as an independent compensator without worry whether the active power exchanged is positive or negative, since the resultant net active power manifesting itself at the common DC link would be balanced by the shunt converter whose control would ensure the maintenance of the DC voltage. Of course, this arrangement would require that the rating of the shunt converter is determined under the expected worst-case condition under which the operation of the IPFC is still expected.

11.5 EQUIPMENT STRUCTURAL AND RATING CONSIDERATIONS

The IPFC basic circuit implementation comprising two back-to-back connected, voltage-sourced converters is similar to that of the UPFC, and thus the general structural and rating consideration detailed in Section 10.7 are also valid for the IPFC. However, the converters of the IPFC are operated differently than the series converter of the UPFC. Each IPFC converter is able to provide independent series *reactive* compensation *for its own transmission line,* same as the series converter of the UPFC, to control the *active power* flow in it. But, through the back-to-back connection of the converters at the common DC bus, they establish a path for active power flow between the two lines, which enables them also to execute *active* power compensation to control *reactive* line power flow without the support of a shunt converter. The operating role of the two converters is the same for reactive power compensation, but different for active power compensation. That is, although each converter can execute the reactive compensation independently of the other, one converter needs the active power support of the other for active power compensation. This means that one converter has the role of providing the desired reactive and active power compensation of the *priority* line (the line in need of power flow optimization), while the other is in *support* role, providing the necessary (positive or negative) active power from its own line to support the compensation of the priority line. The requirement for the IPFC series converters to control self-sufficiently active as well as reactive power makes the rating considerations different from those of the series converter of the UPFC.

Since the capacity of the two lines may be different and, depending on the prevailing transmission requirements and line loading, the roles of the two converters may need to be reversed, the rating and operating considerations for the IPFC can become complex requiring repetitions of load flow studies to define the desired and sustainable operating conditions. However, the principles involved are quite simple and are summarized below for guidance:

1. Determine the series reactive compensation requirements for each line independently, as if only a controllable series reactive compensator was to be applied, to define the necessary rating of each converter to execute the reactive compensation under the stipulated operating conditions. From these studies, the maximum current and voltage rating of each converter, which may not occur simultaneously, for the required reactive compensation can be established. Note that under worst-case contingency conditions, such as partial equipment failure, line failure, or other reasons that prevent active power transfer from one line to the other, *this is* the compensation that will be available for the line.
2. Define the criteria for priority compensation. Determine the rating of the *priority* converter as if it was the series converter of a UPFC, using the process described in Section 10.7.3.1, with the maximum (quadrature) voltage and line current obtained by the process described in 1. (Note that the maximum quadrature voltage will define the radius of the unconstrained circular operating area for the injected voltage.) The required end result of this process is the maximum voltage, line current, and *active power* rating that the priority converter of the IPFC must have for the maximum reactive and active power compensation, which the priority line may require.
3. Determine if the line in *support* role is able to provide the maximum active power for the priority line. The consideration here is that active power usually has to be *supplied* to the priority line to *decrease* the reactive power flow in it. Consequently, the corresponding active power *absorption* from the support line will *increase* the reactive power flow in it. Therefore, the general requirement for sustainable operation of the IPFC is that either the support line is a higher capacity line than the priority line, or that it is less loaded than the priority line.
4. If the support line *has* the capacity to provide the required maximum active power for the priority line, then the rating of the support converter is determined from the appropriate combination of the maximum reactive compensation

required for its own line and the maximum active power need of the priority converter compensating the other line.

5. If the support line does *not* have the capacity to provide the required maximum active power for the priority line, then the process has to be reversed to (a) find first the attainable active power of the support line; (b) determine with that the ratings of both the support and priority converters; (c) assess whether, with the constraint of reduced, or not always available active power transfer capacity, the IPFC still provides enough benefit in terms of power flow optimization, reactive loop flow minimization, damping capability, etc., to justify its practical implementation; and (d) considering the possibility of adding a third converter in shunt connection to support both series converters in active power compensation, and also providing the option for bus voltage regulation. In assessing the possible options and benefits, it should be kept in mind that the extension of a two-line, converter-based series compensating system into an IPFC is practically "free," since it requires no additional power components, only suitable control algorithm, and the addition of a shunt-connected converter for supporting reactive power flow control in the lines can also provide very economical means for shunt reactive compensation.

11.6 PROTECTION CONSIDERATIONS

In its basic form, the IPFC is totally composed of series-connected converters. The converters connected in series with the line require special protection considerations to handle surge currents caused by line faults or other system disturbances, as is discussed in Section 10.8.1 for the UPFC. The main issue is how to handle possible surge currents of the line during faults and abnormal operating conditions. In most practical cases, it would not be economically feasible to rate the series-connected converter to handle the worst-case surge currents of the line. Depending on the magnitude of possible surge currents, generally, there are three possibilities for handling line surge currents, as discussed in Section 10.8.1:

1. If the line is "weak," the converter may be able to handle the possible surge current without making its rating economically impractical.
2. The possible worst line current is too large for the converter to handle economically, so it is "bypassed" during fault conditions by an external device.
3. The converter, either by internal external means, limits the injected voltage to a maximum value for line current higher than a given maximum.

Circuit solution to implement these methods is given for the series converter of the UPFC in Figures 10.21 and 10.22.

Apart from these specific methods and circuit configuration for surge current protection, the control system of the series converter does have additional sophisticated protection functions, which monitor internal circuit parameters, component failures, redundancy conditions, and the overall performance of the system and give alarm indication or, in extreme cases initiate shut downs, if parameter limits are exceeded, or internal component failures or circuit malfunctions prevent normal operation.

11.7 APPLICATION EXAMPLE: IPFC AT NYPA'S MARCY SUBSTATION

The only practical realization of the IPFC to date is within the Convertible Static Compensator (CSC) installation of the New York Power Authority (NYPA) that is a multifunction transmission compensator employing two back-to-back converters, which can be configured and operated as an IPFC controlling the reactive and active compensation of two independent lines [2,4].

This first IPFC in the world, with a total rating of ±200 MVA, was commissioned as one of the several operating configurations of the two-converter CSC in 2002 at the Marcy Substation of the NYPA. The project was jointly sponsored by the Electric Power Research Institute and NYPA, and designed and manufactured by the Westinghouse (subsequently Siemens) FACTS and Custom Power Division.

11.7.1 Background Information, System, and Equipment Requirements

System description. NYPA as a public benefit energy corporation was founded in 1931. It is the largest nonfederal electric utility in the United States and a sale power supplier throughout New York State (26% of the State's electric energy needs) and neighboring states with more than 42.3 billion kWh energy sales. It has over 7000 MW generating capacity, more than 1400 circuit miles with 115, 230, 345, and 765 kV transmission lines. The NYPA power system is shown in Figure 11.7.

System problems. Prior to the CSC project in the late 1990s, the maximum transfer limit of the Central-East interface was *voltage constrained* to 2850 MW (and even to lower under system outage contingency) and operated at 100 MW below this limit 75% of the time. During 1996 and 1997, the system had 18 major emergencies and 192 alarm states. Studies indicated that certain contingencies could ignite oscillatory system conditions. The projected South-East New York load growth of greater than 3% indicated the need to increase the transmission capability of the Central-North transmission corridor.

Objectives of the CSC project. The main objectives of the CSC were to increase the power flow transfer capability of the Central-North transmission corridor, to improve voltage control capability at Marcy, to enhance system damping during oscillatory contingencies, and to provide operational flexibility during system outages. To this end, the CSC was specified as a single, multifunction all solid-state, modular compensating system with expandable hardware structure, capable of providing any one, or any combination of the following compensation and control functions:

- voltage support and regulation by reactive shunt compensation;
- power flow control by reactive series compensation;
- independent active and reactive power flow control;

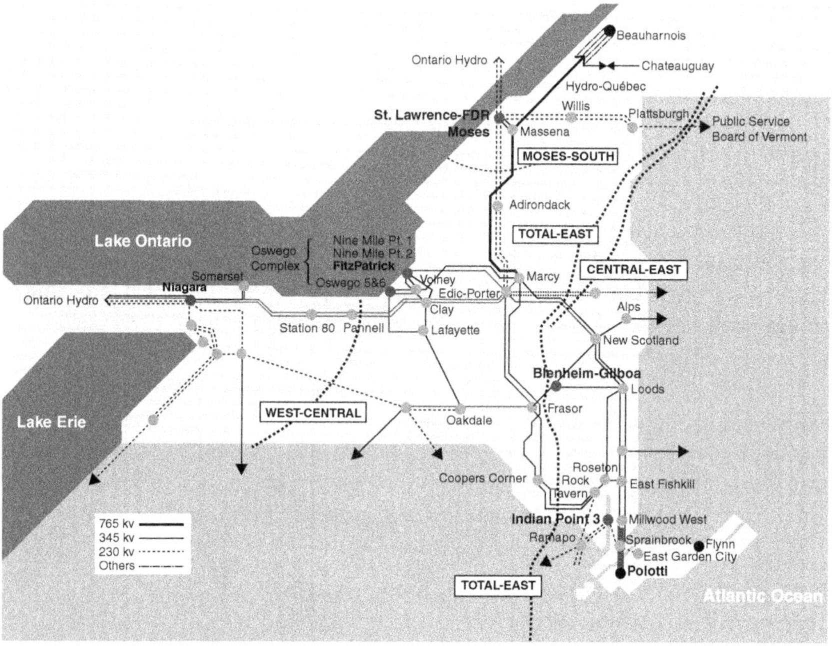

Figure 11.7 NYPA power system.

– active and reactive line load balancing;

– stability improvement.

11.7.2 Description of the CSC/IPFC

The CSC for the Marcy Substation was designed to meet the above defined system requirements. It is structurally similar to the UPFC at AEP's Inez substation, described in Chapter 10, with extended functional and control capability, in particularly, with the important addition of the IPFC operating mode that allows power flow control in two different lines.

Power circuit structure. The 345 kV 200 MVA CSC comprises two identical gate turn-off (GTO) thyristor-based converters, each rated for ±100 MVA. Each converter includes multiple high power, three-level, 48-pulse GTO valve structures feeding an intermediate (low voltage) transformer rated for approximately 50 MVA. The converter and magnetic configuration provides harmonic neutralization, and thus the output is a three-phase voltage set of nearly sinusoidal quality that is coupled to the transmission line via the intermediate transformer by main coupling transformer. The main coupling transformer for shunt connection is rated for 200 MVA and has two sets of secondary windings to accommodate both converters in parallel for reactive shunt compensation. Two coupling transformers, each rated for 100 MVA, are used to connect their corresponding converter to the selected line, to provide the configuration required for UPFC or IPFC operation. Thus, the

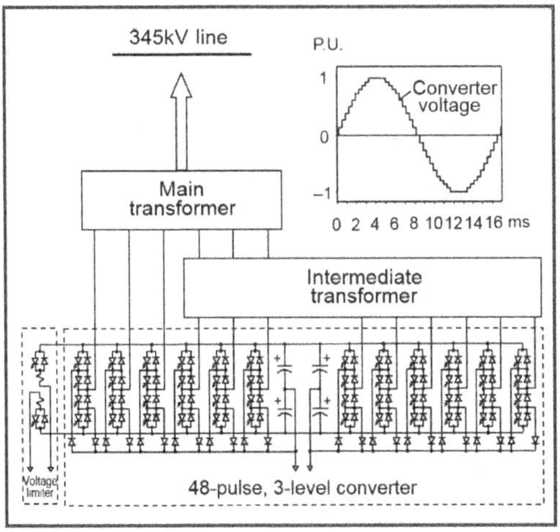

Each Converter

Nominal rating:	±100 MVA
Transmission voltage:	345 kV
Converter output voltage: (to coupling transformer)	22 kV (phase to phase)
Pole output voltage:	5.5 kV
No. of converter pulses:	48
Type of converter pole:	3-level
No. of poles:	12
No. of 3-phase converters:	4
Type of semiconductor:	GTO
Rating of GTO:	4.5 kV, 4 kA
No. of GTOs per valve:	6 (5+1)
DC voltage:	6+6 kV
DC capacitor rating:	72 kJ

Note: All ratings are nominal

Figure 11.8 Basic structure of the CSC converters and the corresponding technical data.

converters can be connected to function as a STATCOM, regulating the voltage at the Marcy bus. The converters can also be connected through the insertion transformers in series with Marcy-New Scotland and Marcy-Coopers Corners 345 kV lines. The series transformers allow the converters to operate as a UPFC from the Marcy bus with series power flow control in either of the two lines. Also, *they can be operated as an IPFC with active power transfer capability between the two lines.* And, of course, the converters can be disconnected at their DC terminal and operated independently as reactive shunt and/or series compensators (i.e., STATCOM and/or SSSC). The converter structure is shown schematically with relevant information in Figure 11.8. The installed converter is shown in Figure 11.9. The one-line diagram in Figure 11.10 shows the connection arrangement of the CSC in the NYPA system at the Marcy Substation, and Figure 11.11 shows the actual installation.

Control system. Both converters comprising the CSC are controlled from a single central control system housed in the control room. The control system is similar to that described for the Inez UPFC in Chapter 10, but facilitates more system configurations, operating modes, and control functions. The CSC basic internal control architecture is shown in Figure 11.12, and the operational and functional options are illustrated in Figure 11.13. The internal control of the CSC is interfaced with NYPA's especially designed master controller that provides the appropriate references, derived from measured system variables and operation requirements, for the CSC according to the overall system optimization control concept illustrated in Figure 11.14.

11.7.3 Importance of the NYPA Installation

This CSC with the full implementation of the IPFC concept extended the functional capability of converter-based modular transmission controllers to the series compensation of multiline systems with active power transfer option for the control of

Figure 11.9 The CSC converter hall.

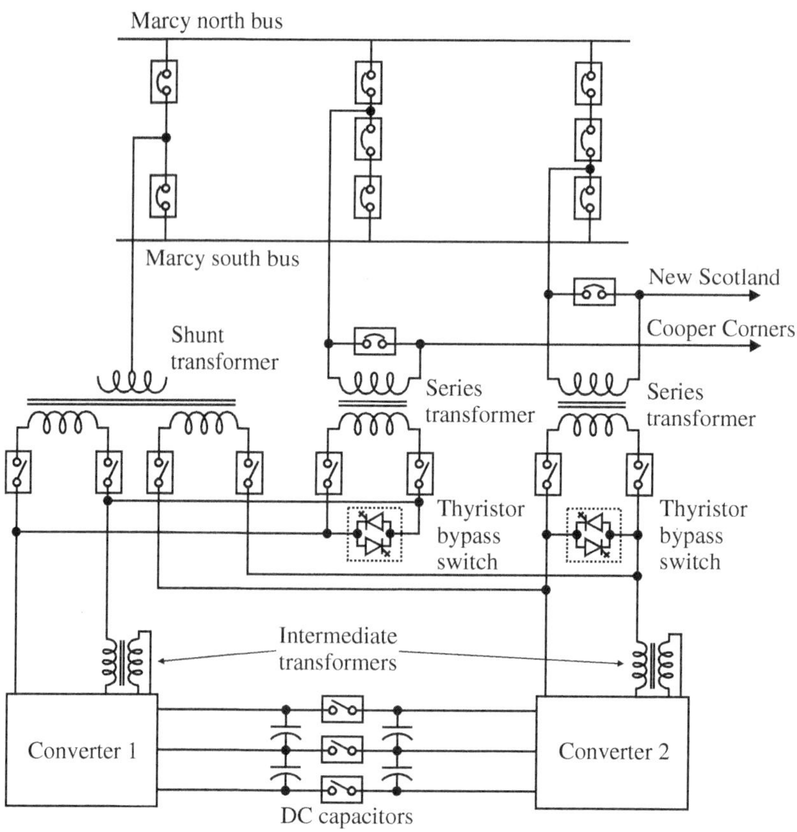

Figure 11.10 Simplified UPFC power circuit in the NYPA system at the Marcy Substation.

Figure 11.11 NYPA CSC installation at the Marcy Substation.

reactive power flow. With this, the practicality of back-to-back converters with not only with shunt-series connection for UPFC, but also with *series-series* AC coupling of two independent lines for IPFC operation has been verified. The practicality of converting a two-converter-based compensator into five different operating modes to provide transmission compensation control functions according to prevailing system

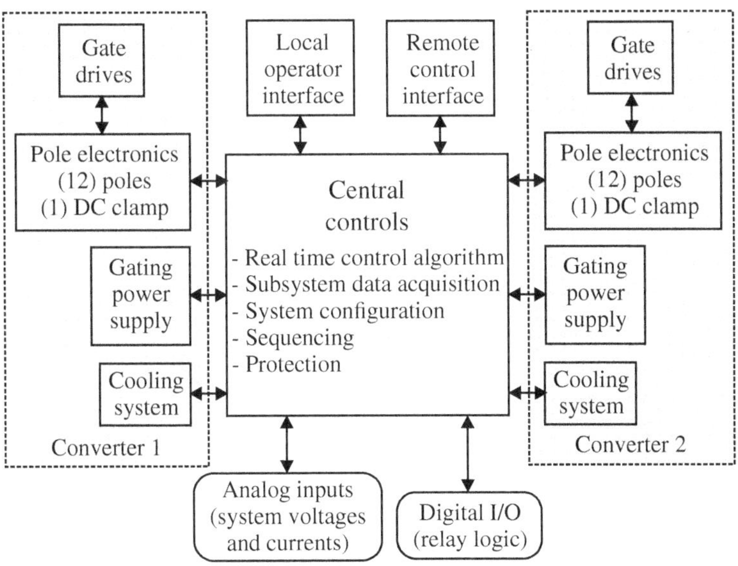

Figure 11.12 Basic functional control structure of the NYPA CSC [4].

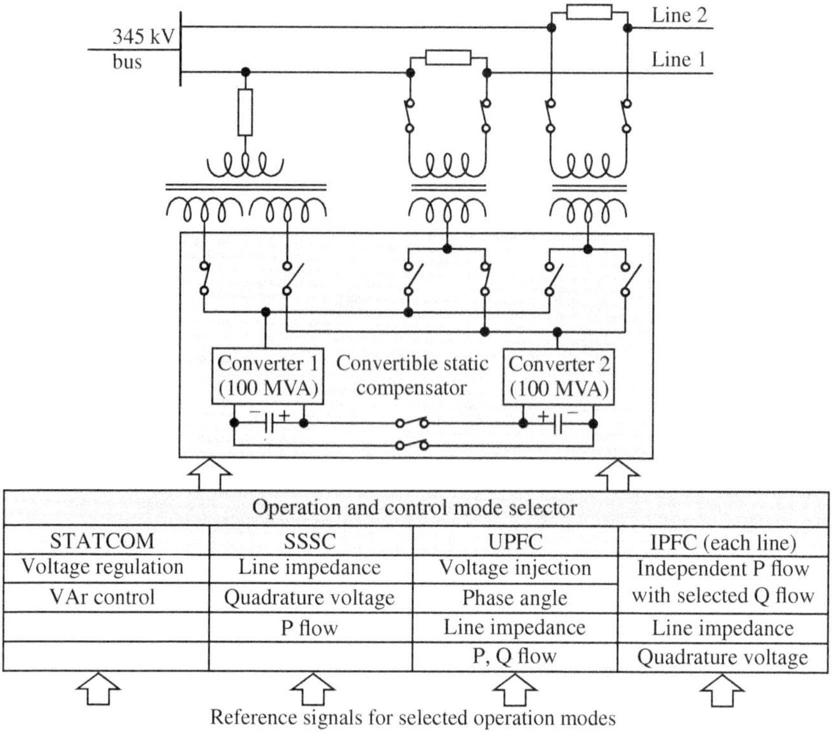

Figure 11.13 Functional and operational capabilities of the NYPA CSC [4].

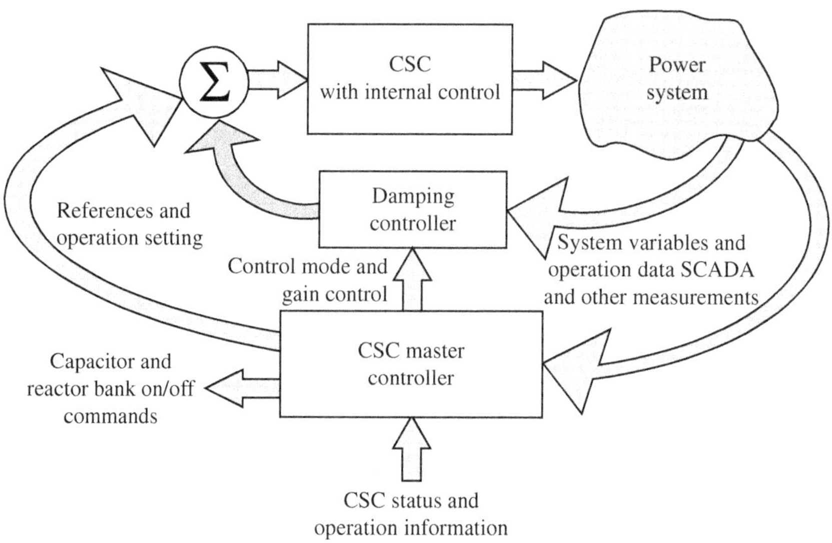

Figure 11.14 The CSC system control structure [4].

requirements has been demonstrated. The NYPA installation also proved the operation of series-coupled converters of the IPFC configuration in the case of high voltage (345 kV) lines.

REFERENCES

[1] Gyugyi, L., Sen, K.K., Schauder, C.D. The interline power flow controller concept: a new approach to power flow management in transmission systems, *IEEE Transactions on Power Delivery*, vol. 14, no. 3, pp. 1115–1123, July 1999.

[2] Fardanesh, B., Gyugyi, L., Henderson, M., Shperling, B., Zelingher, S., Schauder, C., Lam, B., Mountford, J, Adapa, R., Edris, A. Convertible static compensator application to the New York transmission system. CIGRE Paper 14-103, 1998.

[3] Hingorani, N. G., and Gyugyi, L. *Understanding FACTS – Concepts and Technology of Flexible AC Transmission Systems*. Wiley-IEEE Press, New York, 2000.

[4] Fardanesh, B., Gyugyi, L., Shperling, B., Uzunovic, E., Zelingher, S., Kovalsky, L., Macdonald, S., Schauder, C., Edris, A. NYPA convertible static compensator validation of controls and steady-state characteristics. CIGRE Paper, 14–104, 2002.

CHAPTER 12

SEN TRANSFORMER: A POWER REGULATING TRANSFORMER

Kalyan K. Sen

12.1 BACKGROUND

The recently developed flexible alternating current transmission systems (FACTS) controllers have shown that the transmission lines can be utilized to their fullest extent through independent control of active and reactive power flows. The direct benefit of independent control is to optimize both the active and reactive power flows, thus creating lower losses and higher efficiency in electric transmission systems, and lower wholesale electric market costs to loads. FACTS controllers can increase or decrease power flow through desired transmission paths, which improves line utilization. Also, grid congestion is avoided by redirecting the excess power flows from overloaded lines to underloaded lines.

Based on the aforementioned discussion, the world's first unified power flow controller (UPFC) was designed and commissioned at the American Electric Power (AEP)'s Inez substation in 1998. Since the introduction of the power electronics-based UPFC, a great deal has been learnt about the FACTS controllers and what the utility needs for its everyday use. To meet the power industry's pressing need for the most economical ways to transfer bulk power along a desired path, a power flow controller must offer *essential features*, such as:

- high reliability with the lowest number of components that are free from becoming obsolete;
- fast enough response for utility applications;
- portability so that it can be easily relocated to wherever it is needed the most, since the need for power flow control may change with time due to new generation, load, and so on;
- easy maintenance;
- lowest installation and operating costs with the highest efficiency;
- interoperability so that components from various suppliers can be used, resulting in a global manufacturing standard.

Advanced Solutions in Power Systems: HVDC, FACTS, and Artificial Intelligence, First Edition.
Edited by Mircea Eremia, Chen-Ching Liu, and Abdel-Aty Edris.

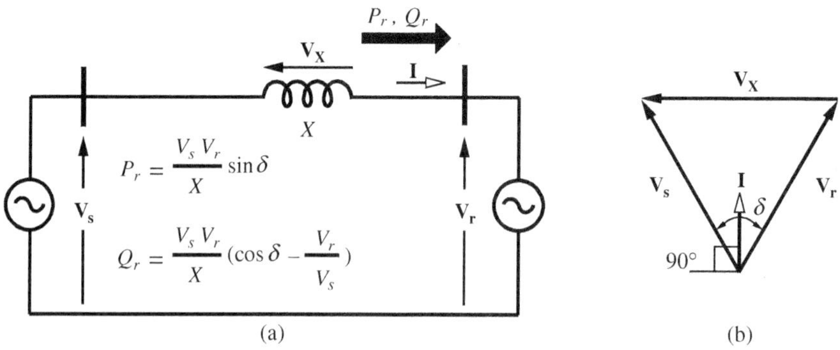

Figure 12.1 (a) Power transmission system and (b) phasor diagram.

The objective of this chapter is to present the Sen transformer (ST) [1–16] that offers a whole slew of power flow control features, such as independent power flow, voltage, phase angle, impedance, fault current, all in one unit, which uses time-tested components proven to be reliable, cost-effective, and portable to meet today's utility needs. In addition, various techniques that are used today to control the power flow in electric transmission lines are briefly reviewed.

12.1.1 Traditional Power Flow Controllers

Transmission of power in a single line with a sending-end voltage, $\mathbf{V_s}$ (of magnitude, V_s, and angle, δ_s), and a receiving-end voltage, $\mathbf{V_r}$ (of magnitude, V_r, and angle, δ_r), connected by a line reactance (X) and the related phasor diagram are shown in Figure 12.1. Neglecting the line resistance to maintain simplicity in the expressions, the voltage, $\mathbf{V_X}$ (i.e., $V_s - V_r$), across the line reactance (X) is the difference between the sending- and receiving-end voltages. The resulting line current (**I**) lags the voltage ($\mathbf{V_X}$) by 90°. The active and reactive power flows, P_r and Q_r, at the receiving end are

$$P_r = A \sin\delta \tag{12.1}$$

$$Q_r = A[\cos\delta - (V_r/V_s)] \tag{12.2}$$

where $A = V_s V_r / X$ and $\delta = \delta_s - \delta_r$.

The power flow control parameters are transmission line voltage magnitudes, V_s and V_r, at its sending and receiving ends, their phase angles' difference (δ), and line reactance (X). Any of these parameters can be controlled individually with the use of the following, now considered conventional, equipment: voltage regulating transformer (VRT), shunt or parallel-connected switched inductor/capacitor, static VAr compensator (SVC), or static synchronous compensator (STATCOM), for *voltage regulation*, as shown in Figure 12.2; phase angle regulator (PAR) or phase shifting transformer, for *phase-angle regulation*, as shown in Figure 12.3; and thyristor-controlled series capacitor (TCSC), for *line reactance regulation*, as shown in Figure 12.4.

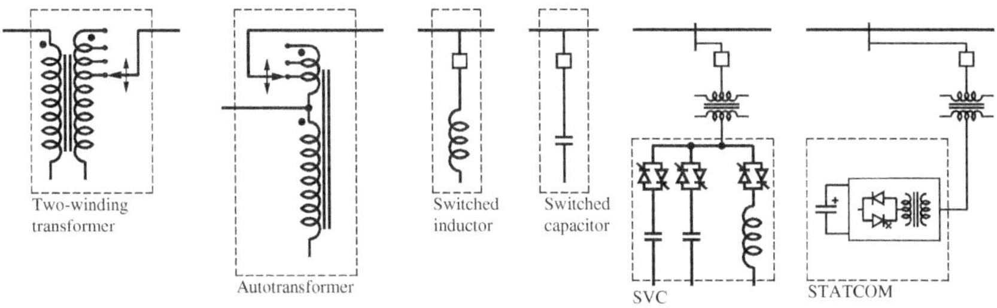

Figure 12.2 Transmission line voltage regulators.

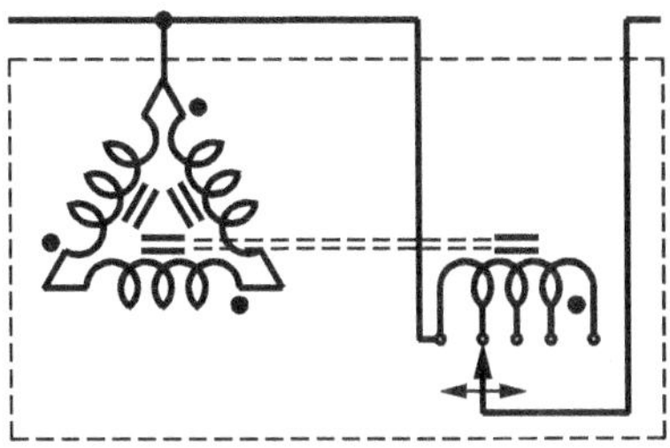

Figure 12.3 Transmission line voltage phase angle regulator.

In a transformer, the exciting voltage is applied to the primary windings and a compensating voltage is induced in the secondary windings. The secondary voltage is varied with the use of load tap changers (LTCs). An LTC can step up/down the voltage without interruption of the load current. In a two-winding transformer, which is referred to as a shunt-shunt configuration, the primary winding is connected in shunt (parallel) to the input terminal and the secondary winding is connected in shunt to the output terminal. Each of the primary and secondary windings is rated for the full line voltage and carries the full line current and, therefore, is rated for its full transmitted power. For example, in a two-winding transformer with input voltage of 1 V and output voltage of 1.1 V, delivering an output current of 1 A, the primary winding is rated at 1 V and 1.1 A; the secondary winding is rated at 1.1 V and 1 A. Therefore, the combined power rating of the two windings is 2.2 VA.

In an autotransformer, which is referred to as a shunt-series configuration, the primary winding is connected in shunt to the input terminal and the secondary winding is connected in series between the input and the output terminals. The primary winding that is rated for the full line voltage carries only a fraction of the line current; but, the secondary winding that is rated for a fraction of the line voltage carries the

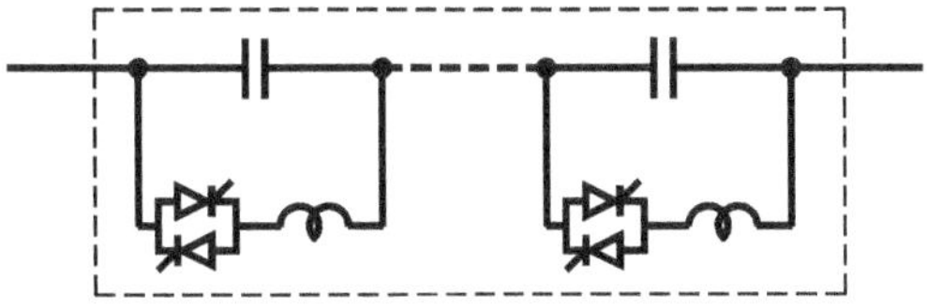

Figure 12.4 Thyristor-controlled series capacitor for transmission line reactance regulation.

full line current. Therefore, each winding is rated for a fractional amount of the full transmitted power. Considering the previous example, an autotransformer with input voltage of 1 V and output voltage of 1.1 V, delivering an output current of 1 A, would require a primary winding that is rated at 1 V and 0.1 A and a secondary winding that is rated at 0.1 V and 1 A. Therefore, the combined power rating of the two windings is 0.2 VA, resulting in a saving of 91% of the power rating over the two-winding transformer. Lower power rating is an important factor in the design of a power flow controller, since the equipment cost is directly related to its power rating.

The indirect way to regulate the line voltage is to connect an inductor or a capacitor in shunt with the transmission line. A shunt-connected inductor absorbs reactive power from the line and lowers the line voltage, whereas a shunt-connected capacitor delivers reactive power to the line and raises the line voltage. The SVC connects fixed capacitors in a step-like manner in shunt with the line through thyristor switches and also connects inductors in shunt with the line through thyristor switches whose duty cycle can be varied, thereby making it function as a variable inductor. The indirect way to regulate the shunt-connected inductor or capacitor is to use a voltage-sourced converter (VSC)-based STATCOM that connects an electronically generated sinusoidal voltage (with some harmonic components) in shunt with the transmission line through a tie inductor. The same concept has been practiced with the use of a synchronous condenser.

The phase angle of a line voltage can be regulated with the use of a PAR. The line voltage is applied to the primary windings and the induced secondary voltage that is varied with the use of LTCs is connected in series with the line. The series-connected compensating voltage is in quadrature with the line voltage. Through the use of a TCSC, a series-connected variable capacitor or a variable inductor can be implemented. As a result, the effective reactance of the transmission line is regulated.

Any traditional power flow controller regulates one of the power flow control parameters and, in turn, affects both the active and reactive power flows in the line simultaneously as shown in Figure 12.5; however, it may be desirable to control both

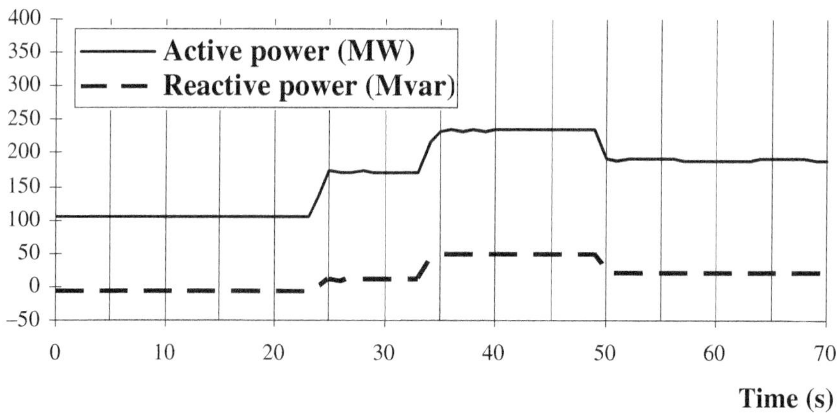

Figure 12.5 Simultaneous power flow control by changing only one control parameter. *Source*: Sen & Keri 2003 [3]. Reproduced with permission of IEEE.

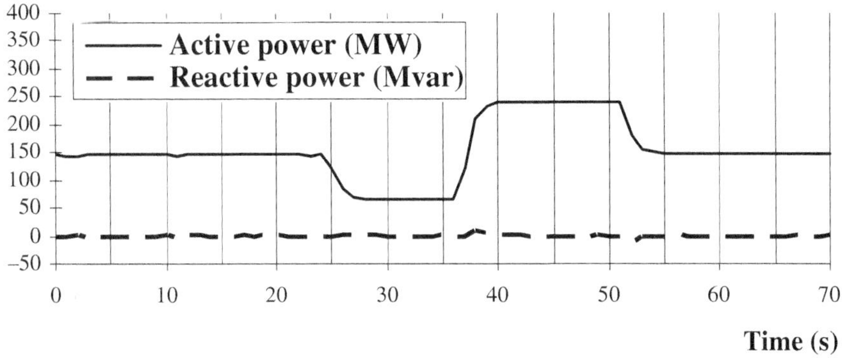

Figure 12.6 Independent power flow control by changing both control parameters; voltage magnitude and phase angle. *Source*: Sen & Keri 2003 [3]. Reproduced with permission of IEEE.

the active and reactive power flows independently as shown in Figure 12.6. As a result, the active and reactive power flows can be optimized so that the revenue-generating active power flow can be maximized. While maintaining the voltage stability, if the reactive power flow is minimized, it results in lower losses in the line and higher overall efficiency. While maintaining unity power factor load, the active power flow in the line is varied at different levels, such as 145, 65, 240, and 145 MW levels, respectively. Note that the response time of the UPFC is slowed down to be in several seconds to meet the utility's needs, even though the power electronics-based UPFC is perfectly capable to provide a faster response in ms.

12.1.2 Essential Control Parameters and Their Implementations

The power flow control parameters are voltage magnitudes, their phase angles, and line reactance. Any of these parameters can be controlled with the use of a power flow controller as described in Section 12.1.1. The functionality of a series reactance compensator (TCSC or VSC-based *static synchronous series compensator*) can be achieved by modifying the magnitude and phase angle of the line voltage. Therefore, the power flow control parameters reduce to only two: the voltage magnitude and its phase angle. The control of either one of these parameters results in changes in active and reactive power flows simultaneously. For optimum operation of a transmission line, independent control of active and reactive power flows in the line is necessary. These require the transmission line voltage to be modified to a specific magnitude and phase angle. In other words, both parameters—voltage magnitude and its phase angle—need to be changed simultaneously in order to achieve an independent control of active and reactive power flows. Then, the next question becomes "*How do we change the voltage magnitude and its phase angle simultaneously*?" The answer is that there are two ways.

Consider that the point of compensation in the transmission line is at its sending end. Assuming that there are no changes in the transmission line's impedance and the voltage at the receiving end, a power flow controller can control the active and

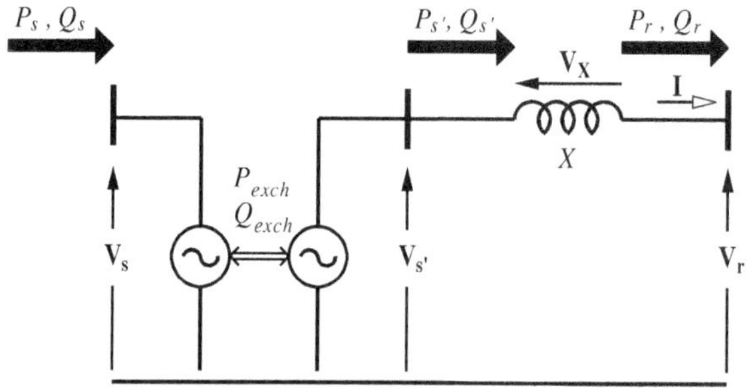

Figure 12.7 Shunt-shunt configuration for an independent power flow control in a transmission line.

reactive power flows (P_r and Q_r), at the receiving end, to be a particular pair of values by modifying the sending-end voltage to a specific magnitude and phase angle. In a shunt-shunt configuration, a shunt-connected voltage source ($\mathbf{V}_{s'}$) that is coupled with another shunt-connected voltage source ($\mathbf{V}_s$) supplies the line with a voltage of desired magnitude and angle as shown in Figure 12.7. In a shunt-series configuration, a series-connected voltage source ($\mathbf{V}_{s's}$) that is coupled with a shunt-connected voltage source ($\mathbf{V}_s$) modifies the line voltage to be $\mathbf{V}_{s'}$ of desired magnitude and angle as shown in Figure 12.8. Then, the next question becomes "*How many different ways can we implement shunt-shunt and shunt-series configurations*?" The answer is that there are three known ways to implement these configurations—by using (a) power electronic inverter, also known as VSC, (b) electrical machine, and (c) transformer/tap changers.

Various features of all shunt-shunt and shunt-series configurations are summarized in Table 12.1.

12.2 THE SEN TRANSFORMER CONCEPT

For many decades, the regulations of voltage magnitude and its phase angle have been accomplished using the VRT and PAR, respectively. Each of the VRT and PAR regulates one of the power flow control parameters–voltage and its phase angle—and, in

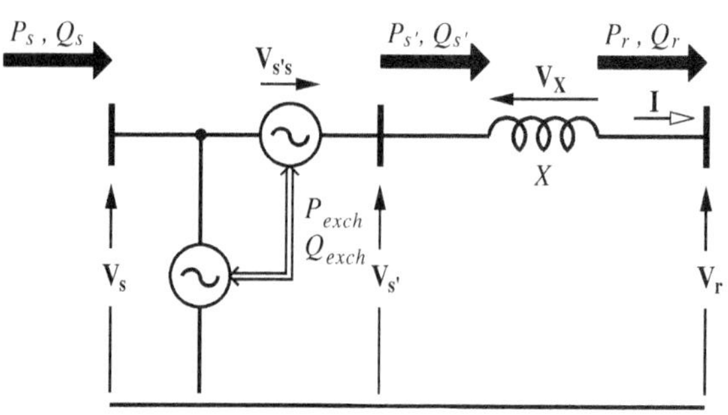

Figure 12.8 Shunt-series configuration for an independent power flow control in a transmission line.

TABLE 12.1 Various features of all shunt-shunt and shunt-series configurations

Features	Power electronics	Electric machine	Transformer/ tap changers
Shunt-shunt configuration			
Independent active and reactive power flow control	Yes	Yes	Yes
Different frequency system	Yes	Yes	No
Different phase system	Yes	Yes	Yes
Intermediate DC transmission	Yes	No	No
Shunt-series configuration			
Independent active and reactive power flow control	Yes	Yes	Yes
One frequency system	Yes	Yes	Yes

turn, controls both active and reactive power simultaneously. An independent power flow controller combines the operations of the VRT and PAR in one unit, which is the basis of ST that is truly a power regulating transformer. The ST uses the time-tested components of transformer and tap changers. If desired, the mechanical taps can be upgraded with suitable electronic switches to achieve a faster dynamic response.

12.2.1 Shunt-Series Configuration for ST

The shunt-shunt and shunt-series configurations for an independent power flow controller, shown in Figures 12.7 and 12.8, can be thought of as a two-winding transformer and an autotransformer, respectively, with one difference. In a transformer, the output voltage ($\mathbf{V}_{s'}$) is different in magnitude than the input voltage ($\mathbf{V}_s$), but their phases are the same; whereas in an independent power flow controller, the output voltage ($\mathbf{V}_{s'}$) is different both in magnitude and phase angle with respect to the input voltage ($\mathbf{V}_s$).

The expressions for active and reactive power flows in equations (12.1) and (12.2) are sine and cosine functions, respectively. The variation of active power as a function of reactive power is shown in Figure 12.9, which is the characteristic of a shunt-shunt configuration. The P_r–Q_r characteristic is not shown in the second and third quadrants, since the power angle (δ) is not used in the range of 90–270°. If the fourth-quadrant operation of a shunt-shunt configuration is needed for reversal of active power flow, note that the shunt-series configuration can also be used for this purpose [10].

The shunt-shunt configuration is capable of controlling the power angle (the phase angle between the voltage at the point of compensation and the voltage at the far end of the transmission line) over its full 360° range. The maximum transfer of active power along a lossless transmission line (with quality factor $Q = X/R = \infty$) between the sending and receiving ends takes place at the 90° power angle. In practice, the actual power angle is significantly lower and it depends on the line length, system characteristics, and load flows. As an example, a compensation range (ψ_{max}) of $\pm$ 11.54° with an uncompensated power angle of 30° is shown in the shaded area, which

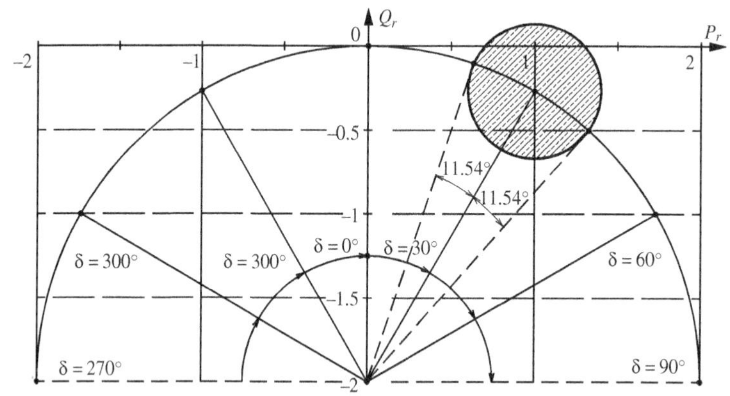

Figure 12.9 Q_r versus P_r at the receiving end of the transmission line for the range of power angle δ (i.e., δ_s–δ_r) from 270° to 90° when $V_s = V_r = 1$, and $X = 0.5$, and $R = 0$ ($X/R = \infty$).

is the characteristic of a shunt-series configuration. In this application, the shunt-shunt configuration is severely restricted to operate within the first quadrant when used as a power flow controller.

In some special cases for point-to-point transfer of power between two isolated networks or interconnection of two power lines with different voltages or phase angles or frequencies, the transmission line voltage can be controlled to a specific magnitude and phase angle by connecting a compensating voltage in shunt with the line using a shunt-shunt configuration.

Regardless of whether the compensating voltage is series- or shunt-connected, it is at any phase angle with the current flowing through it and, therefore, it exchanges active and reactive powers at the point of compensation with the transmission line. It is important to note that in the shunt-series configuration, the series-connected compensating voltage carries the full line current and its terminal voltage rating is a fractional amount of the line voltage. Therefore, the rating of the shunt-series power hardware is only a fraction of the transmitted power of the line. In the shunt-shunt configuration, the shunt-connected compensating voltage also carries the full line current, but its terminal voltage rating is the same as the full line voltage. Therefore, the rating of the shunt-shunt power hardware is the same as the transmitted power of the line. For the same amount of controllable transmitted power, the shunt-series configuration requires lower power-rated hardware than what is required in a shunt-shunt configuration. Therefore, ST uses a shunt-series configuration.

12.2.2 Principle of Operation of ST

Consider the simple power transmission system with a sending-end voltage, $\mathbf{V_s} = V_s\underline{|\delta_s}$, a receiving-end voltage, $\mathbf{V_r} = V_r\underline{|\delta_r}$, the voltage ($\mathbf{V_X}$) across the line reactance (X), and a compensating voltage ($\mathbf{V_{s's}}$) as shown in Figure 12.10a.

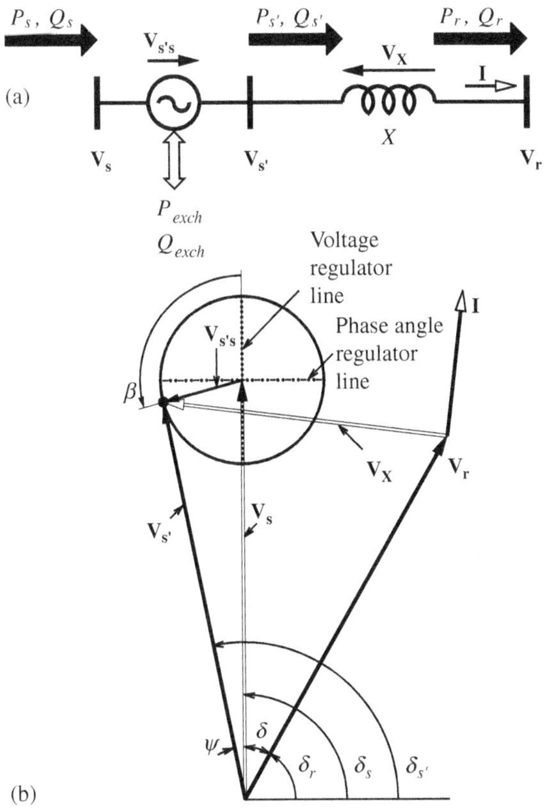

Figure 12.10 The effect of using an ST: (a) power transmission system with a series-connected compensating voltage, $\mathbf{V}_{s's}$ and (b) phasor diagram.

When the compensating voltage, $\mathbf{V}_{s's}$ (magnitude, $V_{s's}$ and phase angle, $\delta_s + \beta$), is added in series with the transmission line at the sending-end voltage ($\mathbf{V}_s$), the modified sending-end voltage becomes $V_{s'}(= \mathrm{V}_s+\mathrm{V}_{s's})$. The magnitude ($V_s$) and phase angle ($\delta_s$) of the transmission line voltage ($\mathbf{V}_s$) are modified simultaneously to be of one particular magnitude ($V_{s'}$) and at a specific phase angle ($\delta_{s'}$). Figure 12.10b shows the phasor diagram related to a series-connected compensating voltage ($\mathbf{V}_{s's}$) with a fixed magnitude of 0.2 p.u. and its entire controllable range of $0° \leq \beta \leq 360°$. The difference voltage ($\mathbf{V}_{s'} - \mathbf{V}_r$) provides the voltage ($\mathbf{V}_X$) across the line reactance (X). As the relative phase angle (β) is varied over its full 360° range, the end of phasor ($\mathbf{V}_{s's}$) moves along a circle with its center located at the end of phasor ($\mathbf{V}_s$). The rotation of phasor ($\mathbf{V}_{s's}$) with an angle (β) modulates both the magnitude and phase angle of the phasor ($\mathbf{V}_X$). In a special case, when $\beta = 0°$ or 180°, the ST controls the magnitude of the line voltage, as in the case of the VRT. In another special case, when $\beta = \pm 90°$, the ST controls mainly the phase angle of the line voltage, as in the case of the PAR.

Therefore, in summary, a series-connected compensating voltage ($\mathbf{V}_{s's}$) modifies:

- voltage magnitude with an autotransformer;
- phase angle with a PAR;
- but, both magnitude and phase angle of the line voltage with an ST.

12.2.3 Reduction in Number of Taps

A series-connected compensating voltage ($\mathbf{V}_{s's}$), shown in Figure 12.11a, modifies the magnitude and phase angle of the transmission line voltage independently. The related phasor diagram is shown in Figure 12.11b. For the desired amount of active and reactive power flows in the line, the compensating voltage has to be a specific magnitude ($V_{s's}$) and phase angle (β) with respect to the line voltage. The desired

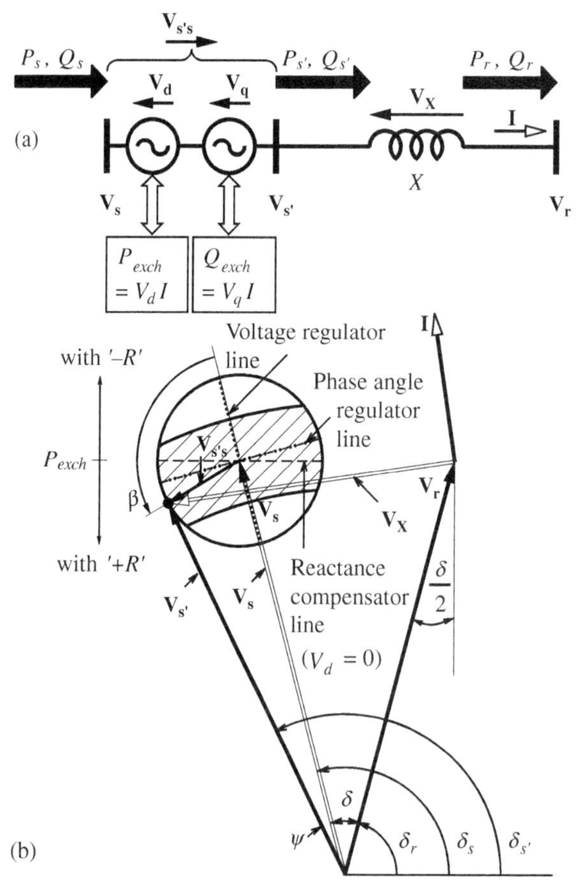

Figure 12.11 The effect of using an ST: (a) power transmission system with a series-connected compensating voltage ($\mathbf{V}_{s's}$) and (b) phasor diagram.

series-connected compensating voltage can be composed of two orthogonal voltages: one regulates the magnitude of the line voltage as in a VRT and another regulates the phase angle of the line voltage as in a PAR. Note that the ST combines the functions of the VRT and PAR in a single unit that results in a reduced amount of hardware from what is required for VRT and PAR individually. The phase-angle-regulating component can be varied through its entire range, since it does not contribute significantly to increase the magnitude of the line voltage. However, the voltage regulating component cannot be varied through its entire range, since the magnitude of the line voltage has to be bounded by the regulatory limit. Therefore, the number of taps used in an ST can also be limited to what is really needed and can be used.

In the transmission system considered above, the magnitude ($V_{s'}$) and phase angle ($\delta_{s'}$) of the modified sending-end voltage ($\mathbf{V_{s'}}$) are modulated between (V_s + $V_{s's}$) and ($V_s - V_{s's}$) and between ($\delta_s + \psi_{max}$) and ($\delta_s - \psi_{max}$), respectively, where $V_{s's} = 0.2$ p.u. and $\psi_{max} = 11.54°$, as shown in Figure 12.12a. Note that the line voltage is allowed to be controlled within a narrow band, for example, 10%, of the nominal value.

The compensating voltage ($\mathbf{V_{s's}}$) is at any phase angle with the prevailing line current ($\mathbf{I}$) and, therefore, exchanges with the line both active power ($P_{exch} = V_d I$) and reactive power ($Q_{exch} = V_q I$), where $\mathbf{V_d}$ and $\mathbf{V_q}$ are the respective active or direct and reactive or quadrature components of the compensating voltage, decomposed with respect to the line current, with load convention. These exchanged active and reactive powers (P_{exch} and Q_{exch}) are also sinusoidal functions of the relative phase angle

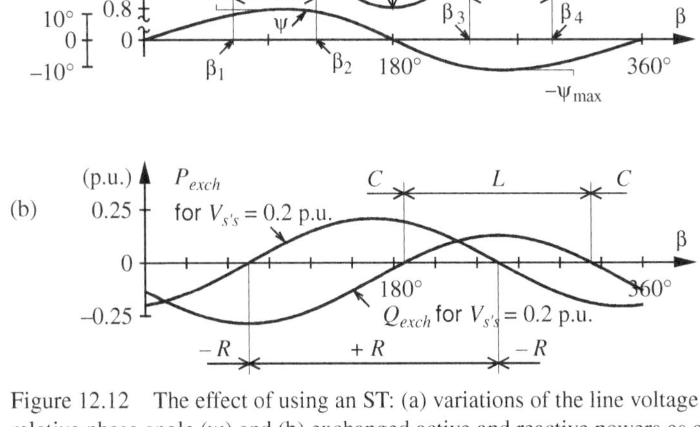

Figure 12.12 The effect of using an ST: (a) variations of the line voltage magnitude and relative phase angle (ψ) and (b) exchanged active and reactive powers as a function of relative phase angle (β).

(β), as shown in Figure 12.12b. For a given magnitude of a compensating voltage, the exchanged capacitive power (Q_{exch}) is larger than its inductive counterpart due to the fact that the capacitive compensation produces a larger line current. The $\mathbf{V_d}$ component of the compensating voltage provides the series resistance emulation, which can be a positive resistor ($+R$) that absorbs active power from the line or a negative resistor ($-R$) that delivers active power to the line. The reactive or quadrature component of the compensating voltage provides the series reactance emulation, which can be a capacitor (C) that delivers reactive power to the line and, in the process, increases the power flow of the line or an inductor (L) that absorbs reactive power from the line and decreases the power flow of the line. Therefore, the desired compensating voltage is actually an impedance emulator. The effect of impedance emulation is such that the active and reactive power flows in a transmission line can be regulated independently.

The active and reactive power flows (P_r and Q_r) at the receiving end vary with the relative phase angle (β) in a sinusoidal manner as shown in Figure 12.13a. The relationship between Q_r and P_r is a circle centered at (P_{rn}, Q_{rn}) with a radius of A_r, where P_{rn} and Q_{rn} are the natural active and reactive power flows at the receiving end of the transmission line without any compensation and A_r is the maximum power flow enhancement. The active and reactive power flows at the receiving end of

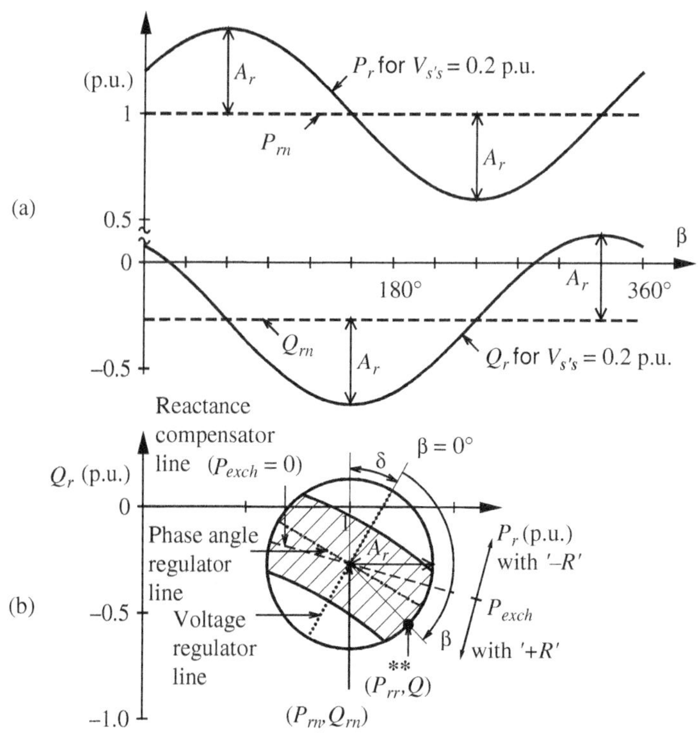

Figure 12.13 The effect of using an ST: (a) variation of the receiving-end active and reactive power flows (P_r and Q_r) as a function of the relative phase angle (β) and (b) receiving-end Q_r versus P_r.

the line can be controlled within the range defined by the *P*–*Q* plot in Figure 12.13b by choosing the magnitude ($V_{s's}$) of the compensating voltage ($\mathbf{V_{s's}}$) within its rated value and the relative phase angle (β) between 0° and 360°. If the magnitude ($V_{s's}$) of the compensating voltage is increased, the controllable range of the power flows in the *P*–*Q* plane is also increased. When the controllable range extends to all four quadrants, bidirectional and independent control of active and reactive power flows are achieved. For the desired amount of active and reactive power flows (P_r^*, Q_r^*), within the allowable control range, the magnitude ($V_{s's}$) of the compensating voltage is related to the distance, $\sqrt{(P_r^* - P_{rn})^2 + (Q_r^* - Q_{rn})^2}$, between ($P_{rn}$, Q_{rn}) and (P_r^*, Q_r^*). The relative phase angle (β) of the compensating voltage ($\mathbf{V_{s's}}$) is the phase angle between the voltage regulator line (i.e., $\beta = 0°$ line) and the line that passes through (P_{rn}, Q_{rn}) and (P_r^*, Q_r^*).

When a series-connected compensating voltage is used to modify the transmission line voltage, the following occurs.

- The compensating voltage can be either in phase or out of phase with the phase-to-neutral voltage of the transmission line to implement the function of the VRT.
- The compensating voltage can be in quadrature with the phase-to-neutral voltage of the transmission line to implement the function of the PAR.
- The compensating voltage can be such that it provides series reactance compensation because it is in quadrature with the prevailing line current. If the circular controllable area is equally divided by the reactance compensator line ($V_d = 0$ or $P_{exch} = 0$), the upper and lower halves represent P_{exch} due to $-R$ and $+R$, respectively.

Any compensator that provides compensation for one power flow control parameter (voltage magnitude, its phase angle, or line reactance) operates on a set of linear operating points inside the *P*–*Q* circle as defined above. Since these operating characteristic lines are neither horizontal nor vertical in the *P*–*Q* plane at the receiving end, controlling one parameter with the use of these compensators affects both the active and reactive power flows in the transmission line simultaneously, as shown in Figure 12.13b. Note the limited power flow control area due to the allowable narrow control band of the line voltage.

12.2.4 ST Circuit

Figure 12.14a shows the ST circuit, which is a three-phase transformer with Y-connected primary windings (A, B, and C), called the exciter unit, and nine secondary windings (a_1, a_2, and a_3 on the core of the A phase; b_1, b_2, and b_3 on the core of the B phase; and c_1, c_2, and c_3 on the core of the C phase), called the compensating voltage unit. This unit can be used to form a series-connected compensating voltage that is at any phase angle with respect to the transmission line voltage. By choosing the number of turns from each of the three windings and, therefore, the magnitudes of the components of the three 120° phase-shifted induced voltages, the compensating

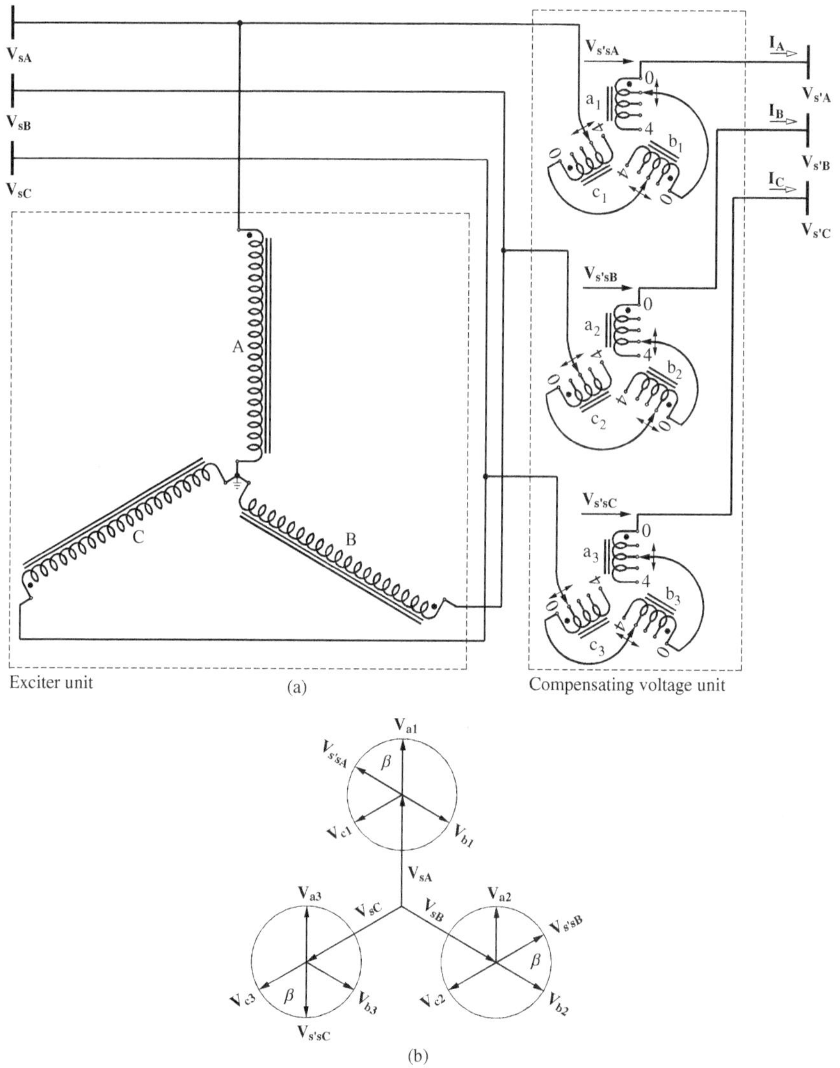

Figure 12.14 (a) ST for series-connected voltage compensation in the entire control range of 0° through 360° and (b) phasor diagram.

voltage ($\mathbf{V}_{s's}$) in any phase is derived from the phasor sum of the voltages induced in a three-phase winding set (a_1, b_1, and c_1 for compensation in the A phase; a_2, b_2, and c_2 for compensation in the B phase; and a_3, b_3, and c_3 for compensation in the C phase). A control algorithm is executed to operate the tap control unit, as discussed later. Figure 12.14b shows the related phasor diagram.

It should be noted that each of a_1, b_2, and c_3 is tapped at the same number of turns; each of b_1, c_2, and a_3 is tapped at the same number of turns; and each of c_1, a_2, and b_3 is tapped at the same number of turns. However, the number of turns in the a_1-b_2-c_3 set, b_1-c_2-a_3 set, and c_1-a_2-b_3 set can be different from each other with one exception—when the ST is used as a voltage regulator to decrease the modified sending-end voltage, there are same number of turns in two windings that are connected to each phase in the case when the relative phase angle (β) is 180°. The same power circuit can be used for both voltage regulation and impedance regulation (for independent control of active and reactive power flows) by proper programming of the tap control unit as discussed later.

12.2.5 Operating Points of ST

The LTCs, used in an ST, change their positions in steps; therefore, the compensating points of an ST are discrete in the allowable control range. As an example, if the allowable magnitude ($V_{s's}$) of the compensating voltage ($\mathbf{V}_{s's}$) is assumed to be 0.2 p.u. and the change of voltage in each step of an LTC is 0.05 p.u., then there are $N = 4$ steps associated with each secondary winding. The total number of possible compensating points (cp) in this example is defined by

$$cp = 3N(N + 1) \tag{12.3}$$

and it is 60. A comprehensive study of the power system network integrated with the power flow controller can only show how many steps is the right number of steps.

Figure 12.15 shows that the theoretical circular control area is actually a hexagon. Shorter steps in the LTCs make the compensating points closer to each other and vice versa.

The theoretically possible compensating points of the modified sending-end voltage with the use of the ST are shown in Figure 12.15a; the corresponding active and reactive power flows at the receiving end within the control range of 0° and 360° are shown in Figure 12.15b. Figures 12.15c and 12.15d show the practical control range of the ST. The filled dots, within the allowable voltage range in Figure 12.15c, show a total of 36 compensating points of the ST lie within the 10% range of the nominal voltage. Figure 12.15d shows that the maximum active power flow enhancement at the receiving end, within the allowable voltage range, is actually 0.35 p.u. instead of the theoretically possible 0.40 p.u., as shown in Figure 12.15b.

12.2.6 Control of ST

The series-connected compensating voltage ($\mathbf{V}_{s's}$) in the transmission line emulates a positive or negative resistance (R_d) and a capacitive or inductive reactance (X_q) in series with the line without any discrete circuit components, but just with a transformer and LTCs. Whether the control algorithm implements a resistance, reactance, or impedance, first the compensating voltage and its relative phase angle (β) with respect to the exciter voltage ($\mathbf{V}_s$) are calculated. Depending on the operating range of β (0° < β < 120°, 120° < β < 240°, or 240° < β < 360°), the two neighboring active compensating windings in each phase are selected. At β = 0°, 120°, and 240°, only one of the three secondary windings in each phase is active. Then, from the measured magnitude of the exciter voltage (V_s), the tap control unit determines the number of

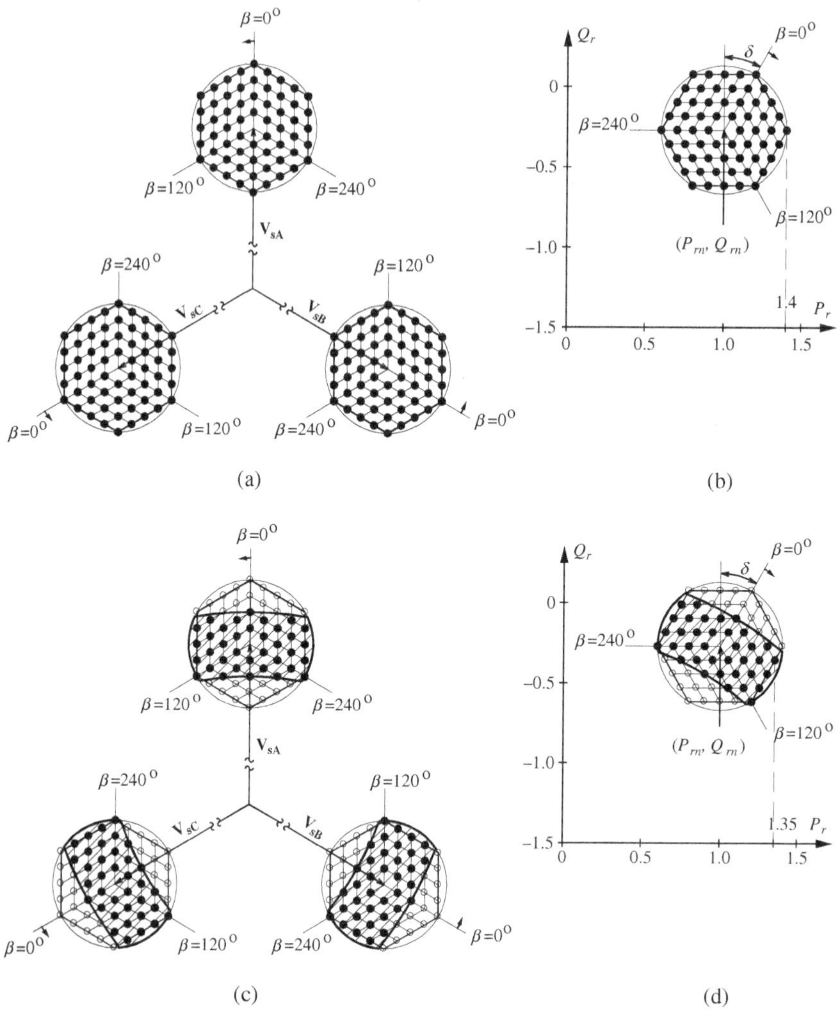

Figure 12.15 Compensating points with the use of the Sen transformer within the entire control range of 0° and 360°. Theoretically possible: (a) modified sending-end voltage and (b) active and reactive power flows at the receiving end. Practically possible: (c) modified sending-end voltage and (d) active and reactive power flows at the receiving end.

turns necessary on each winding of the compensating voltage unit. Based on this calculation, the appropriate taps are switched on to connect the required number of turns in series with the line. A special control algorithm must be implemented to avoid the jittering effect between the neighboring taps.

12.2.6.1 *Impedance Emulation*

Figure 12.16 shows the general control block diagram of the series impedance emulation using the ST. An instantaneous three-phase set of line voltages (v_s) is used to

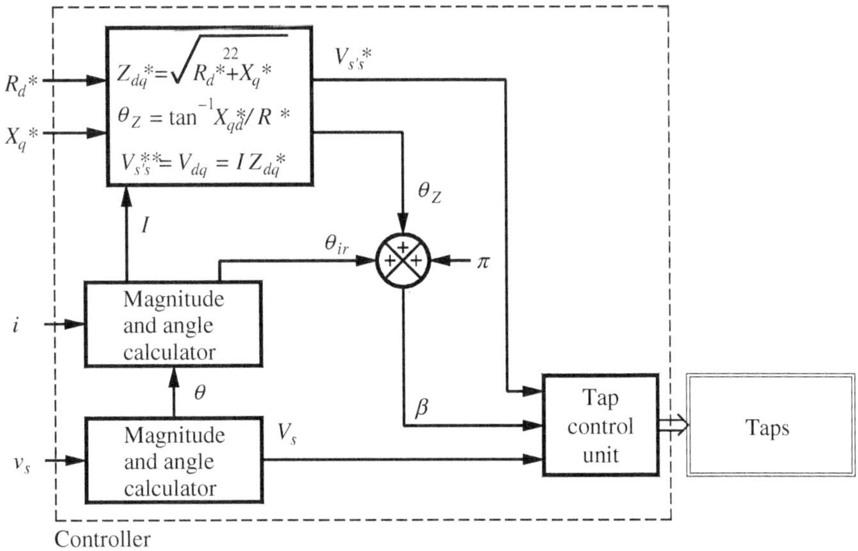

Figure 12.16 Series impedance emulation control block diagram of the ST.

calculate the phase-locked loop (PLL) angle (θ), which is phase locked to the phase a of the line voltage (v_{sa}). From an instantaneous three-phase set of measured line currents (i), the magnitude (I) and its relative phase angle (θ_{ir}) with respect to the PLL angle (θ) are calculated. From the compensating resistance demand ($R_d{}^*$) and the compensating reactance demand ($X_q{}^*$), the magnitude ($Z_{dq}{}^*$) and phase angle (θ_z) of the compensating impedance demand are calculated. The magnitude (I) of the line current multiplied by the magnitude of the compensating impedance ($Z_{dq}{}^*$) demand is the magnitude of the compensating voltage ($V_{s's}{}^*$) demand. The relative phase angle (β) of the compensating voltage demand is $\theta_{ir} + \theta_z + \pi$. Once the desired magnitude of the compensating voltage ($V_{s's}{}^*$) demand and its relative phase angle (β) are defined, the tap control unit figures out the contribution from each winding of a three-phase set (a_1, b_1, and c_1 for compensation in the A phase; a_2, b_2, and c_2 for compensation in the B phase; and a_3, b_3, and c_3 for compensation in the C phase) to constitute ($V_{s's}{}^*$).

12.2.6.2 Closed Loop Power Flow Control

The closed loop automatic power flow control block diagram is shown in Figure 12.17. The automatic power flow control algorithm works in such a way that the desired active and reactive power flows ($P_{s'}{}^*$ and $Q_{s'}{}^*$) are compared with the measured active and reactive power flows ($P_{s'}$ and $Q_{s'}$) and the errors are passed through an error amplifier, which produces, with respect to the exciter voltage (V_s), the required direct and quadrature components (e_d and e_q) of the series-connected compensating voltage. Next, the magnitude ($V_{s's}$) of the compensating voltage, $v_{s's}$ (i.e., $\sqrt{e_d^2 + e_q^2}$), and its relative phase angle (β) with respect to the phase angle of the exciter voltage are calculated. Note that for a particular active and reactive power

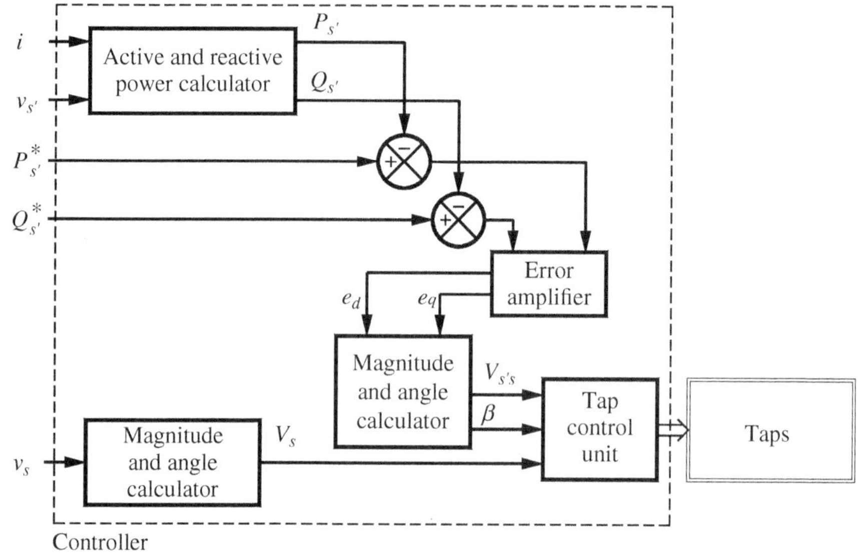

Figure 12.17 Closed loop automatic power flow control block diagram of the ST.

flows ($P_{s'}$* and $Q_{s'}$*) demand, if the magnitude ($V_{s's}$) of the compensating voltage exceeds the rated voltage of the compensating voltage unit, then $V_{s's}$ is limited to its design rating and the active and reactive power flows are regulated to an intermediate value.

12.2.6.3 Open Loop Power Flow Control

The controllable range of the active and reactive power flows can easily be determined with an open loop control by specifying to the tap control unit the rated compensating voltage ($V_{s's}$*) and varying the relative phase angle (β) within its entire 360° range, as shown in Figure 12.18.

Once the compensating voltage ($\mathbf{V}_{s's}$) and its relative phase angle (β) are determined, the tip of phasor ($\mathbf{V}_{s's}$) falls within four possible operating points, as shown in Figure 12.19.

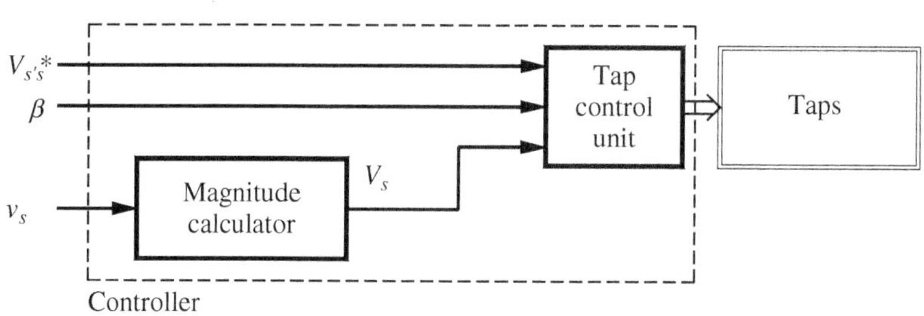

Figure 12.18 Open loop compensating voltage unit control block diagram of the ST.

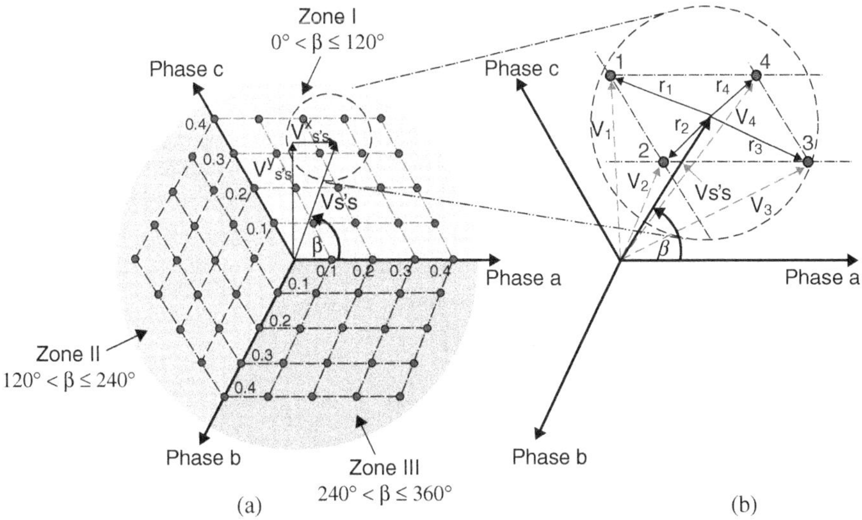

Figure 12.19 Selection of tap positions. Reprinted with permission from [6].

The selected operating point is based on producing the least error given by the following equation:

$$\varepsilon = \|r_k\|_2 = \sqrt{(V_k^x - V_{s's}^x)^2 + (V_k^y - V_{s's}^y)^2} \tag{12.4}$$

where ε is the least error and r_k is the error at a possible operating point for k = 1, 2, 3, and 4. $V_k{}^x$ and $V_{s's}{}^x$ are the projections of voltage of a possible operating point and the compensating voltage on the *a* phase (*x*-axis). $V_k{}^y$ and $V_{s's}{}^y$ are the projections of voltage of a possible operating point and the compensating voltage on the perpendicular to *a* phase (*y*-axis). A hysteresis loop must be implemented to eliminate the jittering effect among the four neighboring taps.

12.2.7 Simulation Results

A basic simulation of the ST was carried out and the results are shown in this section. Consider a power system network with parameters as shown in Table 12.2.

TABLE 12.2 Network parameters

Parameters	Values
Base values	160 MVA and 128 kV
Sending end line-to-line voltage	1∠0° p.u.
Receiving end line-to-line voltage	1∠–20° p.u.
Series impedance for sending-end source	6.25% = 1.0053 Ω and 19.73 mH
Series impedance for receiving-end source	0 Ω and 0 mH
Transmission line impedance	18.75% = 3.0159 Ω and 59.19 mH

The ST consists of three primary windings and nine secondary windings (three for series connection in each of the three phases). The primary and secondary windings are considered to have voltage ratings of 1.0 and 0.2 p.u., respectively. Considering an LTC contact located at every 5% mark, each secondary winding consists of four LTC taps. The leakage reactance of the ST was set to zero to eliminate its secondary effect. The natural active and reactive power flows (P_{rn} and Q_{rn}) at the receiving end are 209.12 MW (1.31 p.u.) and −68.54 MVAr (−0.43 p.u.), respectively.

The control was implemented in an open loop voltage injection mode, traveling from one operating point to the next in one particular hexagon in the entire control range of 0–360°. Figure 12.20 shows the active and reactive power flows (P_r and Q_r)

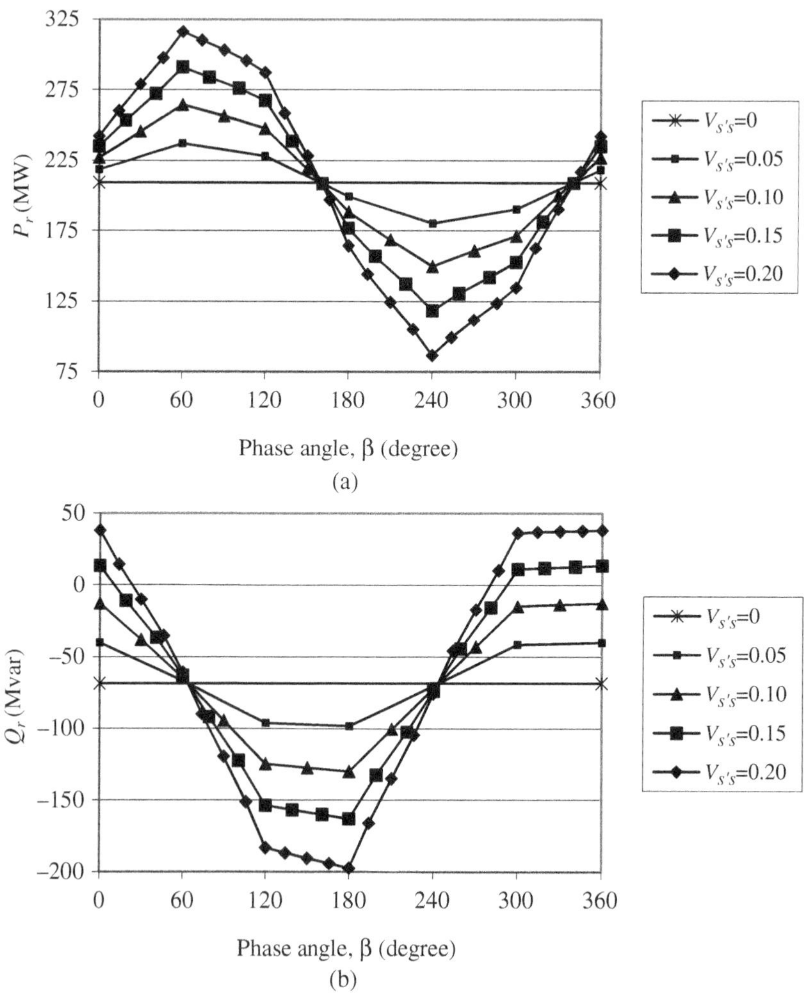

Figure 12.20 (a) Active power (P_r) and (b) reactive power (Q_r) flows at the receiving end during the entire control range of the relative phase angle (β) from 0° to 360°.

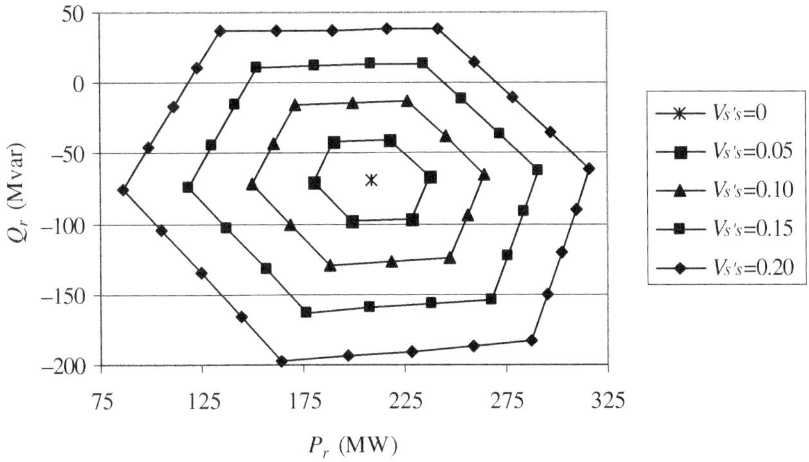

Figure 12.21 Active power (P_r) versus reactive power (Q_r) flows at the receiving end during the entire control range of the relative phase angle (β) from 0° to 360°.

at the receiving end during the entire stepped control range of the relative phase angle (β) from 0° to 360°.

Figure 12.21 shows the active power (P_r) versus reactive power (Q_r) flow at the receiving end during the entire control range of the relative phase angle (β) from 0° to 360°. The *P*–*Q* plot is somewhat different from an ideal hexagon due to the fact that the primary voltage ($\mathbf{V}_s$) keeps moving from its original (uncompensated) location because of the change of load in the line during compensation and the amount of Thèvenin impedance at the point of compensation and the shunt loading due to the exciter unit of the ST. A more ideal hexagon-like characteristic results if the point of compensation is closer to a stiff bus.

12.2.8 Limited Angle Operation of ST

The ST requires the use of a three-phase transformer with three primary windings and nine secondary windings (or three single-phase transformers, each with one primary winding and three secondary windings). The compensating voltage ($\mathbf{V}_{s's}$) is of variable magnitude and at a variable phase angle. However, in many instances, the capability of connecting a compensating voltage in series with a line within its entire range of 360° is not needed. The active power flow can be increased to the maximum possible level within the first 120° of the possible 0–360° range of the relative phase angle (β). The active power flow can be decreased to the minimum possible level within the next 120° of the possible 0–360° range of β. In this case, the circuit configuration for the full 360° range of operation can be simplified to the "limited angle" configuration.

At β = 0°, 120°, and 360°, only one of the three secondary windings in each phase is active. This active winding in each phase carries the full line current at the rated winding voltage that is a fraction of the line voltage. Therefore, during the 360°

range of operation, each of the nine secondary windings takes its turn to provide solely the full compensating voltage and carry the line current. Therefore, each of the nine secondary windings must be designed to carry the rated line current at rated winding voltage. This requires the magnetic rating of the 12 windings (three primary and nine secondary) of the ST to be 12/6 = 2 p.u., considering a transformer with six windings (three primary and three secondary) to be of 1 p.u. rating.

The ST, with 360° range of voltage compensation capability, uses only three primary windings and at the most six secondary windings at any given operating point. The remaining three secondary windings stay inactive. Therefore, it is possible to achieve a 360° operating range just by using six secondary windings instead of nine secondary windings with hardware reconfiguration in every 120° range of operation. In this way, the number of windings required in the ST is reduced at the expense of additional disconnect switches. Considering the three primary windings, the total number of windings required is 3+6 = 9 instead of 3+9 = 12. Therefore, the practical magnetic rating of the ST is only 9/6 = 1.5 p.u. for 120° range of operation, instead of 2 p.u. for 360° range of operation. Note that by using six secondary windings, one of the six operating regions (0–120°, 120–240°, 240–360°, 300–60°, 60–180°, and 180–300°) can be selected. The ST for a limited angle operation between 0° and 120° is shown in Figure 12.22.

Figure 12.22a shows that the line voltage is applied to a shunt-connected, three-phase transformer's primary windings in the exciter unit. A total of six secondary windings (a_1 and a_2 on the core of the A phase; b_2 and b_3 on the core of the B phase; and c_3 and c_1 on the core of the C phase) constitute the limited angle compensating voltage unit. By choosing the number of turns from each of the two windings, the compensating voltage in any phase is derived from the phasor sum of the voltages induced in a two-phase winding set (a_1 and c_1 for connection in the A phase; a_2 and b_2 for connection in the B phase; and b_3 and c_3 for connection in the C phase). The relative phase angle (β) of the series-connected compensating voltage ($\mathbf{V}_{s's}$) with respect to the line voltage ($\mathbf{V}_s$) can vary between 0° and 120°. Figure 12.22b shows the related phasor diagram.

For $N = 4$ in each secondary winding, the total number of possible compensating points (cp) in the 120° control range of operation is given by

$$cp = N(N + 2) \tag{12.5}$$

and it is 24. The filled dots, within the allowable voltage range between $\beta = 0°$ and $\beta = 120°$, in Figure 12.23a, show a total of 16 compensating points in the ST. Figure 12.23b shows the corresponding P–Q diagram for the power flow at the receiving end.

12.2.9 ST Configuration for Very High Voltage Applications

In theory, there is no restriction as to the level of line voltage where the ST configuration can be used for an independent control of active and reactive power flows in the line. However, there must be some practical limitations, such as the availability of high voltage and high current taps. In this case, the taps must be isolated from the high voltage transmission line as shown in Figure 12.24a. In such

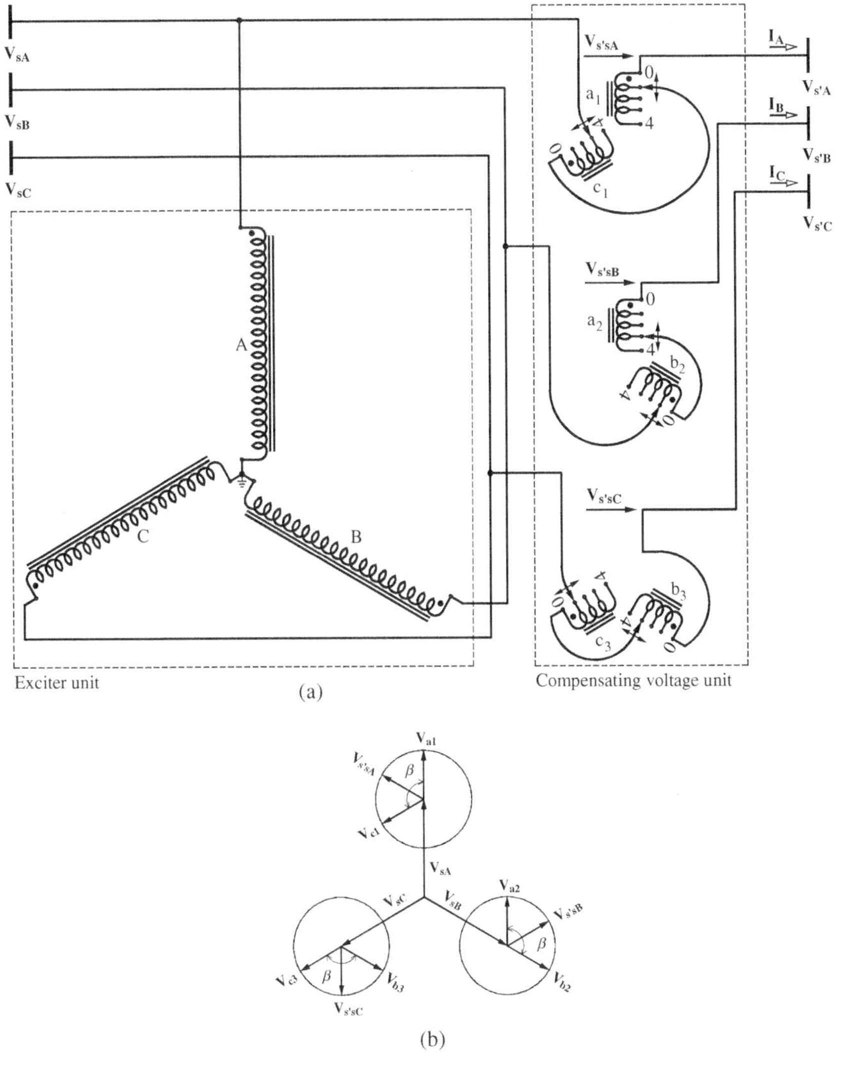

Figure 12.22 (a) ST for voltage compensation in the range of 0° through 120° and (b) phasor diagram.

an application, the ST configuration consists of two cores. The windings from the exciter unit and the compensating voltage unit are placed on the first core. The windings from the series unit are placed on the second core. Three secondary voltages from the compensating voltage unit, each from a three-winding set, are used to excite three single-phase step-down transformers (or one three-phase transformer with three isolated secondary windings) of the series unit. The secondary voltages out of the series unit are the actual compensating voltages that are connected in series with the line, as shown in the figure. Figure 12.24b shows the related phasor diagram.

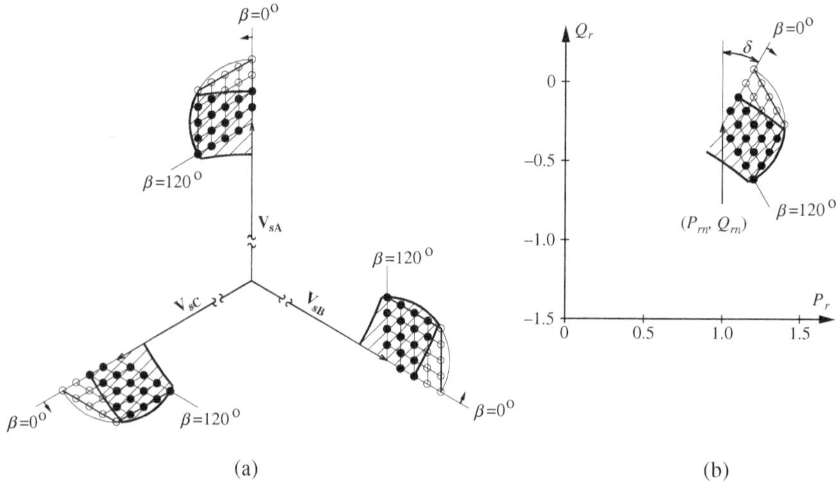

Figure 12.23 (a) Modified sending-end voltage points and (b) corresponding receiving-end active and reactive power points for operation within the control range of 0° and 120°.

The series-connected compensating voltage is only a few percent of the line voltage. Note that in a one-core configuration of the ST, the series winding and the LTCs in a compensating voltage unit carry high line current and even higher fault current. In a two-core configuration, the LTCs normally operate at a low current and ride through a high fault current. In such cases, three secondary voltages of the compensating voltage unit, each from a three-winding set, need to be high voltage and low current rated.

12.2.10 ST Configuration for Voltage Matching and Power Flow Control

The basic scheme of the ST can be modified for interconnection of transmission lines of different voltages while providing independent control of active and reactive power flows. Figure 12.25a shows one such scheme in which two transmission lines, one at 345 kV and the other at 138 kV, can be interfaced with the ST when the compensating voltage unit is connected at the 138 kV line side. The primary windings of the ST are energized with the voltage from the 345 kV line. The magnitude and phase angle of the stepped-down 138 kV voltages are controlled with the use of the compensating voltage unit. Figure 12.25b shows the related phasor diagram. This scheme is particularly attractive when the front-end autotransformer is needed anyway, irrespective of the power flow control requirement. The additional compensating voltage unit can be procured at a marginal cost.

The above scheme can be extended to interconnect the transmission lines of different voltage levels while providing independent control of active and reactive power flows in the higher voltage line. In that case, the primary windings of the ST

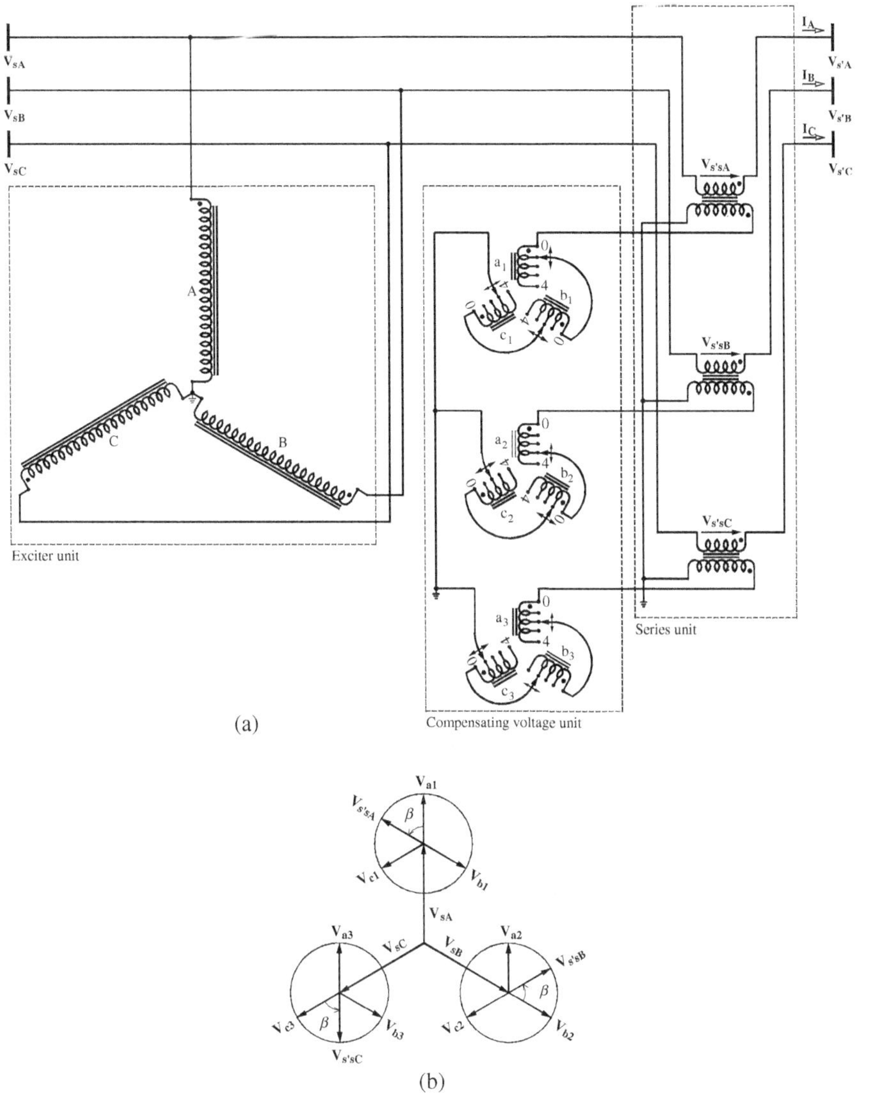

Figure 12.24 (a) ST configuration using taps with lower voltage and current ratings and (b) phasor diagram.

are energized from the lower voltage side. The magnitude and phase angle of the stepped up voltages are modified with the use of the compensating voltage unit.

12.2.11 Case-In Point for Using ST

For more than a century, the transmission line's voltage magnitude and phase angle have been regulated with VRT and PAR that use transformers and LTCs. The power flow controller in a transmission system application is expected to operate reliably

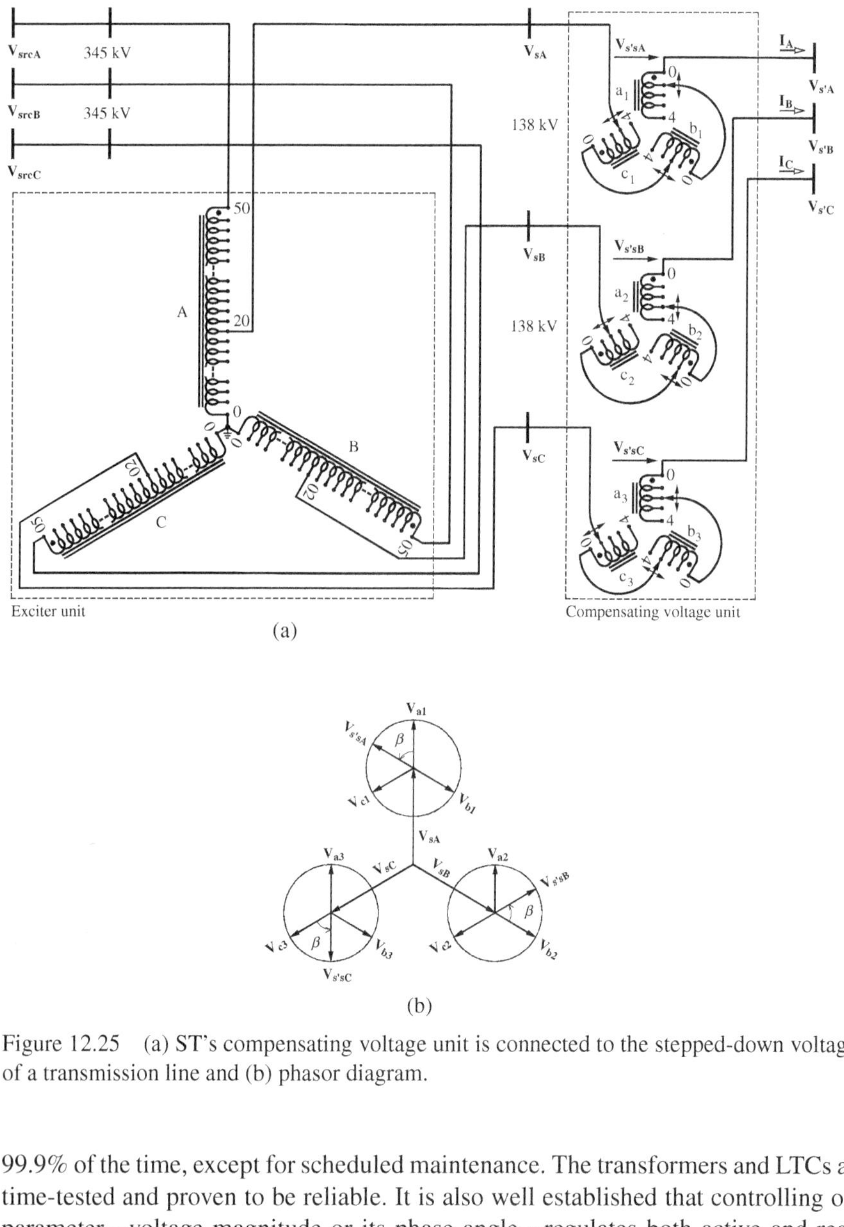

Figure 12.25 (a) ST's compensating voltage unit is connected to the stepped-down voltage of a transmission line and (b) phasor diagram.

99.9% of the time, except for scheduled maintenance. The transformers and LTCs are time-tested and proven to be reliable. It is also well established that controlling one parameter—voltage magnitude or its phase angle—regulates both active and reactive power flows simultaneously. However, the active and reactive power flows need to be optimized for the most efficient operation of the electric transmission lines. Therefore, an independent control of active and reactive power flows is required. The world's first UPFC at AEP showed that an independent control of active and reactive power flows can be achieved, as shown in Figure 12.6, with a series-connected compensating voltage that varies both the magnitude and the phase angle of the line voltage simultaneously. The experience gained from the installation of a total of three

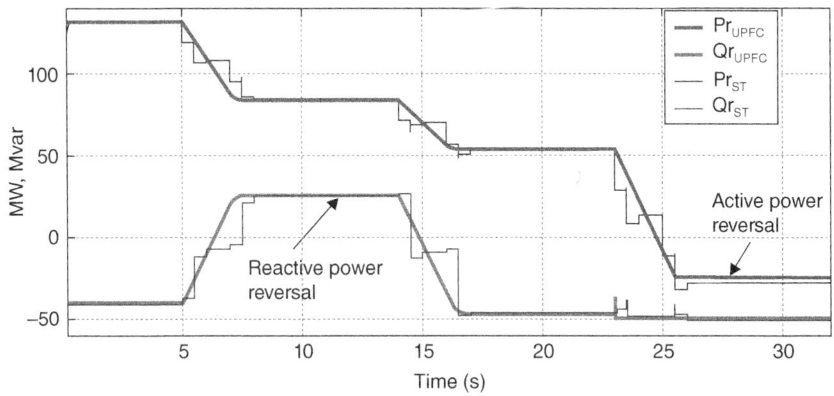

Figure 12.26 ST simulation results for open loop power flow control mode of operation. *Source*: Faruque & Dinavahi 2007 [6]. Reproduced with permission of IEEE.

UPFCs has taught us about the superior capabilities of the power electronics-based UPFC and also what the most utilities need for their everyday use, which is an independent power flow controller that provides a response time of several seconds as shown in Figure 12.6. The UPFC can offer fast responses in ms, but this capability cannot be utilized, as shown in Figure 12.6, in order to assure continued operation under contingencies.

The ST uses the reliable and proven transformer and LTC-based technology at a dynamic rate, which is limited by the speed of operation of the LTCs to create a series-connected compensating voltage that modifies the magnitude and phase angle of the line voltage simultaneously. Since the high speed of operation is not required in most utility applications, the ST becomes an attractive solution to provide an independent control of active and reactive power flows. Over the last one decade, the ST concept has emerged as an alternate low cost and highly efficient independent power flow controller.

The simulated transient responses of P_r and Q_r in a transmission line due to the use of ST and UPFC are superimposed and shown in Figure 12.26. The natural power flows until t = 5 s. A series-connected compensating voltage ($\mathbf{V}_{s's}$) of magnitude 0.2 p.u. is applied at a relative phase angle (β) of 300° at t = 5 s and 240° at t = 14 s. While keeping the relative phase angle at 240°, the magnitude of the compensating voltage is increased to 0.4 p.u. at t = 23 s. The line power due to the application of the UPFC is changed smoothly, whereas that due to the application of the ST is changed in multiple steps. Regardless of how the transitions take place, the steady-state power flow values are the same in both cases.

It should be noted that the series-connected compensating voltage in both the ST and the UPFC exchanges active and reactive powers with the line. In the case of the ST, these exchanged active and reactive powers flow back and forth through the shunt unit to the line. However, in the case of the UPFC, only the exchanged active power flows back and forth through the shunt unit to the line. Each of the series- and shunt-connected VSCs in the UPFC is capable of generating or absorbing reactive power. If this capability of supplying shunt reactive power is really needed, particularly when the short-circuit impedance at the point of compensation is high, a separate capacitor bank as a shunt reactive compensator may be used along with the ST. Ultimately, the life cycle cost of a power flow controller must be the deciding factor for its installation and operation.

12.2.12 Optimization of Transformer and LTC Ratings

The compensating voltage in series with a transmission line has a possible range of operation of 360°. If the rated compensating voltage ($\mathbf{V}_{s's}$) is added to the sending-end voltage ($\mathbf{V}_s$) at 0° or 180°, the modified sending-end voltage, $\mathbf{V}_{s'}$ (i.e., $\mathbf{V}_s + \mathbf{V}_{s's}$), goes out of the allowable range, which may be within the ±10% of the nominal value as shown in Figure 12.23a. Therefore, the winding and the LTCs outside the operating range need not be used as shown by the nonfilled dots in the figure.

Moreover, in the 360° range of operation, the maximum power flow enhancement occurs within the first 120° and, therefore, the rest of the operating region is practically not necessary. The ST, in its simpler, customized configuration with a reduced amount of hardware, provides a limited angle (120°) operation, which is adequate in most utility applications. Consequently, this requirement is satisfied by constructing the ST as shown in Figure 12.22a. This configuration requires only three primary windings and six secondary windings. This results in a 1.5 p.u. rated transformer. In comparison, the autotransformer and the PAR, each with 1 p.u. rating, can jointly perform in the 90° range of operation, requiring a total of 2 p.u. rating. Therefore, it can be concluded that the functions of the autotransformer and PAR are combined in a single unit of ST that results in a reduced amount of hardware from what is required for a separate autotransformer and a PAR.

12.2.13 Concluding Remarks

The ST generates a compensating voltage ($\mathbf{V}_{s's}$) of variable magnitude and variable phase angle through autotransformer action between the shunt-connected primary windings and the series-connected secondary windings. When connected in series with the transmission line, the compensating voltage modifies the magnitude and phase angle of the uncompensated sending-end voltage ($\mathbf{V}_s$) to be the modified sending-end voltage ($\mathbf{V}_{s'}$) and, therefore, independently controls the active and reactive power flows in the transmission line.

The ST offers an area of controllability in the *P*–*Q* plane just like the UPFC (see Figures 12.9 and 12.26). In comparison, it can be noted that the VRT and PAR, which are also made out of transformers and LTCs, offer only linear characteristics in the *P*–*Q* plane (see Figure 12.13) and, therefore, do not regulate the active and reactive power flows independently.

The transformers and LTCs used in the ST are time-tested and proven to be reliable. The dynamic performance of this technology is limited by the operational speed of the LTCs, which is adequate in most utility applications. The dynamic performance of the ST can be improved, when needed, by replacing the mechanical LTCs with faster electronic switches.

The basic scheme of the ST can be modified for special applications, such as limited angle operation (see Figures 12.22 and 12.23) that further reduces the required amount of hardware, power flow control in high voltage transmission lines (see Figure 12.24), and interconnection of transmission lines of different voltages or phase angles while providing independent control of active and reactive power flows (see Figure 12.25).

REFERENCES

[1] Sen, K. K. and Sen, M. L. Versatile power flow transformers for compensating power flow in a transmission line. U.S. Patents 6,335,613, 6,384,581, 6,396,248, and 6,420,856, 2002.

[2] Sen, K. K. and Sen, M. L. Introducing the family of Sen transformers: a set of power flow controlling transformers. *IEEE Transactions on Power Delivery*, vol. 18, no. 1, pp. 149–157, January 2003.

[3] Sen, K. K. and Keri, A. J. Comparison of field results and digital simulation results of voltage-sourced converter-based FACTS controllers. *IEEE Transactions on Power Delivery*, vol. 18, no. 1, pp. 300–306, January 2003.

[4] Sen, K. K. and Sen, M. L. Comparison of the Sen transformer with the unified power flow controller. *IEEE Transactions on Power Delivery*, vol. 18, no. 4, pp. 1523–1533, October 2003.

[5] Faruque, M. O. and Dinavahi, V., Santoso, S., and Adapa R. Review of electromagnetic transient models of non-VSC FACTS. *IEEE Transactions on Power Delivery*, vol. 20, no. 2, pp. 1065–1079, April 2005.

[6] Faruque, M. O. and Dinavahi, V. A tap-changing algorithm for the implementation of Sen transformer. *IEEE Transactions on Power Delivery*, vol. 22, no. 3, pp. 1750–1757, July 2007.

[7] Asghari, B., Faruque, M. O., and Dinavahi, V. Detailed real-time transient model of Sen transformer. *IEEE Transactions on Power Delivery*, vol. 23, no. 3, pp. 1513–1521, July 2008.

[8] Sen, K. K. and Sen, M. L. *Introduction to FACTS Controllers: Theory, Modeling, and Applications*. Chapter 9, IEEE Press and John Wiley & Sons, pp. 307–372, 2009.

[9] Fentie, D. Transient modeling and protection of the Sen Transformer, M.S. Thesis, University of Saskatchewan, 2010.

[10] Sen, M. L. and Sen, K. K. Introducing the SMART Power Flow Controller – an integral part of smart grid. Paper no. 103, *IEEE Electrical Power and Energy Conference*, October 2012.

[11] Fentie, D., Garcia, J. C., Gokaraju, R., and Faried, S. O. EMT Model of the 'Sen Transformer' for Fault Analysis Studies, *International Conference on Power Systems Transients (IPST 2015)*, Cavtat, Croatia, June 2015.

[12] Sen, K. K. and Sen, M. L. SMART Power Flow Controller for Smarter Grid Applications. Paper no. 15PESGM1910, *IEEE Power & Energy Society General Meeting*, Denver, USA, July 2015.

[13] Liu, J. and Dinavahi, V. Nonlinear magnetic equivalent circuit based real-time Sen transformer electromagnetic transient model on FPGA for HIL emulation, *IEEE Transactions on Power Delivery*, ISSN: 0885-8977, DOI: 10.1109/TPWRD.2016.2518676, January 2016.

[14] Liu, J. Real-Time Nonlinear Frequency-Dependent Electromagnetic Transient Power Transformer Model Hardware Emulation, Ph.D. Thesis, University of Alberta, 2016.

[15] Sen, K. K. and Sen, M. L. Comparison of Operational Characteristics between the Sen Transformer and Phase Angle Regulator. Paper no. 16PESGM0959, *IEEE Power & Energy Society General Meeting*, Boston, USA, July 2016.

[16] Sen, K. K. and Sen, M. L. Unique Capabilities of Sen Transformer: a Power Flow Regulating Transformer. Paper no. 16PESGM0961, *IEEE Power & Energy Society General Meeting*, Boston, USA, July 2016.

CHAPTER 13

MEDIUM VOLTAGE POWER ELECTRONICS DEVICES FOR DISTRIBUTION GRIDS

Ion Etxeberria-Otadui, David Frey, Seddik Bacha, and Bertrand Raison

13.1 INTRODUCTION

The use of power electronic devices in the power system is not a new phenomenon, as it started in the 1970s. Today, there are several kinds of power electronic devices connected to T&D grids operating all around the world. These devices are based on different technologies (semiconductors, topologies, etc.) and configurations (series, shunt, or universal) and they carry out a number of functionalities in the grid as shown in Table 13.1.

Obviously, not all the devices are equally spread in the grids: and while some of them are widely installed all around the world (e.g., static VAr compensation [SVC] and high voltage direct current [HVDC] devices), the introduction of other devices has been extremely limited (e.g., UPFC or UPQC devices).

The application of power electronic devices was initiated first in transmission-level applications (in the range of tens of MWs and hundreds of kVs), and has gained then gradually place in distribution-level applications where lower power and voltage ratings are required. This fact can be explained by the different nature and necessities of transmission and distribution grids (power flow and losses optimization in transmission, while power quality in distribution) and the lack of suitable power electronic technology.

The history of power converters in T&D applications goes back to 1954 when the first HVDC transmission system was built, using mercury arc valves [1]. Until the end of the 1960s, mercury arc valves were used for AC to DC or DC to AC conversion, frequency conversion, and many switching applications. Their use became relatively extensive mostly in electric transportation, rolling mills, and other industrial applications.

Nevertheless, since 1960, the mercury arc valve was increasingly replaced by a power electronic device: the thyristor. Started to emerge in the middle of the 1960s,

Advanced Solutions in Power Systems: HVDC, FACTS, and Artificial Intelligence, First Edition.
Edited by Mircea Eremia, Chen-Ching Liu, and Abdel-Aty Edris.

TABLE 13.1 Main T&D power electronic functionalities and devices

			Technology		
		Connection level	Forced commuted	Line-commuted	Functionality
CONFIGURATION	Shunt	Transmission	STATCOM[1]	SVC[2]	Reactive power compensation and voltage regulation
		Distribution	D-STATCOM[3]	SVC	
	Series	Transmission	SSSC[4]	TCSC[5]	Power flow optimization and oscillation damping
			HVDC[6]	HVDC	High voltage DC Transmission
		Distribution	DVR[7]	-	Protection of sensitive loads
	Universal	Transmission	UPFC[8]	-	Combination of series and shunt functionalities
		Distribution	UPQC[9]	-	

[1] STATCOM (Static Synchronous COMpensator), voltage source converter (VSC)-based shunt transmission compensator.
[2] SVC (Static VAr Compensator) thyristor-based shunt compensator.
[3] D-STATCOM (distribution static synchronous compensator), VSC-based shunt distribution compensator.
[4] SSSC (static synchronous series compensator), VSC-based series transmission compensator.
[5] TCSC (thyristor controlled series capacitor), thyristor-based series compensator.
[6] HVDC (high voltage DC transmission).
[7] DVR (dynamic voltage restorer) VSC-based series distribution compensator.
[8] UPFC (universal power flow compensator), VSC-based universal transmission compensator.
[9] UPQC (universal power quality conditioner), VSC-based universal distribution compensator.

it had proven to be very successful in industrial applications and within a decade, the thyristor replaced the mercury arc valve completely (including HVDC applications). It provided a significant improvement in terms of size and complexity of the converters, as well as a higher reliability, lower cost, and lower maintenance requirements. The thyristor indeed revolutionized the power conversion with increasingly higher power ratings becoming available. Indeed, the series and parallel connection of individual thyristors made it possible the development of converters of basically any power rating.

Also, as a result of the appearance of thyristors, a new AC SVC technology was developed in the 1970s—SVC [2]. SVC was initially developed to counteract the random voltage variations caused by electric arc furnaces in steel mills. The first SVC was delivered to a steel facility in 1972 with exceptionally good performances: voltage stabilization, loss reduction, and with a considerable mitigation of flicker effects. Since then, hundreds of SVC devices have been installed all over the world in many different applications such as rolling mills, electric arc furnaces, and utility applications. Thanks to the thyristor technology, HVDC and SVC have been widely accepted in power system applications as an effective way to control active and reactive power flow, respectively.

Nevertheless, the development of semiconductor technology led to the introduction of new components with turn off capability, which opened the way for forced commuted inverters and choppers in the following years. In fact, gate-controlled turn-off was introduced in the late 1970s with the gate turn-off (GTO) thyristor [3]. By making it possible to build efficient converters for controlling the output frequency, the GTO made it possible the development of new devices. On the one hand, variable-speed AC motor drives were introduced both for industrial and transportation (electrical traction) applications. On the other hand, GTOs were also introduced in the power systems field, for building static synchronous COMpensator (STATCOM) compensators, an improved SVC system based on an inverter topology. Even if the first transmission STATCOM (developed by KEPCO and Mitsubishi Motors in 1980) was based on forced-commutated thyristor inverters (20 MVAr), since 1986 most of the installed STATCOM compensators have employed GTO technology, following the path settled by EPRI and Westinghouse Electric Company with their 1 MVAr device [4].

However, a large snubber remained obligatory—an evolutionary link to the fast thyristor commutation circuits of the past, and to the fundamental need to limit the d*v*/d*t* across a thyristor structure. A major improvement in GTO technology occurred in 1994, with the development of the GCT, in which fast unity-gain turn-off (gate commutation of the anode current) allows snubberless operation by suppressing the thyristor structure in all operating modes but conduction [5]. This has resulted in a snubberless switch for high power applications, permitting simple inverter designs with lower losses than previous technologies (even if a current clamp is necessary in order to limit current derivatives, for example, in voltage source converter (VSC) topologies). In the last 10 years GCTs (and especially IGCTs) have gained an important market portion in the range of medium voltage applications, mainly for variable speed drive applications, due to their very high power switching capacity.

In parallel with thyristor-based components, transistor-based devices have also experienced an important evolution toward higher power and voltages since the 1980s. Insulated gate bipolar transistors (IGBTs) were introduced in the 1980s, and were first used in industrial inverters, replacing bipolar transistors. Their high impedance, low power gate input, and increasingly high power switching capability have made them gain significant importance in medium power applications since the end of the 1990s, also in the field of power transmission and distribution. At the end of the 1990s, VSC-based HVDC converters were developed using series-connected press-pack IGBTs [6]. This technology besides being used for several other DC transmission applications has also been employed in distribution voltage levels in order to develop static VAr compensators, also known as D-STATCOMs (commercially called SVC-light by ABB) [7].

13.2 HIGH POWER SWITCHING VALVES: ASSOCIATION OF SEMICONDUCTOR COMPONENTS

Power components available nowadays in the market are based on silicon and are limited to several thousands of volts (Figure 13.1). This is related to the characteristics

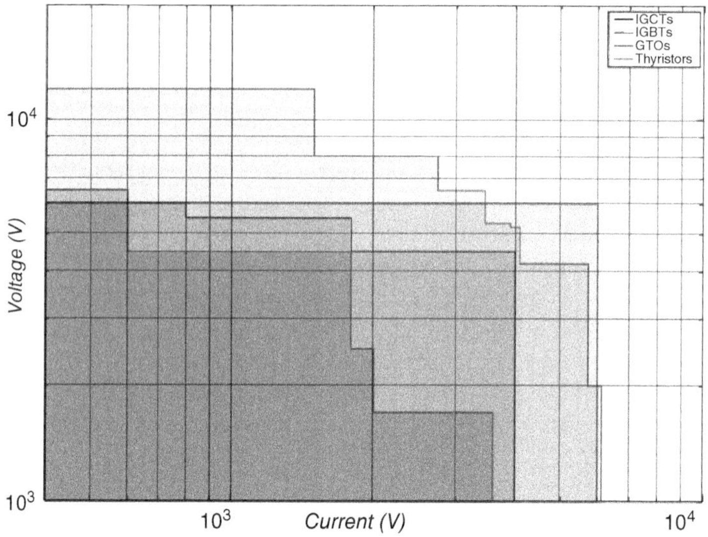

Figure 13.1 Voltage and current ratings of currently commercially available power semiconductors.

of the semiconductor material used (especially its breakdown field) and to the values of the doping concentrations which can be obtained. However, the elevation of the breakdown voltage leads to an elevation of the on-state voltage and thus—the losses elevation.

In order to try to overcome this limitation, new materials are being currently developed [8]. The most advanced seems to be the silicon carbide (SiC) for which commercial components of several hundred volts are already available [9]. They are essentially unipolar components such as Schottky Barrier diodes or JFET. Prototypes of SiC thyristors have breakdown voltages of tens of kilovolts. Other materials, such as diamond, are also studied but their elaboration is less advanced in the high power domain. Their breakdown electric fields being 10 times higher than that of the silicon, they should allow a better breakdown voltage, limiting this way the losses in the semiconductors. In addition, they offer another important advantage which is a maximum operating temperature much higher than in the case of silicon allowing a reduction of cooling necessities.

Figure 13.1 shows the voltage and current ratings of currently commercially available power semiconductors (IGCTs, IGBTs, GTOs, and thyristors). As it can be seen, in most of the cases, they are far from the required ratings in many medium voltage/power applications, and it is therefore necessary to build higher power switching valves by discrete semiconductor series and/or parallel association.

13.2.1 Series Association

The series connection of discrete semiconductors in a unique power switching valve makes it possible to achieve higher voltages, and consequently higher power switching capabilities. The critical point on this kind of valves is the correct voltage sharing between individual semiconductors, in order to avoid overvoltages due to unequal voltage distributions [10, 11]. Unequal voltage distribution may lead to different lifetimes of components (as they operate in different conditions) and even their destruction if a particular threshold is overpassed. This condition must be fulfilled both in static and dynamic (switchings) states. In the first state, voltage is almost constant across semiconductors while in the second state, there is a fast transient variation (several thousands of volts per microsecond).

There may be many causes that lead to an unequal voltage sharing: different characteristics of the semiconductors, of their driver circuits, of the parasitic elements (capacitances and inductances), etc. For example, in the case of transistor-based technologies, one main reason of unbalances is the dispersion between the parasitic capacities (especially the output C_{oss} and reverse C_{rss} capacities, which control the transistor commutation speed).

For the bipolar components, as for example, the IGBTs, the charges recombination at blocking state induces a tail current, which can lead to a significant unbalance. The value of the tail current depends not only on the design characteristics and the type of the IGBT, but also on the electrical parameters applied on it at the commutation (current and voltage between collector and emitter).

The other causes of unbalance are related to the peripheral elements of the component. The most important is the gate driver circuits. The propagation delays between the different control signals or the asymmetry of these circuits (output resistance or stray inductance) may induce important unbalances of the commutation speed and thus, of the final voltage. Another point is that in high voltage circuits, the transistor drivers should be insulated between them. This is necessary in order to ensure that the control signals and the gate driver circuits are insulated. However, insulation circuits provoke common mode capacities. These common mode capacitances can also appear due to a capacitive effect between the layout of the gate driver and the ground plate. During the commutations, because of the existence of dv/dt of several thousands volts per microsecond, very high currents may appear. These currents may interact with the transistor drivers and generate unbalances. In practice, these capacities will induce the deviation of a part of the power current and disturb the transistor commutation.

To reduce the influence of parasitic capacities, a good solution is to use a vertical structure, where components are located one over the other (Figure 13.2). With this solution, the capacitances that support the highest dv/dt are further from the ground plate and therefore are smaller, reducing this way their effect. The semiconductors in press-pack packages are very well suited for this type of geometry. In addition, they present better characteristics versus the occurrence of a short circuit of the components (they have a short circuit fault-mode).

In addition, there are several voltage balancing techniques that are used to improve the series connection of semiconductors. A possible classification of them is presented in Figure 13.3 [12, 13].

13.2.1.1 Load Side Solutions

These solutions act directly on the power side of the semiconductor in order to limit the voltage across the semiconductor, and guarantee a correct voltage sharing

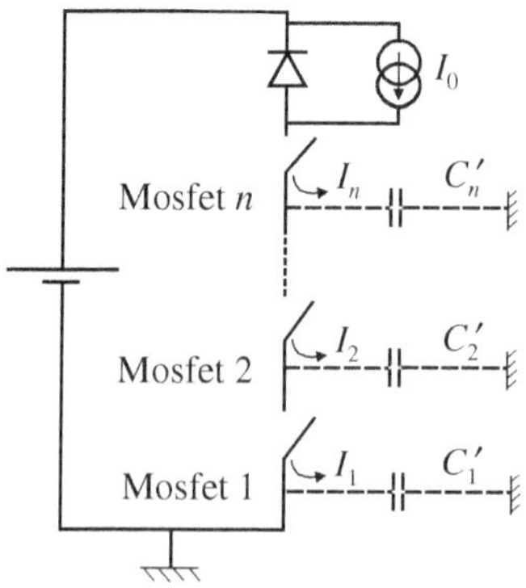

Figure 13.2 Localization of parasitic capacitances in an "n" transistors series association.

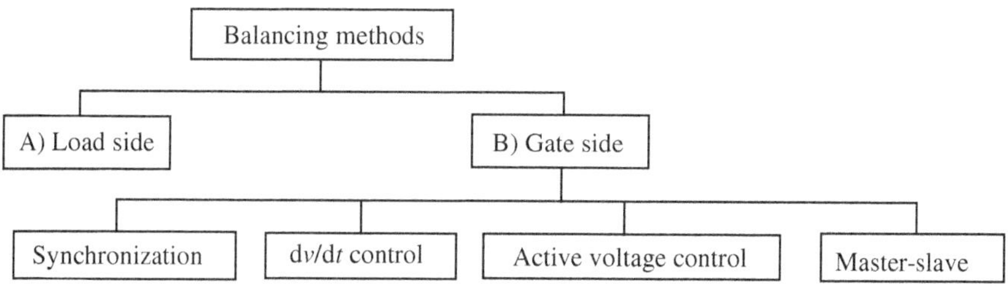

Figure 13.3 Classification of voltage balancing methods for semiconductor series connection.

between the different series-connected devices. They are divided in dynamic and static methods, depending on whether they act during switching transients or only during switched-off operation.

Static methods are generally based on parallel resistors, with a sufficiently high resistive value in order to limit losses during switched-off conditions. Concerning dynamic methods, there are four main techniques [14–17] (see Figure 13.4). In all the cases, an amount of energy is stored on a capacitor during the switch-off of the semiconductor and then dissipated on a resistor, displacing part of the switching losses from the semiconductor itself to the auxiliary balancing circuit.

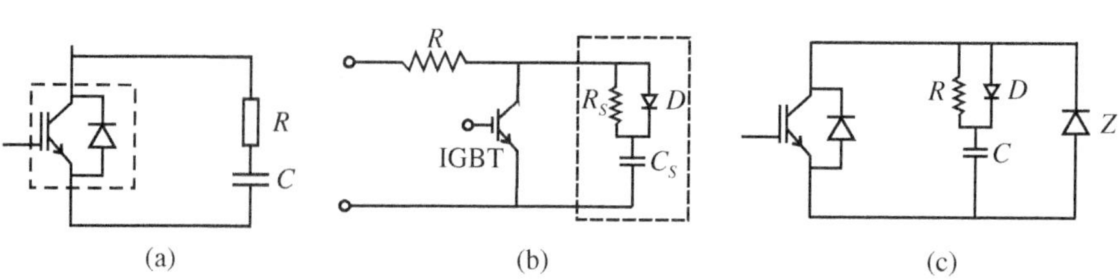

Figure 13.4 Load side voltage balancing solutions: (a) RC snubber, (b) RCD snubber, and (c) RCDZ snubber.

The main components in the first three snubber methods (RC, RCD, and RCDZ) are a capacitor and a resistor connected in parallel with the semiconductor. Their function is to limit the voltage commutation speed.

The simplest structure is the RC snubber. The aim of the resistance is to limit the capacitor discharge current (which will be added to the load current through the semiconductor at the switch-on), but it will also limit the capacitor charge dynamics. In the case of the RCD snubber [14], the capacitor charge and discharge dynamics are decoupled by using a diode to accelerate the charging process. In the case of the RCDZ snubber [16], in addition to limiting the voltage commutation speed, the maximum voltage across the semiconductor is also limited by using a parallel zener diode.

Figure 13.5 shows the voltage waveform evolution during the switching ON/OFF of two series-connected IGBTs with an RC and an RCD snubber. As it can be observed, the voltage derivative during the switching off is imposed in both cases by the capacitor.

13.2.1.2 Gate Side Solutions

These methods are characterized by the fact that they act on the semiconductor gate in order to change their conduction characteristics and to guarantee a correct sharing of the voltage between the different series-connected components. Therefore, they are only valid for transistor-based technology, in which the switching process of the semiconductor can be controlled via a gate signal. They are used in combination with previously mentioned static voltage sharing techniques (parallel resistors), and they permit a faster response and lower power losses. The main principles are:

- *Synchronization*: this method is based on the detection of switching speed deviations (i.e., voltage unbalances) and acting by delaying the fastest semiconductors. This can be achieved by using an FPGA [18] or a DSP unit [19–21], which compares the voltage across the different semiconductors and decides the delay to be introduced to the gate-signals of the devices or the gate impedance to use with each of them.
- *Voltage derivative control* [22,23]: in this case, the power side of the semiconductor is linked to the gate side by using a capacitor. The voltage derivative across the semiconductor acts as a feedback signal, which varies the gate current and therefore the commutation speed. Figure 13.6 presents some possible implementations of this technique.
- *Voltage magnitude control:* in this technique, there is also a feedback from the power side to the gate side. In order to clamp the voltage maximum value, zener diodes or transient voltage suppression diodes are used [24,25] (Figure 13.7).
- *Master-slave* [26]: in this solution, the different series-connected components play different roles. The master components will start the commutation, while slave ones will follow it.

Figure 13.8 shows the voltage waveform evolution during the switching ON/OFF of two series-connected IGBTs with a voltage derivative control method and a voltage magnitude control method. As it can be observed, in the case of voltage

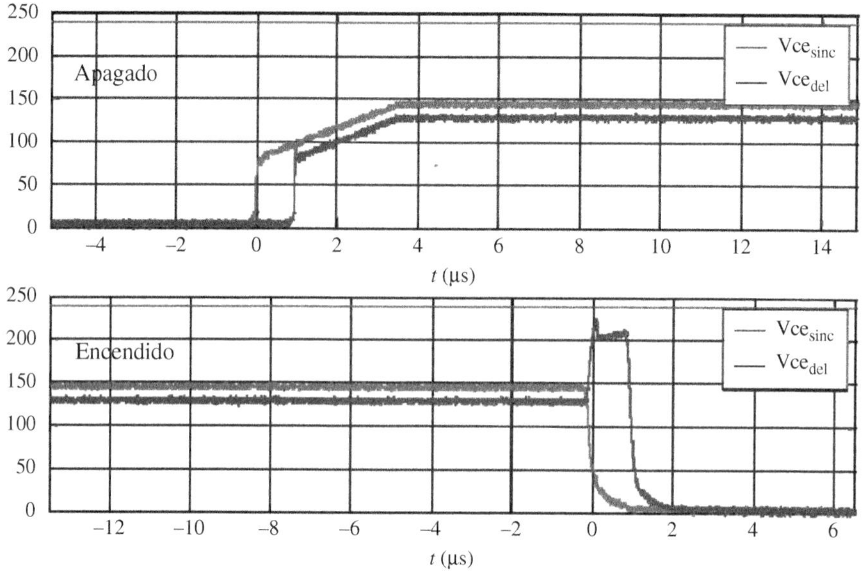

(a)

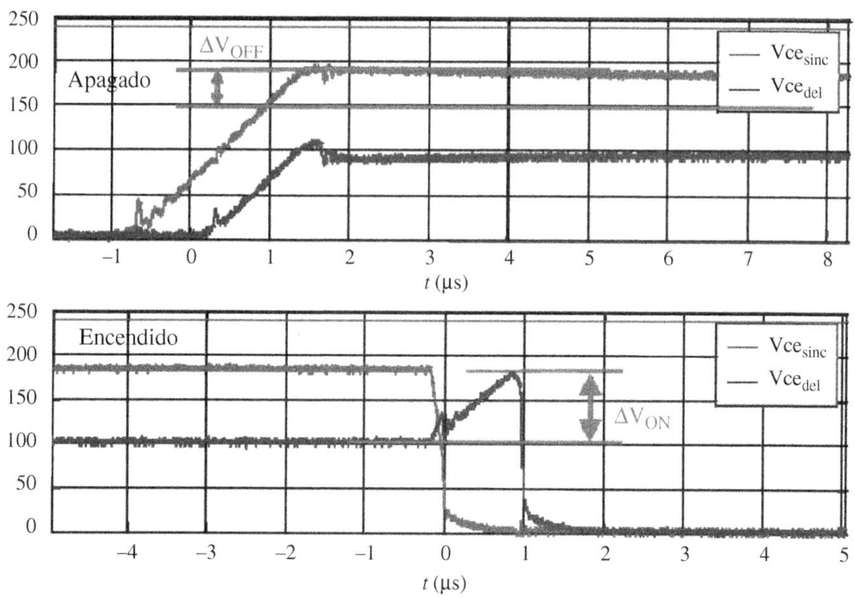

(b)

Figure 13.5 Behavior of series-connected IGBTs: (a) load side solutions with RC and (b) RCD snubber.

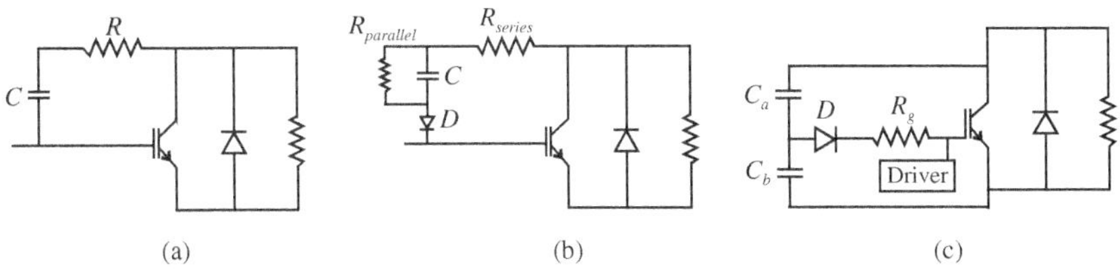

Figure 13.6 Gate side solution: voltage derivative control.

magnitude clamp, there can be some resonance problems with stray inductances if the chosen switching speed is too high.

13.2.2 Parallel Association

The parallel connection of discrete semiconductors in a unique power switching valve makes it possible to achieve higher currents, and consequently higher power switching capabilities. This is especially interesting with IGBT technologies in which simultaneous high voltage and high current switching semiconductors are not currently available (for example, the 6.5 kV IGBT is only able to handle 600 A). The critical point in this case is the correct current sharing between individual semiconductors, in order to avoid overcurrents due to unequal current distributions [27]. This current sharing is strongly dependent on their specific characteristics:

a. *Deviating chip temperatures*
Assuming that junction temperature is identical for each device, a higher amount of current will be conducted through the semiconductor with lowest on-state voltage at a given current.

A higher current means also higher power losses, and as a result its junction temperature tends to become higher than that of the rest of components. In this situation, the behavior of the component depends on its technology:

- *MOSFETs, NPT, and field stop IGBTs* have a positive saturation voltage temperature coefficient. This means that if the temperature increases (due to higher losses induced by a higher current), on-state voltage will also increase (i.e., on-state resistance will increase and therefore as the voltage is fixed, the current through the semiconductor will decrease), and inversely. Consequently, this feedback between temperature and current share ensures

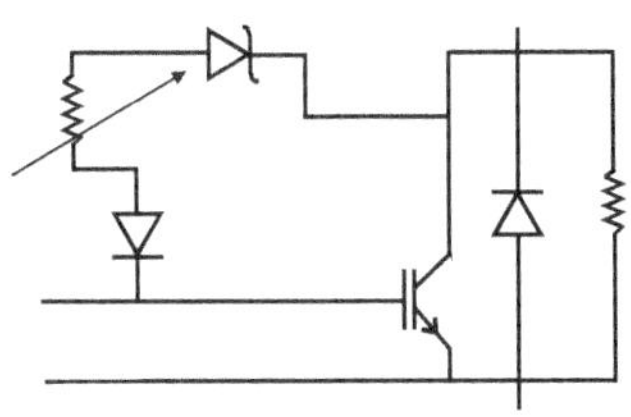

Figure 13.7 Gate side solution: voltage magnitude control with a zener or TVS diode.

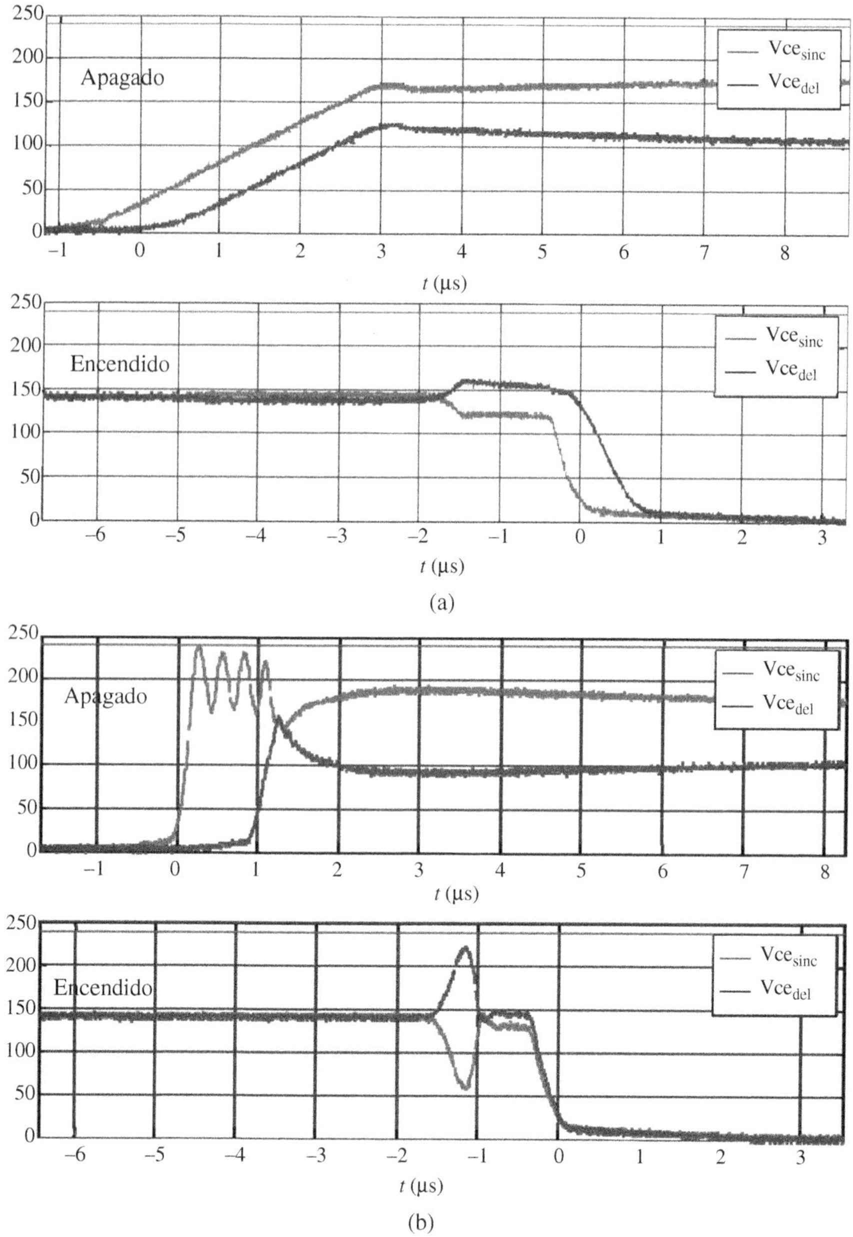

Figure 13.8 Behavior of series-connected IGBTs: gate side solutions voltage derivative (a) and magnitude (b) clamps.

thermal stability of parallel devices, although it does not force an equal current sharing.

- *PT IGBTs* have a positive saturation voltage temperature coefficient only above a threshold point in current. Generally, this threshold point is in the range of the IGBT-rated current or even slightly above. Under this value, the component presents a negative temperature coefficient, which may result in thermal instability when paralleling. Consequently, special care must be taken when using PT IGBT devices in parallel applications. A good cooling system (a common heat sink and minimized thermal resistance between devices) will induce a heating of the adjacent components too, increasing therefore their current share and diminishing the risk of thermal instability.

b. *Deviating impedances in the load circuit*
This deviation should be avoided by a symmetrical design of IGBT current paths (identical stray inductances). An alternative solution is adding an output inductor for each of the IGBT branches. This way, if this inductance is greater than the impedance of each of the branches, the currents will be shared symmetrically through the output chokes.

c. *Variation of the single-chip parameters*
This variation may be due to fabrication tolerances and must be avoided as much as possible. If possible, chosen devices should correspond to a unique production lot (smallest parameter deviations guaranteed).

d. *Different impedances in the gate-drive circuits*

The balanced sharing of the load current between paralleled MOSFETs or IGBTs may be obtained by a suitable control on the gate charge of the devices, as the IGBTs should switch on and off simultaneously. To assure simultaneous switching, two possibilities can be proposed: to use one IGBT driver channel in order to drive all paralleled IGBTs or to use separate driver channels. This has the disadvantage of transition time differences in the signal coupling path and has the advantage that equalizing processes via auxiliary collector and auxiliary emitter will not occur. The more cost-effective option is generally to control all paralleled IGBTs with one driver channel. This can be illustrated on the base of a particular example, around the parallel connection of various IGBTs on an experimental set-up (see Figure 13.9).

The two different paralleling options (various drivers or a unique one) are shown schematically in Figure 13.10.

Figure 13.11 shows the picture of the unique driver solution (courtesy of TRAINELEC Electric Traction Converters).

Above mentioned statements are confirmed experimentally, and it can be observed that the unique driver is the best solution. This is due to the fact the multiple driver solution introduces a higher dispersion on the gate voltages application time (due to the fiber optics path) than if a unique gate driver and multiple gate-circuits are used. Figure 13.10 presents the current measured during the switching on of two parallel IGBTs both with a unique driver (Figure 13.10a) and with multiple drivers (Figure 13.10b).

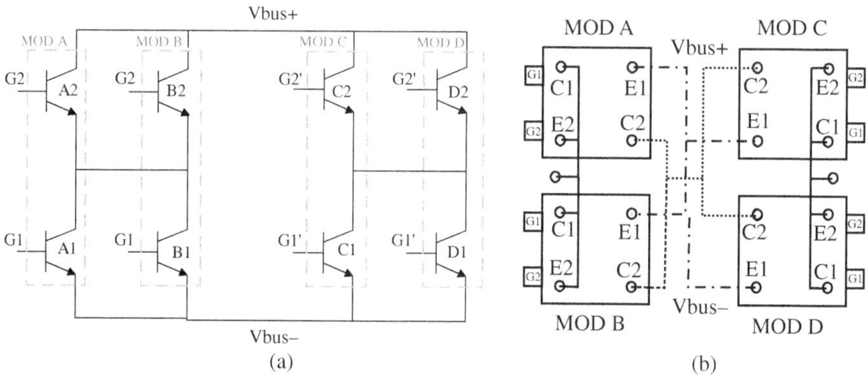

Figure 13.9 Experimental set-up for the test of parallel IGBTs.

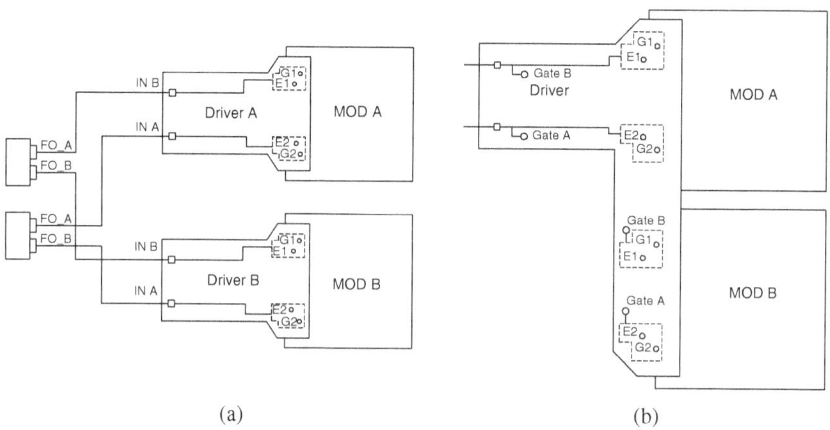

Figure 13.10 Paralleling with several drivers (a) or with a unique driver (b).

Figure 13.11 Detail of the unique driver paralleling solution. Courtesy of TRAINELEC Electric Traction Converters.

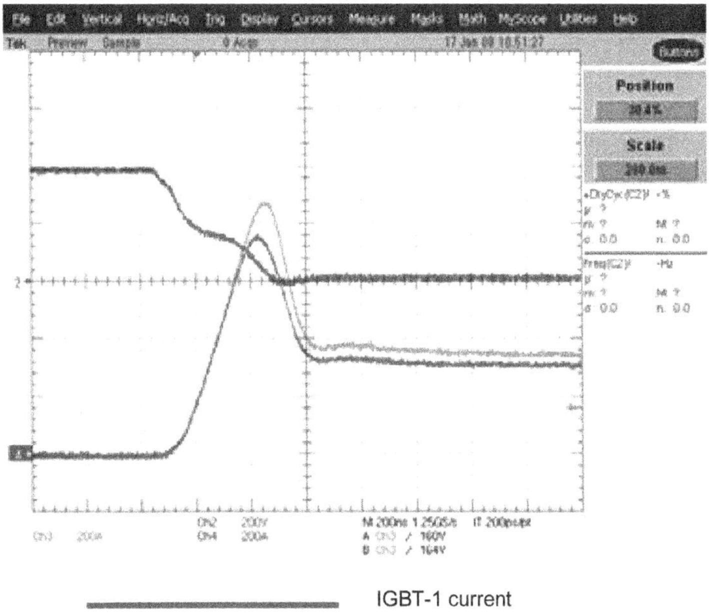

(a)

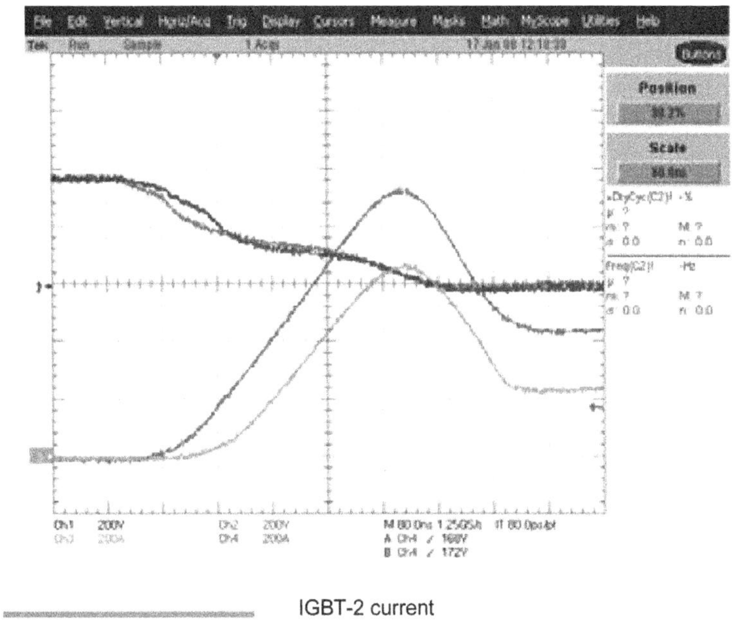

(b)

Figure 13.12 Parallel IGBT switch on currents with a unique driver (a) and with multiple drivers (b). Courtesy of TRAINELEC Electric Traction Converters.

13.3 TOPOLOGIES USED IN HIGH POWER CONVERTERS

In addition to the series or parallel association of discrete semiconductors as power switching valves, the use of topological solutions can also permit to increase the power and voltage rating of converters for utility applications. Furthermore, topological solutions, as will be shown in this chapter, allow increasing the power quality of converter's output waveforms, which is one of the main requirements in distribution grid applications. Traditionally, the main topological solutions used for higher power utility applications have been the multi-pulse and the multilevel configurations [28].

Multi-pulse topologies are based on simple conversion units producing quasi-square-wave voltage (known as 6 pulse units), which are combined via phase-shifting isolation transformers, improving the waveform of the absorbed current and increasing the overall power of the facility. For a high number of pulses, the structure of the phase shift transformer becomes complex and its implementation appears difficult and expensive. However, most existing STATCOMs rated above 80 MVAr use either 24 or 48 pulse converters. In the same way, HVDC converters have traditionally been built using this solution, in order to minimize the harmonic content of the exchanged currents. In the multilevel topology, an array of switches is used to select the output voltage from a number of available DC power supplies. The DC voltage sources are typically implemented using capacitors with a charge balancing scheme used to maintain the voltage constant, which is frequently the main concern on this kind of techniques. Multi-pulse converters are still used in very high voltage and power applications, while multilevel topologies are becoming the standard solution in medium voltage and medium power distribution applications [29]. Consequently, in the following sections, the three main multilevel topologies will be presented.

13.3.1 Series H-Bridge Multilevel Inverter (Cascaded Topology)

This structure was first presented in 1975 [30]. It is composed, as its name indicates, of several bridge inverters associated in series in a cascaded configuration. It is presented in its general form as shown in Figure 13.13. If single phase H-bridge converters are considered, each inverter is able to generate three different voltages ($+V_{DC}$, $-V_{DC}$, and 0).

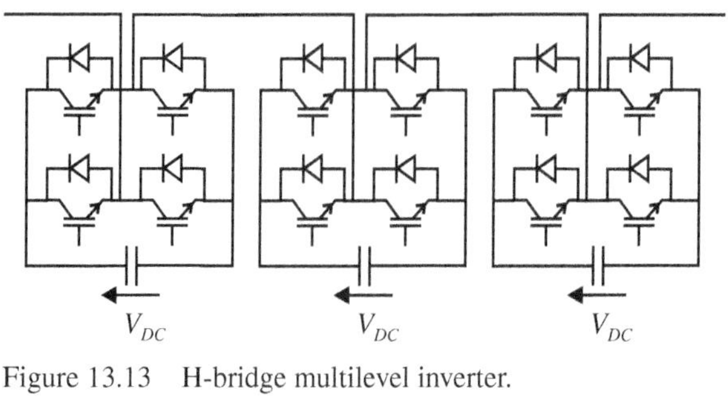

Figure 13.13 H-bridge multilevel inverter.

If we consider that N is the number of associated in series inverters, then we can express the number of voltage levels according to the following expression:

$$\text{Number of voltage levels} = 2N - 1$$

The main interest of this structure is its simplicity and modularity. However, its major inconvenience is that it needs the use of several isolated continuous DC links (as much as converters). This is a drawback in applications in which an active power transfer is required, but not in the case of exclusive reactive power exchange, as simple capacitors may be used as voltage sources. This is the reason why this topology has been widely proposed for reactive power compensation applications (such as the STATCOMs) especially for the distribution network [31].

13.3.2 Structure NPC and MPC (Neutral Point Clamped—Multiple Point Clamped)

The neutral point clamped (NPC) and its generalized version the multiple point clamped (MPC) structures are the most extended multilevel topologies in the industry (Figure 13.14). The NPC structure was first presented in 1980 by Nabae et al. [32], and it is the configuration used by most of the manufacturers in medium voltage and medium power applications (especially not only in drives, but also in utility applications).

The difference between the two structures is that one can generate an odd number of voltage levels from a pair number of capacitances and thus leaves the neutral point accessible, this is the NPC structure. The other structure, the MPC, possesses

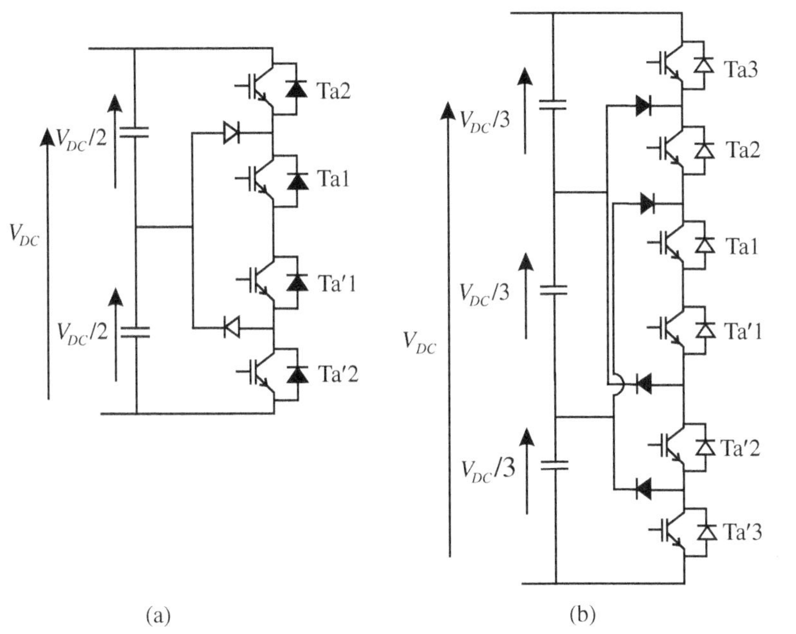

Figure 13.14 NPC with three voltage levels (a) and MPC with four voltage levels (b).

pair voltage levels and is composed of an impair number of capacitances, which makes the neutral point not accessible.

Its principle is based on a continuous voltage source that distributes its voltage equally among the series-connected capacitances. Then by influencing the switches, more or less capacitors are connected to the charge and thus the voltage is modulated.

The principal advantage of this structure is that it only needs one DC voltage source. However, the difficulty to equilibrate the capacitance voltages and the complexity of the modulation techniques limits in practice the number of levels that can be implemented. With three voltage levels for each inverter leg (three-phase NPC structure), it is possible to obtain a total of seven different voltage levels applied on the charge (a classical three-phased bridge inverter delivers five voltage levels), and it is the most widely applied configuration.

13.3.3 Flying Capacitor Structure

This topology was introduced by Meynard and Foch in 1992 [33, 34]. It is formed from the imbrications of commutation cells, which are connected on the extremities to different DC links. The topology for four voltage levels is presented in Figure 13.15.

The principle of this structure is based on the natural balance of the flying capacitors at the kV_{DC}/N voltage, where k is the position of the capacitance and N is the total number of continuous buses when the current circulating through the structure is alternative. However, it is necessary to ensure, via the control strategy, that the voltage of all capacitances remains constant.

This structure presents, like the NPC and MPC structures, the advantage to use only one voltage source at the entrance, the other voltage levels are generated internally. In comparison with the NPC and MPC structures, this solution needs less diodes, but a high number of capacitors (proportional to the number of desired levels) making the structure less reliable, for example, than the cascaded structure.

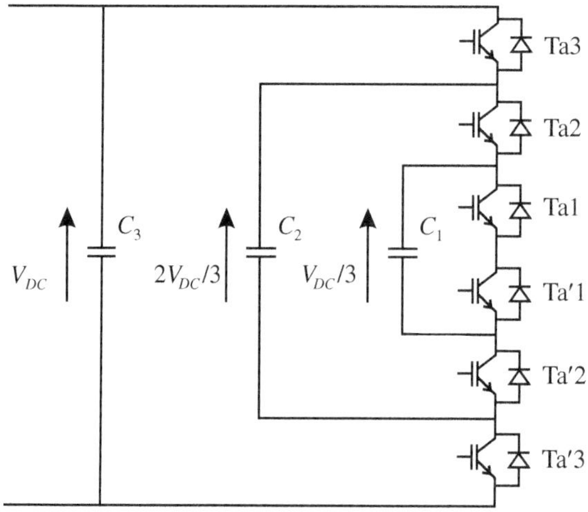

Figure 13.15 Flying capacitor structure with four voltage levels.

13.3.4 Some Industrial Examples

The D-FACTS-related industrial products are many and are presented in almost large manufacturer's offers in Europe, United States, or Japan.

The first applications, which can be called D-FACTS, have been dedicated for MV purpose and are based on thyristor's technology. One can cite various reactive power compensation systems like SVC with thyristor-controlled reactor, for instance: arc furnaces voltage stabilization or catenaries voltage boosting [35]. The rise of IGBTs leaded to more efficient VSC-based structures like D-STATCOM [36] for similar applications concerning the voltage profiles but with other possibilities related to power quality issues, for instance: mitigation of voltage dips effects and voltage harmonics rejection (dynamic voltage restorer [DVR]) [37] or current active filtering. In [38], a wide panel of industrial products is described and commented.

13.4 POWER ELECTRONIC CONVERTER CONTROL

The control of power electronics converters consists essentially of sending opening and closing orders to semiconductors in order to obtain the desired power exchange between the input (source/grid) and the output (grid/load) of the converter, following a hierarchical structure. Figure 13.16 presents this control hierarchy for the case of a generic VSC.

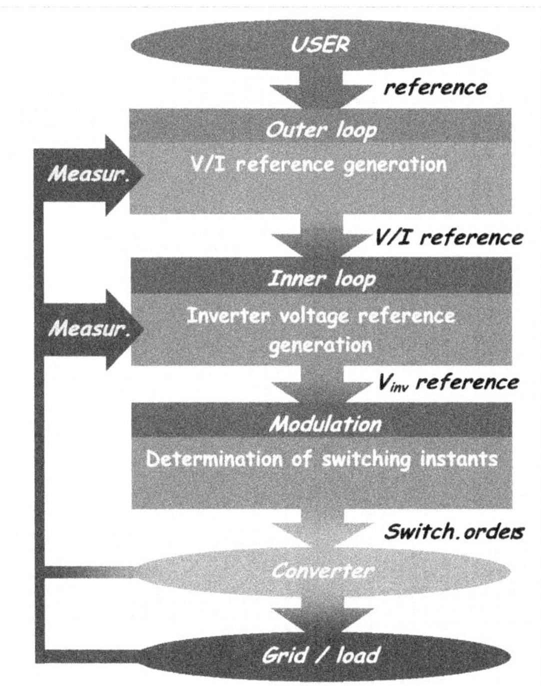

Figure 13.16 Control hierarchy for a generic voltage source converter.

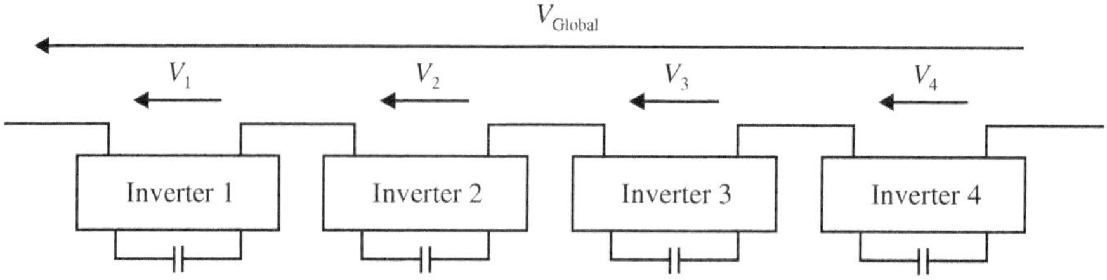

Figure 13.17 H-bridge multilevel converter.

Therefore, main converter control levels for a generic VSC are:

- An outer control loop that from the user's references (sinusoidal waveform, balanced voltages, control of the voltage in the point of common coupling, reactive power injection, voltage sag correction, etc.) and the measurement of network and/or load variables calculates the voltage/current to be exchanged with the grid (in the case of utility applications).
- An inner current/voltage control loop that calculates the voltages to be created by the inverter in order to ensure the desired output voltage/current.
- A modulation level that "translates" the inverter voltage reference into switching orders for the different semiconductors of the converter.

In the following sections, these three control levels will be deeply analyzed, especially focused on grid applications.

13.4.1 Modulation Techniques

The modulation that can be implemented in a converter depends essentially on its topology: number of levels, configuration, etc. In this section, main modulation techniques for H-bridge multilevel converters are presented (Figure 13.17).

13.4.1.1 Full Wave Modulation

In this control principle, the more the reference voltage increases (positive or negative value), the more the inverters will contribute to generate it (Figure 13.8).

In the case of the H-bridge cascaded converters, each inverter is a three-level one delivering +/−V_{DC} or 0 voltages, in a full wave mode, where V_{DC} is the DC bus

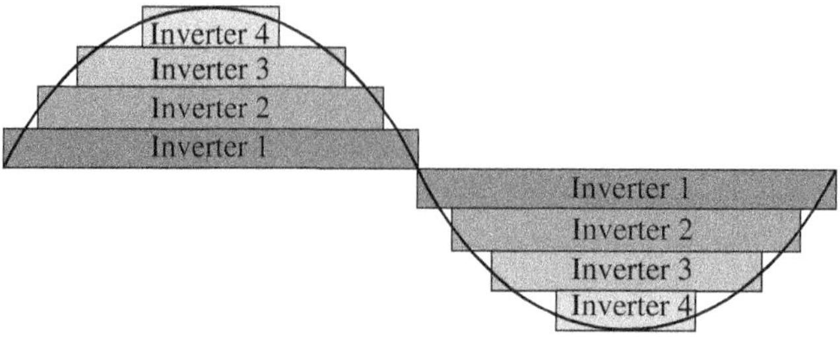

Figure 13.18 Full wave modulation technique.

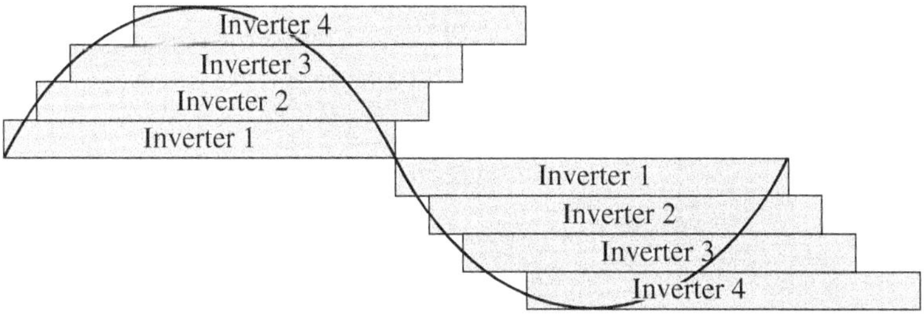

Figure 13.19 Distribution and equilibrium strategy of each cell losses.

voltage level. This modulation is especially interesting for high power applications as each semiconductor switches only 4 times per grid period, in which loss minimization is crucial.

The main drawbacks of this technique are:

- a large low frequency range harmonics contents;
- the low ability to track complex waveforms;
- a recurrent unbalancing operation of the inverters units, due to the different width time conduction.

To mitigate this unbalance, a load distribution method can be used that permits the same solicitation by the phase shifting of conduction times (see Figure 13.19). In this manner, the losses and the constraints between the different modules are equally shared.

13.4.1.2 Vertically Shifted Carrier Signals

Contrary to the previous solution, this control strategy is based on PWM carriers. There are two PWM carriers for each inverter. There is one carrier for the positive output voltage and one carrier for the negative one. All PWM carriers have the same amplitude and phase, but their amplitude and offset depend on the number of inverters. The commutation orders for each converter are given by the comparison between the modulating signal and the different PWM carriers.

With this solution, there is no equilibrium between the RMS output voltages delivered by the different inverters. The inverter driven by the PWM carrier 1 will produce the maximum voltage and the one driven by the PWM carrier 3 the least (Figure 13.20). This induces that the losses will not be equally shared between all the converters.

Its main advantage compared to the previous solution is that it offers a better modulation quality. The output signal switching frequency is the same as the PWM carrier frequency. Therefore, the output signal harmonics have a higher frequency and will be easier to filter.

To improve the output voltage quality, especially for low output voltage values, a solution is to increase the PWM carrier frequency close to 0. Figure 13.21 shows the different PWM carriers and the sinusoid modulation signal.

This kind of modulation is well suited for NPC or MPC converters.

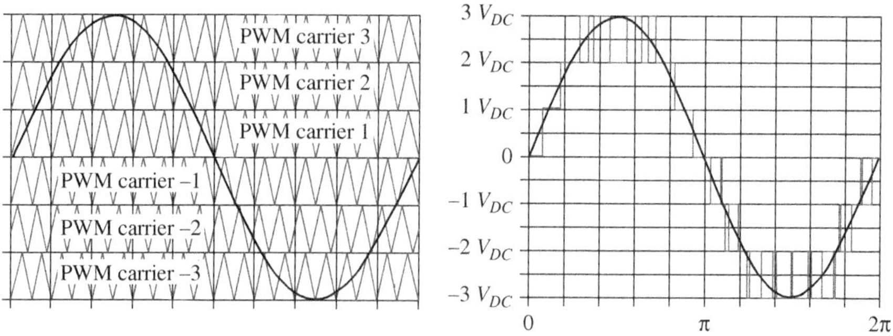

Figure 13.20 PWM carriers and modulation signal vertical shifted carrier signals—case of three modules converter.

13.4.1.3 Horizontal Shifted Carrier Signals

This kind of modulation is also based on the generation of two PWM carriers for each inverter. Each group of carriers is shifted from one another by $2\pi/N$, where 2π corresponds to the PWM carrier period and N corresponds to the total number of modules (Figure 13.22).

In this example, the number of inverters is 2. The output switching frequency is 2 times higher than the PWM carrier frequency.

The main advantages of the HSCS consist of:

- an equal distribution of the constraints among all the inverters;
- an increase in output switching frequency, which is N times the PWM carrier frequency.

These two characteristic features make this strategy particularly interesting. This control strategy is well suited for floating capacitance multilevel converters.

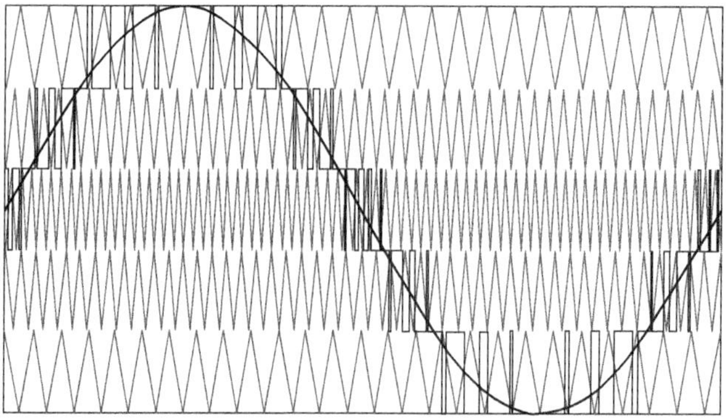

Figure 13.21 Control strategy to improve the harmonic quality for small modulator values.

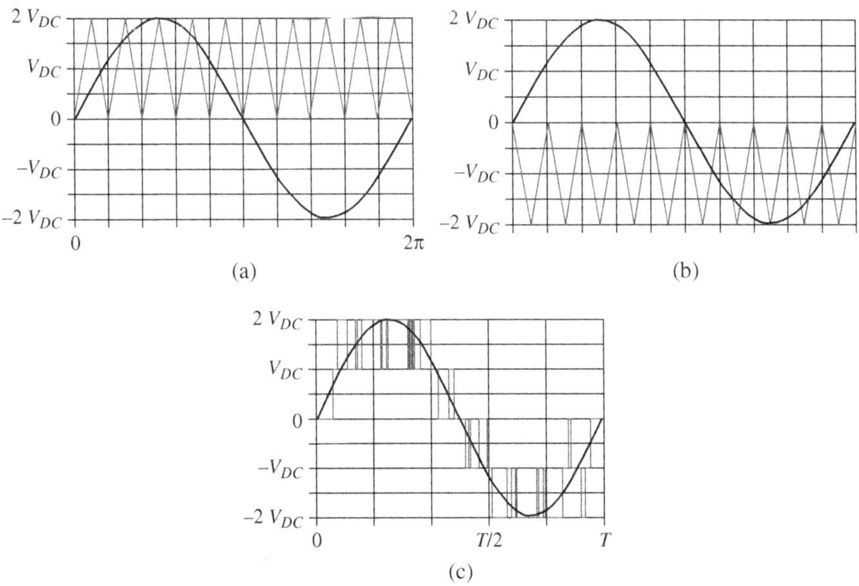

Figure 13.22 PWM carrier and modulation signal for horizontal shifted carrier signals: (a) comparison of the carrier and the modulation signal for a positive value; (b) comparison of the carrier and the modulation signal for a negative value; and (c) output voltage.

13.4.1.4 "Hybrid" Converters Control

The term "hybrid" means that the cascaded converter is composed of cells of different characteristics. For example, the converter can be composed of various H-bridges with different DC link voltages. Figure 13.23 presents two seven-level cascaded converters one based on a symmetrical configuration and the other on a hybrid configuration.

Figure 13.24 shows an example where the high voltage converter has a voltage bus 2 times higher than the low voltage converter, which has a fixed DC bus value of V_{DC}. The output voltage is a sinusoid with 2.5 V_{DC} amplitude.

The output voltage of the high voltage inverter can take three values:

- $2V_{DC}$ when the output voltage is comprised between V_{DC} and $3V_{DC}$;
- V when the output voltage is comprised between $-V_{DC}$ and V_{DC};
- $-2V_{DC}$ when the output voltage is comprised between V_{DC} and $3V_{DC}$.

The low voltage converter will generate the additional signal to obtain the desired output voltage. Therefore, we will subtract to the voltage reference the voltage generated by the high voltage converter. This signal will then be used as reference for the low voltage converter, which is classically controlled through a PWM.

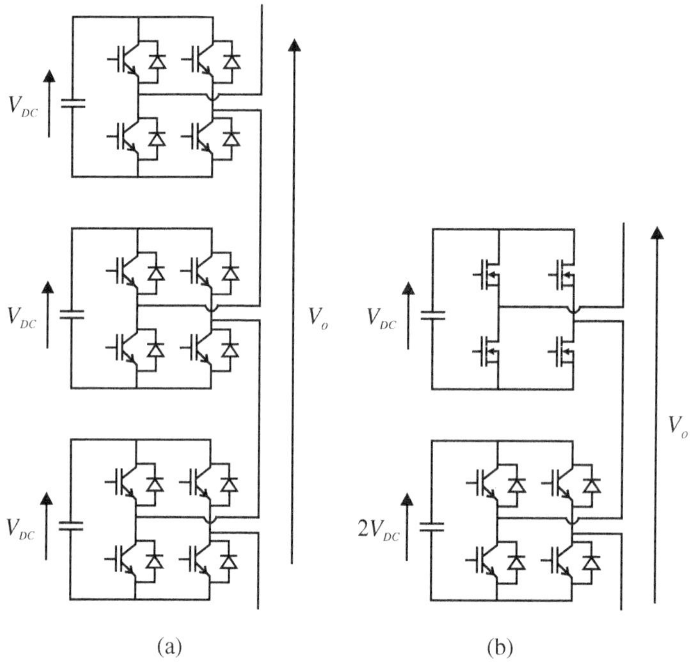

Figure 13.23 Symmetrical seven-level converter (a) and hybrid seven-level converter (b).

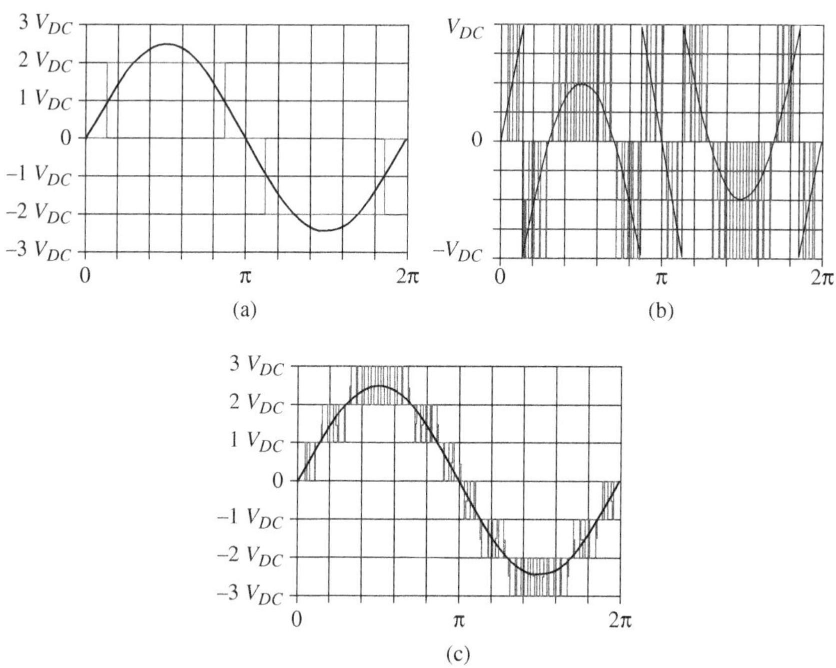

Figure 13.24 Principle of the hybrid inverter modulation: (a) high voltage inverter output voltage; (b) low voltage inverter modulation; and (c) converters output voltage.

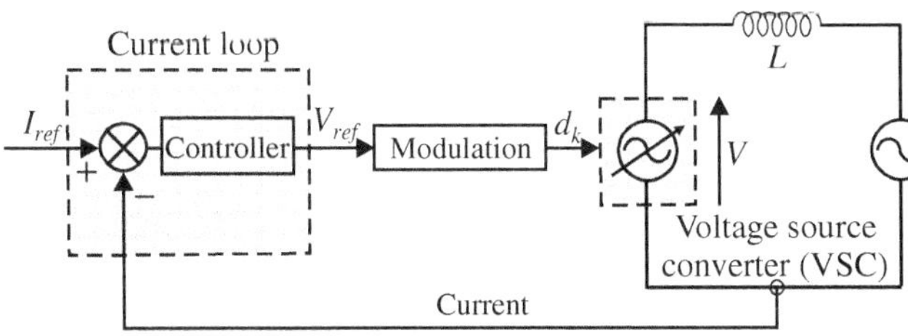

Figure 13.25 Fixed frequency current control.

13.4.2 Inner Current Control Loop

Power converters that are connected in parallel with the grid (STATCOMs, generator interfaces, active filters, rectifiers, etc.) are generally controlled in current-mode. They act as controlled current sources, exchanging reactive/active power with grid according to the objectives of the application.

In a fixed switching frequency configuration control of a VSC (the most common one in high power applications), the current control loop is implemented over a modulation layer. The current loop calculates the voltage to be generated by the inverter in order to obtain the desired exchanged current with the grid, while the modulation layer determines the switching off and on times of semiconductors that are necessary to apply the necessary voltage (Figure 13.25).

In the literature, many kinds of controllers can be found, with various control quality and complexity levels. The choice of the controller depends on the requirements of the application: dynamics, bandwidth, steady-state behavior, etc.

13.4.2.1 Modeling

A VSC converter is connected to the grid through an inductive filter (*L* or LCL), with the equivalent scheme shown in Figure 13.26: a controlled voltage source, *V*, (which represents the VSC neglecting its switching harmonics) connected to the grid, E, through an inductance, *L* (resistance is neglected).

If the coupling filter is a simple inductor, *L*:

$$H(s) = \frac{i}{v} = \frac{1/L}{s + r/L}\,; \; L \cdot \frac{dI}{dt} = v - e$$

The continuous model of this system is presented in Figure 13.27.

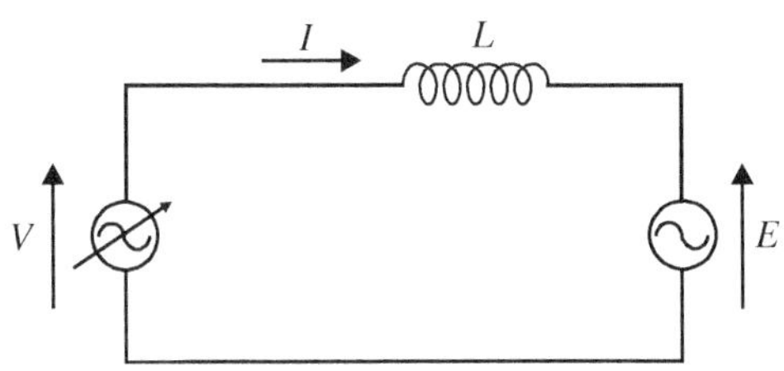

Figure 13.26 Equivalent scheme of a single phase inverter connected to the network.

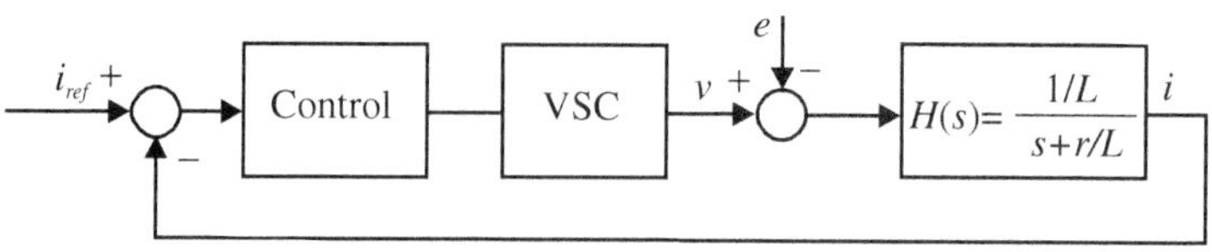

Figure 13.27 Current control scheme of the voltage inverter with L filter.

If we consider that the computational delay is one sampling period (T_s) and a zero-order holder is used, the discrete transfer function of the system with respect to the reference is:

$$H_1(z) = \frac{i(z)}{i_{ref}(z)} = Z\left\{ B_0(s)e^{-sT_s}\frac{1/L}{s+r/L} \right\}$$

$$H_1(z) = \frac{1}{r}\frac{1-a}{z(z-a)} \quad \text{where } a = e^{-\frac{r}{L}T_s}$$

Concerning the transfer function with respect to the disturbance (network voltage e), it must be considered that this is essentially a continuous phenomenon and its effect on the system can only be sampled (Figure 13.28).

As a consequence, to determine the transfer function with respect to the disturbance, an approximation must be carried out by supposing that the dynamics of the disturbance is very low compared to the sampling frequency (Figure 13.29).

Therefore, this approximated transfer function with respect to the perturbation will be expressed as:

$$H_2(z) = \frac{i(z)}{e(z)} = Z\left\{ B_0(s)\frac{1/L}{s+r/L} \right\} = \frac{1}{r}\frac{1-a}{z-a}$$

where: $a = e^{-\frac{r}{L}T_s}$.

We suppose here that the perturbation can be accessed, can be sampled, and that we can evaluate its effect on the discrete model of the system. This hypothesis is normally valid if the grid is not very distorted.

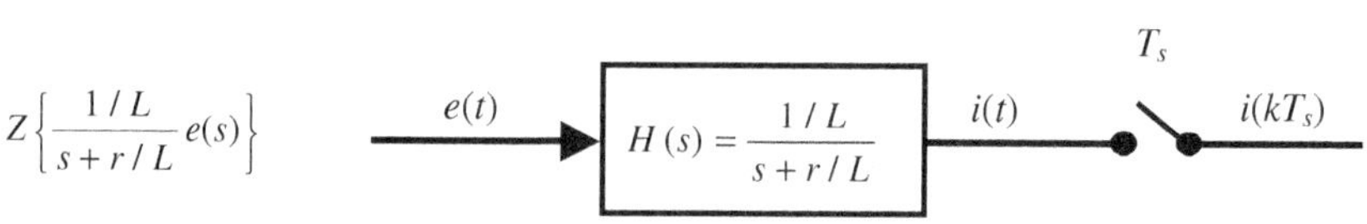

Figure 13.28 Disturbance modeling: real model.

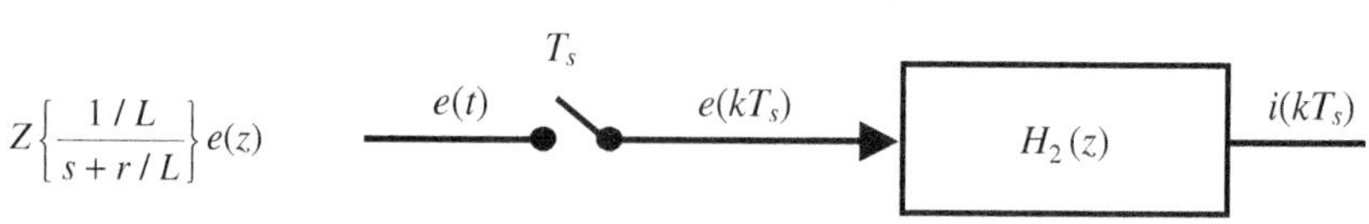

Figure 13.29 Modeling of the disturbance: approximation.

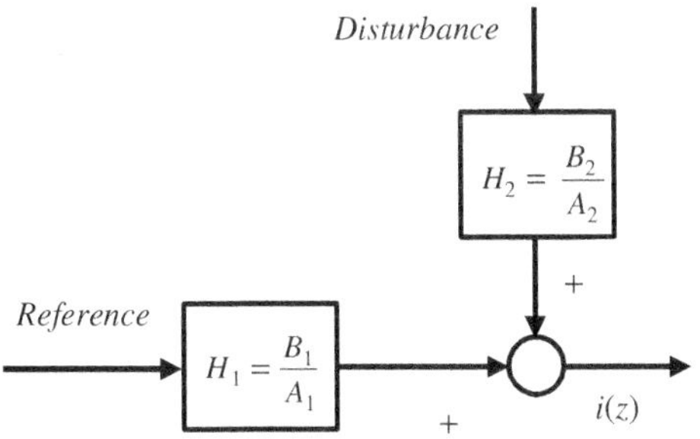

Where:

$$H_1(z) = \frac{B_1}{A_1} = \frac{1}{r}\frac{1-a}{z(z-a)}$$

$$H_2(z) = \frac{B_2}{A_2} = \frac{1}{r}\frac{1-a}{z-a}$$

Figure 13.30 Global sampled model of the voltage inverter connected with an *L* filter.

The global sampled model of the voltage inverter connected with an *L* filter is presented in Figure 13.30.

13.4.2.2 Current Control Techniques

There are several control technologies that can be considered for grid applications. In this chapter, controllers based on integration technologies are considered, as they are widely used in such applications. In addition, instead of considering a tuning for a single frequency, several harmonics are considered, in order to reflect the most demanding situation in the grid: the requirement to inject several harmonics on the grid or to protect from existing harmonic disturbances.

MULTIPLE ROTATING FRAMES. PI controllers implemented in rotating frames permit to control AC magnitudes with a zero steady-state error. Theoretically, if the objective is to control various harmonic components, it is necessary to use as many rotating frames as considered harmonics (taking into account not only their frequency, but also their sequence). Nevertheless, the properties of the most common grid harmonics make it possible to simplify the general structure by taking into account only the following sequences:

Positive: 1, 7, 13, 19, etc.
Negative: 5, 11. 17, etc.

Consequently, these frames can be derived, for example, starting from the fundamental frequency rotary frame as shown in Figure 13.31. The control structure is presented in Figure 13.32.

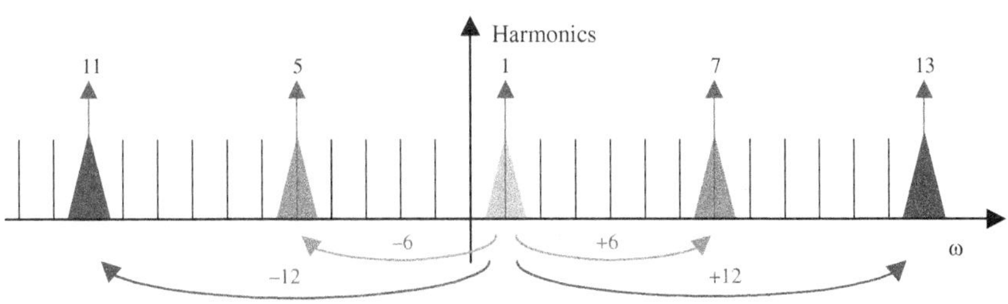

Figure 13.31 Asymptotical spectrum of a multiple rotating frame controller.

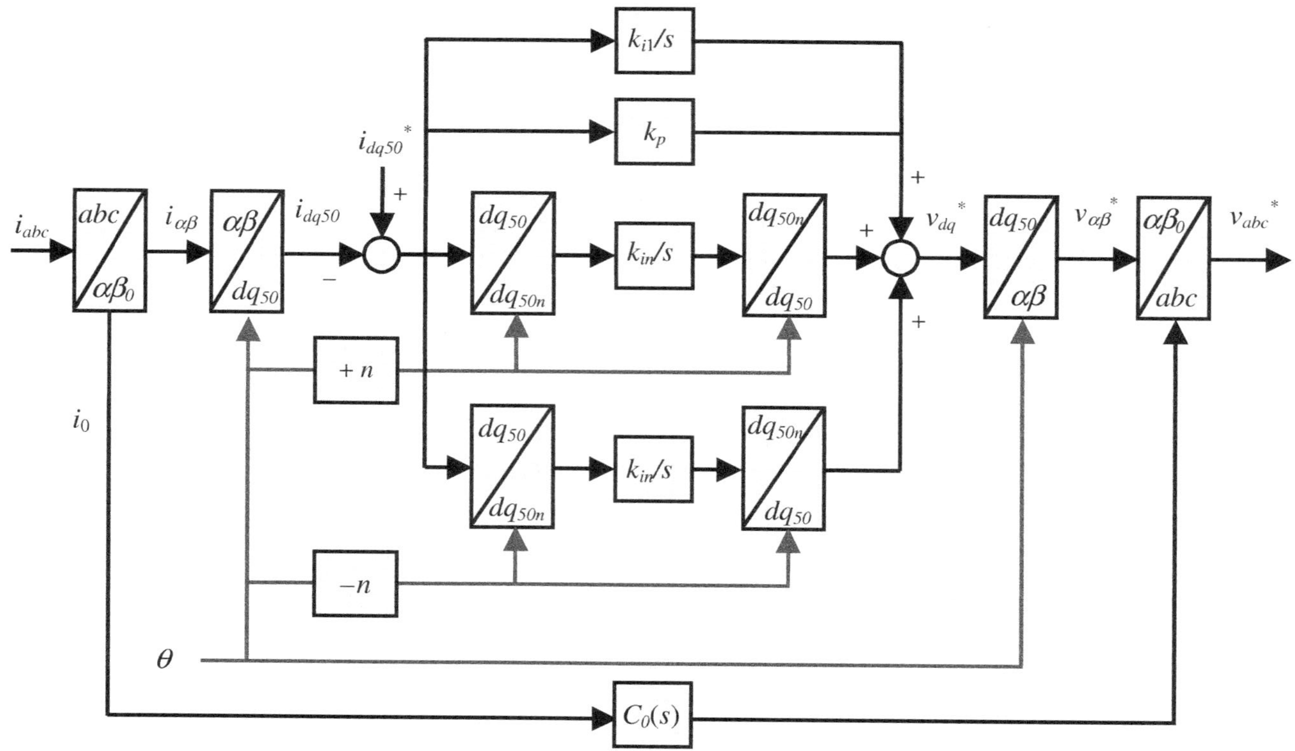

Figure 13.32 Multiple rotating frames: control structure.

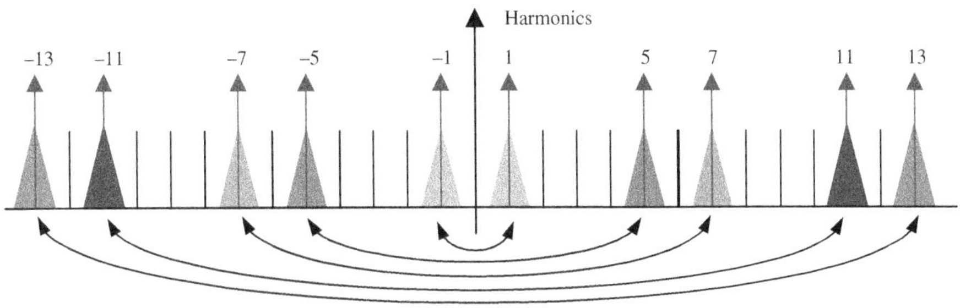

Figure 13.33 Asymptotical spectrum of a multi-resonant controller.

MULTI-RESONANT STRUCTURE. The rotating frame PI controllers are equivalent to resonant controllers in fixed reference frames because they exhibit the same ability to control alternative signals. The discrete transfer function of this kind of controller is the following one (by considering a cosine function):

$$k_p + \frac{k_i s}{s^2 + \omega_0^2} \rightarrow k_p + \frac{k_i T_s[z^2 - \cos(\omega_0 T_s)z]}{z^2 - 2\cos(\omega_0 T_s)z + 1}$$

In this case, the answer in frequency is symmetrical in positive and negative harmonic components and it is not possible to choose the frequencies to be controlled because the two sequences (positive and negative) are always present in the spectrum.

This resonant controller can be connected both in series or parallel. For instance, if a shunt configuration is chosen, the structure of the controller is presented in Figure 13.34.

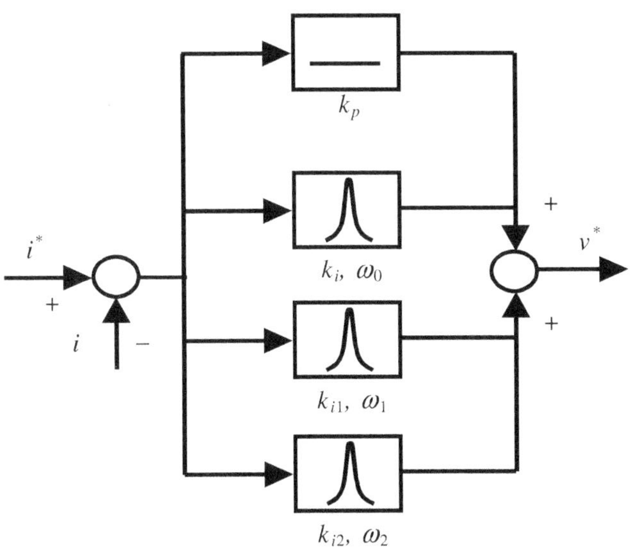

Figure 13.34 Shunt multi-resonant structure.

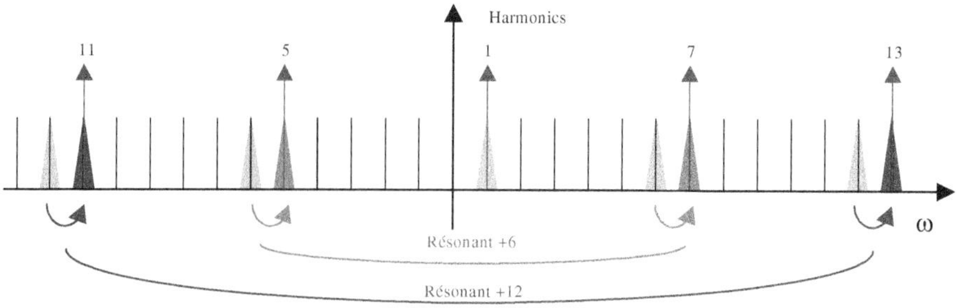

Figure 13.35 Asymptotical spectrum of a multi-frequency hybrid controller.

HYBRID STRUCTURE. A new multi-frequency structure, known as "hybrid" structure, has been proposed in [39,40]. It combines the properties of the two previous presented structures: it uses resonant and PI controllers in a single rotating frame. Thus, the spectrum of the controller is optimized to control the harmonic components that are the most frequent in the grid (considering their sequence of course), while minimizing the required computational load. With only one PI controller and two resonant controllers per frame (tuned for the harmonic components 6 and 12), it is possible to control four harmonics and the fundamental (1, 5, 7, 11, and 13) (Figure 13.35).

Figure 13.36 shows the control structure of the hybrid multi-frequency controller.

COMPARISON OF THE CONTROL TECHNOLOGIES. The three technologies are efficient to control various harmonic components as they have been designed for this purpose, and they achieve similar performances both in terms of dynamic and steady-state behavior. For instance, Figure 13.37 shows the response to a sudden variation of the reference for some harmonic components (5, 7, 11, and 13) while considering the multi-resonant structure (Figure 13.37a) and the hybrid one (Figure 13.37b). The dynamic responses are very similar in both cases. Figure 13.38 shows the results obtained with the same structures at an active filtering application in steady-state conditions (the polluting load is a rectifier with a capacitive filter). In both cases, the grid current is very close to a sine wave.

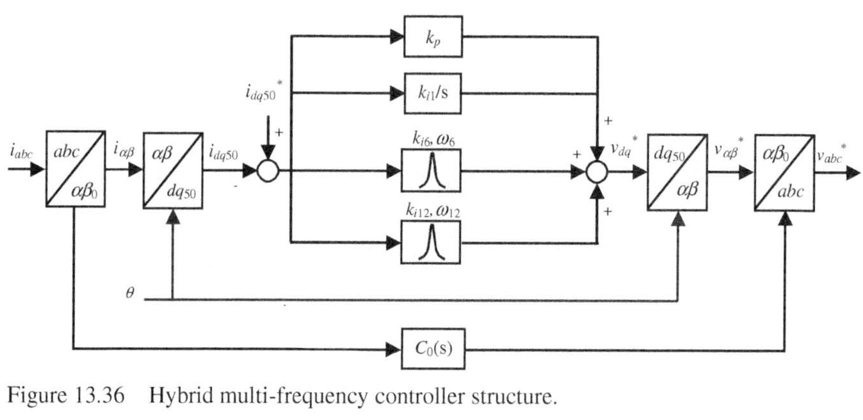

Figure 13.36 Hybrid multi-frequency controller structure.

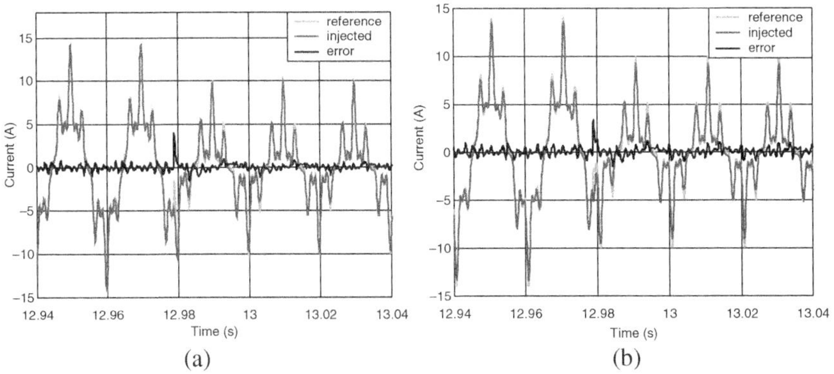

Figure 13.37 Transient response with the multi-resonant structure (a) and the hybrid one (b).

The main difference between these two structures is the computational load necessary for their implementation. The hybrid structure is optimized for such complex applications, due to a lower computational load compared to the two other technologies [40].

13.4.3 Inner Voltage Control Loop

In some grid applications, it is necessary to apply a controlled voltage in order to compensate for grid voltage disturbances, for example. These converters are generally connected in series with the grid, as it is the case of the DVR compensator.

Many types of voltage controllers exist. They can be used not only for compensation purposes, but also for UPS applications. In the less demanding applications, open loop structures can be used, but if a high accuracy is needed it is mandatory to implement closed loop structures, which are able to compensate for

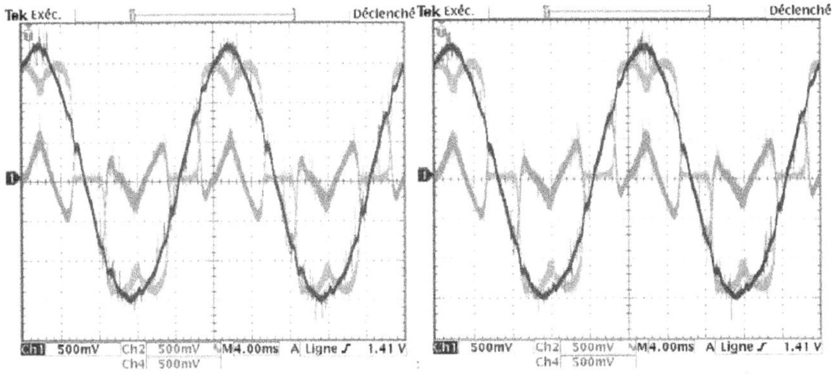

Figure 13.38 Steady-state response (active filter application) with the multi-resonant structure (a) and the hybrid one (b).

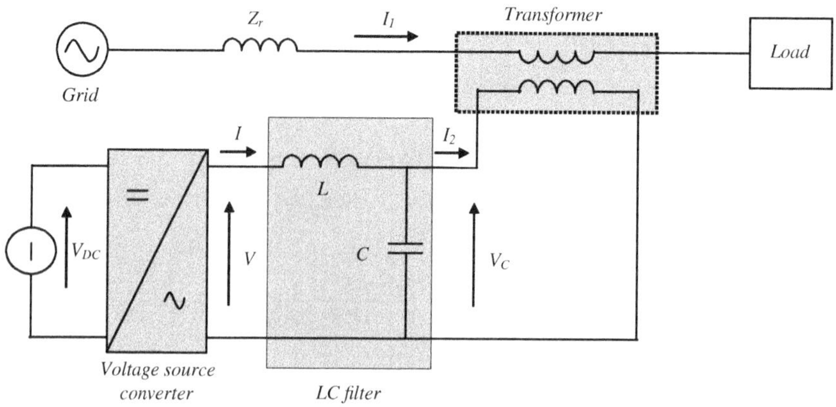

Figure 13.39 Single phase equivalent scheme of a series-connected voltage source converter.

disturbances (such as the effects of the coupling filter) and to control the applied voltage. In this case also, VSC-based fixed switching frequency solutions are the mostly used.

13.4.3.1 Modeling

Voltage-controlled compensators have generally an LC output filter in order to eliminate the switching frequencies and obtain an almost perfect sinusoidal output voltage. Nevertheless, the behavior of this filter is far from being perfect due to the voltage drops that are produced by the circulating currents. Figure 13.39 presents the equivalent single phase scheme of a series-connected VSC (coupled through an LC filter and a transformer).

The LC filter can be modeled as follows:

$$\frac{dx}{dt} = Ax(t) + B_1 \cdot V(t) + B_2 \cdot I_2(t)$$

$$\begin{bmatrix} \dfrac{dI}{dt} \\ \dfrac{dV_C}{dt} \end{bmatrix} = \begin{bmatrix} -\dfrac{R}{L} & -\dfrac{1}{L} \\ \dfrac{1}{C} & 0 \end{bmatrix} \cdot \begin{bmatrix} I \\ V_C \end{bmatrix} + \begin{bmatrix} \dfrac{1}{L} \\ 0 \end{bmatrix} \cdot V + \begin{bmatrix} 0 \\ -\dfrac{1}{C} \end{bmatrix} \cdot I_2$$

where the input of the system is the inverter voltage (V), the output is the compensator voltage (V_c), and the disturbance is the output current (I_2). The continuous transfer function derived from this expression is:

$$V_C(s) = \frac{1}{LCs^2 + RCs + 1} \cdot V(s) - \frac{Ls + R}{LCs^2 + RCs + 1} \cdot I_2(s)$$

This model can be sampled by considering the use of a zero-order holder for the transfer function with respect to the reference and a linear interpolation function

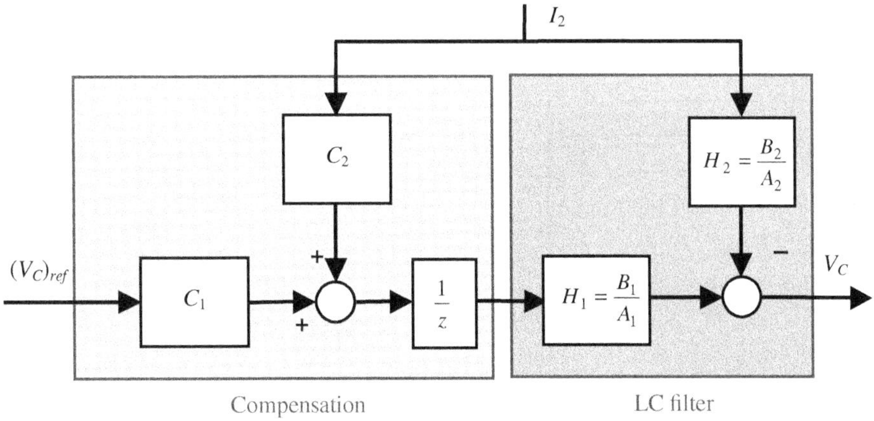

Figure 13.40 Control structure of the inverse model technique.

for the sampling with respect to the perturbation. The discrete transfer function will consequently be:

$$V_C(z) = \frac{az + b}{z^2 + cz + d} V(z) + \frac{ez^2 + fz + g}{z^2 + cz + d} I_2(z)$$

13.4.3.2 Voltage Control Techniques

In this section, two voltage control techniques for VSCs will be presented, the first one in open loop and the second one in closed loop, and both based on the compensation of the discrete model of the system.

OPEN LOOP—INVERTED MODEL COMPENSATION. This controller is based on two terms, C_1 and C_2, and the objective is to compensate with both of them the discrete transfer function of the system (Figure 13.40).

The term C_1 must be designed in order to compensate the transfer function of the system with respect to the reference. As the directly inverted expression is not causal, a delay must be added, and the resulting expression is:

$$C_1 = z^{-1} \frac{A_1}{B_1} = \frac{z^2 + cz + d}{az^2 + bz}$$

The system response with respect to the reference is presented in Figure 13.41.

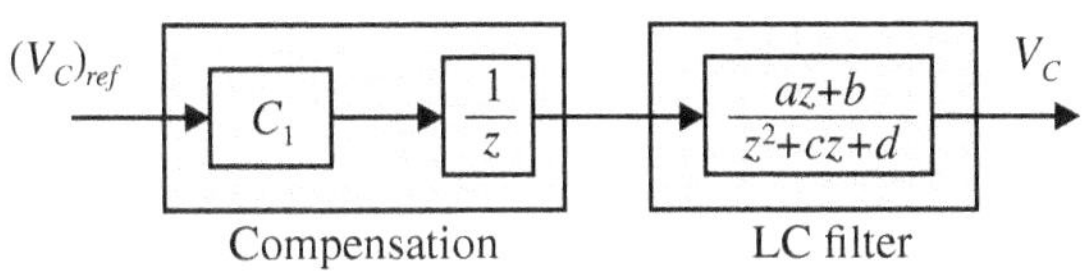

Figure 13.41 Inverse model: compensation on the reference.

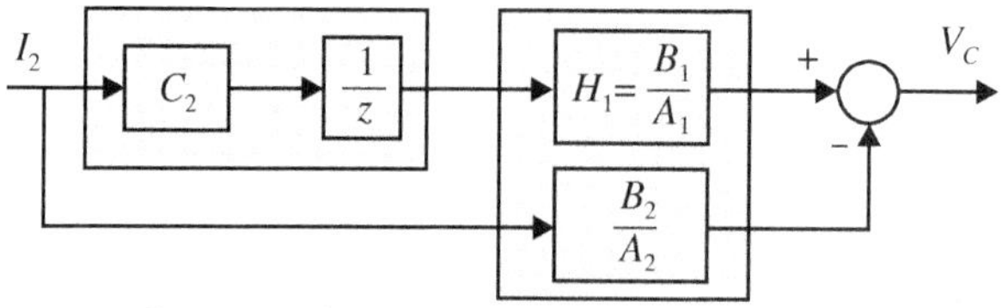

Figure 13.42 Inverse model: compensation of the disturbance.

Concerning the term C_2, it is designed to eliminate the effect of the transfer function with respect to the disturbance:

$$C_2 \frac{B_1}{A_1} - \frac{B_2}{A_2} = 0$$

$$C_2 = \frac{B_2 A_1}{B_1 A_2}$$

As $A_1 = A_2$, then the term becomes:

$$C_2 = \frac{B_2}{B_1}$$

As the degree of B_2 $(ez^2 + fz + g)$ is greater than the degree of B_1 $(az + b)$, a delay has been added to B_1 to guarantee the causality of the term. As a consequence, the controller C_2 is:

$$C_2 = \frac{ez^2 + fz + g}{az^2 + bz}$$

It is possible to compensate this extra delay added to the controller and the computational delay by adding a prediction term to the output of the compensation system in order to improve the behavior of the system in steady state. For instance, a periodicity or resonant prediction could be used (see Figure 13.42) [40].

The global control structure of the system is presented in Figure 13.43.

Closed loop—Deadbeat response controller. The objective of this method is to obtain a deadbeat response of the system, that is, to achieve a double

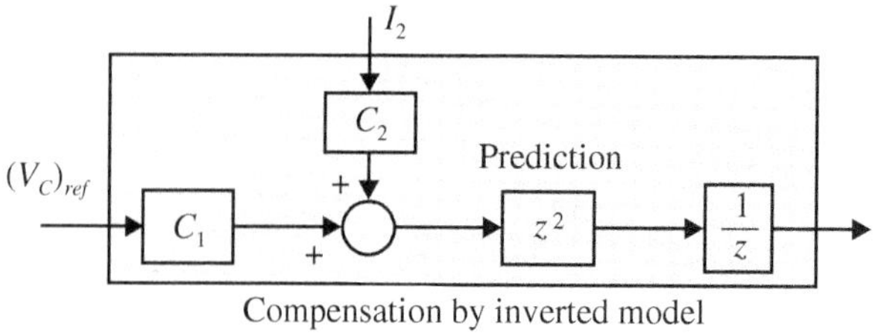

Figure 13.43 Compensation by inverted model and prediction.

delay response. This is accomplished by fixing the appropriate poles to the closed loop transfer function. This transfer function is:

$$V_C = \frac{B_1 T}{zA_1 R + B_1 S}(V_C)_{ref} + \frac{B_2 z A_1 R}{A_2(zA_1 R + B_1 S)} I_2$$

In the studied case, the order of the system to be controlled is 3 (by taking into account the delay) and then a second-order controller is used:

$$S = s_0 z^2 + s_1 z + s_2$$
$$R = z^2 + r_1 z + r_2$$

To get a deadbeat response, all the coefficients of the characteristic polynomial must be zero. Consequently:

$$\begin{cases} s_0 = \dfrac{-c^3 + 2dc + \dfrac{adc^2}{b} - \dfrac{ad^2}{b}}{-ac + \dfrac{a^2 d}{b} + b} \\ r_1 = -c \\ r_2 = c^2 - d - a s_0 \\ s_1 = -\dfrac{d r_2}{b} \\ s_2 = 0 \end{cases}$$

The transfer function in open loop will be:

$$H_{BO} = \frac{B_1 S}{zA_1 R}$$

And the closed loop transfer function will be:

$$H_{BF} = T\frac{B_1}{zA_1 R + B_1 S} = T\frac{B_1}{z^5}$$

It can be noted that the condition to get the desired double delay response answer is:

$$T = \frac{z^3}{B_1}$$

Hence:

$$H_{BF} = T\frac{B_1}{zA_1 R + B_1 S} = \frac{1}{z^2}$$

The deadbeat response controller is designed as shown in Figure 13.44.

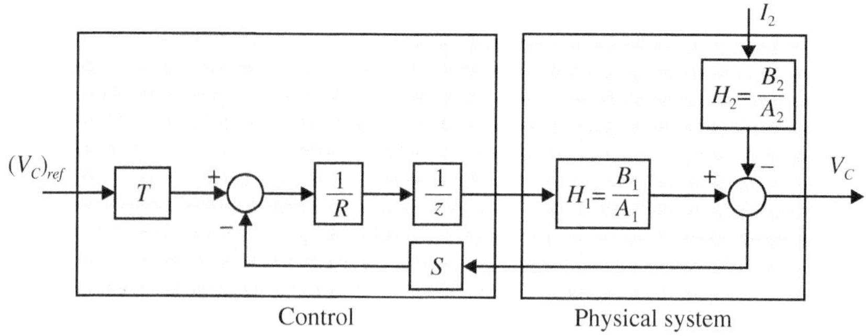

Figure 13.44 Deadbeat response controller.

It is possible to add the compensation term to the system in order to improve its disturbance rejection capability:

$$\frac{V_C}{I_2} = \frac{\frac{H_2'}{H_1'}H_1 - H_2}{1 + \frac{1}{z}\frac{S}{R}H_1}$$

where H_1 and H_2 are the real transfer functions of the physical system, and H_1' and H_2' are the estimated transfer functions. Figure 13.45 shows the scheme of the deadbeat control with disturbance rejection capability.

Comparison of control technologies. Both presented control techniques are able to control the main low frequency harmonic components, besides the fundamental voltage. Therefore, for instance, Figure 13.46 presents the response of a series-connected VSC controlled with these two techniques to a voltage sag (50%) in a heavily disturbed network (harmonic components 5 and 7) with a load that absorbs

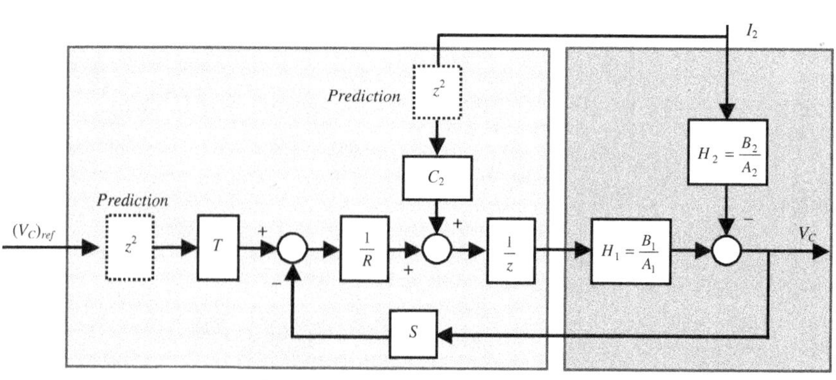

Figure 13.45 Scheme of the deadbeat control with disturbance rejection capability.

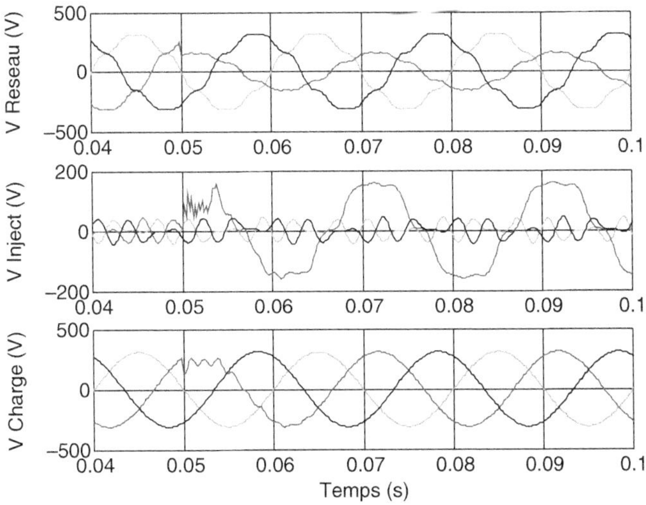

(a)

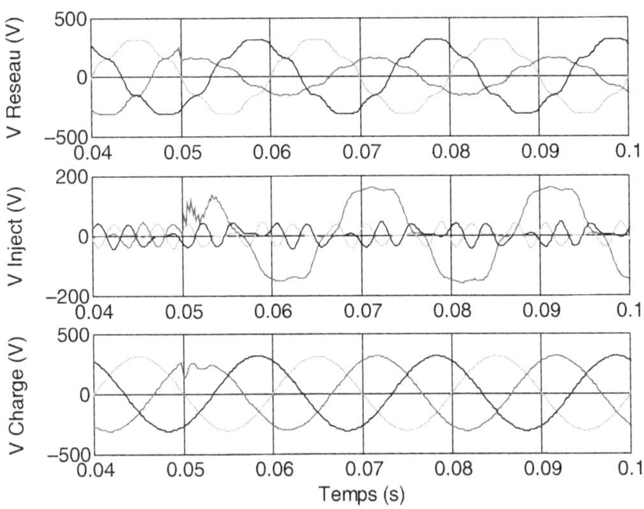

(b)

Figure 13.46 Compensation of single phase voltage sag with a series-connected VSC: open loop inverted (a) and close loop deadbeat (b) controls.

a non-sinusoidal current (with harmonic components 5 and 7)[1]. It can be noticed that in both cases, the network voltage is correctly compensated after a short transient. The main drawback of these methods lies in their sensitivity to parametric variations because they affect not only on the system stability but also in the accuracy (steady-state error).

[1] These results have been simulated with Matlab/Simulink for a switching frequency of 10 kHz and a sampling frequency of 5 kHz.

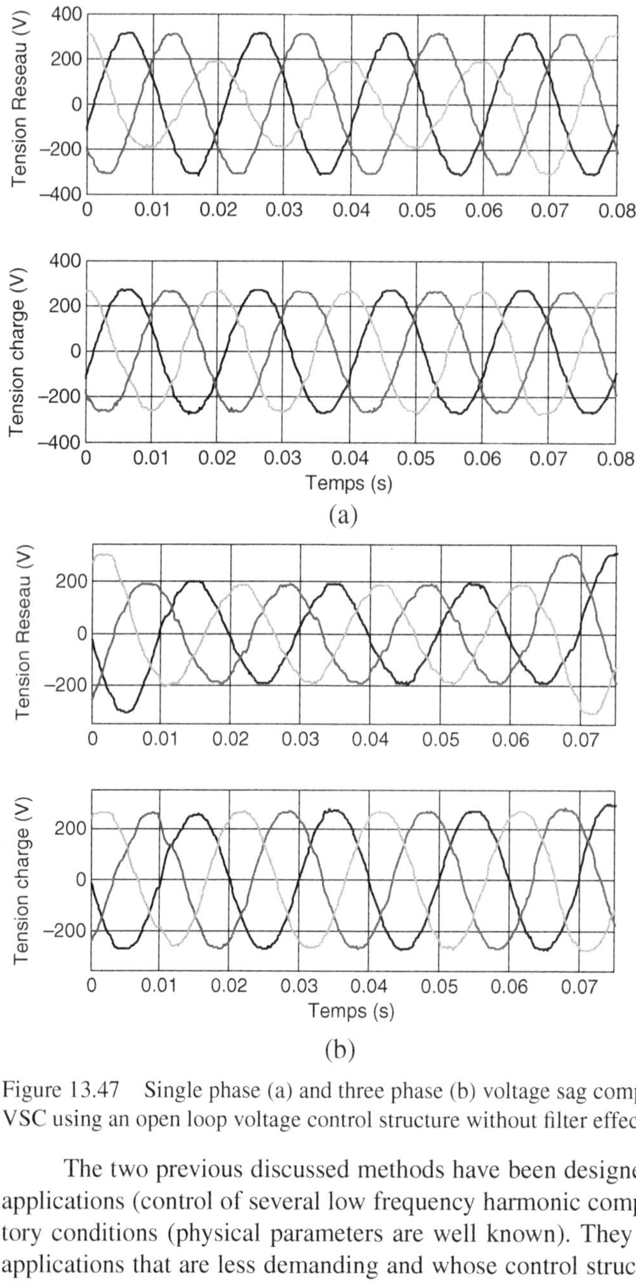

Figure 13.47 Single phase (a) and three phase (b) voltage sag compensation with a series VSC using an open loop voltage control structure without filter effect compensation.

The two previous discussed methods have been designed for very demanding applications (control of several low frequency harmonic components) but in laboratory conditions (physical parameters are well known). They can be used for other applications that are less demanding and whose control structures can be then simpler. For instance, Figure 13.47 shows the compensation of a single phase voltage sag on the upper part (a) and the compensation of a three phase voltage sag on the bottom part, experimentally obtained using an open loop voltage control structure without compensation of the effects of the coupling filter[2]. It can be noticed that even if the

[2] The results have been obtained experimentally with a switching frequency of 10 kHz and a sampling frequency of 5 kHz.

compensator is not able to filter the harmonic components of the voltage, the sags are correctly compensated and therefore even if simple, this method could be suitable for some applications of load protection.

REFERENCES

[1] Chokhawala, R., Danielsson, B., and Angquist, L. Power semiconductors in transmission and distribution applications. *13th International Symposium on Power Semiconductor Devices and ICs, ISPSD'01*, 2001.

[2] Sood, V. K. *HVDC and FACTS Controllers. Applications of Static Converters in Power Systems.* Kluwer Academic Publishers, Boston, 2004.

[3] Stemmler, H. High-power industrial drives. *Proceedings of the IEEE*, vol. 82, no. 8, pp. 1266–1286, August1994.

[4] Yu, Q., Li, P., Liu, W., and Xie, X. Overview of STATCOM technologies. *IEEE International Conference on Electric Utility Deregulation, Restructuring and Power Technologies (DRPT2004)*, 2004.

[5] Steimer, P. K., Gruning, H. E., Werninger, J., Carroll, E., Klaka, S., and Linder, S. IGCT-a new emerging technology for high power, low cost inverters. *Industry Applications Conference*, New Orleans, October 1997.

[6] Axelsson, U., Holm, A., Liljegren, C., Eriksson, K., and Weimers, L. Gotland HVDC LIGHT transmission—world's first commercial small scale DC transmission. *CIRED Conference*, Nice-France, May 1999.

[7] Grunbaum, R. SVC light: a powerful means for dynamic voltage and power quality control in industry and distribution. *8th International Conference on Power Electronics and Variable Speed Drives*, London, September 2000.

[8] Ozpineci, B., and Tolbert, L. M. *Comparison of Wide-Bandgab Semiconductors for Power Electronics Applications.* Oak Ridge National Laboratory, ORNL/TM-2003/257, 2003. Available: http://web.ornl.gov/~webworks/cppr/y2001/rpt/118817.pdf.

[9] Friedrichs, P. Silicon carbide power devices—status and upcoming challenges. *European Conference on Power Electronics and Applications*, Aalborg, September 2007.

[10] Bruckman, M., Sommer, R., Fasching, M., and Sigg, J. Series connection of high voltage IGBT modules. *33rd IEEE-IAS Annual Meeting*, vol. 2, pp. 1067–1072, October 1998.

[11] Githiari, A. N., and Palmer, P. R. Analysis of IGBT modules connected in series. *IEE Proceedings G - Circuits, Devices and Systems*, vol. 145, no. 5, pp. 354–360, October 1998.

[12] Wang, Y., Abu, M., Palmer, K., and Palmer, P. Controlled switching of high voltage IGBTs in series. *IEEE Conference on Electron Devices and Solid-State Circuits*, December 2003.

[13] Palmer, P., and Githiari, A. The series connection of IGBT's with active voltage sharing. *IEEE Transactions on Power Electronics*, vol. 12, no. 4, pp. 637–644, 1997.

[14] Chen, J. F., Lin, J. N., and Ai, T. H. The techniques of the serial and paralleled IGBTs. *Proceedings of the 22nd IEEE International Conference on Industrial Electronics, Control, and Instrumentation – IECON*, Taipei, August 1996.

[15] Schütze, D. Series connection of IGBTs. *EUPEC*, 2001.

[16] Podlesak, T., McMurray, J., and Carter, J. Solid state switch utilizing GTO's in series. *Proceedings of the 24th Intersociety Energy Conversion Engineering Conference*, Washington, August 1989.

[17] Busatto, G., Cascone, B., Fratelli, L., and Luciano, A. Series connection of IGBTs in hard-switching applications. *33rd IEEE-IAS Annual Meeting*, 1998.

[18] Zhou, D., and Braun, D. A practical series connection technique for multiple IGBT devices. *Proceedings of 2001 IEEE Power Electronics Specialists Conference*, Vancouver, Canada, June 17–21, 2001.

[19] Gerster, C. Fast high-power/high-voltage switch using series-connected IGBTs with active gate-controlled voltage-balancing. *IEEE Applied Power Electronics Conference – APEC'94*, Orlando, FL, 1994.

[20] Gerster, C., Hofer, P., and Karrer, N. Gate-control strategies for snubberless operation of series connected IGBTs. *27th IEEE Annual Conference on Power Electronics Specialists Conference*, Baveno, vol. 2, pp. 1739–1742, June 1996.

[21] Belverde, G., Galluzzo, A., Melito, M., Musumeci, S., and Raciti, A. Snubberless voltage sharing of series-connected insulated-gate devices by a novel gate control strategy. *IEEE Transactions on Power Electronics*, vol. 16, no. 1, pp. 132–141, January 2001.

[22] Park, S., and Jahns, T. M. Flexible dv/dt and di/dt control method for insulated gate power switches. *IEEE Industry Applications Conference*, Chicago, 2001.

[23] Abbate, C., Busatto, G., Fratelli, L., Iannuzzo, F., Cascone, B., and Giannini, G. Series connection of high power IGBT modules for traction applications. *2005 European Conference on Power Electronics and Applications*, Dresden, 2005.

[24] Piazzesi, A., and Meysenc, L. Series connection of 3.3 kV IGBTs with active voltage balancing. *35th Annual IEEE Power Electronics Specialists Conference*, Aachen, Germany, 2004.

[25] Raciti, A., Belverde, G., Galluzzo, A., Greco, G., Melito, M., and Musumeci, S. Control of the switching transients of IGBT series strings by high-performance drive units. *IEEE Transactions on Industrial Electronics*, vol. 48, no. 3, pp. 482–490, 2001.

[26] Flury, G., Gollentz, B., and Conilh, C. Fault-tolerant inverter for subsea applications. *European Conference on Power Electronics and Application*, Barcelona, Spain, 2009.

[27] Yang, H., Xuhui, W., Lingyun, G., Li, W., and Feng, W. Investigation of parallel connection of IGBTs. *Proceedings of the 8th International Conference on Electrical Machines and Systems – ICEMS*, 2005.

[28] Soto, D., and Green, T. C. A comparison of high-power converter topologies for the implementation of FACTS controllers. *IEEE Transactions on Industrial Electronics*, vol. 49, no. 5, pp. 1072–1080, October 2002.

[29] Janning, J., and Bocquel, A. Multi-level medium voltage inverters for reactive power and harmonic compensation. *European Conference on Power Electronics and Applications*, Dresden, 2005.

[30] Baker, R. H. *Electric Power Converter*. U.S. Patent 3,867,643, February 1975.

[31] Sirisukprasert, S., Huang, A. Q., and Lai, J. S. Modeling, analysis and control of cascaded-multilevel converter-based STATCOM. *IEEE Power Engineering Society General Meeting*, Toronto-Ontario, Canada, July 2003.

[32] Nabae, A., Takahashi, I., and Akagi, H. A neutral-point-clamped PWM inverter. *IAS Annual Meeting*, Cincinnati, Ohio, USA, pp. 761–766, 1980.

[33] Meynard, T. A., and Foch, H. Multi-level conversion: high voltage choppers and voltage-source inverters. *Power Electronics Specialists Conference – PESC*, Toledo, 1992.

[34] Meynard, T. A., and Foch, H. Multi-level conversion: high voltage choppers and voltage-source inverters. 23rd Annual IEEE Power Electronics Specialists Conference PESC'92, Toledo, vol. 1, pp. 397–403, 1992.

[35] Bacha, S., Etxeberria, I., Crappe, M., Fahe, S., Styczynski, Z. A., Courtois, C., Talibart, A., Fracchia, M., Stuart, M., Consani, T., Merli, E., and Borgignon, P. Using SVC for voltage regulation in railways network High Voltage Booster European project. *9th European Conference on Power Electronics and Applications - EPE 2001*, Graz, Austria, August 2001.

[36] Benchaib, A., Lee-Kwet-Sun, E., Thierry, J. L., and de Preville, G. Industrial D-STATCOM chain link modelling and control. *12th European Conference on Power Electronics and Applications – EPE'07*, Aalborg, Denmark, September 2007.

[37] Affolter, R., and Connell, B. Experience with a dynamic voltage restorer for a critical manufacturing facility. *2003 IEEE PES Transmission and Distribution Conference and Exposition*, vol. 3, pp. 937–939, September 2003.

[38] Davidson, C. C., and De Préville, G. The future of high power electronics in transmission and distribution power systems. *13th European Conference on Power Electronics and Applications – EPE'09*, Barcelona, Spain, September 2009.

[39] Etxeberria-Otadui, I., Lopez de Heredia, A., Gaztanaga, H., Bacha, S., and Reyero, M. R. A single synchronous frame hybrid (SSFH) multifrequency controller for power active filters. *IEEE Transactions on Industrial Electronics*, Vol. 53, no. 5, pp. 1640–1648, October 2006.

[40] Etxeberria-Otadui, I. Sur les Systèmes de l'Electronique de Puissance dédiés à la Distribution Electrique—Application à la Qualité de l'Energie. Ph.D. Thesis, National Polytechnic Institute of Grenoble, France, 2003.

PART III

ARTIFICIAL INTELLIGENCE TECHNIQUES

In the memory of our friend Alain Germond

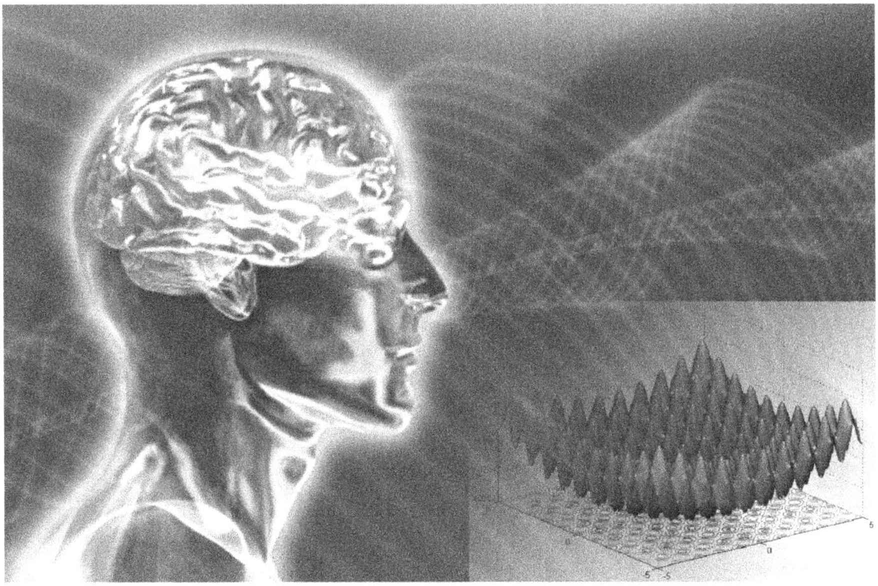

Advanced Solutions in Power Systems: HVDC, FACTS, and Artificial Intelligence, First Edition.
Edited by Mircea Eremia, Chen-Ching Liu, and Abdel-Aty Edris.

14. Artificial Intelligence and Computational Intelligence: A Challenge for Power System Engineers
Chen-Ching Liu, Alexandru Stefanov, and Junho Hong
15. Expert Systems
Mircea Eremia, Kevin Tomsovic, and Gheorghe Cârțină
16. Neural Networks
Dagmar Niebur, Ganesh Kumar Venayagamoorthy, and Ekrem Gursoy
17. Fuzzy Systems
Germano Lambert-Torres, Luiz Eduardo Borges da Silva, Carlos Henrique Valerio de Moraes, and Yvo Marcelo Chiaradia Masselli
18. Decision Trees
Constantin Bulac and Adrian Bulac
19. Genetic Algorithms
Anastasios Bakirtzis and Spyros Kazarlis
20. Multiagent Systems
Nan-Peng Yu and Chen-Ching Liu
21. Heuristic Optimization Techniques
Kwang Y. Lee, Malihe M. Farsangi, Jong-Bae Park, and John G. Vlachogiannis
22. Unsupervised Learning and Hybrid Methods
Nikos Hatziargyriou and Manolis Voumvoulakis

CHAPTER 14

ARTIFICIAL INTELLIGENCE AND COMPUTATIONAL INTELLIGENCE: A CHALLENGE FOR POWER SYSTEM ENGINEERS

Chen-Ching Liu, Alexandru Stefanov, and Junho Hong

AS DISTRIBUTED GENERATION resources increase in the power grid, the inherent uncertainties of wind and solar energy will lead to greater complexity in power system operation, control, forecasting, and scheduling. At the same time, the interest in smart grid has grown significantly. A smart grid environment integrates advanced information and communications technology for grid monitoring and "smart" energy delivery, control, and energy consumption. The integration of plug-in hybrid and electric vehicles further adds new challenges to the control of power grids at all voltage levels. Smart grid is an important application area for artificial intelligence (AI) and computational intelligence (CI), as solutions to complex problems in power system engineering and electric energy markets depend on logic reasoning, heuristic search, perception, and the abilities to handle uncertainties. As higher volumes of information need to be processed, AI and CI offer new solutions by centralized or distributed intelligence. The enormous amount of information in the smart grid environment can be managed by CI that serves as decision support tools.

AI is concerned with decision-making capabilities such as knowledge representation, search methods, inference techniques, heuristic reasoning, and machine learning. CI techniques include expert systems, fuzzy logic, genetic algorithms (GAs), and artificial neural networks (ANNs). CI can further involve adaptive mechanisms for intelligent behaviors in complex environments, such as the ability to adapt, generalize, abstract, discover, and associate. These techniques provide new opportunities for the field of smart grid and other computer applications in power engineering.

AI and CI provide efficient tools for planning, operation, maintenance, market monitoring, and risk management. This new generation of technologies offers new perspectives complementary to other computer solutions. CI tools can be used to combine the power grid operations with trading and ancillary services of the electric energy markets. Moreover, CI tools have demonstrated their feasibility and

Advanced Solutions in Power Systems: HVDC, FACTS, and Artificial Intelligence, First Edition.
Edited by Mircea Eremia, Chen-Ching Liu, and Abdel-Aty Edris.

benefits in key areas, for example, distribution automation, load forecasting, operation and planning, and system diagnosis and restoration. For example, pattern recognition methods and ANNs can be deployed to assess the security of power grids with respect to various contingencies. Neural networks may be used to forecast the levels of load and determine the amounts of energy supply from power plants. Combined with fuzzy logic, they can be utilized in price forecasting for energy markets. The economic study of the power market can be enhanced by the use of multi-agent systems, machine learning, data mining, and game theory techniques. For planning and scheduling, problems such as unit commitment, optimal dispatch, network reconfiguration, and maintenance scheduling can be addressed by the use of CI or bioinspired techniques, for example, ANNs, fuzzy systems, GAs, swarm intelligence, ant colonies, simulated annealing, and tabu search. Expert systems and knowledge-based systems can be used for alarm processing and decision support for system restoration. Fault diagnosis can be accomplished by neural networks based on computer applications, which recognize fault types based on the voltage and current waveforms. Agents and data mining offer useful applications to smart grids for information mining.

Smart grid technologies rely on intelligent systems for handling of uncertainties. These can be provided by both AI and CI techniques. AI uses descriptions and associations to achieve the desired state. The CI techniques, for example, fuzzy logic, ANNs, evolutionary computation, and swarm intelligence, are based on randomized processes consisting of iterative cycles of generation and evaluation. They provide intelligent behaviors under uncertainty conditions, being able to anticipate information that may support the decision making of the power grid.

Distributed and coordinated intelligence is one of the enabling technologies for smart grid. Operation of the existing grid is currently performed by system operators. CI techniques may enhance the smarter operation of power systems by automating appropriate tasks. They can be used in many areas of the grid management. However, CI technologies should not be operated in an isolated manner as the interpretation of data sets by a system operator may be in the benefit of others. CI interoperability and information exchange are essential for the smart grid environment. Various autonomous systems may interoperate, for example, by the use of multi-agent system technology. In this approach, each of the systems meets its own objectives, while exchanging standard messages with other agents. The multi-agent architecture for the smart grid can take advantage of the existing communication standards.

AI and CI have been applied to address the challenges that arise from the increasing complexity and highly nonlinear nature of electric power systems. AI and CI techniques provide effective solutions to the design of nonlinear, adaptive, and optimal controllers for generator excitation systems, High-Voltage Direct Current (HVDC), and Flexible Alternating Current Transmission System (FACTS) devices. AI and CI techniques are used for the identification of faults in HVDC systems. ANNs, fuzzy logic, and GAs have applications to the detection and classification of different AC side faults and DC faults in HVDC. An ANN training algorithm for a standard 12-pulse line-commutated converter HVDC was reported in [1] for the identification and classification of various faults. An online current controller based on ANNs was proposed for HVDC systems in [2]. The inverse model ANN controller has advantages over the traditional Proportional-Integral (PI), HVDC controllers in terms of adaptability, damping, and noise immunity. FACTS devices are mainly used for voltage stability, reactive power compensation, power flow control, and dynamic stability enhancement. AI and CI are used to integrate the FACTS devices into modern power systems, for example, identification of type, number, size, settings, and optimum location. For example, particle swarm optimization was used to identify the optimal location for placement and parameters of Static VAr Compensators (SVCs) [3]. AI and CI methods are applied to solve the reactive power planning and optimal power dispatch problems that incorporate FACTS devices and controllers. The state-of-the-art of various AI and CI methods used to solve the optimal power dispatch for power systems with FACTS was reported in [4]. The major advantages in using AI and CI are the capabilities to handle various qualitative constraints, find solutions to multi-objective optimization problems, and identify the global optimum solution with a high probability.

The theoretical foundation of AI and CI and their applications in power systems are presented in a recently published Wiley-IEEE Press book [5]. The following chapters in this volume provide their comprehensive expansion that help enabling the smart grid, that is, expert systems, ANNs, fuzzy systems, decision trees (DTs), GAs, multi-agent systems, heuristic optimization, and unsupervised learning.

Expert systems are computer solutions that offer expertise in certain areas. The expert knowledge may be represented by rules, models, or DTs. It is gathered from years of experience of operators and system engineers, and used to support solution of complex problems. The embedded knowledge and inference mechanism allow logic reasoning. Expert systems are suitable for applications that must process a large amount of data in a short time period, for example, real-time control, operations, and planning. Control applications were designed for SVC, generator voltage regulators, and plant operations. Expert systems can be used to monitor equipment, diagnose faults, or process alarms. Logic and heuristics can be applied to diagnose abnormal operating conditions, for example, faults. Their applications are also extended to support the decision making of operators during system restoration. They are suitable for operator training as well. Rule-based expert systems can be used to support transmission planning tasks. Other applications are concerned with relay settings.

ANNs are biologically inspired systems of neurons that generate outputs as nonlinear functions of its weight-controlled inputs. During supervised or unsupervised training, the network learns the weights of inputs. *ANNs are characterized by their learning and self-organization capabilities.* ANNs have power grid-related applications, for example, generator controls, intelligent relays, monitoring and diagnosis, and network fault detection. ANNs have been developed for power system security assessment as well as for load forecasting.

Fuzzy systems provide techniques for uncertainty handling. Unlike traditional probabilistic techniques that are based on precise theories, valuable knowledge can be based on subjective judgment. A basic function of fuzzy logic describes the degree of participation with a value between zero and one. By adopting the partial set membership, fuzzy systems are useful in a range of applications where the available information involves uncertainty. Fuzzy sets can be used to translate inputs from numerical to symbolic, and vice versa, for both input and output. A significant difference between ANNs and fuzzy logic is that fuzzy systems do not have learning and adaptation capabilities. A specific application to smart grid is intelligent control. Many fuzzy-based applications are combined with expert systems and ANNs. Fuzzy logic can be used for applications such as real-time control, system restoration, load forecasting, monitoring, and diagnosis.

DTs are classifiers predicting class labels for data items. They are constructed by analyzing a set of training examples and applied to unknown scenarios. DTs are useful tools to support accurate predictions when they are trained with high quality data sets. DT applications include machine-learning tools for power system dynamic performance assessment and control.

GAs are based on heuristic search that solve optimization problems by random generation of new solutions and continue to improve the quality of the solution over iterations. After the initial population of solutions was generated, crossover and mutation techniques are used to obtain a new generation of solutions. The quality of solutions depends on the dimension and coverage of the solution space. GAs have applications in power system design, operation, unit commitment, and maintenance scheduling. GAs are used for large-scale, complex, and nonlinear optimization problems.

Multi-agent systems are a collection of heterogeneous agents with different goals that exchange messages for efficient coordination. An agent is a computer program with autonomous actions to fulfill its objectives. Agents can communicate with each other and cooperate to achieve their common goals. The control functions and decision-making authority are distributed to areas where agents operate. Applications of multi-agent systems to smart grid include monitoring, diagnosis, distributed control, grid restoration, and reconfiguration.

Heuristic optimization consists of techniques that mimic behaviors that can be found in nature, for example, ant colony optimization, simulated annealing, tabu search, evolutionary computing, and particle swarm optimization. Such optimization methods are robust and perform well computationally. They can be widely applied to smart grid, for example, reactive power and voltage control, system security, power system expansion planning, and design.

Unsupervised learning methods explore the structure of data without guidance that can reveal unknown or unexpected features. Applications include clustering, dimensionality reduction, feature extraction, and vector quantization. Unsupervised learning methods can be used for solving various power grid-related problems, for example, security control.

CYBER SECURITY IN A SMART GRID ENVIRONMENT. Smart grid provides intelligent means of managing the energy supply and consumption by taking advantage of two-way communications between energy suppliers and customers. It integrates conventional power grids with their data communication networks to allow fast and efficient data collection and analysis. However, the IP-based communications introduce new threats to cyber security of the smart grid. Different cyber threats, for example, malware, spyware, viruses, and direct intrusions, can exploit the vulnerabilities introduced by new distributed technologies such as smart meters and sensors. Data confidentiality, integrity, and availability are critical issues. New technologies are essential to enhance cyber security of the smart grid. Cyber intrusions that concern both communication networks and power systems must be detected quickly and their impacts mitigated.

Ongoing research is addressing the exposure of a smart grid to cyber threats. In [6], a distributed intrusion detection system (IDS) is proposed applied to a multi-layer network architecture of a smart grid. Multiple intelligent analyzing modules have been developed and deployed in the multiple layers of the smart grid, for example, home area networks, neighboring area networks, and wide area networks. Support vector machines and artificial immune systems (AISs) are applied for malicious data classification and cyber attacks detection. The analytical modules are trained at each level of the smart grid, which is able to exchange relevant information for efficient detection. In [7], the authors present the architecture and design issues of a threat prevention and sensing engine based on Intelligent System (IS) techniques, for example, ANNs and fuzzy logic. Petri net techniques are used in a vulnerability assessment framework to systematically evaluate the vulnerabilities of cyber security for Supervisory Control And Data Acquisition (SCADA) systems [8]. The cyber intrusions and cyber-net embedded with firewall and password models are analyzed by a generalized stochastic Petri net. The impact of the intrusion is evaluated by the potential loss of load in the power grid. Substation automation is highly dependent on computer and communication networks. Thus, its cyber security is important for the smart grid environment. An anomaly inference algorithm considering temporal anomalies has been proposed for early detection of malicious intrusions in the computer network of a substation [9]. Cyber intrusions are detected that may affect the operation of substation automation and damage the power grid.

CI techniques, for example, fuzzy logic, neural networks, agent-based systems, evolutionary computation, AISs, and cellular automata, can be applied in the area of information security providing solutions to different cyber security applications. Defense systems can take the advantage of AI for intelligent decisions in detecting a broad range of cyber attacks. Solutions such as IDSs may deploy the agent-based technology, for example, autonomous, intelligent, or mobile agents. Intelligent agents can be used for classifier algorithms to detect suspicious activities. AI techniques can provide intelligent packet inspection as data flow is continuously increasing and there is limited time for full packet inspections. Intelligent response against malicious activities is essential for minimizing the associated impact.

The difficulty is in programming the needed applications described by invisible and continuous learning processes, while connections between events are enabled. Ongoing research is aimed at offering new solutions to such issues. For example, traditional statistical methods are combined with AI techniques in analyzing the data acquired from sensors, recognizing patterns, correlating events, and detecting malicious behaviors. The intrusion detection models require AI techniques to enhance the learning and adapting capabilities over time. Rule-based applications can be used to automatically develop rules in order to explain the collected historical data and signal anomalies. Threat detection systems based on neural networks can make predictions using known and unknown patterns. A network can be trained to filter incoming network connections and detect abnormal behaviors. Today's communication networks are described by protocols at different communication layers with many services. The many dimensions of a network connection are reduced significantly by deploying the property of dimensionality reduction and data visualization in neural networks. Thus, the connections that are not from the same group of clusters with the trusted ones can be discovered.

It is becoming increasingly difficult to identify cyber security solutions based on conventional fixed algorithms for defending against continuously evolving attacks. The large amount of information must be processed very fast. Events in the cyber system can be correlated and decisions are needed in real-time. These challenges can be overcome by deploying AI techniques that provide processing power, flexibility, and learning capabilities to software. Intelligent cyber attacks can be counteracted by intelligent security mechanisms. AI and knowledge-based applications deploy new security mechanisms, for example, dynamic security perimeters, situation awareness, IDS, and automatic incident response.

ANNs have been applied to intrusion prevention and detection systems, denial of service recognition, malware detection and classification, and forensics. Multi-agent systems have been developed to guard against distributed denial of service attacks. They can assess accurately the state of the cyber system. Combined with ANNs, solutions are identified for the agent-based distributed intrusion detection.

Distributed IDSs for smart grid may rely on AISs to detect and classify malicious data. AISs are intelligent algorithms that model human immune systems involving learning, memory, and optimizing abilities. IDS based on AISs can efficiently detect network intrusions. Existing AI and CI techniques are useful tools to address issues arising from cyber threats in the smart grid environment. However, intelligent solutions are still needed, for example, knowledge handling during incident response and automatic decision making.

Distribution Automation in a Smart Grid Environment. A primary objective of smart grid development is to enhance further automation at all levels of the power grid. Through advanced information and communications technology, the power grid and end users are monitored for efficient and sustainable energy consumption. Distribution systems have special considerations in the smart grid environment as most of the power outages and disturbances occur at these levels. Distribution automation systems (DASs) can provide utilities with flexible control capabilities over their assets and reliable electricity distribution to customers. DAS includes automation at substations, feeders, and customer side facilities. Applications of DASs and distribution management systems are critical parts of the smart grid environment. Some examples of applications are network reconfiguration, fault identification, isolation and service restoration, integrated voltage/reactive power control, automated meter reading and advanced metering infrastructure, network modeling, emergency load shedding, and load management. One of the main features of a DAS for smart grid is the self-reconfiguration capability especially when intervention of site engineers is difficult.

Operational support tools based on rule-based systems are proposed in [10] for restoration and loss reduction of distribution systems. The expert system techniques have been applied to distribution feeder restoration. Several steps may be included such as group restoration, zone restoration, and load transfer. An expert system was developed for online decision support of system operators in a distribution SCADA environment. The capabilities of DAS in providing efficient restoration solutions were extended from single-fault to multiple-fault cases in [11]. A two-step

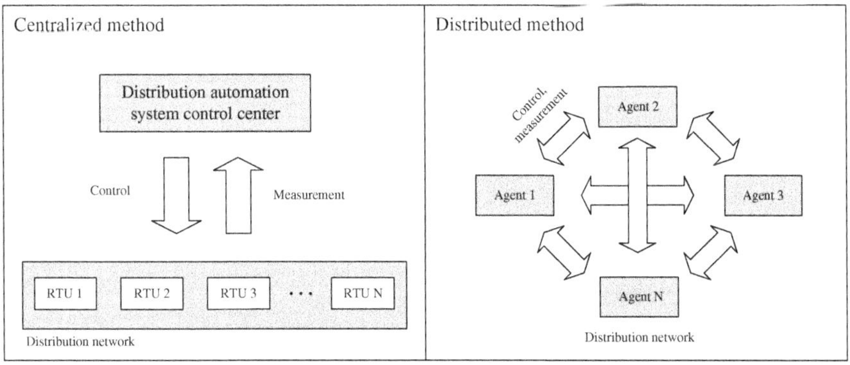

Figure 14.1 Centralized and distributed method in a distribution automation system.

restoration scheme, that is, sequential and simultaneous, has been proposed for multiple fault scenarios. The AI method was extended, implemented, and operated in the real world environment.

A DAS has a complex structure and its operation requires high level knowledge since it is connected to customers directly. The most important role of a DAS is to supply reliable electricity to customers, for example, detecting a fault and restoring the distribution system after the fault is isolated. DAS has evolved from a centralized to a distributed method [12]. The centralized method collects information from distributed remote terminal units (RTUs) and makes a decision at the distribution automation system control center, while distributed method uses a multi-agent system (MAS) that is able to make its own decisions (see Figure 14.1).

The agent and multi-agent system technologies have been applied for enhancing power system stability and reliability [13, 14]. Agents can be used to acquire data, monitor a system, and identify an optimal control action in a DAS. Installation of multi-agent systems is intended to enhance distribution system reliability.

Multi-agent system applications are under development for the distribution system restoration in a smart grid environment. The installation of a multi-agent module to protective devices such as reclosers and relays and RTUs provide the foundation for intelligence, sensors, and actuators in a smart distribution system. The main framework for distribution system restoration has three steps, that is, update restoration plan, fault detection, and MAS-based distribution system restoration.

At the first stage, the restoration plan is updated. After installation of an MAS at a distribution system, an adaptation process is needed. This includes transferring communication information and the reliable restoration plan. The communication information about a central agent and all terminal agents will be transferred from the central agent. This information includes IP address, protocol, agent ID, and so on. The MAS is flexible in that it allows the system to add or remove agents. A terminal agent can share the information about event and measurements with other agents. In a smart grid environment, there are many uncertainties since distributed generators can be added to or removed from a distribution system. A distribution system restoration plan needs to be changed according to the changing conditions of the distribution system. This information is transferred from the central agent to terminal agents since

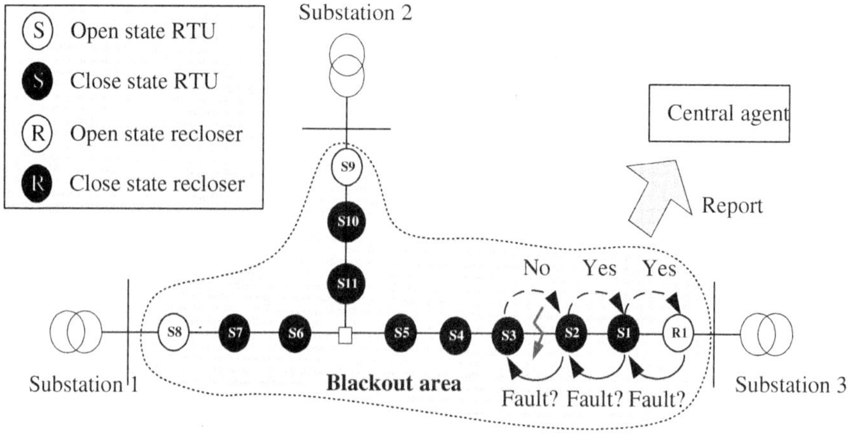

Figure 14.2 Multi-agent-based fault location detection.

the restoration plan is calculated at the central agent for enhancing reliability. The main role of a central agent is to support adaptation of the terminal agents.

Detecting the fault location at stage two is an important part of distribution system restoration. There is an advantage to use a multi-agent-based system for a DAS. The agents will find the fault location with exchanging information and cooperation between agents. Figure 14.2 is depicting the detection process of fault location using a multi-agent system. When a fault occurred between S3 (RTU 3) and S2, a fault current will flow from substation 3 to the faulted section. Recloser 1, R1, will open and will reclose one or more times until the fault is no longer seen by the recloser or until it is locked out. After opening R1, the outage area (R1-S8-S9) will be generated as shown in Figure 14.2. The agents representing the recloser and RTU will send unsolicited fault indicator message to other agents until they receive the “no” message. Therefore, MAS is able to detect the fault location by cooperation among agents.

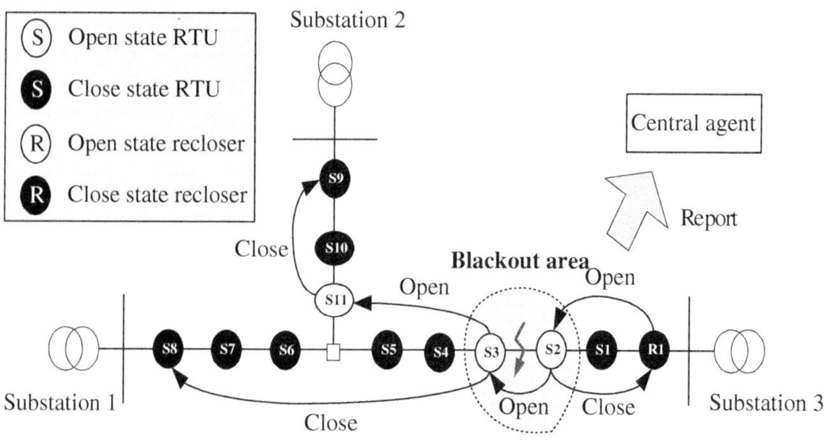

1. R1-S2 (open), S2-R1 (close) 2. S2-S3 (open), S3-S11 (open), S3-S8 (close) 3. S11-S9 (close)

Figure 14.3 Multi-agent-based system restoration.

Distributed restoration using MAS is performed at stage three. After detecting the faulted area, MAS will minimize the fault area and perform restoration actions, which were downloaded at the first stage when updating the restoration plan. As shown in Figure 14.3, R1 sends "open" command to S2 and S2 replies with a "close" command to R1 in order to restore area R1-S1. Then, S2 sends "open" command to S3, and S3 sends "open" command to S11. After confirming that S11 is open, S3 sends a "close" command to S8 to restore area S4-S8. Finally, S11 sends a "close" command to S9. The restoration process is then completed.

REFERENCES

[1] Sanjeevikumar, P., Paily, B., Basu, M., and Conlon, M. Classification of fault analysis of HVDC systems using artificial neural network. *Proceedings of International Universities Power Engineering Conference*, IEEE UPEC, September 2–5, 2014.

[2] Narendra, K. G., Sood, V. K., Khorasani, K., and Patel, R. V. Investigation into an artificial neural network based on-line current controller for an HVDC transmission link. *IEEE Transactions on Power Systems*, vol. 12, no. 4, pp. 1425–1431, November 1997.

[3] Sheeba, R., Jayaraju, M., Mansoor, M., Shanavas, T. N., and Sundareswaran, K. Identification of optimal location of SVC through artificial intelligence techniques. *Proceedings of Innovative Smart Grid Technologies*, IEEE ISGT, December 1–3, 2011.

[4] Burade, P. G., and Helonde, J. B. A novel approach for optimal power dispatch using artificial intelligence methods. *Proceedings of International Conference on Control, Automation, Communication and Energy Conservation*, IEEE INCACEC, June 4–6, 2009.

[5] Lee, K. Y., and El-Sharkawi, M. A. *Modern Heuristic Optimization Techniques: Theory and Applications to Power Systems*. In K. Y. Lee, and M. A. El-Sharkawi, Eds., IEEE Press Book, ISBN: 978-0-471-45711-4, Wiley, New York, 2008.

[6] Zhang, Y., Wang, L., Sun, W., Green II, R. C., and Alam, M. Distributed intrusion detection system in a multi-layer network architecture of smart grids. *IEEE Transactions Smart Grid*, vol. 2, no. 4, pp. 796–808, December 2011.

[7] Parati, N., Malik, L., and Joshi, A. G. Artificial intelligent based threat prevention and sensing engine: architecture and design issues. *Proceedings of International Conference on Emerging Trends in Engineering and Technology*, IEEE ICETET, July 16–18, 2008.

[8] Ten, C.-W., Liu, C. C., and Govindarasu, M. Vulnerability assessment of cybersecurity for SCADA systems. *IEEE Transactions on Power Systems*, vol. 23, no. 4, pp. 1836–1846, November 2008.

[9] Ten, C.-W., Hong, J., and Liu, C. C. Anomaly detection for cybersecurity of the substations. *IEEE Transactions on Smart Grid*, vol. 2, no. 4, pp. 865–873, December 2011.

[10] Liu, C. C., Lee, S. J., and Venkata, S. S. An expert system operational aid for restoration and loss reduction of distribution systems. *IEEE Transactions on Power Systems*, vol. 3, no. 2, pp. 619–626, May 1988.

[11] Lim, S., Lee, S. J., Choi, M. S., Lim, D. J., and Hua, B. N. Service restoration methodology for multiple fault case in distribution systems. *IEEE Transactions on Power Systems*, vol. 21, no. 4, pp. 1638–1644, November 2006.

[12] Lim, I. H., Kim, Y. I., Lim, H. T., Choi, M. S., Hong, S., Lee, S. J., Lim, S. I., Lee, S. W., and Ha, B. N. Distributed restoration system applying multi-agent in distribution automation system. *Proceedings of IEEE PES General Meeting*, July 20–24, 2008.

[13] McArthur, S. D. J., Davidson, E. M., Catterson, V. M., Dimeas, A. L., Hatziargyriou, N. D., Ponci, F., and Funabashi, T. Multi-agent systems for power engineering applications - part I: concepts, approaches, and technical challenges. *IEEE Transactions on Power Systems*, vol. 22, no. 4, pp. 1743–1752, November 2007.

[14] Li, H., Sun, H., Wen, J., Cheng, S., and He, H. A fully decentralized multi-agent system for intelligent restoration of power distribution network incorporating distributed generations. *IEEE Computational Intelligence Magazine*, vol. 7, no. 4, pp. 66–76, November 2012.

CHAPTER 15

EXPERT SYSTEMS

Mircea Eremia, Kevin Tomsovic, and Gheorghe Cârțină

15.1 FUNDAMENTAL CONCEPTS

15.1.1 Definitions

Generally, an "expert" is someone who possesses detailed knowledge in a particular field of interest, a high class specialist, who can perform an expertise on certain issues. An expert must be able to:

- make a precise formulation of the problem given data provided by the user;
- determine an appropriate solution and the particular steps to be taken;
- explain the decision;
- provide support for realization of the solution in a particular field.

In terms of functionality, an "expert system" can be defined as a program that relays on a bank of knowledge to perform certain specific tasks, which can be sometimes difficult, and are usually solved by the human expert in the field. The performances of an intelligent program depend primarily on the "quality" of the knowledge base (KB) and the organization of the knowledge, which should allow the problem to be solved within an acceptable period. An academic definition of the expert system was given by Professor Edward Feigenbaum of Stanford University—-United States: "*... an intelligent program, which utilizes knowledge, facts and reasoning techniques for solving a problem that usually requires the knowledge of a human expert*" [1].

In this respect, an expert system is a data processing system, which, in addition to classical computational elements, contains a KB (consisting of facts and rules relating these facts), and the elements necessary for handling the knowledge (in the form of an inference mechanism). It should also contain components necessary for acquiring new knowledge, representation of results, and the analysis of the result (i.e., the explanation component).

Similar to a human expert, who relies on one's own knowledge and reasoning to reach a conclusion, an expert system relays on its embedded knowledge and on the inference mechanism to support the processes of logical reasoning.

Expertise carried out by a human specialist relies on "long-term memory," made up of facts, structures, and rules, which correspond to the KB of an expert system.

Advanced Solutions in Power Systems: HVDC, FACTS, and Artificial Intelligence, First Edition.
Edited by Mircea Eremia, Chen-Ching Liu, and Abdel-Aty Edris.

The expert system mimics the thinking process of the human expert. When solving a problem, the human expert applies a "reasoning method" to the long-term memory or knowledge of the general problem and the specific case being considered. The part of the expert system, which performs the reasoning function, is called the "inference engine." The same inference engine (IE) can be used for reasoning with different KBs.

In the literature, a number of similar definitions can be found, including:

- "An expert system is series of software programs that rely on current information, a database and a set of inference rules for solving a problem which requires high human competence. The database and the inference rules are shaped by the competence of the best practitioners in solving the problem" [2].
- "An expert system collects the knowledge of a human expert in a particular field, in a form which can be implemented on a computer. It uses them to provide support for a decision at a comparable level with the human expert and to justify its reasoning" [3].

15.1.2 Expert System Characteristics

Expert systems are significantly different from conventional data processing systems. Contrary to traditional data processing programs, which use predetermined data and algorithms for solving problems, the expert system generally implies a more diverse approach. For example, the solution often includes symbolic representation and heuristic search. One often highlighted difference between classical programming and expert systems is constituted by the difference between algorithmic programming languages and declarative programming languages. The algorithmic approach includes, as an undifferentiated ensemble, the logic elements needed to solve a problem, the specific properties of the studied objects (functions), and the specific parameters. An interesting approach in the concept of computational programs was given in 1962 by E. Feigenbaum. He proposed a separation between the logical mechanisms, which defined in a rigorous manner, have a general character and can be applied to any problem, and the properties and parameters of each analyzed object. This form of formulating the problem is called a *declarative approach*. The declarative approach has adopted, in the writing of the application programs, the logical separation of the specific modules of each object from parameters of calculation (physical constants and variables).

In this way, expert systems are identified and separated into three levels of organization:

- The most general level is the formulation of the logic used in solving the problem. The procedures that correspond to this level can be applied to any type of problem.
- The next level contains the specific properties of each class of problems or each type of object.
- The most specific level is the data that individualize the objects or the problems.

The methods used for solving problems in many expert systems are not mathematical procedures or data processes, but techniques of qualitative reasoning[1] that bind the elements by rules of judgment or heuristics, and through laws and theoretical definitions. An algorithm is a procedure that guarantees to find a valid solution to a problem within a given time, or to show that the solution does not exist. Conversely, while the heuristic methods may not be guaranteed to work, they often find solutions in less time than using systematic or exhaustive search algorithms.

In general, the difference between an analytical program and an expert system lies in:

- The expert system uses the data from a facts base, which is filtered through a set of known facts, independent of the algorithm used for solving the problem. The knowledge is stored in a KB.
- Due to a crisp delimitation between the component elements (rules, databases, and IE), an expert system can be modified by simply adding or extracting rules, without the need for rewriting the program. In conventional programs, the knowledge regarding the problem and the solution methods are "mixed," which make subsequent changes to the program difficult. The design of an expert system is a linear process, achieved through repeated iterative discussions with the human experts for the development of the KB.
- All the logical and mathematical operations that contribute to the answer are brought to the user through a user interface, which acts as an independent explanatory module named.

For purposes of this chapter, we consider four main characteristics that qualify a program as an expert system:

- It works at an expert competence level.
- It uses an inference mechanism to achieve deduction.
- The expertise conducted by the expert system is based on specialized knowledge.
- Programming of the expert system involves the description and presentation of the knowledge of experts in the field; this knowledge will be saved in the database for future use.

The performances of expert systems depend on the accuracy and the volume of knowledge that they possess and the quality of the mechanisms, which enable them to operate with that knowledge in order to reach valid conclusions.

[1] Reasoning = a chain of judgment (sentences), starting only from certain knowledge, recorded in a number of sentences called a premise so that new knowledge is obtained and expressed in a proposition called conclusion. Reasoning is correct only if the conclusion is a logical consequence of the premises. Reason can be defined as a chain of symbolic representation, which should lead to the attainment of a purpose (to demonstrate, to persuade, to elucidate, to decide, to justify, to interpret, to explain, etc.).

15.1.3 Artificial Expertise or Human Expertise?

Expert systems offer many advantages over a human expert. Among the fundamental arguments in support of the artificial expertise contained in an expert system is:

- The expertise is not influenced by the physical or mental state of the expert. An expert may take different decisions depending on his/her situation, for example, in stress inducing conditions.
- The reasoning done by an expert system is more consistent and the results are easier to reproduce than those on its human counterpart.
- An expert system may reflect the consensus in a field more than an individual expert.

Other arguments stem from the easiness with which it is possible to transfer or reproduce knowledge existent in an expert system. Usually, the transfer of knowledge from one person to another is a long and laborious operation; electronically stored knowledge is far easier to access. There is also a clear identity between the way in which the expertise is obtained by the expert system and its natural description with the help of language. Finally, another argument is the cost of such a system, consisting only of developing and maintenance costs. This is summarized in Table 15.1 [4,5].

TABLE 15.1 Expert system or human expertise?

	Artificial Expertise	Human Expertise
Advantages	• Permanence (knowledge of an expert system can be stored for an unlimited period) • High availability (easily transferred to any computer) • Explanation facilities (can follow the chain of reason that leads to a conclusion) • Consistency (contains a level of expertise from many experts) • Fast replay	• Creativity (a human expert is capable of producing new knowledge) • Adaptability (to new situations or unpredictable events) • Utilize sensorial capacity (visual, audio, and tactile)
Disadvantages	• Lack of creativity • Limited adaptability (an expert system is not capable to learn by itself new concepts; it needs the assistance of an expert to reorganize its knowledge)	• Hard to transmit expertise • Difficulty in documentation • Unpredictable • Expensive

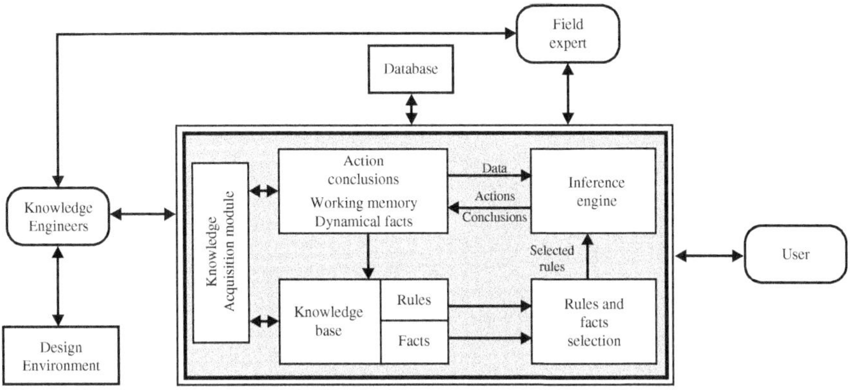

Figure 15.1 The structure of an expert system.

15.2 ARCHITECTURE OF EXPERT SYSTEMS

As discussed in Section 15.1, an expert system contains these modules: the KB, the database, the IE, the knowledge acquisition, and the software interfaces with the expert or application engineer, knowledge engineer, and user (Figure 15.1) [6].

15.2.1 Knowledge Base

The KB contains a collection of relevant knowledge about a certain field in the form of rules and facts. The rules refer to the operations that can be performed upon the objects contained in the database. The rules change relatively infrequently, for example, as a result in changes which occur in technical regulations. In essence, the rules constitute a complete and noncontradictory assembly of knowledge necessary for solving a problem. The facts represent the dynamic part of the KB and show the state of the objects at a certain moment in time.

15.2.1.1 Knowledge Representation

The representation of knowledge has a formal structure, most commonly through rules (or production rules), semantic networks, frames, object-oriented programming, or formal logic. These are described in the following.

PRODUCTION RULES. The representation of knowledge as production rules belongs to the AI category known as production systems. At the foundation of the production system is the concept of a production rule with the following structure

<condition> → <action>

that can be interpreted in the following manner:

IF <condition part> is fulfilled
THEN <action part> can be executed

Example:
IF <breaker A is switched off>
THEN <switch off the disconnecting switch B>

In other words, the condition specifies those criteria that must be fulfilled in order for the rule to be applicable. The mechanism of production rules was borrowed from the formal languages theory. In a production system, knowledge has several components as described below:

- declarative or factual knowledge, representing pieces of knowledge grouped in the form of data structures memorized in a collection termed context;
- procedural knowledge, represented in the form of condition-action type pairs, that is, production rules, and stored in a set called rule base;
- strategic or control knowledge, represented through rules that support the decision upon the sequence of actions to be taken in the process of solving the analyzed problem.

The condition part, also called premise, is made of logical propositions that must be valid in order for the rule to be applied. The action part corresponds to the firing of a conclusion or a new hypothesis. The main advantages of knowledge representation through production rules are:

- Self-modularity—-each rule can be considered an independent structural entity. The fulfillment of the premise is its activation condition; the modularity allows the rule base to be easily maintained. Adding, removing, or modifying a rule does not necessarily imply modifying others.
- Modularity in the problem-solving formalization. The rules are an assembly of elementary constituents (e.g., particles), which are added and combined in order to form a result for the problem at hand. The order in which they are introduced in the KB is not important.
- Natural expression power of rules—-studies show that experts frequently form their opinions in simple rules based on context.
- Simple accessibility of the rule base, which stems from the features and uniformity of the rule structure used during design. It helps in maintaining and manipulating the domain knowledge.

Production systems also present a series of disadvantages, including:

- The difficulty of constructing an optimal sequence of actions. This is a consequence of the modularity given by the rules structure. Systematic testing of all rules in order to identify those that may be used in a particular situation is time-consuming.
- In some situations, depending on the design of the rule base, the order in which rules are applied may influence the conclusion.

Still, production rules have been the most commonly used representation in expert systems.

SEMANTIC NETWORKS are a structure used in knowledge representation in the form of complex graphs consisting of nodes connected by arcs. These were first introduced for natural language understanding but have been applied to a variety of problems. The nodes in the network are used for objects, attributes, events, and states representations. The arcs represent the relationships between the objects in the nodes. These arcs represent relationships or propositions. There are commonly two types of links ISA (is-a) and AKO (a-kind-of).

Example: IF <A-disconnected> THEN <disconnect line>?

Graphs and semantic networks can be organized in hierarchical order: this fact allows for some attributes to be inherited from general types to specialized types.

FRAMES. A frame is a group of attributes that describe a given object. Each attribute is stored in a slot that contains the value/values attached to this attribute. The use of frames allows a form of knowledge representation as data blocks, each block being represented by a slot. Frames can be programmed with inheritance properties, thus hierarchic connections between them can be generated. The connections between different objects may be represented by a tree.

OBJECT-ORIENTED PROGRAMMING is a step forward in data representation. The main characteristic of object-orientated programming is the unification, in the same structure named "object" of both data and the methods that act upon them. From the programming point of view, the structures mentioned above are defined by using the concept of class[2] and the instances of these classes are objects with which the program operates.

LOGIC PROGRAMMING offers another way of expressing assertions and providing a formal context for making inferences. Here, the KB is exclusively made out of a collection of formulas that describe the universe of discourse. Of major interest when using logic is the possibility of obtaining new results by inference, that is, modus ponens, resolution, modus tollens, and so on. Logic allows a simple expression of facts because it has no ambiguities. This strict formalism also proves limiting when uncertainty is inherent to the knowledge.

In practice, most expert systems use a combination of knowledge representation techniques and even traditional programming structures—-rules or frames (or frames integrated in semantic networks) as rules are well suited for reasoning and frames can be used for efficient manipulation and storage of knowledge; objects as properties such as inheritance, encapsulation, and polymorphism, object-oriented representation ensures greater flexibility, and so on.

15.2.1.2 Facts Database

The facts database is the dynamic part of knowledge-based systems. It contains specific information related to the application. In power systems, facts might represent the status of different switchgear types, a malfunction alarm of a transformer, an electrical value outside the admissible limits, and so on. They change as the states of the

[2] Class—-a group of objects with common attributes; each object that belongs to the same group is called an instance.

objects they are associated with change (e.g., as a result of a malfunction or system reconfiguration). Facts have the following formal representation:

<*Object*>, <*Relationship*>, <*Value*>

where:

<Object> represents say the equipment unit (circuit breaker, separator, etc.) whose state may be established through measurement or logical deduction.

<Relationship> corresponds to a relationship between the objects or between objects and their different states.

<Value> gives the state for the object.

Example: <Breaker A> <is in position> <connected>

Depending on the application field, the expert system has a certain number of facts relating to the respective field, facts that describe states or give quantitative information. Part of the data from the facts base is introduced by human–machine dialogue, while others are automatically loaded through network data readings.

15.2.1.3 Quality of Knowledge Base

A good knowledge representation language should be expressive, concise, unambiguous, context-insensitive, and effective. The KB must be clear and correct. All important relations should be defined, and the irrelevant details suppressed, to achieve a compromise between correctness and clarity. The question of efficiency is widely dealt with in the literature. Ideally, the separation between the KB and the inference procedure should be maintained. This allows the creator of the KB to focus only on the context of the knowledge, and not the inference procedure. In practice, some considerations of efficiency are unavoidable. Automatic methods that can eliminate the most obvious sources of inefficiency in a given encoding have been developed. They work the same way that optimizing compilers can speed the execution of a program, but these methods cannot overcome a profligate knowledge engineer who has no concern for efficiency [5]. The knowledge engineer must understand enough about the domain in question to represent the important objects and relationships. He or she must also understand enough about the implementation of the inference procedure to assure that queries can be answered in a reasonable amount of time.

15.2.2 Inference Engine

The IE is a general program that implements the mechanism of reasoning for the KB. By using a variety of different procedures, the IE applies the knowledge and data according to this reasoning mechanism. It has the task of employing the rules and generating answers to questions posed by the users. The IE can be viewed as the processing element of the expert system, which by starting from the facts (problem input data), uses the corresponding rules from the KB to make the necessary associations and connections to offer a solution for the given problem. The basics of these mechanisms are discussed in the following.

15.2.2.1 *Inference Mechanisms*

The purpose of the IE of an expert system is to give meaning to KB, connecting the rules so as to arrive at meaningful conclusions. A control strategy is the set of norms that guide the search for applicable rules in the rules database of the expert system. Without an adequate control strategy, the solution may require excessive computational time. Considering a tree structure for the rules base, the search strategy analysis may be done in two planes:

(i) Rules exploration where three modes emerge:

- The *deductive mode* consists in exploring a rule starting from premise to conclusion. This type of exploration of rules is specific to problems in which the available data may be assimilated to a premise, but no data available directly applicable to a conclusion.
- The *inductive mode* explores each rule starting from conclusion to premise. This procedure is scope–centered. The unknown premises are sought based on conclusions or scope. These premises then become the new scope and lead to a new set of credible premises.
- The *mixed search* procedure corresponds in a larger measure to the practical human reasoning and to conditions in specific problems with a tree broader in shape, in which some initial premises and some conclusions are known, but important intermediary data are unknown.

(ii) Decision graph exploration where search can proceed as:

- *Depth-first search*, which is the most widely used search form in expert systems. This is a search in the interior of a graph where each node is treated according to the rule "the last arriving node is the first taken into consideration." Each arc of the graph is followed to its extremity before the exploration of another arch; it explores each possible track from one starting node up to the node that corresponds to the search scope, before trying any new direction. This is the case of search within a graph, in which every newly found node is explored until no more progress is possible. In Figure 15.2a, an example of depth-first search is presented, for fulfilling the scope E. The order in which the nodes are explored is: A-B-D-B-E.
- *Breadth-first search* explores the nodes on a first come first served basis. The graph is completely explored on each level before proceeding deeper into the graph. Figure 15.2b presents the breadth-first search technique. The order of browsing through the graph from the A node up to the E node is indicated by the arrows. Thus, the nodes are explored in the following order: A-B-C-B-D-E.
- *Heuristic search* procedures may explore a graph in a variety of ways to reduce computational complexity. This may be unavoidable for systems with a large number of rules or a complex dataset. Heuristic procedures avoid the exploration of certain search directions that are considered to be inefficient. Some of these procedures adopt mechanisms of search orientation and others use random search.

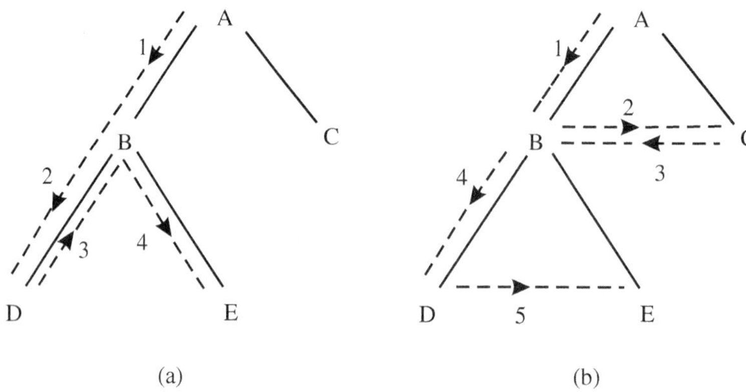

Figure 15.2 Example of search strategies: (a) depth-first search and (b) breadth-first search.

Typically, the best approach to increase the search performance depends upon the problem at hand. For example, let the facts base be constituted of facts *B* and *C*; the search scope labeled as *H* with rules R1 to R9 as in the following:

R1: IF B AND D AND E → F

R2: IF D AND G → A

R3: IF C AND F → A

R4: IF B → X

R5: IF D → E

R6: IF A AND X → H

R7: IF C → D

R8: IF X AND C → A

R9: IF X AND B → D

Figures 15.3a–15.3c illustrate two types of trees that correspond to this rule base and some assumed facts. The continuous lines represent the applied rules and the nodes represent different states of the facts base. The dashed lines correspond to applicable rules, which the IE did not choose to fire.

The process of *forward chaining* (deductive reasoning) (Figure 15.3a) is an inference technique that uses IF – THEN rules for finding the solution (H) of a problem starting from the known facts (B and C). Here, X is deduced by using rule R4 upon B and C, and in the same way, D is deduced by using rule R7 and so on. After one of the rules is applied, new facts (from the conclusion) are added to the facts base.

The following strategies have been applied to establish the order in which the rules are fired: the first encountered rule that can be applied, according to their numbering, R4 (R7), R7 (R8, R9) (Figure 15.3a), or the one with the highest number of conditions (Figure 15.3b). Thus, in the second case, rule R8 is preferred over R7 because it contains a larger number of conditions. "Success" is considered when a new scope is added to the facts base, and "fail" when none of the rules can be applied. The

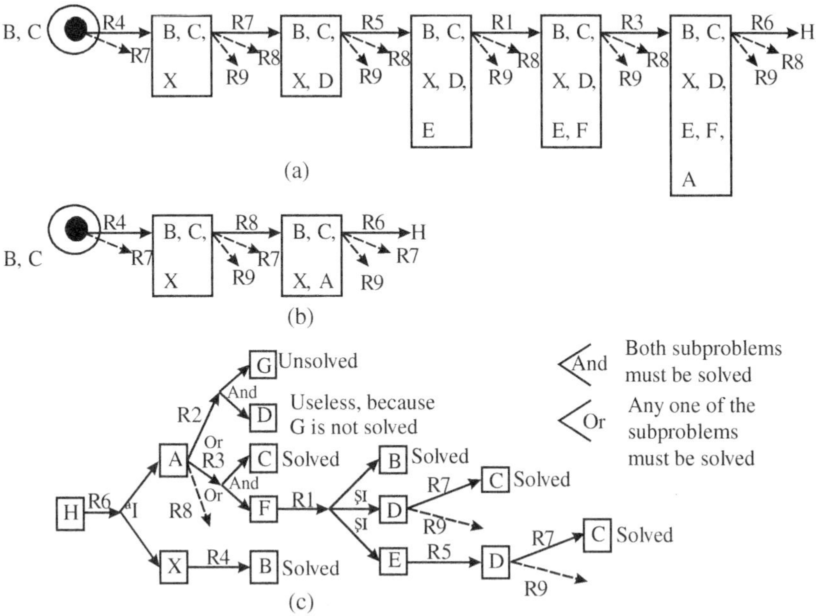

Figure 15.3 Inference procedures: (a) forward chaining with the first encountered applicable rule fired, (b) forward chaining with rules fired according to the highest number of conditions, and (c) backward chaining.

search ends when the scope H is added to the facts base. In the presented example, six inferences have been performed in the first case (a) and three inferences have been performed in the second case (b). The procedure of forward chaining or deductive reasoning presents two major advantages, which are absolutely necessary in most applications. First, it allows the possibility of finding all the possible solutions. Second, it enriches the facts base at each cycle of the IE, which simplifies the subsequent deductions. The main disadvantage of this procedure is the lack of completeness (does not guarantee the finding of the solution if it exists).

Backward chaining (inductive reasoning) (Figure 15.3c) uses reasoning based on the scope. Starting from the scope, a problem is successively decomposed into subproblems, until elementary problems are obtained (known facts with direct solutions). Therefore, it is said that backward chaining is directed by the scope (in comparison with forward chaining), in the sense of finding the facts that allow the scope or conclusion to be fulfilled. The IE uses rules from right to left, that is, in the direction conclusion to premise. The rules that are selected are those containing in the conclusion part (right) the initial specified scope, and their premises become subscopes to be explored. This process of decomposition of a problem in subproblems is repeated until all the subscopes have been explored, or when no applicable rules are found.

Consider the same KB as the previous but now starting with the verification of scope H. When browsing rules R1…R9, it can be observed that only rule R6 contains H in the conclusion part. Therefore, H branches to the subscopes A and X based on rule R6. Subscope A can be verified by using one of the three simple rules R2, R3, or R8. The solution strategy is based on using the first rule that can be applied (according to their numbering). Subscope X is true if rule R4, which has X as conclusion, can be applied. In case of failure (e.g., the G subscope cannot be deduced), the procedure chains backward and the last rule that was applicable but not fired is now applied. Tree exploration is stopped when the initial scope is demonstrated (as in the given example) or when all the possibilities have been explored without success. The major disadvantage of the backward chaining is that a blockage can appear.

Aside from pure forward and backward chaining techniques, *mixed search*, which corresponds better to the human reasoning, can be applied. The main advantage of this kind of reasoning consists on one hand of a reduced internal memory space, because the search tree is usually smaller. On the other hand, the time for finding a solution may be short, because the search is done only for subscopes that have a chance of verification. Still, mixed search strategies can be difficult to implement and can lead to inefficient search. Other inference mechanisms have also been developed, including inexact inference mechanisms, which work with coefficients of trust instead of simple binary TRUE and FALSE values.

15.2.2.2 Inference Engine Components

The IE is the core of an expert system. It creates dynamic reasoning by selecting the rules that are to be used and establishes the order in which those rules are to be fired. Whatever reasoning method, the basic iteration of an IE has four stages: selection, filtering, conflict resolution, and rule firing (Figure 15.4) [7, 8].

- **(i)** *Rule selection*—-during this stage, the subproblems are partitioned to shorten the search time for the subsequent stages. This stage is necessary when the KB is large and covers a broad range of expertise.
- **(ii)** *Filtering or pattern matching*—-compares the premises of the previously selected rules with the facts that characterize the problem to be solved in order to determine the applicable rules. After this stage, no rule, one rule, or more than one rule may be applicable. If there are no applicable rules then either the processing is completed or further input from the user is required in order to continue.
- **(iii)** *Conflict resolution*—-this stage is necessary if after the pattern matching stage more than one rule is applicable. Only one rule will be selected for execution. Many criteria for resolution may be used, including first rule from the list, the rule with the highest number of facts in the premise, the most commonly used rule, rule most recently satisfied, and so on [8]. The performance of the IE depends on the character of this selection, which can, not only find the solution faster or slower, but may also lead to very different solutions. The selection depends critically on the context and the reasoning stage in which the rule is found to be satisfied.
- **(iv)** *Rule firing* —-this results in the addition of one or more new facts in the facts base. It is possible, during this stage, to call the external procedures (access to

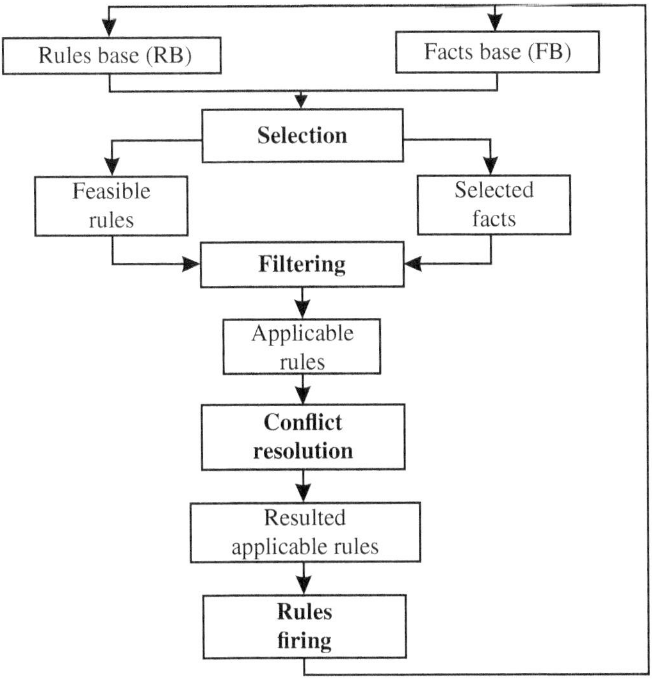

Figure 15.4 Basic iterative steps of an inference engine.

the databases or table processors) or to user queries which in turn would lead to additional facts.

15.2.2.3 Other Modules

In order to understand the design of an expert system, the role of human operators that are interacting with the system must also be known:

- *The field expert* or *the application engineer* represents the person or persons who is/are currently the experts in solving the problem at hand. They provide the knowledge leading to rule base and the actualization of the expert system.
- *The knowledge engineer* is the person who collaborates with the field expert to encode the KB in a declarative form, which can be used by the expert system.
- *The user* is the person who will use the expert system for obtaining useful advice of the same type that would be obtained from the field expert.

The Knowledge Acquisition Module transfers the knowledge from a source to the expert system (Figure 15.5). The expertise from the domain that has to be transferred to the expert system KB is a collection of definitions, relations, specialized facts, heuristics, procedures, strategies, and hypotheses. The knowledge engineer is a specialist in computers with sufficient experience in practical construction of expert systems.

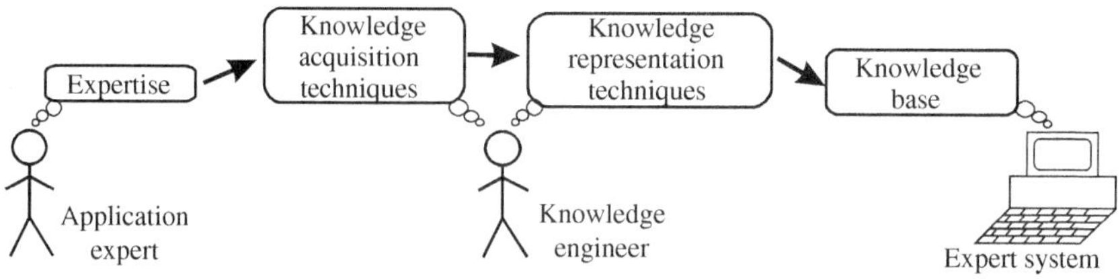

Figure 15.5 Knowledge acquisition module.

Knowledge acquisition is typically based on interviewing experts to obtain information about:

- personal methods and/or types of expertise for the problem at hand;
- personal experience during problem solving in the past;
- personal knowledge of situations that allow the selection of methods or an expertise for a problem at hand.

Extracting the expertise from a human expert with the help of some specific means or through the questionnaire method is a heterogenic process. Knowledge acquisition from human experts is known as *knowledge engineering,* and it can be divided into the following five stages:

(i) *Identification*: the correct definition of the problem and determination of its characteristics.

(ii) *Conceptualization*: determination of the concepts that can support the representation of knowledge.

(iii) *Formalization*: selecting a method of knowledge representation and the inference mechanism.

(iv) *Implementation*: the actual representation of knowledge in the selected formalism (rules, semantic networks, frames, objects, etc).

(v) *Testing*: verification of the knowledge and system validation.

The process is iterative. Testing results may indicate the need for reformulating the problem, redefinition of key concepts, redevelopment of knowledge structures, or refinement of the knowledge. During this process, the knowledge engineer works closely with the field expert.

Knowledge engineering instruments help automate the knowledge acquisition process. If such an instrument is available, an expert in a well-defined field may construct an expert system without the presence of the knowledge engineer. Another advantage of knowledge engineering instruments is a rapid expert system prototype. From the software point of view, the knowledge acquisition module of the expert system represents, in the simplest case, an editor in which knowledge may be introduced or modified, in a form accepted by the programming language in which the expert system is implemented.

The User Interface Module is the part of expert system that allows the user to query the expert system, introduce new information, or receive various recommendations. The user appreciation of an expert system depends in large measure upon the similarity between the mode of information representation within the expert system and the model which the user has "in his mind" concerning the problem that needs to be solved.

The user should be able to interact naturally with the computer without need to adapt to various ways of communication. This clearly causes a problem-solving dilemma between the universe of communication modes and the more complex interpretation of events specific to a particular conclusion. This also raises new cognitive considerations because various modes of communication relate to certain modes of reasoning. These objectives may only be achieved by an intelligent interface that can react in anticipative mode and assist in a cooperative way for both the application and the user.

The Explanation Unit. An expert system must also have the capacity to provide explanations to the operator about the reasoning used in order to arrive at a certain recommendation or decision. The more explicit the representation of knowledge in the knowledge database, the more efficient the explanation process. Each type of explicit knowledge may facilitate the explanation.

15.3 EXPERT SYSTEMS APPLICATION

15.3.1 Expert System for V–Q Control in Power Systems

In the last four decades, a special interest has been given to the planning of compensation means and V–Q regulation in power systems as well as to perfecting the wide area monitoring and control techniques to maintain voltage at desired levels and prevent phenomena of instability and voltage collapse. In this context, and in addition to classical methods (algorithms), several new heuristic approaches have been proposed. These heuristic techniques are based upon the experience and knowledge accumulated by specialists from the power engineering field. The purpose of an expert system for voltage stability evaluation and V–Q control is to assist the area operators or dispatcher to evaluate the risk of instability phenomenon appearance or voltage collapse and to decide the appropriate V–Q control settings. The expert system suggests solutions for control of the voltage level in the network as a consequence of events such as increases in consumption, unavailability of generators or power distribution network branches.

15.3.1.1 The Structure of the Expert System

The role of the EXPERTVQ system is to assist the grid operators in evaluating the voltage stability and applying the most efficient measures of V–Q control imposed by events that occur in the system, such as load increase or network topology changes, carrying out the following steps [9, 10]:

(i) Assess the voltage stability of the electrical power system and identify the most vulnerable nodes in terms of V–Q variation.

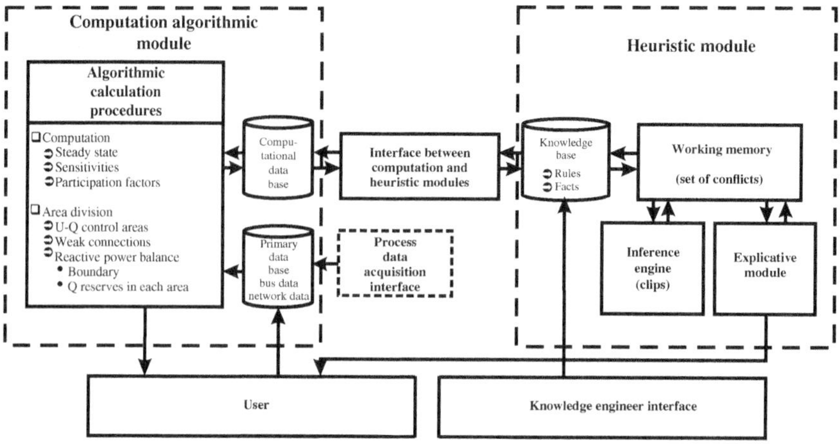

Figure 15.6 The EXPERTVQ system structure.

(ii) Identify, based on the concept of weak connections, the V–Q control areas and establish the reactive power balance and reserve of each zone.

(iii) Detect the nodes with voltage problems and the areas in which they are located.

(iv) Appreciate the voltage fault degree of complexity, and accordingly, launch a specific procedure to solve the problem.

(v) Through a user friendly interface, provide the operator with timely information on the state of the power system, and the necessary control actions needed in order to steer the voltage level to an admissible band.

Given the complex aspects of the tasks that it has to complete, EXPERTVQ has a hybrid structure that includes an algorithmic calculation module, a heuristic module, and an interface with the user and knowledge engineer (Figure 15.6). This hybrid approach is typical of proposed expert systems for voltage control (see, e.g. [11, 12]).

15.3.1.2 The Algorithmic Calculation Module

The algorithmic calculation module, implemented in C++ programming language, is designed to complete the facts in the knowledge database from the heuristic module with system state information at the time of analysis. The steady-state load flow computations are carried out using the Newton–Raphson method on the primary database that contains information entered by the user, nodal data, and network topology obtained directly from the system. This determines:

- the local stability indicators S_{VQ} and S_{QgQc} (discussed in the following) and the nodal voltage sensitivity with the variation of the control values;
- the factors of participation for the PQ nodes to the system V–Q variation modes, by analyzing the reduced Jacobian matrix.

V–Q CONTROL AREAS AND WEAK CONNECTIONS. Static methods were used to establish the V–Q correlations of a specific power system, based on a mathematic model consisting of the linearized load flow equations written in the following form:

$$\begin{bmatrix} J_{P\theta} & J_{PV} \\ J_{Q\theta} & J_{QV} \end{bmatrix}\begin{bmatrix} \Delta\theta \\ \Delta V \end{bmatrix} = \left[\begin{array}{c|c|c} \dfrac{\partial P}{\partial\theta} & \dfrac{\partial P}{\partial V_c} & \dfrac{\partial P}{\partial V_g} \\ \hline \dfrac{\partial Q_c}{\partial\theta} & \dfrac{\partial Q_c}{\partial V_c} & \dfrac{\partial Q_c}{\partial V_g} \\ \hline \dfrac{\partial Q_g}{\partial\theta} & \dfrac{\partial Q_g}{\partial V_c} & \dfrac{\partial Q_g}{\partial V_g} \end{array}\right]\begin{bmatrix} \Delta\theta \\ \Delta V_c \\ \Delta V_g \end{bmatrix} = [J']\begin{bmatrix} \Delta\theta \\ \Delta V_c \\ \Delta V_g \end{bmatrix} = \begin{bmatrix} \Delta P \\ \Delta Q_c \\ \Delta Q_g \end{bmatrix} \tag{15.1}$$

where $[J']$ is the complete Jacobian matrix. The standard model from (15.1) was modified in order to include the influences of loads characteristics, the transformers with control under load, and the limitations imposed on the generator stator and rotor currents in triggering phenomena of instability and voltage collapse [13].

SENSITIVITY COEFFICIENTS. For voltage stability evaluation and selection of the most effective means of control necessary to solve a voltage problem, the following were used:

- Sensitivity S_{VQ} of voltage at a consumption node i as the variation of reactive power consumed in the same node is defined by the relationship:

$$S_{VQ} = \frac{\partial V_i}{\partial Q_i} \cong \frac{\Delta V_{c,i}}{\Delta Q_{c,i}} \tag{15.2}$$

- Sensitivity $S_{Q_gQ_c}$ of the generators output reactive power as the variation of the reactive power consumed in a consumer node i is defined by the relationship:

$$S_{Q_gQ_c} = \partial\left(\sum_{k\in n_g} Q_{g,k}\right)\bigg/\partial Q_{c,i} \cong \sum_{k\in n_g} \Delta Q_{g,k}\bigg/\Delta Q_{c,i} \tag{15.3}$$

- The sensitivity matrix $[H]$, which gives the relationship between the voltage variations $[\Delta V_c]$ in the P–Q nodes and generators output reactive power variations $[\Delta Q_g]$, respectively, and the variation of the quantities associated with the control variables (voltage variations in generator nodes $[\Delta V_g]$, reactive power output variation at capacitor banks or variation in the reactive power absorbed by compensation reactors $[\Delta Q_{inj}]$, and the transformer ratio variation $[\Delta N_{ij}]$) as:

$$\begin{bmatrix} \Delta V_c \\ \Delta Q_g \end{bmatrix} = \begin{bmatrix} H_{VV} & H_{VQ} & H_{VN} \\ H_{QV} & H_{QQ} & H_{QN} \end{bmatrix}\begin{bmatrix} \Delta V_g \\ \Delta Q_{inj} \\ \Delta N_{ij} \end{bmatrix} = [H]\begin{bmatrix} \Delta V_g \\ \Delta Q_{inj} \\ \Delta N_{ij} \end{bmatrix} \tag{15.4}$$

These are calculated by using a numerical derivation procedure based on the linear model defined in (15.1).

PARTICIPATION FACTORS. Given the strong bound between voltage and reactive power in a power system, the V–Q dependency is described by $[\Delta Q] = [J_R][\Delta V]$, where $[J_R]$ is the reduced Jacobian matrix obtained by eliminating the angle variations $[\Delta\theta]$ from equation (15.1) considering $[\Delta P] = 0$.

By applying modal analysis to the $\left[J_R\right]$ matrix, it can be shown that:

$$S_{VQ} = \frac{\partial V_i}{\partial Q_i} = \sum_k \frac{R_{ki}L_{ik}}{\lambda_k} = \sum_k \frac{P_{ki}}{\lambda_k} \tag{15.5}$$

where R_{ki} and L_{ik} are right and left elements of the eigenvectors R_k, respectively, left L_k corresponding to the eigenvalue λ_k, and P_{ki} is the factor of participation of node i to the variation mode k and constitutes an indicator of the node's vulnerability.

V–Q CONTROLLABILITY AND AREAS OF CONTROL. The phenomena of instability and voltage collapse are determined by the inability of the network to transmit an additional load to a demand center as a result of reaching the maximum power transfer capacity. Accordingly, V–Q control areas are defined, consisting of PQ and PV nodes that are similar in terms of voltage behavior and the links between them are identified as vulnerable links. The zone separating algorithm relies on the information provided by the submatrix $[J_{QV}]$ extracted from the complete Jacobian matrix that has the elements [14]:

$$L_{ik} = \frac{\partial Q_i}{\partial V_k} = V_i V_k Y_{ik} \sin(\theta_i - \theta_k - \gamma_{ik}) \tag{15.6}$$

For a vulnerable electric line, the value L_{ik} has a negative, small value because it either has a high impedance (Y_{ik} is close to zero), or it operates near the limit of maximum power transfer capacity, that is, $\sin(\theta_i - \theta_k - \gamma_{ik}) \cong 0$. Therefore, increasing the reactive power transmitted through this electric line leads to L_{ik} becoming zero or even positive, which may lead to a singularity of the $[J']$ matrix and system instability because of the voltage and/or angle bifurcations.

15.3.1.3 The Heuristic Module

The heuristic module was developed in the 'C' Language Integrated Production System (CLIPS) environment [15] that supports building expert system applications, and which, in addition to the fact that it has a high performance IE with forward chaining, also allows a relatively easy interface with algorithmic calculation modules written in C++. Furthermore, the knowledge engineer can customize many different strategies within CLIPS. It is intended for facts-based inference from the knowledge database by using a system of production rules [11, 16] of the form:

IF condition_1 & condition_2 &....& condition_n THEN conclusion (action)

The following specific tasks are implemented:

Task T1: Assessing the stability of the voltage and the risk of triggering the phenomena of instability and voltage collapse. The local indicators of stability are based on the set of rules:

R1.1 IF all the sensibilities $S_{Q_g Q_c}$ are positive
THEN the system is stable and controllable V–Q

R1.2 IF there are $S_{Q_g Q_c}$ values greater than the threshold $S^{\max}_{Q_g Q_c}$ value
THEN the system is in a critical area of operation and the risk of triggering the phenomena of instability and voltage collapse is high.

Task T2: Determination of the number of control areas and their characteristics. The classifying of the importing, or exporting, of reactive power zones and controlled or not controlled areas is determined by the information referring to the balance of reactive power at the boundary, the reserve of reactive power in the area, and the state of the means of V–Q control with the specific rules. An example of such a rule is presented below:

R2.1 IF in an area there are reactive power reserves
THEN the area can be controlled

Task T3: Identifying vulnerable nodes and the mode of V–Q variation. The identification of the vulnerable nodes and their characteristics is made on the basis of the participation of the PQ nodes in the mode of V–Q variation according to the smallest eigenvalue. Among the rules that are used to perform this task are:

R3.1 IF the participation factor P_{ik} of the node i at mode k is higher than $P^{\max}_{ik}$
THEN the node is vulnerable

R3.2 IF there are vulnerable nodes distributed in several areas of V–Q control
THEN the mode of V–Q variation is "dispersed."

Task T4: Identifying and classifying voltage problems. The heuristic module identifies the nodes with voltage problems, their nature, and severity by comparing the values of the nodal voltage with the admissible voltage values of the normal operation mode $V_{\min}$ and $V_{\max}$, and with those of the emergency operation mode $V^{em}_{\min}$ and $V^{em}_{\max}$. In this case, it uses a set of rules from which the following were extracted:

R4.1 IF the voltage of a node is smaller than $V_{\min}$
THEN the node has a "negative deviation" voltage problem

R4.2 IF there are nodes in which the voltage falls outside the $(V_{\min}, V_{\max})$ band but inside the $(V^{em}_{\min}, V^{em}_{\max})$ bandwidth
THEN the network has a voltage problem with a severity degree of 1.

R4.3 IF there is at least one node in which the voltage exceeds the interval of $(V^{em}_{\min}$ and $V^{em}_{\max})$
THEN the network has a voltage problem with a severity degree of 2.

Task T5: Selecting the means of control to resolve the voltage problem. This task is activated if the severity degree of the voltage problem is 1. In order to accomplish this task, a complex set of rules is applied. It is based on the prioritization of activating the means of control, their availability, and effect. Thus, the existing means of control in the node with the most severe problem are analyzed first, then the nodes of the area in which it is located, and finally in the nodes of the neighboring areas.

The voltage-related issues are quantified in the EXPERTVQ system by a heuristic function, and the control mean selected is that for which the function takes the maximum value. After selecting the most effective means of control, based on sensitivity coefficients, the heuristic module estimates the effect of its action and, if necessary, it requests the recalculation of sensitivity coefficients by the algorithmic module.

15.3.1.4 Case Study

In order to test the ability of the EXPERTVQ system, to assess the stability of the voltage and to solve a voltage problem, the standard IEEE 30-bus network as presented in Figure 15.7 was used.

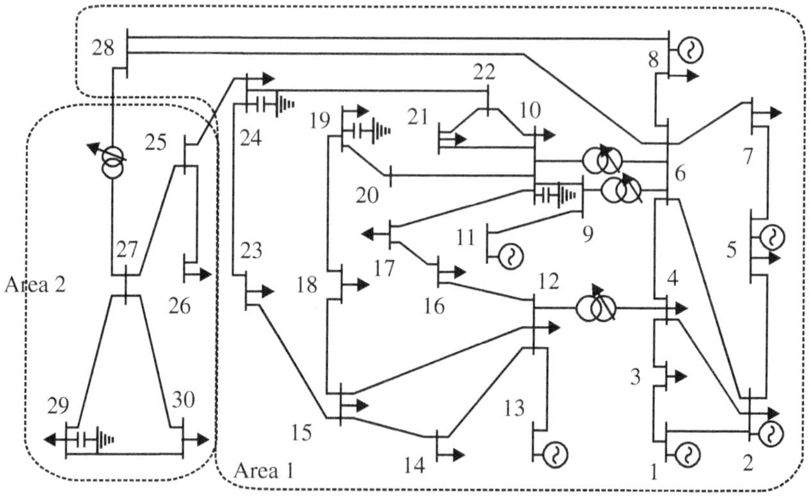

Figure 15.7 The single-line diagram of the IEEE 30-bus test network.

During tests, the following parameters were assumed [9, 10]:

- the vulnerability factor of the branches, $\alpha = 0.075$;
- the normal admissible limits of voltage are $V_{\min} = 0.95 V_n$ and $V_{\max} = 1.05 V_n$ for PQ nodes, $V_{\max} = 1.1\ V_n$ for PV nodes, and the emergency limits are $V^{em}_{\min} = 0.9\ V_n$ and $V^{em}_{\max} = 1.1\ V_n$;
- the vulnerability factor of the nodes $P^{\max}_{ik} = 0.05$ and the coefficient of safe operation $S^{\max}_{Q_g Q_c} = 1.5$.

TABLE 15.2 Synthesis of actions proposed by the expert system

Iter.	The most affected bus	Suggested action	Estimated effect
1	29	Connect the capacitor bank in node 29 at maximum tap position	Voltage-level improvements, and $V_{27} > V_{\min}$
2	29	Modify the transformer ratio $N_{28\text{-}27}$ from 1.032 to 1.055	The voltage problems in area 1 are solved and only three nodes in area 2 remain problematic
3	18	Connect the capacitor bank from bus 19 at maximum tap position	All voltage problems are solved

Note that the values of the vulnerability factors of the branches and nodes are specific to each network, but the admissible values of the voltage have a general character. The specific scenario of interest is a voltage problem due to a load increase at bus 29 correlated with a disconnecting of the line 1–2 and the generator at node 13.

In the first stage, based on the initial facts and on the tasks that it has to complete, the EXPERTVQ program offers the following summary of the state of the system:

Task T1: The system is stable and V–Q controllable, but the risk of voltage instability and voltage collapse phenomena is high.

Task T2: The system is divided into two zones: area 1 is exporting and controllable, and area 2 is importing and controllable.

Task T3: The mode of variation of V–Q is dispersed, the most vulnerable nodes being node 9 in area 1 and node 29 in area 2.

Task T4: There are 11 nodes in the network with voltage problems (14, 15, 18, 19, 23, and 24 in area 1, respectively, 25, 26, 27, 29, and 30 in area 2). The severity degree of the voltage problem is 1, and the most affected node is 29 from area 2.

At this stage, because the severity degree is 1, the heuristic procedure is triggered. For the purpose of solving this problem, three iterations are performed. The suggested actions and the estimated effects are summarized in Table 15.2.

In order to validate the actions suggested by the expert system, they have been input to the primary database and then system continues in algorithmic calculation mode. The obtained results are presented in the histogram from Figure 15.8, alongside with those existing in the initial database. The analysis confirms the estimation made by the heuristic module.

15.3.2 Other Applications of Expert Systems

15.3.2.1 Expert System for Distribution Networks Reconfiguration

The distribution electrical networks are designed in simple loop or meshed configuration, but they are operated radially or in arborescent configuration. Depending on the

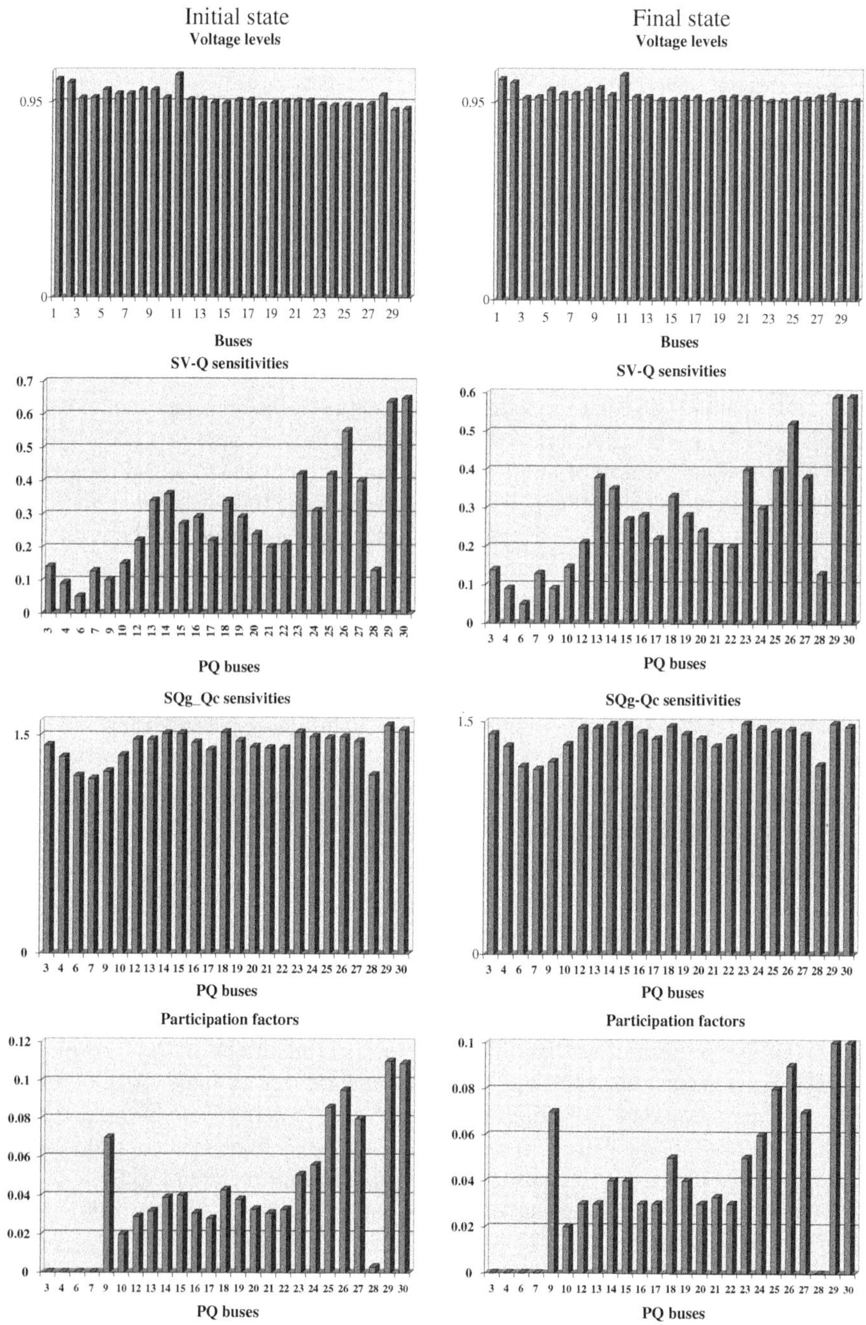

Figure 15.8 The results given by the algorithmic module for the initial and final states.

distribution of loads in different time periods of a day or during the year season, as well as other characteristics, the network configuration can be modified by changing the operating state of some branches with the purpose of achieving technical or economic benefits, such as minimization of the total active power losses or improvement of the voltage level. Various optimization techniques aiming various objectives may be employed to determine the optimal configuration. While the heuristic methods are the most used, the tabu search is an efficient solution used today [17].

The expert system (ES) assists the network operator in a real-time activity to perform the network reconfiguration. The reconfiguration process consists in a series of steps, proposed and verified by calculations, which include connection/disconnection of network branches. In an on-line system, ES uses real-time data available within the SCADA system from the dispatching center, among which the state quantities, the state of branches operation, etc. For the action proposed at each step, ES calls an algorithmic module to calculate the network load flow and checks if the security constraints are satisfied. If no violation is identified, the decision proposed in the current step is sent to the maneuvers module to assist the human operator in the equipment switching. Once the current step is validated by the system operator, a new action is proposed in a new step, and the reconfiguration process continues until the final objective is achieved.

15.3.2.2 Expert System Power System Restoration After Blackouts

In the last two decades, power systems around the world have been subjected to a large number blackouts, of various severity. In order to minimize the effects of a blackout, immediate restoration actions must be taken in order to supply the loads. Specific grid codes regarding the procedures to be followed by the system operators in case of a blackout are available in any power system. However, automatic tools may help the human operator to rapidly perform calculations and equipment energization.

The expert system can be a component of the energy management system (EMS) of a dispatch center, and it can communicate with other tools and systems. The ES can consist of several modules such as database module, reasoning module, data update module, algorithmic computation module, on-line assisting equipment maneuverings module [18]. The ES determines the spread of the blackout, identifies the possible sources to energize the affected areas, and proposes the steps to be followed by the human operator. These steps can be the result of previous procedures predefined by the human expert into the ES modules. The searching strategy is based on two cores: one core aims to synchronize the power plants, and one core aims to reconnect the loads. The human operator may also decide if alternative steps should be considered.

REFERENCES

[1] Barr, A., and Feigenbaum, A. *Le manuel de l'intelligence artificielle*, Tome 1. Editions Eyrolles, Paris, 1986.

[2] Schulte, R. P., Larsen, S. L., Sheble, G. B., Wrubel, J. N., and Wollenberg, B. F. Artificial intelligence solutions to power system operating problems. *IEEE Transactions on Power Systems*, vol. PWRS-2, no. 4, pp. 920–922, November 1987.

[3] Dillon, T. S., and Laughton, M. A. *Expert Systems Applications in Power Systems*. Prentice Hall International, New York, 1990.
[4] Razanamampandry, J. Système expert pour l'optimisation des tensions et des puissances réactive d'un réseau de transport d'énergie électrique. Thèse No. 1392, Ecole Polytechnique Fédérale de Lausanne, 1995.
[5] Song, Y. H., Johns, A., and Aggarwal, R. *Computational Intelligence Applications to Power Systems*. Kluwer Academic Publishers, Dordrecht, Boston, London, 1996.
[6] Tinguely, C. Système expert pour l'analyse de sécurité d'un réseau de transport d'énergie électrique. Thèse No. 1089, Ecole Polytechnique Fédérale de Lausanne, 1992.
[7] Chatain, J. N., and Dussauchoy, A. *Systèmes experts. Méthodes et Outils*. Editions Eyrolles, Paris, 1987.
[8] Voyer, R. *Moteurs De Systèmes Experts*. Editions Eyrolles, Paris, 1987.
[9] Eremia, M., Trecat, J., and Germond, A. *Réseaux électriques. Aspects Actuels*. Editura Tehnică, Bucureşti, 2000.
[10] Eremia, M., Bulac, C., Bulac, A. I., Triştiu, I., and Trecat, J. Voltage stability in power systems: assessment and control by artificial intelligence techniques. Paper 38-188, CIGRE 2000 Session, Paris, 2000.
[11] Liu, C. C., and Tomsovic, K. An expert system assisting decision-making of reactive power/voltage control. *IEEE Transactions on Power Systems*, vol. 1, no. 3, pp. 195–201, August 1986.
[12] Liu, C. C., Marathe, H., and Tomsovic, K. *A Voltage Control Expert System and its Performance Evaluation*, pp. 105–152. In: Expert Systems Application in Power Systems, Dillon, T. S., and Laughton, M. A. (Eds.), Prentice Hall Inc., Englewood Cliffs, 1990.
[13] Eremia, M., Cârţină, G. H., Petricică, D., Bulac, A. I., Bulac, C., Triştiu, I., and Grigoraş, G. H. *Artificial Intelligence Techniques in Power Systems Operation* (in Romanian). Editura AGIR, Bucureşti, 2006.
[14] Schlueter, A., Ben Kilani, K., and Ahn, U. *Impact of Modeling Accuracy on Type, Kind, and Class of Stability Problems in a Power System Model*. Proceedings of the ECC & NSF International Workshop on Bulk Power System Voltage Stability, Security and Control, Phenomena-III, Davos, Switzerland, pp. 117–156, August 1994.
[15] *CLIPS Reference Manual*. Software Technology Branch, 1993.
[16] Germond, A. J., and Niebur, D. Survey of knowledge based systems in power systems: Europe. *Proceedings of the IEEE*, vol. 80, no. 5, pp. 732–744, May 1992.
[17] Tristiu, I. Reconfiguration of medium voltage distribution electrical networks, PhD Thesis, University Politehnica of Bucharest, 1998. (in Romanian)
[18] Kostic, T. Decision aid function for restoration of transmission power systems after a blackout, PhD thesis, École Polytechnique Fédérale de Lausanne, 1997.

CHAPTER 16

NEURAL NETWORKS

Dagmar Niebur, Ganesh Kumar Venayagamoorthy, and Ekrem Gursoy

16.1 INTRODUCTION

Neural networks (NNs), also called artificial neural systems, neurocomputers, parallel distributed processors, or connectionist models, are an attempt to mimic (emulate) at least partially the structure and functions of brains and nervous systems of living creatures. Generally speaking, an NN is an information or signal processing system composed of a large number of a simple processing elements, called artificial neurons or simply nodes, which are interconnected by direct links called connections and which cooperate to perform parallel distributed processing in order to solve a desired computational task. One of the attractive features of NNs is their capability to adapt themselves to changing environmental conditions by changing their connection strengths or structure. NNs are sometimes considered as grossly simplified models of the human brain. This view is somewhat exaggerated and misleading because the human brain is not well understood and indeed its behavior is very complex. For an excellent introduction to the theory of NNs, the interested readers are referred to the comprehensive book by Haykin, [1].

Within the context of electric power engineering, the majority of NNs are used as approximators or classifiers. Neural nets have a long history in power engineering and this chapter cannot thoroughly review all types of neural nets nor all types of power engineering applications. An early overview on neural net applications can be found in [2], [3], or [4]. More recent and specialized reviews of the areas of neural nets for electric load forecasting were published by [5], [6], and [7]. Other reviews include applications to synchronous machine control [8], renewable energy systems [9], and generation scheduling [10].

16.2 NEURAL NETWORK ARCHITECTURES

In the following subsections, several NN models for nonlinear power system monitoring, identification, and control are introduced. While the structure of a physically based model including circuit-based model is determined from physical reasoning,

Advanced Solutions in Power Systems: HVDC, FACTS, and Artificial Intelligence, First Edition.
Edited by Mircea Eremia, Chen-Ching Liu, and Abdel-Aty Edris.

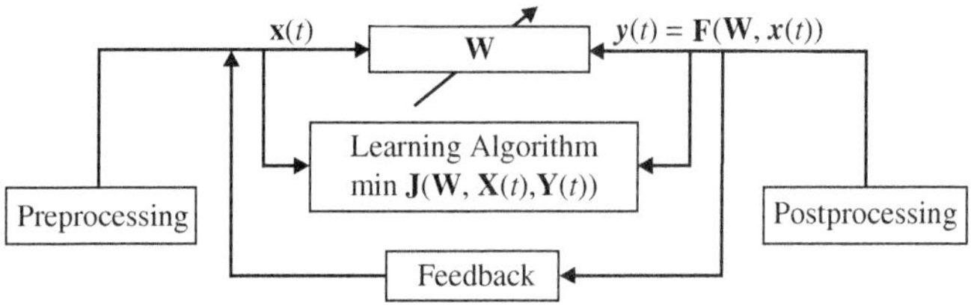

Figure 16.1 A general neural network structure. *Source*: Cichocki et al. 1999 [11]. Adapted with permission of Elsevier.

the NN model is typically characterized as having no (or sparse) information about the physical problem incorporated in the model structure. The model provides an approximation or a classification in a certain function space of the unknown mapping between input (or control) vector $\mathbf{x}(t)$ and output or measurement variables $\mathbf{y}(t)$. These NN models are also referred to as physical model-free. For our purposes, it is not relevant whether any of these NNs have a physiological or neurological relevance.

Let K, M, and N be positive integer numbers. The input or control vector $\mathbf{x}(t_k)$ given at time t_k is an M-dimensional real vector $\mathbf{x}(t_k) = [x_1(t_k), \ldots, x_M(t_k)]^T$, where $k = 1, \ldots K$. Its associated output vector $\mathbf{y}(t_k)$ is an N-dimensional real vector $\mathbf{y}(t_k) = [y_1(t_k), \ldots, y_N(t_k)]^T$. In NN terminology, the set of K pairs of input and output vectors $(\mathbf{x}(t_k), \mathbf{y}(t_k))$ is referred to as the training set of the neural network.

The general NN is of the form $\mathbf{y}(t) = \mathbf{F}(\mathbf{W}, \mathbf{x}(t))$, where the M $\times$ N—dimensional real or complex matrix $\mathbf{W}$ is called the weight matrix, whose values are determined as by a so-called learning algorithm. If the target outputs are known, the learning method is called supervised. It then minimizes an objective function $\mathbf{J}(\mathbf{W}, \mathbf{X}(t), \mathbf{Y}(t))$. Here the input and output matrix $\mathbf{X}(t)$ and $\mathbf{Y}(t)$ have the k input and output vectors of the training set as columns, where $k = 1, \ldots K$.

Figure 16.1 illustrates this concept. For feedforward networks, the feedback loop is disabled. For dynamic NNs, the feedback operator can be a time delay operator acting on one or several output vector elements.

If the target outputs are not known, the learning method is called unsupervised and determines the weight matrix $\mathbf{W}$ by minimizing an objective function J($\mathbf{W}$, $\mathbf{X}(t)$), which does not depend on the output vectors.

If the target outputs are not known but the desired end result is known, the learning method is referred to as reinforcement learning. The weight matrix $\mathbf{W}$ is determined to minimize an objective function J($\mathbf{W}$, $\mathbf{X}(t)$), which does not depend on the output vectors but the final result.

16.2.1 Feedforward Neural Networks

16.2.1.1 The Multi-Layer Perceptron

The multi-layer perceptron (MLP) is a feedforward NN model (Figure 16.2). If the input vector components are normalized to be in the range of [-1,1], the postprocessing function f (called activation function of the hidden NN layer) is chosen to be $f(h) = \tanh(b_h)$.

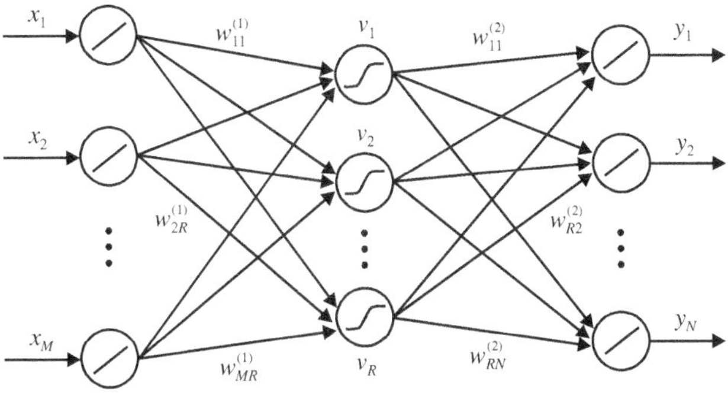

Figure 16.2 Multi-layer perceptron.

The postprocessing function g (activation function of the output layer) is commonly chosen to be the identity. The R hidden units v_r, $r = 1, \ldots, R$ and the output units $y_j j = 1, \ldots, N$ process the input as follows:

$$v_r = \tanh\beta \left(\sum_{m=1}^{M} w_{rm} x_m \right) \tag{16.1}$$

and thus:

$$y_j(t) = g\left(\sum_{r=1}^{R} w_{2jr} v_r(t) \right) = g\left(\sum_{r=1}^{R} w_{2jr} \tanh\beta \left(\sum_{m=1}^{M} w_{1rm} x_m(t) \right) \right); \quad j = 1, \ldots N \tag{16.2}$$

The universal approximation theorem [1] guarantees that a sufficiently good multi-layer perceptron approximator with maximal two hidden layers exists for any L^2 integrable function, but the theorem does not specify how (i.e., with how many hidden neurons and/or layers) it can be constructed.

16.2.1.2 Radial Basis Function Networks

Figure 16.3 shows the radial basis function (RBF) network, which is a nonlinear approximator where the output y is computed as a linear combination of the RBFs f_k centered around a vector c_k that depends on the statistics of the input vectors.

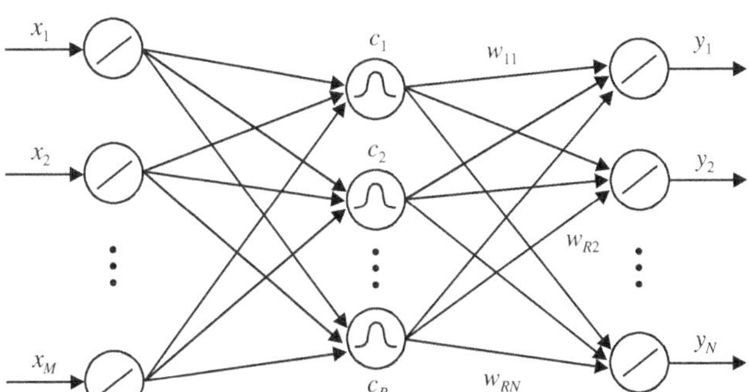

Figure 16.3 Radial basis function network.

The outputs are determined by:

$$\mathbf{y}(t) = \mathbf{b} + \sum_{r=1}^{R} \mathbf{w}_r f_r(\|\mathbf{x}(t) - \mathbf{c}_r\|) \tag{16.3}$$

$$f_k(\|\mathbf{x}(t) - \mathbf{c}_k\|) = exp(-\frac{\|\mathbf{x}(t) - \mathbf{c}_k\|^2}{2\sigma_k^2}) \tag{16.4}$$

where:

$$\sigma_k = \left(\frac{1}{h}\sum_{j=1}^{h} \|\mathbf{c}_k - \mathbf{c}_j\|^2\right)^{1/2} \tag{16.5}$$

where $\|.\|$ denotes the 2-norm or Euclidian norm. One such set of RBFs that is frequently used are Gaussian activation functions centered on the mean value c_j and with a receptive field whose size is proportional to the variance σ_K, called spread. R denotes the number of (hidden) RBF units, and h denotes the number of combination of centers. RBFs present NN implementations of Kernel regression with Gaussian basis functions. The RBF network presents an important approximation tool because like spline functions it provides a quantifiable optimal solution to a multi-dimensional function approximation problem under certain regularization constraints concerning the smoothness of the class of approximating RBF functions.

RBF networks are local approximators by design since every basis function contributes to the overall model in a significant manner only if the Euclidian distance between its center and the input vector is small. This is in contrast to linear regression and MLP that learns function parameters and thus its estimation capability does not depend on the training data range.

16.2.2 Feedback (Recurrent) Neural Networks

The NN shown in Figure 16.4 is commonly referred to as an Elman recurrent neural network (RNN) It has a similar structure as the MLP but instead of a single feedforward path, the hidden layer output is fed back as a time-delayed input. Under certain assumptions about the observability and controllability of the dynamic system given by the input-output training set, these types of networks are well suited to monitor and control discrete systems [1].

Inputs are processed in a similar manner as in the MLP where in addition to the original input vector $x(t)$, the time-delayed outputs are fed back to the system.

$$\begin{aligned} y_j(t) &= g\left(\sum_{r=1}^{R} w_{2jr} v_r(t)\right) = \\ &= g\left(\sum_{r=1}^{R} w_{2jr} \tanh \beta \left(\sum_{m=1}^{M} w_{1rm} x_m(t) + \sum_{m=M+1}^{M_t} w_{1rm} y_m(t-1)\right)\right); \\ &j = 1, \ldots N \end{aligned} \tag{16.6}$$

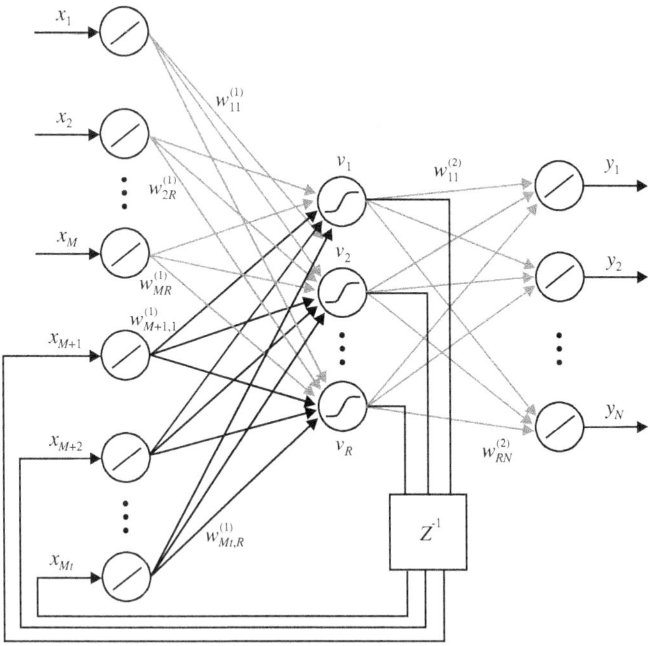

Figure 16.4 Recurrent neural network of Elman type. *Source*: Welch et al. 2009 [12]. Adapted with permission of IEEE.

16.3 ADAPTIVE CRITIC DESIGNS

Adaptive critic designs (ACDs) are NN designs for optimization over time based on combined concepts of reinforcement learning and dynamic programming [13–15]. ACDs use two NNs, the critic and the action network to solve the Hamilton–Jacobi–Bellman equation of optimal control. The critic network approximates the cost-to-go function **J** of Bellman's equation of dynamic programming, given below.

$$\mathbf{J}(t) = \sum_{k=0}^{\infty} \gamma^k U(t+k) \tag{16.7}$$

where γ is a discount factor between 0 and 1, and $U(t)$ is a utility function or a local performance index. The action NN, also referred to as the actor in the ACD literature, provides optimal control to minimize or maximize the cost-to-go function **J**. Several other ACD approaches such as the dual heuristic programming (DHP) and the global dual-heuristic programming exist [13]. These are all model-dependent designs. The model independent designs are referred to as action dependent (AD), and include action-dependent heuristic dynamic programming (ADHDP), action-dependent dual heuristic programming (ADDHP), and action-dependent global dual heuristic programming (ADGDHP).

16.4 INDEPENDENT COMPONENT ANALYSIS

Independent component analysis (ICA) neural networks belong to the class of unsupervised networks where the goal is to estimate the linear model matrix A and the statistically independent source signals **s**(t) based on the known measurements $\mathbf{x}(t)$ for a given time series of measurement [16]. In matrix form, we obtain:

$$\mathbf{X} = \mathbf{AS} \tag{16.8}$$

So, here $\mathbf{S} = [\mathbf{s}_1(t)...\mathbf{s}_N(t)]$ represents the vector of unknown signal sources, $\mathbf{X} = [\mathbf{x}_1(t)...\mathbf{x}_M(t)]$ is the measurement or observation vector, and A is the so-called M×N—dimensional mixing matrix.

A blind source separation technique [17] aims to recover the unknown signals and the mixing matrix from observed mixtures. ICA is a blind source separation technique that exploits the statistical independence between the sources [18]. Statistical properties of signals are a key factor in estimation by ICA, since there is almost no other information available.

The recovery model can be written as:

$$\mathbf{Y} = \mathbf{W}^T\mathbf{X} \tag{16.9}$$

Here, **Y** is an estimate of original sources matrix **S,** and $\mathbf{W}^{\mathrm{T}}$ is an estimate of the so-called separating or demixing matrix, that is, the (pseudo)inverse of matrix **A** in general. In this chapter, **A** is assumed to be a square matrix, representing the case where the number of measurements is equal to the number of sources.

In order to use the ICA for source separation, mixing and demixing models given in (16.8) and (16.9) have to satisfy the following assumptions:

1. The source signals are statistically independent.
2. At most, one of the source signals is Gaussian distributed.
3. The number of observations is greater or equal to the number of sources.

In the complex case, the sources **S**, their estimates **Y**, the measurements **X**, the mixing matrix **A,** and the separating matrix **W** can be complex-valued vectors and matrices. Since many power system measurements are complex valued, the complex demixing model in (16.10) instead of (16.9) is used.

$$\mathbf{Y} = \mathbf{W}^H\mathbf{X} \tag{16.10}$$

Here the columns of **Y** represent the neural outputs, the columns of **X** those of the neural inputs, and **W** the role of the weight matrix.

16.5 LEARNING ALGORITHMS: THE DETERMINATION OF WEIGHTS

The objective of learning algorithm is formulated as an error minimization problem, which can subsequently be solved with nonlinear gradient descent techniques or heuristic techniques including genetic algorithms and particle swarm optimization (PSO) [19]. The detailed algorithms are beyond the scope of this chapter.

16.5.1 Supervised Learning Objectives

For most neural nets including the MLP, the RBF, and the RNN, the mean square error is used as the objective function. The least square error between calculated output vector and target output vector to be minimized is:

$$\mathbf{J}(\mathbf{W},\mathbf{X},\mathbf{Y}) = \frac{1}{2}\sum_{k=1}^{K} \|\mathbf{F}(\mathbf{W},\mathbf{x}(t_k)) - \mathbf{y}(t_k)\|^2 = \text{min!} \tag{16.11}$$

Note that for MLP models, the objective function J is a nonlinear convex function of the combined weight matrix $\mathbf{W} = [\mathbf{w}_1, \mathbf{w}_2]$ whose first columns are the weight matrix $\mathbf{w}_1$ and the subsequent columns are the weight matrix $\mathbf{w}_2$ because the output y_i depends on the weights w_{2ij} and the output of the hidden unit v_j the latter depending on w_{1jk}. A local but not necessarily global minimum is calculated by setting the gradient of the objective function with respect to the weights W_{ij} to zero.

$$\text{grad } \mathbf{J}(\mathbf{W},\mathbf{X},\mathbf{Y}) = 0 \tag{16.12}$$

The error gradient is a nonlinear function of the weights and its minimization requires an iterative procedure. Stochastic gradient descent, referred to as backpropagation, and the Levenberg–Marquard algorithm or other quasi-Newton techniques are popular choices.

For the RBF model, the error function **J** depends only on one set of weights **W**. Furthermore, the gradient of **J** is linear in the weight terms; so after nonlinear preprocessing, the global minimum can be determined in an efficient one-step procedure using the pseudo-inverse, singular value decomposition or, in an on-line application, linear recursive least-square techniques.

Since the activation function of the nonlinear sigmoid units of the MLP saturates at +1 and −1, the input and target data for the MLP net needs to be normalized such that each input vector component falls into this range. Matlab's NN toolbox provides the following data normalization function that centers the data around zero and ensures that its minimum and maximum values are −1 and +1, respectively. Let **x** be row of M data points representing one system parameter. Let $x_{\max}$ and $x_{\min}$ be its maximum and minimum value. Then, each data point $\boldsymbol{x}_i$, i = 1, … M, will be normalized as follows:

$$x_i^{norm} = \frac{2(x_i - x_{\min})}{x_{\max} - x_{\min}} - 1 \tag{16.13}$$

16.5.2 Unsupervised ICA Learning Objectives

For unsupervised ICA learning, the only known variable is the measurement matrix **X**. Therefore, the constraints of the problem are $\mathbf{Y} = \mathbf{W}^H\mathbf{X}$. The other assumptions are

- the statistical independence of the source signals;
- the non-Gaussianity of the source signals.

Thus, the optimization objectives become to maximize signal independence subject to constraints.

16.5.2.1 Off-Line ICA Learning

For off-line learning, we assume that the matrix W, which is an estimate of the inverse of the mixing matrix A, is constant. One example of a function that measures statistical independence is kurtosis, which is the fourth-order statistics. Let E denote the statistical expectation and the objective function to be optimized is:

$$\mathbf{J}(\mathbf{w}) = E\{|\mathbf{w}^{\mathbf{H}}\mathbf{x}|^4\} \tag{16.14}$$

To simplify the ICA algorithm, signals are preprocessed by centering and whitening. Centering transforms the observed signals to zero-mean variables and whitening linearly transforms the observed and centered signals, such that the transformed signals are uncorrelated, have zero mean, and their variances equal unity. This yields the orthogonality condition on the columns of $\mathbf{W}$ in equation (16.15).

The optimization problem using a single unit contrast function subject to the constraint of decorrelation can be defined as:

$$\begin{aligned} &\text{maximize} && \sum_{i=1}^{N} J(\mathbf{w}_i) \\ &\text{under constraint} && E\left\{\mathbf{w}_k^H \mathbf{w}_i\right\} = \delta_{jk} \\ &\text{and} && \mathbf{Y} = \mathbf{W}^H \mathbf{X} \end{aligned} \tag{16.15}$$

where w_i, $i = 1, \ldots N$ are the rows of the matrix $\mathbf{W}$, $\delta_{jk} = 1$ for $j = k$ and $\delta_{jk} = 0$ otherwise.

The optimization task given in (16.15) can be solved using a single unit deflation algorithm, where independent components are estimated one-by-one. To estimate several independent components, this algorithm is executed using several units. After each iteration, vectors are decorrelated to prevent convergence to the same maximum. This algorithm, also called FastICA, is based on a fixed-point iteration scheme presented in [20]. Its complex version is described in [21].

16.5.2.2 On-Line (Adaptive) ICA Learning

In the electric power system examples considered in this chapter, the mixing matrix A represents the equivalent Y-bus matrix, which thus contains topology information. Since the topology may change during the estimation, we relax the condition that the estimate of the inverse of the mixing matrix A, that is the matrix W, remains constant.

The natural gradient algorithm [22], which tries to minimize the mutual information between outputs is well suited for on-line estimation. The objective function then becomes

$$\mathbf{J}(\mathbf{W}) = -\frac{1}{2}\log|\det(\mathbf{W}^T\mathbf{W})| - \sum_{i=1}^{N}\log p_i(y_i) \tag{16.16}$$

Here, $p_i(y_i)$ are the probability density functions of the outputs. Without loss of generality, we assume super-Gaussian distribution of sources. Then the update rule for this algorithm using the objective function is given by

$$\Delta \mathbf{W}_t = \eta_t \left(\mathbf{I} - \tanh(\mathbf{y}_t)\mathbf{y}_t^T\right) \mathbf{W}_t \quad (16.17)$$

A method for on-line adaptation of the learning parameter η_t and an application to the Extended Infomax algorithm [23] is given in [24].

16.6 EXAMPLES OF NEURAL NETWORK APPLICATIONS FOR POWER SYSTEM MONITORING AND CONTROL

The North American electric power grid is the world's largest single machine ever built by man. It is a complex adaptive system under semi-autonomous control. It is spatially and temporally complex, nonconvex, nonlinear, and nonstationary with uncertainties at many levels. Its efficient, reliable, and safe operation and control is a challenge today as seen from the 2003 Northeast American Blackout [25]. The complexity grows with growing demand to integrate of renewable sources of energy such as wind and solar farms. To overcome the challenges brought about complexity, shear size of the network and others, the power system needs intelligence at all levels—horizontally and vertically, be it at component, area, or system level. This means intelligent information processing is needed everywhere a critical decision is made for performance control, locally or globally. For this to be possible, large number of distributed sensors and actuators are required on the network.

16.6.1 On-Line Estimation of Electric Power System Active Loads

This section summarizes the result published also in [26]. The DC power flow equations of the steady-state power flow model are given as:

$$\underbrace{\begin{bmatrix} \Delta P_2 \\ \vdots \\ \Delta P_n \end{bmatrix}}_{\Delta \mathbf{P}} = \underbrace{\begin{bmatrix} -B_{22} & \cdots & -B_{2n} \\ \vdots & \ddots & \vdots \\ -B_{n2} & \cdots & -B_{nn} \end{bmatrix}}_{\mathbf{B}} \underbrace{\begin{bmatrix} \Delta\delta_2 \\ \vdots \\ \Delta\delta_n \end{bmatrix}}_{\Delta\delta} \quad (16.18)$$

where ΔP_i are the net active power deviations entering the network at bus i, $|\delta_i|$ is the voltage angle deviation at bus i, and B_{ik} are the elements of the susceptance matrix between busses i and k. A similar relation can be obtained between active power branch flows and the bus voltage angles:

$$\Delta \mathbf{P}_{branch} = \mathbf{B}_f \Delta\delta \quad (16.19)$$

where B_f is a rectangular matrix. If we solve for $\Delta\delta$ in (16.18) and substitute $\Delta\delta$ in (16.19), we obtain the following vector equation representing the active power branch flows as a linear mixture of the net active bus power injections:

$$\Delta \mathbf{P}_{branch} = \mathbf{B}_f \mathbf{B}^{-1} \Delta \mathbf{P} \quad (16.20)$$

In equation (16.20), the matrix product $\mathbf{A} = \mathbf{B}_f\mathbf{B}^{-1}$ matrix represents the network topology and parameters implicitly.

It is now assumed that the network parameters are partially or completely unknown. Furthermore, if the network topology changes, then the linear mixing model $\mathbf{A}$ in (16.20) will no longer be time-invariant. This is a realistic assumption for electric power systems where the system topology and parameters change during the day. In normal operation, topology change occurs when the circuit breakers are opened and/or closed in order to reconfigure the electric network for more reliable and economic operation. Also system parameters can change by switching series and parallel reactors to keep the power flows and voltages within specified limits. Therefore, the matrix $\mathbf{A}$ changes during the day. In this section, we use an on-line ICA algorithm with an adaptive learning parameter to estimate the active load profiles of electric loads under conditions of topology change. Such an application will provide valuable information to power system operators for real-time operation of power systems.

A sample load profile is shown in Figure 16.5. Electric load profiles can be decomposed into two components. The first one is the slow varying component, a variation representing general characteristics. As an example, residential loads show peaks during the noon and the evening hours, however, industrial loads have more or less constant load profiles throughout the day. Also, rising and falling outside temperatures will contribute to the change of this slow variation. The second part is the fast varying component that represents individual behavior of electric loads at any given point in time. We have shown in [27] for sample loads that the fast varying component represents the independent and non-Gaussian part of the load profile. The slow varying and fast varying components have been separated by a linear filter.

Load profile and measurements are generated using computer simulations. Typical electric load profiles were downloaded from the ERCOT website [14]. To

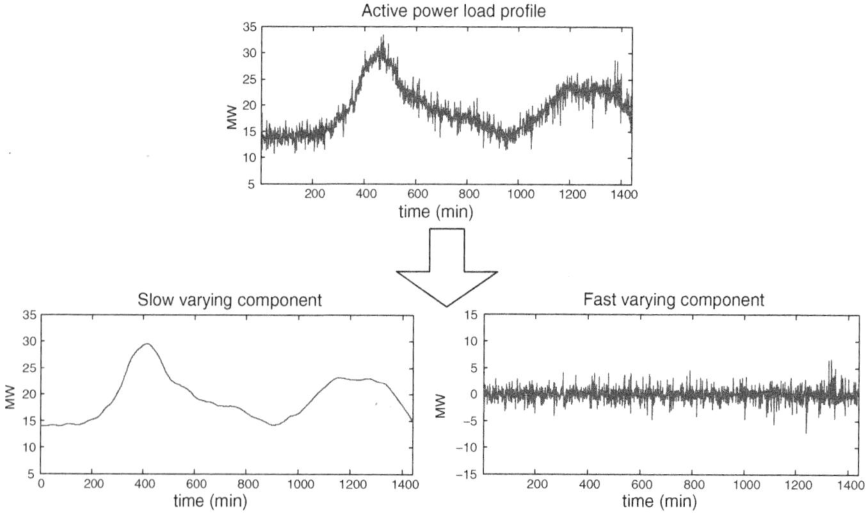

Figure 16.5 Filtered active load profile.

differentiate the fast and slow varying components of the load profiles, a zero mean Laplace distributed random variable with a 0.002 variance is added to the normalized load profiles. Actual load profiles for simulations are obtained by multiplying the load profiles with the active power of each load. The "measurements" are obtained by running a DC power flow and recording the active power flows on selected branches. A public-domain DC power flow program [28] is used for the power flow calculations. Simulations are performed for a 24-h period with 1440 data representing the 1-min reading. A 24-h window data is used for whitening and filtering. When a new measurement arrives, this is included in the window and the oldest data is dropped.

The proposed on-line load profile estimation algorithm is tested on a modified IEEE 14-Bus test system shown in Figure 16.6. There are eight electric loads on busses 3, 4, 5, 6, 9, 10, 13, and 14. Also, there are eight active power flow measurements on branches 2-5, 3-4, 4-5, 6-11, 6-12, 7-9, 9-14, and 13-14.

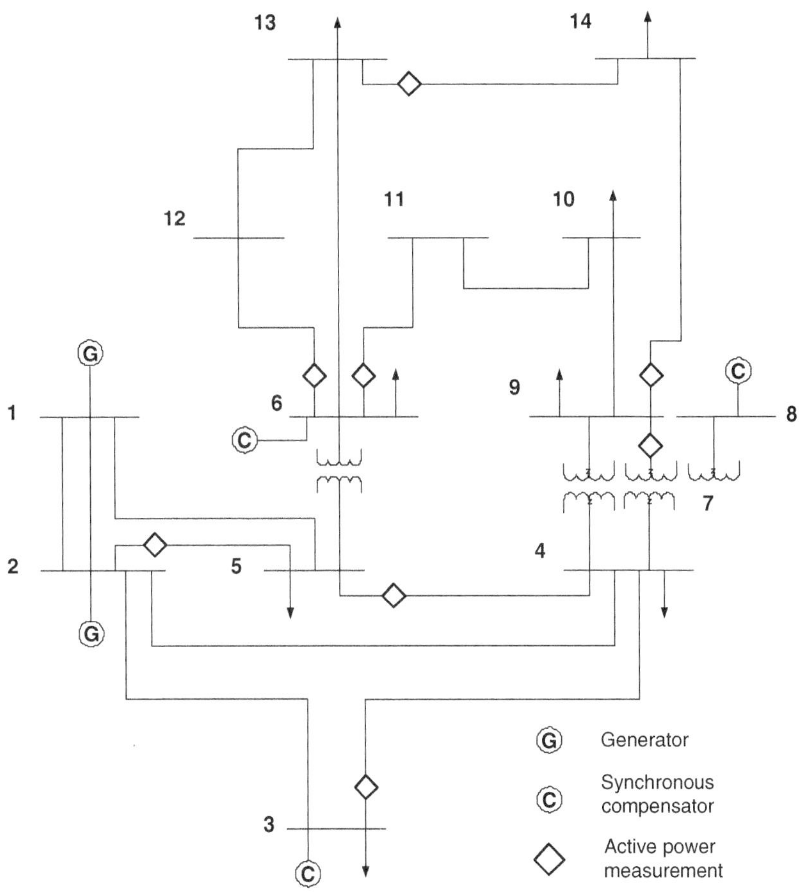

Figure 16.6 Modified IEEE 14-Bus test system.

TABLE 16.1 Errors and correlation coefficients between actual and estimated load profiles without topology changes

Bus number	Correlation coefficient	Maximum percentage error (%)	Mean percentage error (%)
3	0.9555	13.6	4.48
4	0.9576	18.8	5.56
5	0.9940	8.66	3.32
6	0.8754	27.7	11.8
9	0.9865	17.6	3.14
10	0.9948	26.6	4.06
13	0.9646	7.13	2.14
14	0.9255	13.1	3.77

The case study illustrates the performance of the algorithm in a power system where a topology change occurs during the day. In order to simulate the changing environment, the topology of the network is changed by opening the two circuit breakers on both ends of the branch between busses 2 and 4 at time step $t_c = 1100$. The point in time of the topology change can be identified by examining the change in the elements of the separating matrix or the change in the learning parameter. Once this point is determined, the measured data can be divided into two time intervals: the first includes data points before and including the topology change, the second one contains all points after the topology change.

The estimate of the load profile after topology change will be $\mathbf{Y}^a = \mathbf{W}^a\mathbf{X}^a$, where the superscript a denotes variables after the topology change. Since we assume that only the network topology changes and the variations in the loads are small, we can obtain the scaling of the load profile by using the points just before the topology change occurred. The errors and correlation coefficients between actual and estimated load profiles without and with topology changer are given in Tables 16.1 and 16.2, respectively. The estimation error corresponding to the load profile at bus 4 is greater than the others because this is a bus connected to the branch whose status changed. For

TABLE 16.2 Errors and correlation coefficients between actual and estimated load profiles after topology change

Bus number	Correlation coefficient	Maximum percentage error (%)	Mean percentage error (%)
3	0.9824	7.29	1.51
4	0.4283	24.01	5.02
5	0.9108	5.80	1.35
6	0.9983	5.59	3.97
9	0.8130	8.80	2.92
10	0.9935	7.98	1.59
13	0.9968	14.53	3.13
14	0.9695	5.86	2.21

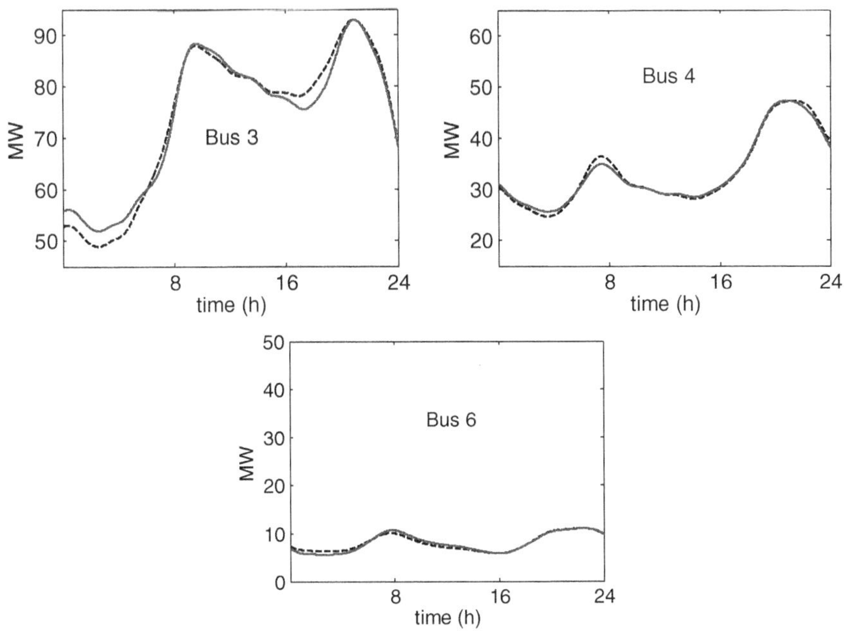

Figure 16.7 Actual (solid line) and estimated (dashed line) load profiles at busses 3, 4, and 10.

those elements of the mixing matrix that changed in greater magnitude, the algorithm requires a larger number of iterations to converge.

Figure 16.7 shows examples of the actual and estimated load profiles without topology change at busses 3, 4, and 10. More detailed results can be found in [26].

16.6.2 Harmonic Source Identification Using Off-Line ICA

This section summarizes the results published in [29]. The harmonic steady-state system equations under nonsinusoidal condition are given by the following linear set of equations:

$$\boldsymbol{I}_h = \boldsymbol{Y}_h \boldsymbol{V}_h \tag{16.21}$$

where h is the harmonic order of the frequency, $\boldsymbol{I}_h$ is the bus current injection vector, $\boldsymbol{V}_h$ is the bus voltage vector, and $\boldsymbol{Y}_h$ is the system admittance matrix at frequency h. The linear equation (16.21) is solved for each frequency of interest.

For electric power systems, it may be difficult to obtain (a) accurate system parameters ($\boldsymbol{Y}_h$), especially for higher harmonics and (b) enough harmonic phasor measurements to guarantee system observability. Using time sequence data of available measurements, that is, complex harmonic voltage measurement sequences at a limited number of busses, preferable substations, the ICA algorithm is able to estimate the load profiles of harmonic current sources without the knowledge of system parameters and topology using the statistical properties of time series data only.

The linear measurement model for the harmonic load flow given in (16.21) can be defined as:

$$\mathbf{V}_h(t_k) = \mathbf{Z}_h \mathbf{I}_h(t_k) k = 1, 2..., K \tag{16.22}$$

Here, $\mathbf{V}_h(t_k) \in C^M$ are the known harmonic voltage measurement vectors, $I_h(t_i) \in C^N$ are the unknown harmonic current source vectors, $\mathbf{Z}_h \in C^{M \times N}$ is the unknown mixing matrix relating measurements to the sources, h is the harmonic order, t_k is the sample index, and K is the number of samples. In the presence of only harmonic voltage measurements, $\mathbf{Z}_h$ is the system impedance matrix at harmonic order h.

The general representation of the linear system equations in (16.21) can be written as:

$$\begin{bmatrix} I_h^1 \\ I_h^2 \\ \vdots \\ I_h^N \end{bmatrix} = \begin{bmatrix} Y_h^{1,1} + Y_h^{L1} & Y_h^{1,2} & \cdots & Y_h^{1,N} \\ Y_h^{1,1} & Y_h^{2,2} + Y_h^{L2} & \cdots & Y_h^{2,1} \\ \vdots & \vdots & \ddots & \vdots \\ Y_h^{N,1} & Y_h^{N,2} & \cdots & Y_h^{N,N} + Y_h^{LN} \end{bmatrix} \begin{bmatrix} V_h^1 \\ V_h^2 \\ \vdots \\ V_h^N \end{bmatrix} \tag{16.23}$$

Here, ${Y_h}^{i,j}$ represents the equivalent admittance at frequency h between node i and j, and ${Y_h}^{Li}$ represents the admittance of linear loads connected to bus i, which are modeled with impedance models. We can separate the admittance matrix into two parts:

$$\begin{bmatrix} I_h^1 \\ I_h^2 \\ \vdots \\ I_h^N \end{bmatrix} = \begin{bmatrix} Y_h^{1,1} & Y_h^{1,2} & \cdots & Y_h^{1,N} \\ Y_h^{1,1} & Y_h^{2,2} & \cdots & Y_h^{2,1} \\ \vdots & \vdots & \ddots & \vdots \\ Y_h^{N,1} & Y_h^{N,2} & \cdots & Y_h^{N,N} \end{bmatrix} \begin{bmatrix} V_h^1 \\ V_h^2 \\ \vdots \\ V_h^N \end{bmatrix} + \begin{bmatrix} Y_h^{L1} & 0 & \cdots & 0 \\ 0 & Y_h^{L2} & \cdots & 0 \\ \vdots & \vdots & \ddots & \vdots \\ 0 & 0 & \cdots & Y_h^{LN} \end{bmatrix} \begin{bmatrix} V_h^1 \\ V_h^2 \\ \vdots \\ V_h^N \end{bmatrix} \tag{16.24}$$

The second term on the right-hand side of equation (16.24) represents the linear loads as a vector of harmonic current sources. Rewriting (16.24), we get

$$-\mathbf{I}_h^L + \mathbf{I}_h = \mathbf{Y}_h \mathbf{V}_h \tag{16.25}$$

Here, ${\mathbf{I}_h}^L$ is the harmonic current source vector corresponding the second term on the right-hand side of (16.24). In the ICA model, the mixing matrix $\mathbf{A}$, which represents the impedance matrix ${\mathbf{Y}_h}^{-1}$ in the harmonic domain, is required to be time-independent for all time steps $t_k, i = 1, 2, \ldots K$. Using this simple manipulation, the load model for linear loads changes from impedance model to current model and the admittance matrix is kept constant. Using current models for linear loads increases the number of sources to be estimated in addition to the nonlinear loads. However, considering that there are no linear loads on harmonic source busses and combining the linear loads, which have similar load profiles, the number of sources to be estimated can be reduced.

16.6.3 ICA-Based Harmonic Source Identification Case Study

For this section, typical load profiles were downloaded from the website of Electric Reliability Council of Texas [30]. To distinguish fast and slow varying components of the load profiles, a zero mean Laplace distributed random variable with 0.002 variance is added to the normalized load profiles. Here the Laplace distributed random variables represent the fast varying components, which are statistically independent, and the load profiles represent the slow varying components mentioned in the previous section. The apparent power of each load is multiplied by one of these load profiles. Harmonic measurement vectors, that is the harmonic bus voltages, are simulated by harmonic power flow. The public-domain MATPOWER [28] program was modified to carry out both fundamental and harmonic power flow calculations. First, the fundamental frequency power flow solution is obtained. Harmonic sources are modeled as constant power loads in this step. Harmonic current source models are obtained for harmonic sources using the power flow solution. Next, we calculate the harmonic bus voltages by solving the linear system equations in (16.21) for each harmonic frequency of interest. In harmonic analysis, a transmission line, a generator, and a transformer are modeled as a π-model, a subtransient reactance, and a short-circuit impedance, respectively. The impedance model #2 given in [31] is used to model linear loads in harmonic domain. The harmonic measurement vector obtained by harmonic power flow is used in ICA for the estimation of harmonic sources.

The harmonic load identification algorithm is tested on a modified IEEE 30-Bus test system shown in Figure 16.8.

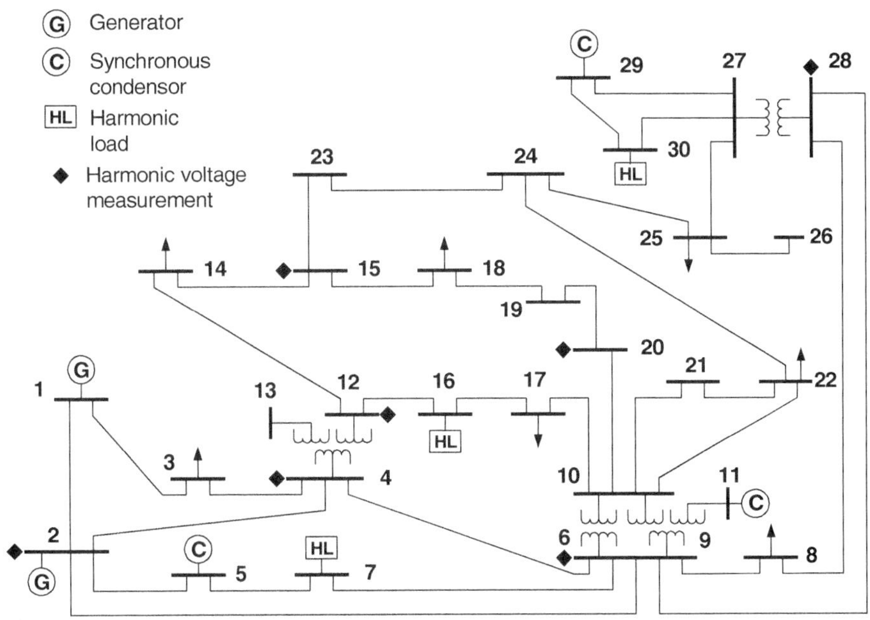

Figure 16.8 Modified IEEE 30-Bus system.

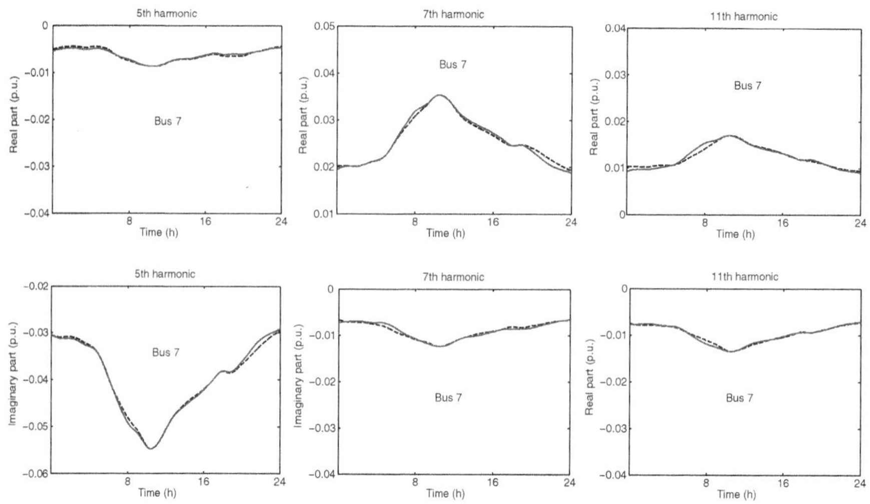

Figure 16.9 Harmonic components of current injection at bus 7.

The system is assumed to be balanced. We placed three harmonic producing loads at busses 7, 16, and 30. These busses are geographically and electrically far from each other, which is usually the case in power systems. These harmonic loads are modeled as harmonic current injection sources. Harmonic current spectrums given in [32] are used for these loads and simulations are obtained up to the 17th harmonic. Harmonic source power ratings are $45 + j20$ MVA, 25 MVar, and $25 + j16$ MVA at busses 7, 16, and 30, respectively. In addition to harmonic loads, there are seven linear loads at busses 3, 8, 14, 17, 18, 22, and 25. Four of these linear loads have a power rating close to harmonic load power ratings. The remaining busses are no-load busses. Both the harmonic loads and linear loads have 1-min varying load shapes, which are obtained by adding Laplace distributed random variables to slow varying load shapes normalized to unity and by multiplying them with the power rating of each load. For the measurements, harmonic voltage measurements are used, since in general these measurements are easier and more reliable than other harmonic measurements, such as power measurements. There are seven harmonic voltage measurements placed at busses 2, 4, 6, 12, 15, 20, and 28. In general, it is easier and less expensive to obtain measurements at substations than at other busses. Therefore, six of these measurements are located at substations. None of the measurements are located at the load busses. This reflects the case of a deregulated network where generation and distribution are unbundled.

The FastICA algorithm referenced in Section 16.5.2.1 was implemented in Matlab 7.0. The calculation time of the algorithm for the test cases is in orders of seconds in a Pentium4, 2.54GHz, 512MB RAM computer. FastICA computes estimates of the real and imaginary part of the load shapes of harmonic current sources shown in Figures 16.9 and 16.10.

Figure 16.9 shows the smoothed current profiles of the estimated harmonic current injection at bus 7. The solid lines represent the actual load shapes and the dashed lines represent the estimated ones. The real and imaginary part of the current

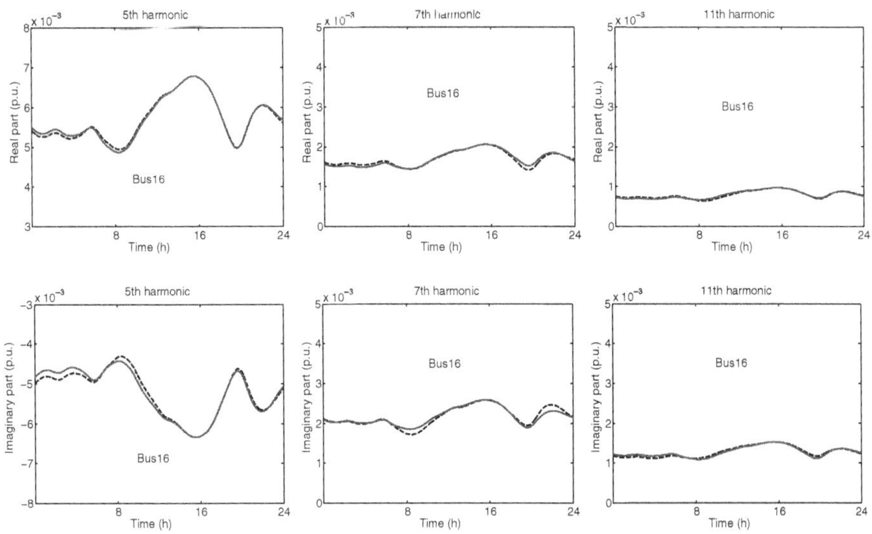

Figure 16.10 Harmonic components of current injection at bus 16.

is estimated by applying the complex ICA algorithm to the complex harmonic voltage measurements. Tables 16.3 and 16.4 show the error and the correlation coefficients between estimated and actual load shapes. The results from Figure 16.9 and Table 16.3 show that the proposed algorithm is capable of estimating the load profiles of harmonic injection within a small error range. Correlation coefficients are close to 1, indicating the matching of high accuracy between estimated and actual profiles.

In Figure 16.10, harmonic current injection profiles of the nonlinear loads at bus 16 are given. The graphs show that there are some slight differences between estimates and actual shapes. However, estimates are still tracking the actual load shapes.

The results given in the tables and our observations with the technique indicate that the signal strength and proximity of measurements to the sources affects in the accuracy of results. In general, these results show that the sources with high magnitude and close to the measurements are estimated more accurately. These issues will be further investigated in future work.

TABLE 16.3 Errors between estimated and actual smoothed harmonic current profiles at bus 7

Harmonic order		Correlation coefficient	Maximum percentage error (%)	Mean percentage error (%)
5	Real	0.9845	8.95	3.97
	Imag	0.9979	4.20	1.06
7	Real	0.9951	6.4	1.67
	Imag	0.9752	11.08	3.04
11	Real	0.9827	11.6	2.81
	Imag	0.9882	6.59	1.93

TABLE 16.4 Errors between estimated and actual smoothed harmonic current profiles at bus 16

Harmonic order		Correlation coefficient	Maximum percentage error (%)	Mean percentage error (%)
5	Real	0.9959	1.88	0.73
	Imag	0.9870	3.63	1.55
7	Real	0.9719	7.16	2.25
	Imag	0.9650	7.27	2.24
11	Real	0.9755	5.95	2.31
	Imag	0.9670	5.15	2.39

In our simulations, measurement noise is ignored. Sensitivities to measurement noise are investigated in [27] and [29].

16.6.4 Wind Speed Forecasting

In order to carry out economical dispatch of wind generated electricity, it is important to have capabilities to make short-term predictions of wind speeds, which directly affects the generation capacity. Without this ability, a wind farm operator is prone to allocate more generation units or supplemental energy reserves than necessary in order to ensure budgeted electricity outputs are met [33], with an end result of increased operating costs. NNs have been extensively studied and shown to provide accurate short-term load predictions [34, 35]. NNs have also shown to be promising tools for wind and solar energy predictions [12, 36, 37].

The short-term wind speed prediction abilities of an MLP, RNN, and simultaneous recurrent neural network (SRN) are investigated in a study carried out by one of the authors [12]. Each network has four input neurons (one for each of the inputs and also a bias value) and one neuron in the output layer. In order to keep NN parameter size identical, the MLP has 10 neurons in its hidden layer while the RNN and SRN each have five neurons in the hidden layer, which also means that the context layers are five neurons wide. The input weight (w_1) size is of 4×10 and output weight (W_2) size is of 10×1 for the MLP case; the RNN and SRN both have a w_1 of size 9×5 and a w_2 of size 5×1; in all cases, the total number of weights is 50.

The training algorithm for all these networks is the PSO. The data used to train the NNs in this study includes three inputs (in addition to a bias value) recorded by instruments on or near the 82 m tall M2 tower of the National Renewable Energy Lab's National Wind Technology Center site [38] located 5 miles south of Boulder, Colorado:

(i) current wind speed measured at 80 m from the ground in m/s;

(ii) current temperature at 2 m in degrees C;

(iii) current percent relative humidity.

The training data includes values for each of these inputs in 1min intervals, with each data point representing the mean of readings taken every 2 s during that

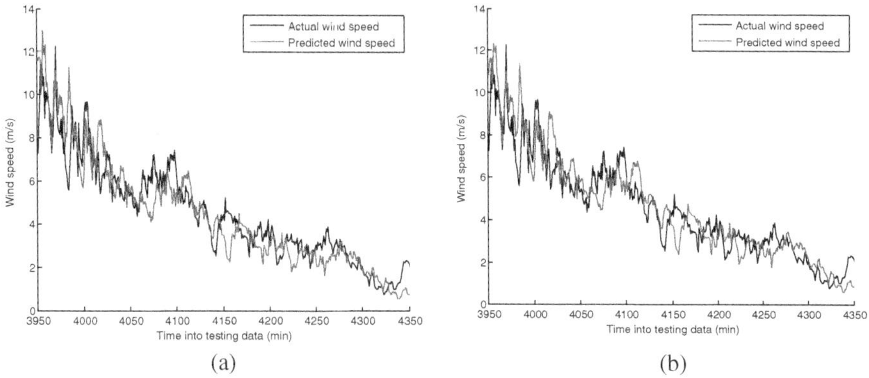

Figure 16.11 The actual performance of the neural networks trained using PSO, showing both actual wind speed and predicted wind speed for a period in a dataset for a week in May, 2008: (a) MLP and (b) SRN.

minute. This training data is used to optimize the values of parameters of each NN in order to provide the best prediction of wind speed 15 min into the future. Figure 16.11 shows the performance of MLP and SRN in predicting the wind speed over 400 min. Table 16.5 shows the performance comparison of each network type in terms of the absolute relative error, calculated using (16.1).

$$ARE = \left| \frac{\text{Actual Wind Speed} - \text{Predicted Wind Speed}}{\text{Actual Wind Speed}} \right| \tag{16.26}$$

The ability to perform accurate short-term wind speed predictions is very useful for the electric energy industry. This provides wind farm operators the ability to anticipate potential wind power, and participate in the electricity market and operation of the power network accordingly. The results obtained in this study show that while all types of NNs investigated have the potential for short-term wind speed prediction, the best results are generally obtained from the feedback architectures, especially on data outside the normal training range. However, with this increase in accuracy comes an increase in training and run time due to the feedback loop of these networks.

16.6.5 Optimal Control of Grid Independent Photovoltaic System

ACDs have been demonstrated to carry out optimal energy dispatch in a grid independent photovoltaic system [39,40]. The objectives of optimal control are threefold:

TABLE 16.5 Mean absolute relative error for each data set

Neural network	Mean ARE on training data	Mean ARE on testing data
MLP	0.3847	0.5038
RNN	0.3892	0.4354
SRN	0.3795	0.4544

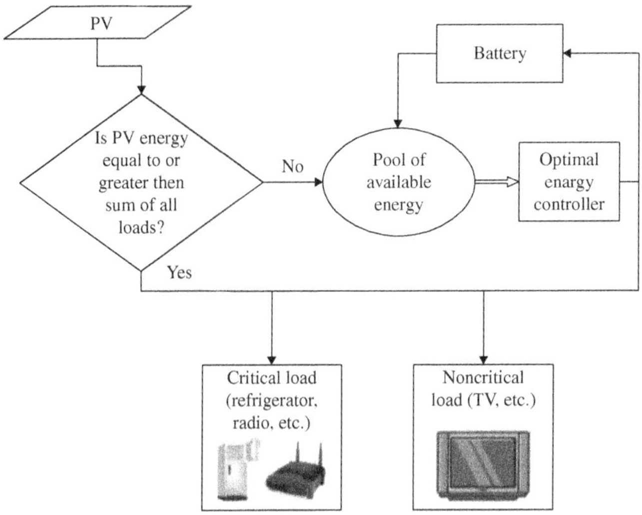

Figure 16.12 Photovoltaic (PV) system model (energy flow shown by arrow direction), shown with optimal controller bypass (for cases with over abundance of energy). Emphasis is placed on meeting the critical load demand.

(1) to maximize or fully dispatch the required energy to the critical loads at all times, (2) to dispatch energy to charge the battery to enable it to power the critical load when the collector cannot meet the critical load demand, and (3) to dispatch energy to the noncritical loads while not compromising on the first two objectives.

The complete PV system considered consists of the PV array (solar cells), maximum power point tracker, battery, critical and noncritical loads, battery charge controller, inverter, and controller. The critical load consists of loads that should never be dropped (such as refrigeration and/or radio communication), and the noncritical load contains items that are nonessential (e.g., television). The minimum charge for the battery was also determined to be 30% (this is consistent with standard deep cycle lead-acid batteries). This arrangement is shown in Figure 16.12.

In Figure 16.12, energy is dispatched in the direction of the arrow. During this simulation, if the energy from the PV array was ever greater than the sum of the loads (both critical and noncritical) and there was enough energy to completely power the battery, then optimal control was not used. Instead, all loads were powered and the battery was completely charged. This case occurred infrequently.

For this case study, an ADHDP adaptive critic approach is adopted for the design of a PV optimal controller [39]. Action-dependent ACDs do not need system models to develop the optimal control policy (action network output). A block diagram of the controller that was developed is shown in Figure 16.13.

The four inputs and three outputs of the ADHDP controller are, respectively:

- Inputs:
  1. Solar energy from the PV array as a fraction of total possible energy from the PV array.

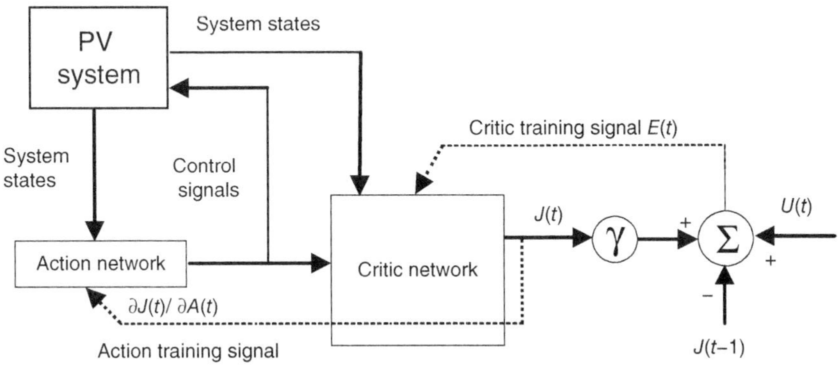

Figure 16.13 Structure of the ADHDP-based optimal PV controller design.

2. Critical load as a fraction of maximum critical load.
3. Noncritical load as a fraction of maximum noncritical load.
4. Current battery charge as a fraction of total charge.

- Outputs:
  1. Energy dispatched to the critical load.
  2. Energy dispatched to the noncritical load.
  3. Energy dispatched to the battery (which can be positive or negative, depending on whether the battery is being charged or being used as a source).

The above controller is compared to a second optimal controller-based Mamdani fuzzy logic and a conventional priority controller. The inputs and outputs of these controllers are very similar to the ADHDP controller discussed above and see [40] for details.

Figure 16.14 shows the battery state of charge for Caribou, Maine, for the entire year using the PV-priority [39], ADHDP based, and optimized fuzzy logic controllers.

Table 16.6 shows the percentage of critical and noncritical loads met by the various controllers for the Caribou, Maine, case study. A metric is developed to compare the performance of the controllers in meeting the three objectives mentioned above [39]. It can be seen that ADHDP controller has the best performance and then followed by the optimized fuzzy controller. The optimized fuzzy controller is one whose parameters are optimized using PSO [19].

The ADHDP controller is most likely able to satisfy more of the loads because of its ability to discount actions into the future, whereas the fuzzy logic controller does not. This property gives the ADHDP controller the ability to keep the average battery charge higher than the PV-priority controller (to be available for use in times of lesser available PV energy in some possible future state), but still allows it to power much more of the critical load and to a lesser extent the noncritical load. Not keeping the batteries fully charged at all times also allows the controller to capture more energy during the day in the summer months when the PV generated energy is higher, while still satisfying the noncritical loads. As the winter months approach, it lessens the

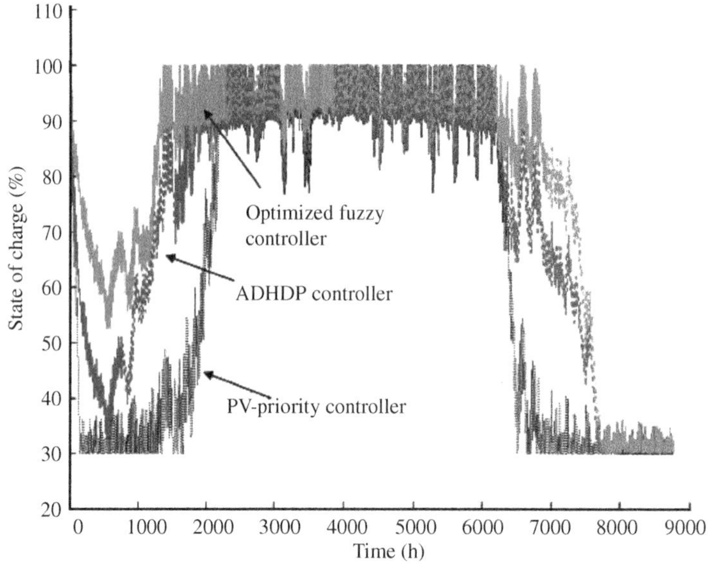

Figure 16.14 State of charge (for the entire year) of the battery using the PV-priority controller (dotted line—bottom curve), the ADHDP-based optimal controller (dashed line—middle curve), and the fuzzy optimal controller (solid line—top curve).

time that the noncritical load is powered so that it can conserve battery charge during times of lower available solar energy. The optimized fuzzy controller does not act as dynamically and cannot maintain the loads for as long as the ADHDP controller, although it is much better than the PV-priority controller.

16.6.6 Adaptive Neurocontrol of a FACTS Device: The Unified Power Flow Controller

With the ever increasing complexities in the electric power grid across the globe and the growing need to provide stable, secure, controlled, economic, and high quality electric power—especially in today's deregulated environment—it is envisaged that flexible AC transmission system (FACTS) controllers are going to play a critical role

TABLE 16.6 Comparison of the performance of three controllers

Controller:	PV-priority	ADHDP	Nonoptimized fuzzy logic	Optimized fuzzy logic
Critical load satisfied	84.22%	96.54%	93.04%	92.97%
Noncritical load satisfied	77.21%	61.87%	32.16%	56.07%
Average battery charge	63.87%	74.35%	76.58%	80.21%
Total score	1.951	2.094	1.895	2.053

in power transmission systems [41]. Transmission congestion results when there is insufficient capacity to transmit power over existing lines and maintain the required safety margins for reliability. FACTS enhance the stability of the power system both with its fast control characteristics and continuous compensating capability. The two main objectives of FACTS technology are to control power flow and increase the transmission capacity over an existing transmission corridor [42].

Gyugyi then proposed the unified power flow controller (UPFC), a new generation of FACTS devices in 1991 [43] as a combination of the static synchronous compensator (STATCOM) and the static synchronous series compensator (SSSC). The two are coupled via a common DC link, to allow bidirectional flow of real power between series output terminals of the SSSC and the shunt terminals of the STATCOM, and are controlled to provide concurrent real and reactive series line compensation without an external electric energy source. The UPFC by means of angularly unconstrained series voltage injection is able to control, concurrently or selectively, the transmission line voltage, impedance, and angle or, alternatively, the real and reactive power flow in the line. The UPFC may also provide independently controllable shunt-reactive compensation.

FACTS devices are nonlinear and when connected to a nonlinear power system with a changing configuration, the performance of the conventionally designed FACTS controllers therefore also degrades as the operating points move away from the point where the controllers were designed. One possible solution to all these problems is the use of nonlinear adaptive intelligent neurocontrollers based on advanced identification and control schemes.

For identifying and controlling the dynamics of a UPFC and a power system, the single machine infinite bus power system in Figure 16.15 is studied. This power

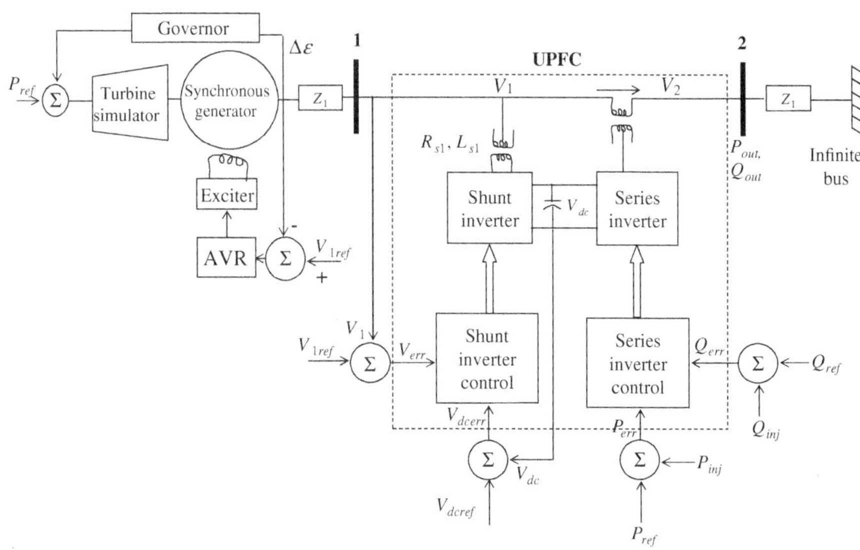

Figure 16.15 Single machine infinite bus system with the UPFC.

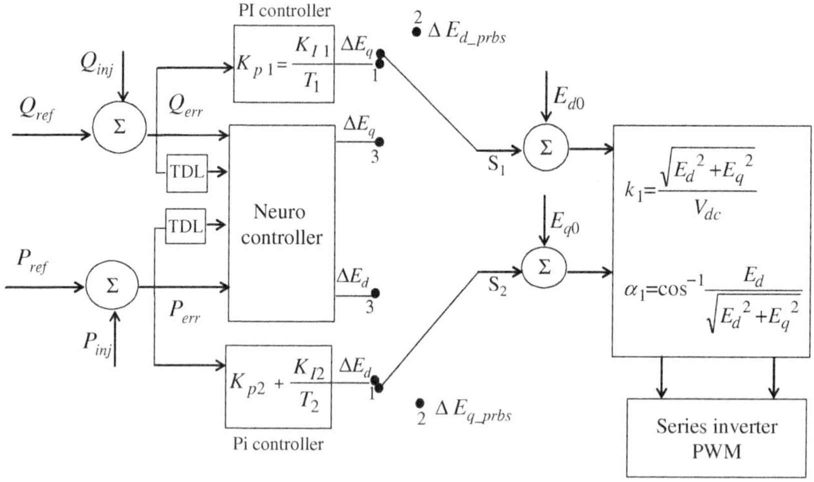

Figure 16.16 Series inverter control with conventional proportional-integral (PI) controllers and neurocontroller.

system comprises of a synchronous generator with exciter-automatic voltage regulator and turbine-governor combinations connected to an infinite bus through two sections of transmission lines. The UPFC is placed between the two sections of the transmission lines, between busses 1 and 2 as shown in Figure 16.15. This simple system is chosen in order to evaluate the performance of the UPFC with two different control strategies—PI and neurocontrol.

The conventional PI control for the series and shunt branches of the UPFC (with the switches S_1 and S_2 at position "1") is shown in Figures 16.16 and 16.17, respectively.

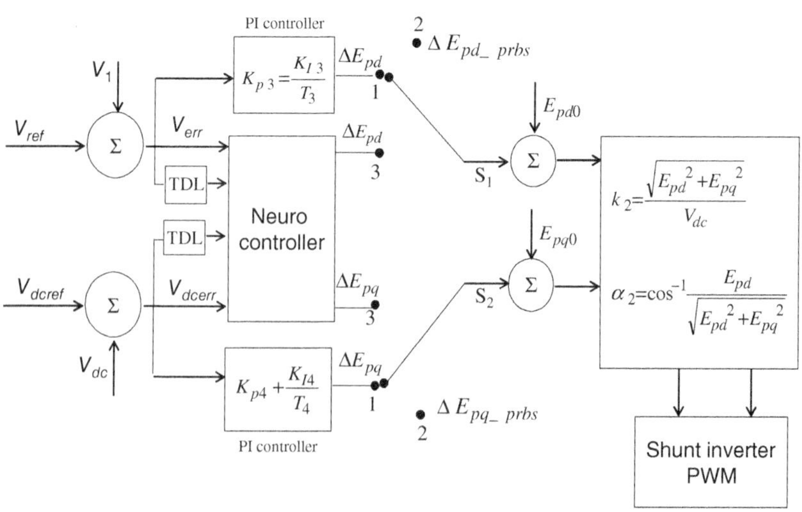

Figure 16.17 Shunt inverter control with conventional PI controllers and neurocontroller.

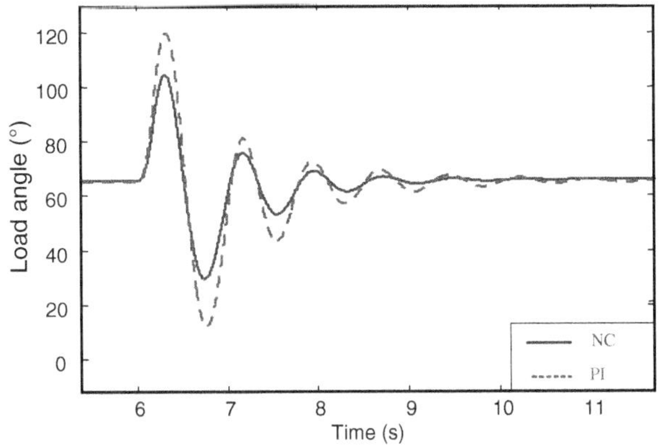

Figure 16.18 Load angle response of the synchronous generator operating ($P = 0.65$ p.u and $Q = 0.12$ p.u) for a 180 ms three-phase short circuit at bus 2.

The shunt and series inverters' PI controllers and neurocontrollers' performances are evaluated by applying a 180 ms three-phase short circuit fault at bus 2 (Figure 16.15) at different operating points given below. Figures 16.18 and 16.19 show the response of the generator with the neurocontrollers (solid lines) and with the PI controllers (dashed lines).

Figure 16.18 shows the load angle response and it can be seen that the response of the neurocontrollers is found to perform better than that of the PI controllers in damping the system oscillations. The maximum overshoot and the settling time are less when compared to that with the PI controllers.

Figure 16.19 shows the load angle response for an operating point, which is much further away from the one at which the PI controllers were fine tuned. For the UPFC with the PI controllers, the load angle increases drastically after the fault,

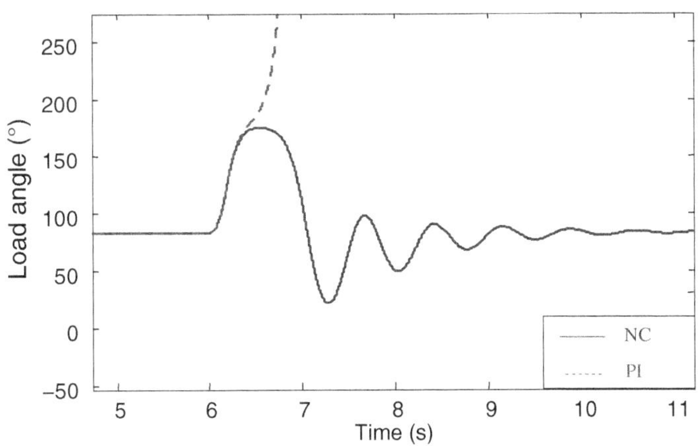

Figure 16.19 Load angle response of the synchronous generator operating ($P = 0.8$ p.u and $Q = 0.15$ p.u) for a 180 ms three-phase short circuit at bus 2.

and the system goes unstable. The UPFC with neurocontrollers on the other hand damps out the oscillations and restores the system to stability. The neurocontrollers give performances similar to those at previous operating points and maintain system stability. This is because the neurocontrollers are trained online and hence they are able to adapt to changes in operating conditions with the use of the neuroidentifiers. More details are provided in [14], [44], [45], and [46].

16.6.7 Wide-Area Monitoring and Control

SRNs are powerful tools for prediction based on internal recurrence and have been shown to overcome the varying time delays in the communication channels for wide-area monitoring (WAM) and control (WAC). The real-time implementation of a WAC on a two-area four machine power system, shown in Figure 16.20, was carried out at one of the author's laboratory [47]. Despite varying time delays, a system damping higher than that provided by local controllers (see Figure 16.21) can be guaranteed with WAC over a significant range of delays [47]. This is possible due to multi-step prediction capabilities of SRNs.

Wide-area coordinating control is becoming an important issue and a challenging problem especially when large wind farms are connected to the power grid. An optimal wide-area monitor based on an RBF NN has been developed in [48] to identify the input-output dynamics of the nonlinear power system. Its parameters are optimized with PSO. Based on the WAM, a WAC is then designed by using the DHP

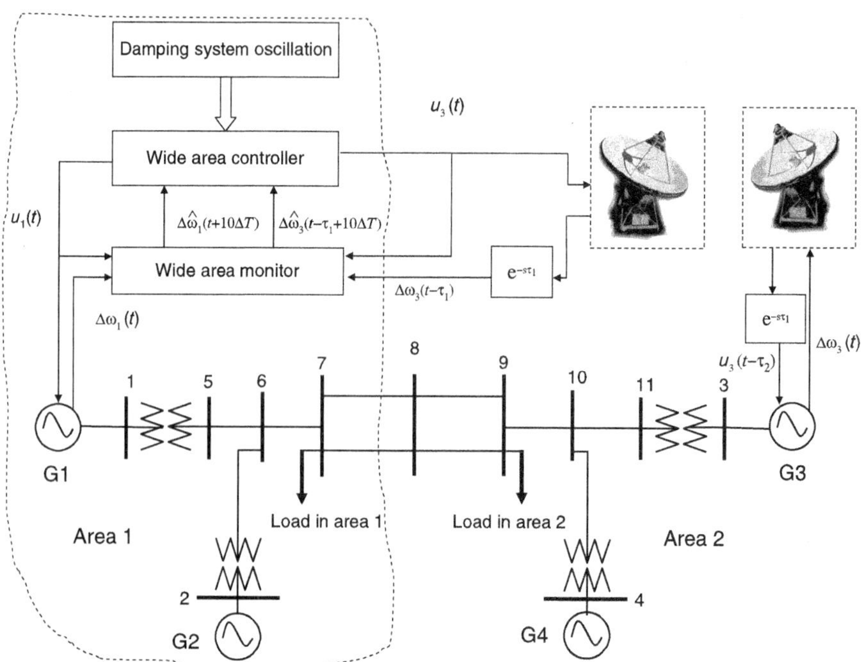

Figure 16.20 Two-area four-machine power system with communication delays in WAC.

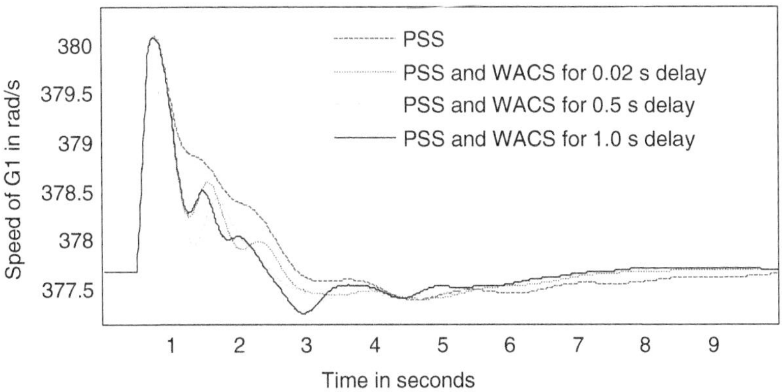

Figure 16.21 Speed of generator G1 during a disturbance with different communication delays.

method and RBFs, while considering the effect of signal transmission delays. The WAC operates at a global level to coordinate the actions of local power system controllers including those on the wind farm. Each local controller communicates with the WAC, receives remote control signals from the WAC to enhance its dynamic performance, and therefore helps improve system-wide dynamic and transient performance.

REFERENCES

[1] Haykin, S. *Neural Networks and Learning Machines*, 3rd edition. Prentice Hall, Upper Saddle River, New Jersey, 2008.

[2] El-Sharkawi, M., and Niebur, D., Eds., *Artificial Neural Networks with Applications to Power Systems*. IEEE-Power Engineering Society, IEEE Neural Network Council and IEEE Educational Activities, Piscataway, NJ, July 1996, ISBN: 0-7803-4010-8, NTSC Product No. HV6957.

[3] Dillon, T., and Niebur, D., Eds., *Neural Net Applications in Power Systems*. CRL Publishing Ltd., Leics, UK, 1996. ISBN 0 9527874 0 7.

[4] Germond, A., Niebur, D., and Eremia, M. Applications des systèmes intelligents dans les réseaux électriques. Invited chapter in: *Réseaux électriques – Aspects Actuels*, M. Eremia, J. Trecat, and A. Germond, Eds., Editura Tehnică, Bucharest, Romania, pp. 239–316, 2000, ISBN 973-31-1526.

[5] Alfares, H. K., and Nazeeruddin, M. Electric load forecasting: literature survey and classification of methods. *International Journal Systems Science*, vol. 33, no. 1, pp. 23–34, 2002.

[6] Hippert, H. S., Pedreira, C. E., and Souza, R. C. Neural networks for short-term load forecasting: a review and evaluation. *IEEE Transactions on Power Systems*, vol. 16, no. 1, pp. 44–55, February 2001.

[7] Singh, D., and Singh, S. P. Self organization and learning methods in short term electric load forecasting: a review. *Electric Power Components and Systems*, vol. 30, no. 10, pp. 1075–1089, 2002.

[8] Farag, W. A., Quintana, V. H., Lambert-Torres G. Applications of artificial intelligence techniques in synchronous machine control: a review. *Engineering Intelligent Systems for Electrical Engineering and Communications*, CRL Publishing Ltd, vol. 6, no. 2, pp. 65–73, 1998.

[9] Kalogirou, S. A. Artificial neural networks in renewable energy systems applications: a review. *Renewable and Sustainable Energy Reviews*, vol. 5, no. 4, pp. 373–401, 2001.

[10] Yamin, H. Y. Review on methods of generation scheduling in electric power systems. *Electric Power Systems Research*, vol. 69, no. 2–3, pp. 227–248, 2004.

[11] Cichocki, A., Karhunen, J., Kasprzak, W., Vigario, R. Neural networks for blind separation with unknown number of sources. *Neurocomputing*, Elsevier, vol. 24, no. 1–3, pp. 55–93, 1999.

[12] Welch, R., Ruffing, S., and Venayagamoorthy, G. K. Comparison of feedforward and feedback neural network architectures for short term wind speed prediction. *IEEE-INNS International Joint Conference on Neural Networks*, Atlanta, GA, USA, June 14–19, 2009.

[13] Prokhorov, D. V., and Wunsch, D. C. Adaptive critic designs. *IEEE Transactions of Neural Networks*, vol. 8, no. 5, pp. 997–1007, September 1997.

[14] Venayagamoorthy, G. K., Harley, R. G., and Wunsch, D. C. Comparison of heuristic dynamic programming and dual heuristic programming adaptive critics for neurocontrol of a turbogenerator. *IEEE Transactions on Neural Networks*, vol. 13, no. 3, pp. 764–773, May 2002.

[15] Werbos, P. J. Approximate dynamic programming for real time control and neural modeling. In: *Handbook of Intelligent Control*, D. A. White and D. A. Sofge, Eds., Van Nostrand Reinhold, New York, pp. 493–525, 1992, ISBN 0-442-30857-4.

[16] Niebur, D., Gursoy, E., and Liao, H. Independent component analysis techniques for power system load estimation- a signal-processing approach. In: *Applied Mathematics For Restructured Electric Power-Optimization, Control, And Computational Intelligence*, Chow, J. H., Wu, F. F., and Momoh, J. A. Eds., Springer Verlag, New York, NY, pp. 287–317, 2004.

[17] Cardoso, J. F. Blind signal separation: statistical principles. *Proceedings of the IEEE*, vol. 86, pp. 2009–2025, 1998.

[18] Karhunen, J., Oja, E., Wang, L., Vigario, R., and Joutsensalo, J. A class of neural networks for independent component analysis. *IEEE Transactions on Neural Networks*, vol. 8, no. 3, pp. 486–504, 1997.

[19] Gudise, V. G., and Venayagamoorthy, G. K. Comparison of particle swarm optimization and backpropagation as training algorithms for neural networks. *IEEE Swarm Intelligence Symposium*, Indianapolis, IN, USA, pp. 110–117, April 24–26, 2003.

[20] Hyvarinen, A. Fast and robust fixed-point algorithms for independent component analysis. *IEEE Transactions on Neural Networks*, vol. 10, pp. 626–634, 1999.

[21] Hyvarinen, A. A fast fixed-point algorithm for independent component analysis of complex valued signals. *International Journal of Neural Systems*, vol. 10, pp. 1–8, 2000.

[22] Amari, S., Cichocki, A., and Yang, H. H. A new learning algorithm for blind signal separation. In: *Advances in Neural Information Processing Systems*, Touretzky, D., Mozer, M., and Hasselmo, M. Eds., pp.757–763, MIT Press, Cambridge MA, 1996.

[23] Lee, T. W., Girolami, M., and Sejnowski, T. J. Independent component analysis using an extended infomax algorithm for mixed subgaussian and supergaussian sources. *Neural Computation*, vol. 11, pp. 417–441, 1999.

[24] Schraudolph, N. N., and Giannakopoulos, X. Online independent component analysis with local learning rate adaptation. In: *Advances in Neural Information Processing Systems*, S. A. Solla, T. K. Leen, and K.-R. Muller, Eds., MIT Press, Cambridge, vol. 12, pp. 789–795, 2000.

[25] Amin, S. M., and Wollenberg, B. F. Toward a smart grid. *IEEE Power and Energy Magazine*, vol. 3, no. 5, pp. 34–41, September–October 2005.

[26] Gursoy, E. and Niebur, D. On-line estimation of electric power system active loads. (Invited paper), *Proceedings of the World Congress on Computational Intelligence*, Vancouver, BC, Canada, 7/16-7/21, 2006.

[27] Liao, H. W., and Niebur, D. Load profile estimation in electric transmission networks using independent component analysis. *IEEE Transactions on Power Systems*, vol. 18, pp. 707–715, 2003.

[28] Zimmerman, R., and Gan, D. MATPOWER - Matlab routines for solving power flow problems, 1997. Available at http://www.pserc.cornell.edu/matpower/matpower.html

[29] Gursoy, E. and Niebur, D. Harmonic load identification using independent component analysis. *IEEE Transactions on Power Delivery*, vol. 24, no. 1, pp. 285–292, January 2009.

[30] Load profiling. The Electric Reliability Council of Texas. Available at http://www.ercot.com/mktinfo/loadprofile

[31] Burch, R., Chang, G., Hatziadoniu, C., Grady, M., Liu, Y., Marz, M., Ortmeyer, T., Ranade, S., Ribeiro, P., and Xu, W. Impact of aggregate linear load modeling on harmonic analysis: a comparison

of common practice and analytical models. *IEEE Transactions on Power Delivery*, vol. 18, pp. 625–630, 2003.

[32] Abu-Hashim, R., Burch, R., Chang, G., Grady, M., Gunther, E., Halpin, M., Hatziadoniu, C., Liu, Y., Marz, M., Ortmeyer, T., Rajagopalan, V., Ranade, S., Ribeiro, P., Sims, T., and Xu, W. Test systems for harmonics modeling and simulation. *IEEE Transactions on Power Delivery*, vol. 14, pp. 579–587, 1999.

[33] Fabbri, A., Roman, T., Abbad, J., and Quezada, V. Assessment of the cost associated with wind generation prediction errors in a liberalized electricity market. *IEEE Transactions on Power Systems*, vol. 20, no. 3, pp. 1440–1446, August 2005.

[34] Bashir, Z. A., and El-Hawary, M. E. Applying wavelets to short-term load forecasting using PSO-Based neural networks. *IEEE Transactions on Power Systems*, vol. 24, no. 1, pp. 20–27, February 2009.

[35] Park, D. C., El-Sharkawi, M. A., Marks, R. J., Atlas, L. E., and Damborg, M. J. Electric load forecasting using an artificial neural network. *IEEE Transactions on Power Systems*, vol. 6, no. 2, pp. 442–449, May 1991.

[36] Barbounis, T. G., Theocharis, J. B., Alexiadis, M. C., and Dokopoulos, P. S. Long-term wind speed and power forecasting using local recurrent neural network models. *IEEE Transactions on Energy Conversion*, vol. 21, no. 1, pp. 273–284, March 2006.

[37] Ruffing, S., and Venayagamoorthy, G. K. Short to medium range time series prediction of solar irradiance using an echo state network. *Intelligent System Applications to Power Systems (ISAP 2009)*, Curitiba, Brazil, November 8–12, 2009.

[38] National Renewable Energy Laboratory's National Wind Technology Center (online). Available at http://www.nrel.gov/wind/

[39] Venayagamoorthy, G. K., and Welch, R. L. Energy dispatch controllers for a photovoltaic system. *Engineering Applications of Artificial Intelligence*, vol. 23, no. 2, pp. 249–261, March 2010.

[40] Welch, R., and Venayagamoorthy, G. K. Energy dispatch fuzzy controller for a grid independent photovoltaic system. *Energy Conversion and Management*, vol. 51, no. 5, pp. 928–937, May 2010.

[41] Venayagamoorthy, G. K., and Harley, R. G. Computational Intelligence for Control of FACTS Devices, in *Applied Mathematics for Restructured Electric Power Systems*, Chow JH, Wu FF, and Momoh JA (Eds.), Springer Power Electronics and Power Systems Series, Verlang, 2005, pp. 201–237, ISBN: 0-387-23470-5.

[42] Chunlei, L., Hongbo, S., and Yu, D. C. A novel method of power flow analysis with unified power flow controller (UPFC). *IEEE Power Engineering Society Winter Meeting*, vol. 4, pp. 2800–2805, July 16–20, 2000.

[43] Hingorani, N. G., and Gyugyi, L. *Understanding FACTS: Concepts and Technology of Flexible AC Transmission Systems*. IEEE Press, 1999, ISBN 0-7803-3455-8.

[44] Venayagamoorthy, G. K., and Kalyani, R. P. Two Separate Continually Online Trained Neurocontrollers for a Unified Power Flow Controller. *IEEE Transactions on Industry Applications*, vol. 41., no. 4, pp. 906–916, July/August 2005.

[45] Ray, S., Venayagamoorthy, G. K., and Watanabe, E. H. A computational approach to optimal damping controller design for a GCSC. *IEEE Transactions on Power Delivery*, vol. 23, no. 3, pp. 1673–1681, July 2008.

[46] Ray, S., and Venayagamoorthy, G. K. Wide-area signal-based optimal neurocontroller for a UPFC. *IEEE Transactions on Power Delivery*, vol. 23, no. 3, pp. 1597–1605, July 2008.

[47] Ray, S., and Venayagamoorthy, G. K. Real-time implementation of a measurement based adaptive wide area control system considering communication delays. *IET Proceedings on Generation, Transmission and Distribution*, vol. 2, no. 1, pp. 62–70, January 2008.

[48] Qiao, W., Venayagamoorthy, G. K., and Harley, R. G. Optimal wide-area monitoring and non-linear adaptive coordinating control of a power system with wind farm integration and multiple FACTS devices. *Neural Networks*, vol. 21, no. 2–3, pp. 466–475, March/April 2008.

CHAPTER 17

FUZZY SYSTEMS

Germano Lambert-Torres, Luiz Eduardo Borges da Silva, Carlos Henrique Valerio de Moraes, and Yvo Marcelo Chiaradia Masselli

17.1 INTRODUCTION

Suppose a man plain of hair in his head. In a classical notation of set (it means, in Cantor sense), we can state that "this man doesn't belong to the set of bald men." But now, you pull out just one hair. Without one hair wire, is this man bald? Or in other words, is this man member of the bald men set? The answer is clearly no. Now, pull out one more hair. Make the same question. The answer continues to be no. And now, continue this process until your answer becomes yes. At this time, return the last hair in the man's head, in this situation he does not belong to the bald men set, but without this last hair he belongs to the bald men set. Conclusion of this experiment: the difference between a bald man and a not bald man is one single hair. Of course, this is not true.

Let us go to another experiment. One characteristic we associate with an old person is wrinkles. How old is an old person for you? 55, 60, or 75 years old? Suppose 65 years old. In a classical sense of set, 65 years old minus one day, the person is not old. Then, this person does not belong to the old people set. In the next day, the person already with 65 years old becomes a member of the old people set. Until here, we do not have a problem with this representation. However, when we see this situation in real world, applying the characteristic "wrinkles," with 65 years old minus one day, the person is not old, and then he/she does not have any wrinkle. But on his/her birthday, he/she will receive suddenly as a "gift" a bunch of wrinkles. Of course, this is also not true.

In these two examples, we are in trouble to represent correctly the problem. It occurs because the real world is complex. And to model this world is a task much more complex.

If we extend this concept to power system problems, the same idea can be applied. For instance, if we try to classify the system load level in light, medium, and heavy. The limits of medium label could be between 30% and 65% of the total expected load. Using this classification, 64.9% of the total load is classified as medium and all operation's actions are taken using this classification. However, the load changes just a little bit to 65.1% as the classification is heavy and the operator

Advanced Solutions in Power Systems: HVDC, FACTS, and Artificial Intelligence, First Edition.
Edited by Mircea Eremia, Chen-Ching Liu, and Abdel-Aty Edris.

must take other actions. And more, if the load level is 31%, the classification continues to be medium and the actions are the same. Of course, this is not a good approach to solve a problem.

The problem we have in both representations is that the classical Cantor's definition of a set is not appropriate to represent a lot of problems in real world. A possible solution for that is to create several sets between the bald men set and the hairy men set or between the young people set and the old people set. However, this solution is not a very good one, because in all cases it is possible to find a step between the sets. And loss of hairs and gain of wrinkles usually are continuous processes.

The solution for this problem was proposed by Lofti Asker Zadeh, Professor at University of California at Berkeley, in 1965 in his seminal paper entitled *Fuzzy Sets* [1]. In this paper, he established new concepts for sets, expanding the concept of membership and creating one possibility to better represent the real world.

Professor Zadeh had this idea when he was studying complex systems. At this time, he already had many important contributions in classical control, such as linear systems, stable systems, and time-invariant systems, but none of these was appropriate to represent a simple biological system or an economic phenomenon. The idea for the fuzzy sets theory came to his mind in early 1960s while he was studying biological systems, when he realized that "we need a radically different kind of mathematics, the mathematics of fuzzy or cloudy quantities which are not describable in terms of probability distributions." The concept and the name of the fuzzy sets theory started with this theory.

Professor Zadeh collaborated and wrote many papers about control with his friend, Professor Richard Bellman, who was enthusiastic about this new theory and was the first to see the new possibilities at least in engineering.

Initially, this concept was not very well understood by his colleagues, and many of them stated that fuzzy sets were inappropriate as mathematical theory [2]. The concept was first exploited by professors and researchers from Europe (especially in France and Poland) and Japan, while in the United States the fuzzy set theory started to be popular only in the 1980s. An example of that was the first IEEE conference about fuzzy logic held in San Diego in 1992 [3].

Nowadays, it is quite easy to understand the first proposed principles of fuzzy sets, and these principles can be applied without big effort. For example, to solve the two problems presented earlier using the Zadeh's fuzzy set ideas, we can just define a set where each time you pull out a hair of person's head, this person becomes more and more a member of the bald men set. Using the same reasoning, every day a person becomes more member of the old people set, and of course, he/she will get continuously more wrinkles.

As Zadeh said, "the closer one looks at a real-world problem, the fuzzier become its solution."

Fuzzy logic was transformed from a "buzz word" to an important technological area, with many publications in conferences and transactions. Many Japanese products applying fuzzy concepts, such as household appliances and electronic equipment, have been manufactured [4].

This chapter is divided into six sections. Section two presents the fundamental notions of the theory. It presents a relation of classical and fuzzy sets and concepts of fuzzy graphs and fuzzy relations. Section three presents the foundation of fuzzy logic, with the fuzzification inference process and defuzzification process. Section four presents a popular fuzzy model to represent complex systems. Section five presents an application of fuzzy control. Finally, in the last section, some application areas of fuzzy techniques have been mentioned.

17.2 FUNDAMENTAL NOTIONS

To Zadeh, "the theory of fuzzy sets is, in effect, a step toward a rapprochement the precision of classical mathematics and the pervasive imprecision of the real world – a rapprochement born of the incessant human quest for a better understanding of mental processes and cognition" [5].

17.2.1 Classical Sets

The principal definitions and concepts of the classical Cantor's sets are well-known and are described here with the only purpose to guide the readers about the notation and a comparison with the similar concepts in fuzzy sets theory. Some numerical examples are presented to illustrate certain places. A traditional definition of a set can be expressed by the following statement.

Let **U** be a reference set and **A** be a subset of **U**, then $\mathbf{A} \subset \mathbf{U}$. It indicates that an element x of **U** can be a member of **A** or not. If x is a member of **A**, we say that "x belongs to **A**," or in other words, we can also say "x is member of **A** with degree equal to 1, $\mu_{\mathbf{A}}(x) = 1$." The symbol $\in$ represents this relation, $x \in \mathbf{A}$.

On the other hand, if x is not a member of **A**, we say that "x doesn't belongs **A**," or in other words, we can also say "x is member of **A** with degree equal to 0, $\mu_{\mathbf{A}}(x) = 0$." The symbol $\notin$ represents this relation, $x \notin \mathbf{A}$. We can see that in classical sets, the membership values are taken in the set $\mathrm{M} = \{0,1\}$.

Using the complementary set of **A** with respect to **U**, denoted by $\neg\mathbf{A}$, we have the following properties of a Boolean binary algebra:

$$\mathrm{A} \cap \neg\mathrm{A} = \emptyset \text{ (intersection) and } \mathrm{A} \cup \neg\mathrm{A} = \cup \text{ (union).}$$

The union and intersection operators can also be expressed by maximum and minimum operations, respectively. Consider the reference set $\mathbf{U} = \{x_1, x_2, x_3, x_4\}$ and the following sets $\mathbf{A} = \{x_1, x_3, x_4\}$ and $\mathbf{B} = \{x_2, x_4\}$. We can also write these sets in the following way:

$$\mathrm{A} = \{(x_1|1), (x_2|0), (x_3|1), (x_4|1)\} \text{ and } \mathrm{B} = \{(x_1|0), (x_2|1), (x_3|0), (x_4|1)\}$$

then:

$$\begin{aligned}\mathrm{A} \cup \mathrm{B} &= \max\{\{(x_1|1), (x_2|0), (x_3|1), (x_4|1)\}, \{(x_1|0), (x_2|1), (x_3|0), (x_4|1)\}\} \\ &= \{(x_1|1), (x_2|1), (x_3|1), (x_4|1)\}\end{aligned}$$

and:

$$\begin{aligned}\mathrm{A} \cap \mathrm{B} &= \min\{\{(x_1|1), (x_2|0), (x_3|1), (x_4|1)\}, \{(x_1|0), (x_2|1), (x_3|0), (x_4|1)\}\} \\ &= \{(x_1|0), (x_2|0), (x_3|0), (x_4|1)\}.\end{aligned}$$

17.2.2 Fuzzy Sets

In the same way, let **U** be a reference set and **A** be a subset of **U**, then **A** ⊂ **U**. It indicates that an element x of **U** can be a member of **A** or not. The difference is that the membership values are taken in the interval M = [0,1]. Let **A** be defined by **A** = $\{x_1, x_3, x_4\}$, but if we use values taken in the interval M, we can define a fuzzy set ~**A** by: ~**A** = $\{(x_1 \mid 0.2), (x_2 \mid 0), (x_3 \mid 0.8), (x_4 \mid 1)\}$. The interpretation of each element of the fuzzy set ~**A** is: x_1 is a member of **A** with degree 0.2, it means, x_1 is a "little" member of **A**, x_2 is a member of **A** with degree 0, it means, x_2 is not a member of **A**, x_3 is a member of **A** with degree 0.8, it means, x_3 is a "strong" member of **A**, and x_4 is a member of **A** with degree 1, it means, x_4 is a "totally" member of **A**.

An interpretation of this fuzzy set ~**A** for the problem described in section one can be the following. We have four people $\{x_1, x_2, x_3,$ and $x_4\}$, and if the set ~**A** represents the bald men set, we can have, for example, x_1 is a man with some failures in his hair (x_1 is a "little" member of **A**), x_2 is a hairy man (x_2 is not a member of **A**), x_3 is a man of little quantity of hair in his head (x_3 is a "strong" member of **A**), and x_4 is a complete bald man (x_4 is a "totally" member of **A**).

A rigorous definition of the Zadeh's concepts can be the following statement.

Let **U** be a set, denumerable or not, and let x be an element of **U**. The fuzzy set ~**A** of **U** is a set of ordered pairs: $\{(x \mid \mu_{\mathbf{A}}(x))\}$, $\forall\, x \in$ **U**, where the values of $\mu_{\mathbf{A}}(x)$ are the degree of the membership of x in **A**, and these values are taken in a set **M**.

The set **M** is any lattice structure. To be a lattice structure (or a reticular set), the set must be an ordered set, it means that for each two pair of elements, there exists one and only one element of **M** as an inferior limit, and there exists one and only one element of **M** as a superior limit.

Usually, this set **M** uses the real number set structure in the interval [0,1].

For the second experiment presented in section one of this chapter, we can represent graphically the classical set (old people set, members who are more than 65 years old) and a fuzzy set ~**B** that expresses the following membership function: $\mu_{\mathbf{B}}(x) = 0$, for $x < 25$; $\mu_{\mathbf{B}}(x) = 0.025\,x - 0.625$, for $25 \leq x < 65$; $\mu_{\mathbf{B}}(x) = 1$, for $x \geq 65$. Figure 17.1 presents these sets.

17.2.2.1 Operations on Fuzzy Sets

Let U be a reference set and M its associated membership set, and let ~A and ~B be two fuzzy sets of U. The following operations can be defined:

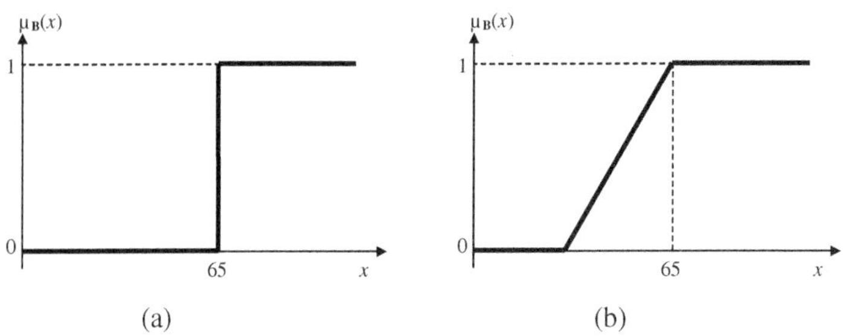

Figure 17.1 Examples of old people set: (a) classical set and (b) fuzzy set.

a) Inclusion: "~**A** is included in ~**B**" if and only if $\forall x \in$ **U**: $\mu_{\mathbf{A}}(x) \leq \mu_{\mathbf{B}}(x)$

b) Equality: "~**A** and ~**B** are equal" if and only if $\forall x \in$ **U**: $\mu_{\mathbf{A}}(x) = \mu_{\mathbf{B}}(x)$

c) Complementation: "~**A** = ¬~**B**" if $\forall x \in$ **U**: $\mu_{\mathbf{B}}(x) = 1 - \mu_{\mathbf{A}}(x)$ for **M** = [0,1]

d) Intersection: $\forall x \in$ **U**: $\mu_{\mathbf{A}\cap\mathbf{B}}(x) = \min(\mu_{\mathbf{A}}(x), \mu_{\mathbf{B}}(x))$

e) Union: $\forall x \in$ **U**: $\mu_{\mathbf{A}\cup\mathbf{B}}(x) = \max(\mu_{\mathbf{A}}(x), \mu_{\mathbf{B}}(x))$

f) Disjunctive sum: $\forall x \in$ **U**: ~**A** ⊕ ~**B** = (~**A** ∩ ¬~**B**) ∪ (¬~**A** ∩ ~**B**)

g) Algebraic sum: $\forall x \in$ **U**: $\mu_{\mathbf{A}\hat{}\mathbf{B}}(x) = \mu_{\mathbf{A}}(x) + \mu_{\mathbf{B}}(x) - \mu_{\mathbf{A}}(x).\mu_{\mathbf{B}}(x)$

h) Algebraic product: $\forall x \in$ **U**: $\mu_{\mathbf{A}\cdot\mathbf{B}}(x) = \mu_{\mathbf{A}}(x).\mu_{\mathbf{B}}(x)$

i) Limited sum: $\forall x \in$ **U**: $\mu_{\mathbf{A}*\mathbf{B}}(x) = \min(1, \mu_{\mathbf{A}}(x)+\mu_{\mathbf{B}}(x))$

j) Limited product: $\forall x \in$ **U**: $\mu_{\mathbf{A}\otimes\mathbf{B}}(x) = \max(0, \mu_{\mathbf{A}}(x)+\mu_{\mathbf{B}}(x) - 1)$

k) Difference: $\forall x \in$ **U**: ~**A** - ~**B** = (~**A** ∩ ¬~**B**)

If we consider:

$$\sim A = \{(x_1|0.2), (x_2|0), (x_3|0.8), (x_4|1)\} \text{ and}$$
$$\sim B = \{(x_1|0.1), (x_2|0.5), (x_3|0.2), (x_4|0.5)\}$$

we can have the following result for the previous operations:

c) Complementation of: ~**A**:¬~**A** = $\{(x_1|0.8), (x_2|1), (x_3|0.2), (x_4|0)\}$

d) Complementation of: ~**B**: ¬~**B** = $\{(x_1|0.9), (x_2|0.5), (x_3|0.8), (x_4|0.5)\}$

e) Intersection: ~**A** ∩ ~**B** = $\{(x_1|0.1), (x_2|0), (x_3|0.2), (x_4|0.5)\}$

f) Union: ~**A** ∪ ~**B** = $\{(x_1|0.2), (x_2|0.5), (x_3|0.8), (x_4|1)\}$

g) Disjunctive sum: ~**A** ⊕ ~**B** = $\{(x_1|0.2), (x_2|0.5), (x_3|0.8), (x_4|0.5)\}$

h) Algebraic sum: ~**A** ^ ~**B** = $\{(x_1|0.28), (x_2|0.5), (x_3|0.84), (x_4|1)\}$

i) Algebraic product: ~**A** • ~**B** = $\{(x_1|0.02), (x_2|0), (x_3|0.16), (x_4|0.5)\}$

j) Limited sum: ~**A** ∗ ~**B** = $\{(x_1|0.3), (x_2|0.5), (x_3|1), (x_4|1)\}$

k) Limited product: ~**A** ⊗ ~**B** = $\{(x_1|0), (x_2|0), (x_3|0), (x_4|0.5)\}$

l) Difference: ~**A** - ~**B** = $\{(x_1|0.2), (x_2|0), (x_3|0.8), (x_4|0.5)\}$

Note that, except in particular cases, the difference has not commutative property, it means ~**A** - ~**B** ≠ ~**B** - ~**A**. Other definitions, such as Hamming distance, Euclidean distance, entropy, cardinality, power set, and so on, are not shown here, but they exist and can be found in [6].

Also, these operations can also be represented graphically. Figure 17.2 presents two fuzzy sets and the results with union and intersection operators.

17.2.2.2 Properties of Fuzzy Sets

Let U be a reference set and M its associated membership set, and let ~A, ~B, and ~C be three fuzzy sets of U. We can present the following properties:

a) Commutativity: ~**A** ∩ ~**B** = ~**B** ∩ ~**A** ~**A** ∪ ~**B** = ~**B** ∪ ~**A**

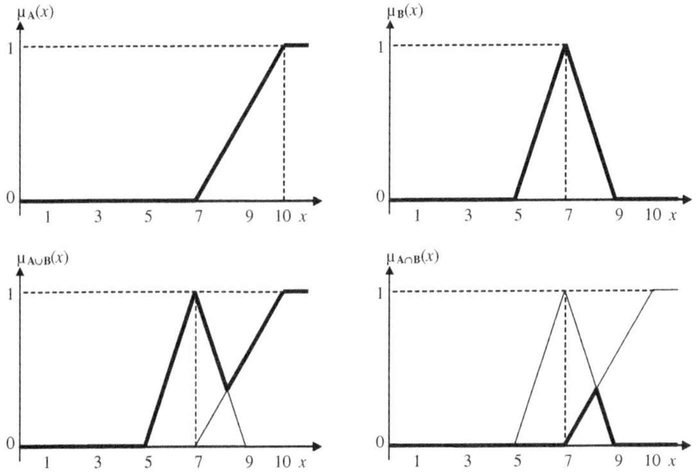

Figure 17.2 Examples of fuzzy sets and the resultant sets after union and intersection operations.

b) Associativity: $(\sim\mathbf{A} \cap \sim\mathbf{B}) \cap \sim\mathbf{C} = \sim\mathbf{A} \cap (\sim\mathbf{B} \cap \sim\mathbf{C})$; $(\sim\mathbf{A} \cup \sim\mathbf{B}) \cup \sim\mathbf{C} = \sim\mathbf{A} \cup (\sim\mathbf{B} \cup \sim\mathbf{C})$

c) Idempotence: $\sim\mathbf{A} \cap \sim\mathbf{A} = \sim\mathbf{A}$ $\quad \mathbf{A} \cup \sim\mathbf{A} = \sim\mathbf{A}$

d) Distributivity: $(\sim\mathbf{A} \cup \sim\mathbf{B}) \cap \sim\mathbf{C} = (\sim\mathbf{A} \cap \sim\mathbf{C}) \cup (\sim\mathbf{B} \cap \sim\mathbf{C})$
$(\sim\mathbf{A} \cap \sim\mathbf{B}) \cup \sim \mathrm{C} = (\sim\mathrm{A} \cup \sim\mathrm{C}) \cap (\sim\mathrm{B} \cup \sim\mathrm{C})$

e) Empty set: $\sim\mathbf{A} \cap \sim\emptyset = \sim\emptyset$ $\quad \sim\mathbf{A} \cup \sim\emptyset = \sim\mathbf{A}$

f) Reference set: $\sim\mathbf{A} \cap \mathrm{U} = \sim\mathbf{A}$ $\quad \sim\mathbf{A} \cup \mathbf{U} = \mathbf{U}$

g) Involution $\quad \neg(\neg\sim\mathbf{A}) = \sim\mathbf{A}$

h) De Morgan's theorems: $\neg(\sim\mathbf{A} \cap \sim\mathbf{B}) = \neg\sim\mathbf{A} \cup \neg\sim\mathbf{B}$ $\neg(\sim\mathbf{A} \cup \sim\mathbf{B}) = $
$= \neg\sim\mathbf{A} \cap \neg\sim\mathbf{B}$

17.2.3 Linguistic Values

Usually, in engineering or in computer science, the mathematic statements are exact, such as "$x = 3$," "$5 \leq x < 9$," and "$y = 2x + 6$." However, the most part of time, information from the real world are imprecise and can only be expressed correctly by linguistic propositions, such as "the transformer load is around 50%," "the maintenance cost is high," and "the student degrees are very low."

Some engineers believe that the exact statements have more information about a fact than the "fuzzy" information, but it is not true. For example, for a statement like "$5 \leq x < 9$," it means that x is true in the interval between 5 and 9. However, usually, the limit values (5 and 9) are obtained by computation from the real world data. These data have some errors, typical of measurement data and the computational approximations. With this aspect, maybe the real value is not 5, but some value takes in the interval [4.9,5.1]. So, for a value of x like $x = 4.99$, if you apply the classical statement, the conclusion is "x is not true," but in practical what is the difference between 4.99 and 5?

And more, have all values of x in the interval [5, 9] the same degree of truth? It can be, and the representation using the classical set definition is good enough. However, if it is not true, it is important to represent the membership degree in this interval using fuzzy set concepts. For example, the set ~**B** represented in Figure 17.2 can express the values of membership degree.

Another point that must be observed is linked with the nature of a linguistic statement. When this kind of statement is provided, a set of conclusions is immediately made for the readers. Let there be two statements "the temperature is 50°F" and "the temperature is low." With the second statement, it is possible to conclude that the people must be dressed with heavy clothes, they have some inconveniences to stay outside for long period, and so on. With the first statement (exact statement), it is not possible to conclude anything because if in Brazil this temperature expresses a cold weather, in the Alaska cities this temperature is "high." This example has been provided in order to conclude that "usually, the linguistic statements contain more information than the exact statements."

Fuzzy sets theory allows representing linguistic values [7, 8] such as "small / big," "low / high," "around," and so on. Let the following statements "the student's degree is high" and "the student's degree is around 7," shown in Figure 17.2, as sets ~**A** and ~**B**, respectively. If we transform in discontinuous sets, we have the following membership degrees for each student degree:

$$\sim \mathrm{A} = \{(0|0),(1|0),(2|0),(3|0),(4|0),(5|0),(6|0),(7|0),(8|0.5),(9|1),(10|1)\},$$

$$\sim \mathrm{B} = \{(0|0),(1|0),(2|0),(3|0),(4|0),(5|0),(6|0.5),(7|1),(8|0.5),(9|0),(10|0)\},$$

Also, other linguistic values can be applied in the previous linguistic values such as "very," "little," "very very," "not," and so on. For example, the statements "the student's degree is very high" and "the student's degree is not high" are shown in Figure 17.3 and can be represented by:

$$\sim \textbf{very A} = \{(0|0),(1|0),(2|0),(3|0),(4|0),(5|0),(6|0),(7|0),(8|0.25),(9|1),(10|1)\},$$

$$\sim \textbf{not A} = \{(0|1),(1|1),(2|1),(3|1),(4|1),(5|1),(6|1),(7|1),(8|0.75),(9|0),(10|0)\},$$

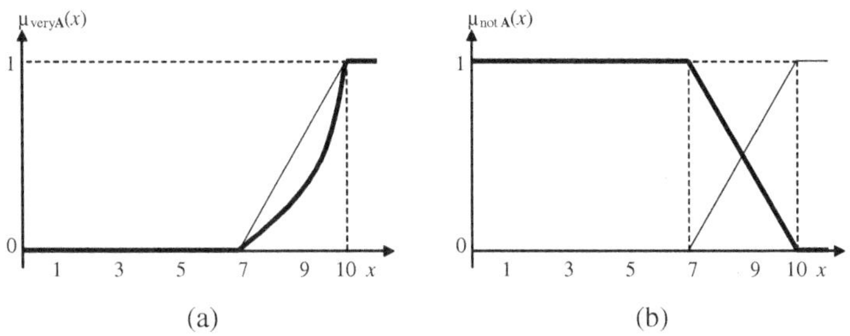

Figure 17.3 Intensifier values: "very" and "not."

Using a concise definition, a linguistic variable is one that contains words or natural language statements. For instance, if "age" is expressed as a linguistic variable x, then the set of linguistic values of x can assume the following values:

$$\mathrm{T}x = \{\text{young, old, very young, very old, not young, not old, little young}, \ldots\}$$

Each one of these terms is a fuzzy sets label. Applying now a rigorous definition, a linguistic variable is characterized by a 5-uple (x, Tx, **U**, G, M) [9], where:

- x is the name of the variable (x = age).
- T is a set of terms of x.
- U is the reference set.
- G is a syntax rule ("Grammar"), which defines the sentences (v) in Tx.
- M is a semantic rule ("Meaning"), which associates each linguistic (v) with its meaning M(v), where M (v) is a fuzzy subset of U.

It exists primitive terms, which have a meaning defined *a priori*, and serves as a base for compose terms. Primate terms can be "young" or "old," while compose terms can be "very young" or "not old." For example, the meaning of the compose terms can be computed by applying the following syntax rule (G(v)):

$$\text{very}(v) = v^2 \quad \text{not}(v) = 1 - v$$

Other syntax rule operations can be conjunctions, such as "and," "or," and "or exclusive." With this, compose terms can be created: "very young or very old" and "young and not old." These conjunctions are computed by union, intersection, and disjunctive sum operators, respectively.

If the primate terms "young" and "old" are defined by:

$$\text{young} = \{(0|1), (10|1), (20|0.75), (30|0.5), (40|0.25), (50|0), (60|0), (70|0), (80|0), \ldots\}$$

$$\text{old} = \{(0|0), (10|0), (20|0), (30|0), (40|0.25), (50|0.5), (60|0.75), (70|1), (80|1), \ldots\}$$

So, the compose terms can be computed:

$$\text{very young} = \{(0|1), (10|1), (20|0.56), (30|0.25), (40|0.06), (50|0), (60|0), (70|0), (80|0), \ldots\}$$

$$\text{very old} = \{(0|0), (10|0), (20|0), (30|0), (40|0.06), (50|0.25), (60|0.56), (70|1), (80|1), \ldots\}$$

$$\text{not young} = \{(0|0), (10|0), (20|0.25), (30|0.5), (40|0.725), (50|1), (60|1), (70|1), (80|1), \ldots\}$$

$$\text{not old} = \{(0|1), (10|1), (20|1), (30|1), (40|0.725), (50|0.5), (60|0.25), (70|0), (80|0), \ldots\}$$

$$\text{very young or very old} = \max(\mu_{\text{very young}}(x), \mu_{\text{very old}}(x)) = \{(0|1), (10|1), (20|0.56), (30|0.25), (40|0.06), (50|0.25), (60|0.75), (70|1), (80|1), \ldots\}$$

$$\text{young and not old} = \min(\mu_{\text{young}}(x), \mu_{\text{not old}}(x)) = \{(0|1), (10|1), (20|0.75), (30|0.5), (40|0.25), (50|0), (60|0), (70|0), (80|0), \ldots\}$$

17.2.4 Fuzzy Statements

Data modeling can be defined as an attempt to represent, in a clear way, the available information from a set of data. A data modeling procedure is important because merging data in algebraic expressions allows the user to better visualize, understand, and interpret its structure. This is very useful, especially in large databases such as the control center databases that are used to help the operator in a decision-making process. On the other hand, algebraic expressions can be easily handled and incorporated into computational programs.

The general form to represent a base, $T(x)$, for a variable x with k fuzzy values, can be described as:

$$T(x) = \left\{T_x^1, T_x^2, \ldots, T_x^k\right\}$$

where T_x^i represents the ith fuzzy value for the variable x.

Thus, an input variable vector, X, can have the following structure:

$$\mathrm{X} = [x_i, \mathrm{U}_i, \left\{T_{x_i}^1, T_{x_i}^2, \ldots, T_{x_i}^n\right\}, \{\mathrm{M}_{x_i}\} | i = 1, \ldots, p]$$

where x_i represents an input variable i, n is a generic number of fuzzy values for this variable, and p is the total number of input variables. Each variable has its associated universe set, U_i, and the domain set of the membership functions, M_{x_i}.

The degree of membership of an element x to a fuzzy set A can be denoted by $\mu_A(x)$. To express specific knowledge, input variables can be combined through linguistic conjunctions such as "and" and "or." Each linguistic conjunction has a meaning in the fuzzy logic theory and represents a specific operation (*and* ↔ minimum and *or* ↔ maximum). Also, in some cases, the complement of a membership can represent its negation; for example, the complement of A is $\neg A$, which means "no *A*." In addition, this theory allows the use of adverbs and hedges to intensify or minimize a linguistic value, such as "very," "quite," and "about." Specific mathematical operations are related to each hedge [10].

According to these considerations, a fuzzy statement can be defined as an attribution of a fuzzy value to a variable. This fuzzy value can be a single value (with or without linguistic hedges) or a composed value (with two or more single values that are combined by linguistic conjunctions). The general form of a fuzzy statement can be written as x_i is T_{x_i}, where T_{x_i} represents the fuzzy value of the variable x_i.

Some examples of fuzzy statements are "The transformer overload is small," "The oil temperature is about 75°C," and "The value of x is big or not very small," where "transformer overload," "oil temperature," and "value of x" are the variables of the statement; "small," "about 75°C," and "big or not very small" are their fuzzy values, respectively.

17.2.5 Fuzzy Conditional Statements

A fuzzy decision system must provide a mapping between input and output variables. In our cases, this mapping is done by using fuzzy conditional statements in "If-then" form. Other kinds of mapping, using artificial neural network techniques or traditional methods, can also handle this problem with advantages and drawbacks [11].

To be a fuzzy conditional statement, an If-then rule must have its premise and/or consequence represented by fuzzy statements. A typical structure of a decision system has multiple input and output variables (MIMO systems). Let us consider a system with p input variables, x, in the premise of the rules, and m output variables, y, in the consequence. Thus, the general form of a fuzzy conditional statement is as follows:

$$\text{If}\{x_1 \text{ is } T_{x_1}, \ldots, x_p \text{ is } T_{x_p}\} \text{ then } \{y_1 \text{ is } T_{y_1}, \ldots, y_m \text{ is } T_{y_m}\}.$$

In this case, the premise fuzzy statements are combined using a comma, to denote the conjunction "and."

Having above elements, it is possible to write complex affirmatives that may represent the real world situations. Through fuzzy rules, it is possible to write the control actions, by using the terms of these imprecise ideas, which comprise the variable's state. As an example, in the ABS break system, it is possible to write:

IF vehicle velocity is *low AND* its location is *near the wall THEN* brake pressure be *medium.*

In one conventional Proportional-Integral (PI), this rule should be far more specific:

IF vehicle velocity *lower than 20 km/h AND* its location *lower than 5 (m) THEN* brake pressure be *20 bar.*

Because the fuzzy rules act over a wide range, usually being more comprehensive than the conventional, the number of fuzzy rules that represent a determined control is smaller than with conventional rules.

In mathematical terms, a production rule may be considered as the charting between different universe sets. Being E_1, E_2, and E_3 the universe sets describing the possible values (or states) for the variables x, y, and z, respectively. If in the rule's premises are founded fuzzy sets from E_1 and E_2 and in the consequence there are fuzzy sets from E_3, it is possible to denote the relation by:

$$E_1 \times E_2 \sim_{\Gamma}\rightarrow E_3$$

The charting Γ allows those values from x and y to be converted for z values, that is to say:

$$z \in \Gamma\{x \times y\}$$

The same concept of fuzziness can also be applied to decision variables, restrictions, and objectives. Usually, decision variables are considered as deterministic and only objectives and restrictions have limits not very well defined. However, objective functions and restrictions can be both characterized by fuzzy statements, and a fuzzy decision is computed as an intersection among the fuzzy restrictions and fuzzy objectives [12]. A general structure of this problem can be expressed by the following statement.

Let P a set of possible decisions (alternatives), G a fuzzy objective function (a fuzzy subset in P, defined by its characteristic function μ_G: $P \rightarrow [0,1]$), and C a fuzzy restriction (a fuzzy subset in P, defined by μ_C: $P \rightarrow [0,1]$), the fuzzy decision D is obtained by the intersection of G and C functions:

$$D = G \cap C$$

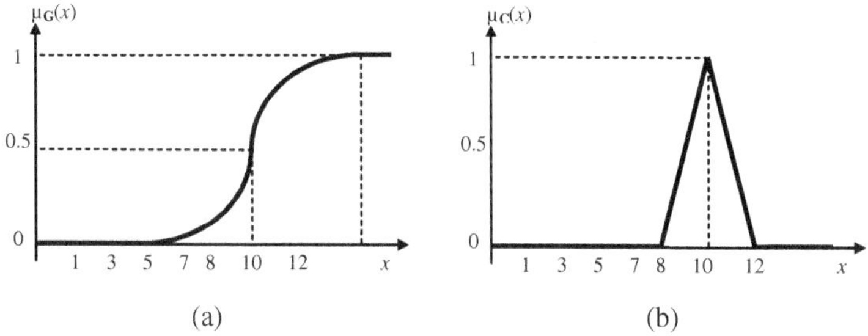

Figure 17.4 Examples of membership functions: (a) "very bigger than 5" and (b) "around 10."

Using the intersection operator, the characteristic function is computed by:

$$\mu_D(x) = \text{MIN}\{\mu_G(x), \mu_C(x)\}$$

The decision D resulted by n fuzzy objective functions $G_1, G_2, \ldots, G_n$, and m fuzzy restriction functions $C_1, C_2, \ldots, C_m$, is obtained by:

$$D = G_1 \cap G_2 \cap \ldots \cap G_n \cap C_1 \cap C_2 \cap \ldots \cap C_m$$

As a numerical example, consider G as an objective function and C as a restriction given by the following linguistic statements: G must be very bigger than 5 and C must be around 10. These statements (characteristic functions or membership functions) are shown in Figure 17.4 and numerically expressed by:

$$\mu_G(x) = \begin{cases} 1/\left(1 + e^{-(x-10)}\right) & \text{for } x > 5 \\ 0 & \text{for } x \leq 5 \end{cases}$$

$$\mu_C(x) = \begin{cases} 0.5x - 4 & \text{for } 8 < x < 10 \\ -0.5x - +6 & \text{for } 10 < x < 12 \\ 0 & \text{for } x \leq 8 \text{ and } x \geq 12 \end{cases}$$

The fuzzy decision function D is obtained by the intersection of previous characteristic functions. The resultant function is shown in Figure 17.5.

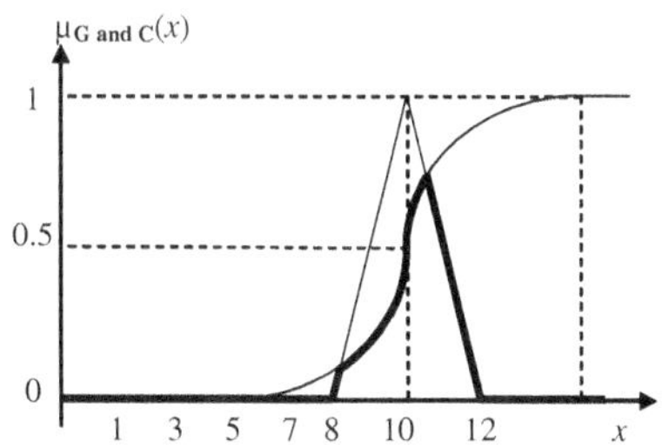

Figure 17.5 Decision membership function between "very bigger than 5" and "around 10."

B

$\mathfrak{R}$		d	e	f	g
	a	0	0	1	1
A	b	0	1	1	1
	c	1	0	0	0

(a)

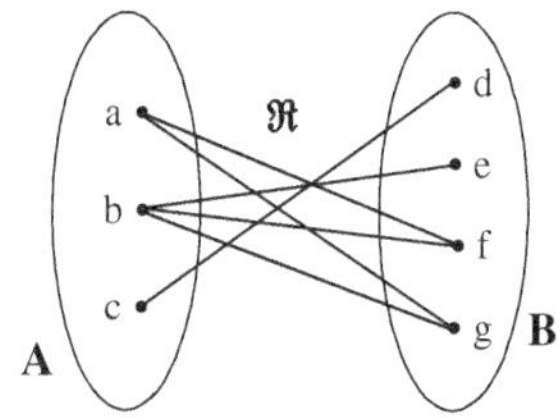

(b)

Figure 17.6 Examples of ordinary relation.

17.2.6 Ordinary and Fuzzy Relations

An ordinary relation expressed links among worlds. For example, let $\mathbf{U}_1 = \{a, b, c\}$ and $\mathbf{U}_2 = \{d, e, f, g\}$, and $\mathbf{M} = [0, 1]$. The relation $\mathfrak{R}$ is expressed by the Cartesian product $\mathbf{U}_1 \times \mathbf{U}_2$. These links can be represented by a matrix or a graph, as shown in Figure 17.6.

Numerically, this relation can be expressed by:

$$\mathfrak{R}(\mathrm{U}_1 \times \mathrm{U}_2) = \{((a,d)|0),((a,e)|0),((a,f)|1),((a,g)|1),((b,d)|0),((b,e)|1),\\ ((b,f)|1),((b,g)|1),((c,d)|1),((c,e)|0),((c,f)|1),((c,g)|0)\}$$

Thus, in general, an ordinary relation can be defined as a set of degree of membership of each n-tuple from a set $\mathrm{U}_1 \times \mathbf{U}_2 \times \ldots \times \mathbf{U}_n$ (Cartesian product).

A fuzzy relation, denoted by $\sim\mathfrak{R}$, has the same structure of ordinary relations but with fuzzy set concepts. Let there be two sets $\mathbf{U}_1$ and $\mathbf{U}_2$, let x an element of $\mathbf{U}_1$ and y an element of $\mathbf{U}_2$. For each element of the set of ordered pairs (x, y), defined by Cartesian product $\mathbf{U}_1 \times \mathbf{U}_2$, a degree of membership taken in a set $\mathbf{M}$ must be associated. For example, Figure 17.7 presents a sample of a fuzzy relation $\sim\mathfrak{R}$.

Numerically, this fuzzy relation can be expressed by:

$$\sim\mathfrak{R}(\mathrm{U}_1 \times \mathrm{U}_2) = \{((a,d)|0.2),((a,e)|0.3),((a,f)|0.6),((a,g)|0.9),((b,d)|0),\\ ((b,e)|0.7),((b,f)|1),((b,g)|0.9),((c,d)|1),((c,e)|0.2),((c,f)|0.1),((c,g)|0)\}$$

B

~$\mathfrak{R}$		d	e	f	g
	a	0.2	0.3	0.6	0.9
A	b	0	0.7	1	0.9
	c	1	0.2	0.1	0

(a)

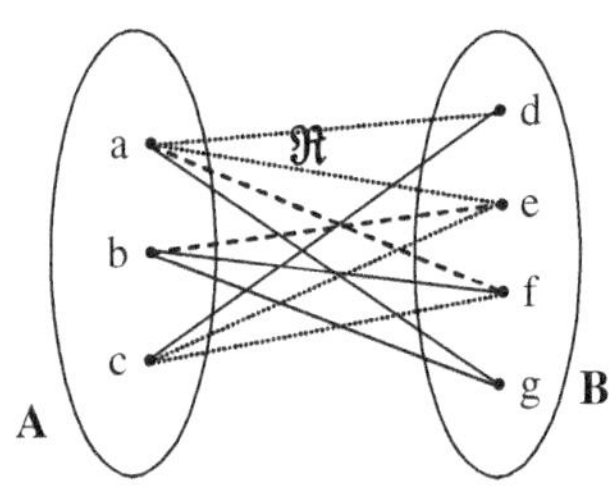

(b)

Figure 17.7 Examples of ordinary relation.

In addition, some operations can be defined. Let $\sim\mathfrak{R}$ and $\sim\mathfrak{I}$ be fuzzy relations defined in $\mathbf{U}_1 \times \mathbf{U}_2$, and (x, y) an ordered pair of $\mathbf{U}_1 \times \mathbf{U}_2$; then, the following basic operations using fuzzy relations exist:

a) Inclusion: $\sim\mathfrak{R} \subset \sim\mathfrak{I}$: "$\sim\mathfrak{R}$ is an enclosure of $\sim\mathfrak{I}$" or "$\sim\mathfrak{I}$ is an envelope of $\sim\mathfrak{R}$" if and only if $\mu_{\mathfrak{R}}(x,y) \leq \mu_{\mathfrak{I}}(x,y)$ for all (x,y)

b) Equality: $\sim\mathfrak{R} = \sim\mathfrak{I}$: "$\sim\mathfrak{R}$ is equal to $\sim\mathfrak{I}$" if and only if $\mu_{\mathfrak{R}}(x,y) = \mu_{\mathfrak{I}}(x,y)$ for all (x,y)

c) Complementation: $\neg\sim\mathfrak{R}$: $\mu_{\neg\mathfrak{R}}(x,y) = 1 - \mu_{\mathfrak{R}}(x,y)$ for all (x,y) belongs to $\mathbf{U}_1 \times \mathbf{U}_2$ and $\mathbf{M} = [0,1]$

d) Intersection: $\sim\mathfrak{R}$ and $\sim\mathfrak{I}$: $\mu_{\mathfrak{R}\cap\mathfrak{I}}(x,y) = \min(\mu_{\mathfrak{R}}(x,y), \mu_{\mathfrak{I}}(x,y))$ for all (x,y)

e) Union: $\sim\mathfrak{R}$ or $\sim\mathfrak{I}$: $\mu_{\mathfrak{R}\cup\mathfrak{I}}(x,y) = \max(\mu_{\mathfrak{R}}(x,y), \mu_{\mathfrak{I}}(x,y))$ for all (x,y)

f) Disjunctive sum: $\sim\mathfrak{R}$ or (exclusive) $\sim\mathfrak{I}$: $\sim\mathfrak{R} \oplus \sim\mathfrak{I} = (\sim\mathfrak{R} \cap \neg\sim\mathfrak{I}) \cup (\neg\sim\mathfrak{R} \cap \sim\mathfrak{I})$

g) Difference: $\sim\mathfrak{R}$ minus $\sim\mathfrak{I}$: $\sim\mathfrak{R} - \sim\mathfrak{I} = \sim\mathfrak{R} \cap \neg\sim\mathfrak{I}$

h) Algebraic product: $\sim\mathfrak{R}$ and-probabilistic $\sim\mathfrak{I}$: $\sim\mathfrak{R} \bullet \sim\mathfrak{I}$: $\mu_{\mathfrak{R}\bullet\mathfrak{I}}(x,y) = \mu_{\mathfrak{R}}(x,y) \,.\, \mu_{\mathfrak{I}}(x,y)$ for all (x,y)

i) Algebraic sum: $\sim\mathfrak{R}$ or-probabilistic $\sim\mathfrak{I}$: $\sim\mathfrak{R} \otimes \sim\mathfrak{I}$: $\mu_{\mathfrak{R}\otimes\mathfrak{I}}(x,y) = \mu_{\mathfrak{R}}(x,y)+\mu_{\mathfrak{I}}(x,y) - \mu_{\mathfrak{R}}(x,y).\mu_{\mathfrak{I}}(x,y)$ for all (x,y).

17.3 FUZZY LOGIC

The name "Fuzzy Logic" in the beginning seems to refer to something confuse (cloudy), even providing scepticism with respect to equipment's control increasing their capacity to perform the work. Fuzzy logic is a way to use inherent analogical process data that shift through a continuous digital computer band, which works with well-defined numeric data, that is, discrete values [13, 14]. For example, considering a brake system directed by a microcontroller, this controller takes the decisions according to the brake temperature, speed, and other system variables.

The temperature variable in this system may be divided in a "states" band: "cold," "fresh," "normal," "tepid," and "hot." Though, it is not easy to set up the transition from one state to the next one; an arbitrary limit should be defined as to divide, for example, the "tepid" from the "hot," but this would drive to a discontinue change when the entry value pass through the limit. The microcontroller should be able to detect it.

The way would be the creation of the "fuzzy states" or more common "fuzzy memberships," which allows the graduated change from a state to the next. You could define the entry temperature by the use of intermediate functions.

In this way, the entry variable state will no longer to jump suddenly from one state to the next one, it looses gradually the value in one state while it gains value in the next state, instead. Up to in a moment, the brake temperature "true value" will almost always be in some point between two consecutive functions: 0.6 normal and 0.4 tepid, or 0.7 normal and 0.3 fresh, and so on. Figure 17.8 shows a representation of such consecutive functions.

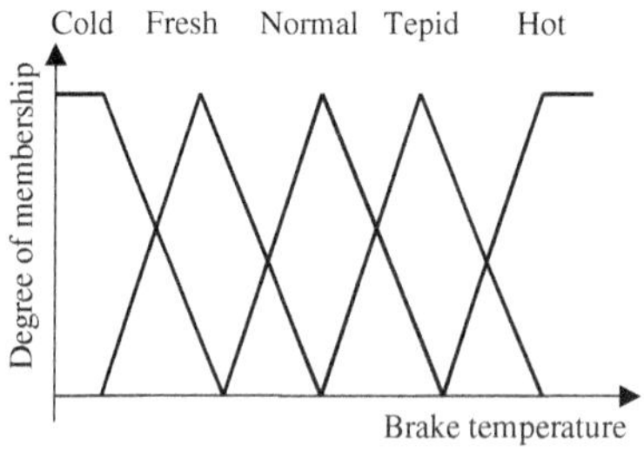

Figure 17.8 Brake temperature memberships.

The entry variables of a fuzzy control system are generally map out inside consecutive function sets—the process for conversion of an intermediate entry value into a fuzzy value is called "fuzzification." We should note that a control system may have entry keys (on/off) together with analogical entries, and such entries (on/off) of the path always will have a true value equal to 1 or 0—but such entries are in fact a simplified case of a fuzzy variable, so the system may work with them with no difficulties.

Setting up the entry variables map out inside the consecutive functions and true values, the microcontroller then takes the decisions for making the actions to follow the rules:

IF the brake temperature is tepid *AND* the speed is not too fast
THEN the brake pressure is slightly reduced

Here, in this case, the two entry variables are "brake temperature" and "speed." The outing variable, "brake pressure," is by the same token, generated from a fuzzy set as a starting point, which may have values as "static," "slightly reduced," "slightly increased," and so on.

The decision is based on a set of rules: All the applied rules are invoked, using the consecutive functions and true values which come from the entries, for the determination of the rule result—that in return will be map out inside the consecutive function and the true value, controlling the outing variable—and after these results are combined as to generate a specific answer, the present brake pressure, a procedure known as "defuzzification" [12]. The combination of fuzzy operations with rules based on "conclusion" describes an "expert fuzzy system."

17.3.1 Fuzzy Control

Fuzzy controls are very simple; they consist of one entry stage, a processing stage, and an outing stage.

The entry stage maps out sensors or another type of entries (numerical data upon which the system will base its decisions) in a proper way the consecutive functions and the true values; the processing stage selects each adequate rule and provides one result for each one of them, and then they combine the results of these rules; and finally, the outing stage converts the combined result from the previous stage into the control [15].

The most usual shape of consecutive functions is the triangular (as in Figure 17.8), even though trapeze-shaped curves, besides other shapes, are also used, the shape is generally less important than the number of curves and where they are located. From 3 up to 7 curves are in general appropriate to cover the required entry values band.

The processing stage is, as already discussed, based upon a collection of logic rules shaped on statements as IF—THEN, where IF is called the "antecedent" and THEN is called "consequence." Typical fuzzy control systems have dozens of rules.

For practical purposes, the set of rules usually has several antecedents that are combined using fuzzy operators, such as AND, OR, and NO, (besides again, the definitions are varying): AND (in an informal definition) uses in a simple way the minimum weight for all the antecedents, whereas OR uses the maximum values [16].

There are several ways for the definition of a rule's result, but one more usually used and simple is called "max-min," in which the outing of the consecutive function is given by the true value generated by the premise.

Rules may be solved in parallel hardware or in sequential software. All rule's results are "defuzzificated" for a concise value by one of the several methods; there are several theoretical ones, all of them with advantages and disadvantages.

The centered method is very popular, in which the result's "mass center" supplies the concise value; in the process of defuzzification by pondered average, the centered of each area is calculated separately and the outing value is calculated through the average of these centered and pondered with the maximum value of the pertinence function. In the process by maximum average of pertinence function, the pertinence function maximum value average is calculated, being this outing value [5].

The fuzzy control systems project is based on empirical methods—basically a methodical approach for try-and-error. The process follows the steps given below:

- The system operational specifications, the entries, and outings are documented.
- The fuzzy sets for the entries are documented.
- The set of rules is documented.
- The defuzzification method is determined.
- It is executed through a test, for checking the system and making the detail adjustments as required.

17.3.2 Fuzzy Controller

In a conventional control strategy, input variables are continuously read and manipulated (computed); thus, the controller continuously generates values for the system output, updating the process. In a fuzzy control scheme, the steps mentioned above are present and, in addition, two more steps are incorporated: fuzzification and defuzzification. The fuzzification process is the first step of the fuzzy controller and is done after reading the input variable values. Its mission is to transform "actual values" of the variables to "fuzzy values" (linguistic values), which are then manipulated by the controller. The defuzzification process is the last step of the fuzzy controller and it transforms "fuzzy values" back to "actual values." Figure 17.9 presents these steps of the control process.

A partial example of the fuzzification process is shown in the first part of Figure 17.10, where inputs $x = 10$ and $y = 8$ generate values for the memberships "small"

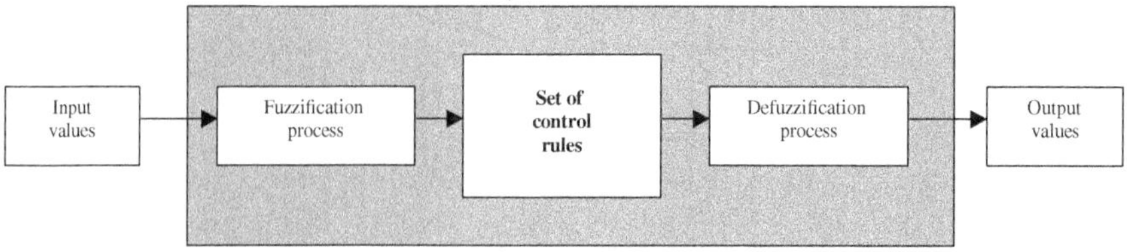

Figure 17.9 Fuzzy control scheme.

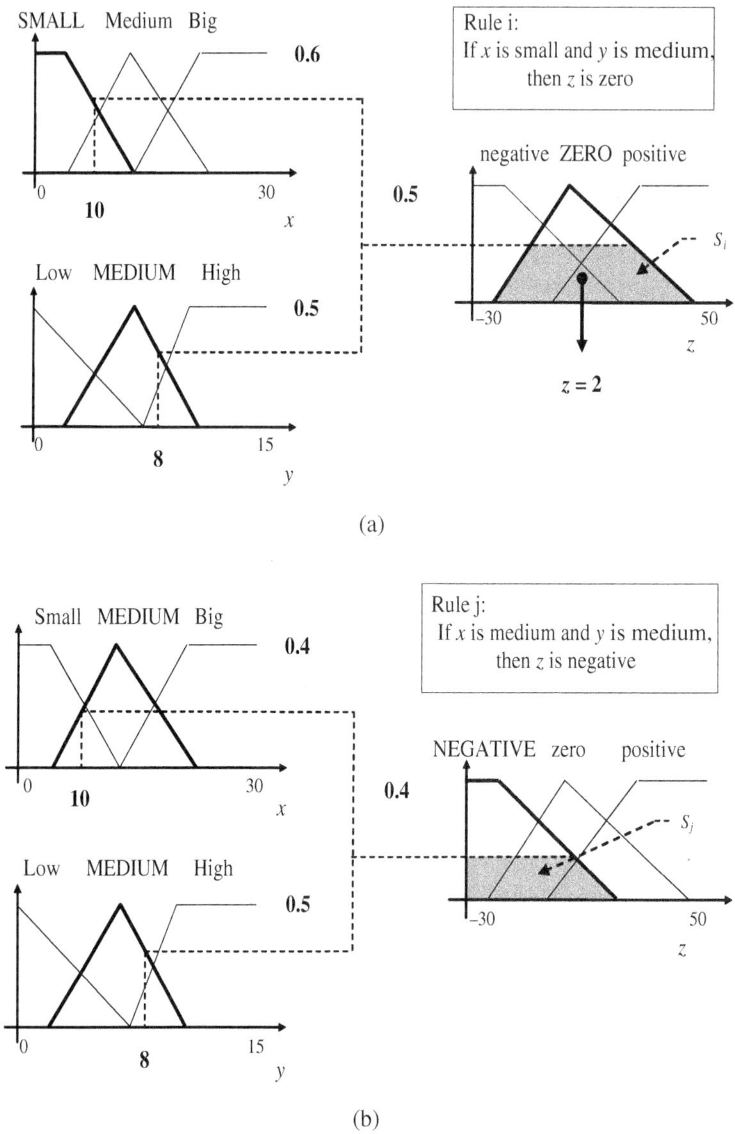

Figure 17.10 Fuzzification inference and defuzzification processes.

and "medium" equal to 0.6 and 0.5, respectively. An example of defuzzification process is presented in the next item.

17.3.3 Fuzzy Inference Process

A fuzzy inference process executes each rule, that is, (a) the input variables are transformed into the fuzzy statements and (b) an output value is computed. The execution of each rule is made using *modus ponens*, which means that the premise of the rule produces the degree of membership for the conclusion of the same rule. This membership degree is a function of the fuzzy values of the input variables, and of the conjunctions used among them. Let us consider the example of Figure 17.10, where the two following fuzzy control rules are shown:

Rule i: If **x** is small and **y** is medium, then **z** is zero.

Rule j: If **x** is medium and **y** is medium, then **z** is negative.

Note that during the fuzzification process for the rule *i*, the values obtained are 0.6 and 0.5, as mentioned above. Since the liaison element used between the fuzzy statement is the conjunction "and" (which represents the "minimum" operator), the conclusion of the rule (value "zero") is limited by the minimum value of the premise, that is, min(0.6, 0.5) = 0.5.

Each rule produces a limited area (as illustrated by the shaded area in Figure 17.5, according to the resultant value from the premise and the output membership function of the conclusion). In the same way, for the rule j, the limitation of the value "negative" is 0.4. Of course, it is possible that the value that comes from the premise is zero. In this case, the rule has no influence on the computation of the final output value.

After the execution of all the rules, the defuzzification process begins. Each produced area is computed. Then, the maximum operation is applied to define the biggest one. In Figure 17.10, S_i is bigger than S_j, that is, Max $(S_i, S_j) = S_i$. The final actual output value is computed using the center of gravity method of S_i.

The value of the abscissa found is the actual value of the output variable. Figure 17.5 shows an example of this process, where the centroid produces a value of $z = 2$.

There are many types of defuzzification processes, such as maximum value, maximum average, center of gravity (centroid), center of mass of the highest region, weighted average method, center of sums, first (or last) of maxima, and so on [15].

17.4 FUZZY MODEL

This section presents an algorithm to build a structure for fuzzy conditional statements [17], where the premise contains fuzzy statements and the consequence is represented by first- or second-order equations. The general form of the proposed fuzzy conditional statement is given by:

$$\text{If } \{x_1 \text{ is } T_{x_1}, \ldots, x_p \text{ is } T_{x_p}\}$$
$$\text{then } \{y_1 = f(x_1, \ldots, x_p), \ldots, y_m = f(x_1, \ldots, x_p)\}.$$

In this case, the premise fuzzy statements are combined using comma, meaning the conjunction "and."

The proposed alternative data model follows a systematic approach suggested by Sugeno-Takagi [11, 17]. The proposed method eliminates some major restrictions in previous work by using an L-fuzzy set approach [18]. In this theory, the domain of the membership function can be taken in any lattice numerical structure. For our purpose, the numerical structure used is $\Re$ (the set of real numbers). Also, the linear and quadratic functions are explored in both fuzzy values and consequence functions. The main idea is to gain flexibility and accuracy in the data representation.

It is worthwhile to mention that many other developments have been recently reported in the literature in this area. Among them, we can mention the one for general data representation purposes of references [19, 20].

17.4.1 Problem Formulation

Let us consider a set of fuzzy conditional statements, with p inputs and a single output variable. The general form of these statements is the following:

$$\text{If } \{x_1 \text{ is } T_{x_1}, \ldots, x_p \text{ is } T_{x_p}\} \text{ then } \{y = f(x_1, \ldots, x_p)\}.$$

where the fuzzy values Tx and the function $f(\bullet)$ can be represented by first- or second-order equations.

For the sake of clarity, the formulation presented in this section uses a system with a single input and a single output (SISO) only. Apparently, this simple formulation seems to lack generalization; however, it is not true because of the existing decoupling among the input variables. The description of the modeling process with multiple input variables is presented in the Section 17.4.3. Thus, the general fuzzy conditional statement form for a SISO system can be represented by:

$$\text{If } \{x \text{ is } T_i\} \text{ then } \{y = a_{0i} + a_{1i}x + a_{2i}x^2\}$$

where:

$$T_i = \begin{cases} 0 & \text{for } x_{i1} > x < x_{i2} \\ m_{0i} + m_{1i}x + m_{2i}x^2 & \text{for } x_{i1} \leq x \leq x_{i2} \end{cases}$$

Thus, the solution vector Θ can be expressed as follows:

$$\Theta = [x_{i1} \quad x_{i2} \quad m_{0i} \quad m_{1i} \quad m_{2i} \quad a_{0i} \quad a_{1i} \quad a_2]^t$$

17.4.2 The Algorithm to Solve for the Vector Θ

Four kinds of solutions for the vector Θ have been studied, named, FF, FS, SF, and SS, where F represents a first-order equation approximation, and S a second-order equation approximation. Thus, an FS rule solution means that the premise fuzzy values have been approximated using a first-order equation (implying $m_{2i} = 0$), and the consequence, by a second-order equation.

The first step in the solution is to choose a maximum average representation error. The proposed method solves for the vector Θ by an algorithm that, initially, computes the possible equations of the consequence by linear and quadratic regressions. This computation starts with the inferior limit of the variable; points are added in this regression until the maximum average representation error cannot be achieved. At this stage, the last regression is the consequence of the first rule. The same procedure is carried out for the superior limit of the variable, given us the equation for the consequence of the second rule.

At this point, two possibilities can occur. The first possibility is a superposition of the intervals of the above regressions. When it occurs, the use of fuzzy conditional statements is not necessary because the two regressions can represent the model with the required maximum average error. In this case, single production rules can be used to represent the model. In the second possibility (more frequently), there is an interval that must still be modeled. Therefore, the algorithm should continue.

Up to this point, the algorithm has determined the following partial solution vector:

$$\Theta' = [x_{i1} \quad x_{i2} \quad a_{0i} \quad a_{1i} \quad a_{2i}]^t$$

where the parameters a_{0i} , a_{1i}, and a_{2i} have been computed in the regression (consequence), and the interval $[x_{i1}, x_{i2}]$ is the one that must be modeled.

The premise computation is achieved by the center of gravity method [9]. Points are randomly chosen in the interval to be modeled and, by the use of a reverse inference process, the coefficients of the fuzzy values (m_{ji}) are computed.

This means that, at this point, the algorithm has computed the second solution vector,

$$\Theta'' = [m_{0i} \quad m_{1i} \quad m_{2i}]^t$$

The above procedure is repeated for each kind of model (FF, FS, SF, and SS); the fuzzy model with the best representation is finally chosen. A performance index is defined to create a common base for the process evaluation. The performance index is computed by the summation of all individual relative errors, without using the signal.

Following, each step of the proposed algorithm is described.

1. Choose the maximum average representation error for the model.
2. Make a list of the input-variable values by rearranging them in an increasing order.
3. Use an auxiliary list to incorporate new points and compute the regression. This additional process continues until a new point does not allow maintaining the maximum average error (step 1). The result is an auxiliary list in the interval $[x_{1\min} \ x_{11}]$.
4. Use the same process of step 3, but start in the superior limit of the list of step 2. The result is an auxiliary list in the interval $[x_{12} \ x_{1\max}]$.
5. Test. Verify if there is a superposition between the computed auxiliary lists (steps 3 and 4), that is, the question is "$x_{11} \geq x_{12}$?" If the answer is yes, to go step 9; otherwise, continue.

6. Choose randomly two (or three) points that do not belong to the previous auxiliary lists, that is, in the interval $[x_{11}\ x_{12}]$. Compute, by fuzzy inverse inference, the fuzzy values. Note: two points are used to compute first-order fuzzy values, and three, for second-order fuzzy values.
7. Repeat step 6, for many different sets of points.
8. Evaluate the results of steps 6 and 7, and choose the model with the better performance index.
9. Repeat steps 3–8 for all models (FF, FS, SF, and SS).
10. Evaluate all models and choose the one with better performance index.

17.4.3 A Multiple Input/Output Decision System

In this section, a system with multiple input variables and a single output variable (MISO) is analyzed. If the system has multiple outputs, it is usually possible to transform the MIMO system in many MISO systems, one for each output variable. When this approach is not possible, a fuzzy decision system must be designed to change information among them.

Let us consider an input vector with p input variables, $x_1, x_2, \ldots, x_p$, and a single output variable y. The main idea in this case is to generate a search tree, where the best local models are chosen. Following, each step of the search tree formation is presented.

Step A:
This search process starts by the representation of the system with only one variable in the premise, while the consequence incorporates all variables. This first step is shown in Figure 17.11 which, for the sake of clarity, contains only three input variables.

Step A comprises four models for each variable (in our example, FF, FS, SF, and SS are computed for each one of the three variables, resulting in 12 models). The best model, according to the performance index, is the *father node* and is used to

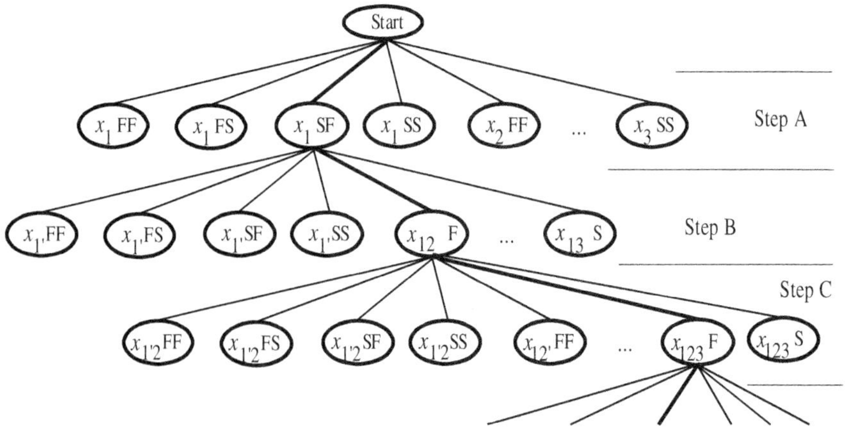

Figure 17.11 Search tree for a multiple input variable space.

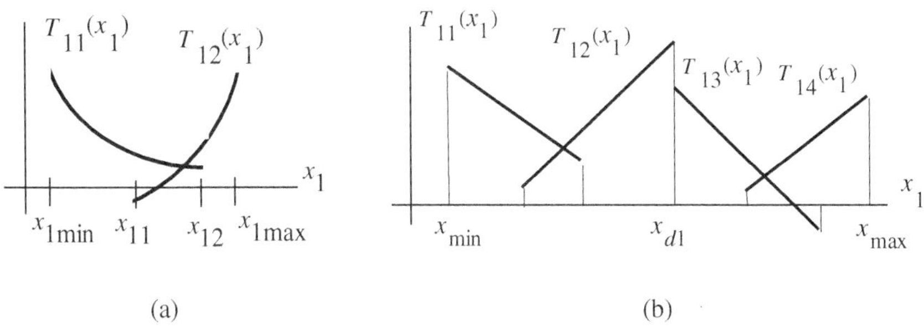

Figure 17.12 Examples of fuzzy values.

continue the search process. In Figure 17.11, the model FS of variable x_1 is chosen, having the following general form:

$$\text{R1: If } x_1 \text{ is } T_{11}, \text{ then } y_{11} = a_{01} + a_{11}x_1 + a_{21}x_2 + a_{31}x_3$$

$$\text{R2: If } x_1 \text{ is } T_{12}, \text{ then } y_{12} = a_{02} + a_{12}x_1 + a_{22}x_2 + a_{32}x_3$$

where the fuzzy values T_{11} and T_{12} are second-order equations in x_1 , which is defined in the intervals $[x_{1min}\ x_{12}]$ and $[x_{11}\ x_{1max}]$, respectively. Figure 17.12 illustrates these values.

Step B:

In the second step, new models are obtained according to the best model of step A. In order to do that, two new types of premise division must be computed: (a) for variables not present in the premise so far and (b) for variables already in the premise. For the sake of clarity, we use the example of Figure 17.1 to describe this approach. Thus, the first type includes the variables x_2 and x_3, while the second one includes the variable x_1.

For instance, let us consider the model x_{12}F, which is a possible example of model that is obtained for the first type of the premise division. The set of resultant fuzzy conditional statements is:

$$\text{R1: If } x_1 \text{ is } T_{11} \text{ and } x_2 \text{ is } T_{21}, \text{ then } y_{11} = a_{01} + a_{11}x_1 + a_{21}x_2 + a_{31}x_3$$

$$\text{R2: If } x_1 \text{ is } T_{11} \text{ and } x_2 \text{ is } T_{22}, \text{ then } y_{13} = a_{03} + a_{13}x_1 + a_{23}x_2 + a_{33}x_3$$

$$\text{R3: If } x_1 \text{ is } T_{12} \text{ and } x_2 \text{ is } T_{21}, \text{ then } y_{12} = a_{02} + a_{12}x_1 + a_{22}x_2 + a_{32}x_3$$

$$\text{R4: If } x_1 \text{ is } T_{12} \text{ and } x_2 \text{ is } T_{22}, \text{ then } y_{14} = a_{04} + a_{14}x_1 + a_{24}x_2 + a_{34}x_3$$

Note that the consequence equations y_{11} and y_{12} remain the same as in step A. The fuzzy values T_{11} and T_{12} are second-order equations, and T_{21} and T_{22} are first-order equations, justifying in this way the letter F in the name of the model (x_{12}F).

The second type of model, with a variable already used in the premise, has its interval divided into two by the use of an inflexion (or a saddle) point or a temporary reduction of the model maximum average error. This division results in intervals, denoted by $[x_{1min}\ x_{d1}]$ and $[x_{d1}\ x_{1max}]$. At this point, and with these new sets of intervals, the search process computes all four kinds of models for each interval independently. For instance, let us consider the model x_{11}FF, for which the resultant fuzzy conditional statements are the following:

$$\text{R1: If } x_1 \text{ is } T_{11}, \text{ then } y_{11} = a_{01} + a_{11}x_1 + a_{21}x_2 + a_{31}x_3$$

$$\text{R2: If } x_1 \text{ is } T_{12}, \text{ then } y_{13} = a_{03} + a_{13}x_1 + a_{23}x_2 + a_{33}x_3$$

$$\text{R3: If } x_1 \text{ is } T_{13}, \text{ then } y_{14} = a_{04} + a_{14}x_1 + a_{24}x_2 + a_{34}x_3$$

$$\text{R4: If } x_1 \text{ is } T_{14}, \text{ then } y_{12} = a_{02} + a_{12}x_1 + a_{22}x_2 + a_{32}x_3$$

where all fuzzy values T_{1k} are recomputed, and two new consequence equations (y_{13} and y_{14}) appear; the other two equations (y_{11} and y_{12}) remain the same. Figure 17.12 b) shows the fuzzy values T_{1k}, which are first-order equations in order to justify the name of model (x_{11}FF).

As shown in Figure 17.11, the best model is x_{12}F, which can be described by:

- fuzzy values T_{11} and T_{12} (second-order equations);
- fuzzy values T_{21} and T_{22} (first-order equations—F);
- first-order equations in the consequences.

Step C:
In this step, following the example of Figure 17.1, the models below are generated:

a) models with variables already used in the premise:

- $x_{1'2}$FF , $x_{1'2}$FS , $x_{1'2}$SF , $x_{1'2}$SS—where the interval of variable x_1 is divided, and T_{21} and T_{22} are the same fuzzy values computed in the step B.
- $x_{12'}$FF , $x_{12'}$FS , $x_{12'}$SF , $x_{12'}$SS—where the interval of variable x_2 is divided, and T_{11} and T_{12} are the same fuzzy values computed in the step B.

b) models with new variables in the premise:

- x_{123}F , x_{123}S—where the fuzzy values T_{11}, T_{12}, T_{21}, and T_{22} are the same as obtained in step B, and first- and second-order equations for fuzzy values of the variable x_3 are computed. All consequences remain the same as in step B.

The model with the best performance index is chosen in order to pursue the search process. From Figure 17.11, this model is x_{123}F.

The algorithm then starts building a search tree until a model is found for which all points satisfy the maximum error (step 1) and/or the maximum performance index value.

The search process above uses local best models to continue building a tree. This process does not give any guarantee that the global optimal model (minimum individual error and minimum number of rules) can be achieved. To obtain an optimal model, it is necessary to build all branches in the tree. However, this approach is execution-time too expensive; also, the gain that can be obtained with the data representation is not evident. For these reasons, a search using a local minimum is used in this work.

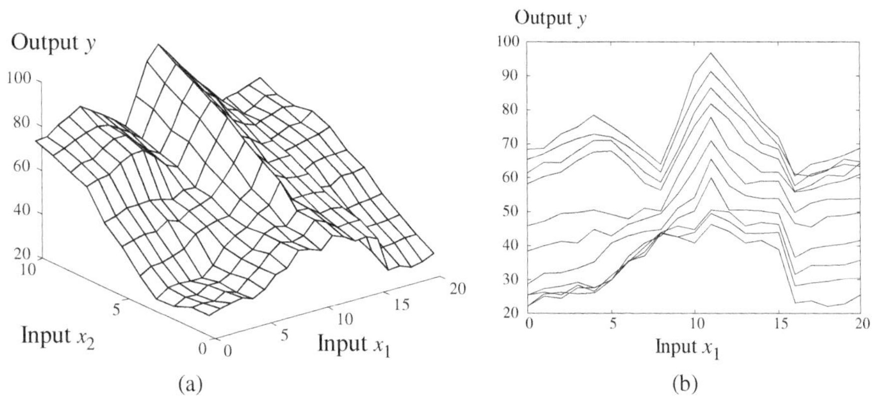

Figure 17.13 Illustrative example: load data: (a) spatial projection and (b) planar projection of the variable x_1.

17.4.4 Illustrative Example

This section presents an example with two input variables and a single output variable. This choice of number of variables allows a graphical treatment of intermediary steps of the solution. The load data to be modeled is shown in Figure 17.13a. These data represent a distribution feeder load for several hours during a day (variable x_1), and for several different days (variable x_2). A projection of the daily load for different days is shown in Figure 17.13b.

For the sake of clarity, only the best two models in each step are presented in this example, indicating their performance indices. The maximum average representation error allowed is 2%.

Step A:
In the first step of the search process, fuzzy conditional statements with only a single premise variable are generated. Thus, eight models are built (four for each variable, corresponding to FF, FS, SF, and SS representations). The best two models obtained, in this step of the example, use the FF representation.

Set 1: for variable x_1—performance index = 51.89

$$\text{R1: If } x_1 \text{ is } T_{11}, \text{ then } y_{11} = 25.4 + 1.02x_1 - 0.84x_2$$

$$\text{R2: If } x_1 \text{ is } T_{12}, \text{ then } y_{12} = 13.3 + 0.58x_1 + 4.978x_2$$

where:

$$T_{11} = 2.6145 - 0.3229x_1 \quad \text{for interval} = [0, 15]$$

$$T_{12} = -15.2476 + 1.8053x_1 \quad \text{for interval} = [4, 20]$$

Set 2: for variable x_2—performance index = 49.12

$$\text{R1: If } x_2 \text{ is } T_{21}, \text{ then } y_{21} = 25.4 + 1.02x_1 - 0.84x_2$$

$$\text{R2: If } x_2 \text{ is } T_{22}, \text{ then } y_{22} = 35.2 + 2.15x_1 + 3.26x_2$$

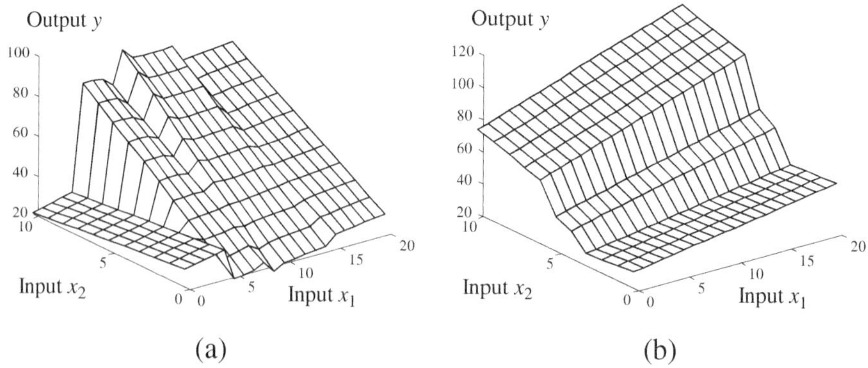

Figure 17.14 Results of the step A of the search process: (a) set 1 and (b) set 2.

where:

$$T_{21} = 3.8009 - 0.5602x_2 \quad \text{for interval} = [0, 6]$$

$$T_{22} = 0.9184 + 0.0272x_2 \quad \text{for interval} = [3, 10]$$

Observing the performance indices, we can verify that between the two best models, the one with better index corresponds to the second set of propositions (using the variable x_2). The surface produced for each set of propositions is shown in Figure 17.14.

In programming the algorithm, we have included an additional step that considers more than one father node when the index difference among the nodes is small. In this example, we can see that the difference between the two indices is very narrow; thus, the program will continue the search tree building using two father nodes.

Step B:

As in step A, we have two father nodes; each of them independently generates a branch in the search tree. In this step, two kinds of propositions are generated. The first one is related to the division of the premise interval and the second one is related to the inclusion of a new variable in the premise. After all cases are computed, those with better performance index are found to be:

- interval division of variable x_1 into [0 , 9] and [9 , 20] intervals;
- inclusion of variables x_1 and x_2 in the premise, using first-order equation for the fuzzy values.

These two cases result in the following sets of propositions:
Set 3: for variable x_1—performance index = 40.17

$$\text{R1: If } x_1 \text{ is } T_{11}, \text{ then } y_{11} = 25.4 + 1.02x_1 - 0.84x_2$$

$$\text{R2: If } x_1 \text{ is } T_{12}, \text{ then } y_{12} = 24.6 + 2.25x_1 + 0.62x_2$$

$$\text{R3: If } x_1 \text{ is } T_{13}, \text{ then } y_{13} = 21.8 + 1.85x_1 + 2.92x_2$$

$$\text{R4: If } x_1 \text{ is } T_{14}, \text{ then } y_{14} = 13.3 + 0.58x_1 + 4.978x_2$$

where:

$$T_{11} = 5.9650 - 1.2413x_1 \quad \text{for interval} = [0, 7]$$

$$T_{12} = -16.2184 + 3.4437x_1 \quad \text{for interval} = [4, 9]$$

$$T_{13} = -225.7657 + 18.8971x_1 \quad \text{for interval} = [9, 15]$$

$$T_{14} = -8.2198 + 0.9220x_1 \quad \text{for interval} = [11, 20]$$

Set 4: for variable x_2, including x_1—performance index = 21.39

R1: If x_1 is T_{11} and x_2 is T_{21}, then $y_{11} = 25.4 + 1.02x_1 - 0.84x_2$

R2: If x_1 is T_{11} and x_2 is T_{22}, then $y_{12} = 35.2 + 2.15x_1 + 3.26x_2$

R3: If x_1 is T_{12} and x_2 is T_{21}, then $y_{13} = 13.3 + 0.58x_1 + 4.978x_2$

R4: If x_1 is T_{12} and x_2 is T_{22}, then $y_{14} = 25.1 + 1.42x_1 + 1.41x_2$

where:

$$T_{11} = 1.40 - 0.3107x_1 \quad \text{for interval} = [0, 15]$$

$$T_{12} = -1.5975 + 0.1352x_1 \quad \text{for interval} = [4, 20]$$

$$T_{21} = 0.8200 - 0.1033x_2 \quad \text{for interval} = [0, 6]$$

$$T_{22} = -0.4800 + 0.1667x_2 \quad \text{for interval} = [3, 10]$$

In verifying the performance indices for these two sets of propositions above, it is encountered that set 4 has the best performance. This is the father node for the next step. The surface produced by each set of the propositions is shown in Figure 17.15.

Step C:

In this step, the interval of each variable is divided, and first- and second-order approximations for fuzzy values are tested. The best two models obtained in this step are:

Set 5: for variable x_1 (FF-FF)—performance index = 5.29

R1: If x_1 is T_{11} and x_2 is T_{21}, then $y_{11} = 25.4 + 1.02x_1 - 0.84x_2$

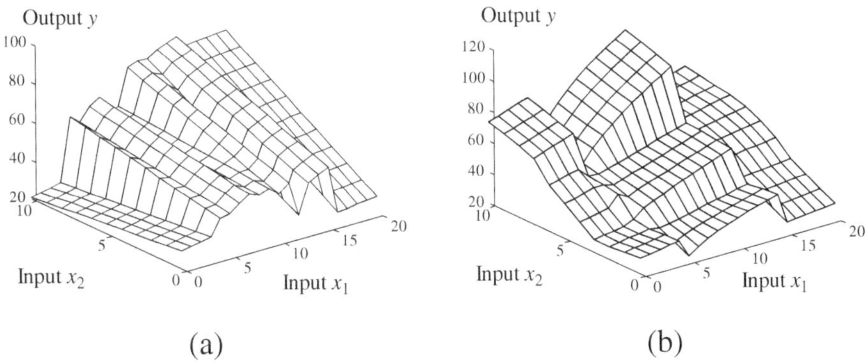

Figure 17.15 Results of the step B of the search process: (a) set 3 and (b) set 4.

R2: If x_1 is T_{11} and x_2 is T_{22}, then $y_{12} = 35.2 + 2.15x_1 + 3.26x_2$

R3: If x_1 is T_{12} and x_2 is T_{21}, then $y_{13} = 24.6 + 2.25x_1 + 0.62x_2$

R4: If x_1 is T_{12} and x_2 is T_{22}, then $y_{14} = 40.7 - 0.14x_1 + 2.34x_2$

R5: If x_1 is T_{13} and x_2 is T_{21}, then $y_{15} = 21.8 + 1.85x_1 + 2.92x_2$

R6: If x_1 is T_{13} and x_2 is T_{22}, then $y_{16} = 40.2 + 0.42x_1 + 4.56x_2$

R7: If x_1 is T_{14} and x_2 is T_{21}, then $y_{17} = 13.3 + 0.58x_1 + 4.978x_2$

R8: If x_1 is T_{14} and x_2 is T_{22}, then $y_{18} = 25.1 + 1.42x_1 + 1.41x_2$

where:

$$T_{11} = 1.2502 - 0.115x_1 \quad \text{for interval} = [0, 7]$$

$$T_{12} = 0.8720 + 0.212x_1 \quad \text{for interval} = [4, 9]$$

$$T_{13} = 2.1750 + 0.125x_1 \quad \text{for interval} = [9, 15]$$

$$T_{14} = -1.5975 + 0.1352x_1 \quad \text{for interval} = [11, 20]$$

$$T_{21} = 0.8200 - 0.1033x_2 \quad \text{for interval} = [0, 6]$$

$$T_{22} = -0.480 + 0.1667x_2 \quad \text{for interval} = [3, 10]$$

Set 6: for variable x_1 (SS-FF)—performance index = 4.04

R1: If x_1 is T_{11} and x_2 is T_{21}, then $y_{11} = 25.4 + 1.02x_1 - 0.84x_2$

R2: If x_1 is T_{11} and x_2 is T_{22}, then $y_{12} = 35.2 + 2.15x_1 + 3.26x_2$

R3: If x_1 is T_{12} and x_2 is T_{21}, then $y_{13} = 24.6 + 2.25x_1 + 0.62x_2$

R4: If x_1 is T_{12} and x_2 is T_{22}, then $y_{14} = 40.7 - 0.14x_1 + 2.34x_2$

R5: If x_1 is T_{13} and x_2 is T_{21}, then $y_{15} = 21.8 + 1.85x_1 + 2.92x_2$

R6: If x_1 is T_{13} and x_2 is T_{22}, then $y_{16} = 40.2 + 0.42x_1 + 4.56x_2$

R7: If x_1 is T_{14} and x_2 is T_{21}, then $y_{17} = 3.3 + 0.58x_1 + 4.978x_2$

R8: If x_1 is T_{14} and x_2 is T_{22}, then $y_{18} = 25.1 + 1.42x_1 + 1.41x_2$

where:

$$T_{11} = 1.400 - 0.3327x_1 + 0.022x_1^2 \quad \text{for interval} = [0, 7]$$

$$T_{12} = 0.770 - 0.421x_1 + 0.055x_1^2 \quad \text{for interval} = [4, 9]$$

$$T_{13} = 2.1750 + 0.125x_1 \quad \text{for interval} = [9, 15]$$

$$T_{14} = -1.5975 + 0.1352x_1 \quad \text{for interval} = [11, 20]$$

$$T_{21} = 0.8200 - 0.1033x_2 \quad \text{for interval} = [0, 6]$$

$$T_{22} = -0.480 + 0.1667x_2 \quad \text{for interval} = [3, 10]$$

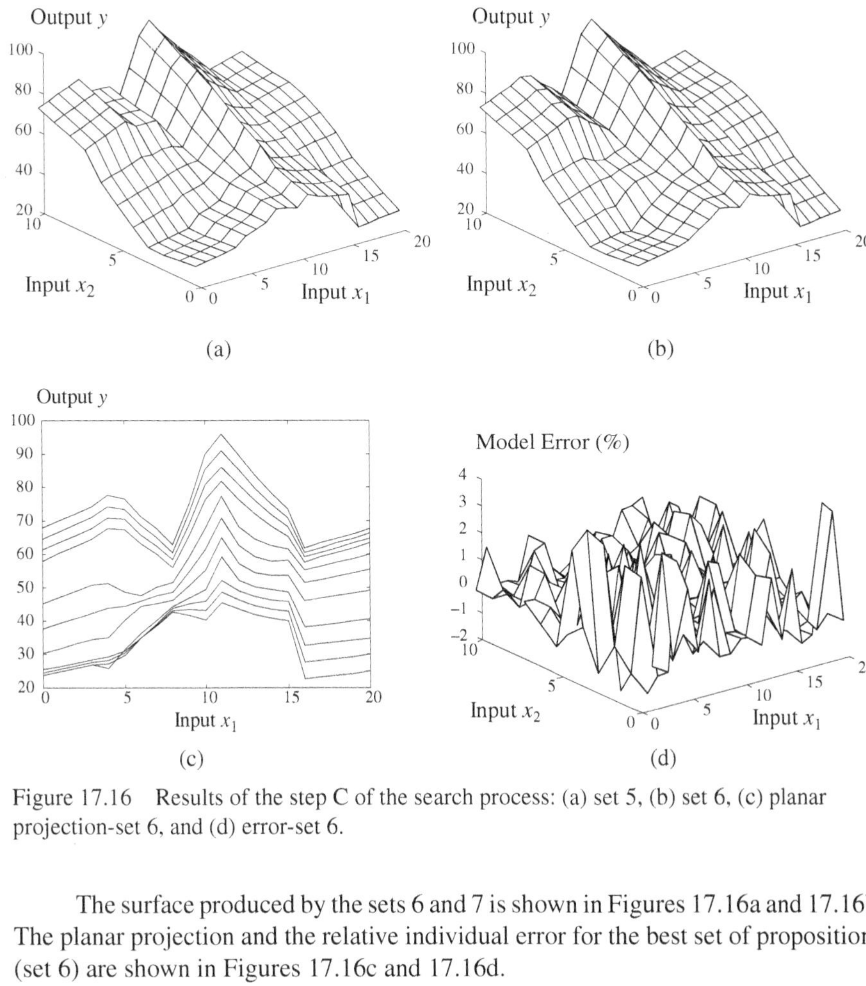

Figure 17.16 Results of the step C of the search process: (a) set 5, (b) set 6, (c) planar projection-set 6, and (d) error-set 6.

The surface produced by the sets 6 and 7 is shown in Figures 17.16a and 17.16b. The planar projection and the relative individual error for the best set of propositions (set 6) are shown in Figures 17.16c and 17.16d.

Comments: Both models obtained in step C have a performance index close to 2% of the medium error (231 points have been used in the initial surface, 11 curves × 21 points). The average representation errors by the sets 5 and 6 are 2.29% and 1.75%, respectively. Therefore, the maximum average error (2%) has remained the same. In addition, by observing Figure 17.16d, we can verify another important characteristic of the model; the errors have no tendency and are distributed around zero.

17.5 AN APPLICATION OF FUZZY LOGIC IN CONTROL SYSTEM

This section presents an application of fuzzy logic to solve an important industrial problem: the power factor, a relation between active and reactive power consumed by industries. Usually, the industries have used capacitor banks to solve this problem, compensating reactive power demand. However, switching large capacitors on a

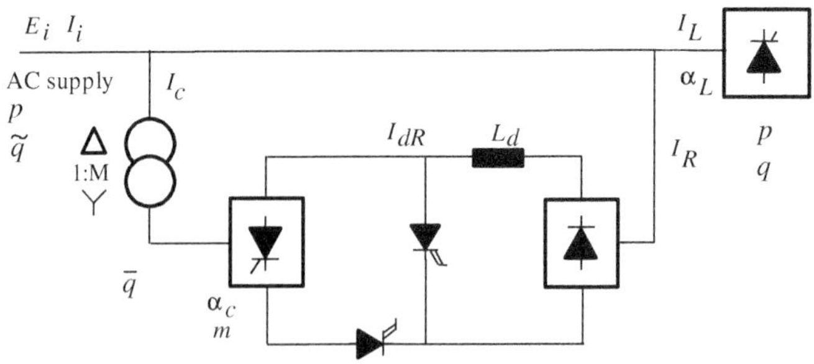

Figure 17.17 Active power factor compensator.

three-phase system is source of several significant problems. These problems may be acceptable if the capacitors are not switched often. The disturbances associated with the voltage transients, during the instant of switching, have forced new solutions to guarantee the power quality of the electrical distribution systems [21, 22].

This application presents a generalized hybrid converter, composed basically of two bridges as shown in Figure 17.17, and also two Insulated Gate Bipolar Transistors (IGBTs) [23]. One bridge, using diodes, works as a rectifier, the other, using thyristors, works as an inverter. The main goal concerning the application of this hybrid converter is to compensate the power factor, since in industrial electrical systems, a poor power factor is one of the main disadvantages of using controllable DC motors and noncontrollable AC motors.

Originally, the hybrid converter was prompted by the desire to improve the performance of existent slip-recovery drives. Experiences with the hybrid converter showed the possibility to apply the same architecture, with a little improvement in the control strategy, as power factor improvement. The basic idea behind this circuit is to provide the line feeder with the necessary reactive power in order to compensate the load displacement factor, and consequently the power factor. In power electronics words, the name of this equipment is static VAr compensator.

The main objective of this section is to present an example of fuzzy control strategy and how it can be implemented. The complex problem to be solved is due to the difficulty to obtain relations between the control variables, and problems involving unstable regions of the hybrid converter operation. The implemented control system has been built with learning characteristics.

The control was implemented using a three-level hierarchical structure (as shown in Figure 17.18a, where the uses of hints, coded in fuzzy conditional statements (the observer), improve the learning of the neural network (control routines) responsible for the proper functioning of the active power factor compensator).

The first level of the control structure comprises the protection routines. These routines are always active. They represent the "self protection instinct" of the system. This level also handles the short-term controller, described by a set of fuzzy implications [24], which automatically reacts to fail in the converter or in the load. This controller represents the "mechanical reflexes" of the system. It constitutes the lower level of the control structure.

The second level is responsible for all parameters that the lower level needs to work properly and closed loop tasks. For example, the adjustment of the system's output, or to inform about the validity of an operating point. The neural network independent structures, defined for this level, have to cope with two different groups of knowledge. One is based on mathematical routines relating to some converter variables. The other is related more to the empirical knowledge about the converter behavior. All the information this level manipulates are directly related to the system operational characteristics. This second level represents the "skills" of the system.

The third level, described in terms of fuzzy conditional statements, concatenates engineering knowledge and designer's expectations named here as hints. The conditional parts of the fuzzy propositions look at the overall behavior of the system and the action parts, instead of acting directly on the system's variables, activate the appropriate lower level autonomous routines to realize a precise task in order to achieve the desired goal, and improve the converter behavior. This level represents the "intelligence" of the system.

This multilevel control structure can be represented mathematically by a hybrid neural fuzzy automaton, where each autonomous routine represents a different state of the system. To edit and run the rules, a rule-based production system was used. This system, called FUZZY-FORTH [25], was designed using FORTH, which is a high level language especially designed for real-time process control, and fast enough to ensure response times of a few milliseconds.

17.5.1 Control Strategy

The developed intelligent control strategy can be divided into two main aspects. The first copes with the exact information about the reactive power needed to be compensated. The second aspect is the training procedure of the control structure.

As mentioned before, the structure for this intelligent control can be divided into three well-defined levels (Figure 17.18a). The third level, the "Observer," is

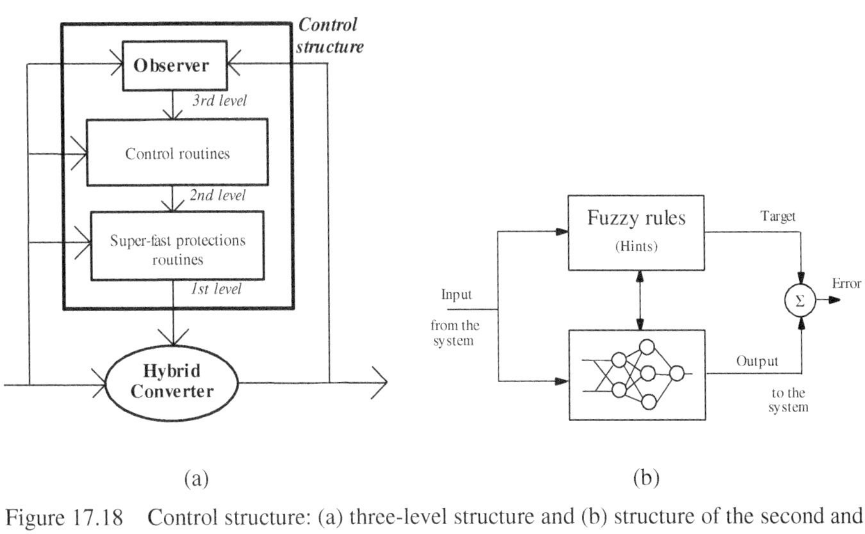

Figure 17.18 Control structure: (a) three-level structure and (b) structure of the second and third levels.

responsible for the training and learning capability of the system, that is, the progressive improvement in the control system behavior. The structure of the third level is a fuzzy rule based containing the hints to be used in order to train the second level. The structure of the second level is basically a neural network, containing the knowledge about the active power factor compensator. Figure 17.18b represents this structure.

The neural network is used to memorize relations between variables in a progressive learning procedure. Instead of developing the neural network learning procedure based only on examples, additional information about the target function can be incorporated in the learning process. Such information is the common sense about the desired electrical system behavior and it is obtained from an expert (for example, bus voltage cannot increase too much, voltage oscillation is not desired, and power factor must be kept between limits).

The use of hints in learning is a powerful idea [26] that boosts the information contained in a set of data, and wherever such a method is implemented, the full benefit of the hint is automatically incorporated in the neural network memory. The idea behind this "Observer" can be summarized as taking a certain set of input data, and the algorithm will decide which hint must be taken into account. Then, using an appropriate training scheme, this operating point is incorporated into the model.

The learning structure can be summarized as: a certain environment A is the set on which the unknown and desired function f is defined. The function f takes values from some set Y. The learning process takes some properties, facts, or pieces of information about f as input. These properties produce a set of hints named hypothesis (g), where g: A => Y. That attempts to approximate f.

The degree to which a hypothesis *g* is considered an approximation of f is measured by the distance or error $\varepsilon(g, f)$. The error ε is basically the disagreement between g and f. A number of points $a_1, \ldots, a_N$ are picked from **A** and the values of f on these points are provided according to a certain hint. So, the input of the learning process is the set of examples given by $(a_1, f(a_1)), \ldots, (a_N, f(a_N))$ that satisfy a set of hints $H_1, \ldots, H_M$. A hint is defined as a fuzzy conditional rule in the form: "If *x* is X and *w* is W then *z* is Z," and a set of fuzzy rules represents the minimal knowledge about the desired system behavior. The defuzzification method used is the composite maximum [27], because the result of interest is only the sensitivity of the truth produced by the single rule that has the highest predicate truth.

17.5.2 Production System

Expert systems are often built around production systems due to the fact that these provide an effective method for coding the engineering knowledge, representing it in the form "If A then B." The rule-based production system described here (FUZZY-FORTH) is a highly efficient, practical, and flexible means for implementing complex real-time control systems. Production systems have several advantages over conventional programming. Firstly, since the knowledge can be represented within the well-defined structure of rules, utilization of such systems is independent of a specific field or application. Secondly, production systems are built in such a way that rules can be easily manipulated, which is very useful in development.

The ordinary meaning of a conditional statement is a Boolean condition-action pair, where, if the conditional part is satisfied then the action should be executed. In our system, the binary nature of the variables is extended to cope with the concept of membership functions, which allows for managing fuzzy variables. The information contained in a set of rules produces an answer through the "Inference Engine". The production system can be divided into two main structures, that is, the rule editor and the inference engine. The FUZZY-FORTH rule editor is contained in a set of FORTH defining words that help construct the rules. It produces rules and places them in the FORTH dictionary adding them to the available knowledge base, which we call the "Passive Rule Set." Only those rules involved in a specific task are placed in what we call the "Active Rule Set," and will be tested by the inference engine. A rule is defined in the form:

RULE: (name) *IF* (variable) *IS* (fuzzy label) OR fuzzy label) *OR* ...*AND*
(variable) *IS* (fuzzy label) *OR* fuzzy label) *OR* ... *AND*
...
(variable) *IS* (fuzzy label) *OR* ...
THEN (actions)
END-RULE

For the sake of simplicity, the action part of the rule will be assumed to be deterministic, although it could involve all kinds of structures: deterministic, fuzzy, or even conditional (IF-THEN). The action part of a rule can add rules to the active rule set, remove some (through "EXCLUDE"), or even reinitialize the entire active rule set. In this way, the rule-base can effectively modify itself as the system evolves, creating the dynamic behavior necessary to implement the multilevel structure of the control system described here.

Another very important function implemented into FUZZY-FORTH is a class of words called "modifiers," which modify fuzzy sets in various ways. These modifiers have names like "very," "not," and "extremely". When they appear before a fuzzy label, they change the characteristics in a specific way. The modifiers are defined by:

MOD: (name) (operation) ... (operation)
... (operation) ...; MOD
Defined in this way, they may be introduced in a rule like
... (variable) IS (modifier) (fuzzy label OR ...

This structure is designed in such a way that several modifiers may be cascaded before a fuzzy label (for instance, IF variable IS not very big THEN ...). Using all these functions, it is possible to write rules very close to the linguistic form of expressing engineering knowledge, making then easy to write and understand. Examples of these rules are:

RULE: Rule.1 **IF** delta (Ed) **IS** zero **OR** Very Small **AND**
delta (Ic) **IS** Very Big **AND**
delta (Id) **IS** Zero
THEN Alpha large-increment **END-RULE**
RULE: Rule.2 **IF** delta(Ed) **IS** pL **OR** pB **AND**

```
                delta(Id) IS nS OR zero OR pS AND
                delta(Ic) IS zero OR pS
             THEN Alpha big-decrement
                Delta big-increment END-RULE
RULE; Rule.3 IF delta(Ed) IS pM AND
                delta(Id) IS nS OR Zero OR pS AND
                delta(Ic) IS Zero OR pS
             THEN Alpha medium-decrement
                Delta medium-increment END-RULE
```

The value of variables and the membership functions for each rule are left on the stack and FUZZY-FORTH system infers the consequence automatically.

17.6 FINAL REMARKS

Fuzzy set techniques have been extensively used to solve problems in different areas of knowledge. The number of applications never stops to grow up. Some of these areas are:

a) Fuzzy pattern recognition [28,29]—identification processes where a set of data is compared to a known structure. Usually, statistical approaches have been used in this area. Fuzzy techniques are used to identify and extract data feature and to characterize feature selection.

b) Fuzzy classification [30–33]—classification processes where the feature space is partitioned (clustering) using fuzzy metrics, such as fuzzy distances, fuzzy similarities, or fuzzy equivalence relations.

c) Fuzzy decision making [34–36]—decision processes where a problem with multi-objectives must be computed to produce a single answer or a ranking of alternatives. Usually, fuzzy techniques provide support to express statements, by fuzzy statements or by fuzzy numbers. Fuzzy operations and fuzzy relations are also continuously used.

d) Fuzzy control [4, 16, 37–41]—a general structure of a fuzzy controller is shown in Figure 17.9. Fuzzy control has been extensively used in many ordinary applications such as autofocus for cameras, image stabilization for camcorders, temperature control for air-conditioning, image quality control of TV set, home heating system, and so on.

e) Fuzzy optimization [26, 27, 42]—to find the global better solution under constraints. The objective function and the constraints can have a fuzzy meaning. Sometimes, the constraints can also incorporate fuzzy linguistic values.

f) Fuzzy hardware [43, 44]—design issues for digital fuzzy logic control circuits. Digital fuzzy logic controller implementations are the most commercially successful in the market. The main advantages of the fuzzy hardware include flexibility, programmability, and compatibility with other digital systems.

Finally, the fuzzy set techniques, in special fuzzy logic, have been incorporated with other techniques, creating hybrid systems. Some examples of these hybrid systems are fuzzy sets and neural networks [45–47], genetic algorithms [48–50], expert systems [51,52], and rough sets [53,54].

ACKNOWLEDGMENTS

The authors would like to thank FAPEMIG, Minas Gerais state research funding agency, CNPq, a Brazilian research funding agency, and CAPES, in the form of research scholarships, for their financial support.

REFERENCES

[1] Zadeh, L. A. Fuzzy sets. *Information and Control*, vol. 8, pp. 338–353, 1965.
[2] McNeill, D., and Freiberger, P. *Fuzzy Logic*. Simon & Schuster, 1993.
[3] Marks II, R. J. *Fuzzy Logic Technology and Applications*. IEEE Press, 1994.
[4] Quail, S., and Adnan, S. State of the art in household appliances using fuzzy logic. *IEEE Proceedings on the 2nd International Workshop on Industrial Fuzzy Control and Intelligent Systems*, IEEE Press, pp. 204–213, 1993.
[5] Kaufmann, A. *Introduction to the Theory of Fuzzy Subsets*. Academic Press, 1975.
[6] Kruse, R., Gebhardt, J. E., and Klawonn, F. *Foundations of Fuzzy Systems*. John Wiley & Sons, 1994.
[7] Berkan, R. C., and Trubatch, S. L. *Fuzzy Systems Design Principles: Building Fuzzy If-Then Rule Bases*. IEEE Press, 1997.
[8] Wolkenhauer, O. *Data Engineering: Fuzzy Mathematics in Systems Theory and Data Analysis*. John Wiley & Sons, 2001.
[9] Lambert-Torres, G., and Mukhedkar, D. On-line fuzzy rule set generator. In: *Advanced Information Processing in Automatic Control*, R. Husson, Ed., Pergamon Press, pp. 113–118, 1990.
[10] Zimmerman, H. J. *Fuzzy Set Theory*. Kluwer Academic, 1990.
[11] Sugeno, M., and Takagi, T. Multi-dimensional fuzzy reasoning. *Fuzzy Sets and Systems*, vol. 9–2, pp. 313–325, 1983.
[12] Ross, T. *Fuzzy Logic with Engineering Applications*. John Wiley & Sons, 2004.
[13] Mamdani, E. H. Application of fuzzy algorithms for control of simple dynamic plant. *Proceedings of IEEE*, vol. 121, no. 12, pp. 1585–1588, 1974.
[14] Ying, H. *Fuzzy Control and Modeling: Analytical Foundations and Applications*. John Wiley & Sons, 2000.
[15] Jantzen, J. *Foundations of Fuzzy Control*. John Wiley & Sons, 2007.
[16] Lee, C. C. Fuzzy logic in control systems: fuzzy controllers – part I and II. *IEEE Transactions on Systems, Man and Cybernetics*, vol. 20, no. 2, pp. 404–435, 1990.
[17] Takagi, T., and Sugeno, M. Fuzzy identification of systems and its applications to modelling and control. *IEEE Transactions on Systems, Man and Cybernetics*, vol. 15, no. 1, pp. 116–132, 1985.
[18] Goguen, J. A. L-fuzzy sets. *Journal of Mathematical Analysis and Applications*, vol. 18, pp. 145–174, 1967.
[19] Takagi, T., and Hayashi, I. NN-driven fuzzy reasoning. *International Journal of Approximate Reasoning*, vol. 5, no. 3, pp. 191–212, 1991.
[20] Hayashi, Y., Buckley, J., and Czogala, E. Fuzzy neural network with fuzzy signals and weights. *International Journal on Intelligent Systems*, vol. 8, pp. 527–537, 1993.
[21] El-Hawary, M. E. *Electric Power Applications of Fuzzy Systems*. IEEE Press, 1998.
[22] Lambert-Torres, G., Abe, J. M., Mucheroni, M. L., and Cruvinel, P. E. *Advances in Intelligent Systems and Robotics*. IOS Press, 2003.

[23] Oliveira, L. E., Borges da Silva, L. E., Silva, V. F., Lambert-Torres, G., and Bonaldi, E. Active power filter for harmonic current compensation using the Id-Iq control method without PLL and with the best transient time. *WSEAS Transactions on Systems*, vol. 2, no. 2, pp. 328–333, 2003.
[24] Dubois, D., and Prade, H. *Fuzzy Sets and Systems: Theory and Applications*. Academic Press, 1980.
[25] Borges da Silva, L. E., Lambert-Torres, G., Silva, V. F., Nakashima, K., April, G. E., and Olivier, G. Adaptive fuzzy techniques for slip-recovery drive control. In: *Fuzzy Logic Technology and Applications*, J. M. Robert Mark II, Ed., IEEE Press, pp. 196–203, 1994.
[26] Kosko, B. *Fuzzy Thinking*. Hyperion Press, 1993.
[27] Klir, G., and Folger, T. *Fuzzy Sets, Uncertainty, And Information*. Prentice Hall, 1988.
[28] Bezdek, J. *Pattern Recognition with Fuzzy Objective Function Algorithms*. Plenum Press, 1981.
[29] Duda, R., and Hart, R. *Pattern Classification and Scene Analysis*. John Wiley & Sons, 1973.
[30] Bezdek, J., and Harris, J. Fuzzy partitions and relations: an axiomatic basis for clustering. *Fuzzy Sets and Systems*, vol. 1, pp. 82–92, 1978.
[31] Zadeh, L. Similarity relations and fuzzy orderings. *Information Sciences*, vol. 3, pp. 177–200, 1971.
[32] Tamura, S., Higuchi, S., and Tanaka, K. Pattern classification based on fuzzy relations. *IEEE Transactions on Systems, Man and Cybernetics*, vol. 1, no. 1, pp. 61–66, 1971.
[33] Ruspini, E. Numerical methods for fuzzy clustering. *Information Sciences*, vol. 2, pp. 319–350, 1970.
[34] Sakawa, M. *Fuzzy Sets and Interactive Multiobjective Optimization*. Plenum Press, 1993.
[35] Terano, T., Asai, K., and Sugeno, M. *Fuzzy System Theory and its Applications*. Academic Press, 1992.
[36] Watson, S., Weiss, J., and Donnell, M. Fuzzy decision analysis. *IEEE Transactions on Systems, Man and Cybernetics*, vol. 9, no. 1, pp. 1–9, 1979.
[37] Cox, E. Adaptive fuzzy systems. *IEEE Spectrum*, vol. 30, no. 2, pp. 27–31, February 1993.
[38] Egusa, Y., Akahori, H., Morimura, A., and Wakami, N. An electronic video camera image stabilizer operated on fuzzy theory. *IEEE International Conference on Fuzzy Systems*, pp. 851–858, 1992.
[39] Mamdani, E. H., and Gaines, R. R. *Fuzzy Reasoning and its Applications*. Academic Press, 1981.
[40] Sugeno, M. *Industrial Application of Fuzzy Control*. North-Holland, 1985.
[41] Höppner, F., Klawonn, F., Kruse, R., and Runkler, T. *Fuzzy Cluster Analysis: Methods for Classification, Data Analysis and Image Recognition*. John Wiley & Sons, 1999.
[42] Sakawa, M. *Fuzzy Sets and Interactive Multiobjective Optimization*. Plenum Press, 1993.
[43] Watanabe, H., Dettloff, W., and Yount, K. A VLSI fuzzy logic controller with reconfigurable, cascadable architecture. *IEEE Journal of Solid-State Circuits*, vol. 25, no. 2, pp. 376–382, 1990.
[44] Patyra, M. J., and Mlynek, D. J. *Fuzzy Logic: Implementation and Applications*. John Wiley & Sons, 1996.
[45] Tsoukalas, L. H., Uhrig, R. E., and Zadeh, L. A. *Fuzzy and Neural Approaches in Engineering*. John Wiley & Sons, 1997.
[46] Kartalopoulos, S. V. *Understanding Neural Networks and Fuzzy Logic: Basic Concepts and Applications*. John Wiley & Sons, 1995.
[47] Badiru, A. B., and Cheung, J. *Fuzzy Engineering Expert Systems with Neural Network Applications*. John Wiley & Sons, 2002.
[48] Sait, S. M., and Youssef, H. *Iterative Computer Algorithms with Applications in Engineering: Solving Combinatorial Optimization Problems*. John Wiley & Sons, 2000.
[49] Adeli, H., and Sarma, K. C. *Cost Optimization of Structures: Fuzzy Logic, Genetic Algorithms, and Parallel Computing*. John Wiley & Sons, 2006.
[50] Deboeck, G. J. *Trading on the Edge: Neural, Genetic, and Fuzzy Systems for Chaotic Financial Markets*. John Wiley & Sons, 1994.
[51] Schneider, M., Langholz, G., Kandel, A., and Chew, G. *Fuzzy Expert System Tools*. John Wiley & Sons, 1996.
[52] Siler, W., and Buckley, J. J. *Fuzzy Expert Systems and Fuzzy Reasoning*. John Wiley & Sons, 2004.
[53] Jensen, R., and Shen, Q. *Computational Intelligence and Feature Selection: Rough and Fuzzy Approaches*. John Wiley & Sons, 2008.
[54] Pedrycz, W., Skowron, A., and Kreinovich, V. *Handbook of Granular Computing*. John Wiley & Sons, 2008.

CHAPTER 18

DECISION TREES

Constantin Bulac and Adrian Bulac

18.1 INTRODUCTION

To make computer do more than what they are programmed to do, to make them acquire knowledge that is not programmed in to them, to make computer reason with the knowledge to produce new knowledge, to teach computer to respond to many situations, and be able to learn how to respond to new situations without supervision. This seems like an impossible task, as everyone knows that computers only do what they are programmed to do. When a computer runs a program, if the program is correct then we would expect the program to be correct every time it is run and to come up with the same results given the same data. In this sense, computers are predictable, unlike human beings who are capable of acquiring knowledge from all around, reasoning with the knowledge to make new knowledge, and also capable of coming up with knowledge that does not seem to come from anywhere.

Therefore, *automatic learning* (*AL*) is the most provocative domain of artificial intelligence and at the same time, the hardest to be implemented in automatic systems. The goal of the AL techniques is to reproduce the processes of human thinking and to construct rules reproducing the thinking of human experts. Unlike the expert systems, the data base of AL systems is automatically modified without the intervention of knowledge engineer.

Automatic learning aims at extracting information in various forms from DBs. The three paradigms of AL are [1]:

- *Statistical methods* are generally characterized by having an explicit underlying probability model, which provides a probability of being in each class rather than simply a classification. In addition, it is usually assumed that the techniques will be used by statisticians, and hence some human intervention is assumed with regard to variable selection and transformation, and overall structuring of the problem.
- *Machine learning (ML)* is a subfield of symbolic artificial intelligence concerned with learning. As most research is artificial intelligence, ML tries to understand the learning capabilities of human beings and to construct algorithms capable to reproduce the human thinking. There are several

Advanced Solutions in Power Systems: HVDC, FACTS, and Artificial Intelligence, First Edition.
Edited by Mircea Eremia, Chen-Ching Liu, and Abdel-Aty Edris.

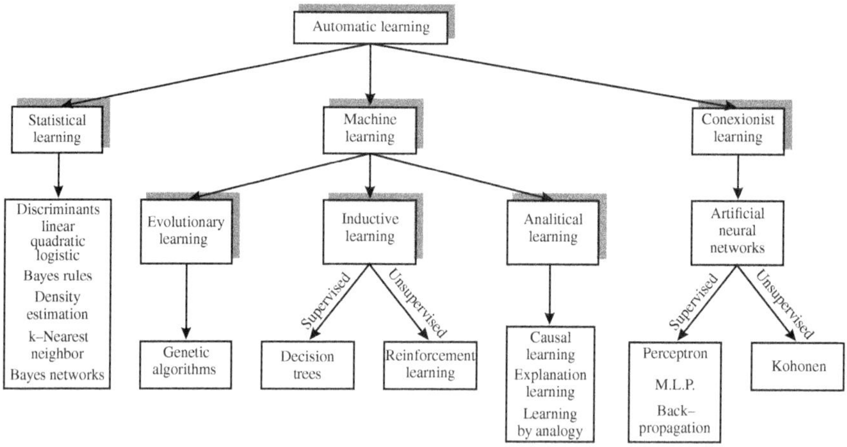

Figure 18.1 Automatic learning techniques.

subareas of ML, such as learning by analogy, concept formation, discovery by experimentation, explanation-based learning, and concept learning.

- *Artificial neural networks* (*ANNs*) are computational systems whose architecture and operation are inspired from our knowledge about biological neural cells (neurons) in the brain. ANNs can be described either as mathematical and computational models for nonlinear function approximation, data classification, clustering, and nonparametric regression or as simulations of the behavior of collections of model biological neurons.

A detailed classification of AL methods is depicted in Figure 18.1.

18.2 DECISION TREES

Decision trees, either classification or regression trees, are especially attractive type of models for three main reasons:

- They have an intuitive representation, the resulting model is easy to understand and assimilate by humans.
- The decision trees are nonparametric models, no intervention being required from the user, and thus they are very suited for exploratory knowledge discovery.
- Performance degrades gracefully with the increase in the size of training data.

Formally, one can define a *decision tree* to be either:

- a leaf node (or *answer node*) that contains a class name, or;
- a nonleaf node (or *decision node*) that contains an attribute test with a branch to another decision tree for each possible value of the attribute.

A decision tree can be seen as a divide-and-conquer strategy for object classification.

18.2.1 Decision Tree Construction

To construct a decision tree from a set *TS* of training cases, let the classes be denoted $\{C_1, C_2, ..., C_k\}$. There are three possibilities:

- The current training set (*TS*) contains one or more cases, all belonging to a single class C_j. The decision tree is a leaf identifying class C_j.
- *TS* contains no cases: The decision tree is again a leaf, but the class to be associated with the leaf must be determined from information other than *TS*. For example, the leaf might be chosen in accordance with some background knowledge of the domain, such as the overall majority class.
- *TS* contains cases that belong to a mixture of classes: In this situation, the idea is to refine *TS* into subsets of cases that are, or seem to be heading toward, single-class collection of cases. A test is chosen based on a single attribute, which has two or more mutually exclusive outcomes. *TS* is partitioned into subsets TS_1, ..., TS_p where TS_i contains all the cases in *TS* that have outcome *i* of the chosen test. The decision tree for *TS* consists of a decision node identifying the test and one branch for each possible outcome. The same tree-building machinery is applied recursively to each subset of training cases, so that the ith branch leads to the decision tree constructed from the subset TS_i of training cases.

The best-known method of decision trees generation is the top-down induction of decision trees (TDIDT) algorithm. The procedure TDIDT is a greedy algorithm, the implicit aim of the iterative search being to generate a tree as simple as possible, which contains maximum of information on the training examples classification variables. An implementation of the up-down induction of the decision trees (TDIDT) method is the ID3 algorithm presented below.

ID3 (Examples, Classes, Attributes)
Creates the node root

If all the examples belong to the same class it labels the root with the name of the respective class.

Otherwise
Beginning

Computes the informative gain for each attribute, using the entropy calculation formula
Selects attribute *A* by means of the lowest medium entropy (the biggest informative gain) and this becomes the test associated with the root
For each possible value v of this attribute

It attaches a new branch to the root corresponding to $A = v$

Let be Examples (v) all those examples for which $A = v$

If Examples (v) is void, the ramification becomes a leaf labeled with the class corresponding to all the examples from the set Examples.

Else the Branch created will be ID3 (Examples (v), Classes, Attributes \A)

End
End Algorithm

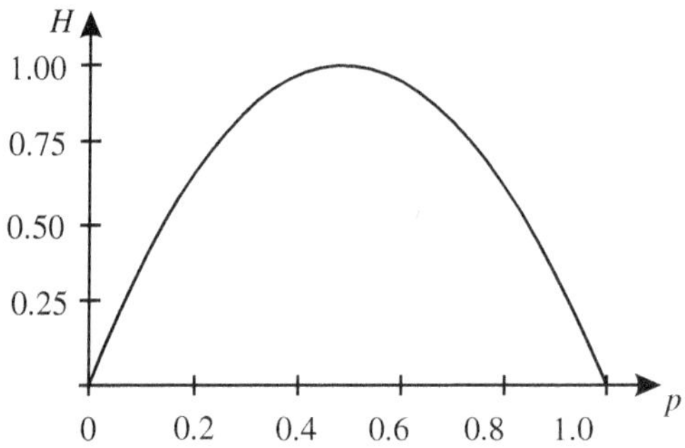

Figure 18.2 The entropy associated with a given training set.

The method of selecting the attribute by means of which the partitioning of the driving set is achieved is very important and sometimes difficult enough to be automated, the order of the attribute consideration influencing the size and the performance of the resulting decision tree.

In order to define information gain precisely, we need to define a measure commonly used in information theory, called entropy, which characterizes the (im)purity of an arbitrary collection of examples.

Given a set S, containing r possible answers to a question and the associated probability for an answer is p_i $(i = 1, 2, \ldots, r)$, the entropy of set S relative to this classification is defined as:

$$H = -\sum_{i=1}^{r} p_i \log p_i \tag{18.1}$$

In all calculations involving entropy, we define 0log0 to be 0. Note that the entropy is 0 if all members of S belong to the same class. Note the entropy is 1 (at its maximum) when the collection contains an equal number of examples from each class. If the collection contains unequal number of examples, the entropy is between 0 and 1.

During the generation process, if we denote by n_i the number of driving examples belonging to class c_i and by n the number of examples to be classified during this step, the classification entropy is defined by formula [2]:

$$H_C \stackrel{\Delta}{=} -\sum_{i=1}^{m} \frac{n_i}{n} \log_2 \frac{n_i}{n} \tag{18.2}$$

The classification entropy represents the degree of impurity of the classes in the driving set. Denoting by n_i^j the number of examples belonging to class c_i and to the result j of the test, and by c^j the overall number of examples that belong to the result j of the test, the entropy test is defined by [2]:

$$H_{C|T} \stackrel{\Delta}{=} -\sum_{j=1}^{p} \frac{n^j}{n} \sum_{i=1}^{m} \frac{n_i^j}{n^j} \log_2 \frac{n_i^j}{n^j} \tag{18.3}$$

The test informative quantity is calculated as [2]:

$$I_C^T \stackrel{\Delta}{=} H_C - H_{C|T} \tag{18.4}$$

The test resultant entropy is defined as [2]:

$$H_T \stackrel{\Delta}{=} -\sum_{j=1}^{p} \frac{n^j}{n} \log_2 \frac{n^j}{n} \tag{18.5}$$

while the Gini impurity measure is [3]:

$$G = 1 - \frac{1}{n} \sum_{j=1}^{p} \sum_{i=1}^{m} \frac{(n_i^j)^2}{n^j} \tag{18.6}$$

There are several methods for the attribute selection. Thus, within algorithm Classification and Regression Trees (CART) [3], the selected test is that which has the lowest Gini impurity measure. Within the method ID3 [2] when the referred gain criterion [4] is used, the test which has the maximum proportion between the informative quantity and the entropy of the test result is selected. The disadvantage of the utilization of the gain or referred gain criterion resides in the generated decision trees being out-of-balance. This disadvantage disappears if method ULg is used, in which case the gain of a test is defined as [1]:

$$gain(T) \stackrel{\Delta}{=} \frac{2 \times I_C^T}{H_C + H_T} \tag{18.7}$$

The logarithm base utilized within the entropy calculation has been considered base two, without any pertinent theoretical explanation. An explanation could be that base two was used because of the binary divisions of the sets. As, generally, an attribute can trigger a p-air partition of the examples set (in p subsets), we consider the logarithm base to be p, for $p \geq 2$. This base must be used for a new attribute selection criteria [5]:

$$normalized\_gain(A, S, n) \stackrel{\Delta}{=} \left(-\sum_{j=1}^{k} p_j \log_n p_j \right) - \left(-\sum_{i=1}^{n} p_i \sum_{j=1}^{k} p_{ij} \log_n p_{ij} \right) \tag{18.8}$$

where

$$p_j = \frac{n^j}{n} \quad \text{and} \quad p_{ij} = \frac{n_i^j}{n^j}$$

The attribute selection criteria disregard the test complexity and cost, all being considered to be performed during the same period and to require the same memory resources. If the cost of a test is $C(T)$, then the test with the largest gain is selected, the latter being calculated as follows:

- method EG2 [6]:

$$\frac{2^{I_C^T} - 1}{C(T) + 1} \tag{18.9}$$

- method CS-ID3 [7]:

$$\frac{(I_C^T)^2}{C(T)} \tag{18.10}$$

- method IDX [8]:

$$\frac{I_C^T}{C(T)} \tag{18.11}$$

The binary splitting rule (BSR) had been proposed by Breiman and can be applied only for binary tests [3]:

$$BSR = \frac{n_s}{n}\frac{n_d}{n}\left(\sum_{i=1}^{k}\left|\frac{n_{si}}{n_s} - \frac{n_{di}}{n_d}\right|\right)^2 \tag{18.12}$$

where

n_s is the number of positive answers.
n_d – the number of negative answers.
n_{si} – the number of positive answers belonging to class C_i.
n_{di} – the number of negative answers belonging to class C_i.
k – the number of classes.

The test with maximum (BSR) value is selected for splitting. The binary splitting rule offers better results than entropy [9].

The impurity measures max minority and sum minority are more recently introduced in the context of decision trees.

We define the maximum minority of *j* result

$$Minority(j) = \sum_{i=1, i \neq \max n_i^j}^{p} n_i^j \tag{18.13}$$

and the maximum minority is [10]:

$$MaxMinority = \max_{j=1}^{p} Minority(j) \tag{18.14}$$

The splitting test with maximum value of *MaxMinority* function is selected to be the best test for splitting.

The sum minority is similar with max minority, but the value used to select the best splitting is [11]:

$$SumMinority = \sum_{j=1}^{p} Minority(j) \tag{18.15}$$

18.2.2 Decision Tree Pruning

A third problem, which Breiman et al. [3] deem the most important, concerns the determination of the leaves. There are two different ways to cope with this: either by prospectively deciding when to stop the growth of a tree (*prepruning*) or by retrospectively reducing the size of a fully expanded tree, $T_{\max}$, by pruning some branches (*postpruning*) [12]. *Prepruning* methods establish stopping rules for preventing the growth of those branches that do not seem to improve the predictive accuracy of the tree. Some rules are [13]:

- All observations reaching a node belong to the same class.
- All observations reaching a node have the same feature vector (but do not necessarily belong to the same class).
- The number of observations in a node is less than a certain threshold.
- There is no rejection for chi-square tests on the independence between a feature Fj and the class attribute C.
- The merit attributed to all possible tests, which partition the set of observations in the node, is too low.

In general, pruning methods aim to simplify those decision trees that overfitted the data. Their beneficial effects have attracted the attention of many researchers, who have proposed a number of methods. Unfortunately, their variety does not encourage the comprehension of both the common and the individual aspects. For instance, while some methods proceed from the root toward the leaves of $T_{\max}$ when they examine the branches to prune (*top-down approach*), other methods follow the opposite direction, starting the analysis from the leaves and finishing it with the root node (*bottom-up approach*). Furthermore, while some methods only use the *TS* to evaluate the accuracy of a decision tree, others exploit an additional *pruning set*, sometimes improperly called test set, which provides less biased estimates of the predictive accuracy of a pruned tree.

Henceforth, I_A will denote the set of *internal* (nonterminal) nodes of a decision tree A, F_A will denote the set of *leaves* in A, and N_A denotes the set of nodes of A. Thus, $N_A = I_A \cup F_A$. The *branch* of A containing a node t and all its descendants will be indicated as A_t. The number of leaves of A will be denoted by the cardinality of F_A, $|F_A|$. The number of training examples of class C_i reaching a node t will be denoted by $n_i(t)$, the total number of examples in t by $n(t)$, and the number of examples not belonging to the majority class by $e(t)$.

In the following subsections, six pruning methods are presented.

18.2.2.1 Reduced Error Pruning

This method is conceptually the simplest and uses the pruning set to evaluate the efficacy of a subtree of $A_{\max}$. It starts with the complete tree $A_{\max}$ and, for each internal node t of $A_{\max}$, it compares the number of classification errors made on the pruning set when the subtree A_t is kept, with the number of classification errors made when t is turned into a leaf and associated with the *best* class. Sometimes, the simplified tree has a better performance than the original one. In this case, it is advisable to prune A_t. This branch pruning operation is repeated on the simplified tree until further pruning increases the misclassification rate.

Quinlan restricts the pruning condition given above with another constraint: A_t can be pruned only if it contains no subtree that results in a lower error rate than A_t itself [12]. This means that nodes to be pruned are examined according to a bottom-up traversal strategy.

The reduced error pruning method finds the smallest version of the most accurate subtree with respect to the pruning set.

Another positive property of this method is its linear computational complexity, since each node is visited only once to evaluate the opportunity of pruning it. On the other hand, a problem with reduced error pruning is its biasness toward overpruning. This is due to the fact that all evidence encapsulated in the TS and exploited to build $A_{\max}$ is neglected during the pruning process. This problem is particularly noticeable when the pruning set is much smaller than the TS, but becomes less relevant as the percentage of cases in the pruning set increases.

18.2.2.2 Pessimistic Error Pruning

This pruning method is characterized by the fact that the same TS is used for both growing and pruning a tree [12]. Obviously, the *apparent* error rate, which is the error rate on the TS, is optimistically biased and cannot be used to choose the best pruned tree. For this reason, Quinlan introduces the continuity correction for the binomial distribution that might provide "a more realistic error rate."

More precisely, let

$$r(t) = \frac{e(t)}{n(t)} \tag{18.16}$$

be the apparent error rate in a single node t when the node is pruned, and

$$r(A_t) = \frac{\sum_{s \in F_{A_t}} e(s)}{\sum_{s \in F_{A_t}} n(s)} \tag{18.17}$$

be the apparent error rate for the whole subtree A_t. Then, the continuity correction for the binomial distribution gives:

$$r'(t) = \frac{e(t) + 1/2}{n(t)} \tag{18.18}$$

By extending the application of the continuity correction to the estimation of the error rate of A_t, we have:

$$r'(t) = \frac{\sum_{s \in F_{A_t}} [e(s) + 1/2]}{\sum_{s \in F_{A_t}} n(s)} = \frac{\sum_{s \in F_{A_t}} e(s) + \frac{|F_{A_t}|}{2}}{\sum_{s \in F_{A_t}} n(s)} \tag{18.19}$$

For simplicity, henceforth, we will refer to the number of errors rather than to the error rate, that is:

$$e'(t) = [e(t) + 1/2] \tag{18.20}$$

for a node t, and:

$$e'(T_t) = \sum_{s \in F_{A_t}} e(s) + \frac{|F_{A_t}|}{2} \tag{18.21}$$

for a subtree A_t.

The subtree A_t makes less errors on the TS than the node t when t becomes a leaf, but sometimes it may happen that $n'(t) \leq n'(A_t)$ due to the continuity correction, in which case the node t is pruned.

The algorithm evaluates each node starting from the root of the tree and, if a branch A_t is pruned then the descendants of t are not examined. This top-down approach gives the pruning technique a high run speed.

18.2.2.3 Minimum Error Pruning

For a k-class problem, the expected probability that an observation reaching the node t belongs to the ith class is the following [14]:

$$p_i(t) = \frac{n_i(t) + p_{ai} \cdot m}{n(t) + m} \tag{18.22}$$

where p_{ai} is the a priori probability of the ith class, and m is a parameter that determines the impact of the a priori probability on the estimation of the a posteriori probability $p_i(t)$. For simplicity, m is assumed to be equal for all the classes. When a new observation reaching t is classified, the expected error rate is given by:

$$EER(t) = \min_i\{1 - p_i(t)\} = \min_i\left\{\frac{n(t) - n_i(t) + m(1 - p_{ai})}{n(t) + m}\right\} \tag{18.23}$$

When the probability distribution is uniform and equal for all classes ($m = k$ and $p_{ai} = \frac{1}{k}$ for $i = 1, \ldots, k$), we get:

$$EER(t) = \min_i\left\{\frac{n(t) - n_i(t) + k - 1}{n(t) + k}\right\} = \frac{e(t) + k - 1}{n(t) + k} \tag{18.24}$$

In the minimum error pruning method, the expected error rate for each internal node $t \in I_A$ is computed. This is called *static error*, ES(t).

Then, the expected error rate of A_t, called *dynamic* (or *backed-up*) *error*, ED(t), is computed as a weighted sum of the expected error rates of the tth children:

$$ED(t) = \sum_{i=1}^{n_f} ps(t_i)ES(t_i) \tag{18.25}$$

where

n_f is the number of successor for node t.

t_i – the successor of node t.

ps – the probability that an example belongs to t_i.

Generally, the higher the m, the more severe the pruning.

18.2.2.4 Critical Value Pruning

This postpruning method is very similar to a prepruning technique. Indeed, a threshold, named *critical value,* is set for the node selection measure [15]. Then, an internal node of the tree is pruned if the value obtained by the selection measure for each test associated with edges coming out of that node does not exceed the critical value.

Nevertheless, it may happen that the pruning condition is met by a node t but not by all its children. In this case, the branch A_t is kept because it contains relevant nodes. This further check is typical of a bottom-up method and represents the main difference from those prepruning methods that prevent a tree from growing even if subsequent tests might turn out to be meaningful.

The degree of pruning clearly changes with the critical value: the choice of a higher critical value results in a more drastic pruning. The method consists of two main steps:

- Prune A_{max} for increasing critical values.
- Choose the best tree among the sequence of pruned trees, by measuring the significance of the tree as a whole and its predictive ability.

If this method is used to prune a decision tree constructed using gain ratio method, the possible pruned nodes are the root and the leaves.

18.2.2.5 Cost-Complexity Pruning

This method is also known as the CART pruning algorithm [3]. It consists of two steps:

1. Selection of a parametric family of subtrees of A_{max}, $\{A_0, A_1, \dots, A_F\}$, according to some heuristics.
2. Choice of the best tree A_i according to an estimate of the true error rates of the trees in the parametric family.

As regards the first step, the basic idea is that A_{i+1} is obtained from A_i by pruning those branches that show the lowest increase in apparent error rate per pruned leaf. Indeed, when a tree A is pruned in a node t, its apparent error rate increases by the amount $r(t) - r(A_t)$, while its number of leaves decreases by $|F_{A_t}| - 1$ units. Thus, the following ratio

$$\alpha = \frac{r(t) - r(A_t)}{|F_{A_t}| - 1} \tag{18.26}$$

measures the increase in apparent error rate per pruned leaf. Then, A_{i+l} in the parametric family is obtained by pruning all nodes in A_i with the lowest value of α. The first tree A_0 is obtained by pruning A_{max} (cutting all braches having a value equal with 0), while the last tree A_F is the root tree. It is possible to prove that each tree A_i is characterized by a distinct value α_i, such that $\alpha_i < \alpha_{i+1}$. Therefore, the set $\{A_0, A_1, \dots, A_F\}$ is actually a *parametric family* of trees that we will denote as $A_{max}(\alpha)$. The parametric family can be built in a time that is quadratic in the number of internal nodes.

In the second phase, the best tree in $A_{max}(\alpha)$ with respect to predictive accuracy is chosen.

In this former proposal, the TS *MA* used to build A_{max} is partitioned into v subsets MA_1, MA_2, ..., MA_v and then v auxiliary decision trees A_1, A_2,..., A_v are induced from the TSs. Following the same approach as before, it is possible to define v distinct parametric families $A_1(\alpha)$, $A_2(\alpha)$,..., $A_v(\alpha)$, which can help to define the accuracy of the decision trees in $A_{max}(\alpha)$. More precisely, the error rate of $A_{max}(\alpha)$ is estimated as the average of the error rates of the trees $A_1(\sqrt{\alpha_i\alpha_{i+1}})$, $A_2(\sqrt{\alpha_i\alpha_{i+1}})$,..., $A_v(\sqrt{\alpha_i\alpha_{i+1}})$.

When an independent pruning set is used, the cost-complexity pruning method is at a disadvantage with respect to the reduced error pruning method because it can only choose a tree in the set $\{A_0, A_1, \dots, A_F\}$ instead of the set of all possible subtrees of $A_{\max}$ [16].

It is demonstrated that the standard deviation of the empirical error count estimator e_C, used with independent sets, is given by [17]:

$$e_C = \sqrt{\frac{e(1-e)}{N}} \tag{18.27}$$

where

e is the true expected error rate of the classifier (in this case, $A_{\max}$).

N – size of the independent set used for computing the error rate estimate.

18.2.2.6 Error-Based Pruning

This is the pruning method implemented in C4.5, the learning system that we employed in our experiments for building the trees [4]. It is considered an improvement on the pessimistic error pruning method, since it is based on a far more pessimistic estimate of the expected error rate. Both use information in the TS for building and simplifying trees.

Unlike pessimistic error pruning, this method visits the nodes of $A_{\max}$ according to a bottom-up postorder traversal strategy instead of a top-down strategy. The true novelty is that error-based pruning simplifies a decision tree A by *grafting* a branch A_t onto the place of the parent of t itself, in addition to pruning nodes.

The sum of the predicted error rates of all the leaves in a branch A_t is considered to be an estimate of the error rate of the branch itself. Thus, by comparing the predicted error rate for t with that of the branch A_t and of the largest subbranch $A_{t'}$ rooted in a child t' of t, we can decide whether it is convenient to prune A_t, to graft $A_{t'}$ onto the place of t or to keep A_t.

18.3 OBLIQUE DECISION TREES

In the previous section were described methods for construction of binary decision trees. For these trees, the border between two neighboring regions of different classes is known as a decision boundary. Since the test condition involves only a single attribute, the decision boundaries are rectilinear (i.e., parallel to the coordinate axes). This limits the expressiveness of the decision tree representation for modeling complex relationships among continuous attributes.

Figure 18.3 illustrates a data set that cannot be classified effectively by a binary decision tree. An oblique decision tree can be used to overcome this limitation because it allows test conditions that involve more than one attribute.

For two-class learning tasks, a multivariate test can be represented by a linear threshold unit (LTU). For multi-class tasks, two possible representations are a LTU or a linear machine (LM). [18, 19]

An LTU is a binary test of the form:

$$W^T X \geq 0 \tag{18.28}$$

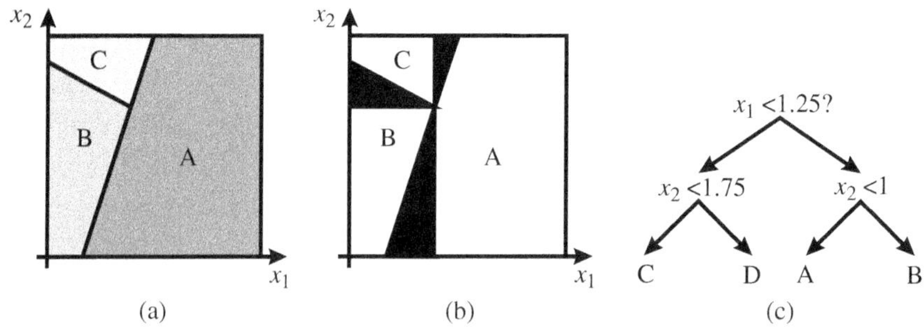

Figure 18.3 Example of data set that cannot be partitioned optimally using conditions involving single attribute.

where $X = (x_1, x_2, \ldots, x_n, 1)^T$ is an instance description (a pattern vector) consisting of a constant 1 and the n features that describe the instance. W is a vector of $n + 1$ coefficients, also known as weights. If $W^T X \geq 0$, then the LTU infers that X belongs to one class C_1, otherwise the LTU infers that X belongs to the other class C_2.

An LM is a set of R linear discriminant functions that are used collectively to assign an instance to one of the R classes. Let $X = (x_1, x_2, \ldots, x_n, 1)$ be an instance description (a pattern vector) consisting of a constant 1 and the n features that describe the instance. Thus, each discriminant function has the form $g_i(X) = W_i^T X$ where W_i is a vector of $n + 1$ coefficients. An LM instance X belongs to class C_i if and only if $g_i(X) > g_j(X)$, $(\forall) i \neq j$, $1 \leq i, j \leq R$. For the rare case in which $g_i(X) = g_j(X)$, some arbitrary decision is made: one implementation of an LM chooses the smaller of i and j in these cases.

18.3.1 Recursive Least Squares Procedure

The recursive least squares (RLS) algorithm minimizes the mean squared error of the training data. Specifically, after instance k is observed, RLS updates the weight vector W, as follows:

$$W_k = W_{k-1} - K_k \left(X_k^T W_k - y_k \right) \tag{18.29}$$

where $X_k = (x_1, x_2, \ldots, x_n, 1)^T$ is the instance vector, y_k is the value of the independent variable (the class), and $K_k = P_k X_k$ is the weight assigned to the update. P, the error-covariance matrix, is a weighting matrix of size $n \times n$, which over time decreases and smoothes the errors made by the linear combination.

After each instance k is observed, P is updated as follows:

$$P_k = P_{k-1} - P_{k-1} X_k \left[1 + X_k^T P_{k-1} X_k \right]^{-1} X_k^T P_{k-1} \tag{18.30}$$

RLS requires an initial set of weights (W_0) and an initial error-covariance matrix. If little is known about the true W and W_0 is set to zero, then the initial values of P should reflect this uncertainty: the diagonal elements should be set to large values to indicate a large initial error variance and little confidence in the initial estimate of W_0.

18.3.2 The Thermal Training Procedure

The thermal training procedure can be applied to an LTU or an LM. A method for training an LM is to adjust the weights:

$$W = W \pm cX \tag{18.31}$$

where the correction

$$c = \frac{W^T X}{2X^T X} + \varepsilon \tag{18.32}$$

causes the updated LM to classify the instance correctly.

A better correction is [20]:

$$c = \frac{\beta}{\beta + k} \tag{18.33}$$

where β is annealed during training and

$$k = \frac{(W_j - W_i)^T X}{2X^T X} + \varepsilon \tag{18.34}$$

As the amount of error k approaches 0, the correction c approaches 1 regardless of β.

Therefore, to ensure that the LM converges, the amount of correction c is annealed regardless of k. This is achieved by multiplying c by β, because it is already annealing, giving the correction [21]:

$$c = \frac{\beta^2}{\beta + k} \tag{18.35}$$

18.3.3 OC1 Algorithm

The previous methods cannot find the optimal linear discriminants for specific goodness measures, the size of the linear program grows very fast with the number of instances and the number of attributes.

OC1 imposes no restrictions on the orientation of the hyperplanes. However, in order to be at least as powerful as standard decision trees (DT) methods, it first finds the best axis-parallel (univariate) split at a node before looking for an oblique split. OC1 uses an oblique split only when it improves over the best axis-parallel split.

OC1 imposes no restrictions on the orientation of the hyperplanes. However, in order to be at least as powerful as standard DT methods, it first finds the best axis-parallel (univariate) split at a node before looking for an oblique split. OC1 uses an oblique split only when it improves over the best axis-parallel split.

Let n be the number of examples in E, n be the number of attributes (or dimensions) for each example, and k be the number of categories (classes). Then, we can write $E_j = (x_{j1}, x_{j2}, \ldots, x_{jn}, C_j)$ for the jth example from the TS E, where x_{ji} is the value of attribute i and C_j is the category label. The equation of the current hyperplane H at a node of the decision tree is written as:

$$\sum_{i=1}^{n} a_i x_i + a_0 = 0 \tag{18.36}$$

If we substitute a point (an example) E_j into the equation for H, we get

$$\sum_{i=1}^{n} a_i x_{ji} + a_{j0} = v_j \tag{18.37}$$

where the sign of V_j tells us whether the point E_j is above or below the hyperplane H.

If H splits the TS E perfectly, then all points belonging to the same category will have the same sign for V_j., that is, $sign(v_i) = sign(v_j)$ if and only if $C_i = C_j$.

OC1 adjusts the coefficients of H individually, finding a locally optimal value for one coefficient at a time. It works as follows. Treat the coefficient a_q as a variable, and treat all other coefficients as constants. Then, V_j can be viewed as a function of a_q. In particular, the condition that E_j is above H is equivalent to:

$$V_j > 0$$

$$a_q > \frac{a_q x_{jq} - v_j}{x_{jq}} \stackrel{def}{=} u_j \tag{18.38}$$

assuming that $x_{jq} > 0$, which we ensure by normalization.

Perturbation algorithm for a single coefficient a_q is:

For $j = 0, \ldots, n$ compute $u_j = \frac{a_q x_{jq} - v_j}{x_{jq}}$
Sort $u_0, u_1, \ldots, u_n$ in nondecreasing order.
For $j = 0, \ldots, n$ compute the values $a_q^{(j)} = \frac{u_j + u_{j+1}}{2}$
Compute $a_q = \max_j \text{Impurity}(x > a_q^{(j)})$
H_1 = result of substituting a_q with previous value
If Impurity(H_1) < Impurity(H) then H = H_1

Now that we have a method for locally improving a coefficient of a hyperplane, we need to decide which of the d+1 coefficients to pick for perturbation. We experimented with three different methods for choosing which coefficient to adjust, namely, sequential, best first and random.

Sequential:

Repeat until none of the coefficient values is modified in the *For* loop:

For $i = 1$ to n, Perturb($H;i$)

Best:

Repeat until coefficient q remains unmodified (q is a coefficient which, when perturbed, results in the maximum improvement of the impurity measure):

Perturb($H;q$)

Random:
Repeat a fixed number of times (q is a random integer between 1 and $n+1$):
Perturb($H;q$)

18.4 APPLICATIONS OF DECISION TREES IN POWER SYSTEMS

Security assessment is a major topic in planning and operation of electric power systems. It consists of evaluating the ability of the power system to face various contingencies and of proposing appropriate remedial actions able to counter its main weaknesses, whenever deemed necessary.

Contingencies may be external or internal events (e.g., faults subsequent to lightning vs. operator-initiated switching sequences) and may consist of small/slower large/fast disturbances (e.g., random behavior of the demand pattern vs. generator or line tripping).

Random sampling techniques are considered to screen all relevant situations in a given context, while existing numerical simulation tools are exploited—if necessary in parallel—to derive detailed security information. The heart of the framework is provided by machine-learning methods used to extract and synthesize relevant information and to reformulate it in a suitable way for decision making. This consists of transforming the DB of case-by-case numerical simulations into a power system security *knowledge base* (KB). As illustrated in Figure 18.4, a large variety of AL methods may be used here in a toolbox fashion, according to the type of information they may exploit and/or produce.

The final step consists of using the extracted synthetic information (decision trees, rules, statistical, or neural network approximates) either in real-time, for fast and effective decision making, or in the off-line study environment, so as to gain new physical insights and to derive better system and/or operation planning strategies [22].

To use the artificial intelligence techniques for voltage stability assessment must be created a hybrid system that contains: a DB, a KB, an algorithmic module (AM), and an intelligent module (IM). The AM can work in series with the IM or in parallel (Figure 18.5) [23]. The first case corresponds to series working and is used when the intelligent system is a knowledge-based system (expert system). The second situation is used when the intelligent system is implemented by a machine-learning technique (a decision tree or an ANN).

AM completes the KB of heuristic module with information on power system state. Using the ***primary DB*** which contains the user's or directly acquired

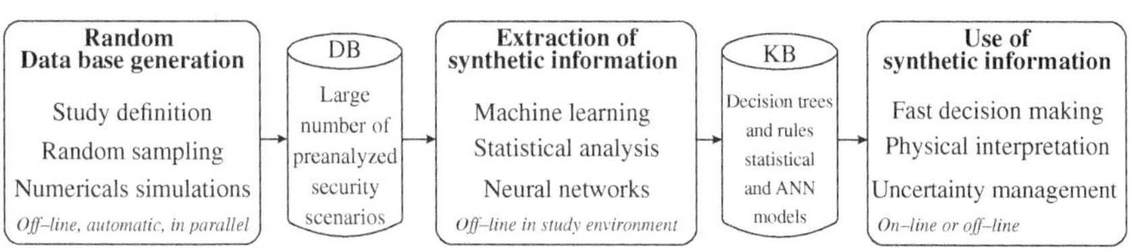

Figure 18.4 Machine learning for security assessment.

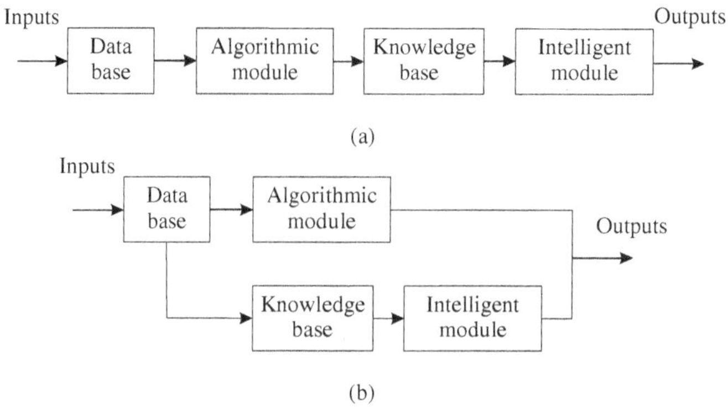

Figure 18.5 The ways for implementing A.I. techniques for voltage stability assessment.

information about busses data and the network topology, the AM calculates the steady-state regime using Newton–Raphson method and afterward computes:

- the local stability indices S_{VQ} and S_{QgQc} and a table of sensitivity factors for each load bus and for the control equipment;
- participation factors of each PQ bus for the V–Q variation modes of the examined system using the modal analysis of reduced Jacobean matrix;
- V–Q control zones (areas) and the weak links.

In order to establish the correlation between reactive power and voltage for each power system, static methods have been used based on mathematical model described by the linear steady-state equations of the power system [24, 25]:

$$\begin{bmatrix} J_{P\theta} & J_{PV} \\ J_{Q\theta} & J_{QV} \end{bmatrix} \begin{bmatrix} \Delta\theta \\ \Delta V \end{bmatrix} = \begin{bmatrix} \frac{\partial P}{\partial \theta} & \frac{\partial P}{\partial V_c} & \frac{\partial P}{\partial V_g} \\ \frac{\partial Q_c}{\partial \theta} & \frac{\partial Q_c}{\partial V_c} & \frac{\partial Q_c}{\partial V_g} \\ \frac{\partial Q_g}{\partial \theta} & \frac{\partial Q_g}{\partial V_c} & \frac{\partial Q_g}{\partial V_g} \end{bmatrix} \begin{bmatrix} \Delta\theta \\ \Delta V_c \\ \Delta V_g \end{bmatrix} = [J'] \begin{bmatrix} \Delta\theta \\ \Delta V_c \\ \Delta V_g \end{bmatrix} = \begin{bmatrix} \Delta P \\ \Delta Q_c \\ \Delta Q_g \end{bmatrix} \tag{18.39}$$

where $[J']$ is the complete Jacobean matrix.

Considering the role of load characteristics, the transformer on-load tap control, and the stator and the rotor power limitations of generators that may trigger the voltage instability and collapse phenomena, the standard model represented by equation (18.39) was modified to include their influence, as well [24].

THE SENSITIVITY FACTORS CALCULATION. To evaluate voltage stability and to select the most efficient control equipment that are needed to solve a voltage problem, the next variables have been employed:

- S_{VQ}—sensitivity factors for a load bus i for a reactive power variation in this bus:

$$S_{VQ} = \frac{\partial V_i}{\partial Q_i} \cong \frac{\Delta V_{c,i}}{\Delta Q_{c,i}} \tag{18.40}$$

- S_{QgQc}—sensitivity factors for generators' injected reactive power for a reactive power variation at a load bus:

$$S_{Q_gQ_c} = \partial\left(\sum_{k\in n_g} Q_{g,k}\right)\Big/\partial Q_{c,i} \cong \sum_{k\in n_g} \Delta Q_{g,k}\Big/\Delta Q_{c,i} \tag{18.41}$$

- $[H]$—sensitivity matrix describing the correlation between the $[\Delta V_c]$ and $[\Delta Q_g]$ variations of the PQ bus voltage and generator-injected reactive power, respectively, and the variations of the control equipment (the variation of generator bus voltage $[\Delta V_g]$, the variation of condenser-injected reactive power or the variation of coils reactive power consumption $[\Delta Q_{inj}]$, and the variation of transformer ratios $[\Delta N_{ij}]$):

$$\begin{bmatrix} \Delta V_c \\ \Delta Q_g \end{bmatrix} = \begin{bmatrix} H_{VV} & H_{VQ} & H_{VN} \\ H_{QV} & H_{QQ} & H_{QN} \end{bmatrix} \begin{bmatrix} \Delta V_g \\ \Delta Q_{inj} \\ \Delta N_{ij} \end{bmatrix} = [H] \begin{bmatrix} \Delta V_g \\ \Delta Q_{inj} \\ \Delta N_{ij} \end{bmatrix} \tag{18.42}$$

These are calculated using a numerical derivation procedure based on the linear model defined by equation (18.39) [24].

The Participation Factors Calculation. Considering the power system strong correlation between voltage and reactive power, the V–Q dependence is described by the equation $[\Delta Q] = [J_R][\Delta V]$, where $[J_R]$ is the reduced Jacobean matrix obtained from equation (18.39) by eliminating the $[\Delta\theta]$ variations.

Applying the modal analysis for the $[J_R]$, one obtains [24]

$$S_{VQ} = \frac{\partial V_i}{\partial Q_i} = \sum_k \frac{R_{ki}L_{ik}}{\lambda_k} = \sum_k \frac{P_{ki}}{\lambda_k} \tag{18.43}$$

where R_{ki} and L_{ik} are the elements of the right eigenvector and the left eigenvector, respectively, which correspond to the eigenvalue λ_k, and P_{ki}—the participation factor of i bus for the k variation mode—is an index of bus vulnerability.

The Voltage—Reactive Power Controllability and Control Areas. The instability and voltage collapse phenomena are determined by the network inability to grant consumers an additional power request because it reached the upper limit of power transfer.

Considering this fact, V–Q control areas are defined, including PQ (active power—reactive power) and PV (active power—voltage) nodes with a similar voltage behavior, and links between these nodes are identified. These links are called ***vulnerable lines***. The area algorithm [25] is based on the information supplied by the $[J_{QV}]$ submatrix. This submatrix is extracted from the complete Jacobean matrix $[J']$, and its elements are:

$$L_{ik} = \frac{\partial Q_i}{\partial V_k} = V_i V_k Y_{ik} \sin(\theta_i - \theta_k - \gamma_{ik}) \tag{18.44}$$

For a vulnerable transmission line, L_{ik} is a negative quantity, but close to zero since such a line has a large impedance (Y_{ik} is nearly zero), or it works close to the upper limit of transmissible power, which is that $\sin(\theta_i - \theta_k - \gamma_{ik}) \cong 0$. When the power transferred on such a line increases, then L_{ik} becomes zero or a positive quantity. This will cause the $[J']$ matrix to be singular and a voltage instability phenomena will be triggered because of voltage or angle bifurcation [24].

18.5 CASE STUDY

The case study was performed on the test network called TEST2 (Figure 18.6) [26]. The bus characteristics are shown in Table 18.1, and the bus steady-state results are shown in Table 18.2.

As much as 500 states were simulated in order to create the driving set and the test set. These states were generated by increasing the active and reactive power values consumed in nodes 1, 6, and 9 at values 80%, 90%, 100%, 110%, 120%, and 130% of the base case values. Having 6 values for 3 nodes we obtain $6^3 = 216$ possible operating conditions. In addition, considering the status of the line L1-10 (possible values: connected and disconnected) we have 432 examples. Other 68 cases where generated by setting random values from selected attributes. The obtained examples were split in two: two-thirds were used to generate the learning set and the last one-third was used to generate the test set.

Using the program MLT2001, eight decision trees were generated [23]:

A1 – by using criterion CART;

A2 – by using criterion gain;

A3 – by using criterion referred gain;

A4 – by using criterion ULg;

A5 – by using criterion normalized gain;

A6 – by using criterion EG2;

A7 – by using criterion CS-ID3;

and A8 – by using criterion IDX.

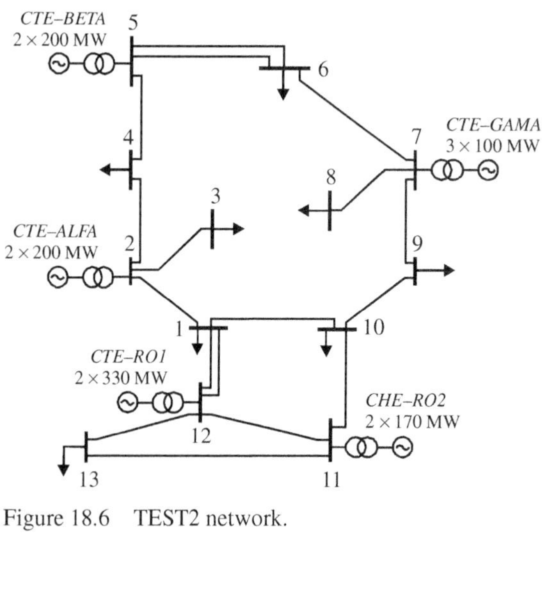

Figure 18.6 TEST2 network.

TABLE 18.1 The bus date of the TEST2 network

Bus	V_n (kV)	P_g (MW)	Q_g (MVAr)	P_c (MW)	Q_c (MVAr)	V_{imp} (u.r)	$Q_{\min}$ (MVAr)	$Q_{\max}$ (MVAr)	Type
1	220	0	0	250	155	—	—	—	PQ
2	220	255	—	0	0	1.045	0	250	PV
3	220	0	0	60	35	—	—	—	PQ
4	220	0	0	190	130	—	—	—	PQ
5	220	240	—	0	0	1.023	0	240	PV
6	220	0	0	220	135	—	—	—	PQ
7	220	240	—	0	0	1.023	0	240	PV
8	220	0	0	65	35	—	—	—	PQ
9	220	0	0	130	70	—	—	—	PQ
10	220	0	0	200	140	—	—	—	PQ
11	220	165	—	0	0	1.050	−60	235	PV
12	220	—	—	—	—	1.080	0	400	Vθ
13	220	0	0	240	145	—	—	—	PQ

TABLE 18.2 Base case steady state

Bus	V (kV)	θ (deg.)	P_g (MW)	Q_g (MVAr)	P_c (MW)	Q_c (MVAr)	The reactive power reserve of generators
1	0.979	−7.028	0	0	250	155	—
2	1.045	−1.924	255	114.08	0	0	135.92 MVAr (54%)
3	1.019	−3.904	0	0	60	35	—
4	0.979	−7.388	0	0	190	130	—
5	1.023	−5.497	240	145.80	0	0	94.20 MVAr (39%)
6	0.978	−9.147	0	0	220	135	—
7	1.023	−7.751	240	157.34	0	0	82.66 MVAr (34%)
8	0.998	−8.723	0	0	65	35	—
9	0.971	−9.568	0	0	130	70	—
10	0.960	−8.753	0	0	200	140	—
11	1.050	−1.475	165	64.57	0	0	170.43 MVAr (73%)
12	1.08	0.000	440.7	273.24	0	0	126.76 MVAr (32%)
13	1.003	−4.717	0	0	240	145	—

Table 18.3 provides the sensitivity factors calculated for various busses of the test network to evaluate the voltage stability.

TABLE 18.3 The values of local voltage stability indices

Bus	1	3	4	6	8	9	10	13
S_{VcQc}	−0.08	−0.15	−0.08	−0.07	−0.13	−0.09	−0.09	−0.08
S_{QgQc}	1.20	1.05	1.11	1.10	1.05	1.14	1.21	1.12

TABLE 18.4 Performances of decision trees

	Complexity		Errors (%)			
Tree	Number of nodes	Height	Training	Testing	Learning	Construction method
A_1	37	8	0.00	3.61	1.20	CART
A_2	37	9	0.00	3.61	1.20	Gain
A_3	39	9	0.00	3.61	1.20	Gain ratio
A_4	33	8	0.00	3.01	1.00	ULg
A_5	39	7	0.00	3.61	1.20	Normalized gain
A_6	33	7	0.00	2.40	0.80	EG2
A_7	33	7	0.00	2.40	0.80	CS-ID3
A_8	31	7	0.00	2.40	0.80	IDX

In Table 18.4 are presented the proprieties and performances of these trees.

Figures 18.7 and 18.8 show the errors and the features (number of nodes and height) of the decision trees generated.

The previous figures reveal that:

1. The best results (the smallest errors) are obtained when using a method that considers the cost involved by the tests (EG2, CS-ID3, or IDX).
2. If the test costs are disregarded, the best results are obviously obtained through the method ULg;

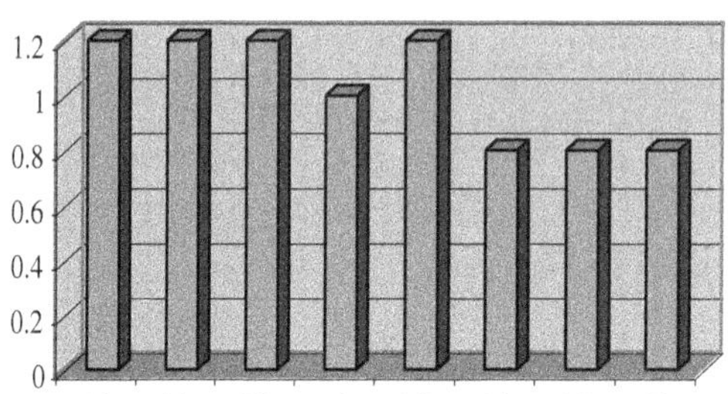

Figure 18.7 Learning errors of the decision trees.

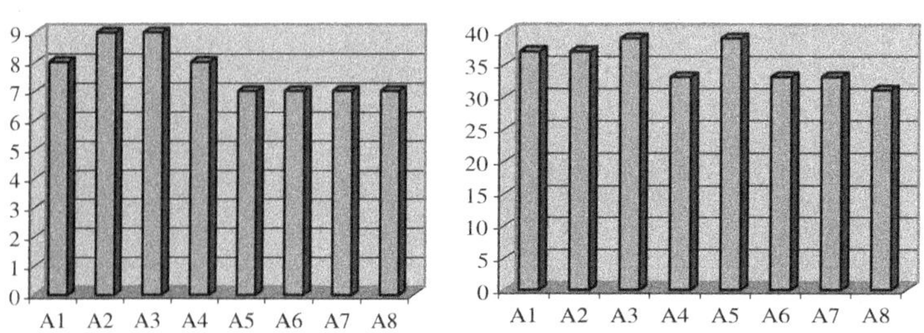

(a) (b)

Figure 18.8 Features of the eight generated decision trees: (a) number of nodes and (b) height.

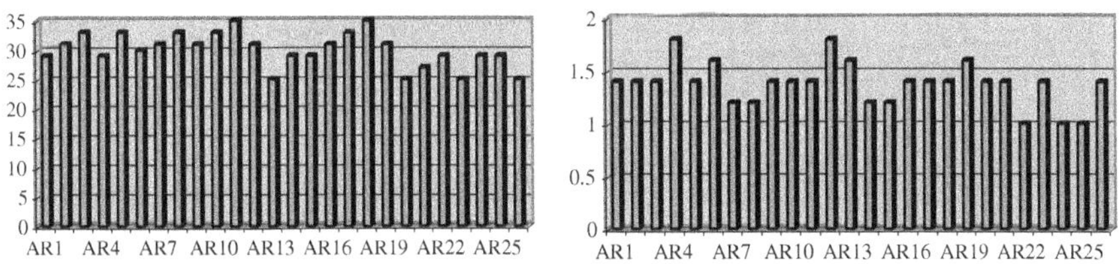

Figure 18.9 Errors of the decision trees obtained through pruning.

3. Out of all algorithms, the simplest structure is obtained by means of the method IDX.

In order to simplify the structure of the eight generated decision trees, a pruning process was used, this being valid if the ratio between the number of examples belonging to a certain class and the overall number of examples is less than 0.05 (existing 26 nodes that can be pruning, which means we have 26 pruned trees—AR_1, …, AR_{26}). The errors and features of the 26 decision trees obtained in this way are shown in Figures 18.9 and 18.10.

The previous figures indicate that through a simple pruning, the performance of the tree obtained by means of the method IDX (A8) cannot be improved too much, while the performances of the tree A4 (method ULg) are greatly improved following this process.

The trees obtained through the pruning process of A1 tree (AR_1, AR_2, and AR_3) present the same error, due to the fact that the three nodes are located on the same branch. In such cases, the pruning should be applied at the level of the node nearest to the root, since the simplest structure (the errors remain stable) is thus obtained.

To compare the performance of obtained decision trees was used a function named *goodness* [27]

$$goodness(T) = \varepsilon_l(T) + C \cdot NN(T) \tag{18.45}$$

where

$NN(T)$	is	the number of nodes.
$\varepsilon_l(T)$	–	the learning error of decision tree.

For different values of C parameter was calculated the goodness of the eight not-pruned decision trees (Figure 18.11) and the goodness of 26 pruned decision trees (Figure 18.12).

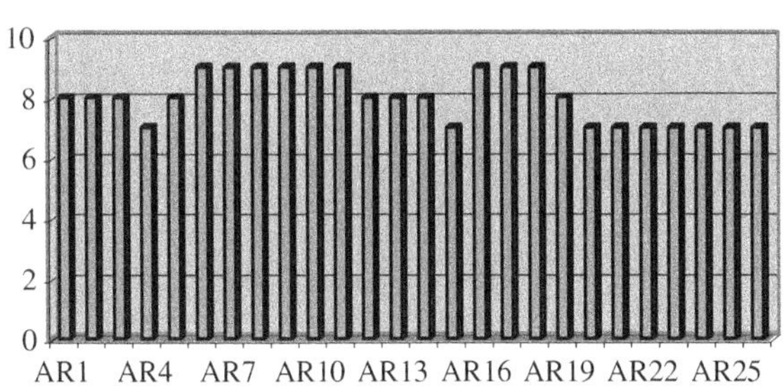

Figure 18.10 Features of the 26 decision trees obtained through pruning.

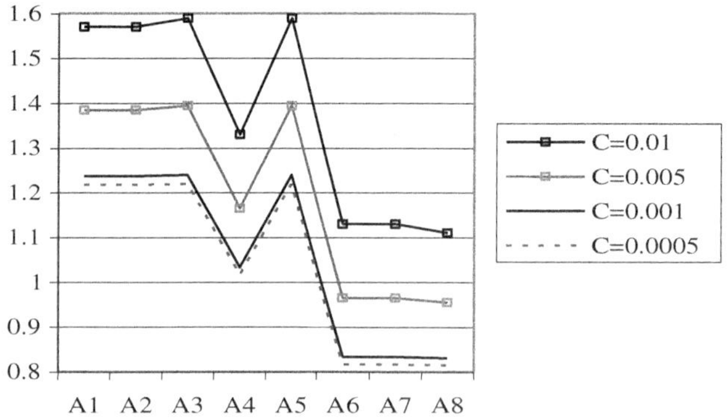

Figure 18.11 The goodness of the eight not-pruned decision trees.

If a pruning algorithm is not used, the best results are obtained by a method based on attribute costs (EG2, CS-ID3, and IDX). The best pruning decision trees are obtained by using EG2, CS-ID3, gain, or ULg method.

The previous relation of goodness does not take into consideration the height of decision tree. It is known that the complexity of validation process depends strongly on the height. From this reason, a better relation for the goodness function is:

$$goodness(T) = \varepsilon_l(T) + \frac{C}{2} \cdot (NN(T) + 2^{h(T)}) \tag{18.46}$$

where $h(T)$ is the high of decision tree.

The results obtained by using this relation are presented in Figures 18.13 and 18.14.

For not-pruned decision trees, the best construction methods are EG2, CS-ID3, and IDX. If CART or ULg method is used, the results are two times better than in the case when gain or gain ratio method is used.

Figure 18.15 shows the decision tree generated by using ULg method.

A decision tree can be written using a sentence of rules. This is very useful for automatic implementation into programming language. Afterward is presented the sentence of rules for the decision tree generated by ULg method.

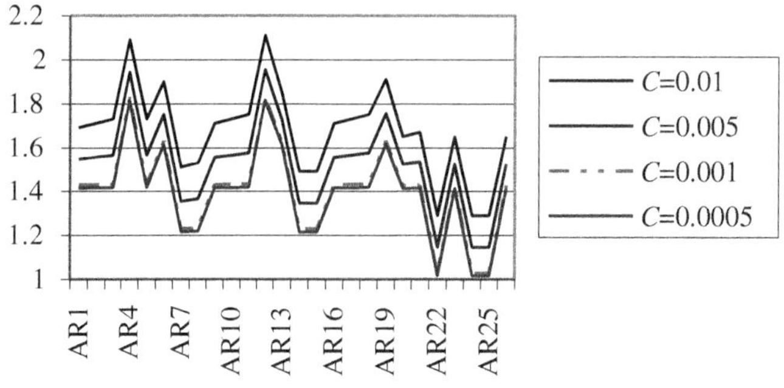

Figure 18.12 The goodness of the 26 pruned decision trees.

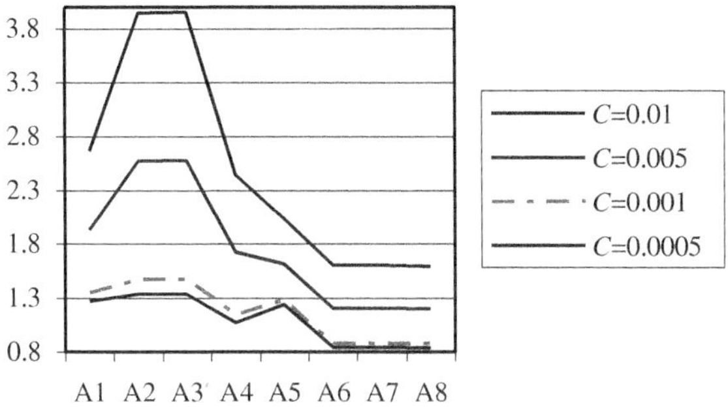

Figure 18.13 The goodness of the eight not-pruned decision trees.

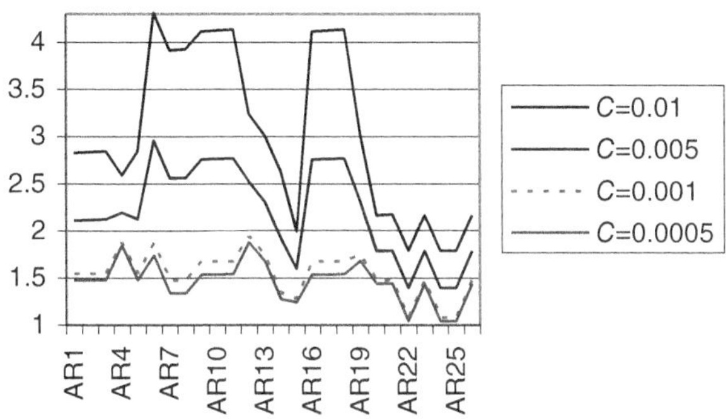

Figure 18.14 The goodness of the 26 pruned decision trees.

```
IF P9<156 THEN
    IF P6<275 THEN
        stable
    ELSE
        IF P9<143 THEN
            stable
        ELSE
            IF p1<275 THEN
                IF L1_10=disconnected THEN
                    stable
                ELSE
                    unstable
            ELSE
                stable
ELSE
    IF P6<250 THEN
        IF P9<200 THEN
 IF L1_10=connected THEN
stable
ELSE
```

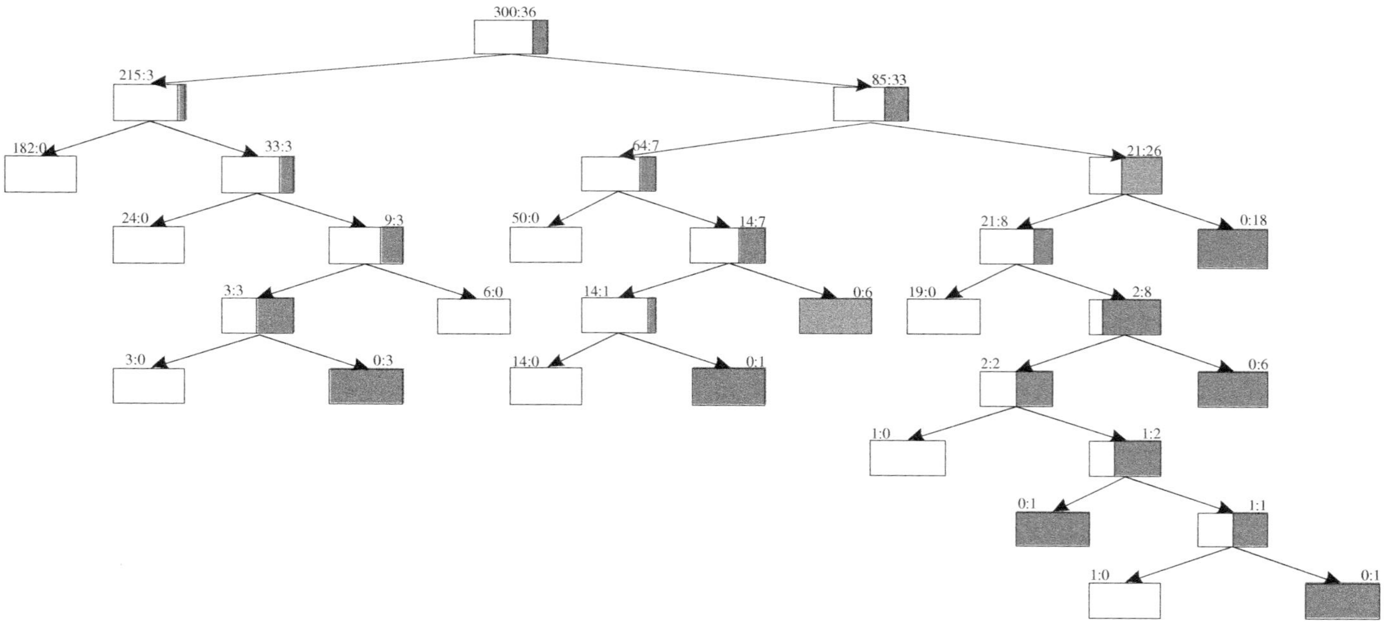

Figure 18.15 Decision tree obtained by using ULg method.

```
IF P1<275
IF Q1<155 THEN
stable
ELSE
unstable
ELSE
stable
        ELSE
            IF L1_10=connected THEN
                IF P9<280 THEN
                    stable
                ELSE
                    unstable
            ELSE
                unstable
    ELSE
        IF L1_10=connected THEN
            IF P9<250 THEN
                stable
            ELSE
                IF Q6<130 THEN
                    IF P9<300 THEN
                        stable
                    ELSE
                        IF P1<220 THEN
                            unstable
                        ELSE
                            IF P9<275 THEN
                                stable
                            ELSE
                                unstable
                ELSE
                    unstable
        ELSE
            unstable
```

We have provided a comparative study of decision trees construction algorithm as regards how their performances are influenced by the selected attribute for the partition of current training set.

We may thus conclude that the decision trees provide a mature, flexible, and effective tool for the analysis of large databases of information about power systems stability.

REFERENCES

[1] Wehenkel, L. *Automatic Learning Techniques in Power Systems*. Kluwer Academic Publishers, New York, 1998.

[2] Quinlan, J. R. Induction of decision trees. *Machine Learning*, Kluwer Academic Publishers, vol. 1, no. 1, pp. 81–106, March 1986.

[3] Breiman, L., Friedman, J. H., Olstren, R. A., and Stone, C. J. *Classification and Regression Trees*. Woodsworth International, Los Angeles, 1984.

[4] Quinlan, J. R. *Programs for Machine Learning*. Morgan Kaufmann, San Mateo, 1993.

[5] Jun, B. H., Kim, C. S., Song, H. Y., and Kim, J. A new criterion in selection and discretization of attributes for the generation of decision tree. *IEEE Transactions on Pattern Analysis and Machine Intelligence*, vol. 19, no. 12, pp. 1371–1375, December 1997.

[6] Nunez, M. The use of background knowledge in decision tree induction. *Machine Learning*, vol. 6, no. 3, pp. 231–250, May 1991.

[7] Tan, M., and Schlimmer, J. Cost-sensitive concept learning of sensor use in approach and recognition. *Proceedings of the Sixth International Workshop on Machine Learning*, New York, 1989.

[8] Norton, S. W. Generating better decision trees. *Proceedings of the Eleventh International Joint Conference on Artificial Intelligence*, Detroit, 1989.

[9] Buntine, W., and Niblett, T. A further comparison of splitting rules for decision tree induction. *Machine Learning*, vol. 8, no. 1, pp. 75–85, January 1992.

[10] Heath, D., Kasif, S., and Salzberg, S. Learning oblique decision trees. *Proceedings of the 13th International Joint Conference on Artificial Intelligence*, Chambery, 1993.

[11] Murthy, S. K., Kasif, K., Salzberg, S., and Beigel, R. OC1: randomized induction of oblique decision trees. *Proceedings of the Eleventh National Conference on Artificial Intelligence,* Washington, 1993.

[12] Quinlan, J. R. Simplifying decision trees. *International Journal of Man-Machine Studies*, vol. 27, no. 3, pp. 221–234, September 1987.

[13] Eremia, M., Petricică, D., Bulac, A. I., Bulac, C., and Triştiu, I. *Artificial Intelligence Techniques. Concepts and Applications in Power Systems (in Romanian: Tehnici de inteligenţă artificială. Concepte şi aplicaţii în sistemele electroenergetice)*. AGIR Publishing House, Bucharest, 2001.

[14] Niblett, T., and Bratko, I. Learning decision rules in noisy domains. *Proceedings of Expert Systems Conference*, Cambridge, 1986.

[15] Mingers, J. Expert systems - rule induction with statistical data. *Journal of Operational Research Society*, vol. 38, no. 1, pp. 39–47, 1987.

[16] Gelfond, S. B., Ravishakar, C. S., and Delp, E. J. An interactive growing an Pruning algorithm for classification tree design. *IEEE Transaction Pattern Analysis and Machine Intelligence*, vol. 13, no. 2, pp. 163–174, February 1991.

[17] Kittler, J., and Devijver, L. E. Statistical properties of error estimators in performance assessment of recognition systems. *IEEE Transactions on Pattern Analysis and Machine Intelligence*, vol. 4, no. 2, pp. 215–220, 1982.

[18] Nilsson, N., Learning Machines: Foundations of Trainable Pattern-Classifying Systems, McGraw-Hill, New York, 1965.

[19] Duda, R. O., and Hart, P. E. *Pattern Classification and Scene Analysis*. John Wiley & Sons, New York, 1973.

[20] Brodley, C. E., and Utgoff, P. E. Multivariate decision trees. COINS Technical Report 92-82, University of Massachusetts, 1992.

[21] Brodley, C. E., and Utgoff, P. E. Multivariate decision trees. *Machine Learning*, vol. 19, no. 1, pp. 45–77, 1995.

[22] Wehenkel, L., Van Cutsem, T., Pavella, M., Jacquemart, Y., Heilbronn, B., and Pruvot, P. Machine learning, neural networks and statistical pattern recognition for voltage security: a comparative study. *International Journal of Engineering Intelligent Systems for Electrical Engineering and Communications*, vol. 2, no. 4, pp. 233–245, December 1994.

[23] Bulac, A. I. Application of artificial intelligence techniques – decision trees in voltage stability assessment. Ph.D. Thesis, University "Politehnica" of Bucharest, 2001.

[24] Bulac, C. Contributions to voltage stability analysis in mixed alternating current – direct current power systems. Ph.D. Thesis, University "Politehnica" of Bucharest, 1998.

[25] Trecat, J., Eremia, M., Bulac, C., and Balan, Gh. Voltage stability evaluation indices. *Proceedings of the IEEE/NTUA Athens PowerTech Conference*, Athens, Greece, 1993.

[26] Eremia, M., Crişciu, H., Ungureanu, B., and Bulac, C. *Computer Aided Analysis of the Electric Power Systems Regimes (in Romanian: Analiza asistată de calculator a regimurilor sistemelor electroenergetice)*. Technical Publishing House, Bucharest, 1985.

[27] Bulac, A. I., Eremia, M., and Bulac, C. Decision trees for voltage stability assessment. *Proceedings of ISAP Conference*, Budapest, 2001.

CHAPTER 19

GENETIC ALGORITHMS

Anastasios Bakirtzis and Spyros Kazarlis

19.1 INTRODUCTION TO EVOLUTIONARY COMPUTATION

When dealing with real-world problems and solution methods one question may come in mind:

> What is the most powerful problem solver in the universe?

To this question, one could give the following answer:

> The human brain of course that created "the wheel, New York, wars and so on. (After Douglas Adams)

But there is also another apparent answer:

> It is the evolution mechanism that evolved the human brain." (After Darwin et al.)

During the past years, scientists and engineers have looked deeper into the structure and mechanisms of natural systems, trying to mimic their dynamical behavior. To this effort, they have a strong-growing ally: the increasing computational power of modern computers. The obvious motivation for biologists is to understand the rules and dynamics of natural systems. However, for the engineers the underlying motivation is different: the perspective to construct abstract digital clones of natural systems in the form of computer software or even computer hardware, with the aim to use them for solving engineering problems more efficiently.

Along this direction, the effort to understand and abstractly simulate the human brain has led to the birth of the science of "neurocomputing." The ambition to understand and utilize the fuzziness of nature has led to the science of "fuzzy systems." And finally, the urge to simulate the mechanism of the "Darwinian evolution" has led to the dawn of the "evolutionary computation" (EC) science.

The EC science is about algorithms and techniques that try to simulate the biological evolution of species in nature as well as evolution-developed biological mechanisms, systems and collective behaviors.

The goal is to construct automated and efficient methods for finding optimal or near-optimal solutions for hard real-world problems using evolutionary mechanics.

Advanced Solutions in Power Systems: HVDC, FACTS, and Artificial Intelligence, First Edition.
Edited by Mircea Eremia, Chen-Ching Liu, and Abdel-Aty Edris.

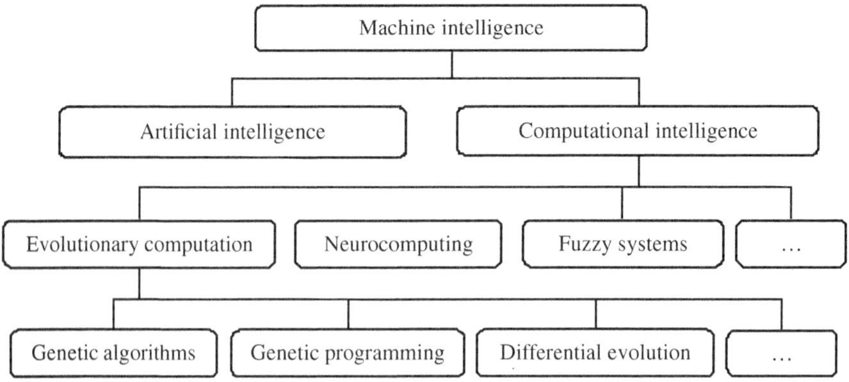

Figure 19.1 Taxonomy of EC.

Another adjacent goal is to create adaptive artificial systems that self-evolve according to their current environment.

Many methods and algorithms are nowadays under the "evolutionary computation" umbrella, but only one family of algorithms clearly outstands from the crowd: the "genetic algorithms."

Genetic algorithms or GAs are probably the most studied and "evolved" class of evolution-inspired methods and are probably the most efficient too. GAs use an abstract flow diagram, an extracted form Darwin's theory that utilizes the most important "key" principles of natural evolution. This is the reason why they manage to inherit the enormous exploration power of their natural counterpart and exhibit such effectiveness in discovering high quality solutions in vast search spaces.

19.1.1 Taxonomy

Genetic algorithms constitute a branch of the science of EC, which itself is a branch of "Computational Intelligence" (or soft computing) together with neurocomputing and fuzzy systems. Computational intelligence is a branch of "machine intelligence" and is a "brother" of "artificial intelligence." This taxonomy is graphically shown in Figure 19.1.

The most important "children" of EC are graphically shown in Figure 19.2.

19.1.2 Initial Inspiration and Basic Principles

GAs and more generally EC methods are inspired by the admirable power of nature to transform over time, the single-cell organisms into organized multicell miracles like the superior mammals and finally the human being. And the last has miraculously managed not only to understand the mechanism that evolved him but also to use this mechanism for solving his problems and advancing his technical abilities.

But what are the key principles of natural evolution that were transferred into the algorithms of these methods?

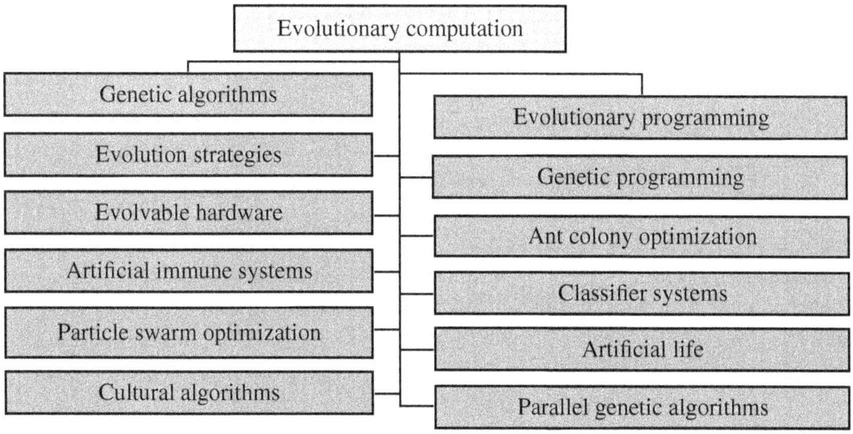

Figure 19.2 Various EC methods.

The first principle is that the evolution is performed on populations of beings and not individual organisms. In nature, species are met in populations of beings that live together, forming communities, with a small or large degree of organization. The members of a population, then, are able to mate and reproduce themselves into offspring that become again members of the same population. With time, offspring members become mature and replace their parents in this reproduction sequence. The presence of a population adds statistical behaviors to the overall evolution dynamics, such as the observed inertia of the population to change, and the parallel exploration of multiple areas of the search space.

The second principle is the "survival of the fittest" rule. This rule tells us that not all population members manage to reproduce themselves, at least not with the same chances. Fitter individuals have higher probability to survive and reach their mature age, and then mate and reproduce, whereas less fit individuals have lower probability to achieve this goal. In this rule, we can observe a "cruel" criterion imposed by the natural environment that promotes the reproduction of organisms with higher abilities and adaptability, concerning the challenges of the surrounding environment. This "elitistic" rule applies a bias on the population dynamics that "pushes" the whole population to gradually better itself. And this is the main lever of the evolution mechanism that makes individuals (solutions) to proceed toward optimality.

The third principle is that the individuals' characteristics are encoded in a manner that makes it easier to recombine them into more fruitful combinations. In nature, this is done by the organisms' DNA whereas in EC methods strings of symbols are used to encode the solutions' "genotype." The "encoding" method is of high importance to the overall performance of an EC method. This is true especially in cases of constrained optimization problems, where a number of constraints may be initially resolved by the encoding method itself that could "encode" only feasible solutions and leave infeasible ones outside of the encoded solutions space.

The fourth principle is the "genetic information recombination" principle. According to this principle, natural beings mate usually in pairs and recombine their

DNA structures, forming thus the DNA of the offspring. This "recombination" procedure favors the mixing of specific characteristics of population members into new blends that are left to be tested by the environment and time. In other words, the population members exchange genetic information to each other, forming new overall fit recombinations from partially fit characteristics. This principle constitutes the basic "search" mechanism of the evolution procedure.

The fifth principle is the principle of "mutation." According to this principle, there is always a small probability, during the reproduction phase that builds a new organism, where a small perturbation may occur in its DNA structure, leading to a new structure that did not exist in either of its parents. This "perturbation" may occur usually due to chemical factors or due to cosmic radiation, at the moment of conception. This mechanism can inject new information into the populations' genotypes and results in a nonzero probability to visit every point in the search space, regardless of the initial random population. This principle comprises a secondary search operator that is based on rare stochastic alterations of the populations' genetic information. Unlike its infrequent action, this "mutation operator" is of great importance to the overall performance of an EC method.

19.1.3 On the Evolution Theory

The first steps were taken by biologists and geneticists who tried to make computer simulations of the Darwin's theory.

The first attempt to simulate evolution in a computer is dated back in 1954, where Baricelli [1, 2] made his first experiments using the computer of Princeton University, NJ. However, his work was not welcome with significant interest by the scientific community.

Meanwhile, another geneticist, Fraser [3, 4], publishes a series of papers concerning the computer simulation of artificial selection of organisms. His work included all the basic elements of modern genetic algorithms and was also described in Crosby [5].

Another series of papers were published in the 1960s by Bremermann [6, 7], reporting work that also used populations of solutions to optimize problems, with recombination, mutation, and selection operators. Other researchers like Box [8], Bledsoe [9], and Friedman [10] also published papers in the same area but attracted minor scientific attention.

Later researchers did better in attracting wide scientific attention, like Ingo Rechenberg and Hans-Paul Schwefel in the 1960s and early 1970s, who introduced a technique called "evolution strategies" (or "Evolutionsstrategie" in German) for solving engineering problems [11–14]. According to many people, these two scientists are considered as the "fathers" of EC.

Another independently developed approach at the same time was the "evolutionary programming" technique of Fogel [15], which was proposed for generating artificial intelligence. Evolutionary programming originally used finite state machines for predicting environments and used variation and selection to optimize the predictive logics

GAs were introduced by Holland [16] in the 1960s and were developed by himself and his students and colleagues at the University of Michigan in the 1960s and the 1970s. In contrast with evolution strategies and evolutionary programming, Holland's original goal was not to design algorithms to solve specific problems, but rather to formally study the phenomenon of adaptation as it occurs in nature and to develop ways in which the mechanisms of natural adaptation might be imported into computer systems. He also proposed a preliminary mathematical theory called "Schema theorem." This theory shows that it is possible to predict population quality at the next generation and originally explains why GAs work. The "Schema theorem" will comprise the basis of almost all subsequent theoretical work on GAs. His book entitled *Adaptation in Natural and Artificial Systems*, published in 1975, is considered as a milestone in the history of EC and crowned him as the "father" of genetic algorithms.

In 1978, Holland and Reitman published their work [17] on learning classifier systems (LCS) where they introduced a prototype LCS called CS-1 (Cognitive System One). In 1989, Koza [18] establishes the notion of genetic programming that was first seen as a paradigm of Gas, but it was quickly recognized as an independent and most important evolutionary method, with many applications and offspring scientific research areas like "evolvable hardware."

These early research efforts where widely inspirational and therefore followed-up by massive research work from all over the world, resulting in thousands of published papers and dozens of conferences and workshops worldwide. This explosive evolution has turned "evolutionary computing" into a well-recognized modern science, with a wide variety of practical applications.

19.1.4 DNA-Like Solution Encoding

Before applying an evolutionary method on an optimization problem, one has to decide about the encoding method (or representation method) to use, in order to encode the candidate solutions of the real problem into DNA-like genotype strings. When choosing a representation, we have to bear in mind how the genotypes will be decoded and evaluated and what the genetic operators might be like. In other words, the encoding method must suit the specific characteristics of the problem, such as the existence of constraints, the type and number of independent problem variables, and problem-specific eccentricities, if any. Moreover, the chosen encoding method must facilitate the application of genetic operators, such as recombination and mutation operators, as well as other specific operators.

In the literature, there is a large number of proposed encodings each of which better suits specific problems or families of problems. The general encoding issue is to find a representation function E: $S \rightarrow G$ that maps the set of candidate solutions S into the set of corresponding encoded genotypes G. Function E encodes each candidate solution vector $\vec{s} \in S$ into a corresponding genotype vector $\vec{g} \in G$.

Binary Encoding: In this case $G = \{0, 1\}^n$, where n is the genotype length, i.e. the length of binary strings that represent solutions. Thus every candidate solution $\vec{s}$ is transformed into a binary string of length n as shown in Figure 19.3. The encoded DNA-like string is called "genotype," whereas the decoded real solution vector is

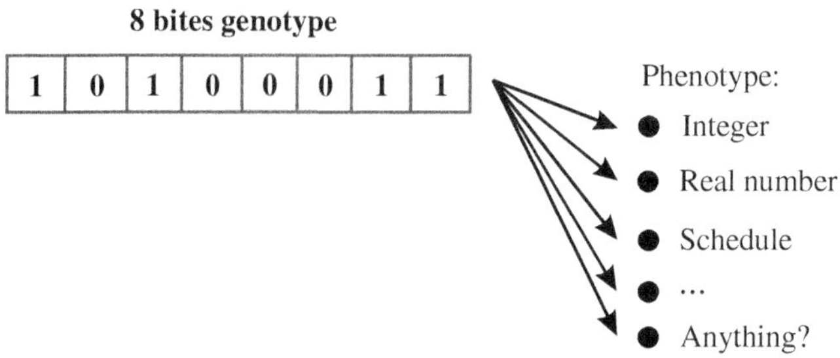

Figure 19.3 Binary encoding.

called the "phenotype." Depending on the specific problem, the phenotype could be an integer, or a vector of integers, a real value (or vector of real values) a schedule, the order of a set of items, etc.

REAL ENCODING: In this case $G = R^n$, which means that the "encoded" solutions are real-valued vectors. This "representation" is also called "natural encoding" when the candidate solutions themselves are also real-valued vectors. In this case, "natural" means in fact that there is no "encoding" at all and that the evolutionary method must cope with the original form of the candidate solutions, i.e. real numbers. As it is easily understood, this "representation" does not allow the application of the usual recombination and permutation operators used in DNA-like encodings. So, one must invent new operators to work with real numbers or vectors of real numbers.

ORDER-BASED (OR PERMUTATION) ENCODING: This encoding is used for ordering problems, i.e. problems where the goal is to find the best order combination of a number of discrete elements that maximizes (or minimizes) a given function. In order to preserve feasibility of solutions in these problems, through the use of genetic operators, one has to adopt either a special variation of the order representation method, or special recombination and mutation operators that preserve feasibility. An example of such a representation for an instance of the "traveling salesman problem" is shown in Figure 19.4.

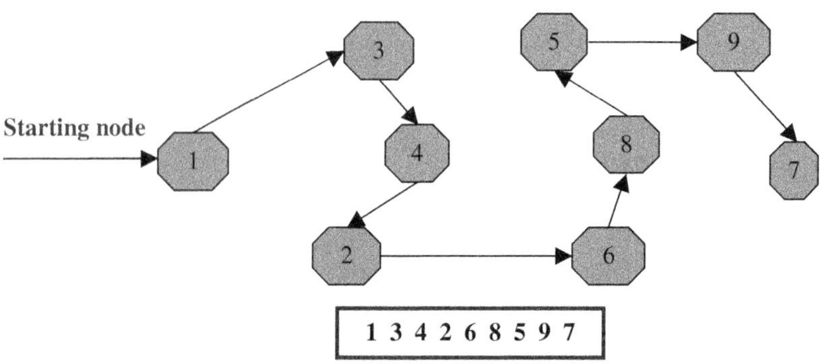

Figure 19.4 Simple order representation for the "traveling salesman problem."

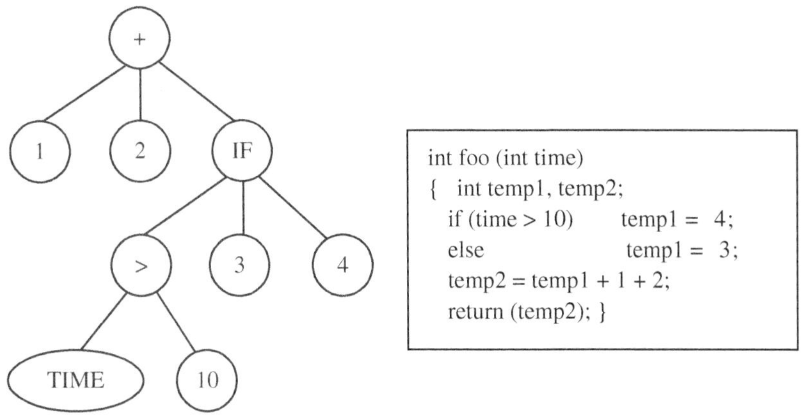

Figure 19.5 Tree representation of a software function foo().

Tree-Based Encoding: This representation is mainly used in "genetic programming" to encode tree-structured solutions like "algorithms," "circuits," "symbolic mathematical expressions," and so on. Tree encoding needs special recombination operators that operate on tree branches, and special mutation operators as well. Usually the tree-formed genotypes produced are of variable length. An example of such an encoding is shown in Figure 19.5.

19.1.5 Solution Evaluation

In order to apply the "survival of the fittest" rule on the population members at each generation, we have to "evaluate" each produced solution by a "fitness function." In general, the fitness function $F : G \rightarrow R$ maps the set of genotypes G into the set of real R. In other words, it assigns a real value that expresses fitness, to every genotype produced by the evolutionary algorithm (Figure 19.6).

In mathematically formulated optimization problems, the fitness function coincides with the function to be optimized. In general, an optimization problem can be formulated as follows: optimize the objective function $F(\vec{x})$, $\vec{x} \in S$, where $\vec{x}$ is a generalized vector of problem parameters and S is the solution search space.

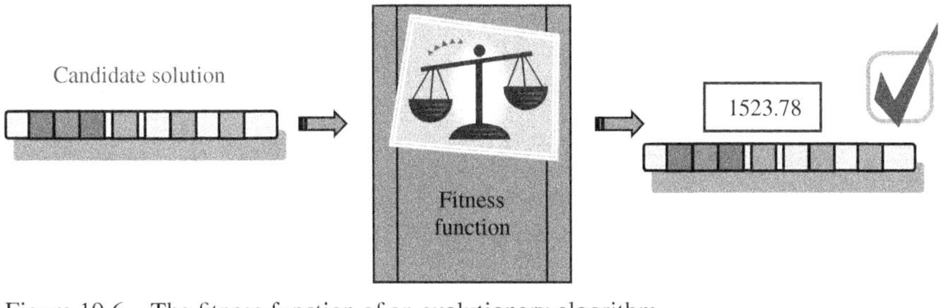

Figure 19.6 The fitness function of an evolutionary algorithm.

So in these cases, the fitness function is $F(\vec{x})$ itself, or is a function of it, especially when other factors are considered like problem constraints.

However, in some real-world problems, the evaluation of solutions cannot be done simply by evaluating a function. In such problems, the fitness may need to be calculated as the outcome of an experiment, the execution of a computer program, or even human judgment. Of course, it is easily understood that it is critical for the fitness evaluation procedure to be fast and accurate, in order to allow the EC algorithm to execute with reliability and within logical time limits.

19.1.6 Genetic Information Recombination

As stated earlier, search is primarily performed by combining the genetic information encapsulated into the DNA-like genotype structures of the population. The operators that perform this recombination procedure are called "recombination operators." These operators exploit the already discovered solutions and combine partially optimized solutions with the hope to form overall optimized ones.

Here it must be noted that although the recombination operator has the ability to rearrange parts of solutions to form new ones, it only recombines information that already exists within the population. In other words, it lacks the ability to inject new information into the pool of genetic material handled by the EC algorithm.

To cover this gap, EC methods usually use an additional operator called "mutation." Mutation is performed with significantly low probability in order not to disrupt the genotypes that have been already formed by the recombination operators. When applied, it focuses on a specific part (loci) of the solution genotype and performs a perturbation to it, e.g., by altering genetic information symbols. In a simple binary encoding, mutation may select bits at random and turn them from zero to one and vice versa.

An example of a recombination and a mutation operator is shown in Figure 19.7.

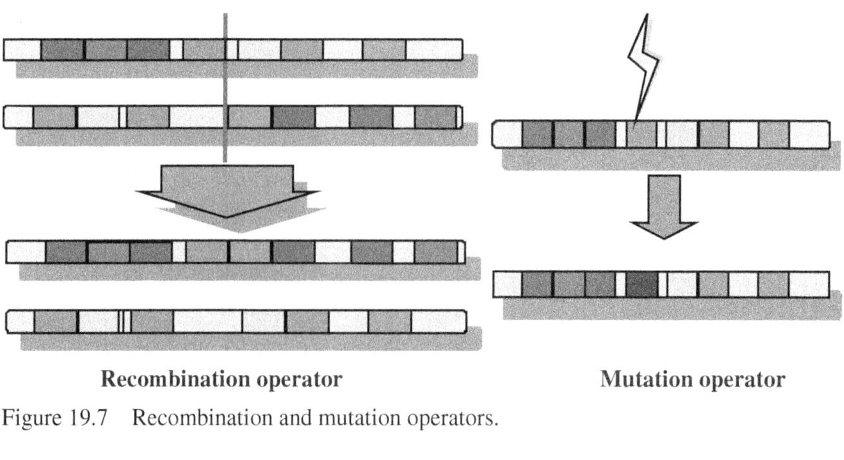

Figure 19.7 Recombination and mutation operators.

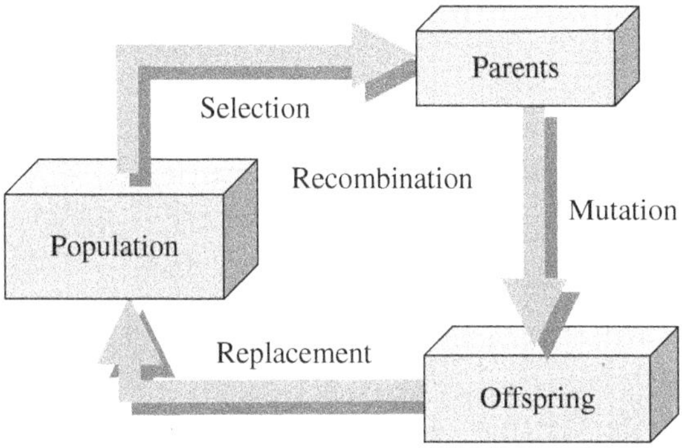

Figure 19.8 The circle of evolution.

19.1.7 The Circle of Evolution

With all the ingredients prepared, the logic of the evolutionary algorithm must be constructed. The simulated evolution mainly comprises a single loop, the "circle of evolution" (Figure 19.8). Within this loop, the following steps take place:

(a) Parents are selected from the population of solutions, with a "survival of the fittest" criterion. The first population is usually produced at random;

(b) parents' genotypes are recombined and probably mutated to form offspring genotypes.

(c) when a sufficient offspring population is produced, it replaces the original "parent" population, and the circle continues.

With minor variations, all evolutionary paradigms follow the above specification. A complete circle is usually referenced as a "generation." A large number of generations are usually needed for the algorithm to converge to an optimal or near optimal solution.

19.1.8 Evolutionary Algorithms as Global Optimizers

Most traditional optimization methods used in science and engineering applications can be divided into two broad classes: (a) direct search methods requiring only the objective function values and (b) gradient search methods requiring gradient information either exactly or numerically. These methods usually work on point-by-point basis. They start with an initial guess and subsequently produce new solutions through the algorithms iterations. The usual stop criterion is when the value of gradient of the objective function becomes close to zero. In cases where the search space is unimodal and smooth, as shown in Figure 19.9, such methods can approach the global optimum with confidence and speed. But most real-world optimization problems are not of this form. They are usually multimodal, discontinuous, and in case of constraints the search space is divided into a number of isolated subareas (see Figure 19.9).

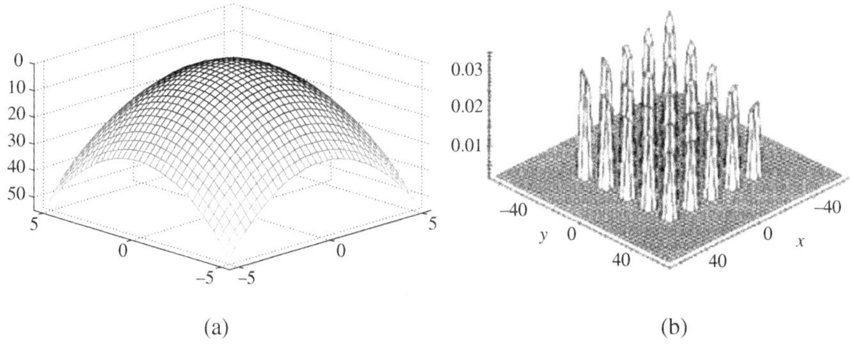

Figure 19.9 A smooth unimodal function (sphere) and a multimodal one (Shekel's foxholes).

In these problem cases, the above methods usually climb-up the slope where the first random solution is located, finding nothing more than a local optimum. Thus they in fact perform local optimization.

On the other hand, population-based search methods, as the EC methods, spread multiple points throughout the whole search space, that behave like autonomous and collaborating agents, that search many areas of the search space simultaneously. Thus it is very likely that the produced "optimal" solution may be the "globally optimal" solution. And although EC methods cannot guarantee that, yet they work with a global perspective. This is supported not only by the population of points but also by the stochastic nature of genotype transition rules that allow for every search direction to be followed and every subarea or point to be visited by the algorithm. Thus they are justifiably characterized as "global optimizers."

19.1.9 Evolutionary Computation Paradigms

EC is a science that embraces a large number of algorithms and methods and an even larger number of their variations. In the following, the most important EC paradigms are quoted:

GENETIC ALGORITHMS [16, 26–30]: They are perhaps the most recognized family of algorithms within the EC community and therefore widely studied and applied. They use a population of solutions, usually encoded in symbol strings (e.g., binary strings), that is randomly initialized at the beginning. They use recombination as well as mutation operators in the reproduction phase. Parents are usually selected in pairs to produce offspring. Parent selection conforms to the "survival of the fittest" rule. Offspring replace parents for the evolution circle to continue.

EVOLUTION STRATEGIES [11–14]: They usually start from a single solution, or a small number of initial solutions, that are not encoded. Performing multiple mutation-like operations on the parents they produce λ new solutions that are evaluated. Form the offspring, or the combined set of offspring and parents, the parents of the next "generation" are chosen. Later a recombination operator was added [14]. Because of the small populations, they use they best suit problems where the evaluation of solutions is extremely time consuming.

Evolutionary Programming [15]: It handles problem solutions in a nonencoded form, using thus the natural form of problem variables, such as real values. It starts with a randomly generated initial population of solutions that are evaluated. It then selects parents from the population with a probability proportional to their relative fitness within the population. It applies only mutation-like operators on the parents to produce offspring. The mutation operators are usually problem dependent. When the offspring population is produced, it replaces that of the parents. It generally resembles a real-coded GA using only the mutation operator, and no recombination.

Genetic Programming [19]: The initial scope of this method was to automatically evolve computer software, for solving algorithmic problems. To achieve this, a new representation was used, the tree-based representation. According to this, computer programs are encoded as tree structures of variable size. Special tree-based recombination and mutation operators have been designed to work with these tree structures. Every tree-encoded program is evaluated by actually "running" the program in a simulated way and evaluating the results. It has been used to propose solutions to other problems as well, where the solutions can be described by tree structures. Some significant applications are the evolutionary design of analog and digital electronic circuits, the evolution of symbolic mathematical expressions, and its contribution to evolvable hardware. Due to its ability to invent structures (like circuits) to fit specific demands, many researchers assign to this method the title of the "invention machine."

Learning Classifier Systems [17]: These systems belong to the category of rule learning systems and especially in the GBML (Genetics Based Machine Learning) family. They have the ability to evolve knowledge in the form of simple "if <condition> then <action>" rules. They use evolutionary algorithms to produce a proper set of rules to describe complex or intelligent systems. The rules are applied to a set of input data, and through rule activation they produce the corresponding output. The rules are trimmed through a training process where the evolutionary algorithm is used to produce new and fitter set of rules. They can simulate input–output black-box functions just like neural networks do.

Evolvable Hardware [19]: It is a special application of genetic programming that due to its high potential has become an independent branch of EC. The aim is to construct self-adaptive hardware in the form of reconfigurable analog or digital circuits that could be able to adapt to different needs and specifications of the current working environment. They are based on the recently developed technology of reconfigurable hardware, like the FPGAs (field programmable gate arrays). Such hardware comprises a number of "cells" each of which can be programmed to simulate a digital gate, or an analog component, in case of analog platforms. Additionally, the interconnections between cells can also be programmed via an interconnection matrix. All programming is done using Electrically Erasable Programmable Read-Only Memory (EEPROM)-like hardware that is nonvolatile and electronically reprogrammable. By combining this reconfigurable hardware together with evolutionary methods, it is possible to develop hardware modules that can self-adapt to given working specifications. And furthermore, if the specifications change in time, the hardware is able to reconfigure itself optimally again, in order to track the changing environment.

PARALLEL GENETIC ALGORITHMS [20]: Parallel genetic algorithms (PGAs) simultaneously maintain a number of different evolving populations, that all have the same goal to achieve to find the optimum solution to a specific problem. PGAs mimic nature's situation where the individuals of a species (like humans) do not live in one unified population, but in a number of partially isolated smaller populations, in the form of countries, cities, islands, etc. Each population may have its own "rules" for mating or reproduction and could contain unique genetic material in its individuals. But due to the ability of subpopulations to communicate, this genetic material, or a possible evolution accomplishment, can spread to adjacent subpopulations, where it will be exploited using the properties of the local subpopulation. PGAs can also apply different evolutionary tactics to each subpopulation, in order to build a polyphony of search methods that search for the optimum.

DIFFERENTIAL EVOLUTION [21]: This can be seen as a variation of a real-coded GA, and it is applied only on continuous variable problems. It uses a population of solutions x_i, $i = 1, \ldots, N$, that is initially produced randomly and evaluated through a fitness function. Then a loop begins for all solutions x_i, building one offspring at a time that immediately replaces a parent, if it is proved to be fitter. The new offspring is produced in two phases: (a) via the formula $v_i = x_{r1} + A(x_{r2} - x_{r3})$, where a perturbation occurs that combines three population members different than x_i and (b) by combining parts of the candidate v_i and the original x_i solution.

With a distant view of this technique, we can recognize the use of all key elements of an evolutionary algorithm, namely a population of solutions, recombination, mutation, and survival of the fittest, but in a different mixture. The interesting and intelligent innovation of this technique that also justifies its title is the adaptive-size mutation it uses, via the formula $v_i = x_{r1} + A(x_{r2} - x_{r3})$. When the population is spread around the search space, the mutation size is large, depending on the large differences between random population members. As the population converges, these differences become smaller and lead to a narrow-size mutation operator that searches only the neighborhood of the area where the population has converged.

SWARM INTELLIGENCE AND ANT COLONY OPTIMIZATION [22,23]: These methods simulate the collective behavior of swarms or ant colonies that exhibits a form of "collective intelligence." They evolve populations of independent individual (autonomous agents) that explore the search space to avoid local optima (obstructions) and find the global optimum (food). The individuals have the ability to communicate with each other and exchange information for the landscape, by secreting substances along their paths that can be tracked and recognized by other individuals. Thus, the sum of individual behaviors forms a collective "mind" that efficiently explores the search space and guides individuals to the optimum.

ARTIFICIAL IMMUNE SYSTEMS [24]: They are inspired from the biological immune systems of superior organisms that protect them from foreign and probably harmful substances and pathogenic organisms that produce foreign proteins called antigens. It does that by first recognizing the foreign proteins among the millions of different self-proteins and second by producing matching killer cells called antibodies to exterminate the foreign cells, without harming the self ones. Moreover, this recognition-reaction behavior is not centrally supervised but is distributed among the organisms' cells. And this mechanism is trained through evolution to efficiently protect the body from external attacks.

Artificial immune systems try to simulate and benefit from the following properties of biological immune systems:

(a) Their remarkable ability for pattern recognition and classification that first recognizes self and non-self cells and the non-self cells are further categorized in order to induce an appropriate type of defensive mechanism;

(b) their ability to learn the pattern of new enemy cells that invade the body, in order to have a quicker response in future infections;

(c) the phenomenon of the distributed memory observed in the system that "remembers" previous attacks and "generalizes" its response over similar antigen patterns;

(d) its unsupervised way of learning, using distributed and parallel training mechanisms.

CULTURAL ALGORITHMS [75]: They comprise another EC paradigm that tries to simulate the evolution of "culture" within organized human communities, as being a knowledge and experience accumulating system, that can be used by individuals for solving problems and confront different challenging situations. Individuals' behavior and abilities are partly determined by their DNA inheritance and partly by their acquired properties that they obtain by growing up and interacting with the communities "culture" or "belief space." Moreover, individuals have the ability to enrich this cultural heritage by adding their individual contributions to it that shapes the "belief space" for the generations to come. This knowledge accumulation and reuse mechanism can be simulated by an evolutionary algorithm, to achieve problem-solving tasks with high efficiency. In this category, we also find "memetic" algorithms that are in fact a variation of GAs using hill climbing operators on every offspring produced by the GA. The hill climbing effect on each individual simulates the atom's growing up progress.

ARTIFICIAL LIFE [25]: This term includes a series of methods that utilize evolutionary algorithms, in order to model, describe, and facilitate the understanding of the evolution mechanisms of natural biological systems. These algorithms can work at the macroscopic level, i.e. modeling the evolution and dynamic behavior of populations but can also work at the microscopic or atomic level, for modeling the evolution and operation of biological mechanisms at the cell level.

19.1.10 Application Areas

Due to their remarkable efficiency for solving hard optimization problems, evolutionary algorithm methods have been already applied on numerous application areas. In the following, we try to quote a fraction of them:

(a) optimization of continuous-variable problems;

(b) optimization of combinatorial problems and order-based problems;

(c) optimization of constrained or multiobjective problems;
(d) system modeling and identification;
(e) planning, scheduling, and control problems;
(f) resource allocation problems;
(g) engineering and industrial design;
(h) rule-based systems and machine learning;
(i) self-adaptive system and hardware;
(j) algorithm and structure invention;
(k) Data Mining;
(l) pattern recognition and classification;
(m) production of art (graphics, music, etc.).

Below we cite a small number of specific applications where evolutionary algorithms have proved to provide "success stories":

(a) solution of optimization problems in power systems;
(b) cutting and packing problems;
(c) job shop scheduling problems;
(d) routing and scheduling problems;
(e) timetabling problems;
(f) as training or design algorithms for neural networks, fuzzy systems, and other techniques;
(g) feature extraction and pattern recognition problems;
(h) network topology problems (communications, transport, supply networks, etc.);
(i) Very-large-scale integration (VLSI) circuit design;
(j) applications in robotics (optimal trajectory calculation, path discovery, obstacle avoidance, etc.);
(k) optimal configuration of reconfigurable hardware;
(l) game playing;
(m) digital image processing;
(n) problems of chemical processes and bioinformatics.

19.1.11 Advantages and Disadvantages

The main advantages of evolutionary algorithms are the following:

(a) They are generally applicable, i.e. they do not impose any restrictions on the form of the problems on which they can be applied, such as linearity, continuity, or differentiability;

(b) they optimally combine "exploration" with "exploitation." This means that they simultaneously search the space for new areas, but also combine information from the space to guide the search;

(c) they exhibit an "implicit parallelism," that is they process in parallel a vast number of solution patterns, much larger than the number of atoms in the population;

(d) they have a low development cost, as they are algorithms with low algorithmic complexity and can be implemented with small programming effort;

(e) they can easily hybridize with other methods to produce hybrid optimization schemas;

(f) they can be supplied with man-produced or previously discovered solutions that can be injected in the initial population. This way the EC method can start from a level higher than total randomness;

(g) they provide a number of alternative solutions in the final population and not just a single proposed solution.

However, they also exhibit some disadvantages. The some of which are

(a) they cannot guarantee the optimal solution in bounded time limits;

(b) they have weak theoretical basis;

(c) they need some parameter tuning, in order to work optimally;

(d) they usually demand a significant amount of computational time and therefore are labeled as "offline" algorithms;

(e) they do not provide sensitivity information around the optimal solution.

19.2 GENETIC ALGORITHMS

GAs constitute the most representative and the most popular and recognizable paradigm of EC. They have achieved that by being a simple enough metaphor of the most important principles of biological evolution, without significant exceptions, and without the true complexity of the biological mechanisms at cell-level. Thus GAs include all the important elements of the "natural evolution" algorithm, a fact that makes them exceptionally effective, and at the same time they maintain a simplicity that makes them easy to understand and to be applied on a wide variety of problems and areas. Their dominance over the other EC paradigms is so clear that many people use their name to refer to the whole EC science.

GAs are mainly used as global optimization algorithms that seek optimal or near optimal solutions in difficult optimization problems. Their use is especially meaningful in cases of multivariable multimodal nonconvex nonsmooth optimization problems with vast search spaces, where other methods show poor performance.

The foundation of GAs is attributed to John Holland and the year 1962 when he publishes his work "Outline for a logical theory of adaptive systems" [26] and sets the basis for the development of GAs. Moreover in 1975 he publishes his ground-breaking book entitled *Adaptation in Natural and Artificial Systems* [16] that inspired and set off the entire scientific area of EC.

GAs have global applicability without imposing any restrictions for the type of the problem or the characteristics of the optimization function. However, they demand a significant amount of time in order to execute, as they evolve through a large number of successive iterations that are called "generations." Thus they are generally classified as "offline" optimization algorithms.

GAs have some kind of a theoretical basis that mostly relies on Holland's work on "schemata" and the "Schema theorem" [16]. But although the theoretical work on GAs may show the right way for aspiring GA practitioners, the problem-specific successful setup of a GA implementation and its parameter-trimming process remain more of an art rather than a pure mathematical process.

19.2.1 Basic GA Principles

The implementation of a genetic algorithm is based on five major principles that are as follows:

MAINTENANCE OF A POPULATION OF ENCODED SOLUTIONS. GAs do not operate on a single solution but on a number of problem solutions that form the evolving "population" of the GA. Thus, in order for the GA to begin, an initial number of problem solutions must be created. These solutions are usually produced at random using a good random number generator. But if there exist some known good solutions to the problem, they can also be part of the initial population, or the population can be built "around" them. In this way, the GA can be facilitated by starting from a good "point" in the solution space, and not from completely random solutions, that most probably have poor qualities.

Moreover, one of nature's important principles employed by GAs is the operation on DNA-like encoded solutions. This means that the solutions in the GA's population do not have their original form (e.g., real numbers or vectors of real numbers) but they are encoded in a form that resembles DNA. The encoding of solutions generally uses an alphabet of symbols with a low cardinality, like the binary alphabet, and defines a mapping function that can transform each candidate solution to a symbol string of specific or even variable length, and vice versa.

The encoded form of a solution is called a "genotype," whereas the decoded solution that is in a more human-understandable form, is called the "phenotype."

BUILDING A FITNESS FUNCTION. Once the population of solutions is created, it is time for evaluating them for their merits. This process corresponds to the natural process of breeding the individual up to a full adult organism that exhibits specific qualities in concern to its environment. In Gas, this is done instantly by calling a specific function called the "fitness function." This function takes an atom's genotype as a parameter and calculates its fitness as the returned outcome. The calling of the fitness function can be seen as taking a sample from the problem's search space. The GA pinpoints a candidate spot on the search space, and the fitness function tells us how good or how bad this solution is. So, successive GA generations of solutions that are combined together and are eventually evaluated can be seen as "sampling" the search space in an "intelligent" manner that drives successive "samplings" toward optimality.

THE PARENT SELECTION MECHANISM THAT IMPLEMENTS "SURVIVAL OF THE FITTEST." After the GA population of solutions has been produced and evaluated, the reproduction phase has to begin. Thus, solutions have to be selected from the "pool" of solutions maintained in the population, in order to combine their characteristics and produce offspring solutions that will inherit parts of their parents.

In nature, this process is not apparent and guaranteed for all individuals, as they have to (a) survive to reach maturity and the ability to reproduce and (b) fight other individuals in order to find their mate for reproduction. This natural law that governs species' reproduction is called the "survival of the fittest." In simple words, this means that fitter individuals have better chances to mate and reproduce, whereas less fit individuals have lesser chances. In a more mathematical form, it means that the probability of an individual to transfer its DNA to the next generation is proportional to its fitness.

In a GA, this "survival of the fittest" law is reflected in the "parent selection" mechanism that is used to select parents from the population of solutions. Parents are selected from the population based on their fitness. Fitter solutions have higher probability to be selected for reproduction, whereas solutions with worse qualities have lower or even zero probability to leave offspring.

REPRODUCTION OPERATORS THAT CREATE OFFSPRING. With the parents selected, it is time to apply the reproduction operators in order to create offspring solutions. The essence of the inner mechanism of reproduction operators is this: The information that is encapsulated into the parents' genotypes has to be combined and interchanged in order to produce a new offspring genotype that inherits parts of genotypic information from both parents.

The usual way to do this is by applying a version of the "crossover" operator. According to the simpler version of this operator, a random "cutting" point (or crossover point) is selected across the parents' genotypes, which are virtually cut at this point. Then the first half is taken from the first parent and the second half from the other and concatenated together to from the offspring genotype. Thus the offspring genotype inherits one part from the first parent and one part from the second one. This is shown in Figure 19.7.

While the crossover operator imitates the combining of parents' DNA observed in nature during a conception, there is another natural operator that simultaneously takes place and this is "mutation." Mutation in nature is triggered by environmental factors like chemical substances, or cosmic radiation, and results in the random alteration of the DNA information of the conceived organism. Of course, this is done with a very small probability.

Although infrequent, the mutation operator is irreplaceable for the effectiveness of the evolution process. This is because the only thing the crossover operator does is to recombine existing information within the population, making thus the process to be highly dependent on the information contained in the initial random population. The mutation operator covers this inefficiency by injecting new information within the population's genotypes, and thus making it possible to reach every possible solution in the search space regardless of the initial population. This is why mutation is a valuable operator to the whole evolution process.

Replacement of Parents with Offspring. When an adequate number of offspring have been produced, (i.e., equal to the number of parents), the production of the new generation can be considered as complete. For the algorithm to go, on the newly produced offspring must take the place of their parents in the evolution process. So the parent population is replaced by that of the offspring and a new reproduction circle begins.

19.2.2 GA Flow Diagram

A flow diagram that shows the basic algorithm of GAs is shown in Figure 19.10.

Observing the flow diagram, it is evident that the basic structure of a GA employs two loops: the inner loop or reproduction loop and the outer loop or generations loop. The inner (reproduction) loop is responsible for producing the offspring of the next population by executing the following sequence:

Parent Selection → Offspring Reproduction → Offspring Evaluation

The outer (generations) loop is responsible for repeating the evolution process for many iterations that are called generations. This loop executes the following sequence:

Offspring production→Replacement of parents→ termination criterion checking

So it is obvious that the basic GA flow is of limited complexity, a fact that favors its wide applicability.

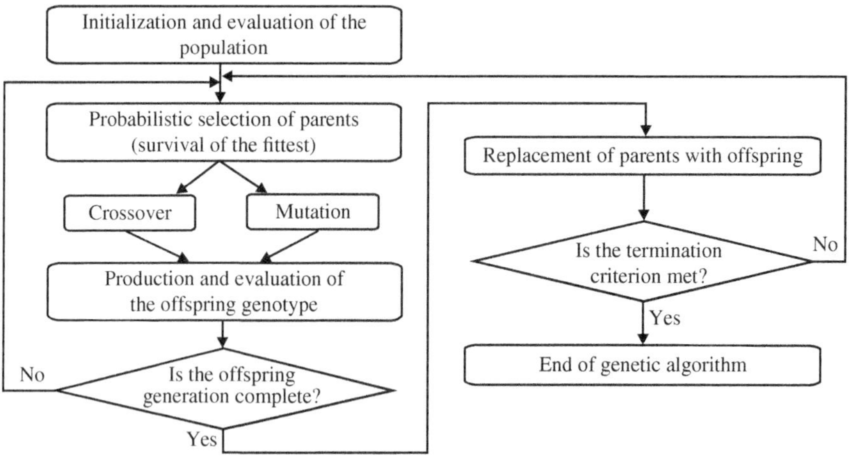

Figure 19.10 Flow diagram of a simple GA.

19.2.3 Solution Encoding

In order for the GA to be applied on a specific problem, one must first decide the encoding method to use, to transform candidate problem solutions into DNA-like strings. This transformation scheme is called a GA "representation" or "encoding" method.

The encoding that will be chosen is of high importance to the efficiency of the whole project. This is because the encoding used is responsible for the "shape" and the characteristics of the search space where the GA will operate. Thus, a suitable and somewhat "clever" encoding may facilitate the optimization process, whereas an inappropriate one may make GA's effort harder.

For the transformed solutions to be close to their natural equivalents, one must use a symbol alphabet of low cardinality to form encoded genotypes, similar to nature's DNA chains that are made of sequences of "nucleotide bases" (cytosine, guanine, adenine, and thymine). As GAs are implemented on computers that natively use the binary alphabet, it is a common practice to use the alphabet of 0s and 1s to encode GA solutions. This choice facilitates all GA operations, by using binary operators and functions that are built-in for most software development languages.

In simpler problem cases, the length of genotypes is constant whereas in some cases variable-length genotypes are used. Thus, for a simple case, if S is the set of all possible solutions to a problem, the encoding function E maps solutions from S to binary strings of length L and vice versa:

$$E : S \leftrightarrow \{0, 1\}^L$$

In case of a single variable problem of the form:

Find $x \in [\min, \max] \subseteq R$: maximize (or minimize) $F(x)$ the single variable x may be encoded to a binary string of length L:

1	0	0	1	0	1	1	0	1	0

Binary string of length L

Consequently, in a multivariable problem of the form:

Find $\vec{x} = (x_1, x_2, \ldots, x_n) \in S \subseteq R^n$: maximize (or minimize) $F(\vec{x})$ where:

$$x_1 \in [x_{1,\min}, x_{1,\max}] \subseteq R$$
$$x_2 \in [x_{2,\min}, x_{2,\max}] \subseteq R$$
$$\ldots$$
$$x_n \in [x_{n,\min}, x_{n,\max}] \subseteq R$$

Each variable may be encoded to a binary string of length L. These substrings (chromosomes) are then concatenated to form the unified solution genotype, as shown

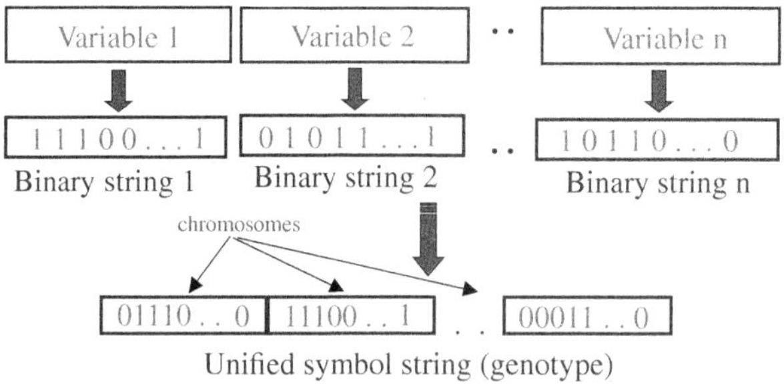

Figure 19.11 Encoding of solution vectors to multipart binary strings in a multi variable problem.

in Figure 19.11. The length of each binary string may not be necessarily uniformly equal, but may vary according to the required precision for each variable.

In the sequel, a number of encoding paradigms are demonstrated for different kinds of problems:

Case 1: Problems with Integer Variables. These problems are of the form:

Find $\vec{x} = (x_1, x_2, \ldots, x_n) \in S \subseteq I^n$: maximize (or minimize) $F(\vec{x})$

where

$$x_1 \in [x_{1,\min}, x_{1,\max}] \subseteq I$$
$$x_2 \in [x_{2,\min}, x_{2,\max}] \subseteq I$$
$$\ldots$$
$$x_n \in [x_{n,\min}, x_{n,\max}] \subseteq I$$

Let us consider, for example, a benchmark problem defined by De Jong [27] called De Jong's *F5* function or Shekel's foxholes:

Maximize

$$f(\bar{x}) = 0.002 + \sum_{j=1}^{25} \frac{1}{j + \sum_{i=1}^{2} (x_i - a_{ij})^6}$$
$$x_n \in [-65536, 65536] \subseteq I, \; i = 1, 2$$

A visualization of the function's search space is shown in Figure 19.12.

In this problem, variables x_1, x_2 are integers and each may have 131,072 different discrete values, from −65,536 to 65,535. Thus, for each variable, the encoding binary string should have a length of $L = \log_2 131,072 = 17$ (2^{17}=131,072). So the genotype will have a total length of 2×17=34 bits. As problem variables have also negative values, a method should be chosen to represent the integer range [−65536, 65535] to binary numbers (biased, 2s complement, etc.). If the "biased" method is chosen, then the encoding sequence is as follows:

Take x_i variable→add the bias 65,536→number is positive→convert to 17 bit binary and the decoding sequence is obviously symmetrical:

Take 17 bit binary→convert to integer→subtract the bias 65,536→take x_i variable

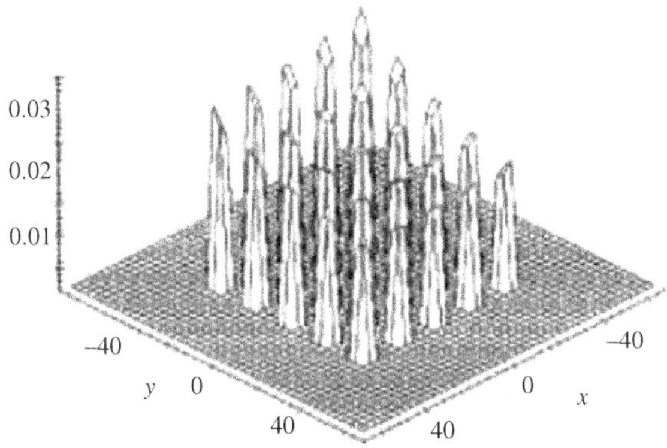

Figure 19.12 De Jong's F5 or Shekel's foxholes.

Some examples of solution vectors and their corresponding encodings for this problem are shown in Table 19.1.

NOTATION ON BINARY AND GRAY CODING. Binary encoding of integer numbers creates the following discontinuity in symbol space: Successive numbers may have a large hamming distance. For example, numbers 01111111_2 (127_{10}) and 10000000_2 (128_{10}) are successive as integers but very distant in binary (symbol) space as they have a hamming distance of 8. To overcome this, one may use Gray coding instead of binary to encode integers. According to Gray coding, two successive integers differ in only 1 bit, thus having a hamming distance of 1. Table 19.2 shows the first eight integers encoded as binary and as Gray-coded binaries.

CASE 2: PROBLEMS WITH REAL VARIABLES. These problems are of the form:

Find $\vec{x} = (x_1, x_2, \ldots, x_n) \in S \subseteq R^n$: maximize (or minimize) $F(\vec{x})$.

where

$$x_1 \in [x_{1,\min}, x_{1,\max}] \subseteq R$$
$$x_2 \in [x_{2,\min}, x_{2,\max}] \subseteq R$$
$$\ldots$$
$$x_n \in [x_{n,\min}, x_{n,\max}] \subseteq R$$

TABLE 19.1 Decoding examples for binary numbers to integer numbers

Genotype			
Chromosome 1	Chromosome 2	x_1	x_2
00000000000000000	00000000000000000	−65,536	−65,536
00000000000000001	00000000000000000	−65,535	−65,536
…		…	…
11111111111111111	00000000000000000	65,536	−65,536
11111111111111111	00000000000000001	65,536	−65,535
…		…	…
11111111111111111	11111111111111111	65536	65536

TABLE 19.2 Coding integer numbers in binary and Gray code

Integer	Binary	Gray
0	0000	0000
1	0001	0001
2	0010	0011
3	0011	0010
4	0100	0110
5	0101	0111
6	0110	0101
7	0111	0100

Let us consider, for example, a benchmark problem defined again by De Jong [27] called De Jong's F_1 function or sphere:

$$\text{Minimize:} f(\bar{x}) = \sum_{i=1}^{n} x_i^2,\ x_i \in [-5.12{,}5.12] \subseteq R,\ i = 1, n$$

A visualization of the function's search space for $n = 2$ is shown in Figure 19.13 (z-axis is inverted).

In this problem, variables x_1, x_2 are real numbers, and take infinite values in the range [−5.12, 5.12]. On the other hand, binary strings (or symbol strings in general) of a given length L have a finite number of different combinations. Thus, we must discretize the range of real variables into a number of distinct values, in order for the last to be encoded as symbol strings. The details of descretization are directly affected by the length of chromosomes L that defines the total number of discrete values that will be encoded.

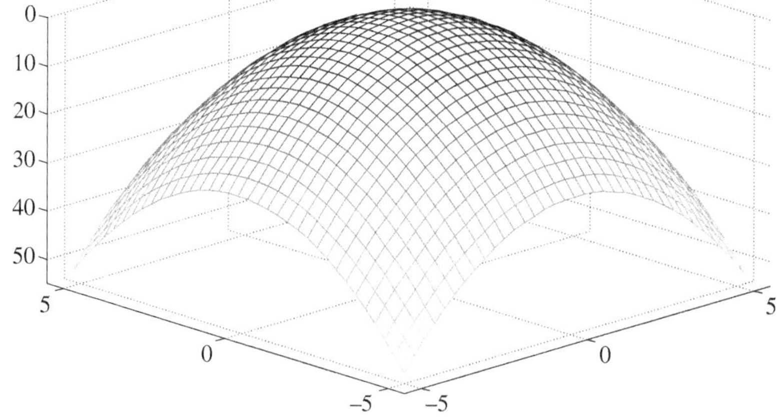

Figure 19.13 De Jong's F1 or sphere.

TABLE 19.3 Discrete values encoded and the corresponding binary strings

Discrete values	Integer	Binary chromosome
−5.12	0	00000000
−5.0798	1	00000001
−5.0396	2	00000010
−4.9995	3	00000011
…	…	…
−0.02007	127	011111111
0.02007	128	100000000
…	…	…
5.0798	254	11111110
5.12	255	11111111

For example, if we decide to use 8 bits for each variable, this will give us $2^8 = 256$ different values in the range [−5.12, 5.12] and result in a discretization step of

$$\begin{aligned} \textit{Discretization step} &= (\max - \min)/(\textit{discrete values} - 1) = \\ &= [5.12 - (-5.12)]/(256 - 1) = 10.24/255 \approx 0.040157 \end{aligned}$$

The discrete values encoded for one variable and the corresponding binary strings are shown in Table 19.3. If the discretization step is too big and the accuracy of discrete values is not adequate, it can be lowered just be increasing the number of bits for the encoded chromosome. For example, if in the former case we choose 10 bits, and thus $2^{10} = 1024$ different values, then the discretization step drops at a value that is one fourth of the original:

$$\begin{aligned} \textit{Discretization step} &= (\max - \min)/(\textit{discrete values} - 1) = \\ &= [5.12 - (-5.12)]/(1024 - 1) = 10.24/1023 \approx 0.0100392 \end{aligned}$$

So for real numbers, the encoding sequence is as follows: take x_i variable→subtract min value→divide by the discretization step →round to the nearest integer→convert to binary

(e.g., for $x = -5.0$ we do the following calculation:

$$\frac{(-5.0 - (-5.12))}{1024 \cdot 255} = 2.988$$

Then we round it to the nearest integer: *round*(2.988) = 3, and convert it to binary: $3_{10} = 00000011_2$. It is obvious that the discretization affects the numbers' accuracy, as $x = -5.0$ is encoded to 00000011 that represents the real value of −4.9995)

The decoding sequence is obviously symmetrical:

Take the binary→ convert to integer→ multiply by the discretization step→ add min value

CASE 3: COMBINATORIAL PROBLEMS. Except from problems that employ continuous variables, there is another category of problems that are diverse in nature and therefore it is difficult to find a common formal description for all of them: the combinatorial problems.

These problems usually demand to find a feasible perturbation of a number of objects in such a way as to optimize a cost function. Some subcategories of combinatorial problems are

(a) scheduling and timetabling problems;
(b) cutting and packing problems;
(c) networking and routing problems;
(d) resource allocation problems;
(e) graph problems (traveling salesman, graph coloring, etc).

In combinatorial problems, special representation schemes are used that are mostly problem specific and best suit the specific GA application.

For example, Let us consider the unit commitment problem that is a common combinatorial problem in power systems. It involves determining the start-up and shutdown schedules of thermal units to be used to meet forecasted demand over a future short-term (24–168 h) period. The objective is to minimize total production costs while observing a large set of operating constraints. This problem is schematically shown in Figure 19.14.

As units may be exclusively on or off, at a certain hour, we can represent this information by a binary digit. Thus H bits can encode the schedule of a single unit and consequently $N \times H$ bits can be used to encode the whole schedule. This encoding is shown in Figure 19.15.

For different combinatorial problems, other problem-specific encodings have been proposed, like "order representation" for ordering problems, "tree-based

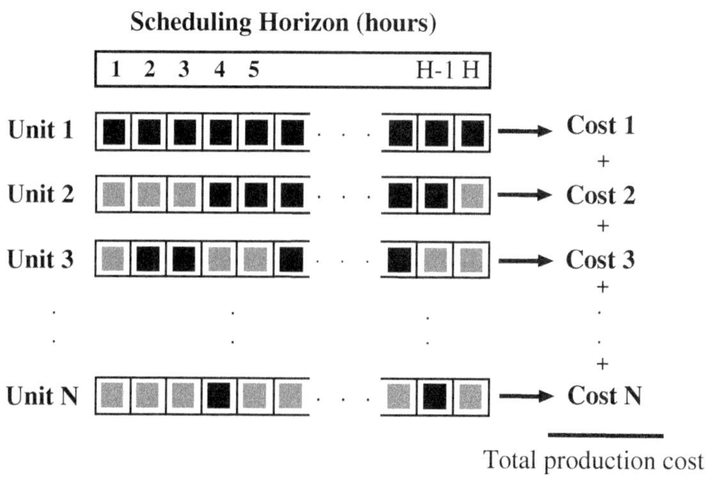

Figure 19.14 Schematic of the unit commitment problem (black squares indicate units are on for that hour).

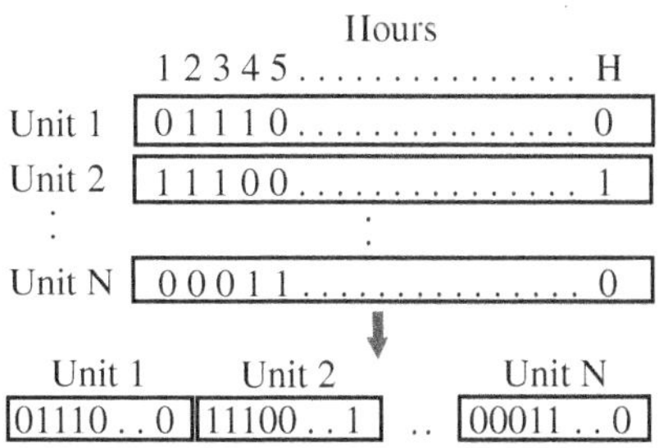

Figure 19.15 Encoding method for the unit commitment problem.

representation" for optimizing complex structures like circuits, and computer programs, etc.

19.2.4 Fitness Function

For the GA to work, one must build a function with the ability to evaluate each genetically produced solution: the fitness function.

The fitness function defines a mapping from the solution space S to the set of reals:

Fitness function $F(x) : S \rightarrow R$

Through this function, the GA is able to evaluate samples from the search space that will be used to further guide the search to new sampling areas and finally to the optimum.

As it is easily conceived, this function is crucial in many ways:

(a) It determines the cost function that the GA has to optimize, representing the real problem to be solved;

(b) it sculpts the shape of the search surface where the GA operates;

(c) it determines GA overall execution time, as it is called very frequently.

GAs do not impose any restrictions as of the characteristics of the fitness function, like continuity, convexity, or differentiability. Also, in some real problem cases the "fitness function" may not be a "function" at all, as the evaluation of solutions may demand real-world experimentation and a fitness that comes as the result of a real practical experiment. However, it has to have a couple of merits that are as follows:

(a) It has to be as fast as possible, because it is the most time-consuming component in the GA and it is called most frequently;

(b) it must accurately represent the real problem's cost function;

(c) it must be consistent and noise-free, that is for the same genotype it must yield the same quality number. Otherwise, it may mislead the GA in the optimization process.

Because of its role of sampling the search space and because of its time-consuming character, GA practitioners compare the performance of different genetic algorithms or other evolutionary methods over the same total number of fitness evaluations, and not over the same number of generations.

19.2.5 Parent Selection Methods

For implementing the "survival of the fittest" rule, GAs must employ proper parent selection methods that will select parents with a probability proportional to their fitness. Thus, high quality solutions should be selected quite often, whereas low quality solutions should be selected rarely or not at all. The probability of reproduction for "bad" solutions is deliberately not zeroed, as the crossover between a good and a bad solution may lead to an even better genotype. Moreover, the existence of bad solutions as well as good ones increases the diversity of information contained within the population and prevents premature convergence.

Many selection methods have been proposed in the literature [28, 29]. In the sequel, four of the most commonly used selection methods are surveyed.

METHOD 1: ROULETTE WHEEL SELECTION. This is probably the most commonly used selection method in GAs. This method selects parents with a probability directly proportional to their fitness. It works as follows: It virtually assigns a sector of a wheel to every parent in the population, according to the parents' relative fitness within the population. Then it spins the wheel. Obviously, the probability of selecting a parent is proportional to its relative fitness within the population ($F_i / \sum F_i$). This method is demonstrated in Figure 19.16.

Although simple to implement and obviously fair, this methods has some drawbacks which are as follows:

(a) It may lead to premature convergence, as extremely fit solutions may dominate the population very quickly;

(b) when the population is small, statistical deviations may occur from the expected selection probabilities for each parent;

(c) in case of a converged population, where the fitness values of all parents are similar, there is no "selection pressure" toward the best genotypes;

(d) The method imposes a restriction: Fitness values must be positive.

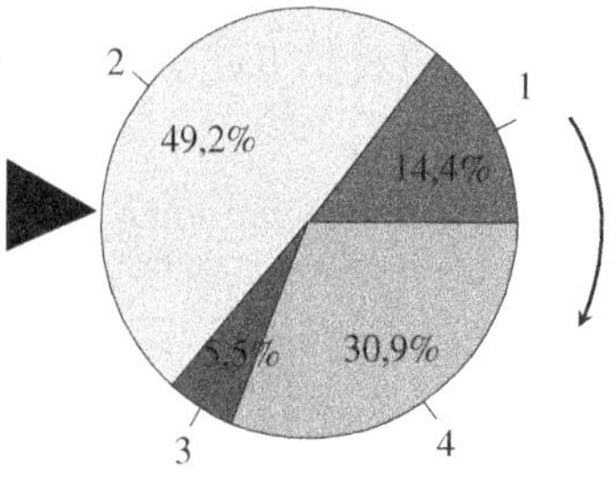

No	Genotype	Fitness	Percent
1	00110110	169	14.4
2	10111001	576	49.2
3	00101111	64	5.5
4	11101100	361	30.9
	Total	1170	100

Figure 19.16 The roulette wheel parent selection method.

METHOD 2: DETERMINISTIC SAMPLING. This method is also proportional, but it precalculates the number of times each parent should be selected, according to its fitness. Thus, it works in a more "deterministic" way, and it does not leave everything to be calculated at random.

This method works like this way: First, it calculates the expected number of times each parent should be selected, according to its relative fitness within the population, as

$$E(x_i) = 2 \times N \times \left(F(x_i) / \sum_{i=1}^{N} F(x_i) \right)$$

where x_i is the ith parent, $E(x_i)$ is the estimated number of times of selection, $F()$ is the fitness function, N is the population size, $2 \times N$ is the total number of selections to perform (two parents for each offspring), and $E(x_i)$ is generally a real number with a fractional part. Thus, initially parent x_i is selected a number of times equal to the integer part of $E(x_i)$:

$$Sel(x_i) = \mathrm{int}(E(x_i))$$

As it may happen that

$$\sum_{i=1}^{N} Sel(x_i) < 2 \times N$$

In that case, and in order to decide the remaining selections

$$r = 2 \times N - \sum_{i=1}^{N} Sel(x_i)$$

the method does the following:

(a) It sorts parents according to the fractional part of $E(x_i)$;

(b) the remaining selections r are made among the parents that are higher in the sort order.

With a distant view, we could state that deterministic sampling is nothing but a "deterministic" version of roulette wheel parent selection that corrects the possible statistical drifts of the second, in cases of small populations.

METHOD 3: TOURNAMENT SELECTION. This method is quite different from the previous methods. Given a population of size N, it selects k parents with uniform random probability. Then the k selected parents participate in a virtual tournament, and the best parent among the k wins, and is selected as a parent for reproduction.

The number of k is called "tournament size" and is critical for the performance of the method.

When $k \rightarrow N$, the method is degraded to simply selecting the best individual of the population, a practice that destroys information diversity and will lead the population into mathematically certain homogenization.

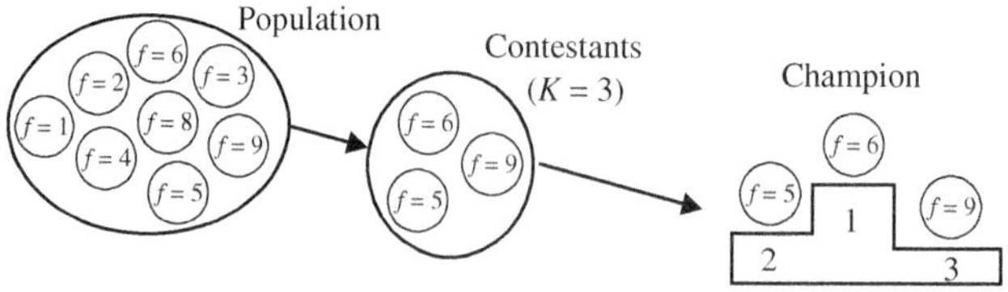

Figure 19.17 Tournament selection method.

When $k \rightarrow 0$, the method is again degraded to uniform random selection that will apply zero "selection pressure" to the best of the genotypes, therefore turning the whole algorithm into undirected random search.

In Figure 19.17, a graphical representation of the tournament selection method is displayed.

METHOD 4: RANKING SELECTION. This method tries to overcome the disadvantages of proportional selection methods like roulette wheel and deterministic sampling. It does that by basing selection probabilities not on the individuals' relative fitness within the population but on a heuristic scale of ranks assigned to each parent in accordance with their fitness-based sort order.

In detail, it works as follows: First, it sorts population individuals according to their fitness, in a descending order, so that best individuals are on top and worst individuals are at the bottom of the array.

The position (1,..., N) of each individual x_i, $i = 1,\ldots, N$ in the sorted array is called the individual's "rank" and denoted by $r(x_i)$. Then a transformation function $h(.)$ is used that takes the individual's rank as an input and calculates the selection probability for this individual.

A commonly used function is the following:

$$h(x_i) = \frac{N - r(x_i) + 1}{\sum_{i=1}^{N} r(x_i)} \times 100\%$$

where N is the population size, x_i is the ith individual in the population, $r(x_i)$ is the associated rank of each individual.

For example, in a population with $N = 3$ that contains solutions A, B, and C with fitness values $F(A) = 5$, $F(B) = 2$, and $F(C) = 19$, the sorted order of individuals is C, A, B, and thus their rank values are $r(A) = 2$, $r(B) = 3$, and $r(C) = 1$. Obviously, $\Sigma r(x_i) = 6$, so after calculations we get

$$h(A) = (3 - 2 + 1)/6 \cdot 100\% = 33.33\%$$

$$h(B) = (3 - 3 + 1)/6 \cdot 100\% = 16.66\%$$

$$h(C) = (3 - 1 + 1)/6 \cdot 100\% = 50.00\%$$

Ranking selection has the advantage of being able to soften and equalize large variances observed in the individual fitness values to more logical levels, but it can also magnify the minor differences of individuals of a population that has converged and homogenized. Its disadvantage is that it requires some computational overhead in order to work.

19.2.6 Basic Genetic Operators

After parent selection, it is time for the reproduction phase, where child solutions will be born form parent ones, by combining the "genetic" information encapsulated inside the selected parents. This is done by genetic recombination and mutation operators that operate on the parents' genotypes in a specific combinatorial and stochastic manner, and finally produce the offspring genotype, that must inherit solution characteristics from both parents. The basic operators used in GAs are essentially two: the "crossover operator" and the "mutation operator."

19.2.6.1 The Crossover Operator

The most well-known and widely used operator for the GA reproduction task is the "crossover operator" [16, 30]. The crossover operator has been proposed and used in a number of variations; some of which will be discussed here. It can be applied not only on binary-coded GAs but also on GAs that use other encoding methods, sometimes with slight adaptations. It generally takes parts of the solution genotypes from both parents and concatenates them into a new genotype that represents the offspring.

The simpler form of the crossover operator is "one-point crossover." This version of crossover simply selects a "cutting point" or "crossover point" within the parents' genotypes with uniform random distribution and takes the first part from one parent and the second part from the other, in order to build the offspring genotype. Obviously, for a genotype of length L, the possible crossover points are $L - 1$. This procedure is shown in Figure 19.18.

As shown, from each operation we may obtain two "conjugate" offspring genotypes that both inherit patterns from their parents. This fact can reduce the required operations in half. Thus for producing an offspring population of 100 genotypes only 50 crossover operations are needed.

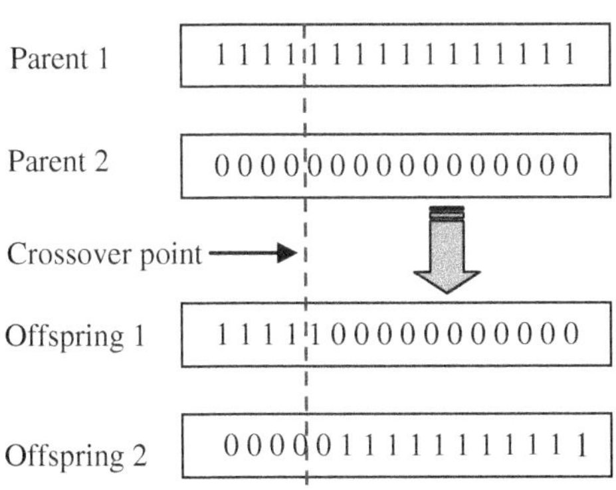

Figure 19.18 One point crossover operator.

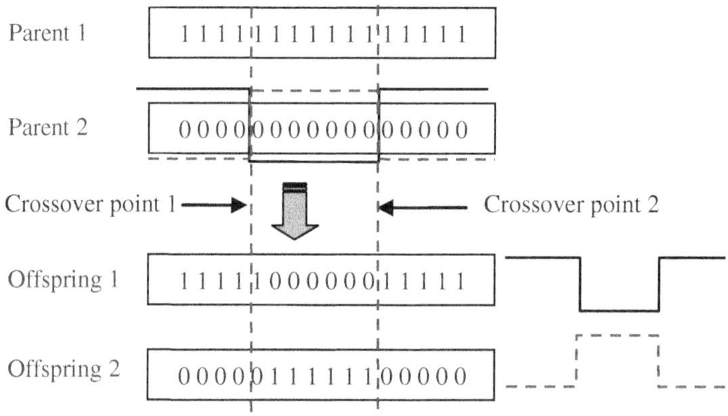

Figure 19.19 Two point crossover operator.

Although sufficient for genotypes with small lengths, the one-point crossover operator can be proven inadequate for large genotypes. Therefore, a more enhanced crossover version has been proposed that improves the ability to combine the parent symbol strings in a more efficient manner.

This version is the "two-point crossover operator." This operator selects two crossover points again with uniform random distribution and alternately select parts between these points from the first and the second parent. This procedure is demonstrated in Figure 19.19.

Again, in this version of crossover, two symmetrically produced offspring can be obtained by each operation.

Along this line, we could upscale the crossover operator, by increasing the number of crossover points, to enhance even more the operator's combinatorial ability. In this case, we have the "multipoint crossover operator" [31], for which the number of crossover points can be problem dependent and must be defined by the GA practitioner for each application, usually after some experimentation. The multipoint crossover operator is shown in Figure 19.20.

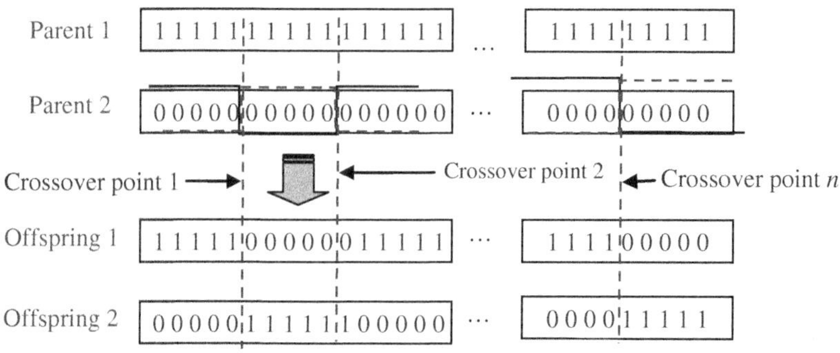

Figure 19.20 The multipoint crossover operator.

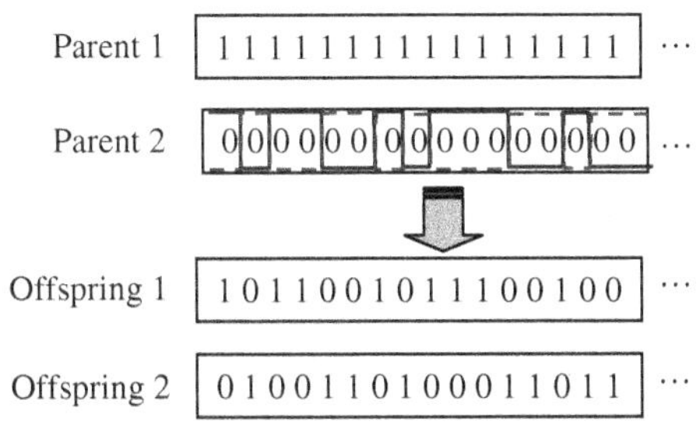

Figure 19.21 Uniform crossover.

A special worth-mentioning version of the crossover operator is "uniform crossover." This operator selects each bit or symbol of the offspring genotype from the first or the second parent with equal probability. It can be seen as a generalization of the multipoint crossover scheme. The uniform crossover operator is shown in Figure 19.21.

The already mentioned crossover operators, although being sufficiently general, they are not always suitable for special encodings like order-based encodings, where they could produce infeasible solutions. Such a case for an instance of the "traveling salesman" problem is shown in Figure 19.22. As shown in this figure, offspring genotypes may be infeasible, by containing the same node number twice, and not "visiting" a specific node. Obviously, offspring 1 contains node no. 2 twice and does not contain node no. 4, whereas offspring 2 contains node 4 twice and does not contain node 2.

For such cases, special recombination operators have been invented in order to preserve feasibility. A paradigm of such operators is the "partially mapped (or matched) crossover" or PMX for brevity [32]. This operator works as follows: It selects two cutting points along the parents' genotypes and then defines an association among the corresponding symbols of the two parents that lie inside the cutting points. Then the symbols of the first parent that are associated with symbols of the second are replaced with the corresponding symbols. The symbol replacements take place through the whole genotype and not only between the cutting points. This procedure produces the first offspring. The same happens for the second parent in order to produce the second "symmetrical" offspring.

The PMX operator is demonstrated in Figure 19.23.

Traveling salesman problem	
Parent 1	1 2 3 ¦ 4 5 6 7 8 9
Parent 2	1 4 3 ¦ 2 8 7 6 5 9
Crossover point	→
Offspring 1	1 2 3 ¦ 2 8 7 6 5 9
Offspring 2	1 4 3 ¦ 4 5 6 7 8 9

Figure 19.22 Simple one point crossover may create infeasible solutions on special encodings.

Parent 1: 9 8 4 | 5 6 7 | 1 3 2
Parent 2: 8 7 1 | 2 3 9 | 5 4 6
Symbol associations:
5 ↔ 2
6 ↔ 3
7 ↔ 9
After symbol replacements:
Offspring 1: 7 8 4 | 2 3 9 | 1 6 5
Offspring 2: 8 9 1 | 5 6 7 | 2 4 3

Figure 19.23 The PMX operator.

When looking at the offspring genotypes, we can confirm that they truly resemble their parents, meeting thus the "inheritance" criterion. However, the PMX operator also guarantees their feasibility and thus it is undoubtedly a most suitable recombination operator for this encoding. Other crossover operators of this kind are the order crossover [33], cycle crossover [34], and edge recombination [35].

In real-coded GAs, where the solutions are not encoded and are usually in the form of real parameter vectors, the crossover operator cannot be used without some customization. In cases of multivariable problems, where each genotype consists of a vector of n real parameters, we can assume that there exist $n - 1$ possible cutting points "between" the vectors' parameters. Thus we can build a crossover operator that can exchange parameter values as a whole, between parent vectors (Figure 19.24).

In the case of a single-variable problem, it is obvious that the above operator cannot be applied. In this marginal case, a different recombination operator can be used: the mean-value recombination operator (Figure 19.25). According to this operator, the value of the offspring chromosome is calculated as the mean value of the corresponding parent chromosomes.

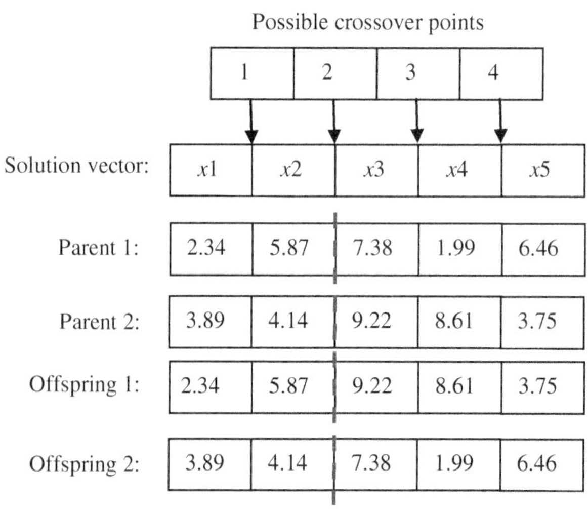

Figure 19.24 Crossover operator for real-coded GAs.

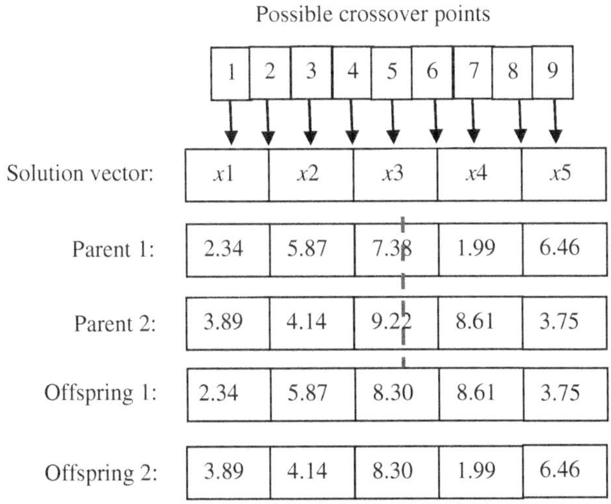

Figure 19.25 Generalized crossover operator for real-coded GAs with averaging capability.

The above two recombination methods can be integrated into a generalized recombination scheme with $2n - 1$ possible cutting points, one for each variable and one for each gap between variables. If the crossover point chosen is one of the "gap" points, then whole variables are exchanged. But if it is a "variable" point then the corresponding parent chromosomes are also averaged.

19.2.6.2 Mutation

Although the reproduction operators like crossover have the power to recombine genetic information of candidate solutions in an effective way, they still cannot do one thing: to produce information that does not already exist within the population.

It is true that recombination operators can only mix the existing information into new blends. Hence, the final outcome of the optimization process is highly depending on the initial population. If a solution subpart, that is needed for the optimum solution, happens to exist within some genotype of the initial population, then there is a nonzero probability that the optimum will be reached. If however this subpart does not exist anywhere in the initial population, then the optimum cannot be reached.

The operator that covers that gap is the "mutation" operator. This operator is assigned the task to inject new information within the population of solutions, by slightly perturbing the genotypes produced by crossover, with a small probability. The small probability is needed in order not to systematically disrupt the genotypes produced by crossover.

So, the mutation operator works like this way: It takes the offspring genotype that has just been produced by crossover and with a small probability (e.g., 1%); it alters the genotype's symbols, producing thus a slightly modified version of the original that now carries new information.

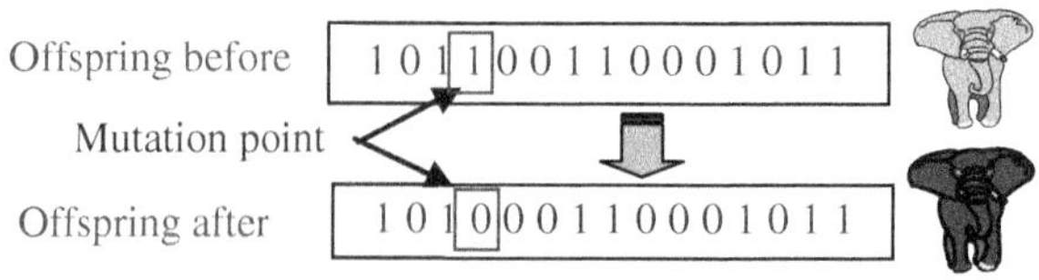

Figure 19.26 The binary mutation operator.

In case of binary-coded Gas, the mutation used is "binary mutation." This version of mutation alters the genotype's bits from 0 to 1 and vice versa with a small "per-bit' probability of about 0.01 or 1%. A mutation probability of 1% per bit means that it is expected that only 1 of 100 bits will be altered by mutation. This operator is shown in Figure 19.26.

The mutation operator adds diversity in the population of solutions and increases the dispersion of solutions throughout the whole search space. Using an extremely high mutation rate (per-bit probability) leads to the destructive randomization of produced solutions and totally neutralizes the effects of selection and recombination. Thus the GA is degraded into a random search algorithm.

On the other hand, too small mutation rates may lead to premature convergence and homogenization of solutions over local optima. Thus, it is easily understood that the appropriate mutation rate is crucial for the global GA performance.

While binary mutation is suitable for binary-coded GAs, other encodings may need different mutation operators. For example, in the case of real-coded GAs, where the genotypes retain their natural representation as (usually) real numbers, the mutation operator has to be an operator that will slightly alter offspring solutions. This can be done by adding a random variable to the offspring solution variables. The random variable may follow any random distribution. So for uniform random distribution and a single variable problem, the mutation operator will perform the following:

$$\text{Mutation}(x) = x + u(-a, +a)$$

where $u(-a, a)$ is a uniform random variable in the range$(-a, a)$. Similarly, for normal distribution, the mutation operator will be

$$\text{Mutation}(x) = x + N(0, \sigma)$$

where $N(0, \sigma)$ is a random variable of normal distribution, with a mean value of 0 and a standard deviation σ (σ^2 is called variance). The above are shown graphically in Figure 19.27.

Similarly, custom mutation operators are required for other encodings like order-based representations or tree representations. Of course in every case, the objective for the mutation operator is to preserve the feasibility of solutions.

19.2.7 Elitism

In traditional GAs, each parent generation is entirely replaced by the offspring generation. Due to the stochastic nature of the selection methods and the reproduction

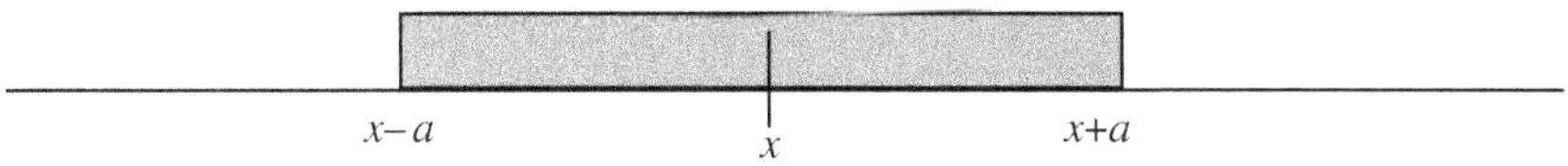

Mutation of a real variable by adding a uniform random variable

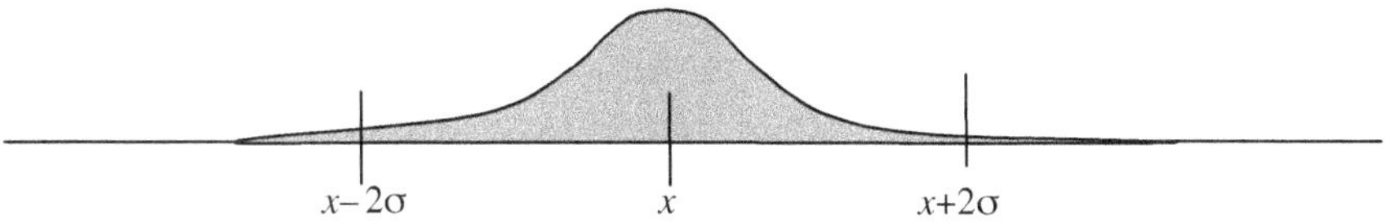

Mutation of a real variable by adding a Gaussian variable

Figure 19.27 Mutation operators for real-coded GAs.

operators, there is always a nonzero possibility that a good solution found at generation k will not survive the process and will not be transferred to generation $k + 1$. This may happen because the good solution:

(a) may not be chosen at all by stochastic selection methods;

(b) may be selected but be disrupted by crossover; or

(c) may survive crossover but degrade by mutation.

Due to this fact, we could observe during a GA run that the best-so-far solution of the GA may not monotonically increase toward the optimum, but it may decrease, and generally vary up and down.

To avoid this, a special technique is applied to GAs, called "elitism." The elitism technique does the following: Before offspring reproduction, it first finds the best solution of the parent generation. Then it copies that solution, intact, to the offspring generation. After that, the reproduction phase begins, but iterates for $N - 1$ times, and not N, as the first offspring has already been produced by elitism.

From another point of view, elitism may be seen as a cloning process of the best parent to the generation of offspring, with the aim to deliberately preserve its good DNA for the generations to come. Thus, a good solution found by the GA at an arbitrary generation cannot be lost through the stochastic evolutionary operators, but is carried to "infinity" with certainty, and can only get overwhelmed by an even better solution. Therefore, the mechanism of "elitism" assures that the best-so-far solution of a GA will monotonically increase, from generation to generation, toward the optimum.

19.2.8 Other Genetic Operators

A number of complementary genetic operators have been proposed in the literature. Some of them are problem specific, and some of them are generally applicable, but with an overall contribution that depends upon the specific problem considered. A small number of them are described in the following:

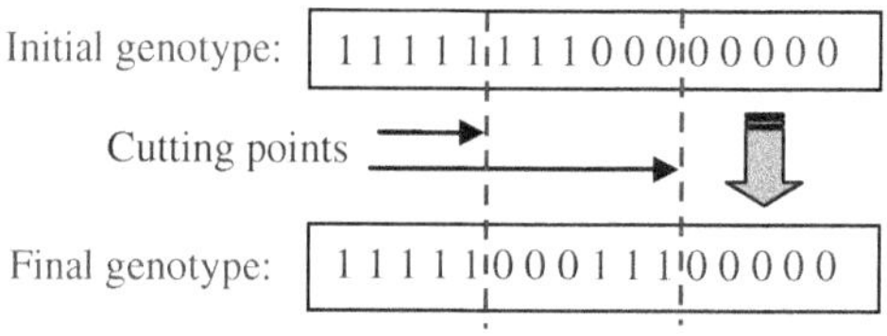

Figure 19.28 The inversion operator.

INVERSION: This operator is applied on a single genotype (like mutation) and was first proposed by Holland [16]. It selects two cutting points within the genotype's length, and the bits within these cutting points are inversed (Figure 19.28).

The inversion operator is more effective in cases of ordering problems, where it causes a middle part of the solution to be considered at a "reversed" order.

SWAP CHROMOSOMES OPERATOR: This operator [36–38] also works on a single genotype that has to comprise multiple chromosomes (e.g., one for each problem variable). It selects two chromosomes at random and swaps their values (Figure 19.29).

This operator may be very efficient in cases of problems where the solution exhibits some symmetry or homogeny. This is caused by the combined effect of this operator together with recombination that tends to copy genetic material from one chromosome to the other.

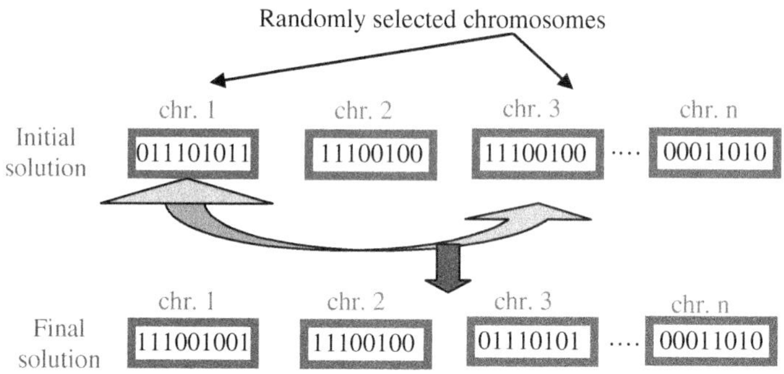

Figure 19.29 The swap chromosomes operator.

19.2.9 Hill-Climbing Operators

GAs can easily integrate other search and optimization methods, like analytical gradient descent methods, operational research methods (like dynamic programming), or other heuristic methods like simulated annealing. This can be done by implementing the different method as an explicit genetic operator that could act solely or additionally to the standard genetic operators on the whole or just a part of the population of solutions.

A common hybridization is with local search or hill-climbing techniques. Such techniques start from a given solution and perform local search in its neighborhood, intending to find even better solutions or a gradient in the search space that can lead to better solutions. These techniques work in an "elitistic" way. This means that they start from a specific solution point and try to find an even better solution point in the neighborhood of the first. If they find one, then the new solution point is taken as a base. Otherwise, the old solution point is retained.

The local search operator can be applied either only to the best solution of every generation, with the ambition to promote this solution even more toward optimality, or to every solution in the population. In the latter case, we have a special version of a GA where each genetically produced solution is locally optimized. This GA version has its own name, which is "memetic algorithms" [39]. Memetic algorithms nowadays constitute an independent scientific area that attracts a large number of researchers.

The hybridization of GAs with local search techniques is very fruitful. This is because GAs, although they are efficient in locating the general area of the global optimum, they are often unable to explore this area exhaustively and effectively to reach the exact optimum with a high degree of accuracy and consistency and within a small number of generations. This is mainly due to the random-based mechanisms for the selection, recombination or even mutation of the evolving solutions, and the homogenization of the population around the area of the global optimum. This problem becomes more serious in cases of constrained optimization problems where the violation of constraints is incorporated in the problem's fitness function as a penalty function that aims in leading the evolving solutions toward the feasible part of the search space. The penalty functions enhance the complexity of the search space and often lead to the creation of narrow "ridges" of random directions, around the feasible/infeasible border areas, that are difficult for the GA to discover and follow toward the exact global optimum.

Many hybrid genetic schemes have been proposed in the literature that combine GAs with hill-climbing or local search techniques [30, 40–43] to solve both continuous variable and combinatorial problems. Most GA–hill climbing hybrids are implemented for continuous variable problems. In such problems, the hill climbers used are often designed to perform independent steps along each axis in the space (one variable at a time) searching for better solutions. One such operator is the PhenoMute (PM) hill-climbing operator suggested in [36, 44, 37].

The PM hill-climbing operator can be applied only to continuous variable problems, where solutions are real valued vectors $(x_1, x_2, \dots, x_n)$. Moreover, it is usually applied only to the best solution of every generation in order to be economical in terms of fitness function evaluations. This operator works as follows:

1. Executes a loop for all solution parameters x_i, $i = 1, \dots, n$;
2. for each variable x_i, it adds and subtracts a small and a large step value (e.g., ± 0.01 and ± 1.0). The magnitude of the step values depends upon the range of the variable;
3. in every one of these four perturbations, the solution is evaluated by a call to the fitness function;
4. if at some step, the new calculated fitness is better than the old one, then the perturbation is adopted. Otherwise, the old solution is restored.

The operator is described in pseudocode in the following lines:
Operator phenotype mutation (S)

```
begin
    Q=evaluate(X=(x_1,x_2,...,x_n))
    X_orig=X
    for i = 1 to n do
          for Step in [+BigStep,+SmallStep,-BigStep,
          -SmallStep] do
            x_i = x_i + Step
            Qtest=evaluate(X)
            if Qtest better than Q then goto EXIT else X
            = X orig
          endfor
          label EXIT:
    endfor
end
```

As seen in the pseudocode, the operator begins with the x_1 parameter of the solution ($i = 1$) and performs four (4) modifications to it by adding the values +BigStep, +SmallStep, −BigStep, −SmallStep to the original parameter value, one by one. After each modification, the resulting solution is evaluated. If it is better than the original one, it replaces it and the modifications of x_i stop; otherwise, it is discarded and the modifications continue. The same procedure is repeated for all parameters in the solution vector, performing thus a local search in the neighborhood of the solution produced by the genetic search (Figure 19.30).

Although, for every generation, the operator adds at most $4n$ fitness evaluations, it accelerates dramatically the search speed toward the final optimum. The values BigStep and SmallStep represent a large and a small modification step for the solution's

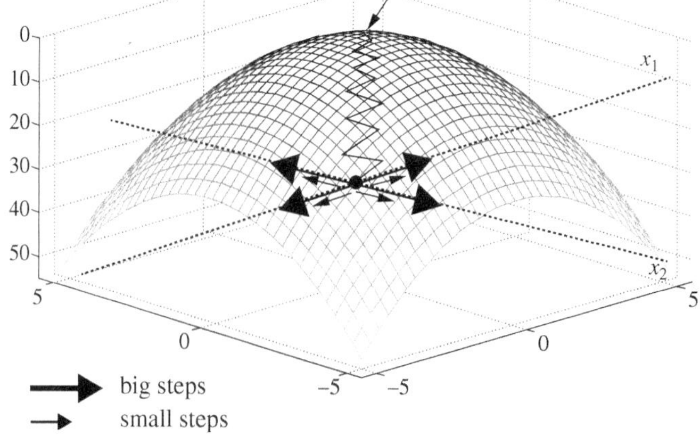

Figure 19.30 Orthogonal steps along each axis, performed by the PM hill-climbing operator.

parameters. The value of BigStep is used to "cover distances," whereas the value of SmallStep is used for fine-grained optimization.

19.2.10 Parent Replacement Methods

Once the offspring population has been produced, it must take the role of parents, in order for the evolution circle to continue. The replacement of parents with offspring signals the beginning of a new generation of solutions in a GA. But is there a single way to perform this replacement? The answer is no: There are more than one ways to replace parents with offspring. Thus, we must decide for one more GA parameter: the parent replacement method to use.

There are two basic variations of parent replacement, namely (a) generational replacement and (b) steady-state reproduction.

GENERATIONAL REPLACEMENT: This is the original parent replacement method conceived by the GA's creator J. H. Holland. According to this method, the whole population of parents is replaced by that of the offspring. This implies that

(a) there are two distinct populations in a GA implementation at any time: the parents and the offspring. Solutions are selected from the population of parents to mate and reproduce and gradually build the population of offspring;

(b) the size of the two populations must be identical, as the one replaces the other;

(c) as the old parent population is completely replaced, and as the offspring population consists of recombinations of the parents, it is not guaranteed that a good solution found in generation g will survive in generation $g + 1$;

(d) there is a "generational gap" between parents and offspring at each generation that can be seen as a discontinuous step in the optimization process.

STEADY-STATE REPRODUCTION: According to this method, when a new offspring is produced it directly replaces the worst of the parents and takes its place in the "mating pool," that is the set of solutions capable of being selected as parents. This implies that

(a) there is only one population at any time: the population of parents. When a new offspring solution is produced from the parents, it replaces one of them and takes its place in the parent population;

(b) the notion of "a generation" does not clearly exist in this parent replacement method, as the offspring reproduction and the immediate parent replacement just goes on and on. However, we can technically introduce such a concept by counting n offspring productions as a generation, where n is the size of the population;

(c) steady-state reproduction leads to a more smooth transition of the GA population toward optimality, as the population is gradually and partially replaced by new solutions. Moreover, highly fit solutions of a population instance are more probable to be retained within the population for a longer time, as the newly produced solutions replace the worst parents and not the best ones.

19.2.11 Fitness Scaling

The parent selection methods in GAs adopt a proportional scheme of matching the selection probabilities of parents with their fitness values. In most methods, the probability of selection of an individual X_i is proportional to its relative fitness within the population, or

$$P_{\text{sel}}(x_i) = \text{Fitness}(x_i) \Big/ \sum_{i=1}^{n} \left(\text{Fitness}(x_i)\right)$$

For example, let us consider the problem of finding the maximum value of the function $\sin(x) + 1$:

$$\text{Find } x_{\text{opt}} \in [0, 2\pi] : \text{maximize } f(x) = \sin(x) + 1$$

Let us consider the population of five solutions shown in Table 19.4:

In Table 19.4, the "percent contribution to the next generation" figures are calculated by each individual's relative fitness within the population ($P_{\text{sel}}(x_i)$), multiplied by 100%. Also, it is assumed that in order to produce the next generation of five individuals, 10 parent selections must be made.

In the above table, the relatively high contribution of good solutions to the next generation is more than obvious. Parent no 2 (genotype = 1.43) is the best solution in the population and has a probability of selection $P_{\text{sel}}(X_2) = 0.372$, or 37.2%, and an expected number of times to be selected is equal to 4.

Similarly, parent no. 1 is the second best and has a probability of selection $P_{\text{sel}}(X_1) = 0.282$, or 28.2% and an expected number of times to be selected is equal to 3. On the other hand, bad solutions like parent no. 5 have a very small probability of selection, and probably they will not be selected at all. Thus, it is expected that the GA will evolve the next generation mostly around the two best solutions (no. 1 and 2).

But if we change the fitness function of the problem to the following form:

$$\text{Find } x_{\text{opt}} \in [0, 2\pi] : \text{maximize } f(x) = \sin(x) + 100$$

Results of what happened in the next generation, we can find in Table 19.5, where the fitness values of solutions do not differ significantly; good solutions are not selected more frequently than bad solutions. This results in the impedance of the evolution progress of the GA.

TABLE 19.4 Population of five individuals

No.	Genotype	Fitness	Percent contribution to the next generation	Expected times to be selected
1	0.54	1.51	28.2	2.82 → 3
2	1.43	1.99	37.2	3.72 → 4
3	5.89	0.61	11.4	1.14 → 1
4	3.02	1.12	20.9	2.09 → 2
5	4.21	012	2.2	0.22 → 0

TABLE 19.5 Resulted population at the next generation

No.	Genotype	Fitness	Percent contribution to the next generation	Expected times to be selected
1	0.54	100.51	20.0879	2.008 → 2
2	1.43	100.99	20.1838	2.018 → 2
3	5.89	99.61	19.9080	1.99 → 2
4	3.02	100.12	20.0099	2.0 → 2
5	4.21	99.12	19.8101	1.981 → 2

This is what happens in a GA when its population converges around a local optimum solution, and the population's genotypes are homogenized. In such cases, the GA alone is not able to apply distinct selection pressure to the best of the genotypes, in order to boost the evolution process toward the exact optimum.

In order to solve this problem, we must apply a "fitness scaling" method to correct the scaling of the solutions' fitness [30]. Two are the most commonly used methods for this: windowing and linear normalization.

WINDOWING: After all solutions are evaluated, the minimum fitness value is found, $F_{\min}$. Then this value is subtracted from all fitness values $F(X_i)$ giving new fitness values $F^*(X_i)$:

$$\text{Find } F_{\min} = \text{minimum } F(X_i),\ i = 1 \ldots N,$$
$$F^*(X_\text{i}) = F(X_i) - F_{\min}, i = 1 \ldots N$$

This method acts like removing the direct current (DC) component that exists in all fitness values and does not contain information, and leaving only the alternating current (AC) variance of the fitness values, which is the important part.

For the above example, $F_{\min} = 99.12$. But in order to avoid to have the new fitness of the worst solution equal to 0, we can agree to a slightly lower value, like $F_{\min} = 99.00$. Thus the windowing method results are presented in Table 19.6.

It is obvious that the distinction between good and bad solutions has been restored.

LINEAR NORMALIZATION: Parent solutions are sorted in an ascending order, according to their original fitness values, such that $\forall i, j \in [1, N]$, if $i < j$, then $F(X_i) < F(X_j)$. Then, new fitness values $F^*(X_i)$, $i = 1, \ldots, N$ are assigned to them in the following manner: First, a "starting" fitness value is assigned to the worst parent (i.e.,

TABLE 19.6 Results for windowing method

No	Genotype	Fitness	New Fitness $F^*(X) = F(X) \rightarrow F_{\min}$	Percent contribution to the next generation	Expected times to be selected
1	0.54	100.51	1.51	28.2	2.82 → 3
2	1.43	100.99	1.99	37.2	3.72 → 4
3	5.89	99.61	0.61	11.4	1.14 → 1
4	3.02	100.12	1.12	20.9	2.09 → 2
5	4.21	99.12	0.12	2.2	0.22 → 0

with the smaller fitness value). Then this fitness value is increased linearly and is assigned to subsequent parents in the sorted list:

$$F^*(x_i) = StVal + i \cdot Incr$$

where *StVal* is the "starting value" for the worst parent, and *Incr* is the amount of linear increment of the new fitness values for subsequent parents.

Thus, for the method proposed and with the above example, we present a numerical example in Table 19.7:

TABLE 19.7 Exemple of linear normalization method

No	Sorted genotypes	Ascending fitness	New fitness $F^*(X) = StVal + i \times Incr$	Percent contribution to the next generation	Expected times to be selected
1	4.21	99.12	20	6.66	0.66 → 1
2	5.89	99.61	40	13.33	1.33 → 1
3	3.02	100.12	60	20.00	2.00 → 2
4	0.54	100.51	80	26.66	2.66 → 3
5	1.43	100.99	100	30.00	3.00 → 3

In the above paradigm, the "starting value" and the "increment" value have been taken equal to 20. It is again obvious that now there is some significant "selection pressure" on the best of the genotypes.

In general, we could transform original fitness values $F(X_i)$, $i = 1,\ldots, N$, in new fitness values $F^*(X_i)$, $i = 1,\ldots, N$, by using a transformation function $\Phi(.)$, such that

$$F^*(X_i) = \Phi(F(X_i)), i = 1 \ldots N$$

where function $\Phi(.)$ may or may not be linear.

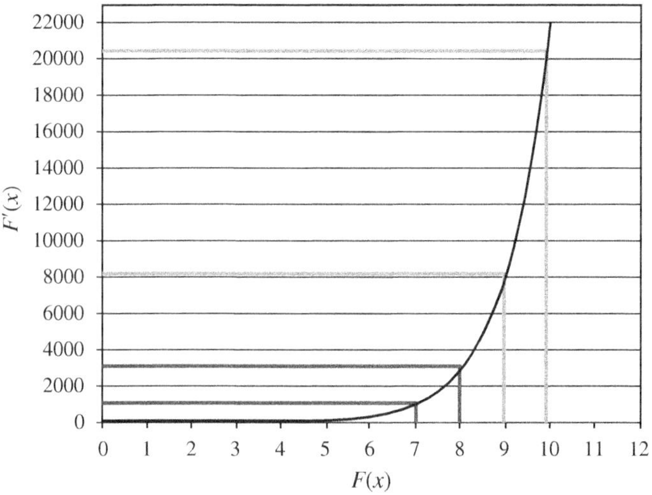

Figure 19.31 Exponential transformation of fitness values.

For example, we can consider an exponential transformation of the fitness function (Figure 19.31), by using the following function $\Phi(.)$:

$$F^*(x_i) = e^{F(x_i)}, \ i = 1 \dots N$$

This exponential transformation causes nonlinear increment of the fitness distance of neighboring solutions, as the fitness values increase toward optimality.

19.2.12 GA Control Parameters Determination

It is clear upto now that the internal mechanism of a GA employs a number of parameters that affect its performance. These parameters can determine the specific nature of the algorithm's search strategy, and the way it will try to approach the optimum. And in conclusion they affect the overall optimization performance of a GA. As such we could mention the following GA parameters:

(a) the population size;
(b) the parameters of the selection method (e.g., the tournament size);
(c) the number of crossover points in a multipoint crossover operator;
(d) the application probability of crossover;
(e) the per-bit probability of the mutation operator;
(f) the application probability of other complementary operators;
(g) the parameters of the fitness scaling function, etc.

Thus, the determination of all these parameters can be considered as an optimization problem itself. Moreover, the application of a GA may be accompanied by three problems:

1. finding the right setup for control parameters, for the specific problem;
2. the fact that for each problem the "right" setup may be different;
3. for a single problem, the optimal GA setup may be different at the various stages of the search process.

Many researchers have tried to address this problem [45–48], proposing several schemes for optimizing the control parameters of GAs, such as adaptive methods and metalevel GAs. The primary goal of these methods is to maintain the population diversity in a suitable level, even after the GA has converged on a suboptimal solution. The population diversity gives the GA a nonzero probability for escaping any local optima or deceptive attractors and discovering the global optimum.

An easy way to do this is by adopting a dynamic method for configuring the GA parameters that affect the population diversity. The latter depends on various parts of a GA, like the selection method used, the probability of mutation and mutation-like operators, and the fitness scaling function.

For example, in a GA that uses binary mutation with a per-bit probability PM, and a fitness scaling function of the form $F^*(X) = F(X)^n$, we could implement the following adaptive mechanism and incorporate it into the GA:

Adaptive method for controlling GA parameters:
At every generation do the following:

Step 1. Extract statistical data from the population that can measure the population's convergence or homogenization;

Step 2. If excessive convergence is detected, do the following corrections;

Step 2.1. Increase the mutation probability PM;

Step 2.2. Make the fitness scaling function less "aggressive," by lowering the exponent n;

*Step 3.*If excessive diversity is detected do the following corrections;

Step 3.1. Decrease the mutation probability PM;

*Step 3.2.*Make the fitness scaling function more "aggressive," by increasing he exponent n.

By an adaptive scheme like the above, the population's diversity can remain in a suitable level, throughout the whole search procedure, increasing the GA's probability of locating the area of the global optimum, and enhancing its overall performance.

19.2.13 Niche and Species

In nature, organisms are arranged into different "species" that are groups of organisms sharing the same basic characteristics that are also significantly different from those of other groups. Moreover, different populations or species occupy different "positions" in the ecology system, exhibiting different responses to the distribution of resources and their competitors. This differentiation is frequently addressed in biology by the phrase "niche and species." This notion of genetic diversity may be very useful to simulated evolutionary algorithms, especially in case of problems like the following:

$$\text{Maximize}\, f(x) = abs(\sin(5x)),\ x \in [0 \ldots \pi]$$

This function has 5 equally optimal peaks, at the points $x = k \cdot \pi/5,\ k = 1 \ldots 5$.

In such a landscape, it is expected that initially the population of GA will be scattered to all peaks. But as the simulation proceeds, the atoms will concentrate only on one peak. This phenomenon is called "genetic drift" [27] and is due to stochastic sampling errors, caused by the use of small finite populations. The right GA behavior should equally distribute the population solutions to the peaks and retain them there through the whole search process. This way the equally optimal peaks will enjoy the same level of exploration (Figure 19.32).

Trying to generalize the former we study another case, the case where the peaks are not equally optimal (Figure 19.33). In such a case, the GA should distribute the populations over the different peaks in proportion to every peak's quality. This means that highly fit peaks that may also be very promising should attract more individuals in order to be explored more exhaustively, whereas peaks with lower qualities should be occupied by lesser individuals.

In order to achieve this behavior, a number of methods have been proposed in the literature.

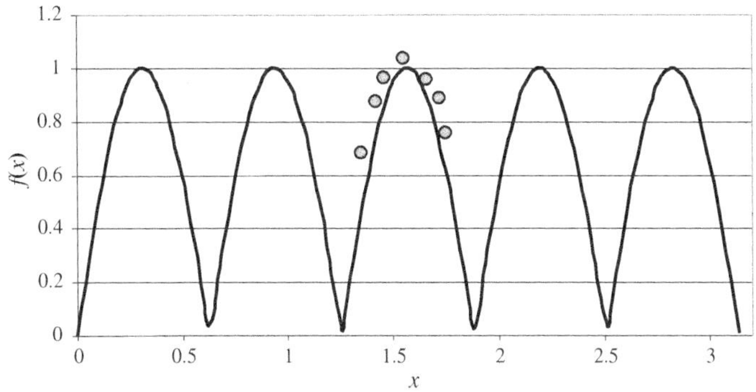

Figure 19.32 The fitness landscape of $f(x) = abs(\sin(5x))$, $x \in [0 \dots \pi]$ and the convergence of GA solutions.

De Jong's "Crowding" Method: This method [27] uses steady-state reproduction for the population renewal. Therefore, there is only one population of parents and when an offspring solution is produced it immediately replaces a parent in the population, keeping thus the population size constant. The novelty of the method is that when a new genotype is produced it replaces a parent with whom it has the largest similarity, or, in other words, the smallest Hamming distance. This way, the newly produced genotypes that happen to lie on a specific region of the search space replace parents that lie within the same region. This mechanism causes the development of "species" or genetically different subpopulations, in diverse areas of the search space, and assures their long-term survival and the long-term exploration of these areas.

Holland's "Fitness Sharing" Method: This method [16] uses a conceptually simple but a brilliant technique to favor the migration of solutions toward the various "peaks" of the search space and the development of local subpopulations with

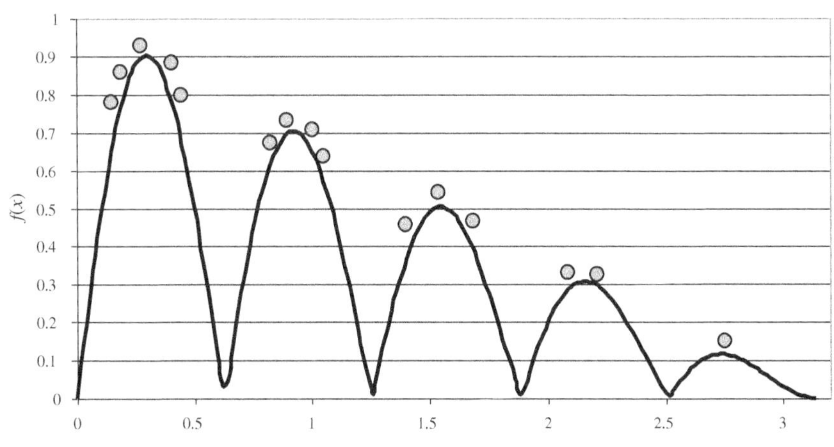

Figure 19.33 Optimal GA population distribution over peaks of variable quality.

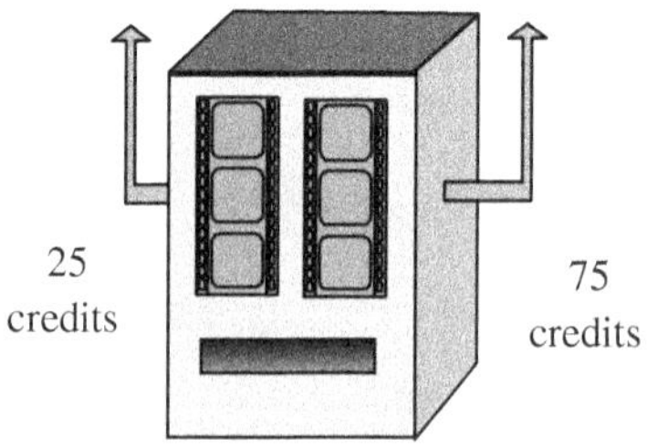

Figure 19.34 The two-armed slot machine for explaining Holland's fitness sharing method.

sizes proportional to the peaks' qualities. When assigning the fitness value to a solution, it first divides that value by the number of individuals that happen to lie in the neighborhood of that solution. Thus, a solution area of a specific quality, where a large number of individuals are placed, produces smaller fitness values for the individuals than a relatively poorer in quality area but with a smaller number of individuals. In order to better understand this method, Let us imagine a two-armed slot machine, as shown in Figure 19.34. The right arm is associated with a payoff of 75 credits, whereas the left one with a payoff of 25 credits. So, the fitness function of the problem can be formulated as

$$F(x) = \begin{cases} 75 \text{ credits, if } x = \text{right} \\ 25 \text{ credits, if } x = \text{left} \end{cases}$$

An initial random population of 100 players, with no prior knowledge of the system's fitness function, is expected to be equally spread among the two alternative solutions. But if we allow the players to reproduce according to their fitness (credit payoff), we will sooner or later observe that the whole population of 100 players will converge on the 75 credit solution.

For the application of the "fitness sharing" method, we have to use a modified fitness function like the following:

$$F(x) = \begin{cases} 75 \text{ credits}/N_{75}, & \text{if } x = \text{right} \\ 25 \text{ credits}/N_{25}, & \text{if } x = \text{left} \end{cases}$$

where N_{75} is the number of solutions that have chosen the right arm and $N_{25} = 100 - N_{75}$ is the number of solutions that have chosen the left arm.

According to this fitness function, if all individuals choose the 75-credit arm, they will be rewarded with the fitness value of $F(\text{right}) = 75$ credits/100 = 0.75 credits, each. At this situation, if a single player randomly migrates to the 25-credit solution, he will be rewarded the value of

$$F(\text{left}) = 25/1 = 25 \text{ credits}$$

whereas the rest 99 players that lie on the 75-credit solution will be rewarded the value of

$$F(\text{right}) = 75/99 \approx 0.7575 \text{ credits}$$

Obviously, now there is a strong motivation for the 99 individuals of the right arm to migrate to the left arm that has a much larger payoff. But as the migration proceeds, the population N_{25} of the players that have selected the left arm will gradually increase, causing the left-arm payoff to gradually decrease. Simultaneously, the population N_{75} of the players that have selected the right arm will gradually decrease, causing the right-arm payoff to increase. This dynamic system will eventually balance, and this will happen when

$$F(\text{right}) = F(\text{left}) \Leftrightarrow 75/N_{75} = 25/N_{25}$$

or when $N_{75} = 75$ and $N_{25} = 25$.

Of course, the above can be generalized for k-arms, with a fitness function of the form

$$F(x_i) = \text{Quality}(x_i)/N_{x_i},\ i = 1 \dots k$$

where Quality(x_i) is the original fitness value for solution x_i and N_{x_i} is the number of individuals occupying solution x_i, $i = 1 \dots k$.

In this case, the equilibrium point is reached when

$$\text{Quality}(x_1)/N_{x_1} = \text{Quality}(x_2)/N_{x_2} = \cdots = \text{Quality}(x_k)/N_{x_k}$$

It is obvious that at the equilibrium point, the size of the subpopulation at each alternative solution is proportional to the solution's relative fitness, as compared to the fitness values of alternative solutions.

Consequently, this method succeeds in creating and maintaining diverse subpopulations at all alternative solutions, of sizes that are proportional to each solution's relative quality in the search space. However, it exhibits a serious drawback: In practical problems that have millions of alternative solutions, the method cannot be applied in the above analytical form.

THE GOLDBERG AND RICHARDSON METHOD: This method [49] introduces a practical mechanism for forming "niche and species" within a genetically evolving population, closely following Holland's ideas that can be applied to any search space. According to this method, the following steps are taken:

(a) all population individuals are first evaluated by the initial fitness function $F(x_i)$, $\forall x_i$, $i = 1,\dots, N$;

(b) then for each individual x_i the following take place;

(c) We calculate a distance metric $d(x_i, x_j)$ for each combination of x_i and x_j, $j = 1,\dots, N$, as the hamming distance of their genotypic representations;

(d) we calculate a sharing measure $s(x_i, x_j) = s(d(x_i, x_j))$ of x_i and x_j based on their distance, and by using a function as the one depicted in Figure 19.35. According to this function, when the distance $d(.)$ between two solutions is close to 0, the sharing value is close to 1; and when $d(.)$ tends to its maximum value, $s(.)$ tends to 0.

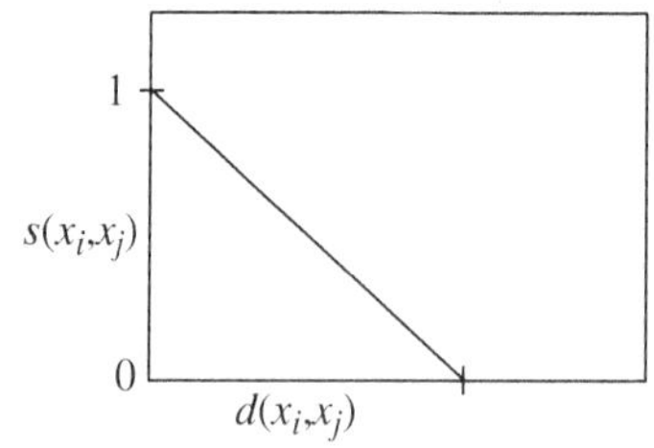

Figure 19.35 Triangular sharing function.

(a) Then a new modified fitness value is calculated for solution x_i by the following formula:

$$F'(x_i) = \frac{F(x_i)}{\sum_{j=1}^{N} s(d(x_i, x_j))}$$

According to this formula, the original quality value of x_i, $F(x_i)$ is divided by the sum of all sharing measures between i and j, $j = 1,\ldots, N$. Note that the sum includes the $j = i$ case, in order for the denominator not to become zero.

Thus when an individual x_i is different from all other individuals x_j, $j = 1,\ldots, N, j \neq i$, which consequently means that it is "alone" in its area, then in the sum of the denominator all sharing values $s(i,j)$, $j = 1,\ldots, N, j \neq i$, will be close to 0, while $s(i,i)$ will be equal to 1. This will cause the denominator to have a final value close to 1, and consequently $F'(x_i)$ to be nearly equal to $F(x_i)$. But if there are some individuals x_j, $j = 1,\ldots, N$, $j \neq i$ that lie within the neighborhood of x_i, then the sharing value $s(i,j)$ for these individuals will not be 0. This will cause the sum in the denominator to increase and $F'(x_i)$ to decrease.

MATING RESTRICTIONS: The formation of "niche and species" can be implemented or enforced by another scheme introduced by Goldberg [30], called as "mating restrictions." According to this scheme, there can be some restrictions imposed on the mating procedure as of who can mate with whom. When an individual A is first selected as a parent, then the second parent B is selected among those individuals that are in the neighborhood of A. In practice, the second parent is selected from a subset of the population, comprising individuals that have a similarity measure with solution A greater than a predefined threshold λ.

This way, it is most likely that the offspring solution produced by these parents that will also eventually replace them will lie at the same area of the search space, as its two parents. This mechanism also enforces the creation and maintenance of distinct subpopulations that stay within specific areas of the search space and reproduce themselves with a local scope. These populations persistently explore these areas to the extreme, offering suboptimal alternative solutions to the overall GA search procedure.

THE "MATING TEMPLATES" OF BOOKER: In his work [50], Booker proposes a method where the mating strategy of the GA is not a priori determined, but it is genetically evolved by the GA together with the solution genotypes. To achieve this, Booker

TABLE 19.8 Matching templates added to genotypes

Individual No	<mating template>	<functional genotype>
1	# 1 0 #	1010
2	# 0 1 #	1100
3	# 0 0 #	0000
...	...	...

adds to the genotype of every solution, a "mating template," which is an additional symbol string of the same length as the genotype, but constructed over a three-symbol alphabet {0, 1, #} as shown in Table 19.8.

The symbols 0 and 1 in a mating template must exactly match corresponding 0s and 1s in the functional genotype, but a "#" is a "don't care" or "wildcard" symbol that matches both 0s and 1s at that position.

So in order for a parent A to mate with parent B, the mating template of A must match the functional genotype of B or vice versa (unidirectional match).

For example, considering the individuals of Table 19.8, individual no. 1 is allowed to mate with no. 2, as the template of no. 1 matches the genotype of no. 2. But it cannot mate with individual no. 3, as the template of no. 1 does not match with the genotype of individual no. 3.

Mating templates are initially created at random, together with the functional genotype parts. Afterward, the mating templates are left to evolve, together with the functional genotype, as the whole hybrid string undergoes crossover and mutation. Thus, the determination of the mating preferences of individuals is genetically evolved by the same mechanism that evolves the solutions, in a second-order adaptation scheme.

19.2.14 Diversity Enhancement

In search spaces that contain a large number of local optima, it is necessary to force the dispersion of the population individuals throughout the whole search space, in order to explore all local optima areas.

To do so, we can apply "niche and species" methods together with mating restriction ones, in order to develop and maintain discrete subpopulations at the areas of local optima. But due to the limited population size, we expect to have sampling errors, especially in case of problems with a large number of local optima. That is, we expect to "lose" some areas of local optima, at the initialization phase, where the first population is randomly initialized. This problem is more severe in cases where the local optima may outnumber the population's individuals. In such cases, even with a perfectly uniform scattering of the initial solutions, it is impossible to cover all areas of local optima.

In order to enhance the possibility of such "orphan" areas to be visited in the future by population individuals, and consequently be explored, in the general effort of finding the global optimum, we must artificially enhance the population diversity. This can be achieved by enhancing the probability of the mutation operator, or by employing additional mutation-like operators, of general or problem-specific nature, that will increase the randomness and the scattering of solutions in the search space [53].

This diversity enhancement mechanism is extremely useful especially in cases of "dynamic optimization problems" [51,52], that is problems with fitness functions that change over time. In such problems, the shape of the fitness landscape and the location of the global optimum may change over time. Thus, for a potential algorithm that is applied to such problems, it is not sufficient just to be able to locate the global optimum, but also to be able to follow it, or rediscover it, through time. To achieve this, a potential genetic algorithm must have artificially enhanced population diversity, in order to be able to dynamically follow the movement of the global optimum, even after its population has already converged.

19.2.15 Constrained Optimization with GAs

Optimization tasks in real-world problems are often subject to a number of problem constraints. A general constrained optimization problem can be formulated as follows:

Optimize the objective function $F(\vec{x})$, $\vec{x} \in S$, where $\vec{x}$ is a generalized vector of problem parameters (continuous or combinatorial variables), and S is the solution search space, subject to the following inequality constraints:

$$IC_i(\vec{x}) \leq 0, i = 1, \ldots, p$$

and the following equality constraints:

$$EC_j(\vec{x}) = 0, j = 1, \ldots, q$$

where $IC_i(\vec{x})$ and $EC_j(\vec{x})$ are functions of $\vec{x}$ that may not be in analytical form.

In the special case where $\vec{x} = (x_1, x_2, \ldots, x_n) \in R^n (S \subseteq \mathrm{R}^n)$, we have the general nonlinear programming problem.

The general constrained optimization problem is intractable [54], i.e. no global algorithm (other than exhaustive search) has been found yet to provide the optimum solution for every conceivable type of the objective function $F(\vec{x})$ and the constraint functions $IC_i(\vec{x})$ and $EC_j(\vec{x})$. This makes room for innovation and application of a great variety of optimization methods and techniques that attempt to solve the above general problem formulation, for specific types of the objective function and constraints.

GAs constitute a class of optimization algorithms that have proved to be able to handle complex real-world optimization problems and have a great number of successful applications as reported in the literature. However, GAs do not incorporate a default mechanism for handling problem constraints. This is the reason why a number of methods have been proposed in the literature for providing GAs with the capability of handling constraints. The proposed methods vary significantly in the way they try to attack the problem constraints.

Category A: Restriction of the Genotype Space. According to this method category, the search space is restricted in such a way that it does not contain infeasible solutions [30]. This is the simplest method for handling elementary constraints and was used in the traditional GA implementations. This is a wise thing to do when it is possible (e.g., in the case of bounded problem variables), since it leads to a smaller search space. However, this method is of little use when dealing with the majority of real-world constrained optimization problems (e.g., with constraints that involve a number of variables).

CATEGORY B: REJECTION OF INFEASIBLE SOLUTIONS (DEATH PENALTY). Methods of this category discard infeasible solutions as soon as they are generated. The offspring reproduction procedure (parent selection—application of genetic operators—offspring production) is repeated until the population of offspring is filled with feasible solutions. During this procedure, any infeasible offspring that is produced is immediately rejected. This method does not utilize the information contained in infeasible solutions. Also in case the GA probability of producing a feasible solution is very small, a lot of CPU time is consumed in the effort of finding feasible solutions (by means of genetic operators) [30, 55], in order to form the offspring population of every generation. In many constrained optimization problems, with low feasible to whole search space ratio, the task of finding a single feasible solution may be very difficult. In such problems, the rejection method may result in endless loops of the offspring reproduction phase. The rejection method can be used in problems with relatively high feasible to whole search space ratios, where the majority of offspring produced genetically are feasible.

CATEGORY C: APPROXIMATION OF INFEASIBLE SOLUTION WITH FEASIBLE ONES, OR REPAIRING OF THE INFEASIBLE SOLUTIONS. Methods of this category approximate an invalid solution by its nearest valid one [57] or repair it to become a valid one [56, 58]. Such approximation (or repair) algorithms can be time consuming. Also, the resulting valid solution may be substantially different from the originally produced solution. Moreover, in certain problems, finding a feasible approximation of an infeasible solution may be as difficult as the optimization problem (constraint satisfaction problems [59]).

CATEGORY D: PENALTY FUNCTION METHODS. Methods of this category add penalty terms to the fitness function that are related to the violation of constraints. In this way, the invalid solutions are considered as valid but they are penalized in proportion to the degree of violation of the constraints. According to this principle, the fitness function has the following general form:

$$Q(\vec{x}) = \begin{cases} F(\vec{x}), & \text{if } \vec{x} \text{ is a feasible solution} \\ F(\vec{x}) + P(\vec{x}), & \text{if } \vec{x} \text{ is infeasible} \end{cases}$$

where $P(\vec{x})$ is a penalty function that usually depends on the degree of constraint violation introduced by solution $\vec{x}$.

This method is probably the most commonly used method for handling constraints and is implemented in many variations. However, it involves the building of a suitable penalty function for the specific problem that will efficiently balance the need for constraint satisfaction and the optimization of the objective function itself. High penalty values generally make the fitness landscape more difficult to search, as they create steep barriers along the feasible/infeasible boundaries of the search space. On the other hand, these steep barriers prevent the GA from converging to infeasible solution areas. Conversely, low penalty values result in a fitness function that is easier to search but they increase the possibility that the GA converges to an infeasible solution.

This is the reason why a large number of approaches have been proposed for building efficient penalty functions. These approaches can be divided in three subcategories: (1) the static penalty methods [61, 60], where penalties are static with relation to time and only depend on the constraint violation; (2) the varying/adaptive penalty methods [36, 38, 62, 63], where penalties are time variant; and (3) enhanced penalty methods [41, 64] that use penalties but also incorporate other techniques such as novel selection schemes or operators.

Category E: Methods That Use Special Phenotype-to-Genotype Representations (Decoders) to Encode Only the Feasible Part of the Search Space. Methods of this category use special phenotype-to-genotype representation schemes (stated also as decoders) that minimize or eliminate the possibility of producing infeasible solutions through the standard genetic operators, crossover and Mutation [65, 55].

Such decoders define a mapping between the set F of feasible solutions (phenotypes) and a set D of decoded solutions (genotypes). It is important that several conditions must be satisfied:

1. For each solution $\vec{x} \in F$, there is an encoded solution $\vec{d} \in D$;
2. each encoded solution $\vec{d}$ corresponds to a feasible solution $\vec{x}$;
3. all solutions in F should be represented by the same number of encodings $\vec{d}$;
4. the transformation $\vec{x} \rightarrow \vec{d}$ must be computationally fast;
5. the transformation must have the locality feature in the sense that small changes in the encoded solution $\vec{d}$ result in small changes in the actual solution $\vec{x}$.

Category F: Methods Using Special Recombination and Permutation Operators That Preserve Feasibility. In this method category, special problem-specific recombination and permutation operators are designed, which are similar to traditional crossover and mutation operators, and produce only feasible solutions. We should mention the Genocop method of Michalewicz [55] and its extensions (Genocop II [66] and Genocop III [56]), where the feasibility of solutions regarding the linear constraints is preserved by the use of special "closed" operators. The special mutation operator first calculates the feasible domain around the specific solution and then performs the mutation within that domain, whereas the crossover operator is an arithmetic crossover of the form: $\vec{o} = a\vec{p}_1 + (1 - a)\vec{p}_2$, where $\vec{o}$ is the offspring and $\vec{p}_1, \vec{p}_2$ the two parents, and $0 \leq \alpha \leq 1$. The above crossover always produces a feasible solution, as, the feasible part of the space, regarding linear constraints, is always convex.

Category G: Methods That Use Two-Phase Strategies for Satisfying the Constraints and Optimizing the Objective Function. Methods of this category [67, 68] first concentrate on the effort of finding feasible solutions, and when this is achieved they try to optimize the objective function preserving the solution feasibility.

Category H: Multi Objective Optimization Methods. The optimization of an objective function $F(\vec{x})$ that is subject to $n = p + q$ constraints can be seen as a multiobjective optimization problem [69], where the objective function $F(\vec{x})$ and the n measures of constraint violation $d_i(\vec{x})$, $I = 1,\ldots, n$, constitute a $n + 1$ dimensional vector $\vec{U} = (F(\vec{x}), d_1(\vec{x}), d_2(\vec{x}), \ldots, d_n(\vec{x}))$, each member of which must be optimized (minimized for $d_i(\vec{x})$, $i = 1,\ldots, n$). This formulation allows the application of any multiobjective optimization method [69] to the area of constrained optimization problems. Some classical multiobjective optimization methods create an aggregating optimization function $Q_{agg}(\vec{x})$ that usually is a combination (linear or not) of the elements of vector $\vec{U}$ (e.g., $Q_{agg}(\vec{x}) = w_o F(\vec{x}) + \sum_{i=1}^{n} w_i d_i(\vec{x})$, where w_i, $i = 0,\ldots, n$ are weight coefficients). This approach resembles the various penalty approaches mentioned in category D. Other multiobjective optimization techniques focus on creating sophisticated parent selection strategies to promote the simultaneous optimization of each of the objectives within the evolving population. Some of these selection strategies aim in producing certain fractions of the offspring population by selecting parents according to each one of the objectives, separately. Other methods use Pareto-based fitness assignment mechanisms that are based on Pareto optimality (or Pareto dominance) such as Pareto ranking [70]. Such methods assign ranks to individuals in a population according to whether they are dominated by other individuals (i.e., if other individuals exist that have better objective values concerning all of the objectives) or not.

Category I: Coevolutionary Methods That Maintain More Than One Population. In this category [56, 71, 72], we have a number of methods that usually use two different evolving populations, in an effort of finding the optimal solution in a constrained environment. Usually one population holds solutions that satisfy the constraints and is used as a pool of ideal solutions that help the second population to avoid the constrained areas. In some other cases, the two populations approach the optimum from different angles: one from the area of feasible solutions and the other from the area of infeasible ones.

Category J: Complex Hybrid Methods. Methods of this class combine one or more of the above methods together with calculus-based or other novel methods to achieve the goal of constraint optimization. In this category, we find the use of simulated annealing [66], mathematical programming techniques (simplex method) [73], fuzzy systems [74], cultural algorithms [75], artificial immune systems [76], and ant colony search models [77].

19.3 ON THE OPTIMAL LOCATION AND OPERATION OF FACTS DEVICES BY GENETIC ALGORITHMS

The GAs have been extensively used to achieve the optimal solution for numerous power system problems, technical and/or economical, by means of flexible alternating current transmission system (FACTS) devices. An optimization problem based on a GA requires large computation time and thereby GAs are not suited for real-time operation systems but for offline calculations only. Power system optimization mathematical problems are formulated either for the design stage—when a FACTS device is planned for installation in the future, or for the operating planning stage—when the use of the existing equipments is scheduled for the next day.

When installation of shunt devices (SVC – static VAr compensator or STATCOM – static synchronous compensator) is intended, the most common objective functions aim to determine the optimal location (a network bus) and size. The two devices are capable of controlling the bus voltage by generating or absorbing reactive power. While any voltage increase causes the active power losses to reduce, the shunt FACTS devices are usually installed in transmission networks to increase the transmission capacity by increasing the static stability reserve.

Series devices, such as thyristor-controlled series capacitor (TCSC) and phase shifting transformer (PST), are installed in series with a transmission line to control the power flow and avoid network congestions, while an indirect effect is to improve the system stability. Thereby, an optimization problem that seeks to install a series device aims to identify the best location (an electrical line) and the appropriate size that provides the full range of the desired control. A procedure for optimal location of phase shifters in the French network by a GA is proposed in [78]. While a classical PST is capable of maintaining only the active power flow on a transmission line to a predefined value, a TCSC can be used, in addition, for power oscillations damping. Forced changes in the power flows can result in increased active power losses, and thus minimization of losses in this case should be removed from the objective function.

Sometimes several FACTS devices may be required to provide the power system control. A solution for such a situation may be found in [79], where four types of FACTS devices were considered for installation. A GA was employed also in [80] to determine the best location and size for six different types of FACTS devices to be placed in the Hydro-Québec network, aiming to improve the steady-state performances from the viewpoints of static loadability and losses; one or two devices of the same type were considered.

The FACTS devices are complementary to the classical equipments that are used in a power system to control the voltage, among which the synchronous generators, synchronous compensators, transformers, shunt reactor, and shunt capacitors. On the other hand, voltage control is performed hierarchically and involves the most effective equipment whenever necessary, while maintaining appropriate reactive power reserves. Therefore, when FACTS devices are included in an optimization problem for day-ahead operation planning, the operating characteristics of classical devices should also be considered for optimization, such as the set-point voltage for the synchronous machines, tap position for transformers, or the on/off position for the shunt devices.

For a national power system however, indentifying the best location for FACTS devices, for both the steady-state and fast acting operation, may be easier than it appears. This is because usually the location of voltage deviation or network overloading problems is identified based on experience, and thus the number of candidate buses or lines for FACTS devices installation can be significantly reduced.

REFERENCES

[1] Barricelli, N. A. Esempi numerici di processi di evoluzione. *Methodos*, vol. 21–22, no. 6, pp. 45–68, 1954.

[2] Barricelli, N. A. Symbiogenetic evolution processes realized by artificial methods. *Methodos*, vol. 35–36, no. 9, pp. 143–182, 1957.
[3] Fraser, A. Simulation of genetic systems by automatic digital computers. I. Introduction. *Australian Journal of Biological Sciences*, vol. 10, pp. 484–491, 1957.
[4] Fraser, A., and Burnell, D. *Computer Models in Genetics*, McGraw-Hill, New York, 1970.
[5] Crosby, J. L. *Computer Simulation in Genetics*, John Wiley & Sons, Ltd, London, 1973.
[6] Bremermann, H. J. *The Evolution of Intelligence. The Nervous System as a Model of Its Environment*, (Tech. Rep. No. 1, Contract No. 477(17)), Department of Mathematics, University of Washington, Seattle, WA, 1958.
[7] Bremermann, H. J. Optimization through evolution and recombination. In: *Self-Organizing Systems*, M. C. Yovits, Jacobi, G. T., and Goldstein, G. D. Eds., Spartan, Washington, DC, 1962.
[8] Box, G. E. P. Evolutionary operation: A method for increasing industrial productivity. *Journal of the Royal Statistical Society C* vol. 6, no. 2, pp. 81–101, 1957.
[9] Bledsoe, W. W. The use of biological concepts in the analytical study of systems. In: ORSA–TIMS National Meeting, San Francisco, CA, 1961.
[10] Friedman, G. J. Digital simulation of an evolutionary process. *General Systems Yearbook 4*, pp. 171–184, 1959.
[11] Schwefel, H. P. Numerische Optimierung von Computer-Modellen. Ph.D. Thesis, Technical University of Berlin, Germany, 1974.
[12] Schwefel, H. P. *Numerische Optimierung von Computor-Modellen mittels der Evolutionsstrategie – mit einer vergleichenden Einführung in die Hill-Climbing- und Zufallsstrategie*, Birkhäuser, Stuttgart, Germany, 1977. ISBN 3764308761.
[13] Schwefel, H. P. *Numerical Optimization of Computer Models (Translation of 1977 Numerische Optimierung von Computor-Modellen mittels der Evolutionsstrategie)*. John Wiley & Sons, Ltd, Chichester, UK, 1981. ISBN 0471099880
[14] Rechenberg, I. Evolutionsstrategie – Optimierung technischer Systeme nach Prinzipien der biologischen Evolution. Ph.D. Thesis, reprinted by Fromman-Holzboog, 1973.
[15] Fogel, L. J., Owens, A. J., and Walsh, M. J. *Artificial Intelligence through Simulated Evolution*. John Wiley & Sons, Inc., New York, 1966.
[16] Holland, J. H. *Adaptation in Natural and Artificial Systems*. University of Michigan Press, Ann Arbor, MI, 1975.
[17] Holland, J. H., and Reitman, J. S. Cognitive systems based on adaptive algorithms. In: *Pattern Directed Inference Systems*, D. A. Waterman and F. Hayes-Roth, Eds., Academic Press, New York, pp. 313–329, 1978.
[18] Koza, J. R. Hierarchical genetic algorithms operating on populations of computer programs. In: *Proceedings of the 11th International Joint Conference on Artificial Intelligence*. Morgan Kaufmann, San Mateo, CA, vol. I., pp. 768–774, 1989.
[19] de Garis, H. *Evolvable Hardware: Genetic Programming of a Darwin Machine*. In: *Artificial Neural Nets and Genetic Algorithms*, Proceedings of the International Conference. Springer-Verlag, Innsbruck, Austria, 1993.
[20] Muhlenbein, H. Evolution in time and space – the parallel genetic algorithm. In: *Foundations of Genetic Algorithms*, G. Rawlins, Ed. Morgan-Kaufmann, pp. 316–337.
[21] Storn, R., and Price, K. Differential evolution—a simple and efficient adaptive scheme for global optimization over continuous spaces. Technical Report TR-95-012, ICSI, March 1995.
[22] Beni, G., and Wang, J. Swarm intelligence in cellular robotic systems. In: Proceedings of NATO Advanced Workshop on Robots and Biological Systems, Tuscany, Italy, June 26–30, 1989.
[23] Colorni, A., Dorigo, M., and Maniezzo, V. Distributed optimization by ant colonies. In: *Proceedings of the First European Conference on Artificial Life*, F. J. Varela and P. Bourgine, Eds., MIT Press, Cambridge, MA, pp. 134–142, 1992.
[24] Farmer, J. D., Packard, N. H., and Perelson, A. S. The immune system, adaptation and machine learning. *Physica D*, vol. 22, pp. 187–204, 1986.
[25] Langton, C. G. Artificial life. In: *Artificial Life*, Langton, Ed., Addison-Wesley, Reading, MA, p. 1, 1989.
[26] Holland, J. H. Outline for a logical theory of adaptive systems. *Journal of ACM*, vol. 9, no. 3, pp. 279–314, 1962.

[27] De Jong, K. A. *An Analysis of the Behaviour of a Class of Genetic Adaptive Systems*, University of Michigan, Ann Arbor, MI (University Micro®lms no. 76-9381), 1975.

[28] Mitchell, M. *An Introduction to Genetic Algorithms*. MIT Press, Cambridge, MA, 1996.

[29] Goldberg, D. E., and Deb, K. A comparative analysis of selection schemes used in genetic algorithms. In: *Foundations of Genetic Algorithms*, Morgan Kaufmann, pp. 69–93, 1991.

[30] Goldberg, D. E. *Genetic Algorithms in Search, Optimisation and Machine Learning*. Addison-Wesley, Reading, MA, 1989.

[31] Spears, W. M., and De Jong, K. A. An analysis of multi-point crossover. In: *Foundations of Genetic Algorithms*, G. Rawlins, Ed., Morgan Kaufman, San Mateo, CA, pp. 301–315, 1991.

[32] Goldberg, D. E., and Lingle, R. Alleles, loci and the traveling salesman problem, In: *Proceedings of an International Conference on Genetic Algorithms*, Morgan Kauffman, pp. 10–19, 1985.

[33] Davis, L. Applying adaptive algorithms to epistatic domains. In: *Proceedings of the Ninth International Joint Conference on Artificial Intelligence*, Morgan Kaufmann, Los Angeles, CA, pp. 162–164, 1985.

[34] Oliver, I. M., Smith, D. J., and Holland, J. R. A study of permutation crossover operators on the traveling salesman problem, In: *Proceedings of the Second International Conference on Genetic Algorithms on Genetic Algorithms and Their Application* (Cambridge, MA), J. J. Grefenstette, Ed., L. Erlbaum Associates, Hillsdale, NJ, pp. 224–230, 1987.

[35] Whitley, D., Starkweather, T., and Fuquay, D'. A. Scheduling problems and traveling salesman: the genetic edge recombination operator. In: *Proceedings of the Third International Conference on Genetic Algorithms*, Morgan Kaufmann, San Francisco, CA, pp. 133–140, 1989. ISBN:1-55860-066-3.

[36] Petridis, V., and Kazarlis, S. Varying quality function in genetic algorithms and the cutting problem. Proceedings of the First IEEE Conference on Evolutionary Computation,1994. *IEEE Service Center*, vol. 1, pp. 166–169, 1994.

[37] Bakirtzis, A., Petridis, V., and Kazarlis, S. A genetic algorithm solution to the economic dispatch problem. *IEE Proceedings—Generation, Transmission, Distribution*, vol. 141, no. 4, pp. 377–382, July 1994.

[38] Kazarlis, S. A., Bakirtzis, A. G., and Petridis, V. A genetic algorithm solution to the unit commitment problem, *IEEE Transactions on Power Systems*, vol. 11, no. 1, pp. 83–92, February 1996.

[39] Moscato, P. On evolution, search, optimization, genetic algorithms and martial arts: towards memetic algorithms. Caltech Concurrent Computation Program, C3P Report 826, 1989.

[40] Renders, J.-M., and Bersini, H. Hybridizing genetic algorithms with hill-climbing methods for global optimization: two possible ways. In: *Proceedings of the First IEEE International Conference on Evolutionary Computation*, IEEE Press, Piscataway, NJ, vol. 1, pp. 312–317, 1994.

[41] Kazarlis, S., Papadakis, S., Theocharis, J., and Petridis, V. Micro-genetic algorithms as generalized hill climbing operators for GA optimization. *IEEE Transactions on Evolutionary Computation*, vol. 5, no. 3, pp. 204–217, June 2001.

[42] Ackley, D. H. Stochastic iterated genetic hill climbing. Ph.D. Thesis, Department of Computer Sciences, Carnegie Mellon University, Pittsburgh, PA, 1987.

[43] Miller, J. A., Potter, W. D., Gandham, R. V., and Lapena, C. N. An evaluation of local improvement operators for genetic algorithms, *IEEE Transactions on Systems, Man, Cybernetics*, vol. 23, no. 5, pp. 1340–1351, 1993.

[44] Petridis, V., Kazarlis, S., and Bakirtzis, A. Varying fitness functions in genetic algorithm constrained optimization: The cutting stock and unit commitment problems. *IEEE Transactions on Systems Man and Cybernetics*, vol. 28, no. 5, pp. 629–640, October 1998.

[45] Grefenstette, J. Optimization of control parameters for genetic algorithms. *IEEE Transactions Systems Man Cybernetics*, vol. 16, no. 1, pp. 122–128, January 1986.

[46] Stephens, C. R., García Olmedo, I., Mora Vargas, J., and Waelbroeck, H. Self-adaptation in evolving systems. *Artificial Life*, vol. 4, no. 2, pp. 183–201, Spring 1998.

[47] Fogel, D. B., Fogel, L. J., and Atmar, J. W. Meta-evolutionary programming. In: *Conference on Signals, Systems and Computers, Conference Record of the Twenty-Fifth Asilomar*, vol. 1, no. 4–6, pp. 540–545, November 1991.

[48] Eiben, A. E., Hinterding, R., and Michalewicz, Z. *Parameter control in evolutionary algorithms. IEEE Transactions on Evolutionary Computation*, vol. 3, no. 2, pp. 124–141, Jul 1999.

[49] Goldberg, D. E., and Richardson, J. *Genetic Algorithms with Sharing for Multimodal Function Optimization*, 2nd International Conference on Genetic Algorithms, 41-49.

[50] Booker, L. B. Improving the Performance of Genetic Algorithms in Classifier Systems, In: *Proceedings of the 1st international Conference on Genetic Algorithms*, J. J. Grefenstette, Ed., L. Erlbaum Associates, Hillsdale, NJ, pp. 80–92, 1985.

[51] Goldberg, D. E., and Smith, R. E. Nonstationary function optimization using genetic algorithms with dominance and diploidy, In: *International Conference on Genetic Algorithms*, J. J. Grefenstette, Ed., Lawrence Erlbaum Associates, pp. 59–68, 1987.

[52] Branke, J., Kaußler, T., Schmidt, C., and Schmeck, H. A multi-population approach to dynamic optimization problems. In: *Adaptive Computing in Design and Manufacturing*, Springer, pp. 299–307, 2000.

[53] Cobb, H. G. An investigation into the use of hypermutation as an adaptive operator in genetic algorithms having continuous, time-dependent nonstationary environments. Technical Report AIC-90-001, Naval Research Laboratory, Washington, DC, 1990.

[54] Michalewicz, Z., and Schoenauer, M. Evolutionary algorithms for constrained parameter optimization problems. *IEEE Transactions on Evolutionary Computation*, vol. 4, no. 1, pp. 1–32, 1996.

[55] Michalewicz, Z. *Genetic Algorithms +Data Structures = Evolution Programs*, 2nd edition. Springer-Verlag, 1992.

[56] Michalewicz, Z., and Nazhiyath, G. Genocop III: a co-evolutionary algorithm for numerical optimization problems with nonlinear constraints. In: *Proceedings of the Second IEEE Conference on Evolutionary Computation*, Perth, Australia, IEEE Press, Piscataway, NJ, vol. 2, pp. 647–651, 1995.

[57] Nakano, R., and Yamada, T. Conventional genetic algorithm for job shop problems. In: *Proceedings of the Fourth International Conference on Genetic Algorithms and Their Applications*, R. K. Belew and. L. B. Booker, Eds., Morgan Kaufmann, San Mateo, CA, pp. 474–479, 1991.

[58] Orvosh, D., and Davis, L. Using a genetic algorithm to optimize problems with feasibility constraints. In: *Proceedings of the First IEEE Conference on Evolutionary Computation*, IEEE Press, Piscataway, NJ, pp. 548–553, 1994.

[59] Eiben, A. E., and Ruttkay, Z. S. Self-adaptivity for constraint satisfaction: learning penalty functions. In: *Proceedings of the Third IEEE Conference on Evolutionary Computation*, IEEE Press, Piscataway, NJ, pp. 258–261, 1996.

[60] Homaifar, A., Lai, S., and Qi, X. Constrained optimization via genetic algorithms. *Simulation*, vol. 62, no. 4, pp. 242–254, 1994.

[61] Richardson, J. T., Palmer, M. R., Liepins, G., and Hilliard, M. Some guidelines for genetic algorithms with penalty functions. In: *Proceedings of the Third International Conference on Genetic Algorithms*, J. D. Schaffer, Ed., Morgan Kaufmann, San Mateo, CA, pp. 191–197, 1989.

[62] Siedlecki, W., and Sklanski, J. Constrained genetic optimization via dynamic reward-penalty balancing and its use in pattern recognition. In: *Proceedings of the Third International Conference on Genetic Algorithms*, J. D. Schaffer, Ed., Morgan Kaufmann, San Mateo, CA, pp. 141–150, 1989.

[63] Coello, C. A. Self-adaptive penalties for GA-based optimization. In: *Proceedings of the 1999 Congress on Evolutionary Computation (CEC 1999)*, July 6–9, 1999, IEEE Press, Washington, DC, vol. 1, pp. 573–580, 1999.

[64] Runarsson, T. P., and Yao, X. Stochastic ranking for constrained evolutionary optimization. in *IEEE Transactions on Evolutionary Computation*, vol. 4, no. 3, pp. 284–294, September 2000.

[65] Koziel, S., and Michalewicz, Z. Evolutionary algorithms, homomorphous mappings, and constrained parameter optimization. *IEEE Transactions on Evolutionary Computation*, vol. 7, no. 1, pp. 19–44, 1999.

[66] Michalewicz, Z., and Attia, N. Evolutionary optimization of constrained problems. In: *Proceedings of the Third Annual Conference on Evolutionary Programming*, A. V. Sebald and L. J. Fogel, Eds., World Scientific, River Edge, NJ, pp. 98–108, 1994.

[67] Hinterding, R., and Michalewicz, Z. Your brains and my beauty: parent matching for constrained optimisation. In: *Proceedings of the Fifth IEEE International Conference on Evolutionary Computation*, Anchorage, AK, May 4–9, 1998, pp. 810–815.

[68] Schoenauer, M., and Xanthakis, S. Constrained GA optimization. In: *Proceedings of the Fifth International Conference on Genetic Algorithms*, S. Forrest, Ed., Morgan Kaufmann, Los Altos, CA, pp. 573–580, 1993.

[69] Fonseca, C. M., and Flemming, P. J. An overview of evolutionary algorithms in multiobjective optimization. *Evolutionary Computation*, vol. 3, no. 1, pp. 1–16, 1995.

[70] Srinivas, N, and Deb, K. Multiobjective optimization using nondominated sorting in genetic algorithms. *Evolutionary Computation*, vol. 2, no. 3, 221–248, 1995.

[71] Leriche, R. G., Knopf-Lenoir, C., and Haftka, R. T. A segregated genetic algorithm for constrained structural optimization. In: *Proceedings of the Sixth International Conference on Genetic Algorithms*, L. J. Eshelman Ed., Morgan Kaufmann, San Francisco, CA, pp. 558–565, 1995.

[72] Paredis, J. Co-evolutionary constraint satisfaction. In: *Proceedings of the Third Conference on Parallel Problem Solving from Nature*, Springer, Berlin, Germany, pp. 46–55, 1994.

[73] Belur, S. V. CORE: Constrained optimization by random evolution. In: *Late Breaking Papers at the 1997 Genetic Programming Conference*, J. R. Koza, Ed., Stanford Bookstore, Stanford University, CA, pp. 280–286, 1997.

[74] Van Le, T. A Fuzzy evolutionary approach to constrained optimisation problems. In: *Proceedings of the Third IEEE International Conference on Evolutionary Computation*, Nagoya, Japan, May 1996, pp. 274—278, 1996.

[75] Reynolds, R. G. An introduction to cultural algorithms. In: *Proceedings of the Third Annual Conference on Evolutionary Programming*, A. V. Sebald and L. J. Fogel, Eds., World Scientific, River Edge, NJ, pp. 131–139, 1994.

[76] Hajela, P., and Lee, J. Constrained genetic search via schema adaptation. An immune network solution. *Structural Optimization*, vol. 12, no. 1, pp. 11–15, 1996.

[77] Bilchev, G., and Parmee, I. C. Constrained and multi-modal optimisation with an ant colony search model. In: *Proceedings of the Second International Conference on Adaptive Computing in Engineering Design and Control* (ACEDC'96), I. C. Parmee and M. J. Denham, Eds., University of Plymouth, Plymouth, UK, pp. 145–151, March 1996.

[78] Paterni, P., Vilet, S., Bena, M., and Yokoyama, A. Optimal location of phase shifters in the French network by genetic algorithm. *IEEE Transactions on Power Systems*, vol. 14, no. 1, February 1999.

[79] Gerbex, S., Cherkaoui, R., and Germond, A. Optimal location of multi-type FACTS devices in a power system by means of genetic algorithms *IEEE Transactions on Power Systems*, vol. 16, no. 3, August 2001.

[80] Ghahremani, E., and Kamwa, I. Analysing the effects of different types of FACTS devices on the steady-state performance of the Hydro-Québec network. *IET Generation, Transmission & Distribution*, vol. 8, no. 2, pp. 233–249, 2014.

CHAPTER **20**

MULTIAGENT SYSTEMS

Nan-Peng Yu and Chen-Ching Liu

20.1 OVERVIEW

When you are watching the movie *Lord of the Rings*, can you imagine that in the ferocious battle sequences, each orc in the enormous crowds is actually modeled as an agent that virtually hears and responds to movements around them, senses proximity and collision with others, and engages in battle animations with each other [1]? When you are driving a GM Daewoo vehicle on freeway, can you imagine that the production process of the exterior components of your automobile might be planned and scheduled by an agent-based scheduling system [2] and the painting of your vehicle might be done by an agent-based automotive paint shop control system [3]? These are two examples for the many applications of multiagent systems (MAS). What is an agent? What are the application domains of MAS? What are the advantages and disadvantages of the multiagent technology? This chapter provides a discussion of the above questions.

20.1.1 What is an Agent? What is a Multiagent System?

Ideas from many disciplines, including computer science, artificial intelligence, economics, psychology, and sociology, contribute to the development of MAS. The core concept of this dynamic field, "*agent*," does not have a single definition that is universally adopted. However, there is a definition that incorporates the key features of an agent. An agent is a virtual entity (e.g., a software program) or physical entity (e.g., a robot) that resides in an environment and is capable of *autonomous* actions in this environment. An agent should be able to perceive and act upon its environment, which may be dynamic and uncertain. The basic property of an agent, "autonomous," means that a software agent must be able to operate without direct intervention of people or other agents and has control over its own action and internal state [4].

As an intelligent entity, an agent should be capable of acting and behaving *flexibly* and *rationally*. The term "flexible" means an agent is *reactive*, *proactive*, and *social* [5]. By reactive, it is meant that the agent perceives its environment and responds in a timely manner to changes in the environment [4]. By proactive, it is meant that the agent does not simply act in response to its environment but is able

Advanced Solutions in Power Systems: HVDC, FACTS, and Artificial Intelligence, First Edition.
Edited by Mircea Eremia, Chen-Ching Liu, and Abdel-Aty Edris.

to achieve a goal by taking the initiative [4]. By social, it is meant that in order to achieve its goals, the agent interacts with people or other agents [4]. To behave rationally, an agent should be equipped with core skills, such as learning, problem solving, planning, and decision making.

An MAS is an organization of possibly heterogeneous and self-motivated agents that interact with each other. Agents in an MAS may have conflicting goals or they could coordinate with each other to accomplish a common mission.

20.1.2 Why Multiagent Systems?

The reasons for researchers from various disciplines to study MAS are quite different. In many subbranches of social science such as sociology, psychology, and economics, the inception of MAS concept facilitates modeling of more realistic cognitive and intelligent agents and interactions among them in a human society context. Therefore, the study of MAS leads to a better understanding of the interactive processes among humans on social, psychological, and economical levels. In economics, the Walrasian equilibrium model devised in late 1880s remains the fundamental paradigm framing the way economists think about modeling economic systems that do not address how production, pricing, and trade take place in the real world through various forms of procurement processes [6]. Retooled with MAS that enables the social communication between agents, the formation of social interaction patterns, and the development of shared perception, economists are able to model the procurement process without significant simplification. Hence, MAS as a tool in economics provides a better understanding of observed global regularities such as employment and growth rate and leads to computational laboratories for discovery of good economic designs [7].

A challenge faced by researchers in computer science and artificial intelligence is how to deal with distributed, large-scale, dynamic, open, and heterogeneous systems such as Internet, multidatabase system, shipboard control, and automation system and transfer line system. Solving those inherently distributed and complex applications in a centralized way is not only counter-intuitive, but also often not possible [8]. MAS, as distributed systems that are composed of autonomous, intelligent, and rational agents, become a natural tool to understand, manage, and use those distributed complex systems. For example, in the domain of manufacturing, transfer lines that move work pieces sequentially through a series of stations where individual mechanisms perform specific tasks are difficult to operate due to complex interactions among their mechanisms [9] (A typical transfer line may be 100 m long and contains dozens of stations with hundreds of mechanisms and more than 1500 degrees of freedom in movement overall [9]). In such a complex system, identifying the mechanism responsible for a particular system failure can be time-consuming. The MAS methodology is suited to tackle the problem.

20.1.3 Applications of Multiagent Technology

The two examples in the beginning of this chapter offered a glimpse of the broad commercial and industrial applications of MAS. Agents are applied to a wide variety of areas, ranging from comparatively small systems such as email filters to large, open, complex, mission critical systems such as air traffic control [4]. This section presents an overview of the main application areas and domains where agent-based methodology is used and provides some example applications.

20.1.3.1 Industrial Applications

Industrial applications of agent technology were among the first to be developed. Today, various agent technologies and MAS paradigms have been deployed in industrial operations.

PRODUCT DESIGN: Collaborative design of complex manufactured products involves teams of designers in various locations and requires an appropriate conflict resolution scheme among them. The key question is how to allocate product characteristics to various design teams. The inherent distribution and complexity of the design process makes multiagent technology a sound alternative to multidimensional optimization that has analytical solutions only in special cases. The exemplar application in this area is the design system responsible agents for product-process integrated design. It is an agent-based software tool and method for using market place dynamics among members of distributed design teams to coordinate set-based design of a manufactured product [8]. Trade-offs are made collectively by designers through buying and selling units of design characteristics in the marketplaces of characteristic agents.

REAL-TIME CONTROL: Highly variable customization of individual products, equipment failure, and time-consuming diagnostic process usually results in low productivity of production lines in automotive industries. These challenges of the real-time control system can be conveniently and flexibly handled by the agent technology. A special class of agent systems, the Holonic Manufacturing System, which consists of autonomous, self-reliant, and cooperative manufacturing units, offers a new approach to organizing and controlling the production process [9]. Semi-real operation of an agent-based engine assembly line at Daimler Chrysler engine assembly plant demonstrated the scalability, robustness, reliability, and increased productivity [9].

LOGISTICS: Frequent changes of customer requirements and operation conditions and temporal information unavailability caused by communication bottlenecks make efficient central planning and scheduling difficult in practice. Therefore, tools such as MAS that are based on distributed decision making satisfy the requirements of a complex logistic problem. A successful field test and commercial application have been reported by Magenta who developed agent-based i-Scheduler system. Road Transportation i-Scheduler tested on logistics provides data with 4000 transportation instructions and 200 trucks operating on the UK business network [10]. The agent-based system demonstrated the capability to significantly improve scheduling effectiveness in comparison with conventional rule-based schedulers [10].

SIMULATION: The uncertain behavior arising from complex interactions among agents is an obstacle to deployment of massive industrial MAS. Currently, the only available method for validating system performance is to simulate the developed multiagent model under all possible operating scenarios [11]. Rockwell Automation designed and developed manufacturing agent simulation tool based on Foundation for Intelligent Physical Agents (FIPA) standard and JADE platform to visualize and simulate handling of materials in flexible manufacturing [12].

TRAINING: Decision making in a complex, dynamic, and critical environment, such as military missions, requires a significant effort. For the purpose of training, multiagent technology has been deployed in several applications. University of Southern California together with Los Angeles fire department and personnel from the Homeland Security Center for Risk and Economic Analysis of Terrorism Events developed a MAS DEFACTO for incident commanders' training [13]. DEFACTO incorporates state-of-the-art artificial intelligence, three-dimensional visualization into a high fidelity system to help commanders gain experience and draw valuable lessons that can be applied in the real world.

DISTRIBUTED DIAGNOSTICS: Advancement can be made in complex onboard aircraft, ship, and manufacturing workshop failure diagnostics with the adoption of multiagent technology. Rockwell Automation demonstrated the potentials of agent-technology through the application of real-time diagnostics and reconfiguration of the shipboard chilled water system for vessels of the US Navy [14]. The deployed agents collectively detect equipment failures and prevent failures from propagating through dynamic reconfiguration.

AIR TRAFFIC CONTROL: The highly dynamic nature of the air traffic control environment places stringent real-time constraints on the system's reasoning process, and requires the system to exhibit both goal-directed and reactive behavior [15]. These requirements make MAS a valuable tool for realizing effective commercial applications. Ljunberg and Lucas developed a sophisticated agent air traffic control system based on the Belief-desire-intention (BDI) agent model known as Optimal Aircraft Sequencing using Intelligent Scheduling (OASIS) [15]. This air traffic control system is successfully deployed at Sydney airport in Australia to help alleviate air traffic congestion.

UNMANNED AERIAL VEHICLE CONTROL: Due to the potential to significantly reduce cost and the risk to aircrew in military operations, unmanned aerial vehicle (UAV) control is of interest to the defense sector. QuinetiQ developed an MAS for controlling teams of uninhabited air vehicles (UAVs) [16]. The decision making to control a UAV fleet is conducted by a human operator together with the developed MAS. The overall system has been evaluated in a number of human-in-the-loop trials within a detailed synthetic environment [16].

20.1.3.2 Commercial Applications

Humans would be easily swamped by the high volume of information available via Internet and World Wide Web. The characteristics of intelligent agents, such as autonomous and persistent, are useful for commercial applications, especially in information management and electronic commerce.

INFORMATION MANAGEMENT: Due to the increasing richness and diversity of information available, the task of locating information source, gathering, filtering, and integrating information in support of decision making has become increasingly difficult. This motivates the development of MAS for information gathering and decision support in an Internet-based information environment. One example project is the WARREN financial portfolio management system [17], which is designed to monitor the ongoing portfolio management process. It is based on multiagent architecture that coordinates three kinds of agents, that is, interface, task, and information agents.

ELECTRONIC COMMERCE: Intelligent agents play a crucial role in electronic commerce where dynamic and heterogeneous interactions between tens of thousands of organizations and tens of millions of individuals are involved [18]. The major advantage of using software agents representing their human owners in e-commerce is that it dramatically reduces certain types of frictional costs and time incurred in the exchange of commodities. A learning, autonomous bidding agent called ATTac-2001 is developed [19]. The intelligent agent uses boosting techniques to learn conditional distributions of auction clearing prices and bids simultaneously for multiple interacting goods.

20.1.3.3 Medical Applications

Many problems in health care domain share a number of similarities, which include spatially distributed knowledge and expertise and the need for coordination among different individuals with various skills and functions. Features and properties of MAS suggest that they are a good option for medical applications.

PATIENT SCHEDULING: There is increasing complexity in interrelated medical procedures and their associated tests and treatments. A multiagent solution using the generalized partial global planning approach that preserves the existing hospital organization and authority structures was proposed [20]. Computer simulation results demonstrated the usefulness of the multiagent solution by both decreasing patient stays and increasing hospital unit throughput.

PATIENT MONITORING: The patient monitoring and care in surgical intensive care unit (SICU) calls for sharing of information and coordination among experts with distinct expertise. The GUARDIAN system was designed [21] to help patient care management in SICU. The function of patient monitoring is distributed to three types of agents that are organized into hierarchies.

20.1.3.4 Entertainment Applications

Intelligent agents play a role in the leisure industry. Development in agent architectures, communication and cooperation, and programming languages and tools has made agents suitable to model and develop characters in computer games and films. In addition, the entertainment industry provides a good platform to demonstrate the capabilities of multiagent technology without putting human life or property at risk.

VIDEO GAMES: As computer games become more complex and consumers demand more sophisticated computer-controlled opponents, game developers are required to place a higher emphasis on artificial intelligence aspects of their games [22]. For example, in most modern video games, the nonplayer characters (NPCs) are hard-coded. Hence, no matter how many times the player exploits a weakness, NPCs never learn to adapt. To solve this problem, the rtNEAT method designed in [23] allows the player to teach a team of agents through a series of customized training exercises. The capability to improve NPCs' performance is illustrated in a robot combat game NERO.

MOVIES: From the hundreds penguin agents that sing as they walk along in the movie Happy Feet to the thousands warrior agents that engage in the large-scale battle scene in the movie Mummy 3, multiagent technology has been used extensively in recent movies. As the industry leader in autonomous agent-driven animation,

Massive has been adopted by many studios to fulfill the creative needs of the blockbuster movies to provide epic scenes such as battle scenes of the Lord of the Rings trilogy [24].

20.1.4 Challenges and Future of Multiagent Technology

Despite significant theoretical development of multiagent technology and numerous successful industrial deployments in the past decade, there are a number of broad technological challenges and potential for future research and applications. These challenges are summarized and categorized into subchallenges [25, 26].

20.1.4.1 Design Methodologies for Software Development of Agent-Based Systems

A bottleneck in the progress of agent technology is the lack of mature software development methodologies for agent-based systems. The meaning of a mature software development methodology for agent-based systems is threefold. From an analysis point of view, core concepts that relate to environments for MAS and individual agents must be clearly defined [25]. From a design point of view, to facilitate agent-based software reusability, a set of generic libraries that include organizational models, agent architectures, communication languages, ontology patterns, and knowledge bases should be provided [25]. At the tool level, a flexible, easy-to-use agent-based programming platform is needed to facilitate developers in all phases of development, including design, testing, maintenance, and visualization [25].

20.1.4.2 Ensure User Confidence and Trust in Agent-Based Systems

To build confidence of users in deploying a group of agents that represent them and act effectively on their behalf in an open system, elaborate verification and extensive reliability testing need to be carried out. In addition, for a user to trust the outcome of an open agent system, they must trust the agents representing other parties [25]. Potential solutions to this problem include the establishment of reputation tracking mechanism, adoption of social rules, the use of electronic contracts, and the study of important legal issues involved.

20.1.4.3 Enable Agent Adaptation in Artificial System

Truly intelligent agents should be equipped with learning tools that help them adapt to a new, open, and dynamic environment. Compared to well-advanced single-agent learning techniques, development of multiagent learning techniques is in its infancy. Although initial work in multiagent learning has shown promising results in a simple environment, it is questionable how well those techniques will perform when they are scaled up to a more complex and realistic environment.

Open topics that are likely to draw future research in this field are scalability of multiagent learning, adaptive dynamics with heterogeneous learning agents, and the task of problem decomposition [27].

20.1.4.4 Promote Interoperability in an Open Environment

The FIPA played a crucial role in promoting the technologies and interoperability specifications that facilitate the end-to-end interworking of intelligent agent system in modern commercial and industrial settings. In the future, to move standards for agents and agent-based systems into the wider context of software development, the agent technology needs to work and integrate with nonagent technologies. To this end, the IEEE Computer Society has adopted FIPA as part of its family of standards committees in 2005 [28].

20.1.4.5 Develop Semantic Infrastructure and Common Ontology for Agent Communication and Information Management

Although FIPA-agent communication language (ACL) has become the de facto standard for agent communications, there is a wide gap between being able to parse and generate messages that conform to the FIPA-ACL standards and being able to perform meaningful conversations [29]. To support a formal description of information and perform a meaningful conversation, a shared semantics of communication process and a set of common ontology or knowledge bases are needed. Hence, there is a growing interest in the use of social concepts such as artificial institution as context to facilitate the creation of objective and external semantics of ACL [19, 30].

20.1.4.6 Enhance Reasoning Capabilities for Agents in Open Environment

Negotiation, argumentation, and deliberation mechanisms are at the heart of agent reasoning. Thus, research on agent reasoning calls for the development of negotiation and argumentation algorithms that are effective. In addition, techniques are needed to enable agents to identity, create, and dissolve coalitions, which in certain conditions function more efficiently than an individual group.

20.2 MULTIAGENT TECHNOLOGY OVERVIEW

This section gives an overview of the techniques on MAS. The section is divided into four subsections. The first subsection briefly reviews the classical architectures of individual intelligent agents. Key issues concerning a society of agents, including communication, negotiation, and coordination, are discussed in subsection two. Then, the programming languages, tools, and frameworks that are crucial to the implementation of MAS are presented. The last subsection is an introduction to the MAS-related standards.

20.2.1 Architectures for Intelligent Agents

The question of how to design a sensible architecture supporting the operation of an intelligent agent with rational behaviors has been of interest to researchers in artificial intelligence. Different architectures have been proposed based on various conceptions of how an agent should produce actions motivated from its goals and perceptions of the environment. Upon realizing the drawbacks of the traditional symbolic approach, a purely reactive architecture that operates in a simple stimulus-response fashion emerges. Later, the deliberative architecture that has deeper philosophical foundations emerged. The idea of combining the strength of both reactive and proactive approaches yields the last class: layered architectures. The presentation of different intelligent agent architectures draws heavily upon the work of [31] and [32].

20.2.1.1 Logic-Based Architectures

Logic-based architectures find their roots in the basic assumption of traditional (symbolic) artificial intelligence: Intelligent behavior can be achieved through manipulation of symbol structures representing bits of knowledge. Although logic-based approaches are elegant and have a clean semantic, they have limited success due to two difficulties. First, within the symbolic framework, it is extraordinarily difficult to actually represent and reason about complex, dynamic, real-world environment. Then, the inherent computational complexity of theorem proving makes the decision-making process slow, which is not acceptable in a time-constrained environment.

20.2.1.2 Reactive Architectures

In contrast to the logic-based architecture, reactive agents do not have a symbolic representation of their environment and act using a stimulus/response type of behavior triggered by sensor data [33]. The subsumption architecture defined by Brooks is the best-known reactive agent architecture among the proposed architectures and theories. The key idea behind the subsumption architecture is that complex behavior may simply be the reflection of a complex environment rather than a product of an extremely complex control system [34]. The subsumption architecture decomposes the problem vertically into a number of levels of competence on the basis of desired external manifestations of the control system (see Figure 20.1 [34]). Each layer is composed of a set of processors that are finite state machines with the ability to monitor input wires and sending messages. In the subsumption architecture, control is layered with higher level layers subsuming the roles of lower level layers when they wish to take control. Despite obvious advantages of reactive approaches, that is, simple, robust, and cheap, they suffer from several fundamental flaws. It is highly questionable whether a reactive agent can make appropriate decisions with limited local information and no model of the global environment. Due to the absence of agent state, it is almost impossible to let a purely reactive agent learn from its past experience. Moreover, the complex interaction and resultant emergent behavior make it very difficult to engineer purely reactive agents to perform complicated tasks.

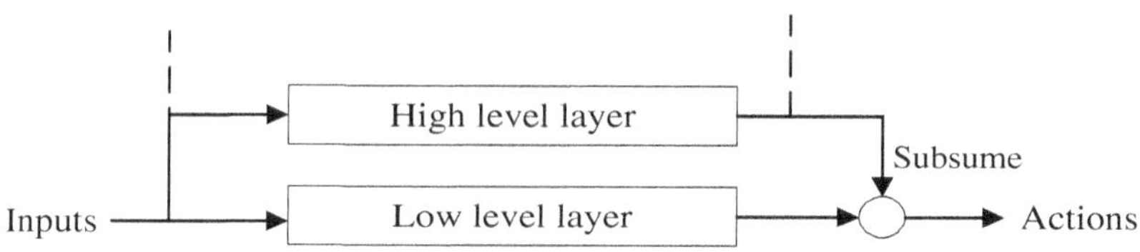

Figure 20.1 Subsumption architecture.

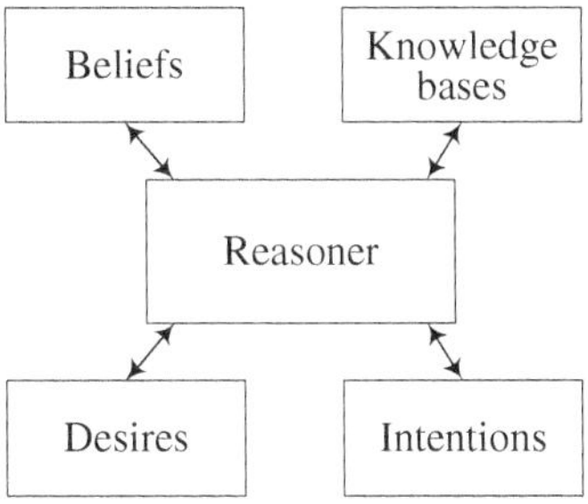

Figure 20.2 The core of procedural reasoning system architecture.

20.2.1.3 Belief-Desire-Intention Architectures

Belief-desire-intention (BDI) architectures draw their foundations from both artificial intelligence and philosophy [35]. The BDI architecture includes direct representations of the agent's beliefs, desires, and intentions. Beliefs represent the information an agent holds about its environment and about itself. Desires represent objectives and goals of the agent, which interacts with its current beliefs and intentions. Intentions represent the desires that the agent has committed to fulfill. One of the most successful agent systems, the procedural reasoning system, was developed with the BDI architecture [36]. The core of the system structure, shown in Figure 20.2, consists of a database containing current beliefs, a set of desires, a stack of current intentions, a set of knowledge bases, and a reasoner for manipulating these components [36]. The reasoner runs the entire system by updating beliefs based on new information from sensors, maintaining and activating new knowledge bases, choosing and placing one procedure specification into the process stack, and executing the plan [36]. Despite attractive features of the BDI architecture such as intuitiveness and clear functional decomposition, it still faces challenging design issues such as how to strike a balance between reconsideration of intention and focusing on the current commitment.

20.2.1.4 Layered (Hybrid) Architectures

A lesson learned from the above discussion of the agent design architectures is that neither a completely deliberative (proactive) nor completely reactive approach is suitable. This idea leads naturally to combining different agent design philosophies and arranging them into a hierarchy of interacting layers. The layered architecture can be categorized into two groups: horizontal layering and vertical layering based on the way information and control flow. In horizontal layering, different layers are operating concurrently, but each layer is independently motivated and is independently connected to the sensors and effectors. A prototypical example of the horizontal layering architecture is the TouringMachine [37]. The main advantage of this structure is the simplicity of the conceptual design process [31]. Due to the fact that different layers are competing with each other to generate action suggestions, the horizontal layering structure usually demands a centralized controller (mediator). However, it is difficult to design a mediator that can properly coordinate complex interactions among various layers. This problem is partly solved by the vertical layering architecture in which sensory input and action output are each dealt with by at most one layer.

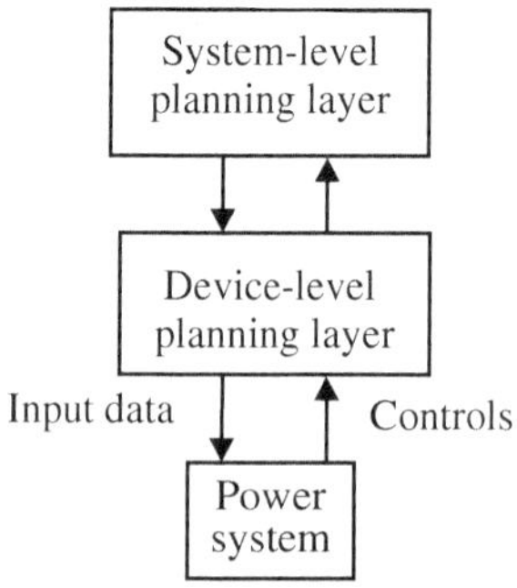

Figure 20.3 Two-pass vertical layered architectures.

The vertical layering architecture is further divided into two categories: one-pass and two-pass control architectures [31]. In a one-pass architecture, information and control flow sequentially from the initial layer that received sensory input to the final layer that generates action output. In a two-pass architecture, information flows up to the highest level layer and then control commands flow down the architecture. An example of a two-pass vertical layered architecture is shown in Figure 20.3 [38]. INTERRAP [39] is a well-known exemplar implementation of this agent architecture. The great advantage of this architecture is that it dramatically reduces the complex interactions among different layers. This main advantage comes with a sacrifice of the system robustness: If one layer fails, the entire system fails.

20.2.2 Multiagent Systems and Societies of Agents

Agents are situated in some physical and/or computational environments. In order to operate effectively and interact with each other productively, agents need an environment that can provide an appropriate infrastructure to facilitate interactions such as communication and negotiation. In addition, the agent itself should be designed in such a way that it could conduct effective communication, carry out rational negotiation, and coherently cooperate with other agents. This section presents three key aspects of societies of agents: communication, negotiation, and coordination.

20.2.2.1 Communication

Communication is important for all kinds of interactions in the human society. It is also fundamental in an environment of autonomous agents that are interacting with each other. Without it, all the interactions between agents such as negotiation, argumentation, and cooperation will not be possible. Research on communication between agents originated from the speech act theory [40], which treats communication as an action. A number of performative verbs such as request, promise, report, and insist are identified. The beauty of speech act theory is that it separates semantics of the domain independent communication protocol from the domain-dependent content language.

Inspired and highly influenced by the speech act theory, the knowledge query and manipulation language (KQML) was developed as a deliverable of the US government's ARPA knowledge sharing effort (KSE). KQML is a protocol for exchanging message and knowledge. A KQML message includes a performative, an envelope information of the message that contains several parameters and the actual content which is left undefined. The knowledge interchange format (KIF), which is also developed as part of KSE, is intended primarily to represent the content of a KQML message. Largely based on first-order predicate calculus, KIF was envisaged as a formal syntax for representing knowledge within computer systems.

Despite the broad uptake of KQML by the MAS community, it suffers from several major drawbacks such as lack of rigorously defined semantics and ad hoc performative set that is not tightly constrained. This led to the development of the popular FIPA-ACL, which is proposed by the FIPA. FIPA-ACL incorporates key features of KQML and made two major improvements. The most important modification is standardizing and defining a new library of normative performatives that ensure interoperability. In addition, a comprehensive formal semantic language (SL) was defined with a concrete syntax. It was suggested as a candidate content language for use in conjunction with the FIPA-ACL [41].

20.2.2.2 Negotiation

The ability of negotiation is crucial for the cooperation between autonomous agents that may not share a common goal. Two issues are of major concern to the study of negotiation: (1) how to design a mechanism or protocol governing the encounter between agents so that a mutually beneficial agreement can be reached efficiently and fairly and (2) how to design a negotiation strategy for an individual agent under a particular protocol such that a "good enough" outcome will be obtained reliably. Generally, there are two approaches for agents to negotiate and reach agreement: game-theoretic negotiation and argumentation-based negotiation.

The game-theoretic negotiation framework usually sets up a formal multiagent game in which agents negotiate according to some specific rational strategy. Despite the desirable attributes, that is, simplicity, robustness, and stability of this approach, it has several limitations. The major limitation is that in the game-theoretic negotiation framework, the agents simply follow a strategy which is made up of a set of equations and rules; hence, it is hard to justify their negotiation position. Due to the lack of a logical understanding of the negotiation outcome, it is difficult for humans to trust the software agents and delegate the negotiation tasks to them. Realizing these limitations, the argumentation-based negotiation approaches were proposed as an attempt to rectify the drawbacks of game-theoretic negotiation approach [42]. In the argumentation-based negotiation process, agents lay out arguments for and against propositions, together with justifications for acceptability of these arguments [43].

20.2.2.3 Coordination

Coordination is the process in which various agents with different subgoals work together harmoniously to achieve a common goal. The traditional approach of coordination tries to predict possible interactions between agents in advance, and design a system to suppress negative interferences and promote potential positive interactions. This approach works well in a simple foreseeable environment; however, when the

MAS is very complex it is impossible to anticipate all possible interactions. This limitation of the traditional approach led to the development of coordination techniques that make agents resolve conflicts and work cooperatively together at run time.

Several approaches emerge for dynamically coordinating agent activities. Partial global planning (PGP) is one of them. The main idea of PGP is that informing local scheduling of nonlocal considerations improves the coordination between agents by avoiding redundant activities, shifting tasks to idle agents, and providing predictive results [44]. Based on the concept that all coordination mechanisms can ultimately be reduced to commitments and their associated conventions [45], Jennings investigated the use of commitments in the coordination of an industrial control system called ARCHON [46]. Coordination can be achieved through mutual modeling in which each agent builds a model of other agents and coordinates its activities according to predictions of the models. Researchers have used norms and social laws that are regulating human activities to coordinate activities of agents.

20.2.3 Programming Languages, Tools, and Frameworks for Multiagent Systems

One of the bottlenecks to advance the theory in MAS and technology transfer to industry is the lack of mature and easy-to-use design and development tools. Therefore, the development of programming languages and tools that can effectively facilitate MAS design and development is critical for bridging the gap between theory and practice. Almost, two decades have passed since the visionary paper "agent-oriented programming language" by Yoav Shoham [47] was published in the early 1990s. There are still heated debates on the key issue of what is the most natural, convenient, and efficient way to program agents. Drastically different approaches have been taken by various researchers to develop programming languages, tools, and frameworks for MAS. A brief survey of the state-of-the-art of programming languages, platforms, and frameworks is presented here. The summary given below is largely based on discussions in [48] and [49].

20.2.3.1 Programming Languages for Multiagent Systems

Initial works in the field of agent-oriented language can be divided into three groups. These range from purely imperative languages at one extreme, to purely declarative ones at the other end and various hybrid approaches in between [48].

DECLARATIVE LANGUAGES: Generally, the declarative paradigm describes what the computer program should accomplish rather than how to accomplish it. Most declarative languages find their roots in theories of formal logic. The declarative nature of agent-oriented design concepts and abstractions naturally leads to the design and development of declarative languages that can effectively support MAS programming. The agent programming languages that can be labeled by declarative languages include: CLAIM [.50], FLUX [51], MINERVA [52], KABUL [53], DALI [54], and ReSpecT [55].

IMPERATIVE LANGUAGES: As opposed to declarative languages, imperative languages describe the exact sequence of commands for the computer to perform. Usually, nonagent-oriented imperative languages are implemented in an ad-hoc manner to develop MAS. JACK Agent Language [56], which extends from an imperative object-oriented language Java, is an example of this class.

HYBRID APPROACHES: Hybrid approaches combine the merits of both declarative and imperative languages. The declarative features specify the theoretical part of the problem, whereas the imperative features specify the low level procedures. Imperative components are usually written in object-oriented languages such as Java. Imperative components are sometimes responsible for interfacing with legacy codes that are imperative in style. Agent languages that can be categorized as hybrid approaches include 3APL [57], IMPACT [58], Jason [59], Go! [60], and AFAPL/AFAPL2 [61].

20.2.3.2 Integrated Development Environment

Most integrated development environments (IDEs) only support a particular abstract agent programming language with the intention to enhance productivity of MAS developers by automating tedious coding tasks. Common services that an IDE provides include project management, online error detection, autocompletion, refactoring, testing, and debugging tools. Representative examples of agent IDEs are 3APL IDE [62], Jason IDE [63], JDE [64], CAFnE [65], Visual Soar [66], AgentBuilder [67], AgentFactory [68], and the Living Systems Developer [69].

20.2.3.3 Frameworks for Multiagent Systems Development

As mentioned before, most agent-based development platforms and IDEs only implement the semantics of a particular agent programming language. However, there is one type of agent development frameworks that is intended to be generic, such as TuCSon [70] and JADE [32]. They facilitate the development of agent-based applications by providing a flexible infrastructure for agent life-cycle support and interagent communication services.

20.2.4 Multiagent System-Related Standards

Two major consortia, FIPA and object management group (OMG), have been working on producing a wide range of standards specifications to cover major aspects of MAS.

20.2.4.1 The Foundation for Intelligent Physical Agents

FIPA was originally formed in 1996 as an international nonprofit organization that is dedicated to promoting the industry of intelligent agents by developing open specifications supporting interoperability among agents and agent-based applications [28]. Through years of open collaboration among its member organizations, a set of standard specifications supporting interagent communication and key middleware services has been set up. Entering the new century, FIPA issued a new mission statement: "To promote technologies and interoperability specifications that facilitate the end-to-end inter networking of intelligent agent systems in modern commercial and industrial settings" [71]. A life-cycle model was adopted to classify FIPA specifications, which contain three major phases: preliminary, experimental, and standard. Up to date, there are 25 specifications that have been promoted to the status of standards, 14 specifications have the experimental status, and three specifications are at a preliminary phase. FIPA standards govern the basics of an agent architecture, including agent life-cycle management, interagent message transport, message structure, interagent interaction protocols, and security. Users are left with the flexibility to design an agent that meets its goals. The standard specifications of FIPA reflect its core concepts and focus on agent communication, agent management, and agent architecture.

AGENT COMMUNICATIONS: The FIPA-ACL finds its roots in speech act theory, which states that message represents actions that are called performatives. There are 22 performatives in communicative act library, which has four basic performatives: request, inform, confirm, and disconfirm. FIPA also standardizes a set of interaction protocols such as requests, and query to coordinate multimessage actions. Different content languages can be employed to express the content of FIPA-ACL. The most popular language FIPA SL is standardized and specified in [41].

AGENT MANAGEMENT: The second fundamental aspect of FIPA is addressed by agent management that establishes the logical reference model for creation, registration, location, communication, migration, and operation of agents. It specifies how a FIPA compliant agent can exist, operate, and be managed. A FIPA compliant agent platform (AP) provides the physical infrastructure that consists of the machines, operating system, FIPA agent management components, agents themselves, and any additional support software [32]. An AP has two utility agents: agent management system (AMS) and directory facilitator (DF). AMS is mandatory, as it allocates agent identifiers (AIDs) to each agent that registered with it, keeps track of the status of an agent, and terminates the life of an agent when it deregistered. DF is optional; it provides yellow page services that allow every agent to advertise its services on a nondiscriminatory basis. AP also provides a message transport service (MTS) to transport FIPA-ACL messages between agents on the same platform or within different platforms.

AGENT ARCHITECTURE: The FIPA abstract agent architecture provides a common, unchanging point of reference for FIPA-compliant implementations that capture most critical and salient features of an agent system [72]. Most important mandatory items specified in the architecture are the ACL message structure, message transport, agent directory services, and service directory services. As mentioned in the agent management section, communication between two agents relies on an MTS that transports FIPA-ACL messages. As mentioned in agent communication section, the structure of a message is a set of key values written in FIPA-ACL. The content of the message is expressed in a content language, such as FIPA-SL and FIPA KIF [72]. Essentially, the two directory services allow agents to register themselves or services that they provide, and to search for specific agents for services.

After standardization of the core specifications, the FIPA Modeling Technical Committee (Modeling TC) was established to develop vendor-neutral common semantics, meta-model, and abstract syntax for agent-based methodologies. A primary goal of the modeling TC is to clarify the definition of the agent unified modeling language and recommend methodology and tools to support it. In mid-2005, FIPA, the standards organization for agents and MAS, was officially accepted by the IEEE as its eleventh standards committee, lending credibility to the use of FIPA as standards for industrial and commercial MAS applications [71]. Since then, four working groups have been established to carry out further standardization projects in the areas of agents and web service interoperability, human-agent communications, mobile agents, and peer-to-peer nomadic agents [71].

20.2.4.2 The Object Management Group

Although not as active as FIPA on producing agent-related standards, as an international computer consortium with hundreds of member organizations, OMG could and will be an important driving force in standardization of the agent technology. The main advantage of OMG in promoting agent technology-related standards is that it could try to leverage and interoperate with other OMG specifications in the agent area. OMG established a special interest group that is focused on developing agent-related standards. The mission of the AP special interest group is to identify and recommend new OMG specifications in the area of agent technology and to recommend agent-related extensions to its existing and emerging specifications. In addition, OMG has two working groups that are also related to agent technology: ontology working group and service-oriented architecture working group.

20.3 APPLICATIONS OF MULTIAGENT SYSTEMS IN POWER ENGINEERING

This section reports on the applications of MAS in power engineering. Although in many application domains multiagent technology has grown out of the laboratory and made it to industry applications, power industry is behind in adopting the technology. Up to date, in power engineering domain, there are only a limited number of demonstrations and deployed MAS in actual operation. Most applications are research concepts or software prototypes. There are two main obstacles of adopting agent-based solutions in power industry: risk that comes with the adoption of new technology and the lack of easy-to-use design and development methodology and tools. Despite many good applications, the reliability and effectiveness of multiagent technology has not been proven in large-scale industry applications. Power systems as one of the most critical infrastructures place emphasis on reliability. As described in Section 2, dozens of multiagent programming languages, tools, and frameworks have been developed. Nevertheless, none of them are tailored for power system applications. The promotion of massive uptake of the technology calls for the provision of flexible design and development tools that are aimed at power system applications.

McArthur et al. [73] provided a comprehensive bibliographical analysis of agent research in power engineering. This section is not intended to repeat their work but to generalize the information and assess various application domains by the level of maturity, applicability of multiagent technology, and the use of agent concepts. Four application domains have been identified: modeling and simulation, monitoring and diagnostics, restoration and reconfiguration, and distributed control [73].

20.3.1 Modeling and Simulation

MAS technologies have been used for modeling and simulation of different aspects of power engineering. A major application is the simulation of restructured electricity market, which serves two main purposes: decision support tools for market participants and market design analysis tools for regulators. Due to complex interactions among strategic behaviors of market players, various layers of market designs, and the underlying physical network, it is difficult to fully evaluate the impact of a new market design using conventional simulation tools. With the capability to simulate inherently complex systems, MAS-based method is well suited to model complex phenomena of an electricity market and provide insights into new market rules and regulatory policies. Regulation and market instruments and market participants also call for a flexible multiagent-based electricity market simulation platform on which they could test their strategies in various dynamic market environments.

Multiagent-based modeling and simulation applications evolved from promising research concepts in early stages to useful agent-based software prototypes that provide insightful ideas to the design of different layers of electricity markets. Some applications are summarized below. An agent-based simulation (ABS) model of the wholesale market for electricity market in England and Wales was developed to compare market prices and bidding strategies of individual generators under pool and bilateral market mechanisms [74]. The new electricity trading arrangement (NETA) in the United Kingdom was tested in an ABS model [75]. In [76], Bunn and Oliveira adapted and extended their previous ABS platform to evaluate two power companies' market power in the framework of NETA. A multiagent-based day-ahead electricity market was modeled in [77] to study the complex interactions of market participants' bidding behaviors and its implications on market power. A wholesale electricity market framework called agent-based modeling of electricity systems (AMES) was used to investigate the effects of demand-bid price sensitivity and supply-offer price caps on prices for bulk electric power [78]. Customers' response under time-of-use electricity pricing is studied in an MAS [79]. A multiagent-based toolbox was constructed in [80] to study the effects of different mechanisms to promote generation expansion.

20.3.2 Monitoring and Diagnostics

Monitoring and diagnostics applications require gathering and interpretation of raw data from a variety of sources that are physically distributed. For example, in the case of transformer condition monitoring, oil temperature measurements, electrical loading, dissolved gas analysis results, partial discharge measurements, tap changer positions, and maintenance and servicing records are critical data for reliable and effective condition monitoring [81]. Moreover, with an increasing amount of information, and more strict time constraints, it becomes difficult even for an experienced engineer to take on laborious processes to retrieve meaningful information for monitoring and diagnostics decision support. For instance, a complete disturbance diagnosis is time-consuming even for an experienced engineer because it involves four labor and knowledge intensive steps: Supervisory control and data acquisition (SCADA) data gathering, incident and event identification, additional data retrieval, and additional data interpretation [82].

Common functionalities that can be found in monitoring and diagnostics applications involve data gathering, interpretation, information retrieval, and integration. In these applications, agents are put into a fixed hieratical structure with different roles and specific interaction protocols. Domain-specific ontology is designed to facilitate the exchange of information between agents.

Monitoring and diagnostics applications are not as critical to the safety and reliability of the entire power system compared to system control and restoration applications. This leads to relatively mature monitoring and diagnostics applications. The representative application of monitoring is the transformer condition monitoring multiagent system [83], which has been developed as an agent-based software prototype. The validity of this application has been demonstrated with the partial discharge data generated from a transformer defect that was set up in the laboratory. The representative application of diagnostics is the protection engineering diagnostic agents [84] that has been successfully transferred from laboratory to deployment at a Scottish utility for automated disturbance diagnosis.

20.3.3 Restoration and Reconfiguration

The automatic reconfiguration of power system components is highly desired in both distribution system restorations and reconfiguration of shipboard power system applications. Mathematical programming, heuristics, expert systems, and meta-heuristic techniques have been proposed to obtain the target configuration of the system [85]. Heuristic, expert systems, and meta-heuristic techniques cannot guarantee optimality of the solution. Traditional optimization is able to obtain the optimal solution. However, since it calls for a powerful computing facility, this approach suffers from the problem of a single-point-of-failure [86]. The complex nature of a combinatorial problem and the geographically distributed system information makes MAS a promising approach to solve the power system restoration and reconfiguration problems. To settle conflicts among various self-interested agents and helping them reach agreements, negotiation, and coordination, concepts are used in the system designs. In addition, sometimes, an organization structure was set up with different layers in which agents were assigned various roles.

Up to date, MAS-based power system restoration and reconfiguration applications are at the stage of research or software prototypes. Although several designs have been demonstrated to be effective in simple test systems based on computer simulations, significant research and development are needed before they can be deployed in actual operations. A good restoration design for distribution systems that achieve a target network was proposed by Nagata et al. [85]. In the proposed MAS, local bus agents formulate a restoration plan through negotiation, and then check the restoration plan with a global facilitator. A navy ship power system reconfiguration design that is aimed at restoring the capacity as much as possible to serve the loads was provided in [87]. Through negotiation among three different types of agents, the system can perform reconfiguration using local information without a central controller.

20.3.4 Distributed Control

With the emergence of distributed generation, an increasing size of the power system, and further restructuring of the electricity market, it becomes harder for traditional centralized control systems to handle certain control operations. Take the example of microgrid control, distributed energy resources with different characteristics and owners need to be coordinated in a cohesive way to perform the tasks: voltage control, black start, transition from grid connected to island mode and vice versa, load-shedding, feeding local critical loads, and producing heat for local installations. In addition, the microgrid control system should be flexible so that it will automatically adjust to the connection and disconnection of distributed generators. Moreover, the microgrid market operation calls for distributed decision making of intelligent distributed generation resources. All those challenging requirements do not match the capabilities of a centralized control approach. Solving such an inherently distributed and complex application in a centralized way is counter-intuitive. The MAS technology provides an alternative, which distributes the solution process across multiple intelligent agents capable of cooperation and coordination.

The following applications have been investigated based on the concepts of MAS: microgrid control [88], power system stability control [89], voltage regulation on the distribution feeders [90], and strategic power infrastructure defense (SPID) [91]. Two major agent concepts, which are negotiation and coordination, have been used in those applications. Various negotiation techniques and auctioning protocols are relied upon, when self-interested end users represented by intelligent agents are trying to reach agreements on daily electricity transactions in the microgrid market operations. Designing a very complex control system such as the SPID system demands a layered structure of the MAS that consists of many autonomous agents assigned with different tasks. Coordination protocols and collaboration enforcement mechanisms are critical to the successful design of such complex systems as they support collective behaviors such as conflict resolution, resources sharing, and collective planning. All those collective behaviors are crucial for the SPID system to perform control actions such as failure analysis and adaptive controls on the complex power system.

Despite innovative research on MAS-based distributed controls, the applications have not reached the stage of practical operation. Most applications are agent-based software prototypes, which have been validated by computer simulations. In a few cases, physical demonstrations have been developed to test the research ideas. For instance, a microgrid is installed in the Power System Laboratory of the National Technical University of Athens to test the proposed multiagent microgrid control system [88].

20.4 ELECTRICITY MARKETS MODELING AND SIMULATION WITH MULTIAGENT SYSTEMS

Restructuring of electric power industry around the world from a vertically integrated monopoly to open competitive markets has been a difficult process. Many challenging issues that do not exist in the design of other commodity markets arise in this transition due to unique physical characteristics of electricity. The requirement of instant balance between supply and demand, very limited storability, and compliance with laws and physics and transmission constraints makes electricity a peculiar product.

Both regulators and various market participants are faced with new challenges. The California energy crisis in 2000–2001 has shown what could happen to an electricity market if emerging market design issues are not thoroughly tested. Although much experience has been gained and costly and valuable lessons have been learned, there is still a lack of systematic methodology to evaluate the impact of a market design from both engineering and economic points of view. In the new market environment, various market participants are striving to develop effective decision support tools that could assist trading in a day-ahead market, negotiation of bilateral contracts, and expansion of generation units and transmission networks.

20.4.1 Why Multiagent System?

Difficulties in design effective market rules and decision support tools arise from complex interactions among strategic behaviors of market players, various layers of market designs, and the complex underlying physical network.

Traditional approach to exploring interactions between strategic bidding and market designs includes game theoretic techniques, stylized oligopoly model of competition such as Bertrand and Cournot, and supply function equilibrium or laboratory game experiments using human participants. Although oligopoly models have yielded useful results, analytical constraints require considerable simplifications in the market structure [76]. While laboratory game experiments using human subjects as market participants can discover market flaws and loopholes, the limited computation capability might prohibit human participants from making reasonable decisions in a complex market environment. Game theoretic models provided many useful insights into gaming in electricity markets. However, potential strategies used by electricity market participants are often too complex to be modeled by standard game theoretic techniques [92].

Unlike the traditional approach, the MAS method does not postulate stylized oligopoly model. Instead, it represents market participants with adaptive agents that have distinct goals, characteristics, decision-making tools, and resources. Those adaptive agents are endowed with capabilities to learn from past experience and project the future. It is observed that the accumulated bidding experience from interacting with other market participants in repeated similar auctions may change the perception of a player has of others [93]. These market dynamics could be well captured by empowering agents with carefully designed learning engines. Well-designed adaptive agents have the potential to develop trading, operating, and planning strategies beyond what can be reasonably hard-coded into heuristic agents or expected of human game participants [94].

20.4.2 Literature on Multiagent-Based Modeling of Electricity Markets

Researchers have proposed multiagent-based simulation to study various issues in electricity markets. A multiagent evolutionary modeling of the proposed NETA in the United Kingdom was presented in [75]. It is shown that ABS provides pricing and strategic insights ahead of NETA's actual introduction. The electricity market complex adaptive systems model was developed [92]. The effects of power plant outages and price setting rules on LMPs are analyzed with the proposed model. The result hinted the potential of using a multiagent approach to electricity market simulation. An ABS was performed [76] to evaluate individual market power in electricity markets. The authors concluded that ABS techniques brought substantial insights into the behaviors of generators in the wholesale electricity market. An MAS that simulates competitive electricity market (MASCEM) was proposed in [95] which is implemented in the open agent architecture. Agent-based computational methods for the study of wholesale and retail electricity markets were discussed in [96]. Distinctions in problem formulation between the two levels are explored. Yu, Liu, and Price presented a flexible and integrative MAS method to assess market designs [97]. Realistic simulation scenarios are constructed for evaluation of the proposed Pennsylvania, Jersey, Maryland Power Pool (PJM)-like market power mitigation rules of the California electricity market. The incorporation of PJM-like local market power mitigation rules is shown to be effective in suppressing the exercise of market power.

20.4.3 Multiagent System Design for Electricity Market Modeling and Simulation

20.4.3.1 Purpose

The main motivation of modeling electricity markets with MAS is that the model serves as a computational platform on top of which regulators and market participants could explore and test potential effects of their actions [96].

From a regulator's point of view, it is of interest to test whether existing market rules and architecture promote efficiency of the market and enhance reliability of the power grid. In addition, the regulator would like to explore the impact of changes to market rules in both engineering and economic aspects before the actual implementation. Testing and exploration on market rules can be conducted for the following purposes:

In a wholesale electricity market, how different auction types (marginal pricing and pay-as-bid) affect the bidding behaviors of market participants and the overall efficiency of the market?

Whether simultaneous or sequential clearance of energy market and ancillary service markets makes trading strategies more complex? Which one will lead to a more efficient market?

Which way of design retail market and combining with the wholesale market will be more beneficial to market participants while incurring less cost in restructuring?

What market power mitigation rules should be implemented to curb the exercise of local market power while sending appropriate economic signals for generation investment?

How to properly design a capacity market that induces sufficient generation investment in the right location and of the right type to satisfy reliability standard at the least cost?

Whether the point-to-point or flowgate paradigm of financial transmission rights serves as a better hedging tool for market participants and leads to more efficient congestion management?

From the market participants' point of view, the multiagent models of electricity markets could be used as a test bed for assessing trading strategies, risk management practices, and generation/transmission investment plans. Testing and assessment on decision support tools can be performed for the following purposes:

Evaluate the performance of portfolio management in short-to medium-term trading. Analyze portfolio management practice under different scenarios. Perform sensitivity analysis to examine how optimal portfolio mix varies as a result of a change in the value of key variables in an MAS model.

Test whether a certain generation investment plan will generate reasonable returns on investment under a preferred level of risk. How the changes in market environment and advancement in generation technology affect the investment plan.

Compare the performance of various electricity price forecasting methodologies (structural approach, purely statistical approach, and hybrid approach) under different scenarios. Test how the data availability, forecast time horizon, and changes in the value of key variable in the model affect the accuracy of the forecasting model.

Potential applications of multiagent-based modeling of electricity markets are not limited to the ones mentioned above. The multiagent-based model is well suited to analyze complex phenomena of an electricity market originated from interactions among strategic behaviors of market players, various layers of market designs, and the complex underlying physical network.

20.4.3.2 MAS Structure

The development of agent-based programming languages, tool, and frameworks mentioned in the previous section accelerates the utilization of more mature and effective tools for applications in electricity market simulations. In the initial stage of agent-based electricity market simulations, Excel and Visual Basic were used to develop the simulation platforms. With advancement in programming languages and tools that are specialized for agent-based applications, researchers in the field of electricity market simulations start to use them for their MAS developments. Open agent architecture, a framework for integrating a community of heterogeneous software agents in a distributed environment, was utilized in [95] to develop MASCEM. RepastJ, a Java-based toolkit designed specifically for agent-based modeling in social sciences, was used in [98] to develop AMES whole power market framework. JADE, the widely used agent-oriented middleware, is chosen to be the platform on which the proposed MAS for electricity day-ahead market was built [97].

Although many tools could be used to develop MAS for electricity market modeling and simulation, JADE is recommended for the following reasons. When using agent-based approach to solve a problem, there are a number of domain independent issues that must be addressed, such as how to allow agents to communicate [32]. JADE provides the domain independent infrastructure, which allows developers to focus on the construction of key logics. In addition, since JADE is written in Java, it benefits from a large set of programming abstractions, which greatly facilitate the development of MAS. More importantly, JADE fully complies with the FIPA specifications, which is maintained by the standards organization for agents and MAS. Based on the above considerations, JADE is chosen to be the middleware on which the proposed MAS was implemented.

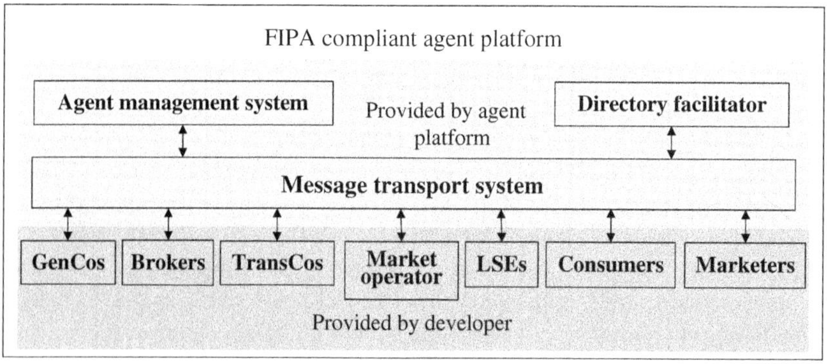

Figure 20.4 Structure of the multiagent platform for electricity market simulation.

The structure of an extensible multiagent platform based on JADE for electricity market simulation is depicted in Figure 20.4. JADE provides two utility agents: AMS and DF, and an interagent messaging system through which agents communicate with each other. AMS allocates AIDs to each agent that registered with it, and provides a "white page" service, where an agent can ask for the address of another. DF provides a "yellow page" service, where agents register the services they provide, and an agent can ask for all agents to provide a particular service. The market operator (MO), generation companies (GenCos), broker, transmission companies (TransCos), load serving entities (LSEs), consumers, and marketers will be designed by the developers. The particular design of these agents depends on the purpose of the electricity market simulation and will be illustrated in the next section.

20.4.3.3 Agents

The MO is responsible for reliable operation of the power grid and efficient management of the wholesale markets. Depending on the specific application, different combinations of functions should be implemented for MO. Key functions of the MO include management of day-ahead market, ancillary service market, real-time market, Financial Transmission Rights (FTR) primary auction and secondary market, capacity market, mitigating of market power, scheduling physical bilateral transactions and scheduling generation and transmission outages, and publishing load forecasts.

Implementation of MO's day-ahead market function in JADE is illustrated here. The sequence of tasks that MO will carry out is as follows. On day *d*-7, MO publishes initial hourly load forecasts for day *d*, and updates it daily as the operating day approaches. On day *d*-1, MO receives supply offers and demand bids from GenCos and LSEs in the morning for day-ahead market for day *d*. In the afternoon, upon receiving all demand bids and supply offers, the MO runs the security constrained unit commitment and security constrained economic dispatch programs to determine the hourly unit dispatches and locational marginal prices (LMPs). At the end of process, MO sends dispatch schedules, LMPs, and settlement information to the corresponding GenCos and LSEs.

Several behaviors need to be implemented for MO in order to perform the task of managing day-ahead market. The first behavior is called publishLoad-Forecast(), which cyclically (everyday) sends out messages to inform all market participants about the load forecasts. The second behavior is called receiveBidsOffers(), which receives messages sent by GenCos and LSEs. These messages contain all the required information for supply offers and demand bids. Upon receiving all supply offers and demand bids, the third behavior clearDayAheadMarket() will be executed, which clears the bid-in supply with bid-in demand and procures ancillary service at the minimum cost. Finally, the fourth behavior sendDayAheadResult() will be carried out, which sends out dispatch schedules, LMPs, and settlement information to the corresponding GenCos and LSEs.

GENCOS: GenCos' long-term goal is to optimize the return on investments from engaging in electric energy supply. GenCos typically own a portfolio of electricity generating facilities located at different busses of the power grid. Core functions of GenCos include plan for generation expansion, purchase fuels, sell electricity power through the wholesale market, engage in physical bilateral contracts, procure various financial instruments to hedge-related risks, and model and forecast load, electricity, and fuel prices. Various functions should be designed and integrated by the portfolio management methodology. In the short term, GenCos' portfolio management's objectives are: (a) to receive a stable cash flow from selling bulk power through multiple markets (pool, bilateral, FTR, and fuel) and (b) to mitigate risks resulting from uncertain load, unexpected outages and fuel, and electricity and emission allowance price volatilities. In the long term, GenCos' portfolio management's objectives are: (a) to receive a reasonable return on the generation investment and (b) to mitigate risks associated with different types of generating plant technologies, changing load pattern and power grid configuration, and changes in market rules and regulations. The GenCo agent will learn from past experience in participating in the electricity market and other related markets and adapt its portfolio management practice.

Implementation of GenCo's day-ahead market activities in JADE is described here. The sequence of tasks that GenCo will carry out is as follows. Suppose on day d-1, GenCo agent receives the load forecast information published by MO. After analyzing the load forecast and other information that it has, GenCo calculates the supply offer according to its bidding strategy. Then, GenCo submits its supply offer for day d to MO. Finally, upon receiving the dispatch schedules, LMPs, and settlement information from MO, GenCo calculates its daily profit and updates its database and bidding strategy.

Several behaviors need to be implemented for GenCo in order to perform the task of participating in the day-ahead market. The first behavior is named receiveLoadForecast(), which receives the load forecast for day d sent out by MO. The second behavior is called makeSupplyOffer(), which calculates its supply offer based on the information that it has. Then, GenCo will send out its supply offer to MO by executing the third behavior sendSupplyOffer(). Afterward, the fourth behavior receiveDayAheadResult() will be carried out, which receives the message sent by MO containing the day-ahead market clearing result. At the end of process, GenCo performs the updateDatabaseStrategy() behavior, which updates its database and bidding strategy.

LSEs: LSEs purchase bulk power from the wholesale electricity market to serve the consumers. Central functions of LSEs include negotiate physical bilateral contracts with GenCos, purchase electric power through wholesale market, perform load forecasting, and offer demand response incentives to customers. All these functions should be unified into the portfolio management process of an LSE. The portfolio management objectives of an LSE are: (a) to maintain low cost of purchasing electricity, (b) to provide reliable electricity service, and (c) to mitigate risks associated with price volatility, and potential changes in market rules and regulations. LSE agents also learn from its past experience in participating in the electricity market.

Implementation of LSE's day-ahead market tasks in JADE is demonstrated here. LSE's sequence of tasks in the day-ahead market is listed. On day *d*-1, LSE performs load forecasting for its customer for day *d*. Then, LSE agent receives the load forecast information in the whole network published by MO. After analyzing the available information, LSE calculates the demand bids and submits it to MO. Finally, upon receiving the LMPs and settlement information from MO, it calculates the daily profit and updates its database and bidding strategy. Behaviors that are similar to the ones listed for GenCo should be programmed to allow LSE to participate in the day-ahead market.

TRANSCOS: TransCos provide electric transmission service with its transmission assets such as overhead and underground transmission lines, and transformers. The duties and responsibilities of a TransCo include: (a) operating its transmission system reliably according to the instructions of MO, (b) planning transmission system upgrades and expansions to meet the future needs, (c) constructing upgrades and expansion projects to meet the future needs, and (d) maintaining the transmission equipment to reduce equipment failures. The revenue of a TransCo comes from two sources: (a) transmission usage charge and (b) congestion rent, which is calculated as differences of marginal congestion component of the day-ahead Locational Marginal Price (LMP) at the FTR delivery point and receipt point.

CONSUMERS: The residential, commercial, and industrial customers' objective is to maximize their utilities in the long term. Large commercial and industrial customers might be able to reach long-term power supply agreement with suppliers directly and negotiate the electricity rates. Smaller residential customers will either pay the fixed rate or choose to participate in various responsive incentive programs provided by LSEs. In order to reduce the transaction cost, the most likely scenario is to employ price responsive smart controller in households. These devices can be used to adjust set-points of home appliances such as air conditioner or refrigerator according to the preference and instructions set by the consumer.

MARKETERS: Marketers are entities that do not own any physical generation asset or have any obligations to serve load. They wish to take a position in the electricity market and other related markets. Although GenCos and LSEs might want to hedge their positions in the market, the marketers might want to assume the risks and pool them together to gain the risk premium.

BROKERS: Brokers are agents that set up market places such as a financial/physical bilateral contract market, which matches offers and bids from GenCos and LSEs automatically. These markets could significantly lower the transaction costs of negotiating short-term bilateral contracts.

REFERENCES

[1] BBC h2g2: Computer animation, 2005. Available at http://www.bbc.co.uk/dna/h2g2/A3421045

[2] Chung, K. T., and Wu, C.-H. *Dynamic Scheduling with Intelligent Agents: An Application Note.* Meta Application Note 105, Metra, Palo Alto, CA, 1997.

[3] Morley, D. Painting trucks at general motors: the effectiveness of a complexity-based approach. In: *Embracing Complexity*, Ernst & Young, Boston, MA, pp. 53–58, 1996.

[4] Wooldridge, M., and Jennings, N. Intelligent agents: theory and practice. *The Knowledge Engineering Review*, vol. 10, pp. 115–152, 1995.

[5] Jennings, N., and Wooldridge, M., Eds., *Agent Technology: Foundations, Applications, and Markets.* Springer-Verlag, 1998.

[6] Tesfatsion, L., and Judd, K. L., Eds., *Handbook of Computational Economics, Volume 2: Agent-Based Computational Economics.* Handbooks in Economics Series, North-Holland, 2006.

[7] Tesfatsion, L. Introduction to the special issue on agent-based computational economics. *Journal of Economic Dynamics and Control*, Elsevier, vol. 25, no. 3–4, pp. 281–293, March 2001.

[8] Parunak, H. V. D., Ward, A., Fleischer, M., and Sauter, J. A marketplace of design agents for distributed concurrent set-based design. *Proceedings of ISPE/CE97: Fourth ISPE International Conference on Concurrent Engineering: Research and Applications*, 1997.

[9] Bussmann, S., and Sieverding, J. Holonic control of an engine assembly plant – an industrial evaluation. *Proceedings of IEEE Conference on SMC*, Tucson, AZ, 2001.

[10] Himoff, J., Rzevski, G., and Skobelev, P. Magenta technology: multi-agent logistics *i*-scheduler for road transportation. *Proceedings of AAMAS-06 – Industry Track*, pp. 1514–1521, 2006.

[11] Mařík, V., and McFarlane, D. Industrial adoption of agent-based technologies. *IEEE Intelligent Systems*, vol. 20, no. 1, pp. 27–35, 2005.

[12] Iglesias, C., González, J., Velasco, J., Martín, P., Hernández, F., Oliveros, E., Toledo, J., Escobero, A., Núñez, M., and García, M. Enhancements of the MIX multiagent platform. Technical report for the ESPRIT project MIX 9119, May 1997.

[13] Schurr, N., Marecki, J., Tambe, M., and Scerri, P. Demonstration of DEFACTO: training tool for incident commanders. *Proceedings of the Fourth International Joint Conference on Autonomous Agents and Multiagent Systems*, AAMAS'05, ACM, New York, pp. 147–148, July 2005.

[14] Maturana, F., Staron, R., Tichý, P., Šlechta, P., and Vrba, P. A strategy to implement and validate industrial applications of holonic systems. *Holonic and Multi-Agent Systems for Manufacturing*, Springer Berlin, Heidelberg, vol. 3593, pp. 111–120, 2005.

[15] Ljunberg, M., and Lucas, A. The OASIS air traffic management system. *Proceedings of the Second Pacific Rim International Conference on AI (PRICAI-92)*, Seoul, Korea, 1992.

[16] Baxter, J., and Horn, G. Controlling teams of uninhabited air vehicles. *Proceedings of AAMAS-05 Industry Track*, ACM Press, Netherlands, pp. 27–33, 2005.

[17] Sycara, K., Decker, K., Pannu, A., Williamson, M., and Zeng, D. Distributed intelligent agents. *IEEE Expert*, vol. 11, no. 6, pp. 36–46, 1996.

[18] Ye, Y., Liu, J., and Moukas, A. Agents in electronic commerce. *Electronic Commerce Research*, vol. 1, pp. 9–14, 2001.

[19] Stone, P., Schapire, R., Csirik, J., Littman, M., and Mcallester, D. *ATTac-2001: A Learning, Autonomous Bidding Agent.* Agent-Mediated Electronic Commerce IV, Designing Mechanisms and Systems, Springer Berlin Heidelberg, pp. 15–35, 2002.

[20] Decker, K., and Li, J. Coordinated hospital patient scheduling. *Proceedings of the 3rd International Conference on Multi-Agent Systems*, ICMAS-98, France, 1998.

[21] Hayes-Roth, B., Hewett, M., Washington, R., Hewett, R., and Seiver, A. *Distributing Intelligence within an Individual.* Distributed Artificial Intelligence, Pitman Publishing, London and Morgan Kaufmann, San Mateo, CA, vol. 2, pp. 385–412, 1989.

[22] Lent, M., Laird, J., Buckman, J., Hartford, J., Houchard, S., Steinkraus, K., and Tedrake, R. Intelligent agents in computer games. *Proceedings of the Sixteenth National Conference on Artificial Intelligence*, Orlando, FL, 1999.

[23] Stanley, K., Bryant, B., and Miikkulainen, R. Evolving neural network agents in the NERO video game. *Proceedings of the IEEE 2005 Symposium on Computational Intelligence and Games*, 2005.

[24] A.I. Driven Character Animation for Film, TV & Game Cinematics. Available at http://www.massivesoftware.com/media-enterainment/

[25] Luck, M., Mcburney, P., and Preist, C. A manifesto for agent technology: towards next generation computing. *Journal of Autonomous Agents and Multi-Agent Systems*, vol. 9, pp. 203–252, 2004.
[26] Pěchouček, M., and Mařík, V. Industrial deployment of multi-agent technologies: review and selected case studies. *Journal of Autonomous Agents and Multi-Agent Systems*, vol. 17, pp. 397–431, 2008.
[27] Panait, L., and Luke, S. Cooperative multi-agent learning: the state of the art. *Journal of Autonomous Agents and Multi-Agent Systems*, vol. 11, pp. 387–434, 2005.
[28] Available at http://www.fipa.org.
[29] Dignum, F., and Eijk, R. Agent communication and social concepts. *Journal of Autonomous Agents and Multi-Agent Systems*, vol. 14, pp. 119–120, 2007.
[30] Fornara, N., Viganò, F., and Colombetti, M. Agent communication and artificial institutions. *Journal of Autonomous Agents and Multi-Agent Systems*, vol. 14, pp. 121–142, 2007.
[31] Wooldridge, M. Intelligent agents. In: G. Weiss, Ed., *Multiagent Systems: A Modern Approach to Distributed Artificial Intelligence*, MIT Press, pp. 42–65, 1999.
[32] Bellifemine, F., Caire, G., and Greenwood, D. *Developing Multi-Agent Systems with JADE*. John Wiley & Sons, February 2007.
[33] Ferber, J. Reactive distributed artificial intelligence. In: G. O'Hare and N. Jennings, Eds., *Foundations of Distributed Artificial Intelligence*, John Wiley & Sons, pp. 287–317, 1996.
[34] Brooks, R. A robust layered control system for a mobile robot. *IEEE Journal of Robotics and Automation*, vol. 2, no. 1, pp. 14–23, 1986.
[35] Bratman, M., Israel, D., and Pollack, M. Plans and resource-bounded practical reasoning. *Computational Intelligence*, vol. 4, pp. 349–355, 1988.
[36] Georgeff, M., and Lansky, A. Reactive reasoning and planning. *Proceedings of the Sixth National Conference on Artificial Intelligence (AAAI-87)*, pp. 677–682, 1987.
[37] Ferguson, I. Touring machines: an architecture for dynamic, rational, mobile agents. Ph.D. Thesis, Clare Hall, University of Cambridge, UK, November 1992.
[38] Liu, C. C., Jung, J., Heydt, G., Vittal, V., and Phadke, A. The strategic power infrastructure defense (SPID) system, a conceptual design. *IEEE Control Systems Magazine*, vol. 20, no. 4, pp. 40–52, August 2000.
[39] Muller, J., Pischel, M., and Thiel, M. Modeling reactive behavior in vertically layered agent architectures. In: M. Wooldridge and N. Jennings, Eds., *Intelligent Agents: Theories, Architectures, and Languages (LNAI 890)*, Spinger-Verlag, Heidelberg, pp. 261–276, 1994.
[40] Searle, J. R. *Speech Acts: An Essay in the Philosophy of Language*. Cambridge University Press, Cambridge, 1969.
[41] FIPA, FIPA 2000. Specification: FIPA SL content language specification, SC00008. Available at http://www.fipa.org/
[42] Sycara, K. P. Multiagent compromise via negotiation. In: *Distributed Artificial Intelligence*, L. Gasser and M. Huhns, Eds., Pitman, London and Morgan Kaufmann, San Mateo, CA, vol. 2, pp. 119–138, 1989.
[43] Wooldridge, M. *An Introduction to Multiagent Systems*. John Wiley & Sons, 2002.
[44] Durfee, E. H., and Lesser, V. R. Partial global planning: a coordination framework for distributed hypothesis formation. *IEEE Transactions on Systems, Man, and Cybernetics*, vol. 21, no. 5, pp. 1167–1183, September 1991.
[45] Jennings, N. R. Commitments and conventions: the foundation of coordination in multi-agent systems. *The Knowledge Engineering Review*, vol. 8, no. 3, pp. 223–250, 1993.
[46] Jennings, N. R., and Wittig, T. ARCHON: theory and practice. In: *Distributed Artificial Intelligence: Theory and Praxis*, N. Avouris and L. Gasser, Eds., Springer Berlin Heidelberg, pp. 179–195, 1992.
[47] Shoham, Y. Agent-oriented programming. *Artificial Intelligence*, vol. 60, pp. 51–92, 1993.
[48] Bordini, R. H., Braubach, L., Dastani, M., Seghrouchni, A. E. F., Gomez-Sanz, J. J., Leit, J., O'Hare, G., Pokahr, A., and Ricci A. A survey of programming languages and platforms for multi-agent systems. *Informatica*, vol. 30, pp. 33–44, 2006.
[49] Bordini, R. H., Dastani, M., Dix, J., and El Fallah Seghrouchni, A., Eds., *Multi-Agent Programming: Languages, Platforms and Applications*. No. 15, Multiagent Systems, Artificial Societies, and Simulation Organizations, Springer, 2005.
[50] Fallah-Seghrouchni, A., and Suna, A. A programming language for autonomous and mobile agents. *IEEE/WIC International Conference on Intelligent Agent Technology*, pp. 474–477, 2003.

[51] Thielscher, M. FLUX: a logic programming method for reasoning agents. *Theory and Practice of Logic Programming*, vol. 5, no. 4–5, pp. 533–565, July 2005.

[52] Leite, J., Alferes, J., and Pereira, L. *MINERVA – A Dynamic Logic Programming Agent Architecture*. Lecture Notes in Computer Science, Springer-Verlag, vol. 2333, pp. 141–157, 2001.

[53] Leite, J. Evolving knowledge bases. Frontiers in Artificial Intelligence and Applications, published in the subseries *Dissertations in Artificial Intelligence,* vol. 81, 2003.

[54] Costantini, S., and Tocchio, A. A logic programming language for multi-agent systems. *Proceedings of the 8th European Conference on Logics in Artificial Intelligence (JELIA'02)*, Springer, vol. 2424 of LNAI, pp. 1–13, 2003.

[55] Omicini, A., and Denti, E. From tuple spaces to tuple centres. *Science of Computer Programming*, vol. 41, no. 3, pp. 277–294, November 2001.

[56] Evertsz, R., Fletcher, M., Jones, R., Jarvis, J., Brusey, J., and Dance, S. Implementing industrial multi-agent systems using JACK^{TM}, *Programming Multiagent Systems, First International Workshop (ProMAS'03)*, Springer-Verlag, vol. 3067 of LNAI, pp. 18–48, 2004.

[57] Dastani, M., van Riemsdijk, M., Dignum, F., and Meyer, J. A programming language for cognitive agents: goal directed 3APL. In: *Programming Multiagent Systems, First International Workshop (ProMAS'03)*, M. Dastani, J. Dix, and A. Fallah-Seghrouchni, Eds., Springer-Verlag, Berlin, vol. 3067 of LNAI, pp. 111–130, 2004.

[58] Subrahmanian, V., Bonatti, P., Dix, J., Eiter, T., Kraus, S., Özcan, F., and Ross, R. *Heterogeneous Active Agents*. MIT-Press, 2000.

[59] Bordini, R., Hübner, J., and Wooldridge, M. *Programming Multi-Agent Systems in AgentSpeak Using Jason*. John Wiley & Sons, 2007.

[60] Clark, K., and McCabe, F. Go!–a multi-paradigm programming language for implementing multi-threaded agents. *Annals of Mathematics and Artificial Intelligence*, vol. 41, no. 2–4, pp. 171–206, 2004.

[61] Muldoon, C. An agent framework for ubiquitous services. Ph.D. Thesis, School of Computer Science and Informatics, University College Dublin, 2008.

[62] Available at http://www.cs.uu.nl/3apl/

[63] Bordini, R. H., and Hübner, J. F., *Jason*, manual, release version 0.9.5, February 2007. Available at http://jason.sourceforge.net/JasonWebSite/Jason%20Home.php

[64] Winikoff, M. JACK^{TM} intelligent agents: an industrial strength platform. Chapter 7 In: *Multi-Agent Programming*, Bordini, R. H., Dastani, M., Dix, J., and El Fallah Seghrouchni, A., Eds., Springer US, 2005.

[65] Jayatilleke, G., Padgham, L., and Winikoff, M. Component agent framework for non-experts (CAFnE) toolkit. In: *Software Agent-Based Applications, Platform and Development Kits*, R. Unland, M. Klusch, and M. Calisti, Eds., Birkhäuser Basel, 2005.

[66] Lehman, J., Laird, J., and Rosenbloom, P. *A gentle introduction to soar, an architecture for human cognition*. In: *An Invitation to Cognitive Science*, S. Sternberg and D. Scarborough, Eds., MIT Press, vol. 4, 1996.

[67] Available at http://www.agentbuilder.com/index.html

[68] Collier, R. Agent factory: a framework for the engineering of agent-oriented applications. Ph.D. Thesis, University College Dublin, 2001.

[69] Available at http://www.whitestein.com/.

[70] Omicini, A., and Zambonelli, F. Coordination for internet application development. *Journal of Autonomous Agents and Multi-Agent Systems*, vol. 2, no. 3, pp. 251–269, September 1999.

[71] History of FIPA. Available at http://www.fipa.org/subgroups/ROFS-SG-docs/History-of-FIPA.htm

[72] FIPA, FIPA 2000 specification: FIPA abstract architecture specification, SC00001. Available at http://www.fipa.org/.

[73] McArthur, S. D. J., Davidson, E. M., Catterson, V. M., Dimeas, A. L., Hatziargyriou, N. D., Ponci, F., and Funabashi, T. Multi-agent systems for power engineering applications – part I: concepts, approaches and technical challenges. *IEEE Transactions on Power Systems*, vol. 22, no. 4, pp. 1743–1752, November 2007.

[74] Bower, J., and Bunn, D. A model-based comparison of pool and bilateral market mechanisms for electricity trading. *The Energy Journal*, vol. 20, no. 3, pp. 2–19, 2000.

[75] Bunn, D. W., and Oliveira, F. S. Agent-based simulation – an application to the new electricity trading arrangements of England and Wales. *IEEE Transactions on Evolutionary Computation*, vol. 5, no. 5, pp. 493–503, October 2001.
[76] Bunn, D. W., and Oliveira, F. S. Evaluating individual market power in electricity markets via agent-based simulation. *Annals of Operations Research*, vol. 121, no. 1–4, pp. 57–77, July 2003.
[77] Yu, N. P., Liu, C. C., and Tesfatsion, L. Modeling of suppliers' learning behaviors in an electricity market environment. *Intelligent Systems Applications to Power Systems, 2007. ISAP 2007. International Conference on.* IEEE, pp. 1–6, November 2007.
[78] Li, H., Sun, J., and Tesfatsion, L. Dynamic LMP response under alternative price-cap and price-sensitive demand scenarios. In *Power and Energy Society General Meeting-Conversion and Delivery of Electrical Energy in the 21st Century, 2008 IEEE*, pp. 1–8. IEEE, 2008.
[79] Yuan, J. H., Wang, J., and Hu, Z. G. Simulation of large customer price response under time-of-use electricity pricing based on multi-agent system, *International Conference on Power System Technology*, 2006.
[80] Ortega-Vazquez, M. A., and Kirschen, D. S. Assessment of generation expansion mechanisms using multi-agent systems. *IEEE Power and Energy Society General Meeting*, July 2008.
[81] McArthur, S. D. J., Strachan, S. M., and Jahn, G. The design of a multi-agent transformer condition monitoring system. *IEEE Transactions on Power Systems*, vol. 19, no. 4, pp. 1845–1852, November 2004.
[82] Hossack, J. A., Menal, J., McArthur, S. D. J., and McDonald, J. R. A multiagent architecture for protection engineering diagnostic assistance. *IEEE Transactions on Power Systems*, vol. 18, no. 2, pp. 639–647, May 2003.
[83] McArthur, S. D. J., and Davidson, E. M. Multi-agent systems for diagnostic and condition monitoring applications. *Power Engineering Society General Meeting, 2004*, vol. 1, pp. 50–54, June 2004.
[84] McArthur, S. D. J., Booth, C. D., McDonald, J. R., and McFadyen, I. T. An agent-based anomaly detection architecture for condition monitoring. *IEEE Transactions on Power Systems*, vol. 20, pp. 1675–1682, November 2005.
[85] Nagata, T., and Sasaki, H. A multi-agent approach to power system restoration. *IEEE Transactions on Power Systems*, vol. 17, pp. 457–462, May 2002.
[86] Solanki, J. M., Khushalani, S., and Schulz, N. N. A multi-agent solution to distribution systems restoration. *IEEE Transactions on Power Systems*, vol. 22, no. 3, pp. 1026–1034, August 2007.
[87] Momoh, J. A. NAVY ship power system restoration using multi-agent approach. *Power Engineering Society General Meeting, 2006. IEEE*, June 2006.
[88] Dimeas, A. L., and Hatziargyriou, N. D. Operation of a multiagent system for microgrid control. *IEEE Transactions on Power Systems*, vol. 20, no. 3, pp. 1447–1455, August 2005.
[89] Ni, H., Heydt, G. T., and Mili, L. Power system stability agents using robust wide area control. *IEEE Transactions on Power Systems*, vol. 17, no. 4, pp. 1123–1131, November 2002.
[90] Baran, M. E., and El-Markabi, I. M. A multiagent-based dispatching scheme for distributed generators for voltage support on distribution feeders. *IEEE Transactions on Power Systems*, vol. 22, no. 1, pp. 52–59, February 2007.
[91] Li, H., Rosenwald, G. W., Jung, J., and Liu, C. C. Strategic power infrastructure defense. *Proceedings of the IEEE*, vol. 93, no. 5, pp. 918–933, May 2005.
[92] North, M., Conzelmann, G., Koritarov, V., Macal, C., Thimmapuram, P., and Veselka, T. E-laboratories: agent-based modeling of electricity markets. *2002 American Power Conference*, April 2002.
[93] Rothkopf, M. H. Daily repetition: a neglected factor in the analysis of electricity auctions. *Electricity Journal*, vol. 12, pp. 61–70, April 1999.
[94] Trotignon, M., and Entriken, R. *Program on Technology Innovation: Optimal Bidding in Power Markets: Q-Learning for Adaptive Agents*. EPRI, Palo Alto, CA and EDF R&D, 1016229, 2007.
[95] Praca, I., Ramos, C., Vale, Z., and Cordeiro, M. MASCEM: a multiagent system that simulates competitive electricity markets. *IEEE Intelligent Systems*, vol. 18, no. 6, pp. 54–60, November 2003.
[96] Widergren, S., Sun, J., and Tesfatsion, L. Market design test environment. *Proceedings, IEEE Power Engineering Society General Meeting*, Montreal, June 2006.
[97] Yu, N., Liu, C.-C., and Price, J. Evaluation of Market Rules Using a Multi-Agent System Method, *IEEE Transactions on Power Systems*, vol. 25, pp. 470–479, 2010.
[98] Sun, J., and Tesfatsion, L. Dynamic testing of wholesale power market designs: an open-source agent-based framework. *Computational Economics*, vol. 30, no. 3, pp. 291–327, October 2007.

CHAPTER 21

HEURISTIC OPTIMIZATION TECHNIQUES

Kwang Y. Lee, Malihe M. Farsangi, Jong-Bae Park, and John G. Vlachogiannis

21.1 INTRODUCTION

There are a large number of different combinatorial optimization problems (COPs) facing electricity utilities, such as optimal load flow, unit commitment, economic dispatch, network reconfiguration, generation planning, transmission planning, reactive power planning, power plant optimization, and integration of renewable energy sources. Deregulation of electricity supply industry worldwide adds ever growing motivations to develop new optimization algorithms. Therefore, best strategies are designed for most effectiveness utilizing the asset under increasing commercial pressure [1, 2]. Various algorithmic and heuristic approaches [3, 4] have been adopted or investigated by power engineers. These include the lambda-iteration method (NM), gradient method, Lagrangian relaxation, benders decomposition, interior point method, linear/nonlinear programming (LP) and dynamic programming, etc.

Several heuristic tools have evolved in the last decades that facilitate solving optimization problems that were previously difficult or impossible to solve. More recently, heuristic techniques such as artificial neural networks, simulated annealing (SA), tabu-search, and evolutionary computing have also been intensively investigated. In particular, for the last 15 years, there has been a growing interest in algorithms inspired by the observation of the natural process and behavior of natural creatures, such as the inheritance of cultural among human generations, quantum mechanics computing, particle swarm optimization (PSO), and ant colony systems (ACS) to help solve complex problems. Reports of applications of each of these tools have been widely published. Recently, these new heuristic tools have been combined among themselves and with knowledge elements, as well as with more traditional approaches such as statistical analysis, to solve extremely challenging problems. Developing solutions with these tools offers two major advantages: (1) development time is much shorter than when using more traditional approaches and (2) the systems are very robust, being relatively insensitive to noisy and/or missing data.

Advanced Solutions in Power Systems: HVDC, FACTS, and Artificial Intelligence, First Edition.
Edited by Mircea Eremia, Chen-Ching Liu, and Abdel-Aty Edris.

To help the power system engineers, the IEEE Power Engineering Society organized the first tutorial course: Tutorial on Modern Heuristic Optimization Techniques with Applications to Power Systems [5]. The second tutorial was organized by the IFAC Symposium on Power Plants and Power Systems Control: A Tutorial Course on Evolutionary Computation Techniques for Power System Optimization [6]. The third tutorial was organized by the ISAP: Tutorial on Intelligent Optimization and Control for Power Systems [7].

21.2 EVOLUTIONARY ALGORITHMS FOR REACTIVE POWER PLANNING

The evolutionary algorithms (EAs), including evolutionary programming (EP), evolutionary strategy (ES), and genetic algorithm (GA), are artificial intelligence methods for optimization based on the mechanics of natural selection, such as mutation, recombination, reproduction, and selection. Mutation randomly perturbs a candidate solution; recombination randomly mixes their parts to form a novel solution; reproduction replicates the most successful solutions found in a population; whereas, selection purges poor solutions from a population. Starting from an initial generation of candidate solutions, this process produces advanced generations with candidates that are successively better suited to their environment.

These methods share many similarities. The EP is introduced first, and followed by ES and GA.

21.2.1 Evolutionary Algorithms

21.2.1.1 Evolutionary Programming

Initialization: The initial population of control variables is selected randomly from the set of uniformly distributed control variables ranging over their upper and lower limits. The fitness score f_i is obtained according to the objective function and the environment.

STATISTICS. The maximum fitness $f_{\max}$, minimum fitness $f_{\min}$, the sum of fitness $\sum f$, and average fitness f_{avg} of this generation are calculated.

MUTATION. Each selected parent, for example, P_i, is mutated and added to its population following the rule:

$$P_{i+m,j} = P_{i,j} + N\left(0, \beta(\bar{x}_j - \underline{x}_j)\frac{f_i}{f_{\max}}\right), \quad j = 1, 2, \ldots, n \tag{21.1}$$

where n is the number of decision variables in an individual, $P_{i,j}$ denotes the jth element of the ith individual, $N(\mu, \sigma^2)$ represents a Gaussian random variable with mean μ and variance σ^2, $f_{\max}$ is the maximum fitness of the old generation, which is obtained in *statistics*, $\bar{x}_j$ and $\underline{x}_j$ are, respectively, maximum and minimum limits of the jth element, and β is the mutation scale, $0 < \beta \leq 1$, which could be adaptively decreased during generations.

If any mutated value exceeds its limit, it will be the given limit value. The mutation process (21.1) allows an individual with larger fitness to produce more offspring for the next generation [8].

COMPETITION. Several individuals (k) that have the best fitness are kept as the parents for the next generation. Other individuals in the combined population of size ($2m - k$) have to compete with each other to get their chances for the next generation, where $2m$ is the population size. A weight value W_i of the ith individual is calculated by the following competition:

$$W_i = \sum_{t=1}^{N} W_{i,t} \tag{21.2}$$

where N is the competition number generated randomly, $W_{i,t}$ is either 0 for loss, or 1 for win as the ith individual competes with a randomly selected (rth) individual in the combined population. The value of $W_{i,t}$ is given in the following equation:

$$W_{i,t} = \begin{cases} 1 & \text{if } U_l < \dfrac{f_r}{f_r + f_i} \\ 0 & \text{otherwise,} \end{cases} \tag{21.3}$$

where f_r is the fitness of the randomly selected rth individual, f_i is the fitness of the ith individual, and U_l is randomly selected from a uniformly distributed set, $U(0,1)$. When all $2m$ individuals get their competition weights, they will be ranked in a descending order according to their corresponding value W_i. The first m individuals are selected along with their corresponding fitness f_i to be the bases for the next generation. The maximum, minimum, and the average fitness and the sum of fitness of current generation are then calculated in the *statistics*.

CONVERGENCE TEST. If the convergence condition is not met, the *mutation* and the *competition* processes will run again. The maximum generation number can be used for convergence condition. Other criteria, such as the ratio of the average and the maximum fitness of the population, are computed and generations are repeated until:

$$\{f_{avg}/f_{\max}\} \geq \delta \tag{21.4}$$

where δ should be very close to 1, which represents the degree of satisfaction. If the convergence has reached a given accuracy, an optimal solution has been found for an optimization problem.

21.2.1.2 Evolutionary Strategy

The ES is very similar to the EP, and the difference is as follows [9, 10]:

In the *mutation* process, each selected parent, for example, P_i, is mutated and added to its population following the rule:

$$P_{i+m,j} = P_{i,j} + N(0, \beta\, \nabla_{dev}), \quad j = 1, 2, \ldots, n \tag{21.5}$$

where n is the number of decision variables, ∇_{dev} is fixed, and its value depends on the size of decision variables.

In the *competition* process, the fitness of individuals of population size $2m$ is sorted in a descending order. The first m individuals are kept as the parents for the next *mutation* process.

21.2.1.3 Genetic Algorithm

GA emphasizes models of DNA selection as observed in nature, such as crossover and mutation, and is applied to abstracted chromosomes. This will be easily realized by a string representation, which costs additional encoding and decoding time. This is in contrast to ES and EP, which emphasize mutational transformations that maintain behavioral linkage between each parent and its offspring. The GA used in this chapter is very similar to the algorithm that can be found in the standard literature on the topic [11–15], also known as the *simple genetic algorithm (SGA)*. We use the three-operator GA with only minor deviations from the original.

Initial Population Generation. In this scheme, an initial population of binary strings is created randomly. Each of these strings represents one feasible solution to the search problem, that is, a point in the search space or a domain satisfying constraints.

Fitness Evaluation. Next, the solution strings are converted into their decimal equivalents and each candidate solution is tested in its environment. The fitness of each candidate is evaluated through some appropriate measures, such as the inverse of the cost function:

$$f = 1/(\alpha + C) \tag{21.6}$$

where C is the cost function to be minimized, and α is the fitness function parameter.

The algorithm is driven toward maximizing this fitness measure. After the fitness of the entire population has been determined, it must be determined whether or not the termination criterion has been satisfied. This criterion can be any number of things. One possibility is to stop the algorithm at some finite number of generations and designate the result as the best fit from the population. Another possibility is to test if the average fitness of the population exceeds some fraction of the best fit in the population. If the criterion is not satisfied, then we continue with the three genetic operations of *reproduction, crossover,* and *mutation*.

Selection And Reproduction. Fitness-proportionate *reproduction* is effected through the simulated spin of a weighted roulette wheel. The roulette wheel is biased with the fitness of each of the solution candidates. The wheel is spun N times, where N is the number of strings in the population. This operation yields a new population of strings that reflect the fitness of the previous generation's fit candidates.

Crossover. The next operation, crossover, is performed on two strings at a time that are selected from the population at random. Crossover involves choosing a random position in the two strings and swapping the bits that occur after this position. In one generation, the crossover operation is performed on a specified percentage of the population. Crossover can occur at a single position (single crossover), or at number of different positions (multiple crossovers). Crossover can also be performed in two different means: tail-tail and head-tail crossovers [12, 13]. The tail-tail crossover is the usual crossover, where the tail ends of the two strings are swapped. In the head-tail crossover, on the other hand, the tail end of one string becomes the head of another string and vice-versa. The tail-tail crossover tends to change the less significant bits, while the head-tail crossover gives more chance of changes by changing the more significant bits. The two crossover methods can be changed during iterations: the head-tail crossover can be used in the earlier generations and then switched to tail-tail crossover in the later generations for fine tuning.

MUTATION. The final genetic operator in the algorithm is mutation. Mutation is performed sparingly, typically after every 100–1000 bit transfers from crossover, and it involves selecting a string at random as well as a bit position at random and changing it from a 1 to a 0 or vice-versa. It is used to escape from a local minimum. After mutation, the new generation is complete and the procedure begins again with the *fitness evaluation* of the population.

21.2.2 Optimal Reactive Power Planning Problem

The optimal reactive power planning (ORPP) problem is to determine the optimal investment of reactive power (VAr) sources over a planning horizon [16]. The cost function to be minimized is the sum of the operation cost and the investment cost. The long-term ORPP is often decomposed into a three-level hierarchical optimization problem using the maximum principle, the Bender's decomposition, and the P–Q decomposition methods [17, 18]. To highlight the use of EAs, however, this chapter considers a short-term ORPP, where the investment is to be performed only once [8–12].

21.2.2.1 Objective Functions

The operation cost is often assumed to have single quadratic cost functions. In reality, the cost function has discontinuities corresponding to change of fuels [10]. Therefore, it is more appropriate to represent the cost function with piecewise quadratic functions. When using piecewise quadratic cost functions, the operation cost is defined as follows:

$$C_F = \sum_{i \in N_g} C_i(P_i) \tag{21.7}$$

$$C_i = \begin{cases} a_{i1} + b_{i1}P_i + c_{i1}P_i^2 & \text{if } \underline{P}_i \leq P_i < P_{i1} \\ a_{i2} + b_{i2}P_i + c_{i2}P_i^2 & \text{if } P_{i1} \leq P_i < P_{i2} \\ \dots & \\ a_{im} + b_{im}P_i + c_{im}P_i^2 & \text{if } P_{im-1} \leq P_i < \overline{P}_i \end{cases}$$

where

N_g	is	the set of generators.
$C_i(P_i)$	–	cost of the ith generator.
a_{ij}, b_{ij}, c_{ij}	–	cost coefficients of the ith generator at the jth power level.
P_i	–	the generated power of the ith generator (MW).
$\underline{P}_i, \overline{P}_i$	–	minimum and maximum real power generations of the ith generator.

Since EAs use the objective function directly, rather than its derivatives, more realistic cost functions can be used for the ORPP problem. Most papers consider the transmission loss in the objective function. However, the minimization of loss does not guarantee the minimization of the operation cost unless all units have the same efficiency. Therefore, the fuel cost has been used for both real and reactive power dispatch [16–19].

The investment cost is simply the installation cost of the VAr sources:

$$C_I = \sum_{i \in N_C} \left\{ C_{fi} + C_{ci} \left| \overline{Q}_{ci} \right| \right\} \tag{21.8}$$

where N_C is the set of compensators, C_{fi} and C_{ci} are, respectively, the fixed and the unit costs for investment, and $\overline{Q}_{ci}$ is the amount of VAr source investment in discrete steps. This cost function assumes the capacitive compensators; however, the reactive compensator can also be included by replacing $\overline{Q}_{ci}$ with $\underline{Q}_{ci}$.

21.2.2.2 P–Q Decomposition

The ORPP problem is decomposed into two subproblems: the Q-optimization module and the P-optimization module.

THE P-MODULE. In this module, the objective is to minimize the sum of the operation cost:

$$C_P = \sum_{l \in N_l} d_l C_F^l \tag{21.9}$$

where C_F^l is the operation cost (21.7) for load level l, d_l is the duration of load level l, and N_l is the set of load levels. The minimization is with respect to the real power generations P_g for each load level subject to the real and reactive power balance of the power system that can be solved by calling the *load flow* program, and the following inequality constraints:

$$\begin{aligned} \underline{P}_{gi} &\le P_{gi} \le \overline{P}_{gi} \\ \underline{Q}_{gi} &\le Q_{gi} \le \overline{Q}_{gi} \\ \underline{T}_k &\le T_k \le \overline{T}_k \\ \underline{V}_i &\le V_i \le \overline{V}_i \\ \underline{Q}_{ci} &\le Q_{ci} \le \overline{Q}_{ci} \end{aligned} \tag{21.10}$$

where P_g and Q_g are generator real and reactive powers, respectively, T are transformer tap-settings, V are bus voltages, and Q_c are the reactive power output of compensators.

The real power generations P_g are the optimization or *decision variables* for the P-module, which are self-constrained. Since the generator bus voltages, transformer tap-settings, and capacitor investments are the optimization variables in the Q-module, they are fixed in the P-module, and thus the constraints are automatically satisfied. The load bus voltages and the generator reactive powers are state variables, which can be constrained by augmenting them as the quadratic penalty terms to the objective function. The P-optimization module is therefore changed to minimize the following generalized objective function:

$$C_P = \sum_{l \in N_l} d_l \left\{ C_F^l + \sum_{i \in N_l} \lambda_{vi}(Vi - Sat(V_i))^2 + \sum_{i \in N_g} \lambda_{gi}(Q_{gi} - Sat(Q_{gi}))^2 \right\} \quad (21.11)$$

where N_l is the set of load busses, λ_{vi} and λ_{gi} are the penalty weights, and $Sat(x)$ is the saturation function defined by:

$$Sat(x) = \begin{cases} \underline{x} & \text{if } x < \underline{x} \\ x & \text{if } \underline{x} \le x \le \bar{x} \\ \bar{x} & \text{if } x > \bar{x} \end{cases} \quad (21.12)$$

THE Q-MODULE. In this module, the objective function is to minimize the sum of the operation cost and the investment cost,

$$C_Q = C_P + C_I, \quad (21.13)$$

with respect to transformer tap-settings T, generator bus voltages, and the VAr source investment $\overline{Q}_c$; subject to the real and reactive power balance equations, which can be solved by calling the *load flow* program, and the inequality constraints (21.10).

The transformer tap-settings, generator bus voltages, and the VAr source investment are the *optimization or decision variables* for the Q-module, which are self-constrained. Since the real power generations are obtained from the last P-module, they should satisfy the constraint. The voltages of load busses and the reactive powers of generators are state variables, which may be out of limits, and penalties are also added to the objective function. The formulation of the penalty parts is the same as the procedure in the P-module. Then the Q-module is to minimize the generalized cost function as follows:

$$C_Q = C_I + \sum_{l \in N_l} d_l \left\{ C_F^l + \sum_{i \in N_l} \lambda_{vi}(Vi - Sat(V_i))^2 + \sum_{i \in N_g} \lambda_{gi}(Q_{gi} - Sat(Q_{gi}))^2 \right\} \quad (21.14)$$

The P–Q decomposition is now complete. The optimization variables for the P-optimization subproblem are generator real power outputs, and those for the Q-optimization subproblem are generator voltage magnitudes, VAr source investments, and transformer tap-settings. The security constraints are the operating limits of these control variables, line flows, and the state variables.

The P- and Q-modules were solved sequentially to obtain local optimal values of the optimization variables for each module. The computation diagram of the ORPP problem is shown in Figure 21.1.

21.2.3 Case Studies

A comparative study is performed for the three EAs and the LP by solving the ORPP problem for the IEEE 30-bus system as a testbed [16–19]. String representations and simulation parameters for EAs are discussed, and the modification of fuel costs with

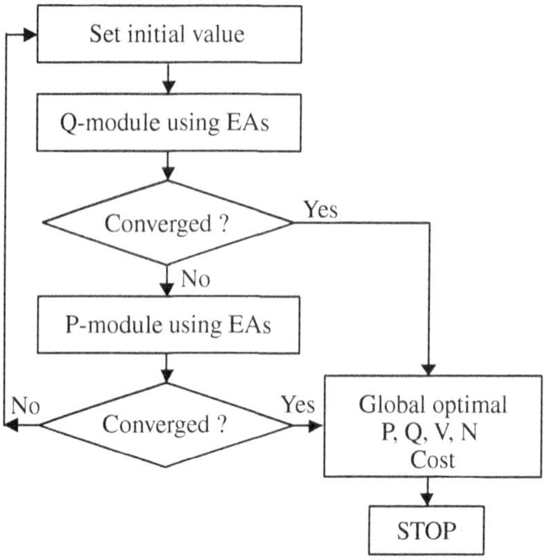

Figure 21.1 Flow chart for ORPP by using EAs.

piecewise quadratic cost functions and the simulation results and the convergence speeds of EAs are given.

1. String representation is an important factor in solving the ORPP problem using the GA. In order to accommodate different decision variables, that is, the investment or operation variables, the following representation method is used.

   A string consists of substrings; the number of substrings is equal to the number of decision or control variables. In the P-optimization problem, the number of substrings is the total number of generators and each substring that represents the generator real power output. In the Q-optimization problem, on the other hand, the decision variables are generator voltage magnitudes, the maximum VAr limits of compensators, and transformer tap-settings. The simulation parameters for the three EAs are given in Tables 21.1 and 21.2.
2. The fuel cost coefficients of the piecewise quadratic cost functions are given in Table 21.3, which are generated from the original single quadratic cost functions [19]. Small units (at busses 5, 8, 11, and 13) are represented by

TABLE 21.1 Simulation parameters in EP, ES, EP+ES, and GA

Method coefficients	ES	EP	EP+ES	GA
Number of parents	25	25	25	25
Number of offspring	25	25	25	25
Standard deviation	0.03	NA	EP(NA), ES(0.03)	NA

NA: Not Applicable.

TABLE 21.2 Simulation parameters in GA

Value parameters	GA
Mutation rate	0.01
Crossover rate	0.8
Abandoning rate	0.9
Maximum generations	100
Population size	25
Control variables	6
Parameter resolution	8 bits per substring
Chromosome length	48
Fitness function parameter α	0.01

two segments, while large units (at busses 1 and 3) are represented by three segments of fuel curves.

As shown in the P- and Q-optimization modules, the formulation is general to include a number of different load levels to optimally balance the savings in the operation cost against the cost of capital investment. However, for simplicity, only the peak load condition [19] is considered for the comparative study. This scenario, obviously, will give an optimistic savings in the operation cost, and therefore could suggest overinvestment.

3. The initial load flow, with generator voltages set to 1.0 p.u., showed that the voltage magnitudes at the load busses 14–30 were all below the lower operating

TABLE 21.3 Cost coefficients of piecewise quadratic function

U	GENERATION							F	Cost coefficients		
	Min	F1	P1	F2	P2	F3	Max		a	b	c
1	50		100		190		200	1	0.000	1.900	0.00355
		1		2		3		2	0.000	2.000	0.00375
								3	0.000	2.200	0.00415
2	20		35		50		80	1	0.000	1.700	0.01700
		1		2		3		2	0.000	1.750	0.01750
								3	0.000	2.050	0.02350
3	15		30		50			1	0.000	1.000	0.06250
		1		2				2	0.000	1.200	0.08250
4	10		25		35			1	0.000	3.250	0.00834
		1		2				2	0.000	3.650	0.01234
5	10		20		30			1	0.000	3.000	0.02500
		1		2				2	0.000	3.300	0.03500
6	12		25		40			1	0.000	3.000	0.02500
		1		2				2	0.000	3.300	0.03500

limit of 0.95 p.u., and the operation cost was as high as 903.31 \$/h. First for optimal power flow, the P-optimization module and the Q-optimization module without the investment were run by using both LP [17] and ES. Real power distribution, voltage magnitudes, and transformer tap-settings were optimized to minimize the operation cost. The results of LP and ES showed that all the control variables and state variables were within their hard limits, but for some load busses, such as busses 18–26, 29, and 30, the voltages were very close to their lower limits. Thus, more reactive power from other VAr sources was needed to improve the voltage profile. As we compared the optimal power flow results, ES gave slightly higher voltage profile at slightly lower operation cost and power loss.

4. For the ORPP problem, the candidate busses for reactive power compensation are 15, 17, 20, 23, 24, and 29. All are shunt capacitors with the investment cost of 0.02 \$/MVAr per unit hour. The fixed cost for installation is neglected for simplicity. However, the site-dependent fixed cost can be incorporated in additional substrings in EAs; the number of additional substrings being equal to the number of candidate busses. The investments are made in discrete steps: 2.5, 5, 7.5, … , 30 MVArs, which become the upper limits $\overline{Q}_{ci}$ in the operational constraints (21.10). In both LP and EAs, when the constraints are violated in any iteration, additional steps of the investments are made in the candidate busses in the Q-optimization module and the iteration is repeated.
5. The results of the ORPP by using LP and EAs are given in Tables 21.4 and 21.5. The results show that all the control variables and state variables such as voltage magnitudes, transformer tap-settings, real power, and reactive power generations are within their hard limits. The results of ES, EP, and EP+ES are very close and comparable with those of LP, and the average differences of voltage magnitudes and real power generations are within 5/1000. The power loss and the operation cost of LP are slightly higher than those of EAs. It is noted

TABLE 21.4 Optimal operation with investment real power and reactive power distribution

Variable	LP	ES	EP	EP+ES
P_1	184.19	179.838	178.67	179.162
P_2	47.77	48.110	49.335	48.345
P_5	20.00	19.822	19.583	19.944
P_8	15.75	21.571	20.733	21.790
P_{11}	13.46	11.402	12.422	11.423
P_{13}	12.00	12.000	12.000	12.071
Q_1	27.71	8.861	13.027	12.409
Q_2	13.35	30.379	25.606	25.827
Q_5	30.12	30.136	30.060	30.753
Q_8	22.49	33.747	23.885	25.890
Q_{11}	29.58	13.826	18.934	17.064
Q_{12}	20.03	9.283	12.845	12.199
Loss	9.8+j44	9.3+j38.3	9.3+j38.7	9.3+j38.5

TABLE 21.5 Bus voltages

Variable (p.u.)	LP	ES	EP	EP+ES
V_1	1.0903	1.084	1.084	1.083
V_2	1.0616	1.064	1.063	1.062
V_3	1.0508	1.050	1.050	1.049
V_4	1.0425	1.043	1.042	1.042
V_5	1.0289	1.032	1.030	1.030
V_6	1.0340	1.037	1.036	1.035
V_7	1.0234	1.027	1.025	1.025
V_8	1.0294	1.038	1.033	1.033
V_9	1.0006	1.028	1.019	1.022
V_{10}	1.0032	1.014	1.010	1.013
V_{11}	1.0584	1.055	1.056	1.056
V_{12}	1.0111	1.019	1.018	1.019
V_{13}	1.0380	1.032	1.035	1.035
V_{14}	1.0000	1.010	1.008	1.010
V_{15}	0.9992	1.011	1.009	1.011
V_{16}	1.0013	1.011	1.009	1.011
V_{17}	0.9988	1.011	1.007	1.010
V_{18}	0.9901	1.003	1.001	1.004
V_{19}	0.9879	1.002	1.000	1.003
V_{20}	0.9924	1.007	1.005	1.008
V_{21}	0.9958	1.006	1.002	1.005
V_{22}	0.9970	1.007	1.002	1.006
V_{23}	0.9981	1.010	1.006	1.009
V_{24}	0.9972	1.002	0.997	1.001
V_{25}	1.0114	1.006	0.996	1.002
V_{26}	0.9936	0.998	0.978	0.984
V_{27}	1.0289	1.018	1.005	1.012
V_{28}	1.0264	1.033	1.032	1.031
V_{29}	1.0183	1.014	1.002	1.009
V_{30}	1.0030	0.995	0.983	0.990

that ES showed the least operation cost at the expense of the highest investment cost, while the combined EP and ES (EP+ES) showed the least total cost (sum of the generation and the investment costs).

Table 21.6 showed that two of the four tap-settings are reversed in direction in LP, while they all are in the same direction in all EAs. This resulted in the fairly even investment of VAr sources in all seven candidate busses for EAs, while it is not the case for LP. It should be noted that the capacitive VArs, Q_{ci}, shown in the table are the *operational* values after the investment, which are in general less than the maximum limits, $\overline{Q}_{ci}$, given by the investment in discrete steps.

The total costs of each *P*- and *Q*-module in each iteration of ORPP are shown in Table 21.7. The iteration process is alternating as *Q*-*P*-*Q*-*P*-, etc. It is shown in the table that only four or five iterations of *P*- and *Q*-modules are required for convergence.

TABLE 21.6 Tap-settings, capacitive VArs, and costs

Variable	LP	ES	EP	EP+ES
N_{11}	1.0155	1.0290	1.0223	1.0165
N_{12}	0.9940	1.0337	1.0199	1.0261
N_{15}	1.0320	1.0298	1.0286	1.0277
N_{36}	0.9930	1.0274	1.0176	1.0152
Q_{c15}(MVAr)	3.05	6.5218	4.7253	4.7541
Q_{c17}(MVAr)	2.89	6.4387	4.4645	4.4418
Q_{c20}(MVAr)	2.79	6.8332	5.0405	5.0644
Q_{c21}(MVAr)	4.73	6.6059	4.7493	4.7540
Q_{c23}(MVAr)	2.96	6.4141	4.4747	4.4762
Q_{c24}(MVAr)	7.47	6.3411	4.2969	4.2865
Q_{c29}(MVAr)	3.09	6.6167	4.7567	4.7358
Gen. cost ($/h)	801.87	801.125	801.58	801.41
Inv. cost ($/h)	0.75	1.05	0.75	0.75
Total cost ($/h)	802.62	802.175	802.33	802.16

Each *P*- or *Q*-module converges in at least 30 generations. The CPU time of VAX for each generation is about 1.85 s, and in the ORPP problem, the CPU time for EAs to reach the *global* minimum is therefore at least 1.85×30×5 s. However, for LP, it needs about 13 iterations to reach a *local* minimum, and the total CPU time is only around 7 s in VAX. Although the EAs take longer than LP, the formulation is straight forward and can handle arbitrary cost functions that may not be convex; and moreover, they can find the global minimum, while LP can only find a local minimum.

6. Within the EA family, the characteristics of ES, EP, and GA were compared. For fair comparison, the same conditions are set for all decision variables of the *Q*-module, and only the *P*-module was run. Sample results are given in Table 21.8. The optimal real power distribution and the operation cost are very close for all methods. However, the number of generations to converge is different for different methods and also different for different runs in the same method [20]. In the EP+ES method, at first, the EP process is called, and the best fitness of these generations decreases rapidly; then, after 10 generations, ES process is called to continue the calculation. It is shown that the ES and ES+EP methods need nearly the same number of generations (on average) to converge, while the EP and GA methods need almost twice as much time to converge.

TABLE 21.7 Total cost during iterations of ES for ORPP

Iter	1	2	3	4	5
Q	961.07	802.12	802.00	802.19	801.90
P	802.56	801.95	802.30	801.90	801.93

TABLE 21.8 Comparison of EAs for *P*-optimization

Method	EP	ES	EP+ES	GA
P1 (MW)	177.92	177.93	177.13	178.46
P2 (MW)	48.587	48.887	49.664	47.891
P3 (MW)	19.975	19.996	19.996	20.879
P4 (MW)	22.739	22.467	22.775	23.086
P5 (MW)	12.613	12.355	12.103	10.703
P6 (MW)	11.745	11.776	11.753	12.328
Loss (MW)	9.995	10.013	9.985	9.944
Total cost	803.87	803.86	803.87	803.64
Generations to converge	60	37	36	63

21.3 GENETIC ALGORITHM FOR GENERATION PLANNING

GA is a search algorithm based on the hypothesis of natural selections and natural genetics [9]. Recently, a global optimization technique using GA has been successfully applied to various areas of power system such as economic dispatch [21, 22], unit commitment [23, 24], reactive power planning [12, 20, 25], and power plant control [13, 26]. GA-based approaches for least-cost generation expansion planning (GEP) have several advantages. Naturally, they can not only treat the discrete variables, but also overcome the dimensionality problem. In addition, they have the capability to search for the global optimum or quasi-optimums within a reasonable computation time. However, there exist some structural problems in the conventional GA, such as premature convergence and duplications among strings in a population as generation progresses [9].

Recently, an improved genetic algorithm (IGA), which can overcome the aforementioned problems of the conventional GA to some extents, was developed [27]. The proposed IGA incorporates the following two main features. First, an artificial creation scheme for an initial population is devised, which also takes the random creation scheme of the conventional GA into account. Second, a stochastic crossover strategy is developed. In this scheme, one of the three different crossover methods is randomly selected from a biased roulette wheel where the weight of each crossover method is determined through preperformed experiments. The stochastic crossover scheme is similar to the stochastic selection of reproduction candidates from a mating pool. The results of the IGA are compared with those of the conventional SGA, the full dynamic programming (DP), and the tunnel-constrained DP employed in WASP.

21.3.1 Generation Expansion Planning Problem

Mathematically, solving a least-cost GEP problem is equivalent to finding a set of optimal decision vectors over a planning horizon that minimizes an objective function under several constraints. The GEP problem to be considered is formulated as follows [28]:

$$\underset{U_1,\ldots,U_T}{\text{Min}} \sum_{t=1}^{T} \{f_t^1(U_t) + f_t^2(X_t) - f_T^3(U_t)\} \tag{21.15}$$

$$s.t.\ X_t = X_{t-1} + U_t \quad (t = 1, \ldots, T) \tag{21.16}$$

$$LOLP(X_t) < \varepsilon \quad t = 1, \ldots, T \tag{21.17}$$

$$\underline{R} \le R(X_t) \le \overline{R} \quad (t = 1, \ldots, T) \tag{21.18}$$

$$\underline{M_t^j} \le \sum_{i \in \Omega_j} x_t^i \le \overline{M_t^j}\ (t = 1, \ldots, T \quad \text{and} \quad j = 1, \ldots, J) \tag{21.19}$$

$$0 \le U_t \le \overline{U_t} \qquad (t = 1, \ldots, T) \tag{21.20}$$

where

T	is	the number of periods (years) in a planning horizon.
J	–	the number of fuel types.
Ω_j	–	the index set for *j-th* fuel-type plant.
X_t	–	the cumulative capacity (MW) vector of plant types in year t
x_t^i	–	the cumulative capacity (MW) of *i-th* plant types in year t.
U_t	–	the capacity addition (MW) vector by plant types in year t.
$\overline{U}_t$	–	the maximum construction capacity (MW) vector by plant types in year t.
u_t^i	–	the capacity addition (MW) of *i-th* plant in year t.
$LOLP(X_t)$	–	the loss of load probability (LOLP) with X_t in year t.
$R(X_t)$	–	the reserve margin with X_t in year t.
ε	–	the reliability criterion expressed in LOLP.
$\overline{R}$, $\underline{R}$	–	the upper and lower bounds of reserve margin.
$\overline{M_t^j}$, $\underline{M_t^j}$	–	the upper and lower bounds of *j-th* fuel type in year t.
$f_t^1(U_t)$	–	discounted construction costs [$] associated with capacity addition U_t in year t.
$f_t^2(X_t)$	–	discounted fuel and O&M costs [$] associated with capacity X_t in year t.
$f_T^3(U_t)$	–	the discounted salvage value [$] associated with capacity addition U_t in year t.

Although the state vector and the decision vector have dimensions of MW, we can easily convert those into vectors, which have information on the number of units in each plant type. This mapping strategy is very useful for GA implementation of a GEP problem such as encoding and treatment of inequality (21.20), and illustrated in the following equations:

$$X_t = \left(x_t^1, \ldots, x_t^N\right)^T \rightarrow X_t' = \left(x_t'^1, \ldots, x_t'^N\right)^T \tag{21.21}$$

$$U_t = \left(u_t^1, \ldots, u_t^N\right)^T \rightarrow U_t' = \left(u_t'^1, \ldots, u_t^N\right)^T \tag{21.22}$$

where

- N is the number of plant types including both existing and candidate plants.
- X_t' – the cumulative number of units by plant types in year t.
- U_t' – the addition number of units by plant types in year t.
- $x_t'^i$ – the ith plant type's cumulative number of units in year t.
- $u_t'^i$ – the ith plant type's addition number of units in year t.

21.3.2 Improved GA for the Least-Cost GEP

21.3.2.1 Overview of Genetic Algorithm

Basically, GA is a search mechanism based on the hypothesis of natural selection [9]. GA is an artificial optimization scheme that emulates the hypothetical adaptive nature of natural genetics. GA provides solutions by generating a set of chromosomes referred to as a generation. Each string (chromosome) has its own fitness measure that reflects how well a creature can survive under surrounding environments. The new generation of strings is provided through three major genetic operations—-reproduction, crossover, and mutation, which provide a powerful global search mechanism. Reproduction is a process in which individual strings are copied into a mating pool according to their fitness values.

Crossover, the most important genetic operator, is a structured recombination operation. In the classical one-point crossover, a random position in a string is chosen and all characters to the right of this position are swapped. Mutation, the secondary operator in GA, is an occasional random alteration of the value of a string position. Variations of the simple GA for power system applications can be found in References [12, 20–26, 29], and [30]. The improvements on the conventional GA will be described in the subsequent sections.

21.3.2.2 String Structure

Since it is convenient to use integer values for GA implementation of a GEP problem, the reordered structure of (21.22) by plant types covering a planning horizon is used for encoding of a string as shown in (21.23).

Here, each element of a string (i.e., $\hat{U}'^n = (u_1'^n, u_2'^n, \ldots, u_T'^n)^T$ for $n = 1,\ldots, N$) corresponds to a substring and its structure is depicted in Figure 21.2.

$$\begin{aligned}\hat{U}' &= \left(u_1'^1, u_2'^1, \ldots, u_T'^1, \ldots, u_1'^n, u_2'^n, \ldots, u_T'^n, \ldots, u_1'^N, u_2'^N, \ldots, u_T'^N\right)^T\\ &= \left(\hat{U}'^1, \ldots, \hat{U}'^n, \ldots, \hat{U}'^N\right)^T \end{aligned} \tag{21.23}$$

21.3.2.3 Fitness Function

The objective function or cost of a candidate plan is calculated through the probabilistic production costing and the direct investment costs calculation [31, 32]. The fitness value of a string can be evaluated using the following equation [9, 13]:

$$f = \alpha/1 + J \tag{21.24}$$

where α is a constant, and J is the objective function of (21.15).

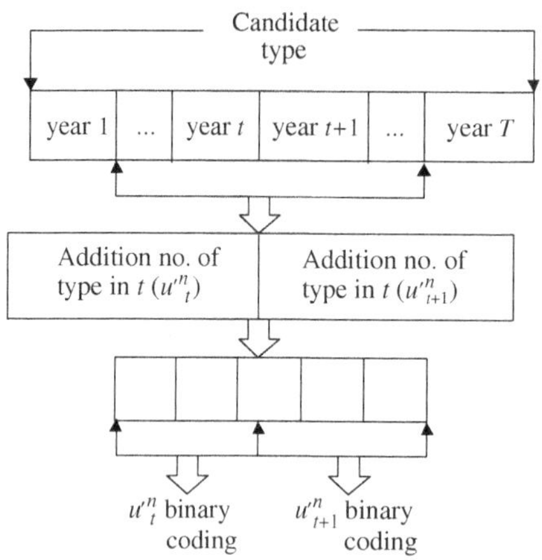

Figure 21.2 A substring structure.

However, this simple mapping occasionally brings about a premature convergence and duplications among strings in a population, since strings with higher fitness values dominate the occupation of a roulette wheel.

To ameliorate these problems, the following modified fitness function, which normalizes the fitness value of strings into real numbers within [0, 1], is used in this chapter [9].

$$f'(i) = \frac{f(i) - f_{\min}}{f_{\max} - f_{\min}} \tag{21.25}$$

where

$f(i)$	is	the fitness value of string I using (21.24).
$f_{\max}$, $f_{\min}$	–	maximum and minimum fitness values in a generation.
$f'(i)$	–	the modified fitness value of string i.

21.3.2.4 Creation of an Artificial Initial Population

It is important to create an initial population of strings spread out throughout the whole solution space, especially in a large-scale problem. One alternative method could be to increase the population size, which yields a high computational burden. This suggests a new artificial initial population (AIP) scheme, which takes the random creation scheme of the conventional GA into account [27]. The procedures are illustrated in the following and in Table 21.9:

Step 1: Generate all possible binary seeds of each plant type considering (21.20). For example, if *i-th* plant type has an upper limit of 3 units per year, then generate four possible binary seeds (i.e., 00,01,10,11).

Step 2: Find the least common multiple m from the numbers of the binary seeds of all types, and fill m binary seeds in a look-up table for all plant types and planning years. For example, if three plant types have upper limits of 3, 3, and 5 units per year, respectively, then the numbers of binary seeds are 4, 4, and 6, and m becomes 12.

TABLE 21.9 Example of look-up table with three plant types for three planning years

	Type 1 (Upper limit: 3 units/year)			Type 2 (Upper limit: 3 units/year)			Type 3 (Upper limit: 5 units/year)		
m	Year 1	Year 2	Year 3	Year 1	Year 2	Year 3	Year 1	Year 2	Year 3
1	00	00	00	00	00	00	000	000	000
2	01	01	01	01	01	01	001	001	001
3	10	10	10	10	10	10	010	010	010
4	11	11	11	11	11	11	011	011	011
5	00	00	00	00	00	00	100	100	100
6	01	01	01	01	01	01	101	101	101
7	10	10	10	10	10	10	000	000	000
8	11	11	11	11	11	11	001	001	001
9	00	00	00	00	00	00	010	010	010
10	01	01	01	01	01	01	011	011	011
11	10	10	10	10	10	10	100	100	100
12	11	11	11	11	11	11	101	101	101
Generated string 1: 011100101011000101010									

Step 3: Select an integer within $[1, m]$ at random for each element $u_t'^n$ of a string in (21.23). Fill the string with the corresponding binary digits, and delete it from the look-up table. Repeat until m different strings are generated.

Step 4: Check the constraints of (21.17)–(21.19). If a string satisfies these constraints for all years, then it becomes a member of an initial population. Otherwise, the only parts of the string that violate the constraints in year t are generated at random until they satisfy the constraints. Go to step 3 n times for $n{\cdot}m$ less than P, where P is the number of strings in a population and n is an arbitrary positive integer.

Step 5: The remaining $P\text{-}n{\cdot}m$ strings are created using uniform random variables with binary number $\{0,1\}$. Go to step 4 to check the constraints and regenerate them if necessary. This process is repeated until all strings, which satisfy the constraints, are generated.

This AIP is based on both artificial and random selection schemes, which allows that all possible string structures can be included in an initial population.

21.3.2.5 *Stochastic Crossover, Elitism, and Mutation*

Most of GA works are based on the Goldberg's SGA framework [9]. In paper [27], two different schemes for genetic operation are proposed: a stochastic crossover technique and the application of elitism. The stochastic crossover scheme covers three different crossover methods; 1-point crossover, 2-point crossover, and 1-point substring crossover as illustrated in Figure 21.3. Each crossover method has its own merits. The 1-point substring crossover can provide diverse bit structures to search solution space, however, it easily destroys the string structure that may have partial information on the optimal structure.

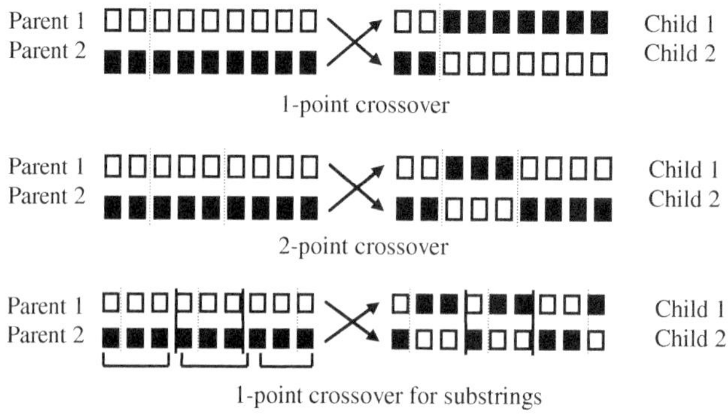

Figure 21.3 Three different crossover methods used.

Although the 1- and 2-point crossovers cannot explore solution space as widely as the above crossover, the probability of destroying an already-found partial optimal structure is very low. The stochastic crossover strategy is similar to the process of stochastic selection of reproduction candidates from a mating pool. That is, one of the three different crossover methods is selected from a biased roulette wheel, where each crossover method has a roulette wheel slot sized according to its performance. The weight for each crossover method has been determined.

The second feature lies in the application of elitism [22]. The roulette wheel selection scheme gives a reproduction opportunity to a set of recessive members and might not give the set of dominant strings (i.e., an elite group) a chance to reproduce. Furthermore, the application of genetic operations changes the string structures of the fittest solutions. Thus, the best solutions in the current generation might not appear in the next generation. To circumvent these problems, an elite group is directly copied into the next generation.

In [27], the conventional mutation scheme [22] is applied in the stochastic crossover operator. However, the mutation procedure is not applied to the set of dominant strings to preserve the elitism.

After genetic operations, all strings are checked whether they satisfy the constraints of (21.17)–(21.19) or not. If any string that violates the constraints of (21.17)–(21.19), only the parts of the string that violate the constraints in year t are generated at random until they satisfy the constraints as described in the AIP scheme.

21.3.3 Case Studies

The IGA, SGA, tunnel-constrained dynamic programming (TCDP) employed in WASP, and full DP were implemented using the FORTRAN77 language on an IBM PC/Pentium (166 MHz) computer.

21.3.3.1 Test Systems Description

The IGA, SGA, TCDP, and DP methods have been applied in two test systems: case 1 for a power system with 15 existing power plants, five types of candidate options,

TABLE 21.10 Forecasted peak demand

Stage (year)	0 (1996)	1 (1998)	2 (2000)	3 (2002)	4 (2004)	5 (2006)	6 (2008)
Peak (MW)	5000	7000	9000	10,000	12,000	13,000	14,000
Stage (year)	—	7 (2010)	8 (2012)	9 (2014)	10 (2016)	11 (2018)	12 (2020)
Peak (MW)	—	15,000	17,000	18,000	20,000	22,000	24,000

and a 14-year study period, and case 2 for a real-scale system with a 24-year study period. The planning horizons of 14 and 24 years are divided into seven and 12 stages (2-year intervals), respectively. The forecasted peak demand over the study period is given in Table 21.10.

Tables 21.10 and 21.11 show the technical and economical data of the existing plants and candidate plant types for future additions, respectively.

21.3.3.2 Parameters for GEP and IGA

There are several parameters to be predetermined, which are related to the GEP problem and GA-based programs. Here, we use 8.5% as a discount rate, 0.01 as LOLP criteria, and 15% and 60% as the lower and upper bounds for reserve margin, respectively. The considered lower and upper bounds of capacity mix are 0% and 30% for oil-fired power plants, 0% and 40% for LNG-fired, 20% and 60% for coal-fired, and 30% and 60% for nuclear, respectively.

Parameters for IGA are selected through experiments as shown in Table 21.13. Especially, the dominant parameters such as crossover probabilities and weights for

TABLE 21.11 Technical and economic data of existing plants

Name (fuel type)	No. of units	Unit capacity (MW)	FOR (%)	Operating cost ($/kWh)	Fixed O&M cost ($/kW-Mon)
Oil #1 (heavy oil)	1	200	7.0	0.024	2.25
Oil #2 (heavy oil)	1	200	6.8	0.027	2.25
Oil #3 (heavy oil)	1	150	6.0	0.030	2.13
LNG G/T #1 (LNG)	3	50	3.0	0.043	4.52
LNG C/C #1 (LNG)	1	400	10.0	0.038	1.63
LNG C/C #2 (LNG)	1	400	10.0	0.040	1.63
LNG C/C #3 (LNG)	1	450	11.0	0.035	2.00
Coal #1 (anthracite)	2	250	15.0	0.023	6.65
Coal #2 (bituminous)	1	500	9.0	0.019	2.81
Coal #3 (bituminous)	1	500	8.5	0.015	2.81
Nuclear #1 (PWR)	1	1,000	9.0	0.005	4.94
Nuclear #2 (PWR)	1	1,000	8.8	0.005	4.63

TABLE 21.12 Technical and economic data of candidate plants

Candidate type	Construction upper limit	Capacity (MW)	FOR (%)	Operating cost ($/kWh)	Fixed O&M cost	Capital cost ($/kW)	Life time (years)
Oil	5	200	7.0	0.021	2.20	812.5	25
LNG C/C	4	450	10.0	0.035	0.90	500.0	20
Coal (Bitum.)	3	500	9.5	0.014	2.75	1062.5	25
Nuc. (PWR)	3	1,000	9.0	0.004	4.60	1625.0	25
Nuc.(PHWR)	3	700	7.0	0.003	5.50	1750.0	25

crossover techniques are determined empirically from a test system with a 6-year planning horizon with other data being the same as cases 1 and 2 [27].

21.3.3.3 Numerical Results

The developed IGA was applied to two test systems, and compared with the results of DP, TCDP, and SGA. Throughout the tests, the solution of the conventional DP is regarded as the global optimum and that of TCDP as a local optimum. Both the global and a local solution can be obtained in case 1; however, only a local solution can be obtained by using TCDP in case 2 since the "curse of dimensionality" prevents the use of the conventional DP.

Figure 21.4 illustrates the convergence characteristics of various GA-based methods in case 1. It also shows the improvement of IGA over SGA. The IGA employing the stochastic crossover scheme (IGA2) has shown better performance than the IGA using the artificial initial population scheme (IGA1). By considering both schemes simultaneously (IGA3), the performance is significantly enhanced.

Table 21.14 summarizes costs of the best solution obtained by each solution method. In case 1, the solution obtained by IGA3 is within 0.18% of the global solution costs, while the solutions by SGA and TCDP are within 1.3% and 0.4%, respectively. In cases 1 and 2, IGA3 has achieved a 0.21% and 0.61% improvement of costs over TCDP, respectively. Although SGA and IGAs have failed in finding the global solution, all IGAs have provided better solution than SGA. Furthermore, solutions of IGA3 are better than that of TCDP in both cases, which imply that it can overcome a local optimal trap in a practical long-term GEP.

TABLE 21.13 Parameters for IGA implementation

Parameters	Values
• Population size	300
• Maximum generation	300
• Probabilities of crossover and mutation	0.6, 0.01
• Number of elite strings	3 (1%)
• Weights of 1-point, 2-point, and 1-point	0.15:0.15:0.70
Crossover for substrings in a biased roulette wheel	

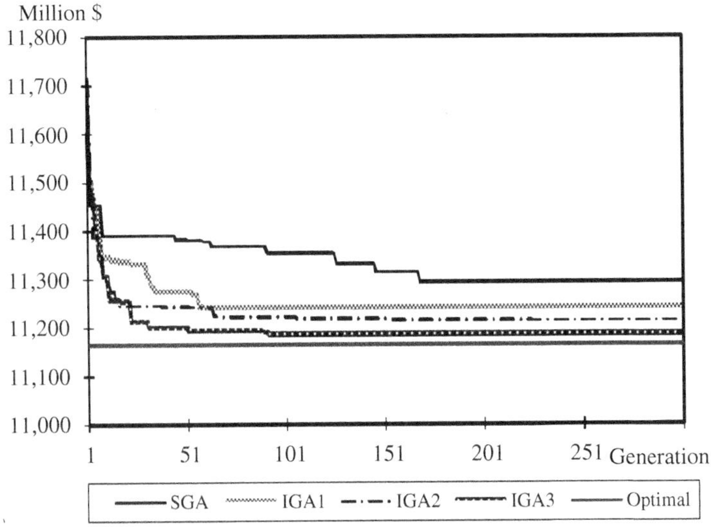

Figure 21.4 Convergence characteristics of IGA method in case 1 system.

Table 21.15 summarizes the generation expansion plans of cases 1 and 2 obtained by IGA3.

21.4 PARTICLE SWARM OPTIMIZATION FOR ECONOMIC DISPATCH

Recently, Eberhart and Kennedy suggested a PSO based on the analogy of swarm of bird and school of fish [33]. The PSO mimics the behavior of individuals in a swarm to maximize the survival of the species. In PSO, each individual makes his decision using his own experience as well as other individuals' experiences [34]. The algorithm, which is based on a metaphor of social interaction, searches a space by

TABLE 21.14 Summary of the best results obtained by each solution method

Solution method		Cumulative discounted cost (10^6 \$) Case 1 (14-year study period)	Case 2 (24-year study period)
DP		11164.2	Unknown
TCDP		11207.7	16746.7
SGA		11310.5	16765.9
	IGA1	11238.3	16759.2
IGA	IGA2	11214.1	16739.2
	IGA3	11184.2	16644.7

TABLE 21.15 Cumulative number of newly introduced plants in cases 1 and 2 by IGA3

Type year	Oil (200 MW)	LNG C/C (450 MW)	Coal (500 MW)	PWR (1000 MW)	PHWR (700 MW)
998	3 (5)[1]	2 (1)	2 (3)	0 (1)	2 (0)
2000	5 (6)	3 (1)	5 (6)	0 (1)	4 (1)
2002	5 (7)	3 (1)	5 (6)	0 (2)	4 (1)
2004	8 (10)	7 (3)	6 (7)	0 (2)	4 (1)
2006	10 (12)	10 (3)	6 (7)	0 (2)	6 (2)
2008	10 (13)	10 (3)	6 (9)	0 (2)	6 (2)
2010	10 (13)	10 (3)	6 (9)	0 (2)	6 (4)
2012	14	11	8	1	7
2014	17	14	8	1	7
2016	19	15	10	1	9
2018	19	17	10	3	9
2020	20	18	12	3	9

1. The figures within parenthesis denote the results of IGA3 in case 1.

adjusting the trajectories of moving points in a multidimensional space. The individual particles are drawn stochastically toward the position of present velocity of each individual, their own previous best performance and the best previous performance of their neighbors [35]. The main advantages of the PSO algorithm are summarized as simple concept, easy implementation, robustness to control parameters, and computational efficiency when compared with mathematical algorithm and other heuristic optimization techniques.

Recently, PSOs have been successfully applied to various fields of power system optimization such as power system stabilizer (PSS) design [36], reactive power and voltage control [34], and dynamic security border identification [37]. The original PSO mechanism is directly applicable to the problems with continuous domain and without any constraints. Therefore, it is necessary to revise the original PSO to reflect the equality/inequality constraints of the variables in the process of modifying each individual's search. Yoshida et al. [34] suggested a modified PSO to control reactive power and voltage considering voltage security assessment. Since the problem was a mixed-integer nonlinear optimization problem with inequality constraints, they applied the classical penalty method to reflect the constraint-violating variables. Abido [36] developed a revised PSO for determining the optimal values of parameters for PSSs. In the study, the velocity of each parameter is limited to a certain value to reflect the inequality constraint problem in the dynamic process. Other variations of PSO on power systems application are given in [73].

21.4.1 Formulation of Economic Dispatch Problem

21.4.1.1 ED Problem with Smooth Cost Functions

The ED problem is to find the optimal combination of power generations that minimizes the total generation cost while satisfying an equality constraint and inequality constraints. The most simplified cost function of each generator can be represented as a quadratic function as given in (21.27) whose solution can be obtained by the conventional mathematical methods [38]:

$$C = \sum_{j \in J} F_j(P_j) \tag{21.26}$$

$$F_j(P_j) = a_j + b_j P_j + c_j P_j^2 \tag{21.27}$$

where

C	is	the total generation cost.
F_j	–	the cost function of generator j.
P_j	–	the electrical output of generator j.
J	–	the set for all generators.
a_j, b_j, c_j	–	cost coefficients of generator j.

While minimizing the total generation cost, the following supply/demand balance should be satisfied. That is, the total generation should be equal to the total system demand plus the transmission network loss. However, the network loss is not considered in this chapter for simplicity:

$$\sum_{j \in J} P_j = D \tag{21.28}$$

where D is the total system demand.

The generation output of each unit should be between its maximum and minimum limits. That is, the following inequality constraint for each generator should be satisfied.

$$P_{j\min} \leq P_j \leq P_{j\max} \tag{21.29}$$

where $P_{j\min}$ and $P_{j\max}$ are minimum and maximum outputs of generator j.

ED Problem with Nonsmooth Cost Functions. In reality, the objective function of an ED problem has discontinuous and nondifferentiable points according to valve-point effects and change of fuels; therefore, the objective function should be composed of a set of nonsmooth cost functions. Here, the two cases with nonsmooth cost functions are considered. One is the case with the valve-point loading problems where the objective function is generally described as the superposition of sinusoidal functions and quadratic functions. The other is the case with the multiple fuel problems where the objective function is expressed as the piecewise quadratic cost functions. In both cases, the problems have multiple minima, therefore, the task to find the global solution is still remains to be tackled [39–41].

Nonsmooth Cost functions with Valve-Point Effects. The generator with multi-valve steam turbines has very different input-output curve compared with the smooth cost function. Typically, the valve point results in, as each steam valve starts to open, the ripples like in Figure 21.5 [41, 42]. To take into account the valve-point effects, the sinusoid functions are added to the quadratic cost functions as follows:

$$F_j(P_j) = a_j + b_j P_j + c_j P_j^2 + |e_j \times \sin(f_j \times (P_{j\min} - P_j))| \tag{21.30}$$

where e_j and f_j are the coefficients of generator j reflecting valve-point effects.

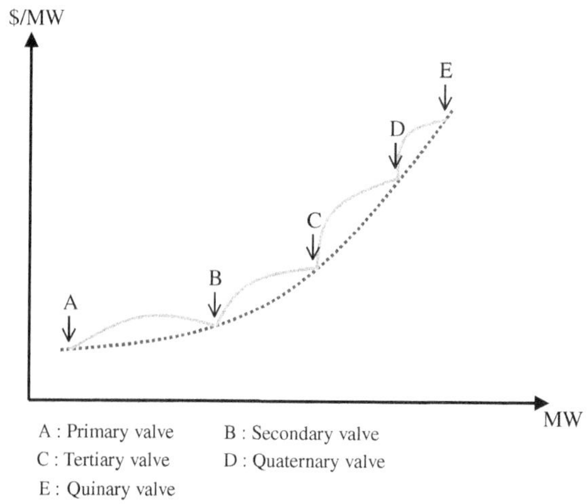

Figure 21.5 Example of cost function with five valves.

21.4.2 Implementation of PSO for ED Problems

21.4.2.1 Overview of the PSO

Kennedy and Eberhart [33] developed a PSO algorithm based on the behavior of individuals (i.e., particles or agents) of a swarm. Its roots are in zoologist's modeling of the movement of individuals (e.g., fish, birds, or insects) within a group. It has been noticed that members within a group seem to share information among them, a fact that leads to increased efficiency of the group [43]. The PSO algorithm searches in parallel using a group of individuals similar to other AI-based heuristic optimization techniques. An individual in a swarm approaches to the optimum through its present velocity, previous experience, and the experience of its neighbors.

In a physical n-dimensional search space, the position and velocity of individual i are represented as the vectors $X_i = (x_{i1}, \dots, x_{in})$ and $V_i = (v_{i1}, \dots v_{in})$ in the PSO algorithm. Let $Pbest_i = (x_{i1}^{Pbest}, \dots, x_{in}^{Pbest})$ and $Gbest_i = (x_{i1}^{Gbest}, \dots, x_{in}^{Gbest})$, respectively, be the best position of individual i and its neighbors' best position so far.

Using the information, the updated velocity of individual i is modified under the following equation in the PSO algorithm:

$$V_i^{k+1} = \omega V_i^k + c_1 rand_1 \times \left(Pbest_i^k - X_i^k\right) + c_2 rand_2 \times \left(Gbest^k - X_i^k\right) \qquad (21.31)$$

where

V_i^k	is	the velocity of individual i at iteration k.
ω	–	the weight parameter.
c_1, c_2	–	the weight factors.
$rand_1, rand_2$	–	the random numbers between 0 and 1.
X_i^k	–	the position of individual i at iteration k.
$Pbest_i^k$	–	the best position of individual i until iteration k.
$Gbest^k$	–	the best position of the group until iteration k.

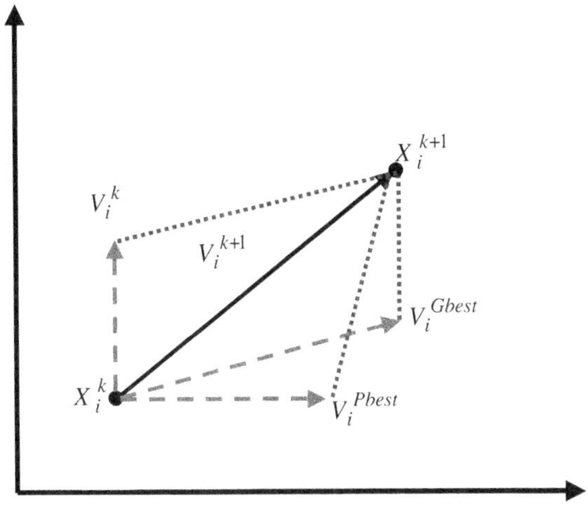

Figure 21.6 The search mechanism of the PSO.

Each individual moves from the current position to the next one by the modified velocity in (21.31) using the following equation:

$$X_i^{k+1} = X_i^k + V_i^{k+1} \tag{21.32}$$

The search mechanism of the PSO using the modified velocity and position of individual i based on (21.31) and (21.32) is illustrated in Figure 21.6.

21.4.2.2 Modified PSO for ED Problems

In [47], a new approach was developed to implement the PSO algorithm in solving the ED problems; especially, with focus on how to deal with the equality and inequality constraints of the ED problems in the process of modifying each individual's search point in the PSO algorithm. Also, to accelerate the convergence speed, the dynamic search space reduction strategy is devised. The dynamic process of the modified PSO algorithm can be summarized as follows:

Step 1: Initialization of a group at random satisfying constraints.

Step 2: Velocity and position update satisfying constraints.

Step 3: Update of *Pbest* and *Gbest.*

Step 4: Activation of space reduction strategy.

Step 5: Go to Step 2 until satisfying stopping criteria.

In the subsequent sections, the detailed implementation strategies of the MPSO are described.

Initialization And Structure Of Individuals. In the initialization process, a set of individuals is created at random. Here, the structure of an individual for ED problem is composed of a set of elements (i.e., generation outputs). Therefore, the position of individual i at iteration 0 can be represented as the vector of $X_i^0 = (P_{i1}^0, \ldots, P_{in}^0)$ where n is the number of generators. The velocity of individual i (i.e., $V_i^0 = (v_{i1}^0, \ldots v_{in}^0)$) corresponds to the generation update quantity covering all generators. The elements of position and velocity have the same dimension, that is, MW in this case. Note that it is very important to create a group of individuals satisfying the equality constraint (21.28) and inequality constraints (21.29). That is, summation of all elements of individual i (i.e., $\sum_{j=1}^{n} P_{ij}^{0}$) should be equal to the total system demand D and the created element j of individual i at random (i.e., P_{ij}^{0}) should be located within its boundary. Although we can create element j of individual i at random satisfying the inequality constraint by mapping [0, 1] into $[P_{j\min}, P_{j\max}]$, it is necessary to develop a new strategy to handle the equality constraint. To do this, the following procedure is suggested for any individual in a group:

Step 1: Set $j = 1$.

Step 2: Select an element (i.e., generator) of an individual at random.

Step 3: Create the value of the element (i.e., generation output) at random satisfying its inequality constraint.

Step 4: If $j = n$-1, then go to *Step 5*; otherwise, $j = j$+1 and go to *Step 2*.

Step 5: The value of the last element of an individual is determined by subtracting $\sum_{j=1}^{n-1} P_{ij}^{0}$ from the total system demand D. If the value is in the range of its operating region, then go to *Step* 6; otherwise, go to *Step* 1.

Step 6: Stop the initialization process.

After creating the initial position of each individual, the velocity of each individual is also created at random. The following strategy is used in creating the initial velocity:

$$(P_{j\min} - \varepsilon) - P_{ij}^{0} \le v_{ij}^{0} \le (P_{j\max} + \varepsilon) - P_{ij}^{0} \tag{21.33}$$

where ε is a small positive real number. The velocity of element j of individual i is generated at random within the boundary. The developed initialization scheme always guarantees to produce individuals satisfying the constraints as well as not to deviate from the concept of the PSO algorithm. The initial $Pbest_i$ of individual i is set as the initial position of individual i and the initial *Gbest* is determined as the position of an individual with minimum payoff of (21.26).

VELOCITY UPDATE. To modify the position of each individual, it is necessary to calculate the velocity of each individual in the next stage, which is obtained from (21.31). In this velocity updating process, the values of parameters such as ω, c_1, and c_2 should be determined in advance. In this chapter, the weighting function is defined as follows [1, 3]:

$$\omega = \omega_{\max} - \frac{\omega_{\max} - \omega_{\min}}{Iter_{\max}} \times Iter \tag{21.34}$$

where

$\omega_{\max}, \omega_{\min}$	are	the initial and final weights.
$Iter_{\max}$	–	the maximum iteration number.
$Iter$	–	the current iteration number.

Also, the parameters in equations (21.31) and (21.34) are selected as follows based on the empirical studies: $c_1 = c_2 = 2.0, \omega_{\max} = 0.5, \omega_{\min} = 0.1$.

Position Modification Considering Constraints. The position of each individual is modified by (21.32). The resulting position of an individual is not always guaranteed to satisfy the inequality constraints due to over/under velocity. If any element of an individual violates its inequality constraint due to over/under speed, then the position of the individual is fixed to its maximum/minimum operating point. Therefore, this can be formulated as follows:

$$P_{ij}^{k+1} = \begin{cases} P_{ij}^{k} + v_{ij}^{k+1} & if \quad P_{ij,\min} \le P_{ij}^{k} + v_{ij}^{k+1} \le P_{ij,\max} \\ P_{ij,\min} & if \quad P_{ij}^{k} + v_{ij}^{k+1} < P_{ij,\min} \\ P_{ij,\max} & if \quad P_{ij}^{k} + v_{ij}^{k+1} > P_{ij,\max} \end{cases} \tag{21.35}$$

Figure 21.7 illustrates how the position of element j of individual i is adjusted to its maximum when overvelocity situation occurs.

Although the aforementioned method always produces the position of each individual satisfying the inequality constraints (21.29), the problem of equality constraint still remains to be resolved. Therefore, it is necessary to develop a new strategy such that the summation of all elements in an individual (i.e., $\sum_{j=1}^{n} P_{ij}^{k}$) is equal to the total system demand.

To resolve the equality constraint problem without intervening the dynamic process inherent in the PSO algorithm, the following heuristic procedure is proposed:

Step 1: Set $j = 1$. Let present iteration be k.

Step 2: Select an element (i.e., generator) of individual i at random and store in an index array $A(n)$.

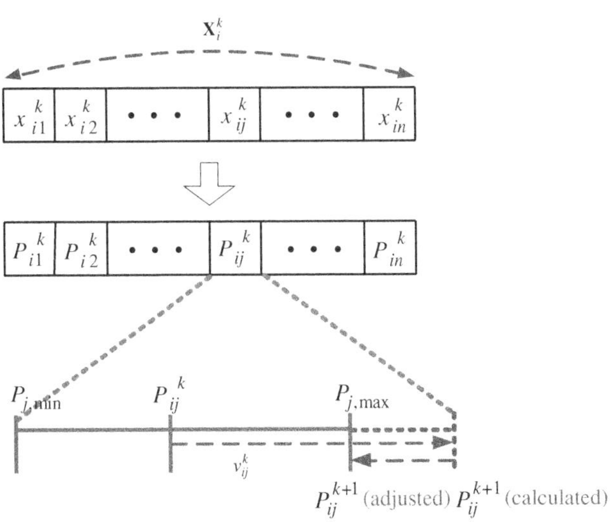

Figure 21.7 Adjustment strategy for an individual's position within boundary.

Step 3: Modify the value of element j using (21.31), (21.32), and (21.35).

Step 4: If $j = n\text{-}1$, then go to *Step* 5, otherwise, $j = j+1$ and go to *Step* 2.

Step 5: The value of the last element of individual i is determined by subtracting $\sum_{j=1}^{n} P_{ij}^{k}$ from D. If the value is not within its boundary, then adjust the value using (21.35) and go to *Step* 6, otherwise go to *Step* 8.

Step 6: Set $l = 1$.

Step 7: Readjust the value of element l in the index array $A(n)$ to the value satisfying equality condition (i.e., $D - \sum_{\substack{j=1 \\ j\neq l}}^{n} P_{ij}^{k}$). If the value is within its boundary, then go to *Step* 8; otherwise, change the value of element-l using (21.35). Set $l = l+1$, and go to *Step* 7. If $l = n+1$, then go to *Step 1*.

Step 8: Stop the modification procedure.

UPDATE OF PBEST AND GBEST. The *Pbest* of each individual at iteration $k+1$ is updated as follows:

$$\begin{aligned} Pbest_i^{k+1} &= X_i^{k+1} \text{ if } TC_i^{k+1} < TC_i^k \\ Pbest_i^{k+1} &= Pbest_i^k \text{ if } TC_i^{k+1} > TC_i^k \end{aligned} \tag{21.36}$$

where TC_i is the object function evaluated at the position of individual i.

Also, G_{best} at iteration $k+1$ is set as the best evaluated position among $Pbest_i^{k+1}$.

SPACE REDUCTION STRATEGY. To accelerate the convergence speed to the solutions, the MPSO has introduced the search space reduction strategy. This strategy is activated in the case when the performance is not increased during a prespecified iteration period. In this case, the search space is dynamically adjusted (i.e., reduced) based on the "distance" between the *Gbest* and the minimum and maximum output of generator j.

To determine the adjusted minimum/maximum output of generator j at iteration k, the distance is multiplied by the predetermined step-size Δ and subtracted (added) from the maximum (minimum) output at iteration k as described in (21.37).

$$\begin{aligned} P_{j\max}^{k+1} &= P_{j\max}^{k} - \left(P_{j\max}^{k} - Gbest_j^k\right) \times \Delta \\ P_{j\min}^{k+1} &= P_{j\min}^{k} + \left(P_{j\min}^{k} - Gbest_j^k\right) \times \Delta \end{aligned} \tag{21.37}$$

Figure 21.8 illustrates how the search space of each generator is dynamically reduced when activated.

Stopping criteria: The MPSO is terminated if the iteration approaches to the predefined maximum iteration.

21.4.2.3 Case Studies

To assess the efficiency and effectiveness of the proposed MPSO, it has been applied to ED problems where the objective functions are either smooth or nonsmooth ones. The results obtained from the MPSO are compared with those of other methods: the NM [38], the hierarchical numerical method [22, 39], the improved evolutionary programming (IEP) [42], the GA [43], the tabu search (TS) [40], the EP [25, 41], the

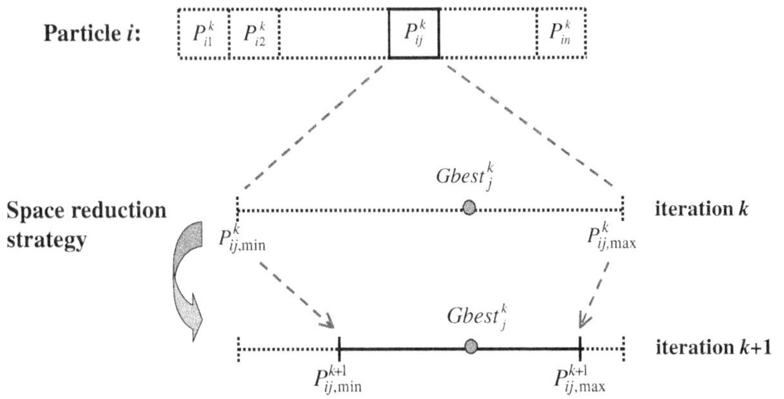

Figure 21.8 Schematic of the dynamic space reduction strategy.

modified Hopfield neural network (MHNN) [45], and the adaptive Hopfield neural network [46].

21.4.2.4 ED Problem with Smooth Cost Functions

The MPSO is applied to an ED problem with three generators with the quadratic cost functions. Table 21.16 shows the input data of the system and the system demand considered is 850 MW.

Table 21.17 shows the comparison of the results from MPSO, NM [38], IEP [42], and MHNN [45].

As seen in Table 21.17, the MPSO has provided the global solution, the same result of the NM, exactly satisfying the equality and inequality constraints. It has been shown that faster convergence is achieved for the proposed MPSO as the number of particles is increased. It has also been shown that a robustness of the MPSO is guaranteed on the random initialization. When the dynamic space reduction strategy is activated, the convergence speed is dramatically increased without trapping in a local optimum [47].

21.4.2.5 ED Problem with Nonsmooth Cost Functions Considering Valve-Point Effects

The MPSO has been applied to the ED problem with three generators where valve-point effects are considered. In Table 21.18, the corresponding coefficients of cost functions and output constraints are given. Here, the total demand considered is

TABLE 21.16 Cost coefficients of test system with three generators

Unit	a_j	b_j	c_j	$P_{j\min}$	$P_{j\max}$
1	561.0	7.92	0.001562	150.0	600.0
2	310.0	7.85	0.00194	100.0	400.0
3	78.0	7.97	0.00482	50.0	200.0

TABLE 21.17 Comparison of simulation results of each method

Unit	NM	MHNN	IEP (pop=10)	MPSO (par=10)
1	393.170	393.800	393.170	393.170
2	334.604	333.100	334.603	334.604
3	122.226	122.300	122.227	122.226
TP	850.000	849.200	850.00000	850.00000
TC	8194.35612	8187.00000	8194.35614	8194.35612

* pop: population size, par: number of particle,
TP: total power (MW), TC: total generation cost [$].

850 MW. It was reported in [40] that the global optimum solution of the problem found is $8234.0.

In Table 21.19, the results from MPSO, GA [43], IEP [42], and EP [41] are compared. It shows that the MPSO has succeeded in finding the global solution as in [40] satisfying the equality and inequality constraints.

It was shown that for different number of particles, the global solution was obtained, and the global solution is also obtained when the dynamic space reduction strategy is applied. However, the obtained value from the PSO with 20 particles without considering the space reduction strategy was observed as 8241.59[$], which verifies the effectiveness of the space reduction strategy [47].

TABLE 21.18 Cost coefficients of test system with valve-point effects

Unit	a_j	b_j	c_j	e_j	f_j	$P_{j\min}$	$P_{j\max}$
1	561.0	7.92	0.001562	300.0	0.0315	100.0	600.0
2	310.0	7.85	0.00194	200.0	0.042	100.0	400.0
3	78.0	7.97	0.00482	150.0	0.063	50.0	200.0

TABLE 21.19 Comparison of simulation results of each method considering valve-point effects

Unit	GA	IEP (pop=20)	EP	MPSO (par=20)
1	300.00	300.23	300.26	300.27
2	400.00	400.00	400.00	400.00
3	150.00	149.77	149.74	149.73
TP	850.00	850.00	850.00	850.00000
TC	8237.60	8234.09	8234.07	8234.07

21.5 ANT COLONY SYSTEM FOR CONSTRAINED LOAD FLOW PROBLEM

The ACS algorithms simulate the behavior of real ants [48, 49]. They are based on the principle that using simple communication mechanisms, an ant group is able to find the shortest path between any two points. During their trips, a chemical trail (pheromone) is left on the ground. The pheromone guides other ants toward the target point. For one ant, the path is chosen according to the quantity of pheromone. The pheromone evaporates over time (i.e., it loses quantity if other ants lay down no more pheromone). If many ants choose a certain path and lay down pheromones, the quantity of the trail increases and thus this trail attracts more and more ants. The artificial ants (or simply, ants) simulate the transitions from one point r to another point s, according to the improved version of ACS, namely, Max-Min AS (MMAS) algorithm [50] as follows:

If the ant k is at point r, has the next point been visited? The ant k maintains a tabu list N_r^k in memory that defines the set of points still to be visited when it is at point r. The ant k chooses to go from point r to point s during a tour with a probability given by [50]:

$$p(r,s) = \frac{\gamma(r,s)}{\sum\limits_{l} \gamma(r,\ell)} \quad s, \ell \in N_r^k \tag{21.38}$$

where matrix $\gamma(r, s)$ represents the amount of the pheromone trail and pheromone intensity, between points r and s.

Then, the pheromone trail on coupling (r, s) is updated according to:

$$\gamma(r,s) = \alpha \cdot \gamma(r,s) + \Delta\gamma^k(r,s) \tag{21.39}$$

where α with $0 < \alpha < 1$ is the persistence of the pheromone trail, so that $(1-\alpha)$ to represent the evaporation and $\Delta\gamma^k(r,s)$ is the amount of pheromone that ant k puts on the trail (r, s). The pheromone update $\Delta\gamma^k(r,s)$ reflects the desirability of the trail (r, s), such as shorter distance and better performance, depending on the application problem. Since the best tour is unknown initially, an ant needs to select a trail randomly, and deposits pheromone in the trail, where the amount of pheromone will depend upon the pheromone update rule (21.39). The randomness implies that pheromone is deposited in all possible trails, not just in the best trail. The trail with favorable update, however, increases the pheromone intensity more than other trails.

After all ants have completed their tours, global pheromone is updated in the trails of the ant with the best tour executed. In the next section, the MMAS algorithm is extended and modified to solve the constrained load flow (CLF) problem.

21.5.1 Formulation of Constrained Load Flow Problem

The load flow problem can be expressed by the next two sets of nonlinear equations:

$$\begin{aligned} Y &= g(X, U) \\ Z &= h(X, U) \end{aligned} \tag{21.40}$$

where

Y is the vector of nodal power injections.

Z – the vector of constrained variables (power flows, reactive powers of PV busses, etc.)

X – the state vector (voltage angles and magnitudes).

U – the control vector (transformer tap-settings, shunt compensation, voltage and power at PV busses, etc.).

The objective of the CLF is to maintain some or all elements of X and Z vectors within given operating limits under the uncertainty of generating units' availabilities and load uncertainties. This can be achieved by selecting appropriate (robust) values of control variables under random variations of loads and generations (noise factors) within their operating range. The next sections of this chapter propose a technique to maintain constrained variables within operating limits over the whole planning period using the ACS algorithm.

21.5.2 Development of Ant Colony System for the Constrained Load Flow Problem

The settings of control variables (tap-settings, VAr compensation blocks, etc.) are combined in order to achieve the power system constraints. In [55], the graph that describes the settings of control variables of the CLF problem is mapped on the AS-graph, which is the space that the artificial ants will walk. Figure 21.9 shows the AS-graph (searching space) for the CLF problem. All possible candidate discrete settings for a control variable are represented by the states r of the AS-graph ($r = 1, \ldots, m$).

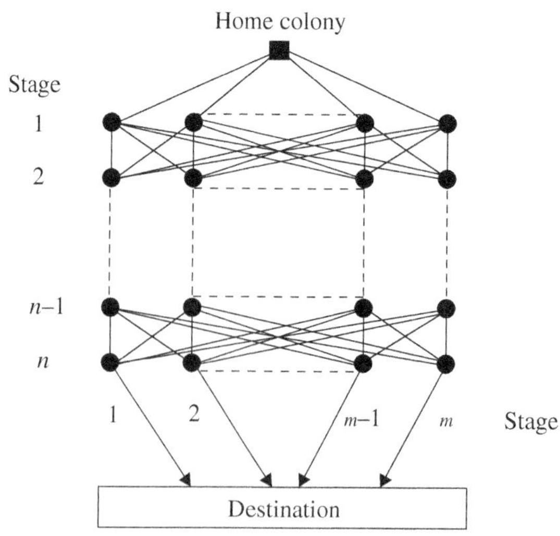

Figure 21.9 Search space for the CLF problem.

The control variables are represented by the stages i ($i = 1, \ldots, n$), where n is the number of the control variables. Each ant will start its tour at the home colony and stop at the destination. The ACS algorithm proceeds as follows:

An operating point comprising a load and generation pattern (operating point of the whole planning period of the system) is randomly created. For this operating point, first of all, AS graph is created and all paths receive an amount of pheromone that corresponds to an estimation of the best solution so that ants test all paths in the initial iterations. Therefore, ACS-algorithm achieves the best exploration of AS-graph in the earlier iterations of convergence and better exploitation at the latest.

Then, ant k chooses the next states to go to in accordance with the transition probability calculated by (21.38). When the ant k moves from one stage to the next, the state of each stage will be recorded in a location list J^k. After the tour of ant k is completed, its location list is used to compute its current solution. Then the pheromone trails composed by nodes of location list J^k are updated in accordance with (21.39) (local update). For the purpose of this research, the pheromone update $\Delta\gamma^k(r,s)$ is chosen as:

$$\Delta\gamma^k(r,s) = 1/Q \cdot f \tag{21.41}$$

where f is the objective function, and Q is a large positive constant.

Application of the ACS algorithm to the CLF problem is linked to the choice of an objective function f, such as the limits of the constrained variables to be satisfied for the whole planning period. An enforced empirical strategy is to consider the variations of constrained variables close to the means of their operating intervals. The objective function f is computed by the average of all constrained variables, normalized in the interval [0, 1], as follows:

$$f = \frac{1}{n} \cdot \sum_{j=1}^{n} \left| \frac{2z_j - z_{j\max} - z_{j\min}}{z_{j\max} - z_{j\min}} \right| \tag{21.42}$$

where n expresses the number of constrained variables, z_j expresses the value of *j-th* constrained variable, and $(z_{j\min}, z_{j\max})$ are its lower and upper limits, respectively.

It must be noticed that $f = -r$, where r is the immediate reward used by the Q-learning algorithm to the CLF problem [51]. The objective function f has this formulation in order for the results provided by the ACS and the Q-learning algorithms to be compared.

In order to exploit the iteration in finding the best solution, the next two steps are considered:

a) When all ants complete their tours, load flow is run and the objective function (21.42) is calculated for each run. Then, the pheromone trails (r,s) of the best ant tour (ant with minimum objective function (21.42)) is updated [global update] as:

$$\gamma(r,s) = \alpha \cdot \gamma(r,s) + \frac{R}{f_{best}} r, s \in J^k_{best} \tag{21.43}$$

where R is a large positive constant. Both Q in (21.41) and R are arbitrarily large numbers. Empirical tests have shown that the ACS-algorithm converges faster when Q is almost equal to R.

b) To avoid search stagnation (the situation where all the ants follow the same path, i.e., they construct the same solution [49]), the allowed range of the pheromone trail strengths is limited to:

$$\gamma(r,s) = \begin{Bmatrix} \tau_{\min} & \text{if} & \gamma(r,s) \leq \tau_{\min} \\ \tau_{\max} & \text{if} & \gamma(r,s) \geq \tau_{\max} \end{Bmatrix} \tag{21.44}$$

For our study, the limits are chosen as:

$$\tau_{\max} = \frac{1}{\alpha f_{gbest}} \tag{21.45}$$

where f_{gbest} is the global best solution (best over the whole past iterations), and

$$\tau_{\min} = \frac{\tau_{\max}}{M^2} \tag{21.46}$$

where M is the number of ants.

The procedure is repeated for a large number of operating states covering the whole planning period. Once we have the set of optimal control settings for a large number of operating points, the one that minimizes the sum of multi-objective function (*mtf*) over the whole planning period is defined as a greedy-optimum control setting:

$$mtf = \min\{\text{total of objective functions}\} = \min\left(\sum_{\substack{\text{over whole}\\ \text{planning period}}} f\right) \tag{21.47}$$

The execution steps of the ACS algorithm applied to the CLF problem are:

(1) Create the AS-graph (search space) that represents the discrete settings (states) of the control variables (stages).

(2) Insert the pheromone matrix $\gamma(m,n)$ according to nodes of AS-graph, where n is the number of stages and m is the number of states.

(3) Initialize the pheromone matrix $\gamma(m,n) = \gamma_0(m,n) = \tau_{\max}$ (in (21.45), in this case, f_{gbest} is an initial estimation of the best solution).

(4) Repeat for a given number of operating points over the whole planning period.

4.1. Repeat until the system convergence or iteration is less than a given maximum number.

4.1.1. Place randomly M ants on the states of the first stage ($i = 1$).

4.1.2. For $k = 1$ to M

4.1.2.1. For $i = 2$ to n

4.1.2.1.1. When the ant-k has selected the r-state of the (i-1)-stage, it currently chooses the s-state of the (i)-stage in which it will move according to transition rule (21.38).

4.1.2.1.2. Move the ant-k to s-state of i–stage.

4.1.2.1.3. Record s to J^k, and set $r = s$.

4.1.3. Run power flow for each ant.

4.1.4. Calculate the objective function (21.42) for each ant.

4.1.5. Update the pheromone of (r,s)-trails for each ant, using the local pheromone update formulae (21.39) and (21.41).

4.1.6. Update the pheromone of (r,s)-trails belonging to best ant tour (f_{best}), using the pheromone update formula (21.43).

4.1.7. In order to avoid the ants' stagnations, enforce the limits (21.44)–(21.46).

4.2. Enforce each of the best control settings over the whole planning period and calculate (21.47).

1. Choose as a greedy-optimum control setting the one that minimizes (21.47).

21.5.3 Results

The ACS algorithm is applied to adjust reactive control variables in the IEEE 14-bus test system shown in Figure 21.10. The test system consists of the slack bus (node 1), three PV (nodes 2, 3, and 6), 10 PQ busses, and 20 branches. It has been used in many probabilistic studies. The network data and load probabilistic data are the same as used in [52]. They comprise six discrete distributions for the active load (at nodes 3, 6, 9, 10, 11, and 14), four discrete distributions for the reactive load (at nodes 9, 10, 11, and 14), with 3–5 impulses each and eight normal distributions for active and reactive loads at the remaining busses.

The total installed capacity is equal to 4.9 p.u. and comprises 14 capacitor banks at node 1, four banks at node 2, two banks at node 3, and two banks at node 6. The voltage at all PV busses is taken equal to 1.0 p.u., and the slack bus voltage is equal

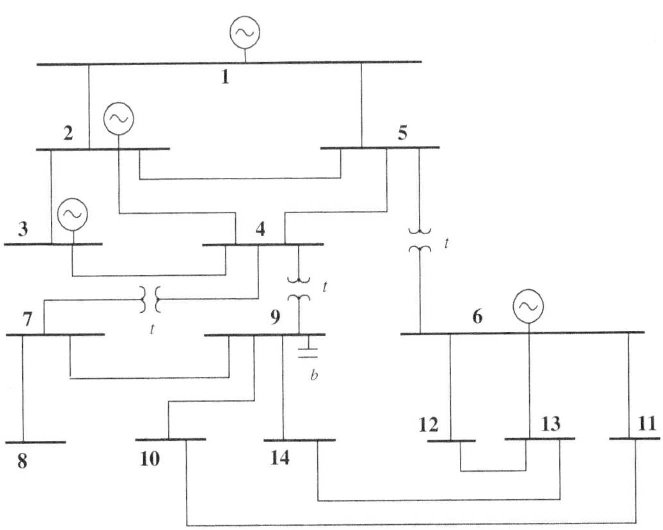

Figure 21.10 Line diagram of the IEEE 14-bus system.

TABLE 21.20 Limits and discretization of actions and limits of constrained variables

Control actions	Umin	Umax	Step
t_{56}	0.90	1.05	0.01
t_{49}	0.90	1.05	0.01
t_{47}	0.90	1.05	0.01
b_9	0.00	0.24	0.03
Constrained variables	Wmin	Wmax	
Q_{g2}	0.00	0.30	
Q_{g3}	0.00	0.70	
Q_{g6}	0.00	0.45	
T_{23}	0.00	0.75	
T_{56}	0.00	0.50	
V_4	0.96	1.05	
V_5	0.96	1.05	
V_7	0.96	1.05	
V_8	0.96	1.05	
V_9	0.96	1.05	
V_{10}	0.96	1.05	
V_{11}	0.96	1.05	
V_{12}	0.96	1.05	
V_{13}	0.96	1.05	
V_{14}	0.96	1.05	

to 1.02 p.u. A fixed network topology is assumed. The control variables comprise all transformer taps (t) and reactive compensation (b) at bus 9.

The upper part of Table 21.20 shows the limits of the control variables U_{min} and U_{max} and the discrete steps in variation. The transformer taps (t_{56}, t_{49}, and t_{47}) are in 16 steps, while the reactive compensation (b_9) is in nine steps. Therefore, the last step of b_9 ($b_9 = 0.24$) is repeated for the next seven steps, for steps 10 through 16.

This makes the pheromone matrix $\gamma(m,n)$ of the AS-graph well defined for all stages and states. In the lower part of Table 21.20, the upper and lower limits of all constrained variables W_{max} and W_{min} are shown.

In our study, the following ACS parameters (Table 21.19) are chosen: $M = 100$, $n = 4$, $m = 16$, $Q = R = 1{,}000{,}000$ and the initial best solution is estimated at 0.0001. The parameter α from our experience shows that any value in the range [0.88, 0.999] works well. In this chapter, it is chosen as $\alpha = 0.9865$. In this study, the search will be terminated if one of the following criteria is satisfied: (a) the number of iterations since the last change of the best solution is greater than 1000 iterations or (b) the number of iterations reaches to 3000 iterations.

The ACS algorithm can be implemented in a large number of load combinations (operating points) selected over the whole planning period.

TABLE 21.21 Comparison of results between ACS, RL, and probabilistic load flow on IEEE 14-bus

	ACS algorithm $mtf = 0.732$		Q-learning algorithm $mtr = -0.804$		Probabilistic load flow —	
Contr. Varib.	Optimal settings		Greedy-optimal settings (a^*)		Optimal settings	
t_{56}	1.01		1.03		0.94	
T_{49}	0.91		0.97		0.97	
T_{47}	0.99		0.90		0.98	
b_9	0.18		0.12		0.12	
Constr.	Wmin	Wmax	Wmin	Wmax	Wmin	Wmax
Q_{G2}	−0.5508*	−0.0795*	−0.5250*	−0.0527*	0.2069	0.3160*
Q_{g3}	0.1554	0.5978	0.1751	0.6183	0.6420	0.6812
Q_{g6}	0.0688	0.3606	0.0891	0.3787	0.3065	0.4161
T_{23}	0.2273	0.6847	0.2284	0.6868	0.7223	0.7799*
T_{56}	0.0731	0.2920	0.1675	0.4528	0.4144	0.4931
V_4	0.9777	0.9984	0.9742	0.9951	0.9654	0.9731
V_5	0.9876	1.0042	0.9864	1.0030	0.9682	0.9744
V_7	0.9864	1.0235	0.9901	1.0274	0.9710	0.9857
V_8	0.9864	1.0235	0.9901	1.0274	0.9710	0.9833
V_9	0.9871	1.0316	0.9840	1.0284	0.9656	0.9833
V_{10}	0.9819	1.0237	0.9775	1.0193	0.9644	0.9803
V_{11}	0.9903	1.0139	0.9831	1.0067	0.9828	0.9912
V_{12}	0.9899	1.0011	0.9802	0.9915	0.9907	0.9936
V_{13}	0.9818	0.9994	0.9725	0.9903	0.9832	0.9878
V_{14}	0.9600	1.0036	0.9541*	0.9977	0.9508*	0.9663

In our study, the algorithm learns the optimum control settings for each of 41 operating points selected from the whole planning period. These full correlated operating points are sampled uniformly from the curves of normal and discrete distribution probabilities as follows:

$$\text{Load step} = \left(\mu \pm k \times \frac{3\sigma}{20}\right) \cdot 100\% \tag{21.48}$$

where $k = 0, 1, 2, 3, \ldots, 20$. The μ and σ are the average values and the standard deviations for normal distributions of loads given in [53]. The μ and σ for the discrete distributions of loads are calculated using the formulae given by [54].

Table 21.21 shows the *mtf*, calculated with (21.47) at 0.732, the greedy-optimum control settings, and the operating space of constrained variables, when these settings are enforced over the whole planning period. It can be seen that even when applying the greedy-optimum control settings, reactive production at node 2 (Q_{g2}) violates its limits. In Table 21.21, these results are also compared to the results of the Q-learning (RL) [51] and the probabilistic CLF method [52], obtained for the same network.

The Q-learning algorithm [13] provides slightly worse results, since Q_{g2} violates its limits and the voltage at node 14 violates its lower limit too. The absolute value of maximum total reward (*mtr*) [51] in this case was calculated at 0.804, which is greater than the corresponding index of the ACS algorithm (*mtf* = 0.732). The Q-learning algorithm [52] took about 38,800 iterations to find the greedy-optimum control settings. In the case of probabilistic CLF [52], the upper limit of reactive production at node 2 (Qg2) and the lower limit of voltage at node 14, as well as the upper limit of the apparent flow (T23) in line 2-3, are violated.

The ACS and Q-learning algorithms provide the optimal results, rather than the near-optimal results given by the probabilistic CLF method, since all constraints including the upper limit of apparent flow (T23) on line 2-3 are satisfied [55].

21.6 IMMUNE ALGORITHM FOR DAMPING OF INTERAREA OSCILLATION

The immune algorithm (IA) has desirable characteristics as an optimization tool and offers significant advantages over traditional methods.

In the IA, *antigen* represents the problem to be solved. An *antibody* set is generated where each member represents a candidate solution. Also, *affinity* is the fit of an antibody to the antigen. In the IA, the role of antibody lies in eliminating the antigen, while the *lymphocyte* helps to produce the antibody [56, 57].

In the immune system, there are two kinds of lymphocyte: T and B; where each of them has its own function. The T lymphocytes develop in bone marrow and travel to thymus to mature. The B lymphocytes develop and mature within the bone marrow. The main purpose of the immune system is to recognize all cells within the body and categorize those cells as self or nonself. Self or self-antigens are those cells that originally belong to the organism and are harmless to its functioning. The disease-causing elements are known as nonself.

Both B cells and T cells have receptors that are responsible for recognizing antigenic patterns by different functions. The attraction between an antigen and a receptor cell (or degree of binding) is known as affinity.

To handle the infection successfully and effectively, both B cells and T cells may be required. After successful recognition, cells capable of binding with nonself antigens are cloned.

In the IA, the elements of the population undergo mutations resulting in a subpopulation of cells that are slightly different. Since the mutation rate is high, this mutation is called hypermutation.

By the above description, the principle of IA can be summarized in Figure 21.11. As the figure shows at the first step, n antibodies are generated randomly and evaluated using a suitable affinity measure. While the affinity of all antibodies is known, new population is generated through three steps: replacement, cloning, and hypermutation. These three steps maintain the diversity and help the algorithm to expand the search space. In the replacement step, the low antibodies are replaced.

Those with the highest affinity are selected to proliferate by cloning where the cloning rate of each immune cell is proportional to its affinity. If the high affinity

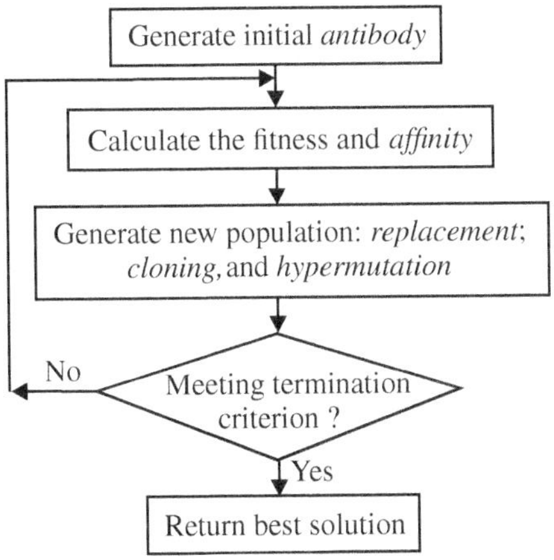

Figure 21.11 General principle of the immune algorithm.

antibody has not been cloned, hypermutation is applied where the mutation rate for each immune cell is inversely proportional to its affinity [56]. When the new population is generated, IA continues with repeated evaluation of the antibodies through replacement, cloning, and hypermutation until the termination criterion is met. The termination criterion could be the number of iterations or when an antibody of maximal affinity is found.

The IA can be implemented through the BIA and RIA. The RIA uses the real codes and the replacement; cloning and hypermutation are applied to the variable directly.

The modified RIA (MRIA) is a hybrid RIA that uses an immune search to explore the search space and a local search to exploit information in the search region.

Local search used in MRIA is a hill-climbing one. Hill-climbing searches the neighborhood of the current solution to find the next solution with more improvement in the value of objective function.

In MRIA, the initial process can produce n antibodies randomly. Each antibody will be evaluated by a fitness function. Then, the first l antibodies with the best fitness function will be selected. Now, the local search is applied on the selected l antibodies. This local search is a hill-climbing one, where a neighborhood is considered for each selected l antibodies. The algorithm searches the neighborhood of the l selected antibodies to find a solution with more improvement in the value of fitness function. If such a solution exists, the old antibody is replaced with the newly obtained antibody. This local search can be repeated m times where m is a real and constant value defined by the user.

21.6.1 Study System and Problem Formulation

A 2-area-4-machine system is used. This test system is illustrated in Figure 21.12. The subtransient model for the generators and the IEEE-type DC1 and DC2 excitation systems are used for machines 1 and 4, respectively. The IEEE-type ST3 compound

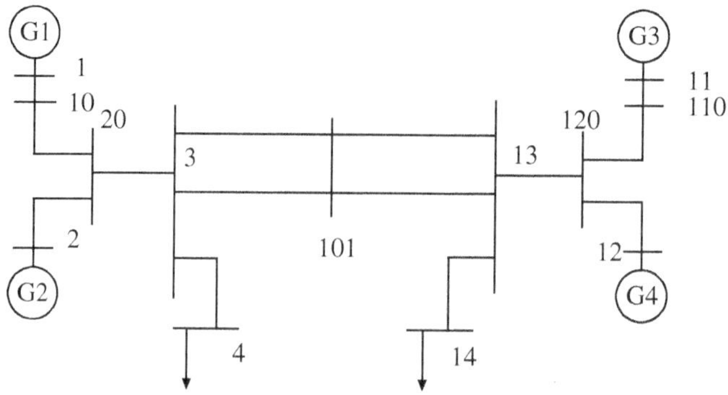

Figure 21.12 Single-line diagram of a 2-area study system.

source rectifier exciter model is used for machine 2, and the first-order simplified model for the excitation systems is used for machine 3.

One SVC is located at bus 101, where voltage swings are the greatest without the SVC. A supplementary controller for SVC is going to be designed using RIA, MRIA, and BIA. The following structure shown by Figure 21.13 is used for the controller where the input to the controller could be the real power of line 13-120 [58].

A damping factor (ς) of around 10–20% for the troublesome low frequency electromemchanical mode is considered adequate. A damping factor of 10% would be acceptable to most utilities and can be adopted as the minimum requirement. Furthermore, having the real part of rotor mode eigenvalue (s) restricted to be less than a value, say α, guarantees a minimum decay rate α. A value $\alpha = -0.5$ is considered adequate for an acceptable settling time. The closed-loop rotor mode location should simultaneously satisfy these two constraints for an acceptable small disturbance response of the controlled system.

Therefore, the parameters of the supplementary controller, T, T_1, T_2, T_3, T_4, are determined by RIA, MRIA, and BIA by optimizing the following objective or cost function:

$$f = \max(real(s) - \min(-\beta \times abs(imag(s)), \alpha)) \tag{21.49}$$

where in this study, β is set to be 1. This fitness function will place the system closed-loop eigenvalues in the D-shaped sector shown in Figure 21.14.

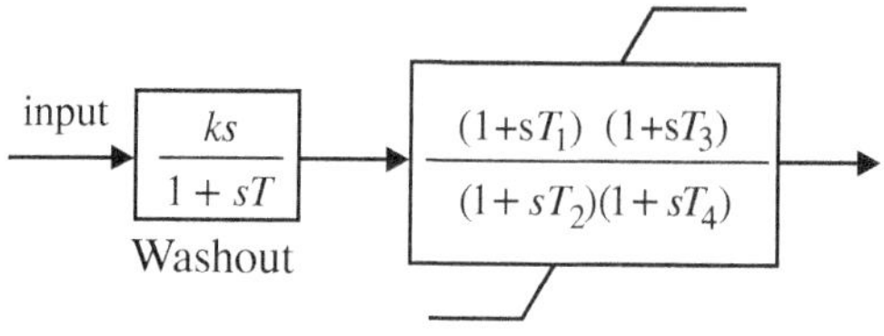

Figure 21.13 A lead-lag structure model block diagram: supplementary controller for SVC.

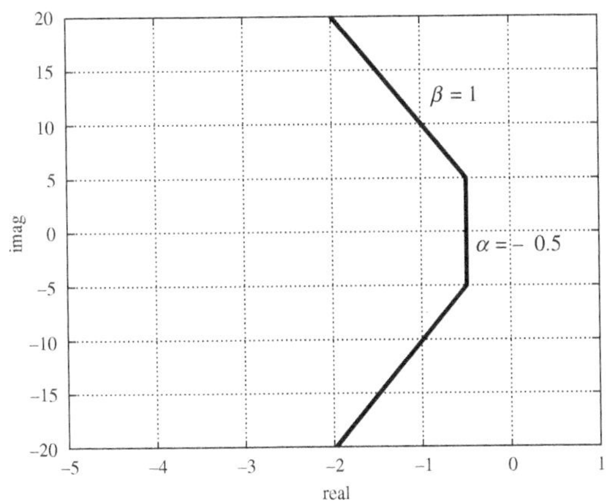

Figure 21.14 A D-shaped sector in the s-plane.

Furthermore, the design problem can be formulated as the following constrained optimization problem, where the constraints are the supplementary controller parameter bounds:

Minimize f

Subject to the constraints:

$$\begin{aligned} &1 \leq T \leq 15 \\ &0 \leq T_i \leq 5, \qquad i = 1, 2, 3, 4 \end{aligned} \tag{21.50}$$

The RIA, MRIA, and BIA are applied to solve this optimization problem and search for optimal or near optimal set of supplementary controller for SVC.

21.6.2 Designing of Supplementary Controller

Population of n antibodies is generated randomly, where n is considered to be 50 in RIA and BIA. The goal of the optimization is to find the best value for the supplementary controller, T, T_1, T_2, T_3, T_4 (Figure 21.13). Therefore, a configuration is considered for each antibody as a vector $[T, T_1, T_2, T_3, T_4]$.

During each generation, the antibodies are evaluated with some measure of fitness, which is calculated from the objective function defined in (21.50). Then the best antibody is chosen. In the current problem, the best antibody is the one that has minimum fitness. This antibody is chosen as antigen and the affinity of other antibodies is calculated with the selected antigen. The affinity of each antibody is calculated by the following equation:

$$affinity = f(antigen)/f(antibody) \tag{21.51}$$

Moving to a new generation is based on the antibodies with the high and low affinity by using cloning and replacement. Also, the hypermutation is applied to each generation in order to recognize not only the antigen itself but also antigens that are similar.

The above procedure continues until the last iteration is met. Here, the number of iterations is set to be 100.

The principle of MRIA applied in this chapter is as follows:

***Step* 1**: *Initialization*. Generate an initial population of n antibodies.

***Step* 2**: *Evaluation*. Evaluate the fitness for each antibody.

***Step* 3**: *Local search*. Apply the local search for the first l antibody with the best fitness. The algorithm searches the neighborhood of the l selected antibodies to find a solution with more improvement in the value of fitness function. If such a solution exists, the old antibody is replaced with the new obtained antibody.

***Step* 4**: *Evaluation operators*. Return the first l antibody from step 3 along with other antibodies (n-l). Generate new population through replacement, cloning, and hypermutation.

***Step* 5**: *Iteration–algorithm*. Steps 2–4 continue iteratively until stopping criteria (the number of iterations) is met.

To find the parameters of supplementary controller, T, T_1, T_2, T_3, T_4 by MRIA, the number of antibodies is set to be 10, for a fare comparison. The local search is applied on the first l antibodies with the best fitness function, where $l = 5$. The local search is repeated 8 times for each antibody. With this setting, the number of evaluations of fitness function in MRIA is 50 ($(5 \times 8) + 10 = 50$), which is the same as RIA. Also, the number of iterations is set to be 100.

To find the best value for the controller; T, T_1, T_2, T_3, T_4; the algorithms are run for 10 independent runs under different random seeds. After 10 independent runs, the following results are observed by the three algorithms:

The 80% of the obtained results by MRIA reach to a minimum cost function (best solution) equal to 6.3769, 60% of the obtained results by RIA reach to a minimum cost function equal to 6.3955, and 10% of the obtained results by BIA reach to a minimum cost function equal to 6.5774. The results obtained by the three algorithms for T, T_1, T_2, T_3, T_4 regarding to the above cost functions are shown in Table 21.22.

For the designed supplementary controller for SVC, the average best-so-far of each run is recorded and averaged over 10 independent runs. To have a better clarity, the convergence characteristics in finding the best values of supplementary controller parameters are given in Figure 21.15.

TABLE 21.22 The results obtained by RIA, MRIA, and BIA

Algorithm	T	T_1	T_2	T_3	T_4	cost function
MRIA	9.6363	0.030514	1.856	0.66318	1.7344	6.3769
RIA	9.081	0.011663	1.609	0.367	1.8104	6.3955
BIA	8.0554	0.28168	1.2889	0.13657	1.1712	6.5774

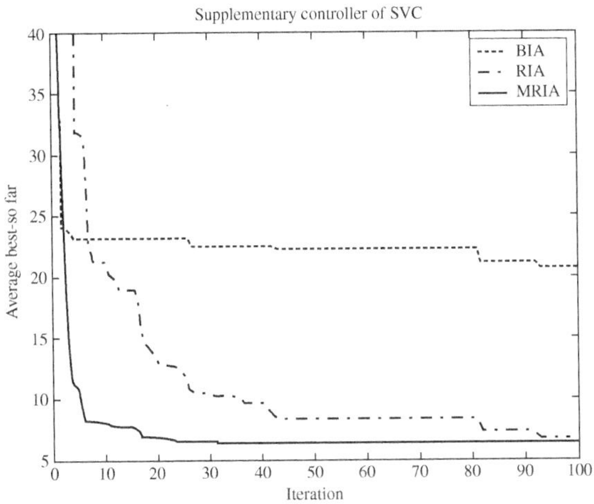

Figure 21.15 Convergence characteristics of RIA, MRIA, and BIA on the average best-so-far function in finding the supplementary controller for SVC.

The obtained supplementary controllers for SVC by RIA, MRIA, and BIA are placed in the study system (Figure 21.12). To show the effectiveness of the designed supplementary controller for SVC, a time-domain analysis is performed for the study system. A three-phase fault is applied in one of the tie circuits at bus 3. The fault persisted for 70.0 ms; following this, the faulted circuit was disconnected by appropriate circuit breaker. The system operated with one tie circuit connecting busses 3 and 101.

The dynamic behavior of the system was evaluated for 15 s. The voltage magnitude at fault bus is shown in Figure 21.16. The machine angle responses were shown to be similarly improved [59].

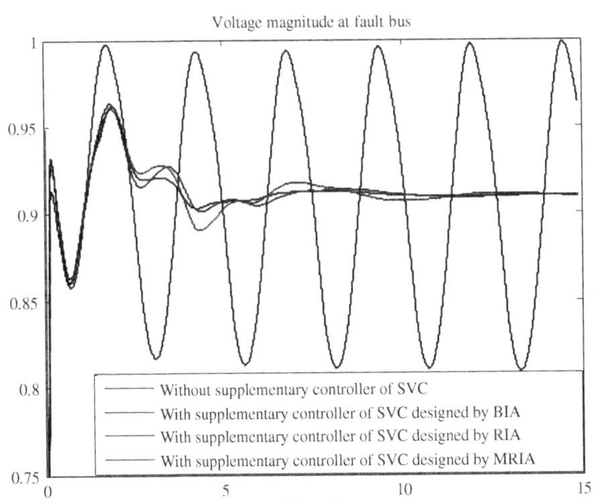

Figure 21.16 The response of the system to a three-phase fault at bus 3.

21.7 SIMULATED ANNEALING AND TABU SEARCH FOR OPTIMAL ALLOCATION OF STATIC VAr COMPENSATORS

TS and SA have been paid attention on different applications in power systems since 1995 and 1990, respectively. Among these applications, a few works have been done on FACTS devices. Bhasaputra and Ongsakul [60] proposed a hybrid TS and SA approach to determine optimal settings of the FACTS parameters for different loading levels, and then find the optimal placement of the multi-type FACTS devices by solving optimal power flow to minimize the total generator fuel cost. Gerbex et al. [61] used three heuristic methods, SA, TS, and GA, to locate FACTS devices in a power system to enhance the power system security. Parallel TS was applied by Mori and Goto [62] for the optimal allocation of UPFCs to maximize the available transfer capability. Mori [63] placed SVC in distribution system using parallel tabu system. TS was used by Kim et al. [64] to tune the input and output gain of a fuzzy logic controller for thyristor-controlled series compensator (TCSC) for a low frequency oscillation in a power system. Similar study was done by Hwang et al. [65].

SA was used by Chen et al. [66] to design output feedback controllers for TCSCs in a power system to damp the critical modes. A two-stage optimization method using an expert system and SA to solve the SVC placement problem is carried out by Jwo et al. [67]. Hsiao et al. [68] developed a computer package based on SA for multiobjective SVC planning of large-scale power systems. In the above methods, the attention has been focused upon IR^2 losses. Relatively little effort has been directly involved with voltage stability improvement. Similar study was carried out by Chang and Huang [69]. SA algorithm was used by Abido [70] for robust PSS design to get optimal settings of the PSS parameters. Also, TS and SA have been applied to several power system problems such as planning, economic dispatch, unit commitment, and network reconfiguration, which can be found in the literature.

21.7.1 Voltage Stability Analysis

Voltage stability is the ability of a power system to maintain acceptable voltages at all busses in the system under normal operation as well as following disturbances. Voltage stability can be categorized to large-disturbance and small-disturbance voltage stabilities. Large-disturbance voltage stability is the ability of the system to control the voltage after being subjected to large disturbances such as system faults and loss of load or generation. Small signal voltage stability is the ability of the system to control voltage after being subjected to small perturbations, such as gradual changes in loads.

Modal or eigenvalue analysis of the system Jacobian matrix near the point of voltage collapse can be used to identify busses vulnerable to voltage collapse [71]. When modal analysis is used, there is no need to drive the system precisely to its "nose point" to ensure that a maximum level of stress is reached. The eigenvector of the critical eigenvalue gives information about the loads responsible for the voltage collapse. The main conclusion from this is that voltage collapse is actually the collapse of a modal voltage. In that case, the power system cannot support a particular combination of reactive power loads.

21.7.2 Simulated Annealing

SA is a derivative-free optimization technique that simulates the physical annealing process in the field of combinatorial optimization. Annealing is the physical process of heating up a solid until it melts, followed by slowly cooling it down by decreasing the temperature of the environment in steps to obtain a perfect structure corresponding to a minimum energy state. The SA is a global search strategy, which tries to avoid local minima by accepting worse solutions with a probability. The probability to accept a solution is defined according to the metropolis distribution:

$$p_i = \begin{cases} 1 & \text{if } f(y) < f(x) \\ \exp\left(\dfrac{f(x) - f(y)}{T}\right) & \text{otherwise} \end{cases} \quad (21.52)$$

where x and y are an initial and new solutions, respectively, and $f(.)$ is the evaluation of the objective function at a solution.

The SA starts from an initial solution x and then a solution y is generated. If y has been improved from x, it will be accepted; otherwise, y will be accepted as the current configuration with a probability proportional to the difference in the value of the objective function, $f(x)$-$f(y)$, on temperature T as shown in (21.52).

21.7.3 Tabu Search

TS algorithm is used to solve COP. This mechanism is somewhat similar to SA, but the main difference between the two search algorithms is that TS uses a flexible memory to store the information and data of the solutions in each iteration. This is to get to the lower objective function values with the help of the information stored, while special features are added to escape from being trapped in the local minima.

The TS searches the neighborhood of the current solution to find the next solution with more improvement in the value of objective function. In order to avoid returning to the local optimum, TS uses a list which is called *Tabu List*. The elements of the tabu list are called tabu moves. Tabu list stores the moves in a data structure such as finite length and restricts the local search algorithm in reusing those moves for some iteration. The number of iterations a move is kept in the list is called *tabu list size*. Since the tabu list may forbid certain worthy or interesting moves found so far, these tabu moves will be accepted as the next move and will be released from the tabu list if they are judged to be worthy. This override of the tabu list is called *aspiration criterion*.

The TS algorithm used in this chapter follows a rather straight forward pattern. As soon as a trial solution is generated, it is checked to see if it is in the tabu list or not. If so, the search algorithm will then search for the frequency counter related to this trial solution. If the frequency counter is smaller than a threshold, it will add up the frequency counter by one, and starts to search for the next move from the neighborhood of the current solution. This mechanism is called *intensification*. And if the frequency counter was more than the threshold, the move will be penalized for some iteration and the search will therefore be directed to unexplored regions of the solution space so far to diversify the search solution. If there is a better solution in the outer spaces of the local search, the search space changes its place to that new region. It should be mentioned that a frequency counter denotes the number of times the solution having been visited throughout the solution process.

21.7.4 Study System and Optimal Allocation of SVCs

21.7.4.1 A 5-Area-16-Machine System

The system shown in Figure 21.17 consists of 16 machines and 68 busses.

This is a reduced order model of the New England (NE) New York (NY) interconnected system. The first nine machines are the simple representation of the NE system generation. Machines 10–13 represent the NY power system. The last three machines are the dynamic equivalents of the three large neighboring areas interconnected to the NY power system.

21.7.4.2 Optimal Allocation of SVCs

Modal analysis [71], SA, and TS are used to locate SVCs optimally in the power system shown in Figure 21.17. Starting from an initial load, the SA is applied. All loads are increased gradually near to the point of collapse. To locate SVC by SA,

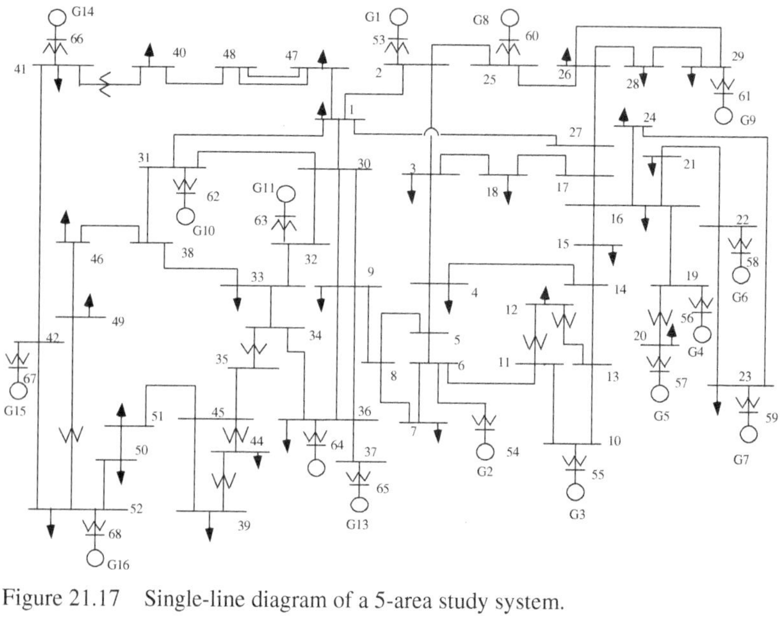

Figure 21.17 Single-line diagram of a 5-area study system.

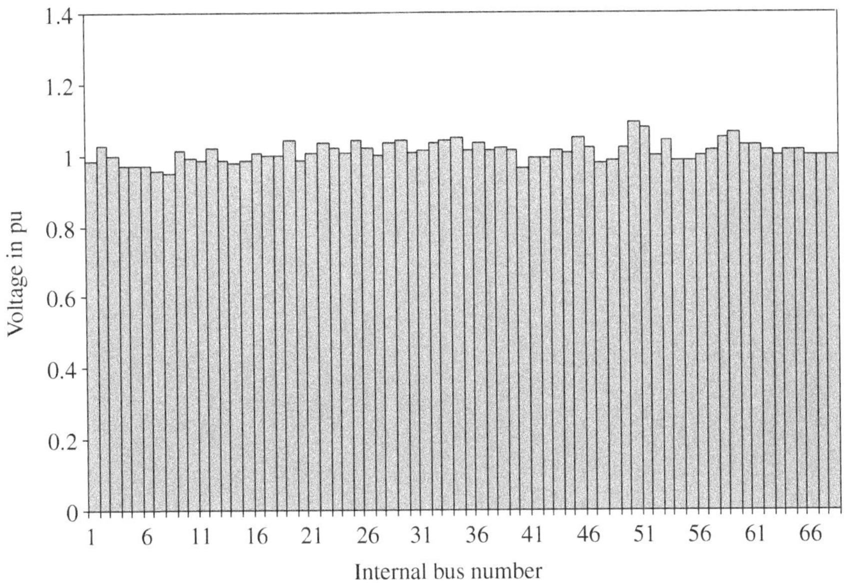

Figure 21.18 Bus voltage magnitude profile of stressed system after placing SVC at bus 48 with the size of 156 MVAr.

suitable busses are selected based on 30 test runs. The obtained results by SA are as follows: 27% of results show that the first SVC should be placed at bus 48 with 156 MVAr (the best-so-far), 13% of results show that a 186 MVAr SVC should be placed at bus 47, and 7% of results show that the first SVC should be placed at bus 40 with the size of 145 MVAr. Bus voltage magnitude profile of stressed system after placing svc at bus 48 is shown in Figure 21.18.

Using modal analysis, it is found that the weakest area in this power system is in the NY system near bus 40, as illustrated in Figure 21.19. Figure 21.20 shows the profile of the voltage when system is heavily stressed and is reached to the point of collapse. Based on modal analysis, bus 40 is a good candidate to place the first SVC. Using the modal analysis, only the weak area can be determined. However, it cannot give any information regarding the size of SVC. Since the size is playing an important role for the SVC to be effective, SA is used to find the MVAr size needed for SVC. To find the optimal compensation level, first, the SVC is placed in the power system at bus 40, and then using SA, the MVAr size of SVC is obtained, which is 145. The obtained level of compensation is the same as the 7% of the results obtained by SA when bus 40 was selected.

After placing the first SVC at bus 40, once again the loads are gradually increased. The best recovery of the voltage profiles is when SVC is located at bus 40. On the other hand, the obtained MVAr size of the SVC when located at bus 40 is less. It means that SA is a good tool to get promising suboptimal results. Therefore, since the SA failed to identify bus 40, which is vulnerable to voltage collapse, only the modal analysis is used to place the second SVC.

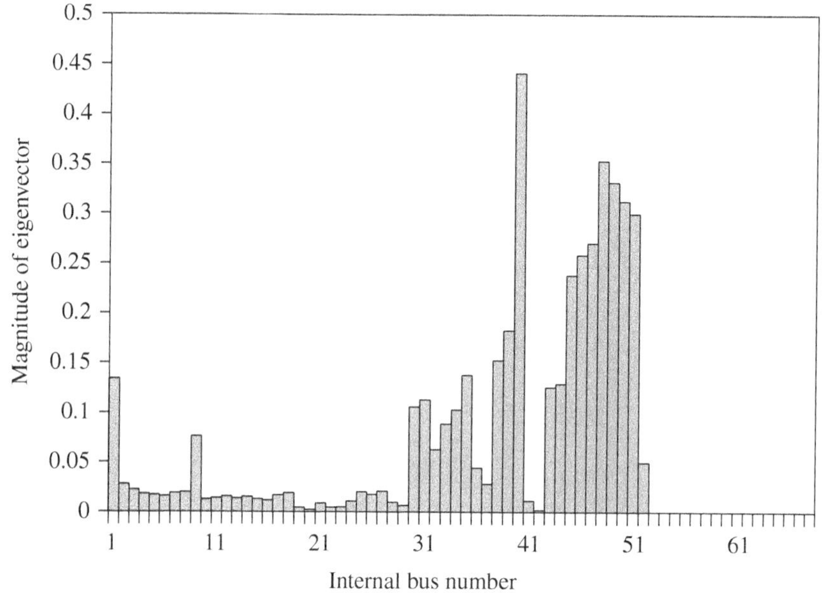

Figure 21.19 The critical eigenvector and the corresponding bus number.

Then, SA is used to find the level of compensation for the SVC. Using the modal analysis, as shown in Figure 21.21, the second worst bus is identified which is bus 50. Figure 21.22 shows the voltage profile when system is under stress. Again SA is used to find the optimal size of the SVC at bus 50. The obtained MVAr size is 155 MVAr.

Now, TS is applied to find the best place of SVC in the power system shown in Figure 21.17. Once again, all loads will increase gradually near to the point of collapse. As in the case of SA, to locate SVC by TS, suitable busses are selected based on 30 test runs. The obtained results by TS are as follows: 86.66% of results show that the first SVC should be placed at bus 40 with 138 MVAr, 6.66% of results

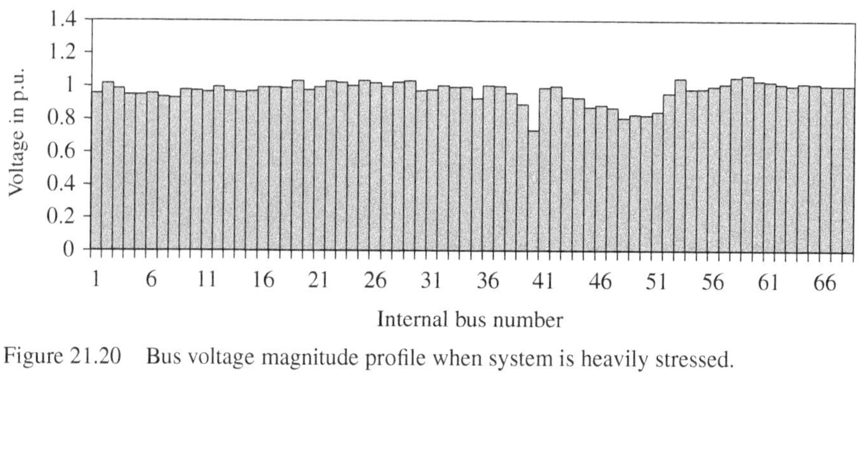

Figure 21.20 Bus voltage magnitude profile when system is heavily stressed.

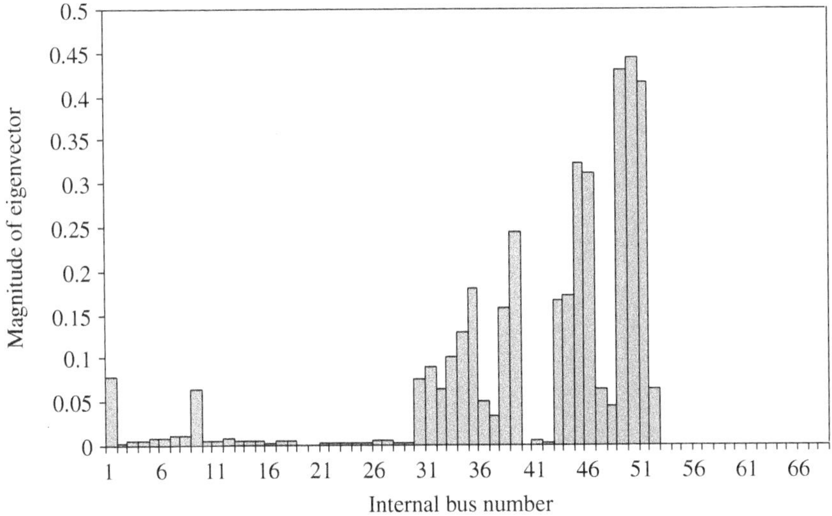

Figure 21.21 The critical eigenvector and the corresponding bus number.

show that a 136.4 MVAr SVC should be placed at bus 48, and 6.66% of results show other busses to place SVC.

The results show that TS has the ability of finding the optimal location better than SA. The fitness and the level of compensation are shown in Figures 21.23 and 21.24. Also, by applying modal analysis and then placing SVC at bus 40, TS finds the level of compensation to be 138 MVAr [72].

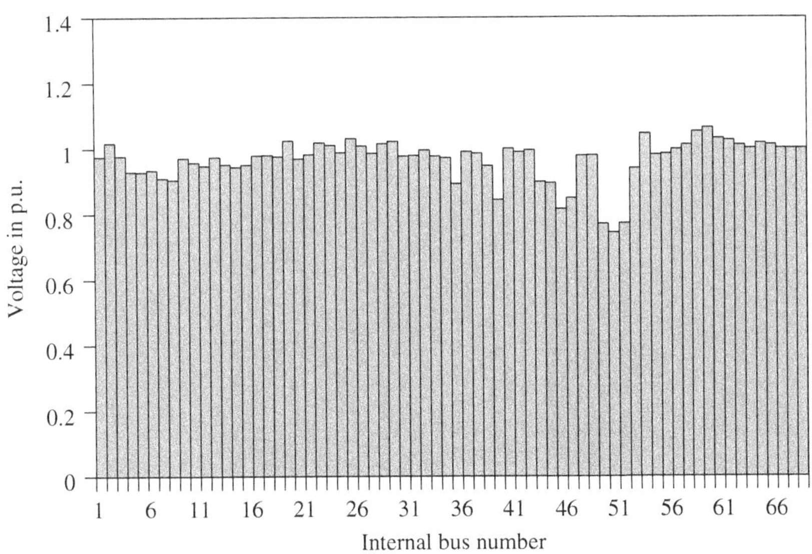

Figure 21.22 Bus voltage magnitude profile when system is heavily stressed.

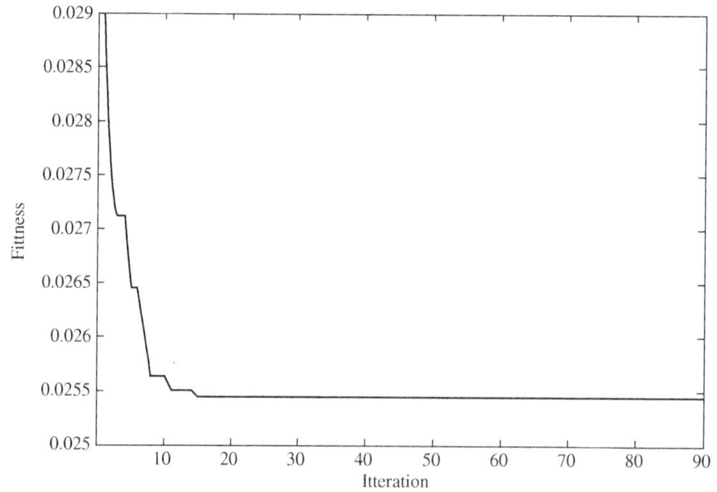

Figure 21.23 Convergence characteristics of TS in finding the solution, 145 MVAr svc at bus 40.

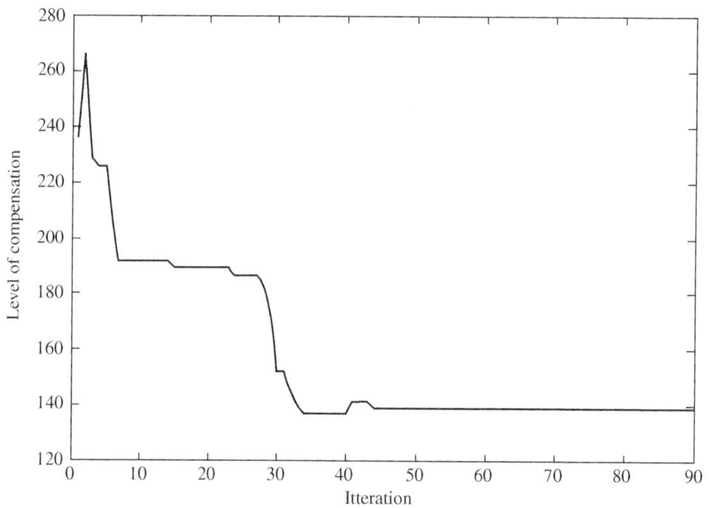

Figure 21.24 Search process of TS for finding the level of compensation.

21.8 CONCLUSIONS

This chapter provides with basic knowledge of recent intelligent optimization and control techniques, and how they are combined with knowledge elements in computational intelligence systems. The chapter is devoted to the application of modern heuristic optimization techniques to power systems, and example applications are presented. The chapter is composed of six optimization techniques applied power systems: EAs, GA, PSO, ant colony search algorithm, IA, SA, and TS. It has been shown that these heuristic techniques can solve very complex large-scale nonlinear optimization problems, which cannot be handled by any analytic approaches.

REFERENCES

[1] Lee, K. Y., and El-Sharkawi, M. A. *Modern Heuristic Optimization Techniques: Theory and Applications to Power Systems*. In: K. Y. Lee and M. A. El-Sharkawi, Eds., IEEE Press Book, Wiley, New York, 2008, ISBN: 978-0-471-45711-4.

[2] Song, Y. H., Ed., *Modern Optimisation Techniques in Power Systems*. Kluwer Academic Publishers, 1999.

[3] Irving, M. R., and Song, Y. H. Optimisation techniques for electrical power systems. *IEEE Power Engineering Journal*, vol. 14, no. 5, pp. 245–254, 2000.

[4] Song, Y. H., and Irving, M. R. Optimisation techniques for electrical power systems. Part 2 heuristic optimisation techniques. *IEEE Power Engineering Journal*, vol. 15, no. 3, pp. 151–160, 2001.

[5] Lee, K. Y., and El-Sharkawi, M. A. Tutorial on modern heuristic optimization techniques with applications to power systems. *IEEE Power Engineering Society*, Publication 02TP160, 2002.

[6] Lee, K. Y., and El-Sharkawi, M. A. A tutorial course on evolutionary computation techniques for power system optimization. *IFAC Symposium on Power Plants and Power Systems Control*, Seoul, Korea, September 15–18, 2003.

[7] Lee, K. Y. *Tutorial on Intelligent Optimization and Control for Power Systems*. ISAP, Arlington, 2005.

[8] Lai, L. L., and Ma, J. T. Application of evolutionary programming to reactive power planning-comparison with nonlinear programming approach. *IEEE/PES Winter Meeting*, #96 WM 248-5 PWRS, January 1996.

[9] Fogal, D. B. An introduction to simulated evolutionary optimization. *IEEE Transactions on Neural Network*, vol. 5, no. 1, pp. 3–14, January 1994.

[10] Park, J. H., Yang, S. O., Lee, H. S., and Park, Y. M. Economic load dispatch using evolutionary algorithm. *Proceedings of International Conference on Intelligent Systems Applications to Power Systems*, Orlando, FL, pp. 441–445, January 1996.

[11] Goldberg, D. E. *Genetic Algorithms in Search, Optimization, and Machine Learning*. Addison-Wesley, 1989.

[12] Lee, K. Y., Bai, X., and Park, Y. M. Optimization method for reactive power planning by using a modified simple genetic algorithm. *IEEE Transactions on Power Systems*, vol. 10, no. 4, pp. 1843–1850, November 1995.

[13] Dimeo, R., and Lee, K. Y. Boiler-turbine control system design using a genetic algorithm. *IEEE Transactions on Energy Conversion*, vol. 10, pp. 752–759, December 1995.

[14] Miranda, V., Ranito, J. V., and Proenica, L. M. Genetic algorithm in optimal multistage distribution network planning. *IEEE PES Winter Meeting*, #94 WM 229-5-PWRS.

[15] Iba, K. Reactive power optimization by genetic algorithm. *IEEE Transactions on Power Systems*, vol. 9, no. 2, pp. 685–692, May 1994.

[16] Lee, K. Y., Ortiz, J. L., Park, Y. M., and Pond, L. G. An optimization technique for reactive power planning of subtransmission network under normal operation. *IEEE Transactions on Power Systems*, vol. PWRS-1, pp. 153–159, May 1986.

[17] Mangoli, M. K., Lee, K. Y., and Park, Y. M. Operational real and reactive power control using linear programming. *Electric Power Systems Research*, vol. 26, pp. 1–10, 1993.

[18] Mangoli, M. K., Lee, K. Y., and Park, Y. M. Optimal long-term reactive power planning using decomposition techniques. *Electric Power System Research*, vol. 26, pp. 41–52, 1993.

[19] Lee, K. Y., Park, Y. M., and Oritz, J. L. A united approach to optimal real and reactive power dispatch. *IEEE Transactions on Power Apparatus and Systems*, vol. PAS-104, pp. 1147–1153, May 1985.

[20] Lee, K. Y., and Yang, F. F. Optimal reactive power planning using evolutionary algorithms: a comparative study for evolutionary programming, evolutionary strategy, genetic algorithm, and linear programming. *IEEE Transactions on Power Systems*, vol. 13, no. 1, pp. 101–108, February 1998.

[21] Walters, D. C., and Sheble, G. B. Genetic algorithm solution of economic dispatch with valve point loading. *IEEE Transactions on PWRS*, vol. 8, no. 3, pp. 1325–1332, 1993.

[22] Chen, P. H., and Chang, H. C. Large-scale economic dispatch by genetic algorithm. *IEEE Transactions on PWRS*, vol. 10, no. 4, pp. 1919–1926, 1995.

[23] Dasgupta, D., and McGregor, D. R. Thermal unit commitment using genetic algorithms. *IEE Proceedings – Generation, Transmission and Distribution*, vol. 141, no. 5, pp.459–465, 1994.

[24] Sheble, G. B., Maifeld, T. T., Brittig, K., and Fahd, G. Unit commitment by genetic algorithm with penalty methods and a comparison of Lagrangian search and genetic algorithm-economic dispatch algorithm. *International Journal of Electric Power & Energy Systems*, vol. 18, no. 6, pp. 339–346, 1996.

[25] Iba, K. Reactive power optimization by genetic algorithm. *IEEE Transactions on PWRS*, vol. 9, no. 2, pp. 685–692, 1994.

[26] Zhao, Y., Edwards, R. M., and Lee, K. Y. Hybrid feedforward and feedback controller design for nuclear steam generators over wide range operation using genetic algorithm. *IEEE Transactions on Energy Conversion*, vol. 12, no. 1, pp. 100–106, 1997.

[27] Park, J. B., Park, Y. M., Won, J. R., and Lee, K. Y. An improved genetic algorithm for generation expansion planning. *IEEE Transactions on Power Systems*, vol. 15, no. 3, pp. 916–922, August 2000.

[28] Park, Y. M., Lee, K. Y., and Youn, L. T. O. New analytical approach for long-term generation expansion planning based maximum principle and Gaussian distribution function. *IEEE Transactions on PAS*, vol. 104, pp. 390–397, 1985.

[29] Fukuyama, Y., and Chiang, H. A parallel genetic algorithm for generation expansion planning. *IEEE Transactions on PWRS*, vol. 11, no. 2, pp. 955–961, 1996.

[30] Park, Y. M., Park, J. B., and Won, J. R. A genetic algorithms approach for generation expansion planning optimization. *Proceedings of the IFAC Symposium on Power Systems and Power Plant Control*, Pergamon, UK, pp. 257–262, 1996.

[31] Jenkins, S. T., and Joy, D. S. *An Electric Utility Optimal Generation Expansion Planning Computer Code*. Wien Automatic System Planning Package (WASP), Oak Ridge National Laboratory, Oak Ridge, Tennessee, ORNL-4945, 1974.

[32] Electric Power Research Institute (EPRI). *Electric Generation Expansion Analysis System (*EGEAS*)*. EPRI EL-2561, Palo Alto, CA, 1982.

[33] Kennedy, J., and Eberhart, R. Particle swarm optimization. *Proceedings of IEEE International Conference on Neural Networks (ICNN'95)*, Perth, Australia, vol. 4, pp. 1942–1948, 1995.

[34] Yoshida, H., Kawata, K., Fukuyama, Y., Takayama, S., and Nakanishi, Y. A particle swarm optimization for reactive power and voltage control considering voltage security assessment. *IEEE Transactions on Power Systems*, vol. 15, no. 4, pp. 1232–1239, November 2000.

[35] Clerc, M., and Kennedy, J. The particle swarm-explosion, stability, and convergence in a multidimensional complex space. *IEEE Transactions on Evolutionary Computation*, vol. 6, no. 1, pp. 58–73, February 2002.

[36] Abido, M. A. Optimal design of power-system stabilizers using particle swarm optimization. *IEEE Transactions on Energy Conversion*, vol. 17, no. 3, pp. 406–413, September 2002.

[37] Kassabalidis, I. N., El-Sharkawi, M. A., Marks, R. J., Moulin, L. S., and Alves da Silva, A. P. Dynamic security border identification using enhanced particle swarm optimization. *IEEE Transactions on Power Systems*, vol. 17, no. 3, pp. 723–729, August 2002.

[38] Wood, A. J., and Wollenberg, B. F. *Power Generation, Operation, and Control*. John Wiley & Sons, Inc. 1984.

[39] Lin, C. E., and Viviani, G. L. Hierarchical economic dispatch for piecewise quadratic cost functions. *IEEE Transactions on Power Apparatus and Systems*, vol. PAS-103, no. 6, June 1984.

[40] Lin, W. M., Cheng, F. S., and Tsay, M. T. An improved tabu search for economic dispatch with multiple minima. *IEEE Transactions on Power Systems*, vol. 17, no. 1, pp. 108–112, February 2002.

[41] Yang, H. T., Yang, P. C., and Huang, C. L. Evolutionary programming based economic dispatch for units with non-smooth fuel cost functions. *IEEE Transactions on Power Systems*, vol. 11, no. 1, pp. 112–118, February 1996.

[42] Park, Y. M., Won, J. R., and Park, J. B. A new approach to economic load dispatch based on improved evolutionary programming. *Engineering Intelligent Systems for Electrical Engineering and Communications*, vol. 6, no. 2, pp. 103–110, June 1998.

[43] Walters, D. C., and Sheble, G. B. Genetic algorithm solution of economic dispatch with the valve point loading. *IEEE Transactions on Power Systems*, vol. 8, no. 3, pp. 1325–1332, August 1993.

[44] Ciuprina, G., Ioan, D., and Munteanu, I. Use of intelligent-particle swarm optimization in electromagnetics. *IEEE Transactions on Magnetics*, vol. 38, no. 2, pp. 1037–1040, March 2002.

[45] Park, J. H., Kim, Y. S., Eom, I. K., and Lee, K. Y. Economic load dispatch for piecewise quadratic cost function using Hopfield neural network. *IEEE Transactions on Power Systems*, vol. 8, no. 3, pp. 1030–1038, August 1993.

[46] Lee, K. Y., Sode-Yome, A., and Park, J. H. Adaptive Hopfield neural network for economic load dispatch. *IEEE Transactions on Power Systems*, vol. 13, no. 2, pp. 519–526, May 1998.

[47] Park, J. B., Lee, K. S., Shin, J. R., and Lee, K. Y. A particle swarm optimization for economic dispatch with non-smooth cost functions. *IEEE Transactions on Power Systems*, vol. 20, no. 1, pp. 34–42, February 2005.

[48] Dorigo, M. Optimization, learning and natural algorithms. Ph.D. Thesis, Politecnico de Milano, Italy, 1992.

[49] Dorigo, M., Maniezzo, V., and Colorni, A. Ant system: optimization by a colony of cooperating agents. *IEEE Transactions Systematics, Man, Cybernetics B*, vol. 26, no. 1, pp. 29–41, February 1996.

[50] Stützle, T., and Hoos, H. H. Improvements on the ant system: introducing the MAX-MIN Ant system. In: *Artificial Neural Networks and Genetic Algorithms*, R. F. Albrecht, G. D. Smith, and N. C. Steele, Eds., Springer Verlag, Wien, New York, pp. 245–249, 1998.

[51] Vlachogiannis, J. G., and Hatziargyriou, N. D. Reinforcement learning for reactive power control. *IEEE Transactions Power Systems*, vol. 19, no. 3, pp. 1317–1325, August 2004.

[52] Karakatsanis, T. S., and Hatziargyriou, N. D. Probabilistic constrained load flow based on sensitivity analysis. *IEEE Transactions Power Systems*, vol. 9, no. 4, pp. 1853–1860, November 1994.

[53] Allan, R. N., and Al-Shakarchi, M. R. G. Probabilistic techniques in a.c. load-flow analysis. *IEE Proceedings – Generation, Transmission and Distribution*, vol. 124, no. 2, pp. 154–160, February 1977.

[54] Kenney, J. F., and Keeping, E. S. *Mathematics in Statistics*. Nostrand, Princeton, NJ, 1962.

[55] Vlachogiannis, J. G., Hatziargyriou, N. D., and Lee, K. Y. Ant colony system based algorithm for constrained load flow problem. *IEEE Transactions on Power Systems*, vol. 20, no. 3, pp. 1241–1249, August 2005.

[56] Musilek, P., Lau, A., Reformat, M., and Wyard-Scott, L. Immune programming. *Information Sciences*, vol. 176, pp. 972–1002, 2006.

[57] Corn, D., Dorigo, M., and Glover, F. *New Ideas in Optimization*. MaGraw-Hill, 1999.

[58] Farsangi, M. M., Nezamabadi-Pour, H., Song, Y. H., and Lee, K. Y. Placement of SVCs and selection of stabilizing signals in power systems. *IEEE Transactions Power Systems*, vol. 22, no. 3, pp. 1061–1071, August 2007.

[59] Farsangi, M., Kyanzadeh, S., Nezamabadi-Pour, H., and Lee, K. Y. Design of a supplementary controller for SVC using hybrid real immune algorithm and local search. *Proceedings of the IEEE Power Engineering Society General Meeting*, Pittsburgh, PA, July 20–24, 2008.

[60] Bhasaputra, P., and Ongsakul, W. Optimal placement of multi-type FACTS devices by hybrid TS/SA approach. *Proceedings of the 2003 International Symposium on Circuits and Systems*, vol. 3, pp. 285–290, 2003.

[61] Gerbex, S., Cherkaoui, R., and Germond, A. J. Optimal location of FACTS devices to enhance power system security. *Power Tech Conference Proceedings*, vol. 3, 2003.

[62] Mori, H., and Goto, Y. A parallel tabu search based method for determining optimal allocation of FACTS in power systems. *International Conference on Power System Technology*, vol. 2, pp. 1077–1082, 2000.

[63] Mori, H. Optimal allocation of FACTS devices in distribution systems. *IEEE Power Engineering Society Winter Meeting*, vol. 2, pp. 936–937, 2001.

[64] Kim, W. G., Hwang, G. H., Kang, H. T., Lee, S. O., and Park, J. H. Design of fuzzy logic controller for firing angle of TCSC using real-type tabu search. *IEEE International Symposium on Industrial Electronics*, vol. 1, pp. 575–580, June 12–16, 2001.

[65] Hwang, G. H., Park, J. H., Mun, K. J., and Kang, H. T. Design of fuzzy logic controller for firing angle of TCSC using real-type tabu search. *The 30th Annual Conference of IEEE Industrial Electronics Society*, vol. 3, pp. 2182–2187, November 2–6, 2004.

[66] Chen, X. R., Pahalawaththa, N. C., Annakkage, U. D., and Kumble, S. C. Design of decentralized output feedback TCSC damping controllers by using simulated annealing. *IEE Proceedings – Generation, Transmission and Distribution*, vol. 145, no. 5, pp. 553–558, 1998.

[67] Jwo, W. S., Liu, C. C., and Hsiao, Y. T. Hybrid expert system and simulated annealing approach to optimal reactive planning. *IEE Proceedings – Generation Transmission and Distribution*, vol. 142, no. 4, pp. 381–385, 1999.

[68] Hsiao, Y. T., Chiang, H. D., Liu, C. C., and Chen, Y. L. A computer package for optimal multi-objective VAR planning in large scale power systems. *IEEE Transactions on Power Systems*, vol. 9, no. 2, pp. 668–676, 1994.

[69] Chang, C. S., and Huang, J. S. Optimal multiobjective SVC planning for voltage stability enhancement. *IEE Proceedings – Generation, Transmission and Distribution*, vol. 145, no. 2, 1998.

[70] Abido, M. A. Robust design of multimachine power system stabilizer using simulated annealing. *IEEE Transactions Power Systems*, vol. 15, no. 3, pp. 297–3043, 2000.

[71] Farsangi, M. M., Nezamabadi-Pour, H., Song, Y. H., and Lee, K. Y. Placement of SVCs and selection of stabilizing signals for in power systems. *IEEE Transactions Power Systems*, vol. 22, no. 3, pp. 1061–1071, August 2007.

[72] Ebrahimi, S., Farsangi, M. M., Nezamabadi-Pour, H., and Lee, K. Y. Optimal allocation of static VAr compensators using modal analysis. Simulated annealing and tabu search. *Proceedings of the IFAC Symposium on Power Plants and Power Systems Control*, Kananaskis (near Calgary), Canada, June 25–28, 2006.

[73] del Valle, Y., Venayagamoorthy, G. K., Mohagheghi, S., Hernandez, J.-C., and Harley, R. G. Particle swarm optimization: basic concepts, variants and applications in power systems. *IEEE Transactions Evolutionary Computation*, vol. 12, no. 2, pp. 171–195, April 2008.

CHAPTER 22

UNSUPERVISED LEARNING AND HYBRID METHODS

Nikos Hatziargyriou and Manolis Voumvoulakis

22.1 GENERALITIES

This chapter deals with the applications of machine-learning techniques in power system dynamic security control with emphasis on unsupervised learning methods and their combination with supervised learning methods. At the beginning, an overview of these methods, which are suitable for power system security control, is provided.

Security is defined as the capability of guaranteeing the continuous operation of a power system under normal operation even following some significant perturbations [1]. Dynamic security assessment (DSA) can deal with transient stability problems and/or voltage stability problems that, respectively, require transient stability assessment and voltage stability assessment [2, 3]. Modern power systems are forced to operate close to their stability limits due to economical and environmental constraints. Under these operating practices, fast and reliable assessment of the system security is necessary in order to avoid a major blackout in the event of a disturbance. Modern energy management systems perform online security analysis to provide the operator valuable information regarding the security status of the system. The main functionalities the control center operators would require, regarding the dynamic security, are [4]:

- evaluation of security status of the current operating point (OP);
- stability margins;
- preventive and/or corrective controls;
- optimum load shedding.

Analytical methods assess the security of a power system, by performing numerical simulation of a contingency in a given state. However, the nonlinear nature of the physical phenomena and the growing complexity of modern power systems make online security assessment a very challenging task. In this aspect, intelligent systems (IS) are seen to have features that can bring benefits in comparison to analytical methods [5]. Although distributed computation is now common place, full

Advanced Solutions in Power Systems: HVDC, FACTS, and Artificial Intelligence, First Edition.
Edited by Mircea Eremia, Chen-Ching Liu, and Abdel-Aty Edris.

simulation methods require some minutes to reach a conclusion. For large power systems in which many contingencies must be assessed, even with multiple Central Processing Unit (CPU) computing, this time may be of concern for online analysis. Once developed, IS technologies such as neural networks (NN) and decision trees (DT) can provide solutions very fast. Furthermore, IS are learning systems, that is, they have the ability to recognize, if a system condition has been seen previously, and predict its security status accordingly. Similarly, if properly designed, an IS can adapt to new conditions by learning from situations previously seen. Finally, IS can provide a high degree of discovery, that is, they have the ability to uncover salient, but previously unknown, characteristics of, or relationships in, a system.

This section gives the general machine-learning framework for DSA. Figure 22.1 shows the framework of the proposed methodology. Three main tasks are distinguished in the procedure:

- knowledge base (KB) generation;
- IS construction;
- online implementation.

KNOWLEDGE BASE GENERATION. For the training of a machine-learning method, a KB is necessary. KB comprises a large number of OPs, each of which is characterized by a vector of predisturbance steady-state variables, called features or attributes, which can be either directly measured (power, voltages, etc.) or indirectly calculated quantities (wind penetration, spinning reserve, etc.). KB is generated offline either by means of random sampling or on the basis of historical data. A set of predefined disturbances $D = \{D_1, \dots, D_n\}$ is simulated for each OP with a time-domain simulation machine, and OPs are classified into secure and insecure, on the basis of a criterion related to the postdisturbance values of system variables. The security criterion takes the form of (22.1):

$$\begin{cases} g(y_t) \geq 0, & OP \text{ is Secure} \\ g(y_t) < 0, & OP \text{ is Insecure} \end{cases} \tag{22.1}$$

where y_t is a vector of postdisturbance values of system variables. The user selects the appropriate security criterion g on the basis of engineering judgment and preliminary system analysis. This criterion may be related to any type of system instability, such as frequency, rotor angle, or voltage instability. The KB is then split into a learning set (LS), used to train the machine-learning algorithms, and a test set (TS) used for evaluating the performance of the developed structures.

MODEL CREATION. The IS is used to extract and synthesize relevant information and to reformulate it in a suitable way for decision making, and it is constructed and trained using the LS. The selection of the suitable machine-learning model is essential for the efficiency of the method. Feature selection techniques, that is, algorithms that output a subset of the input feature set, are applied in order to reduce the dimensionality of the KB. The selected features will characterize properly a variety of power system operating conditions. The noise of the training patterns is reduced by means of data preprocessing such as normalization and scaling of the data. The

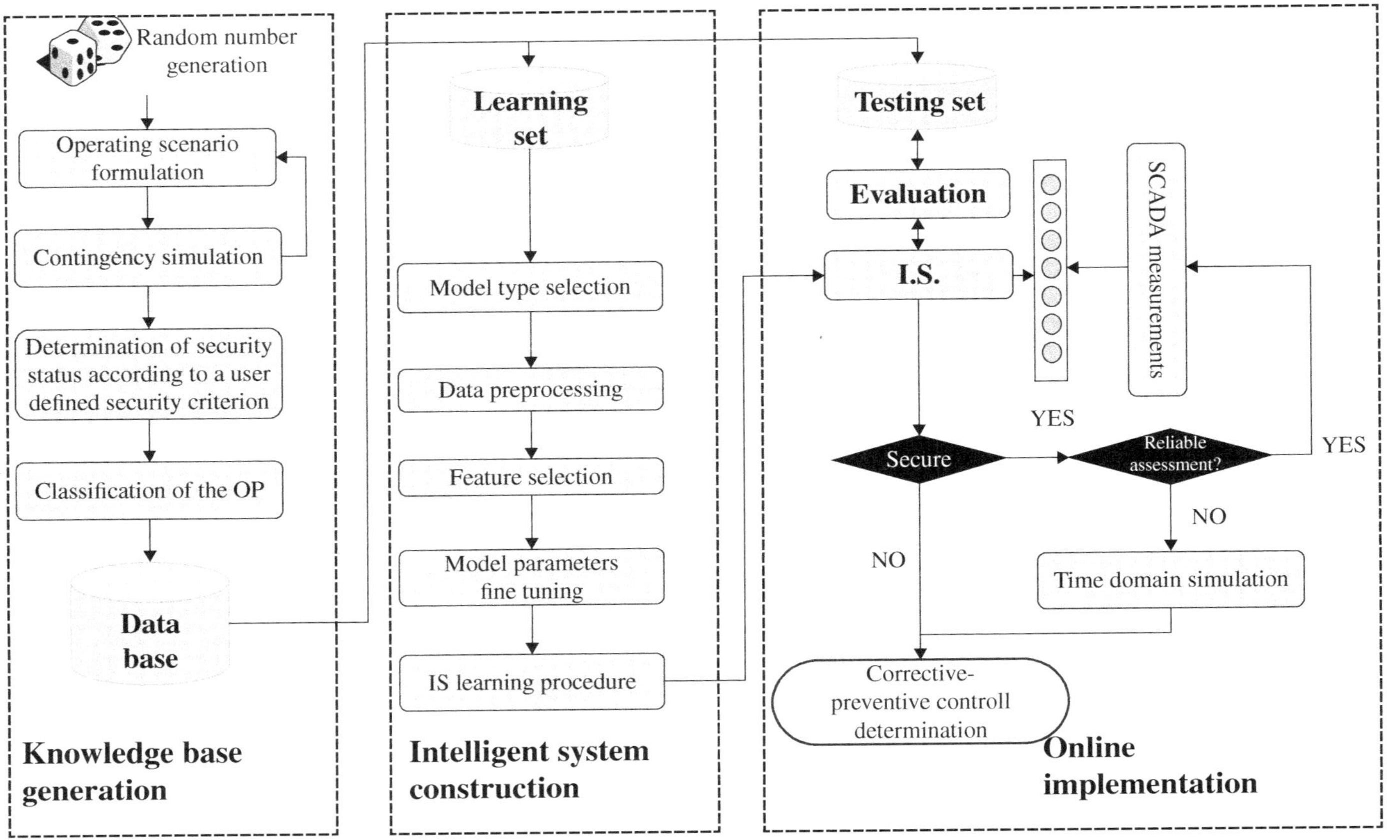

Figure 22.1 Automatic learning framework for dynamic security control.

parameters of the model, such as the number of neurons of an NN, are also tuned, in order to provide an optimized IS.

ONLINE IMPLEMENTATION. During the online implementation of the method, the IS takes as inputs the current OP, in order to estimate the security status of the system. If the IS cannot make a reliable estimate of the security status of the OP, then analytical methods are employed. In case of an insecure OP, preventive or corrective control actions are required. The aim of preventive control is to prevent instability before it actually occurs, whereas the corrective control is to stabilize an unstable power system, shortly after a severe contingency [6].

The heart of the framework illustrated in Figure 22.1 is provided by the IS. A large variety of machine-learning methods may be used in a toolbox fashion, according to the type of information they may exploit and/or produce. Machine-learning methods are organized into three main classifications: (i) supervised learning methods, (ii) unsupervised learning methods, and (iii) reinforcement learning methods.

22.2 SUPERVISED LEARNING METHODS

In supervised learning methods, the training data consist of pairs of input objects and target outputs. The task of the supervised learner is to predict the value of the output for any valid input object. The output values can be either continuous (in which case, the learner performs regression), or class-labeled (classification). Three methods of supervised learning are employed in this chapter: DTs and neuro-fuzzy decision trees (N-FDTs), which perform classification, and radial basis function neural networks (RBFNNs), which can perform both regression and classification.

22.2.1 Decision Trees

The DT is a tree, structured upside down, built on the basis of a KB consisting of a large number of OPs, covering all possible states of the under study power system in order to ensure its representativity. The construction of a DT starts at the root node with the whole LS of two-class OPs (e.g., secure and insecure OPs). At each step, a tip-node of the growing tree is considered and the algorithm decides whether it will be a terminal node or should be further developed. To develop a node, an appropriate attribute is first identified, together with a dichotomy test on its values. The test T is defined as:

$$T : x_j \geq w_* \tag{22.2}$$

The selection of the optimal test is based on maximizing the additional information gained through the test. The information gain is based on the entropy of each subset E_n with respect to the class partition of its elements, defined as

$$H_c(E_n) = -(f_s \log f_s + f_i \log f_i) \tag{22.3}$$

where f_s and f_i are the relative frequencies of the secure and insecure OPs, respectively, in the subset. The selected test is applied to the LS of the node splitting it

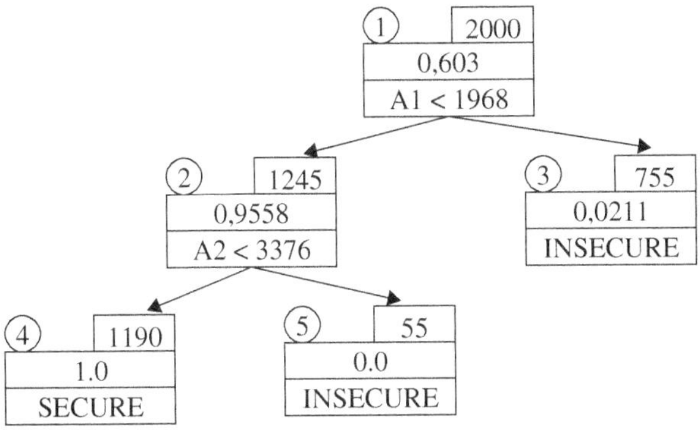

Figure 22.2 A sample decision tree.

into two exclusive subsets, corresponding to the two successor nodes. Every subset (node) is characterized by its security index (SI), defined as the percentage of secure OPs belonging to this subset. The optimal splitting rule is applied recursively to build the corresponding subtrees. In order to detect if one node is terminal, that is, "sufficiently" class pure, the stop splitting rule is used, which checks whether the entropy of the node is lower than a preset minimum value. If it is, the node corresponds to a sufficiently pure subset (states belong to the same class) and is declared a leaf; otherwise, a test T is sought to further split the node. If the node cannot be further split in a statistically significant way, it is termed a dead-end, carrying the two class probabilities estimated on the basis of the corresponding OPs subset. Figure 22.2 shows a sample DT: To infer the output information corresponding to given input attribute values, one traverses the tree, starting at the top-node, and applying sequentially the dichotomous tests encountered to select the appropriate successor. When a terminal node is reached, the output information stored is retrieved.

The nodes of the DT have the following scheme: There are four attributes. In the upper right side is the number of OPs, which belong to the node. In the upper left side is the label of the node. In the middle is the SI, which is the number of OPs belonging to one of the two classes divided to the total OPs of the node. Finally, in the bottom is the dichotomy test of the node or the characterization of the node as dead-end or leaf. The dichotomy test is checked and if it is true the left node is accessed otherwise the right one is selected.

Each leaf of a DT corresponds to a chain of rules, each rule having the form of an inequality (partition test). For example, leaf 4 of the sample DT illustrated in Figure 22.2 corresponds to the chain of rules

IF (ATRR1<1968) AND (ATTR2<3376) THEN LEAF 4.

DTs are evaluated using the testing set (TS). The most important evaluator of the DT reliability and performance is the rate of successful classifications, defined as the ratio of successfully classified OPs to the number of OPs tested.

$$Success\ Rate = \frac{OPs\ Successfully\ classified}{Total\ Number\ of\ OPs}$$

For a two-class partition (secure-insecure), there can be distinguished two types of errors, depending on the actual class of the misclassified OP, false alarms (FAs) and missed alarms (MAs):

$$FA = \frac{OPs\ missclassified\ as\ Insecure}{Total\ Number\ of\ Secure\ OPs}$$

$$MA = \frac{OPs\ missclassified\ as\ Secure}{Total\ Number\ of\ Insecure\ OPs}$$

In the above case, only one input dimension was used at the dichotomy test (univariate tree). An alternative to this is a *multivariate tree,* where at a decision node, all input dimensions can be used and thus it is more general. When all inputs are numeric, a binary linear multivariate node is defined as:

$$T : w_m^T x + w_{m0} \geq 0 \tag{22.4}$$

In the case of a multivariate DT, exhaustive search is no longer possible. Meta-heuristic methods can be used in order to fine tune the weights w_m. A special case is when weights are binary numbers. In that case, the algorithm searches for the optimum aggregation between the candidate attributes, which provide the maximum information gain when it is used to split the node of the tree. In that case, genetic algorithms can be used to optimize the weights w_m.

22.2.2 Neuro-Fuzzy Decision Trees

N-FDTs combine fuzzy logic, DTs, and neural adaptive algorithm methodologies in order to classify given patterns [7]. Once the tree is structured and trained, it can automatically classify new patterns. A pattern is described by a vector consisting of variables-attributes.

A collection of learning patterns is described by input attribute vectors $x^i = \left[x_1^i\ x_2^i \cdots x_p^i\right]$, $(i = 1, \ldots, n)$. Each attribute x_j has been fuzzified into c_j fuzzy sets $F_{jk}, k = 1, \ldots, c_j$. $\mu_{F_{jk}}\left(x_j^i\right)$ is the membership degree of the i-th value of attribute x_j on the fuzzy set F_{jk}. Decision attribute y classifies each pattern to a single class from q classes of concern. Each class l $(l = 1, \ldots, q)$ has been modeled as a crisp set with membership degree of y^i decision attribute value of i-th pattern, defined as follows:

$$\mu_l(y^i) = \begin{cases} 1, & \text{if } y^i \text{ belongs to } l\text{-th class} \\ 0, & \text{otherwise} \end{cases}$$

The tree classifies the patterns according to rules that have to fulfill conditions in every tree-node. Each rule has the following form:

IF x_j^i belongs to k fuzzy class of attribute j and x_j^i belongs to k fuzzy class of attribute j THEN pattern i belongs to the q class.

Index j identifies the number of attributes that describe each pattern $(j = 1,\ldots, N)$, index i identifies the number of the pattern $(i = 1, \ldots, n)$, and index k identifies the number of fuzzy classes in which each attribute is fuzzified $(k = 1, \ldots, c)$. Each vertical level at the tree represents a new attribute that is examined. In each level, there are nodes that separate each attribute in the number of the fuzzy classes. Children-nodes are generated according to the following three tree conditions:

$$b(F_v, l) < b_{def}, \quad E_{qjk} > E_{def}, \quad Index_j < N \tag{22.5}$$

The set F_v is the composite fuzzy set defined over the intersection of several fuzzy sets, for example, $F_v = F_{1k} \cap \cdots \cap F_{pk}$. $\beta(F_v, l)$ is defined as the certainty factor of a rule that measures the degree of which F_V is a subset of class l.

$$b(F_v, l) = \frac{\sum_{i=1}^{n} \min\{\mu_{F_v}(x^i), \mu_l(y^i)\}}{\sum_{i=1}^{n} \mu_{F_v}(x^i)},$$
$$\mu_{F_v}(x^i) = \min\left(\mu_{jk}\left(x_j^i\right), \ldots, \mu_{pk}\left(x_p^i\right)\right) \tag{22.6}$$

Fuzzy ID3 algorithm can be used for the induction of the fuzzy decision tree (FDT) [7–10]. Consider the fuzzy classification problem introduced above. For each fuzzy set $\{F_{jk} \mid k = 1, \ldots, c\}$ of the attribute x_j, the certainty factor concerning the l-th class is defined in (22.7):

$$b_{jk}^l = \frac{\sum_{i=1}^{n} \min\{\mu_{jk}(x_j^i), \mu_l(y^i)\}}{\sum_{i=1}^{n} \mu_{jk}(x_j^i)}, 0 \leq b_{jk}^l \leq 1 \tag{22.7}$$

E_{qjk} shows the equivalent samples that correspond to each path of the tree, and is defined as:

$$E_{qjk} = \sum_{i=1}^{n} \mu_{F_v}(x^i) \tag{22.8}$$

B_{def} and E_{def} are initial values defined in each problem. Every possible combination that corresponds to conditions (22.5) is examined. The output of the structure is calculated for the l class of concern, for each given pattern as:

$$y_l^i = \sum_{m=1}^{M} \mu_{path_m}^i \times b_{ml}, 0 \leq b_{ml} \leq 1$$
$$\mu_{path_m}^i = \prod_j \mu_{G_{jm}}\left(x_j^i\right) \tag{22.9}$$

where m is an index showing the path from the root to each node of the tree. $\mu_{G_{jm}}$ is the value of the Gaussian membership function that has been used to fuzzify the attributes. According to this output, the mean square error of the FDT is calculated as:

$$E = \frac{1}{2n} \sum_{l=1}^{q} \sum_{i=1}^{n} \left(d_l^i - y_l^i\right)^2 \tag{22.10}$$

where d_l^i is the given class in which each pattern belongs. FDT parameters are modified in order to improve the classification accuracy. The necessary condition for the minimization of error is that its differentiations with respect to the Gaussian center locations, Gaussian widths, and certainty factors are all zero. At each epoch (iteration), the complete parameter set $P = \{c_{jk}, \sigma_{jk}, \beta_{mk} \mid k = 1, \ldots, c; l = 1, 2\}$ is moved by a small distance η in the direction in which E decreases most rapidly, that is in the direction of the negative gradient $-\partial E/\partial H$, where **H** is the parameter vector constituted from the set P. This leads to the parameter update rule:

$$H^{t+1} = H^t - \eta \frac{\partial \mathrm{E}}{\partial \mathrm{H}} \tag{22.11}$$

According to (22.11), the parameters are updated in each epoch using (22.12), (22.13), and (22.14):

$$b_{ml}^{t+1} = b_{ml}^t + \frac{\eta}{n} \sum\nolimits_{i=1}^{n} \left(d_l^i - y_l^i\right) \mu_{path_m}^i \tag{22.12}$$

$$c_{ml}^{t+1} = c_{ml}^t + \frac{\eta}{n} \sum\nolimits_{i=1}^{n} \left(\mu_{path_m}^i \frac{\left(x_j^i - c_{jm}^t\right)}{\left(\sigma_{jm}^t\right)^2} \times \sum\nolimits_{l=1}^{q} b_{ml}^{t+1} \left(d_l^i - y_l^i\right) \right) \tag{22.13}$$

$$\sigma_{ml}^{t+1} = \sigma_{ml}^t + \frac{\eta}{n} \sum\nolimits_{i=1}^{n} \left(\mu_{path_m}^i \frac{\left(x_j^i - c_{jm}^t\right)^2}{\left(\sigma_{jm}^t\right)^3} \times \sum\nolimits_{l=1}^{q} b_{ml}^{t+1} \left(d_l^i - y_l^i\right) \right) \tag{22.14}$$

Parameter adaptation continues until error goes below certain small positive error goal ε or the maximum number of training epochs has been completed.

22.2.3 Radial Basis Function Neural Networks

RBFNN is a well-known class of NNs emerged in the late 1980s [11–13]. A typical RBFNN has a feed-forward structure, as illustrated in Figure 22.3, and consists of three layers of nodes: the input layer, the hidden layer, and the output layer.

Each of the d nodes of the input layer corresponds to a feature of the input pattern $X = [x_1, x_2, \dots, x_d]$. The values of x are passed to the hidden layer. Each of the n nodes in the hidden layer is associated with a vector of centers with dimension equal to the number of inputs $C_n = [c_{n1}, c_{n2}, \dots, c_{np}]$ and implements the radial basis function (RBF). Radial functions are a special class of function. Their characteristic feature is that their response decreases (or increases) monotonically with distance from a central point. The center, the distance scale, and the precise shape of the radial function are parameters of the model, all fixed if it is linear. The most general formula for any RBF is

$$h(x) = \varphi\left((x - c)^T R^{-1} (x - c)\right) \tag{22.15}$$

where φ is the function used, c is the center, and R is the metric. The term $(x - c)^T R^{-1} (x - c)$ is the distance between the input vector x and the center c in

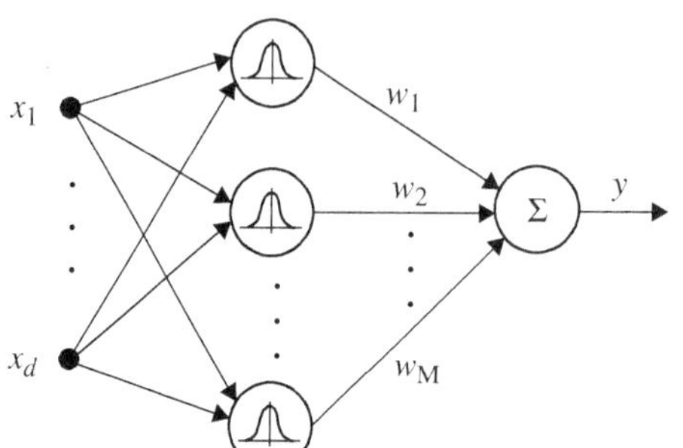

Figure 22.3 Radial basis function neural network.

the metric defined by R. There are several types of functions used, for example, the Gaussian, $\varphi(z) = \exp^{-z}$, the multi-quadric, $\varphi(z) = (1+z)^{\frac{1}{2}}$, the inverse multi-quadric, $\varphi(z) = (1+z)^{-\frac{1}{2}}$, and the Cauchy $\varphi(z) = (1+z)^{-1}$.

Often the metric is Euclidean. In this case, $R = r^2 I$ for some scalar radius r and where I is unity diagonal. If Gaussian is used, then (22.15) simplifies to transfer function (22.16):

$$h(x) = \exp\left(-\frac{(x-c)^T(x-c)}{r^2}\right) \tag{22.16}$$

Its parameters are its center c and its radius r and monotonically decrease with distance from the center. Radial functions could be employed in any sort of model (linear or nonlinear) and any sort of NN (single-layer or multi-layer). However, RBFNs have traditionally been associated with radial functions in a single-layer network such as the one shown in Figure 22.3. Each of d components of the input vector x feeds forward to M basis functions whose outputs are linearly combined with weights $w_j, j = 1, \ldots, M$ into the network output $f(x)$, as:

$$f(x) = \sum_{j=1}^{M} w_j h_j(x) \tag{22.17}$$

and the free variables are the weights w. The model f is expressed as a linear combination of a set of M fixed functions h, which are often called *basis functions*. The flexibility of f, its ability to fit many different functions, derives only from the freedom to choose different values for the weights. The basis functions and any parameters, which they might contain, are fixed. If this is not the case, that is, the basis functions can change during the learning process or if there are more than one hidden layer, then the model is nonlinear. Linear models are simpler to analyze mathematically. In particular, if supervised learning problems are solved by *least squares*, then it is possible to derive and solve a set of equations for the optimal weight values implied by the TS. The same does not apply for nonlinear models, such as multi-layer perceptrons, which require iterative numerical procedures for their optimization.

LEAST SQUARES AND THE OPTIMAL WEIGHT VECTOR (LEARNING PROCESS). When applied to supervised learning with linear models, the least squares principle leads to a particularly easy optimization problem. Since the applied model is like (22.17) and the TS is (x_i, y_i), $i = 1, \ldots, N$, then the least squares recipe is to minimize the *sum-squared-error*,

$$S = \sum_{i=1}^{N} (y_i - f(x_i))^2 \tag{22.18}$$

with respect to the weights of the model. If a weight penalty term is added to the sum-squared-error, as is the case with ridge regression, then the following *cost function* is minimized [12]:

$$C = \sum_{i=1}^{N} (y_i - f(x_i))^2 + \sum_{j=1}^{M} \lambda_j w_j^2 \tag{22.19}$$

where the λ_j, $j = 1, \ldots, M$ are nonnegative regularization parameters. For the optimization of the j-th weight, the differential of the cost function (22.19) is carried out, that is:

$$\frac{\partial C}{\partial w_j} = 2\sum_{i=1}^{N}(f(x_i) - y_i)\frac{\partial f}{\partial w_j}(x_i) + 2\lambda_j w_j \tag{22.20}$$

Substituting the derivative of the model (22.17), $\frac{\partial f}{\partial w_j}(x_i) = h_j(x_i)$, into (22.18) and equating the result to zero leads to:

$$\sum_{i=1}^{N} f(x_i)h_j(x_i) + \lambda_j w_j = \sum_{i=1}^{N} y_i h_j(x_i) \tag{22.21}$$

There are M such equations, for $1 \leq j \leq M$, each representing one constraint on the solution. Since the number of unknowns equals the number of constraints, the system of equations has a unique solution (except under certain pathological conditions). To find that unique solution, linear algebra is employed. Equation (22.21) can be rewritten in vector notation as follows:

$$h_j^T f + \lambda_j w_j = h_j^T y \tag{22.22}$$

The M equations defined in (22.22) are formed in martix form according to (22.23)

$$H^T f + \Lambda w = H^T y \tag{22.23}$$

where:

$$\Lambda = diag[\lambda_1\ \lambda_2\ \cdots\ \lambda_M] \tag{22.24}$$

and where H, which is called the *design matrix*, has the vectors h_j, $j = 1, \ldots, M$ as its columns, $H = [h_1, h_2, \ldots, h_M]$, and has N rows, one for each pattern in the TS. The vector f can be decomposed into the product of two terms, the design matrix and the weight vector, since each of its components is a dot-product between two M-dimensional vectors. For example, the i-th component of f when the weights w are at their optimal values is

$$f_i = f(x_i) = \sum_{j=1}^{M} h_j(x_i)w_j \tag{22.25}$$

and in the matrix form, $f = Hw$.

Finally, substituting this expression for f into (22.23) gives:

$$H^T y = H^T f + \Lambda w = H^T H w + \Lambda w = (H^T H + \Lambda)w \tag{22.26}$$

and the solution to which is

$$w = (H^T H + \Lambda)^{-1} H^T y \tag{22.27}$$

Equation (22.27) is the most general form of the normal equation and is called *local ridge regression*. There are two special cases. In *standard ridge regression* $\lambda_j = \lambda$, $1 \leq j \leq M$, that is:

$$w = \left(H^T H + \lambda I_M\right)^{-1} H^T y \tag{22.28}$$

and *ordinary least squares*, where there is no weight penalty, and is obtained by setting all regularization parameters λ to zero, that is:

$$w = (H^T H)^{-1} H^T y \tag{22.29}$$

The linear equations (22.27–22.29) can all be written more conveniently as the matrix equation:

$$Aw = H^T y \tag{22.30}$$

where H (*design matrix*) is orthogonal of dimensions $N \times M$. The *variance matrix* A^{-1} is defined as:

$$A^{-1} = (H^T H + \Lambda)^{-1} \tag{22.31}$$

and $w = [w_1, w_2, \ldots, w_M]^T$ is the vector of weights, which minimizes the cost function (22.19).

Extrapolation Index. RBFNNs can provide measures of reliability, such as an extrapolation index [14–16]. The extrapolation index determines whether there are enough training data in the vicinity of the test point, in order to consider the output of the RBFNN as reliable. The method of computation of extrapolation index is based on Parzen windows [17], which can be implemented by RBFs. At hidden node n, the Parzen density estimate based on the node width σ_n, and node activation function $a_n(x)$, is:

$$\rho_n = \rho(x_n) = \frac{1}{KV_n} \sum_{k=1}^{K} a_n(x_k) \tag{22.32}$$

where K is the number of training points and:

$$V_n = \int_{-\infty}^{\infty} a_n(x) dx \tag{22.33}$$

which means that ρ_n is the fraction of training patterns with high activation values per set volume. Since the nodes may overlap depending on their centers and widths, the extrapolation index is a weighted average as defined in (22.34):

$$\rho(x) = \frac{\sum_{n=1}^{N} a_n(x)\, \rho_n}{\sum_{n=1}^{N} a_n(x) + 1 - \max(a_n)} \tag{22.34}$$

22.3 UNSUPERVISED LEARNING METHODS

Unsupervised learning methods are not oriented toward a particular prediction task, in contrast to supervised learning methods, where the objective is clearly defined in terms of modeling the underlying correlations between some input variables and some particular output variables. Rather, they try to identify existing underlying relationships among a set of objects characterized by a set of variables or among a set of variables used to characterize a set of objects. Unsupervised learning offers the possibility of exploring the structure of data without guidance in the form of class information and can often reveal features not previously expected and known about. Some examples of unsupervised learning applications are [18].

CLUSTERING. One of the purposes of clustering is to identify homogeneous groups of similar objects, in order to represent a large set of objects by a small number of representative prototypes. The clusters found offer a model of the data in terms of cluster centers, sizes, and shapes which can often be described using less information and in fewer parameters than when required to store the entire training data set. The output of the system should give the cluster label of the input pattern. Graphical, two-dimensional scatter plots may be used as a tool in order to analyze the data and identify clusters.

DIMENSIONALITY REDUCTION. The input data are grouped in a subspace, which has lower dimensionality than the dimensionality of the data. The system has to learn an optimal mapping such that most of the variance in the input data is preserved in the output data.

FEATURE EXTRACTION. The system has to extract features from the input signal. This often means a dimensionality reduction as described above. Another application of the same techniques is to identify similarities (and redundancies) among the different attributes used to characterize objects.

VECTOR QUANTIZATION. This problem occurs when a continuous space has to be discretized. The input of the system is the n-dimensional vector x and the output is a discrete representation of the input space.

In the context of power system security, both applications may be useful as complementary data analysis and preprocessing tools. A representative method of unsupervised learning is self-organized maps (SOMs), proposed by Kohonen [19].

22.3.1 Self-Organized Maps

The Kohonen SOM runs an unsupervised clustering algorithm. It is easily trained and has attractive properties such as topological ordering and good generalization. Kohonen SOMs have been applied successfully in many power system applications, such as DSA and static security assessment ([20–22]). An SOM consists of an array of units (neurons) arranged in a grid and maps a high dimensional input space to a low dimensional output space. The mapping of the SOM is done by feature vectors associated with each unit j:

$$w_j = \left[w_j^1\ w_j^2 \cdots w_j^D\right]$$

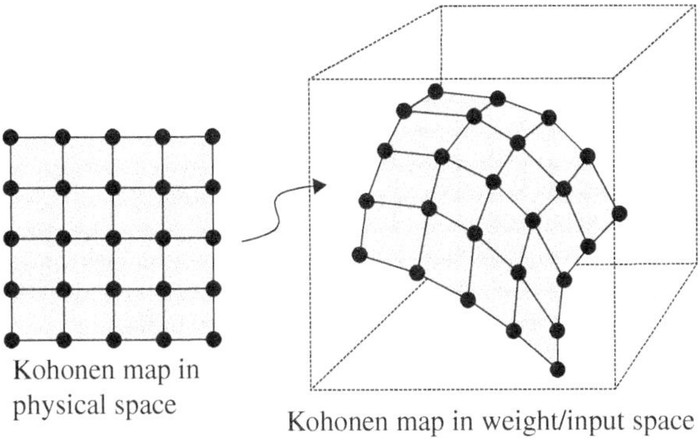

Figure 22.4 Kohonen self-organized map.

where D is the dimensionality of the input data. The feature vectors are structured in a neighborhood grid. If the grid is two-dimensional, the SOM offers the possibility of visualization of its mapping. Figure 22.4 shows a map in physical or topological space on the left and in weight (input) space on the right.

In the SOM, the output units are usually ordered in a two-dimensional grid. The ordering, which is chosen by the user, determines which output neurons are neighbors. During the training process, learning patterns are presented to the network and the weights of neurons are adapted such that the order present in the input space is preserved in the output. This means that learning patterns, which are near to each other in the input space, must be mapped on output units which are also near to each other, that is, the same or neighboring units.

There are various types of neighborhood grids. The most common is the square (Figure 22.5) and the hexagonal (Figure 22.6). Both of them define the way of the map organization and for each grid, the direct neighbors are illustrated in the figures given below.

The hexagonal grid is more uniformly distributed in comparison to the square grid, since all neighbors have the same distance in a hexagonal grid, which is not the case for the square grid.

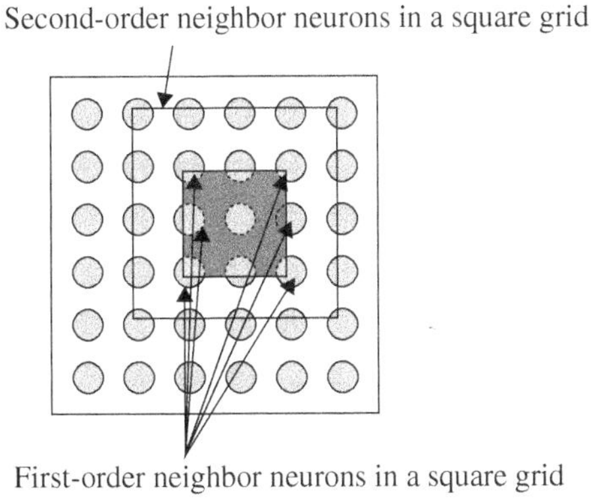

Figure 22.5 SOM neighborhood in a square grid.

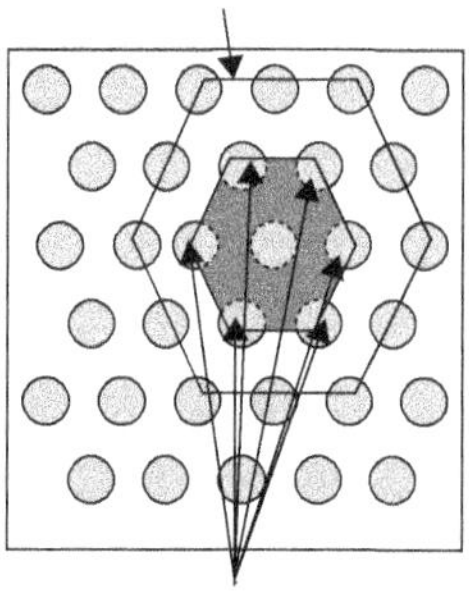

Figure 22.6 SOM neighborhood in a hexagonal grid.

SELF-ORGANIZED MAP TRAINING. During the learning phase, the input vectors are presented randomly. At each presentation of an input vector, $x = [x_1\ x_2 \cdots x_D]$, that is, at each step of the learning process, every neuron of the network calculates a scalar activation function that depends on the input vector and on its own weight vector w_j. This function is chosen to represent distance $\|\ldots\|$ between the input vector and the weight vector of the neuron under consideration. Possible choices are Euclidean distance or the scalar product. The winning unit is considered to be the one with the largest activation. The weight of the winning unit as well as all the weights in a neighborhood around the winning unit is updated. The neighborhood's size generally decreases slowly with each iteration. A sequential description of how to train an SOM is as follows:

1. Initialize all weights. Initial values are given to the weight vectors before training. The SOM training algorithm is robust with regard to initialization of its weight, however a good initialization can contribute to a faster convergence and a better solution. Typically, one of the three following initialization procedures is used:
   - Random initialization, where the weight vectors are initialized with small random values.
   - Sample initialization, where the weight vectors are initialized with random samples drawn from the input data set.
   - Linear initialization, where the weights of all neurons are initialized in a linear way so that the neurons cover the data space. This is achieved by calculating the range of the variable values of the input vector x initializing the weight values using equation (22.35):

$$w_j^i = (j-1)\frac{x_i^{\max} - x_i^{\min}}{N-1} + x_i^{\min} \tag{22.35}$$

   where $x_i^{\min}$ and $x_i^{\max}$ are the minimum and maximum values of input I, respectively, and N is the total number of neurons.
2. Choose input vector randomly in the TS.

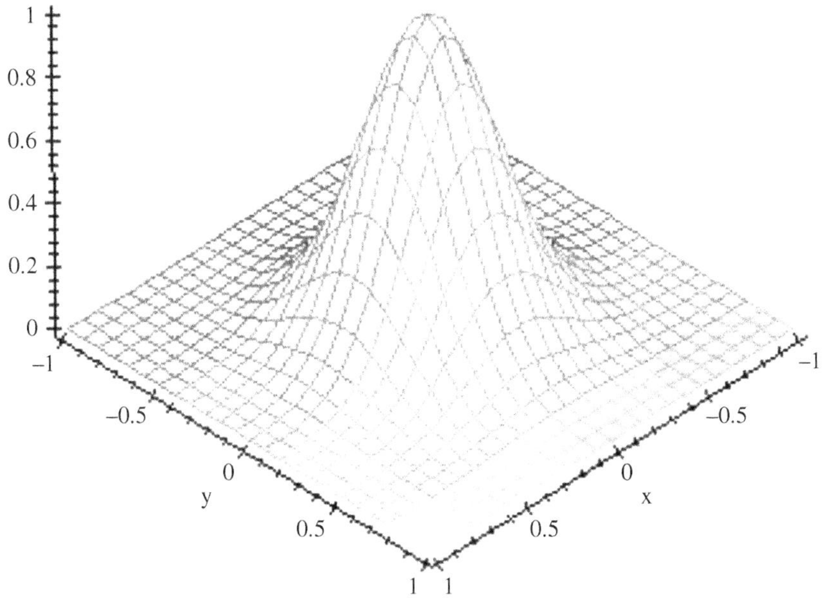

Figure 22.7 Gaussian neighborhood function.

3. Select the winning output unit as the one with the largest similarity measure between all weight vectors w_i and the input vector x. If the Euclidean distance is chosen as the dissimilarity measure, then the winning unit c satisfies equation (22.36):

$$\|x - w_c\| = \min \|x - w_i\|, \quad i = 1, \ldots, N \tag{22.36}$$

4. Define the neighborhood of the winner, for example, by using a neighborhood function $\Omega(i)$ around a winning unit c. For instance, the Gaussian function can be used as the neighborhood function, defined by equation (22.37):

$$\Omega_c(i) = \exp\left(-\frac{\|p_i - p_c\|^2}{2\sigma^2}\right) \tag{22.37}$$

where p_i and p_c are the positions of the output units, i and c, respectively, and σ reflects the scope of the neighborhood. Figure 22.7 illustrates the shape of the Gaussian function.

The scope of the neighborhood declines during the training procedure according to equation (22.38):

$$\sigma(n) = \sigma_0 \exp\left(-\frac{n}{\tau_1}\right) \tag{22.38}$$

where n is a parameter that counts how many times the data have been presented to the SOM (i.e., epochs), σ_0 is the initial value of scope, and τ_1 is a parameter associated with the rate of decline of the scope.

After the definition of the neighborhood function, the weight vector w_c of the selected neuron c and the weight vectors w_i of its neighbors are updated according to the following formula:

$$\Delta w_i = \eta \cdot \Omega_c(i) \cdot (x - w_i) \tag{22.39}$$

where η is a parameter called learning rate. Learning rate also declines with the number of epochs so that the algorithm finally converges.

$$\eta(n) = \eta_0 \exp\left(-\frac{n}{\tau_2}\right) \tag{22.40}$$

where η_0 is the initial value of learning rate, and τ_2 is a parameter that determines the rate of decline of η.

5. If neighborhood function $\geq \varepsilon$, go to 2 else stop.

To achieve convergence, the learning rate and the width of the neighborhood of the winner neuron must shrink to zero with time. A main problem of the SOM algorithm is the fact that the number of training steps of the convergence phase has to be fixed a priori and therefore must be set to a large value in order to ensure convergence. This decelerates the algorithm. The complexity of the algorithm used is O(N, M, D), where N is the number of units of the map, M is the number of samples of the TS, and D is the dimensionality of the OP x.

Different learning processes will be obtained when starting with different initial values w_i and applying different sequences of training vectors $x(t)$ and different learning parameters. It is obvious that some optimal map for the same input data must exist. The mean $\frac{1}{M}\sum_{i=1}^{M} \|x - w_{c(i)}\|$ (quantization error) over the training data is a useful performance index, when comparing maps with the same array structure and definition of $\Omega_c(i)$. An important variant of incremental training described above is the batch training algorithm. In it, the whole TS is gone through at once and only after this, the map is updated with the net effect of all the samples. In fact, the updating is done by simply replacing the prototype vector with a weighted average over the samples, where the weighting factors are the neighborhood function values:

$$w_j(t+1) = \frac{\sum_{i=1}^{N} \Omega_{c(i)}(t) \cdot x_i}{\sum_{i=1}^{N} \Omega_{c(i)}(t)} \tag{22.41}$$

Batch algorithm increases memory requirements, but it assures that training is independent to the sequence of inputs to the SOM.

22.4 SOM VARIANTS

There are many variants to the basic SOM. The goal of these variations is to enable the SOM to follow better the topology of the underlying data set or to achieve better quantization results. Another goal of the variations is to make SOM capable of being online trained, since many real world data analysis and processing tasks require systems with the ability of online self-adaptive learning. This is achieved by using adaptive or flexible neighborhood definitions or even growing map structures. The disadvantage of these methods is that they are computationally heavier and they have less visualization properties since they lead to more complex structures. The methods studied in the next section are the evolving SOM, the growing hierarchical SOM, and the growing neural gas. A variable local topology SOM (VLT-SOM) is also proposed to visualize graphs onto a two-dimensional plane and deal with the Nondeterministic Polynomial Time (NP)-complete problem of graph partitioning. The method can be applied to solve power system adaptive islanding problem.

22.4.1 Evolving SOM

An algorithm of evolving self-organizing map (ESOM) is proposed as a dynamic version of the Kohonen SOM, where network structure is evolved in an online adaptive mode [23]. One of the differences of ESOM compared to the SOM is that the topology of the map is not a priory known and prototype neurons are not organized into one- or two-dimensional grids. The ESOM starts without any neurons. During learning procedure, new neurons are created so that they cover the online incoming data. Each neuron is associated with a weight vector with the same dimension with the input vectors and is connected to other neurons so that neighborhood relationships are maintained. The strength of neighborhood connection is determined by their distance, which has a high value for weak connections and low values for close connections. Weak connections can be pruned, and this can split the ESOM into distinct data structures. The learning procedure of ESOM is summarized as follows:

1. The ESOM is initialized as an empty set of neurons and connections.
2. A new input $x = [x_1\ x_2 \cdots x_D]$ is presented to the ESOM.
3. The distance of the input vector to the weight vectors $W = [w_1\ w_2 \cdots w_N]$ is calculated. The winning unit w_i is the one with the minimum distance. If there is no neuron ($W = \oslash$) or the distance of the winning neuron is above a threshold $\|w_i - x\| \geq \varepsilon$, then a new neuron is inserted in the network as:

$$w_{N+1} = x \tag{22.42}$$

4. The new neuron is inserted representing exactly the poorly matched input vector, resulting in a maximum activation on the new neuron.
5. The two nearest neighbors, that is, the neurons with minimum distance to the newly created neuron, denoted as w_{n1} and w_{n2}, are found and connected with w_{N+1}.
6. Otherwise, update the matching neuron and each of its neighbors denoted as w according to their distances to the data vector x, with each neuron modified as:

$$\Delta w = \gamma e^{-\|w-x\|/\sigma^2}(x - w) \tag{22.43}$$

where γ is a small learning rate, and σ controls the effective neighborhood spread.

7. Reset the strength of connections between the winner (or the newly created node) and its neighbors. The connection strength $s(i,j)$, for the connection $c(w_i, w_j)$, is defined as $s(i,j) = ||w_i - w_j||$.
8. Gradually prune weak connections and inactive neurons.

22.4.2 Growing Hierarchical Self-Organized Map

One of the shortcomings of the SOM lies with its fixed architecture that has to be defined a priori. On the other hand, dynamically growing variants of the SOM such as the ESOM discussed in the previous section tend to produce huge maps that are hard to handle. This leads to the development of the growing hierarchical self-organized map (GHSOM) proposed by Rauber et al. [24]. The GHSOM has a hierarchical structure of multiple layers, where each layer consists of several independent growing SOMs the size of which adapts itself to the requirements of the input space. A graphical representation of a GHSOM is given in Figure 22.8.

For the control of the growth process, two different strategies can be applied, using either the mean quantization error (*mqe*) of a unit or the absolute value namely, the quantization error (*qe*) of a unit. The *mqe* of a unit i is calculated as the mean Euclidean distance between its model vector w_i and the n_C input vectors x_j that are the set of input vectors C_i mapped onto this unit i:

$$mqe_i = \frac{1}{n_C} \sum_{x_j \in C_i} \|w_i - x_j\| \quad n_C = |C_i|,\ C_i \neq \oslash \tag{22.44}$$

As an alternative to the *mq*e, the *qe* can be used as well as a measure to the growth process, since it provides more space for more densely populated regions. The quantization error for a unit i is defined as:

$$qe_i = \sum_{x_j \in C_i} \|w_i - x_j\| \quad C_i \neq \oslash \tag{22.45}$$

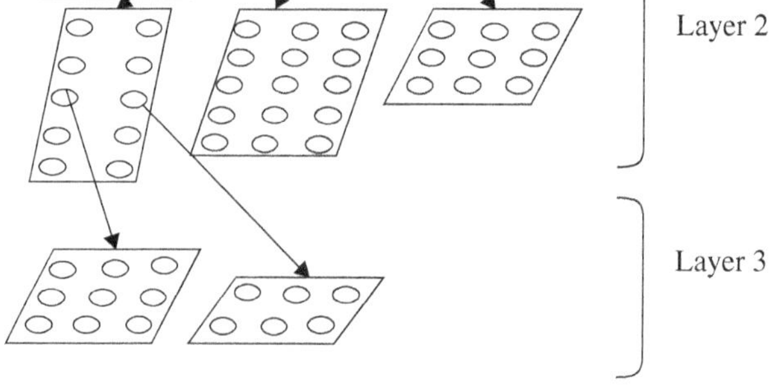

Figure 22.8 GHSOM topology.

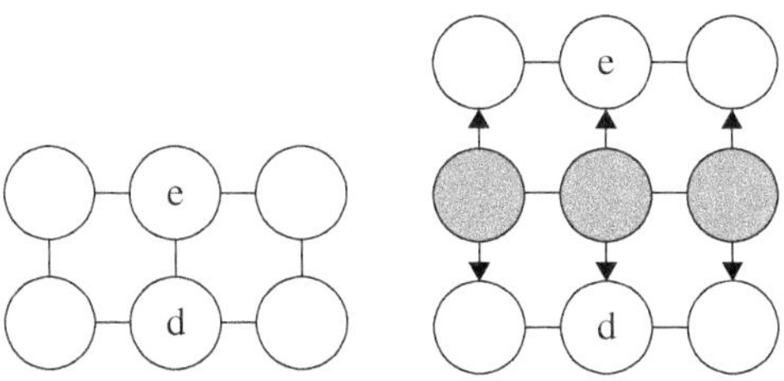

Figure 22.9 Insertion of a row.

The training procedure of the GHSOM starts with the whole data set at layer 0. Quantity mqe_0 (alternatively qe_0) is calculated according to the expressions:

$$mqe_0 = \frac{1}{n_I} \sum_{x_j \in I} \|w_0 - x_j\| \tag{22.46}$$

$$qe_0 = \sum_{x_j \in I} \|w_0 - x_j\| \tag{22.47}$$

where n_I refers to the number of all input vectors x of the input data set I, and w_0 denotes the mean of the input data. The minimum quality of data representation of each unit will be specified as a fraction of mqe_0 (alternatively qe_0), indicated by a parameter τ_2 as defined in the equations below.

$$mqe_i < \tau_2 \cdot mqe_0 \tag{22.48}$$

$$qe_i < \tau_2 \cdot qe_0 \tag{22.49}$$

Equations (22.48) and (22.49) the global termination criterion, this means that all units must represent their respective subsets of data at a mean quantization error smaller than a fraction of τ_2 of mqe_0 (alternatively qe_0). The GHSOM architecture is initialized by creating a new growing SOM beneath the layer 0 map. The initial size of this first layer map is set to 2 × 2 units, with its weight vectors being initialized to random values. The map is trained according to the training procedure described in the previous section. After a fixed number of iterations, the *qe* or *mqe* of all units is calculated and the unit with the highest *qe* (or *mqe*) hereafter called *unit e* is determined.

$$e = \arg \max_i \left(\sum_{x_j \in C_i} \|w_i - x_j\| \right) \tag{22.50}$$

Following the selection of *e*, its most dissimilar neighbor *d* is determined as follows, where N_e is the set of neighboring units of *e*:

$$d = \arg \max_i (\|w_i - x_j\|) \; m_i \in N_e \tag{22.51}$$

A row or column of units is inserted between *d* and *e* as illustrated in Figures 22.9 and 22.10, respectively.

As more units are added to the growing SOM, their *qe* values decrease. For all units not satisfying this condition, a more detailed data representation is required, leading to the addition of further units to provide more map space for data representation. The training process will continue until all units satisfy the global criterion.

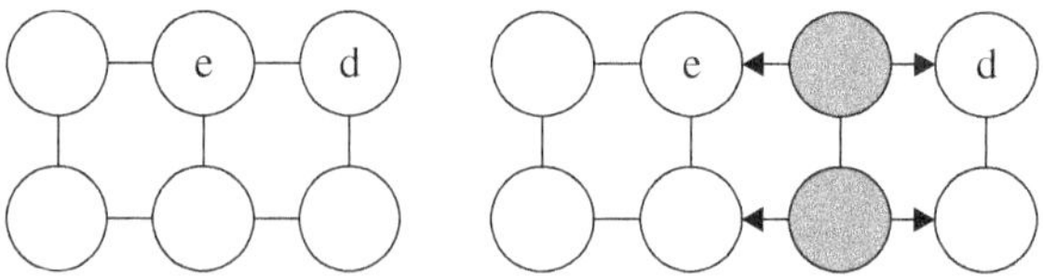

Figure 22.10 Insertion of a column.

The growth process continues until the map's qe referred to as QE_m reaches a certain fraction τ_1 of the qe_u of the corresponding unit u in the upper layer:

$$QE_m < \tau_1 \cdot qe_u \tag{22.52}$$

22.4.3 Growing Neural Gas

The growing neural gas (GNG) was proposed by Fritzke [25]. A GNG network consists of a set A of neurons and a set of connections $C,\ C \subset A \times A$ between the neurons. Each neuron (prototype) is associated with a weight vector as in the case of SOM, and each connection has an age counter variable which is set to zero when the connection is created and incremented by one at each step TS. Connections c_{vw} can be set or deleted between states s_v and s_w and their feature vectors w_v and w_w. These connections represent the topological structure of the data space.

The advantage of GNG is that the number of feature vectors and connections is not fixed. They will be adapted in a continuous process. An adaptation to a changing input data space can be realized easily. The principle accuracy of the input space representation is greater because of a higher degree of freedom in comparison to the SOM. On the other hand, the visualization of the results is not possible. Figure 22.11 shows the structure of the GNG.

The procedure of a GNS construction is described below.

1. The GNG network starts with two prototypes c_1 and c_2 randomly initialized $A = \{c_1, c_2\}$. The connection set is initialized to the empty set $C = \oslash$.
2. A new input vector is presented to the GNG. The distances to all units are calculated and the winner unit (here termed as unit s_1) is determined as the unit with the minimum distance to the input vector. The second winner unit (here

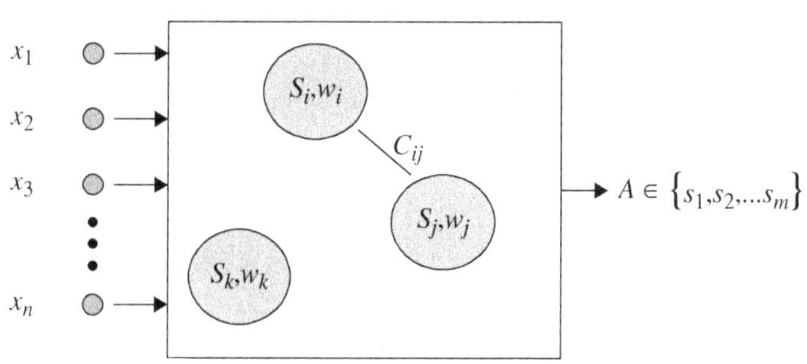

Figure 22.11 Structure of the GNG and an example of connected feature vectors in an input space.

termed as s_2) is also defined as the unit with the minimum distance to the input vector apart from the winner unit.

$$s_1 = \arg\min_{c \in A} \|c - x\| \tag{22.53}$$

$$s_2 = \arg\min_{c \in A \setminus \{s_1\}} \|c - x\| \tag{22.54}$$

3. Create a connection between s_1 and s_2, (if it does not exist already) and set its age counter variable equal to zero

$$C = C \cup \{(s_1, s_2)\} \tag{22.55}$$

$$age(s_1, s_2) = 0 \tag{22.56}$$

4. Update the weights of the winning unit as well as its direct topological neighbors according to the equation

$$\begin{aligned} \Delta w_{s_1} &= \varepsilon_\alpha \cdot (x - w_{s_1}), \\ \Delta w_i &= \varepsilon_b \cdot (x - w_i), \ \forall i \in N_{s_1} \end{aligned} \tag{22.57}$$

ε_a and ε_b are the learning rates for the winner and for its direct topological neighbor.

5. Increment the age variable counter of all connections starting from s_1

$$age(s_1, i) = age(s_1, i) + 1 \ \forall i \in N_{s_1} \tag{22.58}$$

6. When the age counter reaches a threshold value $a_{\max}$, it implies that the edge is redundant and removed.

7. If the number of input signals generated so far is an integer multiple of a parameter λ, insert a new unit as follows:

8. Determine the unit q with the maximum accumulated error:

$$q = \arg\max_{c \in A} E_c \tag{22.59}$$

9. Determine among the neighbors of q the unit f with the maximum accumulated error:

$$f = \arg\max_{c \in N_q} E_c \tag{22.60}$$

10. Add a new unit r to the network and interpolate its reference vector from q and f:

$$A = A \cup \{r\}$$

$$w_r = \frac{w_q + w_f}{2}$$

11. Insert edges connecting the new unit r with units q and f, and remove the original edge between q and f:

$$C = C \cup \{(r, q), (r, f)\}, \quad C = C \setminus \{(q, f)\} \tag{22.61}$$

12. Repeat until the user defined stopping criterion is satisfied.

22.4.4 Variable Local Topology-Self-Organized Map

SOMs with variable local topology proposed in [26] are shown to constitute a reasonably good heuristic to find approximate solutions to the NP-complete graph partitioning problem where a weighted graph has to be partitioned in k subgraphs. The solution to this problem can be applied to power system-controlled islanding. The application of the variable local topology SOM for graph representation can be used as well for power system visualization purposes. In this section, the algorithm (thereafter called variable local topology-SOM [VLT-SOM]) is described. A nondirected graph G is characterized by a set of vertices V and a set of edges E connecting the vertices, which is represented by matrix $[e_{ij}]$:

$$V = \{u_i\}, i = 1, \ldots, n$$

$$[e_{ij}] = \begin{cases} e_{ij} = 0, & \text{if } edge\ (u_i u_j) \text{ does not belong to } E \\ e_{ij} \neq 0, & \text{if } edge\ (u_i u_j) \text{ belongs to } E \end{cases} \tag{22.62}$$

The neighborhood of a vertex u_i is the set of vertices to which u_i is connected. Graph partitioning is an NP-complete problem and heuristic methods are proposed to solve it. In order to partition a graph G = (V;E), vertices are assigned planar coordinates: Let w_i denote the positional vector of vertex v_i in the plane.

The modified algorithm proposes moving the vertices' positional vectors appropriately, starting from an initially random spatial distribution of positional vector. In order to achieve this goal, we use an NN algorithm, the SOM. The idea is that "training vectors" are drawn according to a certain spatial probability density and the graph tries to approximate the distribution of training vectors. Each time a training vector is presented, the closest positional vector is selected and its coordinates are modified so that it gets closer to the presented training vector. In addition, the coordinates of its neighbors are also modified to make them closer to the training vector. After presentation of sufficiently many training vectors, positional vectors should distribute themselves in space to approximate the distribution of training vectors.

1. Vertices of the graph are assigned random planar coordinates:

$$w_i = [w_{i1} \quad w_{i2}], \text{ where } 0 \leq w_{i1}, w_{i2} \leq 1. \tag{22.63}$$

2. A vertice i is randomly selected.
3. For each vertice j with $a_{ij} = 1$, w vector is updated according to the equation

$$w_j^{(t+1)} = w_j^{(t)} + \beta \frac{1}{e\,(i,j)} \left\| w_i^{(t)} - w_j^{(t)} \right\| \tag{22.64}$$

where

$w_i^{(t)}$	is	the weight value at time step t.
β	–	the learning rate.
$\left\| w_i^{(t)} - w_j^{(t)} \right\|$	–	the distance measure between i and j.

The planar can be either segmented beforehand in k rectangles each of them representing a partition of a graph. An alternative to this is the segmentation of the planar after the training procedure, using a heuristic algorithm such as particle swarm optimization [27] and a suitable objective function.

22.5 COMBINED USE OF UNSUPERVISED WITH SUPERVISED LEARNING METHODS

The algorithms discussed above belong to the class of unsupervised learning and are frequently used not only as standalone applications but also as a preprocessing stage. Starting with a high dimensional input space, an unsupervised learning method such as SOM is employed to map the data into a lower dimensional space. In a supervised setting, the discriminant or regression function can be learned in this new space. One advantage of preceding a supervised learner with unsupervised clustering or dimensionality reduction is that the latter does not need labeled data. Labeling the data is costly. A large amount of unlabeled data can be used for learning the cluster parameters and then use a smaller labeled data to learn the second stage of classification or regression.

In the field of DSA of power systems, several supervised learning techniques, such as RBFs, artificial neural networks, and support vector machines, have been applied [28–34] showing satisfactory performance in assessing the dynamic performance of the system. These methods could be combined with unsupervised learning, as a preprocessing step to increase the accuracy of the method. In [35], an automatic learning hybrid method for corrective dynamic security, based on an SOM, is proposed, which classifies the power system's security state according to its load profile. DTs nested in nodes with mixed security states are applied to investigate further their security status. Corrective control is based on the transposition of the system from the insecure nodes of the map to the closest secure ones by changing the load of the system. In this way, load shedding is reduced, which is a very desirable feature of the method.

22.6 APPLICATIONS TO POWER SYSTEMS

22.6.1 Description of the Power System

In this section, the automatic learning framework of Figure 22.1 is applied for the security assessment of the Hellenic power system (Figure 22.12) focusing mainly on voltage stability.

The Hellenic power system comprises steam turbines, which produce the base load and almost 70% of the annual energy, hydro turbines which are used for peak load and frequency regulation and combined cycle units. It is interconnected with the Balkan system with two 400 kV synchronous interconnections, one with Bulgaria and one with FYROM and with a DC interconnection to Italy. In Figure 22.12, the eight load areas in which the system is divided by the Transmission System Operator (TSO) are shown. The derived model includes the generation and transmission system up to 20 kV busses. The total number of busses is 876, and there are 846 lines and

Figure 22.12 The eight areas of the Hellenic power system.

206 transformers. The generating units are represented by 78 machines including their governors, exciters, and over-excitation limiters. The synchronous interconnections at the North of the country were modeled as synchronous generators with high nominal power (20 GW) and inertia coefficient (10 MW·s/MVA). These generators have no over-excitation limiter and their droop is low, so that they can undertake the load, in case of a contingency. The constant impedance (Z), current (I), power (P) (ZIP) model is used for the load assuming 25% constant power, 25% constant current, and 50% constant impedance.

The Hellenic power system is prone to voltage instability, especially during summer [36]. The phenomenon is related to the power transfer from the generating areas in the North and West of Greece to the main load center in the Athens metropolitan area reaching its maximum. The problem is due to the long electrical distance between generation and load, consisting of generator step-up transformers, 400 and 150 kV transmission lines, and 400 kV/150 kV autotransformers and bulk power delivery transformers.

The creation of the KB starts from a base scenario, which is a snapshot of the system at its maximum loading of the year 2005. A coefficient n is considered, which

TABLE 22.1 Two-class partition of the data set

Data set	Learning set	Test set
4000	2000	2000
Two-class partition		
Class A (secure)	1183	1191
Class B (insecure)	817	809

takes values from 0.91 to 1.10 with a step of 0.01 and represents the ratio of the total load to the maximum total load. For each of the 20 values of the coefficient n, 200 OPs are created. The total load of these OPs is normally distributed around the value $nP_{\max}$. The load at each bus of the system is given by equation:

$$P_i^D = N\left(nP_{0,i}^D, \sigma\right) \tag{22.65}$$

where N represents the normal distribution, $P_{0,i}^D$ is the demand of bus i at the base scenario, and σ is the standard deviation ($\sigma = 4\%$). The generated active power of each unit follows a normal distribution with mean value the production of the unit when the system operates at maximum load, multiplied with the coefficient n, as long as it is between its technical minimum and maximum.

$$P_i^G = \min\left(P_{i,\max}^G, N\left(nP_{0,i}^G, \sigma\right)\right) \tag{22.66}$$

where $P_{i,\max}^G$ is the technical maximum of the generator, and $P_{0,i}^G$ is the generation of the unit at base case scenario. The simulated disturbance is the loss of a combined cycle unit at Lavrio with 460 MW nominal active power. This disturbance can lead to dangerously low voltage levels, especially in the region of Athens.

22.6.1.1 RBFNN for DSA

Training of RBFNN requires the determination of the design parameters, that is, the number N of RBF nodes in the hidden layer and the value of their width σ. The method of S-cross folder validation is employed to attain the optimal network structure for better performance. In this procedure, the training data set is first randomly divided into S equal-sized data sets. Then, for a given N and σ, RBFNN is trained using S-1 data sets and the remaining subset is used to test the network's generalization ability. The mean square difference between the target outputs and the predicted outputs is the error associated with the testing subset. This procedure is repeated S times, using different S-1 subsets for training and a different subset for testing at each time. The mean square error over all the testing subsets is the error estimate for the given RBFNN. This procedure is repeated several times with different values of N and σ to obtain the "optimum" network structure with the minimum mean square error.

Parameter N is examined in the range of [1…300], while σ in range of [0…50]. Figure 22.13 shows the average error for all pairs of N and σ. The RBFNN is trained using values $N = 250$ and $\sigma = 3$ and performance results are presented in Table 22.2.

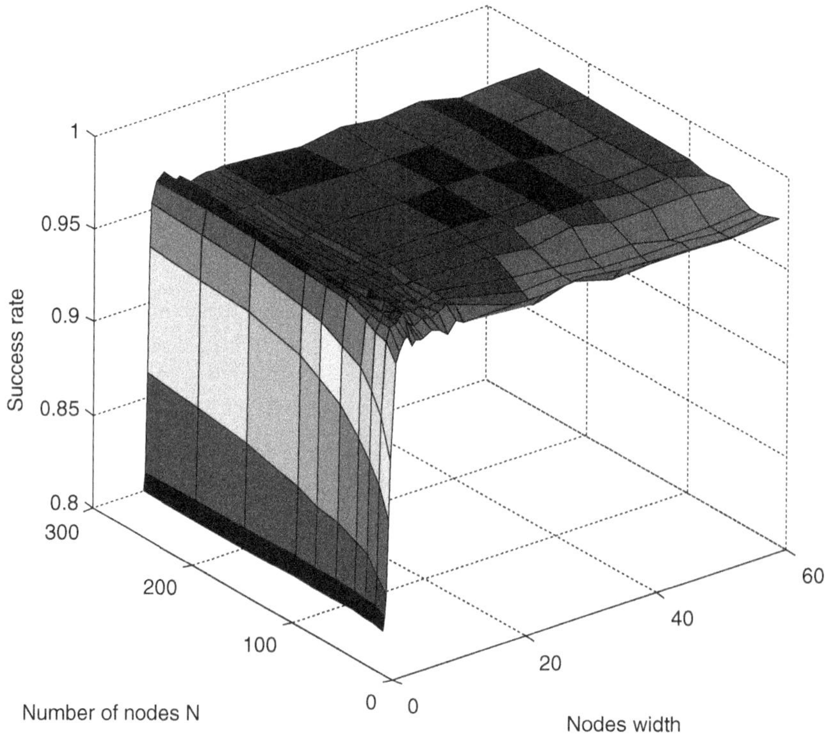

Figure 22.13 Results of S-folder cross validation.

22.6.1.2 *Decision Trees for DSA*

A DT is constructed to perform DSA, on the basis of the LS as illustrated in Figure 22.14. The nodes of the DT have the following scheme: There are four attributes. In the upper right side is the number of OPs, which belong to the node. In the upper left side is the label of the node. In the middle is the safety index, which is the safe OPs of the node divided to the total OPs of the node. Finally, in the bottom is the dichotomy test of the node or the characterization of the node as secure or insecure. In case there is a separation criterion, if it is true we go on the left node, otherwise we go on the right node.

Table 22.3 describes the attributes used in the construction of the DT. The rules derived from the DT are applied to the TS, in order to evaluate the performance of

TABLE 22.2 Evaluation of classification performance

	RBFNN model	DT model
Success rate	99.25% (1985/2000)	98.05% (1985/2000)
False alarms	0.50% (6/1191)	1.51% (18/1191)
Missed alarms	1.11% (9/809)	2.61% (21/809)

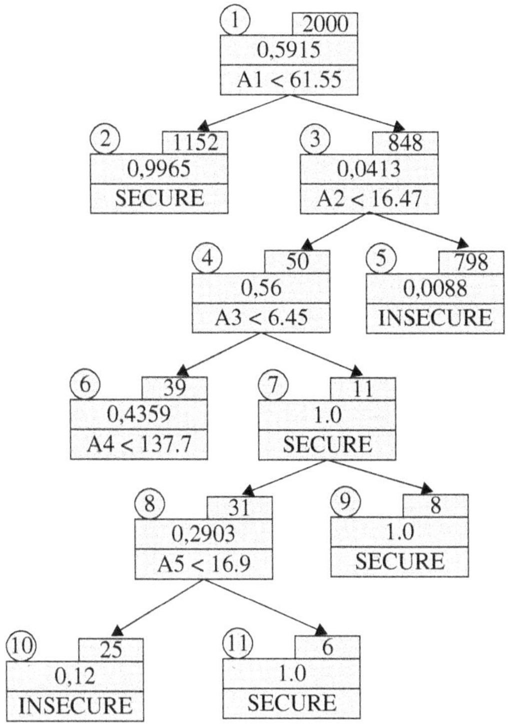

Figure 22.14 Decision tree used for DSA.

the DT to the DSA of the system. The results are illustrated in Table 22.2. From Table 22.2, it is obvious that RBFNNs outperform the DT model in terms of success rate. The advantage of the DTs though is that they provide a clear indication of which features are most important for the classification task.

22.6.1.3 Decision Trees Application for Load Shedding

In order to construct a DT to form a load shedding scheme, the candidate attributes that formulate the rules of the DT must be load-related and controllable variables. Here, the DT is constructed using as candidate attributes the active load of each of the nine areas to which the Greek power system is divided as illustrated in Figure 22.12, and the total load of the Greek power system (10 attributes). Once the DT is constructed, load shedding rules can be derived by inverse reading of it.

TABLE 22.3 Attributes used by the DT

A1	Active losses in area 3
A2	Imported active power from FYROM interconnection
A3	Voltage from generator "STRATOS"
A4	Voltage at substation "N. IONIA"
A5	Voltage of generator "LADONAS"

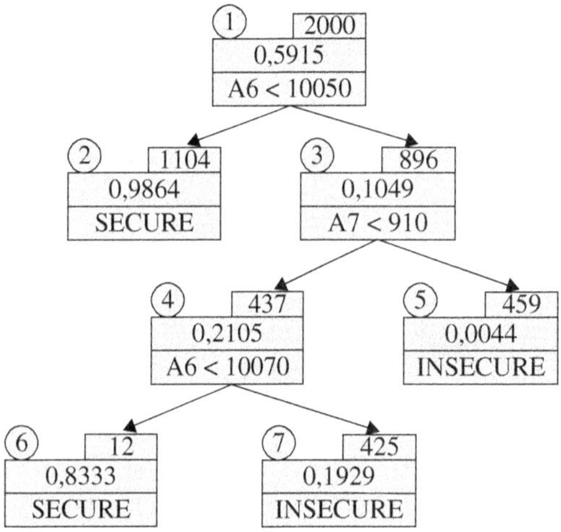

Figure 22.15 Decision tree for load shedding.

Figure 22.15 illustrates the DT, where A6 is the total active load of Greek power system in MW and A7 is the total active load of region 6 in MW. There are two options in order to transpose from the insecure state to a secure one: Curtailment of the load such that the total load is less than 10,050 MW. This action equals to a transposition to leaf 2 of the DT by shedding the appropriate amount of load. Curtailment of the load, such that the load of area 6 is less than 910 MW and the total load is less than 10,070 MW. This action is equivalent to the transposition from an insecure leaf to leaf 6. It should be noted, however, that the SI of leaf 6 is lower than the SI of leaf 5, which means that the first option is more reliable.

22.6.1.4 Genetic Algorithm Aided DTs for Load Shedding

Instead of using the load of the TSO-defined areas of the power system, the aggregate load of suitable groups of load busses can be used to derive more accurate security rules. The objective is to use the aggregate load $\sum_{j \in A(i)} P_D^j$ of a group $A(i)$ of load busses as an attribute for the dichotomy of DT nodes.

The model of the Hellenic power system comprises 241 load busses. Each combination of these busses can formulate a group, which means that there are 2^{241} groups of load busses. The large number of possible solutions indicates that a heuristic search method will be used to provide a suboptimal solution. A genetic algorithm approach is used here to select appropriate groups of load busses, and the process is executed in parallel with the construction of the DT.

The DT is constructed according to Section 22.2.1 with the difference that in this case attributes are not a priori known, but searched via the genetic algorithm method. An initial population of chromosomes is created, each of which represents a group $A(i)$ of load busses. Each chromosome is a vector $V_i : V_i(j) = \{0, 1\}$, $j = 1, \ldots, 241$. If $v(j) = 1$, then the bus j belongs to group $A(i)$ otherwise it does not. The chromosomes are evaluated according to the information gain they provide when the

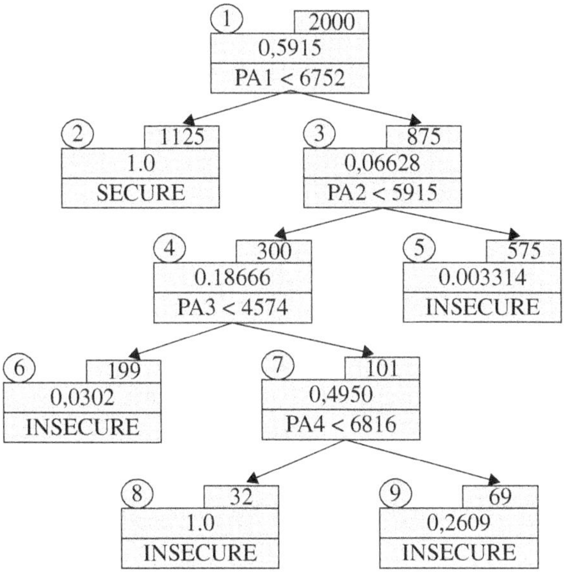

Figure 22.16 Genetic algorithm-aided DT.

aggregate load $\sum_{j \in A(i)} P_D^j$ is used in a dichotomy test of a DT node. After a number of iterations, the chromosome, which provides the maximum information gain, is chosen and an attribute is formulated. This attribute is used to split the node of the DT. The procedure is repeated recursively until the nodes of the DT cannot be further developed.

The initial population comprises 482 chromosomes, and the genetic algorithm is repeated for 1000 iterations. The DT developed by the above procedure is illustrated in Figure 22.16. Table 22.4 provides the attributes used by the GA-aided DT algorithm.

22.6.1.5 Neuro-Fuzzy Decision Trees for DSA

The construction of the N-FDT requires attribute selection. Since the N-FDT is used for corrective control, only controllable, load-related attributes are selected, in order to derive load shedding rules. Instead of using the load of each bus, which would increase the complexity of the method and would provide N-FDTs significantly dependent on the system topology, the aggregate loads of the eight areas are selected. To create the N-FDT using the LS, initialization of the variables k, b_{th}, and E_q is

TABLE 22.4 Attributes used by the GA-aided DT

PA1	Group comprising 157 load busses
PA2	Group comprising 135 load busses
PA3	Group comprising 108 load busses
PA4	Group comprising 159 load busses

TABLE 22.5 Evaluation of two-class classification performance of the N-FDT$_1$ model

Classification of test set by the N-FDT method		
For TS(2000)	Assessed as secure	Assessed as insecure
Secure (1232)	1135	97
Insecure(768)	9	759
Classification performance evaluation		
Success rate		94.7% (1894/2000)
False alarms		7,87% (97/1232)
Missed alarms		1.17% (9/768)

necessary. The values are initialized in a way that both low initial classification error and manageable number of rules are obtained. Different values and combinations were tested increasing each variable in steps. Once the tree is constructed, it is trained for 200 epochs.

The final form of the N-FDT is illustrated in Figure 22.17. The performance of the model is evaluated using the success rate, which is defined as the ratio of successfully classified OPs to the number of OPs tested, the FA rate, which is the ratio of secure OPs classified as insecure to the number of secure OPs, and the MAs rate, which is the ratio of insecure OPs classified as secure to the number of insecure OPs. The classification error is decreasing during the training procedure as illustrated in Figure 22.18. An OP is classified either as secure or insecure according to the certainty factor of the node it belongs to. If $b_{jk} > 0.95$, then the OP is secure, else $b_{jk} < 0.05$ it is insecure, otherwise it is unclassified and combinations with the following attributes are trying. Table 22.5 illustrates the total performance of the structure. The limits between the secure, insecure, and unclassified states are determined by the intersection of their boundary fuzzy classes. These limits were altered during the training as described above.

22.6.1.6 SOM Application for Load Shedding

The proposed method has been applied on the data set of Table 22.1. A 5 × 5 SOM is utilized in order to classify the load profile of the Greek power system. The load level of the eight areas of the Greek power system (Figure 22.12) is used as the input vector for the construction of the SOM. The weight vector of each neuron of the map, $w_i = [w_{i1} \dots w_{i8}]$, represents a typical load profile of the system. An SI, assigned to each neuron of the map, is calculated using the LS after the training of the SOM and equals to the ratio of the secure OPs to the total number of OPs of the neuron.

$$SI_j = \frac{Number\ of\ secure\ OPs\ of\ node\ j}{Total\ number\ of\ OPs\ of\ node\ j} \tag{22.67}$$

The SI of a neuron indicates the degree of security of the OPs that belong to that neuron. SI values close either to 1 or 0 indicate that the corresponding OPs can be characterized with high certainty as secure or insecure, respectively. For this

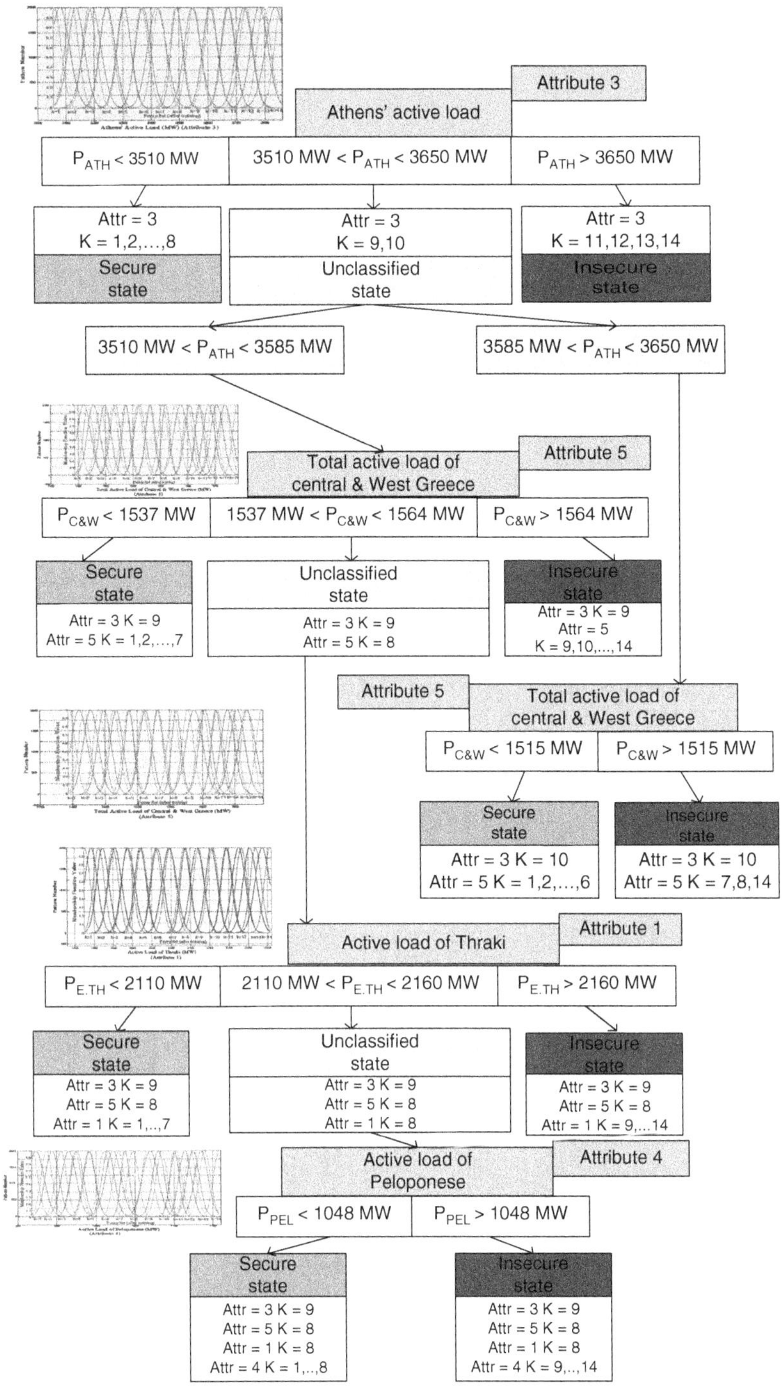

Figure 22.17 N-FDT$_1$ structure.

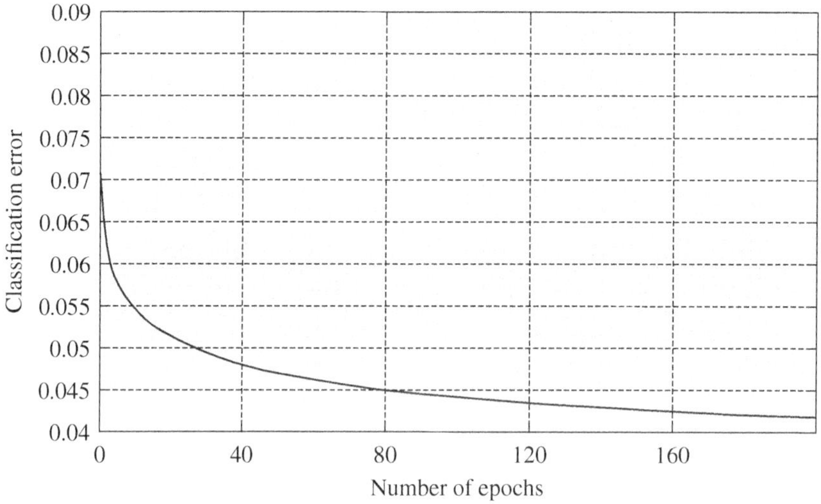

Figure 22.18 Classification error during the training.

assumption to be valid, an adequate number of OPs of the LS must be associated with each neuron of the map. The TS serves for the validation of this assumption.

Figure 22.19 shows the results of the mapping of the TS onto the SOM. At each neuron of the map, the number of OPs that belong to that neuron, along with an SI, and the number of FAs and MAs are illustrated. The size of the SOM is predetermined. As the size of the map increases, a higher resolution of the input data set mapping is provided, however, a fewer number of OPs are associated with each of the neurons of the map, which means that the SI of the neurons cannot be generalized for the OPs of the TS. One of the advantages of the SOM method is the ease of its construction, since it clusters OPs only on the basis of their statistical properties, without information regarding their security status. This is an inherent characteristic of unsupervised learning methods. When the classification criterion is changed, the only modification required is the straightforward recalculation of the numbers shown in Figure 22.17 for each of the neurons of the map. On the contrary, a change of the classification criterion would require, for example, the construction of a new DT.

22.6.1.7 Decision Trees Aided SOM for Load Shedding

In the SOM application for load shedding, there are a number of neurons with intermediate values of SI, indicating a high risk in characterizing the security of the OPs belonging to these neurons. To solve this problem, DTs are embedded at these neurons in order to classify further each OP. As the construction of DTs and the calculation of the SI require the use of postcontingency variables, each contingency is associated with separate SI and DTs.

The inputs of the DTs are the voltage levels of power system busses. Figure 22.20 illustrates the structure of the DTs for each of the positions of the map. Instead of constructing a separate DT for each position of the map, four groups of adjacent positions (position 4 together with 8, 14 with 19, 18 with 23, and 24 with 25) have

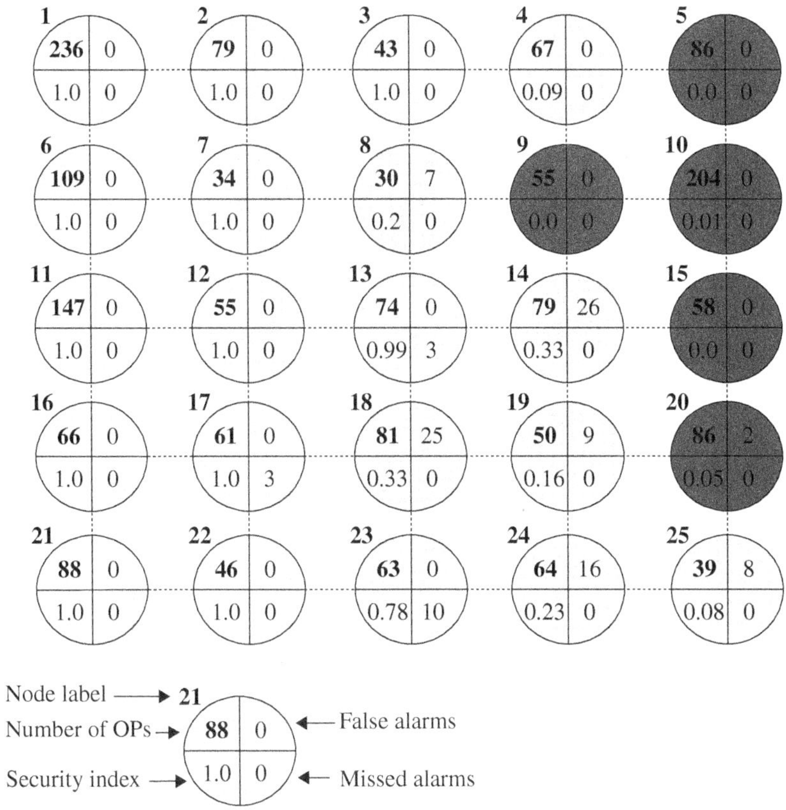

Figure 22.19 SOM for security classification of the power system.

been formed, in order to have enough samples to train the DTs. After having derived the rules from the DT, the model is evaluated using the TS.

Table 22.6 illustrates the results of the DSA method. The number of FAs is reduced significantly in the hybrid SOM–DTs method, which leads to less load being unnecessarily curtailed. On the other hand, the hybrid SOM–DTs lead to a minor increase in the number of MAs (from 16 to 18 MAs). This increase, however, is too small to be statistically significant.

The online application of the DSA method is performed as follows: each OP, characterized by the load profile of the eight areas of the system, is mapped onto the SOM. If the respective neuron has a high SI (SI > 0.95), then no action is taken. Otherwise, the closest secure neuron (e.g., the neuron with SI > 0.95 having minimum distance to the current OP) is sought for, and the load is curtailed so that the system is transposed to that neuron of the map. For neurons with intermediate SI, such as neurons 4, 8, 14, 18, 19, 23, 24, and 25, the SI needs to be recalculated, by applying the rules of the DTs of Figure 22.20. The amount of the curtailed load is given by equation (22.68):

$$P_{curt} = P - \min(\|P - w_i\|) \tag{22.68}$$

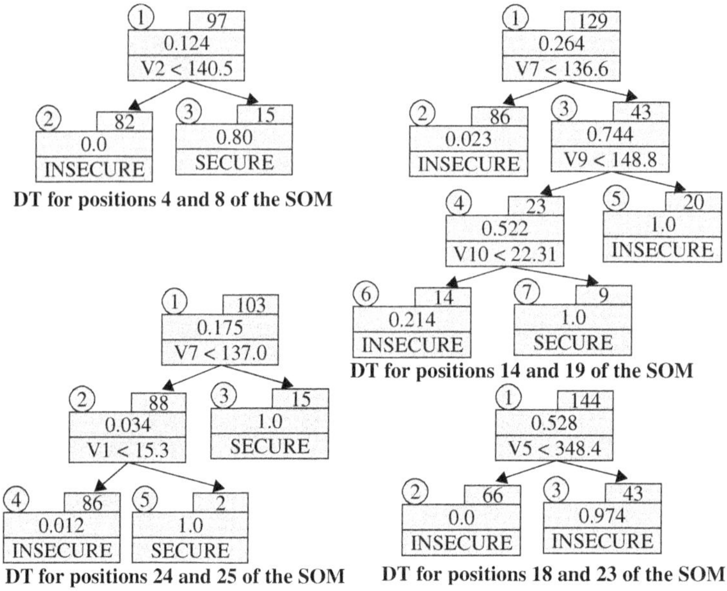

Figure 22.20 Decision trees embedded in the neurons of the SOM.

22.6.2 Preventive Security Control

This section discusses the application of machine-learning methods to the determination of optimal preventive control policy. DTs and reinforcement learning are applied on a realistic model of an autonomous power system considering spinning reserve as a measure for preventive control.

The level of spinning reserve is very important for the security of a power system. When reserve services are inadequate to ensure frequency stability of a power system, emergency control actions, such as under frequency load shedding, are triggered in order to restore the frequency stability. Spinning reserve determination is a very important task, since allocation of a large amount of spinning reserve would lead in uneconomical operation of the system, while inadequate amount of spinning reserve would jeopardize the stability of the system.

TABLE 22.6 Performance indices for each method

	Decision trees	SOM	DT-aided SOM
Success rate	93.65%	93.95%	98.30%
False alarms	8.31% (99/1191)	8.81% (105/1191)	1.34% (16/1191)
Missed alarms	3.46% (28/809)	1.98% (16/809)	2.22% (18/809)
Accumulated curtailed load (TW)	398 TW	487 TW	423 TW
Accumulated unnecessary curtailed load (MW)	18,000	31,655	3760

TABLE 22.7 Generating unit groups

	Vasilikos steam units	Dhekelia steam units	Moni steam units	Moni and Vasilikos gas units
Number of units	3	6	6	5
$P_G^{\min}$ (MW)	60	30	18	4
$P_G^{\max}$ (MW)	130	60	30	28
Forced outage rate	0.589%	3.036%	6.41%	9.788%
Forced outage duration (h)	24	24	24	24
Number of forced outage events	2.13	11.08	23.4	35.72

The classic approach for the determination of the necessary primary frequency control services is based on offline stability analysis of selected extreme conditions. This approach does not take into consideration the varying behavior of the demand and the dynamic characteristics of the generation response [37–39]. Previous publications have faced this problem using online techniques based on energy functions [40–42] and artificial intelligence [43–45]. A successful methodology is given in [45], which takes into account the effect of response time of the offered reserves from the generating units, but it suffers from long execution time. In [33], an efficient, preventive security assessment method, characterized from short execution times, is proposed, providing, however, suboptimal solutions.

Here, two approaches are investigated to solve the problem of spinning reserve allocation. In the first one, a security constrained economic dispatch is performed, with DTs providing the security constraints. In the second approach, a security is considered as a factor of probabilistic cost, which is added to the optimization function and a Reinforcement Learning (RL) framework is employed to provide the spinning reserve policy.

22.6.2.1 *Study Case System*

The application is based on a realistic model of the autonomous power system of Cyprus. There are four groups of generating units as illustrated in Table 22.7, with total installed capacity of 1118 MW.

Under unexpected disturbances, when the spinning reserve of the system fails to control frequency within certain bounds, load shedding automatons are triggered, according to the thresholds described in Table 22.8.

TABLE 22.8 Load shedding automaton thresholds

Frequency threshold	Ratio of load shed to the total load of the system
49	10%
48.5	10%
48	15%

22.6.2.2 Decision Trees for Security Constrained Economic Dispatch

The objective of the proposed methodology is to determine the economic dispatch of the power generation and the primary reserve services required to ensure the secure operating state of the system in the presence of preselected disturbances. The economic dispatch problem can be formulated as follows:

$$Minimize\ C_{gen} = \sum_{n=1}^{N} C_{G,n}(P_{G,n}) \tag{22.69}$$

subject to the following constraints:

- Real power balance constraint

$$\sum_{n=1}^{N} P_{G,n} - P_D - P_{losses} = 0 \tag{22.70}$$

- Real power operating limits

$$P_{G,n}^{\min} \leq P_{G,n} \leq P_{G,n}^{\max} \tag{22.71}$$

- Spinning reserve limits

$$P_n^{\min} \leq P_{G,n} + P_{SR,n} \leq P_n^{\max} \tag{22.72}$$

$$P_{SR,n}^{\min} \leq P_{SR,n} \leq P_{SR,n}^{\max} \tag{22.73}$$

where

$P_G^{\min}$	is the technical minimum of generator.	P_{losses}	– the active power losses.
$P_G^{\max}$	– the technical maximum of generator.	P_D	– an active power demand.
N	– the number of committed generation units.	$P_{G,n}^{\min}, P_{G,n}^{\max}$	– the low and upper limits of power capacity of unit m.
$C_{G,n}$	– the cost of active power provided by unit n.	$P_{SR,n}$	– the spinning reserve of unit n.
$P_{G,n}$	– an active power generation of unit n.	$P_{SR,n}^{\min}, P_{SR,n}^{\max}$	– the minimum and maximum spinning reserves of unit/unit group n.

SECURITY CONSTRAINTS. The security constraints ensure that the system operates above a certain security level. Usually, this security level is defined so that there is no customer interruption under the event of a set of contingencies. A DT is constructed using features related with the spinning reserve or the active power production of the generators of the system. The paths to the secure leafs of the DT represent the security constraints. At least one of these constraints must be satisfied.

For the creation of the KB, a large number of initial OPs are obtained by varying randomly the load at each load busbar and the spinning reserve required. Extensive tests have shown that the development of different DTs, each for a different load range (i.e., low load, medium load, and high load), ensures a better performance of the developed security rules. Thus, the algorithm provides the possibility of considering more than one constraint set (CC) obtained from different DTs suitable for each loading range. The total load of the system is assumed to follow normal distribution around four operating profiles:

1. low load with a total load 300–450 MW;
2. medium load with 450–600 MW;
3. high load with 600–750 MW;
4. peak load with 750–900 MW.

The required spinning reserve of the system also follows normal distribution with mean value the generation of the largest unit and standard deviation $\sigma = 20\%$. For each OP, the outage of the unit with the largest power production, which is a steam unit of Vasilikos station, is simulated using EUROSTAG [46]. The frequency deviation is recorded, and the OPs are labeled as insecure if minimum frequency is below 49 Hz, otherwise they are labeled as secure. Each OP comprises 13 attributes corresponding to the generation and the primary reserve of each unit group, the total primary reserve of the power system, the ratio of the generation of the largest unit to the total generation, and the ratio of total primary reserve to the total generation (Table 22.9).

A total of 6000 OPs are created for each load zone, which means that 24000 OPs are created. These OPs are then divided randomly into two sets, i.e. the Learning Set (LS) and the Training Set (TS) as illustrated in Table 22.10.

Figure 22.21 illustrates the DT for the third load zone. The classification performance of the DTs, when tested with the independent TSs, is illustrated in Table 22.11.

TABLE 22.9 Attributes used

Attribute description	Attribute code
Vasilikos steam units generation	GV
Dhekelia steam units generation	GD
Moni steam units generation	GM
Vasilikos gas units generation	GGV
Moni gas units generation	GGM
Vasilikos steam units PR	PRV
Dhekelia steam units PR	PRD
Moni steam units PR	PRM
Vasilikos gas units PR	PRGV
Moni gas units PR	PRGM
Total PR	TPR
Ratio of generation of the largest unit to the total generation	PPV
Ratio of total PR to the total generation	TPR_G

TABLE 22.10 Knowledge base

	Learning set		Test set	
	Secure	Insecure	Secure	Insecure
Load zone 1	409	2591	451	2549
Load zone 2	719	2281	747	2253
Load zone 3	1388	1612	1440	1560
Load zone 4	2642	358	2624	376

The secure leafs of the DT illustrated in Figure 22.21 are leaf 5, leaf 13, leaf 16, leaf 21, leaf 22, and leaf 25. Each of the secure leafs represents a CC, which determines the dispatch of the generation. Table 22.12 illustrates the extracted CSs for each load zone.

The method is applied on a snapshot of the power system, with total load 650 MW. The following dispatch patterns are examined:

- **DISPATCH PATTERN 1 (DP1):** Dispatch of the total load according to priority list without any spinning reserve requirement
- **DISPATCH PATTERN 2 (DP2):** Dispatch of the load according to priority list, considering the largest active production of the dispatched units as the necessary spinning reserve
- **DISPATCH PATTERN (DP3-DP7):** Dispatch of the load with the Sequential Quadratic Programming (SQP) method and under the constraint set **CS3_1** to **CS3_5**, respectively.

Table 22.13 illustrates the dispatch of the load to the units for each of the examined patterns.

Table 22.14 provides the evaluation indices of the DPs. The system operating cost of each dispatch pattern is computed using the values provided here also.

A time-domain simulation of the tripping of the largest production unit is executed for each of the under study dispatch patterns and the frequency is recorded. The SI of the DPs is the SI of the leaf of the DTs they belong to. The cost of the primary reserve is defined here as the difference between the cost of DP and the cost of DP1, where no primary reserve is required. The values of Table 22.14 indicate that DP4 is the recommended dispatch pattern of the system as it is the lowest cost DP that satisfies the CS imposed by the DT.

22.6.3 Power System-Controlled Islanding

The method of VLT-SOM is applied to map a power system onto a two-dimensional plane in a way that its topological features are preserved. The method is applied for the IEEE 30 bus system, the IEEE 118 bus, and IEEE 17 generators test system. The matrix $[e_{ij}]$ is defined in such a way so as to represent the electrical distance between the two busses i and j. The Thevenin equivalent impedance between busses i and j can be used as a measure of electrical distance in the matrix $[e_{ij}]$. Another option is to consider the dissimilarity of the swing curves of the system, so that generators with

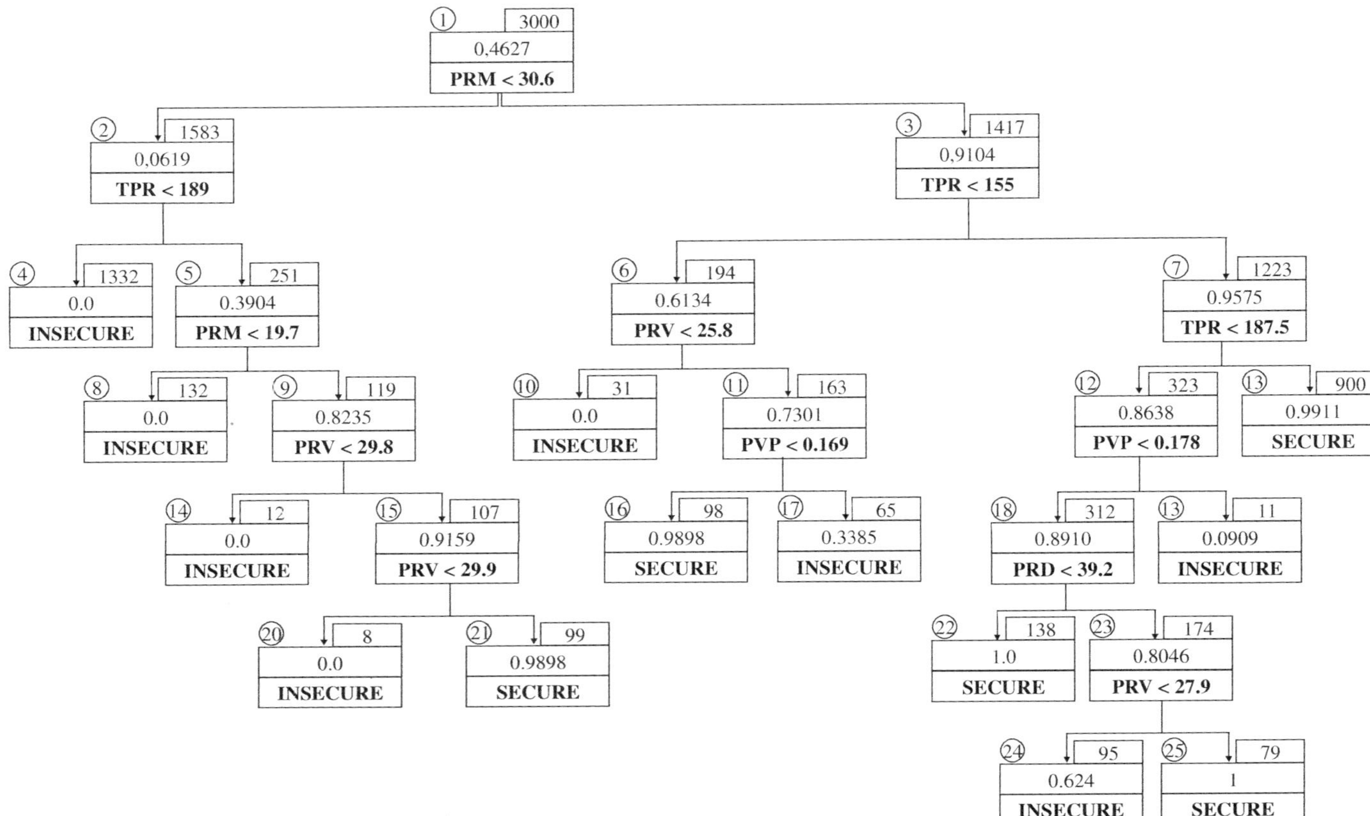

Figure 22.21 Decision tree for load zone 3.

TABLE 22.11 DT classification performance (based on testing set)

	Success rate	Missed alarms	False alarms
Load zone 1	99.27% (2978/3000)	0% (0/2549)	4.88% (22/451)
Load zone 2	99.9% (2997/3000)	0.133% (3/2253)	0% (0/747)
Load zone 3	98.67% (2960/3000)	1.92% (30/1560)	0.694% (10/1440)
Load zone 4	96.76% (2093/3000)	7.71% (29/376)	2.59% (68/2624)

different swing curves are assigned large values of e_{ij}. This can be done by introducing an index W'_{ij}, which is defined by equation:

$$W'_{ij} = \sum_{t_k \in [0,T]} |\omega_i(t_k) - \omega_j(t_k)| \tag{22.74}$$

where [0, T] is the under study time span. Index W'_{ij} is normalized according to equation:

$$W_{ij} = \frac{W'_{ij}}{\max\left(W'_{ij}\right)} \tag{22.75}$$

TABLE 22.12 Security constraint sets (CSs) extracted by the DTs

Load zone 1	
CS1_1	**PVP** < 0,2349
Load zone 2	
CS2_1	GGV>4
CS2_2	(**PVP**<0,2180) ∩ (**GV**<117) ∩ (**PRD**>147,6) ∩ (**GGV** == 0)
CS2_3	(**GV**<108) ∩ (**PVP**>0,218) ∩ (**PVP**<0,219) ∩ (**PRD**>148) ∩ (**GGV** == 0)
Load zone 3	
CS3_1	(**PRM**>30,6) ∩ (**TPR**>187,5)
CS3_2	(**PVP**<0,169) ∩ (**PRV**>25,8) ∩ (**TPR**<155) ∩ (**PRM**>30,6)
CS3_3	(**PRD**<39,2) ∩ (**PVP**<0,178) ∩ (**TPR**<187,48) ∩ (**PRM**>30,6)
CS3_4	(**PRV**>27,9) ∩ (**PRD**>39,2) ∩ (**PVP**<0,178) ∩ (**TPR**<187,48) ∩ (**PRM**>30,6)
CS3_5	(**PRV**>29,9) ∩ (**PRM**>19,7) ∩ (**TPR**>189) ∩ (**PRM**<30,6)
Load zone 4	
CS4_1	(**GGM**>4)
CS4_2	(**TPR**>108) ∩ (**PVP**<0,1588) ∩ (GM>69,5) ∩ (GGM==0)
CS4_3	(**PVP**<0,1647) ∩ (**PRV**>25,3) ∩ (**PVP**>0,159) ∩ (**GM**>69,5) ∩ (**GGM** == 0)
CS4_4	(**TPR_G**<0.2310) ∩ (**PRV**>24,5) ∩ (**PVP**<0,1643) ∩ (**PRV**<25,3) ∩ (**PVP**>0,159) ∩ (**GM**>69,5) ∩ (**GGM** == 0)
CS4_5	(**TPR_G**>0,129) ∩ (**PRV**<23,5) ∩ (**PVP**<0,1643) ∩ (**PRV**<25,3) ∩ (**PVP**>0,159) ∩ (**GM**>69,5) ∩ (**GGM** == 0)
CS4_6	(**PVP**<0,227) ∩ (**PRV**>23,5) ∩ (**PVP**>0,1643) ∩ (**PRV**<25,3) ∩ (**GM**>69,5) ∩ (**GGM** == 0)
CS4_7	(**GM**>125,5) ∩ (**TPR_G**<0.129) ∩ (**PRV**<23,5) ∩ (**PVP**<0,1643) ∩ (**PRV**<25,3) ∩ (**PVP**>0,159) ∩ (**GGM** == 0)
CS4_8	(**PVP**<0,1598) ∩ (**PRV**>24,4) ∩ (**PRV**<24,6) ∩ (**GM**>69,5) ∩ (**GGM** == 0)

TABLE 22.13 Dispatch patterns examined

	DT leaf	Vasilikos steam units (number of units × MW)	Dhekelia steam units (number of units × MW)	Moni steam units (number of units × MW)
DP1	4	3 × 120	5 × 54.4	1 × 18
DP2	4	3 × 120	6 × 45.33	1 × 18
DP3	13	3 × 120	6 × 35	4 × 20
DP4	16	3 × 109.9	5 × 47	4 × 20.4
DP5	22	3 × 105.7	5 × 52	4 × 18
DP6	25	3 × 115.7	6 × 37	4 × 20.25
DP7	21	3 × 120	6 × 34.7	4 × 20.5

A measure of coherence between generators i and j is formulated by equation

$$C_{ij} = 1 - W_{ij}. \tag{22.76}$$

Generators can be divided into clusters according to their coherence. For two generators i, j of the same coherency group $e_{ij} = C_{ij}$, while for two generators in different coherency groups $e_{ij} = -C_{ij}$.

After having mapped the power system onto the two-dimensional plane, the task of power system islanding consists of separating the plane into k segments. In order to achieve this, k points $p_1, p_2, \ldots, p_k$ are considered on the plane with coordinates $p_i = [x_i, y_i]$. These points divide the plane into k segments S_1, S_2, S_k, where segment i includes all points w, for which distance $\|p_i - w\|$ is minimized. Each segment represents an island of the power system comprising the busses that are mapped onto that segment. The task of finding the best positions for points $p_1, p_2, \ldots, p_k$ can be solved using a heuristic technique such as particle swarm optimization [27], given a suitable objective function. This objective function F is defined in order to satisfy the following constraints:

1. *Generation and load balance*

   The load of the formulated islands must be equal to the generation. Quantity F_1 expresses load and generation imbalance

$$F_1 = \sum_{i=1}^{k} \sum_{j=1}^{ni} \left(P_{ij}^{G} - P_{ij}^{D} \right) \tag{22.77}$$

TABLE 22.14 Evaluation of the DPs

	System operating cost £	Minimum frequency	Security index	Total PR.	PR cost £
DP1	18,617	47.75	0.0	68	0
DP2	18,709	48.70	0.0	128	92
DP3	19,309	49.00	0.9911	211	692
DP4	19,141	49.05	0.9898	153	524
DP5	19,252	49.05	1.0	152	635
DP6	19,311	49.0	1.0	183	694
DP7	19,310	48.9	0.9898	211	693

TABLE 22.15 IEEE 30 bus test system load and generation

Bus	Generation (MW)	Load (MW)	Bus	Generation (MW)	Load (MW)
1	50	0	16	0	3.5
2	30	0	17	0	9
3	0	2.4	18	0	3.2
4	0	7.6	19	0	9.5
5	20	0	20	0	2.2
6	0	0	21	0	17.5
7	0	22.8	22	0	0
8	20	0	23	0	3.2
9	0	0	24	0	8.7
10	0	5.8	25	0	0
11	30	0	26	0	3.5
12	0	11.2	27	0	0
13	30	0	28	0	0
14	0	6.2	29	0	2.4
15	0	8.2	30	0	10.6

where k is the number of islands, n_k is the number of busses of island i, P_{ij}^G is the generation of bus j in island i, and P_{ij}^D is the demand of bus j in island i.

2. *Uniform distribution of busses among the islands*

   It is often desirable to split the system into equal-sized islands. Quantity F_2 is a measure of uniform distribution of power system loads to the islands and is defined by equation:

$$F_2 = \sum_{i=1}^{k} \left| P_i^D - \bar{P}^D \right| \tag{22.78}$$

   where P_i^D is the total demand of the i-th island, and $\bar{P}^D$ is the average demand of all the islands.

3. *Number of necessary switching*

   The number of branches that open in order to form the islands must be limited. Otherwise the opening of a large number of branches can cause instability. Quantity F_3 is a measure of the number of opening branches

$$F_2 = a \cdot n_{lines}^2 \tag{22.79}$$

   where α is a weight factor, and n_{lines} is the number of opening branches.

4. *Split the power system to the desirable number of islands*

The number of islands that are created by the proposed methodology is not always equal to the number k of points used to segment the plane. This is because one of the k segments can contain two or more islands. Quantity F_4, which is a modification of F_1, expresses the requirement of creating the desirable number of islands:

$$F_4 = \left(1 + e\frac{(k - k_1)}{q}\right) \sum_{i=1}^{k} \sum_{j=1}^{ni} \left(P_{ij}^G - P_{ij}^D\right) \tag{22.80}$$

where k_1 is the number of the islands, and q is a weight coefficient. The objective function takes the form of equation:

$$F = c_1 F_1 + c_2 F_2 + c_3 F_3 + c_4 F_4 \tag{22.81}$$

where coefficients c_i are user defined. The method is applied for different test power systems, and results are presented in the sections 22.6.3.1 and 22.6.3.2.

22.6.3.1 Application of the Method on the IEEE 30 Bus Test System

The method is applied to the IEEE 30 Bus test system. Table 22.15 provides the load and generation at each bus of the system.

Figure 22.22 illustrates the initial random distribution of system busses onto the plane and Figure 22.23 shows the distribution after the application of VLT-SOM method.

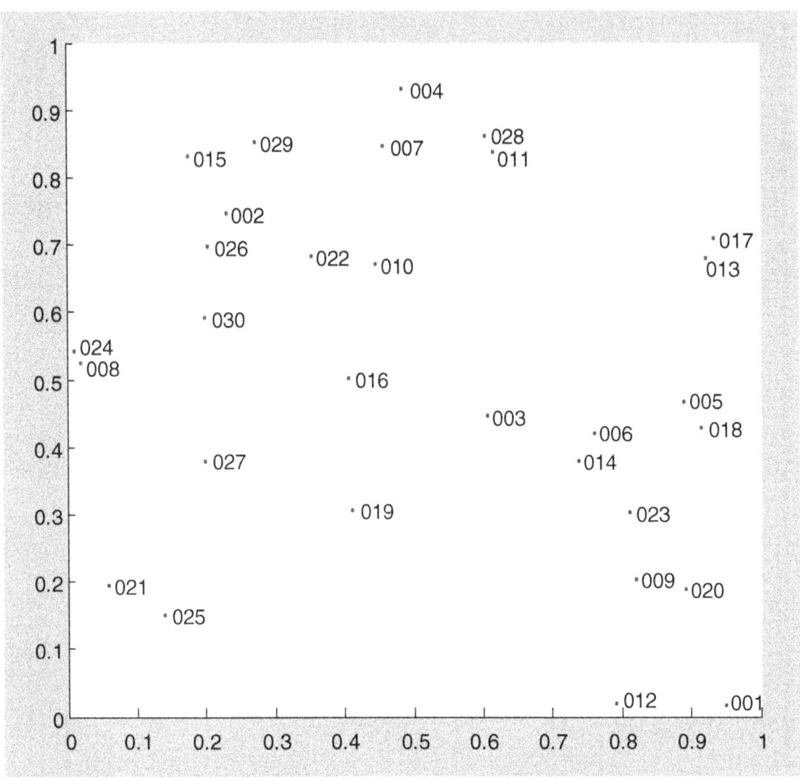

Figure 22.22 Initial random assignment of busses to the plane.

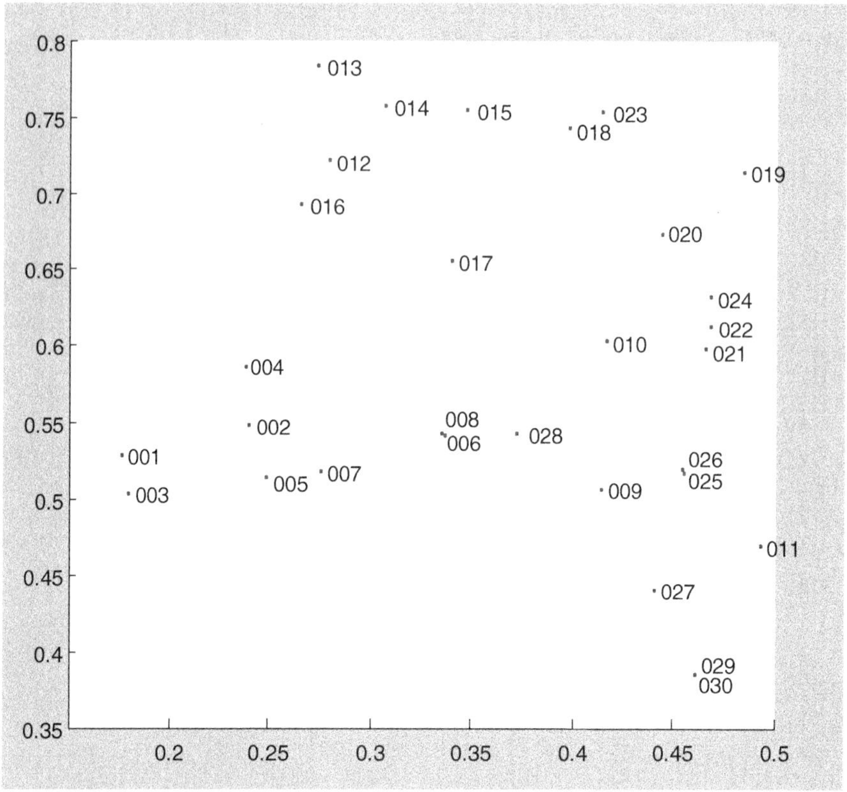

Figure 22.23 Mapping of system busses after the VLT-SOM training.

Table 22.16 presents islanding of the system in two islands, proposed by applying the Particle Swarm Optimization (PSO) method to segment the plane of Figure 22.23.

22.6.3.2 Application of the Method on the IEEE 118 Bus Test System

IEEE 118 busses test system comprises 179 lines, 93 load busses, and 15 generators [47]. Figure 22.24 illustrates the test system, and Tables 22.17 and 22.18 presents islanding of the system in two islands, proposed by applying the PSO method.

TABLE 22.16 Proposed islanding

	Bus	Generation	Load
Island 1	6, 8, 9, 10, 28, 11, 21, 22, 27, 24, 25, 29, 30, 26	50	48,5
Island 2	1, 2, 3 , 4, 5, 12, 7, 13, 14, 15, 16, 18, 23, 17, 19, 20	130	89
Opened branches			
6-2, 6-4, 6-7, 10-17, 10-20, 24-23			

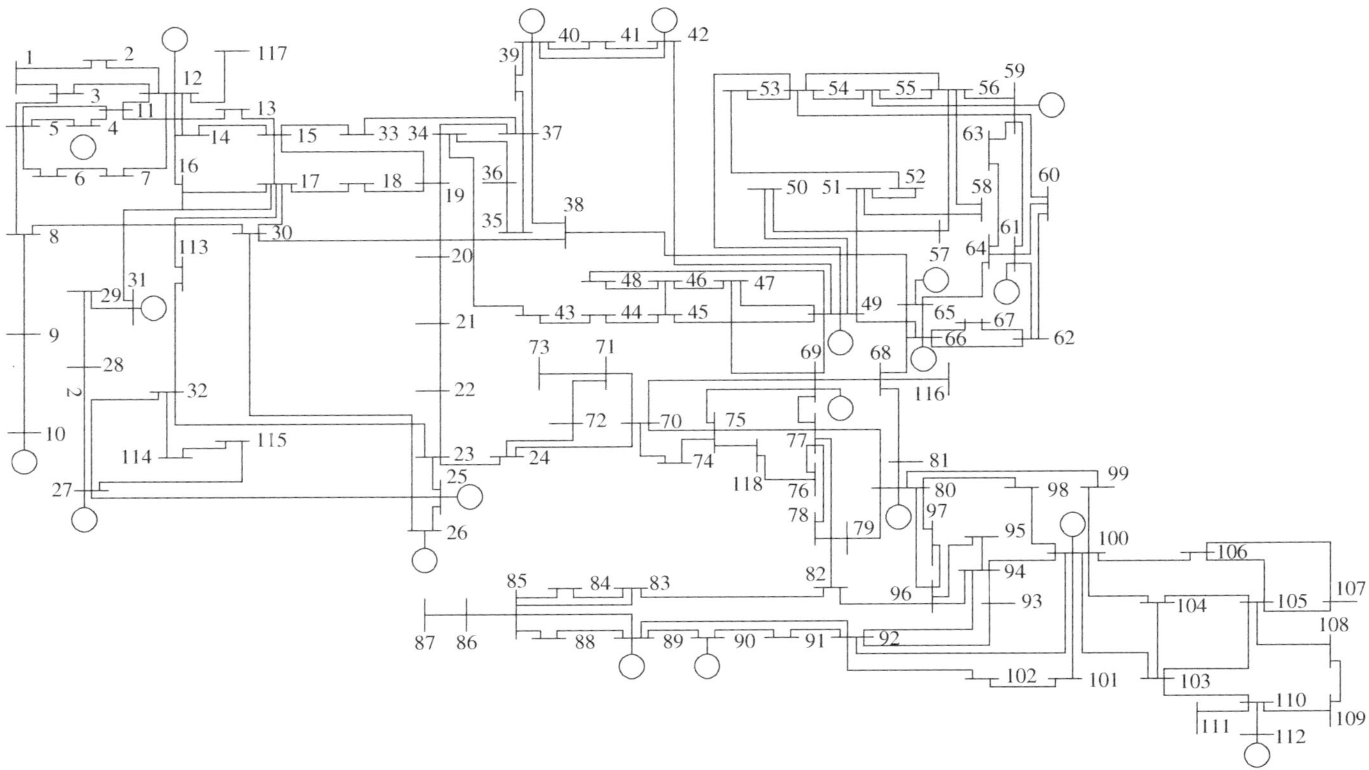

Figure 22.24 IEEE 118 busses test system.

TABLE 22.17 Proposed islanding of IEEE 118 bus test system

Proposed islanding I			
	Bus	Generation	Load
Island 1	23-25, 27-29, 31-32, 68-112, 114-116, 118	1964	1847
Island 2	1-22, 26, 30, 33-67, 113, 117	1823	1803
Opened branches			
17-31, 22-23, 25-26, 47-69, 49-69, 65-68, 113-32			

TABLE 22.18 Proposed islanding of IEEE 118 bus test system (*continued*)

Proposed islanding II			
	Bus	Generation	Load
Island 1	42, 45-68, 80, 81, 92-94, 98-112, 116	1636	1501
Island 2	1-41, 43-44, 69, 70, 72, 74-79, 82-91, 95-97, 113-115, 117-118	2151.6	2143
Disconnected busses	71, 73	0	6
Opened branches			
38-65, 40-42, 41-42, 44-45, 69-47, 69-49, 69-68, 70-71, 72-71, 77-80, 79-80, 89-92, 91-92, 95-94, 96-80, 96-94, 97-80			

REFERENCES

[1] CIGRE TF 38.02.21. Power system security assessment. CIGRE Technical Report, 2004.

[2] Cutsem, T. V., and Vournas, C. *Voltage Stability of Electric Power Systems*. Kluwer Academics, Boston, USA, 1998.

[3] CIGRE T.F. 38.02.13. New trends and requirements for dynamic security assessment. CIGRE Technical Report, *Electra*, 175, 1997.

[4] Kassabalidis, I. N., El-Sharkawi, M. A., Marks II, R. J., Moulin, L. S., and Alves da Silva, A. P. Dynamic security border identification using enhanced particle swarm optimization. *IEEE Transactions on Power Systems*, vol. 17, no. 3, pp. 723–729, August 2002.

[5] Morison, K. On-line dynamic security assessment using intelligent systems. *Proceedings of IEEE-PES General Meeting*, Tampa, Florida, June 2006.

[6] Ajarapu, Q. W. V. A critical review on preventive and corrective control against voltage collapse. *Electric Power Components and Systems*, vol. 29, no. 12, pp. 1133–1144, 2001.

[7] Rajen, B. B., and Gopal, M. On fuzzy-rough sets approach to feature selection. *Pattern Recognition Letters*, vol. 26, pp. 965–975, 2005.

[8] Bhatt, R. B., and Gopal, M. On fuzzy-rough sets approach to feature selection. *Pattern Recognition Letters*, vol. 25, no. 7, pp. 965–975, 2005.

[9] Wang, X.-Z., Yeung, D. S., and Tsang, E. C. C. A comparative study on heuristic algorithms for generating fuzzy decision trees. *IEEE Transactions on SMC-B*, vol. 31, no. 2, pp. 215–226, 2001.

[10] Quinlan, J. R. *Induction of Decision Trees*. Machine Learning 1, Kluwer Academic Publishers, pp. 81–106, Boston, 1986.

[11] Poggio, T., and Girosi, F. Networks for approximation and learning. *Proceedings of the IEEE*, vol. 78, no. 9, pp. 1481–1497, 1990.

[12] Orr, M. J. L. Recent advances in radial basis function networks. Institute for Adaptive and Neural Computation Division of Informatics, Edinburgh University, Scotland, Technical Report, 1999.

[13] Haykin, S. *Neural Networks: A Comprehensive Foundation*. Prentice Hall, Upper Saddle River, NJ, 1994.

[14] Leonard, J. A., Kramer, M. A., and Unger, L. H. A neural network architecture that predicts its own reliability. *Computers and Chemical Engineering*, vol. 16, pp. 819–835, 1992.

[15] Papalexopoulos, A. D., Hubele, N. E., and Ranaweera, D. K. Application of radial basis function neural network model for short term load forecasting. *IEE Proceedings – Generation, Transmission and Distribution*, vol. 142, no. 1, pp. 45–49, January 1995.

[16] Yang, L., Kavli, T., Carlin, M., Clausen, S., and Paul de Groot, F. M. An evaluation of confidence bound estimation methods for neural networks. *Proceedings of 2000 ESIT*, Aachen, Germany, pp. 322–329.

[17] Parzen, E. On estimation of a probability density function and mode. *Annals of Mathematical Statistics*, vol. 33, pp. 1065–1076, 1962.

[18] Kröse, B., and van der Smagt, P. *An Introduction to Neural Networks*, 5th edition. University of Amsterdam, Amsterdam, 1993.

[19] Kohonen, T. *Self Organizing Map*, 2nd edition. Springer Verlag, Berlin, German, 1997.

[20] Niebur, D., and Germond, A. Power system static security assessment using the Kohonen neural network classifier. *IEEE Transactions on Power Systems*, vol. 7, no. 2, pp. 865–871, 1992.

[21] Rehtanz, C., Handschin, E., and Kuhlmann, D. Visualization and analysis of voltage stability using self- organizing neural networks. *Proceedings of ICANN 97*, Lausanne, Switzerland, October 08–10, 1997, Lecture Notes in Computer Science, No. 1327, Springer-Verlag, Berlin, 1997.

[22] Rehtanz, C., and Carsten, L. *Stability assessment of electric power systems using growing neural gas and self-organizing maps*. European Symposium on Artificial Neural Networks Bruges, Belgium, pp. 400–406, 2000.

[23] Deng, D., and Kasabov, N. ESOM: an algorithm to evolve self-organizing maps from on-line data streams. *Proceedings of IJCNN*, IEEE Press, VI: 3–8, 2000.

[24] Rauber, A., Merkl, D., and Dittenbach, M. The growing hierarchical self-organizing map: exploratory analysis of high-dimensional data. *IEEE Transactions on Neural Networks*, vol. 13, no. 6, pp. 1331–1341, 2002.

[25] Fritzke, T. B. *A Growing Neural Gas Network Learns Topologies*. Advances in Neural Information Processing Systems 7, MIT Press, Cambridge, MA, 1995.

[26] Bonabeau, E., and Hnaux, F. Self-organizing maps for drawing large graphs. *Information Processing Letters*, vol. 67, pp. 177–184, 1998.

[27] Kennedy, J., and Eberhart, R. Particle swarm optimization. *Proceedings of the IEEE International Conference on Neural Networks*, Piscataway, NJ, vol. 4, pp. 1942–1948, 1995.

[28] El-Sharkawi, M., and Niebur, D. Artificial neural networks with applications to power systems. IEEE-PES Special Publ. 96 TP-112-0, 1996. *IEEE Service Center, NJ*

[29] Sobajic, D., and Pao, Y. Artificial neural-net based dynamic security assessment for electric power systems. *IEEE Transactions on Power Systems*, vol. 4, pp. 220–228, 1989.

[30] Gavoyiannis, A. E., Vogiatzis, D. G., and Hatziargyriou, N. D. Dynamic security classification using support vector machines. In *Proceedings of the International Conference on Intelligent System Application to Power Systems*, Budapest, Hungary, June 2001.

[31] Moulin, L., da Silva, A., El-Sharkawi, M., and Marks, R. Neural networks and support vector machines applied to power systems transient stability analysis. *Engineering Intelligent System Journal*, vol. 9, no. 4, pp. 205–211, 2001.

[32] Gavoyiannis, A. E., Voumvoulakis, E. M., and Hatziargyriou, N. D. Radial basis functions for dynamic security assessment of power systems with increased wind power penetration. *Proceedings of 15th Power System Computation Conference*, Liege, Belgium, August 2005.

[33] Karapidakis, E., and Hatziargyriou, N. On-line preventive dynamic security of isolated power systems using decision trees. *IEEE Transactions on Power Systems*, vol. 17, no. 2, pp. 297–305, May 2002.

[34] Voumvoulakis, E. M., Gavoyiannis, A. E., and Hatziargyriou, N. D. Decision trees for dynamic security assessment and load shedding scheme. *IEEE PES General Meeting*, Tampa, Florida, June 2006.

[35] Voumvoulakis, E. M., and Hatziargyriou, N. D. Decision trees-aided self organized maps for corrective dynamic security. *IEEE Transactions on Power Systems*, vol. 23, no. 2, pp. 622–630, May 2008.

[36] Vournas, C., Nikolaidis, V., and Tassoulis, A. *Experience from the Athens Blackout of July 12, 2004.* IEEE Saint Petersburg Power Tech, Saint Petersburg, Russia, June 27–30, 2005.

[37] Brown, P. The dispatch of spinning reserve in multiple time scales on an island power system. *Proceedings of University Power Engineering Conference*, Belfast, Ireland, vol. 1, pp. 611–614, 1989.

[38] Wood, A. J., and Wollenberg, B. F. *Power Generation, Operation, and Control.* Wiley, New York, 1996.

[39] Bodo, D. R., Mauzy, D. M., and Trefny, F. J. Economic generation dispatch with response spinning reserve constraints. *IEEE Transactions on Power Systems,* vol. 9, pp. 555–559, February 1994.

[40] Vaahedi, E., Mansour, Y., and Tse, E. A general purpose method for on-line dynamic security assessment. *IEEE Transactions on Power Systems*, vol. 13, pp. 243–249, February 1998.

[41] Li, W., and Bose, A. A coherency based rescheduling method for dynamic security. *IEEE Transactions on Power Systems*, vol. 13, pp. 810–815, August 1998.

[42] Bettiol, A. L., Wehenkel, L., and Pavella, M. Transient stability-constrained maximum allowable transfer. *IEEE Transactions on Power Systems*, vol. 14, pp. 654–659, May 1999.

[43] El-Sharkawi, M. A., and Niebur, D. (Eds.). (1996). Application of artificial neural networks to power systems. *IEEE.* ISO 690.

[44] Wehenkel, L., and Pavella, M. Special issue on automatic learning application to power systems of the Revue-E. *SRBE-Belgium*, Dec. 1996

[45] Wehenkel, L. *Automatic Learning Techniques in Power Systems.* Kluwer Academic Publishers, Norwell, USA, 1998.

[46] Eurostag 3.2, Package, User's Guide.

[47] Available at http://www.ee.washington.edu/research/pstca/

INDEX

Activation function, 756
Agent (see also multiagent systems), 724, 854, 903, 954
  autonomous actions, 724, 903
  flexible, 903
  proactive, 903
  rational, 903
  reactive, 903
  social, 903
Angle
  extinction advance angle, 58
  firing angle, 51, 279, 343
  ignition advance angle, 58
  overlap angle, 52
Anode, 13, 14, 22, 38, 66
Ant colony system (see also evolutionary computation), 931, 961
Artificial Intelligence (AI), 721
  heuristic, 721, 739, 748, 759, 833, 931
  inference, 721, 738, 801
  knowledge representation, 735
  machine learning, 819
  search methods, 853, 880
Artificial Neural Networks (ANN), 755, 820
  adaptive critic design, 759
  back-propagation, 761, 820
  hidden layer, 757, 992
  independent component analysis (ICA), 760
  neural networks, 755
  input layer, 992
  multi-layer perceptron, 756, 820, 993
  off-line ICA learning, 762
  on-line (adaptive) ICA learning, 762
  output layer, 757, 992
  supervised learning, 723, 756
  unsupervised learning, 756, 761, 820, 985
  weights, 760, 830, 937, 956, 993
Automatic learning, 819, 987
  conexionist learning, 820
  machine learning (ML), 819, 855, 985
  statistical methods, 819
Automatic voltage controller, 294
Automatic voltage regulator (AVR), 450, 777

Bipole, 73
Breaker, 142, 188, 545
  backup breaker, 361
  bypass breaker, 216, 350, 394
  DC breaker, 216, 219
  dedicated metallic return (DMR) breaker, 25, 38, 206
  hybrid DC breaker, 216
  metallic return transfer breaker (MRTB), 39, 212
  solid state circuit breaker, 215

Capacitor, 82, 87, 126, 128
  cell capacitor, 149
  DC side capacitor, 146, 473
  floating capacitor, 138
  GTO Controlled Series Capacitor (GCSC), 342
  mechanically switched capacitor (MSC), 271, 521
  thyristor switched capacitor (TSC), 272, 284
Capacitor Commutated Converter (CCC), 26
Cathode, 13, 14, 22, 38, 66
Circle of evolution, 853
  decoding – phenotype, 850, 860
  encoding – genotype, 847, 850

Advanced Solutions in Power Systems: HVDC, FACTS, and Artificial Intelligence, First Edition.
Edited by Mircea Eremia, Chen-Ching Liu, and Abdel-Aty Edris.

Circle of evolution (*Continued*)
  mutation, 852, 862, 877
  recombination, 853, 896
  replacement, 853, 862, 883
  reproduction, 861, 883
  selection, 742, 853, 861
Clustering, 816, 996
Commutation process, 52, 189, 246
  commutating reactance, 83, 100
  commutation failure, 38, 70
  commutation overlap, 55, 83
  commutation time, 18, 52
  commutation voltage, 54, 69, 78
  commutation voltage drop, 54
  overlap time, 52
Compensation, 246, 316, 339
  Advanced Series Compensation (ASC), 389
  series compensation, 339, 351, 382
  shunt compensation, 316, 561, 576
Computational Intelligence (CI), 721, 746
  artificial neural networks, 721, 820, 931
  expert systems, 721, 731,
  fuzzy logic, 721
  genetic algorithms, 721
Converter, 21
  capacitor commutated converter, 26, 246
  converter transformer, 26, 29, 36
  converter valve, 37
  current-sourced converter (CSC), 23
  floating/flying capacitor converter, 138
  full-bridge converter, 143, 144
  half-bridge converter, 139, 168
  H-bridge converter, 128, 694
  line-commutated converter (LCC), 21
  modular multilevel converter (MMC), 138
  multilevel diode-clamped converter, 136
  multilevel converter, 463, 587
  multiple point clamped (MPC), 695
  neutral point clamped (NPC), 126, 554
  voltage-sourced converter (VSC), 13, 206, 527, 549
Constrained load flow, 961
CSC converter, 204, 645
  12-pulse converter, 35, 85, 466
  24-pulse converter, 35, 468
  six pulse converter/bridge, 47
CSC-HVDC configurations
  bipolar HVDC system, 41
  monopolar HVDC system, 41
  point-to-point system, 43
CSC-HVDC control
  compounding point control, 67
  constant current, 62, 67
  constant extinction angle control, 71
  constant valve winding voltage control, 71
  control modes, 67
  current margin order, 75
  DC current control, 69
  emergency power control, 74
  firing angle control, 66
  frequency control, 73
  inverter control, 71
  phase locked loop, 78
  power flow set point, 72
  power modulation, 72
  rectifier control, 69
  tap changer control, 77
  voltage dependent current order limiter (VDCOL), 75
CSC-HVDC hierarchical control
  master control, 72
  pole control, 72
  valve electronics control, 72
Cyber intrusions, 724, 725
Cyber security, 724

Damping of interarea oscillation, 968
Damping circuit, 16, 388, 390, 401
Deadbeat response controller, 712
Decision trees (DTs), 819
  answer node, 820
  attribute selection, 823, 1013
  Classification and Regression Trees (CART), 823
  decision node, 820
  ID3 algorithm, 821, 991
  top-down induction decision trees algorithm (TDIDT), 821
  ULg algorithm, 823, 842
Decision trees applications
  participation factors, 748
  security assessment, 833
  sensitivity factors, 834
  voltage - reactive power controllability, 835
  voltage stability assessment, 833
Decision tree construction, 821, 843
  attributes, 831

classes, 831
entropy, 822
examples, 822
gain, 823
Gini impurity measure, 823
maximum minority, 824
tests, 838
training set, 821
Decision trees pruning, 824
cost-complexity pruning, 828
critical value pruning, 827
error-based pruning, 829
minimum error pruning, 827
pessimistic error pruning, 826
pruning set, 825
reduced error pruning, 825
Defuzzification process, 800
center of gravity, 801
maximum average, 801
maximum value, 801
Diode, 13
diode bridge, 51
diode clamped, 126
free-wheeling diode, 19
Direct current, 53, 54
Direct voltage, 56, 58
Distribution Automation Systems (DAS), 726
automation at substation, 726
fault identification, 726
metering infrastructure, 726
reconfiguration, 726
Distribution STATCOM, 504, 682
Dynamic voltage restorer, 682
Dynamic Security Assessment (DSA)
RBFNN, 1009
decision trees, 1010
neuro-fuzzy decision trees, 1013

Electrode system, 38
electrode line, 38
metallic return, 38, 40
Eurostag software, 494
Evolutionary algorithms for reactive Power planning, 932
constraints, 934
cost function, 935
optimization variables, 936
P-module, 936
P-Q decomposition, 936
Q-module, 937
Evolutionary computation, 847, 932
ant colony system (ACS), 931
differential evolution, 846, 856
evolution strategies, 847, 854
evolutionary programming, 847, 855
evolutionary strategy, 933
evolvable hardware, 847, 855
genetic algorithms, 847, 854
genetic programming, 847, 855
immune algorithm, 968
parallel genetic algorithms, 847, 856
particle swarm optimization (PSO), 847, 951
simulated annealing, 931, 974
tabu search, 931, 974
Evolutionary programming (see also Evolutionary computation), 932
convergence test, 933
competition, 933
Gaussian random variable, 932
initialization, 932
statistics, 932
variance, 932
weights, 933
Evolutionary strategy (see also Evolutionary programming), 933
Expert Systems (ES), 731
artificial expertise, 734
human expertise, 734
reasoning, 733
Expert system architecture, 735
facts database, 737
inference engine, 738
knowledge acquisition, 743
knowledge base, 738
logic programming, 737
object oriented programming, 737
production rules, 735
semantic networks, 737
software interface, 735
Expert system characteristics, 732
expert competence level, 733
inference mechanism, 739
specialized knowledge, 733
Expert system for V-Q control, 745
EXPERTVQ, 745
algorithmic module, 746
heuristic module, 748

Expert system for V-Q control (*Continued*)
participation factors, 748
sensitivity coefficients, 747

Field Programmable Gate Arrays (FPGA), 855
Flexible Alternating Current Transmission Systems (FACTS), 11, 30
Fourier analysis, 52, 83, 275
Fundamental frequency, 52, 130
Fuzzification, 798
Fuzzy control, 798
entry stage, 798
outing stage, 798
processing stage, 798
Fuzzy controller, 799
Fuzzy logic, 797
conditional statements (see also fuzzy model), 801, 805
defuzzification, 798
fuzzification process, 799
inference process, 801
membership, 786
membership function, 795
neuro-fuzzy, 990
rules, 798, 799
statements, 799
search tree, 804
values, 798
Fuzzy model, 801
L-fuzzy Set, 802
Sugeno-Takagi, 802
Fuzzy operations, 788
algebraic product, 789
algebraic sum, 789
complementation, 789
difference, 789
disjunctive sum, 789
equality, 789
inclusion, 789
intersection, 789
union, 789
Fuzzy operators, 799
AND, 799
NO, 799
OR, 799
Fuzzy rule based, 814
production system, 814
Fuzzy set technique, 816
classification, 816
control, 816
decision making, 816
optimization, 816
recognition, 816
Fuzzy sets, 788
operations, 788
properties, 789

Gate, 11
gating command, 130, 286
Gate turn-off (GTO), 11,
Genetic Algorithms (GA), 845
binary encoding, 850
candidate, 849
chromosome, 864
competition, 933
cutting point, 861, 873
De Jong, 889
decoding, 865
elitism, 947
encoding, 864
fitness function, 945
fitness scaling, 884
flow diagram, 862
generation, 853
individual, 847
integer variables, 864
iteration, 954
linear normalization, 885
mating restrictions, 892
minimization, 936
multiobjective optimization, 897
nonlinear programming, 894, 931
offspring, 847, 853, 861
optimal solution, 897, 933
parent, 853, 861, 870
population, 853, 862, 872
population diversity (see also mutation operator), 887
probability of crossover, 887
probability of selection, 884
random variables, 947
real variables, 865
search space, 894
selection, 853, 870
selection and reproduction, 934
Shekel's Foxholes, 854
survival of the fittest, 854
windowing, 885
Genetic algorithm for generation Planning, 943
elite group, 948

fittest solution, 948
mating pool, 948
Genetic operators, 873
crossover operator, 873
hill-climbing operators, 880
inversion operator, 880
mutation, 873
reproduction operators, 861
swap chromosomes operator, 880

Harmonic filter, 37
AC harmonic filter, 87, 191, 540
active filter, 90, 703
band pass (tuned) filter, 83, 87
DC harmonic filter, 90, 188, 191
high frequency filter, 37, 587
high pass (damped) filter, 83, 87
ripple, 20, 173, 474
Harmonics, 37, 56, 78, 83, 157
Heuristic optimization techniques, 931
High Voltage Direct Current Transmission (HVDC), 9, 11, 125, 772
Back-to-back (BTB), 35, 43
CSC-HVDC, 24, 35
VSC-HVDC, 28, 125
HVDC link, 24, 35
HVDC overhead line, 43
HVDC technology, 210, 259

ICA neural networks, 760
Immune algorithm, 968
affinity, 968
antibody, 968
antigen, 968
cloning, 968
hypermutation, 968
lymphocite, 968
Inference engine components, 742
conflict resolution, 742
filtering or pattern matching, 742
rule firing, 742
rule selection, 742
Inference mechanism, 739
backward chaining, 741
breadth-first search, 739
deductive mode, 739
depth-first search, 739
forward chaining, 740
heuristic search, 739
inductive mode, 739
mixed search, 739
Injected voltage, 411, 492
Intelligent cyber attacks, 726
Interaction between AC and DC systems, 96
dynamic overvoltage, 101
effective inertia constant, 99
effective short-circuit ratio, 97
reactive power/strength of the AC system, 100
right of way, 101
short-circuit level, 96
short-circuit ratio, 96
systems stabilization, 96
voltage flickering, 101
Interline Power Flow Controller (IPFC), 629
IPFC at NYPA's Marcy substation, 643
power flow optimization, 641
reactive compensation line, 636
reduce both the active and reactive loop-flow of power, 629
voltage compensation line, 633
Inverter station, 35, 58
equivalent circuit, 57

LC filter, 710
Learning algorithm, 760
Supervised, 988
Unsupervised, 996
Line-to-line voltage, 49
Line-to-neutral voltage, 49
Load shedding, 99
decision trees aided SOM, 1016
genetic algorithm aided SOM application, 1012

Machine intelligence, 846
artificial intelligence, 846
computational intelligence, 846
MAS applications in power engineering (see also multiagent systems), 917
distributed control, 919
electricity markets modeling and simulation, 920
modeling and simulation, 917
monitoring and diagnostics, 918
restoration and reconfiguration, 919
Measurement transducer, 78
capacitive voltage divider, 78
zero flux current transducer, 78
Mechanical phase shifting transformer (see also phase shifter), 410

Metal Oxide Varistor (MOV) (see also varistor), 340
Modulation technique, 698
  carrier wave, 151, 152
  frequency modulation ratio, 153
  hybrid PWM modulation, 164
  level shifted carrier pulse width modulation (LS-PWM), 163
  modulation index, 186
  Nearest Level Control (NLC), 166
  Nearest Vector Control (NVC), 166
  phase opposition disposition (POD-PWM), 164
  phase shifted carrier pulse width modulation (PS-PWM), 163
  phase-disposition modulation (PD-PWM), 146
  Pulse Width modulation (PWM), 20, 125
  selective harmonic elimination method (SHE), 146
  space-vector modulation, 146
  switching frequency, 149
Multiport DC hub, 254
Multiagent systems (MAS), 903
  adaptation, 908
  agent, 903
  agent management, 916
  agent-based simulation, 918
  architectures for intelligent agents, 909
  autonomous, 912
  communication, 912
  communication language, 908
  coordination, 912
  FIPA compliant agent platform, 916
  Foundation of Intelligent Physical Agents (FIPA), 905
  integrated development environment, 915
  intelligent, 903
  JADE, 915
  Java, 915
  negotiation, 909
  object management group, 915
  proactive, 903
  programming languages, 914
  rational agents, 904
  reactive, 903
  real-time control, 905
  social, 903
  training, 906
Multiagent technology applications
  commercial, 912
  entertainment, 907
  industrial, 905
  medical, 907
Multi-layer perceptron (MLP), 756
  feedback (recurrent) neural networks, 758
  feedforward neural networks, 756
  radial basis function (RBF) networks, 757
Multiple input and output variables (MIMO), 794
Multiple input variables and a single output variable (MISO), 804
Multiterminal HVDC installations
  New England, 24
  SACOI, 119
  NEI, 122
Multiterminal HVDC system, 44
  asymmetric monopole system with earth return, 204
  asymmetric monopole system with metallic return, 204
  bipolar HVDC grid, 205
  meshed HVDC grid, 209
  parallel connected system, 45
  series connected system, 45
  symmetrical monopole grid, 204
  system earthing, 206

NGH-SSR damping scheme, 386
Neural networks applications for power systems
  active load estimation, 763
  adaptive neurocontrol of a UPFC device, 776
  harmonic source identification, 767
  optimal control of PV systems, 773
  wide area monitoring and control, 780
  wind speed forecasting, 772
Newton-Raphson, 305, 527, 618

Oblique decision trees, 829
  linear machine, 829
  OC1 algorithm, 831
  recursive least squares (RLS) procedure, 830
  thermal training procedure, 831
On-load tap changer, 36, 417
Optimal location of static VAr compensators, 897

Parent selection mechanism, 861
deterministic sampling, 871
ranking selection, 872
roulette wheel selection, 870
tournament selection, 871
Particle swarm optimization for economic dispatch, 951
ACS algorithm (see also ant colony system), 961
average values, 967
constrained load flow, 961
non-smooth cost function, 952
normal distributions, 965
pheromone trail, 961
Q-Learning algorithm, 963
search mechanism, 955
search space, 955
smooth cost function, 959
standard deviations, 967
velocity, 956
weight factors, 954
Phase-locked loop (PLL), 78, 199, 577, 667
Phase angle regulator (PAR), 409
Phase shifter, 409
Phase shifting transformer (PST), 409
PI controller, 501, 550, 707
Power factor, 52, 56
Power semiconductor devices, 11
not-controllable, 13
fully controllable, 17
reverse-conduction power device, 21
semi-controllable, 13
silicon carbide (SiC), 21
silicon controlled rectifier (SCR), 11
silicon-based, 13
Power semiconductor parameters, 20
blocking voltage, 20
forward voltage drop, 20
surge current, 20
Power semiconductor switching characteristics, 20
peak current turn-off, 20
switching losses, 20
switching speed, 20
turn-off time delay, 20
Power transfer characteristic, 511
PST applications, 439
power flow control, 440
power oscillation damping, 446
transient stability improvement, 444
PST control, 430
quadrature boosting control, 430
quadrature bucking control, 431
PST topology
asymmetrical PST, 412
direct-type PST, 412
exciting transformer, 411
indirect-type PST, 412
quadrature booster (see also phase angle regulator), 409
series reactance, 420
series transformer, 411
shunt reactance, 422
single-core PST, 412
symmetrical PST, 412
two-core PST, 412
Pulse width modulation (PWM), 20, 125, 532

Reactor
arm reactor, 145
mechanically switched reactor (MSR), 271
smoothing reactor, 38, 90, 146
Thyristor-Controlled Reactor (TCR), 272
Rectifier station, 25, 35
equivalent circuit, 56
reactive power requirement, 78
rectifier equations, 47

Self organized maps (SOM) (see also unsupervised learning), 996
Kohonen, 996, 1001
evolving SOM (ESOM), 1001
growing hierarchical SOM (GHSOM), 1001
mean quantization error, 1002
training, 1003
variable local topology (VLT-SOM), 1006
simulated annealing and tabu search applications, 974
aspiration criterion, 975
metropolis distribution, 975
optimal allocation of SVC, 976
Sen transformer (ST), 651
closed loop power flow control, 667
impedance emulation, 666
line reactance regulation, 652
open loop power flow control, 668

Sen transformer (ST) (*Continued*)
optimization of transformer and load tap changer rating, 678
phase angle regulation, 652
power flow enhancement, 662
principle of operation, 658
ST circuit, 663
voltage regulation, 652
Series compensation, 32, 339
Series coupling transformer, 527, 540
Shunt compensation, 316, 398
Single input and a single output (SISO), 802
Smart grid, 721
Snubber circuit, 18
RCD snubber, 686
RCDZ snubber, 686
SSSC applications, 551
compensation of long transmission lines, 551
equalization of power flow in lines, 552
power oscillations damping, 552
receiving-end voltage regulation of a radial line, 552
transient stability improvement, 552
SSSC versus other technologies, 533
fixed series capacitor, 533
fixed series reactor, 533
phase angle regulator, 533
STATCOM applications
fast voltage control, 506
flicker compensation, 506
maintaining voltage level, 506
network transmission capability improvement, 509
oscillations damping, 514
static and transient reserve improvement, 512
STATCOM control, 494
control algorithm, 499
controllable parameters, 499
current controller, 500
rotating d-q frame, 497
voltage controller, 500
STATCOM installations, 515
East Claydon STATCOM, 520
Holly STATCOM, 521
Japan STATCOM, 515
Sullivan STATCOM, 516
Talega STATCOM, 522
STATCOM modeling, 476
basic scheme, 460
current-based steady-state model, 483
detailed dynamic model, 494
equivalent block diagram, 497
equivalent three-phase circuit, 495
simplified dynamic model, 492
unbalanced operation model, 501
V-I characteristic of STATCOM, 475
voltage-based steady-state model, 480
Static Synchronous Compensator (STATCOM), 459
chain link circuit, 463
DC storage capacitor, 474
magnetic circuit, 459
voltage source converter, 459
Static Synchronous Series Compensator (SSSC), 527
control of active power, 547
control of reactance compensation, 531
control of voltage compensation, 531
degree of series compensation, 552
dynamic flow controller, 540
injected voltage in quadrature, 563
operating area, 534
semiconductor switch, 535
voltage source converter, 545
Static VAr Compensator (SVC), 271
gate pulse generator, 294
inductive admittance, 290
linear control domain, 291
shunt compensation, 316
slope reactance, 302
variable susceptance, 311
V-I characteristics, 289
voltage regulator, 294
Storage battery, 527
Static phase shifter (SPS) (see also phase shifter), 428
Subsynchronous resonance (SSR), 386
Supergrid, 249
Supernode, 254
Supervised learning
basis functions, 993
decision trees, 1010
design matrix, 994
epoch, 991
least squares, 993
neuro-fuzzy decision trees, 1013
ridge regression, 993
variance matrix, 995

SVC applications, 314
maintaining the voltage level, 315
oscillations damping, 322
reducing the transient overvoltages, 323
static and transient stability reserve improvement, 317
SVC installations, 324
Auas, Namibia, 330
Channel Tunnel rail link, UK, 333
Forbes, USA, 327
Hagby, Sweden, 326
Harker, UK, 334
Natal, South Africa, 325
Temescal, Mexico, 328
Switch cell, 126

TCSC installations, 387
Imperatriz/Serra da Mesa TCSC, 397
Kanawa River TCSC, 387
Kayenta TCSC, 389
Purnea TCSC, 400
Slatt TCSC, 392
TCSC operation, 351
boost factor, 357
closed loop control, 366
constant angle control, 367
constant current control, 366
constant power control, 369
impedance compensation mode, 345
impedance versus line current, 363
open loop impedance control, 365
voltage compensation mode, 344
Thyristor, 11
Bidirectional control thyristor (BCT), 16
Electrical-Triggered Thyristor (ETT), 15
Emitter Turn-Off Thyristor (ETO), 18
full-wafer thyristor, 17
Gate Turn-Off Thyristor (GTO), 18
Integrated Gate-Commutated Thyristor (GCT), 18
Light Triggered Thyristor (LTT), 15
Phase Control Thyristor (PCT), 12
press-pack thyristor, 140
Super GTO (S-GTO) (see GTO), 18
Thyristor-Controlled Reactor (TCR), 272
Thyristor-Controlled Phase-Angle Regulator (TCPAR), 437
Thyristor-Controlled Phase-Shifting Transformer (TCPST), 410
Thyristor Controlled Series Capacitor (TCSC) (see capacitor), 339
arrester, 350
bypass breaker, 350
bypass switch, 340
damping circuit, 388
spark gap, 340
synchronous voltage reversal, 351
triggered spark gap, 361, 401
ultra-fast switch, 350
Thyristor-Controlled Series Reactor, 31
Thyristor-Controlled Voltage Regulator (TCVR), 31
Transformation ratio, 59, 66, 417, 438
Transient stability, 317, 440, 512
Transistor, 11
Injection Enhanced Gate Transistor (IEGT), 19
Insulated Gate Bipolar Transistor (IGBT), 11
Integrated Gate Controlled Thyristor (IGCT), 11
Metal Oxide Semiconductor Field Effect Transistor (MOSFET), 11

Unified Power Flow Controller (UPFC), 559
ac compensating voltage, 575
circular operating area, 589
controlling the active and reactive power flow, 566
generalized power flow controller, 596
phase-shifting for angle control, 561
reactive shunt compensation for voltage, 561
series compensation for impedance, 561
Universe of discourse, 737
Unsupervised learning, 724, 985
automatic learning framework, 987
clustering, 996
dimensionality reduction, 996
feature extraction, 996
Gaussian neighborhood function, 999
growing neural gas (GNG), 1004
knowledge base generation, 986
model creation, 986
neuron, 996
online implementation, 988
quantization error, 1000
vector quantization, 996

UPFC control system, 575
automatic power flow control mode, 580
automatic voltage control mode, 579
bus voltage regulation and control mode, 579
direct voltage injection mode, 579
functional block diagram, 581
functional (or system) operation control, 575
internal (or converter) control, 575
line impedance compensation mode, 579
phase angle regulation mode, 580
reactive power control mode, 579
system operation control, 578
system optimization control, 577
UPFC at AEP's Inez Station, 600

Varistor, 216
Voltage balancing techniques, 685
gate side solutions, 687
load side solutions, 685
Voltage controlled compensator, 710
Voltage regulation, 62, 82, 403, 439, 517, 575
Voltage stability, 101, 247, 309, 585, 723, 974
VSC-HVDC control, 197
AC voltage control, 197
active power control, 197
coordination of controls, 200
current limiter, 197
DC voltage balance control, 197
DC voltage control, 197
dq frame, 160, 198
firing control, 199
frequency control, 199
inner current controller, 198
outer controller, 197
phase-locked loop control, 199
reactive power control, 197
vector control strategy, 196
VSC converter, 126
12-pulse converter, 247, 446
24-pulse converter, 468
48-pulse converter, 468
cascaded two-level converter, 147
five level converter, 136
multi level converter, 463, 469
multi-pulse two-level, 466
phase leg, 147, 469
pole, 126
P-Q capability diagram, 195
six-pulse converter, 247
switching function, 472
three level converter, 470
two level converter, 471
VSC converter analysis, 166
averaged equivalent circuit, 175
detailed model, 180
duty ratio, 170
dynamic average-value models (AVMs), 166
generalized averaging, 175
simplified model, 180
VSC-HVDC transmission, 125
HVDC Light, 125
HVDC Plus, 125
INELFE project, 144
MonSin, 125

Winding diagram (see also phase shifting transformer), 410
phase shift angle, 413
phasor diagram, 410
quadrature voltage, 412

XLPE cable, 28, 250

Zadeh, Lofti Asker, 786

IEEE Press Series on Power Engineering

Series Editor: M. E. El-Hawary, Dalhousie University, Halifax, Nova Scotia, Canada

The mission of IEEE Press Series on Power Engineering is to publish leadingedge books that cover the broad spectrum of current and forward-looking technologies in this fast-moving area. The series attracts highly acclaimed authors from industry/academia to provide accessible coverage of current and emerging topics in power engineering and allied fields. Our target audience includes the power engineering professional who is interested in enhancing their knowledge and perspective in their areas of interest.

1. *Principles of Electric Machines with Power Electronic Applications, Second Edition*
M. E. El-Hawary

2. *Pulse Width Modulation for Power Converters: Principles and Practice*
D. Grahame Holmes and Thomas Lipo

3. *Analysis of Electric Machinery and Drive Systems, Second Edition*
Paul C. Krause, Oleg Wasynczuk, and Scott D. Sudhoff

4. *Risk Assessment for Power Systems: Models, Methods, and Applications*
Wenyuan Li

5. *Optimization Principles: Practical Applications to the Operations of Markets of the Electric Power Industry*
Narayan S. Rau

6. *Electric Economics: Regulation and Deregulation*
Geoffrey Rothwell and Tomas Gomez

7. *Electric Power Systems: Analysis and Control*
Fabio Saccomanno

8. *Electrical Insulation for Rotating Machines: Design, Evaluation, Aging, Testing, and Repair, Second Edition*
Greg Stone, Edward A. Boulter, Ian Culbert, and Hussein Dhirani

9. *Signal Processing of Power Quality Disturbances*
Math H. J. Bollen and Irene Y. H. Gu

10. *Instantaneous Power Theory and Applications to Power Conditioning*
Hirofumi Akagi, Edson H. Watanabe, and Mauricio Aredes

11. *Maintaining Mission Critical Systems in a 24/7 Environment*
Peter M. Curtis

12. *Elements of Tidal-Electric Engineering*
Robert H. Clark

13. *Handbook of Large Turbo-Generator Operation and Maintenance, Second Edition*
Geoff Klempner and Isidor Kerszenbaum

14. *Introduction to Electrical Power Systems*
Mohamed E. El-Hawary

15. *Modeling and Control of Fuel Cells: Distributed Generation Applications*
M. Hashem Nehrir and Caisheng Wang

16. *Power Distribution System Reliability: Practical Methods and Applications*
Ali A. Chowdhury and Don O. Koval

17. *Introduction to FACTS Controllers: Theory, Modeling, and Applications*
Kalyan K. Sen and Mey Ling Sen

18. *Economic Market Design and Planning for Electric Power Systems*
James Momoh and Lamine Mili

19. *Operation and Control of Electric Energy Processing Systems*
James Momoh and Lamine Mili

20. *Restructured Electric Power Systems: Analysis of Electricity Markets with Equilibrium Models*
Xiao-Ping Zhang

21. *An Introduction to Wavelet Modulated Inverters*
S.A. Saleh and M.A. Rahman

22. *Control of Electric Machine Drive Systems*
Seung-Ki Sul

23. *Probabilistic Transmission System Planning*
Wenyuan Li

24. *Electricity Power Generation: The Changing Dimensions*
Digambar M. Tigare

25. *Electric Distribution Systems*
Abdelhay A. Sallam and Om P. Malik

26. *Practical Lighting Design with LEDs*
Ron Lenk and Carol Lenk

27. *High Voltage and Electrical Insulation Engineering*
Ravindra Arora and Wolfgang Mosch

28. *Maintaining Mission Critical Systems in a 24/7 Environment, Second Edition*
Peter Curtis

29. *Power Conversion and Control of Wind Energy Systems*
Bin Wu, Yongqiang Lang, Navid Zargari, and Samir Kouro

30. *Integration of Distributed Generation in the Power System*
Math H. Bollen and Fainan Hassan

31. *Doubly Fed Induction Machine: Modeling and Control for Wind Energy Generation Applications*
Gonzalo Abad, Jesus Lopez, Miguel Rodrigues, Luis Marroyo, and Grzegorz Iwanski

32. *High Voltage Protection for Telecommunications*
Steven W. Blume

33. *Smart Grid: Fundamentals of Design and Analysis*
James Momoh

34. *Electromechanical Motion Devices, Second Edition*
Paul C. Krause, Oleg Wasynczuk, and Steven D. Pekarek

35. *Electrical Energy Conversion and Transport: An Interactive Computer-Based Approach, Second Edition*
George G. Karady and Keith E. Holbert

36. *ARC Flash Hazard and Analysis and Mitigation*
J. C. Das

37. *Handbook of Electrical Power System Dynamics: Modeling, Stability, and Control*
Mircea Eremia and Mohammad Shahidehpour

38. *Analysis of Electric Machinery and Drive Systems, Third Edition*
Paul Krause, Oleg Wasynczuk, S. D. Sudhoff, and Steven D. Pekarek

39. *Extruded Cables for High-Voltage Direct-Current Transmission: Advances in Research and Development*
Giovanni Mazzanti and Massimo Marzinotto

40. *Power Magnetic Devices: A Multi-Objective Design Approach*
S. D. Sudhoff

41. *Risk Assessment of Power Systems: Models, Methods, and Applications, Second Edition*
Wenyuan Li

42. *Practical Power System Operation*
Ebrahim Vaahedi

43. *The Selection Process of Biomass Materials for the Production of Bio-Fuels and Co-Firing*
Najib Altawell

44. *Electrical Insulation for Rotating Machines: Design, Evaluation, Aging, Testing, and Repair, Second Edition*
Greg C. Stone, Ian Culbert, Edward A. Boulter, and Hussein Dhirani

45. *Principles of Electrical Safety*
Peter E. Sutherland

46. *Advanced Power Electronics Converters: PWM Converters Processing AC Voltages*
Euzeli Cipriano dos Santos Jr. and Edison Roberto Cabral da Silva

47. *Optimization of Power System Operation, Second Edition*
Jizhong Zhu

48. *Power System Harmonics and Passive Filter Designs*
J. C. Das

49. *Digital Control of High-Frequency Switched-Mode Power Converters*
Luca Corradini, Dragan Maksimovic, Paolo Mattavelli, and Regan Zane

50. *Industrial Power Distribution, Second Edition*
Ralph E. Fehr, III

51. *HVDC Grids: For Offshore and Supergrid of the Future*
Dirk Van Hertem, Oriol Gomis-Bellmunt, and Jun Liang

52. *Advanced Solutions in Power Systems: HVDC, FACTS, and Artificial Intelligence*
Mircea Eremia, Chen-Ching Liu, and Abdel-Aty Edris